Studies in Fuzziness and Soft Computing

Volume 337

Series editor

Janusz Kacprzyk, Polish Academy of Sciences, Warsaw, Poland
e-mail: kacprzyk@ibspan.waw.pl

About this Series

The series "Studies in Fuzziness and Soft Computing" contains publications on various topics in the area of soft computing, which include fuzzy sets, rough sets, neural networks, evolutionary computation, probabilistic and evidential reasoning, multi-valued logic, and related fields. The publications within "Studies in Fuzziness and Soft Computing" are primarily monographs and edited volumes. They cover significant recent developments in the field, both of a foundational and applicable character. An important feature of the series is its short publication time and world-wide distribution. This permits a rapid and broad dissemination of research results.

More information about this series at http://www.springer.com/series/2941

Ahmad Taher Azar · Sundarapandian Vaidyanathan
Editors

Advances in Chaos Theory and Intelligent Control

Editors
Ahmad Taher Azar
Faculty of Computers and Information
Benha University
Benha
Egypt

Sundarapandian Vaidyanathan
Research and Development Centre
Vel Tech University
Chennai
India

ISSN 1434-9922 ISSN 1860-0808 (electronic)
Studies in Fuzziness and Soft Computing
ISBN 978-3-319-30338-3 ISBN 978-3-319-30340-6 (eBook)
DOI 10.1007/978-3-319-30340-6

Library of Congress Control Number: 2016933326

Printed on acid-free paper

This Springer imprint is published by Springer Nature
The registered company is Springer International Publishing AG Switzerland

Preface

About the Subject

Chaos theory deals with the behaviour of dynamical systems and maps that are highly sensitive to initial conditions. Sensitivity to initial conditions is usually called the butterfly effect for dynamical systems and maps. Chaotic systems can be observed in many natural systems such as weather and climate. Chaos theory has applications in several areas such as vibration control, electric circuits, chemical reactions, lasers, combustion engines, computers, cryptosystems, encryption, secure communications, biology, medicine, management, finance, etc. Chaotic behaviour can be studied through the analysis of a chaotic mathematical model (discrete or continuous). This book focuses on research problems such as modelling and analysis of chaotic systems, stabilization and control of chaotic systems, chaos synchronization and applications, chaos in communications and cryptosystems, applications of chaos in engineering, etc. Various tools and methodologies have been developed for the analysis of chaotic systems such as Poincaré maps, Lyapunov exponents, active control, adaptive control, sliding mode control, backstepping control, fuzzy logic control, sampled-data feedback control, and in this book we use the latest control techniques for the study of chaotic systems and control.

Intelligent control describes a class of control techniques that use various artificial intelligence (AI) techniques such as neural network control, fuzzy logic control, neuro-fuzzy control, expert systems, genetic control, evolutionary algorithms, and intelligent agents. Intelligent control systems are useful when no mathematical model is available a priori and intelligent control itself develops a system to be controlled. Intelligent control is inspired by the intelligence and genetics of living beings. This book focuses on various applications of intelligent control on nonlinear control systems, including chaotic systems.

About the Book

The new Springer book, *Advances in Chaos Theory and Intelligent Control,* consists of 36 contributed chapters by subject experts who are specialized in the various topics addressed in this book. The special chapters have been brought out in this book after a rigorous review process in the broad areas of Chaos Theory, Control Systems, Computer Science, Information Technology, modelling and engineering applications. Special importance was given to chapters offering practical solutions and novel methods for the recent research problems in the main areas of this book, *viz.* Chaos Theory and Intelligent Control.

Objectives of the Book

The objective of this book makes a modest attempt to cover the framework of Chaos Theory and its applications in a single volume. The book is not only a valuable title on the publishing market, it is also a successful synthesis of intelligent control techniques in the world literature. Several multidisciplinary applications in control, engineering and computational intelligence are discussed inside this volume. This book is recommended to engineers of various specialties, mathematicians, information technology specialists and students of those or related specialties. Both novice and expert readers should find this book a useful reference in the field of chaos theory, soft-computing and intelligent control.

Organization of the Book

This well-structured book consists of 36 full chapters. They are organized into two parts.

Part 1: Advances in Chaos Theory
Part 2: Advances in Intelligent Control

Book Features

- The book chapters deal with the recent research problems in the areas of chaos theory, fuzzy systems, evolutionary algorithms, soft computing, intelligent control, modelling and engineering.
- The book chapters contain a good literature survey with a long list of references.
- The book chapters are well written with a good exposition of the research problem, methodology and block diagrams.

- The book chapters are lucidly illustrated with numerical examples and simulations.
- The book chapters discuss details of engineering applications and future research areas.

Audience

The book is primarily meant for researchers from academia and industry, who are working in the research areas—Computer Science, Information Technology, Engineering, Automation, Chaos and Control Engineering. The book can also be used at the graduate or advanced undergraduate level as a textbook or major reference for courses such as control systems, intelligent control, mathematical modeling, computational science, numerical simulation, applied artificial intelligence, fuzzy logic control, and many others.

Acknowledgements

As the editors, we hope that the chapters in this well-structured book will stimulate further research in chaos theory, control systems, soft computing, intelligent control and utilize them in real-world applications.

We hope sincerely that this book, covering so many different topics, will be very useful for all readers.

We would like to thank all the reviewers for their diligence in reviewing the chapters.

Special thanks go to Springer, especially the book editorial team.

Ahmad Taher Azar
Sundarapandian Vaidyanathan

Contents

Part I
Advances in Chaos Theory

A Novel Design Approach of a Nonlinear Resistor Based on a Memristor Emulator

Ch.K. Volos, S. Vaidyanathan, V.-T. Pham, J.O. Maaita, A. Giakoumis, I.M. Kyprianidis and I.N. Stouboulos

Abstract In this chapter, a novel design method of a nonlinear resistor of type-N consisting of a memristor emulator circuit in parallel with a negative resistor, is presented. The proposed emulator is built with second-generation current conveyors (CCII) and passive elements and its pinched hysteresis loop is holding up to 20 kHz. As an example of using the designed nonlinear resistor, the simple non-autonomous Lacy circuit is chosen. The numerical as well as the simulation results, by using SPICE, reveal the richness of circuit's dynamical behavior confirming the usefulness of the specific design method. The increased complexity that the nonlinear resistor with memristor gives to the circuit is a consequence of the effect of both signals frequency and amplitude to the memristors pinched hysteresis loop. Furthermore, the ease of the specific design method makes this proposal a very attractive option for the design of nonlinear resistors.

Ch.K. Volos (✉) · J.O. Maaita · A. Giakoumis · I.M. Kyprianidis · I.N. Stouboulos
Physics Department, Aristotle University of Thessaloniki,
Thessaloniki 54124, Greece
e-mail: volos@physics.auth.gr

J.O. Maaita
e-mail: jmaay@physics.auth.gr

A. Giakoumis
e-mail: ang1960@el.teithe.gr

I.M. Kyprianidis
e-mail: imkypr@auth.gr

I.N. Stouboulos
e-mail: stouboulos@physics.auth.gr

S. Vaidyanathan
Research and Development Centre, Vel Tech University, Tamil Nadu, India
e-mail: sundar@veltechuniv.edu.in; sundarvtu@gmail.com

V.-T. Pham
School of Electronics and Telecommunications, Hanoi University
of Science and Technology, Hanoi, Vietnam
e-mail: pvt3010@gmail.com

A.T. Azar and S. Vaidyanathan (eds.), *Advances in Chaos Theory and Intelligent Control*, Studies in Fuzziness and Soft Computing 337,
DOI 10.1007/978-3-319-30340-6_1

Keywords Memristor · Emulator · Lacy circuit · Chaos · Bifurcation diagram · Phase portrait

1 Introduction

Chaos is a phenomenon which appears widely and naturally in many dynamical nonlinear systems and its study has led to a vast multidisciplinary research field, ranging from natural sciences (chemistry, biology, ecology, physics, etc.) [11, 44, 63] to social sciences (economics, sociology, etc.) [25, 80, 81] and engineering (electronics, control, communication, cryptography, robotics, etc.) [3, 7, 33, 49, 65–69, 83]. From all these fields, the design of nonlinear circuits seems to attract much interest because of its nature and its rapid development. The easy simulation of chaotic phenomena with nonlinear circuits and the great number of applications, such as in cryptography, in secure communications, robotics and in neuronal networks [4–6, 35, 37, 39, 70, 71, 73], are some of the reasons that appoint the research in this field significant. Also, the discovery of novel nonlinear circuits is not only a fascinating subject, but it also contains several research challenges.

In the last five decades, three major inventions of Professor Leon O. Chua are the landmarks in the evolution of this research field. These are the Chua's circuit, the Cellular Neural/Nonlinear Networks (CNNs), and the memory resistor, which is known as memristor.

The first invention, the Chuas circuit [15], is the first electronic circuit which was designed and constructed in 1983 for displaying chaotic behavior. Its configuration is based on a two-terminal nonlinear resistor (named Chua's diode) having a piecewise-linear $i - v$ characteristic with negative slopes. CNN, the second invention of Chua [17], is a nonlinear system combining the advanced features of neural networks and cellular automata. The classical CNN architecture consists of a number of cells. Each cell includes linear capacitors, linear resistors, linear and nonlinear voltage controlled current sources and independent sources. As a result, CNN architecture is suitable for VLSI implementation [16], which is very important for the industry of electronics. So, CNNs have been applied in various areas, such as in signal processing [21, 50], modeling of complex systems [2], pattern formation [19, 27], bio-inspired robotic visions and biological functions [26]. Finally, his third invention but chronologically the first one, is the memristor [18]. This fourth circuit element presents the missing relation between charge and magnetic flux.

After Chua's paper on memristor in 1971, only a few works appeared in literature for a long time since it was thought that this new circuit element was only a theoretical element and it could not be realized practically. So, until recently, the memristor had received little attention even though a working device made from op-amps and discrete nonlinear resistors had been built and demonstrated in the seminal paper of Chua [18]. The memristor, we can say that it was the Holy Grail of electronics. However, in 2008, Hewlett-Packard scientists, working at their laboratories in Palo Alto-California, announced in Nature [64] that a physical model of memristor has been realized. In their scheme, a memory effect is achieved in solid-state thin film two-terminal device.

Since then, a considerable number of publications have presented noticeable results in models of memristor, fabricated materials and techniques or important applications of memristor such as high-speed low-power processors [76], adaptive filters [23], associative memory [51, 58, 75], neural networks [1, 72], programmable analog integrated circuits [59] and so on.

Especially, the intrinsic nonlinear characteristic of memristor could be exploited in implementing novel chaotic systems with complex dynamics [29, 32]. Memristor can change nonlinearities in conventional systems to create new ones with advanced features. Besides its evident nonlinear characteristic, nanoscopic scale size of memristor promises a revolution in integrated chaotic circuits. In fact, chaotic systems have been designed and realized conveniently using memristors [30, 46].

In this chapter a design method of a nonlinear resistor of type-N, which is used in a variety of nonlinear circuits, consisting of a memristor emulator circuit in parallel with a classical negative resistor, is presented. For this reason, a recently new proposed memristor emulator is built with second-generation current conveyors (CCII) and passive elements and its pinched hysteresis loop is holding up to 20 kHz, which is a very important feature especially in practical applications. For testing the behavior of the proposed nonlinear resistor, the simple non-autonomous Lacy circuit is chosen for using it. The simulation results, by using SPICE, reveals the richness of circuits dynamical behavior confirming the usefulness of the specific design approach. The increased complexity that the nonlinear resistor with memristor gives to the circuit is a consequence of the effect of both signals frequency and amplitude to the memristors pinched hysteresis loop. Furthermore, the ease of the specific design method makes this proposal a very attractive option for the design of non-linear resistors.

This chapter is organized as follows. In the next section, the definition of the memristor as well as its first physical model will be presented. Several existing memristor emulators will be briefly summarized in Sect. 3 as well as the proposed memristor emulator circuit which is the base of the proposed nonlinear resistor. In Sect. 4 the implementation of a nonlinear resistor based on the memristor emulator circuit is described. The simulation results which proved the feasibility of the proposed nonlinear resistor into a nonlinear, non-autonomous circuit, are presented in Sect. 5. Finally, the last section draws the concluding remarks.

2 Memristor—A Brief Review

2.1 The Invention

Until the beginning of 70ies the electronic circuit theory has been spinning around the three known, fundamental two-terminal circuit elements, which are known as: resistor (R), capacitor (C) and inductor (L). These elements reflect the relations between pairs of the four electromagnetic quantities of charge (q), current (i), voltage (v) and magnetic flux (ϕ) that mathematically can be written as:

$$\begin{aligned} dv &= R(i)di \\ dq &= C(v)dv \\ d\phi &= L(i)di \end{aligned} \tag{1}$$

In the case that the factors C, L and R have constant values, the corresponding circuit elements are linear. However, as it can be derived, a relation between the charge (q) and the flux (ϕ) is missing.

At that time (1971), Professor Leon Chua dubbed this missing link by introducing the fourth fundamental element based on the symmetry arguments [18]. This fourth circuit element was named memristor (M), an acronym for memory resistor, which its existence was conjectured due to the following missing relation between the charge (q) and the flux (ϕ).

$$d\phi = M(q)dq \tag{2}$$

The multiplicative term $M(\cdot)$ is called the memristance function. Dividing both sides of (2) by dt one obtains.

$$v = M(q)i \tag{3}$$

If M is constant, Eq. (3) is nothing but the defining relation of a linear resistor (R), as it can be shown from (1). However, Chua has proved theoretically that a memristor is a nonlinear element because its $v - i$ characteristic is similar to that of a Lissajous pattern. So, a memristor with a non-constant M describes a resistor with a memory, more precisely a resistor whose resistance depends on the amount of charge that has passed through the device.

A typical response of a memristor to a sinusoidal input is depicted in Fig. 1. The "pinched hysteresis loop current-voltage characteristic" is an important fingerprint

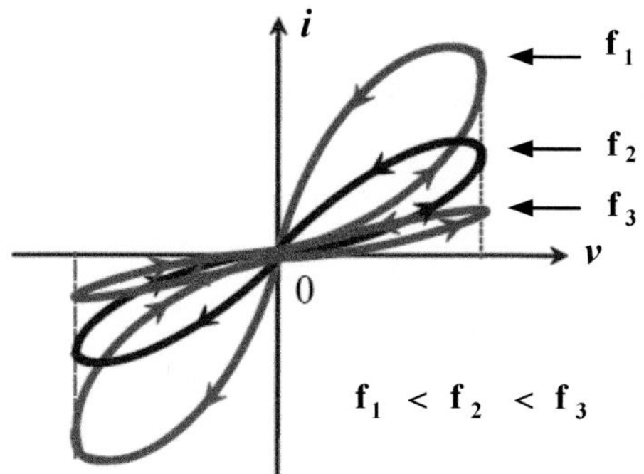

Fig. 1 A typical $v - i$ characteristic curves of a memristor driven by a sinusoidal voltage input

of a memristor. If any device has a current-voltage hysteresis curve, then it is either a memristor or a memristive device. Another signature of the memristor is that the "pinched hysteresis loop" shrinks with the increase in the excitation frequency. The fundamentality of the memristor can also be deduced from this figure, as it is impossible to make a network of capacitors, inductors and resistors with an $v - i$ behavior forming a pinched hysteresis curve [16].

Some of the more interesting properties of memristor are [16]:

- Non-linear relationship between current (i) and voltage (v).
- Does not store energy.
- Similar to classical circuit elements, a system of memristors can also be described as a single memristor.
- Reduces to resistor for large frequencies as evident in the $v - i$ characteristic curve.
- Memory capacities based on different resistances produced by the memristor.
- Non-volatile memory possible if the magnetic flux and charge through the memristor have a positive relationship ($M > 0$).

Furthermore, a more generalized class of systems, in regard to the original definition of a memristor, called memristive systems [20], is introduced. An nth-order current-controlled memristive one-port is represented by

$$\begin{aligned} v &= R(w, i, t)i \\ \frac{dw}{dt} &= f(w, i, t) \end{aligned} \tag{4}$$

where $w \in R^n$ is the n-dimensional state variable of the system.

Also, the nth-order voltage-controlled memristive one-port is defined as:

$$\begin{aligned} i &= G(w, v, t)v \\ \frac{dw}{dt} &= f(w, i, t) \end{aligned} \tag{5}$$

Similarly to memristor, a memristive system has the following properties:

- The memristive system should have a dc characteristic curve passing through the origin.
- For any periodic excitation the $v - i$ characteristic curve should pass through the origin.
- As the excitation frequency increases toward infinity the memristive system has a linear behavior.
- The small signal impedance of a memristive system can be resistive, capacitive, or inductive depending on the operating bias point.

2.2 The First Physical Model of the Memristor

In 2008, Hewlett-Packard scientists, working at their laboratories in Palo Alto-California, announced in Nature [64] that a physical model of memristor has been realized. In their scheme, a memory effect is achieved in solid-state thin film two-terminal device.

This element is passive while the latest realization of a memristor is that of an active one on a base of niobium oxide [55]. The memristor, which is realized by HP researchers, is made of a titanium dioxide layer which is located between two platinum electrodes. This layer is of the dimension of several nanometers and if an oxygen dis-bonding occurs, its conductance will rise instantaneously. However, without doping, the layer behaves as an isolator. The area of oxygen dis-bonding is referred to as space-charge region and changes its dimension if an electrical field is applied. This is done by a drift of the charge carriers. The smaller the insulating layer, the higher the conductance of the memristor. Also, the tunnel effect plays a crucial role. Without an external influence the extension of the space-charge region does not change.

The internal state x is the extent of the space-charge region, which is restricted in the interval [0, 1] and can be described by the equation

$$x = \frac{w}{D}, 0 \leq x \leq 1, x \in R \tag{6}$$

where w is the absolute extent of the space-charge region and D is the absolute extent of the titanium dioxide layer. The memristance can be described by the following equation:

$$M(x) = R_{on}x + R_{off}(1 - x) \tag{7}$$

where R_{on} is the resistance of the maximum conducting state and R_{off} represents the opposite case. So, when $x = 0$, $R = R_{off}$, and when $x = 1$, $R = R_{on}$. The vector containing the internal states of the memristor is one dimensional. For this reason scalar notation is used. The state equation is:

$$\frac{dx}{dt} = \frac{\mu_\nu R_{on}}{D^2} i(t) \tag{8}$$

where μ_ν is the oxygen vacancy mobility and $i(t)$ is the current through the device. By using the Eq. (6) the previous equation can be rewritten as:

$$\frac{dw}{dt} = \frac{\mu_\nu R_{on}}{D} i(t) \tag{9}$$

So, the dynamics of the memristor can therefore be modeled through the time dependence of the width w of the doped region. Integrating Eq. (9) with respect to time,

$$w = w_0 + \frac{\mu_\nu R_{on}}{D} q(t) \tag{10}$$

where w_0 is the initial width of the doped region at $t = 0$ and q is the amount of charges that have passed through the device. Substituting (6), (10) into Eq. (7) gives:

$$M(q) = R_0 - \frac{\mu_\nu R_{on} \Delta R}{D^2} q(t) \tag{11}$$

where

$$R_0 = R_{on} \frac{w_0}{D} + R_{off}(1 - \frac{w_0}{D}) \tag{12}$$

and $\Delta R = R_{off} - R_{on}$. The term R_0 refers to the net resistance at $t = 0$ that serves as the device's memory. This term is associated with the memristive state, which is essentially established through a collective contribution, i.e. it depends directly on the amount of all charges that have flown through the device.

Thats why, we can say that the memristor has the feature to "remember" whether it is on or off after its power is turned on or off. This announcement brought a revolution in various scientific fields, as many phenomena in systems, such as in thermistors whose internal state depends on the temperature [57], spintronic devices whose resistance varies according to their spin polarization [54] and molecules whose resistance changes according to their atomic configuration [13], could be explained now with the use of the memristor. Also, electronic circuits with memory circuit elements could simulate processes typical of biological systems, such as the learning and associative memory [51] and the adaptive behavior of unicellular organisms [53].

3 Memristor's Emulators

3.1 Related Works

After the realization of the first memristor at HP Labs, different types of memristors have been designed [54, 74]. However, there are not any commercial off-the-shelf memristors in the market yet. In order to overcome this difficulty, various emulations of memristors were proposed in literature which use different design methodologies and topologies. Also, smooth continous cubic nonlinear functions, piecewise linear models and SPICE macromodels have been principally used to mimic the memristor behavior, and although some of them are based on those equations reported by HP labs the derived macromodels are in primitive forms and can only be used for high-level simulations, but not to physically build real-world applications.

The first memristor emulator based on active devices was proposed by Chua in [18]. However, this emulator circuit is relatively complex and bulky. Recently, grounded memristor emulator circuits built with op-amps and analog multipliers have been proposed in [9, 10, 47, 48]. Nevertheless, the connectivity of those emulators with other circuit elements in serial or parallel is limited.

In 2010, Muthuswamy [46] developed a flux-controlled memristor with a cubic $q - \phi$ characteristic based on a relatively simple analog circuit. It contains two multipliers and two operational amplifiers as well as an analog integrator, which is used to obtain the magnetic flux across the memristor, while the memductance is obtained by using the multiplier circuit in a feedback loop. This is a first attempt to design such a memristor with cubic $q - \phi$ characteristic, which has been used as a nonlinear element in many works [31, 48, 72].

Many other approaches of analog circuit implementation emulating a memristor by using operational amplifiers and analog multipliers have been presented because of their simplicity and feasibility [8, 40]. In another interesting work a memristive device with a piecewise linear characteristic was realized by a CNN cell [12], based on a circuit which consisted of five operational amplifiers, a capacitor, twelve resistors, a 2N222 diode, a DC voltage source, and a switch ADG201AKN. In this design the voltage v_C across the capacitor is denoted as the internal state variable x of the memristive device.

Unlike the previous memristor emulators using analog electronic circuits, Pershin and Di Ventra [52] implemented a microcontroller-based memristor emulator. The main components of this emulator include a digital potentiometer (AD5206), a control unit (a 16-bit microcontroller dsPIC30F2011), and analog-to-digital converter (an internal 12-bit ADC of the microcontroller). The voltage on the memristor v_M is measured by the ADC. The microcontroller calculates and updates the resistance of the digital potentiometer (x) according to the predefined algorithm. In other words, the resistance is continuously changed and determined by the programmable code in the microcontroller. Depending on the resolution of the selected digital potentiometer, the resistance only varies between the limiting values R_{min} and R_{max}. However, the response time and resolution of its memristances limited due to the limited performance of the ADC converter.

However, memristor emulator which acts as a real memristor device is useful for developing memristor application circuits as well as for the memristor circuit demonstration for educational purpose. For this reason, in the last five years, a number of works, in which memristor emulator circuits for acting as a real memristor fabricated by HP, has been presented.

The first model was published in 2010 by Mutlu and Karakulak [45]. In their work, a memristor emulator via pure analog technology, where the features of the TiO_2 memristor were well initiated, is presented. However, this emulator has the disadvantage that it does not hold the memristance for long enough, while the input signal was not applied due to the leakage current.

In 2012, Kim et al. presented a memristor emulator, which has been designed and built with off-the-shelf solid state devices [36]. The memristor emulating circuit is designed in a way of composing the input resistance as a function of applied voltage or current. In more detail, when an input voltage is applied at the memristor emulator, it is converted into an input current by using an op-amp configuration. Since this current is used at several places of the circuit, its replicas are generated using current mirrors implemented with MOSFETs. One of the distinguished features of a memristor is the capability of keeping the programmed information for a long time until new programming input are presented. The charge stored at a capacitor is the programmed information in this memristor emulator. To avoid discharging during the period when an input signal does not exist, the path to the output terminal is connected to the gate of a MOS type buffer. However, the main weaknesses of this memristor emulator are its nonfloating operation and the small ratio (3.6) between the maximum and the minimum memristance in contrary with that of the real memristor fabricated by HP, which is 160.

Furthermore, the memristor emulator circuit reported in [28] uses a positive second generation current conveyor (CCII+) and a voltage controlled resistor. However, although this emulator is topologically simple, several active devices not only are again necessary for designing the voltage controlled resistor, but the linearity of the transistor operating in the triode region is also limited.

In [24] a memristor emulator based on current feedback operational amplifier was proposed. Although, the emulator circuit is very simple, it cannot be used in complex networks where floating memristor circuits are required.

In 2014, Yu et al. proposed a flux-controlled memristor emulator with floating terminals by making use of four current conveyors, one op-amp, one multiplier, one capacitor and several resistors [79]. This memristor emulator has the ability of being utilized both in floating connections and grounded connections.

In the same year a simple memristor emulator circuit based on differential current conveyor was proposed in [78]. However, numerical simulations were only presented and the pinched hysteresis loops were operated at low frequency.

Furthermore, a memristor emulator should include some important features, such as a sufficiently wide range of memristance, bimodal operability of pulse and continuous signal inputs, a long period of nonvolatility, floating operation, operability with other devices and the ability to be implemented with off-the-shelf devices. As a consequence in 2015, Yang et al. proposes a memristor emulator that contains all of these features [77]. Specifically, in that work, the small variation range of memristance and the nonfloating operation that limit conventional memristor emulators are improved significantly.

3.2 The Proposed Memristor Emulator

In literature many works related with floating memristor emulator circuits have been presented [56, 60, 62]. However, all these emulators require several integrated circuits and, as it is mentioned, they are complex and bulky with the limiting factor that the pinched hysteresis loop operates at low frequency.

For this reason, in 2014, Sanchez-Lopez et al. present a new floating memristor emulator circuit based on second-generation current conveyors (CCII+) and passive elements [43]. The proposed emulator is very simple compared to those topologies reported in [56, 60, 62]. Also, the frequency-dependent pinched hysteresis loop current-voltage characteristic holds up to 20.2 kHz, which is an important advantage of the specific emulator for use in real-world applications.

In this model, CCII+ as an active device, which is widely used to design linear and nonlinear circuits, is chosen. In the CCII+ the following equations are applied

$$\begin{aligned} v_X(t) &= v_Y(t) \\ i_Y(t) &= 0 \\ i_Z(t) &= i_X(t) \end{aligned} \tag{13}$$

where X, Y, Z are the terminals of the AD844AN which is used as CCII+. The schematic diagram of the floating memristor emulator circuit proposed in [43] is shown in Fig. 2. The CCII+ (U_3) and (U_4) are used as current followers, so that the current $i_m = \frac{v_1}{R_1}$ is equal to i_{X4}. Also, the voltage v_3 from the differential voltage amplifier implemented with the CCII+ (U_1) is given by the following equation

$$v_3 = \frac{R_4}{R_2}(v_1 - v_2) \tag{14}$$

where $v_1 = v_{X1}$ and $v_2 = v_5$. Furthermore, the CCII+ (U_2) along with R_3 and C are used to implement an integrator. So,

$$v_4 = \frac{1}{R_3 C}\int_0^t v_3(\tau)d\tau \tag{15}$$

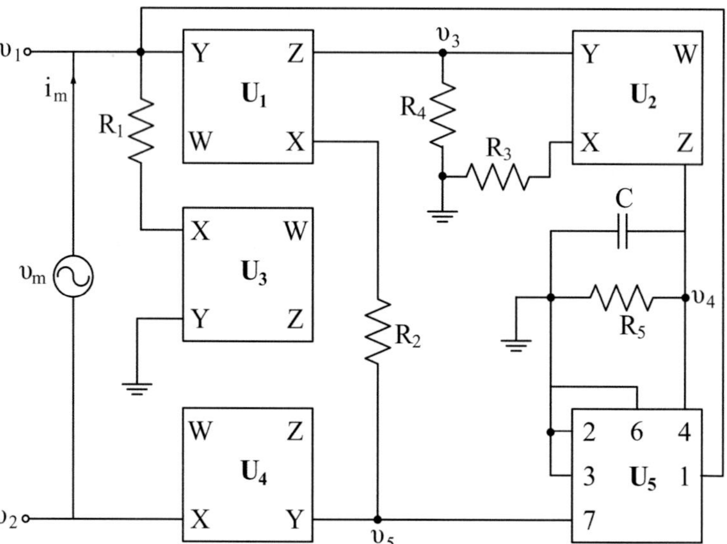

Fig. 2 The schematic of the floating memristor emulator circuit proposed in [43]

Next, a four-quadrant analog multiplier AD644JN (U_5) is used for producing the signal υ_2 as a product of υ_1 and υ_4.

$$\upsilon_2 = \upsilon_5 = -\frac{\upsilon_1 \upsilon_4}{10V} \tag{16}$$

From Eqs. (14) and (15), Eq. (16) can be written as

$$\upsilon_2 = -\frac{R_4 \upsilon_1}{R_2 R_3 C 10V} \int_0^t (\upsilon_1(\tau) - \upsilon_2(\tau)) d\tau \tag{17}$$

or

$$\upsilon_1 - \upsilon_2 = u_1 (1 + \frac{R_4}{R_2 R_3 C 10V} \int_0^t (\upsilon_1(\tau) - \upsilon_2(\tau)) d\tau) \tag{18}$$

However $\upsilon_1 = i_m R_1$, and $\upsilon_m = \upsilon_1 - \upsilon_2$, so the previous equation can be rewritten as

$$\frac{\upsilon_m}{i_m} = R_1 + \frac{R_1 R_4}{R_2 R_3 C 10V} \int_0^t \upsilon_m(\tau) d\tau) \tag{19}$$

or

$$\frac{\upsilon_m}{i_m} = R_1 + \frac{R_1 R_4}{R_2 R_3 C 10V} \phi_m = M(\phi_m) \tag{20}$$

where $\phi_m = \int_0^t \upsilon_m(\tau) d\tau$ is the flux and $M(\phi_m)$ is the memristance, which can be controlled by the voltage signal υ_m. By using a sinusoidal voltage signal ($\upsilon_m = A_m sin(2\pi f t)$) in the input of the memristor, one can study the behavior of the frequency-dependent pinched hysteresis loop. So, for the values of circuit's elements as chosen in Table 1, the pinched hysteresis loops for various values of frequencies of the input signal, by using the Multisim, are displayed in Figs. 3, 4, 5 and 6. In more detail, this figure shows four pinched hysteresis loops, of υ_1 versus υ_m, for frequencies in the range from 5 to 20 kHz. The signal $\upsilon_1 (= i_m R_1)$ is used in order to indirectly plot the current i_m. The feature of working in higher frequencies than

Table 1 Values of memristor emulator circuit's elements

Element	Value
R_1	10 kΩ
R_2	100 kΩ
R_3	1.97 kΩ
R_4	100 kΩ
R_5	10 kΩ
C	5 nF
A_m	1 V
$V_\pm$	±10 V

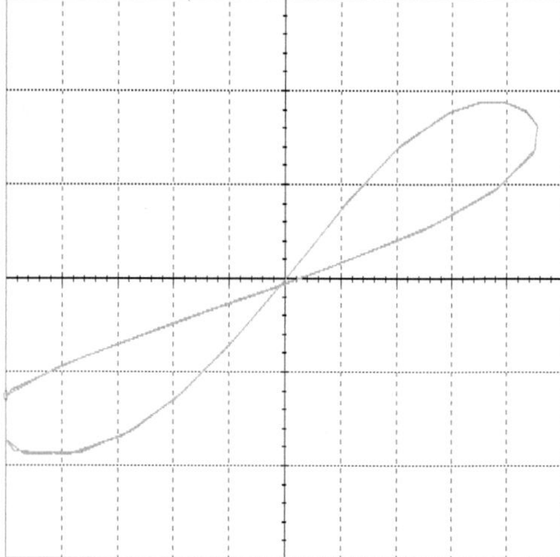

Fig. 3 Frequency-dependent pinched hysteresis loop of v_1 versus v_m, for the memristor emulator circuit, operating at 5 kHz

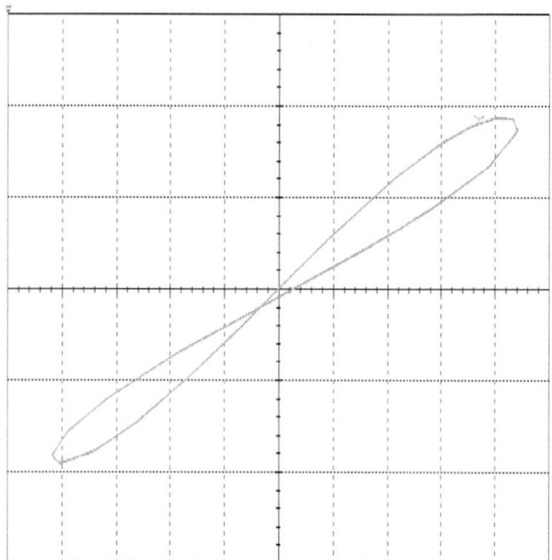

Fig. 4 Frequency-dependent pinched hysteresis loop of v_1 versus v_m, for the memristor emulator circuit, operating at 10 kHz

the real one, makes the specific memristor emulator circuit a suitable candidate for use in real applications. Also, for this figure, one can realize that as the frequency of the voltage source v_m is monotonically increased the memristor behavior becomes a linear time-invariant resistor.

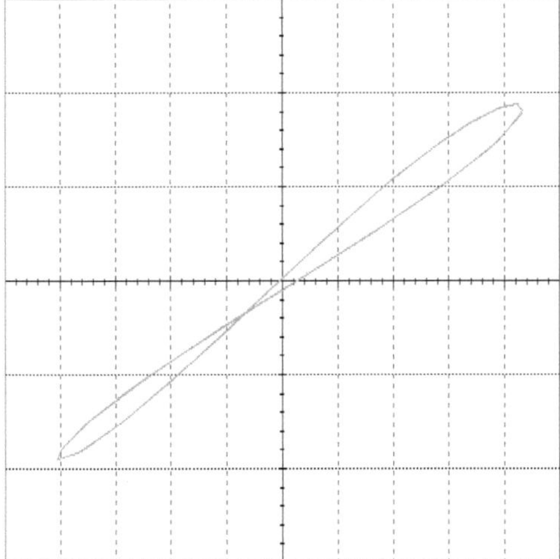

Fig. 5 Frequency-dependent pinched hysteresis loop of υ_1 versus υ_m, for the memristor emulator circuit, operating at 15 kHz

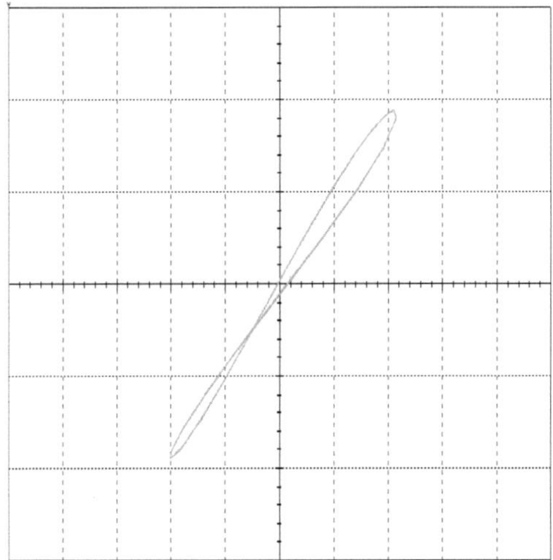

Fig. 6 Frequency-dependent pinched hysteresis loop of υ_1 versus υ_m, for the memristor emulator circuit, operating at 18 kHz

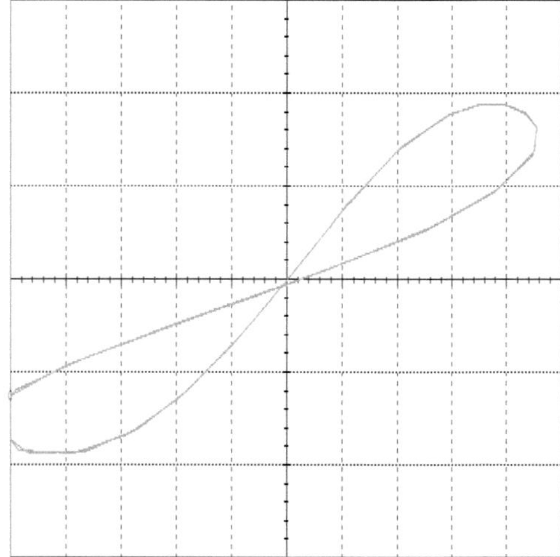

Fig. 7 Frequency-dependent pinched hysteresis loop of v_1 versus v_m, for the memristor emulator circuit, operating at 20 kHz, with $C = 5\,\text{nF}$

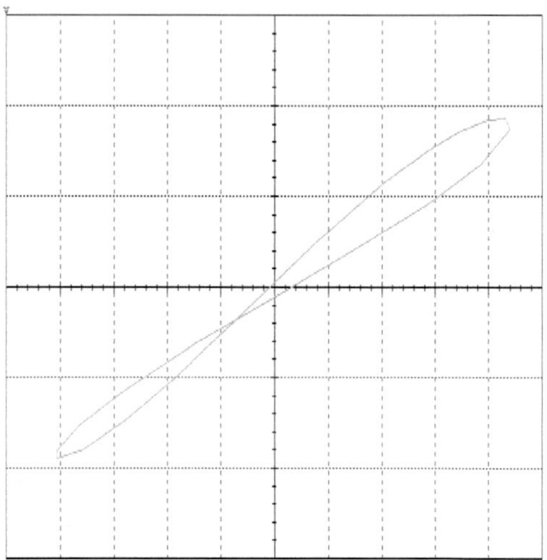

Fig. 8 Frequency-dependent pinched hysteresis loop of v_1 versus v_m, for the memristor emulator circuit, operating at 20 kHz, with $C = 3\,\text{nF}$

Furthermore, by scaling down the value of C, the behavior of hysteresis loop is improved and can be pushed for operating at higher frequencies. This improvement of hysteresis loop as the value of C is decreased is shown in Figs. 7, 8, 9 and 10. So, the proposed memristor emulator circuit, as it is mentioned, is simpler than other topologies reported in literature, but also that, although the frequency-dependent hysteresis loop is not symmetric with regard to the origin, the area enclosed in the first and third quadrants can be relatively equal, if someone adjusts properly the values of the circuit's elelements. The asymmetry is an inherent feature of the circuits topology.

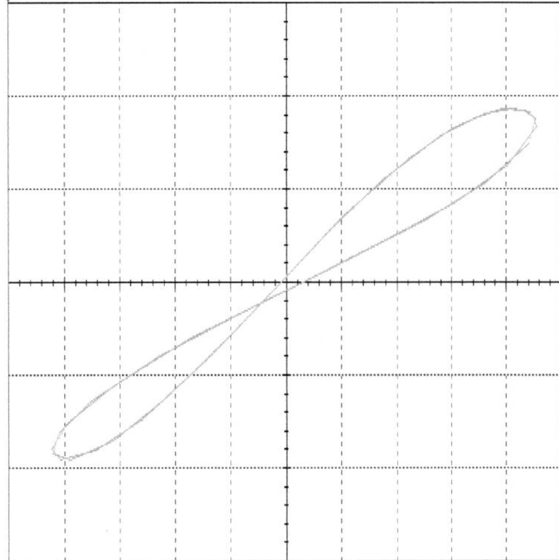

Fig. 9 Frequency-dependent pinched hysteresis loop of v_1 versus v_m, for the memristor emulator circuit, operating at 20 kHz, with $C = 2\,\text{nF}$

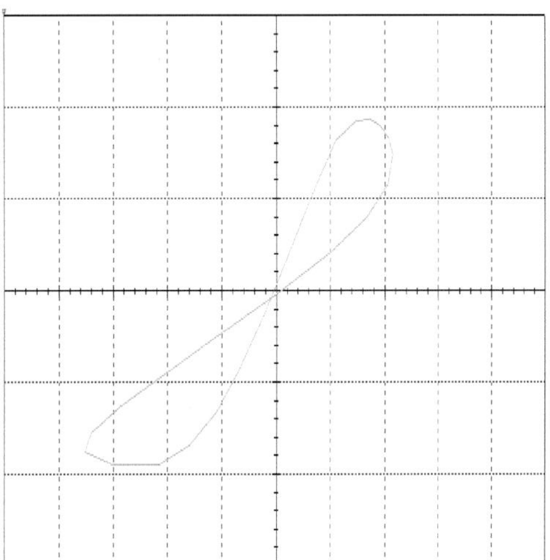

Fig. 10 Frequency-dependent pinched hysteresis loop of v_1 versus v_m, for the memristor emulator circuit, operating at 20 kHz, with $C = 1\,\text{nF}$

4 Implementation of a Nonlinear Resistor Based on the Memristor Emulator

From the beginning of the era of nonlinear circuits, which was basically in 1983 with the design of Chua's circuit [14], its nonlinear element known as the Chuas diode has been used not only in this but also in a variety of nonlinear circuits, due to its simple topology and its easy implementation. Since then, several ways have been proposed in literature to build such an element [22, 34, 41, 42, 61, 82]. Although, many nonlinear functions have been assumed for this element, in its original form it has a 3-segment piecewise-linear $i - u$ characteristic, which is known as nonlinear resistor of type-N.

In general, a two-terminal nonlinear resistor like Chua's diode must be an element that needs to be ad hoc synthesized. Also, its nonlinearity is fundamental for achieving an oscillatory chaotic behavior. That is because the nonlinear element, being a nonlinear locally active resistor, allows the nonlinear circuits to satisfy the criterion of having an element like that in order to present chaotic behavior.

Nowadays, an easy way for implementing nonlinear resistors is based on the op amp approach. The main function of the nonlinear element is to provide an eventually passive negative resistor. The device implementing a nonlinear resistor of type-N is shown in Fig. 11 [34].

If the op-amp has ideal properties, no current flows into the positive input terminal. So, the current is given by

$$i = \frac{\upsilon - \upsilon_0}{R_1} \tag{21}$$

Furthermore, since no current flows in the negative input terminal, the voltage υ can be calculated from

$$\upsilon = \upsilon_d + \frac{R_3}{R_2 + R_3}\upsilon_0 \tag{22}$$

Taking into account that $\upsilon_0 = A_\upsilon \upsilon_d$, the previous equation can be written as

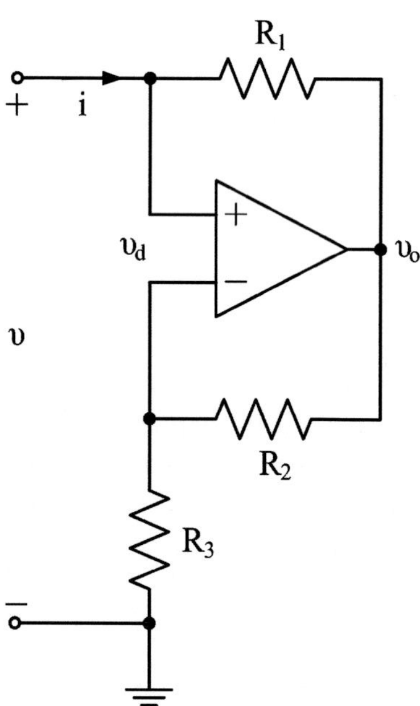

Fig. 11 The schematic of the nonlinear resistor of type-N

$$v = \frac{R_2 + R_3(1 + A_v)}{A_v(R_2 + R_3)} v_0 \tag{23}$$

So, Eq. (21) by using the last equation is transformed to the next one

$$i_v = \frac{(1 + A_v)R_2 + R_3}{R_2 + R_3 A_u + R_3} \tag{24}$$

Under the assumption of a large open loop gain ($A_v \longrightarrow \infty$) and that $R_1 = R_2$, Eq. (24) becomes

$$i_v = -\frac{1}{R_3} \tag{25}$$

which represents the $i - v$ characteristic of the negative resistor. If the voltage takes large values the slope of the $i - v$ characteristic is positive, since the op amp saturates. For example, for the positive saturation ($v_0 = E_{sat}$) the Eq. (21) can be written as

$$i = \frac{v - E_{Sat}}{R_1} \tag{26}$$

which now represents a $i - v$ characteristic which is translated with respect to the origin and has a positive slope. The breakpoint $v = E_1$ of the whole $i - v$ characteristic of the nonlinear resistor can be calculated by substituting $v = E_1$ in Eq. (22) as follows

$$E_1 = \frac{R_2 + R_3(1 + A_v)}{A_v(R_2 + R_3)} E_{sat} \tag{27}$$

In the limit of large A_v, one obtains

$$E_1 = \frac{R_3}{R_2 + R_3} E_{sat} \tag{28}$$

When the op-amp saturates at $v = E_{sat}$ an analogous behavior is obtained. So, the complete $i - v$ characteristic of the nonlinear resistor of type-N implementing with the circuit of Fig. 11, is thus that shown in Fig. 12, which is clearly the characteristic of an eventually passive negative resistor.

By using the Multisim the following values of circuit components of Fig. 11 ($R_1 = R_2 = 4.2\,\mathrm{k}\Omega$, $R_3 = 435\,\Omega$, $V_\pm = 10\,\mathrm{V}$), the values of the slopes (G_a, G_b) and the breakpoints (E, $-E$) of the nonlinear resistor of type-N are given in Table 2.

In this work, a memristor like that of Fig. 2 has been added in parallel with the nonlinear resistor of Fig. 11. Due to the fact that the specific memristor is floating, a small linear resistor ($r = 100\,\Omega$) is added to the terminal (X) of the CCII+ (U_4). With this technique, the complexity that the proposed nonlinear resistor with memristor, gives to a nonlinear circuit is increased, as a consequence of the effect not only of signals amplitude but also of signals frequency to the memristor's pinched hysteresis loop.

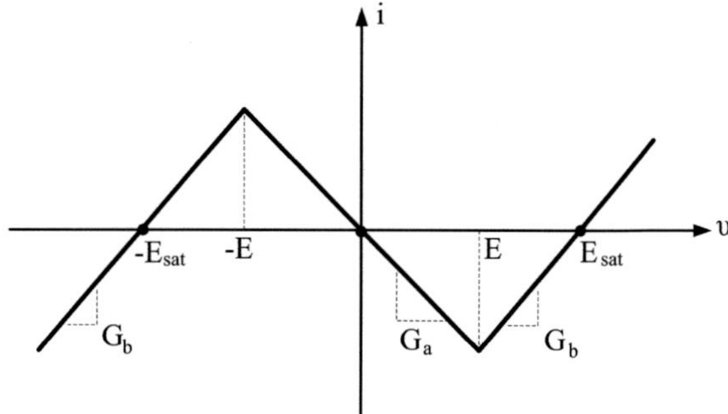

Fig. 12 Nonlinear characteristic of type-N implementing with the circuit of Fig. 11

Table 2 Values of nonlinear resistor circuit's elements

Parameter	Value
G_a	−2.2980 mS
G_b	0.2379 mS
E	0.9001 V
$-E$	−0.9001 V

Table 3 Values of the parameters of the proposed nonlinear element with memristor

Parameter	Value
G_a	−1.6055 mS
G_b	0.9455 mS
E	0.9002 V
$-E$	−0.9745 V

By using the Multisim the values of the slopes (G_a, G_b) and the breakpoints $(E, -E)$ are given in Table 3. As it shown, the absolute values of the breakpoints are different due to the asymmetry of the memristors pinched hysteresis loop. This is one of another factor of complexity that the proposed nonlinear element gives to a circuit. Also, the values of the slopes (G_a, G_b) of the proposed nonlinear element have been changed in comparison with the slopes produced by the nonlinear resistor of type-N, due to the adding in parallel with this of the memristor.

5 Nonlinear Circuit Based on the Proposed Nonlinear Resistor

In this work, a very simple nonlinear, non-autonomous circuit, the well-known Lacy circuit [38] (Fig. 13), is used, for testing the feasibility of this new nonlinear element in such circuits. The choice of a non-autonomous circuit is done so as to examine the effect of the frequency variation to the memristors pinched hysteresis loop and therefore to circuit's behavior.

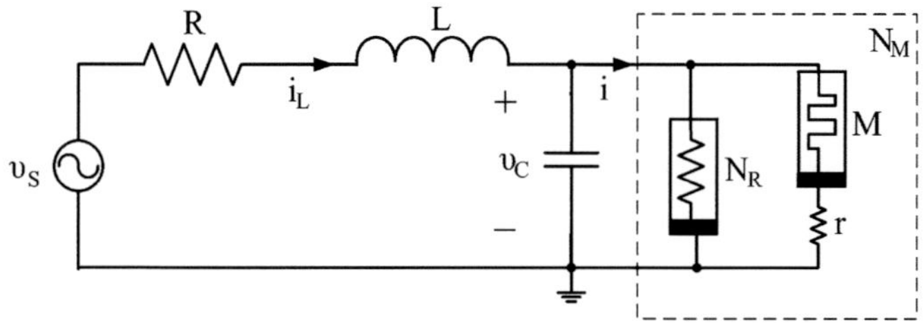

Fig. 13 The schematic of the Lacy circuit implemented with the proposed nonlinear element N_M

The state equations of the Lacy circuit in normalized form are given by:

$$\frac{dx}{d\tau} = = y - f(x)$$
$$\frac{dy}{d\tau} = -\gamma(x + y) + B \sin \Omega\tau \tag{29}$$

where, $f(x)$ describes the $i - v$ characteristic of N-type of the proposed nonlinear element (N_M). This nonlinear element is described by the following equation

$$f(x) = m_b x + 0.5(m_a - m_b)\{|x + 1| - |x - 1|\} \tag{30}$$

In systems equations (29), $x = v_C/E, y = Ri_L/E, \tau = t/RC, \Omega = \omega RC, \gamma = R^2C/L, B = \gamma V_0/E, m_a = RG_a, and\, m_b = RG_b$, are the normalized variables and the systems parameters respectively.

Also, the parameters values of the circuit, which are used in this work, are: $R = 0.7\,\text{k}\Omega, C = 62.8\,\text{nF}$, while V_0 is the amplitude of the sinusoidal voltage source, and $f = \omega/2\pi$, its frequency. The rest of the elements that have been used in the implementation of the proposed nonlinear element have the values which are given in the previous section.

Next, the simulation results of the circuits behavior by solving numerically the system (29) as well as by using the Multisim for emulating the circuit are presented.

5.1 Numerical Simulation Results

The dynamics behavior of the Lacy circuit with the proposed nonlinear element is investigated numerically by employing a fourth order Runge-Kutta algorithm. So, by solving the circuits system (29) the bifurcation diagrams of the signal u_C versus the amplitude of the sinusoidal voltage source V_0, for various values of sources frequencies f, are produced. In detail, the bifurcation diagrams of Figs. 14, 15, 16 and 17 are produced by increasing the sinusoidal voltage source V_0 with a small step, while systems initial conditions at each iteration have different values. This occurs

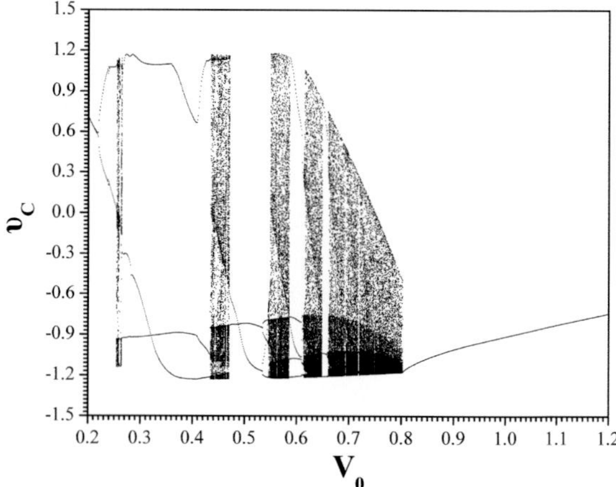

Fig. 14 Bifurcation diagram of u_C versus V_0, for $f = 3\,\text{kHz}$

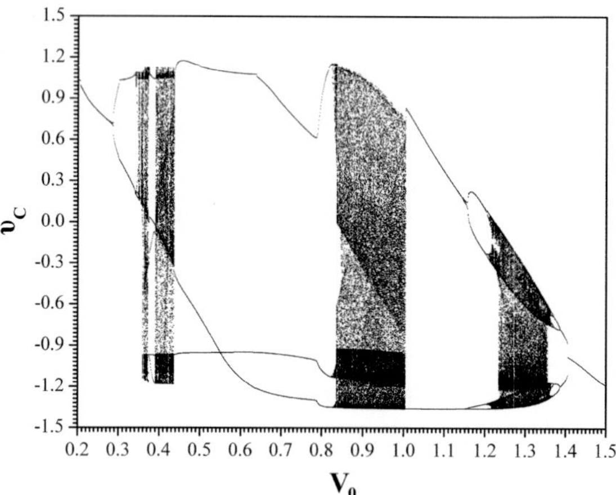

Fig. 15 Bifurcation diagram of u_C versus V_0, for $f = 4\,\text{kHz}$

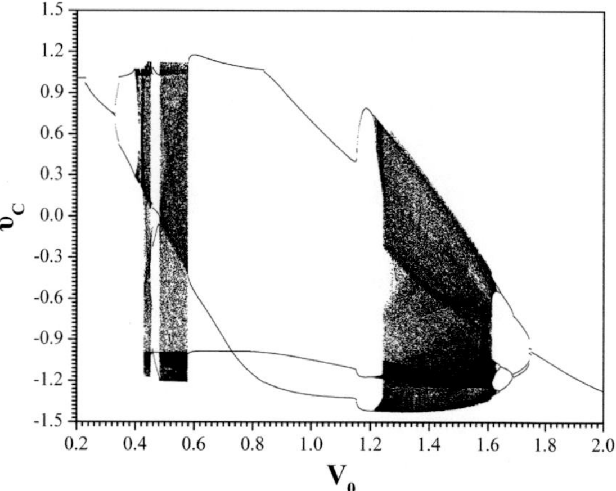

Fig. 16 Bifurcation diagram of u_C versus V_0, for $f = 4.5\,\text{kHz}$

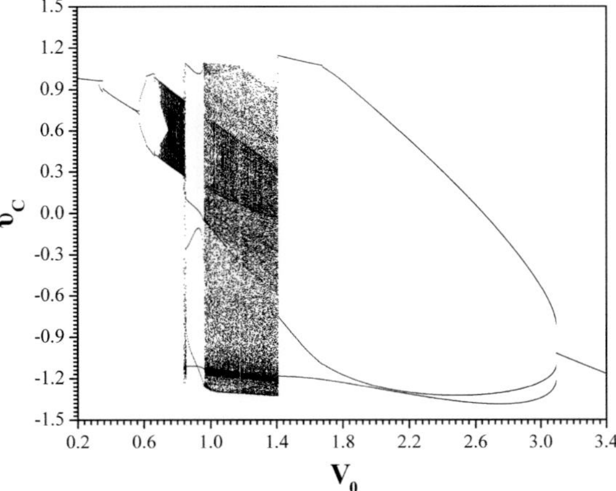

Fig. 17 Bifurcation diagram of u_C versus V_0, for $f = 6\,\text{kHz}$

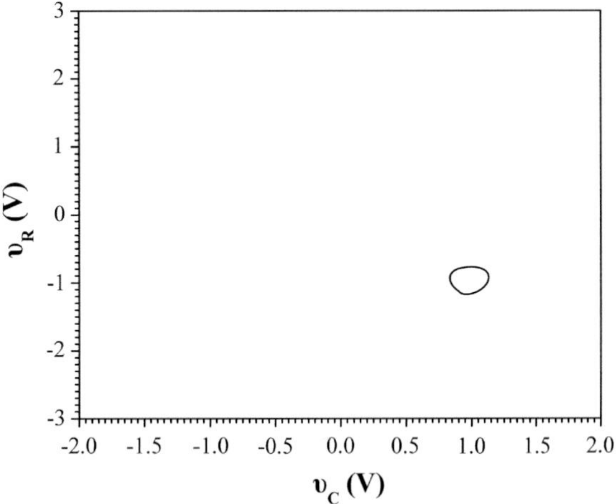

Fig. 18 Phase portrait of the signal υ_R versus the signal υ_C, for $f = 4.5\,\text{kHz}$ with $V_0 = 0.25\,\text{V}$

because the last values of the state variables in the previous iteration become the initial values for the next iteration. This type of a bifurcation diagram is more close to the experimental observation of the system's dynamic behavior as the amplitude of the sinusoidal voltage source V_0 can be increased by the function generator.

From the bifurcation diagrams of Figs. 14, 15, 16 and 17 the rich dynamics behavior of the proposed circuit with the proposed nonlinear resistor has been confirmed. Periodic and chaotic regions are alternated and interesting phenomena related with the nonlinear theory such as the intermittency, period doubling route to chaos and hysteresis in period-3 window are observed. In Figs. 18, 19, 20, 21, 22, 23, 24 and 25 the phase portraits in the $\upsilon_R - \upsilon_C$ plane with a stable frequency of the sinusoidal voltage source ($f = 4.5\,\text{kHz}$) for different values of its amplitude V_0 are displayed. These phase portraits show the route from a periodic attractor, for $V_0 = 0.25\,\text{V}$, around the one of systems equilibria (Fig. 18) to a widen periodic attractor for $V_0 = 2.00\,\text{V}$, through regions of other periodic or chaotic systems behavior.

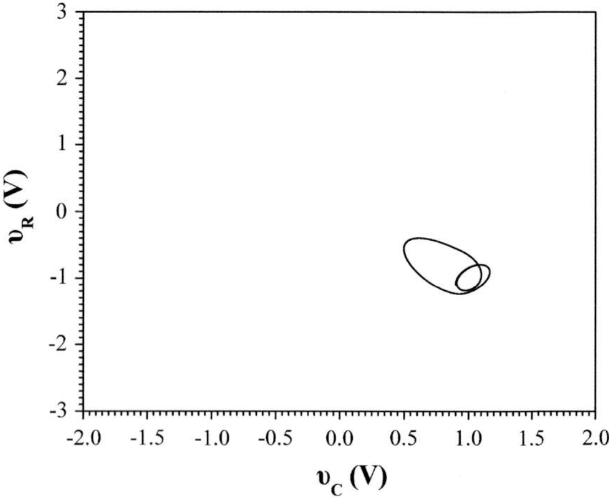

Fig. 19 Phase portrait of the signal υ_R versus the signal υ_C, for $f = 4.5\,\text{kHz}$ with $V_0 = 0.35\,\text{V}$

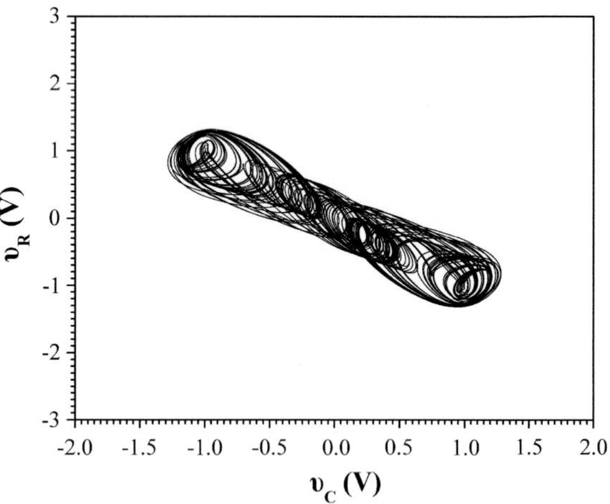

Fig. 20 Phase portrait of the signal υ_R versus the signal υ_C, for $f = 4.5\,\text{kHz}$ with $V_0 = 0.44\,\text{V}$

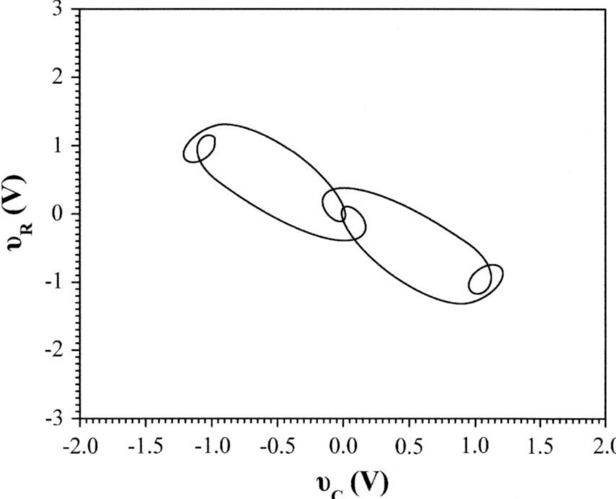

Fig. 21 Phase portrait of the signal υ_R versus the signal υ_C, for $f = 4.5\,\text{kHz}$ with $V_0 = 0.47\,\text{V}$

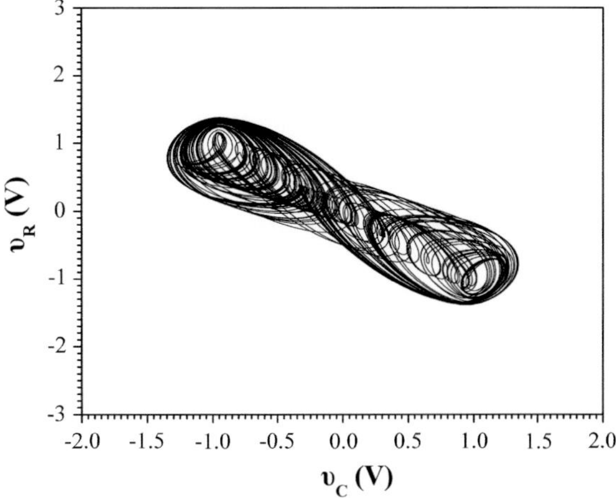

Fig. 22 Phase portrait of the signal υ_R versus the signal υ_C, for $f = 4.5\,\text{kHz}$ with $V_0 = 0.53\,\text{V}$

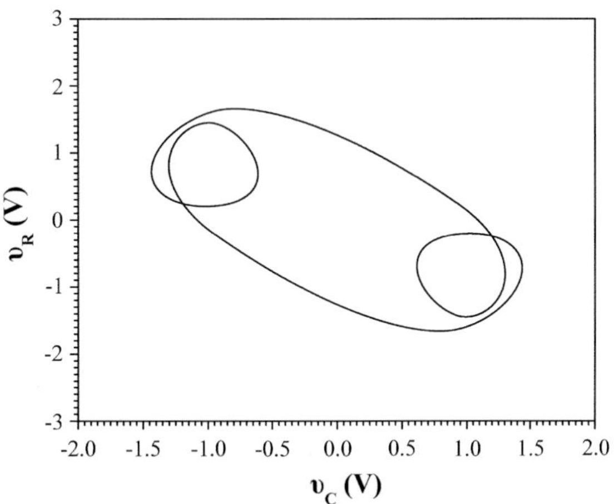

Fig. 23 Phase portrait of the signal υ_R versus the signal υ_C, for $f = 4.5\,\text{kHz}$ with $V_0 = 1.00\,\text{V}$

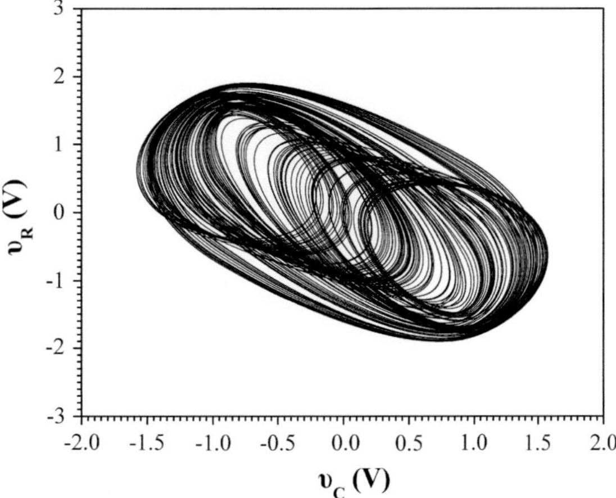

Fig. 24 Phase portrait of the signal υ_R versus the signal υ_C, for $f = 4.5\,\text{kHz}$ with $V_0 = 1.40\,\text{V}$

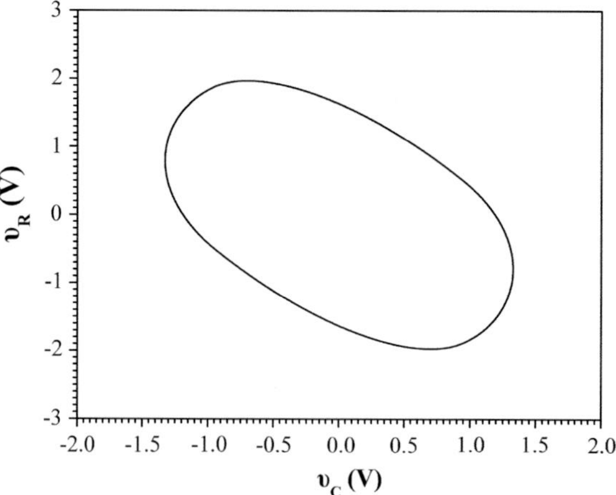

Fig. 25 Phase portrait of the signal v_R versus the signal v_C, for $f = 4.5\,\text{kHz}$ with $V_0 = 2.00\,\text{V}$

5.2 Simulation Results with Multisim

Next, the behavior of Lacy's circuit with the proposed nonlinear element with the memristor emulator is studied through its simulation process in Multisim. The same procedure as for Figs. 18, 19, 20, 21, 22, 23, 24 and 25 cases is followed. So, by using the same values of the circuit's elements ($R = 0.7\,\text{k}\Omega$, $C = 62.8\,\text{nF}$), for $f = 4.5\,\text{kHz}$, and for various values of the amplitude of the sinusoidal voltage source V_0, phase portraits of the signal v_R versus the signal v_C are produced Figs. 26, 27, 28,

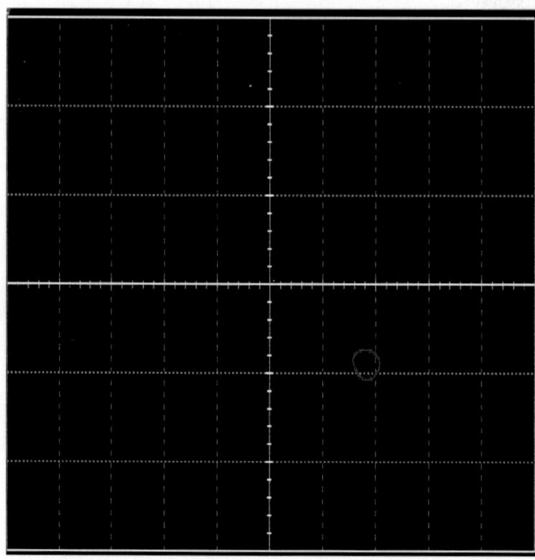

Fig. 26 Phase portrait of the signal v_R versus the signal v_C obtained with Multisim, for $f = 4.5\,\text{kHz}$ with $V_0 = 0.25\,\text{V}$ (X: 0.5 V/div, Y: 1 V/div)

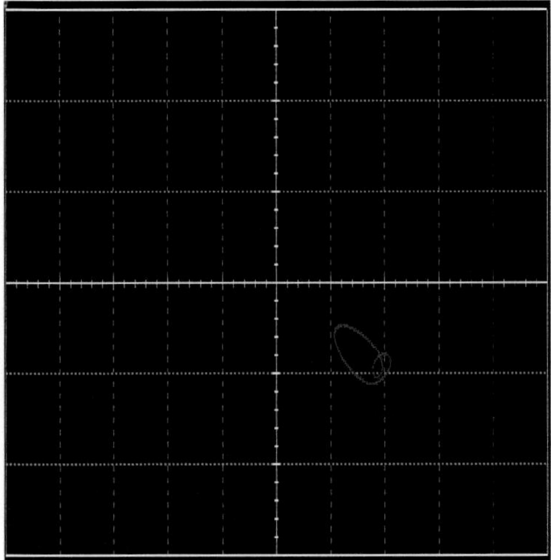

Fig. 27 Phase portrait of the signal υ_R versus the signal υ_C obtained with Multisim, for $f = 4.5\,\text{kHz}$ with $V_0 = 0.35\,\text{V}$ (X: 0.5 V/div, Y: 1 V/div)

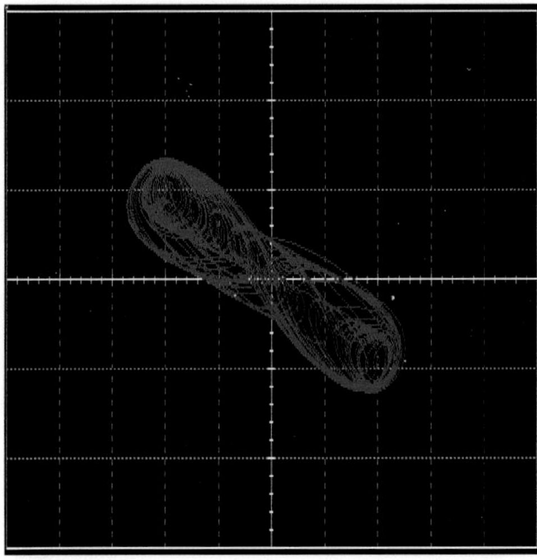

Fig. 28 Phase portrait of the signal υ_R versus the signal υ_C obtained with Multisim, for $f = 4.5\,\text{kHz}$ with $V_0 = 0.44\,\text{V}$ (X: 0.5 V/div, Y: 1 V/div)

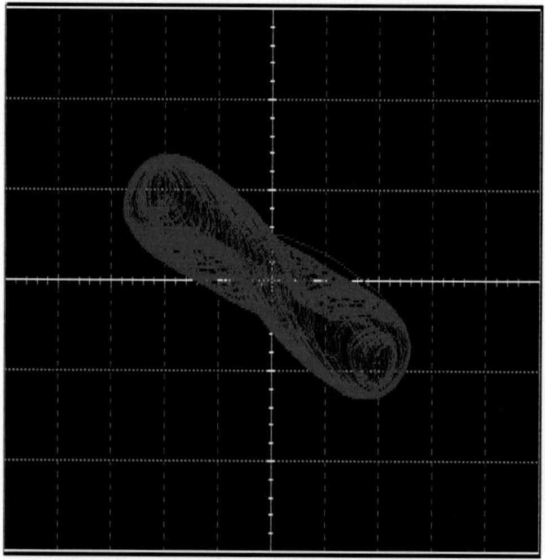

Fig. 29 Phase portrait of the signal v_R versus the signal v_C obtained with Multisim, for $f = 4.5\,\text{kHz}$ with $V_0 = 0.47\,\text{V}$ (X: 0.5 V/div, Y: 1 V/div)

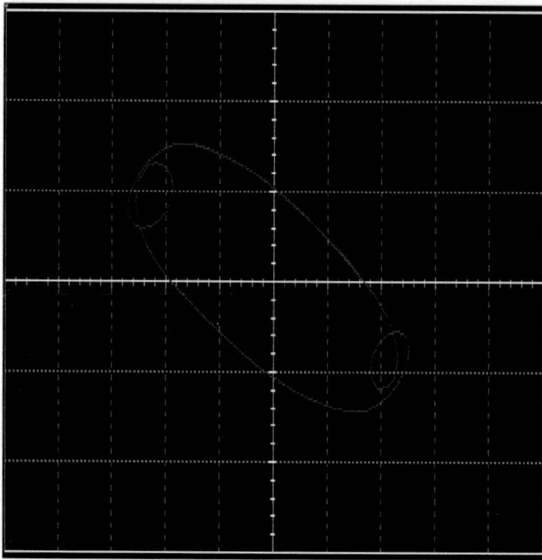

Fig. 30 Phase portrait of the signal v_R versus the signal v_C obtained with Multisim, for $f = 4.5\,\text{kHz}$ with $V_0 = 0.53\,\text{V}$ (X: 0.5 V/div, Y: 1 V/div)

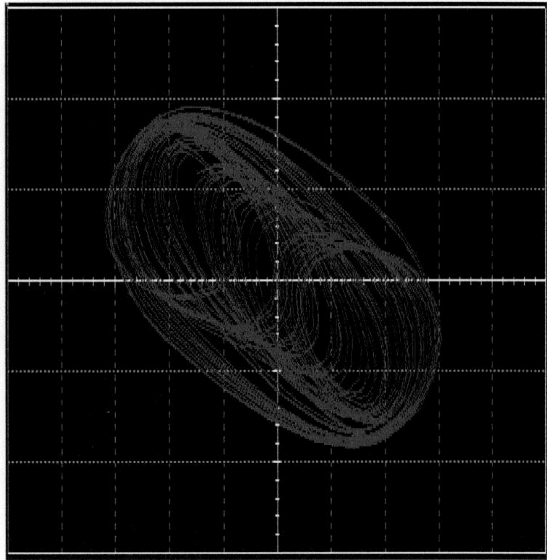

Fig. 31 Phase portrait of the signal v_R versus the signal v_C obtained with Multisim, for $f = 4.5\,\text{kHz}$ with $V_0 = 1.00\,\text{V}$ (X: 0.5 V/div, Y: 1 V/div)

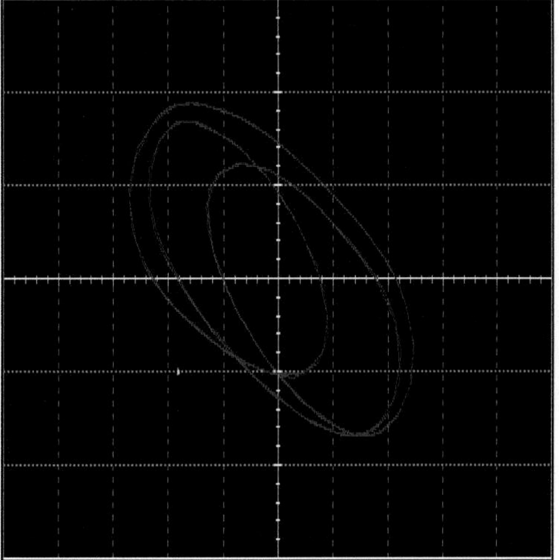

Fig. 32 Phase portrait of the signal v_R versus the signal v_C obtained with Multisim, for $f = 4.5\,\text{kHz}$ with $V_0 = 1.40\,\text{V}$ (X: 0.5 V/div, Y: 1 V/div)

29, 30, 31, 32 and 33. Also, the nonlinear element with the memristor emulator is designed with the procedure and with elements values as described in Sect. 4.

From the comparison of the phase portraits of Figs. 26, 27, 28, 29, 30, 31, 32 and 33 with that of Figs. 18, 19, 20, 21, 22, 23, 24 and 25, a good agreement between the circuit's behavior obtained form numerical simulation and Multisim can be concluded. The only discordance is the small window of periodic behavior for $V_0 = 0.47\,\text{V}$ (Fig. 21), which it can not be detected with Multisim. Therefore, Lacy's circuit, containing the proposed nonlinear element with the memristor emulator, has a rich dynamics behavior, which confirmed the feasibility of the proposed technique for realizing nonlinear element with the specific memristor emulator.

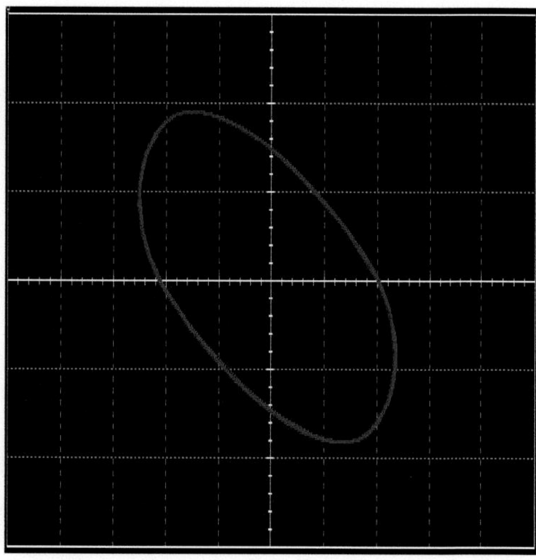

Fig. 33 Phase portrait of the signal v_R versus the signal v_C obtained with Multisim, for $f = 4.5\,\text{kHz}$ with $V_0 = 2.00\,\text{V}$ (X: 0.5 V/div, Y: 1 V/div)

6 Conclusion

In this work, a novel method for designing nonlinear resistors of type-N, consisting of a memristor emulator circuit in parallel with a negative resistor, was presented. The chosen memristor emulator has been built with second-generation current conveyors (CCII) and passive elements and its pinched hysteresis loop is holding up to 20 kHz. The fact, that the proposed memristor emulator circuit is simpler than other topologies as well as it works in higher frequencies that the other emulator circuits reported in literature, makes the proposed nonlinear element a suitable candidate for use in real-world applications.

As an example of use of the designed nonlinear resistor, a simple nonlinear, non-autonomous circuit, the well-known Lacy circuit, was chosen. The numerical as well as the simulation results, by using Multisim, reveals the richness of circuit's dynamics behavior confirming the usefulness of the specific design method. This is due to the increased complexity that the nonlinear resistor with memristor gives to the circuit. This fact was a consequence of the effect of both signals frequency and amplitude to the memristors pinched hysteresis loop.

The ease of the specific design method makes this proposal a very attractive option for the design of nonlinear resistors. So, as a future research steps, the use of the proposed nonlinear element in more nonlinear circuits and its design with other memristor emulators with similar features will be done.

References

1. Adhikari SP, Yang C, Kim H, Chua LO (2012) Memristor bridge synapse-based neural network and its learning. IEEE Trans Neural Netw Learn Syst 23(9):1426–1435
2. Arena P, Caponetto R, Fortuna L, Manganaro G (1997) Cellular neural networks to explore complexity. Soft Comput 1(3):120–136
3. Azar AT, Vaidyanathan S (2015) Chaos modeling and control systems design, vol 581. Springer, Germany
4. Azar AT, Vaidyanathan S (2015) Computational intelligence applications in modeling and control, vol 575. Springer, Germany
5. Azar AT, Vaidyanathan S (2015) Handbook of research on advanced intelligent control engineering and automation. IGI Global, USA
6. Banerjee S (2011) Chaos synchronization and cryptography for secure communications: applications for encryption. IGI Global, USA
7. Banerjee S, Mitra M, Rondoni L (2011) Applications of chaos and nonlinear dynamics in engineering, vol 1. Springer, Germany
8. Bao BC, Xu JP, Zhou GH, Ma ZH, Zou L (2011) Chaotic memristive circuit: equivalent circuit realization and dynamical analysis. Chin Phys B 20(12):120502
9. Bo-Cheng B, Jian-Ping X, Gua-Hua Z, Zheng-Hua M, Ling Z (2011) Chaotic memristive circuit: equivalent circuit realization and dynamical analysis. Chin Phys B 20(12):1–6
10. Bo-Cheng B, Fei F, Wei D, Sai-Hu P (2013) The voltage-current relationship and equivalent circuit implementation of parallel flux-controlled memristive circuits. Chin Phys B 22(6):1–6
11. Bonchev DD, Rouvray D (2005) Complexity in chemistry, biology, and ecology, mathematical and computational chemistry. Springer, USA
12. Buscarino Fortuna L, Frasca M, Gambuzza LV, Sciuto G (2012) Memristive chaotic circuits based on cellular nonlinear network. Int J Bifurc Chaos 22(3):1250070
13. Chen Y, Jung G, Ohlberg D, Li X, Stewart D, Jeppesen J, Nielsen K, Stoddart J, Williams R (2003) Nanoscale molecular-switch crossbar circuits. Nanotechology 14(4):462–468
14. Chua LO (1994) Chua's circuit 10 years later. Int J Circuit Theory Appl 22(4):279–305
15. Chua LO (1994) Chua's circuit: an overview ten year later. J Circuits Syst Comput 4(02): 117–159
16. Chua LO, Yang L (1988) Cellular neural networks: application. IEEE Trans Circuits Syst I: Fundam Theory Appl 35(10):1273–1290
17. Chua LO, Yang L (1988) Cellular neural networks: theory. IEEE Trans Circuits Syst I: Fundam Theory Appl 35(10):1257–1272
18. Chua LO (1971) Memristor-the missing circuit element. IEEE Trans Circuit Theory 18(5): 507–519
19. Chua LO, Hasler M, Moschytz GS, Neirynck J (1995) Autonomous cellular neural networks: a unified paradigm for pattern formation and active wave propagation. IEEE Trans Circuits Syst I: Fundam Theory Appl 42(10):559–577
20. Chua LO, Kang S (1976) Memristive devices and systems. Proc IEEE 64:209–223
21. Chua LO, Roska T (2002) Cellular neural networks and visual computing. Cambridge University Press, Cambridge
22. Cruz JM, Chua LO (1986) A CMOS IC nonlinear resistor for Chua's circuit. IEEE Trans Circuits Syst I 39(12):985–995
23. Driscoll T, Quinn J, Klein S, Kim HT, Kim BJ, Pershin YV, Ventra MD, Basov DN (2010) Memristive adaptive filters. Appl Phys Lett 97(9):093502
24. Elwakil A, Fouda M, Radwan A (2013) A simple model of double-loop hysteresis behavior in memristive elements. IEEE Trans Circuits Syst II: Exp Briefs 60:487–491
25. Eve RA, Horsfall S, Lee M (1997) Chaos, complexity, and sociology: myths, models, and theories. Sage Publications
26. Fortuna L, Arena P, Balya D, Zarandy A (2001) Cellular neural networks: a paradigm for nonlinear spatio-temporal processing. IEEE Circuits Syst Mag 1(4):6–21

27. Goras L, Chua LO, Leenaerts DM (1995) Turing patterns in CNNs—part I: once over lightly. IEEE Trans Circuits Syst I: Fundam Theory Appl 42(10):602–611
28. Hussein A, Fouda M (2013) A simple MOS realization of current controlled memristor emulator. In: Proceedings of international conference on microelectron, pp 1–4
29. Itoh M, Chua LO (2008) Memristor oscillators. Int J Bifurc Chaos 18(11):3183–3206
30. Itoh M, Chua LO (2011) Memristor Hamiltonian circuits. Int J Bifurc Chaos 21(09):2395–2425
31. Itoh M, Chua LO (2013) Duality of memristor circuits. Int J Bifurc Chaos 23(1):1–50
32. Iu HH, Fitch AL (2013) Development of memristor based circuits. World Scientific Publishing, Singapore
33. Kapitaniak T (2000) Chaos for engineers: theory, applications, and control. Springer, Germany
34. Kennedy MP (1992) Robust op amp realization of Chuas circuit. Frequenz 46(3–4):66–80
35. Kennedy M, Rovatti R, Setti G (2000) Chaotic electronics in telecommunications. CRC Press
36. Kim H, Sah MP, Yang C, Cho S, Chua LO (2012) Memristor emulator for memristor circuit applications. IEEE Trans Circuits Syst I: Regular Papers 59(10):2422–2431
37. Kyprianidis IM, Makri AT, Stouboulos IN, Volos CK (2013) Antimonotonicity in a FitzHugh Nagumo type circuit. In: Recent advances in finite differences and applied and computational mathematics, proceedings of 2nd international conference on applied and computational mathematics (ICACM'13), pp 151–156
38. Kyprianidis IM, Stouboulos IN (2003) Chaotic synchronization of three coupled oscillators with ring connection. Chaos Solit Fract 17(2):939–941
39. Larson LL, Liu J-M, Tsimring LS (2006) Digital communications using chaos and nonlinear dynamics. Institute for Nonlinear Science
40. Li Y, Huang X, Guo M (2013) The generation, analysis and circuit implementation of a new memristor based chaotic system. Math Prob Eng
41. Matsumoto T, Chua LO, Komuro M (1985) The double scroll. IEEE Trans Circuits Syst 32(8):798–818
42. Matsumoto T, Chua LO, Tokumasu K (1986) Double scroll via a two-transistor circuit. IEEE Transa Circuits Syst I 33(8):828–835
43. Medoza-Lopez CS-LJ, Carrasco-Aguilar MA, Muniz-Montero C (2014) A floating analog memristor emulator circuit. IEEE Trans Circuits Syst II: Exp Briefs 61(5):309–313
44. Meron E (2015) Nonlinear physics of ecosystems. CRC Press
45. Multu R, Karakulak E (2010) Emulator circuit of TiO_2 memristor with linear dopant drift made using analog multiplier. In: National conference on electrical, electronics and computer engineering (ELECO), pp 380–384
46. Muthuswamy B (2010) Implementing memristor based chaotic circuits. Int J Bifurc Chaos 20(5):1335–1350
47. Muthuswamy B, Chua LO (2010) Simplest chaotic circuit. Int J Bifurc Chaos 20(5):1567–1580
48. Muthuswamy B, Kokate PP (2009) Memristor-based chaotic circuits. IETE Tech Rev 26(6):417–429
49. Nakagawa M (1999) Chaos and fractals in engineering. World Scientific Publishing, Singapore
50. Perez-Munuzuri V, Perez-Villar V, Chua LO (1993) Autowaves for image processing on a two-dimensional CNN array of excitable nonlinear circuits: flat and wrinkled labyrinths. IEEE Trans Circuits Syst I: Fundam Theory Appl 40(3):174–181
51. Pershin YV, Ventra MD (2010) Experimental demonstration of associative memory with memristive neural networks. Neural Netw 23(7):881–886
52. Pershin YV, Ventra MD (2010) Practical approach to programmable analog circuits with memristors. IEEE Trans Circuits Syst I: Regular Papers 57(8):1857–1864
53. Pershin YV, Fontaine SL, Ventra MD (2000) Memristive model of amoeba learning. Phys Rev E 80(2):1–6
54. Pershin YV, Ventra MD (2008) Spin memristive systems: spin memory effects in semiconductor spintronics. Phys Rev B 78(11):1–4
55. Pickett M, Williams R (2012) Sub-100 fJ and sub-nanosecond thermally driven threshold switching in niobium oxide crosspoint nanodevices. Nanotechnology 23(21):215202

56. Sah MP, Yang C, Kim H, Chua LO (2012) A voltage mode memristorbridge synaptic circuit with memristor emulators. Sensors 12(3):3587–3604
57. Sapoff M, Oppenheim R (1963) Theory and application of self-heated thermistors. Proc IEEE 51:1292–1305
58. Shang Y, Fei W, Yu H (2012) Analysis and modeling of internal state variables for dynamic effects of nonvolatile memory devices. IEEE Trans Circuits Syst I: Regular Papers 59(9): 1906–1918
59. Shin S, Kim K, Kang SM (2011) Memristor applications for programmable analog ICs. IEEE Trans Nanotechnol 10(2):266–274
60. Shin S, Zheng L, Weickhardt G, Cho S, Kang S-M (2013) Compact circuit model and hardware emulation for floating memristor devices. IEEE Circuits Syst Mag 13(2):42–55
61. Shi Z, Ran L (2004) Tunnel diode based chuas diode. In: Proceedings of IEEE 6th CAS sympoisum on emerging technologies: mobile and wireless communications, pp 217–220
62. Sodhi A, Gandhi G (2010) Circuit mimicking TiO_2 memristor: a plug and play kit to understand the fourth passive element. Int J Bifurc Chaos 20(8):2537–2545
63. Strogatz SH (1994) Nonlinear dynamics and chaos: with applications to physics, biology, chemistry, and engineering. Perseus Books, Massachutetts
64. Strukov D, Snider G, Stewart G, Williams R (2008) The missing memristor found. Nature 453:80–83
65. Vaidyanathan S, Idowu B, Azar AT (2015) Backstepping controller design for the global chaos synchronization of Sprott's Jerk systems, vol 581. Springer-Verlag GmbH, Berlin
66. Vaidyanathan S, Azar AT (2015) Analysis and control of a 4-D novel hyperchaotic system, vol 581. Springer-Verlag GmbH, Berlin
67. Vaidyanathan S, Azar AT (2015) Analysis, control and synchronization of a nine-term 3-D novel chaotic system, vol 581. Springer-Verlag GmbH, Berlin
68. Vaidyanathan S, Azar AT (2015) Anti-synchronization of identical chaotic systems using sliding mode control and an application to Vaidyanathan-Madhavan chaotic systems, vol 576. Springer-Verlag GmbH, Berlin
69. Vaidyanathan S, Azar AT (2015) Anti-synchronization of identical chaotic systems using sliding mode control and an application to Vaidyanathan-Madhavan chaotic systems, vol 576. Springer-Verlag GmbH, Berlin
70. Volos CK, Kyprianidis IM, Stouboulos IN (2013) Experimental investigation on coverage performance of a chaotic autonomous mobile robot. Robot Auton Syst 61(12):13141322
71. Volos CK, Kyprianidis IM, Stouboulos IN (2013) Image encryption scheme based on continous-time chaotic systems. Nova Science Publishers
72. Volos CK, Kyprianidis I, Stouboulos I (2011) The memristor as an electric synapse-synchronization phenomena. In: Proceedings of international conference DSP2011, pp 1–6
73. Volos CK, Kyprianidis IM, Stouboulos IN (2012) Chaotic path planning generator for autonomous mobile robots. Robot Auton Syst 60(4):651656
74. Wang X, Chen Y, Gu Y, Li H (2010) Spintronic memristor temperature sensor. IEEE Electr Device Lett 31(1):20–22
75. Wang L, Zhang C, Chen L, Lai J, Tong J (2012) A novel memristor-based rSRAM structure for multiple-bit upsets immunity. IEICE Electr Exp 9(9):861–867
76. Yang JJ, Strukov DB, Stewart DR (2013) Memristive devices for computing. Nat Nanotechnol 8(1):13–24
77. Yang C, Choi H, Park S, Kim MPSH, Chua LO (2015) A memristor emulator as a replacement of a real memristor. Semicond Sci Technol 30(1):015007
78. Yesil A, Babacan Y, Kacar F (2014) A new DDCC based memristor emulator circuit and its applications. J Microelectr 45(3):282–287
79. Yu D, Iu HH-C, Fitch AL, Liang Y (2014) Memristor emulator for memristor circuit applications. Float Memristor Emul Based Relax Oscil 61(10):2888–2896
80. Zhang W-B (2005) Differential equations, bifurcations, and chaos in economics, vol 68. World Scientific Publishing, Singapore

81. Zhang W-B (2006) Discrete dynamical systems, bifurcations and chaos in economics, mathematics in science and engineering, vol 204. Elsevier, Netherlands
82. Zhong GQ (1986) Implementation of Chuas circuit with a cubic nonlinearity. IEEE Trans Circuits Syst I 41(12):939–941
83. Zhu Q, Azar AT (2015) Complex system modeling and control through intelligent soft computations, vol 319. Springer, Germany

Dynamics, Synchronization and SPICE Implementation of a Memristive System with Hidden Hyperchaotic Attractor

Viet-Thanh Pham, Sundarapandian Vaidyanathan, Christos K. Volos, Thang Manh Hoang and Vu Van Yem

Abstract The realization of memristor in nanoscale size has received considerate attention recently because memristor can be applied in different potential areas such as spiking neural network, high-speed computing, synapses of biological systems, flexible circuits, nonvolatile memory, artificial intelligence, modeling of complex systems or low power devices and sensing. Interestingly, memristor has been used as a nonlinear element to generate chaos in memristive system. In this chapter, a new memristive system is proposed. The fundamental dynamics properties of such memristive system are discovered through equilibria, Lyapunov exponents, and Kaplan–York dimension. Especially, hidden attractor and hyperchaos can be observed in this new system. Moreover, synchronization for such system is studied and simulation results are presented showing the accuracy of the introduced synchronization scheme. An electronic circuit modelling such hyperchaotic memristive system is also reported to verify its feasibility.

Keywords Chaos · Hyperchaos · Lyapunov exponents · Hidden attractor · No-equilibrium · Memristor · Synchronization · Circuit · SPICE

V.-T. Pham (✉) · T.M. Hoang · V. Van Yem
School of Electronics and Telecommunications,
Hanoi University of Science and Technology, Hanoi, Vietnam
e-mail: pvt3010@gmail.com

T.M. Hoang
e-mail: thang.hoangmanh@hust.edu.vn

V. Van Yem
e-mail: yem.vuvan@hust.edu.vn

S. Vaidyanathan
Research and Development Centre, Vel Tech University,
Tamil Nadu, India
e-mail: sundar@veltechuniv.edu.in

C.K. Volos
Physics Department, Aristotle University of Thessaloniki,
Thessaloniki, Greece
e-mail: volos@physics.auth.gr

A.T. Azar and S. Vaidyanathan (eds.), *Advances in Chaos Theory and Intelligent Control*, Studies in Fuzziness and Soft Computing 337,
DOI 10.1007/978-3-319-30340-6_2

1 Introduction

After the discovery of Lorenz's model for atmospheric convection [1], there has been significant interest in chaotic systems [2–9]. In the past few decades, different chaotic systems have been reported such as Rössler system [10], Arneodo system [11], Chen system [6], Lü system [12], Vaidyanathan system [13], time-delay systems [14] and so on [15, 16]. Chaotic behaviors are useful and have been applied in many fields, for example a double-scroll chaotic attractor has been used to generate true random bits [17], chaotic path planning has been generated for autonomous mobile robots [18], fingerprint images encryption scheme based on chaotic attractors has been implemented, or applications of time delay systems in secure communication have been proposed [19] due to their complex dynamics.

In addition, hyperchaotic system was introduced and studied [20]. Hyperchaotic system is characterized by more than one positive Lyapunov exponent and, thus, presents a higher level of complexity with respect to chaotic system [21]. As a result, hyperchaos is better than conventional chaos in a variety of areas, for instance, hyperchaos increases the security of chaotic-based communication systems significantly [22, 23]. Moreover hyperchaos has used in diverse applications such as cryptosystems [24], neural networks [25], secure communications [22, 23], or laser design [26]. Especially, the intrinsic nonlinear characteristic of memristor has been expointed in designing hyperchaotic oscillators. Some recent researches show that memristor is a potential candidate for generating hyperchaos [27, 28].

In this chapter, our work introduces a memristive system which can exhibit hyperchaotic attractors. Moreover such memristive system does not have equilibrium points. This chapter is organized as follows. Section 2 summarized related works. Section 3 gives a brief representation to the memristive system. Dynamics and properties of such memristive system is introduced in Sect. 4 while the adaptive synchronization scheme is studied in Sect. 5. Section 6 presents circuital implementation of memristive system using SPICE. Finally, conclusions are drawn in Sect. 7.

2 Related Work

Motivated by complex dynamical behaviors of hyperchaotic systems and special features of memristor, some memristor-based hyerchaotic systems have been introduced, recently. Hyperchaos was generated by combining a memristor with cubic nonlinear characteristics and a modified canonical Chua's circuit [28]. This memristor-based modified canonical Chua's circuit is a five-dimensional hyperchaotic oscillator. By extending the HP memristor-based canonical Chua's oscillator, a six-dimensional hyperchaotic oscillator was designed [29]. Authors used a configuration based on two HP memristors in antiparallel [27]. Four-dimensional hyperchaotic memristive systems were discovered by Li et al. [30, 31]. A 4D memristive system with a line of equilibrium was presented in [30] while another memristive system with an uncountable infinite number of stable and unstable equilibria was reported in [31]. A memristor-based hyperchaotic system without equilibrium was introduced in [32]. These memristive systems belong to a new category of chaotic systems with hidden attractors [33, 34].

The terminology "hidden attractor" has been introduced recently although the fact that the problem of analyzing hidden oscillations and to the finding of hidden oscillations in automatic control systems were studied a long time ago. According to a new classification of chaotic dynamics proposed by Leonov and Kuznetsov [33–35], there are two types of attractors: self-excited attractors and hidden attractors. A self-excited attractor has a basin of attraction that is excited from unstable equilibria. In contrast, hidden attractor cannot be found by using a numerical method in which a trajectory started from a point on the unstable manifold in the neighbourhood of an unstable equilibrium [35]. The discovery of dynamical systems with hidden attractors is a great challenge due to their appearance in many research fields such as in mechanics, secure communication and electronics [34, 36–39]. For example, hidden attractor in smooth Chua's system was reported in [40]. Hidden oscillations in mathematical model of drilling system [41] and hidden oscillations in nonlinear control systems [42] were witnessed. Various examples of hidden attractors were summarized in [43–46]. Hidden attractors were observed in a 4-D Rikitake dynamo system [47] or 5-D hyperchaotic Rikitake dynamo system [48]. Hidden attractors in a chaotic system with an exponential nonlinear term were introduced in [49]. In addition, algorithms for searching for hidden oscillations were presented in [50, 51].

Motivated by complex dynamical behaviors of chaotic systems, noticeable characteristics of memristor, and unknown features of hidden attractors, studying memristive hyperchaotic systems with hidden attractors is still an attractive research direction [30, 31].

3 Model of the Memristive System

A flux-controlled memristor is considered in this work. Its memductance is a second degree polynomial function:

$$W(\varphi) = \alpha + 3\beta\varphi^2, \tag{1}$$

with $\alpha = 0.4$ and $\beta = 0.001$. Memductance (1) is similar to known memductance [28, 30, 52, 53]. Using this memristor, a four-dimensinal memristive system is proposed as

$$\begin{cases} \dot{x} = 36y - 36x \\ \dot{y} = -2xz + 20y - axW(\varphi) - b \\ \dot{z} = 2xy - 3z \\ \dot{\varphi} = 2x, \end{cases} \tag{2}$$

where a, b are parameters, and $W(\varphi)$ is the memductance as introduced in (1). It is noting that the memristive system (2) has an uncountable number of equilibrium points when $b = 0$. Moreover, system (2) generate hyperchaos for different values of the parameter a. For example, hyperchaotic attractors is obtained when $a = 30, b = 0$ and the chosen initial conditions are $(x(0), y(0), z(0), \varphi(0)) = (0.5, 0, 0.5, 0)$. In this case, memristive system (2) is similar to the reported one in [31], therefore it will be not discussed in the next sections.

4 Dynamics and Properties of the Memristive System

The memristive system (2) is investigated when $b \neq 0$. It is easy to obtain the equilibrium points for system (2) by solving $\dot{x} = 0$, $\dot{y} = 0$, $\dot{z} = 0$, and $\dot{\varphi} = 0$ that is

$$36y - 36x = 0, \tag{3}$$

$$-2xz + 20y - axW(\varphi) - b = 0, \tag{4}$$

$$2xy - 3z = 0, \tag{5}$$

$$2x = 0, \tag{6}$$

From (3), (5) and (6), we have $x = y = z = 0$. As a results, Eq. (4) reduces to $b = 0$, which is an contradiction. Hence there are not equilibrium points in memristive system (2).

In this work, the parameters are selected as $a = 30$, $b = 0.001$ and the initial conditions are

$$(x(0), y(0), z(0), \varphi(0)) = (0.5, 0, 0.5, 0). \tag{7}$$

Lyapunov exponents, which measure the exponential rates of the divergence and convergence of nearby trajectories in the phase space of the chaotic system [8, 54], are calculated using the well-known algorithm in [55]. The Lyapunov exponents of the system (2) are

$$\lambda_1 = 0.2590, \lambda_2 = 0.0658, \lambda_3 = 0, \lambda_4 = -19.3246. \tag{8}$$

It is noting that the sum of the Lyapunov exponents is negative, and so the novel memristive hyperchaotic system is dissipative. There are two positive Lyapunov exponents, one zero and one negative Lyapunov exponents. Thus, the memtistive system (2) is a four-dimension hyperchaotic system according to [20]. It is worth noting that this memristive system can be classified as a hyperchaotic system with hidden strange attractor because its basin of attractor does not contain neighbourhoods of equilibria [33, 34]. The 3-D and 2-D projections of the hyperchaotic attractors without equilibrium in this case are illustrated in Figs. 1, 2, 3, 4, 5 and 6.

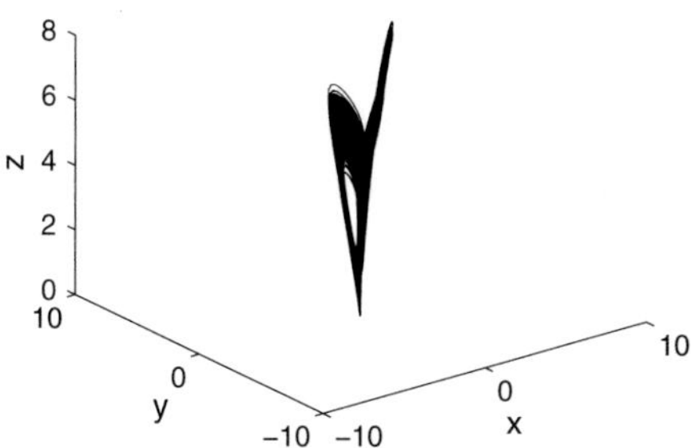

Fig. 1 3-D projection of the hyperchaotic memristive system without equilibrium (2) in the (x, y, z)-space

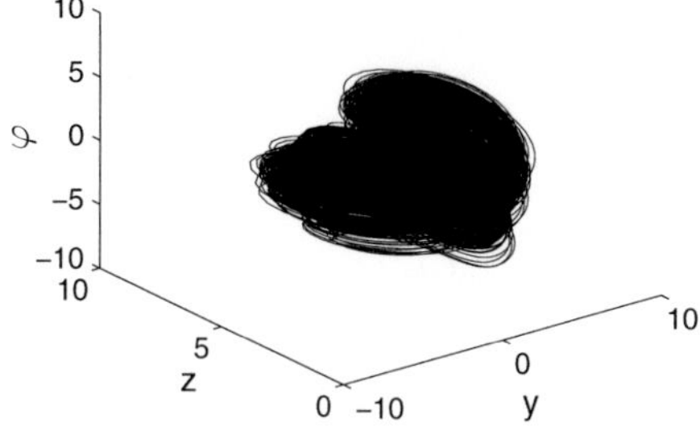

Fig. 2 3-D projection of the hyperchaotic memristive system without equilibrium (2) in the (y, z, φ)-space

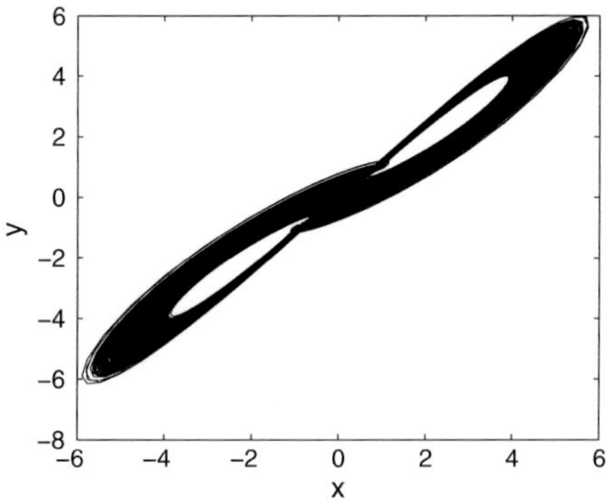

Fig. 3 2-D projection of the hyperchaotic memristive system without equilibrium (2) in the (x, y)-plane

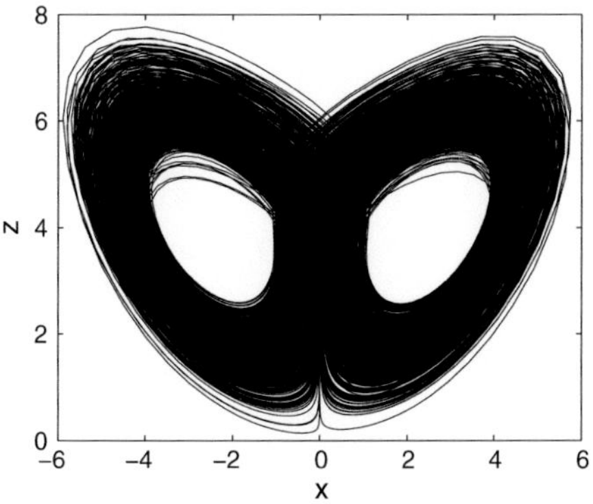

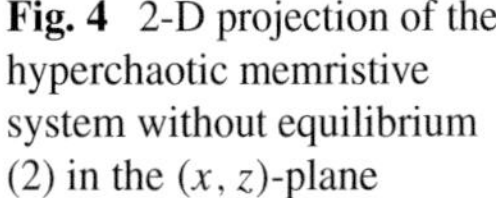

Fig. 4 2-D projection of the hyperchaotic memristive system without equilibrium (2) in the (x, z)-plane

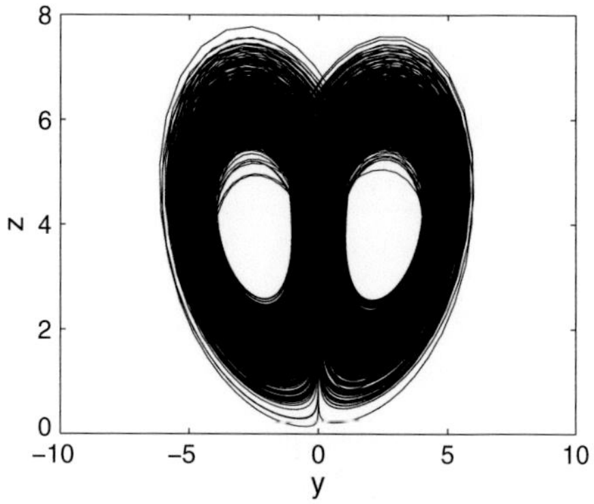

Fig. 5 2-D projection of the hyperchaotic memristive system without equilibrium (2) in the (y, z)-plane

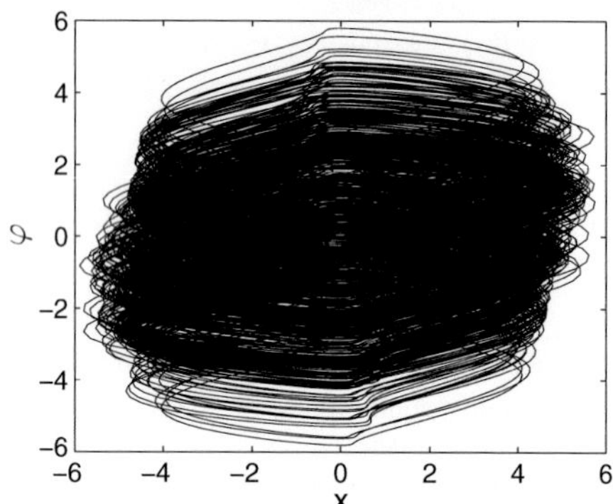

Fig. 6 2-D projection of the hyperchaotic memristive system without equilibrium (2) in the (x, φ)-plane

It is known that the Kaplan–Yorke fractional dimension, which presents the complexity of attractor [56], is defined by

$$D_{KY} = j + \frac{1}{\left|\lambda_{j+1}\right|} \sum_{i=1}^{j} \lambda_i, \tag{9}$$

where j is the largest integer satisfying $\sum_{i=1}^{j} \lambda_i \geq 0$ and $\sum_{i=1}^{j+1} \lambda_i < 0$. The calculated fractional dimension of memristive system (2) when $a = 30$, $b = 0.001$ is

$$D_{KY} = 3 + \frac{\lambda_1 + \lambda_2 + \lambda_3}{|\lambda_4|} = 3.0168. \tag{10}$$

Equation (10) indicates a strange attractor.

5 Adaptive Anti-synchronization of the Memristive System

The possibility of synchronization of two coupled chaotic systems [57–60] is one of the most vital characteristics relating to chaotic systems and their applications. Different research activities using synchronization of nonlinear systems have been investigated in literature [38, 61–69]. For instant, synchronized states in a ring of mutually coupled self-sustained nonlinear electrical oscillators were considered in [70], ragged synchronizability of coupled oscillators was observed in [71], various synchronization phenomena in bidirectionally coupled double-scroll circuits were reported in [72], or observer for synchronization of chaotic systems with application to secure data transmission was studied in [73]. Many synchronization schemes have been introduced such as lag synchronization [74], frequency synchronization [75], projective-anticipating synchronization [76], anti-synchronization [77], adaptive synchronization [78], or hybrid chaos synchronization [63], etc. Here we consider the adaptive synchronization of identical memristive hyperchaotic systems with two unknown parameters.

In this section, we consider the memristive sytem (2) as the master system as follows

$$\begin{cases} \dot{x}_1 = 36y_1 - 36x_1 \\ \dot{y}_1 = -2x_1z_1 + 20y_1 - ax_1W(\varphi_1) - b \\ \dot{z}_1 = 2x_1y_1 - 3z_1 \\ \dot{\varphi}_1 = 2x_1. \end{cases} \tag{11}$$

The states of the master system (11) are x_1, y_1, z_1, φ_1, and $W(\varphi_1)$ is the memductance as given in (1). The slave system is considered as the controlled memristive system and its dynamics is given as

$$\begin{cases} \dot{x}_2 = 36y_2 - 36x_2 + u_x \\ \dot{y}_2 = -2x_2z_2 + 20y_2 - ax_2W(\varphi_2) - b + u_y \\ \dot{z}_2 = 2x_2y_2 - 3z_2 + u_z \\ \dot{\varphi}_2 = 2x_2 + u_\varphi, \end{cases} \tag{12}$$

where x_2, y_2, z_2, φ_2, are the states of the slave system while u_x, u_y, u_z, u_φ are the adaptive controls. These controls will be constructed for the anti-synchronization of the master and slave systems. In order to estimate unknown parameters a and b, $A(t)$ and $B(t)$ are used.

The anti-synchronization error between memristive systems (11) and (12) is described by the following relation

$$\begin{cases} e_x = x_1 + x_2 \\ e_y = y_1 + y_2 \\ e_z = z_1 + z_2 \\ e_\varphi = \varphi_1 + \varphi_2. \end{cases} \tag{13}$$

Therefore, the anti-synchronization error dynamics is determined by

$$\begin{cases} \dot{e}_x = 36e_x - 36e_y + u_x \\ \dot{e}_y = -2\left(x_1 z_1 + x_2 z_2\right) + 20e_y - a\left(x_1 W\left(\varphi_1\right) + x_2 W\left(\varphi_2\right)\right) - 2b + u_y \\ \dot{e}_z = 2\left(x_1 y_1 + x_2 y_2\right) - 3e_z + u_z \\ \dot{e}_\varphi = 2e_x + u_\varphi. \end{cases} \tag{14}$$

Our goal is to find the appropriate controllers u_x, u_y, u_z, u_φ to stabilize the system (14). Thus, we propose the following controllers for system (14):

$$\begin{cases} u_x = -36e_x + 36e_y - k_x e_x \\ u_y = 2\left(x_1 z_1 + x_2 z_2\right) - 20e_y + A\left(t\right)\left(x_1 W\left(\varphi_1\right) + x_2 W\left(\varphi_2\right)\right) + 2B\left(t\right) - k_y e_y \\ u_z = -2\left(x_1 y_1 + x_2 y_2\right) + 3e_z - k_z e_z \\ u_\varphi = -2e_x - k_\varphi e_\varphi, \end{cases} \tag{15}$$

where k_x, k_y, k_z, k_φ are positive gain constants for each controllers and $A(t)$, $B(t)$ are the estimate values for unknown system parameters. The update laws for the unknown parameters are defined as

$$\begin{cases} \dot{A} = -e_y\left(x_1 W\left(\varphi_1\right) + x_2 W\left(\varphi_2\right)\right) \\ \dot{B} = -2e_y. \end{cases} \tag{16}$$

Next, the main result of this section will be presented and proved.

Theorem 5.1 *If the adaptive controller (15) and the updating laws of parameter (16) are chosen, the anti-sychronization between the master system (11) and the slave system (12) is achieved.*

Proof Here $e_a(t)$ and $e_b(t)$ are the parameter estimation errors given as

$$\begin{cases} e_a\left(t\right) = a - A\left(t\right) \\ e_b\left(t\right) = b - B\left(t\right). \end{cases} \tag{17}$$

Differentiating (17) with respect to t, we obtain

$$\begin{cases} \dot{e}_a\left(t\right) = -\dot{A}\left(t\right) \\ \dot{e}_b\left(t\right) = -\dot{B}\left(t\right). \end{cases} \tag{18}$$

Substituting adaptive control law (15) into (14), the closed-loop error dynamics is determined as

$$\begin{cases} \dot{e}_x = -k_x e_x \\ \dot{e}_y = -\left(a - A\left(t\right)\right)\left(x_1 W\left(\varphi_1\right) + x_2 W\left(\varphi_2\right)\right) - 2\left(b - B\left(t\right)\right) - k_y e_y \\ \dot{e}_z = -k_z e_z \\ \dot{e}_\varphi = -k_\varphi e_\varphi \end{cases} \tag{19}$$

Then substituting (17) into (19), we have

$$\begin{cases} \dot{e}_x = -k_x e_x \\ \dot{e}_y = -e_a \left(x_1 W\left(\varphi_1\right) + x_2 W\left(\varphi_2\right)\right) - 2e_b - k_y e_y \\ \dot{e}_z = -k_z e_z \\ \dot{e}_\varphi = -k_\varphi e_\varphi \end{cases} \tag{20}$$

We consider the Lyapunov function as

$$\begin{aligned} V(t) &= V\left(e_x, e_y, e_z, e_\varphi, e_a, e_b\right) \\ &= \tfrac{1}{2}\left(e_x^2 + e_y^2 + e_z^2 + e_\varphi^2 + e_a^2 + e_b^2\right). \end{aligned} \tag{21}$$

The Lyapunov function is clearly definite positive.

Taking time derivative of (21) along the trajectories of (13) and (17) we get

$$\dot{V}(t) = e_x\dot{e}_x + e_y\dot{e}_y + e_z\dot{e}_z + e_\varphi\dot{e}_\varphi + e_a\dot{e}_a + e_b\dot{e}_b. \tag{22}$$

From (18), (20), and (22) we have

$$\begin{aligned} \dot{V}(t) = &-k_x e_x^2 - e_a\left[e_y\left(x_1 W\left(\varphi_1\right) + x_2 W\left(\varphi_2\right)\right) + \dot{A}\right] \\ &-e_b\left(2e_y + \dot{B}\right) - k_y e_y^2 - k_z e_z^2 - k_\varphi e_\varphi^2. \end{aligned} \tag{23}$$

Then by applying the parameter update law (16), Eq. (23) become

$$\dot{V}(t) = -k_x e_x^2 - k_y e_y^2 - k_z e_z^2 - k_\varphi e_\varphi^2. \tag{24}$$

Obviously, derivative of the Lyapunov function is definite negative. According to the Lyapunov stability [79, 80] we obtain $e_x(t) \to 0$, $e_y(t) \to 0$, $e_z(t) \to 0$, $e_\varphi(t) \to 0$, $e_a(t) \to 0$, $e_b(t) \to 0$ exponentially when $t \to 0$ that is, anti-synchronization between master and slave system. This completes the proof. □

We illustrate the proposed anti-synchronization scheme with a numerical example. In the numerical simulations, the fourth-order Runge–Kutta method is used to solve the systems. The parameters of the memristive hyperchaotic systems are selected as $a = 30$, $b = 0.001$ and the positive gain constant as $k = 4$. The initial conditions of the master system (11) and the slave system (12) have been chosen as $x_1(0) = 0.5$, $y_1(0) = 0$, $z_1(0) = 0.5$, $\varphi_1(0) = 0$ and $x_2(0) = -0.9$, $y_2(0) = -0.4$, $z_2(0) = 0.8$, $\varphi_2(0) = 0.5$, respectively. We assumed that the initial values of the parameter estimates are $A(0) = 29$ and $B(0) = 0.5$.

It is easy to see that when adaptive control law (15) and the update law for the parameter estimates (16) are applied, the anti-synchronization of the master (11) and slave system (12) occurred as illustrated in Figs. 7, 8, 9 and 10. It is noting that time series of master states are denoted as blue solid lines while corresponding slave states are plotted as red dash-dot lines in such figures. In addition, the time-history of the anti-synchronization errors e_x, e_y, e_z, and e_φ is presented in Fig. 11.

The anti-synchronization errors converge to the zero, which implies that the chaos anti-synchronization between the memristive systems is realized.

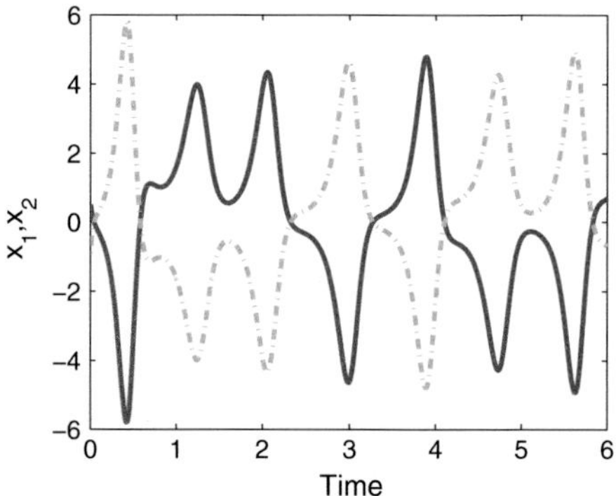

Fig. 7 Anti-synchronization of the states $x_1(t)$ and $x_2(t)$

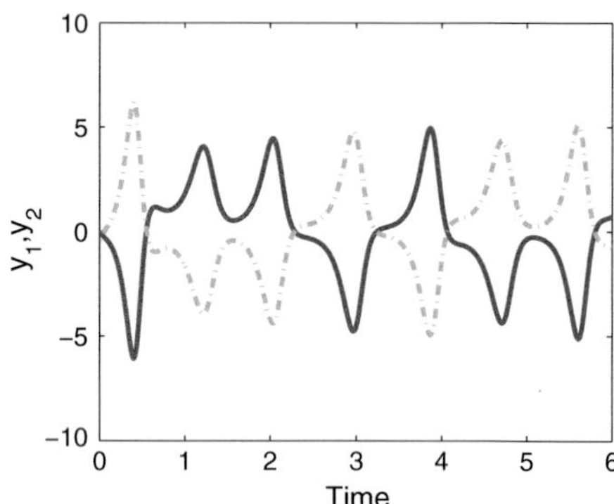

Fig. 8 Anti-synchronization of the states $y_1(t)$ and $y_2(t)$

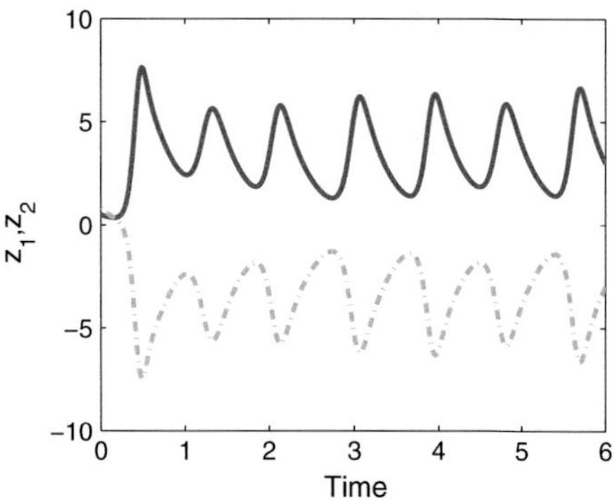

Fig. 9 Anti-synchronization of the states $z_1(t)$ and $z_2(t)$

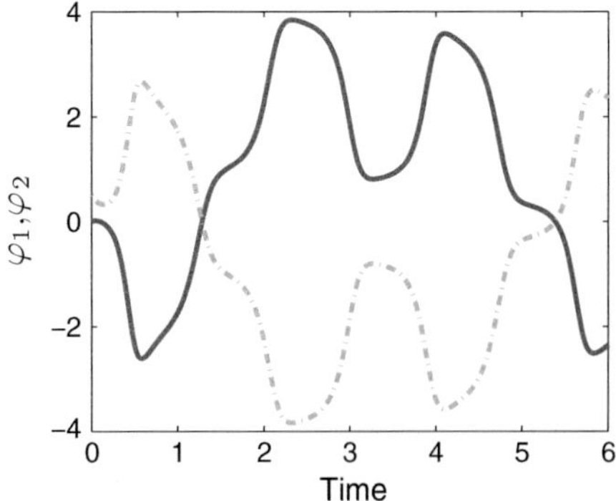

Fig. 10 Anti-synchronization of the states $\varphi_1(t)$ and $\varphi_2(t)$

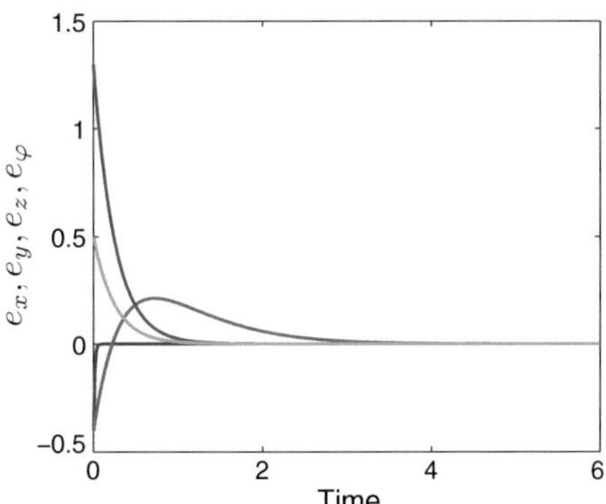

Fig. 11 Time series of the anti-synchronization errors e_x, e_y, e_z, and e_φ

6 Circuit Realization of the Memristive System

Circuital design of chaotic/hyperchaotic systems plays an important role on the field of nonlinear science due to its applications in secure communication, signal processing, random bit generator, or path planning for autonomous mobile robot etc. [17, 18, 22, 62, 81, 82]. In addition, circuital implementation of chaotic/hyperchaotic systems is also provide an effective approach for investigating dynamics of such theoretical models [61, 83]. For example, chaotic attractors can be observed on the oscilloscope easily or experimental bifurcation diagrams can be obtained by varying the values of variable resistors [84, 85].

Therefore, in this work, an electronic circuit is introduced to implement memristive system (2). By using the operational amplifiers approach [85], the circuit is proposed as shown in Fig. 12. Here the variables x, y, z, φ of memristive system (2) are the voltages across the capacitor C_1, C_2, C_3, and C_4, respectively. It is noted

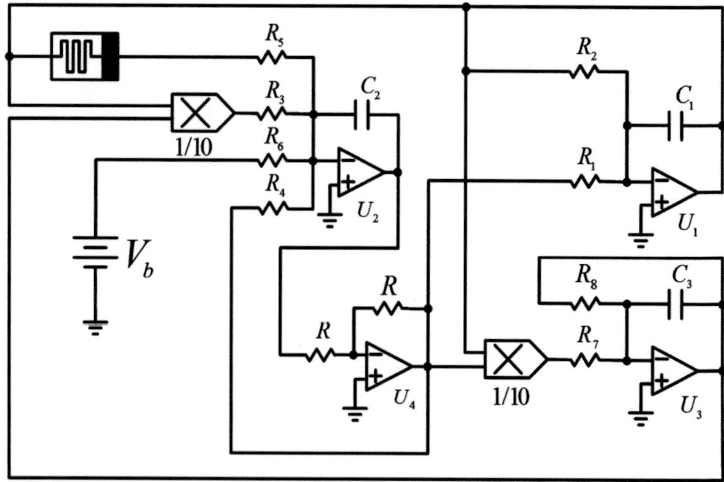

Fig. 12 Schematic of the circuit which modelling hyperchaotic system (2) with the presence of the memristor

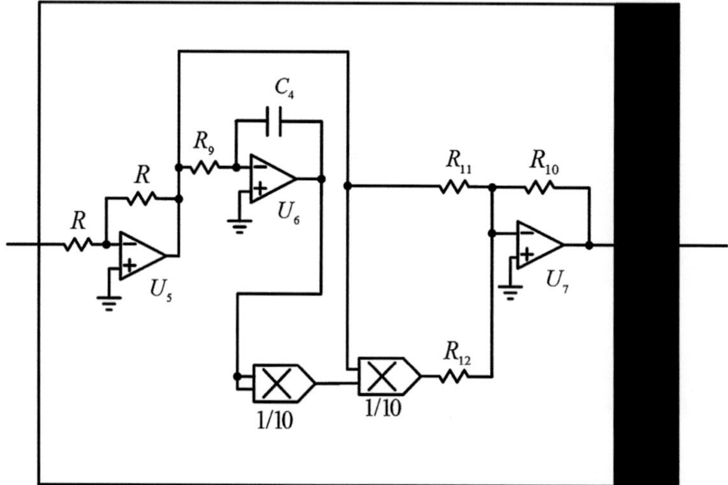

Fig. 13 Schematic of the circuit emulating the memristor

that the detailed schematic of the memristor in Fig. 12 is presented in Fig. 13. This sub-circuit of memristor emulates the memristive device only due to the fact that there are not any commercial off-the-shelf memristive device in the market at the moment [86]. The corresponding circuital equations of circuit can be described as

$$\begin{cases} \frac{dv_{C_1}}{dt} = \frac{1}{R_1C_1}v_{C_2} - \frac{1}{R_2C_1}v_{C_1} \\ \frac{dv_{C_2}}{dt} = -\frac{1}{10R_3C_2}v_{C_1}v_{C_3} + \frac{1}{R_4C_2}v_{C_2} - \frac{1}{R_6C_2}V_b \\ \qquad - \frac{1}{R_5C_2}v_{C_1}\left(\frac{R_{10}}{R_{11}} + \frac{R_{10}}{100R_{12}}v_{C_4}^2\right) \\ \frac{dv_{C_3}}{dt} = \frac{1}{10R_7C_3}v_{C_1}v_{C_2} - \frac{1}{R_8C_3}v_{C_3} \\ \frac{dv_{C_4}}{dt} = \frac{1}{R_9C_4}v_{C_1}, \end{cases} \tag{25}$$

where v_{C_1}, v_{C_2}, v_{C_3}, and v_{C_4} are the voltages across the capacitors C_1, C_2, C_3, and C_4, respectively.

The power supplies of all active devices are $\pm 15V_{DC}$ and the operational amplifiers TL084 are used in this work. The values of components in Figs. 12 and 13 are chosen as follows: $R_1 = R_2 = 1\,\text{k}\Omega$, $R_3 = R_4 = R_7 = 1.8\,\text{k}\Omega$, $R_5 = 1.2\,\text{k}\Omega$, $R_6 = 3.6\,\text{M}\Omega$, $R_8 = 12\,\text{k}\Omega$, $R_9 = 18\,\text{k}\Omega$, $R_{10} = R = 36\,\text{k}\Omega$, $R_{11} = 90\,\text{k}\Omega$, $R_{12} = 120\,\text{k}\Omega$, $V_b = 0.1V_{DC}$, and $C_1 = C_2 = C_3 = C_4 = 4.7\,\text{nF}$.

The designed circuit is implemented in SPICE. The obtained results are displayed in Figs. 14, 15, 16 and 17 which show the hyperchaotic attractors of the designed circuit in different phase planes (v_{C_1}, v_{C_2}), (v_{C_1}, v_{C_3}), (v_{C_2}, v_{C_3}), and (v_{C_1}, v_{C_4}), respectively. Theoretical attractors (see Figs. 3, 4, 5 and 6) are similar with the circuital ones (see Figs. 14, 15, 16 and 17). Moreover, the designed circuit confirms the feasibility of the memristive system.

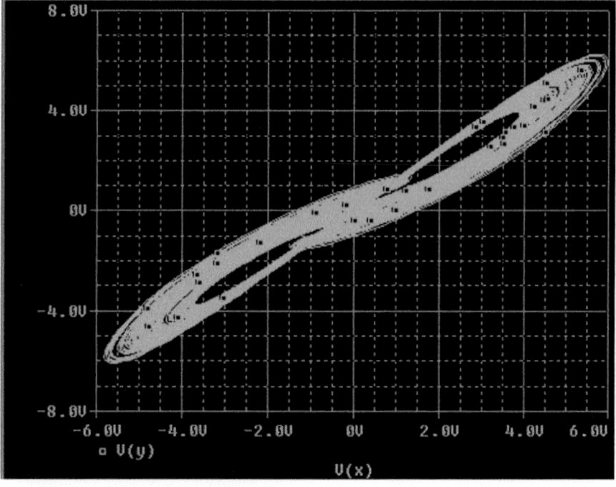

Fig. 14 Hyperchaotic attractor of the designed circuit obtained from SPICE in the (v_{C_1}, v_{C_2}) phase plane

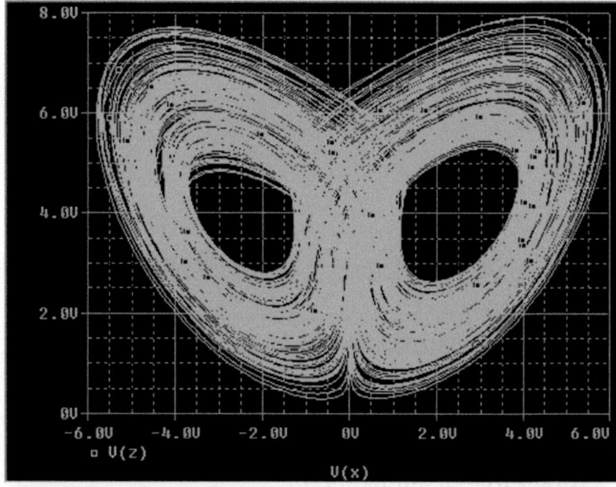

Fig. 15 Hyperchaotic attractor of the designed circuit obtained from SPICE in the (v_{C_1}, v_{C_3}) phase plane

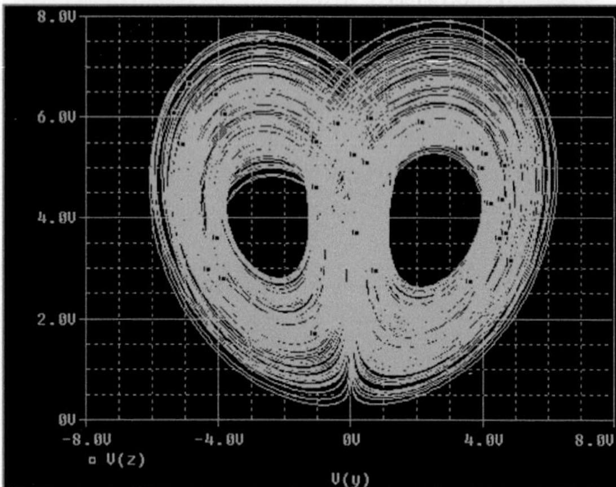

Fig. 16 Hyperchaotic attractor of the designed circuit obtained from SPICE in the (v_{C_2}, v_{C_3}) phase plane

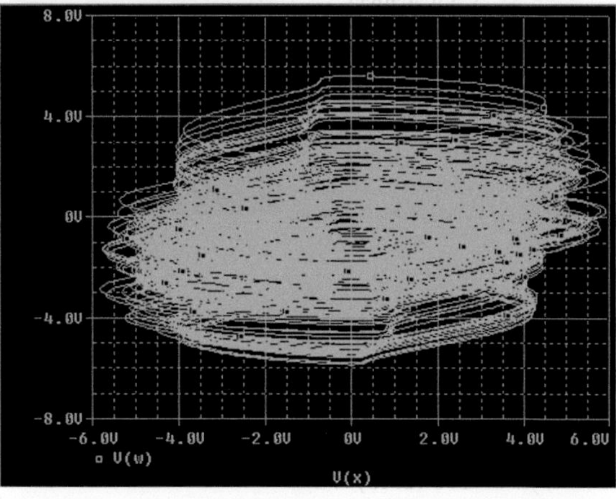

Fig. 17 Hyperchaotic attractor of the designed circuit obtained from SPICE in the (v_{C_1}, v_{C_4}) phase plane

7 Conclusion

A memristive system, which is built by using a memristor, is proposed in this work. The presence of the memristor creates the special features of such as hyperchaos, the absence of equilibrium points, and hidden attractors. Fundamental dynamical behaviors of the memristive hyperchaotic system are investigated through calculating equilibrium points, phase portraits of chaotic attractors, Lyapunov exponents and Kaplan–Yorke dimension. In addition, the capacity of synchronization of memristive systems and the feasibility of such memristive system without equilibrium are verified through anti-synchronization scheme and circuital implementation, respectively.

It is worth noting that the memristive system can exhibit double-scroll hyperchaotic attractor despite the equilibrium points have disappeared. It has been known that equilibrium points of a dynamical system, especially a chaotic one, play an important role when generating multi-scroll attractors. Therefore, investigating no-equilibrium memristive systems with multi-scroll attractors will be studied in future works.

Moreover, the memristive system has potential applications in secure communications and cryptography because of its hyperchaos and feasibility. Further studies in this research direction will be presented in future works.

Acknowledgments This research is funded by Vietnam National Foundation for Science and Technology Development (NAFOSTED) under grant number 102.02-2012.27

References

1. Lorenz EN (1963) Deterministic non-periodic flow. J Atmos Sci 20:130–141
2. Azar AT, Vaidyanathan S (2015) Chaos modeling and control systems design. Springer, Germany
3. Azar AT, Vaidyanathan S (2015) Computational intelligence applications in modeling and control. Springer, Germany
4. Azar AT, Vaidyanathan S (2015) Handbook of research on advanced intelligent control engineering and automation. IGI Global, USA
5. Chen G, Yu X (2003) Chaos control: theory and applications. Springer, Berlin
6. Chen GR (1999) Controlling chaos and bifurcations in engineering systems. CRC Press, Boca Raton
7. Sprott JC (2003) Chaos and times-series analysis. Oxford University Press, Oxford
8. Strogatz SH (1994) Nonlinear dynamics and chaos: with applications to physics, biology, chemistry, and engineering. Perseus Books, Massachusetts
9. Yalcin ME, Suykens JAK, Vandewalle J (2005) Cellular neural networks, multi-scroll chaos and synchronization. World Scientific, Singapore
10. Rössler OE (1976) An equation for continuous chaos. Phys Lett A 57:397–398
11. Arneodo A, Coullet P, Tresser C (1981) Possible new strange attractors with spiral structure. Comm Math Phys 79:573–579
12. Lü J, Chen G (2002) A new chaotic attractor coined. Int J Bif Chaos 12:659–661
13. Vaidyanathan S (2013) A new six-term 3-D chaotic system with an exponential nonlineariry. Far East J Math Sci 79:135–143

14. Barnerjee T, Biswas D, Sarkar BC (2012) Design and analysis of a first order time-delayed chaotic system. Nonlinear Dyn 70:721–734
15. Pham V-T, Volos C, Vaidyanathan S (2015b) Multi-scroll chaotic oscillator based on a first-order delay differential equation. In: Azar AT, Vaidyanathan S (eds) Chaos modelling and control systems design, vol 581., Studies in computational intelligenceSpringer, Germany, pp 59–72
16. Vaidyanathan S, Azar AT (2015b) Analysis, control and synchronization of a nine-term 3-D novel chaotic system. In: Azar AT, Vaidyanathan S (eds) Chaos modelling and control systems design, vol 581., Studies in computational intelligenceSpringer, Germany, pp 19–38
17. Yalcin ME, Suykens JAK, Vandewalle J (2004) True random bit generation from a double-scroll attractor. IEEE Trans Circuits Syst I Regul Papers 51:1395–1404
18. Volos CK, Kyprianidis IM, Stouboulos IN (2012) A chaotic path planning generator for autonomous mobile robots. Robot Auto Syst 60:651–656
19. Hoang TM, Nakagawa M (2008) A secure communication system using projective-lag and/or projective-anticipating synchronizations of coupled multidelay feedback systems. Chaos Solitions Fractals 38:1423–1438
20. Rössler OE (1979) An equation for hyperchaos. Phys Lett A 71:155–157
21. Vaidyanathan S, Azar AT (2015a) Analysis and control of a 4-D novel hyperchaotic system. In: Azar AT, Vaidyanathan S (eds) Chaos modeling and control systems design, vol 581., Studies in computational intelligenceSpringer, Germany, pp 19–38
22. Sadoudi S, Tanougast C, Azzaz MS, Dandache A (2013) Design and FPGA implementation of a wireless hyperchaotic communication system for secure realtime image transmission. EURASIP J Image Video Process 943:1–18
23. Udaltsov VS, Goedgebuer JP, Larger L, Cuenot JB, Levy P, Rhodes WT (2003) Communicating with hyperchaos: the dynamics of a DNLF emitter and recovery of transmitted information. Optics Spectrosc 95:114–118
24. Grassi G, Mascolo S (1999) A system theory approach for designing cryptosystems based on hyperchaos. IEEE Trans Cir Sys I: Fund Theory Appl 46:1135–1138
25. Huang Y, Yang X (2006) Hyperchaos and bifurcation in a new class of four-dimensional hopfield neural networks. Neurocomputing 69:1787–1795
26. Vicente R, Dauden J, Colet P, Toral R (2005) Analysis and characterization of the hyperchaos generated by a semiconductor laser subject to a delayed feedback loop. IEEE J Quantum Electr 41:541–548
27. Buscarino A, Fortuna L, Frasca M, Gambuzza LV (2012) A chaotic circuit based on Hewlett-Packard memristor. Chaos 22:023136
28. Fitch AL, Yu D, Iu HHC, Sreeram V (2012) Hyperchaos in an memristor-based modified canonical chua's circuit. Int J Bif Chaos 22:1250133–1250138
29. Buscarino A, Fortuna L, Frasca M, Gambuzza LV (2012) A gallery of chaotic oscillators based on hp memristor. Int J Bif Chaos 22:1330014–1330015
30. Li Q, Hu S, Tang S, Zeng G (2014) Hyperchaos and horseshoe in a 4D memristive system with a line of equilibria and its implementation. Int J Cir Theory Appl 42:1172–1188
31. Li Q, Zeng H, Li J (2015) Hyperchaos in a 4D memristive circuit with infinitely many stable equilibria. Nonlinear Dyn 79:2295–2308
32. Pham VT, Volos CK, Vaidyanathan S, Le TP, Vu VY (2015c) A memristor-based hyperchaotic system with hidden attractors: dynamics, sychronization and circuital emulating. J Eng Sci Tech Rev 8:205–214
33. Leonov GA, Kuznetsov NV (2013) Hidden attractors in dynamical systems: from hidden oscillation in Hilbert-Kolmogorov, Aizerman and Kalman problems to hidden chaotic attractor in Chua circuits. Int J Bifurc Chaos 23:1330002
34. Leonov GA, Kuznetsov NV, Kuznetsova OA, Seldedzhi SM, Vagaitsev VI (2011) Hidden oscillations in dynamical systems. Trans Syst Contr 6:54–67
35. Jafari S, Sprott JC (2013) Simple chaotic flows with a line equilibrium. Chaos Solitons Fractals 57:79–84

36. Kuznetsov NV, Leonov GA, Seledzhi SM (2011) Hidden oscillations in nonlinear control systems. IFAC Proc 18:2506–2510
37. Pham V-T, Jafari S, Volos C, Wang X, Golpayegani SMRH (2014a) Is that really hidden? The presence of complex fixed-points in chaotic flows with no equilibria. Int J Bifur Chaos 24:1450146
38. Pham V-T, Volos CK, Jafari S, Wei Z, Wang X (2014b) Constructing a novel no-equilibrium chaotic system. Int J Bifurc Chaos 24:1450073
39. Sharma PR, Shrimali MD, Prasad A, Kuznetsov NV, Leonov GA (2015) Control of multistability in hidden attractors. Eur Phys J Special Topics 224:1485–1491
40. Leonov GA, Kuznetsov NV, Vagaitsev VI (2012) Hidden attractor in smooth Chua system. Phys D 241:1482–1486
41. Leonov GA, Kuznetsov NV, Kiseleva MA, Solovyeva EP, Zaretskiy AM (2014) Hidden oscillations in mathematical model of drilling system actuated by induction motor with a wound rotor. Nonlinear Dyn 77:277–288
42. Leonov GA, Kuznetsov NV (2011) Analytical-numerical methods for investigation of hidden oscillations in nonlinear control systems. IFAC Proc 18:2494–2505
43. Brezetskyi S, Dudkowski D, Kapitaniak T (2015) Rare and hidden attractors in van der pol-duffing oscillators. Eur Phys J Special Topics 224:1459–1467
44. Jafari S, Sprott JC, Nazarimehr F (2015) Recent new examples of hidden attractors. Eur Phys J Special Topics 224:1469–1476
45. Shahzad M, Pham VT, Ahmad MA, Jafari S, Hadaeghi F (2015) Synchronization and circuit design of a chaotic system with coexisting hidden attractors. Eur Phys J Special Topics 224:1637–1652
46. Sprott JC (2015) Strange attractors with various equilibrium types. Eur Phys J Special Topics 224:1409–1419
47. Vaidyanathan S, Volos CK, Pham VT (2015c) Analysis, control, synchronization and spice implementation of a novel 4-d hyperchaotic rikitake dynamo system without equilibrium. J Eng Sci Tech Rev 8:232–244
48. Vaidyanathan S, Pham VT, Volos CK (2015b) A 5-d hyperchaotic rikitake dynamo system with hidden attractors. Eur Phys J Special Topics 224:1575–1592
49. Pham VT, Vaidyanathan S, Volos CK, Jafari S (2015a) Hidden attractors in a chaotic system with an exponential nonlinear term. Eur Phys J Special Topics 224:1507–1517
50. Leonov GA, Kuznetsov NV (2011) Algorithms for searching for hidden oscillations in the Aizerman and Kalman problems. Dokl Math 84:475–481
51. Leonov GA, Kuznetsov NV, Vagaitsev VI (2011) Localization of hidden Chua's attractors. Phys Lett A 375:2230–2233
52. Bao B, Liu Z, Xu B (2010) Dynamical analysis of memristor chaotic oscillator. Acta Physica Sinica 59:3785–3793
53. Muthuswamy B (2010) Implementing memristor based chaotic circuits. Int J Bif Chaos 20:1335–1350
54. Sprott JC (2010) Elegant chaos: algebraically simple chaotic flows. World Scientific, Singapore
55. Wolf A, Swift JB, Swinney HL, Vastano JA (1985) Determining Lyapunov exponents from a time series. Phys D 16:285–317
56. Frederickson P, Kaplan JL, Yorke ED, York J (1983) The lyapunov dimension of strange attractors. J Differ Equ 49:185–207
57. Boccaletti S, Kurths J, Osipov G, Valladares DL, Zhou CS (2002) The synchronization of chaotic systems. Phys Rep 366:1–101
58. Fortuna L, Frasca M (2007) Experimental synchronization of single-transistor-based chaotic circuits. Chaos 17:043118-1–043118-5
59. Kapitaniak T (1994) Synchronization of chaos using continuous control. Phys Rev E 50: 1642–1644
60. Pecora LM, Carroll TL (1990) Synchronization in chaotic signals. Phys Rev A 64:821–824
61. Buscarino A, Fortuna L, Frasca M (2009) Experimental robust synchronization of hyperchaotic circuits. Phys D 238:1917–1922

62. Gamez-Guzman L, Cruz-Hernandez C, Lopez-Gutierrez R, Garcia-Guerrero EE (2009) Synchronization of Chua's circuits with multi-scroll attractors: application to communication. Commun Nonlinear Sci Numer Simul 14:2765–2775
63. Karthikeyan R, Vaidyanathan S (2014) Hybrid chaos synchronization of four-scroll systems via active control. J Electr Eng 65:97–103
64. Srinivasan K, Senthilkumar DV, Murali K, Lakshmanan M, Kurths J (2011) Synchronization transitions in coupled time-delay electronic circuits with a threshold nonlinearity. Chaos 21:023119
65. Vaidyanathan S (2014) Analysis and adaptive synchronization of eight-term novel 3-D chaotic system with three quadratic nonlinearities. Eur Phys J Special Topics 223:1519–1529
66. Vaidyanathan S, Azar AT (2015c) Anti-synchronization of identical chaotic systems using sliding mode control and an application to Vaidhyanathan-Madhavan chaotic systems. Stud Comput Intell 576:527–547
67. Vaidyanathan S, Azar AT (2015d) Hybrid synchronization of identical chaotic systems using sliding mode control and an application to Vaidhyanathan chaotic systems. Stud Comput Intell 576:549–569
68. Vaidyanathan S, Idowu BA, Azar AT (2015a) Backstepping controller design for the global chaos synchronization of Sprott's jerk systems. Stud Comput Intell 581:39–58
69. Zhu Q, Azar AT (2015) Complex system modelling and control through intelligent soft computations. Springer, Germany
70. Woafo P, Kadji HGE (2004) Synchronized states in a ring of mutually coupled self-sustained electrical oscillators. Phys Rev E 69:046206
71. Stefanski A, Perlikowski P, Kapitaniak T (2007) Ragged synchronizability of coupled oscillators. Phys Rev E 75:016210
72. Volos CK, Kyprianidis IM, Stouboulos IN (2011) Various synchronization phenomena in bidirectionally coupled double scroll circuits. Commun Nonlinear Sci Numer Simul 71:3356–3366
73. Aguilar-Lopez R, Martinez-Guerra R, Perez-Pinacho C (2014) Nonlinear observer for synchronization of chaotic systems with application to secure data transmission. Eur Phys J Special Topics 223:1541–1548
74. Rosenblum MG, Pikovsky AS, Kurths J (1997) From phase to lag synchronization in coupled chaotic oscillators. Phys Rev Lett 78:4193–4196
75. Akopov A, Astakhov V, Vadiasova T, Shabunin A, Kapitaniak T (2005) Frequency synchronization in clusters in coupled extended systems. Phys Lett A 334:169–172
76. Hoang TM, Nakagawa M (2007) Anticipating and projective–anticipating synchronization of coupled multidelay feedback systems. Phys Lett A 365:407–411
77. Vaidyanathan S (2012) Anti-synchronization of four-wing chaotic systems via sliding mode control. Int J Auto Comput 9:274–279
78. Vaidyanathan S, Volos C, Pham VT, Madhavan K, Idowo BA (2014) Adaptive backstepping control, synchronization and circuit simualtion of a 3-D novel jerk chaotic system with two hyperbolic sinusoidal nonlinearities. Arch Cont Sci 33:257–285
79. Khalil H (2002) Nonlinear systems. Prentice Hall, New Jersey
80. Sastry S (1999) Nonlinear systems: analysis, stability, and control. Springer, USA
81. Barakat M, Mansingka A, Radwan AG, Salama KN (2013) Generalized hardware post processing technique for chaos-based pseudorandom number generators. ETRI J 35:448–458
82. Volos CK, Kyprianidis IM, Stouboulos IN (2013) Image encryption process based on chaotic synchronization phenomena. Signal Process 93:1328–1340
83. Sundarapandian V, Pehlivan I (2012) Analysis, control, synchronization, and circuit design of a novel chaotic system. Math Comp Model 55:1904–1915
84. Bouali S, Buscarino A, Fortuna L, Frasca M, Gambuzza LV (2012) Emulating complex business cycles by using an electronic analogue. Nonlinear Anal Real World Appl 13:2459–2465
85. Fortuna L, Frasca M, Xibilia MG (2009) Chua's circuit implementation: yesterday, today and tomorrow. World Scientific, Singapore
86. Tetzlaff R (2014) Memristors and memristive systems. Springer, New York

Synchronization of Fractional Chaotic and Hyperchaotic Systems Using an Extended Active Control

Sachin Bhalekar

Abstract An extended active control technique is used to synchronize fractional order chaotic and hyperchaotic systems with and without delay. The coupling strength is set to the value less than one to achieve the complete synchronization more easily. Explicit formula for the error matrix is also proposed in this chapter. Numerical examples are given for the fractional order chaotic Liu system, hyperchaotic new system and Ucar delay system. The effect of fractional order and coupling strength on the synchronization time is studied for non-delayed cases. It is observed that the synchronization time decreases with increase in fractional order as well as with increase in coupling strength for the Liu system. For the new system, the synchronization time decreases with increase in fractional order as well as with decrease in coupling strength.

Keywords Chaos · Synchronization · Active control · Extended active control · Fractional order · Caputo derivative

1 Introduction

Natural systems are often modelled by using dynamical systems and most of these equations are nonlinear. A dynamical system with continuous (resp. discrete) time variable is called as differential (resp. discrete) dynamical system. A one dimensional differential system can produce either stable or unstable equations. Further, the oscillatory solutions can be produced by a two dimensional equation. Moreover, in case of third or higher order system, more complex dynamics can be observed. If the solutions are bounded oscillations which are not periodic and are sensitive to initial conditions then these are called chaotic solutions.

S. Bhalekar (✉)
Department of Mathematics, Shivaji University,Vidyanagar, Kolhapur 416004, India
e-mail: sachin.math@yahoo.co.in; sbb_maths@unishivaji.ac.in

A.T. Azar and S. Vaidyanathan (eds.), *Advances in Chaos Theory and Intelligent Control*, Studies in Fuzziness and Soft Computing 337,
DOI 10.1007/978-3-319-30340-6_3

Dynamical systems have a wide range of applications including chaos [1, 2], hyperchaos [3, 4], control theory [5], artificial intelligence [6, 7], population dynamics and so on.

In the pioneering work [8, 9] Pecora and Carroll have shown that chaotic systems can be synchronized by introducing appropriate coupling. Further, the technique of chaos synchronization has been explored in secure communications of analog and digital signals [10] and for developing safe and reliable cryptographic systems [11]. A variety of approaches have been proposed for the chaos synchronization such as nonlinear feedback control [12], adaptive control [13, 14], active control [15–17], sliding mode control [18–22], deadbeat control [23] and backstepping control [24].

In classical calculus, the order of derivative is assumed to be a positive integer. If this integer order is replaced by an arbitrary number (real or complex) then such derivative is called as fractional order derivative. The integer order derivative of a function at a point can be approximated using the values of function in the neighborhood of that point. This local property, however, is not valid for fractional derivative (FD). One has to consider all the values of function starting from initial point to approximate fractional derivatives. This nonlocal nature of FD play an important role while modelling the memory and hereditary properties in the natural systems. The fractional calculus has potential applications in various branches of Science, Engineering and Social Sciences. A well-known fractional diffusion equation is studied by researchers to model numerous phenomena [25–29]. Mainardi in his monograph [30] discussed the application of fractional calculus in linear viscoelasticity. Applications of this branch to bioengineering are discussed by Magin in [31]. Few more applications include signal processing [32, 33], image processing [34], image encryption [35–37], cryptography [38], control theory [39], thermodynamics [40] and nonlinear dynamics [41–43]. Chaotic fractional order systems can be used to generate reliable cryptographic schemes [44–46].

Different inequivalent definitions are provided by the researchers for FD in the literature. Among these definitions, one proposed by Caputo is popular because of its applications in mathematical modelling. The initial conditions arising in Caputo derivative model are more realistic than other derivatives.

Various existence and uniqueness (EU) results for fractional differential equations (FDE) are discussed in the literature. EU result for linear equations of the form

$$^{R}D^{\alpha}y(x) - \lambda y(x) = f(x), \ (n-1 \le Re(\alpha) < n), \tag{1}$$

where $^{R}D^{\alpha}$ is Riemann-Lioville fractional derivative is derived by Barret [47]. Al-Bassam [48] used method of successive approximation to derive EU result for nonlinear problem

$$^{R}D^{\alpha}y(x) = f(x, y(x)), \ (0 < \alpha \le 1). \tag{2}$$

Delbosco and Rodino [49] derived EU results for nonlinear equations using Schauder's fixed point theorem. Cauchy problems of the form (2) with Caputo derivative are investigated by Diethelm and Ford [50]. Existence, uniqueness and stability of solutions of systems of FDEs are proposed by Daftardar-Gejji and Babakhani in [51].

Exact solutions of linear FDEs can be obtained by using transform methods and operational method. Linear FDEs of the form (1) with Caputo fractional derivative are solved by Gorenflo and Mainardi [52] using Laplace transform. Luchko and Gorenflo [53] developed an operational method to solve multi-term linear FDEs of Caputo type. Daftardar-Gejji and coworkers [54, 55] used method of separation of variables to solve these equations.

Nonlinear FDEs can be solved either by numerical methods or approximate analytical methods (AAM). Diethelm, Ford and Freed [56] proposed a predictor-corrector method for solving such equations. A more efficient method is recently derived by Daftardar-Gejji et al. [57]. The solutions provided by AAMs are often in the form of series. Few terms of these series can be used as a good approximation to the exact solution in small interval. Adomian decomposition method (ADM) [58], variational iteration method (VIM) [59], homotopy perturbation method (HPM) [60] and Daftardar-Gejji-Jafari method (DJM) [61] are popular AAMs used by researchers.

In this chapter, we use the method of an extended active control [62] to study the synchronization in case of fractional order chaotic and hyperchaotic systems with and without delay. We show that the coupling strength in the present method is less than that of active control and hence the technique is more efficient.

2 Related Work

Synchronization of fractional order chaotic systems was first studied by Deng and Li [63] who carried out synchronization in the fractional Lü system. Further they have studied synchronization of fractional Chen system [64]. Li and Deng have summarized the theory and techniques of fractional order chaos synchronization in [65]. Li et al. [66] used the Pecora-Carroll (PC) method, the active-passive decomposition method, the one-way coupling and the bidirectional coupling methods to synchronize two identical Chua systems of fractional order. Synchronization in unified system is studied by Wang and Zhang [67] using PC and one way coupling methods. VIM is utilized by Yu and Li [68] to synchronize Rossler hyperchaos system.

Nonlinear control theory is successfully extended by Wang et al. [69] to fractional order Chen systems to achieve synchronization. The same technique is further used by Jun et al. [70] for chaotic synchronization between fractional order financial system and financial system of integer orders.

Mohadeszadeh and Delavari [71] developed adaptive sliding mode technique to synchronize various types of fractional order chaotic systems. The models involving uncertain and unknown parameters were also considered by these authors. Further, they have used particle swarm optimization algorithm for optimizing the controller parameters. Gao et al. [72] proposed a modified sliding mode control scheme to realize complete synchronization of a class of three dimensional fractional order chaotic systems. A modified second order sliding mode control scheme is developed in [73] to synchronize hyperchaotic fractional systems. Guanand and Wang [74] proposed a new chaotic system and investigated the indirect Lyapunov stability by using sliding mode control. Tavazoei and Haeri [75] proposed a controller based on active sliding mode theory to synchronize chaotic fractional order systems in master slave structure. Xu and Wang [76] proposed a fractional integral sliding surface for synchronizing the uncertain fractional-order chaotic systems with Gaussian fluctuation. Uncertain fractional order chaotic systems involving time delay are synchronized using sliding mode control in [77].

Li et al. [78] derived adaptive synchronization of fractional order complex dynamical networks. The adaptive control is further used by Rad, Nikdadian and Bahadorzadeh [79] to synchronize fractional-order Genesio-Tesi chaotic system and by Zhou and Bai [80] for fractional order systems with unknown parameters. An unidirectional adaptive full-state linear error feedback coupling is developed by Leung et al. [81] to synchronize fractional systems. A linear feedback controller and time delayed feedback controller are used by El-Sayed et al. [82] to synchronize novel hyperchaotic fractional order system.

If the amplitudes of the synchronized signals evolve in proportional scale then it is called as a projective synchronization (PS). Xingyuan and Yijie [83] extended the projective synchronization to fractional order systems and designed a scheme to achieve projective synchronization of the fractional order unified system. Fractional order hyperchaotic Lü system and Lorenz system with uncertain parameters are synchronized by Agrawal and Das [84] using PS. A modified generalized projective synchronization (MGPS) of fractional-order chaotic systems is derived by Liu et al. [85]. The MGPS between the hyperchaotic Lorenz system and the Lü system of the base order 0.95 is then discussed by these authors. A hybrid projective synchronization scheme for two identical fractional order chaotic systems is proposed in [86]. In [87], Chen et al. proposed a lag projective synchronization (LPS) where the slave system synchronizes the past state of the driver. LPS between two four-scroll hyperchaotic systems with different fractional orders is discussed and simulated by using a multi-step fractional differential transform method in [88].

The technique of active control (AC) is extended by Bhalekar and Daftardar-Gejji [89, 90] to synchronize and anti-synchronize non-identical commensurate fractional chaotic systems. It is observed that the synchronization is faster as the system order tends to one. Further, Bhalekar utilized AC to synchronize incommensurate order systems [91] and hyperchaotic systems [92] of fractional order.

A new approach named fractional-order dynamics rejection scheme is proposed for fractional-order system based on active disturbance rejection control (ADRC) method in [93]. The ARDC is then used by Gao and Liao [94] to synchronize different fractional order chaotic systems.

Adaptive impulsive synchronization of fractional order chaotic system with uncertain and unknown parameters is discussed in [95]. The finite-time input-to-state stable theory of fractional-order systems proposed by Li and Zhang [96] is used to synchronize fractional-order chaotic system with uncertainties and disturbance. Anti-synchronization of uncertain fractional-order chaotic systems is presented by Ran et al. [97]. Noghredani and Balochian [98] designed sliding mode control for synchronization of fractional order uncertain chaotic systems with input nonlinearity. Balasubramaniam, Muthukumar and Ratnavelu [99] proposed a fuzzy fractional integral sliding mode control for synchronizing fractional-order dynamical systems with mismatched fractional orders.

Chaos synchronization in fractional order delay systems is also studied by researchers [100–102]. Velmurugan et al. [103] discussed the problem of finite-time synchronization of a class of fractional order memristo -based neural networks with time delays. Adaptive pinning synchronization in fractional order uncertain complex dynamical networks with delay is studied in [104].

One of the important applications of chaos synchronization is cryptography. Cryptosystems based on fractional order chaotic systems are more secure than classical ones. In the seminal work, Kiani-B et al. [105] proposed a fractional chaotic communication method using an extended fractional Kalman filter. Sheu et al. [106] presented a modification of the two-channel chaos-based cryptosystems by using fractional chaotic systems. Adaptive synchronization in fractional order PMSM (permanent magnet synchronous motors) system is used by Wang et al. [38] to generate secure communication scheme. In [107], fractional order Chen system is used to obtain a larger key space and thus to improve the level of security of cryptosystem. Muthukumar and co-workers discussed multi-scale synchronization in King Cobra chaotic system [45], sliding mode control [46] and feedback controller to reverse butterfly-shaped chaotic system [44] for image encryption. Image encryption is also done by using delayed fractional order logistic system [108], hyperchaotic Lorenz system [109], Lorenz-like system [110] and so on. A color image encryption scheme is proposed by Huang et al. [111] by applying hyperchaotic systems of fractional order.

3 Preliminaries

In this section, we discuss some basic definitions and analytical results regarding fractional calculus.

Definition 3.1 [112, 113] Riemann-Liouville fractional integration of order α is defined as

$$I^{\alpha} f(t) = \frac{1}{\Gamma(\alpha)} \int_0^t (t-y)^{\alpha-1} f(y)\, dy, \quad t > 0. \tag{3}$$

Definition 3.2 [112, 113] Caputo fractional derivative of order α is defined as

$$D^{\alpha} f(t) = I^{m-\alpha} \left(\frac{d^m f(t)}{dt^m} \right), \quad 0 \leq m-1 < \alpha \leq m. \tag{4}$$

Note that for $0 \leq m-1 < \alpha \leq m,\ a \geq 0$ and $\gamma > -1$

$$I^{\alpha}(t-a)^{\gamma} = \frac{\Gamma(\gamma+1)}{\Gamma(\gamma+\alpha+1)}(t-a)^{\gamma+\alpha}, \tag{5}$$

$$(I^{\alpha}D^{\alpha}f)(t) = f(t) - \sum_{k=0}^{m-1} f^{(k)}(0)\frac{t^k}{k!}. \tag{6}$$

Lemma 3.1 [114] *The fractional order linear system $D^{\alpha}\mathbf{X} = \mathbf{B}\mathbf{X}$ is asymptotically stable if and only if $|arg(\lambda)| > \alpha\pi/2$, for all eigenvalues λ of matrix $\mathbf{B}$. In this case, the component of the state decay towards 0 like $t^{-\alpha}$.*

4 Synchronization Techniques

4.1 Equations Not Involving Delay

Consider the autonomous system

$$D^{\alpha}\mathbf{x} = \mathbf{F}(\mathbf{x}), \tag{7}$$

where $\mathbf{F}$ is function from R^n to R^n and $0 < \alpha \leq 1$ is fractional order. The operator D^{α} represent Caputo fractional derivative [112, 113]. Let us write

$$\mathbf{F}(\mathbf{x}) = \mathbf{L}(\mathbf{x}) + \mathbf{N}(\mathbf{x}), \tag{8}$$

where $\mathbf{L}$ and $\mathbf{N}$ are linear and nonlinear functions of state $\mathbf{x}$ respectively.

The system (7) will be treated as the drive (master) system and the response (slave) system is given by

$$D^{\alpha}\mathbf{y} = \mathbf{F}(\mathbf{y}) + \mathbf{E}\mathbf{u}. \tag{9}$$

The column vector $\mathbf{u}$ in (9) is control term to be chosen and $\mathbf{E} = diag(\epsilon_1, \epsilon_2, \ldots, \epsilon_n)$ is strength matrix. For simplicity, we take $\epsilon_i = \epsilon,\ 1 \leq i \leq n$.

4.1.1 Active Control

For active control, we choose $\epsilon = 1$.

Theorem 4.1 *For fractional order drive system (7) and response system (9), the error system in active synchronization can be written as $D^{\alpha}\mathbf{e} = (\mathbf{L} + \mathbf{A})(\mathbf{e})$.*

proof Defining error vector by $\mathbf{e} = \mathbf{y} - \mathbf{x}$ and subtracting (7) from (9) we get

$$D^{\alpha}\mathbf{e} = \mathbf{L}(\mathbf{e}) + \mathbf{N}(\mathbf{y}) - \mathbf{N}(\mathbf{x}) + \mathbf{u}. \tag{10}$$

We choose active control term as

$$\mathbf{u} = \mathbf{V} - \mathbf{N}(\mathbf{y}) + \mathbf{N}(\mathbf{x}), \tag{11}$$

where $\mathbf{V} = \mathbf{A}\mathbf{e}$ is linear function of $\mathbf{e}$ defined using $n \times n$ real matrix $\mathbf{A}$ called as *active control matrix for given system*. Error equation (10) can now be written as

$$D^{\alpha}\mathbf{e} = (\mathbf{L} + \mathbf{A})(\mathbf{e}). \tag{12}$$

The proof is completed.

Theorem 4.2 *Given fractional order drive system (7) and response system (9), there exists an active control matrix A such that the active control synchronization between (7) and (9) can be achieved if and only if all the eigenvalues μ of $\mathbf{L} + \mathbf{A}$ satisfy*

$$|arg(\mu)| > \alpha\pi/2. \tag{13}$$

Proof is obvious from Lemma 3.1.

4.1.2 Extended Active Control

We generalize the technique of extended active control (EAC) proposed by Tang et al. [62] to synchronize fractional order systems (7) and (9). It is observed that the coupling strength $\epsilon = 1$ used in previous subsection for active control is too large and the stable conditions are too sufficient. The principle of EAC lies in choosing the strength $\epsilon \in (0, 1)$. In other words, it is sufficient to choose some smaller value of coupling strength instead of 1.

With the same choice of $\mathbf{u}$, we write the error system as

$$\begin{aligned} D^{\alpha}\mathbf{e} &= \mathbf{L}(\mathbf{e}) + \mathbf{N}(\mathbf{y}) - \mathbf{N}(\mathbf{x}) + \epsilon\mathbf{u} \\ &= (\mathbf{L} + \epsilon\mathbf{A})(\mathbf{e}) + (1 - \epsilon)(\mathbf{N}(\mathbf{y}) - \mathbf{N}(\mathbf{x})). \end{aligned} \tag{14}$$

For active control, we choose $\epsilon = 1$.

Theorem 4.3 *For fractional order drive system (7) and response system (9), the error system in extended active synchronization can be approximated as $D^{\alpha}\mathbf{e} = \mathbf{B}(\mathbf{x}, \epsilon)\,\mathbf{e}$, where the matrix $\mathbf{B}$ depends on states as well as coupling strength ϵ.*

Using $\mathbf{y} = \mathbf{e} + \mathbf{x}$, first order Taylor's series approximation yields

$$\mathbf{N}(\mathbf{y}) - \mathbf{N}(\mathbf{x}) = \mathbf{J}(\mathbf{x})\mathbf{e}, \tag{15}$$

where **J** is the Jacobian matrix of **N**. Using (15) in (14) we get

$$D^{\alpha}\mathbf{e} = \mathbf{B}(\mathbf{x}, \epsilon)\,\mathbf{e}, \tag{16}$$

where

$$\mathbf{B}(\mathbf{x}, \epsilon) = \mathbf{L} + \epsilon\mathbf{A} + (1 - \epsilon)\mathbf{J}. \tag{17}$$

The proof is completed.

Clearly, the stability of the error system (16) depends on ϵ as well as on the drive system's states. We choose the value of $\epsilon \in (0, 1)$ so that the system (16) becomes stable and the synchronization is achieved.

4.2 Equations Involving Delay

We consider the following system with delay

$$D^{\alpha}\mathbf{x} = \mathbf{F}(\mathbf{x}, \mathbf{x}_{\o}), \tag{18}$$

where, $\tau > 0$ is real number and $\mathbf{x}_{\tau}(t) = \mathbf{x}(t - \tau)$. We write

$$\mathbf{F}(\mathbf{x}) = \mathbf{L_1}(\mathbf{x}) + \mathbf{L_2}(\mathbf{x}_{\tau}) + \mathbf{N}(\mathbf{x}, \mathbf{x}_{\tau}), \tag{19}$$

where $\mathbf{L_1}$ and $\mathbf{L_2}$ are linear functions whereas **N** is a nonlinear function. As in previous section, we consider system (18) as a drive system and the response system is given by

$$D^{\alpha}\mathbf{y} = \mathbf{F}(\mathbf{y}, \mathbf{y}_{\tau}) + \epsilon\mathbf{u}. \tag{20}$$

For active control, we set $\epsilon = 1$ and $\mathbf{u} = \mathbf{A_1}\mathbf{e} + \mathbf{A_2}\mathbf{e}_{\o} - \mathbf{N}(\mathbf{y}, \mathbf{y}_{\tau}) + \mathbf{N}(\mathbf{x}, \mathbf{x}_{\tau})$ where A_1 and A_2 are linear operators to be chosen. The error equation, in this case is

$$D^{\alpha}\mathbf{e} = (\mathbf{L_1} + \mathbf{A_1})(\mathbf{e}) + (\mathbf{L_2} + \mathbf{A_2})(\mathbf{e}_{\tau}). \tag{21}$$

If we choose A_1 and A_2 so that all the roots of characteristic equation

$$det\,(\lambda^{\alpha}I - (\mathbf{L_1} + \mathbf{A_1}) - (\mathbf{L_2} + \mathbf{A_2})\,Exp(-\lambda\tau)) = 0 \tag{22}$$

have negative real parts then the errors will converge to zero as t approaches to infinity [115].

In case of extended active control, we take same A_1 and A_2 as in active control and $0 < \epsilon < 1$. The approximation to the error system is

$$D^{\alpha}\mathbf{e} = \mathbf{B_1}(\mathbf{e}) + \mathbf{B_2}(\mathbf{e}_{\tau}), \tag{23}$$

where $\mathbf{B_1} = \mathbf{L_1} + \epsilon\mathbf{A_1} + (1-\epsilon)\mathbf{J_1}, \mathbf{B_2} = \mathbf{L_2} + \epsilon\mathbf{A_2} + (1-\epsilon)\mathbf{J_2}$ and J_1, J_2 are Jacobian matrices of N with respect to $\mathbf{x}$ and $\mathbf{x}_\tau$ respectively. The coupling strength ϵ can be chosen such that the error system (23) is stable.

5 Examples

5.1 Synchronization of Chaotic Fractional Liu System

Daftardar-Gejji and Bhalekar [116] have studied the chaotic behaviour of the following fractional order Liu system

$$\begin{aligned} D^\alpha x_1 &= -x_1 - y_1^2 \\ D^\alpha y_1 &= 2.5y_1 - 4x_1z_1 \\ D^\alpha z_1 &= -5z_1 + 4x_1y_1, \end{aligned} \tag{24}$$

where $\alpha \in (0, 1)$ is fractional order. The system has three equilibrium points viz. $(0, 0, 0)$, $(-0.8839, 0.9402, -0.6648)$ and $(-0.8839, -0.9402, 0.6648)$ which are all saddle type. Liu system is chaotic for $\alpha \geq 0.91$.

The system (24) is drive system and the response system is given by

$$\begin{aligned} D^\alpha x_2 &= -x_2 - y_2^2 + \epsilon u_1 \\ D^\alpha y_2 &= 2.5y_2 - 4x_2z_2 + \epsilon u_2 \\ D^\alpha z_2 &= -5z_2 + 4x_2y_2 + \epsilon u_3. \end{aligned} \tag{25}$$

Using the notations described in Sect. 4,

$$\mathbf{L} = \begin{pmatrix} -1 & 0 & 0 \\ 0 & 2.5 & 0 \\ 0 & 0 & -5 \end{pmatrix}, \tag{26}$$

and $\mathbf{N}(x_1, y_1, z_1) = \left(-y_1^2, -4x_1z_1, 4x_1y_1\right)^T$. This gives

$$\mathbf{J}(x_1, y_1, z_1) = \begin{pmatrix} 0 & -2y_1 & 0 \\ -4z_1 & 0 & -4x_1 \\ 4y_1 & 4x_1 & 0 \end{pmatrix}. \tag{27}$$

We choose the matrix $\mathbf{A}$ such that the eigenvalues of $\mathbf{L} + \mathbf{A}$ satisfy (13). One of the choices for such $\mathbf{A}$ is

$$\mathbf{A} = \begin{pmatrix} 0 & 0 & 0 \\ 0 & -3.5 & 0 \\ 0 & 0 & 0 \end{pmatrix}. \tag{28}$$

Defining $e_x = x_2 - x_1$, $e_y = y_2 - y_1$, $e_z = z_2 - z_1$, $\mathbf{e} = (e_x, e_y, e_z)$ and using (16)–(17) we can write

$$D^\alpha \mathbf{e} = \mathbf{B}\mathbf{e}, \tag{29}$$

where

$$\mathbf{B} = \begin{pmatrix} -1 & -2(1-\epsilon)y_1 & 0 \\ -4(1-\epsilon)z_1 & 2.5 - 3.5\epsilon & -4(1-\epsilon)x_1 \\ 4(1-\epsilon)y_1 & 4(1-\epsilon)x_1 & -5 \end{pmatrix}. \tag{30}$$

We choose the value of ϵ properly, so as the synchronization is achieved.

5.1.1 Numerical Simulations

We consider different cases of fractional order.

- $\alpha = \mathbf{0.99}$: The local stable region for the zero solution of (29) in the $\epsilon - t$ plane is presented in Fig. 1a. It is clear from Fig. 1a that the synchronization can be achieved with small values of ϵ also. Figure 1b–d show the variation of synchronization errors with time t for $\epsilon = 1$, $\epsilon = 0.9$ and $\epsilon = 0.7$ respectively. In these figures, the

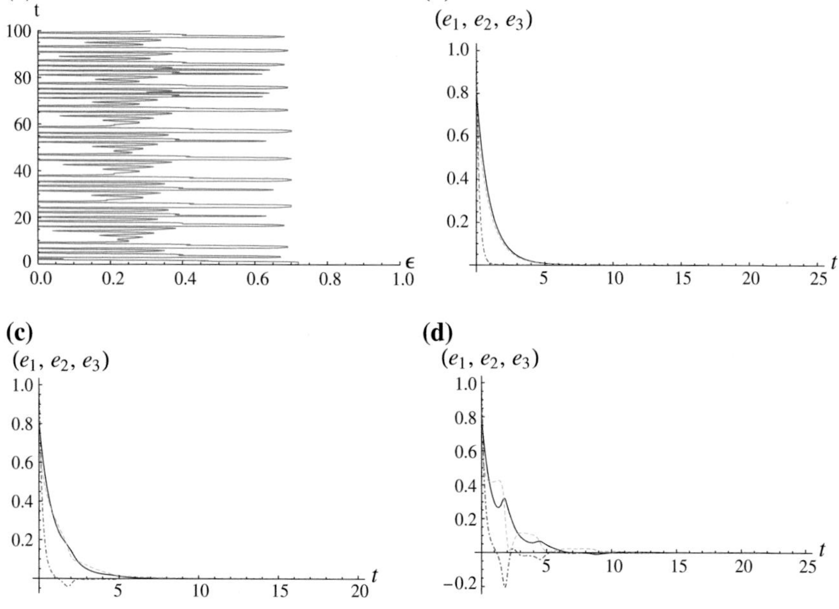

Fig. 1 Simulations for the Liu system with $\alpha = 0.99$. **a** Local stable region, **b** synchronization errors for $\epsilon = 1$, **c** synchronization errors $\epsilon = 0.9$, **d** synchronization errors $\epsilon = 0.7$

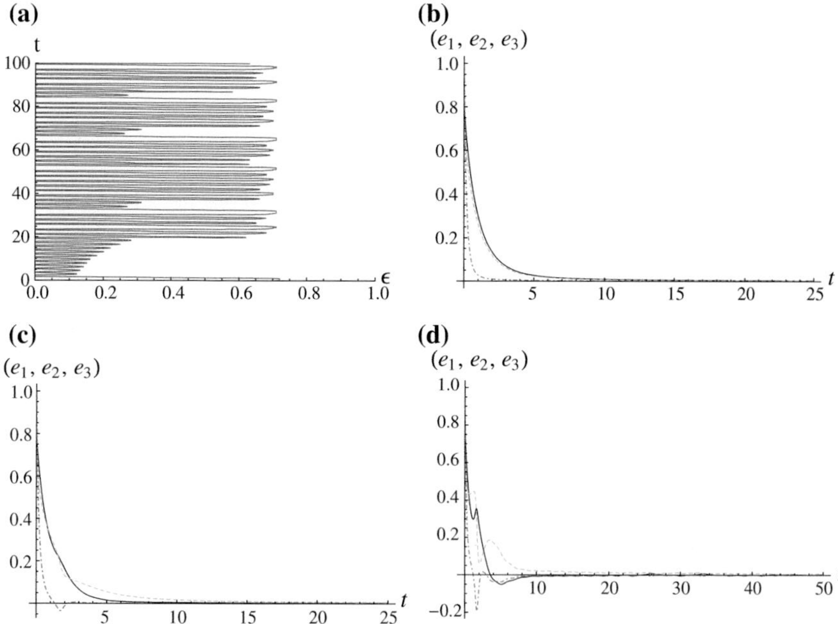

Fig. 2 Simulations for the Liu system with $\alpha = 0.93$. **a** Local stable region, **b** synchronization errors for $\epsilon = 1$, **c** synchronization errors $\epsilon = 0.9$, **d** synchronization errors $\epsilon = 0.7$

error functions e_x, e_y and e_z are given by solid line, dashed line and dot-dashed line respectively.

- $\alpha = \mathbf{0.93}$: Figure 2a shows the local stable region for the zero solution of (29) in the $\epsilon - t$ plane. Figure 2b–d represent the variation of synchronization errors with time t for $\epsilon = 1$, $\epsilon = 0.9$ and $\epsilon = 0.7$ respectively.

We have studied the system for other values of α and ϵ but the results are omitted for brevity.

Observation: It can be observed from figures that the synchronization time *decreases* with *increase* in α as well as with *increase* in ϵ.

5.2 *Synchronization of Hyperchaotic Fractional New System*

Consider the fractional version of the hyperchaotic system proposed by Liang et al. [117]

$$\begin{aligned} D^\alpha x_1 &= 20(y_1 - x_1) \\ D^\alpha y_1 &= x_1 + 10.6y_1 + w_1 - x_1 z_1 \\ D^\alpha z_1 &= -2.8z_1 + x_1^2 \\ D^\alpha w_1 &= -3.7x_1 \end{aligned} \tag{31}$$

as a drive system. The system has two positive, one zero and one negative Lyapunov exponents and the Lyapunove dimension is 3.0719 [117]. The Liang system has unique equilibrium point $(0, 0, 0)$ and the system undergoes Hopf bifurcation at parameter value $b = -c + k/(c - a)$.

The response system is

$$\begin{aligned} D^\alpha x_2 &= 20(y_2 - x_2) + \epsilon u_1 \\ D^\alpha y_2 &= x_2 + 10.6y_2 + w_2 - x_2 z_2 + \epsilon u_2 \\ D^\alpha z_2 &= -2.8z_2 + x_2^2 + \epsilon u_3 \\ D^\alpha w_2 &= -3.7x_2 + \epsilon u_4, \end{aligned} \tag{32}$$

where u_i are defined by (11) and $0 \leq \epsilon \leq 1$. With the notations discussed in Sect. 4,

$$\mathbf{L} = \begin{pmatrix} -20 & 20 & 0 & 0 \\ 1 & 10.6 & 0 & 1 \\ 0 & 0 & -2.8 & 0 \\ -3.7 & 0 & 0 & 0 \end{pmatrix}, \tag{33}$$

and $\mathbf{N}(x_1, y_1, z_1, w_1) = \left(0, -x_1 z_1, x_1^2, 0\right)^T$. This gives

$$\mathbf{J}(x_1, y_1, z_1, w_1) = \begin{pmatrix} 0 & 0 & 0 & 0 \\ -z_1 & 0 & -x_1 & 0 \\ 2x_1 & 0 & 0 & 0 \\ 0 & 0 & 0 & 0 \end{pmatrix}. \tag{34}$$

The matrix $\mathbf{A}$ is chosen as

$$\mathbf{A} = \begin{pmatrix} 0 & 0 & 0 & 0 \\ -1 & -11.6 & 0 & 0 \\ 0 & 0 & 0 & 0 \\ 3.7 & 0 & 0 & -1 \end{pmatrix}. \tag{35}$$

For this choice of $\mathbf{A}$ the eigenvalues of $\mathbf{L} + \mathbf{A}$ satisfy (13).

The synchronization errors are defined by $e_x = x_2 - x_1$, $e_y = y_2 - y_1$, $e_z = z_2 - z_1$, $e_w = w_2 - w_1$, $\mathbf{e} = (e_x, e_y, e_z, e_w)$. Using (16)–(17) we can write

$$D^\alpha \mathbf{e} = \mathbf{B}\mathbf{e}, \tag{36}$$

where

$$\mathbf{B} = \begin{pmatrix} -20 & 20 & 0 & 0 \\ (1-\epsilon)(1-z_1) & 10.6-11.6\epsilon & -(1-\epsilon)x_1 & 1 \\ (1-\epsilon)2x_1 & 0 & -2.8 & 0 \\ -3.7(1-\epsilon) & 0 & 0 & -\epsilon \end{pmatrix}. \tag{37}$$

5.2.1 Numerical Simulations

Now we consider different cases of fractional order and take observations.

- $\alpha = \mathbf{0.99}$: For the zero solution of the error system (36), we have plotted the local stable region in the $\epsilon - t$ plane in Fig. 3a. It is clear from Fig. 3a that the synchronization can be achieved with small values of ϵ. Synchronization errors are plotted in Fig. 3b–d for $\epsilon = 1$, $\epsilon = 0.9$ and $\epsilon = 0.8$ respectively. The error functions e_x, e_y, e_z and e_w are given by solid line, dashed line, dot-dashed line and dotted line respectively.

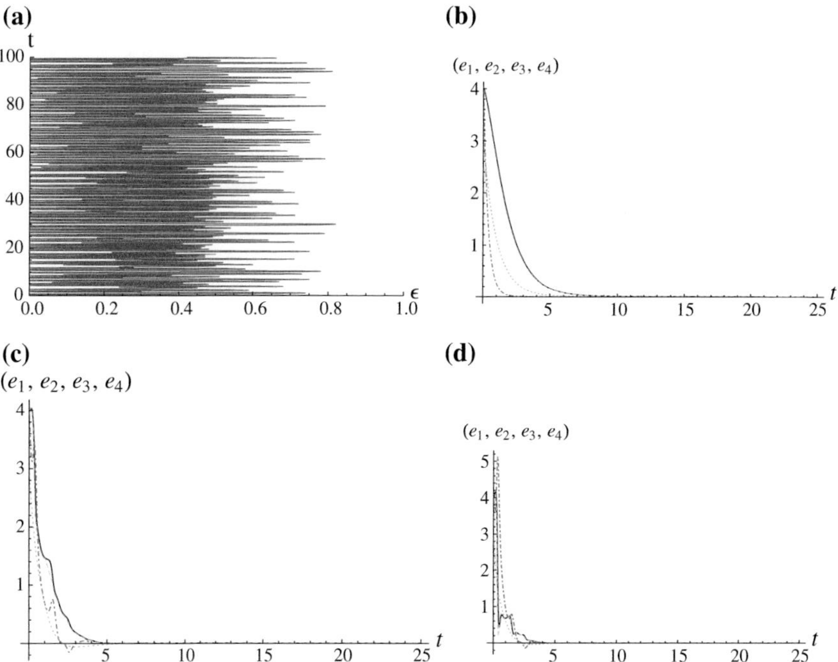

Fig. 3 Simulations for the New system with $\alpha = 0.99$. **a** Local stable region, **b** synchronization errors for $\epsilon = 1$, **c** synchronization errors $\epsilon = 0.9$, **d** synchronization errors $\epsilon = 0.8$

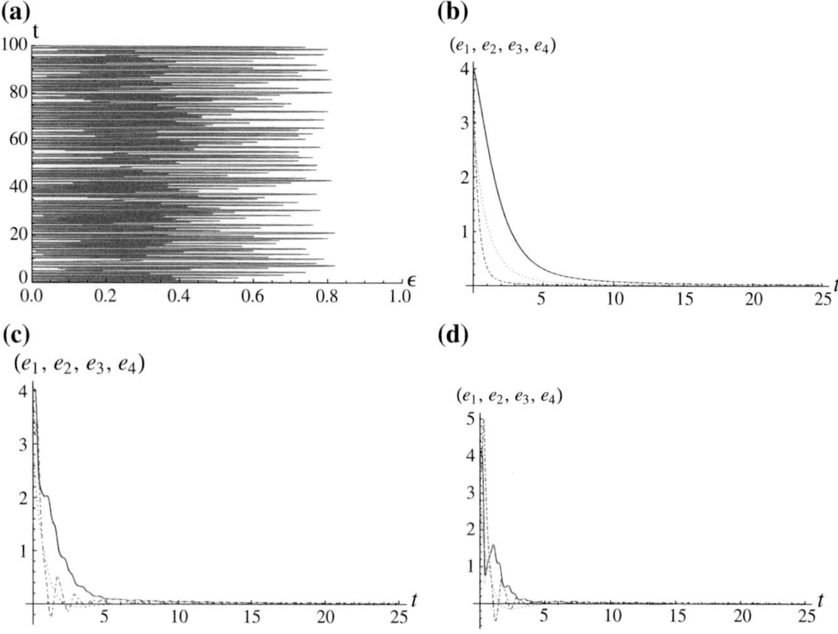

Fig. 4 Simulations for the New system with $\alpha = 0.94$. **a** Local stable region, **b** synchronization errors for $\epsilon = 1$, **c** synchronization errors $\epsilon = 0.9$, **d** synchronization errors $\epsilon = 0.8$

- $\alpha = \mathbf{0.94}$: Figure 4a represents the local stable region for the zero solution of (29) in the $\epsilon - t$ plane. Figure 4b–d show the evolution of synchronization errors with time t for $\epsilon = 1$, $\epsilon = 0.9$ and $\epsilon = 0.8$ respectively.

Observation: It can be observed from figures that the synchronization time *decreases* with *increase* in α as well as with *decrease* in ϵ.

5.3 Synchronization of Fractional Order Delay System

Consider fractional order Ucar system with delay [118]

$$D^{\alpha}x(t) = x(t-\tau) - [x(t-\tau)]^3 = x_\tau - x_\tau^3, \tag{38}$$

$$x_0(t) = 0.5, \; t \leq 0 \tag{39}$$

as a drive system. The system exhibits chaotic motion for different values of fractional order α and delay τ citepucar. Figure 5a shows chaotic attractor shown by the system for $\alpha = 0.99$ and $\tau = 1.7$. We consider the the response system as

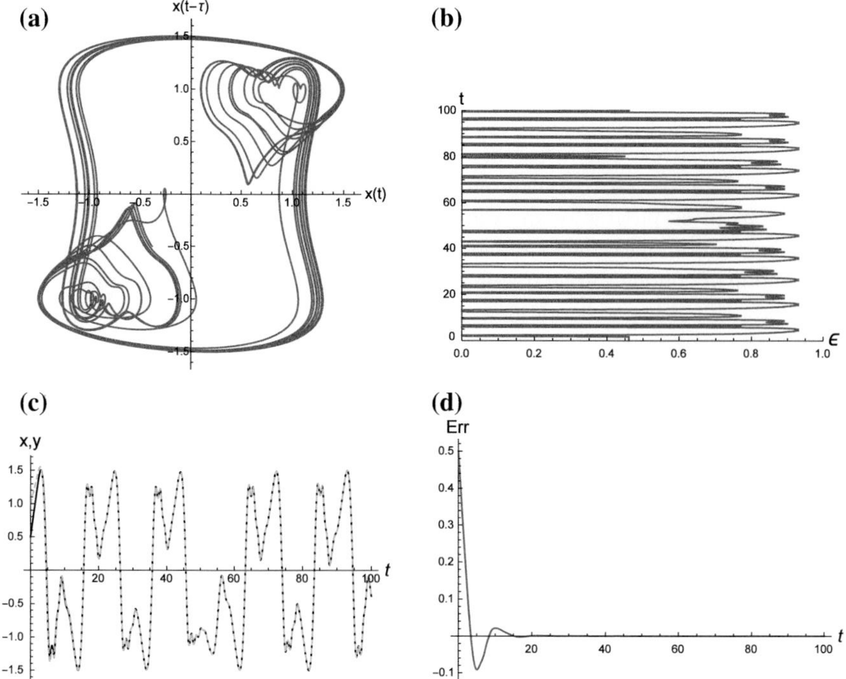

Fig. 5 Simulations for the Ucar delay system with $\alpha = 0.99$ and $\tau = 1.7$. **a** Chaotic attractor, **b** local stable region, **c** synchronized states, **d** synchronization error

$$D^{\alpha} y(t) = y_{\tau} - y_{\tau}^{3} + \epsilon u, \quad y_{0}(t) = 1, \ t \leq 0 \tag{40}$$

where $0 < \alpha < 1$, $\tau > 0$. If we write the control term as

$$u = a(y_{\tau} - x_{\tau}) + y_{\tau}^{3} - x_{\tau}^{3}, \tag{41}$$

then the approximation to the error system (23) becomes

$$D^{\alpha} e = \left(1 + a\epsilon - 3x_{\tau}^{2}(1 - \epsilon)\right) e_{\tau}. \tag{42}$$

If the constant a is set to -1.3 then the system (42) will be stable for $\epsilon = 1$. The local stable region for this case is shown in Fig. 5b. We have shown the synchronized states and error in synchronization in Fig. 5c, d respectively for $\epsilon = 0.95$.

Observations: The extended active control can be used to synchronize delayed systems also. It is clear from the figures that the value $\epsilon = 0.94$ is sufficient for the stability of error system.

6 Discussion

Fractional order systems are now proved to be realistic models for the natural systems. The memory arising in such systems can be better modeled by using (nonlocal) fractional derivatives. The researchers from diverse field are attracting to the relatively new field viz. chaos-synchronization in fractional order systems due to its applications in cryptography and other fields. In this chapter, we have discussed the application of an extended active control (EAC) in chaos and hyperchaos synchronization of commensurate fractional order systems. Fractional order chaotic Liu system, fractional hyperchaotic new system and fractional Ucar delay system are discussed to explain the proposed theory. It is observed that the coupling strength ϵ in active control can be set to the value less than one to get synchronized states. With this modification, the efficiency of the controller can be enhanced remarkably. The EAC will be more useful in practical applications than the classical active control. We hope that this work will be useful to the researchers working in the field of synchronization of fractional systems.

7 Conclusion

Natural systems frequently arise with memory and hereditary properties which cannot be modeled by ordinary derivative models. Due to the nonlocal nature of fractional derivatives, these are capable of modeling memory in the system. Such systems have applications in various fields of science. It is discussed in the literature that the chaotic dynamical systems can be made stable by proper choice of fractional derivative. Further the chaos synchronization in fractional order systems can be used to produce secure communication schemes which are difficult to break. In this article, we have proposed an extended active control in case of commensurate order systems which is useful in practical applications due to small coupling strength. We plan in future to generalize this approach to incommensurate order systems. We also plan to propose new techniques to synchronize fractional order chaotic systems and study their applications.

Acknowledgments Author acknowledges Shivaji University, Kolhapur, India for the research grant provided under the Innovative Research Activities (2014–2016). The author is grateful to Prof. Ahmad Taher Azar for his encouragement and support.

References

1. Vaidyanathan S, Azar AT (2015) Analysis, control and synchronization of a nine-term 3-d novel chaotic system. In: Azar AT, Vaidyanathan S (eds) Chaos modeling and control systems design. Studies in computational intelligence, vol 581. Springer-Verlag GmbH, Berlin, pp 3–17
2. Vaidyanathan S, Azar AT (2015) Analysis, control and synchronization of a nine-term 3-D novel chaotic system. In: Azar AT, Vaidyanathan S (eds) Chaos modeling and control systems design. Studies in computational intelligence, vol 581. Springer, Berlin, pp 3–17
3. Vaidyanathan S, Azar AT (2015) Analysis and control of a 4-D novel hyperchaotic system. In: Azar AT, Vaidyanathan S (eds) Chaos modeling and control systems design. Studies in computational intelligence, vol 581. Springer-Verlag GmbH, Berlin, pp 19–38
4. Vaidyanathan S, Azar AT (2015) Analysis and control of a 4-D novel hyperchaotic system. In: Azar AT, Vaidyanathan S (eds) Chaos modeling and control systems design. Studies in computational intelligence, vol 581. Springer, Berlin, pp 19–38
5. Azar AT, Vaidyanathan S (eds) (2015) Chaos modeling and control systems design. Studies in computational intelligence, vol 581. Springer, New York
6. Azar AT, Vaidyanathan S (2015) Handbook of research on advanced intelligent control engineering and automation. Advances in Computational Intelligence and Robotics (ACIR) Book Series. IGI Global, Hershey PA
7. Azar AT, Vaidyanathan S (2015) Computational intelligence applications in modeling and control. Studies in computational intelligence, vol 575. Springer, New York
8. Pecora LM, Carroll TL (1990) Synchronization in chaotic systems. Phys Rev Lett 64(8):821
9. Pecora LM, Carroll TL (1991) Driving systems with chaotic signals. Phys Rev A 44:2374
10. Hilfer R (ed) (2001) Applications of fractional calculus in physics. World Scientific, Singapore
11. He R, Vaidya PG (1998) Implementation of chaotic cryptography with chaotic synchronization. Phys Rev E 57(2):1532
12. Huang L, Feng R, Wang M (2004) Synchronization of chaotic systems via nonlinear control. Phys Lett A 320:271
13. Liao TL (1998) Adaptive synchronization of two Lorenz systems. Chaos Solitons Fractals 9:1555
14. Yassen MT (2001) Adaptive control and synchronization of a modified Chua's circuit system. Appl Math Comput 135(1):113
15. Bai EW, Lonngre KE (1997) Synchronization of two Lorenz systems using active control. Chaos Solitons Fractals 8:51–58
16. Bai EW, Lonngren KE (2000) Sequential synchronization of two Lorenz systems using active control. Chaos Solitons Fractals 11:1041–1044
17. Vaidyanathan S, Azar AT, Rajagopal K, Alexander P (2015) Design and SPICE implementation of a 12-term novel hyperchaotic system and its synchronization via active control. Int J Model Identif Control 23(3):267–277
18. Azar AT, Serrano FE (2015) Adaptive Sliding mode control of the Furuta pendulum. In: Azar AT, Zhu Q (eds) Advances and applications in sliding mode control systems. Studies in computational intelligence, vol 576. Springer, Berlin, pp 1–42
19. Azar AT, Zhu Q (2015) Advances and applications in sliding mode control systems. Studies in computational intelligence, vol 576. Springer, Germany
20. Vaidyanathan S, Azar AT (2015) Anti-synchronization of identical chaotic systems using sliding mode control and an application to Vaidyanathan-Madhavan chaotic systems. In: Azar AT, Zhu Q (eds) Advances and applications in sliding mode control systems. Studies in computational intelligence book series. Springer, Berlin, pp 527–547
21. Vaidyanathan S, Azar AT (2015) Hybrid synchronization of identical chaotic systems using sliding mode control and an application to Vaidyanathan chaotic systems. In: Azar AT, Zhu Q (eds) (2015) Advances and applications in sliding mode control systems. Studies in computational intelligence book series. Springer, Berlin, pp 549–569

22. Vaidyanathan S, Sampath S, Azar AT (2015) Global chaos synchronisation of identical chaotic systems via novel sliding mode control method and its application to Zhu system. Int J Model Identif Control 23(1):92–100
23. Azar AT, Serrano FE (2015) Deadbeat control for multivariable systems with time varying delays. In: Azar AT, Vaidyanathan S (eds) Chaos modeling and control systems design. Studies in computational intelligence, vol 581. Springer, Berlin, pp 97–132
24. Vaidyanathan S, Idowu BA, Azar AT (2015) Backstepping controller design for the global chaos synchronization of Sprott's jerk systems. In: Azar AT, Vaidyanathan S (eds) Chaos modeling and control systems design. Springer, Berlin, pp 39–58
25. Daftardar-Gejji V, Bhalekar S (2008) Solving multi-term linear and non-linear diffusion-wave equations of fractional order by Adomian decomposition method. Appl Math Comput 202:113–120
26. Ingo C, Magin RL, Parrish TB (2014) New insights into the fractional order diffusion equation using entropy and kurtosis. Entropy 16(11):5838–5852
27. Jesus IS, Machado JAT (2008) Fractional control of heat diffusion systems. Nonlinear Dyn 54(3):263–282
28. Jesus IS, Machado JAT, Barbosa RS (2010) Control of a heat diffusion system through a fractional order nonlinear algorithm. Comput Math Appl 59(5):1687–1694
29. Mainardi F, Luchko Y, Pagnini G (2001) The fundamental solution of the space-time fractional diffusion equation. Fract Calc Appl Anal 4(2):153–192
30. Mainardi F (2010) Fractional calculus and waves in linear viscoelasticity: an introduction to mathematical models. Imperial College Press, London
31. Magin RL (2006) Fractional calculus in bioengineering. Begll House Publishers, Redding
32. Anastasio TJ (1994) The fractional-order dynamics of Brainstem Vestibulo-Oculomotor neurons. Biol Cybern 72:69–79
33. Ortigueira MD, Machado JAT (2006) Fractional calculus applications in signals and systems. Signal Process 86(10):2503–2504
34. Tseng C, Lee SL (2014) Digital image sharpening using Riesz fractional order derivative and discrete hartley transform. In: 2014 IEEE Asia pacific conference on circuits and systems (APCCAS). IEEE, Ishigaki, pp 483–486
35. Ran Q, Yuan L, Zhao T (2015) Image encryption based on nonseparable fractional Fourier transform and chaotic map. Optics Commun 348:43–49
36. Wu GC, Baleanu D, Lin ZX (2015) Image encryption technique based on fractional chaotic time series. J Vibr Control. Article 1077546315574649
37. Zhao J, Chang Y, Li X (2015) A novel image encryption scheme based on an improper fractional-order chaotic system. Nonlinear Dyn 80(4):1721–1729
38. Wang S, Sun W, Ma CY, Wang D, Chen Z (2013) Secure communication based on a fractional order chaotic system. Int J Security Appl 7(5):205–216
39. Sabatier J, Poullain S, Latteux P, Thomas J, Oustaloup A (2004) Robust speed control of a low damped electromechanical system based on CRONE control: application to a four mass experimental test bench. Nonlinear Dyn 38:383–400
40. Meilanov RP, Magomedov RA (2014) Thermodynamics in fractional calculus. J Eng Phys Thermophys 87(6):1521–1531
41. Fu-Hong M, Shu-Yi S, Wen-Di H, En-Rong W (2015) Circuit implementations, bifurcations and chaos of a novel fractional-order dynamical system. Chin Phys Lett 32(3):030503
42. Liao H (2014) Optimization analysis of Duffing oscillator with fractional derivatives. Nonlinear Dyn 79(2):1311–1328
43. Xu B, Chen D, Zhang H, Wang F (2015) Modeling and stability analysis of a fractional-order Francis hydro-turbine governing system. Chaos Solitons Fractals 75:50–61
44. Muthukumar P, Balasubramaniam P (2013) Feedback synchronization of the fractional order reverse butterfly-shaped chaotic system and its application to digital cryptography. Nonlinear Dyn 74:1169–1181
45. Muthukumar P, Balasubramaniam P, Ratnavelu K (2014) Synchronization and an application of a novel fractional order King Cobra chaotic system. Chaos 24(3):033105

46. Muthukumar P, Balasubramaniam P, Ratnavelu K (2015) Sliding mode control design for synchronization of fractional order chaotic systems and its application to a new cryptosystem. Int J Dyn Control. doi:10.1007/s40435-015-0169-y (in press)
47. Barrett JH (1954) Differential equations of non-integer order. Can J Math 64:529–541
48. Al-Bassam MA (1965) Some existence theorems on differential equations of generalized order. Journal fr die reine und angewandte Mathematik 2181:70–78
49. Delbosco D, Rodino L (1996) Existence and uniqueness for a nonlinear fractional differential equation. J Math Anal Appl 2042:609–625
50. Diethelm K, Ford NJ (2002) Analysis of fractional differential equations. J Math Anal Appl 2652:229–248
51. Daftardar-Gejji V, Babakhani A (2004) Analysis of a system of fractional differential equations. J Math Anal Appl 2932:511–522
52. Gorenflo R, Mainardi F (1996) Fractional oscillations and Mittag-Leffler functions. In: International workshop on the recent advances in applied mathematics. Kuwait University, Department of Mathematics and Computer Science, State of Kuwait, pp 193–208
53. Luchko YF, Gorenflo R (1999) An operational method for solving fractional differential equations with the Caputo derivatives. Acta Mathematica Vietnamica 24:207–233
54. Daftardar-Gejji V, Jafari H (2006) Boundary value problems for fractional diffusion-wave equation. Aust J Math Anal Appl 3:1–8
55. Daftardar-Gejji V, Bhalekar S (2008) Boundary value problems for multi-term fractional differential equations. J Math Anal Appl 345:754–765
56. Diethelm K, Ford NJ, Freed AD (2002) A predictor-corrector approach for the numerical solution of fractional differential equations. Nonlinear Dyn 29:3–22
57. Daftardar-Gejji V, Sukale Y, Bhalekar S (2014) A new predictorcorrector method for fractional differential equations. Appl Math Comput 244:158–182
58. Adomian G (1994) Solving Frontier problems of physics: the decomposition method. Kluwer Academic, Dordrecht
59. He JH (1998) Approximate analytical solution for seepage flow with fractional derivatives in porous media. Comput Methods Appl Mech Eng 167:57–68
60. He JH (1999) Homotopy perturbation technique. Comput Methods Appl Mech Eng 178:257–262
61. Daftardar-Gejji V, Jafari H (2006) An iterative method for solving non linear functional equations. J Math Anal Appl 316:753–763
62. Tang RA, Liu YL, Xue JK (2009) An extended active control for chaos synchronization. Phys Lett A 373:1449–1454
63. Deng WH, Li CP (2005) Chaos synchronization of the fractional Lü system. Phys A 353:61–72
64. Deng WH, Li CP (2005) Synchronization of chaotic fractional Chen system. J Phys Soc Jpn 74:1645–1648
65. Li CP, Deng WH (2006) Chaos synchronization of fractional order differential system. Int J Mod Phys B 20(7):791–803
66. Li CP, Deng WH, Xu D (2006) Chaos synchronization of the Chua system with a fractional order. Phys A 360:171–185
67. Wang J, Zhang Y (2006) Designing synchronization schemes for chaotic fractional-order unified systems. Chaos Solitons Fractals 30:1265–1272
68. Yu Y, Li H (2008) The synchronization of fractional-order Rossler hyperchaotic systems. Phys A 387:1393–1403
69. Wang J, Xionga X, Zhang Y (2006) Extending synchronization scheme to chaotic fractional-order Chen systems. Phys A 370:279–285
70. Jun D, Guangjun Z, Shaoying W, Qiongyao L (2014) Chaotic synchronization between fractional-order financial system and financial system of integer orders. In: Control and decision conference (2014 CCDC), the 26th Chinese IEEE. IEEE, Changsha, pp 4924–4928
71. Mohadeszadeh M, Delavari H (2015) Synchronization of fractional-order hyper-chaotic systems based on a new adaptive sliding mode control. Int J Dyn Control. doi:10.1007/s40435-015-0177-y (in press)

72. Gao L, Wang Z, Zhou K, Zhu W, Wu Z, Ma T (2015) Modified sliding mode synchronization of typical three-dimensional fractional-order chaotic systems. Neurocomputing 166:53–58
73. Tian X, Fei S, Chai L (2015) On modified second-order sliding mode synchronization of two different fractional order hyperchaotic systems. Int J Multimed Ubiquitous Eng 10(4): 387–398
74. Guanand J, Wang K (2015) Sliding mode control and modified generalized projective synchronization of a new fractional-order chaotic system. Math Probl Eng. ID 941654
75. Tavazoei MS, Haeri M (2008) Synchronization of chaotic fractional-order systems via active sliding mode controller. Phys A 387:57–70
76. Xu Y, Wang H (2013) Synchronization of fractional-order chaotic systems with Gaussian fluctuation by sliding mode control. Abstr Appl Anal. Article ID 948782, 7 pages
77. Liu H, Yang J (2015) Sliding-mode synchronization control for uncertain fractional-order chaotic systems with time delay. Entropy 17:4202–4214
78. Li J, Guo X, Yao L (2014) Adaptive synchronization of fractional-order general complex dynamical networks. In: 2014 11th world congress on intelligent control and automation (WCICA). IEEE, Shenyang, pp 4367–4372
79. Rad P, Nikdadian M, Bahadorzadeh M (2015) Synchronizing the fractional-order Genesio-Tesi chaotic system using adaptive control. Int J Sci Eng Res 6:1699–1702
80. Zhou P, Bai R (2015) The adaptive synchronization of fractional-order chaotic system with fractional-order $1 < q < 2$ via linear parameter update law. Nonlinear Dyn 80:753–765
81. Leung A, Li X, Chu Y, Rao X (2015) Synchronization of fractional-order chaotic systems using unidirectional adaptive full-state linear error feedback coupling. Nonlinear Dyn 82(1–2):185–199
82. El-Sayed AMA, Nour HM, Elsaid A, Matouk AE, Elsonbaty A (2015) Dynamical behaviors, circuit realization, chaos control and synchronization of a new fractional order hyperchaotic system. Appl Math Model doi:10.1016/j.apm.2015.10.010 (in press)
83. Xingyuan W, Yijie H (2008) Projective synchronization of fractional order chaotic system based on linear separation. Phys Lett A 372:435–441
84. Agrawal S, Das S (2014) Projective synchronization between different fractional-order hyperchaotic systems with uncertain parameters using proposed modified adaptive projective synchronization technique. Math Methods Appl Sci 37:2164–2176
85. Liu J, Liu S, Yuan C (2013) Modified generalized projective synchronization of fractional-order chaotic Lü systems. Adv Diff Equ 2013(1). Article 374
86. Zhou P, Ding R, Cao Y (2012) Hybrid projective synchronization for two identical fractional-order chaotic systems. Discrete Dyn Nat Soc. Article ID 768587, 11 pages
87. Chen L, Chai Y, Wu R (2011) Lag projective synchronization in fractional-order chaotic (hyperchaotic) systems. Phys Lett A 375(21):2099–2110
88. Sun Z (2015) Lag projective synchronization of two chaotic systems with different fractional orders. J Korean Phys Soc 66:1192–1199
89. Bhalekar S, Daftardar-Gejji V (2010) Synchronization of different fractional order chaotic systems using active control. Commun Nonlinear Sci Numer Simul 15(11):3536–3546
90. Bhalekar S, Daftardar-Gejji V (2011) Anti-synchronization of non-identical fractional order chaotic systems using active control. Int J Differ Equ. Article ID 250763
91. Bhalekar S (2014) Synchronization of incommensurate non-identical fractional order chaotic systems using active control. Eur Phys J Special Topics 223(8):1495–1508
92. Bhalekar S (2014) Synchronization of non-identical fractional order hyperchaotic systems using active control. World J Model Simul 10(1):60–68
93. Li M, Li D, Wang J, Zhao C (2013) Active disturbance rejection control for fractional-order system. ISA Trans 52(3):365–374
94. Gao Z, Liao X (2014) Active disturbance rejection control for synchronization of different fractional-order chaotic systems. In: 11th world congress on intelligent control and automation (WCICA). IEEE, Shenyang, pp 2699–2704
95. Li D, Zhang X, Hu Y, Yang Y (2015) Adaptive impulsive synchronization of fractional order chaotic system with uncertain and unknown parameters. Neurocomputing 167:165–171

96. Li C, Zhang J (2015) Synchronisation of a fractional-order chaotic system using finite-time input-to-state stability. Int J Syst Sci doi:10.1080/00207721.2014.998741 (in press)
97. Ran D, Caoyuan M, Yongyi Z, Yanfang L, Jianhua L (2014) Anti-synchronization of a class of fractional-order chaotic system with uncertain parameters. Comput Model New Technol 18(11):108–112
98. Noghredani N, Balochian S (2015) Synchronization of fractional-order uncertain chaotic systems with input nonlinearity. Int J General Syst 44:485–498
99. Balasubramaniam P, Muthukumar P, Ratnavelu K (2015) Theoretical and practical applications of fuzzy fractional integral sliding mode control for fractional-order dynamical system. Nonlinear Dyn 80:249–267
100. Chen L, Chai Y, Wu R, Ma T, Zhai H (2013) Dynamic analysis of a class of fractional-order neural networks with delay. Neurocomputing 111:190–194
101. Dang HG, He WS, Yang XY (2014) Investigation of synchronization for a fractional-order delayed system. Appl Mech Mater 687:447–450
102. Xiaohong Z, Peng C (2015) Different-lags synchronization in time-delay and circuit simulation of fractional-order chaotic system based on parameter identification. Open Electr Electr Eng J 9:117–126
103. Velmurugan G, Rakkiyappan R, Cao J (2015) Finite-time synchronization of fractional-order memristor-based neural networks with time delays. Neural Netw doi:10.1016/j.neunet.2015.09.012 (in press)
104. Liang S, Wu R, Chen L Adaptive pinning synchronization in fractional-order uncertain complex dynamical networks with delay. Phys A: Stat Mech Appl doi:10.1016/j.physa.2015.10.011 (in press)
105. Kiani-B A, Fallahi K, Pariz N, Leung H (2009) A chaotic secure communication scheme using fractional chaotic systems based on an extended fractional Kalman filter. Commun Nonlinear Sci Numer Simul 14(3):863–879
106. Sheu LJ, Chen WC, Chen YC, Weng WT (2010) A two-channel secure communication using fractional chaotic systems. World Acad Sci Eng Technol 65:1057–1061
107. Huang L, Zhang J, Shi S (2015) Circuit simulation on control and synchronization of fractional order switching chaotic system. Math Comput Simul 113:28–39
108. Zhen W, Xia H, Ning L, Xiao-Na S (2012) Image encryption based on a delayed fractional-order chaotic logistic system. Chin Phys B 21. Article ID 050506
109. Zhen W, Xia H, Yu-Xia L, Xiao-Na S (2013) A new image encryption algorithm based on the fractional-order hyperchaotic Lorenz system. Chin Phys B 22(1). Article ID 010504
110. Xu Y, Wang H, Li Y, Pei B (2014) Image encryption based on synchronization of fractional chaotic systems. Commun Nonlinear Sci Numer Simul 19(10):3735–3744
111. Huang X, Sun T, Li Y, Liang J (2015) A color image encryption algorithm based on a fractional-order hyperchaotic system. Entropy 17(1):28–38
112. Kilbas AA, Srivastava HM, Trujillo JJ (2006) Theory and applications of fractional differential equations. Elsevier, Amsterdam
113. Podlubny I (1999) Fractional differential equations. Academic Press, San Diego
114. Matignon D (1996) Stability results for fractional differential equations with applications to control processing. In: Computational engineering in systems and application multiconference. Gerf EC Lille, Villeneuve d'Ascq, Lille, pp 963–968
115. Bhalekar S (2013) Stability analysis of fractional differential systems with delay. In: Daftardar-Gejji V (ed) Fractional calculus: theory and applications. Narosa Publishing House, New Delhi, pp 60–68
116. Daftardar-Gejji V, Bhalekar S (2010) Chaos in fractional ordered Liu system. Comput Math Appl 59(3):1117–1127
117. Liang CG, Song Z, Xin TL (2008) A new hyperchaotic system and its linear feedback control. Chin Phys B 17:4039–4046
118. Bhalekar S (2012) Dynamical analysis of fractional order Ucar prototype delayed system. Signals Image Video Process 6(3):513–519

A Novel 4-D Hyperchaotic Thermal Convection System and Its Adaptive Control

Sundarapandian Vaidyanathan

Abstract In this work, we announce an eleven-term novel 4-D hyperchaotic thermal convection system with two quadratic nonlinearities. The phase portraits of the novel hyperchaotic system are depicted and the qualitative properties of the novel hyperchaotic system are discussed. The novel 4-D hyperchaotic thermal convection system is obtained by introducing a feedback control to the 3-D thermal convection system obtained by Wang et al. (J Fluid Mech 237:479–498, 1992). The Lyapunov exponents of the novel hyperchaotic thermal convection system are obtained as $L_1 = 0.40546$, $L_2 = 0.03583$, $L_3 = 0$ and $L_4 = -6.44038$. Since there are two positive Lyapunov exponents for the novel 4-D thermal convection system, it is hyperchaotic. The Maximal Lyapunov Exponent (MLE) of the novel hyperchaotic system is found as $L_1 = 0.40546$. Also, the Kaplan-Yorke dimension of the novel hyperchaotic system is derived as $D_{KY} = 3.0685$. Since the sum of the Lyapunov exponents is negative, the novel hyperchaotic system is dissipative. Next, an adaptive controller is designed to globally stabilize the novel hyperchaotic thermal convection system with unknown parameters. Finally, an adaptive controller is also designed to achieve global chaos synchronization of the identical novel hyperchaotic thermal convection systems with unknown parameters. MATLAB simulations are depicted to illustrate all the main results derived in this work.

Keywords Chaos · Chaotic systems · Hyperchaos · Hyperchaotic systems · Thermal convection · Adaptive control · Synchronization

---

S. Vaidyanathan (✉)
Research and Development Centre, Vel Tech University, Avadi, Chennai 600062, Tamil Nadu, India
e-mail: sundarvtu@gmail.com

A.T. Azar and S. Vaidyanathan (eds.), *Advances in Chaos Theory and Intelligent Control*, Studies in Fuzziness and Soft Computing 337,
DOI 10.1007/978-3-319-30340-6_4

1 Introduction

In the last few decades, Chaos theory has become a very important and active research field, employing many applications in different disciplines like physics, chemistry, biology, ecology, engineering and economics, among others.

Some classical paradigms of 3-D chaotic systems in the literature are Lorenz system [1], Rössler system [2], ACT system [3], Sprott systems [4], Chen system [5], Lü system [6], Cai system [7], Tigan system [8], etc.

Many new chaotic systems have been discovered in the recent years such as Zhou system [9], Zhu system [10], Li system [11], Wei-Yang system [12], Sundarapandian systems [13, 14], Vaidyanathan systems [15–31], Pehlivan system [32], Sampath system [33], Pham system [34], etc.

The Lyapunov exponent is a measure of the divergence of phase points that are initially very close and can be used to quantify chaotic systems. It is common to refer to the largest Lyapunov exponent as the *Maximal Lyapunov Exponent* (MLE). A positive maximal Lyapunov exponent and phase space compactness are usually taken as defining conditions for a chaotic system.

A hyperchaotic system is defined as a chaotic system with at least two positive Lyapunov exponents. Thus, the dynamics of a hyperchaotic system can expand in several different directions simultaneously. Thus, the hyperchaotic systems have more complex dynamical behaviour and they have miscellaneous applications in engineering such as secure communications [35–37], cryptosystems [38–40], fuzzy logic [41, 42], electrical circuits [43, 44], etc.

The minimum dimension of an autonomous, continuous-time, hyperchaotic system is four. The first 4-D hyperchaotic system was found by Rössler [45]. Many hyperchaotic systems have been reported in the chaos literature such as hyperchaotic Lorenz system [46], hyperchaotic Lü system [47], hyperchaotic Chen system [48], hyperchaotic Wang system [49], hyperchaotic Vaidyanathan systems [50–58], hyperchaotic Pham system [59], etc.

Chaos theory and control systems have many important applications in science and engineering [60–65]. Some commonly known applications are oscillators [66, 67], chemical reactions [68–75], biology [76–85], ecology [86, 87], encryption [88, 89], cryptosystems [90, 91], mechanical systems [92–96], secure communications [97–99], robotics [100–102], cardiology [103, 104], intelligent control [105, 106], neural networks [107–109], memristors [110, 111], etc.

The control of a chaotic or hyperchaotic system aims to stabilize or regulate the system with the help of a feedback control. There are many methods available for controlling a chaotic system such as active control [112–114], adaptive control [115–117], sliding mode control [118, 119], backstepping control [120], etc.

The synchronization of chaotic systems aims to synchronize the states of master and slave systems asymptotically with time. There are many methods available for chaos synchronization such as active control [121–125], adaptive control [126–132], sliding mode control [133–136], backstepping control [137–140], etc.

This work is organized as follows. Section 2 describes the dynamic equations and phase portraits of the eleven-term novel 4-D hyperchaotic thermal convection system. Section 3 details the qualitative properties of the novel hyperchaotic thermal convection system. Our novel hyperchaotic thermal convection system is obtained by introducing a feedback control to the 3-D thermal convection system proposed by Wang et al. [141].

The Lyapunov exponents of the novel hyperchaotic thermal convection system are obtained as $L_1 = 0.40546$, $L_2 = 0.03583$, $L_3 = 0$ and $L_4 = -6.44038$, while the Kaplan-Yorke dimension of the novel hyperchaotic thermal convection system is obtained as $D_{KY} = 3.0685$. Since the sum of the Lyapunov exponents of the novel hyperchaotic thermal convection system is negative, the system is dissipative.

In Sect. 4, we design an adaptive controller to globally stabilize the novel hyperchaotic thermal convection system with unknown parameters. In Sect. 5, an adaptive controller is designed to achieve global chaos synchronization of the identical novel hyperchaotic thermal convection systems with unknown parameters. MATLAB simulations have been shown to illustrate all the main results derived in this research work. Section 6 summarizes the main results of this research work.

2 A Novel 4-D Hyperchaotic Thermal Convection System

In 1992, Wang et al. [141] designed a 3-D model for chaotic flow obtained in a toroidal thermal convection loop heated from below and cooled from above. This 3-D thermal convection system is described by the dynamics

$$\begin{cases} \dot{x}_1 = a(x_2 - x_1) \\ \dot{x}_2 = -x_2 - x_1 x_3 \\ \dot{x}_3 = x_1 x_2 - x_3 - b \end{cases} \tag{1}$$

where x_1, x_2, x_3 are the state variables and a, b are constant, positive, parameters.

In [141], it was shown that the 3-D system (1) is *chaotic* when the system parameters are taken as

$$a = 4, \quad b = 16 \tag{2}$$

For numerical simulations, we take the initial state values as

$$x_1(0) = 0.4, \quad x_2(0) = 0.4, \quad x_3(0) = 0.4 \tag{3}$$

For the parameter values (2) and the initial conditions (3), the Lyapunov exponents of the Wang's thermal convection system (1) are obtained as

$$L_1 = 0.35798, \quad L_2 = 0, \quad L_3 = -6.35501 \tag{4}$$

Also, the Kaplan-Yorke dimension of the Wang's thermal convection system (1) is derived as

$$D_{KY} = 2 + \frac{L_1 + L_2}{|L_3|} = 2.0563 \tag{5}$$

Figure 1 shows the strange attractor of the Wang's thermal convection system (1) in the 3-D space. From Fig. 1, it is clear that Wang's thermal convection system (1) exhibits a *two-scroll* chaotic attractor in the space.

In this section, we describe an eleven-term novel hyperchaotic thermal convection system, which is obtained by introducing a feedback control to Wang's 3-D thermal convection system (1).

This work proposes a novel 4-D hyperchaotic thermal convection system given by the dynamics

$$\begin{cases} \dot{x}_1 = a(x_2 - x_1) + cx_4 \\ \dot{x}_2 = -x_1x_3 - x_2 + cx_4 \\ \dot{x}_3 = x_1x_2 - x_3 - b \\ \dot{x}_4 = -p(x_1 + x_2) \end{cases} \tag{6}$$

where x_1, x_2, x_3, x_4 are the states and a, b, c, p are constant positive parameters.

The system (6) exhibits a *strange hyperchaotic attractor* for the parameter values

$$a = 4, \quad b = 20, \quad c = 0.2, \quad p = 0.5 \tag{7}$$

For numerical simulations, we take the initial conditions as

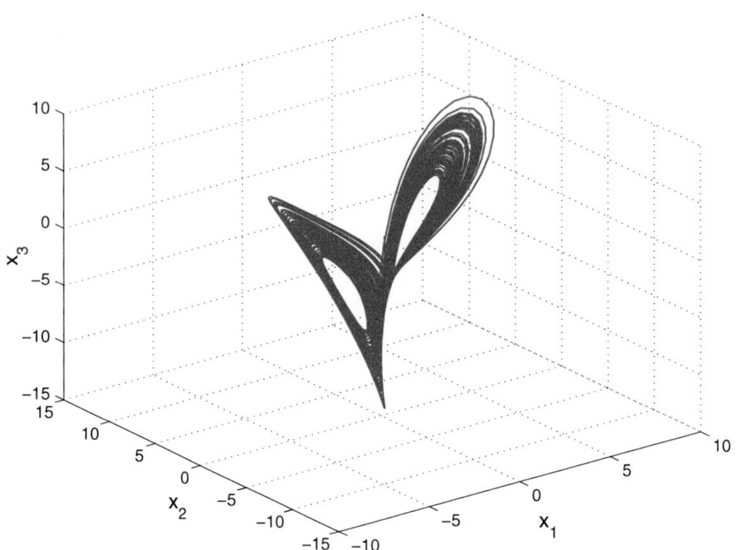

Fig. 1 Strange attractor of the Wang's thermal convection system

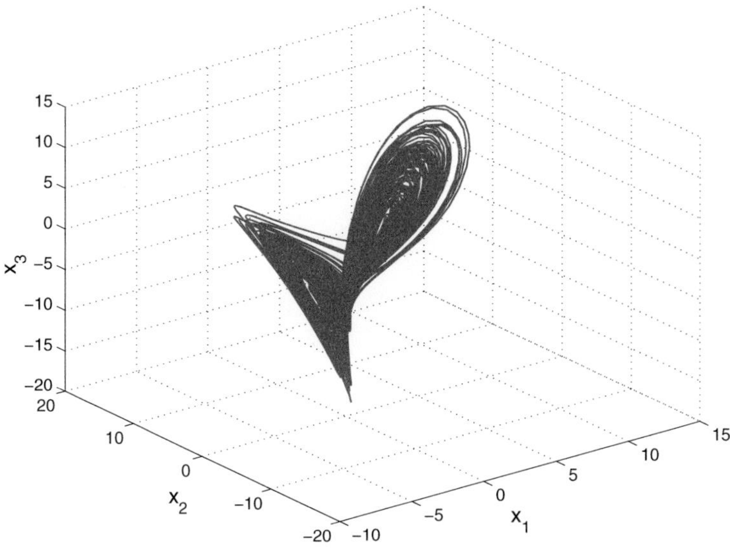

Fig. 2 3-D projection of the hyperchaotic thermal convection system on the (x_1, x_2, x_3) space

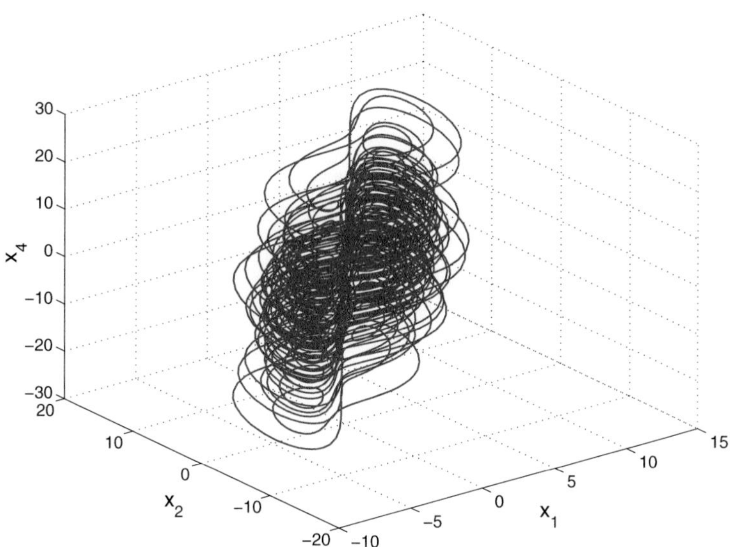

Fig. 3 3-D projection of the hyperchaotic thermal convection system on the (x_1, x_2, x_4) space

$$x_1(0) = 0.4, \quad x_2(0) = 0.4, \quad x_3(0) = 0.4, \quad x_4(0) = 0.4 \tag{8}$$

For the parameter values (7) and the initial conditions (8), the Lyapunov exponents of the 4-D thermal convection system (6) are obtained as

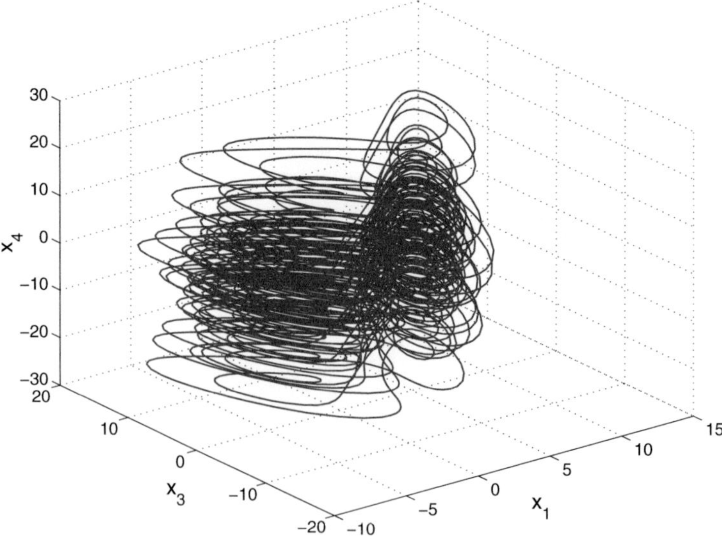

Fig. 4 3-D projection of the hyperchaotic thermal convection system on the (x_1, x_3, x_4) space

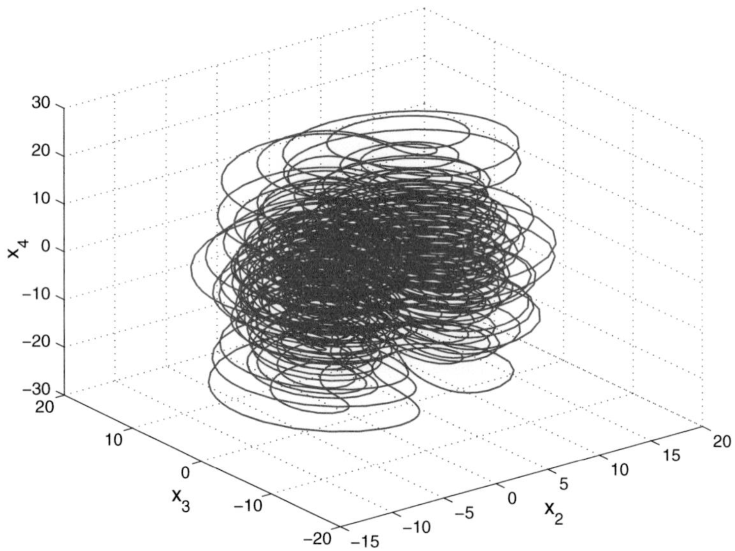

Fig. 5 3-D projection of the hyperchaotic thermal convection system on the (x_2, x_3, x_4) space

$$L_1 = 0.40546, \quad L_2 = 0.03583, \quad L_3 = 0, \quad L_4 = -6.44038 \tag{9}$$

Also, the Kaplan-Yorke dimension of the hyperchaotic 4-D thermal convection system (6) is derived as

$$D_{KY} = 3 + \frac{L_1 + L_2 + L_3}{|L_4|} = 3.0685 \tag{10}$$

Figures 2, 3, 4 and 5 show the 3-D projection of the novel hyperchaotic system (6) on the (x_1, x_2, x_3), (x_1, x_2, x_4), (x_1, x_3, x_4) and (x_2, x_3, x_4) spaces, respectively.

3 Analysis of the Novel 4-D Hyperchaotic Thermal Convection System

In this section, we study the qualitative properties of the novel 4-D hyperchaotic thermal convection system (6). We take the parameter values as in the hyperchaotic case (7).

3.1 Dissipativity

In vector notation, the novel hyperchaotic thermal convection system (6) can be expressed as

$$\dot{\mathbf{x}} = f(\mathbf{x}) = \begin{bmatrix} f_1(x_1, x_2, x_3, x_4) \\ f_2(x_1, x_2, x_3, x_4) \\ f_3(x_1, x_2, x_3, x_4) \\ f_4(x_1, x_2, x_3, x_4) \end{bmatrix}, \tag{11}$$

where

$$\begin{cases} f_1(x_1, x_2, x_3, x_4) = a(x_2 - x_1) + cx_4 \\ f_2(x_1, x_2, x_3, x_4) = -x_1 x_3 - x_2 + cx_4 \\ f_3(x_1, x_2, x_3, x_4) = x_1 x_2 - x_3 - b \\ f_4(x_1, x_2, x_3, x_4) = -p(x_1 + x_2) \end{cases} \tag{12}$$

Let Ω be any region in $\mathbf{R}^4$ with a smooth boundary and also, $\Omega(t) = \Phi_t(\Omega)$, where Φ_t is the flow of f.

Furthermore, let $V(t)$ denote the hypervolume of $\Omega(t)$.

By Liouville's theorem, we know that

$$\dot{V}(t) = \int\limits_{\Omega(t)} (\nabla \cdot f)\, dx_1\, dx_2\, dx_3\, dx_4 \tag{13}$$

The divergence of the novel hyperchaotic system (11) is found as:

$$\nabla \cdot f = \frac{\partial f_1}{\partial x_1} + \frac{\partial f_2}{\partial x_2} + \frac{\partial f_3}{\partial x_3} + \frac{\partial f_4}{\partial x_4} = -a - 2 = -\mu < 0 \tag{14}$$

since $\mu = a + 2 > 0$.

Inserting the value of $\nabla \cdot f$ from (14) into (13), we get

$$\dot{V}(t) = \int\limits_{\Omega(t)} (-\mu)\, dx_1\, dx_2\, dx_3\, dx_4 = -\mu V(t) \tag{15}$$

Integrating the first order linear differential equation (15), we get

$$V(t) = \exp(-\mu t)V(0) \tag{16}$$

Since $\mu > 0$, it follows from Eq. (16) that $V(t) \to 0$ exponentially as $t \to \infty$. This shows that the novel hyperchaotic thermal convection system (6) is dissipative. Hence, the system limit sets are ultimately confined into a specific limit set of zero hypervolume, and the asymptotic motion of the novel hyperchaotic thermal convection system (6) settles onto a strange attractor of the system.

3.2 Equilibrium Points

We take the parameter values as in the hyperchaotic case (7), i.e.

$$a = 4, \quad b = 20, \quad c = 0.2, \quad p = 0.5 \tag{17}$$

The equilibrium points of the novel 4-D hyperchaotic thermal convection system (6) are obtained by solving the system of equations

$$\begin{cases} a(x_2 - x_1) + cx_4 &= 0 \\ -x_1x_3 - x_2 + cx_4 &= 0 \\ x_1x_2 - x_3 - b &= 0 \\ -p(x_1 + x_2) &= 0 \end{cases} \tag{18}$$

A simple calculation shows that the novel 4-D hyperchaotic thermal convection system (6) has a unique equilibrium point given by

$$E_0 = \begin{bmatrix} 0 \\ 0 \\ -b \\ 0 \end{bmatrix} = \begin{bmatrix} 0 \\ 0 \\ -20 \\ 0 \end{bmatrix} \tag{19}$$

The Jacobian matrix of the novel 4-D hyperchaotic thermal convection system (6) at E_0 is obtained as

$$J_0 = J(E_0) = \begin{bmatrix} -4 & 4 & 0 & 0.2 \\ 20 & -1 & 0 & 0.2 \\ 0 & 0 & -1 & 0 \\ -0.5 & -0.5 & 0 & 0 \end{bmatrix} \tag{20}$$

The eigenvalues of the matrix J_0 are obtained as

$$\lambda_1 = -1, \quad \lambda_2 = -11.5720, \quad \lambda_3 = 0.0384, \quad \lambda_4 = 6.5336 \tag{21}$$

This shows that the equilibrium point E_0 is a saddle point, which is unstable.

3.3 Rotation Symmetry About the x_3-axis

It is easy to see that the novel 4-D hyperchaotic thermal convection system (6) is invariant under the change of coordinates

$$(x_1, x_2, x_3, x_4) \mapsto (-x_1, -x_2, x_3, -x_4) \tag{22}$$

Since the transformation (22) persists for all values of the system parameters, it follows that the novel 4-D hyperchaotic system (6) has rotation symmetry about the x_3-axis and that any non-trivial trajectory must have a twin trajectory.

3.4 Invariance

It is easy to see that the x_3-axis is invariant under the flow of the 4-D novel hyperchaotic system (6).

The invariant motion along the x_3-axis is characterized by the scalar dynamics

$$\dot{x}_3 = -b - x_3, \tag{23}$$

which is unstable.

3.5 Lyapunov Exponents and Kaplan-Yorke Dimension

We take the parameter values of the novel system (6) as in the hyperchaotic case (7), i.e.

$$a = 4, \quad b = 20, \quad c = 0.2, \quad p = 0.5 \tag{24}$$

We take the initial state of the novel system (6) as given in (8), i.e.

$$x_1(0) = 0.4, \quad x_2(0) = 0.4, \quad x_3(0) = 0.4, \quad x_4(0) = 0.4 \tag{25}$$

Then the Lyapunov exponents of the system (6) are numerically obtained using MATLAB as

$$L_1 = 0.40546, \quad L_2 = 0.03583, \quad L_3 = 0, \quad L_4 = -6.44038 \tag{26}$$

Since there are two positive Lyapunov exponents in (26), the novel system (6) exhibits *hyperchaotic* behavior.

From the LE spectrum (26), we see that the maximal Lyapunov exponent of the novel hyperchaotic system (6) is $L_1 = 0.40546$.

We find that

$$L_1 + L_2 + L_3 + L_4 = -5.99909 < 0 \tag{27}$$

Thus, it follows that the novel hyperchaotic thermal convection system (6) is dissipative.

Also, the Kaplan-Yorke dimension of the novel hyperchaotic thermal convection system (6) is calculated as

$$D_{KY} = 3 + \frac{L_1 + L_2 + L_3}{|L_4|} = 3.0685, \tag{28}$$

which is fractional.

4 Adaptive Control of the Novel Hyperchaotic Thermal Convection System

In this section, we apply adaptive control method to derive an adaptive feedback control law for globally stabilizing the novel 4-D hyperchaotic thermal convection system with unknown parameters. The main control result in this section is established using Lyapunov stability theory.

Thus, we consider the controlled novel 4-D hyperchaotic thermal convection system given by

$$\begin{cases} \dot{x}_1 = a(x_2 - x_1) + cx_4 + u_1 \\ \dot{x}_2 = -x_1x_3 - x_2 + cx_4 + u_2 \\ \dot{x}_3 = x_1x_2 - x_3 - b + u_3 \\ \dot{x}_4 = -p(x_1 + x_2) + u_4 \end{cases} \tag{29}$$

In (29), x_1, x_2, x_3, x_4 are the states and u_1, u_2, u_3, u_4 are the adaptive controls to be determined using estimates $\hat{a}(t), \hat{b}(t), \hat{c}(t), \hat{p}(t)$ for the unknown parameters a, b, c, p, respectively.

We consider the adaptive feedback control law

$$\begin{cases} u_1 = -\hat{a}(t)(x_2 - x_1) - \hat{c}(t)x_4 - k_1 x_1 \\ u_2 = x_1 x_3 + x_2 - \hat{c}(t)x_4 - k_2 x_2 \\ u_3 = -x_1 x_2 + x_3 + \hat{b}(t) - k_3 x_3 \\ u_4 = \hat{p}(t)(x_1 + x_2) - k_4 x_4 \end{cases} \tag{30}$$

where k_1, k_2, k_3, k_4 are positive gain constants.

Substituting (30) into (29), we get the closed-loop plant dynamics as

$$\begin{cases} \dot{x}_1 = [a - \hat{a}(t)](x_2 - x_1) + [c - \hat{c}(t)]x_4 - k_1 x_1 \\ \dot{x}_2 = [c - \hat{c}(t)]x_4 - k_2 x_2 \\ \dot{x}_3 = -[b - \hat{b}(t)] - k_3 x_3 \\ \dot{x}_4 = -[p - \hat{p}(t)](x_1 + x_2) - k_4 x_4 \end{cases} \tag{31}$$

The parameter estimation errors are defined as

$$\begin{cases} e_a(t) = a - \hat{a}(t) \\ e_b(t) = b - \hat{b}(t) \\ e_c(t) = c - \hat{c}(t) \\ e_p(t) = p - \hat{p}(t) \end{cases} \tag{32}$$

In view of (32), we can simplify the plant dynamics (31) as

$$\begin{cases} \dot{x}_1 = e_a(x_2 - x_1) + e_c x_4 - k_1 x_1 \\ \dot{x}_2 = e_c x_4 - k_2 x_2 \\ \dot{x}_3 = -e_b - k_3 x_3 \\ \dot{x}_4 = -e_p(x_1 + x_2) - k_4 x_4 \end{cases} \tag{33}$$

Differentiating (32) with respect to t, we obtain

$$\begin{cases} \dot{e}_a(t) = -\dot{\hat{a}}(t) \\ \dot{e}_b(t) = -\dot{\hat{b}}(t) \\ \dot{e}_c(t) = -\dot{\hat{c}}(t) \\ \dot{e}_p(t) = -\dot{\hat{p}}(t) \end{cases} \tag{34}$$

We consider the quadratic candidate Lyapunov function defined by

$$V(\mathbf{x}, e_a, e_b, e_c, e_p) = \frac{1}{2}\left(x_1^2 + x_2^2 + x_3^2 + x_4^2\right) + \frac{1}{2}\left(e_a^2 + e_b^2 + e_c^2 + e_p^2\right) \tag{35}$$

Differentiating V along the trajectories of (33) and (34), we obtain

$$\begin{aligned} \dot{V} = & -k_1 x_1^2 - k_2 x_2^2 - k_3 x_3^2 - k_4 x_4^2 + e_a\left[x_1(x_2 - x_1) - \dot{\hat{a}}\right] + e_b\left[-x_3 - \dot{\hat{b}}\right] \\ & + e_c\left[(x_1 + x_2)x_4 - \dot{\hat{c}}\right] + e_p\left[-(x_1 + x_2)x_4 - \dot{\hat{p}}\right] \end{aligned} \tag{36}$$

In view of (36), we take the parameter update law as

$$\begin{cases} \dot{\hat{a}}(t) = x_1(x_2 - x_1) \\ \dot{\hat{b}}(t) = -x_3 \\ \dot{\hat{c}}(t) = (x_1 + x_2)x_4 \\ \dot{\hat{p}}(t) = -(x_1 + x_2)x_4 \end{cases} \tag{37}$$

Next, we state and prove the main result of this section.

Theorem 1 *The novel 4-D hyperchaotic thermal convection system (29) with unknown system parameters is globally and exponentially stabilized for all initial conditions by the adaptive control law (30) and the parameter update law (37), where k_1, k_2, k_3, k_4 are positive gain constants.*

Proof We prove this result by applying Lyapunov stability theory [142].

We consider the quadratic Lyapunov function defined by (35), which is clearly a positive definite function on $\mathbf{R}^8$.

By substituting the parameter update law (37) into (36), we obtain the time-derivative of V as

$$\dot{V} = -k_1x_1^2 - k_2x_2^2 - k_3x_3^2 - k_4x_4^2 \tag{38}$$

From (38), it is clear that $\dot{V}$ is a negative semi-definite function on $\mathbf{R}^8$.

Thus, we can conclude that the state vector $\mathbf{x}(t)$ and the parameter estimation error are globally bounded, i.e.

$$\left[x_1(t)\ x_2(t)\ x_3(t)\ x_4(t)\ e_a(t)\ e_b(t)\ e_c(t)\ e_p(t) \right]^T \in \mathbf{L}_\infty.$$

We define $k = \min\{k_1, k_2, k_3, k_4\}$.

Then it follows from (38) that

$$\dot{V} \le -k\|\mathbf{x}(t)\|^2 \tag{39}$$

Thus, we have

$$k\|\mathbf{x}(t)\|^2 \le -\dot{V} \tag{40}$$

Integrating the inequality (40) from 0 to t, we get

$$k\int_0^t \|\mathbf{x}(\tau)\|^2\, d\tau \ \le\ V(0) - V(t) \tag{41}$$

From (41), it follows that $\mathbf{x} \in \mathbf{L}_2$.

Using (33), we can conclude that $\dot{\mathbf{x}} \in \mathbf{L}_\infty$.

Using Barbalat's lemma [142], we conclude that $\mathbf{x}(t) \to 0$ exponentially as $t \to \infty$ for all initial conditions $\mathbf{x}(0) \in \mathbf{R}^4$.

This completes the proof. ■

For the numerical simulations, the classical fourth-order Runge-Kutta method with step size $h = 10^{-8}$ is used to solve the systems (29) and (37), when the adaptive control law (30) is applied.

The parameter values of the novel 4-D hyperchaotic system (29) are taken as in the hyperchaotic case (7), i.e.

$$a = 4, \quad b = 20, \quad c = 0.2, \quad p = 0.5 \tag{42}$$

We take the positive gain constants as

$$k_1 = 6, \quad k_2 = 6, \quad k_3 = 6, \quad k_4 = 6 \tag{43}$$

Furthermore, as initial conditions of the novel 4-D hyperchaotic thermal convection system (29), we take

$$x_1(0) = 6.2, \quad x_2(0) = 5.7, \quad x_3(0) = 14.3, \quad x_4(0) = 8.9 \tag{44}$$

Also, as initial conditions of the parameter estimates, we take

$$\hat{a}(0) = 14.8, \quad \hat{b}(0) = 13.9, \quad \hat{c}(0) = 11.7, \quad \hat{p}(0) = 7.1 \tag{45}$$

In Fig. 6, the exponential convergence of the controlled states of the novel 4-D hyperchaotic system (29) is shown.

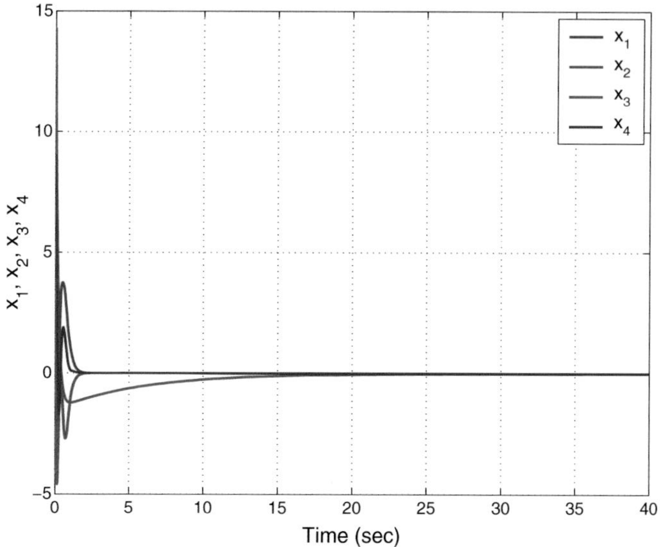

Fig. 6 Time-history of the controlled states x_1, x_2, x_3, x_4

5 Adaptive Synchronization of the Identical Novel Hyperchaotic Systems

In this section, we use adaptive control method to derive an adaptive feedback control law for globally synchronizing identical novel 4-D hyperchaotic thermal convection systems with unknown parameters.

As the master system, we consider the novel 4-D hyperchaotic system given by

$$\begin{cases} \dot{x}_1 = a(x_2 - x_1) + cx_4 \\ \dot{x}_2 = -x_1x_3 - x_2 + cx_4 \\ \dot{x}_3 = x_1x_2 - x_3 - b \\ \dot{x}_4 = -p(x_1 + x_2) \end{cases} \tag{46}$$

In (46), x_1, x_2, x_3, x_4 are the states and a, b, c, p, q are unknown system parameters.

As the slave system, we consider the 4-D novel hyperchaotic system given by

$$\begin{cases} \dot{y}_1 = a(y_2 - y_1) + cy_4 + u_1 \\ \dot{y}_2 = -y_1y_3 - y_2 + cy_4 + u_2 \\ \dot{y}_3 = y_1y_2 - y_3 - b + u_3 \\ \dot{y}_4 = -p(y_1 + y_2) + u_4 \end{cases} \tag{47}$$

In (47), y_1, y_2, y_3, y_4 are the states and u_1, u_2, u_3, u_4 are the adaptive controls to be determined using estimates $\hat{a}(t)$, $\hat{c}(t)$, $\hat{p}(t)$ for the unknown parameters a, c, p, respectively.

The synchronization error between the novel hyperchaotic systems (46) and (47) is defined by

$$\begin{cases} e_1 = y_1 - x_1 \\ e_2 = y_2 - x_2 \\ e_3 = y_3 - x_3 \\ e_4 = y_4 - x_4 \end{cases} \tag{48}$$

Then the error dynamics is obtained as

$$\begin{cases} \dot{e}_1 = a(e_2 - e_1) + ce_4 + u_1 \\ \dot{e}_2 = -y_1y_3 + x_1x_3 - e_2 + ce_4 + u_2 \\ \dot{e}_3 = y_1y_2 - x_1x_2 - e_3 + u_3 \\ \dot{e}_4 = -p(e_1 + e_2) + u_4 \end{cases} \tag{49}$$

We consider the adaptive feedback control law

$$\begin{cases} u_1 = -\hat{a}(t)(e_2 - e_1) - \hat{c}(t)e_4 - k_1e_1 \\ u_2 = y_1y_3 - x_1x_3 + e_2 - \hat{c}(t)e_4 - k_2e_2 \\ u_3 = -y_1y_2 + x_1x_2 + e_3 - k_3e_3 \\ u_4 = \hat{p}(t)(e_1 + e_2) - k_4e_4 \end{cases} \tag{50}$$

where k_1, k_2, k_3, k_4 are positive gain constants.

Substituting (50) into (49), we get the closed-loop error dynamics as

$$\begin{cases} \dot{e}_1 = [a - \hat{a}(t)](e_2 - e_1) + [c - \hat{c}(t)]e_4 - k_1 e_1 \\ \dot{e}_2 = [c - \hat{c}(t)]e_4 - k_2 e_2 \\ \dot{e}_3 = -k_3 e_3 \\ \dot{e}_4 = -[p - \hat{p}(t)](e_1 + e_2) - k_4 e_4 \end{cases} \tag{51}$$

The parameter estimation errors are defined as

$$\begin{cases} e_a(t) = a - \hat{a}(t) \\ e_c(t) = c - \hat{c}(t) \\ e_p(t) = p - \hat{p}(t) \end{cases} \tag{52}$$

In view of (52), we can simplify the error dynamics (51) as

$$\begin{cases} \dot{e}_1 = e_a(e_2 - e_1) + e_c e_4 - k_1 e_1 \\ \dot{e}_2 = e_c e_4 - k_2 e_2 \\ \dot{e}_3 = -k_3 e_3 \\ \dot{e}_4 = -e_p(e_1 + e_2) - k_4 e_4 \end{cases} \tag{53}$$

Differentiating (52) with respect to t, we obtain

$$\begin{cases} \dot{e}_a(t) = -\dot{\hat{a}}(t) \\ \dot{e}_c(t) = -\dot{\hat{c}}(t) \\ \dot{e}_p(t) = -\dot{\hat{p}}(t) \end{cases} \tag{54}$$

We use adaptive control theory to find an update law for the parameter estimates. We consider the quadratic candidate Lyapunov function defined by

$$V(\mathbf{e}, e_a, e_c, e_p) = \frac{1}{2}\left(e_1^2 + e_2^2 + e_3^2 + e_4^2\right) + \frac{1}{2}\left(e_a^2 + e_c^2 + e_p^2\right) \tag{55}$$

Differentiating V along the trajectories of (53) and (54), we obtain

$$\begin{aligned} \dot{V} = & -k_1 e_1^2 - k_2 e_2^2 - k_3 e_3^2 - k_4 e_4^2 + e_a\left[e_1(e_2 - e_1) - \dot{\hat{a}}\right] \\ & + e_c\left[(e_1 + e_2)e_4 - \dot{\hat{c}}\right] + e_p\left[-(e_1 + e_2)e_4 - \dot{\hat{p}}\right] \end{aligned} \tag{56}$$

In view of (56), we take the parameter update law as

$$\begin{cases} \dot{\hat{a}}(t) = e_1(e_2 - e_1) \\ \dot{\hat{c}}(t) = (e_1 + e_2)e_4 \\ \dot{\hat{p}}(t) = -(e_1 + e_2)e_4 \end{cases} \tag{57}$$

Next, we state and prove the main result of this section.

Theorem 2 *The novel hyperchaotic systems (46) and (47) with unknown system parameters are globally and exponentially synchronized for all initial conditions by the adaptive control law (50) and the parameter update law (57), where k_1, k_2, k_3, k_4 are positive gain constants.*

Proof We prove this result by applying Lyapunov stability theory [142].

We consider the quadratic Lyapunov function defined by (55), which is clearly a positive definite function on $\mathbf{R}^8$.

By substituting the parameter update law (57) into (56), we obtain

$$\dot{V} = -k_1 e_1^2 - k_2 e_2^2 - k_3 e_3^2 - k_4 e_4^2 \tag{58}$$

From (58), it is clear that $\dot{V}$ is a negative semi-definite function on $\mathbf{R}^7$.

Thus, we can conclude that the error vector $\mathbf{e}(t)$ and the parameter estimation error are globally bounded, *i.e.*

$$\left[e_1(t)\ e_2(t)\ e_3(t)\ e_4(t)\ e_a(t)\ e_c(t)\ e_p(t) \right]^T \in \mathbf{L}_\infty. \tag{59}$$

We define $k = \min\{k_1, k_2, k_3, k_4\}$.

Then it follows from (58) that

$$\dot{V} \leq -k\|\mathbf{e}(t)\|^2 \tag{60}$$

Thus, we have

$$k\|\mathbf{e}(t)\|^2 \leq -\dot{V} \tag{61}$$

Integrating the inequality (61) from 0 to t, we get

$$k \int_0^t \|\mathbf{e}(\tau)\|^2 \, d\tau \leq V(0) - V(t) \tag{62}$$

From (62), it follows that $\mathbf{e} \in \mathbf{L}_2$.

Using (53), we can conclude that $\dot{\mathbf{x}} \in \mathbf{L}_\infty$.

Using Barbalat's lemma [142], we conclude that $\mathbf{e}(t) \to 0$ exponentially as $t \to \infty$ for all initial conditions $\mathbf{e}(0) \in \mathbf{R}^4$.

This completes the proof. ■

For the numerical simulations, the classical fourth-order Runge-Kutta method with step size $h = 10^{-8}$ is used to solve the systems (46), (47) and (57), when the adaptive control law (50) is applied.

The parameter values of the novel hyperchaotic systems are taken as in the hyperchaotic case (7), i.e.

$$a = 4, \quad b = 20, \quad c = 0.2, \quad p = 0.5 \tag{63}$$

We take the positive gain constants as

$$k_1 = 6, \quad k_2 = 6, \quad k_3 = 6, \quad k_4 = 6 \tag{64}$$

Furthermore, as initial conditions of the master system (46), we take

$$x_1(0) = 3.4, \quad x_2(0) = 16.8, \quad x_3(0) = 12.7, \quad x_4(0) = 21.9 \tag{65}$$

As initial conditions of the slave system (47), we take

$$y_1(0) = 15.1, \quad y_2(0) = -12.5, \quad y_3(0) = 3.8, \quad y_4(0) = 5.8 \tag{66}$$

Also, as initial conditions of the parameter estimates, we take

$$\hat{a}(0) = 2.5, \quad \hat{c}(0) = 3.4, \quad \hat{p}(0) = 4.7 \tag{67}$$

Figures 7, 8, 9 and 10 describe the complete synchronization of the novel hyperchaotic systems (46) and (47), while Fig. 11 describes the time-history of the synchronization errors e_1, e_2, e_3, e_4.

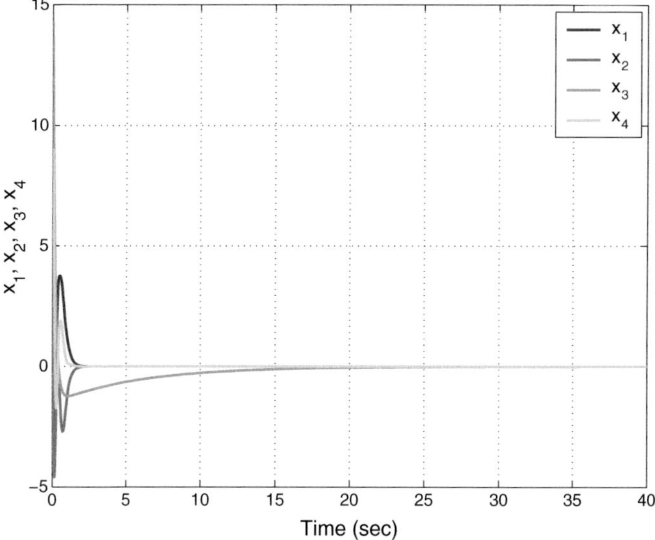

Fig. 7 Synchronization of the states x_1 and y_1

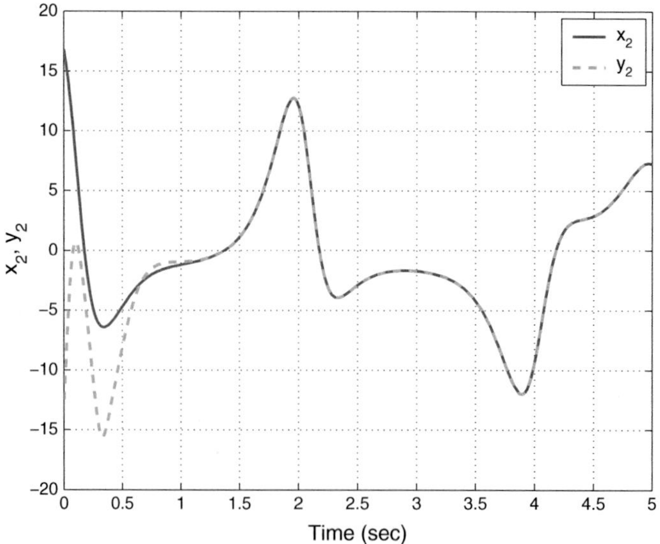

Fig. 8 Synchronization of the states x_2 and y_2

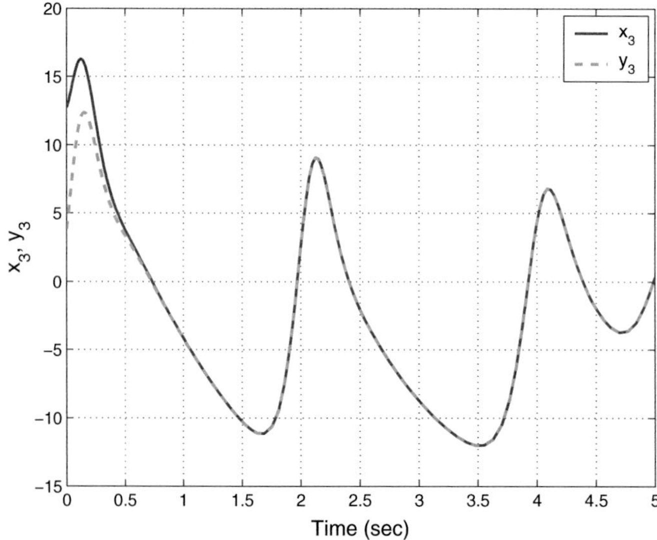

Fig. 9 Synchronization of the states x_3 and y_3

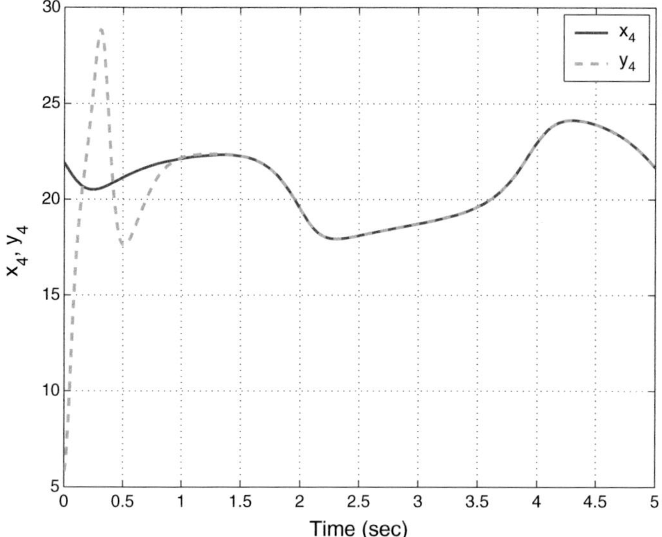

Fig. 10 Synchronization of the states x_4 and y_4

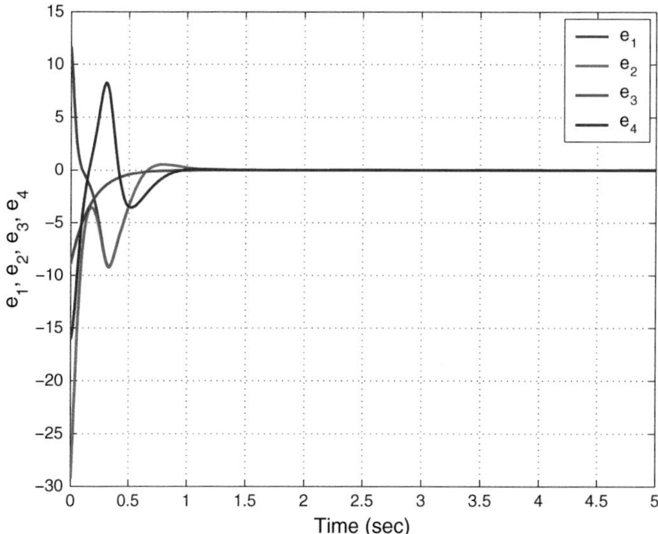

Fig. 11 Time-history of the synchronization errors e_1, e_2, e_3, e_4

6 Conclusions

In this work, we described an eleven-term novel 4-D hyperchaotic thermal convection system with two quadratic nonlinearities. The phase portraits of the novel hyperchaotic system were shown and the qualitative properties of the novel hyperchaotic system were discussed. The novel 4-D hyperchaotic thermal convection system was obtained by introducing a feedback control to the 3-D thermal convection system obtained by Wang et al. [141]. The Lyapunov exponents of the novel hyperchaotic thermal convection system have been obtained as $L_1 = 0.40546$, $L_2 = 0.03583$, $L_3 = 0$ and $L_4 = -6.44038$. Since there are two positive Lyapunov exponents for the novel 4-D thermal convection system, it is hyperchaotic. The Maximal Lyapunov Exponent (MLE) of the novel hyperchaotic system is found as $L_1 = 0.40546$. Also, the Kaplan-Yorke dimension of the novel hyperchaotic system has been derived as $D_{KY} = 3.0685$. Since the sum of the Lyapunov exponents is negative, the novel hyperchaotic system is dissipative. Next, an adaptive controller was designed to globally stabilize the novel hyperchaotic thermal convection system with unknown parameters. Finally, an adaptive controller was also designed to achieve global chaos synchronization of the identical novel hyperchaotic thermal convection systems with unknown parameters. MATLAB simulations have been shown to illustrate and demonstrate all the main results obtained in this work.

References

1. Lorenz EN (1963) Deterministic periodic flow. J Atmos Sci 20(2):130–141
2. Rössler OE (1976) An equation for continuous chaos. Phys Lett A 57(5):397–398
3. Arneodo A, Coullet P, Tresser C (1981) Possible new strange attractors with spiral structure. Common Math Phys 79(4):573–576
4. Sprott JC (1994) Some simple chaotic flows. Phys Rev E 50(2):647–650
5. Chen G, Ueta T (1999) Yet another chaotic attractor. Int J Bifurcat Chaos 9(7):1465–1466
6. Lü J, Chen G (2002) A new chaotic attractor coined. Int J Bifurcat Chaos 12(3):659–661
7. Cai G, Tan Z (2007) Chaos synchronization of a new chaotic system via nonlinear control. J Uncertain Syst 1(3):235–240
8. Tigan G, Opris D (2008) Analysis of a 3D chaotic system. Chaos Solitons Fractals 36: 1315–1319
9. Zhou W, Xu Y, Lu H, Pan L (2008) On dynamics analysis of a new chaotic attractor. Phys Lett A 372(36):5773–5777
10. Zhu C, Liu Y, Guo Y (2010) Theoretic and numerical study of a new chaotic system. Intel Inf Manage 2:104–109
11. Li D (2008) A three-scroll chaotic attractor. Phys Lett A 372(4):387–393
12. Wei Z, Yang Q (2010) Anti-control of Hopf bifurcation in the new chaotic system with two stable node-foci. Appl Math Comput 217(1):422–429
13. Sundarapandian V (2013) Analysis and anti-synchronization of a novel chaotic system via active and adaptive controllers. J Eng Sci Technol Rev 6(4):45–52
14. Sundarapandian V, Pehlivan I (2012) Analysis, control, synchronization, and circuit design of a novel chaotic system. Math Comput Model 55(7–8):1904–1915
15. Vaidyanathan S (2013) A new six-term 3-D chaotic system with an exponential nonlinearity. Far East J Math Sci 79(1):135–143

16. Vaidyanathan S (2013) Analysis and adaptive synchronization of two novel chaotic systems with hyperbolic sinusoidal and cosinusoidal nonlinearity and unknown parameters. J Eng Sci Technol Rev 6(4):53–65
17. Vaidyanathan S (2014) A new eight-term 3-D polynomial chaotic system with three quadratic nonlinearities. Far East J Math Sci 84(2):219–226
18. Vaidyanathan S (2014) Analysis and adaptive synchronization of eight-term 3-D polynomial chaotic systems with three quadratic nonlinearities. Eur Phys J 223(8):1519–1529
19. Vaidyanathan S (2014) Analysis, control and synchronisation of a six-term novel chaotic system with three quadratic nonlinearities. Int J Model Ident Control 22(1):41–53
20. Vaidyanathan S (2014) Generalized projective synchronisation of novel 3-D chaotic systems with an exponential non-linearity via active and adaptive control. Int J Model Ident Control 22(3):207–217
21. Vaidyanathan S (2015) A 3-D novel highly chaotic system with four quadratic nonlinearities, its adaptive control and anti-synchronization with unknown parameters. J Eng Sci Technol Rev 8(2):106–115
22. Vaidyanathan S (2015) Analysis, properties and control of an eight-term 3-D chaotic system with an exponential nonlinearity. Int J Model Ident Control 23(2):164–172
23. Vaidyanathan S, Rajagopal K, Volos CK, Kyprianidis IM, Stouboulos IN (2015c) Analysis, adaptive control and synchronization of a seven-term novel 3-D chaotic system with three quadratic nonlinearities and its digital implementation in LabVIEW. J Eng Sci Technol Rev 8(2):130–141
24. Vaidyanathan S, Azar AT (2015) Analysis, control and synchronization of a nine-term 3-D novel chaotic system. In: Azar AT, Vaidyanathan S (eds) Chaos modelling and control systems design, studies in computational intelligence, vol 581. Springer, Germany, pp 19–38
25. Vaidyanathan S, Madhavan K (2013) Analysis, adaptive control and synchronization of a seven-term novel 3-D chaotic system. Int J Control Theor Appl 6(2):121–137
26. Vaidyanathan S, Pakiriswamy S (2015) A 3-D novel conservative chaotic system and its generalized projective synchronization via adaptive control. J Eng Sci Technol Rev 8(2):52–60
27. Vaidyanathan S, Volos CK, Kyprianidis IM, Stouboulos IN, Pham VT (2015) Analysis, adaptive control and anti-synchronization of a six-term novel jerk chaotic system with two exponential nonlinearities and its circuit simulation. J Eng Sci Technol Rev 8(2):24–36
28. Vaidyanathan S, Volos CK, Pham VT (2015) Analysis, adaptive control and adaptive synchronization of a nine-term novel 3-D chaotic system with four quadratic nonlinearities and its circuit simulation. J Eng Sci Technol Rev 8(2):181–191
29. Vaidyanathan S, Volos CK, Pham VT (2015) Global chaos control of a novel nine-term chaotic system via sliding mode control. In: Azar AT, Zhu Q (eds) Advances and Applications in Sliding Mode Control Systems, Studies in Computational Intelligence, vol 576. Springer, Germany, pp 571–590
30. Vaidyanathan S, Volos C (2015) Analysis and adaptive control of a novel 3-D conservative no-equilibrium chaotic system. Arch Control Sci 25(3):333–353
31. Vaidyanathan S, Volos C, Pham VT, Madhavan K, Idowu BA (2014) Adaptive backstepping control, synchronization and circuit simulation of a 3-D novel jerk chaotic system with two hyperbolic sinusoidal nonlinearities. Arch Control Sci 24(3):375–403
32. Pehlivan I, Moroz IM, Vaidyanathan S (2014) Analysis, synchronization and circuit design of a novel butterfly attractor. J Sound Vib 333(20):5077–5096
33. Sampath S, Vaidyanathan S, Volos CK, Pham VT (2015) An eight-term novel four-scroll chaotic system with cubic nonlinearity and its circuit simulation. J Eng Sci Technol Rev 8(2):1–6
34. Pham VT, Vaidyanathan S, Volos CK, Jafari S (2015a) Hidden attractors in a chaotic system with an exponential nonlinear term. Eur Phys J 224(8):1507–1517
35. Filali RL, Benrejeb M, Borne P (2014) On observer-based secure communication design using discrete-time hyperchaotic systems. Commun Nonlinear Sci Numer Simul 19(5):1424–1432

36. Li C, Liao X, Wong KW (2005) Lag synchronization of hyperchaos with application to secure communications. Chaos Solitons Fractals 23(1):183–193
37. Wu X, Zhu C, Kan H (2015) An improved secure communication scheme based passive synchronization of hyperchaotic complex nonlinear system. Appl Math Comput 252: 201–214
38. Hammami S (2015) State feedback-based secure image cryptosystem using hyperchaotic synchronization. ISA Trans 54:52–59
39. Rhouma R, Belghith S (2008) Cryptanalysis of a new image encryption algorithm based on hyper-chaos. Phys Lett A 372(38):5973–5978
40. Zhu C (2012) A novel image encryption scheme based on improved hyperchaotic sequences. Opt Commun 285(1):29–37
41. Senouci A, Boukabou A (2014) Predictive control and synchronization of chaotic and hyperchaotic systems based on a T-S fuzzy model. Math Comput Simul 105:62–78
42. Zhang H, Liao X, Yu J (2005) Fuzzy modeling and synchronization of hyperchaotic systems. Chaos, Solitons Fractals 26(3):835–843
43. Wei X, Yunfei F, Qiang L (2012) A novel four-wing hyper-chaotic system and its circuit implementation. Procedia Eng 29:1264–1269
44. Yujun N, Xingyuan W, Mingjun W, Huaguang Z (2010) A new hyperchaotic system and its circuit implementation. Commun Nonlinear Sci Numer Simul 15(11):3518–3524
45. Rössler OE (1979) An equation for hyperchaos. Phys Lett A 71:155–157
46. Jia Q (2007) Hyperchaos generated from the Lorenz chaotic system and its control. Phys Lett A 366:217–222
47. Chen A, Lu J, Lü J, Yu S (2006) Generating hyperchaotic Lü attractor via state feedback control. Phys A 364:103–110
48. Li X (2009) Modified projective synchronization of a new hyperchaotic system via nonlinear control. Commun Theoret Phys 52:274–278
49. Wang J, Chen Z (2008) A novel hyperchaotic system and its complex dynamics. Int J Bifurcat Chaos 18:3309–3324
50. Vaidyanathan S (2013) A ten-term novel 4-D hyperchaotic system with three quadratic nonlinearities and its control. Int J Control Theor Appl 6(2):97–109
51. Vaidyanathan S (2014) Qualitative analysis and control of an eleven-term novel 4-D hyperchaotic system with two quadratic nonlinearities. Int J Control Theor Appl 7:35–47
52. Vaidyanathan S (2015) Hyperchaos, qualitative analysis, control and synchronisation of a ten-term 4-D hyperchaotic system with an exponential nonlinearity and three quadratic nonlinearities. Int J Model Ident Control 23(4):380–392
53. Vaidyanathan S, Azar AT, Rajagopal K, Alexander P (2015) Design and SPICE implementation of a 12-term novel hyperchaotic system and its synchronisation via active control. Int J Model Ident Control 23(3):267–277
54. Vaidyanathan S, Azar AT (2015) Analysis and control of a 4-D novel hyperchaotic system. Stud Comput Intell 581:3–17
55. Vaidyanathan S, Volos CK, Pham VT (2015) Analysis, control, synchronization and SPICE implementation of a novel 4-D hyperchaotic Rikitake dynamo system without equilibrium. J Eng Sci Technol Rev 8(2):232–244
56. Vaidyanathan S, Volos C, Pham VT, Madhavan K (2015) Analysis, adaptive control and synchronization of a novel 4-D hyperchaotic hyperjerk system and its SPICE implementation. Arch Control Sci 25(1):135–158
57. Vaidyanathan S, Volos C, Pham VT (2014) Hyperchaos, adaptive control and synchronization of a novel 5-D hyperchaotic system with three positive Lyapunov exponents and its SPICE implementation. Arch Control Sci 24(4):409–446
58. Vaidyanathan S, VTP, Volos CK, (2015) A 5-D hyperchaotic Rikitake dynamo system with hidden attractors. Eur Phys J 224(8):1575–1592
59. Pham VT, Volos C, Jafari S, Wang X, Vaidyanathan S (2014) Hidden hyperchaotic attractor in a novel simple memristive neural network. Optoelectron Adv Mater, Rapid Commun 8(11–12):1157–1163

60. Azar AT, Vaidyanathan S (2015) Chaos modeling and control systems design, studies in computational intelligence, vol 581. Springer, Germany
61. Azar AT, Vaidyanathan S (2015) Computational intelligence applications in modeling and control, studies in computational intelligence, vol 575. Springer, Germany
62. Azar AT, Vaidyanathan S (2015) Handbook of research on advanced intelligent control engineering and automation. Advances in computational intelligence and robotics (ACIR), IGI-Global, USA
63. Azar AT (2010) Fuzzy systems. IN-TECH, Vienna, Austria
64. Azar AT, Zhu Q (2015) Advances and applications in sliding mode control systems, studies in computational intelligence, vol 576. Springer, Germany
65. Zhu Q, Azar AT (2015) Complex system modelling and control through intelligent soft computations, studies in fuzzines and soft computing, vol 319. Springer, Germany
66. Kengne J, Chedjou JC, Kenne G, Kyamakya K (2012) Dynamical properties and chaos synchronization of improved Colpitts oscillators. Commun Nonlinear Sci Numer Simul 17(7):2914–2923
67. Sharma A, Patidar V, Purohit G, Sud KK (2012) Effects on the bifurcation and chaos in forced Duffing oscillator due to nonlinear damping. Commun Nonlinear Sci Numer Simul 17(6):2254–2269
68. Gaspard P (1999) Microscopic chaos and chemical reactions. Phys A 263(1–4):315–328
69. Petrov V, Gaspar V, Masere J, Showalter K (1993) Controlling chaos in Belousov-Zhabotinsky reaction. Nature 361:240–243
70. Vaidyanathan S (2015) Adaptive control of a chemical chaotic reactor. Int J PharmTech Res 8(3):377–382
71. Vaidyanathan S (2015) Adaptive synchronization of chemical chaotic reactors. Int J ChemTech Res 8(2):612–621
72. Vaidyanathan S (2015) Anti-synchronization of Brusselator chemical reaction systems via adaptive control. Int J ChemTech Res 8(6):759–768
73. Vaidyanathan S (2015) Dynamics and control of Brusselator chemical reaction. Int J ChemTech Res 8(6):740–749
74. Vaidyanathan S (2015) Dynamics and control of Tokamak system with symmetric and magnetically confined plasma. Int J ChemTech Res 8(6):795–803
75. Vaidyanathan S (2015) Synchronization of Tokamak systems with symmetric and magnetically confined plasma via adaptive control. Int J ChemTech Res 8(6):818–827
76. Das S, Goswami D, Chatterjee S, Mukherjee S (2014) Stability and chaos analysis of a novel swarm dynamics with applications to multi-agent systems. Eng Appl Artif Intell 30:189–198
77. Kyriazis M (1991) Applications of chaos theory to the molecular biology of aging. Exp Gerontol 26(6):569–572
78. Vaidyanathan S (2015) 3-cells cellular neural network (CNN) attractor and its adaptive biological control. Int J PharmTech Res 8(4):632–640
79. Vaidyanathan S (2015) Adaptive backstepping control of enzymes-substrates system with ferroelectric behaviour in brain waves. Int J PharmTech Res 8(2):256–261
80. Vaidyanathan S (2015) Adaptive biological control of generalized Lotka-Volterra three-species biological system. Int J PharmTech Res 8(4):622–631
81. Vaidyanathan S (2015) Adaptive chaotic synchronization of enzymes-substrates system with ferroelectric behaviour in brain waves. Int J PharmTech Res 8(5):964–973
82. Vaidyanathan S (2015) Adaptive synchronization of generalized Lotka-Volterra three-species biological systems. Int J PharmTech Res 8(5):928–937
83. Vaidyanathan S (2015) Chaos in neurons and adaptive control of Birkhoff-Shaw strange chaotic attractor. Int J PharmTech Res 8(5):956–963
84. Vaidyanathan S (2015) Lotka-Volterra population biology models with negative feedback and their ecological monitoring. Int J PharmTech Res 8(5):974–981
85. Vaidyanathan S (2015) Synchronization of 3-cells cellular neural network (CNN) attractors via adaptive control method. Int J PharmTech Res 8(5):946–955

86. Gibson WT, Wilson WG (2013) Individual-based chaos: extensions of the discrete logistic model. J Theoret Biol 339:84–92
87. Suérez I (1999) Mastering chaos in ecology. Ecol Model 117(2–3):305–314
88. Lang J (2015) Color image encryption based on color blend and chaos permutation in the reality-preserving multiple-parameter fractional Fourier transform domain. Opt Commun 338:181–192
89. Zhang X, Zhao Z, Wang J (2014) Chaotic image encryption based on circular substitution box and key stream buffer. Sign Proces Image Commun 29(8):902–913
90. Rhouma R, Belghith S (2011) Cryptoanalysis of a chaos based cryptosystem on DSP. Commun Nonlinear Sci Numer Simul 16(2):876–884
91. Usama M, Khan MK, Alghatbar K, Lee C (2010) Chaos-based secure satellite imagery cryptosystem. Comput Math Appl 60(2):326–337
92. Azar AT, Serrano FE (2015) Adaptive sliding mode control of the Furuta pendulum. In: Azar AT, Zhu Q (eds) Advances and applications in sliding mode control systems, studies in computational intelligence, vol 576. Springer, Germany, pp 1–42
93. Azar AT, Serrano FE (2015) Deadbeat control for multivariable systems with time varying delays. In: Azar AT, Vaidyanathan S (eds) Chaos Modeling and control systems design, studies in computational intelligence, vol 581. Springer, Germany, pp 97–132
94. Azar AT, Serrano FE (2015) Design and modeling of anti wind up PID controllers. In: Zhu Q, Azar AT (eds) Complex system modelling and control through intelligent soft computations, studies in fuzziness and soft computing, vol 319. Springer, Germany, pp 1–44
95. Azar AT, Serrano FE (2014) Robust IMC-PID tuning for cascade control systems with gain and phase margin specifications. Neural Comput Appl 25(5):983–995
96. Azar AT, Serrano FE (2015d) Stabilizatoin and control of mechanical systems with backlash. In: Azar AT, Vaidyanathan S (eds) Handbook of research on advanced intelligent control engineering and automation. Advances in computational intelligence and robotics (ACIR), IGI-Global, USA, pp 1–60
97. Feki M (2003) An adaptive chaos synchronization scheme applied to secure communication. Chaos Solitons Fractals 18(1):141–148
98. Murali K, Lakshmanan M (1998) Secure communication using a compound signal from generalized chaotic systems. Phys Lett A 241(6):303–310
99. Zaher AA, Abu-Rezq A (2011) On the design of chaos-based secure communication systems. Commun Nonlinear Sci Numer Simul 16(9):3721–3727
100. Mondal S, Mahanta C (2014) Adaptive second order terminal sliding mode controller for robotic manipulators. J Franklin Inst 351(4):2356–2377
101. Nehmzow U, Walker K (2005) Quantitative description of robot-environment interaction using chaos theory. Robot Auton Sys 53(3–4):177–193
102. Volos CK, Kyprianidis IM, Stouboulos IN (2013) Experimental investigation on coverage performance of a chaotic autonomous mobile robot. Robot Auton Syst 61(12):1314–1322
103. Qu Z (2011) Chaos in the genesis and maintenance of cardiac arrhythmias. Prog Biophys Mol Biol 105(3):247–257
104. Witte CL, Witte MH (1991) Chaos and predicting varix hemorrhage. Med Hypotheses 36(4):312–317
105. Azar AT (2012) Overview of type-2 fuzzy logic systems. Int J Fuzzy Syst Appl 2(4):1–28
106. Li Z, Chen G (2006) Integration of fuzzy logic and chaos theory, studies in fuzziness and soft computing, vol 187. Springer, Germany
107. Huang X, Zhao Z, Wang Z, Li Y (2012) Chaos and hyperchaos in fractional-order cellular neural networks. Neurocomputing 94:13–21
108. Kaslik E, Sivasundaram S (2012) Nonlinear dynamics and chaos in fractional-order neural networks. Neural Netw 32:245–256
109. Lian S, Chen X (2011) Traceable content protection based on chaos and neural networks. Appl Soft Comput 11(7):4293–4301
110. Pham VT, Volos CK, Vaidyanathan S, Le TP, Vu VY (2015b) A memristor-based hyperchaotic system with hidden attractors: dynamics, synchronization and circuital emulating. J Eng Sci Technol Rev 8(2):205–214

111. Volos CK, Kyprianidis IM, Stouboulos IN, Tlelo-Cuautle E, Vaidyanathan S (2015) Memristor: A new concept in synchronization of coupled neuromorphic circuits. J Eng Sci Technol Rev 8(2):157–173
112. Sundarapandian V (2010) Output regulation of the Lorenz attractor. Far East J Math Sci 42(2):289–299
113. Vaidyanathan S (2011) Output regulation of Arneodo-Coullet chaotic system. Commun Comput Inf Sci 133:98–107
114. Vaidyanathan S (2011) Output regulation of the unified chaotic system. Commun Comput Inf Sci 198:1–9
115. Sundarapandian V (2013) Adaptive control and synchronization design for the Lu-Xiao chaotic system. Lect Notes Electr Eng 131:319–327
116. Vaidyanathan S (2012) Adaptive controller and syncrhonizer design for the Qi-Chen chaotic system. Lect Notes Inst Comput Sci Social-Informatics Telecommun Eng 84:73–82
117. Vaidyanathan S (2013) Analysis, control and synchronization of hyperchaotic Zhou system via adaptive control. Adv Intell Syst Comput 177:1–10
118. Vaidyanathan S (2012) Global chaos control of hyperchaotic Liu system via sliding control method. Int J Control Theor Appl 5(2):117–123
119. Vaidyanathan S (2012) Sliding mode control based global chaos control of Liu-Liu-Liu-Su chaotic system. Int J Control Theor Appl 5(1):15–20
120. Vaidyanathan S, Idowu BA, Azar AT (2015) Backstepping controller design for the global chaos synchronization of Sprott's jerk systems. Stud Comput Intell 581:39–58
121. Karthikeyan R, Sundarapandian V (2014) Hybrid chaos synchronization of four-scroll systems via active control. J Electr Eng 65(2):97–103
122. Sarasu P, Sundarapandian V (2011) Active controller design for generalized projective synchronization of four-scroll chaotic systems. Int J Syst Sign Control Eng Appl 4(2):26–33
123. Sarasu P, Sundarapandian V (2011) The generalized projective synchronization of hyperchaotic Lorenz and hyperchaotic Qi systems via active control. Int J Soft Comput 6(5):216–223
124. Vaidyanathan S, Rajagopal K (2011) Hybrid synchronization of hyperchaotic Wang-Chen and hyperchaotic Lorenz systems by active non-linear control. Int J Syst Sign Control Eng Appl 4(3):55–61
125. Vaidyanathan S, Rasappan S (2011) Global chaos synchronization of hyperchaotic Bao and Xu systems by active nonlinear control. Commun Comput Inf Sci 198:10–17
126. Sarasu P, Sundarapandian V (2012) Generalized projective synchronization of two-scroll systems via adaptive control. Int J Soft Comput 7(4):146–156
127. Sundarapandian V, Karthikeyan R (2011) Anti-synchronization of hyperchaotic Lorenz and hyperchaotic Chen systems by adaptive control. Int J Syst Sign Control Eng Appl 4(2):18–25
128. Sundarapandian V, Karthikeyan R (2011) Anti-synchronization of Lü and Pan chaotic systems by adaptive nonlinear control. Eur J Sci Res 64(1):94–106
129. Sundarapandian V, Karthikeyan R (2012) Adaptive anti-synchronization of uncertain Tigan and Li systems. J Eng Appl Sci 7(1):45–52
130. Vaidyanathan S (2012) Anti-synchronization of Sprott-L and Sprott-M chaotic systems via adaptive control. Int J Control Theor Appl 5(1):41–59
131. Vaidyanathan S, Pakiriswamy S (2013) Generalized projective synchronization of six-term Sundarapandian chaotic systems by adaptive control. Int J Control Theor Appl 6(2):153–163
132. Vaidyanathan S, Rajagopal K (2012) Global chaos synchronization of hyperchaotic Pang and hyperchaotic Wang systems via adaptive control. Int J Soft Comput 7(1):28–37
133. Sundarapandian V, Sivaperumal S (2011) Sliding controller design of hybrid synchronization of four-wing chaotic systems. Int J Soft Comput 6(5):224–231
134. Vaidyanathan S (2014) Global chaos synchronisation of identical Li-Wu chaotic systems via sliding mode control. Int J Model Ident Control 22(2):170–177
135. Vaidyanathan S, Sampath S (2012) Anti-synchronization of four-wing chaotic systems via sliding mode control. Int J Autom Comput 9(3):274–279

136. Vaidyanathan S, Sampath S, Azar AT (2015) Global chaos synchronisation of identical chaotic systems via novel sliding mode control method and its application to Zhu system. Int J Model Ident Control 23(1):92–100
137. Rasappan S, Vaidyanathan S (2013) Hybrid synchronization of n-scroll Chua circuits using adaptive backstepping control design with recursive feedback. Malays J Math Sci 73(1):73–95
138. Rasappan S, Vaidyanathan S (2014) Global chaos synchronization of WINDMI and Coullet chaotic systems using adaptive backstepping control design. Kyungpook Math J 54(1): 293–320
139. Suresh R, Sundarapandian V (2013) Global chaos synchronization of a family of n-scroll hyperchaotic Chua circuits using backstepping control with recursive feedback. Far East J Math Sci 7(2):219–246
140. Vaidyanathan S, Rasappan S (2014) Global chaos synchronization of n-scroll Chua circuit and Lur'e system using backstepping control design with recursive feedback. Arab J Sci Eng 39(4):3351–3364
141. Wang Y, Singer J, Bau HH (1992) Controlling chaos in a thermal convection loop. J Fluid Mech 237:479–498
142. Khalil HK (2001) Nonlinear systems, 3rd edn. Prentice Hall, New Jersey, USA

Synchronization of Chaotic Dynamical Systems in Discrete-Time

Adel Ouannas and M. Mossa Al-sawalha

Abstract In this study, we investigate the problem of chaos synchronization in discrete-time dynamical systems with different structures and diverse types. Based on Lyapunov stability theory, stability of lineare systems and nonlinear control methods some synchronization criterions are presented in 2D, 3D and N-dimensional discrete-time chaotic systems. Numerical examples and computer simulations are used to show the effectiveness and the feasibility of the proposed synchronization schemes.

1 Introduction

Over the last two decade, control and synchronization of chaos (hyperchaos) have been received a great deal of interest from many fields [1–6]. Various methods have been proposed for the synchronization of chaotic systems [7, 8] and different types of chaos synchronization have been presented [9–12], but most of works have concentrated on continuous-time chaotic systems rather than discrete-time chaotic systems [13–20] and few methods are related to discrete time systems [21]. In practice, discrete-time chaotic systems play a more important role than their continuous-parts [22]. In fact, many mathematical models of physical processes [23, 24], biological phenomena [25], chemical reactions and economic systems [26] were defined using discrete-time chaotic systems. Many chaotic and hyperchaotic maps in 2D and 3D are founded such as Hénon map [27], Lozi map [28], Fold map [29], Lorenz discrete-time system [29], discrete hyperchaotic double scroll [30], Stefanski map [31], Hitzl-Zele map [32], Baier-Klain system [33], Wang system [34], discrete-time Rössler system [35] and Grassi-Miller map, etc.

A. Ouannas (✉)
Laboratory of Mathematics, Informatics and Systems (LAMIS),
University of Larbi Tebessi, Tebessa, Algeria
e-mail: ouannas_adel@yahoo.fr

M.M. Al-sawalha
Mathematics Department, Faculty of Science,
University of Hail, Hail, Kingdom of Saudi Arabia
e-mail: sawalah_moh@yahoo.com

A.T. Azar and S. Vaidyanathan (eds.), *Advances in Chaos Theory and Intelligent Control*, Studies in Fuzziness and Soft Computing 337,
DOI 10.1007/978-3-319-30340-6_5

Recently, synchronization in discrete-time chaotic systems attract more and more attentions and has been extensively studied, due to it's potential applications in secure communication [36–43]. Until now, a variety of approaches have been proposed for the synchronization of discrete chaotic such as hybrid synchronization [44], projective synchronization (PS) [45], adaptive-function projective synchronization (AFPS) [46, 47], lag synchronization (LS) [48], anticipated synchronization (ACS) [49, 50], impulsive synchronization [51], full-state hybrid projective synchronization (FSHPS) [52, 53], function-cascade synchronization (FCS) [54], generalized synchronization (GS) [55, 56] and Q-S synchronization [57]. On the other hand, studying inverse synchronization types such as inverse projective synchronization (IPS) [58], inverse matrix projective synchronization (IMPS) [59], inverse full-state hybrid projective synchronization (IFSHPS) [60], and inverse generalized synchronization (IGS) [61], etc., is an attractive and important idea.

In this chapter, we presente systematic and constructive schemes to study synchronization of chaotic dynamical systems in discrete-time. Based on Lyapunov stability theory, stability of discrete lineare system and nonlinear control methods, some synchronization criterions for arbitrary 3D chaotic systems, general N-dimensional chaotic systems and different dimensional chaotic systems in discrete-time are presented. Diverse synchronization types are investigated such as complete synchronization, matrix projective synchronization, inverse matrix projective synchronization and synchronization with two scaling matrices. Many numerical examples and computer simulations are used to show the effectiveness and the feasibility of the proposed control schemes and the derived synchronization criterions.

This chapter is organized as follow. In Sect. 2, a general approach for complete chaos synchronization between arbitrary quadratic discrete-time systems in 3D is introduced. In Sect. 3, some new synchronization criterions are derived for general N-dimensional chaotic systems in discrete-time. The problem of matrix projective synchronization (MPS) between different dimensional chaotic maps is investigated in Sect. 4. In Sect. 5, the problem of inverse matrix projective synchronization (IMPS) between two general chaotic systems in discrete-time with different dimensions is presented. A new type of synchronization with double scaling matrices is proposed in Sect. 6.

2 Complete Synchronization for Arbitrary 3D Quadratic Maps

Consider the following drive and response chaotic systems:

$$x_i\left(k+1\right)=\sum_{j=1}^{3} a_{ij}x_j\left(k\right)+X^T\left(k\right)C^iX\left(k\right)+\delta_i,\quad 1\leq i\leq 3, \tag{1}$$

$$y_i(k+1) = \sum_{j=1}^{3} b_{ij} y_j(k) + Y^T(k) D^i Y(k) + \gamma_i + u_i, \quad 1 \le i \le 3, \tag{2}$$

where $X(k) = (x_i(k))_{1\le i\le 3}$, $Y(k) = (y_i(k))_{1\le i\le 3}$ are the state vectors of the drive system and the response system, respectively, $(a_{ij}) \in \mathbf{R}^{3\times 3}$, $(b_{ij}) \in \mathbf{R}^{3\times 3}$, $C^i = (c^i_{pq}) \in \mathbf{R}^{3\times 3}$ $(i = 1, 2, 3)$, $D^i = (d^i_{pq}) \in \mathbf{R}^{3\times 3}$ $(i = 1, 2, 3)$, $(\delta_i, \gamma_i)_{1\le i\le 3}$ are real numbers and $(u_i)_{1\le i\le 3}$ are controllers to be determined. Let us define the errors of complete synchronization, between the drive system (1) and the response system (2), as

$$e_i(k) = y_i(k) - x_i(k), \quad 1 \le i \le 3. \tag{3}$$

2.1 General Synchronization Approach

The dynamic of synchronization errors (3), can be derived as follow:

$$e_i(k+1) = \sum_{j=1}^{3} b_{ij} e_j(k) + u_i + L_i + N_i, \quad 1 \le i \le 3, \tag{4}$$

where

$$L_i = \sum_{j=1}^{3} \left(b_{ij} - a_{ij}\right) x_j(k), \quad 1 \le i \le 3, \tag{5}$$

and

$$N_i = Y^T(k) D^i Y(k) - X^T(k) C^i X(k) + \gamma_i - \delta_i, \quad 1 \le i \le 3. \tag{6}$$

To achieve synchronization between the systems (1) and (2), we choose the controllers $(u_i)_{1\le i\le 3}$ as follow:

$$u_i = \sum_{j=1}^{3} \left(\lambda_{ij} - b_{ij}\right) e_j(k) - L_i - N_i, \quad 1 \le i \le 3, \tag{7}$$

where $\lambda_{11} = \frac{1}{b_{11}^2 + l_1}$, $\lambda_{12} = \frac{1}{b_{22}^2 + l_2}$, $\lambda_{13} = -\frac{1}{b_{33}^2 + l_3}$, $\lambda_{21} = \frac{1}{b_{11}^2 + l_1}$, $\lambda_{22} = \frac{1}{b_{22}^2 + l_2}$, $\lambda_{23} = \frac{1}{b_{33}^2 + l_3}$, $\lambda_{31} = \frac{1}{b_{11}^2 + l_1}$, $\lambda_{32} = -\frac{2}{b_{22}^2 + l_2}$, $\lambda_{33} = 0$ and $(l_i)_{1\le i\le 3}$ are unknown constants to be determined later. Hence, we have the following result.

Theorem 1 *If the control constants* $(l_i)_{1\le i\le 3}$ *are chosen such that*

$$l_1 > \sqrt{3}, l_2 > \sqrt{6} \text{ and } l_3 > \sqrt{2}. \tag{8}$$

Then, the drive system (1) and the response system (2) are globally synchronized under the controller law (7).

Proof By substituting Eq. (7) into Eq. (4), the synchronization errors can be written as follow:

$$\begin{aligned} e_1(k+1) &= \frac{1}{b_{11}^2+l_1}e_1(k)+\frac{1}{b_{22}^2+l_2}e_2(k)-\frac{1}{b_{33}^2+l_3}e_3(k), \\ e_2(k+1) &= \frac{1}{b_{11}^2+l_1}e_1(k)+\frac{1}{b_{22}^2+l_2}e_2(k)+\frac{1}{b_{33}^2+l_3}e_3(k), \\ e_3(k+1) &= \frac{1}{b_{11}^2+l_1}e_1(k)-\frac{2}{b_{22}^2+l_2}e_2(k), \end{aligned} \tag{9}$$

For the stability analysis, let us consider the following quadratic Lyapunov function:

$$V(e(k)) = \sum_{i=1}^{3} e_i^2(k), \tag{10}$$

we obtain:

$$\begin{aligned} \Delta V(e(k)) &= V(e(k+1)) - V(e(k)) \\ &= \sum_{i=1}^{3} e_i^2(k+1) - \sum_{i=1}^{3} e_i^2(k) \\ &= \left(\frac{3}{(b_{11}^2+l_1)^2}-1\right)e_1^2(k)+\left(\frac{6}{(b_{22}^2+l_2)^2}-1\right)e_2^2(k)+\left(\frac{2}{(b_{33}^2+l_3)^2}-1\right)e_3^2(k). \end{aligned}$$

Using (8), we can prove that

$$\frac{3}{(b_{11}^2+l_1)^2} < 1, \frac{6}{(b_{22}^2+l_2)^2} < 1 \text{ and } \frac{2}{(b_{33}^2+l_3)^2} < 1, \tag{11}$$

then we get: $\Delta V(e(k)) < 0$. Thus, by Lyapunov stability it is immediate that $\lim_{k\to\infty} e_i(k) = 0$, $(i = 1, 2, 3)$. We conclude that the drive system (1) and the response system (2) are globally synchronized.

2.2 Illustrative Example

In this example, we consider the discrete-time Rössler system as drive system and the controlled Wang system as response system. The discrete-time R össler system [35], is described by

$$\begin{aligned} x_1(k+1) &= \alpha x_1(k)(1-x_1(k)) - \beta(x_3(k)+\gamma)(1-2x_2(k)), \\ x_2(k+1) &= \delta x_2(k)(1-x_2(k)) + \varsigma x_3(k), \\ x_3(k+1) &= \eta((x_3(k)+\gamma)(1-2x_2(k))-1)(1-\theta x_1(k)), \end{aligned} \tag{12}$$

where $\alpha = 3.8$, $\beta = 0.05$, $\gamma = 0.35$, $\delta = 3.78$, $\varsigma = 0.2$, $\eta = 0.1$ and $\theta = 1.9$. The projection of the hyperchaotic attractor of the 3D discrete-time Rössler system is shown in Fig. 1.

The controlled Wang system can be described as

$$\begin{aligned} y_1(k+1) &= a_3 y_2(k) + (a_4+1)\, y_1(k) + u_1, \\ y_2(k+1) &= a_1 y_1(k) + y_2(k) + a_2 y_3(k) + u_2, \\ y_3(k+1) &= (a_7+1)\, y_3(k) + a_6 y_2(k)\, y_3(k) + a_5 + u_3, \end{aligned} \tag{13}$$

where $U = (u_1, u_2, u_3)^T$ is the vector controller. The Wang system (i.e., the system (13) with $u_1 = 0$, $u_2 = 0$, $u_3 = 0$) is hyperchaotic when the parameter values are taken as $(a_1, a_2, a_3, a_4, a_5, a_6, a_7) = (-1.9, 0.2, 0.5, -2.3, 2, -0.6, -1.9)$ [34]. The hyperchaotic attractor of the 3D Wang system is shown in Fig. 2.

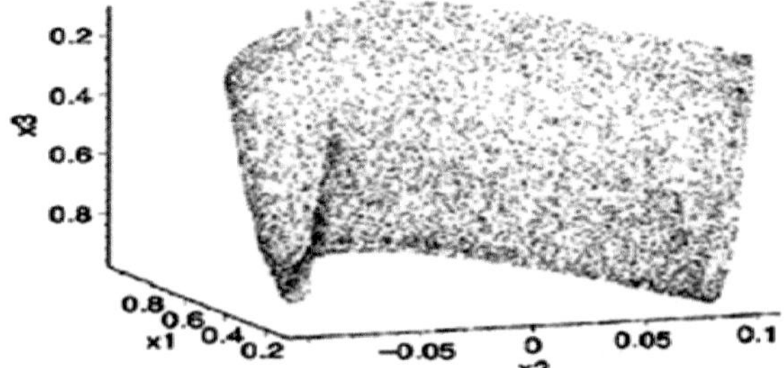

Fig. 1 Hyperchaotic attractor of the discrete-time Rössler when: $\alpha = 3.8$, $\beta = 0.05$, $\gamma = 0.35$, $\delta = 3.78$, $\varsigma = 0.2$, $\eta = 0.1$ and $\theta = 1.9$

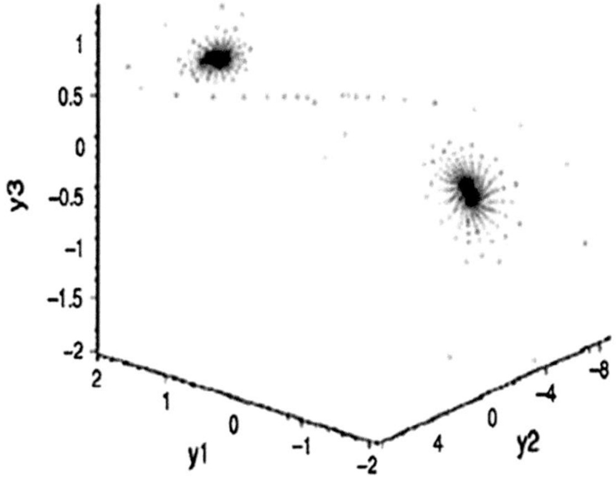

Fig. 2 Hyperchaotic attractor of Wang system when $(a_1, a_2, a_3, a_4, a_5, a_6, a_7, \delta) = (-1.9, 0.2, 0.5, -2.3, 2, -0.6, -1.9, 1)$

To achieve complete synchronization between the discrete-time Rössler system (12) and the controlled Wang system (13), according to our approach presented above, the vector controller $U = (u_1, u_2, u_3)^T$ can be constructed as follow

$$\begin{aligned} u_1 &= \left(\frac{1}{(a_4+1)^2+l_1} - (a_4+1)\right)e_1(k) + \left(\frac{1}{1+l_2} - a_3\right)e_2(k) \\ &\quad + \frac{1}{(a_7+1)^2+l_3}e_3(k) - L_1 - N_1, \\ u_2 &= \left(\frac{1}{(a_4+1)^2+l_1} - a_1\right)e_1(k) + \left(\frac{1}{1+l_2} - 1\right)e_2(k) \\ &\quad + \left(\frac{1}{(a_7+1)^2+l_3} - a_2\right)e_3(k) - L_2 - N_2, \\ u_3 &= \frac{1}{(a_4+1)^2+l_1}e_1(k) - \frac{1}{1+l_2}e_2(k) - (a_7+1)\, e_3(k) - L_3 - N_3, \end{aligned} \tag{14}$$

where

$$\begin{aligned} L_1 &= (a_4+1-\alpha)\, x_1(k) + (a_3 - \beta\gamma 2)\, x_2(k) + \beta x_3(k) + \beta\gamma, \\ L_2 &= a_1 x_1(k) + (1-\beta\gamma 2)\, x_2(k) + (a_2 - \varsigma)\, x_3(k), \\ L_3 &= -\theta(1-\eta\gamma)\, x_1(k) + 2\eta\gamma x_2(k) + (a_7+1-\eta)\, x_3(k) + a_5 - \eta\gamma + 1, \end{aligned} \tag{15}$$

$$\begin{aligned} N_1 &= \alpha x_1^2(k) - 2\beta x_3(k)\, x_2(k), \\ N_2 &= \delta x_2^2(k), \\ N_3 &= a_6 y_2(k)\, y_3(k) - 2\eta\gamma\theta x_1(k)\, x_2(k) + \eta\theta x_1(k)\, x_3(k) \\ &\quad + 2\eta x_2(k)\, x_3(k) - 2\eta\theta x_1(k)\, x_2(k)\, x_3(k), \end{aligned} \tag{16}$$

and $(l_i)_{1\le i\le 3}$ are the control constants. Using the controllers (14), the error system can be described as:

$$\begin{aligned} e_1(k+1) &= \frac{1}{(a_4+1)^2+l_1}e_1(k) + \frac{1}{1+l_2}e_2(k) \\ &\quad + \frac{1}{(a_7+1)^2+l_3}e_3(k), \\ e_2(k+1) &= \frac{1}{(a_4+1)^2+l_1}e_1(k) + \frac{1}{1+l_2}e_2(k) \\ &\quad + \frac{1}{(a_7+1)^2+l_3}e_3(k), \\ e_3(k+1) &= \frac{1}{(a_4+1)^2+l_1}e_1(k) - \frac{1}{1+l_2}e_2(k). \end{aligned} \tag{17}$$

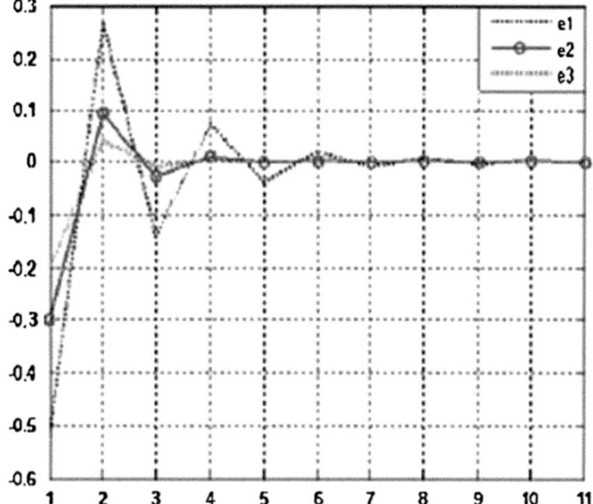

Fig. 3 Time evolution of complete synchronization errors between the drive system (12) and the response system (13)

Corollary 1 *For the two coupled the hyperchaotic discrete-time Rössler system (12) and the controlled hyperchaotic Wang system (13), if we choose the control constants* $(l_i)_{1\leq i\leq 3}$ *such that:* $l_1 = 5$, $l_2 = 4$ *and* $l_3 = 3$. *Then, they are globally complete synchronized as shown in Fig. 3.*

2.3 Conclusion

In this section, a new control scheme was designed to achieve complete synchronization between arbitrary 3-D quadratic chaotic maps. It was shown that the proposed controllers guarantee the asymptotic convergence to zero of the errors between the drive and the response systems. To illustrate the effectiveness of the derived results, the presented approach was applied between the two hyperchaotic maps: discrete-time Rössler system and controlled Wang system.

3 Complete Synchronization Criterions for N-Dimensional Maps

We consider the drive and the response chaotic systems are in the following forms:

$$x_i(k+1) = \sum_{j=1}^{n} a_{ij}x_j(k) + f_i(X(k)), \quad 1 \leq i \leq n, \tag{18}$$

$$y_i(k+1) = \sum_{j=1}^{n} b_{ij}y_j(k) + g_i(Y(k)) + u_i, \quad 1 \leq i \leq n, \tag{19}$$

where $X(k) = (x_1(k), x_2(k), \ldots, x_n(k))^T$, $Y(k) = (y_1(k), y_2(k), \ldots, y_n(k))^T$ are states of the drive system and the response system, respectively, (a_{ij}), $(b_{ij}) \in \mathbb{R}^{n\times n}$, $f_i, g_i : \mathbb{R}^n \to \mathbb{R}$, $1 \le i \le n$, are nonlinear functions and u_i, $1 \le i \le n$, are controllers. Define the state error vector as

$$e(k) = Y(k) - X(k), \tag{20}$$

namely,

$$e_i(k) = y_i(k) - x_i(k), \quad 1 \le i \le n. \tag{21}$$

Then the aim of complete synchronization is to find controllers u_i, $1 \le i \le n$, such that the state of errors satisfy

$$\lim_{k\to\infty} e_i(k) = 0, \quad 1 \le i \le n. \tag{22}$$

3.1 Synchronization Criterion Via Controlling $(b_{ij})_{1\le i,\ j\le n}$

The synchronization errors between the drive system (18) and the response system (19), can be written as follow:

$$e_i(k+1) = \sum_{j=1}^{n} b_{ij} e_j(k) + R_i + u_i, \quad 1 \le i \le n, \tag{23}$$

where

$$R_i = \sum_{j=1}^{n} (b_{ij} - a_{ij}) x_j(k) + g_i(Y(k)) - f_i(X(k)), \quad 1 \le i \le n. \tag{24}$$

To achieve complete synchronization between the systems (18) and (19), we can choose the controllers u_i, $1 \le i \le n$, as follow:

$$u_i = -R_i - \sum_{j=1}^{n} l_{ij} (y_j(k) - x_j(k)) \quad 1 \le i \le n, \tag{25}$$

where $(l_{ij}) \in \mathbb{R}^{n\times n}$ are control constants to be determined later. By substituting Eq. (25) into Eq. (23), the synchronization errors can be written as:

$$e_i(k+1) = \sum_{j=1}^{n} (b_{ij} - l_{ij}) e_j(k), \quad 1 \le i \le n. \tag{26}$$

Theorem 2 *If* $\{l_{ij}\}$ *are chosen such that*

$$\sum_{i=1}^{n}\left(b_{ip}-l_{ip}\right)\left(b_{iq}-l_{iq}\right)=0,\quad p,q=1,2,\ldots n,\ p\neq q \tag{27}$$

and

$$\sum_{i=1}^{n}\left(b_{ij}-l_{ij}\right)^{2}<1,\quad j=1,2,\ldots n, \tag{28}$$

Then, the two systems (18) and (19), are globally synchronized.

Proof Consider the candidate Lyapunov function:

$$V\left(e(k)\right)=\sum_{i=1}^{n}e_i^2\left(k\right), \tag{29}$$

we get:

$$\begin{aligned}\Delta V\left(e(k)\right)&=V\left(e(k+1)\right)-V\left(e(k)\right) \\ &=\sum_{i=1}^{n}e_i^2\left(k+1\right)-\sum_{i=1}^{n}e_i^2\left(k\right) \\ &=\sum_{j=1}^{n}\left(\sum_{i=1}^{n}\left(b_{ij}-l_{ij}\right)^2-1\right) \\ &\quad+\sum_{\substack{p,q=1\\ p\neq q}}^{n}\left(\sum_{i=1}^{n}\left(b_{ip}-l_{ip}\right)\left(b_{iq}-l_{iq}\right)\right)e_p(k)e_q(k)\end{aligned} \tag{30}$$

By using conditions (27 and 28), we obtain $\Delta V\left(e(k)\right)<0$. Then, it is immediate that $\lim_{k\to\infty}e_i\left(k\right)=0,\quad (i=1,\ 2,\ n)$. We conclude that the systems (18) and (19) are globally complete synchronized.

3.2 *Synchronization Criterion Via Controlling* $(b_{ii})_{1\leq i\leq n}$

Now, the error system between the drive system (18) and the response (19), can be written as follow:

$$e_i(k+1)=b_{ii}e_i(k)+R_i+u_i,\quad 1\leq i\leq n, \tag{31}$$

where

$$R_i = \sum_{\substack{j=1 \\ j\neq i}}^{n} b_{ij} y_j(k) + b_{ii} x_i(k) - \sum_{j=1}^{n} a_{ij} x_j(k) + g_i(Y(k)) - f_i(X(k)), \quad 1 \leq i \leq n. \tag{32}$$

To achieve complete synchronization between the systems (18) and (19), we can choose the synchronization controllers u_i $(1 \leq i \leq n)$ as follow:

$$u_i = -R_i - l_i \left(y_i(k) - x_i(k)\right), \quad 1 \leq i \leq n, \tag{33}$$

where $(l_i) \in \mathbb{R}^n$ are control constants to be determined. By substituting Eq. (33) into Eq. (31), the synchronization errors can be written as:

$$e_i(k+1) = (b_{ii} - l_i)\, e_i(k), \quad 1 \leq i \leq n. \tag{34}$$

Theorem 3 *If $\{l_i\}$ are chosen such that*

$$-1 + b_{ii} < l_i < 1 + b_{ii}, \; 1 \leq i \leq n. \tag{35}$$

Then, the two systems (18) and (19) are globally synchronized.

Proof Consider the candidate Lyapunov function: $V(e(k)) = \sum_{i=1}^{n} e_i^2(k)$, we get:

$$\begin{aligned} \Delta V(e(k)) &= V(e(k+1)) - V(e(k)) \\ &= \sum_{i=1}^{n} \left(e_i^2(k+1) - e_i^2(k)\right) \\ &= \sum_{i=1}^{n} \left[(b_{ii} - l_i)^2 - 1\right] e_i^2(k). \end{aligned} \tag{36}$$

By using condition of Theorem 3, we obtain $\Delta V(e(k)) < 0$. Then, it is immediate that: $\lim_{k\to\infty} e_i(k) = 0, \; 1 \leq i \leq n$. We conclude that the systems (18) and (19) are globally complete synchronized.

3.3 Numerical Application

We consider Hitzl-Zele map as the drive system and the controlled Stefanski map as the response system. The Hitzl-Zele map can be described as

$$\begin{aligned} x_1(k+1) &= -bx_2(k), \\ x_2(k+1) &= x_3(k) + 1 - ax_2^2(k), \\ x_3(k+1) &= bx_2(k) + x_1(k), \end{aligned} \tag{37}$$

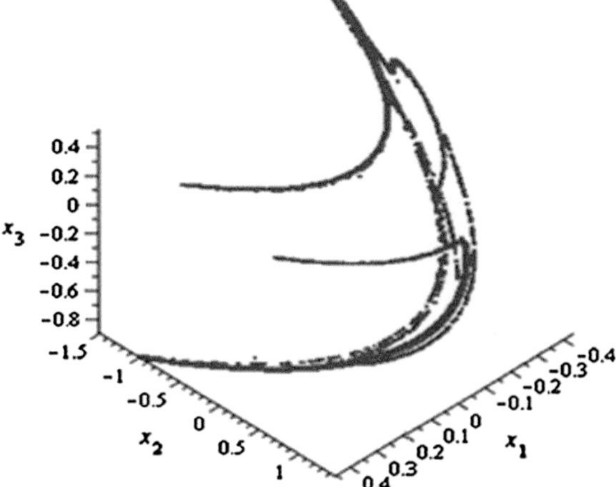

Fig. 4 Chaotic attractor of map (37)

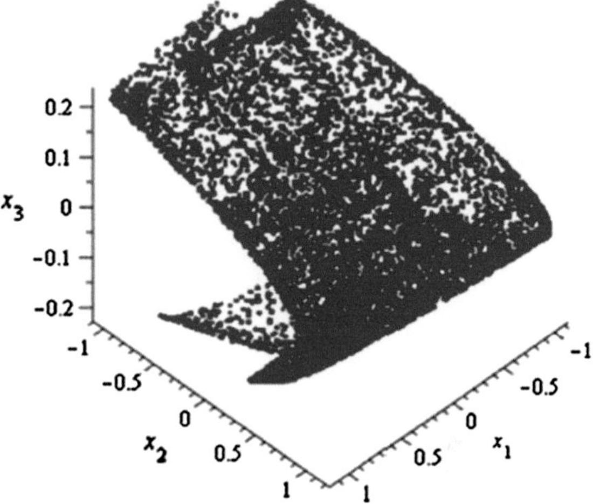

Fig. 5 Hyperchaotic attractor of map (38)

which has a chaotic attractor, for example, when $(a, b) = (1.07, 0.3)$ [32]. The chaotic attractor of the map (37) is shown in Fig. 4.

The controlled Stefanski map can be described as

$$\begin{aligned} y_1(k+1) &= 1 + y_3(k) - \alpha y_2^2(k) + u_1 \\ y_2(k+1) &= 1 + \beta y_2(k) - \alpha y_1^2(k) + u_2 \\ y_3(k+1) &= \beta y_1(k) + u_3 \end{aligned} \tag{38}$$

where $U = (u_1, u_2, u_3)^T$ is the vector controller. Stefanski map has a hyperchaotic attractor when $(\alpha, \beta) = (1.4, 0.2)$ [31]. The chaotic attractor of the map (38) is shown in Fig. 5. First, by applying the control scheme described in subsection 3.1, errors of complete synchronization between the maps (37) and (38) are given as follow:

$$
\begin{aligned}
e_1(k+1) &= e_3(k) + x_3(k) + 1 - \alpha y_2^2(k) + bx_2(k) + u_1, \\
e_2(k+1) &= \beta e_2(k) + \beta x_2(k) - \alpha y_1^2(k) - x_3(k) + ax_2^2(k) + u_2, \\
e_3(k+1) &= \beta e_1(k) + (\beta - 1) x_1(k) - bx_2(k) + u_3,
\end{aligned} \tag{39}
$$

According to Eq. (25), the synchronization controllers errors u_1, u_2 and u_3 can be designed as

$$
\begin{aligned}
u_1(k+1) &= -\sum_{j=1}^{3} l_{1j} e_j(k) - bx_2(k) - 1 + \alpha y_2^2(k), \\
u_2(k+1) &= -\sum_{j=1}^{3} l_{2j} e_j(k) - \beta x_2(k) + x_3(k) + \alpha y_1^2(k) - ax_2^2(k), \\
u_3(k+1) &= -\sum_{j=1}^{3} l_{3j} e_j(k) + bx_2(k) + (1 - \beta) x_1(k),
\end{aligned} \tag{40}
$$

where the control constants $(l_{ij}) \in \mathbb{R}^{3\times 3}$ are chosen as: $l_{11} = l_{12} = l_{21} = l_{23} = l_{32} = l_{33} = 0, 0 < l_{13} < 2$ and $-1 + \beta < l_{22}, l_{31} < 1 + \beta$, then the synchronization errors can be written as: $e_1(k+1) = (1 - l_{13}) e_3(k)$, $e_2(k+1) = (\beta - l_{22}) e_2(k)$ and $e_3(k+1) = (\beta - l_{31}) e_1(k)$. It is easy, by using Lyapunov stability, to show that the zero solution of the error system, in this case, is globally asymptotically stable, and therefore, the maps (37) and (38) are globally complete synchronized. We choose: $l_{13} = 1.5 < 2$ and $l_{22} = l_{31} = 1$, and by using Matlab, we get the numeric result that is shown in Fig. 6.

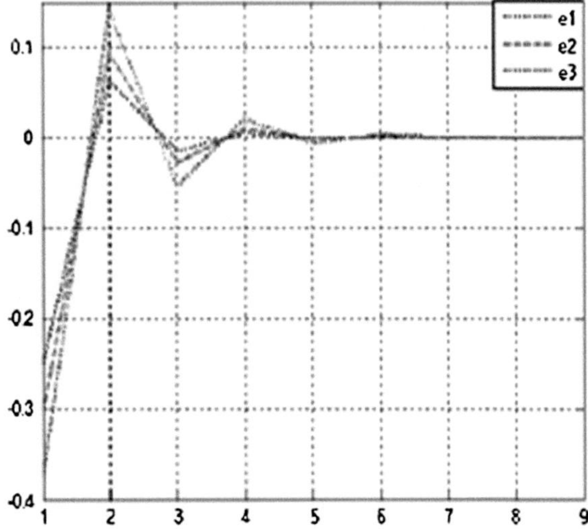

Fig. 6 Complete chaotic synchronization error states between the drive system (37) and the response system (38)

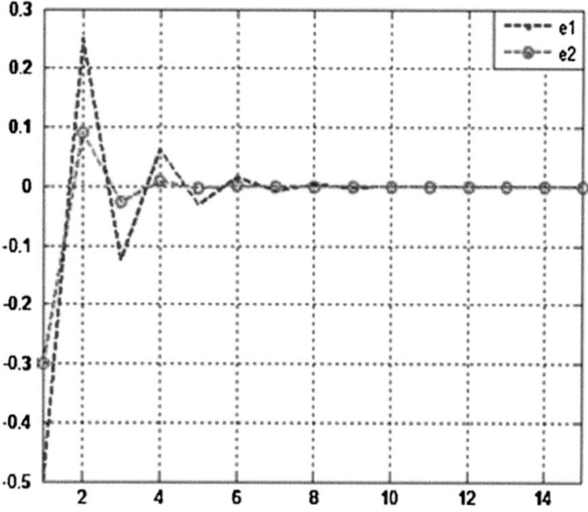

Fig. 7 Complete chaotic synchronization error states between the drive system (37) and the response system (38)

Now, according to the synchronization criterion proposed in subsection 3.2, the errors between the maps (37) and (38) are given as follow:

$$\begin{aligned} e_1(k+1) &= y_3(k) + 1 - \alpha y_2^2(k) + bx_2(k) + u_1, \\ e_2(k+1) &= \beta e_2(k) + \beta x_2(k) - \alpha y_1^2(k) - x_3(k) + ax_2^2(k) + u_2, \\ e_3(k+1) &= \beta y_1(k) - x_1(k) - bx_2(k) + u_3, \end{aligned} \tag{41}$$

According to Eq. (33), the synchronization errors can be designed as

$$\begin{aligned} u_1 &= l_1 e_1(k) - bx_2(k) - 1 + \alpha y_2^2(k), \\ u_2 &= l_2 e_2(k) - \beta x_2(k) + x_3(k) + \alpha y_1^2(k) - ax_2^2(k), \\ u_3 &= -l_3 e_3(k) + bx_2(k) + (1-\beta) x_1(k), \end{aligned} \tag{42}$$

where the control constants $(l_i)_{1\leq i\leq 3}$ are chosen as: $-1 < l_{1,2} < +1$ and $-1+\beta < l_3 < 1+\beta$, then the synchronization errors can be described as follow: $e_1(k+1) = l_1 e_1(k)$, $e_2(k+1) = l_2 e_2(k)$ and $e_3(k+1) = (\beta - l_3) e_3(k)$. It is easy, by using Lyapunov stability, to show that the zero solution of the error system, in this case, is globally asymptotically stable, and therefore, the maps (37) and (38) are globally complete synchronized. We choose: $l_{1,2} = 0.5 < 2$ and $l_3 = 1$, and by using Matlab, we get the numeric result that is shown in Fig. 7.

3.4 Conclusion

In this section, new criterions were derived to achieve complete synchronization between general discrete-time chaotic systems in N-D. It was shown that the proposed method, based on Lyapunov stability theory and controlling the linear part of response system, was theoretically rigorous. Finally, an example of application and numerical simulations were utilized to illustrate the effectiveness of the proposed synchronization criterions.

4 Matrix Projective Synchronization for Different Dimensional Maps

In this section, to study the problem of matrix projective synchronization (MPS), the following drive chaotic system is considered

$$X(k+1) = f(X(k)), \tag{43}$$

where $X(k) \in \mathbb{R}^n$ is the state vector of the drive system (43) and $f : \mathbb{R}^n \to \mathbb{R}^n$. As response system, we consider the following chaotic system

$$Y(k+1) = BY(k) + g(Y(k)) + U, \tag{44}$$

where $Y(k) \in \mathbb{R}^m$, $B = \left(b_{ij}\right)_{m\times m}$, $g : \mathbb{R}^m \to \mathbb{R}^m$ and $U = (u_i)_{1\le i\le m}$ are the state vector of the response system, linear part of the response system, nonlinear part of the response system and a vector controller, respectively. Now, we present the definition of matrix projective synchronization (MPS) between the drive system (43) and the response system (44).

Definition 1 The n-dimensional drive system (43) and the m-dimensional response system (44) are said to be matrix projective synchronized (MPS), if there exists a controller $U = (u_i)_{1\le i\le m}$ and a given matrix $\Lambda = \left(\Lambda_{ij}\right)_{m\times n}$ such that the synchronization error

$$e\left(k\right) = Y\left(k\right) - \Lambda X\left(k\right), \tag{45}$$

satisfies that $\lim_{k\longrightarrow+\infty} \|e\left(k\right)\| = 0$.

4.1 Matrix Projective Synchronization Criterion

The error system between the drive system (43) and the response system (44) can be derived as

$$e(k+1) = BY(k) + g(Y(k)) - \Lambda f(X(k)) + U. \tag{46}$$

Theorem 4 *If there exist a control matrix $L \in \mathbb{R}^{m \times m}$ such that $(B - L)^T (B - L) - I$ is a negative definite matrix, then the drive system (43) and the response system (44) are globally matrix projective synchronized under the following control law*

$$U = -LY(k) - g(Y(k)) + \Lambda f(X(k)) + (L - B)\Lambda X(k), \tag{47}$$

Proof By substituting Eq. (47) into Eq. (46), the error system can be described as

$$e(k + 1) = (B - L)\, e(k). \tag{48}$$

Construct the candidate Lyapunov function in the form

$$V\,(e(k)) = e^T(k)e(k), \tag{49}$$

we obtain

$$\begin{aligned} \Delta V\,(e(k)) &= e^T(k + 1)e(k + 1) - e^T(k)e(k) \\ &= e^T(k)(B - L)^T(B - L)e(k) - e^T(k)e(k) \\ &= e^T(k)\left[(B - L)^T(B - L) - I\right] e(k) < 0. \end{aligned} \tag{50}$$

Thus, from the Lyapunov stability theory, it is immediate that the zero solution of the error system (48) is globally asymptotically stable. Therefore, the systems (43) and (44) are globally matrix projective synchronized.

4.2 Simulation Example

In this example, we consider Hitzl-Zele map as the drive system and the controlled 2D Lorenz discrete-time system as the response system. The Hitzl-Zele map can be described as

$$\begin{aligned} x_1\,(k + 1) &= -bx_2\,(k) \\ x_2\,(k + 1) &= x_3\,(k) + 1 - ax_2^2\,(k) \\ x_3\,(k + 1) &= bx_2\,(k) + x_1\,(k) \end{aligned} \tag{51}$$

where $(a, b) = (1.07, 0.3)$ [32]. The chaotic attractor of the 3D Hitzl-Zele map was shown in Fig. 4.

The controlled 2D Lorenz discrete-time system can be described as

$$\begin{aligned} y_1\,(k + 1) &= (1 + \alpha\beta)\, y_1\,(k) - \beta y_1\,(k)\, y_2\,(k) + u_1 \\ y_2\,(k + 1) &= (1 - \beta)\, y_2\,(k) + \beta y_1^2\,(k) + u_2 \end{aligned} \tag{52}$$

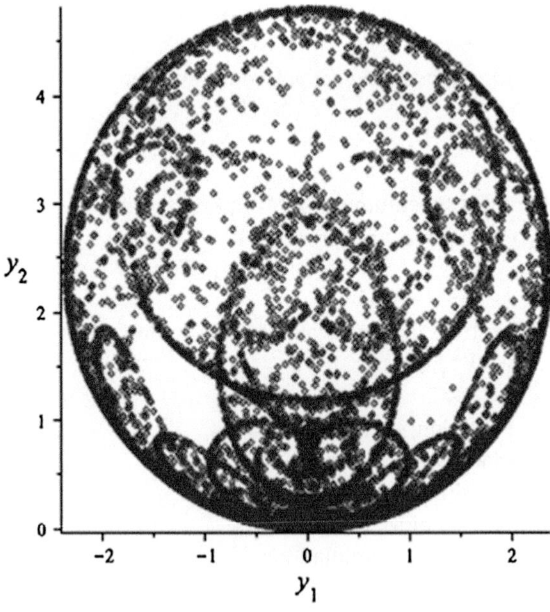

Fig. 8 Chaotic attractor of 2D Lorenz discrete-time system

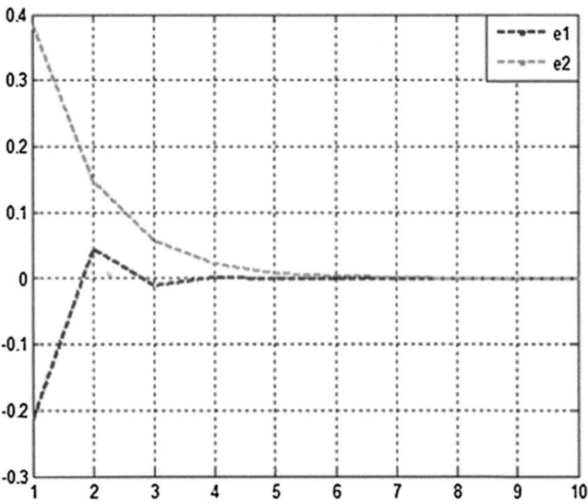

Fig. 9 Time evolution of IMPS errors between the Hitzl-Zele map (51) and the controlled 2D Lorenz discrete-time system (52)

where $U = (u_1, u_2)^T$ is the vector controller. The 2D Lorenz discrete-time system has a chaotic attractor when $(\alpha, \beta) = (1.25, 0.75)$ [29]. The chaotic attractor of Lorenz discrete-time system is shown in Fig. 8.

According to our approach presented above, we obtain

$$B = \begin{pmatrix} 1+\alpha\beta & 0 \\ 0 & 1-\beta \end{pmatrix}, \ and \ g(Y(k)) = \begin{pmatrix} -\beta y_1(k)\, y_2(k) \\ \beta y_1^2(k) \end{pmatrix}.$$

Then, we select the scaling matrix Λ and the control matrix L as

$$\Lambda = \begin{pmatrix} 1 & 0 & 1 \\ 0 & 2 & 0 \end{pmatrix}, L = \begin{pmatrix} 0.5 + \alpha\beta & 0 \\ 0 & 0.5 - \beta \end{pmatrix}.$$

Finally, it is easy to show that $(B - L)^T (B - L) - I$ is a negative definite matrix. Therefore, the systems (51) and (52) are globally matrix projective synchronized. The evolution of the error functions is shown in Fig. 9.

4.3 Conclusion

In this section, the problems of matrix projective synchronization (MPS) in different dimensional discrete-time chaotic systems was analyzed. Based on nonlinear controllers and Lyapunov stability, new conditions for chaos synchronization were derived to achieve MPS behavior between different dimensional drive and response maps. Numerical simulations were used to verify the effectiveness of our approach.

5 Inverse Matrix Projective Synchronization Between n-D and m-D Maps

In this section, to investigate the problem of inverse matrix projective synchronization (IMPS), the drive system and the response system are in the following forms

$$X(k+1) = AX(k) + f(X(k)), \tag{53}$$

$$Y(k+1) = g(Y(k)) + U, \tag{54}$$

where $X(k) = (x_i(k))_{1 \le i \le n}$, $Y(k) = (y_i(k))_{1 \le i \le m}$ are states vector of the drive system and the response system, respectively, A is an $n \times n$ constant matrix, $f = (f_i)_{1 \le i \le n}$ is a nonlinear vector function, $g = (g_i)_{1 \le i \le m}$ and $U = (u_i)_{1 \le i \le m}$ is a vector controller to be determined. Now, we present the definition of inverse matrix projective synchronization (IMPS) between the drive system (53) and the response system (54).

Definition 2 The n-dimensional drive system (53) and the m-dimensional response system (54) are said to be inverse matrix projective synchronized, if there exists a controller $U = (u_i)_{1 \le i \le m}$ and a given matrix $\Lambda = \left(\Lambda_{ij}\right)_{n \times m}$ such that the synchronization error

$$e(k) = X(k) - \Lambda Y(k), \tag{55}$$

satisfies that $\lim_{k \longrightarrow +\infty} \|e(k)\| = 0$.

5.1 Synchronization Results

The error system between the drive system (53) and the response (54) can be derived as

$$e(k+1) = (A - L)\, e(k) + R - \Lambda U, \tag{56}$$

where

$$R = LX(k) + (A - L)\, \Lambda Y(k) + f(X(k)) - \Lambda g(Y(k)), \tag{57}$$

$$\Lambda = \begin{pmatrix} \Lambda_{11} & \Lambda_{12} & \cdots & \Lambda_{1m} \\ \Lambda_{21} & \Lambda_{22} & \cdots & \Lambda_{2m} \\ \vdots & \vdots & \ddots & \vdots \\ \Lambda_{n1} & \Lambda_{n2} & \cdots & \Lambda_{nm} \end{pmatrix}, \tag{58}$$

is a scaling matrix and $L = \left(l_{ij}\right)_{n\times n}$ is an unknown control matrix to be determined. Then, we discuss two kinds of cases: $n < m$ and $m < n$, respectively.

5.1.1 Case 1: $n < m$

In this case, to achieve IMPS between systems (53) and (54), we choose the controller U as

$$U = (u_1, \ldots, u_n, 0, \ldots, 0)^T, \tag{59}$$

by substituting Eq. (59) into Eq. (56), then the error systems between the systems (53) and (54) can be written as

$$e(k+1) = (A - L)\, e(k) + R - \Lambda_1 U_1, \tag{60}$$

where

$$\Lambda_1 = \begin{pmatrix} \Lambda_{11} & \Lambda_{12} & \cdots & \Lambda_{1n} \\ \Lambda_{21} & \Lambda_{22} & \cdots & \Lambda_{2n} \\ \vdots & \vdots & \ddots & \vdots \\ \Lambda_{n1} & \Lambda_{n2} & \cdots & \Lambda_{nn} \end{pmatrix}. \tag{61}$$

and $U_1 = (u_i)_{1\le i\le n}$.

Theorem 5 *For an invertible matrix Λ_1, where inverse matrix projective synchronization between the drive system (53) and the response system (54) will occur if the following conditions are satisfied:*

(i) $U_1 = \Lambda_1^{-1} R$*, where* Λ_1^{-1} *is the inverse of* Λ_1*.*
(ii) $(A - L)^T (A - L) - I$ *is a negative definite matrix.*

Proof By substituting the control law (i) into Eq. (60), the synchronization errors can be written as

$$e(k+1) = (A - L)\, e(k). \tag{62}$$

Construct the candidate Lyapunov function in the form: $V\,(e(k)) = e^T(k)e(k)$, then we obtain

$$\begin{aligned}\Delta V\,(e(k)) &= e^T(k+1)e(k+1) - e^T(k)e(k)\\ &= e^T(k)(A-L)^T(A-L)e(k) - e^T(k)e(k)\\ &= e^T(k)\left[(A-L)^T(A-L) - I\right]e(k).\end{aligned} \tag{63}$$

By using (ii), we get $\Delta V\,(e(k)) < 0$. Thus, from the Lyapunov stability theory, it is immediate that $\lim_{k\to\infty} e_i\,(k) = 0, \quad (i = 1, 2, \ldots, n)$. That is the zero solution of the error system (62) is globally asymptotically stable and therefore, the systems (53) and (54) are globally inverse matrix projective synchronized.

5.1.2 Case 2: $m < n$

Now, the error system between the drive system (53) and the response system (54), can be written as

$$e_i(k+1) = \sum_{j=1}^{n} \left(a_{ij} - l_{ij}\right) e_j(k) + R_i - \sum_{j=1}^{m} \Lambda_{ij} u_j, \quad 1 \le i \le n, \tag{64}$$

where for $1 \le i \le n$

$$\begin{aligned}R_i &= \sum_{j=1}^{n} l_{ij} x_j(k) - \sum_{j=1}^{m} \Lambda_{ij} g_j(Y(k)) \\ &\sum_{j=1}^{n} \left(a_{ij} - l_{ij}\right)\left(\sum_{j=1}^{m} \Lambda_{ij} y_j(k)\right) + f_i(X(k))\end{aligned} \tag{65}$$

To achieve synchronization between the systems (53) and (54), we define new synchronization controllers as

$$\tilde{u}_i = \sum_{j=1}^{m} \Lambda_{ij} u_j, \quad 1 \le i \le n, \tag{66}$$

and we assume that Λ_{ij}, $1 \le j \le m$, are not equal zero. Now, rewriting the error system described in Eq. (64) in the compact form

$$e(k+1) = (A - L)\, e(k) + R - U_2, \tag{67}$$

where $U_2 = (\tilde{u}_i)_{1 \leq i \leq n}$.

Theorem 6 *Inverse matrix projective synchronization between the drive system (53) and the response system (54) will occur if the following conditions are satisfied:*

(i) $U_2 = R$.
(ii) All eigenvalues of $A - L$ *are strictly inside the unit disk.*

Proof By substituting the control law (i) into Eq. (67), the error system can be written as

$$e(k+1) = (A - L)\, e(k). \tag{68}$$

Thus, by asymptotic stability of linear discrete-time systems, if condition (ii) is satisfied, it is immediate that all solutions of error system (68) go to zero as $k \to \infty$. Therefore, the systems (53) and (54) are globally inverse matrix projective synchronized.

5.2 Numerical Results

In order to show the effectiveness of our approach, two examples are used to discuss two kinds of cases: $n < m$ and $m < n$, respectively.

5.2.1 Case: $n < m$

We choose Fold map as drive system and Stefanski map as response system. The Fold map can be described by following:

$$\begin{aligned} x_1(k+1) &= a x_1(k) + x_2(k), \\ x_2(k+1) &= b + x_1^2(k), \end{aligned} \tag{69}$$

which has a chaotic attractor when $(a, b) = (-0.1, -1.7)$ [29]. The chaotic attractor of Fold map is shown in Fig. 10.

The controlled Stefanski map system can be described as

$$\begin{aligned} y_1(k+1) &= 1 + y_3(k) - \alpha y_2^2(k) + u_1, \\ y_2(k+1) &= 1 + \beta y_2(k) - \alpha y_1^2(k) + u_2, \\ y_3(k+1) &= \beta y_1(k) + u_3, \end{aligned} \tag{70}$$

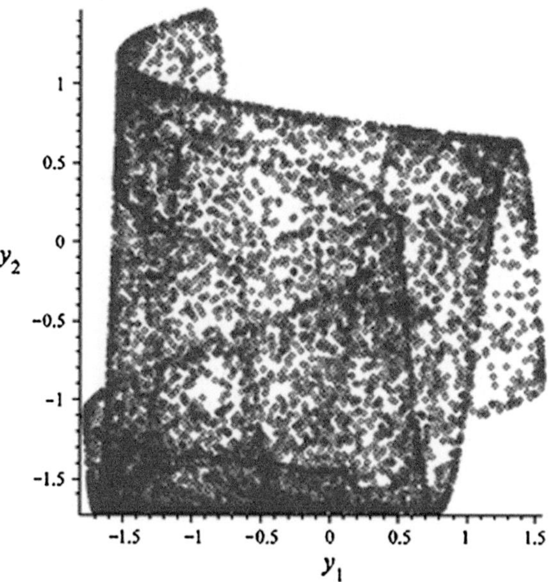

Fig. 10 The chaotic attractor of Fold map

where $(\alpha, \beta) = (1.4, 0.2)$ and $U = (u_1, u_2, u_3)^T$. The hyperchaotic attractor of map (70) was shown in Fig. 5. System (69) can be rewritten as the form $X(k+1) = AX(k) + f(X(k))$, where $X(k) = \begin{pmatrix} x_1(k) \\ x_2(k) \end{pmatrix}$, $A = \begin{pmatrix} a & 1 \\ 0 & 0 \end{pmatrix}$, and $f(X(k)) = \begin{pmatrix} 0 \\ b + x_1^2(k) \end{pmatrix}$.

According to our approach presented in paragraph 5.1.1, the scaling matrix Λ and the control matrix L are selected as

$$\Lambda = \begin{pmatrix} 1 & 0 & 0 \\ 0 & 1 & 0 \end{pmatrix}, L = \begin{pmatrix} a+0.5 & 1 \\ 0 & 0.5 \end{pmatrix}.$$

the vector controller $U = (u_1, u_2, u_3)^T$ is constructed as: $(u_1, u_2)^T = \Lambda_1^{-1} R$ and $u_3 = 0$ where $R = LX(k) + (A - L)\,\Lambda Y(k) + f(X(k)) - \Lambda g(Y(k))$ and $\Lambda_1^{-1} = \begin{pmatrix} 1 & 0 \\ 0 & 1 \end{pmatrix}$.

Simply, we can show that $(A-L)^T(A-L) - I$ is a definite negative matrix. Therefore, in this case, the maps (69) and (70) are globally inverse matrix projective synchronized. The error functions evolution is shown in Fig. 11.

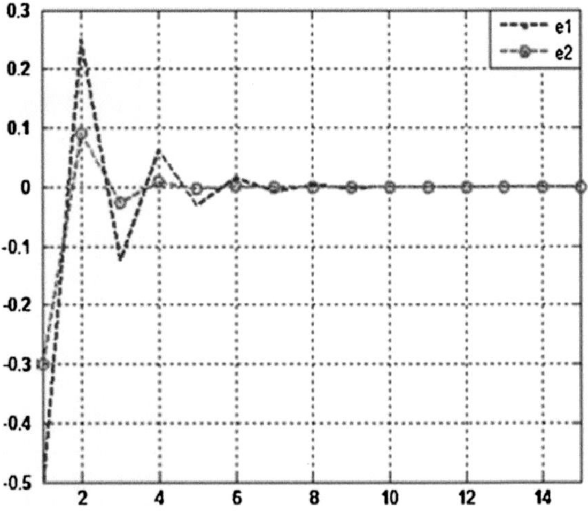

Fig. 11 Time evolution of IMPS errors between the maps (69) and (70)

5.2.2 Case: $m < n$

Now, we choose Stefanski map as drive system and Fold map as response system

$$\begin{aligned} x_1(k+1) &= 1 + x_3(k) - \alpha x_2^2(k), \\ x_2(k+1) &= 1 + \beta x_2(k) - \alpha x_1^2(k), \\ x_3(k+1) &= \beta x_1(k), \end{aligned} \tag{71}$$

$$\begin{aligned} y_1(k+1) &= a y_1(k) + y_2(k) + u_1, \\ y_2(k+1) &= b + y_1^2(k) + u_1. \end{aligned} \tag{72}$$

System (71) can be rewritten as the form $X(k+1) = AX(k) + f(X(k))$, where $X(k) = \begin{pmatrix} x_1(k) \\ x_2(k) \\ x_3(k) \end{pmatrix}$, $A = \begin{pmatrix} 0 & 0 & 1 \\ 0 & \beta & 0 \\ \beta & 0 & 0 \end{pmatrix}$, and $f(X(k)) = \begin{pmatrix} 1 - \alpha x_2^2(k) \\ 1 - \alpha x_1^2(k) \\ 0 \end{pmatrix}$.

By using to the control scheme proposed in paragraph 5.1.2, the scaling matrix Λ and the control matrix L are selected as

$$\Lambda = \begin{pmatrix} 1 & 3 \\ 2 & 1 \\ 1 & -2 \end{pmatrix}, L = \begin{pmatrix} 0.3 & 0 & 1 \\ 0 & \beta + 0.5 & 0 \\ \beta & 0 & -0.3 \end{pmatrix}.$$

So,

$$A - L = \begin{pmatrix} -0.3 & 0 & 1 \\ 0 & -0.5 & 0 \\ 0 & 0 & 0.3 \end{pmatrix}$$

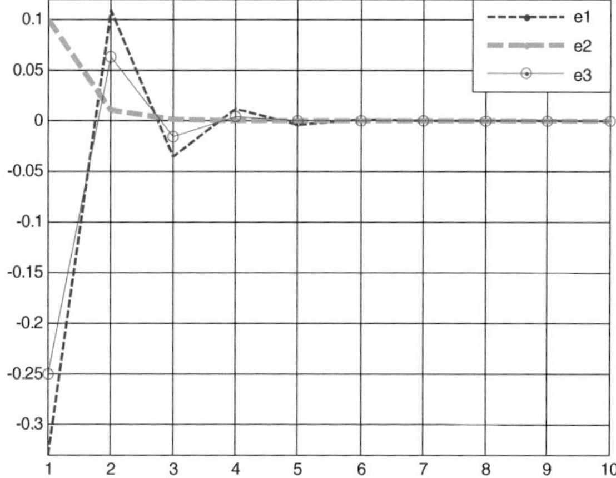

Fig. 12 Complete chaotic synchronization error states between the drive system (71) and the response system (72)

Finally, it is easy to show that all eigenvalues of $A - L$ are strictly inside the unit disk. Therefore, in this case, the maps (71) and (72) are globally inverse matrix projective synchronized. The error functions evolution is shown in Fig. 12.

5.3 Conclusion

In this section, we have developed a systematic and powerful synchronization schemes which were used to study a new synchronization type called inverse matrix projective synchronization (IMPS) between two general $n-$D and $m-$D discrete-time chaoticsystems in two cases $n < m$ and $m < n$,respectively. New synchronization results were derived using Lyapunov stability and stability theory of linear discrete-time dynamical systems. The synchronization criterions were obtained by controlling the linear part of the drive system. Numerical examples and simulations were utilized to verify the effectiveness of the proposed synchronization schemes.

6 Synchronization in Different Dimensions Using Two Scaling Matrices

Consider the following coupled chaotic systems

$$\begin{aligned} X(k+1) &= F(X(k)) \\ Y(k+1) &= G(Y(k)) + U \end{aligned} \tag{73}$$

where $X(k) \in \mathbb{R}^n$, $Y(k) \in \mathbb{R}^m$ are states of drive system and response system, respectively, $F : \mathbb{R}^n \to \mathbb{R}^n$, $G : \mathbb{R}^m \to \mathbb{R}^m$and $U = (u_i)_{1 \le i \le m}$ is a controller.

Definition 3 The n-dimensional drive system $X(k)$ and the m-dimensional response system $Y(k)$ are said to be synchronized in dimension d, with respect to scaling matrices Θ and Φ, respectively, if there exists a controller $U = (u_i)_{1 \le i \le m}$ and given two matrices $\Theta = (\Theta)_{d \times m}$ and $\Phi = (\Phi)_{d \times n}$ such that the synchronization error

$$e(k) = \Theta Y(k) - \Phi X(k), \tag{74}$$

satisfies that $\lim_{k \longrightarrow +\infty} \|e(k)\| = 0$. The number d is called the dimension of synchronization.

Remark 1 When $(\Theta, \Phi) = (I, I)$, $(\Theta, \Phi) = (I, -I)$, $(\Theta, \Phi) = (I, \Phi)$ and $(\Theta, \Phi) = (\Theta, I)$ complete synchronization, anti-synchronization, matrix projective synchronization, and inverse matrix projective synchronization will appear, respectively.

Because in real world all chaotic maps are described by plane equations or space systems, we restrict our study between 2D and 3D discrete chaotic systems in discrete-time and this restriction does'n lose the generality of our main results.

6.1 Synchronization of 2-D Drive System and 3-D Response System in 2D

In this case, the drive and the response chaotic systems are in the following forms

$$X(k+1) = AX(k) + f(X(k)), \tag{75}$$

$$Y(k+1) = g(Y(k)) + U, \tag{76}$$

where $X(k) \in \mathbb{R}^2$, $Y(k) \in \mathbb{R}^3$ are the states of the drive system and the response system, respectively, $A \in \mathbb{R}^{2\times 2}$, $f : \mathbb{R}^2 \to \mathbb{R}^2$ is the nonlinear part of the drive system (75), $g : \mathbb{R}^3 \to \mathbb{R}^3$ and $U \in \mathbb{R}^3$ is a vector controller. The error system, according to definition 3, between the drive system (75) and the response system (76) can be derived as

$$e(k+1) = \Theta g + \Theta U - \Phi AX(k) - \Phi f, \tag{77}$$

where $\Theta = (\Theta_{ij}) \in \mathbb{R}^{2\times 3}$, $\Phi = (\Phi_{ij}) \in \mathbb{R}^{2\times 2}$ are scaling matrices and we assume that:

$$\Phi A = A\Phi, \tag{78}$$

and

$$\Theta = \begin{pmatrix} \Theta_{11} & \Theta_{12} & 0 \\ \Theta_{21} & \Theta_{22} & 0 \end{pmatrix}, \tag{79}$$

Then, the error system (77) can be written as

$$e(k+1) = (A - L_1)\, e(k) + R + \hat{\Theta}\hat{U}, \tag{80}$$

where

$$R = -(A\Theta + L_1\Theta)\, Y(k) + L_1\Phi X(k) + \Theta g - \Phi f, \tag{81}$$

$$\hat{\Theta} = \begin{pmatrix} \Theta_{11} & \Theta_{12} \\ \Theta_{21} & \Theta_{22} \end{pmatrix}, \tag{82}$$

$$\hat{U} = (u_1, u_2)^T, \tag{83}$$

and L_1 is 2×2 control matrix to be determined. To achieve synchronization between the systems (75) and (76), with respect to Θ and Φ, we choose the controller $\hat{U}$ as

$$\hat{U} = -\hat{\Theta}^{-1} R, \tag{84}$$

where $\hat{\Theta}^{-1}$ is the inverse matrix of $\hat{\Theta}$. By substituting Eq. (84) in Eq. (80), the error system can be written as

$$e(k+1) = (A - L_1)\, e(k). \tag{85}$$

By respect to the asymptotic stability property of linear discrete–time dynamical systems, if the control matrix L_1 is chosen such that all eigenvalues of $A - L_1$ are strictly inside the unit disk, it is immediate that all solution of the error system (85) go to zero as $k \to \infty$. Therefore, the systems (75) and (76) are globally synchronized, with respect to scaling matrices Θ and Φ.

6.2 *Synchronization of 2D Drive System and 3D Response System in 3D*

In this case, the drive systems is taken in the following form

$$X(k+1) = f(X(k)), \tag{86}$$

where $X(k) = (x_1(k), x_2(k))^T \in \mathbb{R}^2$, is the state of the drive system, $f : \mathbb{R}^2 \to \mathbb{R}^2$. As the response system we consider the following chaotic system

$$Y(k+1) = BY(k) + g(Y(k)) + U, \tag{87}$$

where $Y(k) = (y_1(k), y_2(k), y_3(k))^T \in \mathbb{R}^3$ is the state of the response system, respectively, $B \in \mathbb{R}^{3\times 3}$, $g : \mathbb{R}^3 \to \mathbb{R}^3$ is a nonlinear function and $U \in \mathbb{R}^3$ is a vector controller. In this case, we assume that $\Theta B = B\Theta$.

The error system between the systems (86) and (87), can be derived as

$$e(k+1) = (B - L_2)\, e(k) + R + \Theta U, \tag{88}$$

where

$$R = L_2 \Theta Y(k) + (B\Phi - L_2\Theta)\, X(k) + \Theta g - \Phi f, \tag{89}$$

and L_2 is 3×3 control matrix to be determined. To achieve synchronization between the drive system (86) and the response system (87), we choose the controller U as

$$U = -\Theta^{-1} R, \tag{90}$$

where Θ^{-1} is the inverse matrix of Θ. By substituting Eq. (90) in Eq. (88), the error system can be written as

$$e(k+1) = (B - L_2)\, e(k). \tag{91}$$

By respect to the asymptotic stability property of linear discrete–time dynamical systems, if the control matrix L_2 is chosen such that all eigenvalues of $B - L_2$ are strictly inside the unit disk, it is immediate that all solutions of the error system (91) go to zero as $k \to \infty$. Therefore, the systems (86) and (87) are globally synchronized, with respect to scaling matrice Θ and Φ.

6.3 Numerical Simulation

We consider Hénon map as the drive system and the controlled 3D map Grassi-Miller as the response system. The Hénon map can be described as

$$\begin{aligned} x_1(k+1) &= x_2(k) + 1 - a x_1^2(k), \\ x_2(k+1) &= b x_1(k), \end{aligned} \tag{92}$$

which has a chaotic attractor when $(\alpha, \beta) = (1.4, 0.2)$ [27]. The chaotic attractor of H, énon map is shown in Fig. 13.

The controlled hyperchaotic Grassi-Miller map can be described as (Grassi and Miller 2012):

$$\begin{aligned} y_1(k+1) &= -0.1 y_3(k) - y_2^2(k) + 1.76 + u_1, \\ y_2(k+1) &= y_1(k) + u_2, \\ y_3(k+1) &= y_2(k) + u_3, \end{aligned} \tag{93}$$

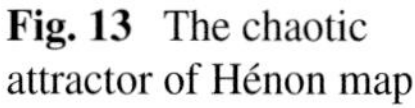

Fig. 13 The chaotic attractor of Hénon map

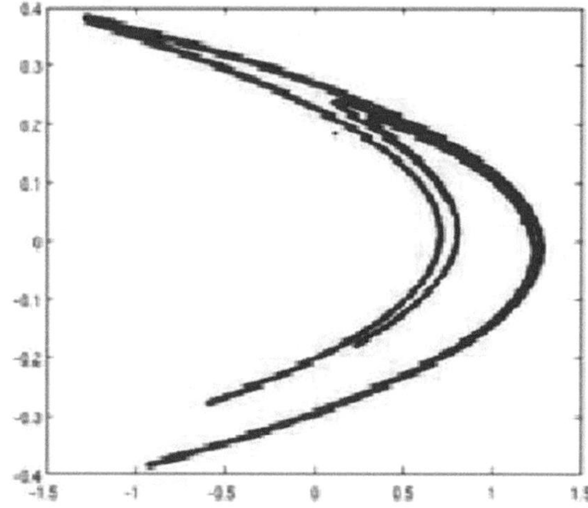

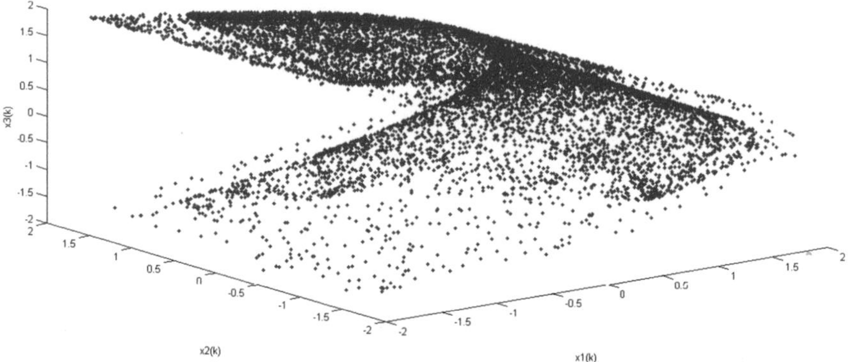

Fig. 14 The hyperchaotic attractors of Grassi-Miller map

where $U = (u_1, u_2, u_3)^T$ is the vector controller. The hyperchaotic attractors of Grassi-Miller map is shown in Fig. 14.

6.3.1 Case 1: Synchronization of Hénon Map and Baier-Klein Map in 2D

In order to observe the synchronization behavior in 2D between the drive system (92) and the response system (93), the quantities A and $f\,(X\,(k))$ are given as

$$A = \begin{pmatrix} 0 & 1 \\ b & 0 \end{pmatrix}, \; f\,(X\,(k)) = \begin{pmatrix} 1 - ax_1^2\,(k) \\ 0 \end{pmatrix}. \tag{94}$$

According to our approach presented in paragraph 6.1, the scaling matric Φ, Θ and the control matrix L_1 are chosen as follows

$$\Phi = \begin{pmatrix} 1 & 1 \\ b & 1 \end{pmatrix}, \Theta = \begin{pmatrix} 2 & 0 & 0 \\ 0 & 2 & 0 \end{pmatrix}, L_1 = \begin{pmatrix} \frac{1}{4} & 1 \\ b & \frac{1}{4} \end{pmatrix}.$$

So,

$$A - L_1 = \begin{pmatrix} -\frac{1}{4} & 0 \\ 0 & -\frac{1}{4} \end{pmatrix},$$

We can show that all eigenvalues of $A - L_1$ are strictly inside the unit disk. Therefore, in this case, the maps (92) and (93) are synchronized in 2D. The error functions evolution is shown in Fig. 15.

6.3.2 Case 2: Synchronization of Hénon Map and Baier-Klein Map in 3D

In order to observe the synchronization behavior in 3D between the drive system (92) and the response system (93), the quantities B and $g(Y(k))$, respectively, are given as

$$B = \begin{pmatrix} 0 & 0 & -0.1 \\ 1 & 0 & 0 \\ 0 & 1 & 0 \end{pmatrix}, g(Y(k)) = \begin{pmatrix} -y_2^2(k) + 1.76 \\ 0 \\ 0 \end{pmatrix}.$$

According to our approach presented in Sect. 1.6.2, the scaling matrices Φ, Θ and the control matrix L_2 are selected as follows

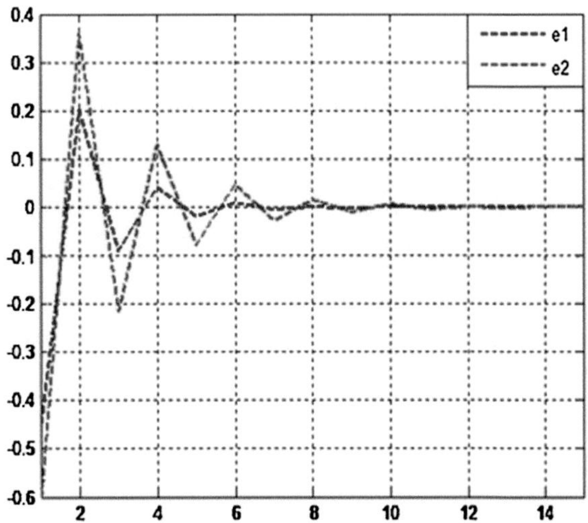

Fig. 15 Time evolution of errors between maps (92) and (93) in 2D

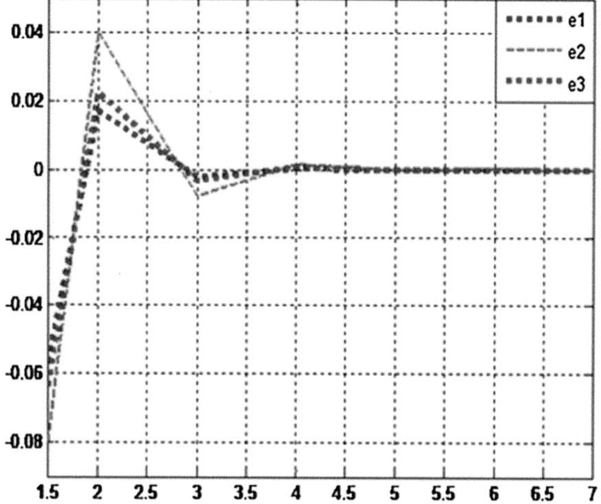

Fig. 16 Time evolution of errors between maps (92) and (93) in 3D

$$\Phi = \begin{pmatrix} 1 & 0 \\ 1 & 0 \\ 0 & 1 \end{pmatrix} .\Theta = \begin{pmatrix} 1 & -0.1 & -0.1 \\ 1 & 1 & -0.1 \\ 1 & 1 & 1 \end{pmatrix}, L_2 = \begin{pmatrix} 0.5 & 0 & -0.1 \\ 1 & 0.5 & 0 \\ 0 & 1 & 0.5 \end{pmatrix}.$$

So,

$$B - L_2 = \begin{pmatrix} -0.5 & 0 & 0 \\ 0 & -0.5 & 0 \\ 0 & 0 & -0.5 \end{pmatrix}.$$

We can see that all eigenvalues of $B - L_2$ are strictly inside the unit disk. Therefore, in this case, the maps (92) and (93) are synchronized in 3D. The error functions evolution is shown in Fig. 16.

6.4 Conclusion

In this section, a new type of synchronization with double scaling matrices was proposed and new synchronization results are derived using new control schemes and stability theory of linear discrete-time dynamical systems. Firstly, when the dimension of synchronization is the same of the response system the synchronization control was achieved by controlling the linear part of the response system. Secondly, if synchronization was made in dimension of the drive system, the linear part of the drive system was controlled. Numerical example and simulations results were used to verify the effectiveness of the proposed schemes.

References

1. Azar AT, Vaidyanathan S (2015) Chaos modeling and control systems design, Studies in computational intelligence. Springer
2. Azar AT, Vaidyanathan S (2015) Computational intelligence applications in modeling and control, Studies in computational intelligence. Springer
3. Azar AT, Vaidyanathan S (2015) Handbook of research on advanced intelligent control engineering and automation, Advances in computational intelligence and robotics (ACIR) Book Series. Springer
4. Blasius B, Stone L (2000) Chaos and phase synchronization in ecological systems. Inter Inter J Bifur Chaos 10:2361–2380
5. Lakshmanan M, Murali K (1996) Chaos in nonlinear oscillators: controlling and synchronization. Singapore
6. Zhu Q, Azar AT (2015) Complex system modelling and control through intelligent soft computations, Studies in fuzziness and soft computing. Springer
7. Boccaletti S, Kurths J, Osipov G, Valladares D, Zhou C (2002) The synchronization of chaotic systems. Phys Rep 1–101
8. Pecora L, Carrol T (1990) Synchronization in chaotic systems. Phys Rev Lett 821–824
9. Grassi G (2013) Continuous-time chaotic systems: arbitrary full-state hybrid projective synchronization via a scalar signal. Chin Phys B 22:080506–6
10. Hui LX, Yu SM (2011) Adaptive full state hybrid function projective lag synchronization in chaotic continuous-time system. Adv Mater Res 383–390
11. Junguo L, Yugeng X (2003) Linear generalized synchronization of continuous-time chaotic systems. Chaos Solitons Fract 825–831
12. Manfeng H, Zhenyuan X, Zhang R (2008) Full state hybrid projective synchronization in continuous-time chaotic (hyperchaotic) systems. Commun Nonlinear Sci Numer Simulat 456–464
13. Ouannas A (2014) Chaos synchronization approach based on new criterion of stability. Nonlinear Dyn Syst Theo 395–401
14. Vaidyanathan S, Azar AT (2015) Analysis and control of a 4-D novel hyperchaotic system. In: Azar AT, Vaidyanathan S (eds) Chaos modeling and control systems design, Studies in computational intelligence. Springer
15. Vaidyanathan S, Azar AT (2015) Analysis, control and synchronization of a nine-term 3-D novel chaotic system. In: Azar AT, Vaidyanathan S (eds) Chaos modeling and control systems design, Studies in computational intelligence. Springer
16. Vaidyanathan S, Azar AT (2015) Anti-synchronization of identical chaotic systems using sliding mode control and an application to vaidyanathan-madhavan chaotic systems. In: Azar AT, Zhu, Q (eds) Advances and applications in sliding mode control systems, Studies in computational intelligence book series. Springer
17. Vaidyanathan S, Azar AT (2015) Hybrid synchronization of identical chaotic systems using sliding mode control and an application to vaidyanathan chaotic systems. In: Azar, AT, Zhu Q (eds) Advances and applications in sliding mode control systems, Studies in computational intelligence book series. Springer
18. Vaidyanathan S, Azar AT, Rajagopal K, Alexander P (2015) Design and SPICE implementation of a 12-term novel hyperchaotic system and its synchronization via active control. Int J Model Identication Control, 267–277
19. Vaidyanathan S, Idowu B, Azar AT (2015) Backstepping controller design for the global chaos synchronization of sprott's jerk systems. In: Azar, AT, Vaidyanathan S (eds) Chaos modeling and control systems design, Studies in computational intelligence. Springer
20. Vaidyanathan S, Sampath S, Azar AT (2015) Global chaos synchronisation of identical chaotic systems via novel sliding mode control method and its application to Zhu system. Springer
21. Ju H (2007) A New approach to synchronization of discrete-time chaotic systems. J Phys Soc Jpn 093,002–093,013

22. Strogatz S (2001) Nonlinear dynamics and chaos: with applications to physics, biology, chemistry, and engineering. Westview Press, Studies In Nonlinearity
23. Matsumoto T, Chua L, Kobayashi K (1986) Hyperchaos: laboratory experiment and numerical confirmation. IEEE Trans Circuits Syst 11:1143–1147
24. Stoop R, Peinke J, Röhricht B, übener RH (1989) A p-Ge semiconductor experiment showing chaos and hyperchaos. Phys D 35:4352–4425
25. Eduardo L, Ruiz-Herrera A (2012) Chaos in discrete structured population models. SIAM J Appl Dynam Syst 11:1200–1214
26. Zhang W (2006) Discrete dynamical systems, bifurcations, and chaos in economics. Elsevier, Boston
27. Hénon M (1976) A two dimentionnal mapping with a strange attractor. Comm Math Phys 50:69–76
28. Lozi R (1978) Un Attracteur etrange du Type Attracteur de Henon. J Phys 9:9–10
29. Itoh M, Yang T, Chua L (2001) Conditions for impulsive synchronization of chaotic and hyperchaotic systems. Int J Bifurcat Chaos 551–560
30. Zeraoulia E, Sprott JC (2009) The discrete hyperchaotic double scroll. Int J Bifurcat Chaos 19:1023–1027
31. Stefanski K (1998) Modelling chaos and hyperchaos with 3D maps. Chaos Solitons Fractals 9:83–93
32. Hitzl D, Zele F (1985) An exploration of the Hénon quadratic map. Phys D 305–326
33. Baier G, Klein M (1990) Maximum hyperchaos in generalized Hénon maps. Phys Lett A 51:281–284
34. Wang X (2003) Chaos in complex nonlinear systems publishing house of electronics industry. Beijing
35. Yan Z (2006) Q-S (Complete or Anticipated) Synchronization backstepping scheme in a class of discrete-time chaotic (Hyperchaotic) systems: a symbolic-numeric computation approach. Chaos 16:013,119–11
36. Bustos A, Hernández C (2009) Synchronization of discrete-time hyperchaotic systems: an apllication in communications. Chaos Solitons Fract 41:1301–1310
37. Bustos A, Hernández C, Gutiérrez CL, Castillo P (2008) Synchronization of different hyperchaotic maps for encryption. Nonlinear Dyn Sys Theo 8:221–236
38. FilaliR., BenrejebM., BorneP. (2014) On observer-based secure communication design using discrete-time hyperchaotic systems. Commun Nonlinear Sci Numer Simulat 13:1424–1432
39. González I, Hernandez C (2013) Double hyperchaotic encryption for security in biometric systems. Nonlinear Dyn Sys Theo 13:55–68
40. Hernández C, Gutierrez L, Bustos A, Castillo P (2010) Communicating encrypted information based on synchronized hyperchaotic maps. Commun Nonlinear Sci Numer Simulat 11:337–349
41. Liu W, Wang Z, Zhang W (2012) Controlled synchronization of discrete-time chaotic systems under communication constraints. Nonlinear Dyn 69:223–230
42. Lu J, Xi Y (2005) Chaos communication based on synchronization of discrete-time chaotic systems. Chin Phys 14:274–278
43. Solak E (2005) Cryptanalysis of observer based discrete-time chaotic encryption schemes. Inter J Bifurcat Chaos 15:653–658
44. Filali R, Hammami S, Benrejeb M, Borne P (2012) On synchronization, anti-synchronization and hybrid synchronization of 3D discrete generalized Hénon map. Nonlinear Dyn Sys Theo 12:81–95
45. Jin Y, Xin L, Chen Y (2008) Function projective synchronization of discrete-time chaotic and hyperchaotic systems using backstepping method. Commun Theor Phys 50:111–116
46. Li Y, Chen Y, Li B (2009) Adaptive control and function projective synchronization in 2D discrete-time chaotic systems. Commun Theor Phys 51:270–278
47. Li Y, Chen Y, Li B (2009) Adaptive Function Projective Synchronization of Discrete-Time Chaotic Systems. Chin Phys Lett 26:040, 504–4
48. Chai Y, Lü L, Zhao H (2010) Lag Synchronization between discrete chaotic systems with diverse structure. Math Mech-Engl 31:733–738

49. Jianfeng L (2008) Generalized (complete, lag, anticipated) synchronization of discrete-time chaotic systems. Commun Nonlinear Sci Numer Simulat 13:1851–1859
50. Yin L, Tianyan D (2010) Adaptive control for anticipated function projective synchronization of 2d discrete-time chaotic systems with uncertain parameters. J Uncertain Syst 13:195–205
51. Yanbo G, Xiaomei Z, Guoping L, Yufan Z (2011) Impulsive synchronization of discrete-time chaotic systems under communication constraints. Commun Nonlinear Sci Numer Simulat 16:1580–1588
52. Grassi G, Miller DA (2012) Dead-beat full state hybrid projective synchronization for chaotic maps using a scalar synchronizing signal. Chin Phys B 17:1824–1830
53. Ouannas A (2014) On full-state hybrid projective synchronization of general discrete chaotic systems. J Nonlinear Dyn 17:1–6
54. An H, Chen Y (2009) The function cascade synchronization scheme for discrete-time hyperchaotic systems. Commun Nonlinear Sci Numer Simulat 14:1494–1501
55. Grassi G (2012) Generalized synchronization between different chaotic maps via dead-beat control. Chin Phys B 21:050505–6
56. Ma Z, Liu Z, Zhang G (2007) Generalized synchronization of discrete systems. Appl Math Mech 28:609–614
57. Yan Z (2005) Q-S synchronization in 3D Hénon-like map and generalized Hénon map via a scalar controller. Phys Lett A 342:309–317
58. Chai Y, Chen L, Wu R (2012) Inverse projective synchronization between two different hyperchaotic systems with fractional order. J Appl Math 1–18
59. Ouannas A, Mahmoud E (2014) Inverse matrix projective synchro-nization for discrete chaotic systems with different dimensions. Intell Electron Syst 3:188–192
60. Ouannas A, Al-sawalha M (2015) On inverse full state hybrid pro- jective synchronization of chaotic dynamical systems in discrete-Time. Inter J Dyn Control 1–7
61. Ouannas A, Odibat Z (2015) Generalized synchronization of different dimensional chaotic dynamical systems in discrete time. Nonlinear Dyn 81:765–771

Mathematical Modelling of Chaotic Jerk Circuit and Its Application in Secure Communication System

Aceng Sambas, Mada Sanjaya WS, Mustafa Mamat and Rizki Putra Prastio

Abstract In chaos-based secure communication schemes, a message signal is modulated to the chaotic signal at transmitter and at receiver the masking signals regenerated and subtracted from the receiver signal. In order to show some interesting phenomena of three dimensional autonomous ordinary differential equations, the chaotic behavior as a function of a variable control parameter, has been studied. The initial study in this chapter is to analyze the eigenvalue structures, various attractors, bifurcation diagram, Lyapunov exponent spectrum, FFT analysis, Poincare maps, while the analysis of the synchronization in the case of bidirectional coupling between two identical generated chaotic systems, has been presented. Moreover, some appropriate comparisons are made to contrast some of the existing results. Finally, the effectiveness of the bidirectional coupling method scheme between two identical Jerk circuits in a secure communication system is presented in details.

Keywords Jerk circuit · Synchronization · Bifurcation diagram · Secure communication system

---

A. Sambas (✉)
Department of Mechanical Engineering, Universitas Muhammadiyah Tasikmalaya, Tasikmalaya, Indonesia
e-mail: acenx.btts@gmail.com

M. Sanjaya WS
Department of Physics, Universitas Islam Negeri Sunan Gunung Djati Bandung, Bandung, Indonesia
e-mail: madasws@gmail.com

M. Mamat
Faculty of Informatics and Computing, Universiti Sultan Zainal Abidin, Kuala Terengganu, Malaysia
e-mail: must@unisza.edu.my

R. Putra Prastio
School of Electrical Engineering and Informatics, Bandung Institute of Technology, Jalan Ganeca No 10, Bandung, Indonesia
e-mail: rppprastio@gmail.com

A.T. Azar and S. Vaidyanathan (eds.), *Advances in Chaos Theory and Intelligent Control*, Studies in Fuzziness and Soft Computing 337,
DOI 10.1007/978-3-319-30340-6_6

1 Introduction

Chaos is used to describe the behavior of certain dynamical nonlinear systems, i.e., systems which state variables evolve with time, exhibiting complex dynamics that are highly sensitive on initial conditions. As a result of this sensitivity, which manifests it self as an exponential growth of perturbations in the initial conditions, the behavior of chaotic systems appears to be almost random [47]. Chaotic behavior has been found in physics [24], control system [1, 35], engineering [1], artificial intelligence [1], Complex system [48], robotics [44], text encryption [44], psychology [26], ecology [19] and economy [45].

Lorenz [9] published an analysis of a simple system with a third-order differential equation that he had extracted from a model of atmospheric convection. In Lorenz meteorological computer model he discovered the underlying mechanism of deterministic chaos: simply-formulated system with only a few variables can display highly complicated behavior that is unpredictable. Using his digital computer, through the printed numbers and simple strip chart plots of the variables, he saw that slight differences in one variable had profound effects on the outcome of the whole system. This was one of the first clear demonstrations of sensitive dependence on initial conditions in chaotic systems.

Rossler [17] constructed several three-dimensional quadratic autonomous chaotic systems, which also have seven terms on the right-hand side, but with only one quadratic nonlinearity. Sprott [25] suggests 19 cases of chaotic systems: case A-S with five linear terms and two nonlinear terms. Sprott [25] gives a new class of chaotic circuit defined by jerk equation. Sprott [25] gives Simple Autonomous chaotic circuit which employs an op-amp as a comparator to provide signum nonlinearity. Pandey et al. [12] modifies the system of Jerk equations into a system of simple quadratic equations. Vaidyanathan et al. [40] found a 3-D novel jerk chaotic system with two hyperbolic sinusoidal nonlinearities and Vaidyanathan et al. [35] found a novel 4-D hyperchaotic hyperjerk system.

Synchronization of chaotic oscillators in particular became popular when Pecora and Carroll [14] published their observations of synchronization in unidirectionally coupled chaotic systems. Till now, various types of synchronization have been reported, namely complete synchronization [16], Hybrid synchronization [35], global synchronization [35], phase synchronization [18], generalized synchronization [37] , anti-synchronization [38], [35], anti-phase synchronization [4], lag synchronization [35], anticipating synchronization [44], active control synchronization [35] and projective synchronization [10]. So, the phenomenon of chaotic synchronization has been intensively and extensively investigated due to its potential applications in a variety of areas, such as in secure communications [19–21], chemical reactions [11], neuronal systems [23] and economic models [42].

In general, this study is focused on the development of chaos and non-linear dynamical system behavior in chaotic electrical oscillator. To study the non-linear dynamics and chaotic behavior, we investigate and analyze some basic properties, such as eigenvalues structure, phase plane, Lyapunov exponent, Poincare map, FFT analysis and diagram bifurcation analysis, while the analysis of the synchronization in the case of bidirectional coupling between two identical generated chaotic systems, has been presented. Moreover, some appropriate comparisons are made to contrast some of the existing results. Finally, the effectiveness of the bidirectional coupling between two identical Jerk circuits in a secure communication system is presented in details.

The plan of the chapter is as follows. In Sects. 2 and 3, the details of the proposed autonomous Jerk circuits simulation using MATLAB 2010 and the analog circuit simulation by using MultiSIM 10.0, are presented respectively. In Sect. 4, a typical chaotic attractor is experimentally demonstrated. In Sect. 5, the Bidirectional coupling method is applied in order to synchronize two identical autonomous Jerk circuits. The chaotic masking communication scheme by using the above mentioned synchronization technique is presented in Sect. 6. Finally, in Sect. 7, the concluding remarks are given.

2 Mathematical Model of Jerk Circuit

In this work, the Jerk circuit, which was firstly presented by Pandey [13], is used. This is a three-dimensional autonomous nonlinear system that is described by the following system of ordinary differential equations:

$$\left.\begin{aligned} \dot{x} &= y \\ \dot{y} &= z \\ \dot{z} &= -x - y - az - bx^2 \end{aligned}\right\} \tag{1}$$

The system has one quadratic term and two positive real constants a and b. The parameters and initial conditions of the Jerk system (1) are chosen as: $a = 0.51$, $b = 0.125$ and $(x_0; y_0; z_0) = (0.001, 0.010, 0.100)$, so that the system shows the expected chaotic behavior.

2.1 Equilibrium Point Analysis

The equilibrium points of (1) denote by $E\,(\bar{x}, \bar{y}, \bar{z})$, are the zeros of its non-linear algebraic system which can be written as:

$$\left.\begin{aligned} 0 &= y \\ 0 &= z \\ 0 &= -x - y - az - bx^2 \end{aligned}\right\} \tag{2}$$

The Jerk system has two equilibrium points E_0 (0, 0, 0) and E_1 (−8, 0, 0). The dynamical behavior of equilibrium points can be studied by computing the eigenvalues of the Jacobian matrix J of system (1) where:

$$J(\bar{x}, \bar{y}, \bar{z}) = \begin{bmatrix} 0 & 1 & 0 \\ 0 & 0 & 1 \\ -1 - 0.250^{*}x & -1 & -0.51 \end{bmatrix} \quad (3)$$

For equilibrium points E_0 (0, 0, 0), the Jacobian becomes:

$$J(0, 0, 0) = \begin{bmatrix} 0 & 1 & 0 \\ 0 & 0 & 1 \\ -1 & -1 & -0.5 \end{bmatrix} \quad (4)$$

The eigenvalues are obtained by solving the characteristic equation, $\det[\lambda \mathrm{I} - \mathrm{J}_1] = 0$ which is:

$$\lambda^3 + 0.51\lambda^2 + \lambda + 1 \quad (5)$$

Yielding eigenvalues of $\lambda_1 = -0.806802$, $\lambda_2 = 0.148401 + 1.103375\mathrm{i}$, $\lambda_3 = 0.148401 - 1.103375\ \mathrm{i}$, for a = 0.51, b = 0.125. For equilibrium points (−8, 0, 0), the Jacobian becomes:

$$J(-8, 0, 0) = \begin{bmatrix} 0 & 1 & 0 \\ 0 & 0 & 1 \\ 1 & -1 & -0.51 \end{bmatrix} \quad (6)$$

The eigenvalues are obtained by solving the characteristic equation, $\det[\lambda \mathrm{I} - J_1] = 0$ which is:

$$\lambda^3 + 0.51\lambda^2 + \lambda - 1 \quad (7)$$

Yielding eigenvalues of $\lambda_1 = 0.600148$, $\lambda_2 = -0.555074 + 1.165395i$, $\lambda_3 = -0.555074 - 1.165395\mathrm{i}$, for a = 0.51, b = 0.125. The above eigenvalues show that the system has an unstable spiral behavior. In this case, the phenomenon of chaos is presented.

2.2 Numerical Simulation Using MATLAB

In this section, numerical simulations are carried out by using MATLAB 2010. The fourth-order Runge-Kutta method is used to solve the system of differential equations (1). Figure 1a–c show the projections of the phase space orbit on to the *xy* plane, the *yz* plane and the *xz* plane, respectively. The time series of Jerk circuit is shown in Fig. 1d. As it is shown, for the chosen set of parameters and initial conditions, the Jerk system presents chaotic attractors of Rossler type.

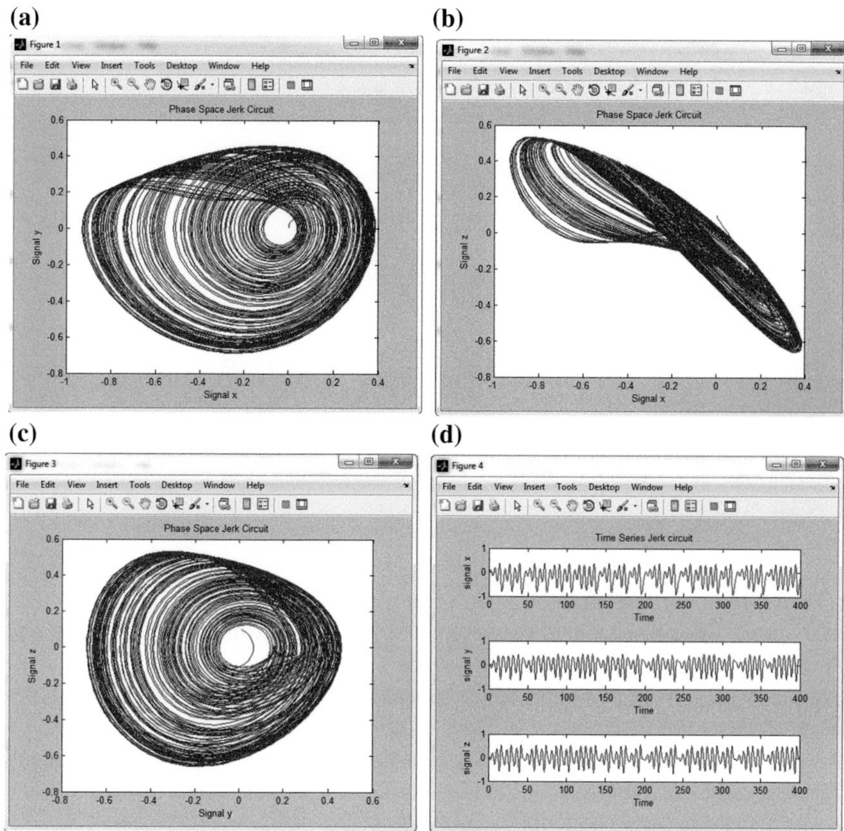

Fig. 1 Numerical simulation results using MATLAB 2010, for a = 0.51, b = 0.125, in **a** $x - y$ plane, **b** $x - z$ plane, **c** $y - z$ plane, **d** time series versus x, y, z signal

2.3 *Lyapunov Exponent Analysis*

It is also known from the theory of nonlinear dynamics that for a three dimensional system, like this, there has been three Lyapunov exponents (λ_1, λ_2 , λ_3). In more details, for a 3D continuous dissipative system the values of the Lyapunov exponents are useful for distinguishing among the various types of orbits. So, the possible spectra of attractors, of this class of dynamical systems, can be classified in four groups, based on Lyapunov exponents [7, 46].

- $(\lambda_1, \lambda_2, \lambda_3) \rightarrow (-, -, -)$: a fixed point
- $(\lambda_1, \lambda_2, \lambda_3) \rightarrow (0, -, -)$: a limit point
- $(\lambda_1, \lambda_2, \lambda_3) \rightarrow (0, 0, -)$: a 2-torus
- $(\lambda_1, \lambda_2, \lambda_3) \rightarrow (+, 0, -)$: a strange attractor

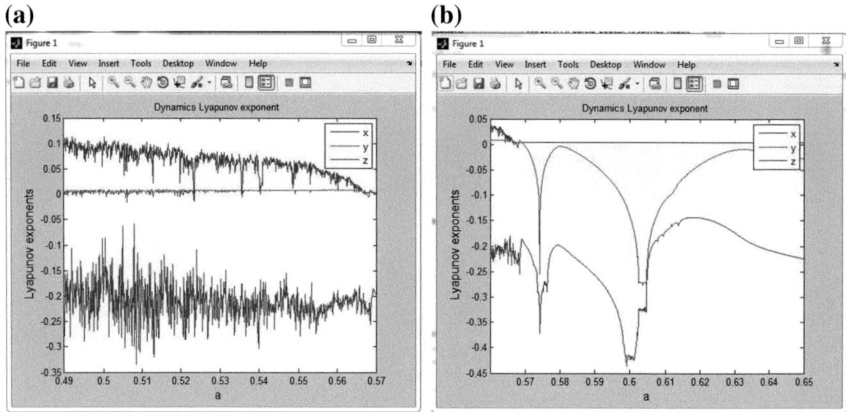

Fig. 2 Nonlinear dynamics of system (1) for b = 0.125. **a** Lyapunov exponents versus the parameter control a [0.49 − 0.57]. **b** Lyapunov exponents versus the parameter control a [0.56 − 0.65], with MATLAB 2010

Therefore, the last configuration is the only possible third-order chaotic system. In this case, a positive Lyapunov exponent reflects a "direction" of stretching and folding and therefore determines chaos in the system.

So, in Fig. 2a, b the dynamics of the proposed system's Lyapunov exponents for the variation of the parameter $a \in [0.49 - 0.65]$, is shown. For $0.49 \leq a \leq 0.57$ a strange attractor is displayed as the system has one positive Lyapunov exponent, while for values of $0.57 < a \leq 0.65$ is a transition to limit point behavior as the system has two negative Lyapunov exponents.

2.4 Bifurcation Diagram Analysis

The word bifurcation denotes a situation in which the solutions of a nonlinear system of differential equations alter their character with a change of a parameter on which the solutions depend. Bifurcation theory studies these changes (e.g. appearance and disappearance of the stationary points, dependence of their stability on the parameter etc.). A MATLAB program was written to obtain the bifurcation diagrams for Jerk circuit of Fig. 3a, b. So, in this diagram a possible bifurcation diagram for system (1), in the range of $0.49 \leq a \leq 0.65$, is shown. For the chosen value of $0.49 \leq a < 0.554$ the system displays the expected chaotic behavior. Also, for $0.554 \leq a < 0.65$, a reverse period doubling route is presented.

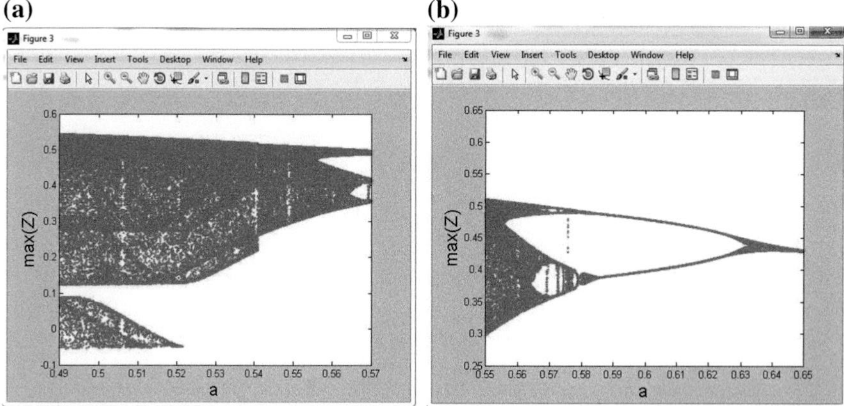

Fig. 3 Nonlinear dynamics of system (1) for specific values set b = 0.125. **a** Bifurcation diagram of z versus the control parameter $a\varepsilon$ [0.49 − 0.57], **b** Bifurcation diagram of z versus the control parameter $a\varepsilon$ [0.55 − 0.65], with MATLAB 2010

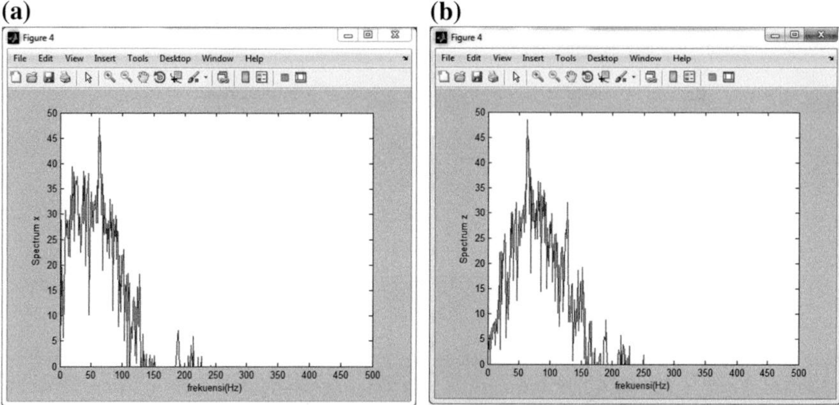

Fig. 4 The frequency spectrum generated numerically from chaotic system. **a** Spectrum x versus frequency. **b** Spectrum z versus frequency, with MATLAB 2010

2.5 Frequency Spectrum of Chaotic System

Many proposed chaos-based encryption schemes have been totally or partially broken by different attacks. One reason is narrow bandwidths of those signals [5, 15]. The frequency spectra of signals, which are generated numerically from the Jerk system (1) proposed in this chapter, are shown in Fig. 4a, b, respectively. One can note that the bandwidths of signals x and z are in about 0–200 Hz and 0–250 Hz.

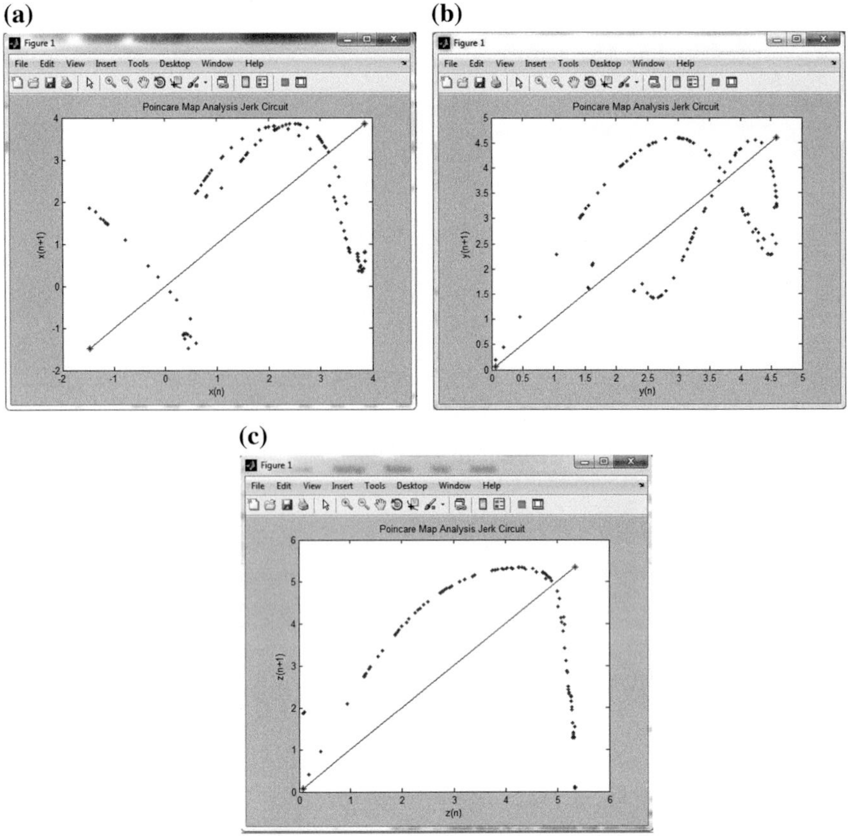

Fig. 5 A gallery of Poincare maps for system (1), for a = 0.51, b = 0. 125. The plots give the maxima of **a** x $(n + 1)$ versus those of x (n); **b** y $(n + 1)$ versus those of y (n); **c** z $(n + 1)$ versus those of $z(n)$, obtained with MATLAB 2010

2.6 Poincare Map Analysis

A Poincare section is often used to reduce a three-dimensional continuous system to a lower-dimensional discrete map. The strength behinds this tool is that these sections have the same topological properties as their continuous counterparts [8]. In the chaotic state the phase portrait is very dense, in the sense that the trajectories of the motion are very close to each other. It can be only indicative of the minima and maxima of the motion. Any other characterization of the motion is difficult to be interpreted. So, one way to capture the qualitative features of the strange attractor is to obtain the Poincare map [6, 45]. Figure 5a–c shows the Poincare section map by using MATLAB, for $a = 0.51$ and $b = 0.125$.

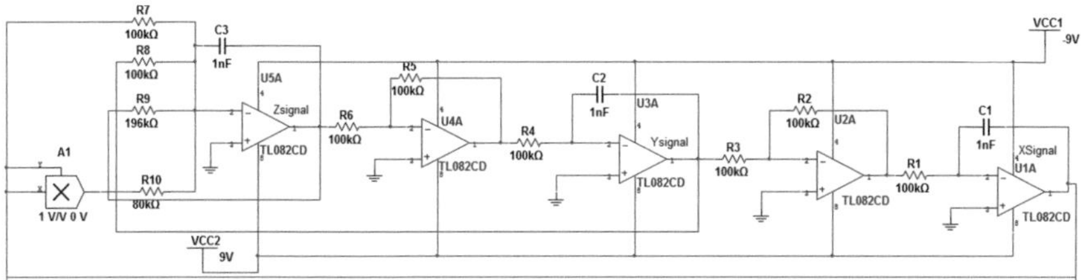

Fig. 6 Schematic of the proposed Jerk circuit by using MultiSIM

3 Analog Circuit Simulation Using MultiSIM

A simple electronic circuit is designed that can be used to study chaotic phenomena. The circuit employs simple electronic elements, such as resistors, capacitors, multiplier and operational amplifiers. In Fig. 6, the voltages of C_1, C_2, C_3 are used as x, y and z, respectively. The nonlinear term of system (1) are implemented with the analog multiplier. The corresponding circuit equation can be described as:

$$\left.\begin{aligned} \dot{x} &= \frac{1}{C_1R_1}y \\ \dot{y} &= \frac{1}{C_2R_4}z \\ \dot{z} &= -\frac{1}{C_3R_7}x - \frac{1}{C_3R_8}y - \frac{1}{C_3R_9}z - \frac{1}{10C_3R_{10}}x^2 \end{aligned}\right\} \tag{8}$$

We choose $R_1 = R_2 = R_3 = R_4 = R_5 = R_6 = R_7 = R_8 = 100\ \text{k}\Omega$, $R_9 = 196\ \text{k}\Omega$, $R_{10} = 80\ \text{k}\Omega$. $C_1 = C_2 = C_3 = C_4 = 1$ nF. The circuit has three integrators (by using Op-amp TL082CD) in a feedback loop and a multiplier (IC AD633). The supplies of all active devices are $\pm$ 9 V. With MultiSIM 10.0, we obtain the experimental observations of system (1) as shown in Fig. 7. As compared with Fig. 1 a good qualitative agreement between the numerical simulation and the MultiSIM 10.0 results of the Jerk circuit is confirmed. The parameter variable a of system (1) is changed by adjusting the resistor R_9, and obeys the following relation:

$$a = \frac{1}{C_3R_9} \tag{9}$$

4 Circuit Design and Implementation for the Jerk Circuit

The chaotic dynamics of system (1) have also been realized by an electronic circuit. The designed circuitry realizing system (1) is shown in Fig. 8. It consists of three channels to conduct the integration of the three state variable x, y and z, respectively. The nonlinear term of system (1) are implemented with the analog multipliers AD633. We obtain the experimental observations of system (1) as shown in Fig. 9.

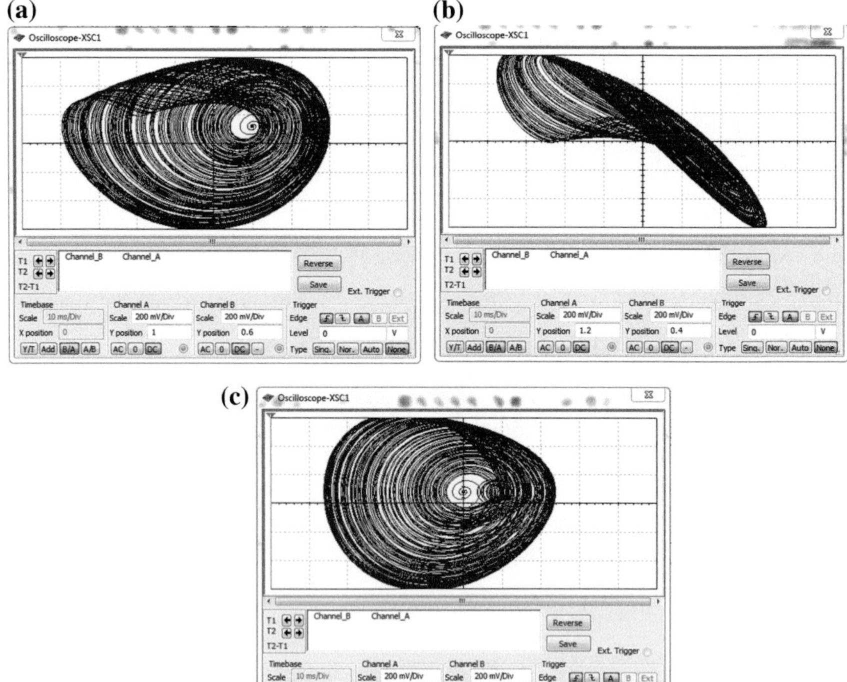

Fig. 7 Various projections of the chaotic attractor using MultiSIM, for $a = 0.51, b = 0.125$. **a** $x - y$ plane. **b** $x - z$ plane. **c** $y - z$ plane

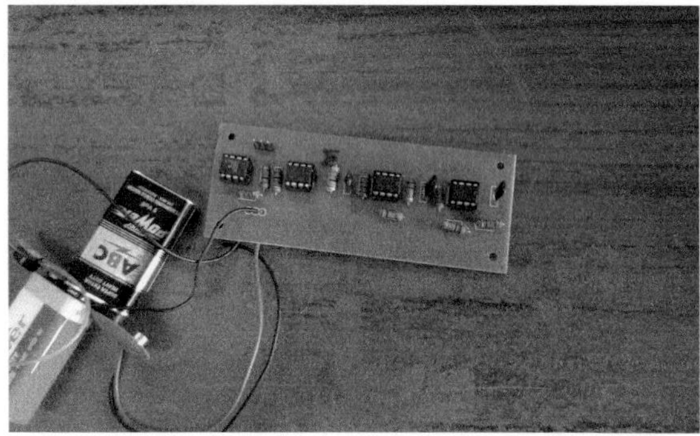

Fig. 8 Electronic Circuit Realization for a = 0.51 and b = 0.125

As compared with Figs. 1 and 7, a good qualitative agreement between the system's numerical simulation by using MATLAB 2010, and its circuit simulation by using MultiSIM is confirmed.

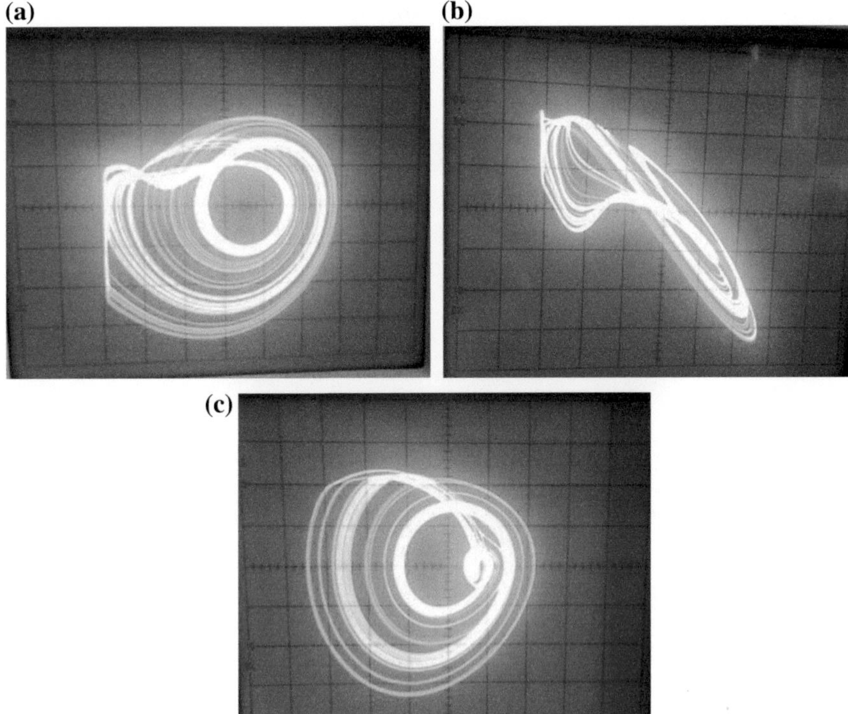

Fig. 9 Experimental observations of the Chaotic attractor in different planes. **a** $x - y$ plane. **b** $x - z$ plane. **c** $y - z$ plane for a = 0.51, b = 0.125

5 Bidirectional Chaotic Synchronization

5.1 Mathematical Model of Bidirectional Coupling

In the case of bidirectional coupling two systems interact and are coupled with each other creating a mutual synchronization. The following master slave (bidirectional coupling) configuration, as described below:

$$\left.\begin{aligned} \dot{x}_1 &= y_1 \\ \dot{y}_1 &= z_1 + g_c(y_2 - y_1) \\ \dot{z}_1 &= -x_1 - y_1 - az_1 - bx_1^2 \\ \dot{x}_2 &= y_2 \\ \dot{y}_2 &= z_2 + g_c(y_1 - y_2) \\ \dot{z}_2 &= -x_2 - y_2 - az_2 - bx_2^2 \end{aligned}\right\} \tag{10}$$

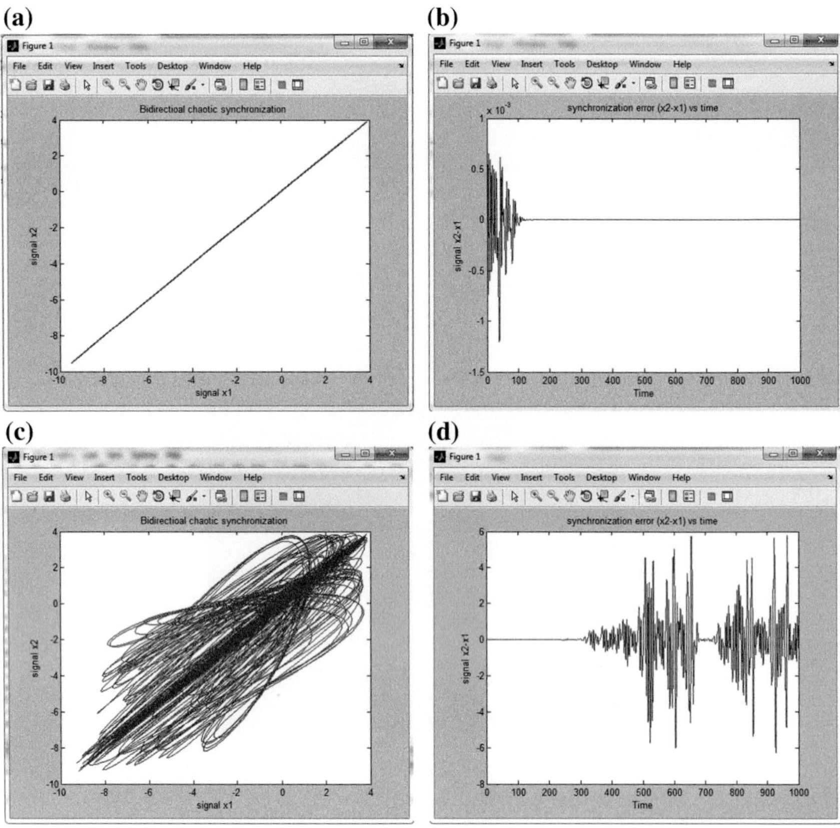

Fig. 10 Phase portrait of $x_2$versus x_1 and error $x_2 - x_1$ in the case of bidirectionally coupled Jerk circuits, for (**a**) $g_c = 0.2$ (full synchronization) and (**c**) $g_c = 0.1$ (full desynchronization), for a = 0.51, b = 0.125

The coupling coefficient g_c is present in the equations of both systems, since the coupling between them is mutual. Numerical simulations of system (10), by using the fourth-order Runge-Kutta method, are used to describe the dynamics of chaotic synchronization of bidirectionally coupled Jerk circuits. In bidirectional (mutual) coupling, both coupled systems are connected in such a way that they mutually influence each other's behavior. Synchronization numerically appears for a coupling factor $g_c \geq 0.2$ as shown in Fig. 10a, with error $e_x = x_1 - x_2 \rightarrow 0$, which implies the complete synchronization.

5.2 Analog Circuit Simulation Using MultiSIM

Synchronization of chaotic motions among the coupled dynamical systems is an important generalization for the phenomenon of synchronization of linear system, which is useful and indispensable in communications. Simulation results show that the two systems synchronize well. Figure 11 shows the circuit schematic for implementing the bidirectional synchronization of coupled Jerk systems. Chaotic synchronization appears for a coupling strength $R_{21} \leq 100\ \text{m}\Omega$, as shown in Fig. 12a. For different initial conditions or resistance coupling strength $R_{21} > 100\ \text{m}\Omega$, the synchronization cannot occur as shown in Fig. 12b.

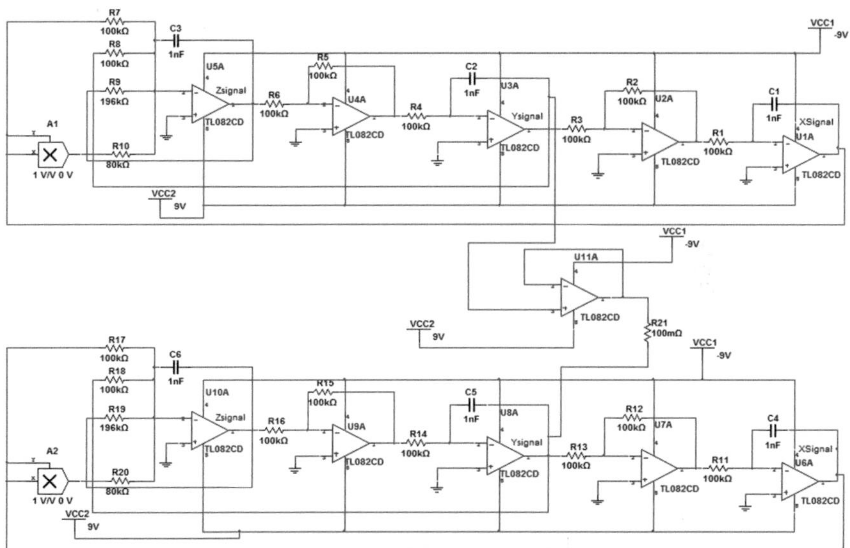

Fig. 11 Bidirectional chaotic synchronization Jerk circuit by using MultiSIM

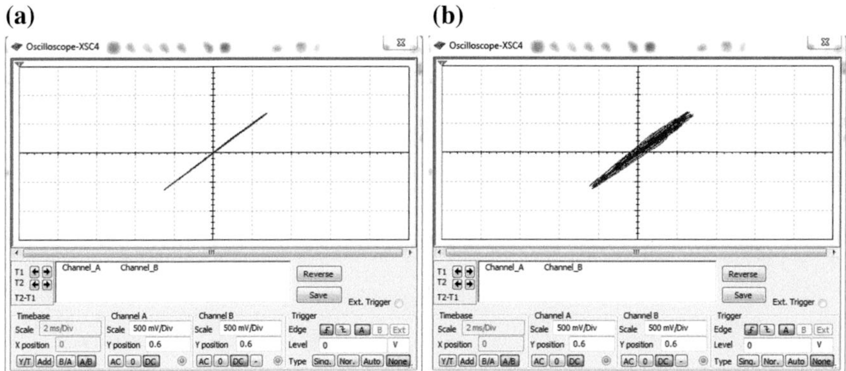

Fig. 12 Synchronization phase portrait of x_2 versus x_1, for **a** $R_{21} = 100\ \text{m}\Omega$ and **b** $R_{21} = 1\ \Omega$, with MultiSIM 10.0

5.3 Circuit Design and Implementation for the Jerk Circuit

The overall circuit was built according to the scheme shown in Fig. 13 by using TL082CD operational Amplifiers. Chaotic synchronization appears for a coupling strength $R_{21} \leq 100$ mΩ, as shown in Fig. 14a. Figure 14b shows the phase portrait (x_1, x_2) when there is no synchronization for $R_{21} > 100$ mΩ.

Fig. 13 Practical circuit of the Bidirectional chaotic synchronization

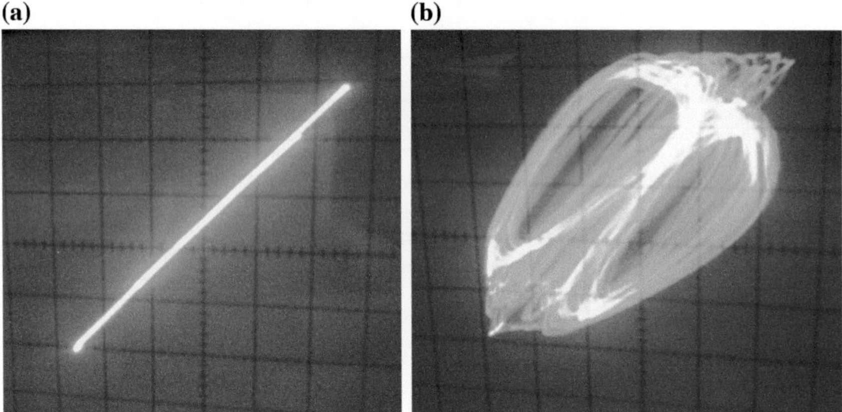

Fig. 14 Bidirectional synchronization phase portrait of x_2 versus x_1, for **a** $R_{21} = 100$ mΩ and **b** $R_{21} = 1$ Ω, with experimental result

6 Applications in Secure Communication System

6.1 Mathematical Model of Secure Communication System

To study the effectiveness of signal masking approach in the Jerk system, we first set the information-bearing signal $m_s(t)$ in the form of sinusoidal wave:

$$m_s(t) = A\sin(2\pi f)t \tag{11}$$

where A and f are the amplitude and the frequency of the sinusoidal wave signal respectively.

The sum of the signal $m_s(t)$ and the chaotic signal $m_{Jerkcircuit}(t)$, produced by the Jerk circuit, is the new encryption signal $m_{encryption}(t)$, which is given by Eq. (12).

$$m_{encryption}(t) = m_s(t) + m_{\mathrm{Jerk\_circuit}}(t) \tag{12}$$

The signal $m_{Jerkcircuit}$ (t) is one of the parameters of equation (1). After finishing the encryption process the original signal can be recovered with the following procedure.

$$m_{decryption}(t) = m_{encryption}(t) - m_{Jerk\_circuit}(t) \tag{13}$$

So, $m_{decryption}(t)$is the original signal and must be the same with $m_s(t)$, Due to the fact that the input signal can be recovered from the output signal, it turns out that it is possible to implement a secure communication system using the proposed chaotic system.

6.2 Numerical Simulation of Secure Communication Systems

In chaos-based secure communication schemes, information signals are masked (encrypted) by chaotic signals at the transmitter and the resulting encrypted signals are sent to the corresponding receiver across a public channel. Perfect chaos synchronization is usually expected to recover the original information signals. In other words, the recovery of the information signals requires the receivers own copy of the chaotic signals which are synchronized with the transmitter ones. Thus, chaos synchronization is the key technique throughout this whole process [47]. Figure 15a–c shows the MATLAB 2010 numerical simulation results for the proposed chaotic masking communication scheme.

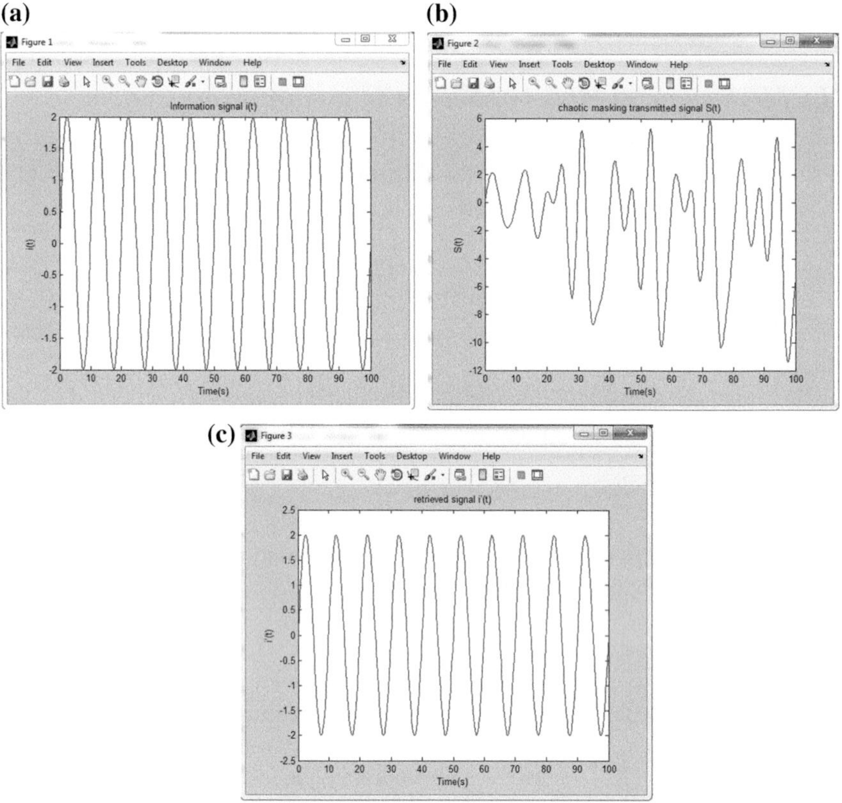

Fig. 15 MATLAB 2010 simulation of Jerk circuit masking communication system when amplitude is 2 V and frequency 1 KHz: **a** Information signal, **b** Chaotic masking transmitted signal, **c** Retrieved signal

6.3 Analog Circuit Simulation of Secure Communication System

The sinusoidal wave signal of amplitude 2 V and frequency 1 KHz is added to the synchronizing driving chaotic signal in order to regenerate the original driving signal at the receiver. Thus, as it can be shown from Fig. 16c, the message signal has been perfectly recovered by using the signal masking approach through the synchronization of chaotic Jerk circuits. Furthermore, Simulation results with MultiSIM 10.0 have shown that the performance of chaotic Jerk circuits in chaotic masking and message recovery is very satisfactory (Fig. 16). Finally, Fig. 17 shows the circuit schematic of implementing the Jerk circuit chaotic masking communication scheme.

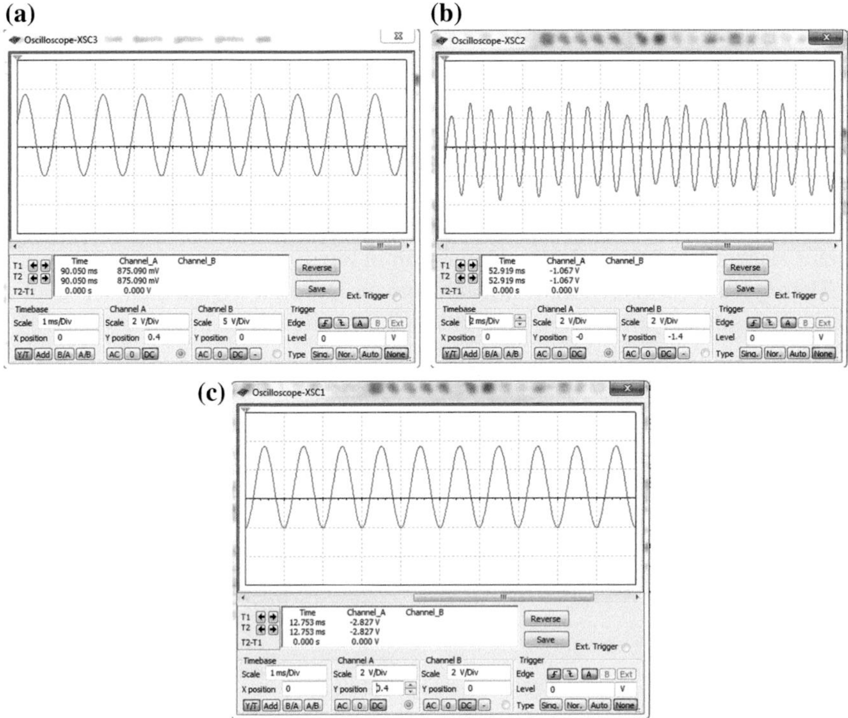

Fig. 16 MultiSIM 10.0 outputs of Jerk circuit masking communication systems, when amplitude is 2 V and frequency 1 KHz: **a** Information signal, **b** Chaotic masking transmitted signal, **c** Retrieved signal

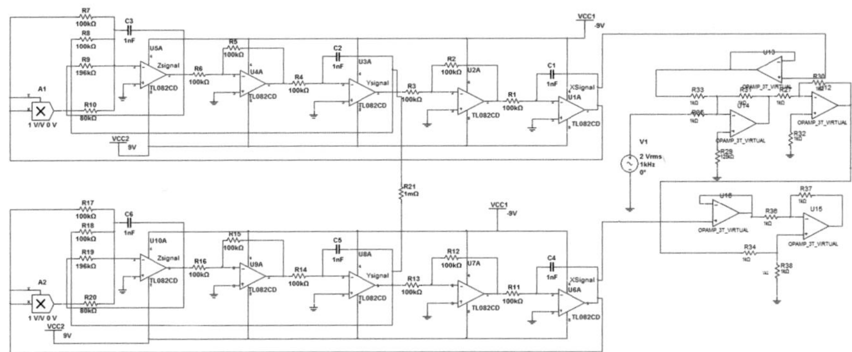

Fig. 17 Jerk circuit masking communication system for sinusoidal wave

6.4 Analog Experimental Results of Secure Communication Systems

We use TL082CD op amps, the analog device AD633JN multiplier, appropriate valued resistors and capacitors for realization circuit. The circuit is supplied with ±9 V power supply. Figure 18 shows the experimental results for masking signal communication system by varying the input signals voltage. The experimental electronic circuit realization of the Jerk system is shown in Fig. 19.

The experimental results show that autonomous Jerk circuit is excellent for chaotic masking communication when the voltage information is at intervals of 1–10 V.

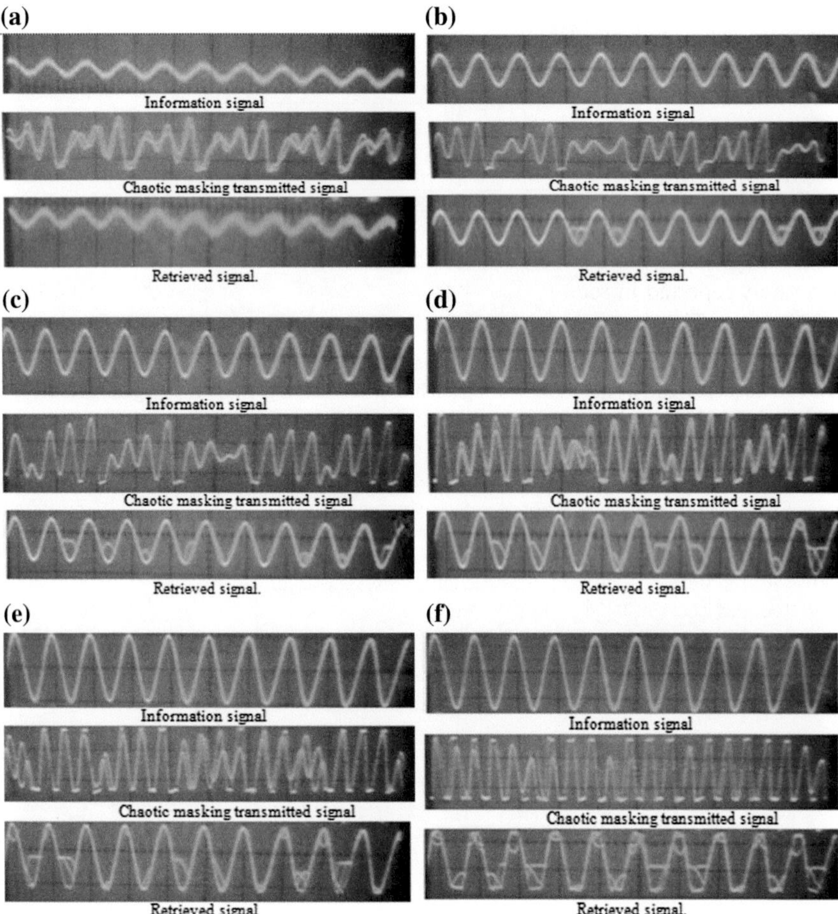

Fig. 18 Experimental results of Jerk circuit masking communication systems. **a** 2 V, **b** 4 V, **c** 6 V, **d** 8 V, **e** 10 V, **f** 14 V

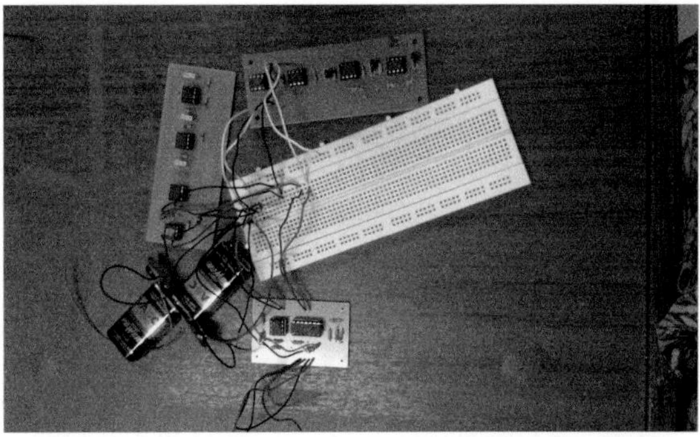

Fig. 19 Practical circuit of the secure communication system

Otherwise, when the voltage information is more than 10 V or less than 1 V, the chaotic masking communication is not occur.

7 Conclusions

In this chapter, we construct a three-dimensional autonomous system which have been rarely reported in reported in literature. The system has rich chaotic dynamics behaviors. The complex dynamics have also been explored in detail, including various periodic and chaotic motions, by means of Lyapunov exponent spectrum, diagram bifurcation analysis, FFT analysis and Poincare map analysis. Moreover, it is implemented via a designed circuit with MultiSIM and tested experimentally in laboratory, showing very good agreement with the simulation result.

The chaotic synchronization of two identical Jerk circuits system has been investigated by implementing bidirectional method technique. The proposed method of synchronization between chaotic circuits can be applied successfully to a secure communication scheme. Chaos synchronization, realization circuit and chaos masking were realized by using MATLAB 2010 and MultiSIM 10.0 programs. The comparison between MATLAB 2010 and MultiSIM simulation results demonstrate the effectiveness of the proposed secure communication scheme. Finally, the effectiveness of the bidirectional coupling scheme between two identical Jerk circuits in a secure communication system is presented in details. Integration of theoretical physics, the numerical simulation by using MATLAB 2010, as well as the implementation of circuit simulations by using MultiSIM and realization circuit have been performed in this study. It is possible to use the jerk system in application to secure communications using voice signal in the future researches.

References

1. Azar AT, Vaidyanathan S (2015) Chaos modeling and control systems design. Studies in computational intelligence, vol 581. Springer, Berlin. ISBN 978-3-319-13131-3
2. Azar A. T and Vaidyanathan S., 2015. Computational intelligence applications in modeling and control. Studies in computational intelligence, vol 575. Springer, Berlin. ISBN 978-3-319-11016-5
3. Azar AT, Vaidyanathan S (2015) Handbook of research on advanced intelligent control engineering and automation. Advances in computational intelligence and robotics (ACIR) Book Series. IGI Global, USA
4. Cao LY, Lai YC (1998) Antiphase synchronism in chaotic systems. Phys Rev E 58:382–386
5. Dong G, Zheng S, Tian L, Du R, Sun M, Shi Z (2009) The analysis of a novel 3-D autonomous system and circuit implementation. Phys Lett A 373:4227–4238
6. Han F (2004) Multi-Scroll Chaos generation via linear systems and hysteresis function series. Dissertation. PhD thesis, Royal Melbourne Institute of Technology, Australia
7. Li C, Wang J, Hu W (2012) Absolute term introduced to rebuild the chaotic attractor with constant Lyapunov exponent spectrum. Nonlinear Dyn. 68(4):575–587
8. Li XF, Chlouverakis KE, Xu D-L (2009) Nonlinear dynamics and circuit realization of a new chaotic flow: a variant of Lorentz. Chen and L. Nonlinear Anal: Real World Appl 10:2357–2368
9. Lorenz EN (1963) Deterministic nonperiodic flow. J Atmos Sci 20:130141
10. Mainieri R, Rehacek J (1999) Projective synchronization in three dimensional Chaotic systems. Phys Rev Lett 82:3042 3045
11. Nakajima K, Sawada Y (1979) Experimental studies on the weak coupling of oscillatory chemical reaction systems. J Chem Phys 72(4):2231–2234
12. Pandey A, Baghel RK, Singh RP (2012) Analysis and circuit realization of a new autonomous Chaotic System. Int J Electron Commun Eng 5:487–496
13. Pandey A, Baghel RK, Singh RP (2013) An autonomous Chaotic circuit for wideband secure communication. Int J Eng Bus Enterp Appl 4(1):44–47
14. Pecora LM, Carroll TL (1990) Synchronization in Chaotic systems. Phys Rev Lett 64:821825
15. Peng JH, Ding EJ, Ding M, Yang W (1996) Synchronizing hyperchaos with a scalar transmitted signal. Phys Rev Lett 76:904
16. Pikovski AS (1984) On the interaction of strange attractors. Z Phys B: Condens Matter 55: 149–154
17. Rossler OE (1976) An equation for continuous chaos. Phys Lett A 57:397398
18. Rulkov NF, Sushchik MM, Tsimring LS, Abarbanel HDI (1995) Generalized synchronization of Chaos in directionally coupled Chaotic systems. Phys Rev E 51:980–994
19. Sambas A, Sanjaya WS, Halimatussadiyah M (2012) Unidirectional Chaotic synchronization of Rossler circuit and its application for secure communication. WSEAS Trans Syst 11:506–515
20. Sambas A, Sanjaya WSM, Mamat M, Halimatussadiyah M (2013) Design and analysis bidirectional Chaotic synchronization of Rossler circuit and its application for secure communication. Appl Math Sci 7(1):11–21
21. Sambas A, Sanjaya WSM, Mamat M (2015) Bidirectional coupling scheme of Chaotic systems and its application in secure communication system. J Eng Sci Technol Rev 8(2):89–95
22. Sanjaya WSM, Mamat M, Salleh Z, Mohd I (2011) Bidirectional Chaotic synchronization of Hindmarsh-Rose neuron model. Appl Math Sci 5(54):2685–2695
23. Sanjaya WSM, Mamat Mohd IM, Salleh Z (2012) Mathematical model of three species food chain interaction with mixed functional response. Int J Mod Phys: Conf Ser 9:334340
24. Shinbrot T, Grebogi C, Wisdom J, Yorke JA (1992) Chaos in a double pendulum. Am J Phys 60:491–499
25. Sprott JC (1994) Some simple chaotic flows. Phys Rev E 50:647–650
26. Sprott JC (2000) Simple Chaotic systems and circuits. Am J Phys 68:758–763
27. Sprott JC (2010) Elegant Chaos algebraically simple chaotic flows. World Scientific, Singapore
28. Sprott JC (2004) Dynamical models of love. Nonlinear Dyn Psych Life Sci 8:303–314

29. Vaidyanathan S, Azar AT (2015) Anti-synchronization of identical Chaotic systems using sliding mode control and an application to Vaidyanathan-Madhavan Chaotic systems. In: Azar AT, Zhu Q (eds) Advances and applications in sliding mode control systems. Studies in computational intelligence book series, vol 576. Springer, Berlin, pp 527–547
30. Vaidyanathan S, Azar AT (2015) Hybrid synchronization of identical Chaotic systems using sliding mode control and an application to Vaidyanathan Chaotic systems. In: Azar AT, Zhu Q (eds) Advances and applications in sliding mode control systems. Studies in computational intelligence book series, vol 576. Springer, Berlin, pp 549–569
31. Vaidyanathan S, Azar AT (2015) Analysis, control and synchronization of a nine-Term 3-D novel Chaotic system. In: Azar AT, Vaidyanathan S (eds) Chaos modeling and control systems design. Studies in computational intelligence book series, vol 581. Springer, Berlin, pp 3–17
32. Vaidyanathan S, Azar AT (2015) Analysis and control of a 4-D novel hyperchaotic system. In: Azar AT, Vaidyanathan S (eds) Chaos modeling and control systems design. Studies in computational intelligence, vol 581. Springer, Berlin, pp 19–38
33. Vaidyanathan S, Azar AT, Rajagobal K, Alexander P (2015) Design and SPICE implementation of a 12-term novel hyperchaotic system and its synchronization via active control. Int J Modell Ident Control (IJMIC) (Forthcoming)
34. Vaidyanathan S, Idowu BA, Azar AT (2015) Backstepping controller design for the global Chaos synchronization of Sprott's Jerk systems. In: Azar AT, Vaidyanathan S (eds) Chaos modeling and control systems design. Studies in computational intelligence, vol 581. Springer, Berlin, pp 19–58
35. Vaidyanathan S, Karthikeyan R (2013) Anti-synchronization of hyperchaotic Lorenz and hyperchaotic Chen systems by adaptive control. Int J Signal Syst Control Eng Appl 4:18–25
36. Vaidyanathan S, Pakiriswamy S (2013) Generalized projective synchronization of six-term sundarapandian chaotic systems by adaptive control. Int J Control Theory Appl 6:153–163
37. Vaidyanathan S, Sampath S, Azar AT (2015) Global chaos synchronization of identical chaotic systems via novel sliding mode control method and its application to Zhu system. Int J Modell Ident Control 23(1):92–100
38. Vaidyanathan S, Volos ChK, Pham VT, Madhavan K, Idowu BA (2014) adaptive backstepping control, synchronization and circuit simulation of a 3-D novel Jerk Chaotic system with two hyperbolic sinusoidal nonlinearities. Arch Control Sci 24(3):257–285
39. Vaidyanathan S, Volos ChK, Pham VT, Madhavan K (2015) Analysis, adaptive control and synchronization of a novel 4-D hyperchaotic hyperjerk system and its SPICE implementation. Arch Control Sci 25(1):135–158
40. Volos ChK, Kyprianidis IM, Stouboulos IN (2012) Synchronization phenomena in coupled nonlinear systems applied in economic cycles. WSEAS Trans Syst 11(12):681–690
41. Volos ChK, Kyprianidis IM, Stouboulos IN (2013) Experimental investigation on coverage performance of a chaotic autonomous mobile robot. Robot Auton Syst 61:1314–1322
42. Volos ChK, Kyprianidis IM, Stouboulos IN (2013) Text encryption scheme realized with a Chaotic Pseudo-random bit generator. J Eng Sci Technol Rev 6:914
43. Volos ChK, Kyprianidis IM, Stavrinides SG, Stouboulos IN, Magafas I, Anagnostopoulos AN (2011) Nonlinear dynamics of a financial system from an engineers point of view. J Eng Sci Technol Rev 4:281–285
44. Voss HU (2000) Anticipating Chaotic synchronization. Phys Rev E 61:5115–5119
45. Wang SX (1998) Simulation of Chaos synchronization, PhD thesis, University Western Ontario, London
46. Wolf A (1986) Quantity Chaos with Lyapunov exponents. Princeton University Press, Chaos
47. Zhang H (2010) Chaos synchronization and its application to secure communication. PhD thesis, University of Waterloo, Canada
48. Zhu Q, Azar AT (2015) Complex system modelling and control through intelligent soft computations. Studies in fuzziness and soft computing, vol 319. Springer, Berlin. ISBN: 978-3-319-12882-5

Dynamic Analysis, Adaptive Feedback Control and Synchronization of An Eight-Term 3-D Novel Chaotic System with Three Quadratic Nonlinearities

Sundarapandian Vaidyanathan and Ahmad Taher Azar

Abstract In this research work, we describe an eight-term 3-D novel chaotic system with three quadratic nonlinearities. First, this work describes the dynamic analysis of the novel chaotic system. The Lyapunov exponents of the eight-term novel chaotic system are obtained as $L_1 = 4.0359$, $L_2 = 0$ and $L_3 = -29.1071$. The Kaplan-Yorke dimension of the novel chaotic system is obtained as $D_{KY} = 2.1384$. Next, this work describes the adaptive feedback control of the novel chaotic system with unknown parameters. Also, this work describes the adaptive feedback synchronization of the identical novel chaotic systems with unknown parameters. The adaptive feedback control and synchronization results are proved using Lyapunov stability theory. MATLAB simulations are depicted to illustrate all the main results for the eight-term 3-D novel chaotic system.

Keywords Chaos · Chaotic systems · Adaptive control · Feedback control · Synchronization

1 Introduction

Chaotic systems are defined as nonlinear dynamical systems which are sensitive to initial conditions, topologically mixing and with dense periodic orbits. Sensitivity to initial conditions of chaotic systems is popularly known as the *butterfly effect*. Small changes in an initial state will make a very large difference in the behavior of the system at future states. Chaotic behaviour was suspected well over hundred

S. Vaidyanathan (✉)
Research and Development Centre, Vel Tech University,
Avadi, Chennai 600062, Tamil Nadu, India
e-mail: sundarvtu@gmail.com

A.T. Azar
Faculty of Computers and Information, Benha University, Banha, Egypt
e-mail: ahmad_t_azar@ieee.org; ahmad.azar@fci.bu.edu.eg

A.T. Azar and S. Vaidyanathan (eds.), *Advances in Chaos Theory and Intelligent Control*, Studies in Fuzziness and Soft Computing 337,
DOI 10.1007/978-3-319-30340-6_7

years ago in the study of three bodies problem by Poincaré [13], but chaos was experimentally established by Lorenz [37] only a few decades ago in the study of 3-D weather models.

The Lyapunov exponent is a measure of the divergence of phase points that are initially very close and can be used to quantify chaotic systems. It is common to refer to the largest Lyapunov exponent as the *Maximal Lyapunov Exponent* (MLE). A positive maximal Lyapunov exponent and phase space compactness are usually taken as defining conditions for a chaotic system.

Since the discovery of Lorenz system in 1963, there is a great deal of interest in the chaos literature in finding new chaotic systems. Some classical paradigms of 3-D chaotic systems in the literature are Rössler system [53], ACT system [1], Sprott systems [63], Chen system [16], Lü system [38], Liu system [36], Cai system [14], Chen-Lee system [17], Tigan system [74], etc.

Many new chaotic systems have been discovered in the recent years such as Zhou system [140], Zhu system [141], Li system [31], Wei-Yang system [133], Sundarapandian systems [66, 71], Vaidyanathan systems [82, 83, 85–88, 91, 98, 108, 111, 113, 122, 124, 126, 128–130], Pehlivan system [43], Sampath system [55], Pham system [44], etc.

Chaos theory and control systems have many important applications in science and engineering [2, 9–12, 142]. Some commonly known applications are oscillators [27, 62], lasers [33, 136], chemical reactions [95, 96, 99, 101, 102, 106], biology [19, 29, 90, 92–94, 97, 100, 104, 105], ecology [22, 64], encryption [30, 139], cryptosystems [52, 75], mechanical systems [4–8], secure communications [20, 40, 137], robotics [39, 41, 131], cardiology [46, 134], intelligent control [3, 34], neural networks [23, 26, 35], memristors [45, 132], etc.

Synchronization of chaotic systems is a phenomenon that occurs when two or more chaotic systems are coupled or when a chaotic system drives another chaotic system. Because of the butterfly effect which causes exponential divergence of the trajectories of two identical chaotic systems started with nearly the same initial conditions, the synchronization of chaotic systems is a challenging research problem in the chaos literature.

Major works on synchronization of chaotic systems deal with the complete synchronization of a pair of chaotic systems called the *master* and *slave* systems. The design goal of the complete synchronization is to apply the output of the master system to control the slave system so that the output of the slave system tracks the output of the master system asymptotically with time.

Pecora and Carroll pioneered the research on synchronization of chaotic systems with their seminal papers [15, 42]. The active control method [25, 56, 57, 65, 70, 76, 80, 114, 115, 118] is typically used when the system parameters are available for measurement. Adaptive control method [54, 58–60, 67–69, 78, 84, 103, 107, 112, 116, 117, 123, 127] is typically used when some or all the system parameters are not available for measurement and estimates for the uncertain parameters of the systems.

Sampled-data feedback control method [21, 32, 135, 138] and time-delay feedback control method [18, 24, 61] are also used for synchronization of chaotic systems.

Backstepping control method [47–51, 73, 119, 125] is also used for the synchronization of chaotic systems, which is a recursive method for stabilizing the origin of a control system in strict-feedback form. Another popular method for the synchronization of chaotic systems is the sliding mode control method [72, 77, 79, 81, 89, 109, 110, 120, 121], which is a nonlinear control method that alters the dynamics of a nonlinear system by application of a discontinuous control signal that forces the system to "slide" along a cross-section of the system's normal behavior.

In this research work, we describe an eight-term 3-D novel polynomial chaotic system with three quadratic nonlinearities. Section 2 describes the 3-D dynamical model and phase portraits of the novel chaotic system. Section 3 describes the dynamic analysis of the novel chaotic system. The Lyapunov exponents of the eight-term novel chaotic system are obtained as $L_1 = 4.0359$, $L_2 = 0$ and $L_3 = -29.1071$. The Kaplan-Yorke dimension of the novel chaotic system is obtained as $D_{KY} = 2.1384$.

Section 4 describes the adaptive feedback control of the novel chaotic system with unknown parameters. Section 5 describes the adaptive feedback synchronization of the identical novel chaotic systems with unknown parameters. The adaptive feedback control and synchronization results are proved using Lyapunov stability theory [28]. MATLAB simulations are depicted to illustrate all the main results for the eight-term 3-D novel chaotic system. Finally, Sect. 6 gives a summary of the main results derived in this work.

2 A Novel 3-D Chaotic System

In this research work, we announce a novel 3-D chaotic system described by

$$\begin{aligned} \dot{x}_1 &= a(x_2 - x_1) + x_2 x_3 \\ \dot{x}_2 &= b x_1 + c x_1 x_3 \\ \dot{x}_3 &= -p(x_1 + x_3) - x_1 x_2 \end{aligned} \tag{1}$$

where x_1, x_2, x_3 are the states and a, b, c, p are constant, positive parameters.

The 3-D system (1) is *chaotic* when the parameter values are taken as

$$a = 24, \quad b = 950, \quad c = 75, \quad p = 1.2 \tag{2}$$

For numerical simulations, we take the initial state of the chaotic system (1) as

$$x_1(0) = 0.6, \quad x_2(0) = 0.8, \quad x_3(0) = 0.4 \tag{3}$$

The novel 3-D chaotic system (1) exhibits a 2-scroll chaotic attractor. Figure 1 describes the 2-scroll chaotic attractor of the novel chaotic system (1) in 3-D view.

Figure 2 describes the 2-D projection of the strange chaotic attractor of the novel chaotic system (1) on (x_1, x_2)-plane. In the projection on the (x_1, x_2)-plane, a 2-scroll chaotic attractor is clearly seen.

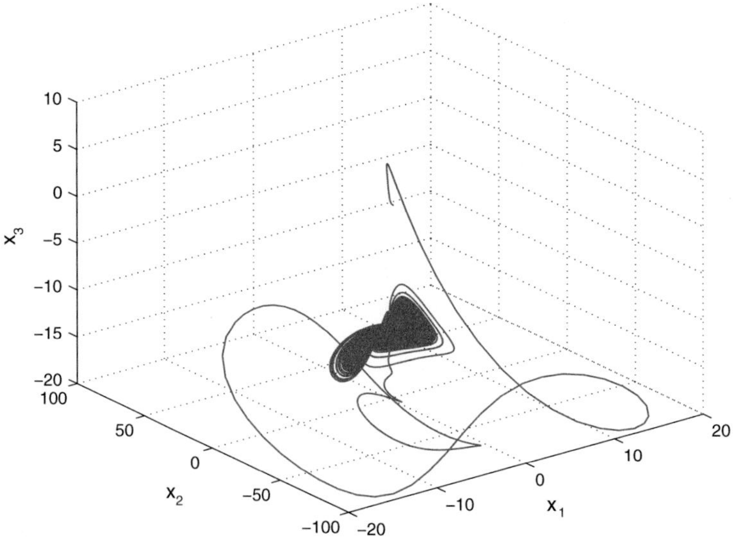

Fig. 1 Strange attractor of the novel chaotic system in $\mathbf{R}^3$

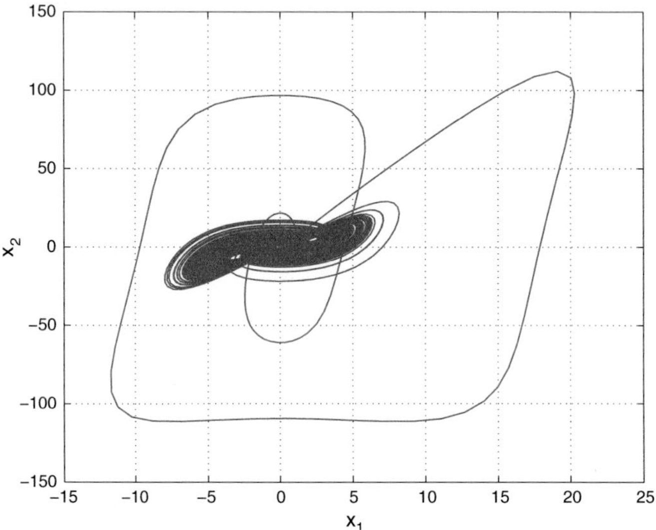

Fig. 2 2-D projection of the novel chaotic system on (x_1, x_2)-plane

Figure 3 describes the 2-D projection of the strange chaotic attractor of the novel chaotic system (1) on (x_2, x_3)-plane. In the projection on the (x_2, x_3)-plane, a 2-scroll chaotic attractor is clearly seen.

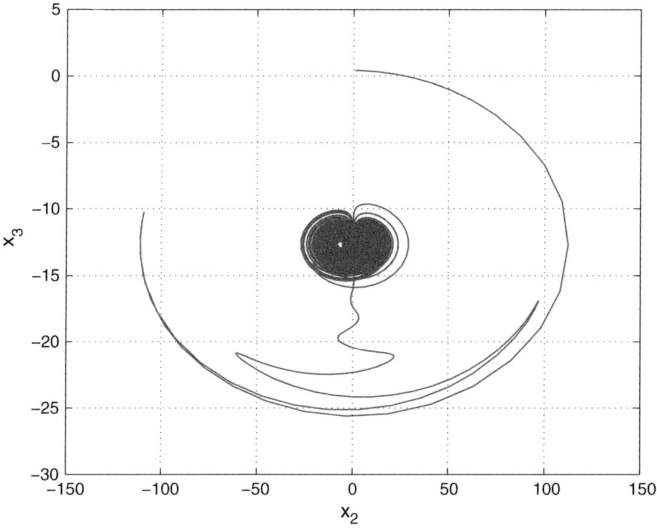

Fig. 3 2-D projection of the novel chaotic system on (x_2, x_3)-plane

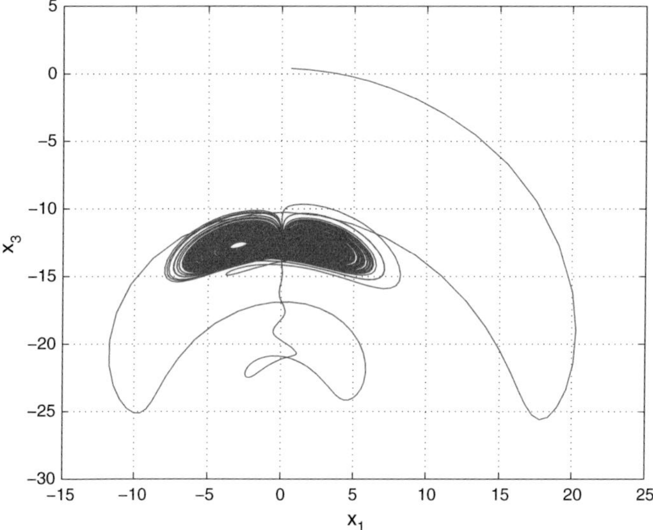

Fig. 4 2-D projection of the novel chaotic system on (x_1, x_3)-plane

Figure 4 describes the 2-D projection of the strange chaotic attractor of the novel chaotic system (1) on (x_1, x_3)-plane. In the projection on the (x_1, x_3)-plane, a 2-scroll chaotic attractor is clearly seen.

3 Analysis of the 3-D Novel Chaotic System

This section gives the qualitative properties of the eight-term novel 3-D chaotic system (1) proposed in this research work.

3.1 Dissipativity

We write the system (1) in vector notation as

$$\dot{x} = f(x) = \begin{bmatrix} f_1(x) \\ f_2(x) \\ f_3(x) \end{bmatrix}, \tag{4}$$

where

$$\begin{aligned} f_1(x) &= a(x_2 - x_1) + x_2 x_3 \\ f_2(x) &= b x_1 + c x_1 x_3 \\ f_3(x) &= -p(x_1 + x_3) - x_1 x_2 \end{aligned} \tag{5}$$

We take the parameter values as

$$a = 24, \quad b = 950, \quad c = 75, \quad p = 1.2 \tag{6}$$

The divergence of the vector field f on $\mathbb{R}^3$ is obtained as

$$\text{div}\, f = \frac{\partial f_1(x)}{\partial x_1} + \frac{\partial f_2(x)}{\partial x_2} + \frac{\partial f_3(x)}{\partial x_3} = -(a + p) = -\mu \tag{7}$$

where

$$\mu = a + p = 25.2 > 0 \tag{8}$$

Let Ω be any region in $\mathbb{R}^3$ with a smooth boundary. Let $\Omega(t) = \Phi_t(\Omega)$, where Φ_t is the flow of the vector field f.

Let $V(t)$ denote the volume of $\Omega(t)$.

By Liouville's theorem, it follows that

$$\frac{dV(t)}{dt} = \int\limits_{\Omega(t)} (\text{div} f) dx_1\, dx_2\, dx_3 \tag{9}$$

Substituting the value of div f in (9) leads to

$$\frac{dV(t)}{dt} = -\mu \int\limits_{\Omega(t)} dx_1\, dx_2\, dx_3 = -\mu V(t) \tag{10}$$

Integrating the linear differential equation (10), $V(t)$ is obtained as

$$V(t) = V(0)\exp(-\mu t), \quad \text{where } \mu = 25.2 > 0. \tag{11}$$

From Eq. (11), it follows that the volume $V(t)$ shrinks to zero exponentially as $t \to \infty$.

Thus, the novel chaotic system (1) is dissipative. Hence, any asymptotic motion of the system (1) settles onto a set of measure zero, *i.e.* a strange attractor.

3.2 Invariance

It is easily seen that the x_3-axis is invariant for the flow of the novel chaotic system (1). The invariant motion along the x_3-axis is characterized by the scalar dynamics

$$\dot{x}_3 = -px_3, \quad (p > 0) \tag{12}$$

which is globally exponentially stable.

3.3 Equilibria

The equilibrium points of the novel chaotic system (1) are obtained by solving the nonlinear equations

$$\begin{aligned} f_1(x) &= a(x_2 - x_1) + x_2x_3 &&= 0 \\ f_2(x) &= bx_1 + cx_1x_3 &&= 0 \\ f_3(x) &= -p(x_1 + x_3) - x_1x_2 &&= 0 \end{aligned} \tag{13}$$

We take the parameter values as in the chaotic case, viz.

$$a = 24, \quad b = 950, \quad c = 75, \quad p = 1.2 \tag{14}$$

Solving the nonlinear system of equations (13) with the parameter values (14), we obtain three equilibrium points of the novel chaotic system (1) as

$$E_0 = \begin{bmatrix} 0 \\ 0 \\ 0 \end{bmatrix}, \quad E_1 = \begin{bmatrix} 2.4107 \\ 5.1051 \\ -12.6667 \end{bmatrix} \text{ and } E_2 = \begin{bmatrix} -2.9774 \\ -6.3051 \\ -12.6667 \end{bmatrix}. \tag{15}$$

The Jacobian matrix of the novel chaotic system (1) at $(x_1^\star, x_2^\star, x_3^\star)$ is obtained as

$$J(x^\star) = \begin{bmatrix} -a & a + x_3^\star & x_2^\star \\ b + cx_3^\star & 0 & cx_1^\star \\ -p - x_2^\star & -x_1^\star & -p \end{bmatrix} \tag{16}$$

The Jacobian matrix at E_0 is obtained as

$$J_0 = J(E_0) = \begin{bmatrix} -24 & 24 & 0 \\ 950 & 0 & 0 \\ -1.2 & 0 & -1.2 \end{bmatrix} \tag{17}$$

The matrix J_0 has the eigenvalues

$$\lambda_1 = -1.2, \quad \lambda_2 = 139.4728, \quad \lambda_3 = -163.4728 \tag{18}$$

This shows that the equilibrium point E_0 is a saddle-point, which is unstable. The Jacobian matrix at E_1 is obtained as

$$J_1 = J(E_1) = \begin{bmatrix} -24.0000 & 11.3333 & 5.1051 \\ -0.0025 & 0 & 180.8025 \\ -6.3051 & -2.4107 & -1.2000 \end{bmatrix} \tag{19}$$

The matrix J_1 has the eigenvalues

$$\lambda_1 = -32.2619, \quad \lambda_{2,3} = 3.5309 \pm 26.6878i \tag{20}$$

This shows that the equilibrium point E_1 is a saddle-focus, which is unstable. The Jacobian matrix at E_2 is obtained as

$$J_2 = J(E_2) = \begin{bmatrix} -24.0000 & 11.3333 & -6.3051 \\ -0.0025 & 0 & -223.3050 \\ 5.1051 & 2.9774 & -1.2000 \end{bmatrix} \tag{21}$$

The matrix J_2 has the eigenvalues

$$\lambda_1 = -31.3862, \quad \lambda_{2,3} = 3.0931 \pm 30.1741i \tag{22}$$

This shows that the equilibrium point E_2 is a saddle-focus, which is unstable.

Hence, E_0, E_1, E_2 are all unstable equilibrium points of the 3-D novel chaotic system (1), where E_0 is a saddle point and E_1, E_2 are saddle-focus nodes.

3.4 Lyapunov Exponents and Kaplan-Yorke Dimension

We take the initial values of the novel chaotic system (1) as in (3) and the parameter values of the novel chaotic system (1) as in (2).

Then the Lyapunov exponents of the novel chaotic system (1) are numerically obtained as

$$L_1 = 4.0359, \quad L_2 = 0, \quad L_3 = -29.1701 \tag{23}$$

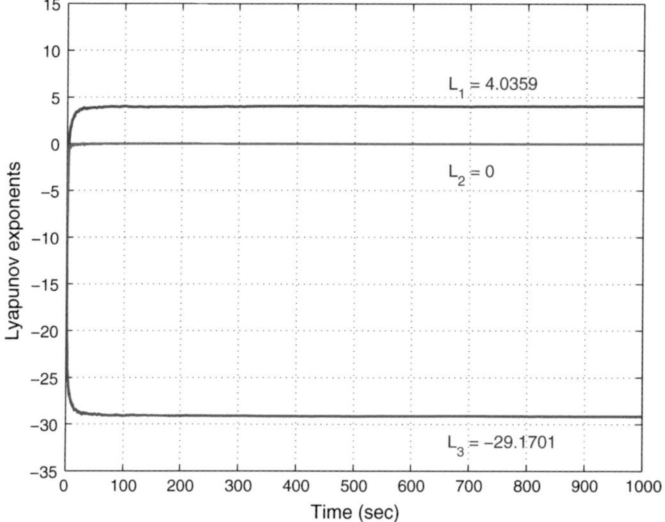

Fig. 5 Lyapunov exponents of the novel chaotic system

Since $L_1 + L_2 + L_3 = -25.1342 < 0$, the system (1) is dissipative. Also, the Kaplan-Yorke dimension of the system (1) is obtained as

$$D_{KY} = 2 + \frac{L_1 + L_2}{|L_3|} = 2.1384 \tag{24}$$

Figure 5 depicts the Lyapunov exponents of the novel chaotic system (1). From this figure, it is seen that the Maximal Lyapunov Exponent (MLE) of the novel chaotic system (1) is $L_1 = 4.0359$. Thus, the novel chaotic system (1) exhibits strong chaotic properties.

4 Adaptive Feedback Control of the 3-D Novel Chaotic System

This section derives new results for adaptive feedback controller design in order to stabilize the unstable novel chaotic system with unknown parameters for all initial conditions.

The controlled novel 3-D chaotic system is given by

$$\begin{aligned} \dot{x}_1 &= a(x_2 - x_1) + x_2 x_3 + u_1 \\ \dot{x}_2 &= b x_1 + c x_1 x_3 + u_2 \\ \dot{x}_3 &= -p(x_1 + x_3) - x_1 x_2 + u_3 \end{aligned} \tag{25}$$

where x_1, x_2, x_3 are state variables, a, b, c, p are constant, unknown, parameters of the system and u_1, u_2, u_3 are adaptive feedback controls to be designed.

An adaptive feedback control law is taken as

$$\begin{aligned} u_1 &= -\hat{a}(t)(x_2 - x_1) - x_2 x_3 - k_1 x_1 \\ u_2 &= -\hat{b}(t)x_1 - \hat{c}(t)x_1 x_3 - k_2 x_2 \\ u_3 &= \hat{p}(t)(x_1 + x_3) + x_1 x_2 - k_3 x_3 \end{aligned} \tag{26}$$

In (26), $\hat{a}(t), \hat{b}(t), \hat{c}(t), \hat{p}(t)$ are estimates for the unknown parameters a, b, c, p, respectively, and k_1, k_2, k_3 are positive gain constants.

The closed-loop control system is obtained by substituting (26) into (25) as

$$\begin{aligned} \dot{x}_1 &= \left[a - \hat{a}(t)\right](x_2 - x_1) - k_1 x_1 \\ \dot{x}_2 &= \left[b - \hat{b}(t)\right] x_1 + \left[c - \hat{c}(t)\right] x_1 x_3 - k_2 x_2 \\ \dot{x}_3 &= -\left[p - \hat{p}(t)\right](x_1 + x_3) - k_3 x_3 \end{aligned} \tag{27}$$

To simplify (27), we define the parameter estimation error as

$$\begin{aligned} e_a(t) &= a - \hat{a}(t) \\ e_b(t) &= b - \hat{b}(t) \\ e_c(t) &= c - \hat{c}(t) \\ e_p(t) &= d - \hat{p}(t) \end{aligned} \tag{28}$$

Using (28), the closed-loop system (27) can be simplified as

$$\begin{aligned} \dot{x}_1 &= e_a(x_2 - x_1) - k_1 x_1 \\ \dot{x}_2 &= e_b x_1 + e_c x_1 x_3 - k_2 x_2 \\ \dot{x}_3 &= -e_p(x_1 + x_3) - k_3 x_3 \end{aligned} \tag{29}$$

Differentiating the parameter estimation error (28) with respect to t, we get

$$\begin{aligned} \dot{e}_a &= -\dot{\hat{a}} \\ \dot{e}_b &= -\dot{\hat{b}} \\ \dot{e}_c &= -\dot{\hat{c}} \\ \dot{e}_p &= -\dot{\hat{p}} \end{aligned} \tag{30}$$

Next, we find an update law for parameter estimates using Lyapunov stability theory.

Consider the quadratic Lyapunov function defined by

$$V(x_1, x_2, x_3, e_a, e_b, e_c, e_p) = \frac{1}{2}\left(x_1^2 + x_2^2 + x_3^2 + e_a^2 + e_b^2 + e_c^2 + e_p^2\right), \tag{31}$$

which is positive definite on $\mathbb{R}^7$.

Differentiating V along the trajectories of (29) and (30), we get

$$\begin{aligned}\dot{V} &= -k_1 x_1^2 - k_2 x_2^2 - k_3 x_3^2 + e_a[x_1(x_2 - x_1) - \dot{\hat{a}}] + e_b[x_1 x_2 - \dot{\hat{b}}] \\ &\quad + e_c[x_1 x_2 x_3 - \dot{\hat{c}}] + e_p[-x_3(x_1 + x_3) - \dot{\hat{p}}]\end{aligned} \tag{32}$$

In view of Eq. (32), an update law for the parameter estimates is taken as

$$\begin{aligned}\dot{\hat{a}} &= x_1(x_2 - x_1) \\ \dot{\hat{b}} &= x_1 x_2 \\ \dot{\hat{c}} &= x_1 x_2 x_3 \\ \dot{\hat{p}} &= -x_3(x_1 + x_3)\end{aligned} \tag{33}$$

Theorem 1 *The novel chaotic system (25) with unknown system parameters is globally and exponentially stabilized for all initial conditions $x(0) \in \mathbb{R}^3$ by the adaptive control law (26) and the parameter update law (33), where k_i, $(i = 1, 2, 3)$ are positive constants.*

Proof The result is proved using Lyapunov stability theory [28]. We consider the quadratic Lyapunov function V defined by (31), which is positive definite on $\mathbb{R}^7$.

Substitution of the parameter update law (33) into (32) yields

$$\dot{V} = -k_1 x_1^2 - k_2 x_2^2 - k_3 x_3^2, \tag{34}$$

which is a negative semi-definite function on $\mathbb{R}^7$.

Therefore, it can be concluded that the state vector $x(t)$ and the parameter estimation error are globally bounded, *i.e.*

$$\left[x_1(t)\ x_2(t)\ x_3(t)\ e_a(t)\ e_b(t)\ e_c(t)\ e_p(t) \right]^T \in \mathbf{L}_\infty. \tag{35}$$

Define

$$k = \min\{k_1, k_2, k_3\} \tag{36}$$

Then it follows from (34) that

$$\dot{V} \le -k\|\mathbf{x}\|^2 \text{ or } k\|\mathbf{x}\|^2 \le -\dot{V} \tag{37}$$

Integrating the inequality (37) from 0 to t, we get

$$k \int_0^t \|\mathbf{x}(\tau)\|^2 \, d\tau \le -\int_0^t \dot{V}(\tau)\, d\tau = V(0) - V(t) \tag{38}$$

From (38), it follows that $\mathbf{x}(t) \in \mathbf{L}_2$.

Using (29), it can be deduced that $\dot{x}(t) \in \mathbf{L}_\infty$.

Hence, using Barbalat's lemma, we can conclude that $\mathbf{x}(t) \to 0$ exponentially as $t \to \infty$ for all initial conditions $\mathbf{x}(0) \in \mathbb{R}^3$.

This completes the proof. □

For numerical simulations, the parameter values of the novel system (25) are taken as in the chaotic case, *viz.*

$$a = 24, \quad b = 950, \quad c = 75, \quad p = 1.2 \tag{39}$$

The gain constants are taken as $k_i = 5$, $(i = 1, 2, 3)$.

The initial values of the parameter estimates are taken as

$$\hat{a}(0) = 3.2, \quad \hat{b}(0) = 20.5, \quad \hat{c}(0) = 12.3, \quad \hat{p}(0) = 6.4 \tag{40}$$

The initial values of the novel system (25) are taken as

$$x_1(0) = 2.5, \quad x_2(0) = -7.6, \quad x_3(0) = 12.7 \tag{41}$$

Figure 6 shows the time-history of the controlled states $x_1(t), x_2(t), x_3(t)$.

Figure 6 depicts the exponential convergence of the controlled states and the efficiency of the adaptive controller defined by (26).

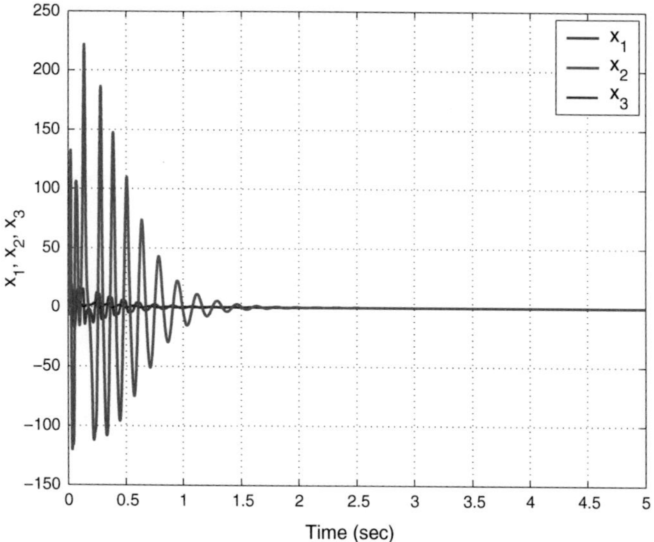

Fig. 6 Time-history of the states $x_1(t), x_2(t), x_3(t)$

5 Adaptive Synchronization of the Identical 3-D Novel Chaotic Systems

This section derives new results for the adaptive synchronization of the identical novel chaotic systems with unknown parameters.

The master system is given by the novel chaotic system

$$\begin{aligned} \dot{x}_1 &= a(x_2 - x_1) + x_2 x_3 \\ \dot{x}_2 &= b x_1 + c x_1 x_3 \\ \dot{x}_3 &= -p(x_1 + x_3) - x_1 x_2 \end{aligned} \tag{42}$$

where x_1, x_2, x_3 are state variables and a, b, c, p are constant, unknown, parameters of the system.

The slave system is given by the controlled novel chaotic system

$$\begin{aligned} \dot{y}_1 &= a(y_2 - y_1) + y_2 y_3 + u_1 \\ \dot{y}_2 &= b y_1 + c y_1 y_3 + u_2 \\ \dot{y}_3 &= -p(y_1 + y_3) - y_1 y_2 + u_3 \end{aligned} \tag{43}$$

where y_1, y_2, y_3 are state variables and u_1, u_2, u_3 are adaptive controls to be designed.

The synchronization error is defined as

$$\begin{aligned} e_1 &= y_1 - x_1 \\ e_2 &= y_2 - x_2 \\ e_3 &= y_3 - x_3 \end{aligned} \tag{44}$$

The error dynamics is easily obtained as

$$\begin{aligned} \dot{e}_1 &= a(e_2 - e_1) + y_2 y_3 - x_2 x_3 + u_1 \\ \dot{e}_2 &= b e_1 + c(y_1 y_3 - x_1 x_3) + u_2 \\ \dot{e}_3 &= -p(e_1 + e_3) - y_1 y_2 + x_1 x_2 + u_3 \end{aligned} \tag{45}$$

An adaptive control law is taken as

$$\begin{aligned} u_1 &= -\hat{a}(t)(e_2 - e_1) - y_2 y_3 + x_2 x_3 - k_1 e_1 \\ u_2 &= -\hat{b}(t) e_1 - \hat{c}(t)(y_1 y_3 - x_1 x_3) - k_2 e_2 \\ u_3 &= \hat{p}(t)(e_1 + e_3) + y_1 y_2 - x_1 x_2 - k_3 e_3 \end{aligned} \tag{46}$$

where $\hat{a}(t), \hat{b}(t), \hat{c}(t), \hat{p}(t)$ are estimates for the unknown parameters a, b, c, p, respectively, and k_1, k_2, k_3 are positive gain constants.

The closed-loop control system is obtained by substituting (46) into (45) as

$$
\begin{aligned}
\dot{e}_1 &= [a - \hat{a}(t)](e_2 - e_1) - k_1 e_1 \\
\dot{e}_2 &= [b - \hat{b}(t)]e_1 + [c - \hat{c}(t)](y_1 y_3 - x_1 x_3) - k_2 e_2 \\
\dot{e}_3 &= -[p - \hat{p}(t)](e_1 + e_3) - k_3 e_3
\end{aligned}
\tag{47}
$$

To simplify (47), we define the parameter estimation error as

$$
\begin{aligned}
e_a(t) &= a - \hat{a}(t) \\
e_b(t) &= b - \hat{b}(t) \\
e_c(t) &= c - \hat{c}(t) \\
e_p(t) &= p - \hat{p}(t)
\end{aligned}
\tag{48}
$$

Using (48), the closed-loop system (47) can be simplified as

$$
\begin{aligned}
\dot{e}_1 &= e_a(e_2 - e_1) - k_1 e_1 \\
\dot{e}_2 &= e_b e_1 + e_c(y_1 y_3 - x_1 x_3) - k_2 e_2 \\
\dot{e}_3 &= -e_p(e_1 + e_3) - k_3 e_3
\end{aligned}
\tag{49}
$$

Differentiating the parameter estimation error (48) with respect to t, we get

$$
\begin{aligned}
\dot{e}_a &= -\dot{\hat{a}} \\
\dot{e}_b &= -\dot{\hat{b}} \\
\dot{e}_c &= -\dot{\hat{c}} \\
\dot{e}_p &= -\dot{\hat{p}}
\end{aligned}
\tag{50}
$$

Next, we find an update law for parameter estimates using Lyapunov stability theory.

Consider the quadratic Lyapunov function defined by

$$
V(e_1, e_2, e_3, e_a, e_b, e_c, e_p) = \frac{1}{2}\left(e_1^2 + e_2^2 + e_3^2 + e_a^2 + e_b^2 + e_c^2 + e_p^2\right), \tag{51}
$$

which is positive definite on $\mathbb{R}^7$.

Differentiating V along the trajectories of (49) and (50), we get

$$
\begin{aligned}
\dot{V} = -k_1 e_1^2 - k_2 e_2^2 - k_3 e_3^2 &+ e_a\left[e_1(e_2 - e_1) - \dot{\hat{a}}\right] + e_b\left[e_1 e_2 - \dot{\hat{b}}\right] \\
&+ e_c\left[e_2(y_1 y_3 - x_1 x_3) - \dot{\hat{c}}\right] + e_p\left[-e_3(e_1 + e_3) - \dot{\hat{p}}\right]
\end{aligned}
\tag{52}
$$

In view of Eq. (52), an update law for the parameter estimates is taken as

$$
\begin{aligned}
\dot{\hat{a}} &= e_1(e_2 - e_1) \\
\dot{\hat{b}} &= e_1 e_2 \\
\dot{\hat{c}} &= e_2(y_1 y_3 - x_1 x_3) \\
\dot{\hat{p}} &= -e_3(e_1 + e_3)
\end{aligned}
\tag{53}
$$

Theorem 2 *The identical novel chaotic systems (42) and (43) with unknown system parameters are globally and exponentially synchronized for all initial conditions* $x(0), y(0) \in \mathbb{R}^3$ *by the adaptive control law (46) and the parameter update law (53), where* k_i, $(i = 1, 2, 3)$ *are positive constants.*

Proof The result is proved using Lyapunov stability theory [28].

We consider the quadratic Lyapunov function V defined by (51), which is positive definite on $\mathbb{R}^7$.

Substitution of the parameter update law (53) into (52) yields

$$\dot{V} = -k_1 e_1^2 - k_2 e_2^2 - k_3 e_3^2, \tag{54}$$

which is a negative semi-definite function on $\mathbb{R}^7$.

Therefore, it can be concluded that the synchronization error vector $e(t)$ and the parameter estimation error are globally bounded, *i.e.*

$$\left[e_1(t)\ e_2(t)\ e_3(t)\ e_a(t)\ e_b(t)\ e_c(t)\ e_p(t) \right]^T \in \mathbf{L}_\infty. \tag{55}$$

Define

$$k = \min\{k_1, k_2, k_3\} \tag{56}$$

Then it follows from (54) that

$$\dot{V} \le -k\|e\|^2 \text{ or } k\|\mathbf{e}\|^2 \le -\dot{V} \tag{57}$$

Integrating the inequality (57) from 0 to t, we get

$$k \int_0^t \|\mathbf{e}(\tau)\|^2 \, d\tau \le -\int_0^t \dot{V}(\tau)\, d\tau = V(0) - V(t) \tag{58}$$

From (58), it follows that $\mathbf{e}(t) \in \mathbf{L}_2$.

Using (49), it can be deduced that $\dot{\mathbf{e}}(t) \in \mathbf{L}_\infty$.

Hence, using Barbalat's lemma, we can conclude that $\mathbf{e}(t) \to 0$ exponentially as $t \to \infty$ for all initial conditions $\mathbf{e}(0) \in \mathbb{R}^3$.

This completes the proof. □

For numerical simulations, the parameter values of the novel systems (42) and (43) are taken as in the chaotic case, *viz.*

$$a = 24, \quad b = 950, \quad c = 75, \quad p = 1.2 \tag{59}$$

The gain constants are taken as $k_i = 5$ for $i = 1, 2, 3$.

The initial values of the parameter estimates are taken as

$$\hat{a}(0) = 10.2, \quad \hat{b}(0) = 24.9, \quad \hat{c}(0) = 8.5, \quad \hat{p}(0) = 7.3 \tag{60}$$

The initial values of the master system (42) are taken as

$$x_1(0) = 2.8, \quad x_2(0) = 2.3, \quad x_3(0) = -2.1 \tag{61}$$

The initial values of the slave system (43) are taken as

$$y_1(0) = 6.4, \quad y_2(0) = -1.5, \quad y_3(0) = 3.8 \tag{62}$$

Figures 7, 8 and 9 show the complete synchronization of the identical chaotic systems (42) and (43).

Figure 7 shows that the states $x_1(t)$ and $y_1(t)$ are synchronized in two seconds (MATLAB).

Figure 8 shows that the states $x_2(t)$ and $y_2(t)$ are synchronized in two seconds (MATLAB).

Figure 9 shows that the states $x_3(t)$ and $y_3(t)$ are synchronized in two seconds (MATLAB).

Figure 10 shows the time-history of the synchronization errors $e_1(t), e_2(t), e_3(t)$. From Fig. 10, it is seen that the errors $e_1(t), e_2(t)$ and $e_3(t)$ are stabilized in two seconds (MATLAB).

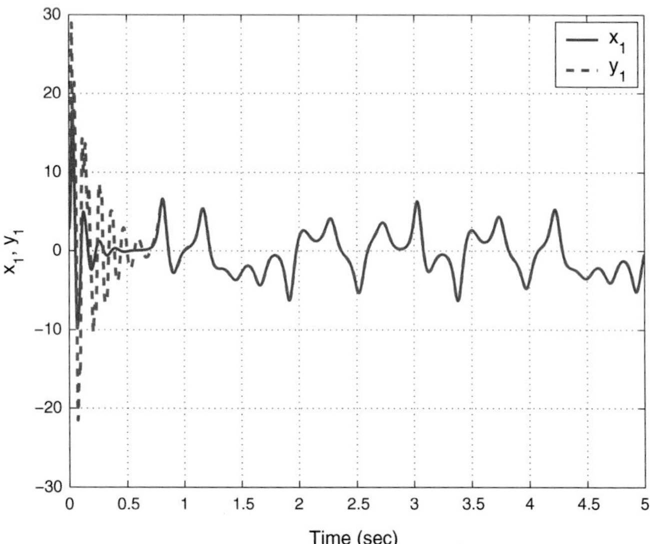

Fig. 7 Synchronization of the states x_1 and y_1

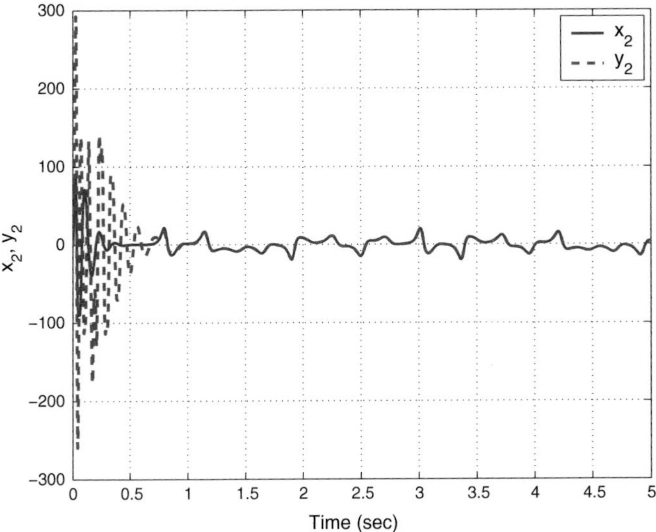

Fig. 8 Synchronization of the states x_2 and y_2

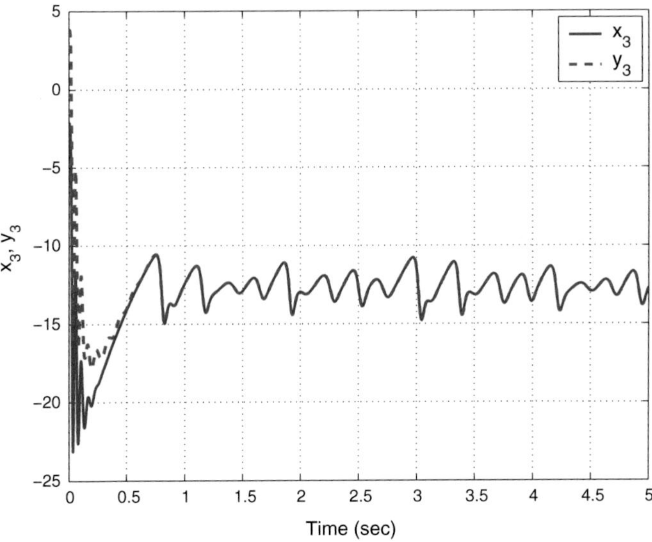

Fig. 9 Synchronization of the states x_3 and y_3

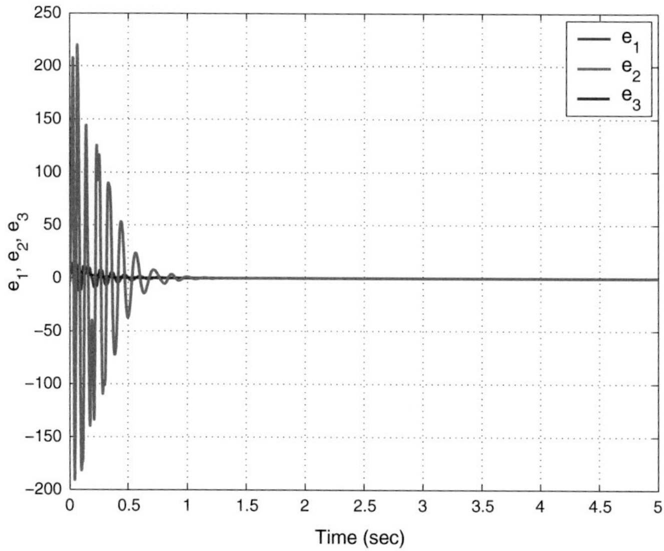

Fig. 10 Time-history of the synchronization errors e_1, e_2, e_3

6 Conclusions

In this research work, an eight-term 3-D novel chaotic system with three quadratic nonlinearities has been proposed and its dynamic analysis has been derived. The Lyapunov exponents of the nine-term novel chaotic system have been obtained as $L_1 = 4.0359$, $L_2 = 0$ and $L_3 = -29.1701$. Since the maximal Lyapunov exponent (MLE) of the novel chaotic system is $L_1 = 4.0359$, which is above four, the novel chaotic system exhibits strong chaotic properties. The novel chaotic system has three unstable equilibrium points. Next, an adaptive controller has been derived for globally stabilizing the novel chaotic system with unknown system parameters. Furthermore, an adaptive synchronizer has been derived for globally and exponentially synchronizing the identical novel chaotic systems with unknown system parameters. The adaptive control and synchronization results were proved using Lyapunov stability theory. MATLAB simulations were shown to demonstrate and validate all the main results derived in this work for the nine-term 3-D novel chaotic system. As future research directions, new control techniques like sliding mode control or backstepping control may be considered for stabilizing the novel chaotic system with three unstable equilibrium points or synchronizing the identical novel chaotic systems for all initial conditions.

References

1. Arneodo A, Coullet P, Tresser C (1981) Possible new strange attractors with spiral structure. Commun Math Phys 79(4):573–576
2. Azar AT (2010) Fuzzy systems. IN-TECH, Vienna, Austria
3. Azar AT (2012) Overview of type-2 fuzzy logic systems. Int J Fuzzy Syst Appl 2(4):1–28
4. Azar AT, Serrano FE (2014) Robust IMC-PID tuning for cascade control systems with gain and phase margin specifications. Neural Comput Appl 25(5):983–995
5. Azar AT, Serrano FE (2015) Adaptive sliding mode control of the Furuta pendulum. In: Azar AT, Zhu Q (eds) Advances and applications in sliding mode control systems. Studies in computational intelligence, vol 576. Springer, Berlin, pp 1–42
6. Azar AT, Serrano FE (2015) Deadbeat control for multivariable systems with time varying delays. In: Azar AT, Vaidyanathan S (eds) Chaos modeling and control systems design. Studies in computational intelligence, vol 581. Springer, Berlin, pp 97–132
7. Azar AT, Serrano FE (2015) Design and modeling of anti wind up PID controllers. In: Zhu Q, Azar AT (eds) Complex system modelling and control through intelligent soft computations. Studies in fuzziness and soft computing, vol 319. Springer, Berlin, pp 1–44
8. Azar AT, Serrano FE (2015) Stabilizatoin and control of mechanical systems with backlash. In: Azar AT, Vaidyanathan S (eds) Handbook of research on advanced intelligent control engineering and automation. Advances in computational intelligence and robotics (ACIR), IGI-Global, USA, pp 1–60
9. Azar AT, Vaidyanathan S (2015) Chaos modeling and control systems design. Studies in computational intelligence, vol 581. Springer, Berlin
10. Azar AT, Vaidyanathan S (2015) Computational intelligence applications in modeling and control. Studies in computational intelligence, vol 575. Springer, Berlin
11. Azar AT, Vaidyanathan S (2015) Handbook of research on advanced intelligent control engineering and automation. Advances in computational intelligence and robotics (ACIR), IGI-Global, USA
12. Azar AT, Zhu Q (2015) Advances and applications in sliding mode control systems. Studies in computational intelligence, vol 576. Springer, Berlin
13. Barrow-Green J (1997) Poincaré and the three body problem. American Mathematical Society, Providence
14. Cai G, Tan Z (2007) Chaos synchronization of a new chaotic system via nonlinear control. J Uncertain Syst 1(3):235–240
15. Carroll TL, Pecora LM (1991) Synchronizing chaotic circuits. IEEE Trans Circuits Syst 38(4):453–456
16. Chen G, Ueta T (1999) Yet another chaotic attractor. Int J Bifurc Chaos 9(7):1465–1466
17. Chen HK, Lee CI (2004) Anti-control of chaos in rigid body motion. Chaos, Solitons Fractals 21(4):957–965
18. Chen WH, Wei D, Lu X (2014) Global exponential synchronization of nonlinear time-delay Lur'e systems via delayed impulsive control. Commun Nonlinear Sci Numer Simul 19(9):3298–3312
19. Das S, Goswami D, Chatterjee S, Mukherjee S (2014) Stability and chaos analysis of a novel swarm dynamics with applications to multi-agent systems. Eng Appl Artif Intell 30:189–198
20. Feki M (2003) An adaptive chaos synchronization scheme applied to secure communication. Chaos, Solitons Fractals 18(1):141–148
21. Gan Q, Liang Y (2012) Synchronization of chaotic neural networks with time delay in the leakage term and parametric uncertainties based on sampled-data control. J Franklin Inst 349(6): 1955–1971
22. Gibson WT, Wilson WG (2013) Individual-based chaos: Extensions of the discrete logistic model. J Theor Biol 339:84–92
23. Huang X, Zhao Z, Wang Z, Li Y (2012) Chaos and hyperchaos in fractional-order cellular neural networks. Neurocomputing 94:13–21

24. Jiang GP, Zheng WX, Chen G (2004) Global chaos synchronization with channel time-delay. Chaos, Solitons Fractals 20(2):267–275
25. Karthikeyan R, Sundarapandian V (2014) Hybrid chaos synchronization of four-scroll systems via active control. J Electr Eng 65(2):97–103
26. Kaslik E, Sivasundaram S (2012) Nonlinear dynamics and chaos in fractional-order neural networks. Neural Netw 32:245–256
27. Kengne J, Chedjou JC, Kenne G, Kyamakya K (2012) Dynamical properties and chaos synchronization of improved Colpitts oscillators. Commun Nonlinear Sci Numer Simul 17(7):2914–2923
28. Khalil HK (2001) Nonlinear systems. Prentice Hall, New Jersey
29. Kyriazis M (1991) Applications of chaos theory to the molecular biology of aging. Exp Gerontol 26(6):569–572
30. Lang J (2015) Color image encryption based on color blend and chaos permutation in the reality-preserving multiple-parameter fractional Fourier transform domain. Optics Commun 338:181–192
31. Li D (2008) A three-scroll chaotic attractor. Phys Lett A 372(4):387–393
32. Li N, Zhang Y, Nie Z (2011) Synchronization for general complex dynamical networks with sampled-data. Neurocomputing 74(5):805–811
33. Li N, Pan W, Yan L, Luo B, Zou X (2014) Enhanced chaos synchronization and communication in cascade-coupled semiconductor ring lasers. Commun Nonlinear Sci Numer Simul 19(6): 1874–1883
34. Li Z, Chen G (2006) Integration of fuzzy logic and chaos theory. Studies in fuzziness and soft computing, vol 187. Springer, Berlin
35. Lian S, Chen X (2011) Traceable content protection based on chaos and neural networks. Appl Soft Comput 11(7):4293–4301
36. Liu C, Liu T, Liu L, Liu K (2004) A new chaotic attractor. Chaos Solitions Fractals 22(5): 1031–1038
37. Lorenz EN (1963) Deterministic periodic flow. J Atmos Sci 20(2):130–141
38. Lü J, Chen G (2002) A new chaotic attractor coined. Int J Bifurc Chaos 12(3):659–661
39. Mondal S, Mahanta C (2014) Adaptive second order terminal sliding mode controller for robotic manipulators. J Franklin Inst 351(4):2356–2377
40. Murali K, Lakshmanan M (1998) Secure communication using a compound signal from generalized chaotic systems. Phys Lett A 241(6):303–310
41. Nehmzow U, Walker K (2005) Quantitative description of robot-environment interaction using chaos theory. Robot Auton Syst 53(3–4):177–193
42. Pecora LM, Carroll TL (1990) Synchronization in chaotic systems. Phys Rev Lett 64(8): 821–824
43. Pehlivan I, Moroz IM, Vaidyanathan S (2014) Analysis, synchronization and circuit design of a novel butterfly attractor. J Sound Vib 333(20):5077–5096
44. Pham VT, Vaidyanathan S, Volos CK, Jafari S (2015) Hidden attractors in a chaotic system with an exponential nonlinear term. Eur Phys J: Spec Topics 224(8):1507–1517
45. Pham VT, Volos CK, Vaidyanathan S, Le TP, Vu VY (2015) A memristor-based hyperchaotic system with hidden attractors: dynamics, synchronization and circuital emulating. J Eng Sci Technol Rev 8(2):205–214
46. Qu Z (2011) Chaos in the genesis and maintenance of cardiac arrhythmias. Prog Biophys Mol Biol 105(3):247–257
47. Rasappan S, Vaidyanathan S (2012) Global chaos synchronization of WINDMI and Coullet chaotic systems by backstepping control. Far East J Math Sci 67(2):265–287
48. Rasappan S, Vaidyanathan S (2012) Hybrid synchronization of n-scroll Chua and Lur'e chaotic systems via backstepping control with novel feedback. Arch Control Sci 22(3): 343–365
49. Rasappan S, Vaidyanathan S (2012) Synchronization of hyperchaotic Liu system via backstepping control with recursive feedback. Commun Comput Inf Sci 305:212–221

50. Rasappan S, Vaidyanathan S (2013) Hybrid synchronization of n-scroll chaotic Chua circuits using adaptive backstepping control design with recursive feedback. Malays J Math Sci 7(2): 219–246
51. Rasappan S, Vaidyanathan S (2014) Global chaos synchronization of WINDMI and Coullet chaotic systems using adaptive backstepping control design. Kyungpook Math J 54(1): 293–320
52. Rhouma R, Belghith S (2011) Cryptoanalysis of a chaos based cryptosystem on DSP. Commun Nonlinear Sci Numer Simul 16(2):876–884
53. Rössler OE (1976) An equation for continuous chaos. Phys Lett A 57(5):397–398
54. Vaidyanathan S, VTP, Volos CK, (2015) A 5-D hyperchaotic Rikitake dynamo system with hidden attractors. Eur Phys J: Spec Topics 224(8):1575–1592
55. Sampath S, Vaidyanathan S, Volos CK, Pham VT (2015) An eight-term novel four-scroll chaotic system with cubic nonlinearity and its circuit simulation. J Eng Sci Technol Rev 8(2):1–6
56. Sarasu P, Sundarapandian V (2011) Active controller design for the generalized projective synchronization of four-scroll chaotic systems. Int J Syst Signal Control Eng Appl 4(2):26–33
57. Sarasu P, Sundarapandian V (2011) The generalized projective synchronization of hyperchaotic Lorenz and hyperchaotic Qi systems via active control. Int J Soft Comput 6(5): 216–223
58. Sarasu P, Sundarapandian V (2012) Adaptive controller design for the generalized projective synchronization of 4-scroll systems. Int J Syst Signal Control Eng Appl 5(2):21–30
59. Sarasu P, Sundarapandian V (2012) Generalized projective synchronization of three-scroll chaotic systems via adaptive control. Eur J Sci Res 72(4):504–522
60. Sarasu P, Sundarapandian V (2012) Generalized projective synchronization of two-scroll systems via adaptive control. Int J Soft Comput 7(4):146–156
61. Shahverdiev EM, Shore KA (2009) Impact of modulated multiple optical feedback time delays on laser diode chaos synchronization. Optics Commun 282(17):3568–3572
62. Sharma A, Patidar V, Purohit G, Sud KK (2012) Effects on the bifurcation and chaos in forced Duffing oscillator due to nonlinear damping. Commun Nonlinear Sci Numer Simul 17(6): 2254–2269
63. Sprott JC (1994) Some simple chaotic flows. Phys Rev E 50(2):647–650
64. Suérez I (1999) Mastering chaos in ecology. Ecol Model 117(2–3):305–314
65. Sundarapandian V (2010) Output regulation of the Lorenz attractor. Far East J Math Sci 42(2):289–299
66. Sundarapandian V (2013) Analysis and anti-synchronization of a novel chaotic system via active and adaptive controllers. J Eng Sci Technol Rev 6(4):45–52
67. Sundarapandian V, Karthikeyan R (2011) Anti-synchronization of hyperchaotic Lorenz and hyperchaotic Chen systems by adaptive control. Int J Syst Signal Control Eng Appl 4(2):18–25
68. Sundarapandian V, Karthikeyan R (2011) Anti-synchronization of Lü and Pan chaotic systems by adaptive nonlinear control. Eur J Sci Res 64(1):94–106
69. Sundarapandian V, Karthikeyan R (2012) Adaptive anti-synchronization of uncertain Tigan and Li systems. J Eng Appl Sci 7(1):45–52
70. Sundarapandian V, Karthikeyan R (2012) Hybrid synchronization of hyperchaotic Lorenz and hyperchaotic Chen systems via active control. J Eng Appl Sci 7(3):254–264
71. Sundarapandian V, Pehlivan I (2012) Analysis, control, synchronization, and circuit design of a novel chaotic system. Math Comput Model 55(7–8):1904–1915
72. Sundarapandian V, Sivaperumal S (2011) Sliding controller design of hybrid synchronization of four-wing chaotic systems. Int J Soft Comput 6(5):224–231
73. Suresh R, Sundarapandian V (2013) Global chaos synchronization of a family of n-scroll hyperchaotic Chua circuits using backstepping control with recursive feedback. Far East J Math Sci 73(1):73–95
74. Tigan G, Opris D (2008) Analysis of a 3D chaotic system. Chaos, Solitons Fractals 36: 1315–1319

75. Usama M, Khan MK, Alghatbar K, Lee C (2010) Chaos-based secure satellite imagery cryptosystem. Comput Math Appl 60(2):326–337
76. Vaidyanathan S (2011) Hybrid chaos synchronization of Liu and Lü systems by active nonlinear control. Commun Comput Inf Sci 204
77. Vaidyanathan S (2012) Analysis and synchronization of the hyperchaotic Yujun systems via sliding mode control. Adv Intell Syst Comput 176:329–337
78. Vaidyanathan S (2012) Anti-synchronization of Sprott-L and Sprott-M chaotic systems via adaptive control. Int J Control Theory Appl 5(1):41–59
79. Vaidyanathan S (2012) Global chaos control of hyperchaotic Liu system via sliding control method. Int J Control Theory Appl 5(2):117–123
80. Vaidyanathan S (2012) Output regulation of the Liu chaotic system. Appl Mech Materbreak 110–116:3982–3989
81. Vaidyanathan S (2012) Sliding mode control based global chaos control of Liu-Liu-Liu-Su chaotic system. Int J Control Theory Appl 5(1):15–20
82. Vaidyanathan S (2013) A new six-term 3-D chaotic system with an exponential nonlinearity. Far East J Math Sci 79(1):135–143
83. Vaidyanathan S (2013) Analysis and adaptive synchronization of two novel chaotic systems with hyperbolic sinusoidal and cosinusoidal nonlinearity and unknown parameters. J Eng Sci Technol Rev 6(4):53–65
84. Vaidyanathan S (2013) Analysis, control and synchronization of hyperchaotic Zhou system via adaptive control. Adv Intell Syst Comput 177:1–10
85. Vaidyanathan S (2014) A new eight-term 3-D polynomial chaotic system with three quadratic nonlinearities. Far East J Math Sci 84(2):219–226
86. Vaidyanathan S (2014) Analysis and adaptive synchronization of eight-term 3-D polynomial chaotic systems with three quadratic nonlinearities. Eur Phys J: Spec Topics 223(8): 1519–1529
87. Vaidyanathan S (2014) Analysis, control and synchronisation of a six-term novel chaotic system with three quadratic nonlinearities. Int J Model Ident Control 22(1):41–53
88. Vaidyanathan S (2014) Generalized projective synchronisation of novel 3-D chaotic systems with an exponential non-linearity via active and adaptive control. Int J Model Ident Control 22(3):207–217
89. Vaidyanathan S (2014) Global chaos synchronization of identical Li-Wu chaotic systems via sliding mode control. Int J Model Ident Control 22(2):170–177
90. Vaidyanathan S (2015) 3-cells cellular neural network (CNN) attractor and its adaptive biological control. Int J PharmTech Res 8(4):632–640
91. Vaidyanathan S (2015) A 3-D novel highly chaotic system with four quadratic nonlinearities, its adaptive control and anti-synchronization with unknown parameters. J Eng Sci Technol Rev 8(2):106–115
92. Vaidyanathan S (2015) Adaptive backstepping control of enzymes-substrates system with ferroelectric behaviour in brain waves. Int J PharmTech Res 8(2):256–261
93. Vaidyanathan S (2015) Adaptive biological control of generalized Lotka-Volterra three-species biological system. Int J PharmTech Res 8(4):622–631
94. Vaidyanathan S (2015) Adaptive chaotic synchronization of enzymes-substrates system with ferroelectric behaviour in brain waves. Int J PharmTech Res 8(5):964–973
95. Vaidyanathan S (2015) Adaptive control of a chemical chaotic reactor. Int J PharmTech Res 8(3):377–382
96. Vaidyanathan S (2015) Adaptive synchronization of chemical chaotic reactors. Int J Chem Tech Res 8(2):612–621
97. Vaidyanathan S (2015) Adaptive synchronization of generalized Lotka-Volterra three-species biological systems. Int J PharmTech Res 8(5):928–937
98. Vaidyanathan S (2015) Analysis, properties and control of an eight-term 3-D chaotic system with an exponential nonlinearity. Int J Model Ident Control 23(2):164–172
99. Vaidyanathan S (2015) Anti-synchronization of Brusselator chemical reaction systems via adaptive control. Int J ChemTech Res 8(6):759–768

100. Vaidyanathan S (2015) Chaos in neurons and adaptive control of Birkhoff-Shaw strange chaotic attractor. Int J PharmTech Res 8(5):956–963
101. Vaidyanathan S (2015) Dynamics and control of Brusselator chemical reaction. Int J ChemTech Res 8(6):740–749
102. Vaidyanathan S (2015) Dynamics and control of Tokamak system with symmetric and magnetically confined plasma. Int J ChemTech Res 8(6):795–803
103. Vaidyanathan S (2015) Hyperchaos, qualitative analysis, control and synchronisation of a ten-term 4-D hyperchaotic system with an exponential nonlinearity and three quadratic nonlinearities. Int J Model Ident Control 23(4):380–392
104. Vaidyanathan S (2015) Lotka-Volterra population biology models with negative feedback and their ecological monitoring. Int J PharmTech Res 8(5):974–981
105. Vaidyanathan S (2015) Synchronization of 3-cells cellular neural network (CNN) attractors via adaptive control method. Int J PharmTech Res 8(5):946–955
106. Vaidyanathan S (2015) Synchronization of Tokamak systems with symmetric and magnetically confined plasma via adaptive control. Int J ChemTech Res 8(6):818–827
107. Vaidyanathan S, Azar AT (2015) Analysis and control of a 4-D novel hyperchaotic system. In: Azar AT, Vaidyanathan S (eds) Chaos modeling and control systems design. Studies in computational intelligence, vol 581. Springer, Berlin, pp 19–38
108. Vaidyanathan S, Azar AT (2015) Analysis, control and synchronization of a nine-term 3-D novel chaotic system. In: Azar AT, Vaidyanathan S (eds) Chaos modelling and control systems design. Studies in computational intelligence, vol 581. Springer, Berlin, pp 19–38
109. Vaidyanathan S, Azar AT (2015) Anti-synchronization of identical chaotic systems using sliding mode control and an application to Vaidhyanathan-Madhavan chaotic systems. Studies in computational intelligence, vol 576, pp 527–547
110. Vaidyanathan S, Azar AT (2015) Hybrid synchronization of identical chaotic systems using sliding mode control and an application to Vaidhyanathan chaotic systems. Studies in computational intelligence, vol 576, pp 549–569
111. Vaidyanathan S, Madhavan K (2013) Analysis, adaptive control and synchronization of a seven-term novel 3-D chaotic system. Int J Control Theory Appl 6(2):121–137
112. Vaidyanathan S, Pakiriswamy S (2013) Generalized projective synchronization of six-term Sundarapandian chaotic systems by adaptive control. Int J Control Theory Appl 6(2):153–163
113. Vaidyanathan S, Pakiriswamy S (2015) A 3-D novel conservative chaotic system and its generalized projective synchronization via adaptive control. J Eng Sci Technol Rev 8(2): 52–60
114. Vaidyanathan S, Rajagopal K (2011) Anti-synchronization of Li and T chaotic systems by active nonlinear control. Commun Comput Inf Sci 198:175–184
115. Vaidyanathan S, Rajagopal K (2011) Global chaos synchronization of hyperchaotic Pang and Wang systems by active nonlinear control. Commun Comput Inf Sci 204:84–93
116. Vaidyanathan S, Rajagopal K (2011) Global chaos synchronization of Lü and Pan systems by adaptive nonlinear control. Commun Comput Inf Sci 205:193–202
117. Vaidyanathan S, Rajagopal K (2012) Global chaos synchronization of hyperchaotic Pang and hyperchaotic Wang systems via adaptive control. Int J Soft Comput 7(1):28–37
118. Vaidyanathan S, Rasappan S (2011) Global chaos synchronization of hyperchaotic Bao and Xu systems by active nonlinear control. Commun Comput Inf Sci 198:10–17
119. Vaidyanathan S, Rasappan S (2014) Global chaos synchronization of n-scroll Chua circuit and Lur'e system using backstepping control design with recursive feedback. Arab J Sci Eng 39(4):3351–3364
120. Vaidyanathan S, Sampath S (2011) Global chaos synchronization of hyperchaotic Lorenz systems by sliding mode control. Commun Comput Inf Sci 205:156–164
121. Vaidyanathan S, Sampath S (2012) Anti-synchronization of four-wing chaotic systems via sliding mode control. Int J Autom Comput 9(3):274–279
122. Vaidyanathan S, Volos C (2015) Analysis and adaptive control of a novel 3-D conservative no-equilibrium chaotic system. Arch Control Sci 25(3):333–353

123. Vaidyanathan S, Volos C, Pham VT (2014) Hyperchaos, adaptive control and synchronization of a novel 5-D hyperchaotic system with three positive Lyapunov exponents and its SPICE implementation. Arch Control Sci 24(4):409–446
124. Vaidyanathan S, Volos C, Pham VT, Madhavan K, Idowu BA (2014) Adaptive backstepping control, synchronization and circuit simulation of a 3-D novel jerk chaotic system with two hyperbolic sinusoidal nonlinearities. Arch Control Sci 24(3):375–403
125. Vaidyanathan S, Idowu BA, Azar AT (2015) Backstepping controller design for the global chaos synchronization of Sprott's jerk systems. Studies in computational intelligence, vol 581, pp 39–58
126. Vaidyanathan S, Rajagopal K, Volos CK, Kyprianidis IM, Stouboulos IN (2015) Analysis, adaptive control and synchronization of a seven-term novel 3-D chaotic system with three quadratic nonlinearities and its digital implementation in LabVIEW. J Eng Sci Technol Rev 8(2):130–141
127. Vaidyanathan S, Volos C, Pham VT, Madhavan K (2015) Analysis, adaptive control and synchronization of a novel 4-D hyperchaotic hyperjerk system and its SPICE implementation. Arch Control Sci 25(1):5–28
128. Vaidyanathan S, Volos CK, Kyprianidis IM, Stouboulos IN, Pham VT (2015) Analysis, adaptive control and anti-synchronization of a six-term novel jerk chaotic system with two exponential nonlinearities and its circuit simulation. J Eng Sci Technol Rev 8(2):24–36
129. Vaidyanathan S, Volos CK, Pham VT (2015) Analysis, adaptive control and adaptive synchronization of a nine-term novel 3-D chaotic system with four quadratic nonlinearities and its circuit simulation. J Eng Sci Technol Rev 8(2):174–184
130. Vaidyanathan S, Volos CK, Pham VT (2015) Global chaos control of a novel nine-term chaotic system via sliding mode control. In: Azar AT, Zhu Q (eds) Advances and applications in sliding mode control systems. Studies in computational intelligence, vol 576. Springer, Berlin, pp 571–590
131. Volos CK, Kyprianidis IM, Stouboulos IN (2013) Experimental investigation on coverage performance of a chaotic autonomous mobile robot. Robot Auton Syst 61(12):1314–1322
132. Volos CK, Kyprianidis IM, Stouboulos IN, Tlelo-Cuautle E, Vaidyanathan S (2015) Memristor: A new concept in synchronization of coupled neuromorphic circuits. J Eng Sci Technol Rev 8(2):157–173
133. Wei Z, Yang Q (2010) Anti-control of Hopf bifurcation in the new chaotic system with two stable node-foci. Appl Math Comput 217(1):422–429
134. Witte CL, Witte MH (1991) Chaos and predicting varix hemorrhage. Med Hypotheses 36(4):312–317
135. Xiao X, Zhou L, Zhang Z (2014) Synchronization of chaotic Lur'e systems with quantized sampled-data controller. Commun Nonlinear Sci Numer Simul 19(6):2039–2047
136. Yuan G, Zhang X, Wang Z (2014) Generation and synchronization of feedback-induced chaos in semiconductor ring lasers by injection-locking. Optik: Int J Light Electron Optics 125(8): 1950–1953
137. Zaher AA, Abu-Rezq A (2011) On the design of chaos-based secure communication systems. Commun Nonlinear Syst Numer Simul 16(9):3721–3727
138. Zhang H, Zhou J (2012) Synchronization of sampled-data coupled harmonic oscillators with control inputs missing. Syst Control Lett 61(12):1277–1285
139. Zhang X, Zhao Z, Wang J (2014) Chaotic image encryption based on circular substitution box and key stream buffer. Sig Process Image Commun 29(8):902–913
140. Zhou W, Xu Y, Lu H, Pan L (2008) On dynamics analysis of a new chaotic attractor. Phys Lett A 372(36):5773–5777
141. Zhu C, Liu Y, Guo Y (2010) Theoretic and numerical study of a new chaotic system. Intell Inf Manag 2:104–109
142. Zhu Q, Azar AT (2015) Complex system modelling and control through intelligent soft computations. Studies in fuzzines and soft computing, vol 319. Springer, Berlin

Qualitative Study and Adaptive Control of a Novel 4-D Hyperchaotic System with Three Quadratic Nonlinearities

Sundarapandian Vaidyanathan and Ahmad Taher Azar

Abstract In this work, we announce an eleven-term novel 4-D hyperchaotic system with three quadratic nonlinearities. The phase portraits of the eleven-term novel hyperchaotic system are depicted and the qualitative properties of the novel hyperchaotic system are discussed. The novel hyperchaotic system has a unique equilibrium at the origin, which is a saddle point. The Lyapunov exponents of the novel hyperchaotic system are obtained as $L_1 = 2.0836$, $L_2 = 0.1707$, $L_3 = 0$ and $L_4 = -26.6499$. The maximal Lyapunov exponent of the novel hyperchaotic system is found as $L_1 = 2.0836$. Also, the Kaplan-Yorke dimension of the novel hyperchaotic system is derived as $D_{KY} = 3.0846$. Since the sum of the Lyapunov exponents is negative, the novel hyperchaotic system is dissipative. Next, an adaptive controller is designed to globally stabilize the novel hyperchaotic system with unknown parameters. Finally, an adaptive controller is also designed to achieve global chaos synchronization of the identical hyperchaotic systems with unknown parameters. MATLAB simulations are depicted to illustrate all the main results derived in this work.

Keywords Chaos · Chaotic systems · Hyperchaos · Hyperchaotic systems · Adaptive control · Synchronization

1 Introduction

In the last few decades, Chaos theory has become a very important and active research field, employing many applications in different disciplines like physics, chemistry, biology, ecology, engineering and economics, among others.

S. Vaidyanathan (✉)
Research and Development Centre, Vel Tech University,
Avadi, Chennai 600062, Tamil Nadu, India
e-mail: sundarvtu@gmail.com

A.T. Azar
Faculty of Computers and Information, Benha University, Banha, Egypt
e-mail: ahmad_t_azar@ieee.org; ahmad.azar@fci.bu.edu.eg

A.T. Azar and S. Vaidyanathan (eds.), *Advances in Chaos Theory and Intelligent Control*, Studies in Fuzziness and Soft Computing 337,
DOI 10.1007/978-3-319-30340-6_8

Some classical paradigms of 3-D chaotic systems in the literature are Lorenz system [1], Rössler system [2], ACT system [3], Sprott systems [4], Chen system [5], Lü system [6], Cai system [7], Tigan system [8], etc.

Many new chaotic systems have been discovered in the recent years such as Zhou system [9], Zhu system [10], Li system [11], Wei-Yang system [12], Sundarapandian systems [13, 14], Vaidyanathan systems [15–31], Pehlivan system [32], Sampath system [33], Pham system [34], etc.

The Lyapunov exponent is a measure of the divergence of phase points that are initially very close and can be used to quantify chaotic systems. It is common to refer to the largest Lyapunov exponent as the *Maximal Lyapunov Exponent* (MLE).

A hyperchaotic system is defined as a chaotic system with at least two positive Lyapunov exponents. Thus, the dynamics of a hyperchaotic system can expand in several different directions simultaneously. Thus, the hyperchaotic systems have more complex dynamical behaviour and they have miscellaneous applications in engineering such as secure communications [35–37], cryptosystems [38–40], fuzzy logic [41, 42], electrical circuits [43, 44], etc.

The minimum dimension of an autonomous, continuous-time, hyperchaotic system is four. The first 4-D hyperchaotic system was found by Rössler [45]. Many hyperchaotic systems have been reported in the chaos literature such as hyperchaotic Lorenz system [46], hyperchaotic Lü system [47], hyperchaotic Chen system [48], hyperchaotic Wang system [49], hyperchaotic Vaidyanathan systems [50–58], hyperchaotic Pham system [59], etc.

Chaos theory and control systems have many important applications in science and engineering [60–65]. Some commonly known applications are oscillators [66, 67], chemical reactions [68–75], biology [76–85], ecology [86, 87], encryption [88, 89], cryptosystems [90, 91], mechanical systems [92–96], secure communications [97–99], robotics [100–102], cardiology [103, 104], intelligent control [105, 106], neural networks [107–109], memristors [110, 111], etc.

The control of a chaotic or hyperchaotic system aims to stabilize or regulate the system with the help of a feedback control. There are many methods available for controlling a chaotic system such as active control [112–114], adaptive control [115–117], sliding mode control [118, 119], backstepping control [120], etc.

The synchronization of chaotic systems aims to synchronize the states of master and slave systems asymptotically with time. There are many methods available for chaos synchronization such as active control [121–125], adaptive control [126–132], sliding mode control [133–136], backstepping control [137–140], etc.

In this research work, we announce an eleven-term novel 4-D hyperchaotic system with three quadratic nonlinearities. We have also designed adaptive controllers for stabilization and synchronization of the novel hyperchaotic systems when the system parameters are unknown.

This work is organized as follows. Section 2 describes the dynamic equations and phase portraits of the eleven-term novel 4-D hyperchaotic system. Section 3 details the qualitative properties of the novel hyperchaotic system. The Lyapunov exponents of the novel hyperchaotic system are obtained as $L_1 = 2.0836, L_2 = 0.1707, L_3 = 0$ and $L_4 = -26.6449$, while the Kaplan-Yorke dimension of the novel hyperchaotic system is obtained as $D_{KY} = 3.0846$.

In Sect. 4, we design an adaptive controller to globally stabilize the novel hyperchaotic system with unknown parameters. In Sect. 5, an adaptive controller is designed to achieve global chaos synchronization of the identical novel hyperchaotic systems with unknown parameters. Section 6 summarizes the main results.

2 A Novel 4-D Hyperchaotic System

In this section, we describe an eleven-term novel hyperchaotic system, which is given by the 4-D dynamics

$$\begin{cases} \dot{x}_1 = a(x_2 - x_1) + x_2 x_3 - x_4 \\ \dot{x}_2 = b x_2 - x_1 x_3 + x_4 \\ \dot{x}_3 = x_1 x_2 - c x_3 \\ \dot{x}_4 = -p(x_1 + x_2) \end{cases} \tag{1}$$

where x_1, x_2, x_3, x_4 are the states and a, b, c, p are constant positive parameters.

The system (1) exhibits a *strange hyperchaotic attractor* for the parameter values

$$a = 40, \quad b = 20.5, \quad c = 5, \quad p = 2.5 \tag{2}$$

For numerical simulations, we take the initial conditions as

$$x_1(0) = 0.5, \quad x_2(0) = 0.8, \quad x_3(0) = 0.6, \quad x_4(0) = 0.2 \tag{3}$$

Figures 1, 2, 3 and 4 show the 3-D projection of the novel hyperchaotic system (1) on the (x_1, x_2, x_3), (x_1, x_2, x_4), (x_1, x_3, x_4) and (x_2, x_3, x_4) spaces, respectively. Figure 1 shows that the the 3-D projection of the strange attractor on the (x_1, x_2, x_3) space of the novel hyperchaotic system (1) has the shape of a *two-scroll* attractor or a *butterfly* attractor.

3 Analysis of the Novel 4-D Hyperchaotic System

In this section, we study the qualitative properties of the novel 4-D hyperchaotic system (1). We take the parameter values as in the hyperchaotic case (2).

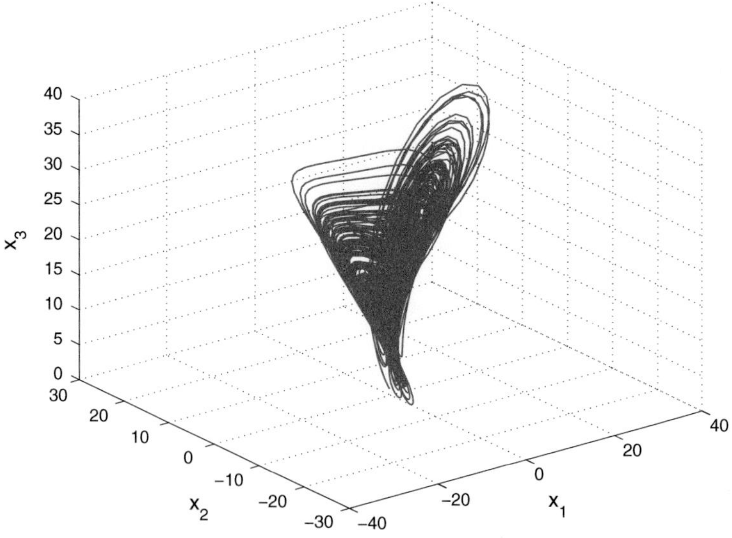

Fig. 1 3-D projection of the novel hyperchaotic system on the (x_1, x_2, x_3) space

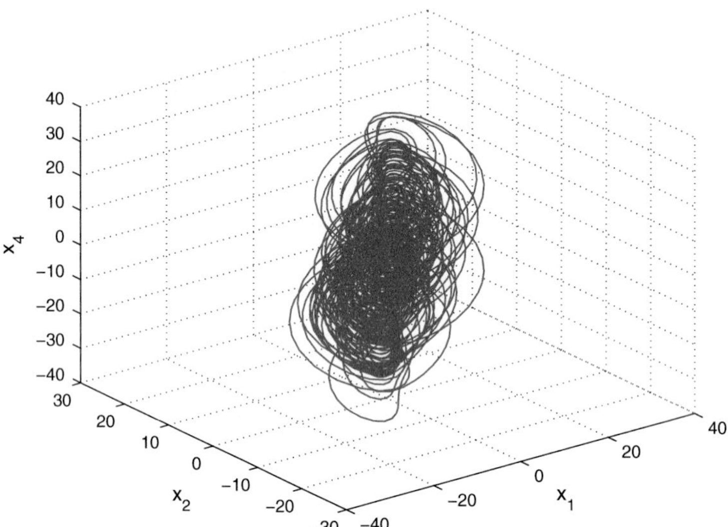

Fig. 2 3-D projection of the novel hyperchaotic system on the (x_1, x_2, x_4) space

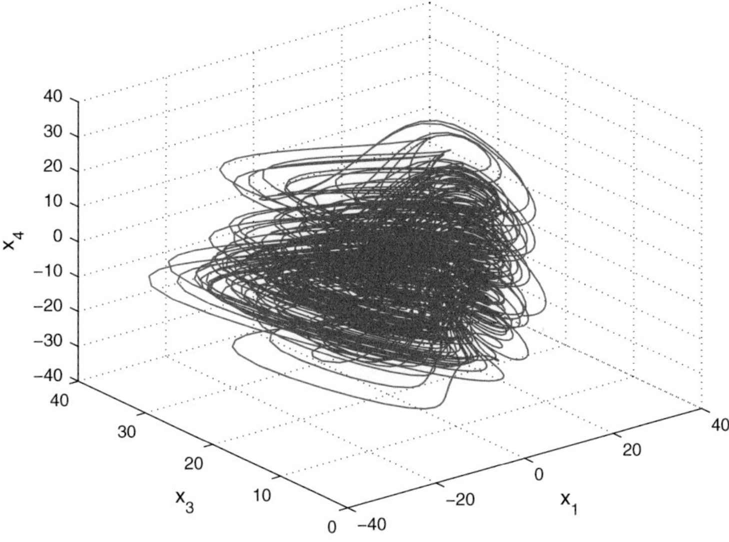

Fig. 3 3-D projection of the novel hyperchaotic system on the (x_1, x_3, x_4) space

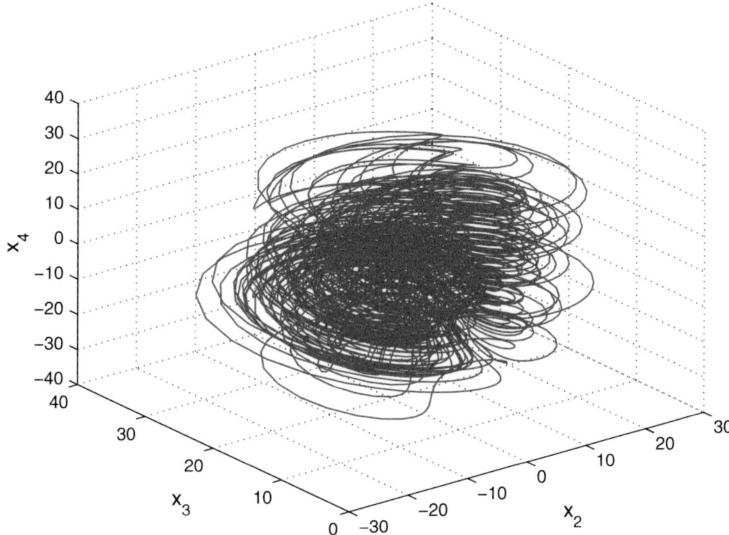

Fig. 4 3-D projection of the novel hyperchaotic system on the (x_2, x_3, x_4) space

3.1 Dissipativity

In vector notation, the novel hyperchaotic system (1) can be expressed as

$$\dot{\mathbf{x}} = f(\mathbf{x}) = \begin{bmatrix} f_1(x_1, x_2, x_3, x_4) \\ f_2(x_1, x_2, x_3, x_4) \\ f_3(x_1, x_2, x_3, x_4) \\ f_4(x_1, x_2, x_3, x_4) \end{bmatrix}, \tag{4}$$

where

$$\begin{cases} f_1(x_1, x_2, x_3, x_4) = a(x_2 - x_1) + x_2 x_3 - x_4 \\ f_2(x_1, x_2, x_3, x_4) = b x_2 - x_1 x_3 + x_4 \\ f_3(x_1, x_2, x_3, x_4) = x_1 x_2 - c x_3 \\ f_4(x_1, x_2, x_3, x_4) = -p(x_1 + x_2) \end{cases} \tag{5}$$

Let Ω be any region in $\mathbf{R}^4$ with a smooth boundary and also, $\Omega(t) = \Phi_t(\Omega)$, where Φ_t is the flow of f.

Furthermore, let $V(t)$ denote the hypervolume of $\Omega(t)$.

By Liouville's theorem, we know that

$$\dot{V}(t) = \int\limits_{\Omega(t)} (\nabla \cdot f)\, dx_1\, dx_2\, dx_3\, dx_4 \tag{6}$$

The divergence of the novel hyperchaotic system (4) is found as:

$$\nabla \cdot f = \frac{\partial f_1}{\partial x_1} + \frac{\partial f_2}{\partial x_2} + \frac{\partial f_3}{\partial x_3} + \frac{\partial f_4}{\partial x_4} = -(a - b + c) = -\mu < 0 \tag{7}$$

since $\mu = a - b + c = 40 - 20.5 + 5 = 24.5 > 0$.

Inserting the value of $\nabla \cdot f$ from (7) into (6), we get

$$\dot{V}(t) = \int\limits_{\Omega(t)} (-\mu)\, dx_1\, dx_2\, dx_3\, dx_4 = -\mu V(t) \tag{8}$$

Integrating the first order linear differential equation (8), we get

$$V(t) = \exp(-\mu t) V(0) \tag{9}$$

Since $\mu > 0$, it follows from Eq. (9) that $V(t) \to 0$ exponentially as $t \to \infty$. This shows that the novel hyperchaotic system (1) is dissipative. Hence, the system limit sets are ultimately confined into a specific limit set of zero hypervolume, and the asymptotic motion of the novel hyperchaotic system (1) settles onto a strange attractor of the system.

3.2 Equilibrium Points

We take the parameter values as in the hyperchaotic case (2).

It is easy to see that the system (1) has a unique equilibrium at the origin.

To test the stability type of the equilibrium point $E_0 = \mathbf{0}$, we calculate the Jacobian matrix of the novel hyperchaotic system (1) at $\mathbf{x} = \mathbf{0}$:

We find that

$$J \triangleq J(E_0) = \begin{bmatrix} -40 & 40 & 0 & -1 \\ 0 & 20.5 & 0 & 1 \\ 0 & 0 & -5 & 0 \\ -2.5 & -2.5 & 0 & 0 \end{bmatrix} \tag{10}$$

The matrix J has the eigenvalues

$$\lambda_1 = -40.1034, \;\; \lambda_2 = -5, \;\; \lambda_3 = 0.3087, \;\; \lambda_4 = 20.2947 \tag{11}$$

This shows that the equilibrium point $E_0 = \mathbf{0}$ is a saddle-point, which is unstable.

3.3 Rotation Symmetry About the x_3-axis

It is easy to see that the novel 4-D hyperchaotic system (1) is invariant under the change of coordinates

$$(x_1, x_2, x_3, x_4) \mapsto (-x_1, -x_2, x_3, -x_4) \tag{12}$$

Since the transformation (12) persists for all values of the system parameters, it follows that the novel 4-D hyperchaotic system (1) has rotation symmetry about the x_3-axis and that any non-trivial trajectory must have a twin trajectory.

3.4 Invariance

It is easy to see that the x_3-axis is invariant under the flow of the 4-D novel hyperchaotic system (1).

The invariant motion along the x_3-axis is characterized by the scalar dynamics

$$\dot{x}_3 = -cx_3, \quad (c > 0) \tag{13}$$

which is globally exponentially stable.

3.5 Lyapunov Exponents and Kaplan-Yorke Dimension

We take the parameter values of the novel system (1) as in the hyperchaotic case (2), *i.e.*

$$a = 40, \quad b = 20.5, \quad c = 5, \quad p = 2.5 \tag{14}$$

We take the initial state of the novel system (1) as given in (3), i.e.

$$x_1(0) = 0.5, \quad x_2(0) = 0.8, \quad x_3(0) = 0.6, \quad x_4(0) = 0.2 \tag{15}$$

Then the Lyapunov exponents of the system (1) are numerically obtained using MATLAB as

$$L_1 = 2.0836, \quad L_2 = 0.1707, \quad L_3 = 0, \quad L_4 = -26.6449 \tag{16}$$

Since there are two positive Lyapunov exponents in (16), the novel system (1) exhibits *hyperchaotic* behavior.

The maximal Lyapunov exponent of the novel hyperchaotic system (1) is easily seen as $L_1 = 2.0836$.

Since $L_1 + L_2 + L_3 + L_4 = -24.3906 < 0$, it follows that the novel hyperchaotic system (1) is dissipative.

Also, the Kaplan-Yorke dimension of the novel hyperchaotic system (1) is calculated as

$$D_{KY} = 3 + \frac{L_1 + L_2 + L_3}{|L_4|} = 3.0846, \tag{17}$$

which is fractional.

4 Adaptive Control of the Novel Hyperchaotic System

In this section, we apply adaptive control method to derive an adaptive feedback control law for globally stabilizing the novel 4-D hyperchaotic system with unknown parameters. The main control result in this section is established using Lyapunov stability theory.

Thus, we consider the controlled novel 4-D hyperchaotic system given by

$$\begin{cases} \dot{x}_1 = a(x_2 - x_1) + x_2 x_3 - x_4 + u_1 \\ \dot{x}_2 = b x_2 - x_1 x_3 + x_4 + u_2 \\ \dot{x}_3 = x_1 x_2 - c x_3 + u_3 \\ \dot{x}_4 = -p(x_1 + x_2) + u_4 \end{cases} \tag{18}$$

In (18), x_1, x_2, x_3, x_4 are the states and u_1, u_2, u_3, u_4 are the adaptive controls to be determined using estimates $\hat{a}(t), \hat{b}(t), \hat{c}(t), \hat{p}(t)$ for the unknown parameters a, b, c, p, respectively.

We consider the adaptive feedback control law

$$\begin{cases} u_1 = -\hat{a}(t)(x_2 - x_1) - x_2 x_3 + x_4 - k_1 x_1 \\ u_2 = -\hat{b}(t)x_2 + x_1 x_3 - x_4 - k_2 x_2 \\ u_3 = -x_1 x_2 + \hat{c}(t)x_3 - k_3 x_3 \\ u_4 = \hat{p}(t)(x_1 + x_2) - k_4 x_4 \end{cases} \tag{19}$$

where k_1, k_2, k_3, k_4 are positive gain constants.

Substituting (19) into (18), we get the closed-loop plant dynamics as

$$\begin{cases} \dot{x}_1 = [a - \hat{a}(t)](x_2 - x_1) - k_1 x_1 \\ \dot{x}_2 = [b - \hat{b}(t)]x_2 - k_2 x_2 \\ \dot{x}_3 = -[c - \hat{c}(t)]x_3 - k_3 x_3 \\ \dot{x}_4 = -[p - \hat{p}(t)](x_1 + x_2) - k_4 x_4 \end{cases} \tag{20}$$

The parameter estimation errors are defined as

$$\begin{cases} e_a(t) = a - \hat{a}(t) \\ e_b(t) = b - \hat{b}(t) \\ e_c(t) = c - \hat{c}(t) \\ e_p(t) = p - \hat{p}(t) \end{cases} \tag{21}$$

In view of (21), we can simplify the plant dynamics (20) as

$$\begin{cases} \dot{x}_1 = e_a(x_2 - x_1) - k_1 x_1 \\ \dot{x}_2 = e_b x_2 - k_2 x_2 \\ \dot{x}_3 = -e_c x_3 - k_3 x_3 \\ \dot{x}_4 = -e_p(x_1 + x_2) - k_4 x_4 \end{cases} \tag{22}$$

Differentiating (21) with respect to t, we obtain

$$\begin{cases} \dot{e}_a(t) = -\dot{\hat{a}}(t) \\ \dot{e}_b(t) = -\dot{\hat{b}}(t) \\ \dot{e}_c(t) = -\dot{\hat{c}}(t) \\ \dot{e}_p(t) = -\dot{\hat{p}}(t) \end{cases} \tag{23}$$

We consider the quadratic candidate Lyapunov function defined by

$$V(\mathbf{x}, e_a, e_b, e_c, e_p) = \frac{1}{2}\left(x_1^2 + x_2^2 + x_3^2 + x_4^2\right) + \frac{1}{2}\left(e_a^2 + e_b^2 + e_c^2 + e_p^2\right) \tag{24}$$

Differentiating V along the trajectories of (22) and (23), we obtain

$$\begin{aligned} \dot{V} &= -k_1x_1^2 - k_2x_2^2 - k_3x_3^2 - k_4x_4^2 + e_a\left[x_1(x_2 - x_1) - \dot{\hat{a}}\right] \\ &\quad + e_b\left[x_2^2 - \dot{\hat{b}}\right] + e_c\left[-x_3^2 - \dot{\hat{c}}\right] + e_p\left[-x_4(x_1 + x_2) - \dot{\hat{p}}\right] \end{aligned} \tag{25}$$

In view of (25), we take the parameter update law as

$$\begin{cases} \dot{\hat{a}}(t) = x_1(x_2 - x_1) \\ \dot{\hat{b}}(t) = x_2^2 \\ \dot{\hat{c}}(t) = -x_3^2 \\ \dot{\hat{p}}(t) = -x_4(x_1 + x_2) \end{cases} \tag{26}$$

Next, we state and prove the main result of this section.

Theorem 1 *The novel 4-D hyperchaotic system (18) with unknown system parameters is globally and exponentially stabilized for all initial conditions by the adaptive control law (19) and the parameter update law (26), where k_1, k_2, k_3, k_4 are positive gain constants.*

Proof We prove this result by applying Lyapunov stability theory [141].

We consider the quadratic Lyapunov function defined by (24), which is clearly a positive definite function on $\mathbf{R}^8$.

By substituting the parameter update law (26) into (25), we obtain the time-derivative of V as

$$\dot{V} = -k_1x_1^2 - k_2x_2^2 - k_3x_3^2 - k_4x_4^2 \tag{27}$$

From (27), it is clear that $\dot{V}$ is a negative semi-definite function on $\mathbf{R}^8$.

Thus, we can conclude that the state vector $\mathbf{x}(t)$ and the parameter estimation error are globally bounded, i.e.

$$\left[x_1(t)\ x_2(t)\ x_3(t)\ x_4(t)\ e_a(t)\ e_b(t)\ e_c(t)\ e_p(t)\right]^T \in \mathbf{L}_\infty.$$

We define $k = \min\{k_1, k_2, k_3, k_4\}$.

Then it follows from (27) that

$$\dot{V} \le -k\|\mathbf{x}(t)\|^2 \tag{28}$$

Thus, we have

$$k\|\mathbf{x}(t)\|^2 \leq -\dot{V} \tag{29}$$

Integrating the inequality (29) from 0 to t, we get

$$k \int_0^t \|\mathbf{x}(\tau)\|^2 \, d\tau \;\leq\; V(0) - V(t) \tag{30}$$

From (30), it follows that $\mathbf{x} \in \mathbf{L}_2$.

Using (22), we can conclude that $\dot{\mathbf{x}} \in \mathbf{L}_\infty$.

Using Barbalat's lemma [141], we conclude that $\mathbf{x}(t) \to 0$ exponentially as $t \to \infty$ for all initial conditions $\mathbf{x}(0) \in \mathbf{R}^4$.

Thus, the novel 4-D hyperchaotic system (18) with unknown system parameters is globally and exponentially stabilized for all initial conditions by the adaptive control law (19) and the parameter update law (26).

This completes the proof. ■

For the numerical simulations, the classical fourth-order Runge-Kutta method with step size $h = 10^{-8}$ is used to solve the systems (18) and (26), when the adaptive control law (19) is applied.

The parameter values of the novel 4-D hyperchaotic system (18) are taken as in the hyperchaotic case (2), i.e.

$$a = 40, \quad b = 20.5, \quad c = 5, \quad p = 2.5 \tag{31}$$

We take the positive gain constants as

$$k_1 = 5, \;\; k_2 = 5, \;\; k_3 = 5, \;\; k_4 = 5 \tag{32}$$

Furthermore, as initial conditions of the novel 4-D hyperchaotic system (18), we take

$$x_1(0) = 18.2, \;\; x_2(0) = 16.7, \;\; x_3(0) = -19.3, \;\; x_4(0) = -20.9 \tag{33}$$

Also, as initial conditions of the parameter estimates, we take

$$\hat{a}(0) = 4.2, \;\; \hat{b}(0) = 10.3, \;\; \hat{c}(0) = 3.7, \;\; \hat{p}(0) = 25.8 \tag{34}$$

In Fig. 5, the exponential convergence of the controlled states of the novel 4-D hyperchaotic system (18) is shown.

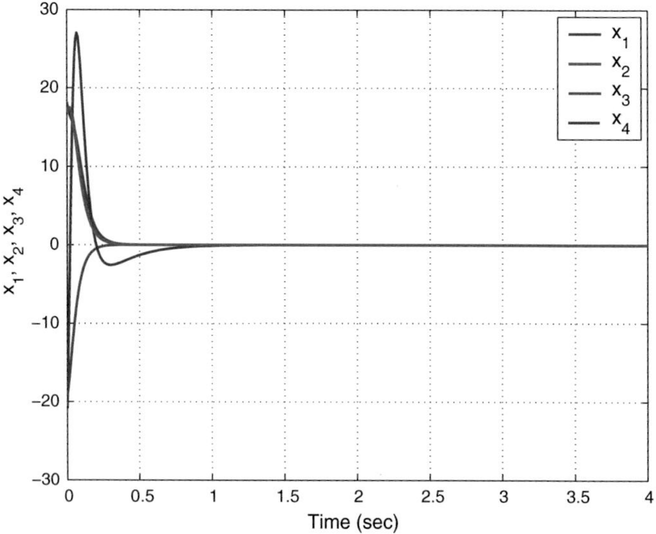

Fig. 5 Time-history of the controlled states x_1, x_2, x_3, x_4

5 Adaptive Synchronization of the Identical Novel Hyperchaotic Systems

In this section, we use adaptive control method to derive an adaptive feedback control law for globally synchronizing identical novel 4-D hyperchaotic systems with unknown parameters.

As the master system, we consider the novel 4-D hyperchaotic system given by

$$\begin{cases} \dot{x}_1 = a(x_2 - x_1) + x_2 x_3 - x_4 \\ \dot{x}_2 = b x_2 - x_1 x_3 + x_4 \\ \dot{x}_3 = x_1 x_2 - c x_3 \\ \dot{x}_4 = -p(x_1 + x_2) \end{cases} \tag{35}$$

In (35), x_1, x_2, x_3, x_4 are the states and a, b, c, p are unknown system parameters. As the slave system, we consider the 4-D novel hyperchaotic system given by

$$\begin{cases} \dot{y}_1 = a(y_2 - y_1) + y_2 y_3 - y_4 + u_1 \\ \dot{y}_2 = b y_2 - y_1 y_3 + y_4 + u_2 \\ \dot{y}_3 = y_1 y_2 - c y_3 + u_3 \\ \dot{y}_4 = -p(y_1 + y_2) + u_4 \end{cases} \tag{36}$$

In (36), y_1, y_2, y_3, y_4 are the states and u_1, u_2, u_3, u_4 are the adaptive controls to be determined using estimates $\hat{a}(t), \hat{b}(t), \hat{c}(t), \hat{p}(t)$ for the unknown parameters a, b, c, p, respectively.

The synchronization error between the novel hyperchaotic systems (35) and (36) is defined by

$$\begin{cases} e_1 = y_1 - x_1 \\ e_2 = y_2 - x_2 \\ e_3 = y_3 - x_3 \\ e_4 = y_4 - x_4 \end{cases} \tag{37}$$

Then the error dynamics is obtained as

$$\begin{cases} \dot{e}_1 = a(e_2 - e_1) - e_4 + y_2 y_3 - x_2 x_3 + u_1 \\ \dot{e}_2 = b e_2 + e_4 - y_1 y_3 + x_1 x_3 + u_2 \\ \dot{e}_3 = -c e_3 + y_1 y_2 - x_1 x_2 + u_3 \\ \dot{e}_4 = -p(e_1 + e_2) + u_4 \end{cases} \tag{38}$$

We consider the adaptive feedback control law

$$\begin{cases} u_1 = -\hat{a}(t)(e_2 - e_1) + e_4 - y_2 y_3 + x_2 x_3 - k_1 e_1 \\ u_2 = -\hat{b}(t) e_2 - e_4 + y_1 y_3 - x_1 x_3 - k_2 e_2 \\ u_3 = \hat{c}(t) e_3 - y_1 y_2 + x_1 x_2 - k_3 e_3 \\ u_4 = \hat{p}(t)(e_1 + e_2) - k_4 e_4 \end{cases} \tag{39}$$

where k_1, k_2, k_3, k_4 are positive gain constants.

Substituting (39) into (38), we get the closed-loop error dynamics as

$$\begin{cases} \dot{e}_1 = \left[a - \hat{a}(t)\right](e_2 - e_1) - k_1 e_1 \\ \dot{e}_2 = \left[b - \hat{b}(t)\right] e_2 - k_2 e_2 \\ \dot{e}_3 = -\left[c - \hat{c}(t)\right] e_3 - k_3 e_3 \\ \dot{e}_4 = -\left[p - \hat{p}(t)\right](e_1 + e_2) - k_4 e_4 \end{cases} \tag{40}$$

The parameter estimation errors are defined as

$$\begin{cases} e_a(t) = a - \hat{a}(t) \\ e_b(t) = b - \hat{b}(t) \\ e_c(t) = c - \hat{c}(t) \\ e_p(t) = p - \hat{p}(t) \end{cases} \tag{41}$$

In view of (41), we can simplify the error dynamics (40) as

$$\begin{cases} \dot{e}_1 = e_a(e_2 - e_1) - k_1 e_1 \\ \dot{e}_2 = e_b e_2 - k_2 e_2 \\ \dot{e}_3 = -e_c e_3 - k_3 e_3 \\ \dot{e}_4 = -e_p(e_1 + e_2) - k_4 e_4 \end{cases} \tag{42}$$

Differentiating (41) with respect to t, we obtain

$$\begin{cases} \dot{e}_a(t) = -\dot{\hat{a}}(t) \\ \dot{e}_b(t) = -\dot{\hat{b}}(t) \\ \dot{e}_c(t) = -\dot{\hat{c}}(t) \\ \dot{e}_p(t) = -\dot{\hat{p}}(t) \end{cases} \tag{43}$$

We use adaptive control theory to find an update law for the parameter estimates. We consider the quadratic candidate Lyapunov function defined by

$$V(\mathbf{e}, e_a, e_b, e_c, e_p) = \frac{1}{2}\left(e_1^2 + e_2^2 + e_3^2 + e_4^2\right) + \frac{1}{2}\left(e_a^2 + e_b^2 + e_c^2 + e_p^2\right) \tag{44}$$

Differentiating V along the trajectories of (42) and (43), we obtain

$$\begin{aligned} \dot{V} &= -k_1 e_1^2 - k_2 e_2^2 - k_3 e_3^2 - k_4 e_4^2 + e_a\left[e_1(e_2 - e_1) - \dot{\hat{a}}\right] \\ &\quad + e_b\left[e_2^2 - \dot{\hat{b}}\right] + e_c\left[-e_3^2 - \dot{\hat{c}}\right] + e_p\left[-e_4(e_1 + e_2) - \dot{\hat{p}}\right] \end{aligned} \tag{45}$$

In view of (45), we take the parameter update law as

$$\begin{cases} \dot{\hat{a}}(t) = e_1(e_2 - e_1) \\ \dot{\hat{b}}(t) = e_2^2 \\ \dot{\hat{c}}(t) = -e_3^2 \\ \dot{\hat{p}}(t) = -e_4(e_1 + e_2) \end{cases} \tag{46}$$

Next, we state and prove the main result of this section.

Theorem 2 *The novel hyperchaotic systems (35) and (36) with unknown system parameters are globally and exponentially synchronized for all initial conditions by the adaptive control law (39) and the parameter update law (46), where* k_1, k_2, k_3, k_4 *are positive gain constants.*

Proof We prove this result by applying Lyapunov stability theory [141].

We consider the quadratic Lyapunov function defined by (44), which is clearly a positive definite function on $\mathbf{R}^8$.

By substituting the parameter update law (46) into (45), we obtain

$$\dot{V} = -k_1 e_1^2 - k_2 e_2^2 - k_3 e_3^2 - k_4 e_4^2 \tag{47}$$

From (47), it is clear that $\dot{V}$ is a negative semi-definite function on $\mathbf{R}^8$.

Thus, we can conclude that the error vector $\mathbf{e}(t)$ and the parameter estimation error are globally bounded, i.e.

$$\left[\, e_1(t)\; e_2(t)\; e_3(t)\; e_4(t)\; e_a(t)\; e_b(t)\; e_c(t)\; e_p(t) \,\right]^T \in \mathbf{L}_\infty. \tag{48}$$

We define $k = \min\{k_1, k_2, k_3, k_4\}$.

Then it follows from (47) that

$$\dot{V} \leq -k \|\mathbf{e}(t)\|^2 \tag{49}$$

Thus, we have

$$k \|\mathbf{e}(t)\|^2 \leq -\dot{V} \tag{50}$$

Integrating the inequality (50) from 0 to t, we get

$$k \int_0^t \|\mathbf{e}(\tau)\|^2 \, d\tau \;\leq\; V(0) - V(t) \tag{51}$$

From (51), it follows that $\mathbf{e} \in \mathbf{L}_2$.

Using (42), we can conclude that $\dot{\mathbf{e}} \in \mathbf{L}_\infty$.

Using Barbalat's lemma [141], we conclude that $\mathbf{e}(t) \to 0$ exponentially as $t \to \infty$ for all initial conditions $\mathbf{e}(0) \in \mathbf{R}^4$.

This completes the proof. ■

For the numerical simulations, the classical fourth-order Runge-Kutta method with step size $h = 10^{-8}$ is used to solve the systems (35), (36) and (46), when the adaptive control law (39) is applied.

The parameter values of the novel hyperchaotic systems are taken as in the hyperchaotic case (2), i.e. $a = 40$, $b = 20.5$, $c = 5$ and $p = 2.5$.

We take the positive gain constants as $k_i = 5$ for $i = 1, \ldots, 4$.

Furthermore, as initial conditions of the master system (35), we take

$$x_1(0) = -6.3, \quad x_2(0) = 12.4, \quad x_3(0) = 15.7, \quad x_4(0) = -22.9 \tag{52}$$

As initial conditions of the slave system (36), we take

$$y_1(0) = 5.1, \quad y_2(0) = -18.5, \quad y_3(0) = 4.8, \quad y_4(0) = 3.4 \tag{53}$$

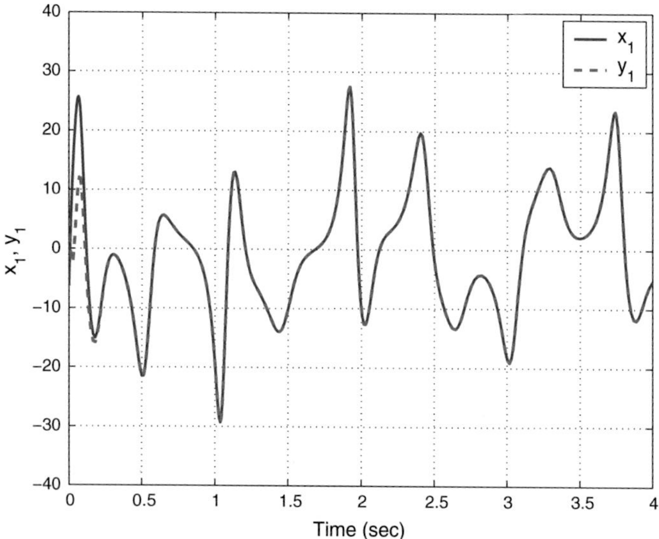

Fig. 6 Synchronization of the states x_1 and y_1

Also, as initial conditions of the parameter estimates, we take

$$\hat{a}(0) = 2.3, \quad \hat{b}(0) = 5.4, \quad \hat{c}(0) = 3.7, \quad \hat{p}(0) = 15.2 \tag{54}$$

Figures 6, 7, 8 and 9 describe the complete synchronization of the novel hyperchaotic systems (35) and (36), while Fig. 10 describes the time-history of the synchronization errors e_1, e_2, e_3, e_4.

6 Conclusions

In this work, we described an eleven-term novel 4-D hyperchaotic system with three quadratic nonlinearities. We discussed the qualitative properties of the novel hyperchaotic system in detail. The novel hyperchaotic system has a unique equilibrium at the origin, which is a saddle point and unstable. The Lyapunov exponents of the novel hyperchaotic system have been obtained as $L_1 = 2.0836, L_2 = 0.1707, L_3 = 0$ and $L_4 = -26.6499$. Also, the Kaplan-Yorke dimension of the novel hyperchaotic system has been derived as $D_{KY} = 3.0846$. Since the sum of the Lyapunov exponents is negative, the novel hyperchaotic system is dissipative. Next, an adaptive controller was designed to globally stabilize the novel hyperchaotic system with unknown parameters. Finally, an adaptive controller was also designed to achieve global chaos synchronization of the identical hyperchaotic systems with unknown parameters. MATLAB simulations are depicted to illustrate all the main results derived in this work.

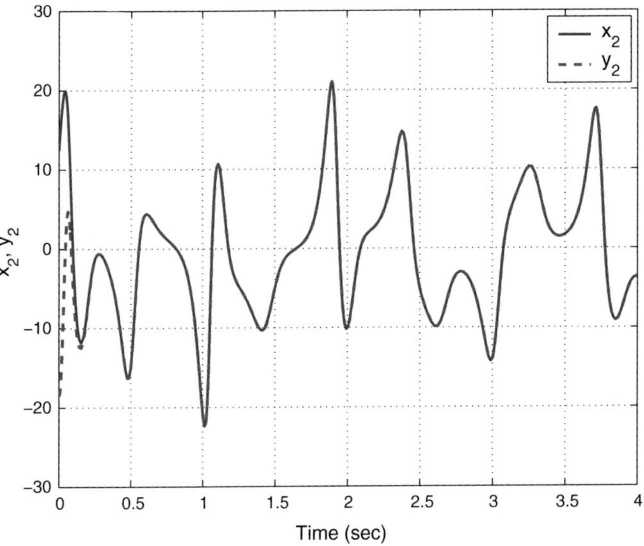

Fig. 7 Synchronization of the states x_2 and y_2

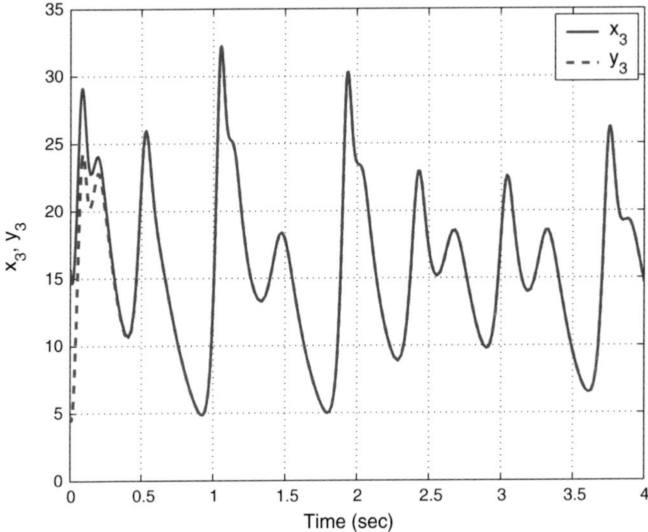

Fig. 8 Synchronization of the states x_3 and y_3

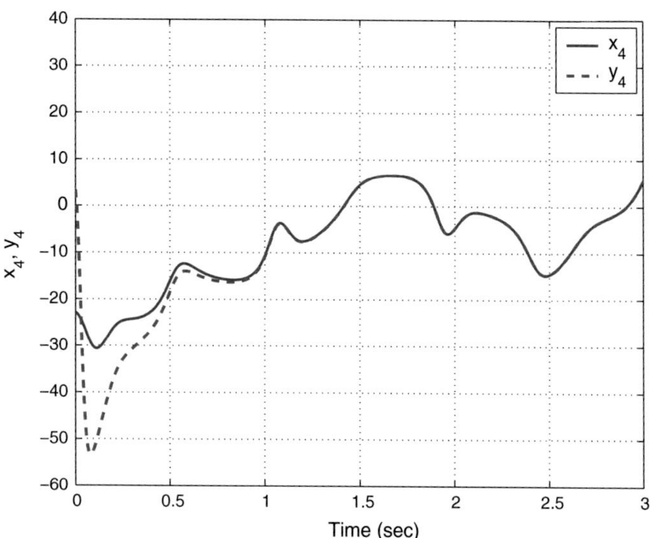

Fig. 9 Synchronization of the states x_4 and y_4

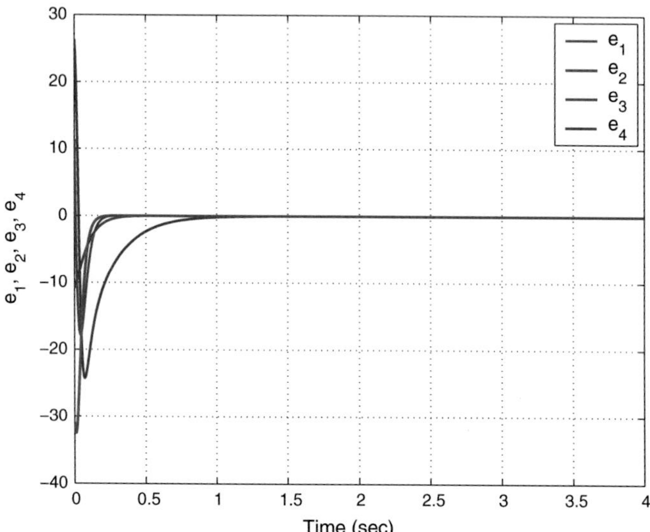

Fig. 10 Time-history of the synchronization errors e_1, e_2, e_3, e_4

References

1. Lorenz EN (1963) Deterministic periodic flow. J Atmos Sci 20(2):130–141
2. Rössler OE (1976) An equation for continuous chaos. Phys Lett A 57(5):397–398
3. Arneodo A, Coullet P, Tresser C (1981) Possible new strange attractors with spiral structure. Commun Math Phys 79(4):573–576
4. Sprott JC (1994) Some simple chaotic flows. Phys Rev E 50(2):647–650
5. Chen G, Ueta T (1999) Yet another chaotic attractor. Int J Bifurcat Chaos 9(7):1465–1466
6. Lü J, Chen G (2002) A new chaotic attractor coined. Int J Bifurcat Chaos 12(3):659–661
7. Cai G, Tan Z (2007) Chaos synchronization of a new chaotic system via nonlinear control. J Uncertain Syst 1(3):235–240
8. Tigan G, Opris D (2008) Analysis of a 3D chaotic system. Chaos Solitons Fractals 36: 1315–1319
9. Zhou W, Xu Y, Lu H, Pan L (2008) On dynamics analysis of a new chaotic attractor. Phys Lett A 372(36):5773–5777
10. Zhu C, Liu Y, Guo Y (2010) Theoretic and numerical study of a new chaotic system. Intell Inf Manage 2:104–109
11. Li D (2008) A three-scroll chaotic attractor. Phys Lett A 372(4):387–393
12. Wei Z, Yang Q (2010) Anti-control of Hopf bifurcation in the new chaotic system with two stable node-foci. Appl Math Comput 217(1):422–429
13. Sundarapandian V (2013) Analysis and anti-synchronization of a novel chaotic system via active and adaptive controllers. J Eng Sci Technol Rev 6(4):45–52
14. Sundarapandian V, Pehlivan I (2012) Analysis, control, synchronization, and circuit design of a novel chaotic system. Math Comput Model 55(7–8):1904–1915
15. Vaidyanathan S (2013) A new six-term 3-D chaotic system with an exponential nonlinearity. Far East J Math Sci 79(1):135–143
16. Vaidyanathan S (2013) Analysis and adaptive synchronization of two novel chaotic systems with hyperbolic sinusoidal and cosinusoidal nonlinearity and unknown parameters. J Eng Sci Technol Rev 6(4):53–65
17. Vaidyanathan S (2014) A new eight-term 3-D polynomial chaotic system with three quadratic nonlinearities. Far East J Math Sci 84(2):219–226
18. Vaidyanathan S (2014) Analysis and adaptive synchronization of eight-term 3-D polynomial chaotic systems with three quadratic nonlinearities. Eur Phys J 223(8):1519–1529
19. Vaidyanathan S (2014) Analysis, control and synchronisation of a six-term novel chaotic system with three quadratic nonlinearities. Int J Model, Ident Control 22(1):41–53
20. Vaidyanathan S (2014) Generalized projective synchronisation of novel 3-D chaotic systems with an exponential non-linearity via active and adaptive control. Int J Model, Ident Control 22(3):207–217
21. Vaidyanathan S (2015) A 3-D novel highly chaotic system with four quadratic nonlinearities, its adaptive control and anti-synchronization with unknown parameters. J Eng Sci Technol Rev 8(2):106–115
22. Vaidyanathan S (2015) Analysis, properties and control of an eight-term 3-D chaotic system with an exponential nonlinearity. Int J Model, Ident Control 23(2):164–172
23. Vaidyanathan S, Azar AT (2015) Analysis, control and synchronization of a nine-term 3-D novel chaotic system. In: Azar AT, Vaidyanathan S (eds) Chaos modelling and control systems design, studies in computational intelligence, vol 581. Springer, Germany, pp 19–38
24. Vaidyanathan S, Madhavan K (2013) Analysis, adaptive control and synchronization of a seven-term novel 3-D chaotic system. Int J Control Theor Appl 6(2):121–137
25. Vaidyanathan S, Pakiriswamy S (2015) A 3-D novel conservative chaotic system and its generalized projective synchronization via adaptive control. J Eng Sci Technol Rev 8(2): 52–60
26. Vaidyanathan S, Volos C (2015) Analysis and adaptive control of a novel 3-D conservative no-equilibrium chaotic system. Arch Control Sci 25(3):333–353

27. Vaidyanathan S, Volos C, Pham VT, Madhavan K, Idowu BA (2014) Adaptive backstepping control, synchronization and circuit simulation of a 3-D novel jerk chaotic system with two hyperbolic sinusoidal nonlinearities. Arch Control Sci 24(3):375–403
28. Vaidyanathan S, Rajagopal K, Volos CK, Kyprianidis IM, Stouboulos IN (2015) Analysis, adaptive control and synchronization of a seven-term novel 3-D chaotic system with three quadratic nonlinearities and its digital implementation in LabVIEW. J Eng Sci Technol Rev 8(2):130–141
29. Vaidyanathan S, Volos CK, Kyprianidis IM, Stouboulos IN, Pham VT (2015) Analysis, adaptive control and anti-synchronization of a six-term novel jerk chaotic system with two exponential nonlinearities and its circuit simulation. J Eng Sci Technol Rev 8(2):24–36
30. Vaidyanathan S, Volos CK, Pham VT (2015) Analysis, adaptive control and adaptive synchronization of a nine-term novel 3-D chaotic system with four quadratic nonlinearities and its circuit simulation. J Eng Sci Technol Rev 8(2):181–191
31. Vaidyanathan S, Volos CK, Pham VT (2015) Global chaos control of a novel nine-term chaotic system via sliding mode control. In: Azar AT, Zhu Q (eds) Advances and applications in sliding mode control systems, studies in computational intelligence, vol 576. Springer, Germany, pp 571–590
32. Pehlivan I, Moroz IM, Vaidyanathan S (2014) Analysis, synchronization and circuit design of a novel butterfly attractor. J Sound Vib 333(20):5077–5096
33. Sampath S, Vaidyanathan S, Volos CK, Pham VT (2015) An eight-term novel four-scroll chaotic System with cubic nonlinearity and its circuit simulation. J Eng Sci Technol Rev 8(2):1–6
34. Pham VT, Vaidyanathan S, Volos CK, Jafari S (2015) Hidden attractors in a chaotic system with an exponential nonlinear term. Eur Phys J 224(8):1507–1517
35. Filali RL, Benrejeb M, Borne P (2014) On observer-based secure communication design using discrete-time hyperchaotic systems. Commun Nonlinear Sci Numer Simul 19(5):1424–1432
36. Li C, Liao X, Wong KW (2005) Lag synchronization of hyperchaos with application to secure communications. Chaos Solitons Fractals 23(1):183–193
37. Wu X, Zhu C, Kan H (2015) An improved secure communication scheme based passive synchronization of hyperchaotic complex nonlinear system. Appl Math Comput 252: 201–214
38. Hammami S (2015) State feedback-based secure image cryptosystem using hyperchaotic synchronization. ISA Trans 54:52–59
39. Rhouma R, Belghith S (2008) Cryptanalysis of a new image encryption algorithm based on hyper-chaos. Phys Lett A 372(38):5973–5978
40. Zhu C (2012) A novel image encryption scheme based on improved hyperchaotic sequences. Opt Commun 285(1):29–37
41. Senouci A, Boukabou A (2014) Predictive control and synchronization of chaotic and hyperchaotic systems based on a T-S fuzzy model. Math Comput Simul 105:62–78
42. Zhang H, Liao X, Yu J (2005) Fuzzy modeling and synchronization of hyperchaotic systems. Chaos Solitons Fractals 26(3):835–843
43. Wei X, Yunfei F, Qiang L (2012) A novel four-wing hyper-chaotic system and its circuit implementation. Procedia Eng 29:1264–1269
44. Yujun N, Xingyuan W, Mingjun W, Huaguang Z (2010) A new hyperchaotic system and its circuit implementation. Commun Nonlinear Sci Numer Simul 15(11):3518–3524
45. Rössler OE (1979) An equation for hyperchaos. Phys Lett A 71:155–157
46. Jia Q (2007) Hyperchaos generated from the Lorenz chaotic system and its control. Phys Lett A 366:217–222
47. Chen A, Lu J, Lü J, Yu S (2006) Generating hyperchaotic Lü attractor via state feedback control. Phys A 364:103–110
48. Li X (2009) Modified projective synchronization of a new hyperchaotic system via nonlinear control. Commun Theoret Phys 52:274–278
49. Wang J, Chen Z (2008) A novel hyperchaotic system and its complex dynamics. Int J Bifurcat Chaos 18:3309–3324

50. Vaidyanathan S, VTP, Volos CK, (2015) A 5-D hyperchaotic Rikitake dynamo system with hidden attractors. Eur Phys J 224(8):1575–1592
51. Vaidyanathan S (2013) A ten-term novel 4-D hyperchaotic system with three quadratic nonlinearities and its control. Int J Control Theor Appl 6(2):97–109
52. Vaidyanathan S (2014) Qualitative analysis and control of an eleven-term novel 4-D hyperchaotic system with two quadratic nonlinearities. Int J Control Theor Appl 7:35–47
53. Vaidyanathan S (2015) Hyperchaos, qualitative analysis, control and synchronisation of a ten-term 4-D hyperchaotic system with an exponential nonlinearity and three quadratic nonlinearities. Int J Model, Ident Control 23(4):380–392
54. Vaidyanathan S, Azar AT (2015) Analysis and control of a 4-D novel hyperchaotic system. Stud Comput Intell 581:3–17
55. Vaidyanathan S, Volos C, Pham VT (2014) Hyperchaos, adaptive control and synchronization of a novel 5-D hyperchaotic system with three positive Lyapunov exponents and its SPICE implementation. Arch Control Sci 24(4):409–446
56. Vaidyanathan S, Azar AT, Rajagopal K, Alexander P (2015) Design and SPICE implementation of a 12-term novel hyperchaotic system and its synchronisation via active control. Int J Model, Ident Control 23(3):267–277
57. Vaidyanathan S, Volos C, Pham VT, Madhavan K (2015) Analysis, adaptive control and synchronization of a novel 4-D hyperchaotic hyperjerk system and its SPICE implementation. Arch Control Sci 25(1):135–158
58. Vaidyanathan S, Volos CK, Pham VT (2015) Analysis, control, synchronization and SPICE implementation of a novel 4-D hyperchaotic Rikitake dynamo system without equilibrium. J Eng Sci Technol Rev 8(2):232–244
59. Pham VT, Volos C, Jafari S, Wang X, Vaidyanathan S (2014) Hidden hyperchaotic attractor in a novel simple memristive neural network. Optoelectron Adv Mater, Rapid Commun 8(11–12):1157–1163
60. Azar AT (2010) Fuzzy systems. IN-TECH, Vienna, Austria
61. Azar AT, Vaidyanathan S (2015) Chaos modeling and control systems design, studies in computational intelligence, vol 581. Springer, Germany
62. Azar AT, Vaidyanathan S (2015) Computational intelligence applications in modeling and control, studies in computational intelligence, vol 575. Springer, Germany
63. Azar AT, Vaidyanathan S (2015) Handbook of research on advanced intelligent control engineering and automation. Advances in computational intelligence and robotics (ACIR), IGI-Global, USA
64. Azar AT, Zhu Q (2015) Advances and applications in sliding mode control systems, studies in computational intelligence, vol 576. Springer, Germany
65. Zhu Q, Azar AT (2015) Complex system modelling and control through intelligent soft computations, studies in fuzzines and soft computing, vol 319. Springer, Germany
66. Kengne J, Chedjou JC, Kenne G, Kyamakya K (2012) Dynamical properties and chaos synchronization of improved Colpitts oscillators. Commun Nonlinear Sci Numer Simul 17(7):2914–2923
67. Sharma A, Patidar V, Purohit G, Sud KK (2012) Effects on the bifurcation and chaos in forced Duffing oscillator due to nonlinear damping. Commun Nonlinear Sci Numer Simul 17(6):2254–2269
68. Gaspard P (1999) Microscopic chaos and chemical reactions. Physica A 263(1–4):315–328
69. Petrov V, Gaspar V, Masere J, Showalter K (1993) Controlling chaos in Belousov-Zhabotinsky reaction. Nature 361:240–243
70. Vaidyanathan S (2015) Adaptive control of a chemical chaotic reactor. Int J PharmTech Res 8(3):377–382
71. Vaidyanathan S (2015) Adaptive synchronization of chemical chaotic reactors. Int J ChemTech Res 8(2):612–621
72. Vaidyanathan S (2015) Anti-synchronization of Brusselator chemical reaction systems via adaptive control. Int J ChemTech Res 8(6):759–768

73. Vaidyanathan S (2015) Dynamics and control of Brusselator chemical reaction. Int J ChemTech Res 8(6):740–749
74. Vaidyanathan S (2015) Dynamics and control of Tokamak system with symmetric and magnetically confined plasma. Int J ChemTech Res 8(6):795–803
75. Vaidyanathan S (2015) Synchronization of Tokamak systems with symmetric and magnetically confined plasma via adaptive control. Int J ChemTech Res 8(6):818–827
76. Das S, Goswami D, Chatterjee S, Mukherjee S (2014) Stability and chaos analysis of a novel swarm dynamics with applications to multi-agent systems. Eng Appl Artif Intell 30:189–198
77. Kyriazis M (1991) Applications of chaos theory to the molecular biology of aging. Exp Gerontol 26(6):569–572
78. Vaidyanathan S (2015) 3-cells cellular neural network (CNN) attractor and its adaptive biological control. Int J PharmTech Res 8(4):632–640
79. Vaidyanathan S (2015) Adaptive backstepping control of enzymes-substrates system with ferroelectric behaviour in brain waves. Int J PharmTech Res 8(2):256–261
80. Vaidyanathan S (2015) Adaptive biological control of generalized Lotka-Volterra three-species biological system. Int J PharmTech Res 8(4):622–631
81. Vaidyanathan S (2015) Adaptive chaotic synchronization of enzymes-substrates system with ferroelectric behaviour in brain waves. Int J PharmTech Res 8(5):964–973
82. Vaidyanathan S (2015) Adaptive synchronization of generalized Lotka-Volterra three-species biological systems. Int J PharmTech Res 8(5):928–937
83. Vaidyanathan S (2015) Chaos in neurons and adaptive control of Birkhoff-Shaw strange chaotic attractor. Int J PharmTech Res 8(5):956–963
84. Vaidyanathan S (2015) Lotka-Volterra population biology models with negative feedback and their ecological monitoring. Int J PharmTech Res 8(5):974–981
85. Vaidyanathan S (2015) Synchronization of 3-cells cellular neural network (CNN) attractors via adaptive control method. Int J PharmTech Res 8(5):946–955
86. Gibson WT, Wilson WG (2013) Individual-based chaos: extensions of the discrete logistic model. J Theoret Biol 339:84–92
87. Suérez I (1999) Mastering chaos in ecology. Ecol Model 117(2–3):305–314
88. Lang J (2015) Color image encryption based on color blend and chaos permutation in the reality-preserving multiple-parameter fractional Fourier transform domain. Opt Commun 338:181–192
89. Zhang X, Zhao Z, Wang J (2014) Chaotic image encryption based on circular substitution box and key stream buffer. Sig Process Image Commun 29(8):902–913
90. Rhouma R, Belghith S (2011) Cryptoanalysis of a chaos based cryptosystem on DSP. Commun Nonlinear Sci Numer Simul 16(2):876–884
91. Usama M, Khan MK, Alghatbar K, Lee C (2010) Chaos-based secure satellite imagery cryptosystem. Comput Math Appl 60(2):326–337
92. Azar AT, Serrano FE (2014) Robust IMC-PID tuning for cascade control systems with gain and phase margin specifications. Neural Comput Appl 25(5):983–995
93. Azar AT, Serrano FE (2015) Adaptive sliding mode control of the Furuta pendulum. In: Azar AT, Zhu Q (eds) Advances and applications in sliding mode control systems, studies in computational intelligence, vol 576. Springer, Germany, pp 1–42
94. Azar AT, Serrano FE (2015) Deadbeat control for multivariable systems with time varying delays. In: Azar AT, Vaidyanathan S (eds) Chaos modeling and control systems design, studies in computational intelligence, vol 581. Springer, Germany, pp 97–132
95. Azar AT, Serrano FE (2015) Design and modeling of anti wind up PID controllers. In: Zhu Q, Azar AT (eds) Complex system modelling and control through intelligent soft computations, studies in fuzziness and soft computing, vol 319. Springer, Germany, pp 1–44
96. Azar AT, Serrano FE (2015) Stabilizatoin and control of mechanical systems with backlash. In: Azar AT, Vaidyanathan S (eds) Handbook of research on advanced intelligent control engineering and automation. Advances in computational intelligence and robotics (ACIR), IGI-Global, USA, pp 1–60

97. Feki M (2003) An adaptive chaos synchronization scheme applied to secure communication. Chaos Solitons Fractals 18(1):141–148
98. Murali K, Lakshmanan M (1998) Secure communication using a compound signal from generalized chaotic systems. Phys Lett A 241(6):303–310
99. Zaher AA, Abu-Rezq A (2011) On the design of chaos-based secure communication systems. Commun Nonlinear Sci Numer Simul 16(9):3721–3727
100. Mondal S, Mahanta C (2014) Adaptive second order terminal sliding mode controller for robotic manipulators. J Franklin Inst 351(4):2356–2377
101. Nehmzow U, Walker K (2005) Quantitative description of robot-environment interaction using chaos theory. Robot Auton Syst 53(3–4):177–193
102. Volos CK, Kyprianidis IM, Stouboulos IN (2013) Experimental investigation on coverage performance of a chaotic autonomous mobile robot. Robot Auton Syst 61(12):1314–1322
103. Qu Z (2011) Chaos in the genesis and maintenance of cardiac arrhythmias. Prog Biophys Mol Biol 105(3):247–257
104. Witte CL, Witte MH (1991) Chaos and predicting varix hemorrhage. Med Hypotheses 36(4):312–317
105. Azar AT (2012) Overview of type-2 fuzzy logic systems. Int J Fuzzy Syst Appl 2(4):1–28
106. Li Z, Chen G (2006) Integration of fuzzy logic and chaos theory, studies in fuzziness and soft computing, vol 187. Springer, Germany
107. Huang X, Zhao Z, Wang Z, Li Y (2012) Chaos and hyperchaos in fractional-order cellular neural networks. Neurocomputing 94:13–21
108. Kaslik E, Sivasundaram S (2012) Nonlinear dynamics and chaos in fractional-order neural networks. Neural Netw 32:245–256
109. Lian S, Chen X (2011) Traceable content protection based on chaos and neural networks. Appl Soft Comput 11(7):4293–4301
110. Pham VT, Volos CK, Vaidyanathan S, Le TP, Vu VY (2015) A memristor-based hyperchaotic system with hidden attractors: dynamics, synchronization and circuital emulating. J Eng Sci Technol Rev 8(2):205–214
111. Volos CK, Kyprianidis IM, Stouboulos IN, Tlelo-Cuautle E, Vaidyanathan S (2015) Memristor: a new concept in synchronization of coupled neuromorphic circuits. J Eng Sci Technol Rev 8(2):157–173
112. Sundarapandian V (2010) Output regulation of the Lorenz attractor. Far East J Math Sci 42(2):289–299
113. Vaidyanathan S (2011) Output regulation of Arneodo-Coullet chaotic system. Commun Comput Inf Sci 133:98–107
114. Vaidyanathan S (2011) Output regulation of the unified chaotic system. Commun Comput Inf Sci 198:1–9
115. Sundarapandian V (2013) Adaptive control and synchronization design for the Lu-Xiao chaotic system. Lect Notes Electr Eng 131:319–327
116. Vaidyanathan S (2012) Adaptive controller and syncrhonizer design for the Qi-Chen chaotic system. Lect Notes Inst Comput Sci, Social-Inform Telecommun Eng 84:73–82
117. Vaidyanathan S (2013) Analysis, control and synchronization of hyperchaotic Zhou system via adaptive control. Adv Intell Syst Comput 177:1–10
118. Vaidyanathan S (2012) Global chaos control of hyperchaotic Liu system via sliding control method. Int J Control Theor Appl 5(2):117–123
119. Vaidyanathan S (2012) Sliding mode control based global chaos control of Liu-Liu-Liu-Su chaotic system. Int J Control Theor Appl 5(1):15–20
120. Vaidyanathan S, Idowu BA, Azar AT (2015) Backstepping controller design for the global chaos synchronization of Sprott's jerk systems. Stud Comput Intell 581:39–58
121. Karthikeyan R, Sundarapandian V (2014) Hybrid chaos synchronization of four-scroll systems via active control. J Electr Eng 65(2):97–103
122. Sarasu P, Sundarapandian V (2011) Active controller design for generalized projective synchronization of four-scroll chaotic systems. Int J Syst Sig Control Eng Appl 4(2):26–33

123. Sarasu P, Sundarapandian V (2011) The generalized projective synchronization of hyperchaotic Lorenz and hyperchaotic Qi systems via active control. Int J Soft Comput 6(5): 216–223
124. Vaidyanathan S, Rajagopal K (2011) Hybrid synchronization of hyperchaotic Wang-Chen and hyperchaotic Lorenz systems by active non-linear control. Int J Syst Sig Control Eng Appl 4(3):55–61
125. Vaidyanathan S, Rasappan S (2011) Global chaos synchronization of hyperchaotic Bao and Xu systems by active nonlinear control. Commun Comput Inf Sci 198:10–17
126. Sarasu P, Sundarapandian V (2012) Generalized projective synchronization of two-scroll systems via adaptive control. Int J Soft Comput 7(4):146–156
127. Sundarapandian V, Karthikeyan R (2011) Anti-synchronization of hyperchaotic Lorenz and hyperchaotic Chen systems by adaptive control. Int J Syst Sig Control Eng Appl 4(2):18–25
128. Sundarapandian V, Karthikeyan R (2011) Anti-synchronization of Lü and Pan chaotic systems by adaptive nonlinear control. Eur J Sci Res 64(1):94–106
129. Sundarapandian V, Karthikeyan R (2012) Adaptive anti-synchronization of uncertain Tigan and Li systems. J Eng Appl Sci 7(1):45–52
130. Vaidyanathan S (2012) Anti-synchronization of Sprott-L and Sprott-M chaotic systems via adaptive control. Int J Control Theor Appl 5(1):41–59
131. Vaidyanathan S, Pakiriswamy S (2013) Generalized projective synchronization of six-term Sundarapandian chaotic systems by adaptive control. Int J Control Theor Appl 6(2):153–163
132. Vaidyanathan S, Rajagopal K (2012) Global chaos synchronization of hyperchaotic Pang and hyperchaotic Wang systems via adaptive control. Int J Soft Comput 7(1):28–37
133. Sundarapandian V, Sivaperumal S (2011) Sliding controller design of hybrid synchronization of four-wing chaotic systems. Int J Soft Comput 6(5):224–231
134. Vaidyanathan S (2014) Global chaos synchronisation of identical Li-Wu chaotic systems via sliding mode control. Int J Model, Ident Control 22(2):170–177
135. Vaidyanathan S, Sampath S (2012) Anti-synchronization of four-wing chaotic systems via sliding mode control. Int J Autom Comput 9(3):274–279
136. Vaidyanathan S, Sampath S, Azar AT (2015) Global chaos synchronisation of identical chaotic systems via novel sliding mode control method and its application to Zhu system. Int J Model, Ident Control 23(1):92–100
137. Rasappan S, Vaidyanathan S (2013) Hybrid synchronization of n-scroll Chua circuits using adaptive backstepping control design with recursive feedback. Malays J Math Sci 73(1):73–95
138. Rasappan S, Vaidyanathan S (2014) Global chaos synchronization of WINDMI and Coullet chaotic systems using adaptive backstepping control design. Kyungpook Math J 54(1): 293–320
139. Suresh R, Sundarapandian V (2013) Global chaos synchronization of a family of n-scroll hyperchaotic Chua circuits using backstepping control with recursive feedback. Far East J Math Sci 7(2):219–246
140. Vaidyanathan S, Rasappan S (2014) Global chaos synchronization of n-scroll Chua circuit and Lur'e system using backstepping control design with recursive feedback. Arab J Sci Eng 39(4):3351–3364
141. Khalil HK (2001) Nonlinear systems, 3rd edn. Prentice Hall, USA

A Novel 4-D Four-Wing Chaotic System with Four Quadratic Nonlinearities and Its Synchronization via Adaptive Control Method

Sundarapandian Vaidyanathan and Ahmad Taher Azar

Abstract In this research work, we describe a ten-term novel 4-D four-wing chaotic system with four quadratic nonlinearities. First, this work describes the qualitative analysis of the novel 4-D four-wing chaotic system. We show that the novel four-wing chaotic system has a unique equilibrium point at the origin, which is a saddle-point. Thus, origin is an unstable equilibrium of the novel chaotic system. We also show that the novel four-wing chaotic system has a rotation symmetry about the x_3 axis. Thus, it follows that every non-trivial trajectory of the novel four-wing chaotic system must have a twin trajectory. The Lyapunov exponents of the novel 4-D four-wing chaotic system are obtained as $L_1 = 5.6253$, $L_2 = 0$, $L_3 = -5.4212$ and $L_4 = -53.0373$. Thus, the maximal Lyapunov exponent of the novel four-wing chaotic system is obtained as $L_1 = 5.6253$. The large value of L_1 indicates that the novel four-wing system is highly chaotic. Since the sum of the Lyapunov exponents of the novel chaotic system is negative, it follows that the novel chaotic system is dissipative. Also, the Kaplan-Yorke dimension of the novel four-wing chaotic system is obtained as $D_{KY} = 3.0038$. Finally, this work describes the adaptive synchronization of the identical novel 4-D four-wing chaotic systems with unknown parameters. The adaptive synchronization result is proved using Lyapunov stability theory. MATLAB simulations are depicted to illustrate all the main results for the novel 4-D four-wing chaotic system.

Keywords Chaos · Chaotic systems · Four-wing systems · Adaptive control · Synchronization

S. Vaidyanathan (✉)
Research and Development Centre, Vel Tech University,
Avadi, Chennai 600062, Tamil Nadu, India
e-mail: sundarvtu@gmail.com

A.T. Azar
Faculty of Computers and Information, Benha University, Banha, Egypt
e-mail: ahmad_t_azar@ieee.org; ahmad.azar@fci.bu.edu.eg

A.T. Azar and S. Vaidyanathan (eds.), *Advances in Chaos Theory and Intelligent Control*, Studies in Fuzziness and Soft Computing 337,
DOI 10.1007/978-3-319-30340-6_9

1 Introduction

Chaotic systems are defined as nonlinear dynamical systems which are sensitive to initial conditions, topologically mixing and with dense periodic orbits. Sensitivity to initial conditions of chaotic systems is popularly known as the *butterfly effect*. Small changes in an initial state will make a very large difference in the behavior of the system at future states.

The Lyapunov exponent is a measure of the divergence of phase points that are initially very close and can be used to quantify chaotic systems. It is common to refer to the largest Lyapunov exponent as the *Maximal Lyapunov Exponent* (MLE). A positive maximal Lyapunov exponent and phase space compactness are usually taken as defining conditions for a chaotic system.

Some classical paradigms of 3-D chaotic systems in chaos literature are Lorenz system [1], Rössler system [2], ACT system [3], Sprott systems [4], Chen system [5], Lü system [6], Liu system [7], Cai system [8], Chen-Lee system [9], Tigan system [10], etc.

Many new chaotic systems have been discovered in the recent years such as Zhou system [11], Zhu system [12], Li system [13], Wei-Yang system [14], Sundarapandian systems [15, 16], Vaidyanathan systems [17–32], Pehlivan system [33], Sampath system [34], Pham system [35], etc.

Chaos theory and control systems have many important applications in science and engineering [36–41]. Some commonly known applications are oscillators [42, 43], lasers [44, 45], chemical reactions [46–48, 48–50], biology [51–58], ecology [59, 60], encryption [61, 62], cryptosystems [63, 64], mechanical systems [65–69], secure communications [70–72], robotics [73–75], cardiology [76, 77], intelligent control [78, 79], neural networks [80–82], finance [83, 84], memristors [85, 86], etc.

Synchronization of chaotic systems is a phenomenon that occurs when two or more chaotic systems are coupled or when a chaotic system drives another chaotic system. Because of the butterfly effect which causes exponential divergence of the trajectories of two identical chaotic systems started with nearly the same initial conditions, the synchronization of chaotic systems is a challenging research problem in the chaos literature.

Major works on synchronization of chaotic systems deal with the complete synchronization of a pair of chaotic systems called the *master* and *slave* systems. The design goal of the complete synchronization is to apply the output of the master system to control the slave system so that the output of the slave system tracks the output of the master system asymptotically with time.

Pecora and Carroll pioneered the research on synchronization of chaotic systems with their seminal papers [87, 88]. The active control method [89–99] is typically used when the system parameters are available for measurement.

Adaptive control method [100–115] is typically used when some or all the system parameters are not available for measurement and estimates for the uncertain parameters of the systems.

Sampled-data feedback control method [116–119] and time-delay feedback control method [120–122] are also used for synchronization of chaotic systems. Backstepping control method [123–130] is also used for the synchronization of chaotic systems, which is a recursive method for stabilizing the origin of a control system in strict-feedback form.

Another popular method for the synchronization of chaotic systems is the sliding mode control method [131–140], which is a nonlinear control method that alters the dynamics of a nonlinear system by application of a discontinuous control signal that forces the system to "slide" along a cross-section of the system's normal behavior.

In this research work, we describe a ten-term novel 4-D four-wing chaotic system with four quadratic nonlinearities. Section 2 describes the 4-D dynamical model and phase portraits of the novel four-wing chaotic system. Section 3 describes the dynamic analysis of the novel four-wing chaotic system. We shall show that the novel four-wing chaotic system has a unique equilibrium at the origin, which is a saddle-point. Thus, the origin is an unstable equilibrium of the novel four-wing chaotic system.

The Lyapunov exponents of the novel 4-D four-wing chaotic system are obtained as $L_1 = 5.6253$, $L_2 = 0$, $L_3 = -5.4212$ and $L_4 = -53.0373$. Thus, the maximal Lyapunov exponent of the novel four-wing chaotic system is obtained as $L_1 = 5.6253$. The large value of L_1 indicates that the novel four-wing system is highly chaotic. Since the sum of the Lyapunov exponents of the novel chaotic system is negative, it follows that the novel chaotic system is dissipative. Also, the Kaplan-Yorke dimension of the novel four-wing chaotic system is obtained as $D_{KY} = 3.0038$.

Section 4 describes the adaptive synchronization of the identical novel chaotic systems with unknown parameters. The adaptive feedback control and synchronization results are proved using Lyapunov stability theory [141]. MATLAB simulations are depicted to illustrate all the main results for the 4-D novel four-wing chaotic system. Finally, Sect. 5 gives a summary of the main results derived in this work.

2 A Novel 4-D Four-Wing Chaotic System

In this work, we announce a novel 4-D four-wing chaotic system described by

$$\begin{aligned}
\dot{x}_1 &= -ax_1 + x_2x_3 + px_4 \\
\dot{x}_2 &= ax_2 - x_1x_3 - px_4 \\
\dot{x}_3 &= -bx_3 + x_1x_2 \\
\dot{x}_4 &= -cx_4 + x_1x_3
\end{aligned} \tag{1}$$

In (1), x_1, x_2, x_3, x_4 are the states and a, b, c, p are constant, positive parameters.

The 4-D system (1) is *chaotic* when the parameter values are taken as

$$a = 17, \quad b = 50, \quad c = 5, \quad p = 2 \tag{2}$$

For numerical simulations, we take the initial state of the chaotic system (1) as

$$x_1(0) = 1.5, \quad x_2(0) = 1.8, \quad x_3(0) = 1.2, \quad x_4(0) = 1.4 \tag{3}$$

The novel 4-D chaotic system (1) exhibits a strange, four-wing chaotic attractor. Figure 1 describes the 3-D projection of the four-wing chaotic attractor of the novel 4-D chaotic system (1) on (x_1, x_2, x_3) space. Figure 2 describes the 3-D projection of the four-wing chaotic attractor of the novel 4-D chaotic system (1) on (x_1, x_2, x_4) space.

Figure 3 describes the 3-D projection of the four-wing chaotic attractor of the novel 4-D chaotic system (1) on (x_1, x_3, x_4) space. Figure 2 describes the 3-D projection of the four-wing chaotic attractor of the novel 4-D chaotic system (1) on (x_2, x_3, x_4) space.

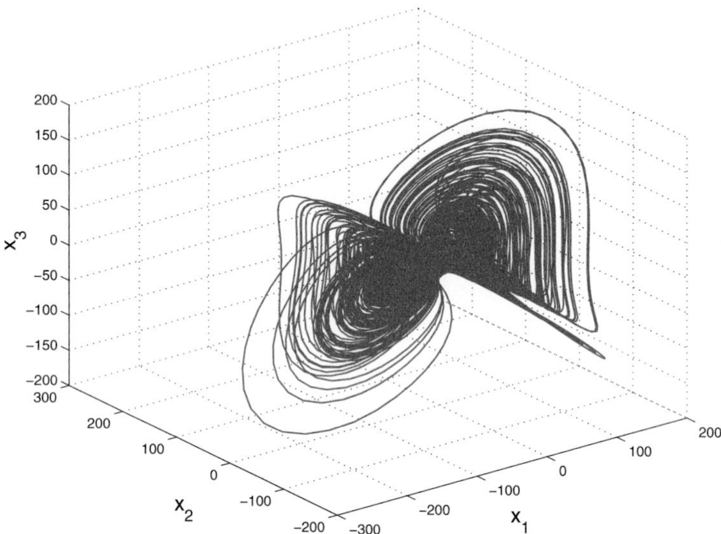

Fig. 1 3-D projection of the novel four-wing chaotic system on (x_1, x_2, x_3) space

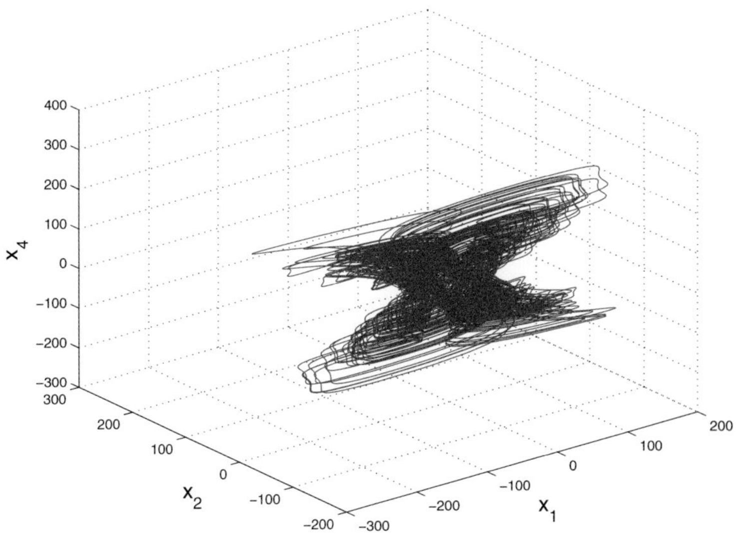

Fig. 2 3-D projection of the novel four-wing chaotic system on (x_1, x_2, x_4) space

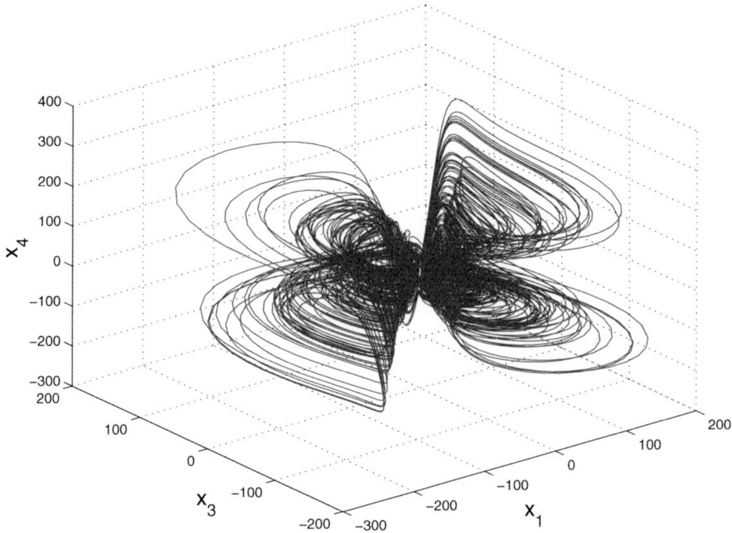

Fig. 3 3-D projection of the novel four-wing chaotic system on (x_1, x_3, x_4) space

3 Analysis of the Novel 4-D Four-Wing Chaotic System

This section gives the qualitative properties of the novel 4-D four-wing chaotic system (1) proposed in this research work.

3.1 Dissipativity

We write the system (1) in vector notation as

$$\dot{x} = f(x) = \begin{bmatrix} f_1(x) \\ f_2(x) \\ f_3(x) \\ f_4(x) \end{bmatrix}, \tag{4}$$

where

$$\begin{aligned} f_1(x) &= -ax_1 + x_2x_3 + px_4 \\ f_2(x) &= ax_2 - x_1x_3 - px_4 \\ f_3(x) &= -bx_3 + x_1x_2 \\ f_4(x) &= -cx_4 + x_1x_3 \end{aligned} \tag{5}$$

We take the parameter values as in the chaotic case, viz.

$$a = 17, \quad b = 50, \quad c = 5, \quad p = 2 \tag{6}$$

The divergence of the vector field f on $\mathbb{R}^4$ is obtained as

$$\text{div}\, f = \frac{\partial f_1(x)}{\partial x_1} + \frac{\partial f_2(x)}{\partial x_2} + \frac{\partial f_3(x)}{\partial x_3} + \frac{\partial f_4(x)}{\partial x_4} = -(b + c) = -\mu \tag{7}$$

where

$$\mu = b + c = 55 > 0 \tag{8}$$

Let Ω be any region in $\mathbb{R}^4$ with a smooth boundary. Let $\Omega(t) = \Phi_t(\Omega)$, where Φ_t is the flow of the vector field f.

Let $V(t)$ denote the hypervolume of $\Omega(t)$.

By Liouville's theorem, it follows that

$$\frac{dV(t)}{dt} = \int\limits_{\Omega(t)} (\text{div} f) dx_1\, dx_2\, dx_3\, dx_4 \tag{9}$$

Substituting the value of div$f$ in (9) leads to

$$\frac{dV(t)}{dt} = -\mu \int\limits_{\Omega(t)} dx_1\, dx_2\, dx_3\, dx_4 = -\mu V(t) \tag{10}$$

Integrating the linear differential equation (10), $V(t)$ is obtained as

$$V(t) = V(0)\exp(-\mu t), \quad \text{where} \;\; \mu = 55 > 0. \tag{11}$$

From Eq. (11), it follows that the hypervolume $V(t)$ shrinks to zero exponentially as $t \to \infty$.

Thus, the novel chaotic system (1) is dissipative. Hence, any asymptotic motion of the system (1) settles onto a set of measure zero, i.e. a strange attractor.

3.2 *Rotation Symmetry*

It is easy to see that the novel 4-D chaotic system (1) is invariant under the change of coordinates

$$(x_1, x_2, x_3, x_4) \mapsto (-x_1, -x_2, x_3, -x_4) \tag{12}$$

Since the transformation (12) persists for all values of the system parameters, it follows that the novel 4-D chaotic system (1) has rotation symmetry about the x_3-axis and that any non-trivial trajectory must have a twin trajectory.

3.3 *Equilibria*

The equilibrium points of the novel chaotic system (1) are obtained by solving the nonlinear equations

$$\begin{aligned} f_1(x) &= -ax_1 + x_2x_3 + px_4 &= 0 \\ f_2(x) &= ax_2 - x_1x_3 - px_4 &= 0 \\ f_3(x) &= -bx_3 + x_1x_2 &= 0 \\ f_4(x) &= -cx_4 + x_1x_3 &= 0 \end{aligned} \tag{13}$$

We take the parameter values as in the chaotic case, viz.

$$a = 17, \quad b = 50, \quad c = 5, \quad p = 2 \tag{14}$$

Solving the nonlinear system of Eq. (13) with the parameter values (14), we obtain a unique equilibrium point at the origin, i.e.

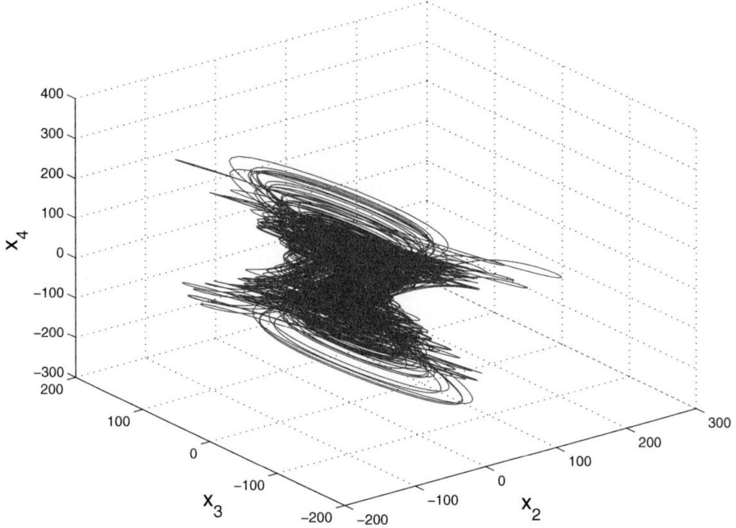

Fig. 4 3-D projection of the novel four-wing chaotic system on (x_2, x_3, x_4) space

$$E_0 = \begin{bmatrix} 0 \\ 0 \\ 0 \\ 0 \end{bmatrix} \tag{15}$$

The Jacobian matrix of the novel chaotic system (1) at E_0 is obtained as

$$J_0 = J(E_0) = \begin{bmatrix} -17 & 0 & 0 & 2 \\ 0 & 17 & 0 & -2 \\ 0 & 0 & -50 & 0 \\ 0 & 0 & 0 & -5 \end{bmatrix} \tag{16}$$

The matrix J_0 has the eigenvalues

$$\lambda_1 = -50, \quad \lambda_2 = -17, \quad \lambda_3 = -5, \quad \lambda_4 = 17 \tag{17}$$

This shows that the equilibrium point E_0 is a saddle-point, which is unstable (Fig. 4).

3.4 Lyapunov Exponents and Kaplan-Yorke Dimension

We take the initial values of the novel four-wing system (1) as in (3), viz.

$$x_1(0) = 1.5, \quad x_2(0) = 1.8, \quad x_3(0) = 1.2, \quad x_4(0) = 1.4 \tag{18}$$

We also take the parameter values of the novel four-wing system (1) as in the chaotic case (2), viz.

$$a = 17, \quad b = 50, \quad c = 5, \quad p = 2 \tag{19}$$

Then the Lyapunov exponents of the novel four-wing system (1) are numerically obtained as

$$L_1 = 5.6253, \quad L_2 = 0, \quad L_3 = -5.4212, \quad L_4 = -53.0373 \tag{20}$$

Since $L_1 + L_2 + L_3 + L_4 = -52.8332 < 0$, the system (1) is dissipative. Also, the Kaplan-Yorke dimension of the system (1) is obtained as

$$D_{KY} = 3 + \frac{L_1 + L_2 + L_3}{|L_4|} = 3.0038 \tag{21}$$

Figure 5 depicts the dynamics of the Lyapunov exponents of the novel 4-D four-wing chaotic system (1).

From Fig. 5, it is seen that the Maximal Lyapunov Exponent (MLE) of the novel 4-D four-wing chaotic system (1) is $L_1 = 5.5623$, which is a large value. Thus, the novel 4-D four-wing chaotic system (1) exhibits strong chaotic properties.

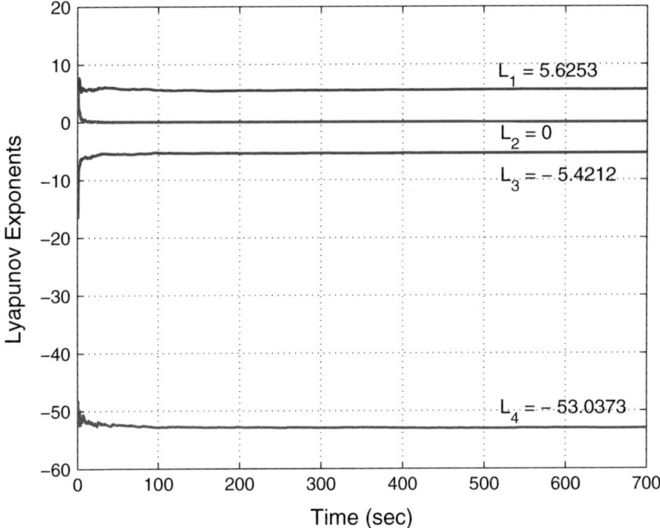

Fig. 5 Dynamics of the Lyapunov exponents of the novel four-wing chaotic system

4 Adaptive Synchronization of the Identical Novel Four-Wing Chaotic Systems

This section derives new results for the adaptive synchronization of the identical novel four-wing chaotic systems with unknown parameters.

The master system is given by the novel four-wing chaotic system

$$\begin{aligned}\dot{x}_1 &= -ax_1 + x_2x_3 + px_4\\ \dot{x}_2 &= ax_2 - x_1x_3 - px_4\\ \dot{x}_3 &= -bx_3 + x_1x_2\\ \dot{x}_4 &= -cx_4 + x_1x_3\end{aligned} \tag{22}$$

where x_1, x_2, x_3, x_4 are state variables and a, b, c, p are constant, unknown, parameters of the system.

The slave system is given by the controlled novel chaotic system

$$\begin{aligned}\dot{y}_1 &= -ay_1 + y_2y_3 + py_4 + u_1\\ \dot{y}_2 &= ay_2 - y_1y_3 - py_4 + u_2\\ \dot{y}_3 &= -by_3 + y_1y_2 + u_3\\ \dot{y}_4 &= -cy_4 + y_1y_3 + u_4\end{aligned} \tag{23}$$

where y_1, y_2, y_3, y_4 are state variables and u_1, u_2, u_3, u_4 are adaptive controls to be designed.

The synchronization error is defined as

$$\begin{aligned}e_1 &= y_1 - x_1\\ e_2 &= y_2 - x_2\\ e_3 &= y_3 - x_3\\ e_4 &= y_4 - x_4\end{aligned} \tag{24}$$

A simple calculation yields the error dynamics

$$\begin{aligned}\dot{e}_1 &= -ae_1 + pe_4 + y_2y_3 - x_2x_3 + u_1\\ \dot{e}_2 &= ae_2 - pe_4 - y_1y_3 + x_1x_3 + u_2\\ \dot{e}_3 &= -be_3 + y_1y_2 - x_1x_2 + u_3\\ \dot{e}_4 &= -ce_4 + y_1y_3 - x_1x_3 + u_4\end{aligned} \tag{25}$$

We consider the adaptive control law given by

$$\begin{aligned}u_1 &= \hat{a}(t)e_1 - \hat{p}(t)e_4 - y_2y_3 + x_2x_3 - k_1e_1\\ u_2 &= -\hat{a}(t)e_2 + \hat{p}(t)e_4 + y_1y_3 - x_1x_3 - k_2e_2\\ u_3 &= \hat{b}(t)e_3 - y_1y_2 + x_1x_2 - k_3e_3\\ u_4 &= \hat{c}(t)e_4 - y_1y_3 + x_1x_3 - k_4e_4\end{aligned} \tag{26}$$

where $\hat{a}(t), \hat{b}(t), \hat{c}(t), \hat{p}(t)$ are estimates for the unknown parameters a, b, c, p, respectively, and k_1, k_2, k_3, k_4 are positive gain constants.

The closed-loop control system is obtained by substituting (26) into (25) as

$$\begin{aligned} \dot{e}_1 &= -[a - \hat{a}(t)]e_1 + [p - \hat{p}(t)]e_4 - k_1 e_1 \\ \dot{e}_2 &= [a - \hat{a}(t)]e_2 - [p - \hat{p}(t)]e_4 - k_2 e_2 \\ \dot{e}_3 &= -[b - \hat{b}(t)]e_3 - k_3 e_3 \\ \dot{e}_4 &= -[c - \hat{c}(t)]e_4 - k_4 e_4 \end{aligned} \tag{27}$$

To simplify (27), we define the parameter estimation error as

$$\begin{aligned} e_a(t) &= a - \hat{a}(t) \\ e_b(t) &= b - \hat{b}(t) \\ e_c(t) &= c - \hat{c}(t) \\ e_p(t) &= p - \hat{p}(t) \end{aligned} \tag{28}$$

Using (28), the closed-loop system (27) can be simplified as

$$\begin{aligned} \dot{e}_1 &= -e_a e_1 + e_p e_4 - k_1 e_1 \\ \dot{e}_2 &= e_a e_2 - e_p e_4 - k_2 e_2 \\ \dot{e}_3 &= -e_b e_3 - k_3 e_3 \\ \dot{e}_4 &= -e_c e_4 - k_4 e_4 \end{aligned} \tag{29}$$

Differentiating the parameter estimation error (28) with respect to t, we get

$$\begin{aligned} \dot{e}_a &= -\dot{\hat{a}} \\ \dot{e}_b &= -\dot{\hat{b}} \\ \dot{e}_c &= -\dot{\hat{c}} \\ \dot{e}_p &= -\dot{\hat{p}} \end{aligned} \tag{30}$$

Next, we find an update law for parameter estimates using Lyapunov stability theory.

Consider the quadratic Lyapunov function defined by

$$V(e_1, e_2, e_3, e_4, e_a, e_b, e_c, e_p) = \frac{1}{2} \sum_{i=1}^{4} e_i^2 + \frac{1}{2} \left(e_a^2 + e_b^2 + e_c^2 + e_p^2\right) \tag{31}$$

Differentiating V along the trajectories of (29) and (30), we get

$$\begin{aligned} \dot{V} = &-k_1 e_1^2 - k_2 e_2^2 - k_3 e_3^2 - k_4 e_4^2 + e_a \left[e_2^2 - e_1^2 - \dot{\hat{a}}\right] \\ &+ e_b \left[-e_3^2 - \dot{\hat{b}}\right] + e_c \left[-e_4^2 - \dot{\hat{c}}\right] + e_p \left[(e_1 - e_2)e_4 - \dot{\hat{p}}\right] \end{aligned} \tag{32}$$

In view of Eq. (32), an update law for the parameter estimates is taken as

$$\begin{aligned} \dot{\hat{a}} &= e_2^2 - e_1^2 \\ \dot{\hat{b}} &= -e_3^2 \\ \dot{\hat{c}} &= -e_4^2 \\ \dot{\hat{p}} &= (e_1 - e_2)e_4 \end{aligned} \tag{33}$$

Theorem 1 *The identical novel 4-D four-wing chaotic systems (22) and (23) with unknown system parameters are globally and exponentially synchronized for all initial conditions* $x(0),\ y(0) \in \mathbb{R}^4$ *by the adaptive control law (26) and the parameter update law (33), where* k_i, $(i = 1, 2, 3, 4)$ *are positive constants.*

Proof The result is proved using Lyapunov stability theory [141].

We consider the quadratic Lyapunov function V defined by (31), which is positive definite on $\mathbb{R}^8$.

Substitution of the parameter update law (33) into (32) yields

$$\dot{V} = -k_1 e_1^2 - k_2 e_2^2 - k_3 e_3^2 - k_4 e_4^2, \tag{34}$$

which is a negative semi-definite function on $\mathbb{R}^8$.

Therefore, it can be concluded that the synchronization error vector $e(t)$ and the parameter estimation error are globally bounded, i.e.

$$\left[e_1(t)\ e_2(t)\ e_3(t)\ e_4(t)\ e_a(t)\ e_b(t)\ e_c(t)\ e_p(t) \right]^T \in \mathbf{L}_\infty. \tag{35}$$

Define

$$k = \min\{k_1, k_2, k_3, k_4\} \tag{36}$$

Then it follows from (34) that

$$\dot{V} \le -k\|e\|^2 \ \text{ or } \ k\|\mathbf{e}\|^2 \le -\dot{V} \tag{37}$$

Integrating the inequality (37) from 0 to t, we get

$$k \int_0^t \|\mathbf{e}(\tau)\|^2\, d\tau \le -\int_0^t \dot{V}(\tau)\, d\tau = V(0) - V(t) \tag{38}$$

From (38), it follows that $\mathbf{e}(t) \in \mathbf{L}_2$.

Using (29), it can be deduced that $\dot{\mathbf{e}}(t) \in \mathbf{L}_\infty$.

Thus, using Barbalat's lemma [141], we can conclude that $\mathbf{e}(t) \to 0$ exponentially as $t \to \infty$ for all initial conditions $\mathbf{e}(0) \in \mathbb{R}^4$.

Hence, we have proved that the identical novel 4-D four-wing chaotic systems (22) and (23) with unknown system parameters are globally and exponentially synchronized for all initial conditions $x(0), y(0) \in \mathbb{R}^4$ by the adaptive control law (26) and the parameter update law (33).

This completes the proof. □

For numerical simulations, the parameter values of the novel systems (22) and (23) are taken as in the chaotic case, viz.

$$a = 17, \quad b = 50, \quad c = 5, \quad p = 2 \tag{39}$$

The gain constants are taken as

$$k_1 = 5, \quad k_2 = 5, \quad k_3 = 5, \quad k_4 = 5 \tag{40}$$

The initial values of the parameter estimates are taken as

$$\hat{a}(0) = 5.3, \quad \hat{b}(0) = 14.9, \quad \hat{c}(0) = 20.1, \quad \hat{p}(0) = 17.8 \tag{41}$$

The initial values of the master system (22) are taken as

$$x_1(0) = 12.4, \quad x_2(0) = -21.3, \quad x_3(0) = 6.1, \quad x_4(0) = -7.3 \tag{42}$$

The initial values of the slave system (23) are taken as

$$y_1(0) = 1.5, \quad y_2(0) = 12.8, \quad y_3(0) = 23.9, \quad y_4(0) = -18.5 \tag{43}$$

Figures 6, 7, 8 and 9 show the complete synchronization of the identical chaotic systems (22) and (23).

Figure 6 shows that the states $x_1(t)$ and $y_1(t)$ are synchronized in two seconds (MATLAB).

Figure 7 shows that the states $x_2(t)$ and $y_2(t)$ are synchronized in two seconds (MATLAB).

Figure 8 shows that the states $x_3(t)$ and $y_3(t)$ are synchronized in two seconds (MATLAB).

Figure 9 shows that the states $x_4(t)$ and $y_4(t)$ are synchronized in two seconds (MATLAB).

Figure 10 shows the time-history of the synchronization errors $e_1(t)$, $e_2(t)$, $e_3(t)$, $e_4(t)$. From Fig. 10, it is seen that the errors $e_1(t)$, $e_2(t)$, $e_3(t)$ and $e_4(t)$ are stabilized in two seconds (MATLAB).

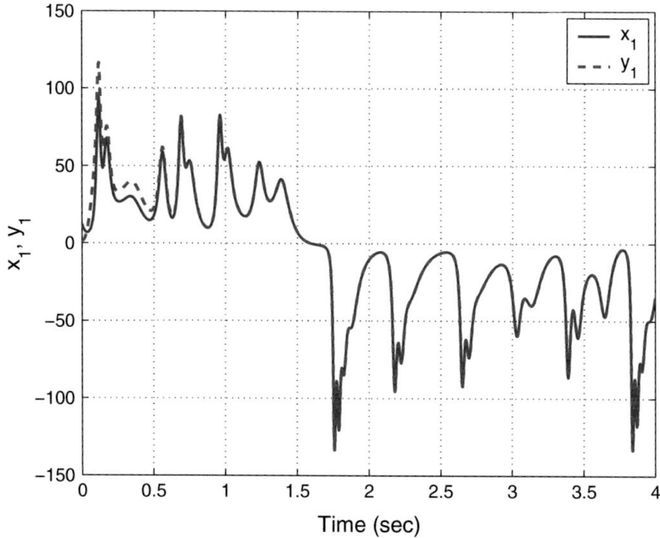

Fig. 6 Synchronization of the states x_1 and y_1

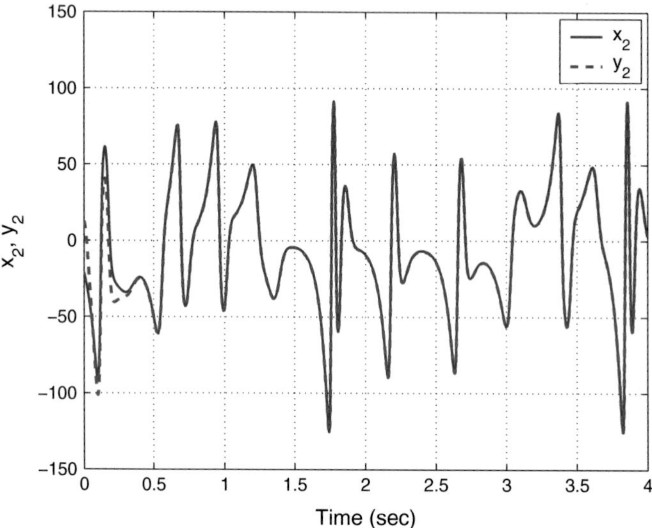

Fig. 7 Synchronization of the states x_2 and y_2

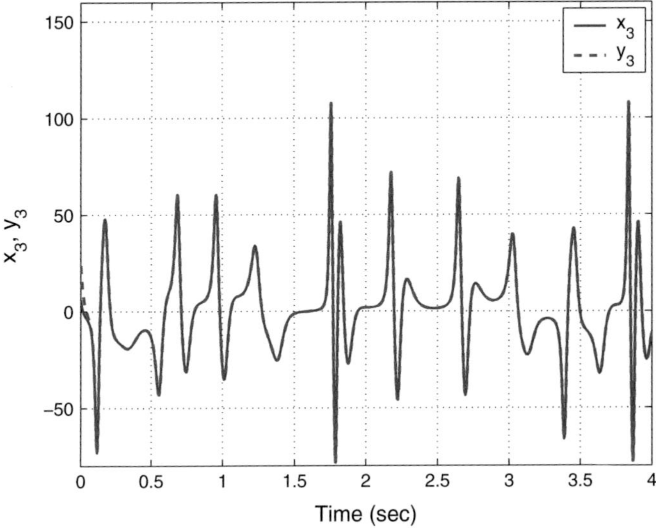

Fig. 8 Synchronization of the states x_3 and y_3

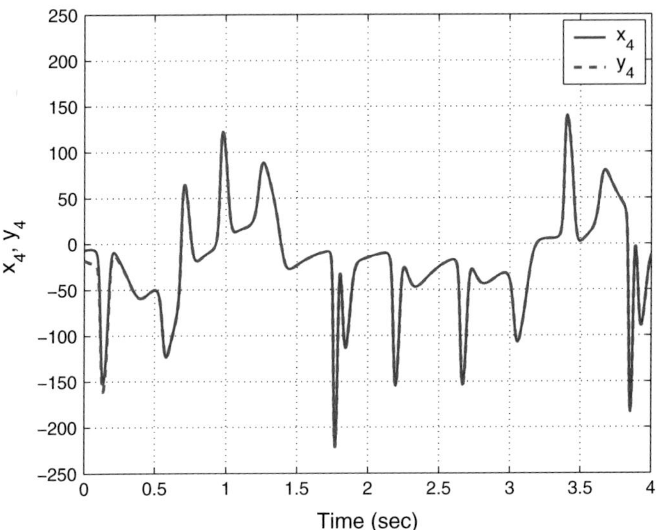

Fig. 9 Synchronization of the states x_4 and y_4

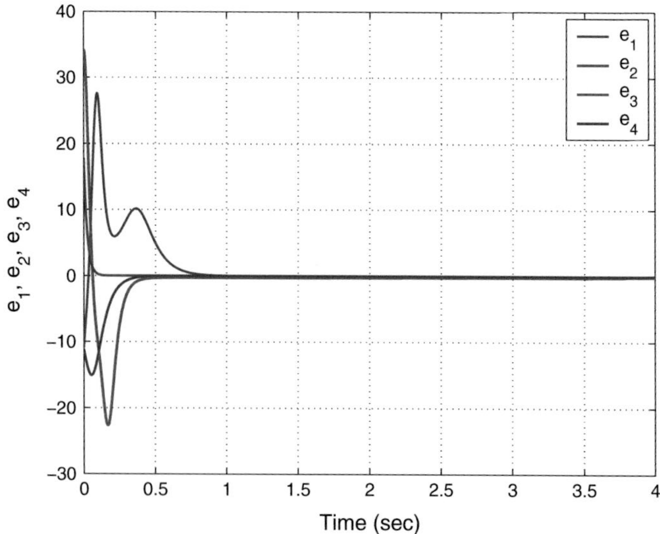

Fig. 10 Time-history of the synchronization errors e_1, e_2, e_3, e_4

5 Conclusions

In this research work, we announced a ten-term novel 4-D four-wing chaotic system with four quadratic nonlinearities. We described the qualitative analysis of the novel 4-D four-wing chaotic system. We showed that the novel four-wing chaotic system has a unique equilibrium point at the origin, which is a saddle-point. Thus, origin is an unstable equilibrium of the novel chaotic system. We also showed that the novel four-wing chaotic system has a rotation symmetry about the x_3 axis. Thus, it follows that every non-trivial trajectory of the novel four-wing chaotic system must have a twin trajectory. The Lyapunov exponents of the novel 4-D four-wing system were obtained as $L_1 = 5.6253$, $L_2 = 0$, $L_3 = -5.4212$ and $L_4 = -53.0373$. Thus, the maximal Lyapunov exponent of the novel four-wing chaotic system is seen as $L_1 = 5.6253$. The large value of L_1 indicates that the novel four-wing system is highly chaotic. Since the sum of the Lyapunov exponents of the novel chaotic system is negative, the novel chaotic system is dissipative. Also, the Kaplan-Yorke dimension of the novel four-wing chaotic system was obtained as $D_{KY} = 3.0038$. Finally, we derived new results for the adaptive synchronization of the identical novel 4-D four-wing chaotic systems with unknown parameters. The adaptive synchronization result was proved using Lyapunov stability theory. MATLAB simulations were shown to illustrate all the main results for the novel 4-D four-wing chaotic system.

References

1. Lorenz EN (1963) Deterministic periodic flow. J Atmos Sci 20(2):130–141
2. Rössler OE (1976) An equation for continuous chaos. Phys Lett A 57(5):397–398
3. Arneodo A, Coullet P, Tresser C (1981) Possible new strange attractors with spiral structure. Commun Math Phys 79(4):573–576
4. Sprott JC (1994) Some simple chaotic flows. Phys Rev E 50(2):647–650
5. Chen G, Ueta T (1999) Yet another chaotic attractor. Int Bifurcat Chaos 9(7):1465–1466
6. Lü J, Chen G (2002) A new chaotic attractor coined. Int J Bifurcat Chaos 12(3):659–661
7. Liu C, Liu T, Liu L, Liu K (2004) A new chaotic attractor. Chaos Solitions Fractals 22(5): 1031–1038
8. Cai G, Tan Z (2007) Chaos synchronization of a new chaotic system via nonlinear control. J Uncertain Syst 1(3):235–240
9. Chen HK, Lee CI (2004) Anti-control of chaos in rigid body motion. Chaos Solitons Fractals 21(4):957–965
10. Tigan G, Opris D (2008) Analysis of a 3D chaotic system. Chaos Solitons Fractals 36: 1315–1319
11. Zhou W, Xu Y, Lu H, Pan L (2008) On dynamics analysis of a new chaotic attractor. Phys Lett A 372(36):5773–5777
12. Zhu C, Liu Y, Guo Y (2010) Theoretic and numerical study of a new chaotic system. Intell Inf Manage 2:104–109
13. Li D (2008) A three-scroll chaotic attractor. Phys Lett A 372(4):387–393
14. Wei Z, Yang Q (2010) Anti-control of Hopf bifurcation in the new chaotic system with two stable node-foci. Appl Math Comput 217(1):422–429
15. Sundarapandian V (2013) Analysis and anti-synchronization of a novel chaotic system via active and adaptive controllers. J Eng Sci Technol Rev 6(4):45–52
16. Sundarapandian V, Pehlivan I (2012) Analysis, control, synchronization, and circuit design of a novel chaotic system. Math Comput Model 55(7–8):1904–1915
17. Vaidyanathan S (2013) A new six-term 3-D chaotic system with an exponential nonlinearity. Far East J Math Sci 79(1):135–143
18. Vaidyanathan S (2013) Analysis and adaptive synchronization of two novel chaotic systems with hyperbolic sinusoidal and cosinusoidal nonlinearity and unknown parameters. J Eng Sci Technol Rev 6(4):53–65
19. Vaidyanathan S (2014) A new eight-term 3-D polynomial chaotic system with three quadratic nonlinearities. Far East J Math Sci 84(2):219–226
20. Vaidyanathan S (2014) Analysis and adaptive synchronization of eight-term 3-D polynomial chaotic systems with three quadratic nonlinearities. Eur Phys J 223(8):1519–1529
21. Vaidyanathan S (2014) Analysis, control and synchronisation of a six-term novel chaotic system with three quadratic nonlinearities. Int J Model Ident Control 22(1):41–53
22. Vaidyanathan S (2014) Generalized projective synchronisation of novel 3-D chaotic systems with an exponential non-linearity via active and adaptive control. Int J Model Ident Control 22(3):207–217
23. Vaidyanathan S (2015) A 3-D novel highly chaotic system with four quadratic nonlinearities, its adaptive control and anti-synchronization with unknown parameters. J Eng Sci Technol Rev 8(2):106–115
24. Vaidyanathan S (2015) Analysis, properties and control of an eight-term 3-D chaotic system with an exponential nonlinearity. Int J Model Ident Control 23(2):164–172
25. Vaidyanathan S, Azar AT (2015) Analysis, control and synchronization of a nine-term 3-D novel chaotic system. In: Azar AT, Vaidyanathan S (eds) Chaos modelling and control systems design, studies in computational intelligence, vol 581. Springer, Germany, pp 19–38
26. Vaidyanathan S, Madhavan K (2013) Analysis, adaptive control and synchronization of a seven-term novel 3-D chaotic system. Int J Control Theor Appl 6(2):121–137

27. Vaidyanathan S, Pakiriswamy S (2015) A 3-D novel conservative chaotic system and its generalized projective synchronization via adaptive control. J Eng Sci Technol Rev 8(2): 52–60
28. Vaidyanathan S, Volos C (2015) Analysis and adaptive control of a novel 3-D conservative no-equilibrium chaotic system. Arch Control Sci 25(3):333–353
29. Vaidyanathan S, Volos C, Pham VT, Madhavan K, Idowu BA (2014) Adaptive backstepping control, synchronization and circuit simulation of a 3-D novel jerk chaotic system with two hyperbolic sinusoidal nonlinearities. Arch Control Sci 24(3):375–403
30. Vaidyanathan S, Volos CK, Kyprianidis IM, Stouboulos IN, Pham VT (2015) Analysis, adaptive control and anti-synchronization of a six-term novel jerk chaotic system with two exponential nonlinearities and its circuit simulation. J Eng Sci Technol Rev 8(2):24–36
31. Vaidyanathan S, Volos CK, Pham VT (2015) Analysis, adaptive control and adaptive synchronization of a nine-term novel 3-D chaotic system with four quadratic nonlinearities and its circuit simulation. J Eng Sci Technol Rev 8(2):174–184
32. Vaidyanathan S, Volos CK, Pham VT (2015) Global chaos control of a novel nine-term chaotic system via sliding mode control. In: Azar AT, Zhu Q (eds) Advances and applications in sliding mode control systems, studies in computational intelligence, vol 576. Springer, Germany, pp 571–590
33. Pehlivan I, Moroz IM, Vaidyanathan S (2014) Analysis, synchronization and circuit design of a novel butterfly attractor. J Sound Vib 333(20):5077–5096
34. Sampath S, Vaidyanathan S, Volos CK, Pham VT (2015) An eight-term novel four-scroll chaotic system with cubic nonlinearity and its circuit simulation. J Eng Sci Technol Rev 8(2):1–6
35. Pham VT, Vaidyanathan S, Volos CK, Jafari S (2015) Hidden attractors in a chaotic system with an exponential nonlinear term. Eur Phys J 224(8):1507–1517
36. Azar AT (2010) Fuzzy systems. IN-TECH, Vienna, Austria
37. Azar AT, Vaidyanathan S (2015) Chaos modeling and control systems design, studies in computational intelligence, vol 581. Springer, Germany
38. Azar AT, Vaidyanathan S (2015) Computational intelligence applications in modeling and control, studies in computational intelligence, vol 575. Springer, Germany
39. Azar AT, Vaidyanathan S (2015) Handbook of research on advanced intelligent control engineering and automation. Advances in computational intelligence and robotics (ACIR), IGI-Global, USA
40. Azar AT, Zhu Q (2015) Advances and applications in sliding mode control systems, studies in computational intelligence, vol 576. Springer, Germany
41. Zhu Q, Azar AT (2015) Complex system modelling and control through intelligent soft computations, studies in fuzzines and soft computing, vol 319. Springer, Germany
42. Kengne J, Chedjou JC, Kenne G, Kyamakya K (2012) Dynamical properties and chaos synchronization of improved Colpitts oscillators. Commun Nonlinear Sci Numer Simul 17(7):2914–2923
43. Sharma A, Patidar V, Purohit G, Sud KK (2012) Effects on the bifurcation and chaos in forced Duffing oscillator due to nonlinear damping. Commun Nonlinear Sci Numer Simul 17(6):2254–2269
44. Li N, Pan W, Yan L, Luo B, Zou X (2014) Enhanced chaos synchronization and communication in cascade-coupled semiconductor ring lasers. Commun Nonlinear Sci Numer Simul 19(6):1874–1883
45. Yuan G, Zhang X, Wang Z (2014) Generation and synchronization of feedback-induced chaos in semiconductor ring lasers by injection-locking. Optik—Int J Light Electron Opt 125(8):1950–1953
46. Vaidyanathan S (2015) Adaptive control of a chemical chaotic reactor. Int J PharmTech Res 8(3):377–382
47. Vaidyanathan S (2015) Anti-synchronization of Brusselator chemical reaction systems via adaptive control. Int J ChemTech Res 8(6):759–768

48. Vaidyanathan S (2015) Dynamics and control of Brusselator chemical reaction. Int J ChemTech Res 8(6):740–749
49. Vaidyanathan S (2015l) Dynamics and control of Tokamak system with symmetric and magnetically confined plasma. Int J ChemTech Res 8(6):795–803
50. Vaidyanathan S (2015) Synchronization of Tokamak systems with symmetric and magnetically confined plasma via adaptive control. Int J ChemTech Res 8(6):818–827
51. Vaidyanathan S (2015) 3-cells cellular neural network (CNN) attractor and its adaptive biological control. Int J PharmTech Res 8(4):632–640
52. Vaidyanathan S (2015) Adaptive backstepping control of enzymes-substrates system with ferroelectric behaviour in brain waves. Int J PharmTech Res 8(2):256–261
53. Vaidyanathan S (2015) Adaptive biological control of generalized Lotka-Volterra three-species biological system. Int J PharmTech Res 8(4):622–631
54. Vaidyanathan S (2015) Adaptive chaotic synchronization of enzymes-substrates system with ferroelectric behaviour in brain waves. Int J PharmTech Res 8(5):964–973
55. Vaidyanathan S (2015) Adaptive synchronization of generalized Lotka-Volterra three-species biological systems. Int J PharmTech Res 8(5):928–937
56. Vaidyanathan S (2015) Chaos in neurons and adaptive control of Birkhoff-Shaw strange chaotic attractor. Int J PharmTech Res 8(5):956–963
57. Vaidyanathan S (2015) Lotka-Volterra population biology models with negative feedback and their ecological monitoring. Int J PharmTech Res 8(5):974–981
58. Vaidyanathan S (2015) Synchronization of 3-cells cellular neural network (CNN) attractors via adaptive control method. Int J PharmTech Res 8(5):946–955
59. Gibson WT, Wilson WG (2013) Individual-based chaos: extensions of the discrete logistic model. J Theor Biol 339:84–92
60. Suérez I (1999) Mastering chaos in ecology. Ecol Model 117(2–3):305–314
61. Lang J (2015) Color image encryption based on color blend and chaos permutation in the reality-preserving multiple-parameter fractional Fourier transform domain. Opt Commun 338:181–192
62. Zhang X, Zhao Z, Wang J (2014) Chaotic image encryption based on circular substitution box and key stream buffer. Sig Process Image Commun 29(8):902–913
63. Rhouma R, Belghith S (2011) Cryptoanalysis of a chaos based cryptosystem on DSP. Commun Nonlinear Sci Numer Simul 16(2):876–884
64. Usama M, Khan MK, Alghatbar K, Lee C (2010) Chaos-based secure satellite imagery cryptosystem. Comput Math Appl 60(2):326–337
65. Azar AT, Serrano FE (2014) Robust IMC-PID tuning for cascade control systems with gain and phase margin specifications. Neural Comput Appl 25(5):983–995
66. Azar AT, Serrano FE (2015) Adaptive sliding mode control of the Furuta pendulum. In: Azar AT, Zhu Q (eds) Advances and applications in sliding mode control systems, studies in computational intelligence, vol 576. Springer, Germany, pp 1–42
67. Azar AT, Serrano FE (2015) Deadbeat control for multivariable systems with time varying delays. In: Azar AT, Vaidyanathan S (eds) Chaos modeling and control systems design, studies in computational intelligence, vol 581. Springer, Germany, pp 97–132
68. Azar AT, Serrano FE (2015) Design and modeling of anti wind up PID controllers. In: Zhu Q, Azar AT (eds) Complex system modelling and control through intelligent soft computations, studies in fuzziness and soft computing, vol 319. Springer, Germany, pp 1–44
69. Azar AT, Serrano FE (2015) Stabilizatoin and control of mechanical systems with backlash. In: Azar AT, Vaidyanathan S (eds) Handbook of research on advanced intelligent control engineering and automation. Advances in computational intelligence and robotics (ACIR), IGI-Global, USA, pp 1–60
70. Feki M (2003) An adaptive chaos synchronization scheme applied to secure communication. Chaos Solitons Fractals 18(1):141–148
71. Murali K, Lakshmanan M (1998) Secure communication using a compound signal from generalized chaotic systems. Phys Lett A 241(6):303–310

72. Zaher AA, Abu-Rezq A (2011) On the design of chaos-based secure communication systems. Commun Nonlinear Syst Numer Simul 16(9):3721–3727
73. Mondal S, Mahanta C (2014) Adaptive second order terminal sliding mode controller for robotic manipulators. J Franklin Inst 351(4):2356–2377
74. Nehmzow U, Walker K (2005) Quantitative description of robot-environment interaction using chaos theory. Robot Auton Syst 53(3–4):177–193
75. Volos CK, Kyprianidis IM, Stouboulos IN (2013) Experimental investigation on coverage performance of a chaotic autonomous mobile robot. Robot Auton Syst 61(12):1314–1322
76. Qu Z (2011) Chaos in the genesis and maintenance of cardiac arrhythmias. Prog Biophys Mol Biol 105(3):247–257
77. Witte CL, Witte MH (1991) Chaos and predicting varix hemorrhage. Med Hypotheses 36(4):312–317
78. Azar AT (2012) Overview of type-2 fuzzy logic systems. Int J Fuzzy Syst Appl 2(4):1–28
79. Li Z, Chen G (2006) Integration of fuzzy logic and chaos theory, studies in fuzziness and soft computing, vol 187. Springer, Germany
80. Huang X, Zhao Z, Wang Z, Li Y (2012) Chaos and hyperchaos in fractional-order cellular neural networks. Neurocomputing 94:13–21
81. Kaslik E, Sivasundaram S (2012) Nonlinear dynamics and chaos in fractional-order neural networks. Neural Netw 32:245–256
82. Lian S, Chen X (2011) Traceable content protection based on chaos and neural networks. Appl Soft Comput 11(7):4293–4301
83. Guégan D (2009) Chaos in economics and finance. Annu Rev Control 33(1):89–93
84. Sprott JC (2004) Competition with evolution in ecology and finance. Phys Lett A 325(5–6):329–333
85. Pham VT, Volos CK, Vaidyanathan S, Le TP, Vu VY (2015) A memristor-based hyperchaotic system with hidden attractors: dynamics, synchronization and circuital emulating. J Eng Sci Technol Rev 8(2):205–214
86. Volos CK, Kyprianidis IM, Stouboulos IN, Tlelo-Cuautle E, Vaidyanathan S (2015) Memristor: a new concept in synchronization of coupled neuromorphic circuits. J Eng Sci Technol Rev 8(2):157–173
87. Carroll TL, Pecora LM (1991) Synchronizing chaotic circuits. IEEE Trans Circ Syst 38(4):453–456
88. Pecora LM, Carroll TL (1990) Synchronization in chaotic systems. Phys Rev Lett 64(8): 821–824
89. Karthikeyan R, Sundarapandian V (2014) Hybrid chaos synchronization of four-scroll systems via active control. J Electr Eng 65(2):97–103
90. Sarasu P, Sundarapandian V (2011) Active controller design for the generalized projective synchronization of four-scroll chaotic systems. Int J Syst Sign Control Eng Appl 4(2):26–33
91. Sarasu P, Sundarapandian V (2011) The generalized projective synchronization of hyperchaotic Lorenz and hyperchaotic Qi systems via active control. Int J Soft Comput 6(5): 216–223
92. Sundarapandian V (2010) Output regulation of the Lorenz attractor. Far East J Math Sci 42(2):289–299
93. Sundarapandian V, Karthikeyan R (2012) Hybrid synchronization of hyperchaotic Lorenz and hyperchaotic Chen systems via active control. J Eng Appl Sci 7(3):254–264
94. Vaidyanathan S (2011) Hybrid chaos synchronization of Liu and Lü systems by active nonlinear control. Commun Comput Inf Sci 204
95. Vaidyanathan S (2012) Output regulation of the Liu chaotic system. Appl Mech Mater 110–116:3982–3989
96. Vaidyanathan S, Rajagopal K (2011) Anti-synchronization of Li and T chaotic systems by active nonlinear control. Commun Comput Inf Sci 198:175–184
97. Vaidyanathan S, Rajagopal K (2011) Global chaos synchronization of hyperchaotic Pang and Wang systems by active nonlinear control. Commun Comput Inf Sci 204:84–93

98. Vaidyanathan S, Rasappan S (2011) Global chaos synchronization of hyperchaotic Bao and Xu systems by active nonlinear control. Commun Comput Inf Sci 198:10–17
99. Vaidyanathan S, Azar AT, Rajagopal K, Alexander P (2015) Design and SPICE implementation of a 12-term novel hyperchaotic system and its synchronisation via active control. Int J Model Ident Control 23(3):267–277
100. Vaidyanathan S, VTP, Volos CK, (2015) A 5-D hyperchaotic Rikitake dynamo system with hidden attractors. Eur Phys J 224(8):1575–1592
101. Sarasu P, Sundarapandian V (2012) Adaptive controller design for the generalized projective synchronization of 4-scroll systems. Int J Syst Sign Control Eng Appl 5(2):21–30
102. Sarasu P, Sundarapandian V (2012b) Generalized projective synchronization of three-scroll chaotic systems via adaptive control. Eur J Sci Res 72(4):504–522
103. Sarasu P, Sundarapandian V (2012) Generalized projective synchronization of two-scroll systems via adaptive control. Int J Soft Comput 7(4):146–156
104. Sundarapandian V, Karthikeyan R (2011) Anti-synchronization of hyperchaotic Lorenz and hyperchaotic Chen systems by adaptive control. Int J Syst Sign Control Eng Appl 4(2):18–25
105. Sundarapandian V, Karthikeyan R (2011) Anti-synchronization of Lü and Pan chaotic systems by adaptive nonlinear control. Eur J Sci Res 64(1):94–106
106. Sundarapandian V, Karthikeyan R (2012) Adaptive anti-synchronization of uncertain Tigan and Li systems. J Eng Appl Sci 7(1):45–52
107. Vaidyanathan S (2012) Anti-synchronization of Sprott-L and Sprott-M chaotic systems via adaptive control. Int J Control Theor Appl 5(1):41–59
108. Vaidyanathan S (2013) Analysis, control and synchronization of hyperchaotic Zhou system via adaptive control. Adv Intell Syst Comput 177:1–10
109. Vaidyanathan S (2015) Hyperchaos, qualitative analysis, control and synchronisation of a ten-term 4-D hyperchaotic system with an exponential nonlinearity and three quadratic nonlinearities. Int J Model Ident Control 23(4):380–392
110. Vaidyanathan S, Azar AT (2015) Analysis and control of a 4-D novel hyperchaotic system. In: Azar AT, Vaidyanathan S (eds) Chaos modeling and control systems design, studies in computational intelligence, vol 581. Springer, Germany, pp 19–38
111. Vaidyanathan S, Pakiriswamy S (2013) Generalized projective synchronization of six-term Sundarapandian chaotic systems by adaptive control. Int J Control Theor Appl 6(2):153–163
112. Vaidyanathan S, Rajagopal K (2011) Global chaos synchronization of Lü and Pan systems by adaptive nonlinear control. Commun Comput Inf Sci 205:193–202
113. Vaidyanathan S, Rajagopal K (2012) Global chaos synchronization of hyperchaotic Pang and hyperchaotic Wang systems via adaptive control. Int J Soft Comput 7(1):28–37
114. Vaidyanathan S, Volos C, Pham VT (2014) Hyperchaos, adaptive control and synchronization of a novel 5-D hyperchaotic system with three positive Lyapunov exponents and its SPICE implementation. Arch Control Sci 24(4):409–446
115. Vaidyanathan S, Volos C, Pham VT, Madhavan K (2015) Analysis, adaptive control and synchronization of a novel 4-D hyperchaotic hyperjerk system and its SPICE implementation. Arch Control Sci 25(1):5–28
116. Gan Q, Liang Y (2012) Synchronization of chaotic neural networks with time delay in the leakage term and parametric uncertainties based on sampled-data control. J Franklin Inst 349(6):1955–1971
117. Li N, Zhang Y, Nie Z (2011) Synchronization for general complex dynamical networks with sampled-data. Neurocomputing 74(5):805–811
118. Xiao X, Zhou L, Zhang Z (2014) Synchronization of chaotic Lur'e systems with quantized sampled-data controller. Commun Nonlinear Sci Numer Simul 19(6):2039–2047
119. Zhang H, Zhou J (2012) Synchronization of sampled-data coupled harmonic oscillators with control inputs missing. Syst Control Lett 61(12):1277–1285
120. Chen WH, Wei D, Lu X (2014) Global exponential synchronization of nonlinear time-delay Lur'e systems via delayed impulsive control. Commun Nonlinear Sci Numer Simul 19(9):3298–3312

121. Jiang GP, Zheng WX, Chen G (2004) Global chaos synchronization with channel time-delay. Chaos Solitons Fractals 20(2):267–275
122. Shahverdiev EM, Shore KA (2009) Impact of modulated multiple optical feedback time delays on laser diode chaos synchronization. Opt Commun 282(17):2572–3568
123. Rasappan S, Vaidyanathan S (2012) Global chaos synchronization of WINDMI and Coullet chaotic systems by backstepping control. Far East J Math Sci 67(2):265–287
124. Rasappan S, Vaidyanathan S (2012) Hybrid synchronization of n-scroll Chua and Lur'e chaotic systems via backstepping control with novel feedback. Arch Control Sci 22(3):343–365
125. Rasappan S, Vaidyanathan S (2012c) Synchronization of hyperchaotic Liu system via backstepping control with recursive feedback. Commun Comput Inf Sci 305:212–221
126. Rasappan S, Vaidyanathan S (2013) Hybrid synchronization of n-scroll chaotic Chua circuits using adaptive backstepping control design with recursive feedback. Malays J Math Sci 7(2):219–246
127. Rasappan S, Vaidyanathan S (2014) Global chaos synchronization of WINDMI and Coullet chaotic systems using adaptive backstepping control design. Kyungpook Math J 54(1): 293–320
128. Suresh R, Sundarapandian V (2013) Global chaos synchronization of a family of n-scroll hyperchaotic Chua circuits using backstepping control with recursive feedback. Far East J Math Sci 73(1):73–95
129. Vaidyanathan S, Rasappan S (2014) Global chaos synchronization of n-scroll Chua circuit and Lur'e system using backstepping control design with recursive feedback. Arab J Sci Eng 39(4):3351–3364
130. Vaidyanathan S, Idowu BA, Azar AT (2015) Backstepping controller design for the global chaos synchronization of Sprott's jerk systems. Stud Comput Intell 581:39–58
131. Sundarapandian V, Sivaperumal S (2011) Sliding controller design of hybrid synchronization of four-wing chaotic systems. Int J Soft Comput 6(5):224–231
132. Vaidyanathan S (2012) Analysis and synchronization of the hyperchaotic Yujun systems via sliding mode control. Adv Intell Syst Comput 176:329–337
133. Vaidyanathan S (2012) Global chaos control of hyperchaotic Liu system via sliding control method. Int J Control Theor Appl 5(2):117–123
134. Vaidyanathan S (2012) Sliding mode control based global chaos control of Liu-Liu-Liu-Su chaotic system. Int J Control Theor Appl 5(1):15–20
135. Vaidyanathan S (2014) Global chaos synchronization of identical Li-Wu chaotic systems via sliding mode control. Int J Model Ident Control 22(2):170–177
136. Vaidyanathan S, Azar AT (2015) Anti-synchronization of identical chaotic systems using sliding mode control and an application to Vaidhyanathan-Madhavan chaotic systems. Stud Comput Intell 576:527–547
137. Vaidyanathan S, Azar AT (2015) Hybrid synchronization of identical chaotic systems using sliding mode control and an application to Vaidhyanathan chaotic systems. Stud Comput Intell 576:549–569
138. Vaidyanathan S, Sampath S (2011) Global chaos synchronization of hyperchaotic Lorenz systems by sliding mode control. Commun Comput Inf Sci 205:156–164
139. Vaidyanathan S, Sampath S (2012) Anti-synchronization of four-wing chaotic systems via sliding mode control. Int J Autom Comput 9(3):274–279
140. Vaidyanathan S, Sampath S, Azar AT (2015) Global chaos synchronisation of identical chaotic systems via novel sliding mode control method and its application to Zhu system. Int J Model Ident Control 23(1):92–100
141. Khalil HK (2001) Nonlinear systems. Prentice Hall, USA

Adaptive Control and Synchronization of Halvorsen Circulant Chaotic Systems

Sundarapandian Vaidyanathan and Ahmad Taher Azar

Abstract In this research work, we describe Halvorsen circulant chaotic systems and its qualitative properties. We show that Halvorsen circulant chaotic system is dissipative and that it has an unstable equilibrium at the origin. The Lyapunov exponents of Halvorsen circulant chaotic system are obtained as $L_1 = 0.8109$, $L_2 = 0$ and $L_3 = -4.6255$. The Kaplan-Yorke dimension of the Halvorsen circulant chaotic system is obtained as $D_{KY} = 2.1753$. Next, this work describes the adaptive control of the Halvorsen circulant chaotic system with unknown parameters. Also, this work describes the adaptive synchronization of the identical Halvorsen circulant chaotic systems with unknown parameters. The adaptive feedback control and synchronization results are proved using Lyapunov stability theory. MATLAB simulations are depicted to illustrate all the main results for the Halvorsen circulant chaotic system.

Keywords Chaos · Chaotic systems · Circulant systems · Adaptive control · Synchronization

1 Introduction

Chaotic systems are defined as nonlinear dynamical systems which are sensitive to initial conditions, topologically mixing and with dense periodic orbits. Sensitivity to initial conditions of chaotic systems is popularly known as the *butterfly effect*. Small changes in an initial state will make a very large difference in the behavior of the system at future states.

S. Vaidyanathan (✉)
Research and Development Centre, Vel Tech University,
Avadi, Chennai 600062, Tamil Nadu, India
e-mail: sundarvtu@gmail.com

A.T. Azar
Faculty of Computers and Information, Benha University, Benha, Egypt
e-mail: ahmad_t_azar@ieee.org; ahmad.azar@fci.bu.edu.eg

A.T. Azar and S. Vaidyanathan (eds.), *Advances in Chaos Theory and Intelligent Control*, Studies in Fuzziness and Soft Computing 337,
DOI 10.1007/978-3-319-30340-6_10

The Lyapunov exponent is a measure of the divergence of phase points that are initially very close and can be used to quantify chaotic systems. It is common to refer to the largest Lyapunov exponent as the *Maximal Lyapunov Exponent* (MLE). A positive maximal Lyapunov exponent and phase space compactness are usually taken as defining conditions for a chaotic system.

Some classical paradigms of 3-D chaotic systems in the literature are Lorenz system [1], Rössler system [2], ACT system [3], Sprott systems [4], Chen system [5], Lü system [6], Liu system [7], Cai system [8], Chen-Lee system [9], Tigan system [10], etc.

Many new chaotic systems have been discovered in the recent years such as Zhou system [11], Zhu system [12], Li system [13], Wei-Yang system [14], Sundarapandian systems [15, 16], Vaidyanathan systems [17–33], Pehlivan system [34], Sampath system [35], Pham system [36], etc.

Chaos theory and control systems have many important applications in science and engineering [37–42]. Some commonly known applications are oscillators [43, 44], lasers [45, 46], chemical reactions [47–54], biology [55–64], ecology [65, 66], encryption [67, 68], cryptosystems [69, 70], mechanical systems [71–75], secure communications [76–78], robotics [79–81], cardiology [82, 83], intelligent control [84, 85], neural networks [86–88], finance [89, 90], memristors [91, 92], etc.

Control of chaotic systems aims to regulate or stabilize the chaotic system about its equilibrium points. There are many methods available in the literature such as active control method [93–96], adaptive control method [97, 98], sliding mode control method [99, 100], backstepping control method [101], etc.

Synchronization of chaotic systems is a phenomenon that occurs when two or more chaotic systems are coupled or when a chaotic system drives another chaotic system. Major works on synchronization of chaotic systems deal with the complete synchronization of a pair of chaotic systems called the *master* and *slave* systems. The design goal of the complete synchronization is to apply the output of the master system to control the slave system so that the output of the slave system tracks the output of the master system asymptotically with time.

Pecora and Carroll pioneered the research on synchronization of chaotic systems with their seminal papers [102, 103]. The active control method [93, 104–112] is typically used when the system parameters are available for measurement. Adaptive control method [113–128] is typically used when some or all the system parameters are not available for measurement and estimates for the uncertain parameters of the systems. Sampled-data feedback control method [129–132] and time-delay feedback control method [133–135] are also used for synchronization of chaotic systems.

Backstepping control method [136–143] is also used for the synchronization of chaotic systems, which is a recursive method for stabilizing the origin of a control system in strict-feedback form. Another popular method for the synchronization of chaotic systems is the sliding mode control method [144–150], which is a nonlinear control method that alters the dynamics of a nonlinear system by application of a discontinuous control signal that forces the system to "slide" along a cross-section of the system's normal behavior.

In this research work, we derive new results for the adaptive control and synchronization of Halvorsen circulant chaotic system [151] with unknown parameters. Halvorsen circulant chaotic system is a famous dissipative circulant chaotic system. The Lyapunov exponents of the Halvorsen circulant chaotic system are obtained as $L_1 = 0.8109$, $L_2 = 0$ and $L_3 = -4.6255$. Also, the Kaplan-Yorke dimension of the Halvorsen circulant chaotic system is derived as $D_{KY} = 2.1753$.

Section 2 describes the dynamic equations and phase portraits of the Halvorsen circulant chaotic system. Section 3 describes the qualitative properties of the Halvorsen circulant chaotic system. Section 4 describes the adaptive control of the Halvorsen circulant chaotic system with unknown parameters. Section 5 describes the adaptive synchronization of the identical Halvorsen circulant chaotic systems with unknown parameters. Finally, Sect. 6 contains the conclusions of this work.

2 Halvorsen Circulant Chaotic System

In this section, we discuss the Halvorsen circulant chaotic system [151], which is described by the 3-D dynamics

$$\begin{cases} \dot{x}_1 = -ax_1 - bx_2 - bx_3 - x_2^2 \\ \dot{x}_2 = -ax_2 - bx_3 - bx_1 - x_3^2 \\ \dot{x}_3 = -ax_3 - bx_1 - bx_2 - x_1^2 \end{cases} \tag{1}$$

which is symmetric with respect to cyclic interchanges of x_1, x_2 and x_3.

Halvorsen showed the circulant chaotic system (1) is *chaotic* when the parameter values are taken as

$$a = 1.27, \quad b = 4 \tag{2}$$

For numerical simulations, we take the initial values of the Halvorsen chaotic system (1) as

$$x_1(0) = 0.2, \quad x_2(0) = 0.6, \quad x_3(0) = 0.2 \tag{3}$$

Figure 1 describes the *strange attractor* of the Halvorsen chaotic system (1) in $\mathbb{R}^3$. Figures 2, 3 and 4 describe the 2-D projection of the strange chaotic attractor of the Halvorsen chaotic system (1) on (x_1, x_2), (x_2, x_3) and (x_1, x_3) coordinate planes respectively.

3 Properties of the Halvorsen Circulant Chaotic System

This section describes the qualitative properties of the Halvorsen chaotic system (1).

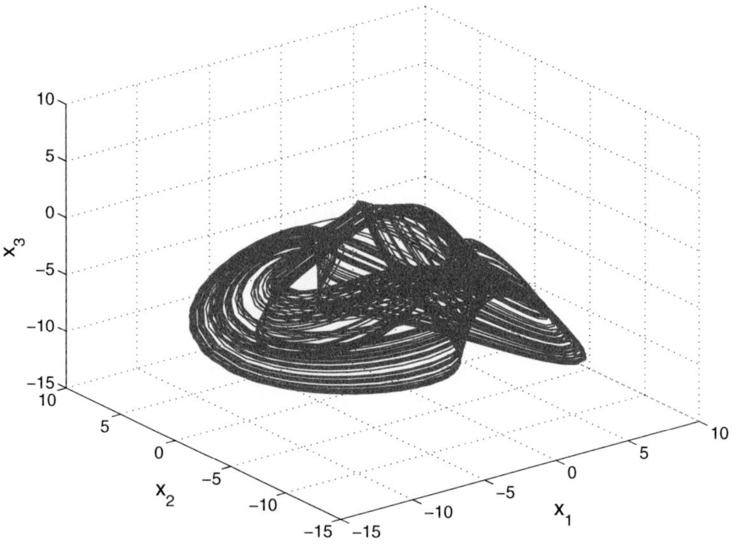

Fig. 1 Strange attractor of the Halvorsen chaotic system in $\mathbf{R}^3$

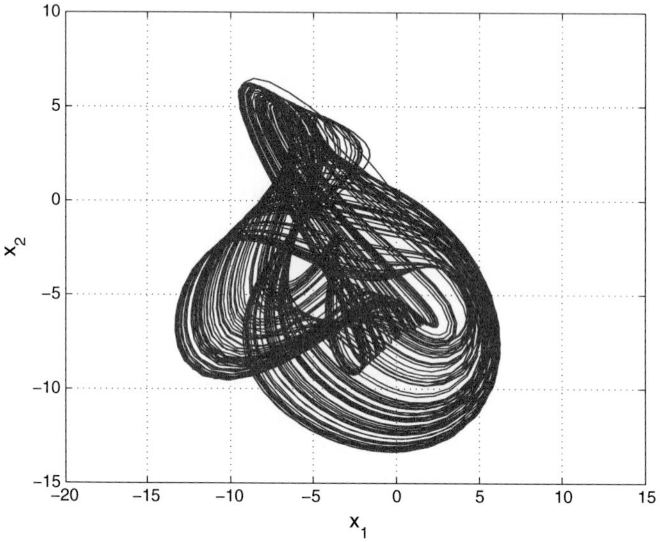

Fig. 2 2-D projection of the Halvorsen chaotic system on (x_1, x_2)-plane

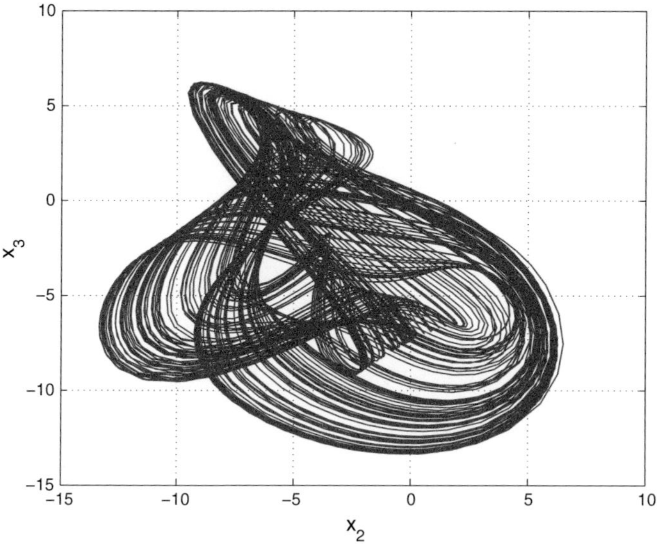

Fig. 3 2-D projection of the Halvorsen chaotic system on (x_2, x_3)-plane

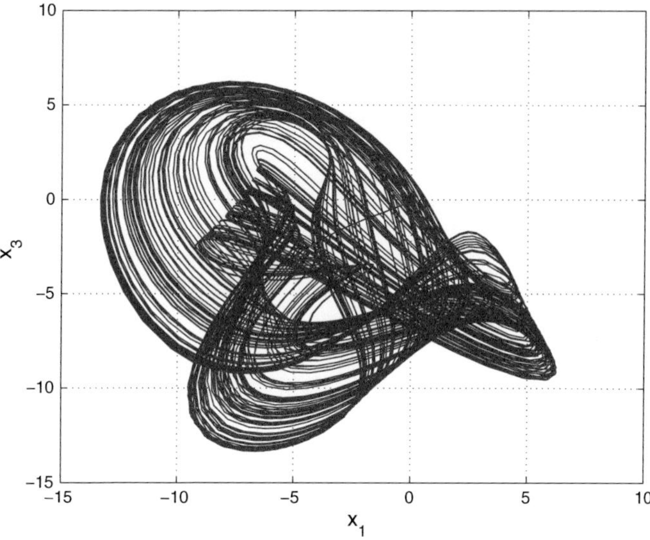

Fig. 4 2-D projection of the Halvorsen chaotic system on (x_1, x_3)-plane

3.1 Dissipativity

We write the Halvorsen system (1) in vector notation as

$$\dot{x} = f(x) = \begin{bmatrix} f_1(x) \\ f_2(x) \\ f_3(x) \end{bmatrix}, \tag{4}$$

where

$$\begin{aligned} f_1(x) &= -ax_1 - bx_2 - bx_3 - x_2^2 \\ f_2(x) &= -ax_2 - bx_3 - bx_1 - x_3^2 \\ f_3(x) &= -ax_3 - bx_1 - bx_2 - x_1^2 \end{aligned} \tag{5}$$

We take the parameter values as in the chaotic case, *viz.*

$$a = 1.27, \quad b = 4. \tag{6}$$

The divergence of the vector field f on $\mathbb{R}^3$ is obtained as

$$\operatorname{div} f = \frac{\partial f_1(x)}{\partial x_1} + \frac{\partial f_2(x)}{\partial x_2} + \frac{\partial f_3(x)}{\partial x_3} = -(a + a + a) = -\mu \tag{7}$$

where $\mu = 3a = 3.81 > 0$.

Let Ω be any region in $\mathbb{R}^3$ with a smooth boundary. Let $\Omega(t) = \Phi_t(\Omega)$, where Φ_t is the flow of the vector field f.

Let $V(t)$ denote the volume of $\Omega(t)$.

By Liouville's theorem, it follows that

$$\frac{dV(t)}{dt} = \int\limits_{\Omega(t)} (\operatorname{div} f)\, dx_1\, dx_2\, dx_3 \tag{8}$$

Substituting the value of div$f$ in (8) leads to

$$\frac{dV(t)}{dt} = -\mu \int\limits_{\Omega(t)} dx_1\, dx_2\, dx_3 = -\mu V(t) \tag{9}$$

Integrating the linear differential equation (9), $V(t)$ is obtained as

$$V(t) = V(0)\exp(-\mu t), \quad \text{where } \mu = 3.81 > 0. \tag{10}$$

From Eq. (10), it follows that the volume $V(t)$ shrinks to zero exponentially as $t \to \infty$.

Thus, the Halvorsen circulant chaotic system (1) is dissipative. Hence, any asymptotic motion of the system (1) settles onto a set of measure zero, i.e. a strange attractor.

3.2 Equilibrium Point

For the chaotic case (2), the Halvorsen circulant chaotic system (1) has a unique equilibrium at the origin, i.e.

$$E_0 = \begin{bmatrix} 0 \\ 0 \\ 0 \end{bmatrix} \tag{11}$$

The Jacobian matrix of the Halvorsen circulant chaotic system (1) at E_0 is obtained as

$$J_0 = J(E_0) = \begin{bmatrix} -1.27 & -4 & -4 \\ -4 & -1.27 & -4 \\ -4 & -4 & -1.27 \end{bmatrix} \tag{12}$$

The matrix J_0 has the eigenvalues

$$\lambda_1 = -9.27, \quad \lambda_2 = 2.73, \quad \lambda_3 = 2.73 \tag{13}$$

This shows that the equilibrium point E_0 is a saddle-point, which is unstable.

3.3 Lyapunov Exponents and Kaplan-Yorke Dimension

We take the parameter values of the Halvorsen chaotic system (1) as in the chaotic case, i.e. $a = 1.27$ and $b = 4$. We take the initial values of the Halvorsen chaotic system (1) as in (3).

Then the Lyapunov exponents of the Halvorsen chaotic system (1) are numerically obtained as

$$L_1 = 0.8109, \quad L_2 = 0, \quad L_3 = -4.6255 \tag{14}$$

Since $L_1 + L_2 + L_3 = -3.8146 < 0$, the Halvorsen chaotic system (1) is dissipative.

Also, the Kaplan-Yorke dimension of the Halvorsen chaotic system (1) is obtained as

$$D_{KY} = 2 + \frac{L_1 + L_2}{|L_3|} = 2.1753 \tag{15}$$

which is fractional.

4 Adaptive Control of Halvorsen Circulant Chaotic System

This section derives new results for the adaptive control of Halvorsen circulant chaotic system.

The controlled Halvorsen circulant chaotic system is given by

$$\begin{aligned} \dot{x}_1 &= -ax_1 - bx_2 - bx_3 - x_2^2 + u_1 \\ \dot{x}_2 &= -ax_2 - bx_3 - bx_1 - x_3^2 + u_2 \\ \dot{x}_3 &= -ax_3 - bx_1 - bx_2 - x_1^2 + u_3 \end{aligned} \tag{16}$$

where x_1, x_2, x_3 are state variables, a, b are constant, unknown, parameters of the system and u_1, u_2, u_3 are adaptive feedback controls to be designed.

We consider the adaptive feedback control law given by

$$\begin{aligned} u_1 &= \hat{a}(t)x_1 + \hat{b}(t)x_2 + \hat{b}(t)x_3 + x_2^2 - k_1x_1 \\ u_2 &= \hat{a}(t)x_2 + \hat{b}(t)x_3 + \hat{b}(t)x_1 + x_3^2 - k_2x_2 \\ u_3 &= \hat{a}(t)x_3 + \hat{b}(t)x_1 + \hat{b}(t)x_2 + x_1^2 - k_3x_3 \end{aligned} \tag{17}$$

In (17), $\hat{a}(t)$, $\hat{b}(t)$ are estimates for the unknown parameters a, b, respectively, and k_1, k_2, k_3 are positive gain constants.

We obtain the closed-loop system by substituting (17) into (16) as

$$\begin{aligned} \dot{x}_1 &= -\left[a - \hat{a}(t)\right]x_1 - \left[b - \hat{b}(t)\right]x_2 - \left[b - \hat{b}(t)\right]x_3 - k_1x_1 \\ \dot{x}_2 &= -\left[a - \hat{a}(t)\right]x_2 - \left[b - \hat{b}(t)\right]x_3 - \left[b - \hat{b}(t)\right]x_1 - k_2x_2 \\ \dot{x}_3 &= -\left[a - \hat{a}(t)\right]x_3 - \left[b - \hat{b}(t)\right]x_1 - \left[b - \hat{b}(t)\right]x_2 - k_3x_3 \end{aligned} \tag{18}$$

To simplify (18), we define the parameter estimation error as

$$\begin{aligned} e_a(t) &= a - \hat{a}(t) \\ e_b(t) &= b - \hat{b}(t) \end{aligned} \tag{19}$$

Using (19), the closed-loop system (18) can be simplified as

$$
\begin{aligned}
\dot{x}_1 &= -e_a x_1 - e_b x_2 - e_b x_3 - k_1 x_1 \\
\dot{x}_2 &= -e_a x_2 - e_b x_3 - e_b x_1 - k_2 x_2 \\
\dot{x}_3 &= -e_a x_3 - e_b x_1 - e_b x_2 - k_3 x_3
\end{aligned}
\tag{20}
$$

Differentiating the parameter estimation error (19) with respect to t, we get

$$
\begin{aligned}
\dot{e}_a &= -\dot{\hat{a}} \\
\dot{e}_b &= -\dot{\hat{b}}
\end{aligned}
\tag{21}
$$

We consider the quadratic Lyapunov function defined by

$$
V(x_1, x_2, x_3, e_a, e_b) = \frac{1}{2}\left(x_1^2 + x_2^2 + x_3^2 + e_a^2 + e_b^2\right), \tag{22}
$$

which is positive definite on $\mathbb{R}^5$.

Differentiating V along the trajectories of (20) and (21), we get

$$
\begin{aligned}
\dot{V} = &-k_1 x_1^2 - k_2 x_2^2 - k_3 x_3^2 + e_a[-\left(x_1^2 + x_2^2 + x_3^2\right) - \dot{\hat{a}}] \\
&+ e_b[-2(x_1 x_2 + x_2 x_3 + x_1 x_3) - \dot{\hat{b}}]
\end{aligned}
\tag{23}
$$

In view of Eq. (23), an update law for the parameter estimates is taken as

$$
\begin{aligned}
\dot{\hat{a}} &= -\left(x_1^2 + x_2^2 + x_3^2\right) \\
\dot{\hat{b}} &= -2(x_1 x_2 + x_2 x_3 + x_1 x_3)
\end{aligned}
\tag{24}
$$

Theorem 1 *The Halvorsen circulant chaotic system (16) with unknown system parameters is globally and exponentially stabilized for all initial conditions* $x(0) \in \mathbb{R}^3$ *by the adaptive control law (17) and the parameter update law (24), where* k_i, $(i = 1, 2, 3)$ *are positive constants.*

Proof The result is proved using Lyapunov stability theory [152].

We consider the quadratic Lyapunov function V defined by (22), which is positive definite on $\mathbb{R}^5$.

Substitution of the parameter update law (24) into (23) yields

$$
\dot{V} = -k_1 x_1^2 - k_2 x_2^2 - k_3 x_3^2, \tag{25}
$$

which is a negative semi-definite function on $\mathbb{R}^5$.

Therefore, it can be concluded that the state vector $x(t)$ and the parameter estimation error are globally bounded, i.e.

$$
\left[x_1(t)\ x_2(t)\ x_3(t)\ e_a(t)\ e_b(t)\right]^T \in \mathbf{L}_\infty. \tag{26}
$$

Define

$$k = \min\{k_1, k_2, k_3\} \tag{27}$$

Then it follows from (25) that

$$\dot{V} \le -k\|\mathbf{x}\|^2 \ \text{ or } \ k\|\mathbf{x}\|^2 \le -\dot{V} \tag{28}$$

Integrating the inequality (28) from 0 to t, we get

$$k \int_0^t \|\mathbf{x}(\tau)\|^2 \, d\tau \ \le \ -\int_0^t \dot{V}(\tau) \, d\tau = V(0) - V(t) \tag{29}$$

From (29), it follows that $\mathbf{x}(t) \in \mathbf{L}_2$.

Using (20), it can be deduced that $\dot{x}(t) \in \mathbf{L}_\infty$.

Hence, using Barbalat's lemma, we can conclude that $\mathbf{x}(t) \to 0$ exponentially as $t \to \infty$ for all initial conditions $\mathbf{x}(0) \in \mathbf{R}^3$.

Thus, we conclude that the Halvorsen circulant chaotic system (16) with unknown system parameters is globally and exponentially stabilized for all initial conditions $x(0) \in \mathbf{R}^3$ by the adaptive control law (17) and the parameter update law (24).

This completes the proof. □

For numerical simulations, we use the classical fourth-order Runge-Kutta method with step-size $h = 10^{-6}$ for finding numerical solutions of systems of ordinary differential equations.

The parameter values of the novel system (16) are taken as in the chaotic case, *viz.*

$$a = 1.27, \quad b = 4 \tag{30}$$

The gain constants are taken as

$$k_1 = 6, \quad k_2 = 6, \quad k_3 = 6 \tag{31}$$

The initial values of the parameter estimates are taken as

$$\hat{a}(0) = 17.2, \quad \hat{b}(0) = 10.3 \tag{32}$$

The initial values of the Halvorsen circulant chaotic system (16) are taken as

$$x_1(0) = 7.2, \quad x_2(0) = 3.5, \quad x_3(0) = -12.9 \tag{33}$$

Figure 5 shows the time-history of the controlled states $x_1(t), x_2(t), x_3(t)$.

From Fig. 5, it is clear that the controlled states $x_1(t), x_2(t), x_3(t)$ are stabilized in just 1 s (MATLAB).

Thus, Fig. 5 depicts the efficiency of the adaptive controller defined by (17).

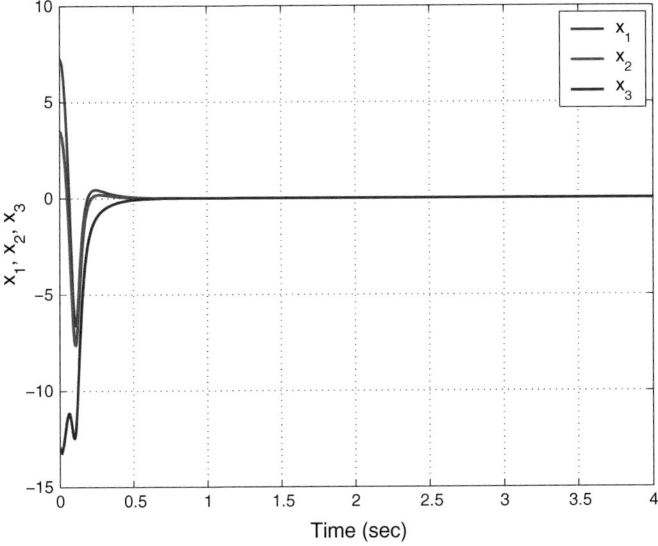

Fig. 5 Time-history of the controlled states $x_1(t), x_2(t), x_3(t)$

5 Adaptive Synchronization of the Identical Halvorsen Circulant Chaotic Systems

This section derives a new result for the adaptive synchronization of the identical Halvorsen circulant chaotic systems with unknown parameters. The main result is established using Lyapunov stability theory.

The master system is given by the Halvorsen chaotic system

$$\begin{aligned} \dot{x}_1 &= -ax_1 - bx_2 - bx_3 - x_2^2 \\ \dot{x}_2 &= -ax_2 - bx_3 - bx_1 - x_3^2 \\ \dot{x}_3 &= -ax_3 - bx_1 - bx_2 - x_1^2 \end{aligned} \tag{34}$$

where x_1, x_2, x_3 are state variables and a, b are constant, unknown, parameters of the system.

The slave system is given by the controlled Halvorsen chaotic system

$$\begin{aligned} \dot{y}_1 &= -ay_1 - by_2 - by_3 - y_2^2 + u_1 \\ \dot{y}_2 &= -ay_2 - by_3 - by_1 - y_3^2 + u_2 \\ \dot{y}_3 &= -ay_3 - by_1 - by_2 - y_1^2 + u_3 \end{aligned} \tag{35}$$

where y_1, y_2, y_3 are state variables and u_1, u_2, u_3 are adaptive controls to be designed.

The synchronization error is defined as

$$\begin{aligned} e_1 &= y_1 - x_1 \\ e_2 &= y_2 - x_2 \\ e_3 &= y_3 - x_3 \end{aligned} \tag{36}$$

The error dynamics is easily obtained as

$$\begin{aligned} \dot{e}_1 &= -ae_1 - be_2 - be_3 - y_2^2 + x_2^2 + u_1 \\ \dot{e}_2 &= -ae_2 - be_3 - be_1 - y_3^2 + x_3^2 + u_2 \\ \dot{e}_3 &= -ae_3 - be_1 - be_2 - y_1^2 + x_1^2 + u_3 \end{aligned} \tag{37}$$

An adaptive control law is taken as

$$\begin{aligned} u_1 &= \hat{a}(t)e_1 + \hat{b}(t)e_2 + \hat{b}(t)e_3 + y_2^2 - x_2^2 - k_1 e_1 \\ u_2 &= \hat{a}(t)e_2 + \hat{b}(t)e_3 + \hat{b}(t)e_1 + y_3^2 - x_3^2 - k_2 e_2 \\ u_3 &= \hat{a}(t)e_3 + \hat{b}(t)e_1 + \hat{b}(t)e_2 + y_1^2 - x_1^2 - k_3 e_3 \end{aligned} \tag{38}$$

where $\hat{a}(t), \hat{b}(t)$ are estimates for the unknown parameters a, b, respectively, and k_1, k_2, k_3 are positive gain constants.

The closed-loop control system is obtained by substituting (38) into (37) as

$$\begin{aligned} \dot{e}_1 &= -\left[a - \hat{a}(t)\right] e_1 - \left[b - \hat{b}(t)\right] e_2 - \left[b - \hat{b}(t)\right] e_3 - k_1 e_1 \\ \dot{e}_2 &= -\left[a - \hat{a}(t)\right] e_2 - \left[b - \hat{b}(t)\right] e_3 - \left[b - \hat{b}(t)\right] e_1 - k_2 e_2 \\ \dot{e}_3 &= -\left[a - \hat{a}(t)\right] e_3 - \left[b - \hat{b}(t)\right] e_1 - \left[b - \hat{b}(t)\right] e_2 - k_3 e_3 \end{aligned} \tag{39}$$

To simplify (39), we define the parameter estimation error as

$$\begin{aligned} e_a(t) &= a - \hat{a}(t) \\ e_b(t) &= b - \hat{b}(t) \end{aligned} \tag{40}$$

Using (40), the closed-loop system (39) can be simplified as

$$\begin{aligned} \dot{e}_1 &= -e_a e_1 - e_b e_2 - e_b e_3 - k_1 e_1 \\ \dot{e}_2 &= -e_a e_2 - e_b e_3 - e_b e_1 - k_1 e_2 \\ \dot{e}_3 &= -e_a e_3 - e_b e_1 - e_b e_2 - k_3 e_3 \end{aligned} \tag{41}$$

Differentiating the parameter estimation error (40) with respect to t, we get

$$\begin{aligned} \dot{e}_a &= -\dot{\hat{a}} \\ \dot{e}_b &= -\dot{\hat{b}} \end{aligned} \tag{42}$$

We consider the quadratic Lyapunov function defined by

$$V(e_1, e_2, e_3, e_a, e_b) = \frac{1}{2}\left(e_1^2 + e_2^2 + e_3^2 + e_a^2 + e_b^2\right), \tag{43}$$

which is positive definite on $\mathbb{R}^5$.

Differentiating V along the trajectories of (41) and (42), we get

$$\begin{aligned} \dot{V} = &-k_1e_1^2 - k_2e_2^2 - k_3e_3^2 + e_a[-\left(e_1^2 + e_2^2 + e_3^2\right) - \dot{\hat{a}}] \\ &+ e_b[-2(e_1e_2 + e_2e_3 + e_1e_3) - \dot{\hat{b}}] \end{aligned} \tag{44}$$

In view of Eq. (44), an update law for the parameter estimates is taken as

$$\begin{aligned} \dot{\hat{a}} &= -\left(e_1^2 + e_2^2 + e_3^2\right) \\ \dot{\hat{b}} &= -2(e_1e_2 + e_2e_3 + e_1e_3) \end{aligned} \tag{45}$$

Theorem 2 *The identical Halvorsen circulant chaotic systems (34) and (35) with unknown system parameters are globally and exponentially synchronized for all initial conditions* $x(0), y(0) \in \mathbb{R}^3$ *by the adaptive control law (38) and the parameter update law (45), where* k_i, $(i = 1, 2, 3)$ *are positive constants.*

Proof The result is proved using Lyapunov stability theory [152].

We consider the quadratic Lyapunov function V defined by (43), which is positive definite on $\mathbb{R}^5$.

Substitution of the parameter update law (45) into (44) yields

$$\dot{V} = -k_1e_1^2 - k_2e_2^2 - k_3e_3^2, \tag{46}$$

which is a negative semi-definite function on $\mathbb{R}^5$.

Therefore, it can be concluded that the synchronization error vector $e(t)$ and the parameter estimation error are globally bounded, i.e.

$$\left[\, e_1(t)\ e_2(t)\ e_3(t)\ e_a(t)\ e_b(t) \,\right]^T \in \mathbf{L}_\infty. \tag{47}$$

Define

$$k = \min\{k_1, k_2, k_3\} \tag{48}$$

Then it follows from (46) that

$$\dot{V} \leq -k\|e\|^2 \ \text{ or } \ k\|\mathbf{e}\|^2 \leq -\dot{V} \tag{49}$$

Integrating the inequality (49) from 0 to t, we get

$$k \int_0^t \|\mathbf{e}(\tau)\|^2 \, d\tau \leq -\int_0^t \dot{V}(\tau)\, d\tau = V(0) - V(t) \tag{50}$$

From (50), it follows that $\mathbf{e}(t) \in \mathbf{L}_2$.

Using (41), it can be deduced that $\dot{\mathbf{e}}(t) \in \mathbf{L}_\infty$.

Hence, using Barbalat's lemma, we can conclude that $\mathbf{e}(t) \to 0$ exponentially as $t \to \infty$ for all initial conditions $\mathbf{e}(0) \in \mathbb{R}^3$.

Since $\mathbf{e}(t) \to 0$ exponentially as $t \to \infty$ for all $\mathbf{e}(0) \in \mathbb{R}^3$, it is immediate that $\mathbf{x}(t) \to \mathbf{y}(t)$ as $t \to \infty$ for all $\mathbf{x}(0), \mathbf{y}(0) \in \mathbb{R}^3$.

Hence, we have shown that the identical Halvorsen circulant chaotic systems (34) and (35) with unknown system parameters are globally and exponentially synchronized for all initial conditions $x(0), y(0) \in \mathbb{R}^3$ by the adaptive control law (38) and the parameter update law (45)

This completes the proof. □

For numerical simulations, the parameter values of the novel systems (34) and (35) are taken as in the chaotic case, *viz.*

$$a = 1.27, \quad b = 4 \tag{51}$$

The gain constants are taken as

$$k_1 = 8, \quad k_2 = 8, \quad k_3 = 8 \tag{52}$$

The initial values of the parameter estimates are taken as

$$\hat{a}(0) = 4.2, \quad \hat{b}(0) = 1.6 \tag{53}$$

The initial values of the master system (34) are taken as

$$x_1(0) = 1.2, \quad x_2(0) = 1.8, \quad x_3(0) = 2.1 \tag{54}$$

The initial values of the slave system (35) are taken as

$$y_1(0) = 1.9, \quad y_2(0) = 2.8, \quad y_3(0) = 0.7 \tag{55}$$

Figures 6, 7 and 8 show the complete synchronization of the identical chaotic systems (34) and (35).

Figure 6 shows that the states $x_1(t)$ and $y_1(t)$ are synchronized in 2 s (MATLAB).

Figure 7 shows that the states $x_2(t)$ and $y_2(t)$ are synchronized in 2 s (MATLAB).

Figure 8 shows that the states $x_3(t)$ and $y_3(t)$ are synchronized in 2 s (MATLAB).

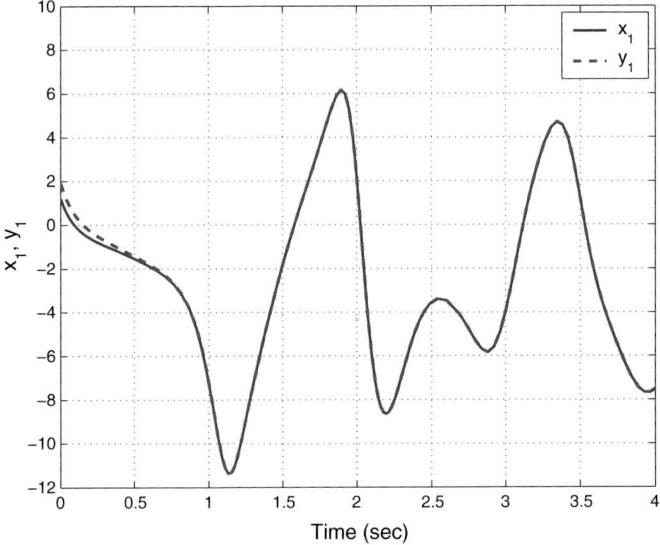

Fig. 6 Synchronization of the states x_1 and y_1

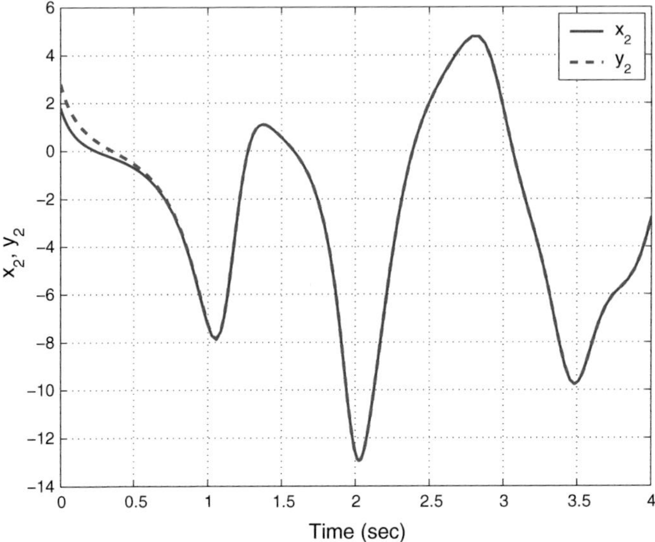

Fig. 7 Synchronization of the states x_2 and y_2

Figure 9 shows the time-history of the synchronization errors $e_1(t), e_2(t), e_3(t)$. From Fig. 9, it is seen that the errors $e_1(t)$, $e_2(t)$ and $e_3(t)$ are stabilized in 2 s (MATLAB).

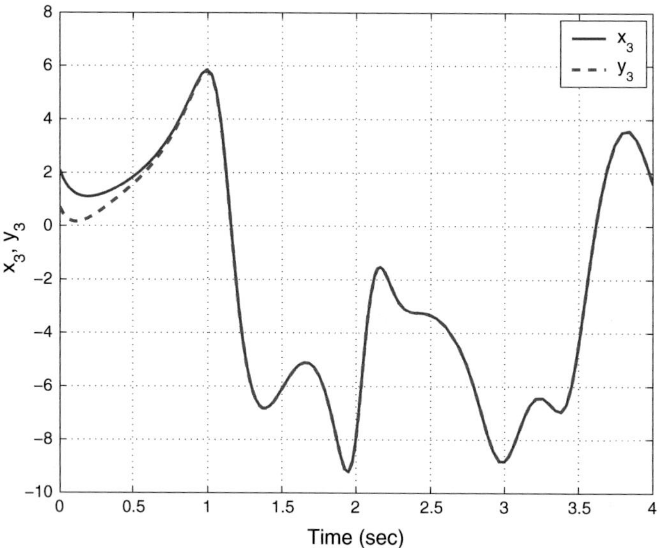

Fig. 8 Synchronization of the states x_3 and y_3

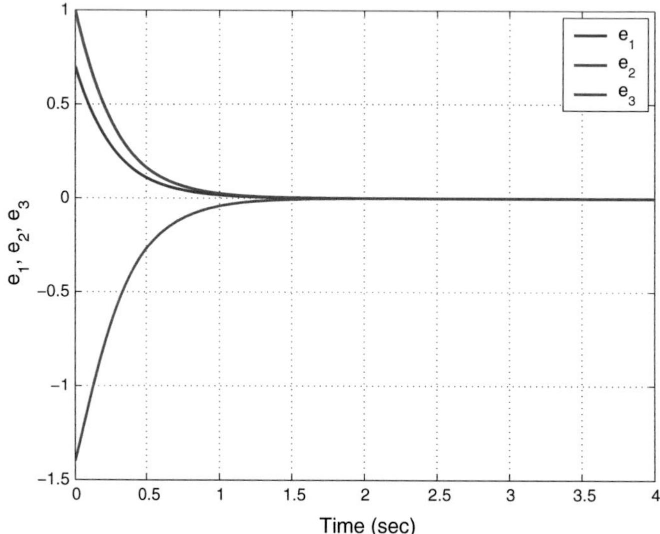

Fig. 9 Time-history of the synchronization errors e_1, e_2, e_3

This shows the efficiency of the adaptive control law (38) and the parameter update law (45) designed in this section.

6 Conclusions

In this research work, we derived new results for the adaptive controller and adaptive synchronizer design for Halvorsen circulant chaotic system. We first discussed the qualitative properties of the Halvorsen circulant chaotic system. Explicitly, we showed that Halvorsen circulant chaotic system is dissipative and that it has an unstable equilibrium at the origin. The Lyapunov exponents of Halvorsen circulant chaotic system have been obtained as $L_1 = 0.8109$, $L_2 = 0$ and $L_3 = -4.6255$, while the Kaplan-Yorke dimension of the Halvorsen circulant chaotic system has been obtained as $D_{KY} = 2.1753$. The adaptive feedback control and synchronization results were proved using Lyapunov stability theory. MATLAB simulations have been depicted to illustrate all the main results for the Halvorsen circulant chaotic system.

References

1. Lorenz EN (1963) Deterministic periodic flow. J Atmos Sci 20(2):130–141
2. Rössler OE (1976) An equation for continuous chaos. Phys Lett A 57(5):397–398
3. Arneodo A, Coullet P, Tresser C (1981) Possible new strange attractors with spiral structure. Commun Math Phys 79(4):573–576
4. Sprott JC (1994) Some simple chaotic flows. Phys Rev E 50(2):647–650
5. Chen G, Ueta T (1999) Yet another chaotic attractor. Int J Bifurcat Chaos 9(7):1465–1466
6. Lü J, Chen G (2002) A new chaotic attractor coined. Int J Bifurcat Chaos 12(3):659–661
7. Liu C, Liu T, Liu L, Liu K (2004) A new chaotic attractor. Chaos, Solitions Fractals 22(5):1031–1038
8. Cai G, Tan Z (2007) Chaos synchronization of a new chaotic system via nonlinear control. J Uncertain Syst 1(3):235–240
9. Chen HK, Lee CI (2004) Anti-control of chaos in rigid body motion. Chaos, Solitons Fractals 21(4):957–965
10. Tigan G, Opris D (2008) Analysis of a 3D chaotic system. Chaos, Solitons Fractals 36: 1315–1319
11. Zhou W, Xu Y, Lu H, Pan L (2008) On dynamics analysis of a new chaotic attractor. Phys Lett A 372(36):5773–5777
12. Zhu C, Liu Y, Guo Y (2010) Theoretic and numerical study of a new chaotic system. Intell Inf Manage 2:104–109
13. Li D (2008) A three-scroll chaotic attractor. Phys Lett A 372(4):387–393
14. Wei Z, Yang Q (2010) Anti-control of Hopf bifurcation in the new chaotic system with two stable node-foci. Appl Math Comput 217(1):422–429
15. Sundarapandian V (2013) Analysis and anti-synchronization of a novel chaotic system via active and adaptive controllers. J Eng Sci Technol Rev 6(4):45–52
16. Sundarapandian V, Pehlivan I (2012) Analysis, control, synchronization, and circuit design of a novel chaotic system. Math Comput Model 55(7–8):1904–1915
17. Vaidyanathan S (2013) A new six-term 3-D chaotic system with an exponential nonlinearity. Far East J Math Sci 79(1):135–143
18. Vaidyanathan S (2013) Analysis and adaptive synchronization of two novel chaotic systems with hyperbolic sinusoidal and cosinusoidal nonlinearity and unknown parameters. J Eng Sci Technol Rev 6(4):53–65

19. Vaidyanathan S (2014) A new eight-term 3-D polynomial chaotic system with three quadratic nonlinearities. Far East J Math Sci 84(2):219–226
20. Vaidyanathan S (2014) Analysis and adaptive synchronization of eight-term 3-D polynomial chaotic systems with three quadratic nonlinearities. Eur Phys J Special Topics 223(8): 1519–1529
21. Vaidyanathan S (2014) Analysis, control and synchronisation of a six-term novel chaotic system with three quadratic nonlinearities. Int J Model Ident Control 22(1):41–53
22. Vaidyanathan S (2014) Generalized projective synchronisation of novel 3-D chaotic systems with an exponential non-linearity via active and adaptive control. Int J Model Ident Control 22(3):207–217
23. Vaidyanathan S (2015) A 3-D novel highly chaotic system with four quadratic nonlinearities, its adaptive control and anti-synchronization with unknown parameters. J Eng Sci Technol Rev 8(2):106–115
24. Vaidyanathan S (2015) Analysis, properties and control of an eight-term 3-D chaotic system with an exponential nonlinearity. Int J Model Ident Control 23(2):164–172
25. Vaidyanathan S, Azar AT (2015) Analysis, control and synchronization of a nine-term 3-D novel chaotic system. In: Azar AT, Vaidyanathan S (eds) Chaos modelling and control systems design, studies in computational intelligence, vol 581. Springer, Germany, pp 19–38
26. Vaidyanathan S, Madhavan K (2013) Analysis, adaptive control and synchronization of a seven-term novel 3-D chaotic system. Int J Control Theory Appl 6(2):121–137
27. Vaidyanathan S, Pakiriswamy S (2015) A 3-D novel conservative chaotic system and its generalized projective synchronization via adaptive control. J Eng Sci Technol Rev 8(2): 52–60
28. Vaidyanathan S, Volos C (2015) Analysis and adaptive control of a novel 3-D conservative no-equilibrium chaotic system. Arch Control Sci 25(3):333–353
29. Vaidyanathan S, Volos C, Pham VT, Madhavan K, Idowu BA (2014) Adaptive backstepping control, synchronization and circuit simulation of a 3-D novel jerk chaotic system with two hyperbolic sinusoidal nonlinearities. Arch Control Sci 24(3):375–403
30. Vaidyanathan S, Rajagopal K, Volos CK, Kyprianidis IM, Stouboulos IN (2015) Analysis, adaptive control and synchronization of a seven-term novel 3-D chaotic system with three quadratic nonlinearities and its digital implementation in LabVIEW. J Eng Sci Technol Rev 8(2):130–141
31. Vaidyanathan S, Volos CK, Kyprianidis IM, Stouboulos IN, Pham VT (2015) Analysis, adaptive control and anti-synchronization of a six-term novel jerk chaotic system with two exponential nonlinearities and its circuit simulation. J Eng Sci Technol Rev 8(2):24–36
32. Vaidyanathan S, Volos CK, Pham VT (2015) Analysis, adaptive control and adaptive synchronization of a nine-term novel 3-D chaotic system with four quadratic nonlinearities and its circuit simulation. J Eng Sci Technol Rev 8(2):174–184
33. Vaidyanathan S, Volos CK, Pham VT (2015) Global chaos control of a novel nine-term chaotic system via sliding mode control. In: Azar AT, Zhu Q (eds) Advances and applications in sliding mode control systems, studies in computational intelligence, vol 576. Springer, Germany, pp 571–590
34. Pehlivan I, Moroz IM, Vaidyanathan S (2014) Analysis, synchronization and circuit design of a novel butterfly attractor. J Sound Vib 333(20):5077–5096
35. Sampath S, Vaidyanathan S, Volos CK, Pham VT (2015) An eight-term novel four-scroll chaotic system with cubic nonlinearity and its circuit simulation. J Eng Sci Technol Rev 8(2):1–6
36. Pham VT, Vaidyanathan S, Volos CK, Jafari S (2015) Hidden attractors in a chaotic system with an exponential nonlinear term. Eur Phys J Special Topics 224(8):1507–1517
37. Azar AT (2010) Fuzzy systems. IN-TECH, Vienna
38. Azar AT, Vaidyanathan S (2015) Chaos modeling and control systems design, studies in computational intelligence, vol 581. Springer, Germany
39. Azar AT, Vaidyanathan S (2015) Computational intelligence applications in modeling and control, studies in computational intelligence, vol 575. Springer, Germany

40. Azar AT, Vaidyanathan S (2015) Handbook of research on advanced intelligent control engineering and automation. advances in computational intelligence and robotics (ACIR), IGI-Global, USA
41. Azar AT, Zhu Q (2015) Advances and applications in sliding mode control systems, studies in computational intelligence, vol 576. Springer, Germany
42. Zhu Q, Azar AT (2015) Complex system modelling and control through intelligent soft computations, Studies in fuzzines and soft computing, vol 319. Springer, Germany
43. Kengne J, Chedjou JC, Kenne G, Kyamakya K (2012) Dynamical properties and chaos synchronization of improved Colpitts oscillators. Commun Nonlinear Sci Numer Simul 17(7):2914–2923
44. Sharma A, Patidar V, Purohit G, Sud KK (2012) Effects on the bifurcation and chaos in forced Duffing oscillator due to nonlinear damping. Commun Nonlinear Sci Numer Simul 17(6):2254–2269
45. Li N, Pan W, Yan L, Luo B, Zou X (2014) Enhanced chaos synchronization and communication in cascade-coupled semiconductor ring lasers. Commun Nonlinear Sci Numer Simul 19(6):1874–1883
46. Yuan G, Zhang X, Wang Z (2014) Generation and synchronization of feedback-induced chaos in semiconductor ring lasers by injection-locking. Optik Int J Light Electron Opt 125(8): 1950–1953
47. Gaspard P (1999) Microscopic chaos and chemical reactions. Physica A: Stat Mech Appl 263(1–4):315–328
48. Petrov V, Gaspar V, Masere J, Showalter K (1993) Controlling chaos in Belousov-Zhabotinsky reaction. Nature 361:240–243
49. Vaidyanathan S (2015) Adaptive control of a chemical chaotic reactor. Int J PharmTech Res 8(3):377–382
50. Vaidyanathan S (2015) Adaptive synchronization of chemical chaotic reactors. Int J ChemTech Res 8(2):612–621
51. Vaidyanathan S (2015) Anti-synchronization of Brusselator chemical reaction systems via adaptive control. Int J ChemTech Res 8(6):759–768
52. Vaidyanathan S (2015) Dynamics and control of Brusselator chemical reaction. Int J ChemTech Res 8(6):740–749
53. Vaidyanathan S (2015) Dynamics and control of Tokamak system with symmetric and magnetically confined plasma. Int J ChemTech Res 8(6):795–803
54. Vaidyanathan S (2015) Synchronization of Tokamak systems with symmetric and magnetically confined plasma via adaptive control. Int J ChemTech Res 8(6):818–827
55. Das S, Goswami D, Chatterjee S, Mukherjee S (2014) Stability and chaos analysis of a novel swarm dynamics with applications to multi-agent systems. Eng Appl Artif Intell 30:189–198
56. Kyriazis M (1991) Applications of chaos theory to the molecular biology of aging. Exp Gerontol 26(6):569–572
57. Vaidyanathan S (2015) 3-cells cellular neural network (CNN) attractor and its adaptive biological control. Int J PharmTech Res 8(4):632–640
58. Vaidyanathan S (2015) Adaptive backstepping control of eEnzymes-substrates system with ferroelectric behaviour in brain waves. Int J PharmTech Res 8(2):256–261
59. Vaidyanathan S (2015) Adaptive biological control of generalized Lotka-Volterra three-species biological system. Int J PharmTech Res 8(4):622–631
60. Vaidyanathan S (2015) Adaptive chaotic synchronization of enzymes-substrates system with ferroelectric behaviour in brain waves. Int J PharmTech Res 8(5):964–973
61. Vaidyanathan S (2015) Adaptive synchronization of generalized Lotka-Volterra three-species biological systems. Int J PharmTech Res 8(5):928–937
62. Vaidyanathan S (2015) Chaos in neurons and adaptive control of Birkhoff-Shaw strange chaotic attractor. Int J PharmTech Res 8(5):956–963
63. Vaidyanathan S (2015) Lotka-Volterra population biology models with negative feedback and their ecological monitoring. Int J PharmTech Res 8(5):974–981

64. Vaidyanathan S (2015) Synchronization of 3-cells cellular neural network (CNN) attractors via adaptive control method. Int J PharmTech Res 8(5):946–955
65. Gibson WT, Wilson WG (2013) Individual-based chaos: extensions of the discrete logistic model. J Theoret Biol 339:84–92
66. Suérez I (1999) Mastering chaos in ecology. Ecol Model 117(2–3):305–314
67. Lang J (2015) Color image encryption based on color blend and chaos permutation in the reality-preserving multiple-parameter fractional Fourier transform domain. Opt Commun 338:181–192
68. Zhang X, Zhao Z, Wang J (2014) Chaotic image encryption based on circular substitution box and key stream buffer. Signal Process Image Commun 29(8):902–913
69. Rhouma R, Belghith S (2011) Cryptoanalysis of a chaos based cryptosystem on DSP. Commun Nonlinear Sci Numer Simul 16(2):876–884
70. Usama M, Khan MK, Alghatbar K, Lee C (2010) Chaos-based secure satellite imagery cryptosystem. Comput Math Appl 60(2):326–337
71. Azar AT, Serrano FE (2014) Robust IMC-PID tuning for cascade control systems with gain and phase margin specifications. Neural Comput Appl 25(5):983–995
72. Azar AT, Serrano FE (2015) Adaptive sliding mode control of the Furuta pendulum. In: Azar AT, Zhu Q (eds) Advances and applications in sliding mode control systems, studies in computational intelligence, vol 576. Springer, Germany, pp 1–42
73. Azar AT, Serrano FE (2015) Deadbeat control for multivariable systems with time varying delays. In: Azar AT, Vaidyanathan S (eds) Chaos modeling and control systems design, studies in computational intelligence, vol 581. Springer, Germany, pp 97–132
74. Azar AT, Serrano FE (2015) Design and modeling of anti wind up PID controllers. In: Zhu Q, Azar AT (eds) Complex system modelling and control through intelligent soft computations, studies in fuzziness and soft computing, vol 319. Springer, Germany, pp 1–44
75. Azar AT, Serrano FE (2015) Stabilizatoin and control of mechanical systems with backlash. In: Azar AT, Vaidyanathan S (eds) Handbook of research on advanced intelligent control engineering and automation., Advances in computational intelligence and robotics (ACIR)IGI-Global, USA, pp 1–60
76. Feki M (2003) An adaptive chaos synchronization scheme applied to secure communication. Chaos, Solitons Fractals 18(1):141–148
77. Murali K, Lakshmanan M (1998) Secure communication using a compound signal from generalized chaotic systems. Phys Lett A 241(6):303–310
78. Zaher AA, Abu-Rezq A (2011) On the design of chaos-based secure communication systems. Commun Nonlinear Syst Numer Simul 16(9):3721–3727
79. Mondal S, Mahanta C (2014) Adaptive second order terminal sliding mode controller for robotic manipulators. J Franklin Inst 351(4):2356–2377
80. Nehmzow U, Walker K (2005) Quantitative description of robot-environment interaction using chaos theory. Robot Auton Syst 53(3–4):177–193
81. Volos CK, Kyprianidis IM, Stouboulos IN (2013) Experimental investigation on coverage performance of a chaotic autonomous mobile robot. Robot Auton Syst 61(12):1314–1322
82. Qu Z (2011) Chaos in the genesis and maintenance of cardiac arrhythmias. Prog Biophys Mol Biol 105(3):247–257
83. Witte CL, Witte MH (1991) Chaos and predicting varix hemorrhage. Med Hypotheses 36(4):312–317
84. Azar AT (2012) Overview of type-2 fuzzy logic systems. Int J Fuzzy Syst Appl 2(4):1–28
85. Li Z, Chen G (2006) Integration of fuzzy logic and chaos theory, studies in fuzziness and soft computing, vol 187. Springer, Germany
86. Huang X, Zhao Z, Wang Z, Li Y (2012) Chaos and hyperchaos in fractional-order cellular neural networks. Neurocomputing 94:13–21
87. Kaslik E, Sivasundaram S (2012) Nonlinear dynamics and chaos in fractional-order neural networks. Neural Netw 32:245–256
88. Lian S, Chen X (2011) Traceable content protection based on chaos and neural networks. Appl Soft Comput 11(7):4293–4301

89. Guégan D (2009) Chaos in economics and finance. Ann Rev Control 33(1):89–93
90. Sprott JC (2004) Competition with evolution in ecology and finance. Phys Lett A 325 (5–6):329–333
91. Pham VT, Volos CK, Vaidyanathan S, Le TP, Vu VY (2015) A memristor-based hyperchaotic system with hidden attractors: dynamics, synchronization and circuital emulating. J Eng Sci Technol Rev 8(2):205–214
92. Volos CK, Kyprianidis IM, Stouboulos IN, Tlelo-Cuautle E, Vaidyanathan S (2015) Memristor: a new concept in synchronization of coupled neuromorphic circuits. J Eng Sci Technol Rev 8(2):157–173
93. Sundarapandian V (2010) Output regulation of the Lorenz attractor. Far East J Math Sci 42(2):289–299
94. Vaidyanathan S (2011) Output regulation of Arneodo-Coullet chaotic system. Commun Comput Inf Sci 133:98–107
95. Vaidyanathan S (2011) Output regulation of the unified chaotic system. Commun Comput Inf Sci 198:10–17
96. Vaidyanathan S, Azar AT, Rajagopal K, Alexander P (2015) Design and SPICE implementation of a 12-term novel hyperchaotic system and its synchronisation via active control. Int J Model Ident Control 23(3):267–277
97. Noroozi N, Roopaei M, Karimaghaee P, Safavi AA (2010) Simple adaptive variable structure control for unknown chaotic systems. Commun Nonlinear Sci Numer Simul 15(3):707–727
98. Vaidyanathan S, Volos CK, Pham VT (2015) Analysis, control, synchronization and SPICE implementation of a novel 4-D hyperchaotic Rikitake dynamo System without equilibrium. J Eng Sci Technol Rev 8(2):232–244
99. Vaidyanathan S (2012) Global chaos control of hyperchaotic Liu system via sliding control method. Int J Control Theory Appl 5(2):117–123
100. Vaidyanathan S (2012) Sliding mode control based global chaos control of Liu-Liu-Liu-Su chaotic system. Int J Control Theory Appl 1(2):15–20
101. Vaidyanathan S, Volos CK, Rajagopal K, Kyprianidis IM, Stouboulos IN (2015) Adaptive backstepping controller design for the anti-synchronization of identical WINDMI chaotic systems with unknown parameters and its SPICE implementation. J Eng Sci Technol Rev 8(2):74–82
102. Carroll TL, Pecora LM (1991) Synchronizing chaotic circuits. IEEE Trans Circuits Syst 38(4):453–456
103. Pecora LM, Carroll TL (1990) Synchronization in chaotic systems. Phys Rev Lett 64(8): 821–824
104. Karthikeyan R, Sundarapandian V (2014) Hybrid chaos synchronization of four-scroll systems via active control. J Electr Eng 65(2):97–103
105. Sarasu P, Sundarapandian V (2011) Active controller design for the generalized projective synchronization of four-scroll chaotic systems. Int J Syst Signal Control Eng Appl 4(2):26–33
106. Sarasu P, Sundarapandian V (2011) The generalized projective synchronization of hyperchaotic Lorenz and hyperchaotic Qi systems via active control. Int J Soft Comput 6(5): 216–223
107. Sundarapandian V, Karthikeyan R (2012) Hybrid synchronization of hyperchaotic Lorenz and hyperchaotic Chen systems via active control. J Eng Appl Sci 7(3):254–264
108. Vaidyanathan S (2011) Hybrid chaos synchronization of Liu and Lü systems by active nonlinear control. Commun Comput Inf Sci 204
109. Vaidyanathan S (2012) Output regulation of the Liu chaotic system. Appl Mech Mater 110–116:3982–3989
110. Vaidyanathan S, Rajagopal K (2011) Anti-synchronization of Li and T chaotic systems by active nonlinear control. Commun Comput Inf Sci 198:175–184
111. Vaidyanathan S, Rajagopal K (2011) Global chaos synchronization of hyperchaotic Pang and Wang systems by active nonlinear control. Commun Comput Inf Sci 204:84–93
112. Vaidyanathan S, Rasappan S (2011) Global chaos synchronization of hyperchaotic Bao and Xu systems by active nonlinear control. Commun Comput Inf Sci 198:10–17

113. Vaidyanathan S, VTP, Volos CK, (2015) A 5-D hyperchaotic Rikitake dynamo system with hidden attractors. Eur Phys J Special Topics 224(8):1575–1592
114. Sarasu P, Sundarapandian V (2012) Adaptive controller design for the generalized projective synchronization of 4-scroll systems. Int J Syst Signal Control Eng Appl 5(2):21–30
115. Sarasu P, Sundarapandian V (2012) Generalized projective synchronization of three-scroll chaotic systems via adaptive control. Eur J Sci Res 72(4):504–522
116. Sarasu P, Sundarapandian V (2012) Generalized projective synchronization of two-scroll systems via adaptive control. Int J Soft Comput 7(4):146–156
117. Sundarapandian V, Karthikeyan R (2011) Anti-synchronization of hyperchaotic Lorenz and hyperchaotic Chen systems by adaptive control. Int J Syst Signal Control Eng Appl 4(2):18–25
118. Sundarapandian V, Karthikeyan R (2011) Anti-synchronization of Lü and Pan chaotic systems by adaptive nonlinear control. Eur J Sci Res 64(1):94–106
119. Sundarapandian V, Karthikeyan R (2012) Adaptive anti-synchronization of uncertain Tigan and Li systems. J Eng Appl Sci 7(1):45–52
120. Vaidyanathan S (2012) Anti-synchronization of Sprott-L and Sprott-M chaotic systems via adaptive control. Int J Control Theory Appl 5(1):41–59
121. Vaidyanathan S (2013) Analysis, control and synchronization of hyperchaotic Zhou system via adaptive control. Adv Intell Syst Comput 177:1–10
122. Vaidyanathan S (2015) Hyperchaos, qualitative analysis, control and synchronisation of a ten-term 4-D hyperchaotic system with an exponential nonlinearity and three quadratic nonlinearities. Int J Model Ident Control 23(4):380–392
123. Vaidyanathan S, Azar AT (2015) Analysis and control of a 4-D novel hyperchaotic system. In: Azar AT, Vaidyanathan S (eds) Chaos modeling and control systems design, studies in computational intelligence, vol 581. Springer, Germany, pp 19–38
124. Vaidyanathan S, Pakiriswamy S (2013) Generalized projective synchronization of six-term Sundarapandian chaotic systems by adaptive control. Int J Control Theory Appl 6(2):153–163
125. Vaidyanathan S, Rajagopal K (2011) Global chaos synchronization of Lü and Pan systems by adaptive nonlinear control. Commun Comput Inf Sci 205:193–202
126. Vaidyanathan S, Rajagopal K (2012) Global chaos synchronization of hyperchaotic Pang and hyperchaotic Wang systems via adaptive control. Int J Soft Comput 7(1):28–37
127. Vaidyanathan S, Volos C, Pham VT (2014) Hyperchaos, adaptive control and synchronization of a novel 5-D hyperchaotic system with three positive Lyapunov exponents and its SPICE implementation. Arch Control Sci 24(4):409–446
128. Vaidyanathan S, Volos C, Pham VT, Madhavan K (2015) Analysis, adaptive control and synchronization of a novel 4-D hyperchaotic hyperjerk system and its SPICE implementation. Arch Control Sci 25(1):5–28
129. Gan Q, Liang Y (2012) Synchronization of chaotic neural networks with time delay in the leakage term and parametric uncertainties based on sampled-data control. J Franklin Inst 349(6):1955–1971
130. Li N, Zhang Y, Nie Z (2011) Synchronization for general complex dynamical networks with sampled-data. Neurocomputing 74(5):805–811
131. Xiao X, Zhou L, Zhang Z (2014) Synchronization of chaotic Lur'e systems with quantized sampled-data controller. Commun Nonlinear Sci Numer Simul 19(6):2039–2047
132. Zhang H, Zhou J (2012) Synchronization of sampled-data coupled harmonic oscillators with control inputs missing. Syst Control Lett 61(12):1277–1285
133. Chen WH, Wei D, Lu X (2014) Global exponential synchronization of nonlinear time-delay Lur'e systems via delayed impulsive control. Commun Nonlinear Sci Numer Simul 19(9):3298–3312
134. Jiang GP, Zheng WX, Chen G (2004) Global chaos synchronization with channel time-delay. Chaos, Solitons Fractals 20(2):267–275
135. Shahverdiev EM, Shore KA (2009) Impact of modulated multiple optical feedback time delays on laser diode chaos synchronization. Opt Commun 282(17):3568–3572
136. Rasappan S, Vaidyanathan S (2012) Global chaos synchronization of WINDMI and Coullet chaotic systems by backstepping control. Far East J Math Sci 67(2):265–287

137. Rasappan S, Vaidyanathan S (2012) Hybrid synchronization of n-scroll Chua and Lur'e chaotic systems via backstepping control with novel feedback. Arch Control Sci 22(3):343–365
138. Rasappan S, Vaidyanathan S (2012) Synchronization of hyperchaotic Liu system via backstepping control with recursive feedback. Commun Comput Inf Sci 305:212–221
139. Rasappan S, Vaidyanathan S (2013) Hybrid synchronization of n-scroll chaotic Chua circuits using adaptive backstepping control design with recursive feedback. Malays J Math Sci 7(2):219–246
140. Rasappan S, Vaidyanathan S (2014) Global chaos synchronization of WINDMI and Coullet chaotic systems using adaptive backstepping control design. Kyungpook Math J 54(1): 293–320
141. Suresh R, Sundarapandian V (2013) Global chaos synchronization of a family of n-scroll hyperchaotic Chua circuits using backstepping control with recursive feedback. Far East J Math Sci 73(1):73–95
142. Vaidyanathan S, Rasappan S (2014) Global chaos synchronization of n-scroll Chua circuit and Lur'e system using backstepping control design with recursive feedback. Arab J Sci Eng 39(4):3351–3364
143. Vaidyanathan S, Idowu BA, Azar AT (2015) Backstepping controller design for the global chaos synchronization of Sprott's jerk systems. Stud Comput Intell 581:39–58
144. Sundarapandian V, Sivaperumal S (2011) Sliding controller design of hybrid synchronization of four-wing chaotic systems. Int J Soft Comput 6(5):224–231
145. Vaidyanathan S (2012) Analysis and synchronization of the hyperchaotic Yujun systems via sliding mode control. Adv Intell Syst Comput 176:329–337
146. Vaidyanathan S (2014) Global chaos synchronization of identical Li-Wu chaotic systems via sliding mode control. Int J Model Ident Control 22(2):170–177
147. Vaidyanathan S, Azar AT (2015) Anti-synchronization of identical chaotic systems using sliding mode control and an application to Vaidhyanathan-Madhavan chaotic systems. Stud Comput Intell 576:527–547
148. Vaidyanathan S, Azar AT (2015) Hybrid synchronization of identical chaotic systems using sliding mode control and an application to Vaidhyanathan chaotic systems. Stud Comput Intell 576:549–569
149. Vaidyanathan S, Sampath S (2011) Global chaos synchronization of hyperchaotic Lorenz systems by sliding mode control. Commun Comput Inf Sci 205:156–164
150. Vaidyanathan S, Sampath S (2012) Anti-synchronization of four-wing chaotic systems via sliding mode control. Int J Autom Comput 9(3):274–279
151. Sprott JC (2010) Elegant chaos: algebraically simple chaotic flows. World Scientific, Singapore
152. Khalil HK (2001) Nonlinear systems. Prentice Hall, New Jersey

Adaptive Backstepping Control and Synchronization of a Novel 3-D Jerk System with an Exponential Nonlinearity

Sundarapandian Vaidyanathan and Ahmad Taher Azar

Abstract In this research work, we announce a seven-term novel 3-D jerk chaotic system with an exponential nonlinearity. First, we discuss the qualitative properties of the novel jerk chaotic system. The novel jerk chaotic system has a unique equilibrium point, which is a saddle-focus. Thus, the unique equilibrium point is unstable. We obtain the Lyapunov exponents of the novel jerk chaotic system as $L_1 = 0.1066$, $L_2 = 0$ and $L_3 = -1.1047$. Also, the Kaplan-Yorke dimension of the novel jerk chaotic system is obtained as $D_{KY} = 2.0965$. Next, an adaptive backstepping controller is designed to stabilize the novel jerk chaotic system with unknown system parameters. Moreover, an adaptive backstepping controller is designed to achieve complete chaos synchronization of the identical novel jerk chaotic systems with unknown system parameters. The main control results are established using Lyapunov stability theory. MATLAB simulations are shown to illustrate all the main results developed in this work.

Keywords Chaos · Chaotic systems · Jerk systems · Backstepping control · Adaptive control · Synchronization

1 Introduction

Chaotic systems are defined as nonlinear dynamical systems which are sensitive to initial conditions, topologically mixing and with dense periodic orbits. Sensitivity to initial conditions of chaotic systems is popularly known as the *butterfly effect*.

S. Vaidyanathan (✉)
Research and Development Centre, Vel Tech University,
Avadi, Chennai 600062, Tamil Nadu, India
e-mail: sundarvtu@gmail.com

A.T. Azar
Faculty of Computers and Information, Benha University, Banha, Egypt
e-mail: ahmad_t_azar@ieee.org; ahmad.azar@fci.bu.edu.eg

A.T. Azar and S. Vaidyanathan (eds.), *Advances in Chaos Theory and Intelligent Control*, Studies in Fuzziness and Soft Computing 337,
DOI 10.1007/978-3-319-30340-6_11

The Lyapunov exponent is a measure of the divergence of phase points that are initially very close and can be used to quantify chaotic systems. It is common to refer to the largest Lyapunov exponent as the *Maximal Lyapunov Exponent* (MLE). A positive maximal Lyapunov exponent and phase space compactness are usually taken as defining conditions for a chaotic system.

Some classical paradigms of 3-D chaotic systems in the literature are Lorenz system [38], Rössler system [56], ACT system [1], Sprott systems [66], Chen system [15], Lü system [39], Liu system [37], Cai system [13], Chen-Lee system [16], Tigan system [80], etc.

Many new chaotic systems have been discovered in the recent years such as Zhou system [151], Zhu system [152], Li system [32], Wei-Yang system [144], Sundarapandian systems [72, 77], Vaidyanathan systems [90, 91, 93–96, 99, 106, 116, 119, 121, 130, 132, 135, 137, 138, 140], Pehlivan system [45], Sampath system [58], Pham system [47], etc.

Chaos theory and control systems have many important applications in science and engineering [2, 9–12, 153]. Some commonly known applications are oscillators [28, 65], lasers [34, 147], chemical reactions [21, 46, 103, 104, 107, 109, 110, 114], biology [18, 30, 98, 100–102, 105, 108, 112, 113], ecology [22, 70], encryption [31, 150], cryptosystems [55, 81], mechanical systems [4–8], secure communications [19, 41, 148], robotics [40, 42, 142], cardiology [49, 145], intelligent control [3, 35], neural networks [24, 27, 36], finance [23, 68], memristors [48, 143], etc.

Control of chaotic systems aims to regulate or stabilize the chaotic system about its equilibrium points. There are many methods available in the literature such as active control method [71, 83, 84, 133], adaptive control method [43, 139], sliding mode control method [87, 89], backstepping control method [141], etc.

Synchronization of chaotic systems is a phenomenon that occurs when two or more chaotic systems are coupled or when a chaotic system drives another chaotic system. Major works on synchronization of chaotic systems deal with the complete synchronization of a pair of chaotic systems called the *master* and *slave* systems. The design goal of the complete synchronization is to apply the output of the master system to control the slave system so that the output of the slave system tracks the output of the master system asymptotically with time.

Thus, if $x(t)$ and $y(t)$ denote the states of the master and slave systems, then the design goal of complete synchronization is to satisfy the condition

$$\lim_{t \to \infty} \|x(t) - y(t)\| = 0, \quad \forall x(0), y(0) \in \mathbb{R}^n \tag{1}$$

Pecora and Carroll pioneered the research on synchronization of chaotic systems with their seminal papers [14, 44]. The active control method [26, 59, 60, 71, 76, 82, 88, 122, 123, 126] is typically used when the system parameters are available for measurement. Adaptive control method [57, 61–63, 73–75, 86, 92, 111, 115, 120, 124, 125, 131, 136] is typically used when some or all the system parameters are not available for measurement and estimates for the uncertain parameters of the systems. Sampled-data feedback control method [20, 33, 146, 149] and time-delay feedback control method [17, 25, 64] are also used for synchronization of chaotic systems.

Backstepping control method [50–54, 79, 127, 134] is also used for the synchronization of chaotic systems. Another popular method for the synchronization of chaotic systems is the sliding mode control method [78, 85, 97, 117, 118, 128, 129].

In the recent decades, there is some good interest in finding novel chaotic systems, which are of the jerk type [69]. Jerk chaotic systems are chaotic systems that can be expressed by an explicit third-order differential equation describing the time evolution of the single scalar variable x given by

$$\dddot{x} = j\left(x, \dot{x}, \ddot{x}\right). \tag{2}$$

The differential equation (2) is called *jerk system*, because the third order time derivative in mechanical systems is called *jerk*. Thus, in order to study different aspects of chaos, the ODE (2) can be considered instead of a 3-D system.

The simplest known jerk system was discovered by Sprott [67] and Sprott's jerk function contains just three terms with a quadratic nonlinearity:

$$j\left(x, \dot{x}, \ddot{x}\right) = -A\ddot{x} + \dot{x}^2 - x \quad (\text{with } A = 2.017) \tag{3}$$

In this research work, we consider jerk systems described by 3-D systems of differential equations.

Defining the phase variables

$$x_1 = x, \quad x_2 = \dot{x}, \quad x_3 = \ddot{x}, \tag{4}$$

we can rewrite the jerk differential equation (2) in system form as

$$\begin{cases} \dot{x}_1 = x_2 \\ \dot{x}_2 = x_3 \\ \dot{x}_3 = j(x_1, x_2, x_3) \end{cases} \tag{5}$$

In this research work, we propose a seven-term novel jerk chaotic system with an exponential nonlinearity. We also derive new results for the adaptive control and synchronization of the novel jerk chaotic system with unknown parameters via backstepping control. The Lyapunov exponents of the novel jerk chaotic system are obtained as $L_1 = 0.1066$, $L_2 = 0$ and $L_3 = -1.1047$. Also, the Kaplan-Yorke dimension of the novel jerk chaotic system is derived as $D_{KY} = 2.0965$.

Section 2 describes the dynamic equations and phase portraits of the novel jerk chaotic system. Section 3 describes the qualitative properties of the novel jerk chaotic system. Section 4 describes the adaptive control of the novel jerk chaotic system with unknown parameters. Section 5 describes the adaptive synchronization of the identical novel jerk chaotic system with unknown parameters. Finally, Sect. 6 contains the conclusions of this work.

2 A Novel 3-D Jerk Chaotic System

In this section, we announce a novel 3-D jerk chaotic system with an exponential nonlinearity, which is described by the 3-D dynamics

$$\begin{cases} \dot{x}_1 = x_2 \\ \dot{x}_2 = x_3 \\ \dot{x}_3 = 5 - a(x_1 + x_2) + bx_1x_2 - \exp(x_1) \end{cases} \tag{6}$$

where x_1, x_2, x_3 are the states and a, b are constant, positive, parameters.

The jerk system (6) is *chaotic* when the parameter values are taken as

$$a = 1, \quad b = 0.4 \tag{7}$$

For numerical simulations, we take the initial values of the novel jerk chaotic system (6) as

$$x_1(0) = 0.1, \quad x_2(0) = 0.4, \quad x_3(0) = 0.6 \tag{8}$$

Figure 1 describes the *strange chaotic attractor* of the novel jerk chaotic system (6) in $\mathbb{R}^3$.

Figures 2, 3 and 4 describe the 2-D projection of the strange chaotic attractor of the novel jerk chaotic system (6) on (x_1, x_2), (x_2, x_3) and (x_1, x_3) coordinate planes respectively.

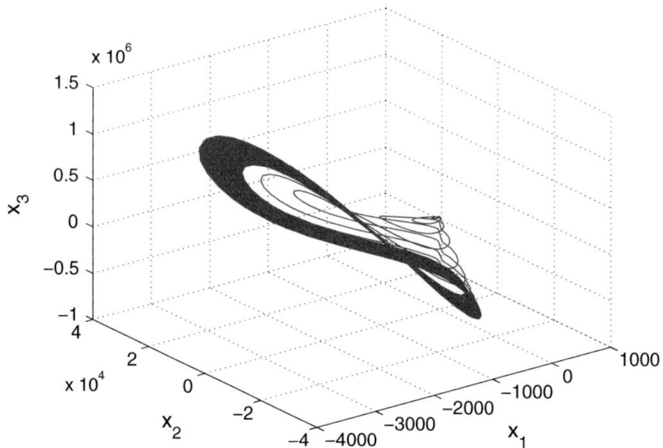

Fig. 1 Strange attractor of the novel jerk chaotic system in $\mathbf{R}^3$

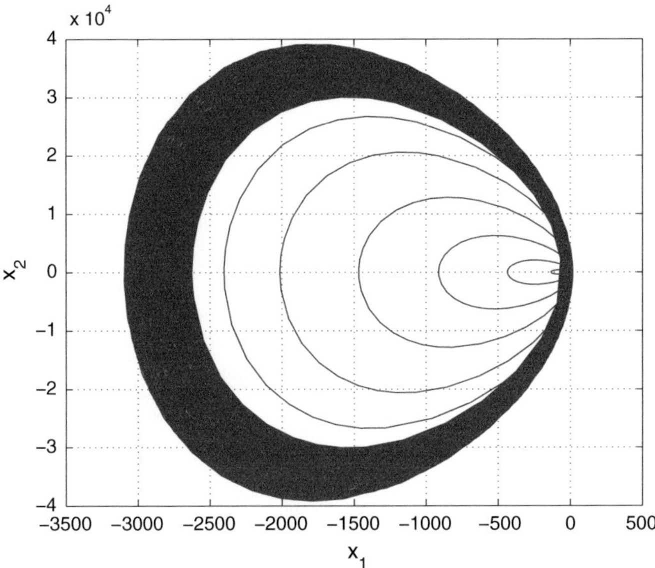

Fig. 2 2-D projection of the novel jerk chaotic system on (x_1, x_2)-plane

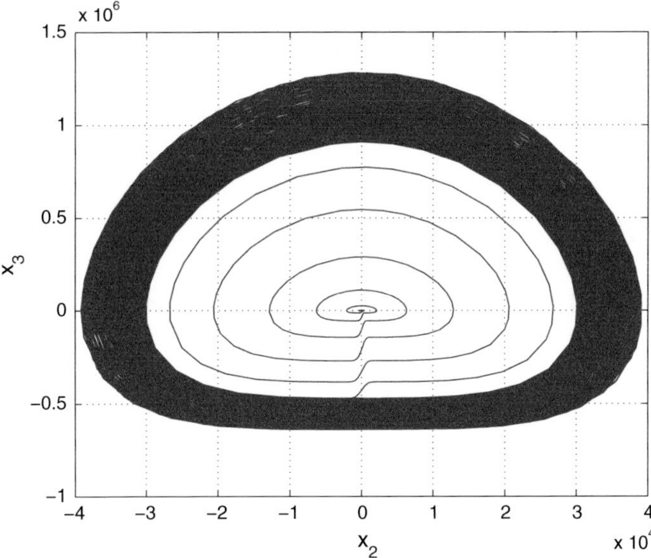

Fig. 3 2-D projection of the novel jerk chaotic system on (x_2, x_3)-plane

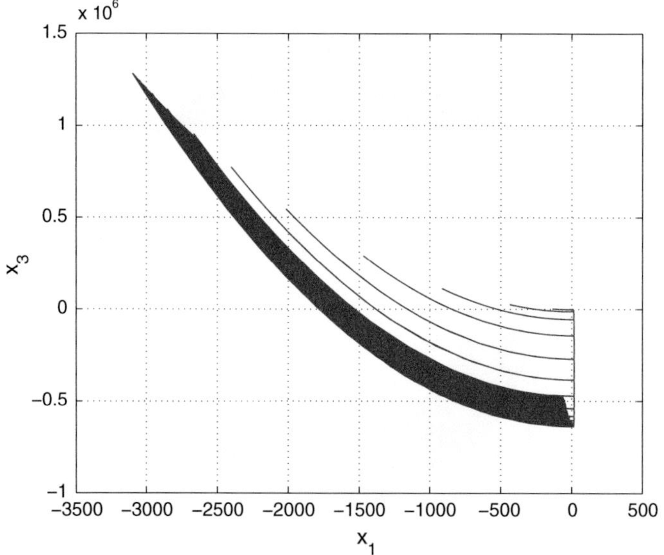

Fig. 4 2-D projection of the novel jerk chaotic system on (x_1, x_3)-plane

3 Properties of the Novel Jerk Chaotic System

This section describes the qualitative properties of the novel jerk chaotic system (6).

3.1 Dissipativity

We write the novel jerk chaotic system (6) in vector notation as

$$\dot{x} = f(x) = \begin{bmatrix} f_1(x) \\ f_2(x) \\ f_3(x) \end{bmatrix}, \tag{9}$$

where

$$\begin{aligned} f_1(x) &= x_2 \\ f_2(x) &= x_3 \\ f_3(x) &= 5 - a(x_2 + x_3) + bx_1x_2 - \exp(x_1) \end{aligned} \tag{10}$$

We take the parameter values as in the chaotic case, viz. $a = 1$ and $b = 0.4$. The divergence of the vector field f on $\mathbb{R}^3$ is obtained as

$$\text{div } f = \frac{\partial f_1(x)}{\partial x_1} + \frac{\partial f_2(x)}{\partial x_2} + \frac{\partial f_3(x)}{\partial x_3} = 0 + 0 - a = -1 < 0 \tag{11}$$

Let Ω be any region in $\mathbb{R}^3$ with a smooth boundary. Let $\Omega(t) = \Phi_t(\Omega)$, where Φ_t is the flow of the vector field f.

Let $V(t)$ denote the volume of $\Omega(t)$.

By Liouville's theorem, it follows that

$$\frac{dV(t)}{dt} = \int\limits_{\Omega(t)} (\operatorname{div} f) dx_1\, dx_2\, dx_3 \tag{12}$$

Substituting the value of $\operatorname{div} f$ in (12) leads to

$$\frac{dV(t)}{dt} = -\int\limits_{\Omega(t)} dx_1\, dx_2\, dx_3 = -V(t) \tag{13}$$

Integrating the linear differential equation (13), $V(t)$ is obtained as

$$V(t) = V(0)\exp(-t) \tag{14}$$

From Eq. (14), it follows that the volume $V(t)$ shrinks to zero exponentially as $t \to \infty$.

Thus, the novel jerk chaotic system (6) is dissipative. Hence, any asymptotic motion of the system (6) settles onto a set of measure zero, i.e. a strange attractor.

3.2 *Equilibrium Point*

The equilibrium points of the novel jerk chaotic system (6) are obtained by solving the system of equations

$$\begin{cases} x_2 = 0 \\ x_3 = 0 \\ 5 - a(x_2 + x_3) + bx_1x_2 - \exp(x_1) = 0 \end{cases} \tag{15}$$

Solving the system (15), we obtain the unique equilibrium point of the novel jerk chaotic system (6) as

$$E_1 = \begin{bmatrix} \ln(5) \\ 0 \\ 0 \end{bmatrix} = \begin{bmatrix} 1.6094 \\ 0 \\ 0 \end{bmatrix} \tag{16}$$

The Jacobian matrix of the novel jerk chaotic system (6) at E_1 is obtained as

$$J_1 = J(E_1) = \begin{bmatrix} 0 & 1 & 0 \\ 0 & 0 & 1 \\ -5 & -0.3562 & -1 \end{bmatrix} \tag{17}$$

The matrix J_1 has the eigenvalues

$$\lambda_1 = -2.0337, \quad \lambda_{2,3} = 0.5167 \pm 1.4803i \tag{18}$$

This shows that the equilibrium point E_1 is a saddle-focus, which is unstable.

3.3 Lyapunov Exponents and Kaplan-Yorke Dimension

We take the parameter values of the novel jerk chaotic system (6) as in the chaotic case (7), i.e.

$$a = 1, \quad b = 0.4 \tag{19}$$

We take the initial values of the novel jerk chaotic system (6) as in (8), i.e.

$$x_1(0) = 0.1, \quad x_2(0) = 0.4, \quad x_3(0) = 0.6 \tag{20}$$

Then the Lyapunov exponents of the novel jerk chaotic system (6) are numerically obtained as

$$L_1 = 0.1066, \quad L_2 = 0, \quad L_3 = -1.1047 \tag{21}$$

Since $L_1 + L_2 + L_3 = -0.9981 < 0$, the novel jerk chaotic system (6) is dissipative.

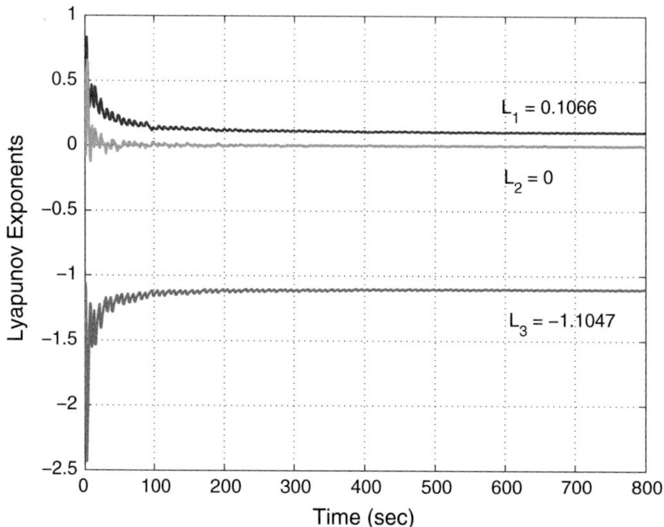

Fig. 5 Dynamics of the Lyapunov exponents of the novel jerk chaotic system

Figure 5 shows the dynamics of the Lyapunov exponents of the novel jerk chaotic system (6). From Fig. 5, the Maximal Lyapunov Exponent (MLE) of the novel jerk chaotic system (6) is obtained as $L_1 = 0.1066$, which is a small value. Thus, the novel jerk chaotic system (6) is *mildly chaotic*.

Also, the Kaplan-Yorke dimension of the novel jerk chaotic system (6) is obtained as

$$D_{KY} = 2 + \frac{L_1 + L_2}{|L_3|} = 2.0965 \tag{22}$$

which is fractional.

4 Adaptive Control of Novel Jerk Chaotic System

In this section, we use backstepping control method to derive an adaptive control law for globally stabilizing the 3-D novel jerk chaotic system with unknown parameters.

Thus, we consider the 3-D novel jerk chaotic system given by

$$\begin{cases} \dot{x}_1 = x_2 \\ \dot{x}_2 = x_3 \\ \dot{x}_3 = 5 - a(x_2 + x_3) + bx_1x_2 - \exp(x_1) + u \end{cases} \tag{23}$$

where a and b are unknown constant parameters, and u is a backstepping control law to be determined using estimates $\hat{a}(t)$ and $\hat{b}(t)$ for a and b, respectively.

The parameter estimation errors are defined as:

$$\begin{cases} e_a(t) = a - \hat{a}(t) \\ e_b(t) = b - \hat{b}(t) \end{cases} \tag{24}$$

Differentiating (24) with respect to t, we obtain the following equations:

$$\begin{cases} \dot{e}_a(t) = -\dot{\hat{a}}(t) \\ \dot{e}_b(t) = -\dot{\hat{b}}(t) \end{cases} \tag{25}$$

Theorem 1 *The 3-D novel jerk chaotic system (23), with unknown parameters a and b, is globally and exponentially stabilized by the adaptive feedback control law,*

$$u(t) = -5 - 3x_1 - (5 - \hat{a}(t))x_2 - (3 - \hat{a}(t))x_3 - \hat{b}(t)x_1x_2 + \exp(x_1) - kz_3 \tag{26}$$

where $k > 0$ is a gain constant,

$$z_3 = 2x_1 + 2x_2 + x_3, \tag{27}$$

and the update law for the parameter estimates $\hat{a}(t), \hat{b}(t)$ *is given by*

$$\begin{cases} \dot{\hat{a}}(t) = -(x_2 + x_3)z_3 \\ \dot{\hat{b}}(t) = x_1 x_2 z_3 \end{cases} \tag{28}$$

Proof We prove this result via backstepping control method and Lyapunov stability theory [29].

First, we define a quadratic Lyapunov function

$$V_1(z_1) = \frac{1}{2} z_1^2 \tag{29}$$

where

$$z_1 = x_1 \tag{30}$$

Differentiating V_1 along the dynamics (23), we get

$$\dot{V}_1 = z_1 \dot{z}_1 = x_1 x_2 = -z_1^2 + z_1(x_1 + x_2) \tag{31}$$

Now, we define

$$z_2 = x_1 + x_2 \tag{32}$$

Using (32), we can simplify the equation (31) as

$$\dot{V}_1 = -z_1^2 + z_1 z_2 \tag{33}$$

Secondly, we define a quadratic Lyapunov function

$$V_2(z_1, z_2) = V_1(z_1) + \frac{1}{2} z_2^2 = \frac{1}{2}\left(z_1^2 + z_2^2\right) \tag{34}$$

Differentiating V_2 along the dynamics (23), we get

$$\dot{V}_2 = -z_1^2 - z_2^2 + z_2(2x_1 + 2x_2 + x_3) \tag{35}$$

Now, we define

$$z_3 = 2x_1 + 2x_2 + x_3 \tag{36}$$

Using (36), we can simplify the equation (35) as

$$\dot{V}_2 = -z_1^2 - z_2^2 + z_2 z_3 \tag{37}$$

Finally, we define a quadratic Lyapunov function

$$V(z_1, z_2, z_3, e_a, e_b) = V_2(z_1, z_2) + \frac{1}{2}z_3^2 + \frac{1}{2}e_a^2 + \frac{1}{2}e_b^2 \tag{38}$$

which is a positive definite function on $\mathbb{R}^5$.

Differentiating V along the dynamics (23), we get

$$\dot{V} = -z_1^2 - z_2^2 - z_3^2 + z_3(z_3 + z_2 + \dot{z}_3) - e_a\dot{\hat{a}} - e_b\dot{\hat{b}} \tag{39}$$

Equation (39) can be written compactly as

$$\dot{V} = -z_1^2 - z_2^2 - z_3^2 + z_3 S - e_a\dot{\hat{a}} - e_b\dot{\hat{b}} \tag{40}$$

where

$$S = z_3 + z_2 + \dot{z}_3 = z_3 + z_2 + 2\dot{x}_1 + 2\dot{x}_2 + \dot{x}_3 \tag{41}$$

A simple calculation gives

$$S = 5 + 3x_1 + (5 - a)x_2 + (3 - a)x_3 + bx_1x_2 - \exp(x_1) + u \tag{42}$$

Substituting the adaptive control law (26) into (42), we obtain

$$S = -[a - \hat{a}(t)]x_2 - [a - \hat{a}(t)]x_3 + [b - \hat{b}(t)]x_1x_2 - kz_3 \tag{43}$$

Using the definitions (25), we can simplify (43) as

$$S = -e_a(x_2 + x_3) + e_b x_1x_2 - kz_3 \tag{44}$$

Substituting the value of S from (44) into (40), we obtain

$$\dot{V} = -z_1^2 - z_2^2 - (1 + k)z_3^2 + e_a\left[-(x_2 + x_3)z_3 - \dot{\hat{a}}\right] + e_b\left[x_1x_2z_3 - \dot{\hat{b}}\right] \tag{45}$$

Substituting the update law (28) into (45), we get

$$\dot{V} = -z_1^2 - z_2^2 - (1 + k)z_3^2, \tag{46}$$

which is a negative semi-definite function on $\mathbb{R}^5$.

From (46), it follows that the vector $\mathbf{z}(t) = (z_1(t), z_2(t), z_3(t))$ and the parameter estimation error $(e_a(t), e_b(t))$ are globally bounded, i.e.

$$\left[\, z_1(t) \; z_2(t) \; z_3(t) \; e_a(t) \; e_b(t) \,\right] \in \mathbf{L}_\infty \tag{47}$$

Also, it follows from (46) that

$$\dot{V} \leq -z_1^2 - z_2^2 - z_3^2 = -\|\mathbf{z}\|^2 \tag{48}$$

That is,

$$\|\mathbf{z}\|^2 \leq -\dot{V} \tag{49}$$

Integrating the inequality (49) from 0 to t, we get

$$\int_0^t |\mathbf{z}(\tau)|^2 \, d\tau \leq V(0) - V(t) \tag{50}$$

From (50), it follows that $\mathbf{z}(t) \in \mathbf{L}_2$.

From Eq. (23), it can be deduced that $\dot{\mathbf{z}}(t) \in \mathbf{L}_\infty$.

Thus, using Barbalat's lemma, we conclude that $\mathbf{z}(t) \to \mathbf{0}$ exponentially as $t \to \infty$ for all initial conditions $\mathbf{z}(0) \in \mathbb{R}^3$.

Hence, it is immediate that $\mathbf{x}(t) \to \mathbf{0}$ exponentially as $t \to \infty$ for all initial conditions $\mathbf{x}(0) \in \mathbb{R}^3$.

This completes the proof.

For the numerical simulations, the classical fourth-order Runge-Kutta method with step size $h = 10^{-8}$ is used to solve the system of differential equations (23) and (28), when the adaptive control law (26) is applied.

The parameter values of the novel jerk chaotic system (23) are taken as in the chaotic case, i.e.

$$a = 1, \quad b = 0.4 \tag{51}$$

We take the positive gain constant as

$$k = 10 \tag{52}$$

Furthermore, as initial conditions of the novel jerk chaotic system (23), we take

$$x_1(0) = 4.7, \quad x_2(0) = 8.3, \quad x_3(0) = -2.9 \tag{53}$$

Also, as initial conditions of the parameter estimates, we take

$$\hat{a}(0) = 10.5, \quad \hat{b}(0) = 7.4 \tag{54}$$

In Fig. 6, the exponential convergence of the controlled states $x_1(t), x_2(t), x_3(t)$ is depicted, when the adaptive control law (26) and (28) are implemented.

From Fig. 6, it is clear that the controlled states $x_1(t), x_2(t), x_3(t)$ converge to zero exponentially in 6 s. This shows the efficiency of the adaptive backstepping controller derived in this section for the novel jerk chaotic system (23).

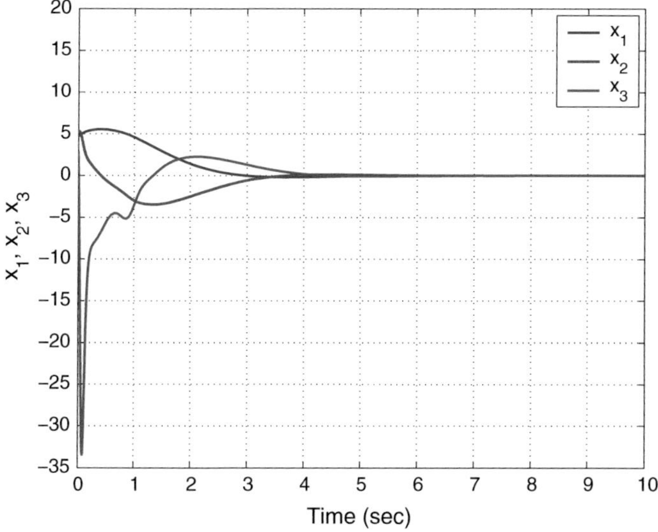

Fig. 6 Time-history of the controlled states $x_1(t), x_2(t), x_3(t)$

5 Adaptive Synchronization of Identical Novel Jerk Chaotic Systems

In this section, we use backstepping control method to derive an adaptive control law for globally and exponentially synchronizing the identical 3-D novel jerk chaotic systems with unknown parameters.

As the master system, we consider the 3-D novel jerk chaotic system given by

$$\begin{cases} \dot{x}_1 = x_2 \\ \dot{x}_2 = x_3 \\ \dot{x}_3 = 5 - a(x_2 + x_3) + bx_1x_2 - \exp(x_1) \end{cases} \tag{55}$$

where x_1, x_2, x_3 are the states of the system, and a and b are unknown constant parameters.

As the slave system, we consider the 3-D novel jerk chaotic system given by

$$\begin{cases} \dot{y}_1 = y_2 \\ \dot{y}_2 = y_3 \\ \dot{y}_3 = 5 - a(y_2 + y_3) + by_1y_2 - \exp(y_1) + u \end{cases} \tag{56}$$

where y_1, y_2, y_3 are the states of the system, and u is a backstepping control to be determined using estimates $\hat{a}(t)$ and $\hat{b}(t)$ for a and b, respectively.

We define the synchronization errors between the states of the master system (55) and the slave system (56) as

$$\begin{cases} e_1 = y_1 - x_1 \\ e_2 = y_2 - x_2 \\ e_3 = y_3 - x_3 \end{cases} \tag{57}$$

Then the error dynamics is easily obtained as

$$\begin{cases} \dot{e}_1 = e_2 \\ \dot{e}_2 = e_3 \\ \dot{e}_3 = -a(e_2 + e_3) + b(y_1 y_2 - x_1 x_2) - \exp(y_1) + \exp(x_1) + u \end{cases} \tag{58}$$

The parameter estimation errors are defined as:

$$\begin{cases} e_a(t) = a - \hat{a}(t) \\ e_b(t) = b - \hat{b}(t) \end{cases} \tag{59}$$

Differentiating (59) with respect to t, we obtain the following equations:

$$\begin{cases} \dot{e}_a(t) = -\dot{\hat{a}}(t) \\ \dot{e}_b(t) = -\dot{\hat{b}}(t) \end{cases} \tag{60}$$

Next, we shall state and prove the main result of this section.

Theorem 2 *The identical 3-D novel jerk chaotic systems (55) and (56) with unknown parameters a and b are globally and exponentially synchronized by the adaptive control law*

$$\begin{cases} u(t) = -3e_1 - [5 - \hat{a}(t)]e_2 - [3 - \hat{a}(t)]e_3 - \hat{b}(t)(y_1 y_2 - x_1 x_2) \\ \qquad\qquad + \exp(y_1) - \exp(x_1) - kz_3 \end{cases} \tag{61}$$

where $k > 0$ is a gain constant,

$$z_3 = 2e_1 + 2e_2 + e_3, \tag{62}$$

and the update law for the parameter estimates $\hat{a}(t)$, $\hat{b}(t)$ is given by

$$\begin{cases} \dot{\hat{a}}(t) = -(e_2 + e_3)z_3 \\ \dot{\hat{b}}(t) = (y_1 y_2 - x_1 x_2)z_3 \end{cases} \tag{63}$$

Proof We prove this result via backstepping control method and Lyapunov stability theory [29].

First, we define a quadratic Lyapunov function

$$V_1(z_1) = \frac{1}{2} z_1^2 \tag{64}$$

where

$$z_1 = e_1 \tag{65}$$

Differentiating V_1 along the error dynamics (58), we get

$$\dot{V}_1 = z_1\dot{z}_1 = e_1e_2 = -z_1^2 + z_1(e_1 + e_2) \tag{66}$$

Now, we define

$$z_2 = e_1 + e_2 \tag{67}$$

Using (67), we can simplify the equation (66) as

$$\dot{V}_1 = -z_1^2 + z_1z_2 \tag{68}$$

Secondly, we define a quadratic Lyapunov function

$$V_2(z_1, z_2) = V_1(z_1) + \frac{1}{2}z_2^2 = \frac{1}{2}\left(z_1^2 + z_2^2\right) \tag{69}$$

Differentiating V_2 along the error dynamics (58), we get

$$\dot{V}_2 = -z_1^2 - z_2^2 + z_2(2e_1 + 2e_2 + e_3) \tag{70}$$

Now, we define

$$z_3 = 2e_1 + 2e_2 + e_3 \tag{71}$$

Using (71), we can simplify the equation (70) as

$$\dot{V}_2 = -z_1^2 - z_2^2 + z_2z_3 \tag{72}$$

Finally, we define a quadratic Lyapunov function

$$V(z_1, z_2, z_3, e_a, e_b) = V_2(z_1, z_2) + \frac{1}{2}z_3^2 + \frac{1}{2}e_a^2 + \frac{1}{2}e_b^2 \tag{73}$$

Differentiating V along the error dynamics (58), we get

$$\dot{V} = -z_1^2 - z_2^2 - z_3^2 + z_3(z_3 + z_2 + \dot{z}_3) - e_a\dot{\hat{a}} - e_b\dot{\hat{b}} \tag{74}$$

Equation (74) can be written compactly as

$$\dot{V} = -z_1^2 - z_2^2 - z_3^2 + z_3S - e_a\dot{\hat{a}} - e_b\dot{\hat{b}} \tag{75}$$

where

$$S = z_3 + z_2 + \dot{z}_3 = z_3 + z_2 + 2\dot{e}_1 + 2\dot{e}_2 + \dot{e}_3 \tag{76}$$

A simple calculation gives

$$S = 3e_1 + (5 - a)e_2 + (3 - a)e_3 + b(y_1 y_2 - x_1 x_2) - \exp(y_1) + \exp(x_1) + u \tag{77}$$

Substituting the adaptive control law (61) into (77), we obtain

$$S = -\left[a - \hat{a}(t)\right](e_2 + e_3) + \left[b - \hat{b}(t)\right](y_1 y_2 - x_1 x_2) - kz_3 \tag{78}$$

Using the definitions (60), we can simplify (78) as

$$S = -e_a(e_2 + e_3) + e_b(y_1 y_2 - x_1 x_2) - kz_3 \tag{79}$$

Substituting the value of S from (79) into (75), we obtain

$$\begin{cases} \dot{V} = -z_1 - z_2 - (1 + k)z_3^2 + e_a\left[-(e_2 + e_3)z_3 - \dot{\hat{a}}\right] \\ \qquad + e_b\left[(y_1 y_2 - x_1 x_2)z_3 - \dot{\hat{b}}\right] \end{cases} \tag{80}$$

Substituting the update law (63) into (80), we get

$$\dot{V} = -z_1^2 - z_2^2 - (1 + k)z_3^2, \tag{81}$$

which is a negative semi-definite function on $\mathbf{R}^5$.

From (81), it follows that the vector $\mathbf{z}(t) = (z_1(t), z_2(t), z_3(t))$ and the parameter estimation error $(e_a(t), e_b(t))$ are globally bounded, i.e.

$$\left[\, z_1(t) \; z_2(t) \; z_3(t) \; e_a(t) \; e_b(t) \,\right] \in \mathbf{L}_\infty \tag{82}$$

Also, it follows from (81) that

$$\dot{V} \leq -z_1^2 - z_2^2 - z_3^2 = -\|\mathbf{z}\|^2 \tag{83}$$

That is,

$$\|\mathbf{z}\|^2 \leq -\dot{V} \tag{84}$$

Integrating the inequality (84) from 0 to t, we get

$$\int_0^t |\mathbf{z}(\tau)|^2 \, d\tau \leq V(0) - V(t) \tag{85}$$

From (85), it follows that $\mathbf{z}(t) \in \mathbf{L}_2$.

From Eq. (58), it can be deduced that $\dot{\mathbf{z}}(t) \in \mathbf{L}_\infty$.

Thus, using Barbalat's lemma, we conclude that $\mathbf{z}(t) \to \mathbf{0}$ exponentially as $t \to \infty$ for all initial conditions $\mathbf{z}(0) \in \mathbb{R}^3$.

Hence, it is immediate that $\mathbf{e}(t) \to \mathbf{0}$ exponentially as $t \to \infty$ for all initial conditions $\mathbf{e}(0) \in \mathbb{R}^3$.

This completes the proof. □

For the numerical simulations, the classical fourth-order Runge-Kutta method with step size $h = 10^{-8}$ is used to solve the system of differential equations (55) and (56).

The parameter values of the novel jerk chaotic systems are taken as in the chaotic case, *i.e.* $a = 1$ and $b = 0.4$.

We take the positive gain constant as $k = 10$.

Furthermore, as initial conditions of the master chaotic system (55), we take

$$x_1(0) = 2.5, \quad x_2(0) = 0.7, \quad x_3(0) = -4.9 \tag{86}$$

As initial conditions of the slave chaotic system (56), we take

$$y_1(0) = -1.3, \quad y_2(0) = 4.2, \quad y_3(0) = 2.5 \tag{87}$$

Also, as initial conditions of the parameter estimates $\hat{a}(t)$ and $\hat{b}(t)$, we take $\hat{a}(0) = 6.2$ and $\hat{b}(0) = 11.5$.

In Figs. 7, 8 and 9, the complete synchronization of the identical 3-D jerk chaotic systems (55) and (56) is shown, when the adaptive control law and the parameter update law are implemented.

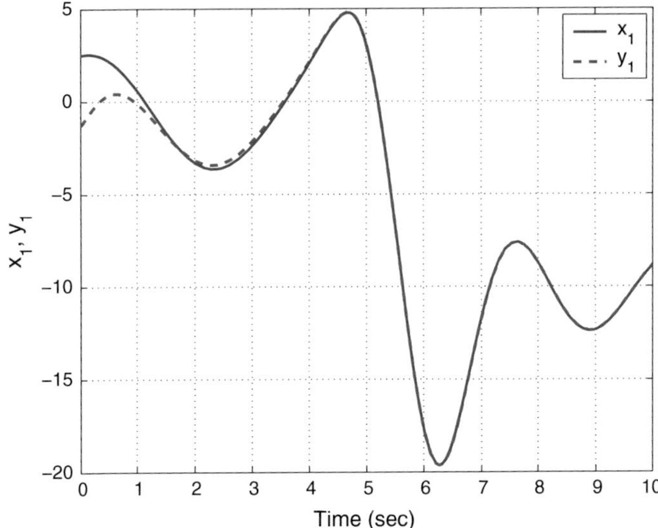

Fig. 7 Synchronization of the states $x_1(t)$ and $y_1(t)$

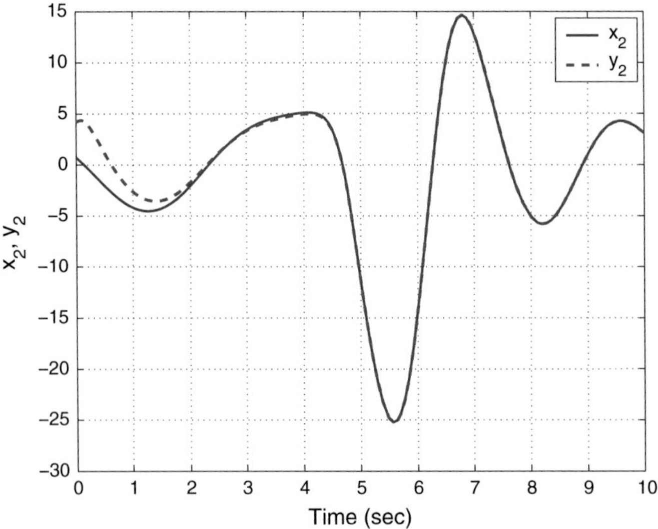

Fig. 8 Synchronization of the states $x_2(t)$ and $y_2(t)$

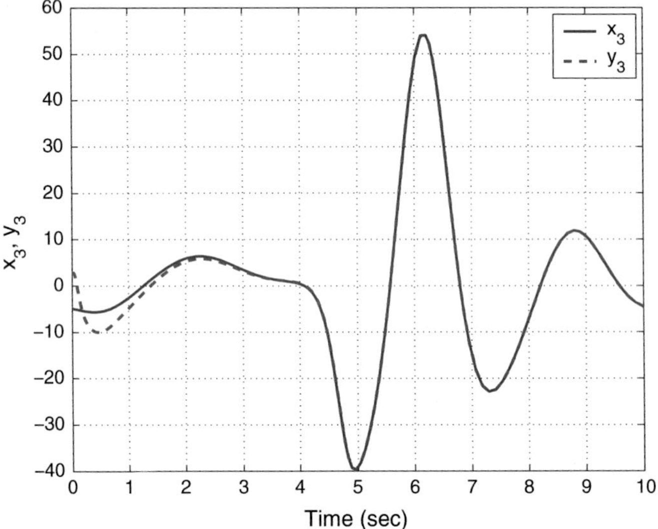

Fig. 9 Synchronization of the states $x_3(t)$ and $y_3(t)$

From Fig. 7, it is clear that the states $x_1(t)$ and $y_1(t)$ are completely synchronized in 6 s.

From Fig. 8, it is clear that the states $x_2(t)$ and $y_2(t)$ are completely synchronized in 6 s.

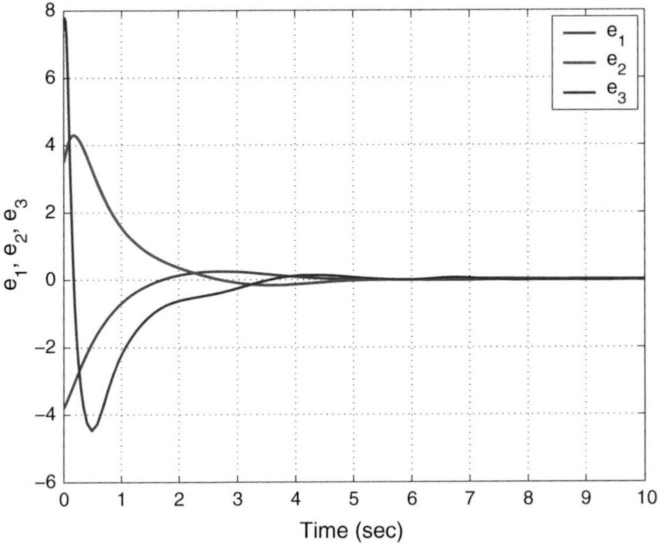

Fig. 10 Time-history of the synchronization errors $e_1(t), e_2(t), e_3(t)$

From Fig. 9, it is clear that the states $x_3(t)$ and $y_3(t)$ are completely synchronized in 6 s.

Also, in Fig. 10, the time-history of the synchronization errors $e_1(t), e_2(t), e_3(t)$, is shown.

From Fig. 10, it is clear that the synchronization errors converge to zero in 6 s. This shows the efficiency of the adaptive backstepping controller designed in this section.

6 Conclusions

In this research work, we announced a seven-term novel jerk chaotic system with an exponential nonlinearity. We derived new results for the adaptive controller and adaptive synchronizer design for the novel jerk chaotic system via backstepping control method. We first discussed the qualitative properties of the novel jerk chaotic system. Explicitly, we showed that the novel jerk chaotic system is dissipative and that it has an unstable equilibrium at the origin. The Lyapunov exponents of novel jerk chaotic system have been obtained as $L_1 = 0.1066$, $L_2 = 0$ and $L_3 = -1.1047$, while the Kaplan-Yorke dimension of the novel jerk chaotic system has been obtained as $D_{KY} = 2.0965$. The adaptive feedback control and synchronization results were proved using Lyapunov stability theory. MATLAB simulations have been depicted to illustrate all the main results for the novel jerk chaotic system.

References

1. Arneodo A, Coullet P, Tresser C (1981) Possible new strange attractors with spiral structure. Commun Math Phys 79(4):573–576
2. Azar AT (2010) Fuzzy systems. IN-TECH, Vienna, Austria
3. Azar AT (2012) Overview of type-2 fuzzy logic systems. Int J Fuzzy Syst Appl 2(4):1–28
4. Azar AT, Serrano FE (2014) Robust IMC-PID tuning for cascade control systems with gain and phase margin specifications. Neural Comput Appl 25(5):983–995
5. Azar AT, Serrano FE (2015) Adaptive sliding mode control of the Furuta pendulum. In: Azar AT, Zhu Q (eds) Advances and applications in sliding mode control systems. Studies in computational intelligence, vol 576. Springer, Berlin, pp 1–42
6. Azar AT, Serrano FE (2015) Deadbeat control for multivariable systems with time varying delays. In: Azar AT, Vaidyanathan S (eds) Chaos modeling and control systems design. Studies in computational intelligence, vol 581. Springer, Berlin, pp 97–132
7. Azar AT, Serrano FE (2015) Design and modeling of anti wind up PID controllers. In: Zhu Q, Azar AT (eds) Complex system modelling and control through intelligent soft computations. Studies in computational intelligence, vol 319. Springer, Berlin, pp 1–44
8. Azar AT, Serrano FE (2015) Stabilizatoin and control of mechanical systems with backlash. In: Azar AT, Vaidyanathan S (eds) Handbook of research on advanced intelligent control engineering and automation. Advances in computational intelligence and robotics (ACIR), IGI-Global, USA, pp 1–60
9. Azar AT, Vaidyanathan S (2015) Chaos modeling and control systems design. Studies in computational intelligence, vol 581. Springer, Berlin
10. Azar AT, Vaidyanathan S (2015) Computational intelligence applications in modeling and control. Studies in computational intelligence, vol 575. Springer, Berlin
11. Azar AT, Vaidyanathan S (2015) Handbook of research on advanced intelligent control engineering and automation. Advances in computational intelligence and robotics (ACIR), IGI-Global, USA
12. Azar AT, Zhu Q (2015) Advances and applications in sliding mode control systems. Studies in computational intelligence, vol 576. Springer, Berlin
13. Cai G, Tan Z (2007) Chaos synchronization of a new chaotic system via nonlinear control. J Uncertain Syst 1(3):235–240
14. Carroll TL, Pecora LM (1991) Synchronizing chaotic circuits. IEEE Trans Circuits Syst 38(4):453–456
15. Chen G, Ueta T (1999) Yet another chaotic attractor. Int J Bifurc Chaos 9(7):1465–1466
16. Chen HK, Lee CI (2004) Anti-control of chaos in rigid body motion. Chaos, Solitons Fractals 21(4):957–965
17. Chen WH, Wei D, Lu X (2014) Global exponential synchronization of nonlinear time-delay Lur'e systems via delayed impulsive control. Commun Nonlinear Sci Numer Simul 19(9):3298–3312
18. Das S, Goswami D, Chatterjee S, Mukherjee S (2014) Stability and chaos analysis of a novel swarm dynamics with applications to multi-agent systems. Eng Appl Artif Intell 30:189–198
19. Feki M (2003) An adaptive chaos synchronization scheme applied to secure communication. Chaos, Solitons Fractals 18(1):141–148
20. Gan Q, Liang Y (2012) Synchronization of chaotic neural networks with time delay in the leakage term and parametric uncertainties based on sampled-data control. J Franklin Inst 349(6):1955–1971
21. Gaspard P (1999) Microscopic chaos and chemical reactions. Physica A 263(1–4):315–328
22. Gibson WT, Wilson WG (2013) Individual-based chaos: Extensions of the discrete logistic model. J Theor Biol 339:84–92
23. Guégan D (2009) Chaos in economics and finance. Annu Rev Control 33(1):89–93
24. Huang X, Zhao Z, Wang Z, Li Y (2012) Chaos and hyperchaos in fractional-order cellular neural networks. Neurocomputing 94:13–21

25. Jiang GP, Zheng WX, Chen G (2004) Global chaos synchronization with channel time-delay. Chaos, Solitons Fractals 20(2):267–275
26. Karthikeyan R, Sundarapandian V (2014) Hybrid chaos synchronization of four-scroll systems via active control. J Electr Eng 65(2):97–103
27. Kaslik E, Sivasundaram S (2012) Nonlinear dynamics and chaos in fractional-order neural networks. Neural Netw 32:245–256
28. Kengne J, Chedjou JC, Kenne G, Kyamakya K (2012) Dynamical properties and chaos synchronization of improved Colpitts oscillators. Commun Nonlinear Sci Numer Simul 17(7):2914–2923
29. Khalil HK (2001) Nonlinear systems. Prentice Hall, New Jersey
30. Kyriazis M (1991) Applications of chaos theory to the molecular biology of aging. Exp Gerontol 26(6):569–572
31. Lang J (2015) Color image encryption based on color blend and chaos permutation in the reality-preserving multiple-parameter fractional Fourier transform domain. Optics Commun 338:181–192
32. Li D (2008) A three-scroll chaotic attractor. Phys Lett A 372(4):387–393
33. Li N, Zhang Y, Nie Z (2011) Synchronization for general complex dynamical networks with sampled-data. Neurocomputing 74(5):805–811
34. Li N, Pan W, Yan L, Luo B, Zou X (2014) Enhanced chaos synchronization and communication in cascade-coupled semiconductor ring lasers. Commun Nonlinear Sci Numer Simul 19(6):1874–1883
35. Li Z, Chen G (2006) Integration of fuzzy logic and chaos theory. Studies in computational intelligence, vol 187. Springer, Berlin
36. Lian S, Chen X (2011) Traceable content protection based on chaos and neural networks. Appl Soft Comput 11(7):4293–4301
37. Liu C, Liu T, Liu L, Liu K (2004) A new chaotic attractor. Chaos, Solitions Fractals 22(5):1031–1038
38. Lorenz EN (1963) Deterministic periodic flow. J Atmos Sci 20(2):130–141
39. Lü J, Chen G (2002) A new chaotic attractor coined. Int J Bifurc Chaos 12(3):659–661
40. Mondal S, Mahanta C (2014) Adaptive second order terminal sliding mode controller for robotic manipulators. J Franklin Inst 351(4):2356–2377
41. Murali K, Lakshmanan M (1998) Secure communication using a compound signal from generalized chaotic systems. Phys Lett A 241(6):303–310
42. Nehmzow U, Walker K (2005) Quantitative description of robot-environment interaction using chaos theory. Robot Auton Syst 53(3–4):177–193
43. Noroozi N, Roopaei M, Karimaghaee P, Safavi AA (2010) Simple adaptive variable structure control for unknown chaotic systems. Commun Nonlinear Sci Numer Simul 15(3):707–727
44. Pecora LM, Carroll TL (1990) Synchronization in chaotic systems. Phys Rev Lett 64(8): 821–824
45. Pehlivan I, Moroz IM, Vaidyanathan S (2014) Analysis, synchronization and circuit design of a novel butterfly attractor. J Sound Vib 333(20):5077–5096
46. Petrov V, Gaspar V, Masere J, Showalter K (1993) Controlling chaos in Belousov-Zhabotinsky reaction. Nature 361:240–243
47. Pham VT, Vaidyanathan S, Volos CK, Jafari S (2015) Hidden attractors in a chaotic system with an exponential nonlinear term. Eur Phys J: Spec Topics 224(8):1507–1517
48. Pham VT, Volos CK, Vaidyanathan S, Le TP, Vu VY (2015) A memristor-based hyperchaotic system with hidden attractors: Dynamics, synchronization and circuital emulating. J Eng Sci Technol Rev 8(2):205–214
49. Qu Z (2011) Chaos in the genesis and maintenance of cardiac arrhythmias. Prog Biophys Mol Biol 105(3):247–257
50. Rasappan S, Vaidyanathan S (2012) Global chaos synchronization of WINDMI and Coullet chaotic systems by backstepping control. Far East J Math Sci 67(2):265–287
51. Rasappan S, Vaidyanathan S (2012) Hybrid synchronization of n-scroll Chua and Lur'e chaotic systems via backstepping control with novel feedback. Arch Control Sci 22(3):343–365

52. Rasappan S, Vaidyanathan S (2012) Synchronization of hyperchaotic Liu system via backstepping control with recursive feedback. Commun Comput Inf Sci 305:212–221
53. Rasappan S, Vaidyanathan S (2013) Hybrid synchronization of n-scroll chaotic Chua circuits using adaptive backstepping control design with recursive feedback. Malays J Math Sci 7(2):219–246
54. Rasappan S, Vaidyanathan S (2014) Global chaos synchronization of WINDMI and Coullet chaotic systems using adaptive backstepping control design. Kyungpook Math J 54(1):293–320
55. Rhouma R, Belghith S (2011) Cryptoanalysis of a chaos based cryptosystem on DSP. Commun Nonlinear Sci Numer Simul 16(2):876–884
56. Rössler OE (1976) An equation for continuous chaos. Phys Lett A 57(5):397–398
57. Vaidyanathan S, VTP, Volos CK, (2015) A 5-D hyperchaotic Rikitake dynamo system with hidden attractors. Eur Phys J: Spec Topics 224(8):1575–1592
58. Sampath S, Vaidyanathan S, Volos CK, Pham VT (2015) An eight-term novel four-scroll chaotic system with cubic nonlinearity and its circuit simulation. J Eng Sci Technol Rev 8(2):1–6
59. Sarasu P, Sundarapandian V (2011) Active controller design for the generalized projective synchronization of four-scroll chaotic systems. Int J Syst Signal Control Eng Appl 4(2):26–33
60. Sarasu P, Sundarapandian V (2011) The generalized projective synchronization of hyperchaotic Lorenz and hyperchaotic Qi systems via active control. Int J Soft Comput 6(5): 216–223
61. Sarasu P, Sundarapandian V (2012) Adaptive controller design for the generalized projective synchronization of 4-scroll systems. Int J Syst Signal Control Eng Appl 5(2):21–30
62. Sarasu P, Sundarapandian V (2012) Generalized projective synchronization of three-scroll chaotic systems via adaptive control. Eur J Sci Res 72(4):504–522
63. Sarasu P, Sundarapandian V (2012) Generalized projective synchronization of two-scroll systems via adaptive control. Int J Soft Comput 7(4):146–156
64. Shahverdiev EM, Shore KA (2009) Impact of modulated multiple optical feedback time delays on laser diode chaos synchronization. Optics Commun 282(17):3568–3572
65. Sharma A, Patidar V, Purohit G, Sud KK (2012) Effects on the bifurcation and chaos in forced Duffing oscillator due to nonlinear damping. Commun Nonlinear Sci Numer Simul 17(6):2254–2269
66. Sprott JC (1994) Some simple chaotic flows. Phys Rev E 50(2):647–650
67. Sprott JC (1997) Some simple chaotic jerk functions. Am J Phys 65(6):537–543
68. Sprott JC (2004) Competition with evolution in ecology and finance. Phys Lett A 325(5–6):329–333
69. Sprott JC (2010) Elegant Chaos: algebraically simple chaotic flows. World Scientific, Singapore
70. Suérez I (1999) Mastering chaos in ecology. Ecol Model 117(2–3):305–314
71. Sundarapandian V (2010) Output regulation of the Lorenz attractor. Far East J Math Sci 42(2):289–299
72. Sundarapandian V (2013) Analysis and anti-synchronization of a novel chaotic system via active and adaptive controllers. J Eng Sci Technol Rev 6(4):45–52
73. Sundarapandian V, Karthikeyan R (2011) Anti-synchronization of hyperchaotic Lorenz and hyperchaotic Chen systems by adaptive control. Int J Syst Signal Control Eng Appl 4(2):18–25
74. Sundarapandian V, Karthikeyan R (2011) Anti-synchronization of Lü and Pan chaotic systems by adaptive nonlinear control. Eur J Sci Res 64(1):94–106
75. Sundarapandian V, Karthikeyan R (2012) Adaptive anti-synchronization of uncertain Tigan and Li systems. J Eng Appl Sci 7(1):45–52
76. Sundarapandian V, Karthikeyan R (2012) Hybrid synchronization of hyperchaotic Lorenz and hyperchaotic Chen systems via active control. J Eng Appl Sci 7(3):254–264
77. Sundarapandian V, Pehlivan I (2012) Analysis, control, synchronization, and circuit design of a novel chaotic system. Math Comput Model 55(7–8):1904–1915

78. Sundarapandian V, Sivaperumal S (2011) Sliding controller design of hybrid synchronization of four-wing chaotic systems. Int J Soft Comput 6(5):224–231
79. Suresh R, Sundarapandian V (2013) Global chaos synchronization of a family of n-scroll hyperchaotic Chua circuits using backstepping control with recursive feedback. Far East J Math Sci 73(1):73–95
80. Tigan G, Opris D (2008) Analysis of a 3D chaotic system. Chaos, Solitons Fractals 36: 1315–1319
81. Usama M, Khan MK, Alghatbar K, Lee C (2010) Chaos-based secure satellite imagery cryptosystem. Comput Math Appl 60(2):326–337
82. Vaidyanathan S (2011) Hybrid chaos synchronization of Liu and Lü systems by active nonlinear control. Commun Comput Inf Sci 204:1–10
83. Vaidyanathan S (2011) Output regulation of Arneodo-Coullet chaotic system. Commun Comput Inf Sci 133:98–107
84. Vaidyanathan S (2011) Output regulation of the unified chaotic system. Commun Comput Inf Sci 198:10–17
85. Vaidyanathan S (2012) Analysis and synchronization of the hyperchaotic Yujun systems via sliding mode control. Adv Intell Syst Comput 176:329–337
86. Vaidyanathan S (2012) Anti-synchronization of Sprott-L and Sprott-M chaotic systems via adaptive control. Int J Control Theory Appl 5(1):41–59
87. Vaidyanathan S (2012) Global chaos control of hyperchaotic Liu system via sliding control method. Int J Control Theory Appl 5(2):117–123
88. Vaidyanathan S (2012) Output regulation of the Liu chaotic system. Appl Mech Mater 110–116:3982–3989
89. Vaidyanathan S (2012) Sliding mode control based global chaos control of Liu-Liu-Liu-Su chaotic system. Int J Control Theory Appl 1(2):15–20
90. Vaidyanathan S (2013) A new six-term 3-D chaotic system with an exponential nonlinearity. Far East J Math Sci 79(1):135–143
91. Vaidyanathan S (2013) Analysis and adaptive synchronization of two novel chaotic systems with hyperbolic sinusoidal and cosinusoidal nonlinearity and unknown parameters. J Eng Sci Technol Rev 6(4):53–65
92. Vaidyanathan S (2013) Analysis, control and synchronization of hyperchaotic Zhou system via adaptive control. Adv Intell Syst Comput 177:1–10
93. Vaidyanathan S (2014) A new eight-term 3-D polynomial chaotic system with three quadratic nonlinearities. Far East J Math Sci 84(2):219–226
94. Vaidyanathan S (2014) Analysis and adaptive synchronization of eight-term 3-D polynomial chaotic systems with three quadratic nonlinearities. Eur Phys J: Spec Topics 223(8): 1519–1529
95. Vaidyanathan S (2014) Analysis, control and synchronisation of a six-term novel chaotic system with three quadratic nonlinearities. Int J Model Ident Control 22(1):41–53
96. Vaidyanathan S (2014) Generalized projective synchronisation of novel 3-D chaotic systems with an exponential non-linearity via active and adaptive control. Int J Model Ident Control 22(3):207–217
97. Vaidyanathan S (2014) Global chaos synchronization of identical Li-Wu chaotic systems via sliding mode control. Int J Model Ident Control 22(2):170–177
98. Vaidyanathan S (2015) 3-cells cellular neural network (CNN) attractor and its adaptive biological control. Int J PharmTech Res 8(4):632–640
99. Vaidyanathan S (2015) A 3-D novel highly chaotic system with four quadratic nonlinearities, its adaptive control and anti-synchronization with unknown parameters. J Eng Sci Technol Rev 8(2):106–115
100. Vaidyanathan S (2015) Adaptive backstepping control of enzymes-substrates system with ferroelectric behaviour in brain waves. Int J PharmTech Res 8(2):256–261
101. Vaidyanathan S (2015) Adaptive biological control of generalized Lotka-Volterra three-species biological system. Int J PharmTech Res 8(4):622–631

102. Vaidyanathan S (2015) Adaptive chaotic synchronization of enzymes-substrates system with ferroelectric behaviour in brain waves. Int J PharmTech Res 8(5):964–973
103. Vaidyanathan S (2015) Adaptive control of a chemical chaotic reactor. Int J PharmTech Res 8(3):377–382
104. Vaidyanathan S (2015) Adaptive synchronization of chemical chaotic reactors. Int J ChemTech Res 8(2):612–621
105. Vaidyanathan S (2015) Adaptive synchronization of generalized Lotka-Volterra three-species biological systems. Int J PharmTech Res 8(5):928–937
106. Vaidyanathan S (2015) Analysis, properties and control of an eight-term 3-D chaotic system with an exponential nonlinearity. Int J Model Ident Control 23(2):164–172
107. Vaidyanathan S (2015) Anti-synchronization of Brusselator chemical reaction systems via adaptive control. Int J ChemTech Res 8(6):759–768
108. Vaidyanathan S (2015) Chaos in neurons and adaptive control of Birkhoff-Shaw strange chaotic attractor. Int J PharmTech Res 8(5):956–963
109. Vaidyanathan S (2015) Dynamics and control of Brusselator chemical reaction. Int J ChemTech Res 8(6):740–749
110. Vaidyanathan S (2015) Dynamics and control of Tokamak system with symmetric and magnetically confined plasma. Int J ChemTech Res 8(6):795–803
111. Vaidyanathan S (2015) Hyperchaos, qualitative analysis, control and synchronisation of a ten-term 4-D hyperchaotic system with an exponential nonlinearity and three quadratic nonlinearities. Int J Model Ident Control 23(4):380–392
112. Vaidyanathan S (2015) Lotka-Volterra population biology models with negative feedback and their ecological monitoring. Int J PharmTech Res 8(5):974–981
113. Vaidyanathan S (2015) Synchronization of 3-cells cellular neural network (CNN) attractors via adaptive control method. Int J PharmTech Res 8(5):946–955
114. Vaidyanathan S (2015) Synchronization of Tokamak systems with symmetric and magnetically confined plasma via adaptive control. Int J ChemTech Res 8(6):818–827
115. Vaidyanathan S, Azar AT (2015) Analysis and control of a 4-D novel hyperchaotic system. In: Azar AT, Vaidyanathan S (eds) Chaos modeling and control systems design. Studies in computational intelligence, vol 581. Springer, Berlin, pp 19–38
116. Vaidyanathan S, Azar AT (2015) Analysis, control and synchronization of a nine-term 3-D novel chaotic system. In: Azar AT, Vaidyanathan S (eds) Chaos modelling and control systems design. Studies in computational intelligence, vol 581. Springer, Berlin, pp 19–38
117. Vaidyanathan S, Azar AT (2015) Anti-synchronization of identical chaotic systems using sliding mode control and an application to Vaidhyanathan-Madhavan chaotic systems. Studies in computational intelligence, vol 576, pp 527–547
118. Vaidyanathan S, Azar AT (2015) Hybrid synchronization of identical chaotic systems using sliding mode control and an application to Vaidhyanathan chaotic systems. Studies in computational intelligence, vol 576, pp 549–569
119. Vaidyanathan S, Madhavan K (2013) Analysis, adaptive control and synchronization of a seven-term novel 3-D chaotic system. Int J Control Theory Appl 6(2):121–137
120. Vaidyanathan S, Pakiriswamy S (2013) Generalized projective synchronization of six-term Sundarapandian chaotic systems by adaptive control. Int J Control Theory Appl 6(2):153–163
121. Vaidyanathan S, Pakiriswamy S (2015) A 3-D novel conservative chaotic system and its generalized projective synchronization via adaptive control. J Eng Sci Technol Rev 8(2):52–60
122. Vaidyanathan S, Rajagopal K (2011) Anti-synchronization of Li and T chaotic systems by active nonlinear control. Commun Comput Inf Sci 198:175–184
123. Vaidyanathan S, Rajagopal K (2011) Global chaos synchronization of hyperchaotic Pang and Wang systems by active nonlinear control. Commun Comput Inf Sci 204:84–93
124. Vaidyanathan S, Rajagopal K (2011) Global chaos synchronization of Lü and Pan systems by adaptive nonlinear control. Commun Comput Inf Sci 205:193–202
125. Vaidyanathan S, Rajagopal K (2012) Global chaos synchronization of hyperchaotic Pang and hyperchaotic Wang systems via adaptive control. Int J Soft Comput 7(1):28–37

126. Vaidyanathan S, Rasappan S (2011) Global chaos synchronization of hyperchaotic Bao and Xu systems by active nonlinear control. Commun Comput Inf Sci 198:10–17
127. Vaidyanathan S, Rasappan S (2014) Global chaos synchronization of n-scroll Chua circuit and Lur'e system using backstepping control design with recursive feedback. Arab J Sci Eng 39(4):3351–3364
128. Vaidyanathan S, Sampath S (2011) Global chaos synchronization of hyperchaotic Lorenz systems by sliding mode control. Commun Comput Inf Sci 205:156–164
129. Vaidyanathan S, Sampath S (2012) Anti-synchronization of four-wing chaotic systems via sliding mode control. Int J Autom Comput 9(3):274–279
130. Vaidyanathan S, Volos C (2015) Analysis and adaptive control of a novel 3-D conservative no-equilibrium chaotic system. Arch Control Sci 25(3):333–353
131. Vaidyanathan S, Volos C, Pham VT (2014) Hyperchaos, adaptive control and synchronization of a novel 5-D hyperchaotic system with three positive Lyapunov exponents and its SPICE implementation. Arch Control Sci 24(4):409–446
132. Vaidyanathan S, Volos C, Pham VT, Madhavan K, Idowu BA (2014) Adaptive backstepping control, synchronization and circuit simulation of a 3-D novel jerk chaotic system with two hyperbolic sinusoidal nonlinearities. Arch Control Sci 24(3):375–403
133. Vaidyanathan S, Azar AT, Rajagopal K, Alexander P (2015) Design and SPICE implementation of a 12-term novel hyperchaotic system and its synchronisation via active control. Int J Model Ident Control 23(3):267–277
134. Vaidyanathan S, Idowu BA, Azar AT (2015) Backstepping controller design for the global chaos synchronization of Sprott's jerk systems. Studies in computational intelligence, vol 581, pp 39–58
135. Vaidyanathan S, Rajagopal K, Volos CK, Kyprianidis IM, Stouboulos IN (2015) Analysis, adaptive control and synchronization of a seven-term novel 3-D chaotic system with three quadratic nonlinearities and its digital implementation in LabVIEW. J Eng Sci Technol Rev 8(2):130–141
136. Vaidyanathan S, Volos C, Pham VT, Madhavan K (2015) Analysis, adaptive control and synchronization of a novel 4-D hyperchaotic hyperjerk system and its SPICE implementation. Arch Control Sci 25(1):5–28
137. Vaidyanathan S, Volos CK, Kyprianidis IM, Stouboulos IN, Pham VT (2015) Analysis, adaptive control and anti-synchronization of a six-term novel jerk chaotic system with two exponential nonlinearities and its circuit simulation. J Eng Sci Technol Rev 8(2):24–36
138. Vaidyanathan S, Volos CK, Pham VT (2015) Analysis, adaptive control and adaptive synchronization of a nine-term novel 3-D chaotic system with four quadratic nonlinearities and its circuit simulation. J Eng Sci Technol Rev 8(2):174–184
139. Vaidyanathan S, Volos CK, Pham VT (2015) Analysis, control, synchronization and SPICE implementation of a novel 4-D hyperchaotic Rikitake dynamo System without equilibrium. J Eng Sci Technol Rev 8(2):232–244
140. Vaidyanathan S, Volos CK, Pham VT (2015) Global chaos control of a novel nine-term chaotic system via sliding mode control. In: Azar AT, Zhu Q (eds) Advances and applications in sliding mode control systems. Studies in computational intelligence, vol 576. Springer, Berlin, pp 571–590
141. Vaidyanathan S, Volos CK, Rajagopal K, Kyprianidis IM, Stouboulos IN (2015) Adaptive backstepping controller design for the anti-synchronization of identical WINDMI chaotic systems with unknown parameters and its SPICE implementation. J Eng Sci Technol Rev 8(2):74–82
142. Volos CK, Kyprianidis IM, Stouboulos IN (2013) Experimental investigation on coverage performance of a chaotic autonomous mobile robot. Robot Auton Syst 61(12):1314–1322
143. Volos CK, Kyprianidis IM, Stouboulos IN, Tlelo-Cuautle E, Vaidyanathan S (2015) Memristor: A new concept in synchronization of coupled neuromorphic circuits. J Eng Sci Technol Rev 8(2):157–173
144. Wei Z, Yang Q (2010) Anti-control of Hopf bifurcation in the new chaotic system with two stable node-foci. Appl Math Comput 217(1):422–429

145. Witte CL, Witte MH (1991) Chaos and predicting varix hemorrhage. Med Hypotheses 36(4):312–317
146. Xiao X, Zhou L, Zhang Z (2014) Synchronization of chaotic Lur'e systems with quantized sampled-data controller. Commun Nonlinear Sci Numer Simul 19(6):2039–2047
147. Yuan G, Zhang X, Wang Z (2014) Generation and synchronization of feedback-induced chaos in semiconductor ring lasers by injection-locking. Optik: Int J Light Electron Optics 125(8):1950–1953
148. Zaher AA, Abu-Rezq A (2011) On the design of chaos-based secure communication systems. Commun Nonlinear Syst Numer Simul 16(9):3721–3727
149. Zhang H, Zhou J (2012) Synchronization of sampled-data coupled harmonic oscillators with control inputs missing. Syst Control Lett 61(12):1277–1285
150. Zhang X, Zhao Z, Wang J (2014) Chaotic image encryption based on circular substitution box and key stream buffer. Signal Process Image Commun 29(8):902–913
151. Zhou W, Xu Y, Lu H, Pan L (2008) On dynamics analysis of a new chaotic attractor. Phys Lett A 372(36):5773–5777
152. Zhu C, Liu Y, Guo Y (2010) Theoretic and numerical study of a new chaotic system. Intell Inf Manag 2:104–109
153. Zhu Q, Azar AT (2015) Complex system modelling and control through intelligent soft computations. Studies in fuzzines and soft computing, vol 319. Springer, Berlin

Generalized Projective Synchronization of a Novel Hyperchaotic Four-Wing System via Adaptive Control Method

Sundarapandian Vaidyanathan and Ahmad Taher Azar

Abstract In this research work, we announce a novel 4-D hyperchaotic four-wing system with three quadratic nonlinearities. First, this work describes the qualitative analysis of the novel 4-D hyperchaotic four-wing system. We show that the novel hyperchaotic four-wing system has a unique equilibrium point at the origin, which is a saddle-point. Thus, origin is an unstable equilibrium of the novel hyperchaotic system. The Lyapunov exponents of the novel hyperchaotic four-wing system are obtained as $L_1 = 2.5266$, $L_2 = 0.1053$, $L_3 = 0$ and $L_4 = -43.0194$. Thus, the maximal Lyapunov exponent (MLE) of the novel hyperchaotic four-wing system is obtained as $L_1 = 2.5266$. Since the sum of the Lyapunov exponents of the novel hyperchaotic system is negative, it follows that the novel hyperchaotic system is dissipative. Also, the Kaplan-Yorke dimension of the novel four-wing chaotic system is obtained as $D_{KY} = 3.0612$. Finally, this work describes the generalized projective synchronization (GPS) of the identical novel hyperchaotic four-wing systems with unknown parasmeters. The GPS is a general type of synchronization, which generalizes known types of synchronization such as complete synchronization, anti-synchronization, hybrid synchronization, etc. The main GPS result via adaptive control method is proved using Lyapunov stability theory. MATLAB simulations are depicted to illustrate all the main results for the novel 4-D hyperchaotic four-wing system.

Keywords Chaos · Chaotic systems · Hyperchaos · Hyperchaotic systems · Four-wing systems · Adaptive control · Generalized projective synchronization

S. Vaidyanathan (✉)
Research and Development Centre, Vel Tech University,
Avadi, Chennai 600062, Tamil Nadu, India
e-mail: sundarvtu@gmail.com

A.T. Azar
Faculty of Computers and Information, Benha University, Benha, Egypt
e-mail: ahmad_t_azar@ieee.org; ahmad.azar@fci.bu.edu.eg

A.T. Azar and S. Vaidyanathan (eds.), *Advances in Chaos Theory and Intelligent Control*, Studies in Fuzziness and Soft Computing 337,
DOI 10.1007/978-3-319-30340-6_12

1 Introduction

Chaotic systems are defined as nonlinear dynamical systems which are sensitive to initial conditions, topologically mixing and with dense periodic orbits.Sensitivity to initial conditions of chaotic systems is popularly known as the *butterfly effect*. Small changes in an initial state will make a very large difference in the behavior of the system at future states.

The Lyapunov exponent is a measure of the divergence of phase points that are initially very close and can be used to quantify chaotic systems. It is common to refer to the largest Lyapunov exponent as the *Maximal Lyapunov Exponent* (MLE). A positive maximal Lyapunov exponent and phase space compactness are usually taken as defining conditions for a chaotic system.

Some classical paradigms of 3-D chaotic systems in chaos literature are Lorenz system [34], Rössler system [50], ACT system [1], Sprott systems [61], Chen system [15], Lü system [35], Liu system [33], Cai system [13], Chen-Lee system [16], Tigan system [72], etc.

Many new chaotic systems have been discovered in the recent years such as Zhou system [140], Zhu system [141], Li system [28], Wei-Yang system [132], Sundarapandian systems [64, 69], Vaidyanathan systems [80, 81, 83–86, 89, 96, 106, 109, 111, 120, 122, 127–129], Pehlivan system [40], Sampath system [52], Pham system [41], etc.

Chaos theory and control systems have important applications in science and engineering [2, 9–12, 142]. Some commonly known applications are oscillators [25, 59], lasers [30, 136], chemical reactions [93, 94, 97, 99, 100, 104], biology [88, 90–92, 95, 98, 102, 103], ecology [20, 62], encryption [27, 139], cryptosystems [49, 73], mechanical systems [4–8], secure communications [18, 37, 137], robotics [36, 38, 130], cardiology [43, 133], intelligent control [3, 31], neural networks [21, 24, 32], memristors [42, 131], etc.

In most of the chaos synchronization approaches, the master-slave or drive-response formalism is used. If a particular chaotic system is called the *master* or *drive* system and another chaotic system is called the *slave* or *response* system, then the idea of synchronization is to use the output of the master system to control the slave system so that the states of the slave system track the states of the master system asymptotically.

In the complete synchronization of chaotic systems with states $\mathbf{x}(t)$ and $\mathbf{y}(t)$, the synchronization error is defined by

$$\mathbf{e}(t) = \mathbf{y}(t) - \mathbf{x}(t) \tag{1}$$

In the anti-synchronization of chaotic systems with states $\mathbf{x}(t)$ and $\mathbf{y}(t)$, the synchronization error is defined by

$$\mathbf{e}(t) = \mathbf{y}(t) + \mathbf{x}(t) \tag{2}$$

In the hybrid synchronization of chaotic systems with states $\mathbf{x}(t)$ and $\mathbf{y}(t)$, the synchronization error is defined by

$$e_i(t) = \begin{cases} y_i(t) - x_i(t) & \text{if } i \text{ is odd} \\ y_i(t) + x_i(t) & \text{if } i \text{ is even} \end{cases} \tag{3}$$

In the generalized projective synchronization (GPS) of chaotic systems with states $\mathbf{x}(t)$ and $\mathbf{y}(t)$, the synchronization error is defined by

$$\mathbf{e}(t) = \mathbf{y}(t) - S\mathbf{x}(t), \quad \text{where } S = \text{diag}[\alpha_1, \alpha_2, \ldots, \alpha_n] \tag{4}$$

or equivalently

$$e_i(t) = y_i(t) - \alpha_i x_i(t), \quad \text{where } \alpha_i \text{ are real scaling constants} \tag{5}$$

In the generalized projective synchronization [53, 54, 86, 111] of chaotic systems, chaotic systems can synchronize up to a constant scaling matrix. Complete synchronization [14, 39, 82, 115], anti-synchronization [65, 66, 76, 112], hybrid synchronization [23, 68, 74], projective synchronization [60] and generalized synchronization [135] are special cases of the generalized projective synchronization of chaotic systems. GPS has important applications in areas like secure communications and secure data encryption.

Pecora and Carroll pioneered the research on synchronization of chaotic systems with their seminal papers [14, 39]. The active control method [23, 53, 54, 63, 68, 74, 78, 112, 113, 116, 123] is typically used when the system parameters are available for measurement.

Adaptive control method [51, 55–57, 65–67, 76, 82, 101, 105, 110, 114, 115, 121, 126] is typically used when some or all the system parameters are not available for measurement and estimates for the uncertain parameters of the systems.

Sampled-data feedback control method [19, 29, 134, 138] and time-delay feedback control method [17, 22, 58] are also used for synchronization of chaotic systems. Backstepping control method [44–48, 71, 117, 124] is also used for the synchronization of chaotic systems, which is a recursive method for stabilizing the origin of a control system in strict-feedback form.

Another popular method for the synchronization of chaotic systems is the sliding mode control method [70, 75, 77, 79, 87, 107, 108, 118, 119, 125], which is a nonlinear control method that alters the dynamics of a nonlinear system by application of a discontinuous control signal that forces the system to "slide" along a cross-section of the system's normal behavior.

In this research work, we describe a ten-term novel 4-D hyperchaotic four-wing system with three quadratic nonlinearities. Section 2 describes the 4-D dynamical model and phase portraits of the novel hyperchaotic four-wing system. Section 3 describes the qualitative properties of the novel hyperchaotic four-wing system.

The Lyapunov exponents of the novel 4-D hyperchaotic four-wing system are obtained as $L_1 = 2.5266$, $L_2 = 0.1053$, $L_3 = 0$ and $L_4 = -43.0194$. Thus, the maximal Lyapunov exponent of the novel hyperchaotic four-wing system is obtained as $L_1 = 2.5266$. Also, the Kaplan-Yorke dimension of the novel hyperchaotic system is obtained as $D_{KY} = 3.0612$.

Section 4 describes the generalized projective synchronization (GPS) of the identical novel hyperchaotic systems with unknown parameters via adaptive control method. The main GPS results for the novel hyperchaotic four-wing systems are established using Lyapunov stability theory [26]. Finally, Sect. 5 gives a summary of the main results derived in this work.

2 A Novel 4-D Hyperchaotic Four-Wing System

In this work, we announce a novel 4-D hyperchaotic four-wing system described by

$$\begin{cases} \dot{x}_1 = ax_1 - x_2x_3 + x_4 \\ \dot{x}_2 = -bx_2 + x_1x_3 + x_4 \\ \dot{x}_3 = -cx_3 + x_1x_2 + x_1 \\ \dot{x}_4 = -px_1 \end{cases} \tag{6}$$

In (6), x_1, x_2, x_3, x_4 are the states and a, b, c, p are constant, positive parameters. The 4-D system (6) is *hyperchaotic* when the parameter values are taken as

$$a = 10, \quad b = 35, \quad c = 16, \quad p = 1.2 \tag{7}$$

For numerical simulations, we take the initial state of the novel 4-D hyperchaotic system (6) as

$$x_1(0) = 0.2, \quad x_2(0) = 0.4, \quad x_3(0) = 0.2, \quad x_4(0) = 0.8 \tag{8}$$

The novel 4-D hyperchaotic system (6) exhibits a strange, four-wing hyperchaotic attractor. Figure 1 shows the 3-D projection of the hyperchaotic attractor of the novel 4-D system (6) on (x_1, x_2, x_3) space. This 3-D projection has the shape of a *four-wing attractor*.

Figure 2 shows the 2-D projection of the hyperchaotic attractor of the novel 4-D system on the (x_1, x_2) plane. This 2-D projection also has the shape of a *four-wing attractor*.

Figure 3 describes the 3-D projection of the hyperchaotic attractor of the novel 4-D system (6) on (x_1, x_3, x_4) space.

Figure 4 describes the 3-D projection of the hyperchaotic attractor of the novel 4-D system (6) on (x_2, x_3, x_4) space.

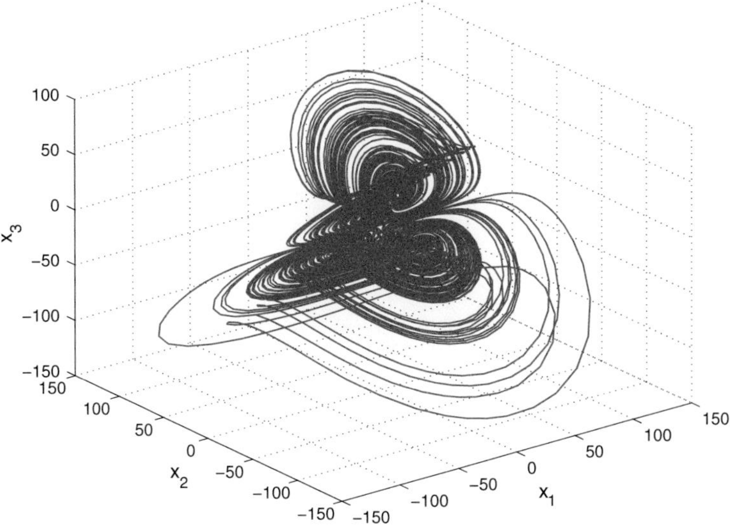

Fig. 1 3-D projection of the novel hyperchaotic system on (x_1, x_2, x_3) space

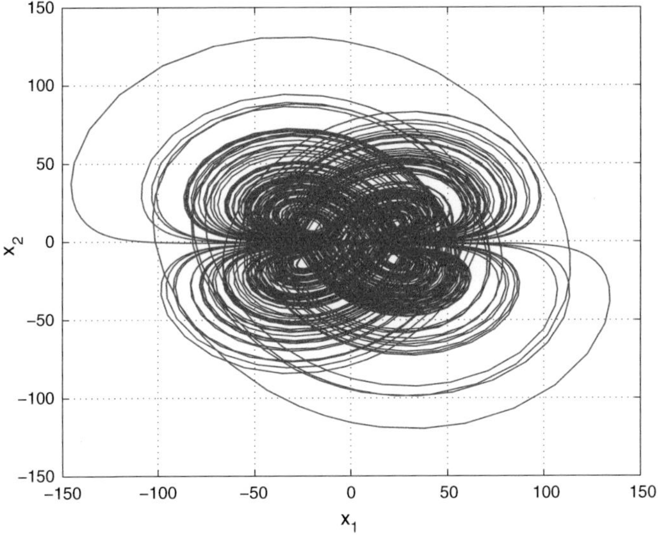

Fig. 2 3-D projection of the novel hyperchaotic system on (x_1, x_2, x_4) space

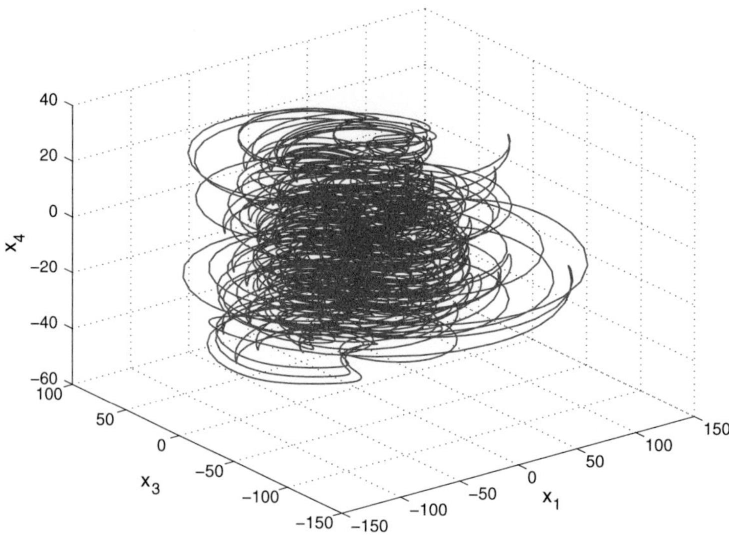

Fig. 3 3-D projection of the novel hyperchaotic system on (x_1, x_3, x_4) space

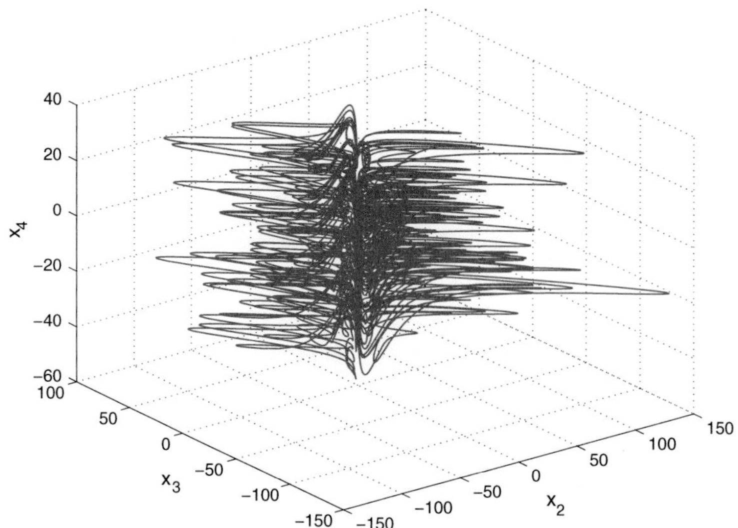

Fig. 4 3-D projection of the novel hyperchaotic system on (x_2, x_3, x_4) space

3 Analysis of the Novel 4-D Hyperchaotic Four-Wing System

This section gives the qualitative properties of the novel 4-D hyperchaotic four-wing system (6) proposed in this research work.

3.1 Dissipativity

We write the system (6) in vector notation as

$$\dot{x} = f(x) = \begin{bmatrix} f_1(x) \\ f_2(x) \\ f_3(x) \\ f_4(x) \end{bmatrix}, \tag{9}$$

where

$$\begin{aligned} f_1(x) &= ax_1 - x_2x_3 + x_4 \\ f_2(x) &= -bx_2 + x_1x_3 + x_4 \\ f_3(x) &= -cx_3 + x_1x_2 + x_1 \\ f_4(x) &= -px_1 \end{aligned} \tag{10}$$

We take the parameter values as in the hyperchaotic case (7), viz.

$$a = 10, \quad b = 35, \quad c = 16, \quad p = 1.2 \tag{11}$$

The divergence of the vector field f on $\mathbb{R}^4$ is obtained as

$$\operatorname{div} f = \frac{\partial f_1(x)}{\partial x_1} + \frac{\partial f_2(x)}{\partial x_2} + \frac{\partial f_3(x)}{\partial x_3} + \frac{\partial f_4(x)}{\partial x_4} = a - b - c = -\mu \tag{12}$$

where

$$\mu = -a + b + c = -10 + 35 + 16 = 41 > 0 \tag{13}$$

Let Ω be any region in $\mathbb{R}^4$ with a smooth boundary. Let $\Omega(t) = \Phi_t(\Omega)$, where Φ_t is the flow of the vector field f.

Let $V(t)$ denote the hypervolume of $\Omega(t)$.

By Liouville's theorem, it follows that

$$\frac{dV(t)}{dt} = \int\limits_{\Omega(t)} (\operatorname{div} f)\, dx_1\, dx_2\, dx_3\, dx_4 \tag{14}$$

Substituting the value of div$f$ in (14) leads to

$$\frac{dV(t)}{dt} = -\mu \int\limits_{\Omega(t)} dx_1\, dx_2\, dx_3\, dx_4 = -\mu V(t) \tag{15}$$

Integrating the linear differential equation (15), $V(t)$ is obtained as

$$V(t) = V(0)\exp(-\mu t), \quad \text{where } \mu = 41 > 0. \tag{16}$$

From Eq. (16), it follows that the hypervolume $V(t)$ shrinks to zero exponentially as $t \to \infty$.

Thus, the novel chaotic system (6) is dissipative. Hence, any asymptotic motion of the system (6) settles onto a set of measure zero, i.e. a strange attractor.

3.2 Equilibria

The equilibrium points of the novel hyperchaotic system (6) are obtained by solving the nonlinear equations

$$\begin{aligned} f_1(x) &= ax_1 - x_2x_3 + x_4 &&= 0 \\ f_2(x) &= -bx_2 + x_1x_3 + x_4 &&= 0 \\ f_3(x) &= -cx_3 + x_1x_2 + x_1 &&= 0 \\ f_4(x) &= -px_1 &&= 0 \end{aligned} \tag{17}$$

We take the parameter values as in the hyperchaotic case (7), viz.

$$a = 10, \quad b = 35, \quad c = 16, \quad p = 1.2 \tag{18}$$

Solving the nonlinear system of equations (17) with the parameter values as in (18), we obtain a unique equilibrium point at the origin, i.e.

$$E_0 = \begin{bmatrix} 0 \\ 0 \\ 0 \\ 0 \end{bmatrix} \tag{19}$$

The Jacobian matrix of the novel hyperchaotic system (6) at E_0 is obtained as

$$J_0 = J(E_0) = \begin{bmatrix} 10 & 0 & 0 & 1 \\ 0 & -35 & 0 & 1 \\ 1 & 0 & -16 & 0 \\ -1.2 & 0 & 0 & 0 \end{bmatrix} \tag{20}$$

The matrix J_0 has the eigenvalues

$$\lambda_1 = -35, \quad \lambda_2 = -16, \quad \lambda_3 = 0.1215, \quad \lambda_4 = 9.8785 \tag{21}$$

This shows that the equilibrium point E_0 is a saddle-point, which is unstable.

3.3 Lyapunov Exponents and Kaplan-Yorke Dimension

We also take the parameter values of the novel hyperchaotic system (6) as in the hyperchaotic case (7). We take the initial values of the novel hyperchaotic system (6) as in (8).

Then the Lyapunov exponents of the novel hyperchaotic system (6) are numerically obtained as

$$L_1 = 2.5266, \quad L_2 = 0.1053, \quad L_3 = 0, \quad L_4 = -43.0194 \tag{22}$$

Since $L_1 + L_2 + L_3 + L_4 = -40.3875 < 0$, the system (6) is dissipative.

Also, the Kaplan-Yorke dimension of the system (6) is obtained as

$$D_{KY} = 3 + \frac{L_1 + L_2 + L_3}{|L_4|} = 3.0612, \tag{23}$$

which is fractional.

4 Generalized Projective Synchronization of the Identical Novel Hyperchaotic Systems

This section derives new results for the generalized projective synchronization of the identical novel hyperchaotic systems with unknown parameters via adaptive control method.

The master system is given by the novel four-wing chaotic system

$$\begin{aligned} \dot{x}_1 &= ax_1 - x_2x_3 + x_4 \\ \dot{x}_2 &= -bx_2 + x_1x_3 + x_4 \\ \dot{x}_3 &= -cx_3 + x_1x_2 + x_1 \\ \dot{x}_4 &= -px_1 \end{aligned} \tag{24}$$

where x_1, x_2, x_3, x_4 are state variables and a, b, c, p are constant, unknown, parameters of the system.

The slave system is given by the controlled novel chaotic system

$$\begin{aligned}\dot{y}_1 &= ay_1 - y_2y_3 + y_4 + u_1\\ \dot{y}_2 &= -by_2 + y_1y_3 + y_4 + u_2\\ \dot{y}_3 &= -cy_3 + y_1y_2 + y_1 + u_3\\ \dot{y}_4 &= -py_1 + u_4\end{aligned} \tag{25}$$

where y_1, y_2, y_3, y_4 are state variables and u_1, u_2, u_3, u_4 are adaptive controls to be designed using estimates of the unknown parameters.

The generalized projective synchronization (GPS) error is defined as

$$\begin{aligned}e_1 &= y_1 - \alpha_1 x_1\\ e_2 &= y_2 - \alpha_2 x_2\\ e_3 &= y_3 - \alpha_3 x_3\\ e_4 &= y_4 - \alpha_4 x_4\end{aligned} \tag{26}$$

where $\alpha_1, \alpha_2, \alpha_3, \alpha_4$ are real GPS scales.

A simple calculation yields the GPS error dynamics

$$\begin{aligned}\dot{e}_1 &= ae_1 - y_2y_3 + \alpha_1 x_2x_3 + y_4 - \alpha_1 x_4 + u_1\\ \dot{e}_2 &= -be_2 + y_1y_3 - \alpha_2 x_1x_3 + y_4 - \alpha_2 x_4 + u_2\\ \dot{e}_3 &= -ce_3 + y_1y_2 - \alpha_3 x_1x_2 + y_1 - \alpha_3 x_1 + u_3\\ \dot{e}_4 &= -p(y_1 - \alpha_4 x_1) + u_4\end{aligned} \tag{27}$$

We consider the adaptive control law given by

$$\begin{aligned}u_1 &= -\hat{a}(t)e_1 + y_2y_3 - \alpha_1 x_2x_3 - y_4 + \alpha_1 x_4 - k_1e_1\\ u_2 &= \hat{b}(t)e_2 - y_1y_3 + \alpha_2 x_1x_3 - y_4 + \alpha_2 x_4 - k_2e_2\\ u_3 &= \hat{c}(t)e_3 - y_1y_2 + \alpha_3 x_1x_2 - y_1 + \alpha_3 x_1 - k_3e_3\\ u_4 &= \hat{p}(t)(y_1 - \alpha_4 x_1) - k_4e_4\end{aligned} \tag{28}$$

where $\hat{a}(t), \hat{b}(t), \hat{c}(t), \hat{p}(t)$ are estimates for the unknown parameters a, b, c, p, respectively, and k_1, k_2, k_3, k_4 are positive gain constants.

The closed-loop control system is obtained by substituting (28) into (27) as

$$\begin{aligned}\dot{e}_1 &= [a - \hat{a}(t)]e_1 - k_1e_1\\ \dot{e}_2 &= -[b - \hat{b}(t)]e_2 - k_2e_2\\ \dot{e}_3 &= -[c - \hat{c}(t)]e_3 - k_3e_3\\ \dot{e}_4 &= -[p - \hat{p}(t)](y_1 - \alpha_4 x_1) - k_4e_4\end{aligned} \tag{29}$$

To simplify (29), we define the parameter estimation error as

$$\begin{aligned} e_a(t) &= a - \hat{a}(t) \\ e_b(t) &= b - \hat{b}(t) \\ e_c(t) &= c - \hat{c}(t) \\ e_p(t) &= p - \hat{p}(t) \end{aligned} \tag{30}$$

Using (30), the closed-loop system (29) can be simplified as

$$\begin{aligned} \dot{e}_1 &= e_a e_1 - k_1 e_1 \\ \dot{e}_2 &= -e_b e_2 - k_2 e_2 \\ \dot{e}_3 &= -e_c e_3 - k_3 e_3 \\ \dot{e}_4 &= -e_p(y_1 - \alpha_4 x_1) - k_4 e_4 \end{aligned} \tag{31}$$

Differentiating the parameter estimation error (30) with respect to t, we get

$$\begin{aligned} \dot{e}_a &= -\dot{\hat{a}} \\ \dot{e}_b &= -\dot{\hat{b}} \\ \dot{e}_c &= -\dot{\hat{c}} \\ \dot{e}_p &= -\dot{\hat{p}} \end{aligned} \tag{32}$$

Next, we find an update law for parameter estimates using Lyapunov stability theory.

Consider the quadratic Lyapunov function defined by

$$V(e_1, e_2, e_3, e_4, e_a, e_b, e_c, e_p) = \frac{1}{2}\sum_{i=1}^{4} e_i^2 + \frac{1}{2}\left(e_a^2 + e_b^2 + e_c^2 + e_p^2\right) \tag{33}$$

Differentiating V along the trajectories of (31) and (32), we get

$$\begin{aligned} \dot{V} = &-k_1 e_1^2 - k_2 e_2^2 - k_3 e_3^2 - k_4 e_4^2 + e_a\left[e_1^2 - \dot{\hat{a}}\right] + e_b\left[-e_2^2 - \dot{\hat{b}}\right] \\ &+ e_c\left[-e_3^2 - \dot{\hat{c}}\right] + e_p\left[-e_4(y_1 - \alpha_4 x_1) - \dot{\hat{p}}\right] \end{aligned} \tag{34}$$

In view of Eq. (34), an update law for the parameter estimates is taken as

$$\begin{aligned} \dot{\hat{a}} &= e_1^2 \\ \dot{\hat{b}} &= -e_2^2 \\ \dot{\hat{c}} &= -e_3^2 \\ \dot{\hat{p}} &= -e_4(y_1 - \alpha_4 x_1) \end{aligned} \tag{35}$$

Theorem 1 *The adaptive control law (28) and the parameter update law (35) achieve exponential and generalized projective synchronization (GPS) between the identical novel 4-D hyperchaotic four-wing systems (24) and (25) with unknown* *system parameters for all initial conditions* $x(0), y(0) \in \mathbb{R}^4$, *where* k_i, $(i = 1, 2, 3, 4)$ *are positive constants.*

Proof The result is proved using Lyapunov stability theory [26].

We consider the quadratic Lyapunov function V defined by (33), which is positive definite on $\mathbb{R}^8$.

Substitution of the parameter update law (35) into (34) yields

$$\dot{V} = -k_1 e_1^2 - k_2 e_2^2 - k_3 e_3^2 - k_4 e_4^2, \tag{36}$$

which is a negative semi-definite function on $\mathbb{R}^8$.

Therefore, it can be concluded that the synchronization error vector $e(t)$ and the parameter estimation error are globally bounded, i.e.

$$\left[e_1(t)\ e_2(t)\ e_3(t)\ e_4(t)\ e_a(t)\ e_b(t)\ e_c(t)\ e_p(t) \right]^T \in \mathbf{L}_\infty. \tag{37}$$

Define

$$k = \min \{k_1, k_2, k_3, k_4\} \tag{38}$$

Then it follows from (36) that

$$\dot{V} \leq -k\|e\|^2 \ \text{ or } \ k\|\mathbf{e}\|^2 \leq -\dot{V} \tag{39}$$

Integrating the inequality (39) from 0 to t, we get

$$k \int_0^t \|\mathbf{e}(\tau)\|^2 \, d\tau \leq - \int_0^t \dot{V}(\tau) \, d\tau = V(0) - V(t) \tag{40}$$

From (40), it follows that $\mathbf{e}(t) \in \mathbf{L}_2$.

Using (31), it can be deduced that $\dot{\mathbf{e}}(t) \in \mathbf{L}_\infty$.

Thus, using Barbalat's lemma [26], we can conclude that $\mathbf{e}(t) \to 0$ exponentially as $t \to \infty$ for all initial conditions $\mathbf{e}(0) \in \mathbb{R}^4$.

Hence, we have proved that the adaptive control law (28) and the parameter update law (35) achieve exponential and generalized projective synchronization (GPS) between the identical novel 4-D hyperchaotic four-wing systems (24) and (25) with unknown system parameters for all initial conditions $x(0), y(0) \in \mathbb{R}^4$.

This completes the proof. □

For numerical simulations, the parameter values of the novel systems (24) and (25) are taken as in the hyperchaotic case, viz.

$$a = 10, \quad b = 35, \quad c = 16, \quad p = 1.2 \tag{41}$$

The gain constants are taken as

$$k_1 = 6, \quad k_2 = 6, \quad k_3 = 6, \quad k_4 = 6 \tag{42}$$

The GPS scales are taken as

$$\alpha_1 = -5.8, \quad \alpha_2 = 3.7, \quad \alpha_3 = 1.2, \quad \alpha_4 = -0.5 \tag{43}$$

The initial values of the parameter estimates are taken as

$$\hat{a}(0) = 15.7, \quad \hat{b}(0) = 13.1, \quad \hat{c}(0) = 17.5, \quad \hat{p}(0) = 12.8 \tag{44}$$

The initial values of the master system (24) are taken as

$$x_1(0) = 2.9, \quad x_2(0) = -11.7, \quad x_3(0) = 16.4, \quad x_4(0) = 7.3 \tag{45}$$

The initial values of the slave system (25) are taken as

$$y_1(0) = -11.5, \quad y_2(0) = 3.8, \quad y_3(0) = -5.8, \quad y_4(0) = -1.2 \tag{46}$$

Figures 5, 6, 7 and 8 show the generalized projective synchronization (GPS) of the identical chaotic systems (24) and (25).

Figure 5 shows that the states $y_1(t)$ and $\alpha_1 x_1(t)$ are synchronized in 1 s (MATLAB).
Figure 6 shows that the states $y_2(t)$ and $\alpha_2 x_2(t)$ are synchronized in 1 s (MATLAB).
Figure 7 shows that the states $y_3(t)$ and $\alpha_3 x_3(t)$ are synchronized in 1 s (MATLAB).
Figure 8 shows that the states $y_4(t)$ and $\alpha_4 x_4(t)$ are synchronized in 2 s (MATLAB).

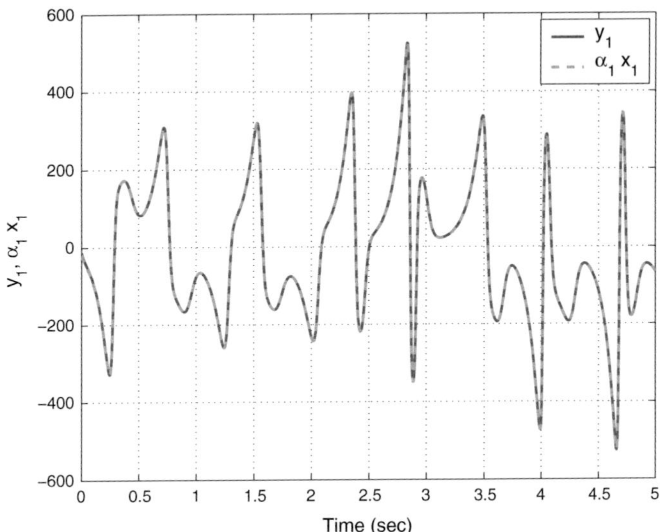

Fig. 5 Synchronization of the states y_1 and αx_1

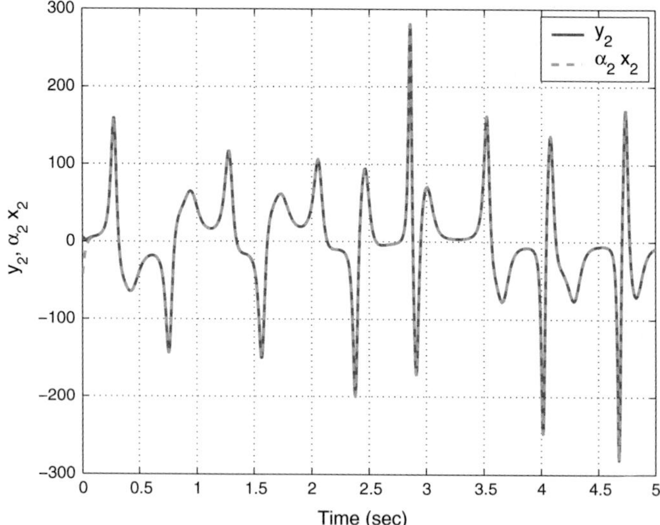

Fig. 6 Synchronization of the states y_2 and αx_2

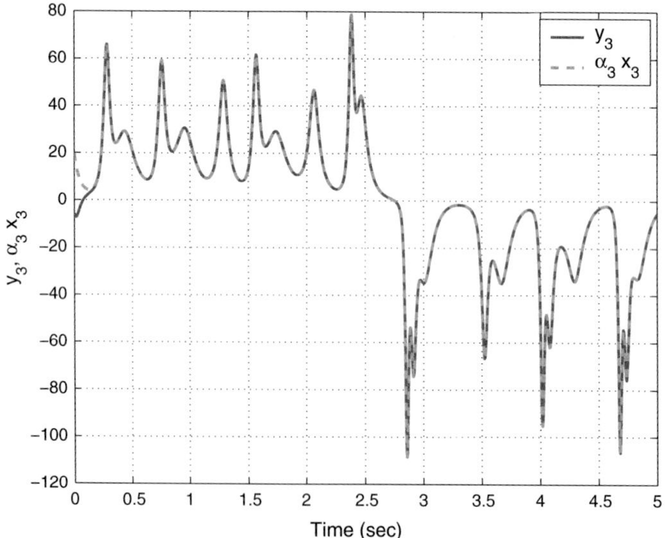

Fig. 7 Synchronization of the states y_3 and αx_3

Figure 9 shows the time-history of the synchronization errors $e_1(t)$, $e_2(t)$, $e_3(t)$, $e_4(t)$.

From Fig. 9, it is seen that the errors $e_1(t)$, $e_2(t)$, $e_3(t)$ and $e_4(t)$ are stabilized in 2 s (MATLAB).

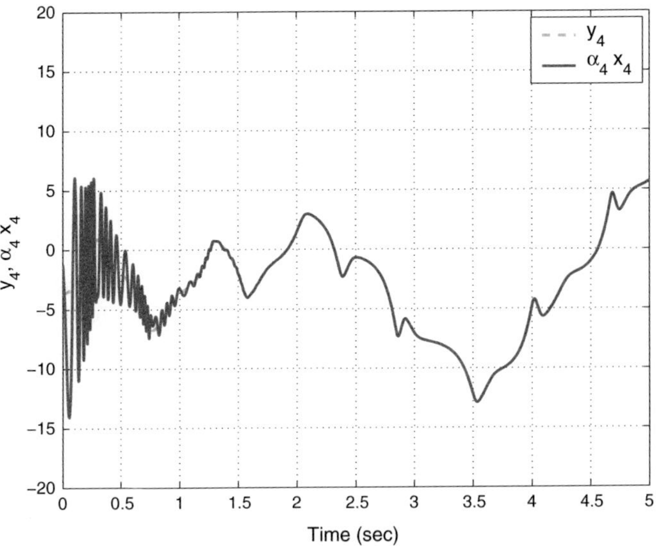

Fig. 8 Synchronization of the states y_4 and αx_4

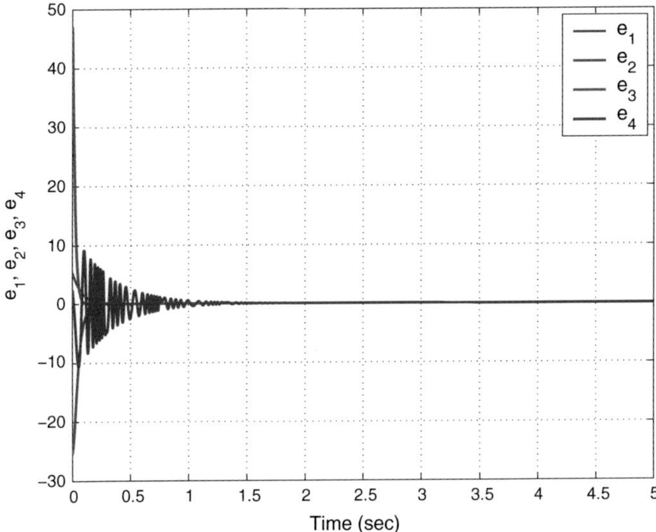

Fig. 9 Time-history of the synchronization errors e_1, e_2, e_3, e_4

5 Conclusions

In this research work, we described a novel 4-D hyperchaotic four-wing system with three quadratic nonlinearities. First, we discussed the qualitative properties of the novel 4-D hyperchaotic four-wing system. We showed that the novel hyperchaotic four-wing system has a unique equilibrium point at the origin, which is a saddle-point. Thus, origin is an unstable equilibrium of the novel hyperchaotic system. The Lyapunov exponents of the novel hyperchaotic four-wing system have been obtained as $L_1 = 2.5266$, $L_2 = 0.1053$, $L_3 = 0$ and $L_4 = -43.0194$. Thus, the maximal Lyapunov exponent (MLE) of the novel hyperchaotic four-wing system is obtained as $L_1 = 2.5266$. Since the sum of the Lyapunov exponents of the novel hyperchaotic system is negative, we showed that the novel hyperchaotic system is dissipative. Also, the Kaplan-Yorke dimension of the novel four-wing chaotic system has been obtained as $D_{KY} = 3.0612$. Finally, this work described the generalized projective synchronization (GPS) of the identical novel hyperchaotic four-wing systems with unknown parameters. The main GPS result via adaptive control method was proved using Lyapunov stability theory. MATLAB simulations have been shown to demonstrate all the main results for the novel 4-D hyperchaotic four-wing system.

References

1. Arneodo A, Coullet P, Tresser C (1981) Possible new strange attractors with spiral structure. Commun Math Phys 79(4):573–576
2. Azar AT (2010) Fuzzy systems. IN-TECH, Vienna
3. Azar AT (2012) Overview of type-2 fuzzy logic systems. Int J Fuzzy Syst Appl 2(4):1–28
4. Azar AT, Serrano FE (2014) Robust IMC-PID tuning for cascade control systems with gain and phase margin specifications. Neural Comput Appl 25(5):983–995
5. Azar AT, Serrano FE (2015) Adaptive sliding mode control of the Furuta pendulum. In: Azar AT, Zhu Q (eds) Advances and applications in sliding mode control systems, studies in computational intelligence, vol 576. Springer, Germany, pp 1–42
6. Azar AT, Serrano FE (2015) Deadbeat control for multivariable systems with time varying delays. In: Azar AT, Vaidyanathan S (eds) Chaos modeling and control systems design, studies in computational intelligence, vol 581. Springer, Germany, pp 97–132
7. Azar AT, Serrano FE (2015) Design and modeling of anti wind up PID controllers. In: Zhu Q, Azar AT (eds) Complex system modelling and control through intelligent soft computations, studies in fuzziness and soft computing, vol 319. Springer, Germany, pp 1–44
8. Azar AT, Serrano FE (2015) Stabilizatoin and control of mechanical systems with backlash. In: Azar AT, Vaidyanathan S (eds) Handbook of research on advanced intelligent control engineering and automation. Advances in computational intelligence and robotics (ACIR). IGI-Global, USA, pp 1–60
9. Azar AT, Vaidyanathan S (2015) Chaos modeling and control systems design, studies in computational intelligence, vol 581. Springer, Germany
10. Azar AT, Vaidyanathan S (2015) Computational intelligence applications in modeling and control, studies in computational intelligence, vol 575. Springer, Germany
11. Azar AT, Vaidyanathan S (2015) Handbook of research on advanced intelligent control engineering and automation. Advances in computational intelligence and robotics (ACIR). IGI-Global, USA

12. Azar AT, Zhu Q (2015) Advances and applications in sliding mode control systems, studies in computational intelligence, vol 576. Springer, Germany
13. Cai G, Tan Z (2007) Chaos synchronization of a new chaotic system via nonlinear control. J Uncertain Syst 1(3):235–240
14. Carroll TL, Pecora LM (1991) Synchronizing chaotic circuits. IEEE Trans Circuits Syst 38(4):453–456
15. Chen G, Ueta T (1999) Yet another chaotic attractor. Int J Bifurcat Chaos 9(7):1465–1466
16. Chen HK, Lee CI (2004) Anti-control of chaos in rigid body motion. Chaos, Solitons Fractals 21(4):957–965
17. Chen WH, Wei D, Lu X (2014) Global exponential synchronization of nonlinear time-delay Lur'e systems via delayed impulsive control. Commun Nonlinear Sci Numer Simul 19(9):3298–3312
18. Feki M (2003) An adaptive chaos synchronization scheme applied to secure communication. Chaos, Solitons Fractals 18(1):141–148
19. Gan Q, Liang Y (2012) Synchronization of chaotic neural networks with time delay in the leakage term and parametric uncertainties based on sampled-data control. J Franklin Inst 349(6):1955–1971
20. Gibson WT, Wilson WG (2013) Individual-based chaos: extensions of the discrete logistic model. J Theoret Biol 339:84–92
21. Huang X, Zhao Z, Wang Z, Li Y (2012) Chaos and hyperchaos in fractional-order cellular neural networks. Neurocomputing 94:13–21
22. Jiang GP, Zheng WX, Chen G (2004) Global chaos synchronization with channel time-delay. Chaos, Solitons Fractals 20(2):267–275
23. Karthikeyan R, Sundarapandian V (2014) Hybrid chaos synchronization of four-scroll systems via active control. J Electr Eng 65(2):97–103
24. Kaslik E, Sivasundaram S (2012) Nonlinear dynamics and chaos in fractional-order neural networks. Neural Netw 32:245–256
25. Kengne J, Chedjou JC, Kenne G, Kyamakya K (2012) Dynamical properties and chaos synchronization of improved Colpitts oscillators. Commun Nonlinear Sci Numer Simul 17(7):2914–2923
26. Khalil HK (2001) Nonlinear systems. Prentice Hall, New Jersey
27. Lang J (2015) Color image encryption based on color blend and chaos permutation in the reality-preserving multiple-parameter fractional Fourier transform domain. Opt Commun 338:181–192
28. Li D (2008) A three-scroll chaotic attractor. Phys Lett A 372(4):387–393
29. Li N, Zhang Y, Nie Z (2011) Synchronization for general complex dynamical networks with sampled-data. Neurocomputing 74(5):805–811
30. Li N, Pan W, Yan L, Luo B, Zou X (2014) Enhanced chaos synchronization and communication in cascade-coupled semiconductor ring lasers. Commun Nonlinear Sci Numer Simul 19(6):1874–1883
31. Li Z, Chen G (2006) Integration of fuzzy logic and chaos theory, studies in fuzziness and soft computing, vol 187. Springer, Germany
32. Lian S, Chen X (2011) Traceable content protection based on chaos and neural networks. Appl Soft Comput 11(7):4293–4301
33. Liu C, Liu T, Liu L, Liu K (2004) A new chaotic attractor. Chaos, Solitions Fractals 22(5):1031–1038
34. Lorenz EN (1963) Deterministic periodic flow. J Atmos Sci 20(2):130–141
35. Lü J, Chen G (2002) A new chaotic attractor coined. Int J Bifurcat Chaos 12(3):659–661
36. Mondal S, Mahanta C (2014) Adaptive second order terminal sliding mode controller for robotic manipulators. J Franklin Inst 351(4):2356–2377
37. Murali K, Lakshmanan M (1998) Secure communication using a compound signal from generalized chaotic systems. Phys Lett A 241(6):303–310
38. Nehmzow U, Walker K (2005) Quantitative description of robot-environment interaction using chaos theory. Robot Auton Syst 53(3–4):177–193

39. Pecora LM, Carroll TL (1990) Synchronization in chaotic systems. Phys Rev Lett 64(8): 821–824
40. Pehlivan I, Moroz IM, Vaidyanathan S (2014) Analysis, synchronization and circuit design of a novel butterfly attractor. J Sound Vib 333(20):5077–5096
41. Pham VT, Vaidyanathan S, Volos CK, Jafari S (2015) Hidden attractors in a chaotic system with an exponential nonlinear term. Eur Phys J Special Topics 224(8):1507–1517
42. Pham VT, Volos CK, Vaidyanathan S, Le TP, Vu VY (2015) A memristor-based hyperchaotic system with hidden attractors: dynamics, synchronization and circuital emulating. J Eng Sci Technol Rev 8(2):205–214
43. Qu Z (2011) Chaos in the genesis and maintenance of cardiac arrhythmias. Prog Biophys Mol Biol 105(3):247–257
44. Rasappan S, Vaidyanathan S (2012) Global chaos synchronization of WINDMI and Coullet chaotic systems by backstepping control. Far East J Math Sci 67(2):265–287
45. Rasappan S, Vaidyanathan S (2012) Hybrid synchronization of n-scroll Chua and Lur'e chaotic systems via backstepping control with novel feedback. Arch Control Sci 22(3):343–365
46. Rasappan S, Vaidyanathan S (2012) Synchronization of hyperchaotic Liu system via backstepping control with recursive feedback. Commun Comput Inf Sci 305:212–221
47. Rasappan S, Vaidyanathan S (2013) Hybrid synchronization of n-scroll chaotic Chua circuits using adaptive backstepping control design with recursive feedback. Malays J Math Sci 7(2):219–246
48. Rasappan S, Vaidyanathan S (2014) Global chaos synchronization of WINDMI and Coullet chaotic systems using adaptive backstepping control design. Kyungpook Math J 54(1): 293–320
49. Rhouma R, Belghith S (2011) Cryptoanalysis of a chaos based cryptosystem on DSP. Commun Nonlinear Sci Numer Simul 16(2):876–884
50. Rössler OE (1976) An equation for continuous chaos. Phys Lett A 57(5):397–398
51. Vaidyanathan S, VTP, Volos CK, (2015) A 5-D hyperchaotic Rikitake dynamo system with hidden attractors. Eur Phys J Special Topics 224(8):1575–1592
52. Sampath S, Vaidyanathan S, Volos CK, Pham VT (2015) An eight-term novel four-scroll chaotic system with cubic nonlinearity and its circuit simulation. J Eng Sci Technol Rev 8(2):1–6
53. Sarasu P, Sundarapandian V (2011) Active controller design for the generalized projective synchronization of four-scroll chaotic systems. Int J Syst Signal Control Eng Appl 4(2):26–33
54. Sarasu P, Sundarapandian V (2011) The generalized projective synchronization of hyperchaotic Lorenz and hyperchaotic Qi systems via active control. Int J Soft Comput 6(5): 216–223
55. Sarasu P, Sundarapandian V (2012) Adaptive controller design for the generalized projective synchronization of 4-scroll systems. Int J Syst Signal Control Eng Appl 5(2):21–30
56. Sarasu P, Sundarapandian V (2012) Generalized projective synchronization of three-scroll chaotic systems via adaptive control. Eur J Sci Res 72(4):504–522
57. Sarasu P, Sundarapandian V (2012) Generalized projective synchronization of two-scroll systems via adaptive control. Int J Soft Comput 7(4):146–156
58. Shahverdiev EM, Shore KA (2009) Impact of modulated multiple optical feedback time delays on laser diode chaos synchronization. Opt Commun 282(17):3568–3572
59. Sharma A, Patidar V, Purohit G, Sud KK (2012) Effects on the bifurcation and chaos in forced Duffing oscillator due to nonlinear damping. Commun Nonlinear Sci Numer Simul 17(6):2254–2269
60. Shi Y, Zhu P, Qin K (2014) Projective synchronization of different chaotic neural networks with mixed time delays based on an integral sliding mode controller. Neurocomputing 123:443–449
61. Sprott JC (1994) Some simple chaotic flows. Phys Rev E 50(2):647–650
62. Suérez I (1999) Mastering chaos in ecology. Ecol Model 117(2–3):305–314
63. Sundarapandian V (2010) Output regulation of the Lorenz attractor. Far East J Math Sci 42(2):289–299

64. Sundarapandian V (2013) Analysis and anti-synchronization of a novel chaotic system via active and adaptive controllers. J Eng Sci Technol Rev 6(4):45–52
65. Sundarapandian V, Karthikeyan R (2011) Anti-synchronization of hyperchaotic Lorenz and hyperchaotic Chen systems by adaptive control. Int J Syst Signal Control Eng Appl 4(2):18–25
66. Sundarapandian V, Karthikeyan R (2011) Anti-synchronization of Lü and Pan chaotic systems by adaptive nonlinear control. Eur J Sci Res 64(1):94–106
67. Sundarapandian V, Karthikeyan R (2012) Adaptive anti-synchronization of uncertain Tigan and Li systems. J Eng Appl Sci 7(1):45–52
68. Sundarapandian V, Karthikeyan R (2012) Hybrid synchronization of hyperchaotic Lorenz and hyperchaotic Chen systems via active control. J Eng Appl Sci 7(3):254–264
69. Sundarapandian V, Pehlivan I (2012) Analysis, control, synchronization, and circuit design of a novel chaotic system. Math Comput Model 55(7–8):1904–1915
70. Sundarapandian V, Sivaperumal S (2011) Sliding controller design of hybrid synchronization of four-wing chaotic systems. Int J Soft Comput 6(5):224–231
71. Suresh R, Sundarapandian V (2013) Global chaos synchronization of a family of *n*-scroll hyperchaotic Chua circuits using backstepping control with recursive feedback. Far East J Math Sci 73(1):73–95
72. Tigan G, Opris D (2008) Analysis of a 3D chaotic system. Chaos, Solitons Fractals 36: 1315–1319
73. Usama M, Khan MK, Alghatbar K, Lee C (2010) Chaos-based secure satellite imagery cryptosystem. Comput Math Appl 60(2):326–337
74. Vaidyanathan S (2011) Hybrid chaos synchronization of Liu and Lü systems by active nonlinear control. Commun Comput Inf Sci 204
75. Vaidyanathan S (2012) Analysis and synchronization of the hyperchaotic Yujun systems via sliding mode control. Adv Intell Syst Comput 176:329–337
76. Vaidyanathan S (2012) Anti-synchronization of Sprott-L and Sprott-M chaotic systems via adaptive control. Int J Control Theory Appl 5(1):41–59
77. Vaidyanathan S (2012) Global chaos control of hyperchaotic Liu system via sliding control method. Int J Control Theory Appl 5(2):117–123
78. Vaidyanathan S (2012) Output regulation of the Liu chaotic system. Appl Mech Mater 110–116:3982–3989
79. Vaidyanathan S (2012) Sliding mode control based global chaos control of Liu-Liu-Liu-Su chaotic system. Int J Control Theory Appl 5(1):15–20
80. Vaidyanathan S (2013) A new six-term 3-D chaotic system with an exponential nonlinearity. Far East J Math Sci 79(1):135–143
81. Vaidyanathan S (2013) Analysis and adaptive synchronization of two novel chaotic systems with hyperbolic sinusoidal and cosinusoidal nonlinearity and unknown parameters. J Eng Sci Technol Rev 6(4):53–65
82. Vaidyanathan S (2013) Analysis, control and synchronization of hyperchaotic Zhou system via adaptive control. Adv Intell Syst Comput 177:1–10
83. Vaidyanathan S (2014) A new eight-term 3-D polynomial chaotic system with three quadratic nonlinearities. Far East J Math Sci 84(2):219–226
84. Vaidyanathan S (2014) Analysis and adaptive synchronization of eight-term 3-D polynomial chaotic systems with three quadratic nonlinearities. Eur Phys J Special Topics 223(8): 1519–1529
85. Vaidyanathan S (2014) Analysis, control and synchronisation of a six-term novel chaotic system with three quadratic nonlinearities. Int J Model Ident Control 22(1):41–53
86. Vaidyanathan S (2014) Generalized projective synchronisation of novel 3-D chaotic systems with an exponential non-linearity via active and adaptive control. Int J Model Ident Control 22(3):207–217
87. Vaidyanathan S (2014) Global chaos synchronization of identical Li-Wu chaotic systems via sliding mode control. Int J Model Ident Control 22(2):170–177
88. Vaidyanathan S (2015) 3-cells cellular neural network (CNN) attractor and its adaptive biological control. Int J PharmTech Res 8(4):632–640

89. Vaidyanathan S (2015) A 3-D novel highly chaotic system with four quadratic nonlinearities, its adaptive control and anti-synchronization with unknown parameters. J Eng Sci Technol Rev 8(2):106–115
90. Vaidyanathan S (2015) Adaptive backstepping control of enzymes-substrates system with ferroelectric behaviour in brain waves. Int J PharmTech Res 8(2):256–261
91. Vaidyanathan S (2015) Adaptive biological control of generalized Lotka-Volterra three-species biological system. Int J PharmTech Res 8(4):622–631
92. Vaidyanathan S (2015) Adaptive chaotic synchronization of enzymes-substrates system with ferroelectric behaviour in brain waves. Int J PharmTech Res 8(5):964–973
93. Vaidyanathan S (2015) Adaptive control of a chemical chaotic reactor. Int J PharmTech Res 8(3):377–382
94. Vaidyanathan S (2015) Adaptive synchronization of chemical chaotic reactors. Int J ChemTech Res 8(2):612–621
95. Vaidyanathan S (2015) Adaptive synchronization of generalized Lotka-Volterra three-species biological systems. Int J PharmTech Res 8(5):928–937
96. Vaidyanathan S (2015) Analysis, properties and control of an eight-term 3-D chaotic system with an exponential nonlinearity. Int J Model Ident Control 23(2):164–172
97. Vaidyanathan S (2015) Anti-synchronization of Brusselator chemical reaction systems via adaptive control. Int J ChemTech Res 8(6):759–768
98. Vaidyanathan S (2015) Chaos in neurons and adaptive control of Birkhoff-Shaw strange chaotic attractor. Int J PharmTech Res 8(5):956–963
99. Vaidyanathan S (2015) Dynamics and control of Brusselator chemical reaction. Int J ChemTech Res 8(6):740–749
100. Vaidyanathan S (2015) Dynamics and control of Tokamak system with symmetric and magnetically confined plasma. Int J ChemTech Res 8(6):795–803
101. Vaidyanathan S (2015) Hyperchaos, qualitative analysis, control and synchronisation of a ten-term 4-D hyperchaotic system with an exponential nonlinearity and three quadratic nonlinearities. Int J Model Ident Control 23(4):380–392
102. Vaidyanathan S (2015) Lotka-Volterra population biology models with negative feedback and their ecological monitoring. Int J PharmTech Res 8(5):974–981
103. Vaidyanathan S (2015) Synchronization of 3-cells cellular neural network (CNN) attractors via adaptive control method. Int J PharmTech Res 8(5):946–955
104. Vaidyanathan S (2015) Synchronization of Tokamak systems with symmetric and magnetically confined plasma via adaptive control. Int J ChemTech Res 8(6):818–827
105. Vaidyanathan S, Azar AT (2015) Analysis and control of a 4-D novel hyperchaotic system. In: Azar AT, Vaidyanathan S (eds) Chaos modeling and control systems design, studies in computational intelligence, vol 581. Springer, Germany, pp 19–38
106. Vaidyanathan S, Azar AT (2015) Analysis, control and synchronization of a nine-term 3-D novel chaotic system. In: Azar AT, Vaidyanathan S (eds) Chaos modelling and control systems design, studies in computational intelligence, vol 581. Springer, Germany, pp 19–38
107. Vaidyanathan S, Azar AT (2015) Anti-synchronization of identical chaotic systems using sliding mode control and an application to Vaidhyanathan-Madhavan chaotic systems. Stud Comput Intell 576:527–547
108. Vaidyanathan S, Azar AT (2015) Hybrid synchronization of identical chaotic systems using sliding mode control and an application to Vaidhyanathan chaotic systems. Stud Comput Intell 576:549–569
109. Vaidyanathan S, Madhavan K (2013) Analysis, adaptive control and synchronization of a seven-term novel 3-D chaotic system. Int J Control Theory Appl 6(2):121–137
110. Vaidyanathan S, Pakiriswamy S (2013) Generalized projective synchronization of six-term Sundarapandian chaotic systems by adaptive control. Int J Control Theory Appl 6(2):153–163
111. Vaidyanathan S, Pakiriswamy S (2015) A 3-D novel conservative chaotic system and its generalized projective synchronization via adaptive control. J Eng Sci Technol Rev 8(2): 52–60

112. Vaidyanathan S, Rajagopal K (2011) Anti-synchronization of Li and T chaotic systems by active nonlinear control. Commun Comput Inf Sci 198:175–184
113. Vaidyanathan S, Rajagopal K (2011) Global chaos synchronization of hyperchaotic Pang and Wang systems by active nonlinear control. Commun Comput Inf Sci 204:84–93
114. Vaidyanathan S, Rajagopal K (2011) Global chaos synchronization of Lü and Pan systems by adaptive nonlinear control. Commun Comput Inf Sci 205:193–202
115. Vaidyanathan S, Rajagopal K (2012) Global chaos synchronization of hyperchaotic Pang and hyperchaotic Wang systems via adaptive control. Int J Soft Comput 7(1):28–37
116. Vaidyanathan S, Rasappan S (2011) Global chaos synchronization of hyperchaotic Bao and Xu systems by active nonlinear control. Commun Comput Inf Sci 198:10–17
117. Vaidyanathan S, Rasappan S (2014) Global chaos synchronization of n-scroll Chua circuit and Lur'e system using backstepping control design with recursive feedback. Arab J Sci Eng 39(4):3351–3364
118. Vaidyanathan S, Sampath S (2011) Global chaos synchronization of hyperchaotic Lorenz systems by sliding mode control. Commun Comput Inf Sci 205:156–164
119. Vaidyanathan S, Sampath S (2012) Anti-synchronization of four-wing chaotic systems via sliding mode control. Int J Autom Comput 9(3):274–279
120. Vaidyanathan S, Volos C (2015) Analysis and adaptive control of a novel 3-D conservative no-equilibrium chaotic system. Arch Control Sci 25(3):333–353
121. Vaidyanathan S, Volos C, Pham VT (2014) Hyperchaos, adaptive control and synchronization of a novel 5-D hyperchaotic system with three positive Lyapunov exponents and its SPICE implementation. Arch Control Sci 24(4):409–446
122. Vaidyanathan S, Volos C, Pham VT, Madhavan K, Idowu BA (2014) Adaptive backstepping control, synchronization and circuit simulation of a 3-D novel jerk chaotic system with two hyperbolic sinusoidal nonlinearities. Arch Control Sci 24(3):375–403
123. Vaidyanathan S, Azar AT, Rajagopal K, Alexander P (2015) Design and SPICE implementation of a 12-term novel hyperchaotic system and its synchronisation via active control. Int J Model Ident Control 23(3):267–277
124. Vaidyanathan S, Idowu BA, Azar AT (2015) Backstepping controller design for the global chaos synchronization of Sprott's jerk systems. Stud Comput Intell 581:39–58
125. Vaidyanathan S, Sampath S, Azar AT (2015) Global chaos synchronisation of identical chaotic systems via novel sliding mode control method and its application to Zhu system. Int J Model Ident Control 23(1):92–100
126. Vaidyanathan S, Volos C, Pham VT, Madhavan K (2015) Analysis, adaptive control and synchronization of a novel 4-D hyperchaotic hyperjerk system and its SPICE implementation. Arch Control Sci 25(1):5–28
127. Vaidyanathan S, Volos CK, Kyprianidis IM, Stouboulos IN, Pham VT (2015) Analysis, adaptive control and anti-synchronization of a six-term novel jerk chaotic system with two exponential nonlinearities and its circuit simulation. J Eng Sci Technol Rev 8(2):24–36
128. Vaidyanathan S, Volos CK, Pham VT (2015) Analysis, adaptive control and adaptive synchronization of a nine-term novel 3-D chaotic system with four quadratic nonlinearities and its circuit simulation. J Engi Sci Technol Rev 8(2):174–184
129. Vaidyanathan S, Volos CK, Pham VT (2015) Global chaos control of a novel nine-term chaotic system via sliding mode control. In: Azar AT, Zhu Q (eds) Advances and applications in sliding mode control systems, studies in computational intelligence, vol 576. Springer, Germany, pp 571–590
130. Volos CK, Kyprianidis IM, Stouboulos IN (2013) Experimental investigation on coverage performance of a chaotic autonomous mobile robot. Robot Auton Syst 61(12):1314–1322
131. Volos CK, Kyprianidis IM, Stouboulos IN, Tlelo-Cuautle E, Vaidyanathan S (2015) Memristor: a new concept in synchronization of coupled neuromorphic circuits. J Eng Sci Technol Rev 8(2):157–173
132. Wei Z, Yang Q (2010) Anti-control of Hopf bifurcation in the new chaotic system with two stable node-foci. Appl Math Comput 217(1):422–429

133. Witte CL, Witte MH (1991) Chaos and predicting varix hemorrhage. Med Hypotheses 36(4):312–317
134. Xiao X, Zhou L, Zhang Z (2014) Synchronization of chaotic Lur'e systems with quantized sampled-data controller. Commun Nonlinear Sci Numer Simul 19(6):2039–2047
135. Yang ZQ, Zhang Q, Chen ZQ (2012) Adaptive linear generalized synchronization between two nonidentical networks. Commun Nonlinear Sci Numer Simul 17(6):2628–2636
136. Yuan G, Zhang X, Wang Z (2014) Generation and synchronization of feedback-induced chaos in semiconductor ring lasers by injection-locking. Optik Int J Light Electron Opt 125(8): 1950–1953
137. Zaher AA, Abu-Rezq A (2011) On the design of chaos-based secure communication systems. Commun Nonlinear Syst Numer Simul 16(9):3721–3727
138. Zhang H, Zhou J (2012) Synchronization of sampled-data coupled harmonic oscillators with control inputs missing. Syst Control Lett 61(12):1277–1285
139. Zhang X, Zhao Z, Wang J (2014) Chaotic image encryption based on circular substitution box and key stream buffer. Signal Process Image Commun 29(8):902–913
140. Zhou W, Xu Y, Lu H, Pan L (2008) On dynamics analysis of a new chaotic attractor. Phys Lett A 372(36):5773–5777
141. Zhu C, Liu Y, Guo Y (2010) Theoretic and numerical study of a new chaotic system. Intell Inf Manage 2:104–109
142. Zhu Q, Azar AT (2015) Complex system modelling and control through intelligent soft computations, Studies in fuzzines and soft computing, vol 319. Springer, Germany

Hyperchaos, Control, Synchronization and Circuit Simulation of a Novel 4-D Hyperchaotic System with Three Quadratic Nonlinearities

Sundarapandian Vaidyanathan, Christos K. Volos and Viet-Thanh Pham

Abstract In this work, we announce an eleven-term novel 4-D hyperchaotic system with three quadratic nonlinearities. The novel 4-D hyperchaotic system has been derived by adding a feedback control to the seven term 3-D Lu-Xiao chaotic system [1]. The phase portraits of the eleven-term novel hyperchaotic system are depicted and the qualitative properties of the novel hyperchaotic system are discussed. The novel hyperchaotic system has a unique equilibrium at the origin, which is a saddle point. Thus, the origin is an unstable equilibrium of the novel hyperchaotic system. The Lyapunov exponents of the novel hyperchaotic system are obtained as $L_1 = 1.6023$, $L_2 = 0.1123$, $L_3 = 0$ and $L_4 = -22.6467$. Also, the Kaplan-Yorke dimension of the novel hyperchaotic system is obtained as $D_{KY} = 3.0757$. Since the sum of the Lyapunov exponents is negative, the novel hyperchaotic system is dissipative. Next, an adaptive controller is designed to globally stabilize the novel hyperchaotic system with unknown parameters. Moreover, an adaptive controller is also designed to achieve global chaos synchronization of the identical hyperchaotic systems with unknown parameters. Finally, an electronic circuit realization of the novel 4-D hyperchaotic system using SPICE is described in detail to confirm the feasibility of the theoretical model.

Keywords Chaos · Chaotic systems · Hyperchaos · Hyperchaotic systems · Adaptive control · Synchronization · Circuit simulation

---

S. Vaidyanathan (✉)
Research and Development Centre, Vel Tech University,
Avadi, Chennai 600062, Tamil Nadu, India
e-mail: sundarvtu@gmail.com

C.K. Volos
Physics Department, Aristotle University of Thessaloniki,
Thessaloniki GR-54124, Greece
e-mail: chvolos@gmail.com

V.-T. Pham
School of Electronics and Telecommunications,
Hanoi University of Science and Technology, 01 Dai Co Viet, Hanoi, Vietnam
e-mail: pvt3010@gmail.com

A.T. Azar and S. Vaidyanathan (eds.), *Advances in Chaos Theory and Intelligent Control*, Studies in Fuzziness and Soft Computing 337,
DOI 10.1007/978-3-319-30340-6_13

1 Introduction

In the last few decades, Chaos theory has become a very important and active research field, employing many applications in different disciplines like physics, chemistry, biology, ecology, engineering and economics, among others.

Some classical paradigms of 3-D chaotic systems in the literature are Lorenz system [2], Rössler system [3], ACT system [4], Sprott systems [5], Chen system [6], Lü system [7], Cai system [8], Tigan system [9], etc.

Many new chaotic systems have been discovered in the recent years such as Zhou system [10], Zhu system [11], Li system [12], Wei-Yang system [13], Sundarapandian systems [14, 15], Vaidyanathan systems [16–32], Pehlivan system [33], Sampath system [34], Pham system [35], etc.

Chaos theory and control systems have many important applications in science and engineering [36–41]. Some commonly known applications are oscillators [42, 43], lasers [44, 45], chemical reactions [46–53], biology [54–63], ecology [64, 65], encryption [66, 67], cryptosystems [68, 69], mechanical systems [70–74], secure communications [75–77], robotics [78–80], cardiology [81, 82], intelligent control [83, 84], neural networks [85–87], finance [88, 89], memristors [90, 91], etc.

A hyperchaotic system is defined as a chaotic system with at least two positive Lyapunov exponents [37]. Thus, the dynamics of a hyperchaotic system can expand in several different directions simultaneously. Thus, the hyperchaotic systems have more complex dynamical behaviour and they have miscellaneous applications in engineering such as secure communications [92–94], cryptosystems [95–97], fuzzy logic [98, 99], electrical circuits [100, 101], etc.

The minimum dimension of an autonomous, continuous-time, hyperchaotic system is four. The first 4-D hyperchaotic system was found by Rössler [102]. Many hyperchaotic systems have been reported in the chaos literature such as hyperchaotic Lorenz system [103], hyperchaotic Lü system [104], hyperchaotic Chen system [105], hyperchaotic Wang system [106], hyperchaotic Vaidyanathan systems [107–112], hyperchaotic Pham system [113], etc.

The control of a chaotic or hyperchaotic system aims to stabilize or regulate the system with the help of a feedback control. There are many methods available for controlling a chaotic system such as active control [114–116], adaptive control [117–119], sliding mode control [120, 121], backstepping control [122], etc.

The synchronization of chaotic systems aims to synchronize the states of master and slave systems asymptotically with time. There are many methods available for chaos synchronization such as active control [123–127], adaptive control [128–136], sliding mode control [137–140], backstepping control [141–144], etc.

In this research work, we announce an eleven-term novel 4-D hyperchaotic system with three quadratic nonlinearities. The novel 4-D hyperchaotic system is derived by adding a feedback controller to the 3-D Lu-Xiao chaotic system [1], which is a seven-term polynomial chaotic system with three quadratic nonlinearities.

This work is organized as follows. Section 2 describes the dynamic equations and phase portraits of the eleven-term novel 4-D hyperchaotic system. Section 3 details the qualitative properties of the novel hyperchaotic system. The Lyapunov exponents of the novel hyperchaotic system are obtained as $L_1 = 1.6023$, $L_2 = 0.1123$, $L_3 = 0$ and $L_4 = -22.6467$, while the Kaplan-Yorke dimension of the novel hyperchaotic system is obtained as $D_{KY} = 3.0757$.

In Sect. 4, we design an adaptive controller to globally stabilize the novel hyperchaotic system with unknown parameters. In Sect. 5, an adaptive controller is designed to achieve global chaos synchronization of the identical novel hyperchaotic systems with unknown parameters. In Sect. 6, an electronic circuit realization of the novel 4-D hyperchaotic system using SPICE is described in detail to confirm the feasibility of the theoretical model. Section 7 summarizes the main results derived in this work.

2 A Novel 4-D Hyperchaotic System

The Lu-Xiao chaotic system [1] is a seven-term polynomial chaotic system, which is described by the 3-D dynamics

$$\begin{cases} \dot{x}_1 = a(x_2 - x_1) + x_2 x_3 \\ \dot{x}_2 = -b x_1 x_3 + c x_1 \\ \dot{x}_3 = d x_1 x_2 - p x_3 \end{cases} \tag{1}$$

where x_1, x_2, x_3 are the states and a, b, c, d, p are positive, constant, parameters.

In [1], it was shown that the Lu-Xiao system (1) is *chaotic,* when the parameter values are taken as

$$a = 20, \quad b = 5, \quad c = 40, \quad d = 4, \quad p = 3 \tag{2}$$

For numerical simulations, we take the initial values as

$$x_1(0) = 0.5, \quad x_2(0) = 0.8, \quad x_3(0) = 0.2 \tag{3}$$

The Lyapunov exponents of the Lu-Xiao system (1) for the parameter values (2) and the initial state (3) can be calculated as

$$L_1 = 1.5202, \quad L_2 = 0, \quad L_3 = -24.4132 \tag{4}$$

and the Kaplan-Yorke dimension of the Lu-Xiao system can be derived as

$$D_{KY} = 2 + \frac{L_1 + L_2}{|L_3|} = 2.0623 \tag{5}$$

From (4), the Maximal Lyapunov Exponent (MLE) of the Lu-Xiao system (1) is obtained as $L_1 = 1.5202$.

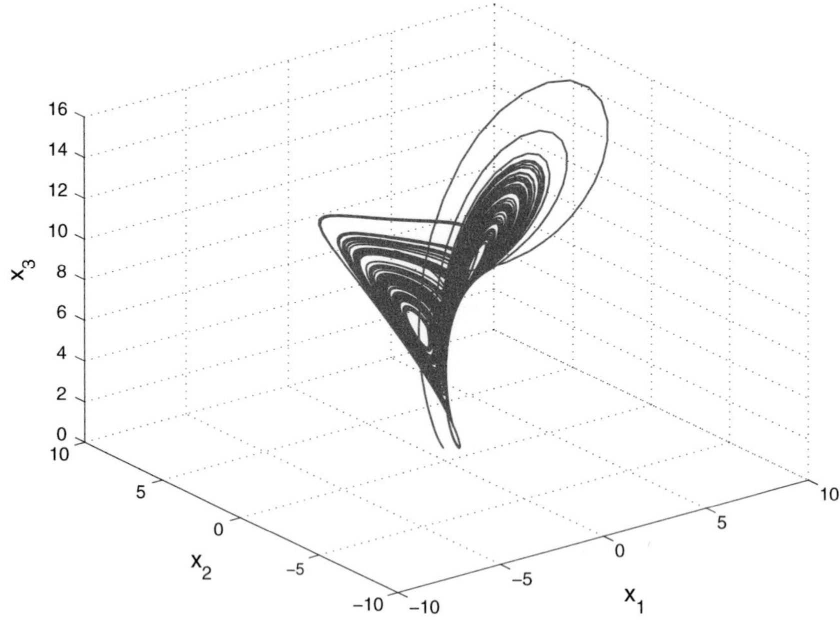

Fig. 1 3-D phase portrait of the Lu-Xiao chaotic system

Figure 1 shows the 3-D phase portrait of the Lu-Xiao chaotic system (1), which has the shape of a *two-scroll attractor*.

In this section, we describe an eleven-term novel hyperchaotic system, which is obtained by adding a feedback control to the Lu-Xiao chaotic system (1) and given by the 4-D dynamics

$$\begin{cases} \dot{x}_1 = a(x_2 - x_1) + x_2x_3 + x_4 \\ \dot{x}_2 = -bx_1x_3 + cx_1 + x_4 \\ \dot{x}_3 = dx_1x_2 - px_3 \\ \dot{x}_4 = -x_1 - x_2 \end{cases} \tag{6}$$

where x_1, x_2, x_3, x_4 are the states and a, b, c, d, p are constant, positive, parameters.

The novel 4-D system (6) has three quadratic nonlinearities.

The system (6) exhibits a *strange hyperchaotic attractor* for the parameter values

$$a = 18, \quad b = 5, \quad c = 40, \quad d = 4, \quad p = 3 \tag{7}$$

For numerical simulations, we take the initial conditions as

$$x_1(0) = 0.5, \quad x_2(0) = 0.8, \quad x_3(0) = 0.2, \quad x_4(0) = 1.3 \tag{8}$$

The Lyapunov exponents of the novel hyperchaotic system (6) for the parameter values (7) and the initial state (8) can be calculated as

$$L_1 = 1.6023, \quad L_2 = 0.1123, \quad L_3 = 0, \quad L_4 = -22.6467 \tag{9}$$

and the Kaplan-Yorke dimension of the novel hyperchaotic system (6) can be derived as

$$D_{KY} = 3 + \frac{L_1 + L_2 + L_3}{|L_4|} = 3.0757 \tag{10}$$

From (9), the Maximal Lyapunov Exponent (MLE) of the novel hyperchaotic system (6) is obtained as $L_1 = 1.6023$, which is greater than the MLE of the Lu-Xiao chaotic system (1), viz. $L_1 = 1.5202$.

Also, the novel hyperchaotic system (6) has two positive Lyapunov exponents and the Kaplan-Yorke dimension of the novel hyperchaotic system (6) is obviously greater than the Kaplan-Yorke dimension of the Lu-Xiao chaotic system (1). Thus, the novel 4-D hyperchaotic system (6) displays more chaotic behaviour and complexity than the 3-D Lu-Xiao chaotic system (1).

Figure 2 shows the 3-D projection of the novel hyperchaotic system (6) on the (x_1, x_2, x_3) space. We note that this has the form a *two-scroll attractor* or a *butterfly attractor*.

Figure 3 shows the 3-D projection of the novel hyperchaotic system (6) on the (x_1, x_2, x_4) space.

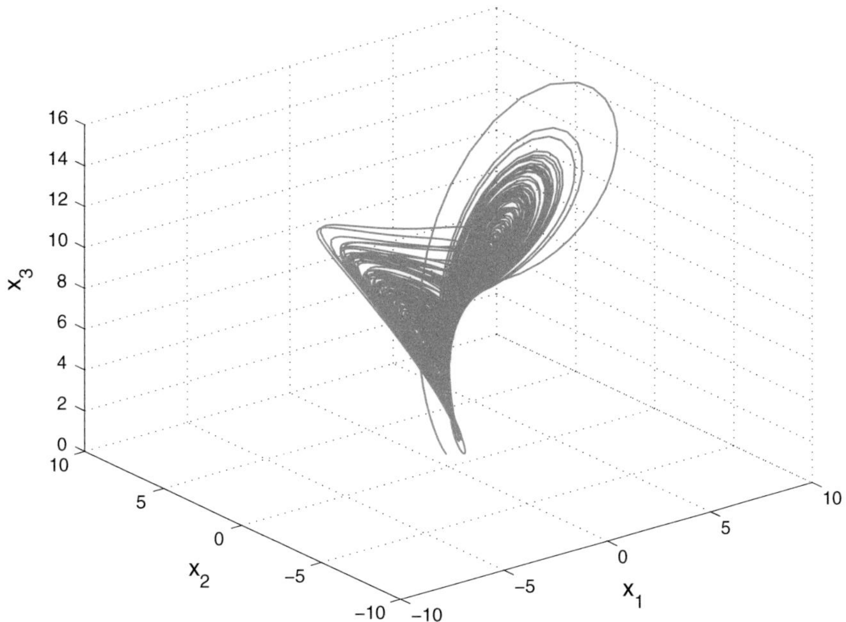

Fig. 2 3-D projection of the novel hyperchaotic system on the (x_1, x_2, x_3) space

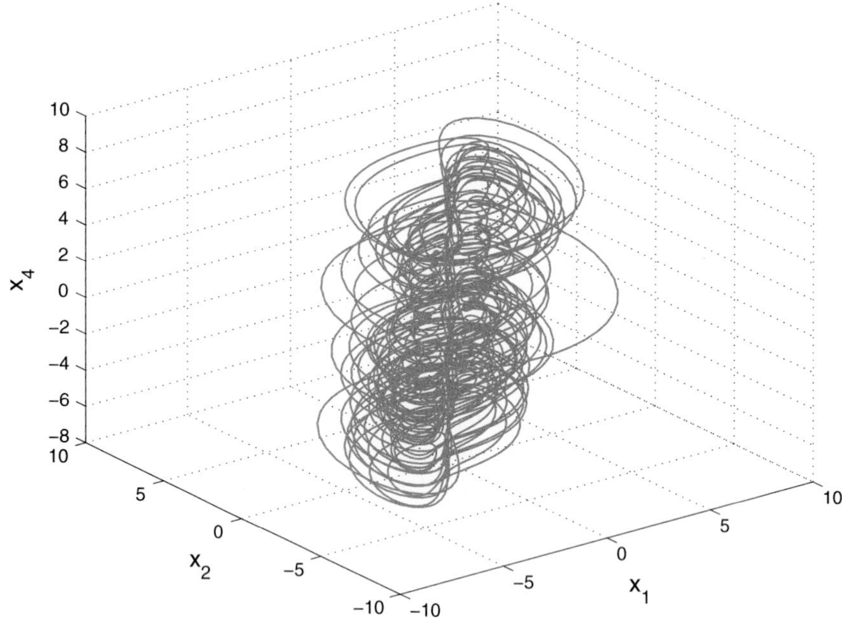

Fig. 3 3-D projection of the novel hyperchaotic system on the (x_1, x_2, x_4) space

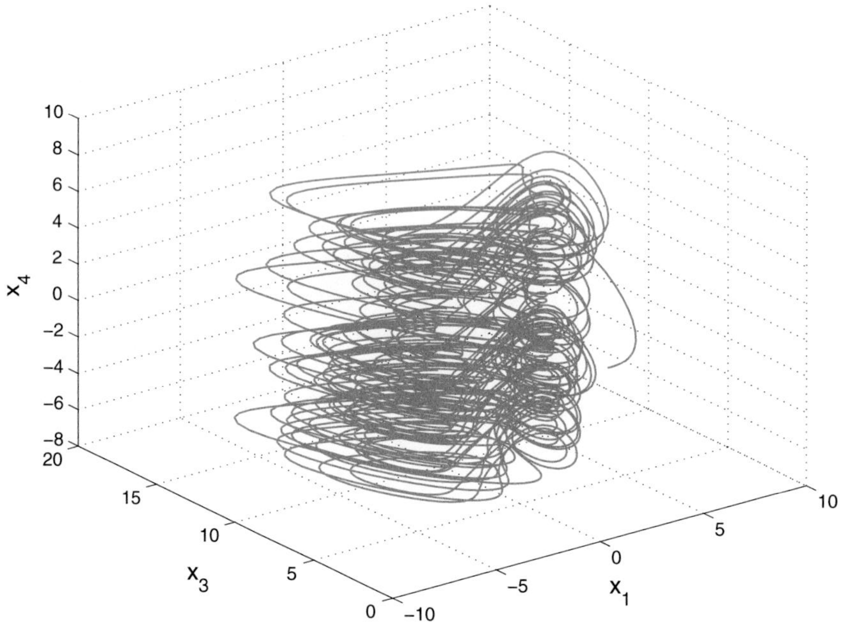

Fig. 4 3-D projection of the novel hyperchaotic system on the (x_1, x_3, x_4) space

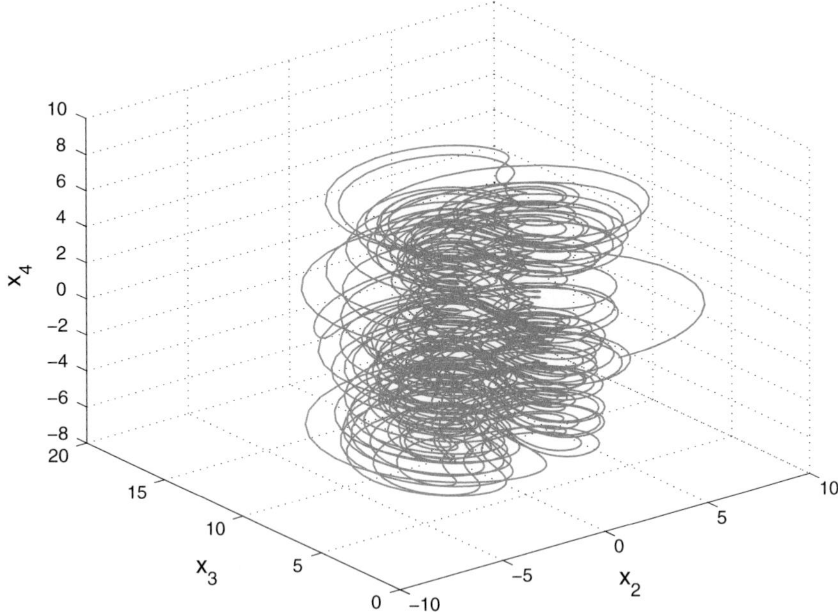

Fig. 5 3-D projection of the novel hyperchaotic system on the (x_2, x_3, x_4) space

Figure 4 shows the 3-D projection of the novel hyperchaotic system (6) on the (x_1, x_3, x_4) space.

Figure 5 shows the 3-D projection of the novel hyperchaotic system (6) on the (x_2, x_3, x_4) space.

3 Analysis of the Novel 4-D Hyperchaotic System

In this section, we give a dynamic analysis of the 4-D novel hyperchaotic system (6). We take the parameter values as in the hyperchaotic case (7).

3.1 Dissipativity

In vector notation, the novel hyperchaotic system (6) can be expressed as

$$\dot{\mathbf{x}} = f(\mathbf{x}) = \begin{bmatrix} f_1(x_1, x_2, x_3, x_4) \\ f_2(x_1, x_2, x_3, x_4) \\ f_3(x_1, x_2, x_3, x_4) \\ f_4(x_1, x_2, x_3, x_4) \end{bmatrix}, \tag{11}$$

where

$$\begin{cases} f_1(x_1, x_2, x_3, x_4) = a(x_2 - x_1) + x_2 x_3 + x_4 \\ f_2(x_1, x_2, x_3, x_4) = -b x_1 x_3 + c x_1 + x_4 \\ f_3(x_1, x_2, x_3, x_4) = d x_1 x_2 - p x_3 \\ f_4(x_1, x_2, x_3, x_4) = -x_1 - x_2 \end{cases} \tag{12}$$

Let Ω be any region in $\mathbf{R}^4$ with a smooth boundary and also, $\Omega(t) = \Phi_t(\Omega)$, where Φ_t is the flow of f. Furthermore, let $V(t)$ denote the hypervolume of $\Omega(t)$.

By Liouville's theorem, we know that

$$\dot{V}(t) = \int\limits_{\Omega(t)} (\nabla \cdot f)\, dx_1\, dx_2\, dx_3\, dx_4 \tag{13}$$

The divergence of the novel hyperchaotic system (11) is found as:

$$\nabla \cdot f = \frac{\partial f_1}{\partial x_1} + \frac{\partial f_2}{\partial x_2} + \frac{\partial f_3}{\partial x_3} + \frac{\partial f_4}{\partial x_4} = -(a + p) = -\mu < 0 \tag{14}$$

where $\mu = a + p = 21 > 0$.

Inserting the value of $\nabla \cdot f$ from (14) into (13), we get

$$\dot{V}(t) = \int\limits_{\Omega(t)} (-\mu)\, dx_1\, dx_2\, dx_3\, dx_4 = -\mu V(t) \tag{15}$$

Integrating the first order linear differential equation (15), we get

$$V(t) = \exp(-\mu t) V(0) \tag{16}$$

Since $\mu > 0$, it follows from Eq. (16) that $V(t) \to 0$ exponentially as $t \to \infty$. This shows that the novel hyperchaotic system (6) is dissipative. Hence, the system limit sets are ultimately confined into a specific limit set of zero hypervolume, and the asymptotic motion of the novel hyperchaotic system (6) settles onto a strange attractor of the system.

3.2 Equilibrium Points

We take the parameter values as in the hyperchaotic case (7), i.e.

$$a = 18, \quad b = 5, \quad c = 40, \quad d = 4, \quad p = 3 \tag{17}$$

It is easy to see that the novel hyperchaotic system (6) has a unique equilibrium at the origin.

To test the stability type of the equilibrium point $E_0 = \mathbf{0}$, we calculate the Jacobian matrix of the novel hyperchaotic system (6) at $\mathbf{x} = \mathbf{0}$.

We find that

$$J_0 \triangleq J(E_0) = \begin{bmatrix} -18 & 18 & 0 & 1 \\ 40 & 0 & 0 & 1 \\ 0 & 0 & -3 & 0 \\ -1 & -1 & 0 & 0 \end{bmatrix} \tag{18}$$

The matrix J_0 has the eigenvalues

$$\lambda_1 = -37.3026, \ \ \lambda_2 = -3, \ \ \lambda_3 = 0.1061, \ \ \lambda_4 = 19.1965 \tag{19}$$

This shows that the equilibrium point $E_0 = \mathbf{0}$ is a saddle-point, which is unstable.

3.3 Rotation Symmetry About the x_3-axis

It is easy to see that the novel 4-D hyperchaotic system (6) is invariant under the change of coordinates

$$(x_1, x_2, x_3, x_4) \mapsto (-x_1, -x_2, x_3, -x_4) \tag{20}$$

Since the transformation (20) persists for all values of the system parameters, it follows that the novel 4-D hyperchaotic system (6) has rotation symmetry about the x_3-axis and that any non-trivial trajectory must have a twin trajectory.

3.4 Invariance

It is easy to see that the x_3-axis is invariant under the flow of the 4-D novel hyperchaotic system (6). The invariant motion along the x_3-axis is characterized by the scalar dynamics

$$\dot{x}_3 = -px_3, \quad (p > 0) \tag{21}$$

which is globally exponentially stable.

3.5 Lyapunov Exponents and Kaplan-Yorke Dimension

We take the parameter values of the novel system (6) as in the hyperchaotic case (7), i.e.

$$a = 18, \ \ b = 5, \ \ c = 40, \ \ d = 4, \ \ p = 3 \tag{22}$$

We take the initial state of the novel system (6) as given in (8), i.e.

$$x_1(0) = 0.5, \quad x_2(0) = 0.8, \quad x_3(0) = 0.2, \quad x_4(0) = 1.3 \tag{23}$$

Then the Lyapunov exponents of the system (6) are numerically obtained using MATLAB as

$$L_1 = 1.6023, \quad L_2 = 0.1123, \quad L_3 = 0, \quad L_4 = -22.6467 \tag{24}$$

Since there are two positive Lyapunov exponents in (24), the novel system (6) exhibits *hyperchaotic* behavior.

From the LE spectrum (24), the maximal Lyapunov exponent (MLE) of the system (6) is obtained as $L_1 = 1.6023$.

We find that

$$L_1 + L_2 + L_3 + L_4 = -20.9321 < 0 \tag{25}$$

Thus, it follows that the novel hyperchaotic system (6) is dissipative.

Also, the Kaplan-Yorke dimension of the novel hyperchaotic system (6) is calculated as

$$D_{KY} = 3 + \frac{L_1 + L_2 + L_3}{|L_4|} = 3.0757, \tag{26}$$

which is fractional.

4 Adaptive Control of the Novel Hyperchaotic System

In this section, we use adaptive control method to derive an adaptive feedback control law for globally stabilizing the novel 4-D hyperchaotic system with unknown parameters.

Thus, we consider the novel 4-D hyperchaotic system given by

$$\begin{cases} \dot{x}_1 = a(x_2 - x_1) + x_2 x_3 + x_4 + u_1 \\ \dot{x}_2 = -b x_1 x_3 + c x_1 + x_4 + u_2 \\ \dot{x}_3 = d x_1 x_2 - p x_3 + u_3 \\ \dot{x}_4 = -x_1 - x_2 + u_4 \end{cases} \tag{27}$$

In (27), x_1, x_2, x_3, x_4 are the states and u_1, u_2, u_3, u_4 are the adaptive controls to be determined using estimates $\hat{a}(t), \hat{b}(t), \hat{c}(t), \hat{d}(t), \hat{p}(t)$ for the unknown parameters a, b, c, d, p, respectively.

We consider the adaptive feedback control law

$$\begin{cases} u_1 = -\hat{a}(t)(x_2 - x_1) - x_2 x_3 - x_4 - k_1 x_1 \\ u_2 = \hat{b}(t) x_1 x_3 - \hat{c}(t) x_1 - x_4 - k_2 x_2 \\ u_3 = -\hat{d}(t) x_1 x_2 + \hat{p}(t) x_3 - k_3 x_3 \\ u_4 = x_1 + x_2 - k_4 x_4 \end{cases} \tag{28}$$

where k_1, k_2, k_3, k_4 are positive gain constants.

Substituting (28) into (27), we get the closed-loop plant dynamics as

$$\begin{cases} \dot{x}_1 = [a - \hat{a}(t)](x_2 - x_1) - k_1 x_1 \\ \dot{x}_2 = -[b - \hat{b}(t)] x_1 x_3 + [c - \hat{c}(t)] x_1 - k_2 x_2 \\ \dot{x}_3 = [d - \hat{d}(t)] x_1 x_2 - [p - \hat{p}(t)] x_3 - k_3 x_3 \\ \dot{x}_4 = -k_4 x_4 \end{cases} \tag{29}$$

The parameter estimation errors are defined as

$$\begin{cases} e_a(t) = a - \hat{a}(t) \\ e_b(t) = b - \hat{b}(t) \\ e_c(t) = c - \hat{c}(t) \\ e_d(t) = d - \hat{d}(t) \\ e_p(t) = p - \hat{p}(t) \end{cases} \tag{30}$$

In view of (30), we can simplify the closed-loop plant dynamics (29) as

$$\begin{cases} \dot{x}_1 = e_a (x_2 - x_1) - k_1 x_1 \\ \dot{x}_2 = -e_b x_1 x_3 + e_c x_1 - k_2 x_2 \\ \dot{x}_3 = e_d x_1 x_2 - e_p x_3 - k_3 x_3 \\ \dot{x}_4 = -k_4 x_4 \end{cases} \tag{31}$$

Differentiating (30) with respect to t, we obtain

$$\begin{cases} \dot{e}_a(t) = -\dot{\hat{a}}(t) \\ \dot{e}_b(t) = -\dot{\hat{b}}(t) \\ \dot{e}_c(t) = -\dot{\hat{c}}(t) \\ \dot{e}_d(t) = -\dot{\hat{d}}(t) \\ \dot{e}_p(t) = -\dot{\hat{p}}(t) \end{cases} \tag{32}$$

We consider the quadratic candidate Lyapunov function defined by

$$V(\mathbf{x}, e_a, e_b, e_c, e_d, e_p) = \frac{1}{2}\left(x_1^2 + x_2^2 + x_3^2 + x_4^2\right) + \frac{1}{2}\left(e_a^2 + e_b^2 + e_c^2 + e_d^2 + e_p^2\right) \tag{33}$$

Differentiating V along the trajectories of (31) and (32), we obtain

$$\dot{V} = -k_1x_1^2 - k_2x_2^2 - k_3x_3^2 - k_4x_4^2 + e_a\left[x_1(x_2 - x_1) - \dot{\hat{a}}\right] + e_b\left[-x_1x_2x_3 - \dot{\hat{b}}\right] + e_c\left[x_1x_2 - \dot{\hat{c}}\right] + e_d\left[x_1x_2x_3 - \dot{\hat{d}}\right] + e_p\left[-x_3^2 - \dot{\hat{p}}\right] \tag{34}$$

In view of (34), we take the parameter update law as

$$\begin{cases} \dot{\hat{a}}(t) = x_1(x_2 - x_1) \\ \dot{\hat{b}}(t) = -x_1x_2x_3 \\ \dot{\hat{c}}(t) = x_1x_2 \\ \dot{\hat{d}}(t) = x_1x_2x_3 \\ \dot{\hat{p}}(t) = -x_3^2 \end{cases} \tag{35}$$

Next, we state and prove the main result of this section.

Theorem 1 *The novel 4-D hyperchaotic system (27) with unknown system parameters is globally and exponentially stabilized for all initial conditions by the adaptive control law (28) and the parameter update law (35), where k_1, k_2, k_3, k_4 are positive gain constants.*

Proof We prove this result by applying Lyapunov stability theory [145].

We consider the quadratic Lyapunov function defined by (33), which is clearly a positive definite function on $\mathbf{R}^9$.

By substituting the parameter update law (35) into (34), we obtain the time-derivative of V as

$$\dot{V} = -k_1x_1^2 - k_2x_2^2 - k_3x_3^2 - k_4x_4^2 \tag{36}$$

From (36), it is clear that $\dot{V}$ is a negative semi-definite function on $\mathbf{R}^9$.

Thus, we can conclude that the state vector $\mathbf{x}(t)$ and the parameter estimation error are globally bounded, i.e.

$$\left[x_1(t)\ x_2(t)\ x_3(t)\ x_4(t)\ e_a(t)\ e_b(t)\ e_c(t)\ e_d(t)\ e_p(t)\right]^T \in \mathbf{L}_\infty.$$

We define $k = \min\{k_1, k_2, k_3, k_4\}$.

Then it follows from (36) that

$$\dot{V} \le -k\|\mathbf{x}(t)\|^2 \tag{37}$$

Thus, we have

$$k\|\mathbf{x}(t)\|^2 \le -\dot{V} \tag{38}$$

Integrating the inequality (38) from 0 to t, we get

$$k \int_0^t \|\mathbf{x}(\tau)\|^2 \, d\tau \le V(0) - V(t) \tag{39}$$

From (39), it follows that $\mathbf{x} \in \mathbf{L}_2$.

Using (31), we can conclude that $\dot{\mathbf{x}} \in \mathbf{L}_\infty$.

Using Barbalat's lemma [145], we conclude that $\mathbf{x}(t) \to 0$ exponentially as $t \to \infty$ for all initial conditions $\mathbf{x}(0) \in \mathbf{R}^4$.

This completes the proof. ■

For the numerical simulations, the classical fourth-order Runge-Kutta method with step size $h = 10^{-8}$ is used to solve the systems (27) and (35), when the adaptive control law (28) is applied.

The parameter values of the novel 4-D hyperchaotic system (27) are taken as in the hyperchaotic case (7), i.e.

$$a = 18, \quad b = 5, \quad c = 40, \quad d = 4, \quad p = 3 \tag{40}$$

We take the positive gain constants as $k_i = 8$ for $i = 1, \ldots, 4$.

Furthermore, as initial conditions of the novel 4-D hyperchaotic system (27), we take

$$x_1(0) = 2.1, \quad x_2(0) = -6.4, \quad x_3(0) = 3.7, \quad x_4(0) = -5.3 \tag{41}$$

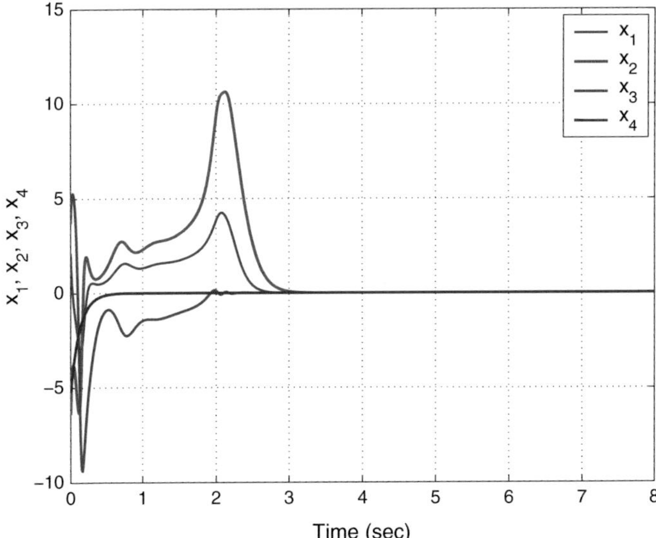

Fig. 6 Time-history of the controlled states x_1, x_2, x_3, x_4

Also, as initial conditions of the parameter estimates, we take

$$\hat{a}(0) = 5.4, \quad \hat{b}(0) = 10.3, \quad \hat{c}(0) = 3.5, \quad \hat{d}(0) = 15.7, \quad \hat{p}(t) = 12.4 \tag{42}$$

In Fig. 6, the exponential convergence of the controlled states of the novel 4-D hyperchaotic system (27) is shown. From Fig. 6, it is clear that the controlled states $x_1(t)$, $x_2(t)$, $x_3(t)$ and $x_4(t)$ converge to zero in less than four seconds.

5 Adaptive Synchronization of the Identical Novel Hyperchaotic Systems

In this section, we use adaptive control method to derive an adaptive feedback control law for globally synchronizing identical novel 4-D hyperchaotic systems with unknown parameters.

As the master system, we consider the novel 4-D hyperchaotic system given by

$$\begin{cases} \dot{x}_1 = a(x_2 - x_1) + x_2 x_3 + x_4 \\ \dot{x}_2 = -b x_1 x_3 + c x_1 + x_4 \\ \dot{x}_3 = d x_1 x_2 - p x_3 \\ \dot{x}_4 = -x_1 - x_2 \end{cases} \tag{43}$$

In (43), x_1, x_2, x_3, x_4 are the states and a, b, c, d, p are unknown parameters.

As the slave system, we consider the 4-D novel hyperchaotic system given by

$$\begin{cases} \dot{y}_1 = a(y_2 - y_1) + y_2 y_3 + y_4 + u_1 \\ \dot{y}_2 = -b y_1 y_3 + c y_1 + y_4 + u_2 \\ \dot{y}_3 = d y_1 y_2 - p y_3 + u_3 \\ \dot{y}_4 = -y_1 - y_2 + u_4 \end{cases} \tag{44}$$

In (44), y_1, y_2, y_3, y_4 are the states and u_1, u_2, u_3, u_4 are the adaptive controls to be determined using estimates $\hat{a}(t), \hat{b}(t), \hat{c}(t), \hat{d}(t), \hat{p}(t)$ for the unknown parameters a, b, c, d, p, respectively.

The synchronization error between the novel hyperchaotic systems (43) and (44) is defined by

$$\begin{cases} e_1 = y_1 - x_1 \\ e_2 = y_2 - x_2 \\ e_3 = y_3 - x_3 \\ e_4 = y_4 - x_4 \end{cases} \tag{45}$$

Then the synchronization error dynamics is obtained as

$$\begin{cases} \dot{e}_1 = a(e_2 - e_1) + e_4 + y_2 y_3 - x_2 x_3 + u_1 \\ \dot{e}_2 = -b(y_1 y_3 - x_1 x_3) + c e_1 + e_4 + u_2 \\ \dot{e}_3 = d(y_1 y_2 - x_1 x_2) - p e_3 + u_3 \\ \dot{e}_4 = -e_1 - e_2 + u_4 \end{cases} \tag{46}$$

We consider the adaptive feedback control law

$$\begin{cases} u_1 = -\hat{a}(t)(e_2 - e_1) - e_4 - y_2 y_3 + x_2 x_3 - k_1 e_1 \\ u_2 = \hat{b}(t)(y_1 y_3 - x_1 x_3) - \hat{c}(t) e_1 - e_4 - k_2 e_2 \\ u_3 = -\hat{d}(t)(y_1 y_2 - x_1 x_2) + \hat{p}(t) e_3 - k_3 e_3 \\ u_4 = e_1 + e_2 - k_4 e_4 \end{cases} \tag{47}$$

where k_1, k_2, k_3, k_4 are positive gain constants.

Substituting (47) into (46), we get the closed-loop error dynamics as

$$\begin{cases} \dot{e}_1 = \left[a - \hat{a}(t)\right](e_2 - e_1) - k_1 e_1 \\ \dot{e}_2 = \left[b - \hat{b}(t)\right](y_1 y_3 - x_1 x_3) + \left[c - \hat{c}(t)\right] e_1 - k_2 e_2 \\ \dot{e}_3 = \left[d - \hat{d}(t)\right](y_1 y_2 - x_1 x_2) - \left[p - \hat{p}(t)\right] e_3 - k_3 e_3 \\ \dot{e}_4 = -k_4 e_4 \end{cases} \tag{48}$$

The parameter estimation errors are defined as

$$\begin{cases} e_a(t) = a - \hat{a}(t) \\ e_b(t) = b - \hat{b}(t) \\ e_c(t) = c - \hat{c}(t) \\ e_d(t) = d - \hat{d}(t) \\ e_p(t) = p - \hat{p}(t) \end{cases} \tag{49}$$

In view of (49), we can simplify the error dynamics (48) as

$$\begin{cases} \dot{e}_1 = e_a(e_2 - e_1) - k_1 e_1 \\ \dot{e}_2 = e_b(y_1 y_3 - x_1 x_3) + e_c e_1 - k_2 e_2 \\ \dot{e}_3 = e_d(y_1 y_2 - x_1 x_2) - e_p e_3 - k_3 e_3 \\ \dot{e}_4 = -k_4 e_4 \end{cases} \tag{50}$$

Differentiating (49) with respect to t, we obtain

$$\begin{cases} \dot{e}_a(t) = -\dot{\hat{a}}(t) \\ \dot{e}_b(t) = -\dot{\hat{b}}(t) \\ \dot{e}_c(t) = -\dot{\hat{c}}(t) \\ \dot{e}_d(t) = -\dot{\hat{d}}(t) \\ \dot{e}_p(t) = -\dot{\hat{p}}(t) \end{cases} \tag{51}$$

We consider the quadratic candidate Lyapunov function defined by

$$V(\mathbf{e}, e_a, e_b, e_c, e_d, e_p) = \frac{1}{2}\left(e_1^2 + e_2^2 + e_3^2 + e_4^2\right) + \frac{1}{2}\left(e_a^2 + e_b^2 + e_c^2 + e_d^2 + e_p^2\right) \tag{52}$$

Differentiating V along the trajectories of (50) and (51), we obtain

$$\begin{aligned} \dot{V} = & -k_1 e_1^2 - k_2 e_2^2 - k_3 e_3^2 - k_4 e_4^2 + e_a \left[e_1(e_2 - e_1) - \dot{\hat{a}} \right] \\ & + e_b \left[-e_2(y_1 y_3 - x_1 x_3) - \dot{\hat{b}} \right] + e_c \left[e_1 e_2 - \dot{\hat{c}} \right] \\ & + e_d \left[e_3(y_1 y_2 - x_1 x_2) - \dot{\hat{d}} \right] + e_p \left[-e_3^2 - \dot{\hat{p}} \right] \end{aligned} \tag{53}$$

In view of (53), we take the parameter update law as

$$\begin{cases} \dot{\hat{a}}(t) = e_1(e_2 - e_1) \\ \dot{\hat{b}}(t) = -e_2(y_1 y_3 - x_1 x_3) \\ \dot{\hat{c}}(t) = e_1 e_2 \\ \dot{\hat{d}}(t) = e_3(y_1 y_2 - x_1 x_2) \\ \dot{\hat{p}}(t) = -e_3^2 \end{cases} \tag{54}$$

Next, we state and prove the main result of this section.

Theorem 2 *The novel hyperchaotic systems (43) and (44) with unknown system parameters are globally and exponentially synchronized for all initial conditions by the adaptive control law (47) and the parameter update law (54), where k_1, k_2, k_3, k_4 are positive gain constants.*

Proof We prove this result by applying Lyapunov stability theory [145].

We consider the quadratic Lyapunov function defined by (52), which is clearly a positive definite function on $\mathbf{R}^9$.

By substituting the parameter update law (54) into (53), we obtain

$$\dot{V} = -k_1 e_1^2 - k_2 e_2^2 - k_3 e_3^2 - k_4 e_4^2 \tag{55}$$

From (55), it is clear that $\dot{V}$ is a negative semi-definite function on $\mathbf{R}^9$.

Thus, we can conclude that the error vector $\mathbf{e}(t)$ and the parameter estimation error are globally bounded, i.e.

$$\left[e_1(t)\ e_2(t)\ e_3(t)\ e_4(t)\ e_a(t)\ e_b(t)\ e_c(t)\ e_d(t)\ e_p(t) \right]^T \in \mathbf{L}_\infty. \tag{56}$$

We define $k = \min\{k_1, k_2, k_3, k_4\}$.

Then it follows from (55) that

$$\dot{V} \leq -k\|\mathbf{e}(t)\|^2 \tag{57}$$

Thus, we have

$$k\|\mathbf{e}(t)\|^2 \leq -\dot{V} \tag{58}$$

Integrating the inequality (58) from 0 to t, we get

$$k\int_0^t \|\mathbf{e}(\tau)\|^2\, d\tau \;\leq\; V(0) - V(t) \tag{59}$$

From (59), it follows that $\mathbf{e} \in \mathbf{L}_2$.

Using (50), we can conclude that $\dot{\mathbf{e}} \in \mathbf{L}_\infty$.

Using Barbalat's lemma [145], we conclude that $\mathbf{e}(t) \to 0$ exponentially as $t \to \infty$ for all initial conditions $\mathbf{e}(0) \in \mathbf{R}^4$.

Thus, we conclude that the novel hyperchaotic systems (43) and (44) with unknown system parameters are globally and exponentially synchronized for all initial conditions by the adaptive control law (47) and the parameter update law (54).

This completes the proof. ■

For the numerical simulations, the classical fourth-order Runge-Kutta method with step size $h = 10^{-8}$ is used to solve the systems (43), (44) and (54), when the adaptive control law (47) is applied.

The parameter values of the novel hyperchaotic systems are taken as in the hyperchaotic case (7), i.e.

$$a = 18, \quad b = 5, \quad c = 40, \quad d = 4, \quad p = 3 \tag{60}$$

We take the positive gain constants as

$$k_1 = 8, \quad k_2 = 8, \quad k_3 = 8, \quad k_4 = 8 \tag{61}$$

Furthermore, as initial conditions of the master system (43), we take

$$x_1(0) = 5.3, \quad x_2(0) = 1.8, \quad x_3(0) = 4.7, \quad x_4(0) = -2.9 \tag{62}$$

As initial conditions of the slave system (44), we take

$$y_1(0) = 10.4, \quad y_2(0) = -4.5, \quad y_3(0) = 1.2, \quad y_4(0) = -15.4 \tag{63}$$

Also, as initial conditions of the parameter estimates, we take

$$\hat{a}(0) = 2.1, \quad \hat{b}(0) = 8.4, \quad \hat{c}(0) = 9.7, \quad \hat{d}(0) = 10.3, \quad \hat{p}(0) = 5.2 \tag{64}$$

Figures 7, 8, 9 and 10 describe the complete synchronization of the novel hyperchaotic systems (43) and (44), while Fig. 11 describes the time-history of the synchronization errors e_1, e_2, e_3, e_4.

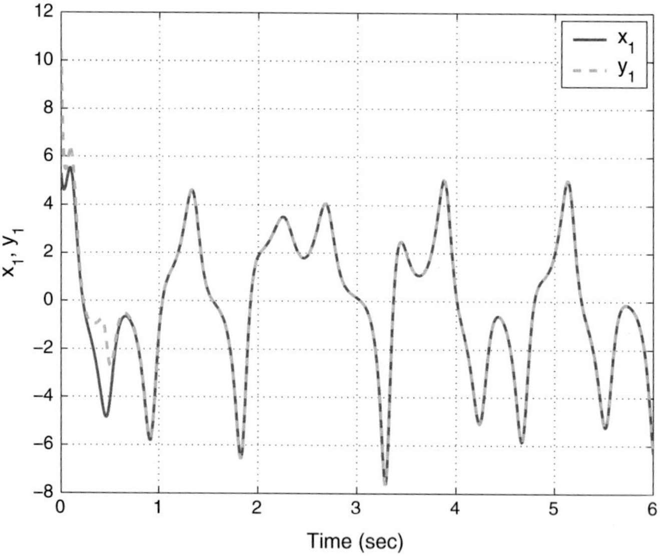

Fig. 7 Synchronization of the states x_1 and y_1

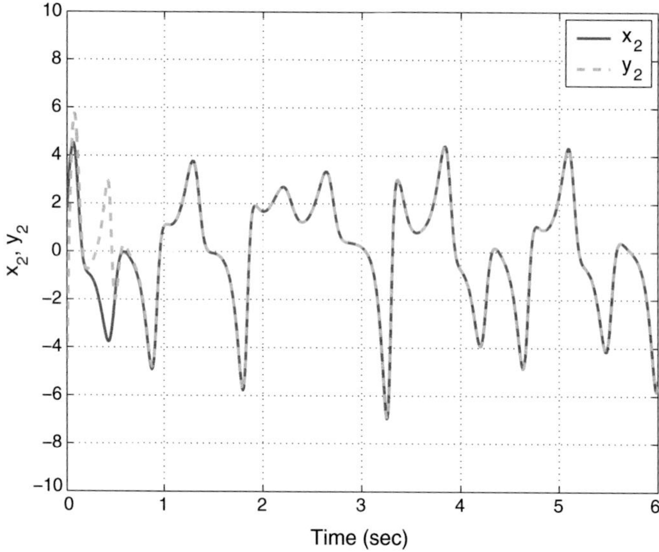

Fig. 8 Synchronization of the states x_2 and y_2

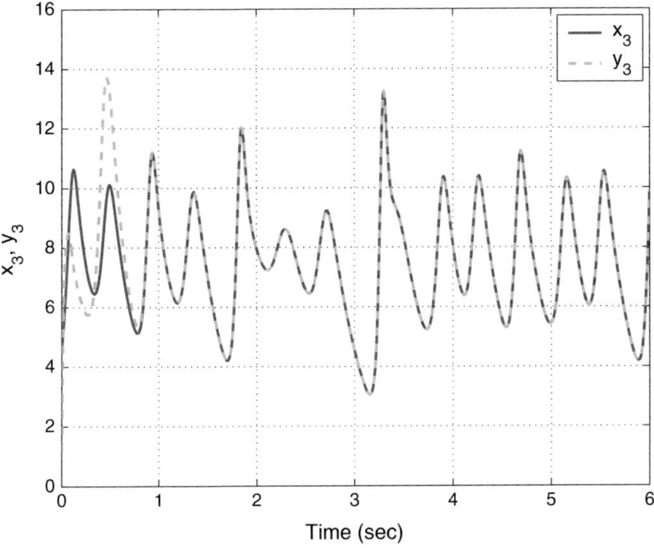

Fig. 9 Synchronization of the states x_3 and y_3

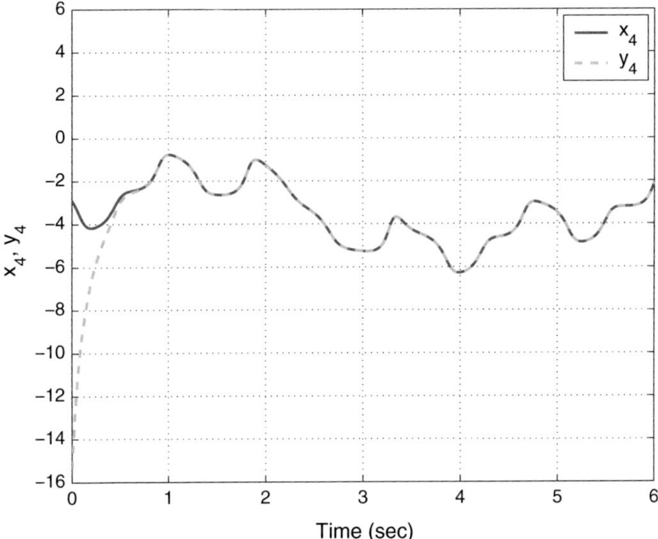

Fig. 10 Synchronization of the states x_4 and y_4

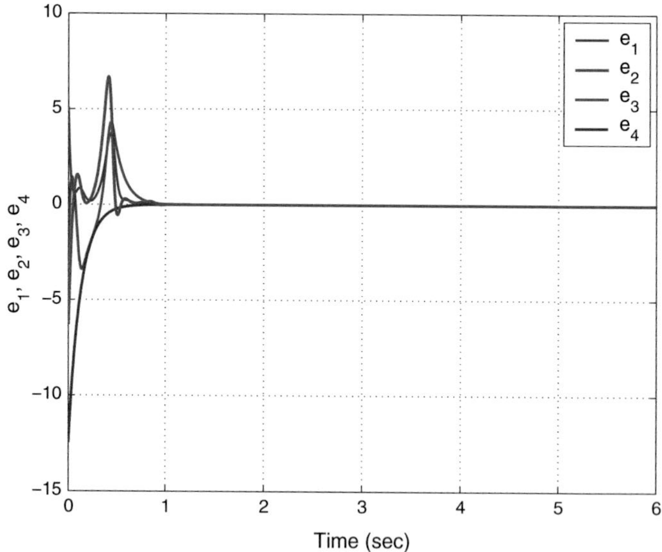

Fig. 11 Time-history of the synchronization errors e_1, e_2, e_3, e_4

Figure 7 shows that the states x_1 and y_1 are synchronized in 1 s.

Figure 8 shows that the states x_2 and y_2 are synchronized in 1 s.

Figure 9 shows that the states x_3 and y_3 are synchronized in 2 s.

Figure 10 shows that the states x_4 and y_4 are synchronized in 2 s.

Figure 11 shows that the complete synchronization errors $e_1(t)$, $e_2(t)$, $e_3(t)$ and $e_4(t)$ converge to zero in 2 s.

6 Circuit Simulation of the Novel Hyperchaotic System

The electronic circuit modelling the new hyperchaotic system (6) is realized by using off-the-shelf components such as resistors, capacitors, operational amplifiers and analog multipliers. Applying the design approach based on the operational amplifiers [26, 28, 33], the circuit has proposed as shown in Fig. 12 where each state variable of system (6), i.e. x_1, x_2, x_3, x_4 is implemented as the voltage across the corresponding capacitors C_1, C_2, C_3, and C_4, respectively.

The circuital equations of the circuit in Fig. 12 are given in the following form

$$\left\{\begin{aligned}\frac{dv_{C_1}}{dt} &= \frac{1}{R_1C_1}v_{C_2} - \frac{1}{R_2C_1}v_{C_1} + \frac{1}{10R_3C_1}v_{C_2}v_{C_3} + \frac{1}{R_4C_1}v_{C_4}\\ \frac{dv_{C_2}}{dt} &= -\frac{1}{10R_5C_2}v_{C_1}v_{C_3} + \frac{1}{R_6C_2}v_{C_1} + \frac{1}{R_7C_2}v_{C_4}\\ \frac{dv_{C_3}}{dt} &= \frac{1}{10R_8C_3}v_{C_1}v_{C_2} - \frac{1}{R_9C_3}v_{C_3}\\ \frac{dv_{C_4}}{dt} &= -\frac{1}{R_{10}C_4}v_{C_1} - \frac{1}{R_{11}C_4}v_{C_3}\end{aligned}\right. \tag{65}$$

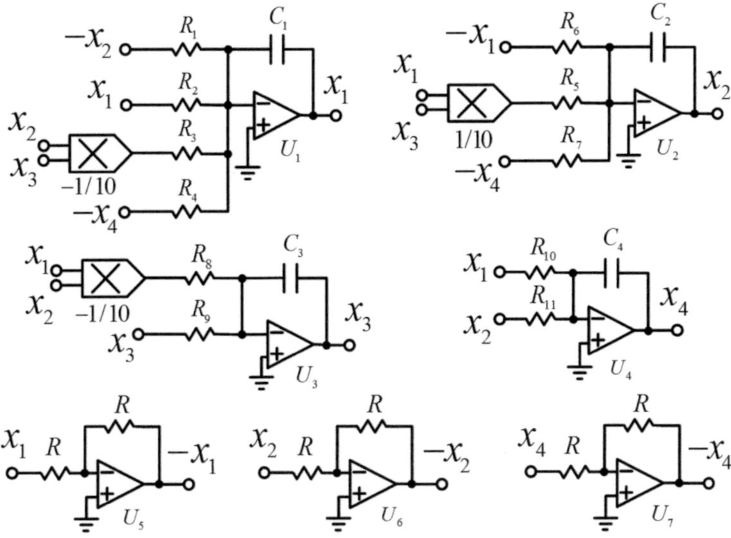

Fig. 12 The designed electronic circuit schematic of the eleven-term novel 4-D hyperchaotic system with three quadratic nonlinearities (6)

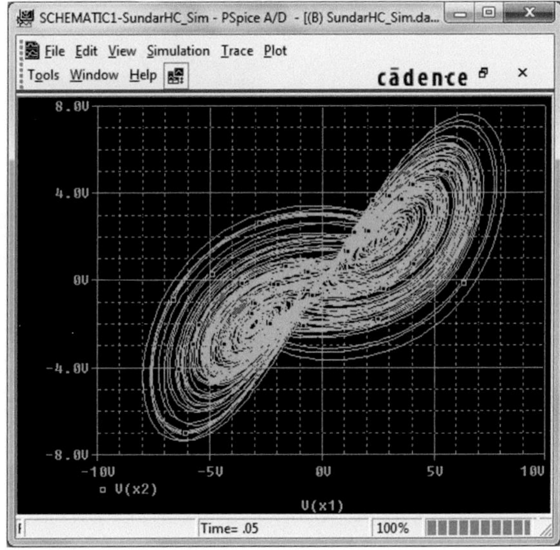

Fig. 13 Phase portrait of the designed electronic circuit obtained from OrCAD in $v_{C_1} - v_{C_2}$ plane

where v_{C_1}, v_{C_2}, v_{C_3}, and v_{C_4} are the voltages across the capacitors C_1, C_2, C_3, and C_4, respectively.

The power supplies of all active devices are $\pm 15V_{DC}$. The TL084 operational amplifiers are used in this work. The values of components in Fig. 12 are chosen to match the parameters of system (6) as follows: $R_1 = R_2 = 20\,\text{k}\Omega$, $R_3 = 36\,\text{k}\Omega$, $R_4 =$

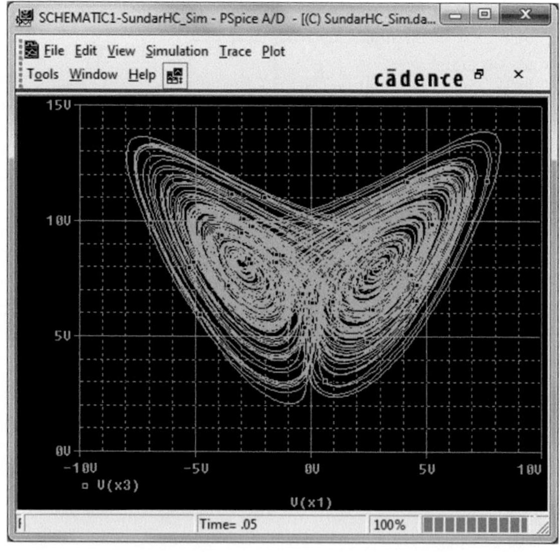

Fig. 14 Phase portrait of the designed electronic circuit obtained from OrCAD in $v_{C_1} - v_{C_3}$ plane

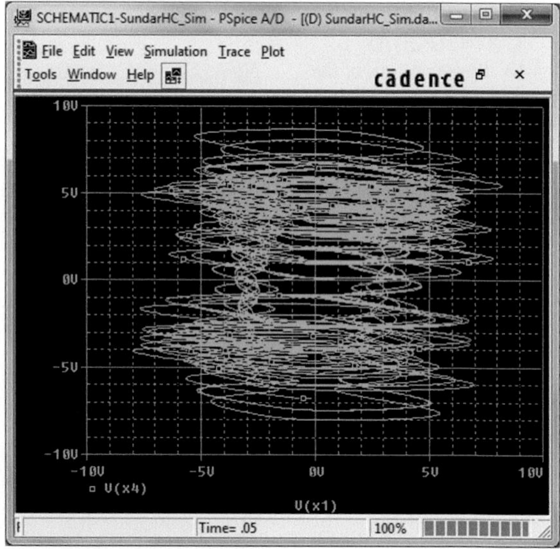

Fig. 15 Phase portrait of the designed electronic circuit obtained from OrCAD in $v_{C_1} - v_{C_4}$ plane

$R_7 = R_{10} = R_{11} = R = 360\,\text{k}\Omega$, $R_5 = 7.2\,\text{k}\Omega$, $R_6 = R_8 = 9\,\text{k}\Omega$, $R_9 = 120\,\text{k}\Omega$, and $C_1 = C_2 = C_3 = C_4 = 1\,\text{nF}$.

The designed circuit is implemented in the electronic simulation package Cadence OrCAD. The SPICE results are displayed in Figs. 13, 14, 15 and 16, which indicate the hyperchaotic attractors in $v_{C_1} - v_{C_2}$, $v_{C_1} - v_{C_3}$, $v_{C_1} - v_{C_4}$and $v_{C_2} - v_{C_3}$ planes.

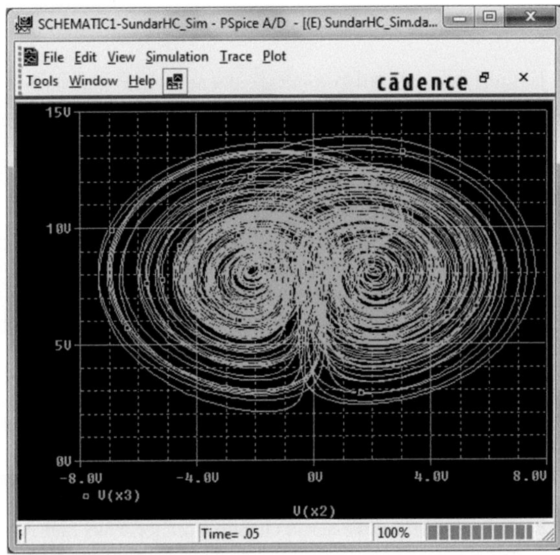

Fig. 16 Phase portrait of the designed electronic circuit obtained from OrCAD in $v_{C_2} - v_{C_3}$ plane

7 Conclusions

In this work, we described an eleven-term novel 4-D hyperchaotic system with three quadratic nonlinearities. We discussed the qualitative properties of the novel hyperchaotic system in detail. The novel hyperchaotic system has a unique equilibrium at the origin, which is a saddle-point and unstable. The novel hyperchaotic system has a rotation symmetry about the x_3-axis. The x_3 axis is an invariant manifold for the novel hyperchaotic system. The Lyapunov exponents of the novel hyperchaotic system have been obtained as $L_1 = 1.6023$, $L_2 = 0.1123$, $L_3 = 0$ and $L_4 = -22.6467$. Since the maximal Lyapunov exponent of the novel hyperchaotic system has a high value, viz. $L_1 = 1.6023$, the system shows highly hyperchaotic behavior. Also, the Kaplan-Yorke dimension of the novel hyperchaotic system is obtained as $D_{KY} = 3.0757$. Since the sum of the Lyapunov exponents is negative, the novel hyperchaotic system is dissipative. Next, an adaptive controller was designed to globally stabilize the novel hyperchaotic system with unknown parameters. Also, an adaptive controller was also designed to achieve global chaos synchronization of the identical hyperchaotic systems with unknown parameters. Finally, an electronic circuit realization of the novel 4-D hyperchaotic system using SPICE has been designed in detail to confirm the feasibility of the theoretical model.

References

1. Lu H, Xiao X (2012) Analysis of a novel autonomous 3-D chaotic system. Int J Adv Comput Technol 4(1):248–255
2. Lorenz EN (1963) Deterministic periodic flow. J Atmos Sci 20(2):130–141
3. Rössler OE (1976) An equation for continuous chaos. Phys Lett A 57(5):397–398
4. Arneodo A, Coullet P, Tresser C (1981) Possible new strange attractors with spiral structure. Commun Math Phys 79(4):573–576
5. Sprott JC (1994) Some simple chaotic flows. Phys Rev E 50(2):647–650
6. Chen G, Ueta T (1999) Yet another chaotic attractor. Int J Bifurcat Chaos 9(7):1465–1466
7. Lü J, Chen G (2002) A new chaotic attractor coined. Int J Bifurcat Chaos 12(3):659–661
8. Cai G, Tan Z (2007) Chaos synchronization of a new chaotic system via nonlinear control. J Uncertain Syst 1(3):235–240
9. Tigan G, Opris D (2008) Analysis of a 3D chaotic system. Chaos, Solitons Fractals 36: 1315–1319
10. Zhou W, Xu Y, Lu H, Pan L (2008) On dynamics analysis of a new chaotic attractor. Phys Lett A 372(36):5773–5777
11. Zhu C, Liu Y, Guo Y (2010) Theoretic and numerical study of a new chaotic system. Intell Inform Manage 2:104–109
12. Li D (2008) A three-scroll chaotic attractor. Phys Lett A 372(4):387–393
13. Wei Z, Yang Q (2010) Anti-control of Hopf bifurcation in the new chaotic system with two stable node-foci. Appl Math Comput 217(1):422–429
14. Sundarapandian V (2013) Analysis and anti-synchronization of a novel chaotic system via active and adaptive controllers. J Eng Sci Technol Rev 6(4):45–52
15. Sundarapandian V, Pehlivan I (2012) Analysis, control, synchronization, and circuit design of a novel chaotic system. Math Comput Model 55(7–8):1904–1915
16. Vaidyanathan S (2013) A new six-term 3-D chaotic system with an exponential nonlinearity. Far East J Math Sci 79(1):135–143
17. Vaidyanathan S (2013) Analysis and adaptive synchronization of two novel chaotic systems with hyperbolic sinusoidal and cosinusoidal nonlinearity and unknown parameters. J Eng Sci Technol Rev 6(4):53–65
18. Vaidyanathan S (2014) A new eight-term 3-D polynomial chaotic system with three quadratic nonlinearities. Far East J Math Sci 84(2):219–226
19. Vaidyanathan S (2014) Analysis and adaptive synchronization of eight-term 3-D polynomial chaotic systems with three quadratic nonlinearities. Eur Phys J: Special Topics 223(8): 1519–1529
20. Vaidyanathan S (2014) Analysis, control and synchronisation of a six-term novel chaotic system with three quadratic nonlinearities. Int J Model Ident Control 22(1):41–53
21. Vaidyanathan S (2014) Generalized projective synchronisation of novel 3-D chaotic systems with an exponential non-linearity via active and adaptive control. Int J Model Ident Control 22(3):207–217
22. Vaidyanathan S (2015) A 3-D novel highly chaotic system with four quadratic nonlinearities, its adaptive control and anti-synchronization with unknown parameters. J Eng Sci Technol Rev 8(2):106–115
23. Vaidyanathan S (2015) Analysis, properties and control of an eight-term 3-D chaotic system with an exponential nonlinearity. Int J Model Ident Control 23(2):164–172
24. Vaidyanathan S, Azar AT (2015) Analysis, control and synchronization of a nine-term 3-D novel chaotic system. In: Azar AT, Vaidyanathan S (eds) Chaos modelling and control systems design, Studies in Computational Intelligence, vol 581. Springer, Germany, pp 19–38
25. Vaidyanathan S, Madhavan K (2013) Analysis, adaptive control and synchronization of a seven-term novel 3-D chaotic system. Int J Control Theory Appl 6(2):121–137
26. Vaidyanathan S, Pakiriswamy S (2015) A 3-D novel conservative chaotic system and its generalized projective synchronization via adaptive control. J Eng Sci Technol Rev 8(2):52–60

27. Vaidyanathan S, Volos C (2015) Analysis and adaptive control of a novel 3-D conservative no-equilibrium chaotic system. Arch Control Sci 25(3):333–353
28. Vaidyanathan S, Volos C, Pham VT, Madhavan K, Idowu BA (2014b) Adaptive backstepping control, synchronization and circuit simulation of a 3-D novel jerk chaotic system with two hyperbolic sinusoidal nonlinearities. Arch Control Sci 24(3):375–403
29. Vaidyanathan S, Rajagopal K, Volos CK, Kyprianidis IM, Stouboulos IN (2015) Analysis, adaptive control and synchronization of a seven-term novel 3-D chaotic system with three quadratic nonlinearities and its digital implementation in LabVIEW. J Eng Sci Technol Rev 8(2):130–141
30. Vaidyanathan S, Volos CK, Kyprianidis IM, Stouboulos IN, Pham VT (2015) Analysis, adaptive control and anti-synchronization of a six-term novel jerk chaotic system with two exponential nonlinearities and its circuit simulation. J Eng Sci Technol Rev 8(2):24–36
31. Vaidyanathan S, Volos CK, Pham VT (2015) Analysis, adaptive control and adaptive synchronization of a nine-term novel 3-D chaotic system with four quadratic nonlinearities and its circuit simulation. J Eng Sci Technol Rev 8(2):181–191
32. Vaidyanathan S, Volos CK, Pham VT (2015) Global chaos control of a novel nine-term chaotic system via sliding mode control. in: Azar AT, Zhu Q Advances and applications in sliding mode control systems, Studies in Computational Intelligence, vol 576. Springer, Germany, pp 571–590
33. Pehlivan I, Moroz IM, Vaidyanathan S (2014) Analysis, synchronization and circuit design of a novel butterfly attractor. J Sound Vib 333(20):5077–5096
34. Sampath S, Vaidyanathan S, Volos CK, Pham VT (2015) An eight-term novel four-scroll chaotic system with cubic nonlinearity and its circuit simulation. J Eng Sci Technol Rev 8(2):1–6
35. Pham VT, Vaidyanathan S, Volos CK, Jafari S (2015) Hidden attractors in a chaotic system with an exponential nonlinear term. Eur Phys J—Special Topics 224(8):1507–1517
36. Azar AT (2010) Fuzzy Syst. IN-TECH, Vienna
37. Azar AT, Vaidyanathan S (2015) Chaos modeling and control systems design, Studies in Computational Intelligence, vol 581. Springer, Germany
38. Azar AT, Vaidyanathan S (2015) Computational intelligence applications in modeling and control, Studies in Computational Intelligence, vol 575. Springer, Germany
39. Azar AT, Vaidyanathan S (2015) Handbook of research on advanced intelligent control engineering and automation. Advances in Computational Intelligence and Robotics (ACIR), IGI-Global, USA
40. Azar AT, Zhu Q (2015) Advances and applications in sliding mode control systems, Studies in Computational Intelligence, vol 576. Springer, Germany
41. Zhu Q, Azar AT (2015) Complex system modelling and control through intelligent soft computations, Studies in Fuzzines and Soft Computing, vol 319. Springer, Germany
42. Kengne J, Chedjou JC, Kenne G, Kyamakya K (2012) Dynamical properties and chaos synchronization of improved Colpitts oscillators. Commun Nonlinear Sci Numer Simul 17(7):2914–2923
43. Sharma A, Patidar V, Purohit G, Sud KK (2012) Effects on the bifurcation and chaos in forced Duffing oscillator due to nonlinear damping. Commun Nonlinear Sci Numer Simul 17(6):2254–2269
44. Li N, Pan W, Yan L, Luo B, Zou X (2014) Enhanced chaos synchronization and communication in cascade-coupled semiconductor ring lasers. Commun Nonlinear Sci Numer Simul 19(6):1874–1883
45. Yuan G, Zhang X, Wang Z (2014) Generation and synchronization of feedback-induced chaos in semiconductor ring lasers by injection-locking. Optik—Int J Light Electron Optics 125(8):1950–1953
46. Gaspard P (1999) Microscopic chaos and chemical reactions. Phys A 263(1–4):315–328
47. Petrov V, Gaspar V, Masere J, Showalter K (1993) Controlling chaos in Belousov-Zhabotinsky reaction. Nature 361:240–243

48. Vaidyanathan S (2015) Adaptive control of a chemical chaotic reactor. Int J PharmTech Res 8(3):377–382
49. Vaidyanathan S (2015) Adaptive synchronization of chemical chaotic reactors. Int J ChemTech Res 8(2):612–621
50. Vaidyanathan S (2015) Anti-synchronization of Brusselator chemical reaction systems via adaptive control. Int J ChemTech Res 8(6):759–768
51. Vaidyanathan S (2015) Dynamics and control of Brusselator chemical reaction. Int J ChemTech Res 8(6):740–749
52. Vaidyanathan S (2015) Dynamics and control of Tokamak system with symmetric and magnetically confined plasma. Int J ChemTech Res 8(6):795–803
53. Vaidyanathan S (2015) Synchronization of Tokamak systems with symmetric and magnetically confined plasma via adaptive control. Int J ChemTech Res 8(6):818–827
54. Das S, Goswami D, Chatterjee S, Mukherjee S (2014) Stability and chaos analysis of a novel swarm dynamics with applications to multi-agent systems. Eng Appl Artif Intell 30:189–198
55. Kyriazis M (1991) Applications of chaos theory to the molecular biology of aging. Exp Gerontol 26(6):569–572
56. Vaidyanathan S (2015) 3-cells cellular neural network (CNN) attractor and its adaptive biological control. Int J PharmTech Res 8(4):632–640
57. Vaidyanathan S (2015) Adaptive backstepping control of enzymes-substrates system with ferroelectric behaviour in brain waves. Int J PharmTech Res 8(2):256–261
58. Vaidyanathan S (2015) Adaptive biological control of generalized Lotka-Volterra three-species biological system. Int J PharmTech Res 8(4):622–631
59. Vaidyanathan S (2015) Adaptive chaotic synchronization of enzymes-substrates system with ferroelectric behaviour in brain waves. Int J PharmTech Res 8(5):964–973
60. Vaidyanathan S (2015) Adaptive synchronization of generalized Lotka-Volterra three-species biological systems. Int J PharmTech Res 8(5):928–937
61. Vaidyanathan S (2015) Chaos in neurons and adaptive control of Birkhoff-Shaw strange chaotic attractor. Int J PharmTech Res 8(5):956–963
62. Vaidyanathan S (2015) Lotka-Volterra population biology models with negative feedback and their ecological monitoring. Int J PharmTech Res 8(5):974–981
63. Vaidyanathan S (2015) Synchronization of 3-cells cellular neural network (CNN) attractors via adaptive control method. Int J PharmTech Res 8(5):946–955
64. Gibson WT, Wilson WG (2013) Individual-based chaos: Extensions of the discrete logistic model. J Theor Biol 339:84–92
65. Suérez I (1999) Mastering chaos in ecology. Ecol Model 117(2–3):305–314
66. Lang J (2015) Color image encryption based on color blend and chaos permutation in the reality-preserving multiple-parameter fractional Fourier transform domain. Optics Commun 338:181–192
67. Zhang X, Zhao Z, Wang J (2014) Chaotic image encryption based on circular substitution box and key stream buffer. Sig Process Image Commun 29(8):902–913
68. Rhouma R, Belghith S (2011) Cryptoanalysis of a chaos based cryptosystem on DSP. Commun Nonlinear Sci Numer Simul 16(2):876–884
69. Usama M, Khan MK, Alghatbar K, Lee C (2010) Chaos-based secure satellite imagery cryptosystem. Comput Math Appl 60(2):326–337
70. Azar AT, Serrano FE (2014) Robust IMC-PID tuning for cascade control systems with gain and phase margin specifications. Neural Comput Appl 25(5):983–995
71. Azar AT, Serrano FE (2015) Adaptive sliding mode control of the Furuta pendulum. In: Azar AT, Zhu Q (eds) Advances and applications in sliding mode control systems, Studies in computational intelligence, vol 576. Springer, Germany, pp 1–42
72. Azar AT, Serrano FE (2015) Deadbeat control for multivariable systems with time varying delays. In: Azar AT, Vaidyanathan S (eds) Chaos modeling and control systems design, Studies in computational intelligence, vol 581. Springer, Germany, pp 97–132
73. Azar AT, Serrano FE (2015) Design and modeling of anti wind up PID controllers. In: Zhu Q, Azar AT (eds) Complex system modelling and control through intelligent soft computations, Studies in fuzziness and soft computing, vol 319. Springer, Germany, pp 1–44

74. Azar AT, Serrano FE (2015) Stabilizatoin and control of mechanical systems with backlash. In: Azar AT, Vaidyanathan S (eds) Handbook of Research on advanced intelligent control engineering and automation. Advances in Computational Intelligence and Robotics (ACIR), IGI-Global, USA, pp 1–60
75. Feki M (2003) An adaptive chaos synchronization scheme applied to secure communication. Chaos, Solitons Fractals 18(1):141–148
76. Murali K, Lakshmanan M (1998) Secure communication using a compound signal from generalized chaotic systems. Phys Lett A 241(6):303–310
77. Zaher AA, Abu-Rezq A (2011) On the design of chaos-based secure communication systems. Commun Nonlinear Sci Numer Simul 16(9):3721–3727
78. Mondal S, Mahanta C (2014) Adaptive second order terminal sliding mode controller for robotic manipulators. J Franklin Inst 351(4):2356–2377
79. Nehmzow U, Walker K (2005) Quantitative description of robot-environment interaction using chaos theory. Robot Auton Syst 53(3–4):177–193
80. Volos CK, Kyprianidis IM, Stouboulos IN (2013) Experimental investigation on coverage performance of a chaotic autonomous mobile robot. Robot Auton Syst 61(12):1314–1322
81. Qu Z (2011) Chaos in the genesis and maintenance of cardiac arrhythmias. Prog Biophys Mol Biol 105(3):247–257
82. Witte CL, Witte MH (1991) Chaos and predicting varix hemorrhage. Med Hypotheses 36(4):312–317
83. Azar AT (2012) Overview of type-2 fuzzy logic systems. Int J Fuzzy Syst Appl 2(4):1–28
84. Li Z, Chen G (2006) Integration of fuzzy logic and chaos theory, Studies in Fuzziness and Soft Computing, vol 187. Springer, Germany
85. Huang X, Zhao Z, Wang Z, Li Y (2012) Chaos and hyperchaos in fractional-order cellular neural networks. Neurocomputing 94:13–21
86. Kaslik E, Sivasundaram S (2012) Nonlinear dynamics and chaos in fractional-order neural networks. Neural Networks 32:245–256
87. Lian S, Chen X (2011) Traceable content protection based on chaos and neural networks. Appl Soft Comput 11(7):4293–4301
88. Guégan D (2009) Chaos in economics and finance. Ann Rev Control 33(1):89–93
89. Sprott JC (2004) Competition with evolution in ecology and finance. Phys Lett A 325(5–6):329–333
90. Pham VT, Volos CK, Vaidyanathan S, Le TP, Vu VY (2015) A memristor-based hyperchaotic system with hidden attractors: dynamics, synchronization and circuital emulating. J Eng Sci Technol Rev 8(2):205–214
91. Volos CK, Kyprianidis IM, Stouboulos IN, Tlelo-Cuautle E, Vaidyanathan S (2015) Memristor: a new concept in synchronization of coupled neuromorphic circuits. J Eng Sci Technol Rev 8(2):157–173
92. Filali RL, Benrejeb M, Borne P (2014) On observer-based secure communication design using discrete-time hyperchaotic systems. Commun Nonlinear Sci Numer Simul 19(5):1424–1432
93. Li C, Liao X, Wong KW (2005) Lag synchronization of hyperchaos with application to secure communications. Chaos, Solitons Fractals 23(1):183–193
94. Wu X, Zhu C, Kan H (2015) An improved secure communication scheme based passive synchronization of hyperchaotic complex nonlinear system. Appl Math Comput 252: 201–214
95. Hammami S (2015) State feedback-based secure image cryptosystem using hyperchaotic synchronization. ISA Trans 54:52–59
96. Rhouma R, Belghith S (2008) Cryptanalysis of a new image encryption algorithm based on hyper-chaos. Phys Lett A 372(38):5973–5978
97. Zhu C (2012) A novel image encryption scheme based on improved hyperchaotic sequences. Optics Commun 285(1):29–37
98. Senouci A, Boukabou A (2014) Predictive control and synchronization of chaotic and hyperchaotic systems based on a $T-S$ fuzzy model. Math Comput Simul 105:62–78

99. Zhang H, Liao X, Yu J (2005) Fuzzy modeling and synchronization of hyperchaotic systems. Chaos, Solitons Fractals 26(3):835–843
100. Wei X, Yunfei F, Qiang L (2012) A novel four-wing hyper-chaotic system and its circuit implementation. Procedia Eng 29:1264–1269
101. Yujun N, Xingyuan W, Mingjun W, Huaguang Z (2010) A new hyperchaotic system and its circuit implementation. Commun Nonlinear Sci Numer Simul 15(11):3518–3524
102. Rössler OE (1979) An equation for hyperchaos. Phys Lett A 71:155–157
103. Jia Q (2007) Hyperchaos generated from the Lorenz chaotic system and its control. Phys Lett A 366:217–222
104. Chen A, Lu J, Lü J, Yu S (2006) Generating hyperchaotic Lü attractor via state feedback control. Phys A 364:103–110
105. Li X (2009) Modified projective synchronization of a new hyperchaotic system via nonlinear control. Commun Theor Phys 52:274–278
106. Wang J, Chen Z (2008) A novel hyperchaotic system and its complex dynamics. Int J Bifurcat Chaos 18:3309–3324
107. Vaidyanathan S (2013) A ten-term novel 4-D hyperchaotic system with three quadratic nonlinearities and its control. Int J Control Theory Appl 6(2):97–109
108. Vaidyanathan S (2014) Qualitative analysis and control of an eleven-term novel 4-D hyperchaotic system with two quadratic nonlinearities. Int J Control Theory Appl 7:35–47
109. Vaidyanathan S, Azar AT (2015) Analysis and control of a 4-D novel hyperchaotic system. Studies in Computational Intelligence, vol 581, pp 3–17
110. Vaidyanathan S, Volos C, Pham VT (2014a) Hyperchaos, adaptive control and synchronization of a novel 5-D hyperchaotic system with three positive Lyapunov exponents and its SPICE implementation. Arch Control Sci 24(4):409–446
111. Vaidyanathan S, Volos C, Pham VT, Madhavan K (2015) Analysis, adaptive control and synchronization of a novel 4-D hyperchaotic hyperjerk system and its SPICE implementation. Nonlinear Dyn 25(1):135–158
112. Vaidyanathan S, Volos CK, Pham VT (2015) Analysis, control, synchronization and SPICE implementation of a novel 4-D hyperchaotic Rikitake dynamo system without equilibrium. J Eng Sci Technol Rev 8(2):232–244
113. Pham VT, Volos C, Jafari S, Wang X, Vaidyanathan S (2014) Hidden hyperchaotic attractor in a novel simple memristive neural network. Optoelectron Adv Mater Rapid Commun 8(11–12):1157–1163
114. Sundarapandian V (2010) Output regulation of the Lorenz attractor. Far East J Math Sci 42(2):289–299
115. Vaidyanathan S (2011) Output regulation of Arneodo-Coullet chaotic system. Commun Comput Inform Sci 133:98–107
116. Vaidyanathan S (2011) Output regulation of the unified chaotic system. Commun Comput Inform Sci 198:1–9
117. Sundarapandian V (2013) Adaptive control and synchronization design for the Lu-Xiao chaotic system. Lecture Notes in Electrical Engineering, vol 131, pp 319–327
118. Vaidyanathan S (2012) Adaptive controller and syncrhonizer design for the Qi-Chen chaotic system. Lecture Notes of the Institute for Computer Sciences, Social-Informatics and Telecommunications Engineering vol. 84, pp 73–82
119. Vaidyanathan S (2013) Analysis, control and synchronization of hyperchaotic Zhou system via adaptive control. Adv Intell Syst Comput 177:1–10
120. Vaidyanathan S (2012) Global chaos control of hyperchaotic Liu system via sliding control method. Int J Control Theory Appl 5(2):117–123
121. Vaidyanathan S (2012) Sliding mode control based global chaos control of Liu-Liu-Liu-Su chaotic system. Int J Control Theory Appl 5(1):15–20
122. Vaidyanathan S, Idowu BA, Azar AT (2015) Backstepping controller design for the global chaos synchronization of Sprott's jerk systems. Studies in Computational Intelligence, vol 581, pp 39–58

123. Karthikeyan R, Sundarapandian V (2014) Hybrid chaos synchronization of four-scroll systems via active control. J Electr Eng 65(2):97–103
124. Sarasu P, Sundarapandian V (2011) Active controller design for generalized projective synchronization of four-scroll chaotic systems. Int J Syst Signal Control Eng Appl 4(2):26–33
125. Sarasu P, Sundarapandian V (2011) The generalized projective synchronization of hyperchaotic Lorenz and hyperchaotic Qi systems via active control. Int J Soft Comput 6(5): 216–223
126. Vaidyanathan S, Rajagopal K (2011) Hybrid synchronization of hyperchaotic Wang-Chen and hyperchaotic Lorenz systems by active non-linear control. Int J Syst Signal Control Eng Appl 4(3):55–61
127. Vaidyanathan S, Rasappan S (2011) Global chaos synchronization of hyperchaotic Bao and Xu systems by active nonlinear control. Commun Comput Inform Sci 198:10–17
128. Vaidyanathan S, Pham VT, Volos CK, (2015) A 5-D hyperchaotic Rikitake dynamo system with hidden attractors. Eur Phys J: Special Topics 224(8):1575–1592
129. Sarasu P, Sundarapandian V (2012) Generalized projective synchronization of two-scroll systems via adaptive control. Int J Soft Comput 7(4):146–156
130. Sundarapandian V, Karthikeyan R (2011) Anti-synchronization of hyperchaotic Lorenz and hyperchaotic Chen systems by adaptive control. Int J Syst Signal Control Eng Appl 4(2):18–25
131. Sundarapandian V, Karthikeyan R (2011) Anti-synchronization of Lü and Pan chaotic systems by adaptive nonlinear control. Eur J Sci Res 64(1):94–106
132. Sundarapandian V, Karthikeyan R (2012) Adaptive anti-synchronization of uncertain Tigan and Li systems. J Eng Appl Sci 7(1):45–52
133. Vaidyanathan S (2012) Anti-synchronization of Sprott-L and Sprott-M chaotic systems via adaptive control. Int J Control Theory Appl 5(1):41–59
134. Vaidyanathan S (2015) Hyperchaos, qualitative analysis, control and synchronisation of a ten-term 4-D hyperchaotic system with an exponential nonlinearity and three quadratic nonlinearities. Int J Model Ident Control 23(4):380–392
135. Vaidyanathan S, Pakiriswamy S (2013) Generalized projective synchronization of six-term Sundarapandian chaotic systems by adaptive control. Int J Control Theory Appl 6(2):153–163
136. Vaidyanathan S, Rajagopal K (2012) Global chaos synchronization of hyperchaotic Pang and hyperchaotic Wang systems via adaptive control. Int J Soft Comput 7(1):28–37
137. Sundarapandian V, Sivaperumal S (2011) Sliding controller design of hybrid synchronization of four-wing chaotic systems. Int J Soft Comput 6(5):224–231
138. Vaidyanathan S (2014) Global chaos synchronisation of identical Li-Wu chaotic systems via sliding mode control. Int J Model Ident Control 22(2):170–177
139. Vaidyanathan S, Sampath S (2012) Anti-synchronization of four-wing chaotic systems via sliding mode control. Int J Autom Comput 9(3):274–279
140. Vaidyanathan S, Sampath S, Azar AT (2015) Global chaos synchronisation of identical chaotic systems via novel sliding mode control method and its application to Zhu system. Int J Model Ident Control 23(1):92–100
141. Rasappan S, Vaidyanathan S (2013) Hybrid synchronization of n-scroll Chua circuits using adaptive backstepping control design with recursive feedback. Malays J Math Sci 73(1):73–95
142. Rasappan S, Vaidyanathan S (2014) Global chaos synchronization of WINDMI and Coullet chaotic systems using adaptive backstepping control design. Kyungpook Math J 54(1): 293–320
143. Suresh R, Sundarapandian V (2013) Global chaos synchronization of a family of n-scroll hyperchaotic Chua circuits using backstepping control with recursive feedback. Far East J Math Sci 7(2):219–246
144. Vaidyanathan S, Rasappan S (2014) Global chaos synchronization of n-scroll Chua circuit and Lur'e system using backstepping control design with recursive feedback. Arab J Sci Eng 39(4):3351–3364
145. Khalil HK (2001) Nonlinear Systems, 3rd edn. Prentice Hall, New Jersey

Complete Synchronization of Hyperchaotic Systems via Novel Sliding Mode Control

Sundarapandian Vaidyanathan and Sivaperumal Sampath

Abstract Chaos in nonlinear dynamics occurs widely in physics, chemistry, biology, ecology, secure communications, cryptosystems and many scientific branches. Synchronization of chaotic systems is an important research problem in chaos theory. Sliding mode control is an important method used to solve various problems in control systems engineering. In robust control systems, the sliding mode control is often adopted due to its inherent advantages of easy realization, fast response and good transient performance as well as insensitivity to parameter uncertainties and disturbance. In this work, we derive a novel sliding mode control method for the complete synchronization of identical chaotic or hyperchaotic systems. The general result is derived using novel sliding mode control method. The general result is established using Lyapunov stability theory. As an application of the general result, the problem of complete synchronization of identical hyperchaotic Vaidyanathan systems (2014) is studied and a new sliding mode controller is derived. The Lyapunov exponents of the hyperchaotic Vaidyanathan system are obtained as $L_1 = 1.4252$, $L_2 = 0.2445$, $L_3 = 0$ and $L_4 = -17.6549$. Since the Vaidyanathan hyperjerk system has two positive Lyapunov exponents, it is hyperchaotic. Also, the Kaplan-Yorke dimension of the Vaidyanathan hyperjerk system is obtained as $D_{KY} = 3.0946$. Numerical simulations using MATLAB have been shown to depict the phase portraits of the hyperchaotic Vaidyanathan system and the sliding mode controller design for the anti-synchronization of identical hyperchaotic Vaidyanathan systems.

Keywords Chaos · Chaotic systems · Hyperchaos · Hyperchaotic systems · Sliding mode control · Synchronization

S. Vaidyanathan (✉)
Research and Development Centre, Vel Tech University, Avadi, Chennai 600062, Tamil Nadu, India
e-mail: sundarvtu@gmail.com

S. Sampath
School of Electrical and Computing, Vel Tech University, Avadi, Chennai 600062, Tamil Nadu, India
e-mail: sivaperumals@gmail.com

A.T. Azar and S. Vaidyanathan (eds.), *Advances in Chaos Theory and Intelligent Control*, Studies in Fuzziness and Soft Computing 337,
DOI 10.1007/978-3-319-30340-6_14

1 Introduction

Chaos theory describes the quantitative study of unstable aperiodic dynamic behaviour in deterministic nonlinear dynamical systems. For the motion of a dynamical system to be chaotic, the system variables should contain some nonlinear terms and the system must satisfy three properties: boundedness, infinite recurrence and sensitive dependence on initial conditions.

Chaos theory and control systems have many important applications in science and engineering [1, 8–11, 102]. Some commonly known applications are oscillators [23, 50], lasers [28, 96], chemical reactions [16, 35, 68, 69, 71, 73, 74, 78], biology [13, 25, 64–67, 70, 72, 76, 77], ecology [17, 51], encryption [26, 100], cryptosystems [43, 57], mechanical systems [3–7], secure communications [14, 33, 98], robotics [32, 34, 90], cardiology [39, 94], intelligent control [2, 30], neural networks [19, 22, 31], memristors [38, 91], etc.

A hyperchaotic system is defined as a chaotic system with at least two positive Lyapunov exponents [8]. Thus, the dynamics of a hyperchaotic system can expand in several different directions simultaneously. Thus, the hyperchaotic systems have more complex dynamical behaviour and they have miscellaneous applications in engineering such as secure communications [15, 27, 95], cryptosystems [18, 42, 101], fuzzy logic [49, 99], electrical circuits [93, 97], etc.

The minimum dimension of an autonomous, continuous-time, hyperchaotic system is four. The first 4-D hyperchaotic system was found by Rössler [44]. Many hyperchaotic systems have been reported in the chaos literature such as hyperchaotic Lorenz system [20], hyperchaotic Lü system [12], hyperchaotic Chen system [29], hyperchaotic Wang system [92], hyperchaotic Vaidyanathan systems [61, 63, 79, 86, 88, 89], hyperchaotic Pham system [36], etc.

The synchronization of chaotic systems aims to synchronize the states of master and slave systems asymptotically with time. There are many methods available for chaos synchronization such as active control [21, 46, 47, 81, 83], adaptive control [37, 45, 48, 52–54, 60, 75, 80, 82], sliding mode control [55, 62, 85, 87], backstepping control [40, 41, 56, 84], etc.

The design goal of complete synchronization of chaotic systems is to use the output of the master system to control the slave system so that the states of the slave system coincide with the states of the master system asymptotically, i.e.

$$\lim_{t \to \infty} \|\mathbf{x}(t) - \mathbf{y}(t)\| = 0, \quad \forall \mathbf{x}(0), \mathbf{y}(0) \in \mathbf{R}^n \tag{1}$$

In this research work, we derive a general result for the complete synchronization of chaotic systems using sliding mode control (SMC) theory [58, 59]. The sliding mode control approach is recognized as an efficient tool for designing robust controllers for linear or nonlinear control systems operating under uncertainty conditions. A major advantage of sliding mode control is low sensitivity to parameter variations in the plant and disturbances affecting the plant, which eliminates the necessity of exact modeling of the plant.

In the sliding mode control theory, the control dynamics will have two sequential modes, viz. the reaching mode and the sliding mode. Basically, a sliding mode controller (SMC) design consists of two parts: hyperplane design and controller design. A hyperplane is first designed via the pole-placement approach in the modern control theory and a controller is then designed based on the sliding condition. The stability of the overall system is guaranteed by the sliding condition and by a stable hyperplane.

This work is organized as follows. In Sect. 2, we discuss the problem statement for the complete synchronization of identical chaotic or hyperchaotic systems. In Sect. 3, we derive a general result for the complete synchronization of identical chaotic or hyperchaotic systems using novel sliding mode control. In Sect. 4, we describe the hyperchaotic Vaidyanathan system and its phase portraits. In Sect. 5, we describe the qualitative properties of the hyperchaotic Vaidyanathan system. The Lyapunov exponents of the hyper system are obtained as $L_1 = 1.4252$, $L_2 = 0.2445$, $L_3 = 0$ and $L_4 = -17.6549$, which shows that the hyperchaotic Vaidyanathan system is hyperchaotic.

In Sect. 6, we describe the sliding mode controller design for the complete synchronization of identical hyperchaotic Vaidyanathan systems using novel sliding mode control and its numerical simulations using MATLAB. Section 7 contains the conclusions of this work.

2 Problem Statement

As the *master* system, we consider the chaotic or hyperchaotic system given by

$$\dot{\mathbf{x}} = A\mathbf{x} + f(\mathbf{x}) \tag{2}$$

where $\mathbf{x} \in \mathbf{R}^n$ denotes the state of the system, $A \in \mathbf{R}^{n\times n}$ denotes the matrix of system parameters and $f(\mathbf{x}) \in \mathbf{R}^n$ contains the nonlinear parts of the system.

As the *slave* system, we consider the controlled identical system given by

$$\dot{\mathbf{y}} = A\mathbf{y} + f(\mathbf{y}) + \mathbf{u} \tag{3}$$

where $\mathbf{y} \in \mathbf{R}^n$ denotes the state of the system and $\mathbf{u}$ is the control.

The complete synchronization error is defined as

$$\mathbf{e} = \mathbf{y} - \mathbf{x} \tag{4}$$

The error dynamics is easily obtained as

$$\dot{\mathbf{e}} = A\mathbf{e} + \psi(\mathbf{x}, \mathbf{y}) + \mathbf{u}, \tag{5}$$

where

$$\psi(\mathbf{x}, \mathbf{y}) = f(\mathbf{x}) - f(\mathbf{y}) \tag{6}$$

Thus, the complete synchronization problem between the systems (2) and (3) can be stated as follows: Find a controller $\mathbf{u}(\mathbf{x}, \mathbf{y})$ so as to render the anti-synchronization error $\mathbf{e}(t)$ to be globally asymptotically stable for all values of $\mathbf{e}(0) \in \mathbf{R}^n$, i.e.

$$\lim_{t \to \infty} \|\mathbf{e}(t)\| = 0 \text{ for all } \mathbf{e}(0) \in \mathbf{R}^n \tag{7}$$

3 A Novel Sliding Mode Control Method for Solving Complete Synchronization Problem

This section details the main results of this work, viz. novel sliding mode controller design for achieving complete synchronization of chaotic or hyperchaotic systems.

First, we start the design by setting the control as

$$\mathbf{u}(t) = -\psi(\mathbf{x}, \mathbf{y}) + Bv(t) \tag{8}$$

In Eq. (8), $B \in \mathbf{R}^n$ is chosen such that (A, B) is completely controllable.

By substituting (8) into (5), we get the closed-loop error dynamics

$$\dot{\mathbf{e}} = A\mathbf{e} + Bv \tag{9}$$

The system (9) is a linear time-invariant control system with single input v.

Next, we start the sliding controller design by defining the sliding variable as

$$s(\mathbf{e}) = C\mathbf{e} = c_1 e_1 + c_2 e_2 + \cdots + c_n e_n, \tag{10}$$

where $C \in \mathbf{R}^{1 \times n}$ is a constant vector to be determined.

The sliding manifold S is defined as the hyperplane

$$S = \{\mathbf{e} \in \mathbf{R}^n \; : \; s(\mathbf{e}) = C\mathbf{e} = 0\} \tag{11}$$

We shall assume that a sliding motion occurs on the hyperplane S.

In sliding mode, the following equations must be satisfied:

$$s = 0 \tag{12a}$$

$$\dot{s} = CA\mathbf{e} + CBv = 0 \tag{12b}$$

We assume that

$$CB \neq 0 \tag{13}$$

The sliding motion is influenced by equivalent control derived from (12b) as

$$v_{\text{eq}}(t) = -(CB)^{-1}\, CA\mathbf{e}(t) \tag{14}$$

By substituting (14) into (9), we obtain the equivalent error dynamics in the sliding phase as follows:

$$\dot{\mathbf{e}} = A\mathbf{e} - (CB)^{-1}CA\mathbf{e} = E\mathbf{e}, \tag{15}$$

where

$$E = \left[I - B(CB)^{-1}C\right]A \tag{16}$$

We note that E is independent of the control and has at most $(n-1)$ non-zero eigenvalues, depending on the chosen switching surface, while the associated eigenvectors belong to $\ker(C)$.

Since (A, B) is controllable, we can use sliding control theory [58, 59] to choose B and C so that E has any desired $(n-1)$ stable eigenvalues.

This shows that the dynamics (15) is globally asympotically stable.

Finally, for the sliding controller design, we apply a novel sliding control law, viz.

$$\dot{s} = -ks - qs^2\,\text{sgn}(s) \tag{17}$$

In (17), $\text{sgn}(\cdot)$ denotes the *sign* function and the SMC constants $k > 0, q > 0$ are found in such a way that the sliding condition is satisfied and that the sliding motion will occur.

By combining Eqs. (12b), (14) and (17), we finally obtain the sliding mode controller $v(t)$ as

$$v(t) = -(CB)^{-1}\left[C(kI + A)\mathbf{e} + qs^2\,\text{sgn}(s)\right] \tag{18}$$

Next, we establish the main result of this section.

Theorem 1 *The sliding mode controller defined by (8) achieves complete synchronization between the identical chaotic systems (2) and (3) for all initial conditions* $\mathbf{x}(0), \mathbf{y}(0)$ *in* $\mathbf{R}^n$*, where* v *is defined by the novel sliding mode control law (18),* $B \in \mathbf{R}^{n\times 1}$ *is such that* (A, B) *is controllable,* $C \in \mathbf{R}^{1\times n}$ *is such that* $CB \neq 0$ *and the matrix* E *defined by (16) has* $(n-1)$ *stable eigenvalues.*

Proof Upon substitution of the control laws (8) and (18) into the error dynamics (5), we obtain the closed-loop error dynamics as

$$\dot{\mathbf{e}} = A\mathbf{e} - B(CB)^{-1}\left[C(kI + A)\mathbf{e} + qs^2\,\text{sgn}(s)\right] \tag{19}$$

We shall show that the error dynamics (19) is globally asymptotically stable by considering the quadratic Lyapunov function

$$V(\mathbf{e}) = \frac{1}{2}\, s^2(\mathbf{e}) \tag{20}$$

The sliding mode motion is characterized by the equations

$$s(\mathbf{e}) = 0 \text{ and } \dot{s}(\mathbf{e}) = 0 \tag{21}$$

By the choice of E, the dynamics in the sliding mode given by Eq. (15) is globally asymptotically stable.

When $s(\mathbf{e}) \neq 0$, $V(\mathbf{e}) > 0$.

Also, when $s(\mathbf{e}) \neq 0$, differentiating V along the error dynamics (19) or the equivalent dynamics (17), we get

$$\dot{V}(\mathbf{e}) = s\dot{s} = -ks^2 - qs^3 \operatorname{sgn}(s) < 0 \tag{22}$$

Hence, by Lyapunov stability theory [24], the error dynamics (19) is globally asymptotically stable for all $\mathbf{e}(0) \in \mathbf{R}^n$.

This completes the proof. ■

4 Hyperchaotic Vaidyanathan System

The hyperchaotic Vaidyanathan system [63] is described by the 4-D dynamics

$$\begin{aligned} \dot{x}_1 &= a(x_2 - x_1) + x_3 + x_4 \\ \dot{x}_2 &= cx_1 - x_1x_3 + x_4 \\ \dot{x}_3 &= -bx_3 + x_1x_2 \\ \dot{x}_4 &= -d(x_1 + x_2) \end{aligned} \tag{23}$$

where x_1, x_2, x_3, x_4 are the states and a, b, c, d are constant, positive, parameters.

In [63], it was shown that the system (23) is hyperchaotic when the parameters take the values

$$a = 12, \quad b = 4, \quad c = 100, \quad d = 5 \tag{24}$$

For numerical simulations, we take the initial values of the hyperchaotic Vaidyanathan system (23) as

$$x_1(0) = 1.5, \quad x_2(0) = 0.6, \quad x_3(0) = 1.8, \quad x_4(0) = 2.5 \tag{25}$$

Figures 1, 2, 3 and 4 show the 3-D projections of the hyperchaotic Vaidyanathan system (23) on (x_1, x_2, x_3), (x_1, x_2, x_4), (x_1, x_3, x_4) and (x_2, x_3, x_4) spaces, respectively.

The 3-D projection of the hyperchaotic Vaidyanathan system (23) on (x_1, x_2, x_3) space has the shape of a *two-scroll attractor* or *butterfly attractor*. Thus, we may also call the hyperchaotic Vaidyanathan system (23) as *hyperchaotic butterfly attractor*.

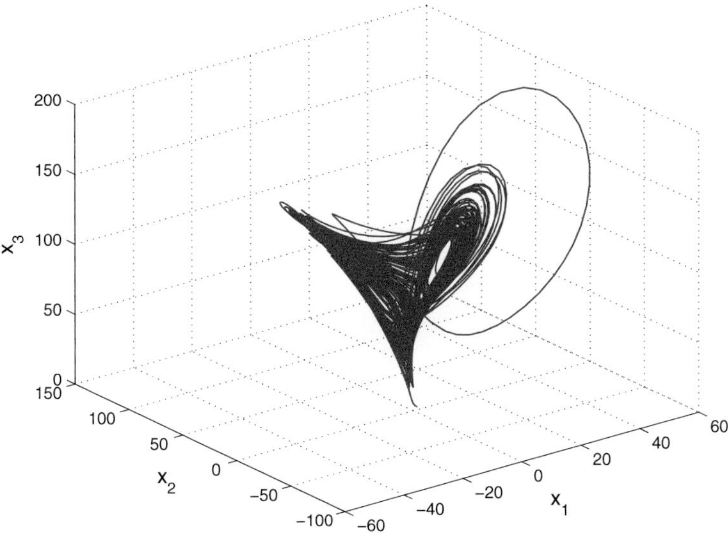

Fig. 1 3-D projection of the hyperchaotic Vaidyanathan system on the (x_1, x_2, x_3) space

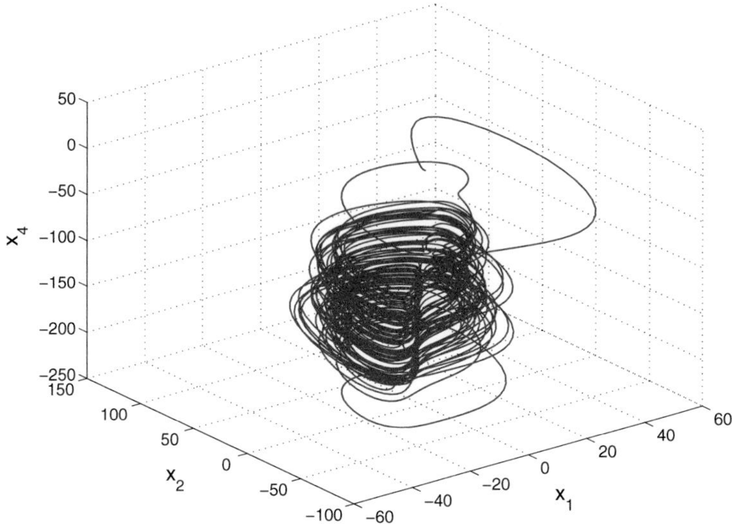

Fig. 2 3-D projection of the hyperchaotic Vaidyanathan system on the (x_1, x_2, x_4) space

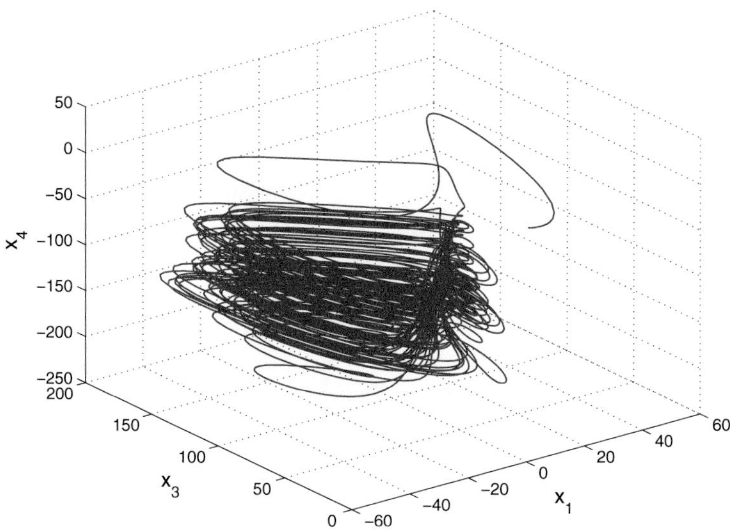

Fig. 3 3-D projection of the hyperchaotic Vaidyanathan system on the (x_1, x_3, x_4) space

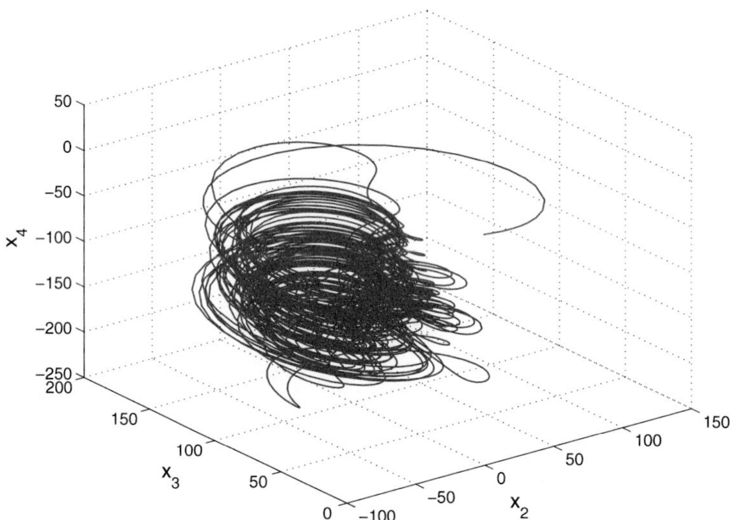

Fig. 4 3-D projection of the hyperchaotic Vaidyanathan system on the (x_2, x_3, x_4) space

5 Qualitative Properties of the Hyperchaotic Vaidyanathan System

5.1 Dissipativity

In vector notation, the hyperchaotic Vaidyanathan system (23) can be expressed as

$$\dot{\mathbf{x}} = f(\mathbf{x}) = \begin{bmatrix} f_1(x_1, x_2, x_3, x_4) \\ f_2(x_1, x_2, x_3, x_4) \\ f_3(x_1, x_2, x_3, x_4) \\ f_4(x_1, x_2, x_3, x_4) \end{bmatrix}, \tag{26}$$

where

$$\begin{cases} f_1(x_1, x_2, x_3, x_4) = a(x_2 - x_1) + x_3 + x_4 \\ f_2(x_1, x_2, x_3, x_4) = cx_1 - x_1x_3 + x_4 \\ f_3(x_1, x_2, x_3, x_4) = -bx_3 + x_1x_2 \\ f_4(x_1, x_2, x_3, x_4) = -d(x_1 + x_2) \end{cases} \tag{27}$$

We take the parameter values as in the hyperchaotic case (24).

Let Ω be any region in $\mathbf{R}^4$ with a smooth boundary and also, $\Omega(t) = \Phi_t(\Omega)$, where Φ_t is the flow of f. Furthermore, let $V(t)$ denote the hypervolume of $\Omega(t)$.

By Liouville's theorem, we know that

$$\dot{V}(t) = \int\limits_{\Omega(t)} (\nabla \cdot f)\, dx_1\, dx_2\, dx_3\, dx_4 \tag{28}$$

The divergence of the hyperchaotic Vaidyanathan system (26) is found as:

$$\nabla \cdot f = \frac{\partial f_1}{\partial x_1} + \frac{\partial f_2}{\partial x_2} + \frac{\partial f_3}{\partial x_3} + \frac{\partial f_4}{\partial x_4} = -(a + b) = -\mu < 0 \tag{29}$$

where $\mu = a + b = 12 + 4 = 16 > 0$.

Inserting the value of $\nabla \cdot f$ from (29) into (28), we get

$$\dot{V}(t) = \int\limits_{\Omega(t)} (-\mu)\, dx_1\, dx_2\, dx_3\, dx_4 = -\mu V(t) \tag{30}$$

Integrating the first order linear differential equation (30), we get

$$V(t) = \exp(-\mu t) V(0) \tag{31}$$

Since $\mu > 0$, it follows from Eq. (31) that $V(t) \to 0$ exponentially as $t \to \infty$. This shows that the hyperchaotic Vaidyanathan system (23) is dissipative. Hence, the system limit sets are ultimately confined into a specific limit set of zero volume, and the asymptotic motion of the hyperchaotic Vaidyanathan system (23) settles onto a strange attractor of the system.

5.2 *Equilibrium Points*

The equilibrium points of the hyperchaotic Vaidyanathan system (23) are obtained by solving the equations

$$\left.\begin{array}{rcll} f_1(x_1,x_2,x_3,x_4) &=& a(x_2-x_1)+x_3+x_4 &= 0 \\ f_2(x_1,x_2,x_3,x_4) &=& cx_1 - x_1x_3 + x_4 &= 0 \\ f_3(x_1,x_2,x_3,x_4) &=& -bx_3 + x_1x_2 &= 0 \\ f_1(x_1,x_2,x_3,x_4) &=& -d(x_1+x_2) &= 0 \end{array}\right\} \tag{32}$$

We take the parameter values as in the hyperchaotic case (24), viz. $a = 12, b = 4$, $c = 100$ and $d = 5$.

Solving the system (32), we see that the system (23) has a unique equilibrium point at the origin, i.e.

$$E_0 = \begin{bmatrix} 0 \\ 0 \\ 0 \\ 0 \end{bmatrix} \tag{33}$$

To test the stability type of the equilibrium point E_0, we calculate the Jacobian of the system (23) at E_0 as

$$J_0 = J(E_0) = \begin{bmatrix} -a & a & 1 & 1 \\ c & 0 & 0 & 1 \\ 0 & 0 & -b & 0 \\ -d & -d & 0 & 0 \end{bmatrix} = \begin{bmatrix} -12 & 12 & 1 & 1 \\ 100 & 0 & 0 & 1 \\ 0 & 0 & -4 & 0 \\ -5 & -5 & 0 & 0 \end{bmatrix} \tag{34}$$

The matrix J_0 has the eigenvalues

$$\lambda_1 = -41.2284, \quad \lambda_2 = -4, \quad \lambda_3 = 0.5239 \text{ and } \lambda_4 = 28.7045 \tag{35}$$

This shows that the equilibrium point E_0 is a saddle point, which is unstable.

5.3 *Lyapunov Exponents and Kaplan-Yorke Dimension*

We take the parameter values of the 4-D system (23) as

$$a = 12, \quad b = 4, \quad c = 100, \quad d = 5 \tag{36}$$

We take the initial values of the 4-D system (23) as

$$x_1(0) = 1.5, \;\; x_2(0) = 0.6, \;\; x_3(0) = 1.8, \;\; x_4(0) = 2.5 \tag{37}$$

Then the Lyapunov exponents of the 4-D system (23) are numerically obtained using MATLAB as

$$L_1 = 1.4252, \;\; L_2 = 0.2445, \;\; L_3 = 0, \;\; L_4 = -17.6549 \tag{38}$$

Equation (38) shows that the 4-D system (23) is hyperchaotic, since it has two positive Lyapunov exponents.

The dynamics of the Lyapunov exponents is depicted in Fig. 5. From Fig. 5, we see that the maximal Lyapunov exponent of the hyperchaotic Vaidyanathan system is given by $L_1 = 1.4252$. Since the sum of the Lyapunov exponents is negative, the system (23) is a dissipative hyperchaotic system.

Also, the Kaplan-Yorke dimension of the hyperchaotic Vaidyanathan system (23) is obtained as

$$D_{KY} = 3 + \frac{L_1 + L_2 + L_3}{|L_4|} = 3.0946, \tag{39}$$

which is fractional.

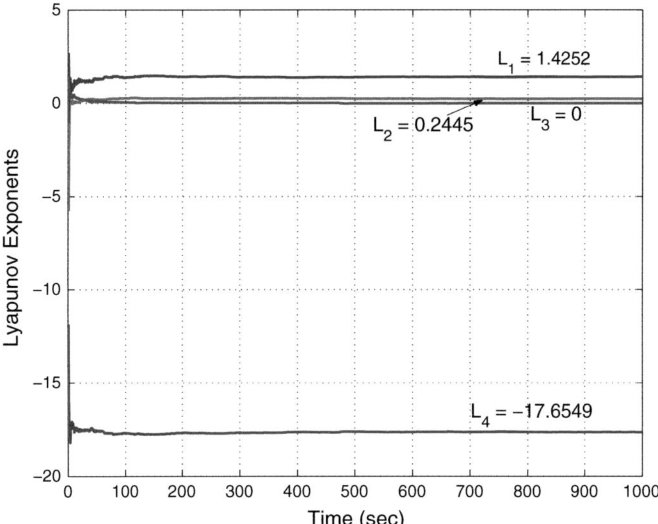

Fig. 5 Dynamics of the Lyapunov exponents of the hyperchaotic Vaidyanathan system

6 Sliding Mode Controller Design for the Complete Synchronization of Hyperchaotic Vaidyanathan Systems

In this section, we describe the sliding mode controller design for the complete synchronization of identical hyperchaotic Vaidyanathan systems [63] by applying the novel method described by Theorem 1 in Sect. 3.

As the master system, we consider the hyperchaotic Vaidyanathan system given by

$$\begin{aligned} \dot{x}_1 &= a(x_2 - x_1) + x_3 + x_4 \\ \dot{x}_2 &= cx_1 - x_1 x_3 + x_4 \\ \dot{x}_3 &= -bx_3 + x_1 x_2 \\ \dot{x}_4 &= -d(x_1 + x_2) \end{aligned} \tag{40}$$

where x_1, x_2, x_3, x_4 are the state variables and a, b, c, d are positive parameters.

As the slave system, we consider the controlled hyperchaotic Vaidyanathan system given by

$$\begin{aligned} \dot{y}_1 &= a(y_2 - y_1) + y_3 + y_4 + u_1 \\ \dot{y}_2 &= cy_1 - y_1 y_3 + y_4 + u_2 \\ \dot{y}_3 &= -by_3 + y_1 y_2 + u_3 \\ \dot{y}_4 &= -d(y_1 + y_2) + u_4 \end{aligned} \tag{41}$$

where y_1, y_2, y_3, y_4 are the state variables and u_1, u_2, u_3, u_4 are the controls.

The complete synchronization error between (40) and (41) is defined as

$$\begin{aligned} e_1 &= y_1 - x_1 \\ e_2 &= y_2 - x_2 \\ e_3 &= y_3 - x_3 \\ e_4 &= y_4 - x_4 \end{aligned} \tag{42}$$

Then the error dynamics is obtained as

$$\begin{aligned} \dot{e}_1 &= a(e_2 - e_1) + e_3 + e_4 + u_1 \\ \dot{e}_2 &= ce_1 + e_4 - y_1 y_3 + x_1 x_3 + u_2 \\ \dot{e}_3 &= -be_3 + y_1 y_2 - x_1 x_2 + u_3 \\ \dot{e}_4 &= -d(e_1 + e_2) + u_4 \end{aligned} \tag{43}$$

In matrix form, we can write the error dynamics (43) as

$$\dot{\mathbf{e}} = A\mathbf{e} + \psi(\mathbf{x}, \mathbf{y}) + \mathbf{u} \tag{44}$$

The matrices in (44) are given by

$$A = \begin{bmatrix} -a & a & 1 & 1 \\ c & 0 & 0 & 1 \\ 0 & 0 & -b & 0 \\ -d & -d & 0 & 0 \end{bmatrix} \quad \text{and} \quad \psi(\mathbf{x}, \mathbf{y}) = \begin{bmatrix} 0 \\ -y_1 y_3 + x_1 x_3 \\ y_1 y_2 - x_1 x_2 \\ 0 \end{bmatrix} \tag{45}$$

We follow the procedure given in Sect. 3 for the construction of the novel sliding controller to achieve complete synchronization of the identical hyperchaotic Vaidyanathan systems (40) and (41).

First, we set $\mathbf{u}$ as

$$\mathbf{u}(t) = -\psi(\mathbf{x}, \mathbf{y}) + Bv(t) \tag{46}$$

where B is selected such that (A, B) is completely controllable.

A simple choice of B is

$$B = \begin{bmatrix} 1 \\ 1 \\ 1 \\ 1 \end{bmatrix} \tag{47}$$

It can be easily checked that (A, B) is completely controllable.

The hyperchaotic Vaidyanathan system (40) displays a strange attractor when the parameter values are selected as

$$a = 12, \quad b = 4, \quad c = 100, \quad d = 5 \tag{48}$$

Next, we take the sliding variable as

$$s(\mathbf{e}) = C\mathbf{e} = \begin{bmatrix} 10 & 8 & -1 & -2 \end{bmatrix} \mathbf{e} = 10e_1 + 8e_2 - e_3 - 2e_4 \tag{49}$$

Next, we take the sliding mode gains as

$$k = 6, \quad q = 0.2 \tag{50}$$

From Eq. (18) in Sect. 3, we obtain the novel sliding control v as

$$v(t) = -50e_1 - 11.8667e_2 - 0.5333e_3 - 0.4e_4 - 0.0133s^2 \operatorname{sgn}(s) \tag{51}$$

As an application of Theorem 1 to the identical hyperchaotic Vaidyanathan systems, we obtain the following main result of this section.

Theorem 2 *The identical hyperchaotic Vaidyanathan systems (40) and (41) are globally and asymptotically synchronized for all initial conditions* $\mathbf{x}(0), \mathbf{y}(0) \in \mathbf{R}^4$ *with the sliding controller* $\mathbf{u}$ *defined by (46), where* $\psi(\mathbf{x}, \mathbf{y})$ *is defined by (45), B is defined by (47) and v is defined by (51).* ■

For numerical simulations, we use MATLAB for solving the systems of differential equations using the classical fourth-order Runge-Kutta method with step size $h = 10^{-8}$.

The parameter values of the hyperchaotic Vaidyanathan systems are taken as in the hyperchaotic case, viz. $a = 12$, $b = 4$, $c = 100$ and $d = 5$.

The sliding mode gains are taken as $k = 6$ and $q = 0.2$.

As an initial condition for the master system (40), we take

$$x_1(0) = 4.9, \quad x_2(0) = -2.5, \quad x_3(0) = 6.9, \quad x_4(0) = 12.4 \tag{52}$$

As an initial condition for the slave system (41), we take

$$y_1(0) = 8.1, \quad y_2(0) = 5.3, \quad y_3(0) = 1.3, \quad y_4(0) = -5.1 \tag{53}$$

Figures 6, 7, 8 and 9 show the complete synchronization of the states of the identical hyperchaotic Vaidyanathan systems (40) and (41).

From Fig. 6, it is clear that the states x_1 and y_1 are synchronized in 1 s.

From Fig. 7, it is clear that the states x_2 and y_2 are synchronized in 2 s.

From Fig. 8, it is clear that the states x_3 and y_3 are synchronized in 1 s.

From Fig. 9, it is clear that the states x_4 and y_4 are synchronized in 2 s.

Figure 10 shows the time-history of the complete synchronization errors e_1, e_2, e_3, e_4.

From Fig. 10, it is clear that all the synchronization errors converge to zero in 2 s.

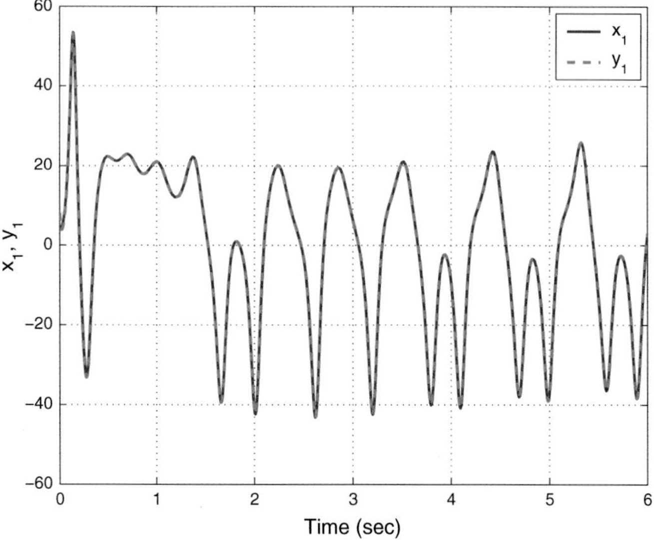

Fig. 6 Complete synchronization of the states x_1 and y_1

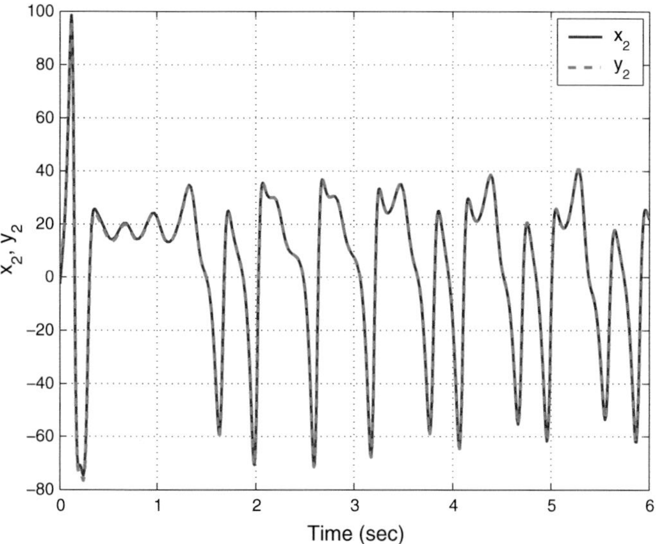

Fig. 7 Complete synchronization of the states x_2 and y_2

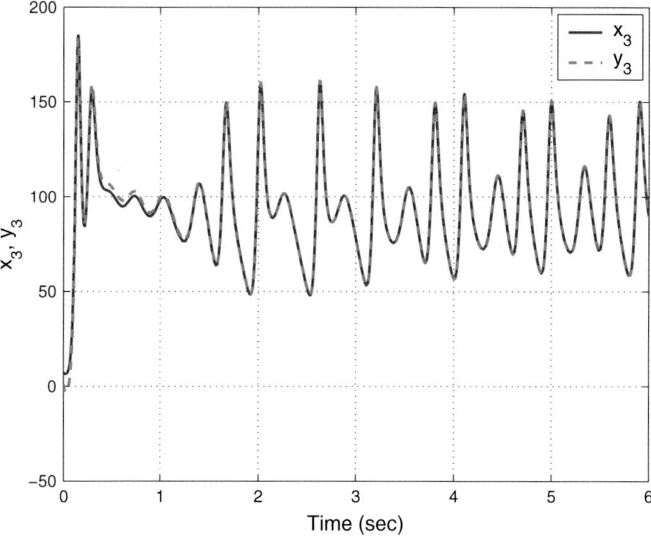

Fig. 8 Complete synchronization of the states x_3 and y_3

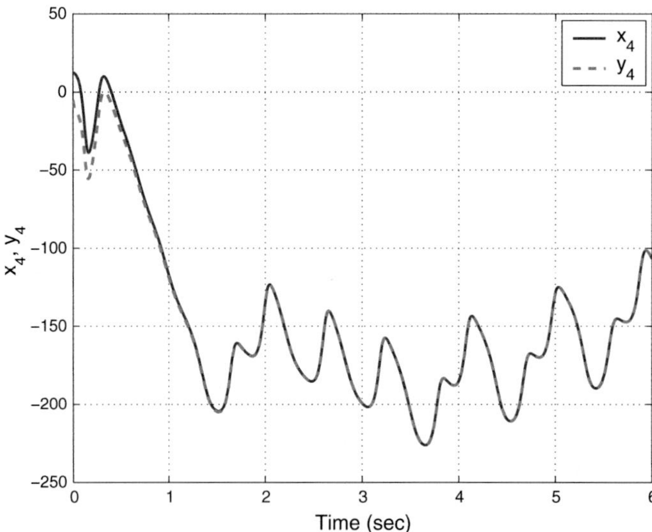

Fig. 9 Complete synchronization of the states x_4 and y_4

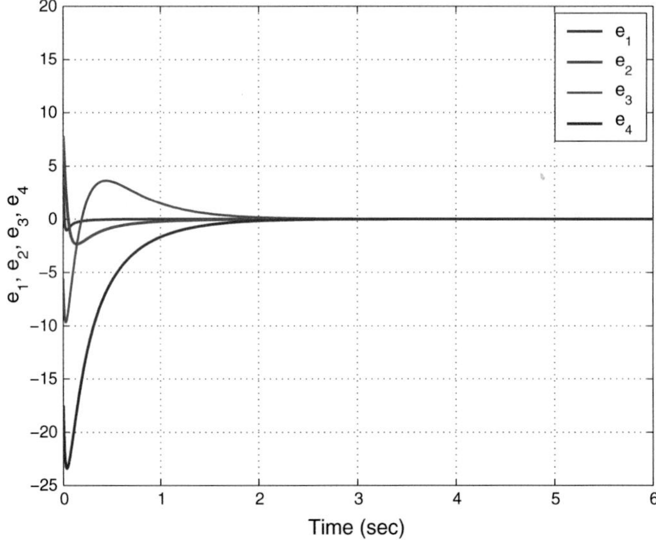

Fig. 10 Time-history of the complete synchronization errors e_1, e_2, e_3, e_4

7 Conclusions

Chaos and hyperchaos have important applications in science and engineering. Hyperchaotic systems have more complex behaviour than chaotic systems and they have miscellaneous applications in areas like secure communications, cryptosystems, etc. In robust control systems, the sliding mode control is commonly used due to its inherent advantages of easy realization, fast response and good transient performance as well as insensitivity to parameter uncertainties and disturbance. In this work, we derived a novel sliding mode control method for the complete synchronization of identical chaotic or hyperchaotic systems. We proved the main result using Lyapunov stability theory. As an application of the general result, the problem of complete synchronization of identical hyperchaotic Vaidyanathan systems (2014) was studied and a new sliding mode controller has been derived. Numerical simulations using MATLAB were shown to depict the phase portraits of the hyperchaotic Vaidyanathan system and the sliding mode controller design for the complete synchronization of identical hyperchaotic Vaidyanathan systems.

References

1. Azar AT (2010) Fuzzy Systems. IN-TECH, Vienna, Austria
2. Azar AT (2012) Overview of type-2 fuzzy logic systems. Int J Fuzzy Syst Appl 2(4):1–28
3. Azar AT, Serrano FE (2014) Robust IMC-PID tuning for cascade control systems with gain and phase margin specifications. Neural Comput Appl 25(5):983–995
4. Azar AT, Serrano FE (2015) Adaptive sliding mode control of the Furuta pendulum. In: Azar AT, Zhu Q (eds) Advances and applications in sliding mode control systems, studies in computational intelligence, vol 576. Springer, Germany, pp 1–42
5. Azar AT, Serrano FE (2015) Deadbeat control for multivariable systems with time varying delays. In: Azar AT, Vaidyanathan S (eds) Chaos modeling and control systems design, studies in computational intelligence, vol 581. Springer, Germany, pp 97–132
6. Azar AT, Serrano FE (2015) Design and modeling of anti wind up PID controllers. In: Zhu Q, Azar AT (eds) Complex system modelling and control through intelligent soft computations, studies in fuzziness and soft computing, vol 319. Springer, Germany, pp 1–44
7. Azar AT, Serrano FE (2015) Stabilizatoin and control of mechanical systems with backlash. In: Azar AT, Vaidyanathan S (eds) Handbook of research on advanced intelligent control engineering and automation. Advances in computational intelligence and robotics (ACIR), IGI-Global, USA, pp 1–60
8. Azar AT, Vaidyanathan S (2015) Chaos modeling and control systems design, studies in computational intelligence, vol 581. Springer, Germany
9. Azar AT, Vaidyanathan S (2015b) Computational intelligence applications in modeling and control, studies in computational intelligence, vol 575. Springer, Germany
10. Azar AT, Vaidyanathan S (2015) Handbook of research on advanced intelligent control engineering and automation. Advances in computational intelligence and robotics (ACIR), IGI-Global, USA
11. Azar AT, Zhu Q (2015) Advances and applications in sliding mode control systems, studies in computational intelligence, vol 576. Springer, Germany
12. Chen A, Lu J, Lü J, Yu S (2006) Generating hyperchaotic Lü attractor via state feedback control. Phys A 364:103–110

13. Das S, Goswami D, Chatterjee S, Mukherjee S (2014) Stability and chaos analysis of a novel swarm dynamics with applications to multi-agent systems. Eng Appl Artif Intell 30:189–198
14. Feki M (2003) An adaptive chaos synchronization scheme applied to secure communication. Chaos Solitons Fractals 18(1):141–148
15. Filali RL, Benrejeb M, Borne P (2014) On observer-based secure communication design using discrete-time hyperchaotic systems. Commun Nonlinear Sci Numer Simul 19(5):1424–1432
16. Gaspard P (1999) Microscopic chaos and chemical reactions. Physica A 263(1–4):315–328
17. Gibson WT, Wilson WG (2013) Individual-based chaos: extensions of the discrete logistic model. J Theoret Biol 339:84–92
18. Hammami S (2015) State feedback-based secure image cryptosystem using hyperchaotic synchronization. ISA Trans 54:52–59
19. Huang X, Zhao Z, Wang Z, Li Y (2012) Chaos and hyperchaos in fractional-order cellular neural networks. Neurocomputing 94:13–21
20. Jia Q (2007) Hyperchaos generated from the Lorenz chaotic system and its control. Phys Lett A 366:217–222
21. Karthikeyan R, Sundarapandian V (2014) Hybrid chaos synchronization of four-scroll systems via active control. J Electr Eng 65(2):97–103
22. Kaslik E, Sivasundaram S (2012) Nonlinear dynamics and chaos in fractional-order neural networks. Neural Netw 32:245–256
23. Kengne J, Chedjou JC, Kenne G, Kyamakya K (2012) Dynamical properties and chaos synchronization of improved Colpitts oscillators. Commun Nonlinear Sci Numer Simul 17(7):2914–2923
24. Khalil HK (2001) Nonlinear systems, 3rd edn. Prentice Hall, USA
25. Kyriazis M (1991) Applications of chaos theory to the molecular biology of aging. Exp Gerontol 26(6):569–572
26. Lang J (2015) Color image encryption based on color blend and chaos permutation in the reality-preserving multiple-parameter fractional Fourier transform domain. Opt Commun 338:181–192
27. Li C, Liao X, Wong KW (2005) Lag synchronization of hyperchaos with application to secure communications. Chaos Solitons Fractals 23(1):183–193
28. Li N, Pan W, Yan L, Luo B, Zou X (2014) Enhanced chaos synchronization and communication in cascade-coupled semiconductor ring lasers. Commun Nonlinear Sci Numer Simul 19(6):1874–1883
29. Li X (2009) Modified projective synchronization of a new hyperchaotic system via nonlinear control. Commun Theoret Phys 52:274–278
30. Li Z, Chen G (2006) Integration of fuzzy logic and chaos theory, studies in fuzziness and soft computing, vol 187. Springer, Germany
31. Lian S, Chen X (2011) Traceable content protection based on chaos and neural networks. Appl Soft Comput 11(7):4293–4301
32. Mondal S, Mahanta C (2014) Adaptive second order terminal sliding mode controller for robotic manipulators. J Franklin Inst 351(4):2356–2377
33. Murali K, Lakshmanan M (1998) Secure communication using a compound signal from generalized chaotic systems. Phys Lett A 241(6):303–310
34. Nehmzow U, Walker K (2005) Quantitative description of robot-environment interaction using chaos theory. Robot Auton Syst 53(3–4):177–193
35. Petrov V, Gaspar V, Masere J, Showalter K (1993) Controlling chaos in Belousov-Zhabotinsky reaction. Nature 361:240–243
36. Pham VT, Volos C, Jafari S, Wang X, Vaidyanathan S (2014) Hidden hyperchaotic attractor in a novel simple memristive neural network. Optoelectron Adv Mater, Rapid Commun 8(11–12):1157–1163
37. Pham VT, Vaidyanathan S, Volos CK, Jafari S (2015) Hidden attractors in a chaotic system with an exponential nonlinear term. Eur Phys J 224(8):1507–1517
38. Pham VT, Volos CK, Vaidyanathan S, Le TP, Vu VY (2015) A memristor-based hyperchaotic system with hidden attractors: dynamics, synchronization and circuital emulating. J Eng Sci Technol Rev 8(2):205–214

39. Qu Z (2011) Chaos in the genesis and maintenance of cardiac arrhythmias. Prog Biophys Mol Biol 105(3):247–257
40. Rasappan S, Vaidyanathan S (2013) Hybrid synchronization of n-scroll Chua circuits using adaptive backstepping control design with recursive feedback. Malays J Math Sci 73(1):73–95
41. Rasappan S, Vaidyanathan S (2014) Global chaos synchronization of WINDMI and Coullet chaotic systems using adaptive backstepping control design. Kyungpook Math J 54(1): 293–320
42. Rhouma R, Belghith S (2008) Cryptanalysis of a new image encryption algorithm based on hyper-chaos. Phys Lett A 372(38):5973–5978
43. Rhouma R, Belghith S (2011) Cryptoanalysis of a chaos based cryptosystem on DSP. Commun Nonlinear Sci Numer Simul 16(2):876–884
44. Rössler OE (1979) An equation for hyperchaos. Phys Lett A 71:155–157
45. Vaidyanathan S, VTP, Volos CK (2015) A 5-D hyperchaotic Rikitake dynamo system with hidden attractors. Eur Phys J 224(8):1575–1592
46. Sarasu P, Sundarapandian V (2011) Active controller design for generalized projective synchronization of four-scroll chaotic systems. Int J Syst Sig Control Eng Appl 4(2):26–33
47. Sarasu P, Sundarapandian V (2011) The generalized projective synchronization of hyperchaotic Lorenz and hyperchaotic Qi systems via active control. Int J Soft Comput 6(5): 216–223
48. Sarasu P, Sundarapandian V (2012) Generalized projective synchronization of two-scroll systems via adaptive control. Int J Soft Comput 7(4):146–156
49. Senouci A, Boukabou A (2014) Predictive control and synchronization of chaotic and hyperchaotic systems based on a T-S fuzzy model. Math Comput Simul 105:62–78
50. Sharma A, Patidar V, Purohit G, Sud KK (2012) Effects on the bifurcation and chaos in forced Duffing oscillator due to nonlinear damping. Commun Nonlinear Sci Numer Simul 17(6):2254–2269
51. Suérez I (1999) Mastering chaos in ecology. Ecol Model 117(2–3):305–314
52. Sundarapandian V, Karthikeyan R (2011) Anti-synchronization of hyperchaotic Lorenz and hyperchaotic Chen systems by adaptive control. Int J Syst Sig Control Eng Appl 4(2):18–25
53. Sundarapandian V, Karthikeyan R (2011) Anti-synchronization of Lü and Pan chaotic systems by adaptive nonlinear control. Eur J Sci Res 64(1):94–106
54. Sundarapandian V, Karthikeyan R (2012) Adaptive anti-synchronization of uncertain Tigan and Li systems. J Eng Appl Sci 7(1):45–52
55. Sundarapandian V, Sivaperumal S (2011) Sliding controller design of hybrid synchronization of four-wing chaotic systems. Int J Soft Comput 6(5):224–231
56. Suresh R, Sundarapandian V (2013) Global chaos synchronization of a family of n-scroll hyperchaotic Chua circuits using backstepping control with recursive feedback. Far East J Math Sci 7(2):219–246
57. Usama M, Khan MK, Alghatbar K, Lee C (2010) Chaos-based secure satellite imagery cryptosystem. Comput Math Appl 60(2):326–337
58. Utkin VI (1977) Variable structure systems with sliding modes. IEEE Trans Autom Control 22(2):212–222
59. Utkin VI (1993) Sliding mode control design principles and applications to electric drives. IEEE Trans Industr Electron 40(1):23–36
60. Vaidyanathan S (2012) Anti-synchronization of Sprott-L and Sprott-M chaotic systems via adaptive control. Int J Control Theor Appl 5(1):41–59
61. Vaidyanathan S (2013) A ten-term novel 4-D hyperchaotic system with three quadratic nonlinearities and its control. Int J Control Theor Appl 6(2):97–109
62. Vaidyanathan S (2014) Global chaos synchronisation of identical Li-Wu chaotic systems via sliding mode control. Int J Model, Ident Control 22(2):170–177
63. Vaidyanathan S (2014) Qualitative analysis and control of an eleven-term novel 4-D hyperchaotic system with two quadratic nonlinearities. Int J Control Theor Appl 7:35–47
64. Vaidyanathan S (2015) 3-cells cellular neural network (CNN) attractor and its adaptive biological control. Int J PharmTech Res 8(4):632–640

65. Vaidyanathan S (2015) Adaptive backstepping control of enzymes-substrates system with ferroelectric behaviour in brain waves. Int J PharmTech Res 8(2):256–261
66. Vaidyanathan S (2015) Adaptive biological control of generalized Lotka-Volterra three-species biological system. Int J PharmTech Res 8(4):622–631
67. Vaidyanathan S (2015) Adaptive chaotic synchronization of enzymes-substrates system with ferroelectric behaviour in brain waves. Int J PharmTech Res 8(5):964–973
68. Vaidyanathan S (2015) Adaptive control of a chemical chaotic reactor. Int J PharmTech Res 8(3):377–382
69. Vaidyanathan S (2015) Adaptive synchronization of chemical chaotic reactors. Int J ChemTech Res 8(2):612–621
70. Vaidyanathan S (2015) Adaptive synchronization of generalized Lotka-Volterra three-species biological systems. Int J PharmTech Res 8(5):928–937
71. Vaidyanathan S (2015) Anti-synchronization of Brusselator chemical reaction systems via adaptive control. Int J ChemTech Res 8(6):759–768
72. Vaidyanathan S (2015) Chaos in neurons and adaptive control of Birkhoff-Shaw strange chaotic attractor. Int J PharmTech Res 8(5):956–963
73. Vaidyanathan S (2015) Dynamics and control of Brusselator chemical reaction. Int J ChemTech Res 8(6):740–749
74. Vaidyanathan S (2015) Dynamics and control of Tokamak system with symmetric and magnetically confined plasma. Int J ChemTech Res 8(6):795–803
75. Vaidyanathan S (2015) Hyperchaos, qualitative analysis, control and synchronisation of a ten-term 4-D hyperchaotic system with an exponential nonlinearity and three quadratic nonlinearities. Int J Model, Ident and Control 23(4):380–392
76. Vaidyanathan S (2015) Lotka-Volterra population biology models with negative feedback and their ecological monitoring. Int PharmTech Res 8(5):974–981
77. Vaidyanathan S (2015) Synchronization of 3-cells cellular neural network (CNN) attractors via adaptive control method. Int J PharmTech Res 8(5):946–955
78. Vaidyanathan S (2015) Synchronization of Tokamak systems with symmetric and magnetically confined plasma via adaptive control. Int J ChemTech Res 8(6):818–827
79. Vaidyanathan S, Azar AT (2015) Analysis and control of a 4-D novel hyperchaotic system. Stud Comput Intell 581:3–17
80. Vaidyanathan S, Pakiriswamy S (2013) Generalized projective synchronization of six-term Sundarapandian chaotic systems by adaptive control. Int J Control Theor Appl 6(2):153–163
81. Vaidyanathan S, Rajagopal K (2011) Hybrid synchronization of hyperchaotic Wang-Chen and hyperchaotic Lorenz systems by active non-linear control. Int J Syst Sig Control Eng Appl 4(3):55–61
82. Vaidyanathan S, Rajagopal K (2012) Global chaos synchronization of hyperchaotic Pang and hyperchaotic Wang systems via adaptive control. Int J Soft Comput 7(1):28–37
83. Vaidyanathan S, Rasappan S (2011) Global chaos synchronization of hyperchaotic Bao and Xu systems by active nonlinear control. Commun Comput Inf Sci 198:10–17
84. Vaidyanathan S, Rasappan S (2014) Global chaos synchronization of n-scroll Chua circuit and Lur'e system using backstepping control design with recursive feedback. Arab J Sci Eng 39(4):3351–3364
85. Vaidyanathan S, Sampath S (2012) Anti-synchronization of four-wing chaotic systems via sliding mode control. Int J Autom Comput 9(3):274–279
86. Vaidyanathan S, Volos C, Pham VT (2014) Hyperchaos, adaptive control and synchronization of a novel 5-D hyperchaotic system with three positive Lyapunov exponents and its SPICE implementation. Arch Control Sci 24(4):409–446
87. Vaidyanathan S, Sampath S, Azar AT (2015) Global chaos synchronisation of identical chaotic systems via novel sliding mode control method and its application to Zhu system. Int J Model, Ident Control 23(1):92–100
88. Vaidyanathan S, Volos C, Pham VT, Madhavan K (2015) Analysis, adaptive control and synchronization of a novel 4-D hyperchaotic hyperjerk system and its SPICE implementation. Arch Control Sci 25(1):135–158

89. Vaidyanathan S, Volos CK, Pham VT (2015) Analysis, control, synchronization and SPICE implementation of a novel 4-D hyperchaotic Rikitake dynamo system without equilibrium. J Eng Sci Technol Rev 8(2):232–244
90. Volos CK, Kyprianidis IM, Stouboulos IN (2013) Experimental investigation on coverage performance of a chaotic autonomous mobile robot. Robot Auton Syst 61(12):1314–1322
91. Volos CK, Kyprianidis IM, Stouboulos IN, Tlelo-Cuautle E, Vaidyanathan S (2015) Memristor: a new concept in synchronization of coupled neuromorphic circuits. J Eng Sci Technol Rev 8(2):157–173
92. Wang J, Chen Z (2008) A novel hyperchaotic system and its complex dynamics. Int J Bifurcat Chaos 18:3309–3324
93. Wei X, Yunfei F, Qiang L (2012) A novel four-wing hyper-chaotic system and its circuit implementation. Procedia Eng 29:1264–1269
94. Witte CL, Witte MH (1991) Chaos and predicting varix hemorrhage. Med Hypotheses 36(4):312–317
95. Wu X, Zhu C, Kan H (2015) An improved secure communication scheme based passive synchronization of hyperchaotic complex nonlinear system. Appl Math Comput 252: 201–214
96. Yuan G, Zhang X, Wang Z (2014) Generation and synchronization of feedback-induced chaos in semiconductor ring lasers by injection-locking. Optik—Int J Light Electron Opt 125(8):1950–1953
97. Yujun N, Xingyuan W, Mingjun W, Huaguang Z (2010) A new hyperchaotic system and its circuit implementation. Commun Nonlinear Sci Numer Simu 15(11):3518–3524
98. Zaher AA, Abu-Rezq A (2011) On the design of chaos-based secure communication systems. Commun Nonlinear Syst Numer Simul 16(9):3721–3727
99. Zhang H, Liao X, Yu J (2005) Fuzzy modeling and synchronization of hyperchaotic systems. Chaos Solitons Fractals 26(3):835–843
100. Zhang X, Zhao Z, Wang J (2014) Chaotic image encryption based on circular substitution box and key stream buffer. Sig Process Image Commun 29(8):902–913
101. Zhu C (2012) A novel image encryption scheme based on improved hyperchaotic sequences. Opt Commun 285(1):29–37
102. Zhu Q, Azar AT (2015) Complex system modelling and control through intelligent soft computations, studies in fuzzines and soft computing, vol 319. Springer, Germany

A Novel 3-D Conservative Jerk Chaotic System with Two Quadratic Nonlinearities and Its Adaptive Control

Sundarapandian Vaidyanathan

Abstract In this research work, we announce a six-term novel 3-D conservative jerk chaotic system with two quadratic nonlinearities. The novel conservative jerk chaotic system is obtained by adding a quadratic nonlinearity to Sprott's 3-D conservative jerk chaotic system (1997). In this work, we first discuss the qualitative properties of the novel 3-D conservative jerk chaotic system. Conservative chaotic systems are characterized by the property that they are *volume conserving*. The novel conservative jerk chaotic system has two saddle-foci equilibrium points. Thus, both equilibrium points of the novel conservative jerk chaotic system are unstable. We obtain the Lyapunov exponents of the novel conservative jerk chaotic system as $L_1 = 0.0452$, $L_2 = 0$ and $L_3 = -0.0452$. Also, the Kaplan-Yorke dimension of the conservative novel jerk chaotic system is obtained as $D_{KY} = 3$. The high value of the Kaplan-Yorke dimension indicates the complexity of the novel conservative jerk chaotic system. Next, an adaptive backstepping controller is designed to stabilize the novel conservative jerk chaotic system with unknown system parameters. Moreover, an adaptive backstepping controller is designed to achieve complete chaos synchronization of the identical novel conservative jerk chaotic systems with unknown system parameters. The main control results are established using Lyapunov stability theory. MATLAB simulations are shown to illustrate all the main results on the novel 3-D conservative jerk chaotic system.

Keywords Chaos · Chaotic systems · Jerk systems · Backstepping control · Adaptive control · Synchronization

S. Vaidyanathan (✉)
Research and Development Centre, Vel Tech University,
Avadi, Chennai 600062, Tamil Nadu, India
e-mail: sundarvtu@gmail.com

A.T. Azar and S. Vaidyanathan (eds.), *Advances in Chaos Theory and Intelligent Control*, Studies in Fuzziness and Soft Computing 337,
DOI 10.1007/978-3-319-30340-6_15

1 Introduction

Chaotic systems are defined as nonlinear dynamical systems which are sensitive to initial conditions, topologically mixing and with dense periodic orbits. Sensitivity to initial conditions of chaotic systems is popularly known as the *butterfly effect*.

Chaotic systems are either conservative or dissipative. The conservative chaotic systems are characterized by the property that they are *volume conserving*. The dissipative chaotic systems are characterized by the property that any asymptotic motion of the chaotic system settles onto a set of measure zero, i.e. a strange attractor. In this research work, we shall announce and discuss a novel conservative jerk chaotic system with two quadratic nonlinearities.

The Lyapunov exponent of a chaotic system is a measure of the divergence of points which are initially very close and this can be used to quantify chaotic systems. Each nonlinear dynamical system has a spectrum of Lyapunov exponents, which are equal in number to the dimension of the state space. The largest Lyapunov exponent of a nonlinear dynamical system is called the *maximal Lyapunov exponent* (MLE).

In the last few decades, many 3-D polynomial chaotic systems (both conservative and dissipative) have been found with varying qualitative properties.

Some classical paradigms of 3-D chaotic systems in the literature are Lorenz system [1], Rössler system [2], ACT system [3], Sprott systems [4], Chen system [5], Lü system [6], Liu system [7], Cai system [8], Chen-Lee system [9], Tigan system [10], etc.

Many new chaotic systems have been discovered in the recent years such as Zhou system [11], Zhu system [12], Li system [13], Wei-Yang system [14], Sundarapandian systems [15, 16], Vaidyanathan systems [17–33], Pehlivan system [34], Sampath system [35], Pham system [36], etc.

Chaos theory and control systems have many important applications in science and engineering [37–42]. Some commonly known applications are oscillators [43, 44], lasers [45, 46], chemical reactions [47–54], biology [55–64], ecology [65, 66], encryption [67, 68], cryptosystems [69, 70], mechanical systems [71–75], secure communications [76–78], robotics [79–81], cardiology [82, 83], intelligent control [84, 85], neural networks [86–88], finance [89, 90], memristors [91, 92], etc.

Control of chaotic systems aims to regulate or stabilize the chaotic system about its equilibrium points. There are many methods available in the literature such as active control method [93–96], adaptive control method [97, 98], sliding mode control method [99, 100], backstepping control method [101], etc.

Synchronization of chaotic systems is a phenomenon that occurs when two or more chaotic systems are coupled or when a chaotic system drives another chaotic system. Major works on synchronization of chaotic systems deal with the complete synchronization of a pair of chaotic systems called the *master* and *slave* systems. The design goal of the complete synchronization is to apply the output of the master system to control the slave system so that the output of the slave system tracks the output of the master system asymptotically with time.

Pecora and Carroll pioneered the research on synchronization of chaotic systems with their seminal papers [102, 103]. The active control method [104–113] is typically used when the system parameters are available for measurement. Adaptive control method [114–129] is typically used when some or all the system parameters are not available for measurement and estimates for the uncertain parameters of the systems. Sampled-data feedback control method [130–133] and time-delay feedback control method [134–136] are also used for synchronization of chaotic systems.

Backstepping control method [137–144] is also used for the synchronization of chaotic systems, which is a recursive method for stabilizing the origin of a control system in strict-feedback form. Another popular method for the synchronization of chaotic systems is the sliding mode control method [145–151], which is a nonlinear control method that alters the dynamics of a nonlinear system by application of a discontinuous control signal that forces the system to "slide" along a cross-section of the system's normal behavior.

In the recent decades, there is some good interest in finding novel chaotic systems, which are of the jerk type [152]. Jerk chaotic systems are chaotic systems that can be expressed by an explicit third-order differential equation describing the time evolution of the single scalar variable x given by

$$\dddot{x} = j\left(x, \dot{x}, \ddot{x}\right). \tag{1}$$

The differential equation (1) is called *jerk system*, because the third order time derivative in mechanical systems is called *jerk*. Thus, in order to study different aspects of chaos, the ODE (1) can be considered instead of a 3-D system.

The simplest known jerk system was discovered by Sprott [153] and Sprott's jerk function contains just three terms with a quadratic nonlinearity:

$$j\left(x, \dot{x}, \ddot{x}\right) = -A\ddot{x} + \dot{x}^2 - x \quad (\text{with } A = 2.017) \tag{2}$$

The jerk system (2) is a dissipative jerk chaotic system. In this research work, we consider conservative jerk chaotic systems described by 3-D systems of differential equations. The conservative chaotic systems are characterized by the property that they are *volume-conserving*.

For this purpose, we define the phase variables

$$\begin{cases} x_1 = x \\ x_2 = \dot{x} \\ x_3 = \ddot{x} \end{cases} \tag{3}$$

Using (3), we can rewrite the jerk differential equation (1) in the system form as

$$\begin{cases} \dot{x}_1 = x_2 \\ \dot{x}_2 = x_3 \\ \dot{x}_3 = j(x_1, x_2, x_3) \end{cases} \tag{4}$$

In this research work, we propose a six-term novel conservative jerk chaotic system with two quadratic nonlinearities. Our novel conservative jerk chaotic system is obtained by adding a quadratic nonlinearity to Sprott's five-term novel conservative jerk chaotic system [153] with one quadratic nonlinearity. In this work, we show that the maximal Lyapunov exponent for the novel conservative chaotic system is greater than the maximal Lyapunov exponent for the Sprott's conservative chaotic system [153].

The Lyapunov exponents of the novel 3-D conservative jerk chaotic system are obtained as $L_1 = 0.0452, L_2 = 0$ and $L_3 = -0.0452$. Also, the Kaplan-Yorke dimension of the novel conservative jerk chaotic system is derived as $D_{KY} = 3$.

We apply backstepping control method to derive new results for the adaptive control and synchronization of the novel conservative jerk chaotic system with unknown parameters.

Section 2 describes the dynamic equations and phase portraits of the novel conservative jerk chaotic system. Section 3 describes the qualitative properties of the novel jerk chaotic system. Section 4 describes the adaptive backstepping control of the novel conservative jerk chaotic system with unknown parameters. Section 5 describes the adaptive backstepping synchronization of the identical novel conservative jerk chaotic systems with unknown parameters. Finally, Sect. 6 contains the conclusions of this work.

2 A Novel 3-D Jerk Chaotic System

First, we consider the five-term Sprott's conservative jerk chaotic system [153], which is described by the 3-D dynamics

$$\begin{cases} \dot{x}_1 = x_2 \\ \dot{x}_2 = x_3 \\ \dot{x}_3 = -ax_1(1 - x_1) - x_2 \end{cases} \tag{5}$$

where x_1, x_2, x_3 are the states and a is a positive, constant, parameter.

We note that Sprott's conservative jerk chaotic system (5) contains just one quadratic nonlinearity.

Sprott showed that the 3-D jerk system (5) is *chaotic*, when we take

$$a = 0.2 \tag{6}$$

For numerical simulations, we take the initial values of the Sprott's conservative jerk chaotic system (5) as

$$x_1(0) = 0.01, \quad x_2(0) = -0.01, \quad x_3(0) = 0.01 \tag{7}$$

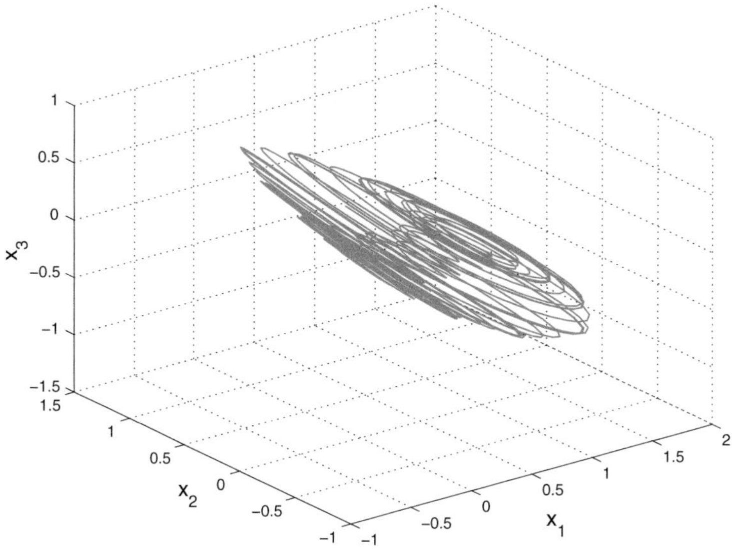

Fig. 1 3-D phase portrait of Sprott's conservative jerk chaotic system in $\mathbf{R}^3$

Figure 1 describes the chaotic phase portrait of the Sprott's conservative jerk chaotic system (5) in $\mathbf{R}^3$.

Also, for the parameter value $a = 0.2$ and the initial state (7), the Lyapunov exponents of the Sprott's conservative jerk chaotic system (5) are calculated as

$$L_1 = 0.0163, \quad L_2 = 0, \quad L_3 = -0.0163 \tag{8}$$

Thus, the maximal Lyapunov exponent (MLE) of the Sprott's conservative jerk chaotic system (5) is obtained as $L_1 = 0.0163$.

In this work, we announce a new conservative jerk chaotic system by adding a quadratic nonlinearity to Sprott's conservative jerk chaotic system (5).

Our novel 3-D conservative jerk chaotic system is described by

$$\begin{cases} \dot{x}_1 = x_2 \\ \dot{x}_2 = x_3 \\ \dot{x}_3 = -ax_1(1 - x_1) - x_2 + bx_2^2 \end{cases} \tag{9}$$

where x_1, x_2, x_3 are the states and a, b are positive, constant, parameters.

The novel jerk system (9) is *chaotic*, when we take the parameter values as

$$a = 0.2, \quad b = 0.01 \tag{10}$$

For numerical simulations, we take the initial state as in (7).

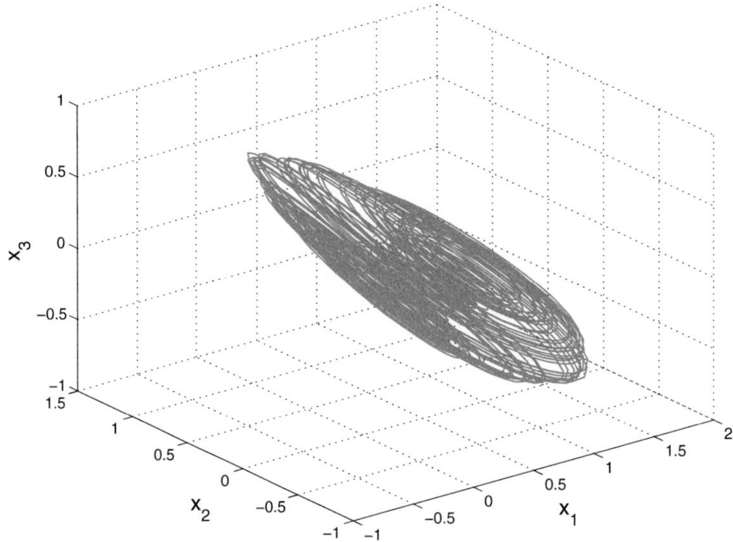

Fig. 2 3-D phase portrait of the novel conservative jerk chaotic system in $\mathbf{R}^3$

For the parameter values (10) and initial state (7), the Lyapunov exponents of the novel conservative jerk chaotic system (9) are obtained as

$$L_1 = 0.0452, \quad L_2 = 0, \quad L_3 = -0.0452 \tag{11}$$

Thus, the maximal Lyapunov exponent (MLE) of the novel 3-D conservative jerk chaotic system (9) is obtained as $L_1 = 0.0452$, which is greater than the maximal Lyapunov exponent (MLE) of the Sprott's conservative jerk chaotic system, *viz.* $L_1 = 0.0163$.

Figure 2 describes the chaotic phase portrait of the novel conservative jerk chaotic system (9) in $\mathbb{R}^3$.

Figures 3, 4 and 5 describe the 2-D projections of the novel 3-D conservative jerk chaotic system (9) on (x_1, x_2), (x_2, x_3) and (x_1, x_3) coordinate planes respectively.

3 Properties of the Novel Jerk Chaotic System

This section describes the qualitative properties of the novel jerk chaotic system (9).

3.1 Volume Conservation of the Flow

We write the novel jerk chaotic system (9) in vector notation as

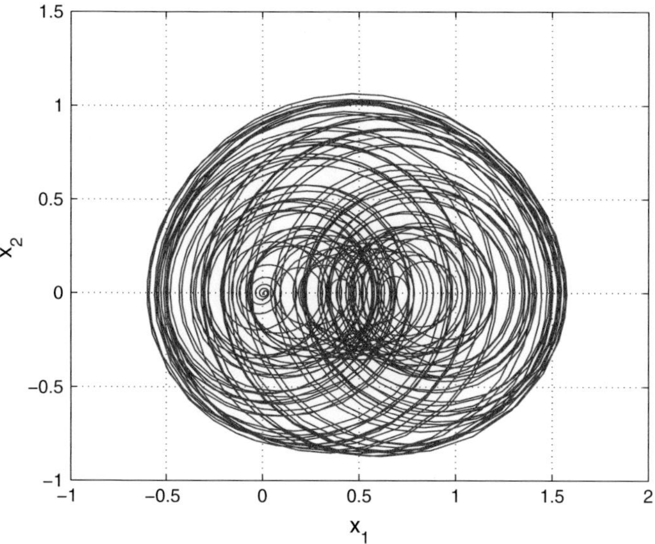

Fig. 3 2-D projection of the novel jerk chaotic system on (x_1, x_2)-plane

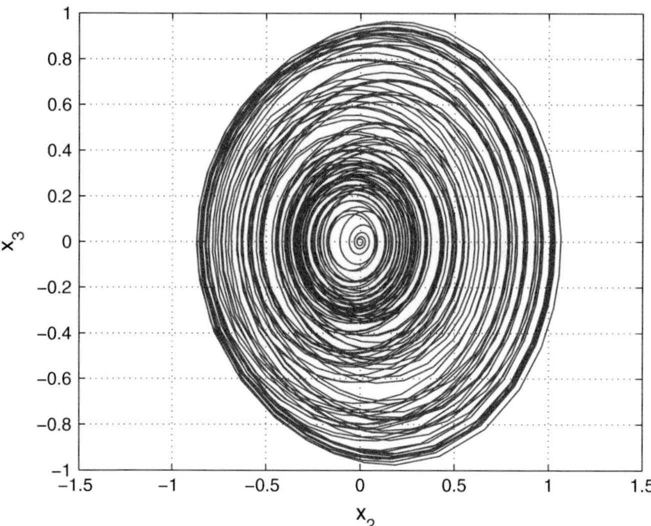

Fig. 4 2-D projection of the novel jerk chaotic system on (x_2, x_3)-plane

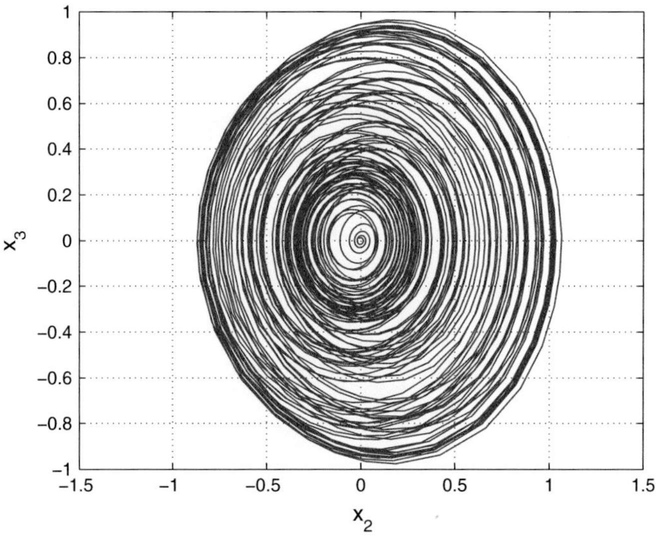

Fig. 5 2-D projection of the novel jerk chaotic system on (x_1, x_3)-plane

$$\dot{x} = f(x) = \begin{bmatrix} f_1(x) \\ f_2(x) \\ f_3(x) \end{bmatrix}, \tag{12}$$

where

$$\begin{aligned} f_1(x) &= x_2 \\ f_2(x) &= x_3 \\ f_3(x) &= -ax_1(1 - x_1) - x_2 + bx_2^2 \end{aligned} \tag{13}$$

We take the parameter values of the novel conservative jerk chaotic system (9) as in the chaotic case (10), *viz.* $a = 2$ and $b = 0.01$.

The divergence of the vector field f on $\mathbb{R}^3$ is obtained as

$$\text{div}\, f = \frac{\partial f_1(x)}{\partial x_1} + \frac{\partial f_2(x)}{\partial x_2} + \frac{\partial f_3(x)}{\partial x_3} = 0 \tag{14}$$

Let Ω be any region in $\mathbb{R}^3$ with a smooth boundary. Let $\Omega(t) = \Phi_t(\Omega)$, where Φ_t is the flow of the vector field f.

Let $V(t)$ denote the volume of $\Omega(t)$.

By Liouville's theorem, it follows that

$$\frac{dV(t)}{dt} = \int\limits_{\Omega(t)} (\text{div} f)dx_1\, dx_2\, dx_3 \tag{15}$$

Substituting the value of div$f$ in (15) leads to

$$\frac{dV(t)}{dt} = 0 \tag{16}$$

Integrating the linear differential equation (16), $V(t)$ is obtained as

$$V(t) = V(0) \text{ for all } t \tag{17}$$

This shows that novel jerk chaotic system (9) is volume-conserving.
Hence, the novel jerk chaotic system (9) is a conservative system.

3.2 Equilibrium Points

The equilibrium points of the novel jerk chaotic system (9) are obtained by solving the system of equations

$$\begin{cases} x_2 = 0 \\ x_3 = 0 \\ -ax_1(1 - x_1) - x_2 + bx_2^2 = 0 \end{cases} \tag{18}$$

Solving the system (18), we obtain two equilibrium points of the novel conservative jerk chaotic system (9) as

$$E_0 = \begin{bmatrix} 0 \\ 0 \\ 0 \end{bmatrix} \text{ and } E_1 = \begin{bmatrix} 1 \\ 0 \\ 0 \end{bmatrix} \tag{19}$$

Next, we determine the stability type of the equilibrium points E_0 and E_1 of the conservative jerk chaotic system (9).

The Jacobian matrix of the novel jerk chaotic system (9) at any $x \in \mathbb{R}^3$ is obtained as

$$J(x) = \begin{bmatrix} 0 & 1 & 0 \\ 0 & 0 & 1 \\ -a + 2ax_1 & -1 + 2bx_2 & 0 \end{bmatrix} \tag{20}$$

Then it follows that the matrix $J_0 = J(E_0)$ is given by

$$J_0 = J(E_0) = \begin{bmatrix} 0 & 1 & 0 \\ 0 & 0 & 1 \\ -0.2 & -1 & 0 \end{bmatrix} \tag{21}$$

The matrix J_0 has the eigenvalues

$$\lambda_1 = -0.1928, \quad \lambda_{2,3} = 0.0964 \pm 1.0138i \tag{22}$$

This shows that the equilibrium point E_0 is a saddle-focus, which is unstable.

Next, the matrix $J_1 = J(E_1)$ is given by

$$J_1 = J(E_1) = \begin{bmatrix} 0 & 1 & 0 \\ 0 & 0 & 1 \\ 0.2 & -1 & 0 \end{bmatrix} \tag{23}$$

The matrix J_0 has the eigenvalues

$$\lambda_1 = 0.1928, \quad \lambda_{2,3} = -0.0964 \pm 1.0138i \tag{24}$$

This shows that the equilibrium point E_1 is a saddle-focus, which is unstable.

Hence, both equilibrium points of the conservative jerk chaotic system (9) are saddle-foci, which are unstable.

3.3 Lyapunov Exponents and Kaplan-Yorke Dimension

We take the parameter values of the novel conservative jerk chaotic system (9) as in the chaotic case (10), i.e.

$$a = 0.2, \quad b = 0.01 \tag{25}$$

We take the initial values of the novel conservative jerk chaotic system (9) as in (7), i.e.

$$x_1(0) = 0.01, \quad x_2(0) = -0.01, \quad x_3(0) = 0.01 \tag{26}$$

Then the Lyapunov exponents of the novel jerk chaotic system (9) are numerically obtained as

$$L_1 = 0.0452, \quad L_2 = 0, \quad L_3 = -0.0452 \tag{27}$$

Since $L_1 + L_2 + L_3 = 0$, the novel jerk chaotic system (9) is conservative.

Also, the Kaplan-Yorke dimension of the novel conservative jerk chaotic system (9) is obtained as

$$D_{KY} = 2 + \frac{L_1 + L_2}{|L_3|} = 3 \tag{28}$$

Since $D_{KY} = 3$, the novel conservative jerk chaotic system (9) exhibits complex chaotic behaviour.

4 Adaptive Backstepping Control of Novel Conservative Jerk Chaotic System

In this section, we use adaptive backstepping control method to derive a feedback control law for globally stabilizing the 3-D novel conservative jerk chaotic system with unknown parameters.

Thus, we consider the 3-D novel conservative jerk chaotic system given by

$$\begin{cases} \dot{x}_1 = x_2 \\ \dot{x}_2 = x_3 \\ \dot{x}_3 = -ax_1(1 - x_1) - x_2 + bx_2^2 + u \end{cases} \tag{29}$$

where a and b are unknown constant parameters, and u is a backstepping control law to be determined using the estimates $\hat{a}(t)$ and $\hat{b}(t)$ for a and b, respectively.

The parameter estimation errors are defined as:

$$\begin{cases} e_a(t) = a - \hat{a}(t) \\ e_b(t) = b - \hat{b}(t) \end{cases} \tag{30}$$

Differentiating (30) with respect to t, we obtain the following equations:

$$\begin{cases} \dot{e}_a(t) = -\dot{\hat{a}}(t) \\ \dot{e}_b(t) = -\dot{\hat{b}}(t) \end{cases} \tag{31}$$

Theorem 1 *The 3-D novel conservative jerk chaotic system (29), with unknown parameters a and b, is globally and exponentially stabilized by the adaptive feedback control law,*

$$u(t) = -[3 - \hat{a}(t)]x_1 - 4x_2 - 3x_3 - \hat{a}(t)x_1^2 - \hat{b}(t)x_2^2 - kz_3, \tag{32}$$

where $k > 0$ is a gain constant,

$$z_3 = 2x_1 + 2x_2 + x_3, \tag{33}$$

and the update law for the parameter estimates $\hat{a}(t), \hat{b}(t)$ is given by

$$\begin{cases} \dot{\hat{a}}(t) = (-x_1 + x_1^2)z_3 \\ \dot{\hat{b}}(t) = x_2^2 z_3 \end{cases} \tag{34}$$

Proof We prove this result via backstepping control method and Lyapunov stability theory [154].

First, we define a quadratic Lyapunov function

$$V_1(z_1) = \frac{1}{2}\, z_1^2 \tag{35}$$

where

$$z_1 = x_1 \tag{36}$$

Differentiating V_1 along the dynamics (29), we get

$$\dot{V}_1 = z_1 \dot{z}_1 = x_1 x_2 = -z_1^2 + z_1(x_1 + x_2) \tag{37}$$

Now, we define

$$z_2 = x_1 + x_2 \tag{38}$$

Using (38), we can simplify the Eq. (37) as

$$\dot{V}_1 = -z_1^2 + z_1 z_2 \tag{39}$$

Secondly, we define a quadratic Lyapunov function

$$V_2(z_1, z_2) = V_1(z_1) + \frac{1}{2}\, z_2^2 = \frac{1}{2}\left(z_1^2 + z_2^2\right) \tag{40}$$

Differentiating V_2 along the dynamics (29), we get

$$\dot{V}_2 = -z_1^2 - z_2^2 + z_2(2x_1 + 2x_2 + x_3) \tag{41}$$

Now, we define

$$z_3 = 2x_1 + 2x_2 + x_3 \tag{42}$$

Using (42), we can simplify the Eq. (41) as

$$\dot{V}_2 = -z_1^2 - z_2^2 + z_2 z_3 \tag{43}$$

Finally, we define a quadratic Lyapunov function

$$V(z_1, z_2, z_3, e_a, e_b) = V_2(z_1, z_2) + \frac{1}{2} z_3^2 + \frac{1}{2} e_a^2 + \frac{1}{2} e_b^2 \tag{44}$$

which is a positive definite function on $\mathbb{R}^5$.

Differentiating V along the dynamics (29), we get

$$\dot{V} = -z_1^2 - z_2^2 - z_3^2 + z_3(z_3 + z_2 + \dot{z}_3) - e_a \dot{\hat{a}} - e_b \dot{\hat{b}} \tag{45}$$

Equation (45) can be written compactly as

$$\dot{V} = -z_1^2 - z_2^2 - z_3^2 + z_3 S - e_a \dot{\hat{a}} - e_b \dot{\hat{b}} \tag{46}$$

where

$$S = z_3 + z_2 + \dot{z}_3 = z_3 + z_2 + 2\dot{x}_1 + 2\dot{x}_2 + \dot{x}_3 \tag{47}$$

A simple calculation gives

$$S = (3 - a)x_1 + 4x_2 + 3x_3 + ax_1^2 + bx_2^2 + u \tag{48}$$

Substituting the adaptive control law (32) into (48), we obtain

$$S = [a - \hat{a}(t)](-x_1 + x_1^2) + [b - \hat{b}(t)]x_2^2 - kz_3 \tag{49}$$

Using the definitions (31), we can simplify (49) as

$$S = e_a(-x_1 + x_1^2) + e_b x_2^2 - kz_3 \tag{50}$$

Substituting the value of S from (50) into (46), we obtain

$$\dot{V} = -z_1^2 - z_2^2 - (1 + k)z_3^2 + e_a \left[(-x_1 + x_1^2)z_3 - \dot{\hat{a}} \right] + e_b \left[x_2^2 z_3 - \dot{\hat{b}} \right] \tag{51}$$

Substituting the update law (34) into (51), we get

$$\dot{V} = -z_1^2 - z_2^2 - (1 + k)z_3^2, \tag{52}$$

which is a negative semi-definite function on $\mathbf{R}^5$.

From (52), it follows that the vector $\mathbf{z}(t) = (z_1(t), z_2(t), z_3(t))$ and the parameter estimation error $(e_a(t), e_b(t))$ are globally bounded, i.e.

$$\left[z_1(t)\ z_2(t)\ z_3(t)\ e_a(t)\ e_b(t) \right] \in \mathbf{L}_\infty \tag{53}$$

Also, it follows from (52) that

$$\dot{V} \le -z_1^2 - z_2^2 - z_3^2 = -\|\mathbf{z}\|^2 \tag{54}$$

That is,

$$\|\mathbf{z}\|^2 \le -\dot{V} \tag{55}$$

Integrating the inequality (55) from 0 to t, we get

$$\int_0^t |\mathbf{z}(\tau)|^2 \, d\tau \leq V(0) - V(t) \tag{56}$$

From (56), it follows that $\mathbf{z}(t) \in \mathbf{L}_2$.

From Eq. (29), it can be deduced that $\dot{\mathbf{z}}(t) \in Ł_\infty$.

Thus, using Barbalat's lemma, we conclude that $\mathbf{z}(t) \to \mathbf{0}$ exponentially as $t \to \infty$ for all initial conditions $\mathbf{z}(0) \in \mathbb{R}^3$.

Hence, it is immediate that $\mathbf{x}(t) \to \mathbf{0}$ exponentially as $t \to \infty$ for all initial conditions $\mathbf{x}(0) \in \mathbb{R}^3$.

This completes the proof. □

For the numerical simulations, the classical fourth-order Runge-Kutta method with step size $h = 10^{-8}$ is used to solve the system of differential equations (29) and (34), when the adaptive control law (32) is applied.

The parameter values of the novel conservative jerk chaotic system (29) are taken as in the chaotic case, i.e. $a = 0.2$ and $b = 0.01$.

We take the positive gain constant as $k = 30$.

Furthermore, as initial conditions of the novel jerk chaotic system (29), we take

$$x_1(0) = 4.7, \quad x_2(0) = 6.3, \quad x_3(0) = -5.2 \tag{57}$$

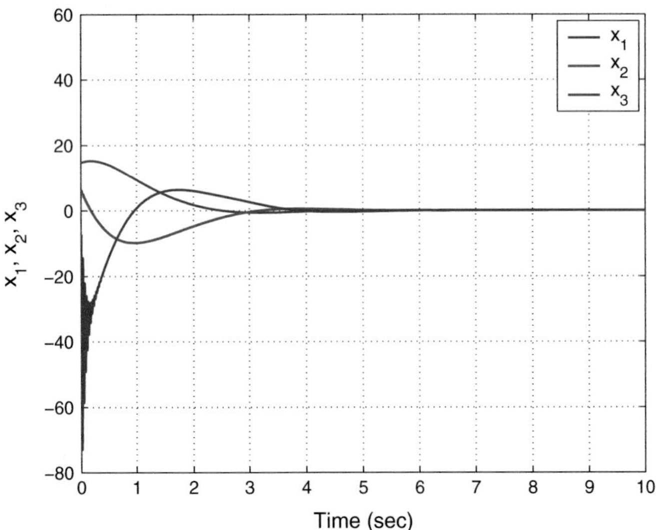

Fig. 6 Time-history of the controlled states $x_1(t), x_2(t), x_3(t)$

Also, as initial conditions of the parameter estimates $\hat{a}(t)$ and $\hat{b}(t)$, we take

$$\hat{a}(0) = 4.8, \quad \hat{b}(0) = 12.1 \tag{58}$$

In Fig. 6, the exponential convergence of the controlled states $x_1(t), x_2(t), x_3(t)$ is depicted, when the adaptive control law (32) and (34) are implemented.

From Fig. 6, it is clear that the controlled states $x_1(t), x_2(t), x_3(t)$ converge to zero exponentially in 6 s. This shows the efficiency of the adaptive backstepping controller derived in this section for the novel conservative jerk chaotic system (29).

5 Adaptive Synchronization of Identical Novel Jerk Chaotic Systems

In this section, we use adaptive backstepping control method to derive a feedback control law for globally and exponentially synchronizing the identical 3-D novel conservative jerk chaotic systems with unknown parameters.

The main adaptive control of this section is established using adaptive control theory and Lyapunov stability theory.

As the master system, we consider the 3-D novel jerk chaotic system given by

$$\begin{cases} \dot{x}_1 = x_2 \\ \dot{x}_2 = x_3 \\ \dot{x}_3 = -ax_1(1 - x_1) - x_2 + bx_2^2 \end{cases} \tag{59}$$

In (59), x_1, x_2, x_3 are the states of the system, and a and b are unknown constant parameters.

As the slave system, we consider the 3-D novel jerk chaotic system given by

$$\begin{cases} \dot{y}_1 = y_2 \\ \dot{y}_2 = y_3 \\ \dot{y}_3 = -ay_1(1 - y_1) - y_2 + by_2^2 + u \end{cases} \tag{60}$$

In (60), y_1, y_2, y_3 are the states of the system, and u is a backstepping control to be determined using estimates $\hat{a}(t)$ and $\hat{b}(t)$ for a and b, respectively.

We define the synchronization errors between the states of the master system (59) and the slave system (60) as

$$\begin{cases} e_1 = y_1 - x_1 \\ e_2 = y_2 - x_2 \\ e_3 = y_3 - x_3 \end{cases} \tag{61}$$

Then the error dynamics is easily obtained as

$$\begin{cases} \dot{e}_1 = e_2 \\ \dot{e}_2 = e_3 \\ \dot{e}_3 = -a(e_1 - y_1^2 + x_1^2) - e_2 + b(y_2^2 - x_2^2) + u \end{cases} \tag{62}$$

The parameter estimation errors are defined as:

$$\begin{cases} e_a(t) = a - \hat{a}(t) \\ e_b(t) = b - \hat{b}(t) \end{cases} \tag{63}$$

Differentiating (63) with respect to t, we obtain the following equations:

$$\begin{cases} \dot{e}_a(t) = -\dot{\hat{a}}(t) \\ \dot{e}_b(t) = -\dot{\hat{b}}(t) \end{cases} \tag{64}$$

Next, we shall state and prove the main result of this section.

Theorem 2 *The identical 3-D novel jerk chaotic systems (59) and (60) with unknown parameters a and b are globally and exponentially synchronized by the adaptive control law*

$$u(t) = -[3 - \hat{a}(t)]e_1 - 4e_2 - 3e_3 - \hat{a}(t)[y_1^2 - x_1^2] - \hat{b}(t)[y_2^2 - x_2^2] - kz_3 \tag{65}$$

where $k > 0$ *is a gain constant,*

$$z_3 = 2e_1 + 2e_2 + e_3, \tag{66}$$

and the update law for the parameter estimates $\hat{a}(t), \hat{b}(t)$ *is given by*

$$\begin{cases} \dot{\hat{a}}(t) = (-e_1 + y_1^2 - x_1^2)z_3 \\ \dot{\hat{b}}(t) = (y_2^2 - x_2^2)z_3 \end{cases} \tag{67}$$

Proof We prove this result via backstepping control method and Lyapunov stability theory [154].

First, we define a quadratic Lyapunov function

$$V_1(z_1) = \frac{1}{2} z_1^2 \tag{68}$$

where

$$z_1 = e_1 \tag{69}$$

Differentiating V_1 along the error dynamics (62), we get

$$\dot{V}_1 = z_1\dot{z}_1 = e_1e_2 = -z_1^2 + z_1(e_1 + e_2) \tag{70}$$

Now, we define

$$z_2 = e_1 + e_2 \tag{71}$$

Using (71), we can simplify the Eq. (70) as

$$\dot{V}_1 = -z_1^2 + z_1z_2 \tag{72}$$

Secondly, we define a quadratic Lyapunov function

$$V_2(z_1, z_2) = V_1(z_1) + \frac{1}{2}z_2^2 = \frac{1}{2}\left(z_1^2 + z_2^2\right) \tag{73}$$

Differentiating V_2 along the error dynamics (62), we get

$$\dot{V}_2 = -z_1^2 - z_2^2 + z_2(2e_1 + 2e_2 + e_3) \tag{74}$$

Now, we define

$$z_3 = 2e_1 + 2e_2 + e_3 \tag{75}$$

Using (75), we can simplify the Eq. (74) as

$$\dot{V}_2 = -z_1^2 - z_2^2 + z_2z_3 \tag{76}$$

Finally, we define a quadratic Lyapunov function

$$V(z_1, z_2, z_3, e_a, e_b) = V_2(z_1, z_2) + \frac{1}{2}z_3^2 + \frac{1}{2}e_a^2 + \frac{1}{2}e_b^2 \tag{77}$$

Differentiating V along the error dynamics (62), we get

$$\dot{V} = -z_1^2 - z_2^2 - z_3^2 + z_3(z_3 + z_2 + \dot{z}_3) - e_a\dot{\hat{a}} - e_b\dot{\hat{b}} \tag{78}$$

Equation (78) can be written compactly as

$$\dot{V} = -z_1^2 - z_2^2 - z_3^2 + z_3S - e_a\dot{\hat{a}} - e_b\dot{\hat{b}} \tag{79}$$

where

$$S = z_3 + z_2 + \dot{z}_3 = z_3 + z_2 + 2\dot{e}_1 + 2\dot{e}_2 + \dot{e}_3 \tag{80}$$

A simple calculation gives

$$S = (3 - a)e_1 + 4e_2 + 3e_3 + a(y_1^2 - x_1^2) + b(y_2^2 - x_2^2) + u \tag{81}$$

Substituting the adaptive control law (65) into (81), we obtain

$$S = [a - \hat{a}(t)]\left(-e_1 + y_1^2 - x_1^2\right) + [b - \hat{b}(t)]\left(y_2^2 - x_2^2\right) - kz_3 \tag{82}$$

Using the definitions (64), we can simplify (82) as

$$S = e_a\left(-e_1 + y_1^2 - x_1^2\right) + e_b\left(y_2^2 - x_2^2\right) - kz_3 \tag{83}$$

Substituting the value of S from (83) into (79), we obtain

$$\begin{cases} \dot{V} = -z_1 - z_2 - (1+k)z_3^2 + e_a\left[(-e_1 + y_1^2 - x_1^2)z_3 - \dot{\hat{a}}\right] \\ \quad + e_b\left[(y_2^2 - x_2^2)z_3 - \dot{\hat{b}}\right] \end{cases} \tag{84}$$

Substituting the update law (67) into (84), we get

$$\dot{V} = -z_1^2 - z_2^2 - (1+k)z_3^2, \tag{85}$$

which is a negative semi-definite function on $\mathbb{R}^5$.

From (85), it follows that the vector $\mathbf{z}(t) = (z_1(t), z_2(t), z_3(t))$ and the parameter estimation error $(e_a(t), e_b(t))$ are globally bounded, i.e.

$$\left[\, z_1(t)\ z_2(t)\ z_3(t)\ e_a(t)\ e_b(t) \,\right] \in \mathbf{L}_\infty \tag{86}$$

Also, it follows from (85) that

$$\dot{V} \le -z_1^2 - z_2^2 - z_3^2 = -\|\mathbf{z}\|^2 \tag{87}$$

That is,

$$\|\mathbf{z}\|^2 \le -\dot{V} \tag{88}$$

Integrating the inequality (88) from 0 to t, we get

$$\int_0^t |\mathbf{z}(\tau)|^2\, d\tau \le V(0) - V(t) \tag{89}$$

From (89), it follows that $\mathbf{z}(t) \in \mathbf{L}_2$.

From Eq. (62), it can be deduced that $\dot{\mathbf{z}}(t) \in \mathbf{L}_\infty$.

Thus, using Barbalat's lemma, we conclude that $\mathbf{z}(t) \to \mathbf{0}$ exponentially as $t \to \infty$ for all initial conditions $\mathbf{z}(0) \in \mathbb{R}^3$.

Hence, it is immediate that $\mathbf{e}(t) \to \mathbf{0}$ exponentially as $t \to \infty$ for all initial conditions $\mathbf{e}(0) \in \mathbb{R}^3$.

This completes the proof. □

For the numerical simulations, the classical fourth-order Runge-Kutta method with step size $h = 10^{-8}$ is used to solve the system of differential equations (59) and (60).

The parameter values of the novel jerk chaotic systems are taken as in the chaotic case, i.e. $a = 0.2$ and $b = 0.01$.

We take the positive gain constant as $k = 30$.

Furthermore, as initial conditions of the master chaotic system (59), we take

$$x_1(0) = 1.5, \quad x_2(0) = 0.7, \quad x_3(0) = -2.9 \tag{90}$$

As initial conditions of the slave chaotic system (60), we take

$$y_1(0) = -2.8, \quad x_2(0) = 6.2, \quad x_3(0) = 1.5 \tag{91}$$

Also, as initial conditions of the parameter estimates $\hat{a}(t)$ and $\hat{b}(t)$, we take $\hat{a}(0) = 8.2$ and $\hat{b}(0) = 12.1$.

In Figs. 7, 8 and 9, the complete synchronization of the identical 3-D conservative jerk chaotic systems (59) and (60) is shown, when the adaptive control law and the parameter update law are implemented.

From Fig. 7, it is clear that the states $x_1(t)$ and $y_1(t)$ are completely synchronized in 6 s.

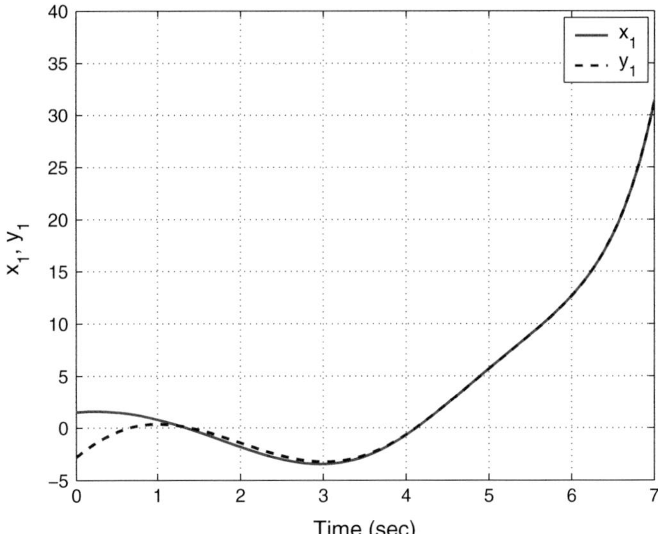

Fig. 7 Synchronization of the states $x_1(t)$ and $y_1(t)$

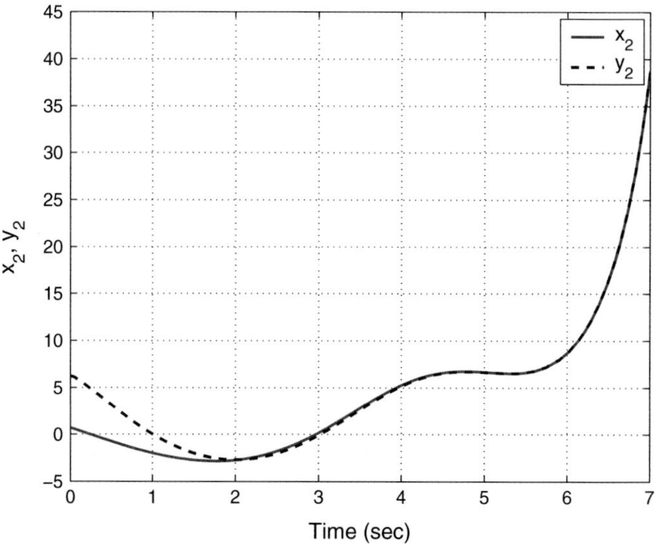

Fig. 8 Synchronization of the states $x_2(t)$ and $y_2(t)$

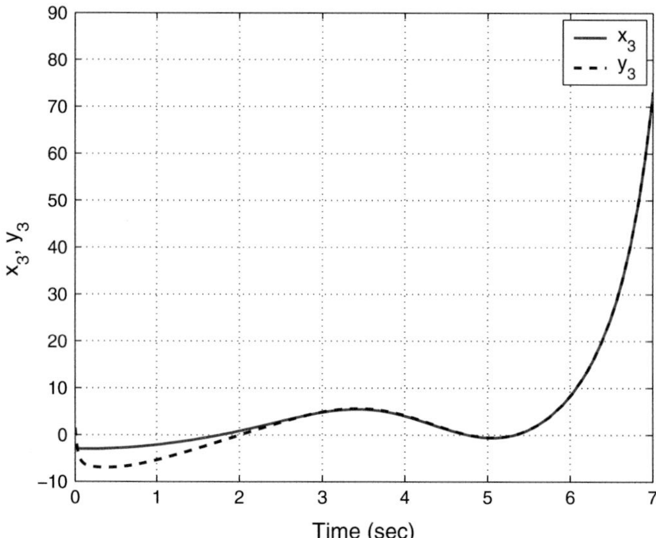

Fig. 9 Synchronization of the states $x_3(t)$ and $y_3(t)$

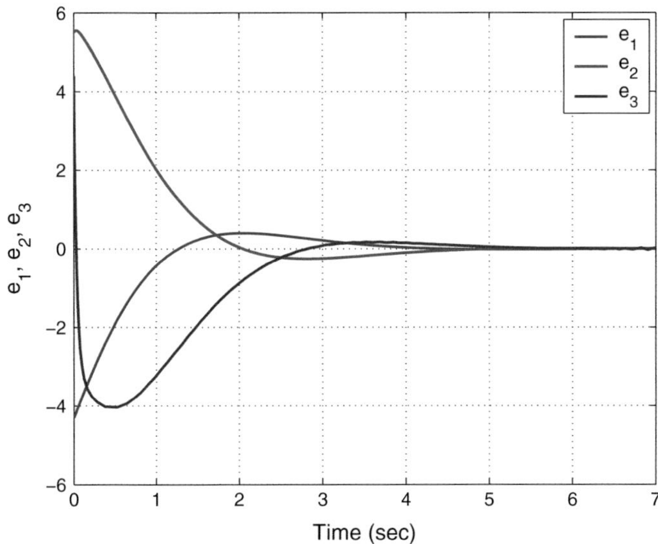

Fig. 10 Time-history of the synchronization errors $e_1(t), e_2(t), e_3(t)$

From Fig. 8, it is clear that the states $x_2(t)$ and $y_2(t)$ are completely synchronized in 6 s.

From Fig. 9, it is clear that the states $x_3(t)$ and $y_3(t)$ are completely synchronized in 6 s.

Also, in Fig. 10, the time-history of the synchronization errors $e_1(t), e_2(t), e_3(t)$, is shown.

From Fig. 10, it is clear that the synchronization errors converge to zero in 6 s. This shows the efficiency of the adaptive backstepping controller designed in this section.

6 Conclusions

In this research work, we announced a six-term novel 3-D conservative jerk chaotic system with two quadratic nonlinearities. We derived new backstepping control results for the adaptive controller and adaptive synchronizer design for the novel conservative jerk chaotic system. We first discussed the qualitative properties of the novel conservative jerk chaotic system. Explicitly, we showed that the novel jerk chaotic system is conservative and that it has two unstable equilibrium points. The Lyapunov exponents of novel jerk chaotic system have been obtained as $L_1 = 0.0452$, $L_2 = 0$ and $L_3 = -0.0452$, while the Kaplan-Yorke dimension of the novel conservative jerk chaotic system has been obtained as $D_{KY} = 3$. The adaptive backstepping control and synchronization results were proved using Lyapunov stability theory. MATLAB simulations have been depicted to illustrate all the main results for the novel conservative jerk chaotic system.

References

1. Lorenz EN (1963) Deterministic periodic flow. J Atmos Sci 20(2):130–141
2. Rössler OE (1976) An equation for continuous chaos. Phys Lett A 57(5):397–398
3. Arneodo A, Coullet P, Tresser C (1981) Possible new strange attractors with spiral structure. Commun Math Phys 79(4):573–576
4. Sprott JC (1994) Some simple chaotic flows. Phys Rev E 50(2):647–650
5. Chen G, Ueta T (1999) Yet another chaotic attractor. Int J Bifurcat Chaos 9(7):1465–1466
6. Lü J, Chen G (2002) A new chaotic attractor coined. Int J Bifurc Chaos 12(3):659–661
7. Liu C, Liu T, Liu L, Liu K (2004) A new chaotic attractor. Chaos, Solitions Fractals 22(5):1031–1038
8. Cai G, Tan Z (2007) Chaos synchronization of a new chaotic system via nonlinear control. J Uncertain Syst 1(3):235–240
9. Chen HK, Lee CI (2004) Anti-control of chaos in rigid body motion. Chaos, Solitons Fractals 21(4):957–965
10. Tigan G, Opris D (2008) Analysis of a 3D chaotic system. Chaos, Solitons Fractals 36: 1315–1319
11. Zhou W, Xu Y, Lu H, Pan L (2008) On dynamics analysis of a new chaotic attractor. Phys Lett A 372(36):5773–5777
12. Zhu C, Liu Y, Guo Y (2010) Theoretic and numerical study of a new chaotic system. Intell Inf Manage 2:104–109
13. Li D (2008) A three-scroll chaotic attractor. Phys Lett A 372(4):387–393
14. Wei Z, Yang Q (2010) Anti-control of Hopf bifurcation in the new chaotic system with two stable node-foci. Appl Math Comput 217(1):422–429
15. Sundarapandian V (2013) Analysis and anti-synchronization of a novel chaotic system via active and adaptive controllers. J Eng Sci Technol Rev 6(4):45–52
16. Sundarapandian V, Pehlivan I (2012) Analysis, control, synchronization, and circuit design of a novel chaotic system. Math Comput Model 55(7–8):1904–1915
17. Vaidyanathan S (2013) A new six-term 3-D chaotic system with an exponential nonlinearity. Far East J Math Sci 79(1):135–143
18. Vaidyanathan S (2013) Analysis and adaptive synchronization of two novel chaotic systems with hyperbolic sinusoidal and cosinusoidal nonlinearity and unknown parameters. J Eng Sci Technol Rev 6(4):53–65
19. Vaidyanathan S (2014) A new eight-term 3-D polynomial chaotic system with three quadratic nonlinearities. Far East J Math Sci 84(2):219–226
20. Vaidyanathan S (2014) Analysis and adaptive synchronization of eight-term 3-D polynomial chaotic systems with three quadratic nonlinearities. Eur Phys J Spec Top 223(8):1519–1529
21. Vaidyanathan S (2014) Analysis, control and synchronisation of a six-term novel chaotic system with three quadratic nonlinearities. Int J Model Ident Control 22(1):41–53
22. Vaidyanathan S (2014) Generalized projective synchronisation of novel 3-D chaotic systems with an exponential non-linearity via active and adaptive control. Int J Model Ident Control 22(3):207–217
23. Vaidyanathan S (2015) A 3-D novel highly chaotic system with four quadratic nonlinearities, its adaptive control and anti-synchronization with unknown parameters. J Eng Sci Technol Rev 8(2):106–115
24. Vaidyanathan S (2015) Analysis, properties and control of an eight-term 3-D chaotic system with an exponential nonlinearity. Int J Model Ident Control 23(2):164–172

25. Vaidyanathan S, Azar AT (2015) Analysis, control and synchronization of a nine-term 3-D novel chaotic system. In: Azar AT, Vaidyanathan S (eds) Chaos modelling and control systems design, studies in computational intelligence, vol 581. Springer, Germany, pp 19–38
26. Vaidyanathan S, Madhavan K (2013) Analysis, adaptive control and synchronization of a seven-term novel 3-D chaotic system. Int J Control Theory Appl 6(2):121–137
27. Vaidyanathan S, Pakiriswamy S (2015) A 3-D novel conservative chaotic system and its generalized projective synchronization via adaptive control. J Eng Sci Technol Rev 8(2): 52–60
28. Vaidyanathan S, Volos C (2015) Analysis and adaptive control of a novel 3-D conservative no-equilibrium chaotic system. Arch Control Sci 25(3):333–353
29. Vaidyanathan S, Volos C, Pham VT, Madhavan K, Idowu BA (2014) Adaptive backstepping control, synchronization and circuit simulation of a 3-D novel jerk chaotic system with two hyperbolic sinusoidal nonlinearities. Arch Control Sci 24(3):375–403
30. Vaidyanathan S, Rajagopal K, Volos CK, Kyprianidis IM, Stouboulos IN (2015) Analysis, adaptive control and synchronization of a seven-term novel 3-D chaotic system with three quadratic nonlinearities and its digital implementation in LabVIEW. J Eng Sci Technol Rev 8(2):130–141
31. Vaidyanathan S, Volos CK, Kyprianidis IM, Stouboulos IN, Pham VT (2015) Analysis, adaptive control and anti-synchronization of a six-term novel jerk chaotic system with two exponential nonlinearities and its circuit simulation. J Eng Sci Technol Rev 8(2):24–36
32. Vaidyanathan S, Volos CK, Pham VT (2015) Analysis, adaptive control and adaptive synchronization of a nine-term novel 3-D chaotic system with four quadratic nonlinearities and its circuit simulation. J Eng Sci Technol Rev 8(2):174–184
33. Vaidyanathan S, Volos CK, Pham VT (2015) Global chaos control of a novel nine-term chaotic system via sliding mode control. In: Azar AT, Zhu Q (eds) Advances and applications in sliding mode control systems, studies in computational intelligence, vol 576. Springer, Germany, pp 571–590
34. Pehlivan I, Moroz IM, Vaidyanathan S (2014) Analysis, synchronization and circuit design of a novel butterfly attractor. J Sound Vib 333(20):5077–5096
35. Sampath S, Vaidyanathan S, Volos CK, Pham VT (2015) An eight-term novel four-scroll chaotic system with cubic nonlinearity and its circuit simulation. J Eng Sci Technol Rev 8(2):1–6
36. Pham VT, Vaidyanathan S, Volos CK, Jafari S (2015) Hidden attractors in a chaotic system with an exponential nonlinear term. Eur Phys J Spec Top 224(8):1507–1517
37. Azar AT (2010) Fuzzy systems. IN-TECH, Vienna
38. Azar AT, Vaidyanathan S (2015) Chaos modeling and control systems design, studies in computational intelligence, vol 581. Springer, Germany
39. Azar AT, Vaidyanathan S (2015) Computational intelligence applications in modeling and control, studies in computational intelligence, vol 575. Springer, Germany
40. Azar AT, Vaidyanathan S (2015) Handbook of research on advanced intelligent control engineering and automation. Advances in computational intelligence and robotics (ACIR). IGI-Global, USA
41. Azar AT, Zhu Q (2015) Advances and applications in sliding mode control systems, studies in computational intelligence, vol 576. Springer, Germany
42. Zhu Q, Azar AT (2015) Complex system modelling and control through intelligent soft computations, studies in fuzzines and soft computing, vol 319. Springer, Germany
43. Kengne J, Chedjou JC, Kenne G, Kyamakya K (2012) Dynamical properties and chaos synchronization of improved Colpitts oscillators. Commun Nonlinear Sci Numer Simul 17(7):2914–2923
44. Sharma A, Patidar V, Purohit G, Sud KK (2012) Effects on the bifurcation and chaos in forced Duffing oscillator due to nonlinear damping. Commun Nonlinear Sci Numer Simul 17(6):2254–2269
45. Li N, Pan W, Yan L, Luo B, Zou X (2014) Enhanced chaos synchronization and communication in cascade-coupled semiconductor ring lasers. Commun Nonlinear Sci Numer Simul 19(6):1874–1883

46. Yuan G, Zhang X, Wang Z (2014) Generation and synchronization of feedback-induced chaos in semiconductor ring lasers by injection-locking. Opt Int J Light Electron Opt 125(8): 1950–1953
47. Gaspard P (1999) Microscopic chaos and chemical reactions. Physica A: Stat Mech Appl 263(1–4):315–328
48. Petrov V, Gaspar V, Masere J, Showalter K (1993) Controlling chaos in Belousov-Zhabotinsky reaction. Nature 361:240–243
49. Vaidyanathan S (2015) Adaptive control of a chemical chaotic reactor. Int J PharmTech Res 8(3):377–382
50. Vaidyanathan S (2015) Adaptive synchronization of chemical chaotic reactors. Int J ChemTech Res 8(2):612–621
51. Vaidyanathan S (2015) Anti-synchronization of Brusselator chemical reaction systems via adaptive control. Int J ChemTech Res 8(6):759–768
52. Vaidyanathan S (2015) Dynamics and control of Brusselator chemical reaction. Int J ChemTech Res 8(6):740–749
53. Vaidyanathan S (2015) Dynamics and control of Tokamak system with symmetric and magnetically confined plasma. Int J ChemTech Res 8(6):795–803
54. Vaidyanathan S (2015) Synchronization of Tokamak systems with symmetric and magnetically confined plasma via adaptive control. Int J ChemTech Res 8(6):818–827
55. Das S, Goswami D, Chatterjee S, Mukherjee S (2014) Stability and chaos analysis of a novel swarm dynamics with applications to multi-agent systems. Eng Appl Artif Intell 30:189–198
56. Kyriazis M (1991) Applications of chaos theory to the molecular biology of aging. Exp Gerontol 26(6):569–572
57. Vaidyanathan S (2015) 3-cells cellular neural network (CNN) attractor and its adaptive biological control. Int J PharmTech Res 8(4):632–640
58. Vaidyanathan S (2015) Adaptive backstepping control of eEnzymes-substrates system with ferroelectric behaviour in brain waves. Int J PharmTech Res 8(2):256–261
59. Vaidyanathan S (2015) Adaptive biological control of generalized Lotka-Volterra three-species biological system. Int J PharmTech Res 8(4):622–631
60. Vaidyanathan S (2015) Adaptive chaotic synchronization of enzymes-substrates system with ferroelectric behaviour in brain waves. Int J PharmTech Res 8(5):964–973
61. Vaidyanathan S (2015) Adaptive synchronization of generalized Lotka-Volterra three-species biological systems. Int J PharmTech Res 8(5):928–937
62. Vaidyanathan S (2015) Chaos in neurons and adaptive control of Birkhoff-Shaw strange chaotic attractor. Int J PharmTech Res 8(5):956–963
63. Vaidyanathan S (2015) Lotka-Volterra population biology models with negative feedback and their ecological monitoring. Int J PharmTech Res 8(5):974–981
64. Vaidyanathan S (2015) Synchronization of 3-cells cellular neural network (CNN) attractors via adaptive control method. Int J PharmTech Res 8(5):946–955
65. Gibson WT, Wilson WG (2013) Individual-based chaos: extensions of the discrete logistic model. J Theoret Biol 339:84–92
66. Suérez I (1999) Mastering chaos in ecology. Ecol Model 117(2–3):305–314
67. Lang J (2015) Color image encryption based on color blend and chaos permutation in the reality-preserving multiple-parameter fractional Fourier transform domain. Opt Commun 338:181–192
68. Zhang X, Zhao Z, Wang J (2014) Chaotic image encryption based on circular substitution box and key stream buffer. Signal Process Image Commun 29(8):902–913
69. Rhouma R, Belghith S (2011) Cryptoanalysis of a chaos based cryptosystem on DSP. Commun Nonlinear Sci Numer Simul 16(2):876–884
70. Usama M, Khan MK, Alghatbar K, Lee C (2010) Chaos-based secure satellite imagery cryptosystem. Comput Math Appl 60(2):326–337
71. Azar AT, Serrano FE (2014) Robust IMC-PID tuning for cascade control systems with gain and phase margin specifications. Neural Comput Appl 25(5):983–995

72. Azar AT, Serrano FE (2015) Adaptive sliding mode control of the Furuta pendulum. In: Azar AT, Zhu Q (eds) Advances and applications in sliding mode control systems, studies in computational intelligence, vol 576. Springer, Germany, pp 1–42
73. Azar AT, Serrano FE (2015) Deadbeat control for multivariable systems with time varying delays. In: Azar AT, Vaidyanathan S (eds) Chaos modeling and control systems design, studies in computational intelligence, vol 581. Springer, Germany, pp 97–132
74. Azar AT, Serrano FE (2015) Design and modeling of anti wind up PID controllers. In: Zhu Q, Azar AT (eds) Complex system modelling and control through intelligent soft computations, studies in fuzziness and soft computing, vol 319. Springer, Germany, pp 1–44
75. Azar AT, Serrano FE (2015) Stabilizatoin and control of mechanical systems with backlash. In: Azar AT, Vaidyanathan S (eds) Handbook of research on advanced intelligent control engineering and automation., Advances in computational intelligence and robotics (ACIR)IGI-Global, USA, pp 1–60
76. Feki M (2003) An adaptive chaos synchronization scheme applied to secure communication. Chaos, Solitons Fractals 18(1):141–148
77. Murali K, Lakshmanan M (1998) Secure communication using a compound signal from generalized chaotic systems. Phys Lett A 241(6):303–310
78. Zaher AA, Abu-Rezq A (2011) On the design of chaos-based secure communication systems. Commun Nonlinear Syst Numer Simul 16(9):3721–3727
79. Mondal S, Mahanta C (2014) Adaptive second order terminal sliding mode controller for robotic manipulators. J Franklin Inst 351(4):2356–2377
80. Nehmzow U, Walker K (2005) Quantitative description of robot-environment interaction using chaos theory. Robot Auton Syst 53(3–4):177–193
81. Volos CK, Kyprianidis IM, Stouboulos IN (2013) Experimental investigation on coverage performance of a chaotic autonomous mobile robot. Robot Auton Syst 61(12):1314–1322
82. Qu Z (2011) Chaos in the genesis and maintenance of cardiac arrhythmias. Prog Biophys Mol Biol 105(3):247–257
83. Witte CL, Witte MH (1991) Chaos and predicting varix hemorrhage. Med Hypotheses 36(4):312–317
84. Azar AT (2012) Overview of type-2 fuzzy logic systems. Int J Fuzzy Syst Appl 2(4):1–28
85. Li Z, Chen G (2006) Integration of fuzzy logic and chaos theory, studies in fuzziness and soft computing, vol 187. Springer, Germany
86. Huang X, Zhao Z, Wang Z, Li Y (2012) Chaos and hyperchaos in fractional-order cellular neural networks. Neurocomputing 94:13–21
87. Kaslik E, Sivasundaram S (2012) Nonlinear dynamics and chaos in fractional-order neural networks. Neural Netw 32:245–256
88. Lian S, Chen X (2011) Traceable content protection based on chaos and neural networks. Appl Soft Comput 11(7):4293–4301
89. Guégan D (2009) Chaos in economics and finance. Ann Rev Control 33(1):89–93
90. Sprott JC (2004) Competition with evolution in ecology and finance. Phys Lett A 325(5–6):329–333
91. Pham VT, Volos CK, Vaidyanathan S, Le TP, Vu VY (2015) A memristor-based hyperchaotic system with hidden attractors: dynamics, synchronization and circuital emulating. J Eng Sci Technol Rev 8(2):205–214
92. Volos CK, Kyprianidis IM, Stouboulos IN, Tlelo-Cuautle E, Vaidyanathan S (2015) Memristor: a new concept in synchronization of coupled neuromorphic circuits. J Eng Sci Technol Rev 8(2):157–173
93. Sundarapandian V (2010) Output regulation of the Lorenz attractor. Far East J Math Sci 42(2):289–299
94. Vaidyanathan S (2011) Output regulation of Arneodo-Coullet chaotic system. Communications in Computer and Information Science 133:98–107
95. Vaidyanathan S (2011) Output regulation of the unified chaotic system. Commun Comput Inf Sci 198:10–17

96. Vaidyanathan S, Azar AT, Rajagopal K, Alexander P (2015) Design and SPICE implementation of a 12-term novel hyperchaotic system and its synchronisation via active control. Int J Model Ident Control 23(3):267–277
97. Noroozi N, Roopaei M, Karimaghaee P, Safavi AA (2010) Simple adaptive variable structure control for unknown chaotic systems. Commun Nonlinear Sci Numer Simul 15(3):707–727
98. Vaidyanathan S, Volos CK, Pham VT (2015) Analysis, control, synchronization and SPICE implementation of a novel 4-D hyperchaotic Rikitake dynamo System without equilibrium. J Eng Sci Technol Rev 8(2):232–244
99. Vaidyanathan S (2012) Global chaos control of hyperchaotic Liu system via sliding control method. Int J Control Theory Appl 5(2):117–123
100. Vaidyanathan S (2012) Sliding mode control based global chaos control of Liu-Liu-Liu-Su chaotic system. Int J Control Theory Appl 1(2):15–20
101. Vaidyanathan S, Volos CK, Rajagopal K, Kyprianidis IM, Stouboulos IN (2015) Adaptive backstepping controller design for the anti-synchronization of identical WINDMI chaotic systems with unknown parameters and its SPICE implementation. J Eng Sci Technol Rev 8(2):74–82
102. Carroll TL, Pecora LM (1991) Synchronizing chaotic circuits. IEEE Trans Circuits Syst 38(4):453–456
103. Pecora LM, Carroll TL (1990) Synchronization in chaotic systems. Phys Rev Lett 64(8): 821–824
104. Karthikeyan R, Sundarapandian V (2014) Hybrid chaos synchronization of four-scroll systems via active control. J Electr Eng 65(2):97–103
105. Sarasu P, Sundarapandian V (2011) Active controller design for the generalized projective synchronization of four-scroll chaotic systems. Int J Syst Signal Control Eng Appl 4(2):26–33
106. Sarasu P, Sundarapandian V (2011) The generalized projective synchronization of hyperchaotic Lorenz and hyperchaotic Qi systems via active control. Int J Soft Comput 6(5): 216–223
107. Sundarapandian V (2010) Output regulation of the Lorenz attractor. Far East J Math Sci 42(2):289–299
108. Sundarapandian V, Karthikeyan R (2012) Hybrid synchronization of hyperchaotic Lorenz and hyperchaotic Chen systems via active control. J Eng Appl Sci 7(3):254–264
109. Vaidyanathan S (2011) Hybrid chaos synchronization of Liu and Lü systems by active nonlinear control. Commun Comput Inf Sci 204:1–10
110. Vaidyanathan S (2012) Output regulation of the Liu chaotic system. Appl Mech Mater 110–116:3982–3989
111. Vaidyanathan S, Rajagopal K (2011) Anti-synchronization of Li and T chaotic systems by active nonlinear control. Commun Comput Inf Sci 198:175–184
112. Vaidyanathan S, Rajagopal K (2011) Global chaos synchronization of hyperchaotic Pang and Wang systems by active nonlinear control. Commun Comput Inf Sci 204:84–93
113. Vaidyanathan S, Rasappan S (2011) Global chaos synchronization of hyperchaotic Bao and Xu systems by active nonlinear control. Commun Comput Inf Sci 198:10–17
114. Vaidyanathan S, Pham VT, Volos CK (2015) A 5-D hyperchaotic Rikitake dynamo system with hidden attractors. Eur Phys J Spec Top 224(8):1575–1592
115. Sarasu P, Sundarapandian V (2012) Adaptive controller design for the generalized projective synchronization of 4-scroll systems. Int J Syst Signal Control Eng Appl 5(2):21–30
116. Sarasu P, Sundarapandian V (2012) Generalized projective synchronization of three-scroll chaotic systems via adaptive control. Eur J Sci Res 72(4):504–522
117. Sarasu P, Sundarapandian V (2012) Generalized projective synchronization of two-scroll systems via adaptive control. Int J Soft Comput 7(4):146–156
118. Sundarapandian V, Karthikeyan R (2011) Anti-synchronization of hyperchaotic Lorenz and hyperchaotic Chen systems by adaptive control. Int J Syst Signal Control Eng Appl 4(2):18–25
119. Sundarapandian V, Karthikeyan R (2011) Anti-synchronization of Lü and Pan chaotic systems by adaptive nonlinear control. Eur J Sci Res 64(1):94–106

120. Sundarapandian V, Karthikeyan R (2012) Adaptive anti-synchronization of uncertain Tigan and Li systems. J Eng Appl Sci 7(1):45–52
121. Vaidyanathan S (2012) Anti-synchronization of Sprott-L and Sprott-M chaotic systems via adaptive control. Int J Control Theory Appl 5(1):41–59
122. Vaidyanathan S (2013) Analysis, control and synchronization of hyperchaotic Zhou system via adaptive control. Adv Intell Syst Comput 177:1–10
123. Vaidyanathan S (2015) Hyperchaos, qualitative analysis, control and synchronisation of a ten-term 4-D hyperchaotic system with an exponential nonlinearity and three quadratic nonlinearities. Int J Model Ident Control 23(4):380–392
124. Vaidyanathan S, Azar AT (2015) Analysis and control of a 4-D novel hyperchaotic system. In: Azar AT, Vaidyanathan S (eds) Chaos modeling and control systems design, studies in computational intelligence, vol 581. Springer, Germany, pp 19–38
125. Vaidyanathan S, Pakiriswamy S (2013) Generalized projective synchronization of six-term Sundarapandian chaotic systems by adaptive control. Int J Control Theory Appl 6(2):153–163
126. Vaidyanathan S, Rajagopal K (2011) Global chaos synchronization of Lü and Pan systems by adaptive nonlinear control. Commun Comput Inf Sci 205:193–202
127. Vaidyanathan S, Rajagopal K (2012) Global chaos synchronization of hyperchaotic Pang and hyperchaotic Wang systems via adaptive control. Int J Soft Comput 7(1):28–37
128. Vaidyanathan S, Volos C, Pham VT (2014a) Hyperchaos, adaptive control and synchronization of a novel 5-D hyperchaotic system with three positive Lyapunov exponents and its SPICE implementation. Arch Control Sci 24(4):409–446
129. Vaidyanathan S, Volos C, Pham VT, Madhavan K (2015) Analysis, adaptive control and synchronization of a novel 4-D hyperchaotic hyperjerk system and its SPICE implementation. Arch Control Sci 25(1):5–28
130. Gan Q, Liang Y (2012) Synchronization of chaotic neural networks with time delay in the leakage term and parametric uncertainties based on sampled-data control. J Franklin Inst 349(6):1955–1971
131. Li N, Zhang Y, Nie Z (2011) Synchronization for general complex dynamical networks with sampled-data. Neurocomputing 74(5):805–811
132. Xiao X, Zhou L, Zhang Z (2014) Synchronization of chaotic Lur'e systems with quantized sampled-data controller. Commun Nonlinear Sci Numer Simul 19(6):2039–2047
133. Zhang H, Zhou J (2012) Synchronization of sampled-data coupled harmonic oscillators with control inputs missing. Syst Control Lett 61(12):1277–1285
134. Chen WH, Wei D, Lu X (2014) Global exponential synchronization of nonlinear time-delay Lur'e systems via delayed impulsive control. Commun Nonlinear Sci Numer Simul 19(9):3298–3312
135. Jiang GP, Zheng WX, Chen G (2004) Global chaos synchronization with channel time-delay. Chaos, Solitons Fractals 20(2):267–275
136. Shahverdiev EM, Shore KA (2009) Impact of modulated multiple optical feedback time delays on laser diode chaos synchronization. Opt Commun 282(17):3568–3572
137. Rasappan S, Vaidyanathan S (2012a) Global chaos synchronization of WINDMI and Coullet chaotic systems by backstepping control. Far East J Math Sci 67(2):265–287
138. Rasappan S, Vaidyanathan S (2012) Hybrid synchronization of n-scroll Chua and Lur'e chaotic systems via backstepping control with novel feedback. Arch Control Sci 22(3):343–365
139. Rasappan S, Vaidyanathan S (2012) Synchronization of hyperchaotic Liu system via backstepping control with recursive feedback. Commun Comput Inf Sci 305:212–221
140. Rasappan S, Vaidyanathan S (2013) Hybrid synchronization of n-scroll chaotic Chua circuits using adaptive backstepping control design with recursive feedback. Malays J Math Sci 7(2):219–246
141. Rasappan S, Vaidyanathan S (2014) Global chaos synchronization of WINDMI and Coullet chaotic systems using adaptive backstepping control design. Kyungpook Math J 54(1): 293–320
142. Suresh R, Sundarapandian V (2013) Global chaos synchronization of a family of n-scroll hyperchaotic Chua circuits using backstepping control with recursive feedback. Far East J Math Sci 73(1):73–95

143. Vaidyanathan S, Rasappan S (2014) Global chaos synchronization of n-scroll Chua circuit and Lur'e system using backstepping control design with recursive feedback. Arab J Sci Eng 39(4):3351–3364
144. Vaidyanathan S, Idowu BA, Azar AT (2015) Backstepping controller design for the global chaos synchronization of Sprott's jerk systems. Stud Comput Intell 581:39–58
145. Sundarapandian V, Sivaperumal S (2011) Sliding controller design of hybrid synchronization of four-wing chaotic systems. Int J Soft Comput 6(5):224–231
146. Vaidyanathan S (2012) Analysis and synchronization of the hyperchaotic Yujun systems via sliding mode control. Adv Intell Syst Comput 176:329–337
147. Vaidyanathan S (2014) Global chaos synchronization of identical Li-Wu chaotic systems via sliding mode control. Int J Model Ident Control 22(2):170–177
148. Vaidyanathan S, Azar AT (2015) Anti-synchronization of identical chaotic systems using sliding mode control and an application to Vaidhyanathan-Madhavan chaotic systems. Stud. Comput Intell 576:527–547
149. Vaidyanathan S, Azar AT (2015) Hybrid synchronization of identical chaotic systems using sliding mode control and an application to Vaidhyanathan chaotic systems. Stud Comput Intell 576:549–569
150. Vaidyanathan S, Sampath S (2011) Global chaos synchronization of hyperchaotic Lorenz systems by sliding mode control. Commun Comput Inf Sci 205:156–164
151. Vaidyanathan S, Sampath S (2012) Anti-synchronization of four-wing chaotic systems via sliding mode control. Int J Autom Comput 9(3):274–279
152. Sprott JC (2010) Elegant chaos: algebraically simple chaotic flows. World Scientific, Singapore
153. Sprott JC (1997) Some simple chaotic jerk functions. Am J Phys 65(6):537–543
154. Khalil HK (2001) Nonlinear Syst. Prentice Hall, New Jersey

A Novel 3-D Circulant Highly Chaotic System with Labyrinth Chaos

Sundarapandian Vaidyanathan

Abstract In this work, we describe a novel 3-D circulant highly chaotic system with labyrinth chaos. The novel chaotic system is a nine-term polynomial system with six sinusoidal nonlinearities. The phase portraits of the novel circulant chaotic system are illustrated and the dynamic properties of the novel circulant chaotic system are discussed. The novel circulant chaotic system has infinitely many equilibrium points and it exhibits labyrinth chaos. We show that all the equilibrium points of the novel circulant chaotic system are saddle-foci and hence they are unstable. The Lyapunov exponents of the novel circulant chaotic system are obtained as $L_1 = 10.3755$, $L_2 = 0$ and $L_3 = -10.4113$. Thus, the Maximal Lyapunov Exponent (MLE) of the novel chaotic system is obtained as $L_1 = 10.3755$, which is a large value. This shows that the novel 3-D circulant chaotic system is highly chaotic. Also, the Kaplan-Yorke dimension of the novel circulant highly chaotic system is obtained as $D_{KY} = 2.9966$. Since the Kaplan-Yorke dimension of the the novel circulant chaotic system has a large value and close to three, the novel circulant chaotic system with labyrinth chaos exhibits highly complex behaviour. Since the sum of the Lyapunov exponents is negative, the novel chaotic system is dissipative. Next, we derive new results for the global chaos control of the novel circulant highly chaotic system with unknown parameters using adaptive control method. We also derive new results for the global chaos synchronization of the identical novel circulant highly chaotic systems with unknown parameters using adaptive control method. The main control results are established using Lyapunov stability theory. MATLAB simulations are depicted to illustrate the phase portraits of the novel circulant highly chaotic system and also the adaptive control results derived in this work.

Keywords Chaos · Chaotic systems · Circulant chaotic system · Chaos control · Adaptive control · Synchronization

---

S. Vaidyanathan (✉)
Research and Development Centre, Vel Tech University, Avadi, Chennai 600062, Tamil Nadu, India
e-mail: sundarvtu@gmail.com

A.T. Azar and S. Vaidyanathan (eds.), *Advances in Chaos Theory and Intelligent Control*, Studies in Fuzziness and Soft Computing 337,
DOI 10.1007/978-3-319-30340-6_16

1 Introduction

Chaotic systems are defined as nonlinear dynamical systems which are sensitive to initial conditions, topologically mixing and with dense periodic orbits. Sensitivity to initial conditions of chaotic systems is popularly known as the *butterfly effect*.

Chaotic systems are either conservative or dissipative. The conservative chaotic systems are characterized by the property that they are *volume conserving*. The dissipative chaotic systems are characterized by the property that any asymptotic motion of the chaotic system settles onto a set of measure zero, i.e. a strange attractor. In this research work, we shall announce and discuss a novel 3-D dissipative highly chaotic circulant chaotic system with six sinusoidal nonlinearities.

The Lyapunov exponent of a chaotic system is a measure of the divergence of points which are initially very close and this can be used to quantify chaotic systems. Each nonlinear dynamical system has a spectrum of Lyapunov exponents, which are equal in number to the dimension of the state space. The largest Lyapunov exponent of a nonlinear dynamical system is called the *maximal Lyapunov exponent* (MLE).

In the last few decades, Chaos theory has become a very important and active research field, employing many applications in different disciplines like physics, chemistry, biology, ecology, engineering and economics, among others.

Some classical paradigms of 3-D chaotic systems in the literature are Lorenz system [1], Rössler system [2], ACT system [3], Sprott systems [4], Chen system [5], Lü system [6], Cai system [7], Tigan system [8], etc.

Many new chaotic systems have been discovered in the recent years such as Zhou system [9], Zhu system [10], Li system [11], Sundarapandian systems [12, 13], Vaidyanathan systems [14–30], Pehlivan system [31], Sampath system [32], Pham system [33], etc.

Chaos theory and control systems have many important applications in science and engineering [34–39]. Some commonly known applications are oscillators [40, 41], lasers [42, 43], chemical reactions [44–51], biology [52–55, 55–61], ecology [62, 63], encryption [64, 65], cryptosystems [66, 67], mechanical systems [68–72], secure communications [73–75], robotics [76–78], cardiology [79, 80], intelligent control [81, 82], neural networks [83–85], finance [86, 87], memristors [88, 89],etc.

The control of a chaotic system aims to stabilize or regulate the system with the help of a feedback control. There are many methods available for controlling a chaotic system such as active control [90–92], adaptive control [93–101], sliding mode control [102, 103], backstepping control [104–106], etc.

Major works on synchronization of chaotic systems deal with the complete synchronization (CS) which has the design goal of using the output of the master system to control the slave system so that the output of the slave system tracks the output of the master system asymptotically with time.

There are many methods available for chaos synchronization such as active control [107–112], adaptive control [113–119], sliding mode control [120–123], backstepping control [124–127], etc.

In this research work, we announce a novel circulant highly chaotic system with Labyrinth chaos. The novel chaotic system is a nine-term polynomial system with six sinusoidal nonlinearities. Using adaptive control method, we have also derived new results for the global chaos control of the novel circulant chaotic system and global chaos synchronization of the identical novel highly chaotic systems when the system parameters are unknown.

This work is organized as follows. Section 2 describes the dynamic equations and phase portraits of the novel circulant highly chaotic system. Section 3 details the qualitative properties of the novel circulant highly chaotic system.

The novel circulant chaotic system has infinitely many equilibrium points and it exhibits labyrinth chaos. The Lyapunov exponents of the novel chaotic system are obtained as $L_1 = 10.3755$, $L_2 = 0$ and $L_3 = -10.4113$. Thus, the Maximal Lyapunov Exponent (MLE) of the novel chaotic system is obtained as $L_1 = 10.3755$, which is large value. This shows that the novel 3-D circulant chaotic system is highly chaotic. Also, the Kaplan-Yorke dimension of the novel circulant highly chaotic system is obtained as $D_{KY} = 2.9966$. Since the Kaplan-Yorke dimension of the the novel circulant chaotic system has a large value and close to three, the novel circulant chaotic system with labyrinth chaos exhibits highly complex behaviour. Since the sum of the Lyapunov exponents is negative, the novel chaotic system is also dissipative.

In Sect. 4, we derive new results for the global chaos control of the novel circulant highly chaotic system with unknown parameters. In Sect. 5, we derive new results for the global chaos synchronization of the identical novel circulant highly chaotic systems with unknown parameters. Section 6 contains the conclusions of this work.

2 A Novel 3-D Circulant Highly Chaotic System with Labyrinth Chaos

A circulant chaotic system is an elegant chaotic system in which the variables are cyclically symmetric [128].

Thus, a 3-D circulant chaotic system has the form

$$\begin{cases} \dot{x}_1 = \varphi(x_1, x_2, x_3) \\ \dot{x}_2 = \varphi(x_2, x_3, x_1) \\ \dot{x}_3 = \varphi(x_3, x_1, x_2) \end{cases} \tag{1}$$

where all the functions are the same except the state variables which are rotated.

A famous circulant chaotic system is the Thomas system [129], which can be expressed as

$$\begin{cases} \dot{x}_1 = \sin x_2 - bx_1 \\ \dot{x}_2 = \sin x_3 - bx_2 \\ \dot{x}_3 = \sin x_1 - bx_3 \end{cases} \tag{2}$$

In the system (2), b is a constant that corresponds to how *dissipative* the system is, and acts as a bifurcation parameter. The Thomas circulant system (2) is found to be *chaotic* when

$$b = 0.2082 \tag{3}$$

For numerical simulations, we take the initial state of the Thomas system (2) as

$$x_1(0) = 0.4, \quad x_2(0) = 0, \quad x_3(0) = 0 \tag{4}$$

The Lyapunov exponents of the Thomas circulant system (2) for the initial state (4) and the parameter value (3) are numerically found as

$$L_1 = 0.0179, \quad L_2 = 0, \quad L_3 = -0.6376 \tag{5}$$

Thus, the Kaplan-Yorke dimension of the Thomas circulant system (2) is derived as

$$D_{KY} = 2 + \frac{L_1 + L_2}{|L_3|} = 2.0281 \tag{6}$$

Figure 1 depicts the 3-D phase portrait of the Thomas circulant chaotic system (2) with labyrinth chaos.

It is also easy to see that the Thomas circulant system (2) has infinitely many equilibrium points given by

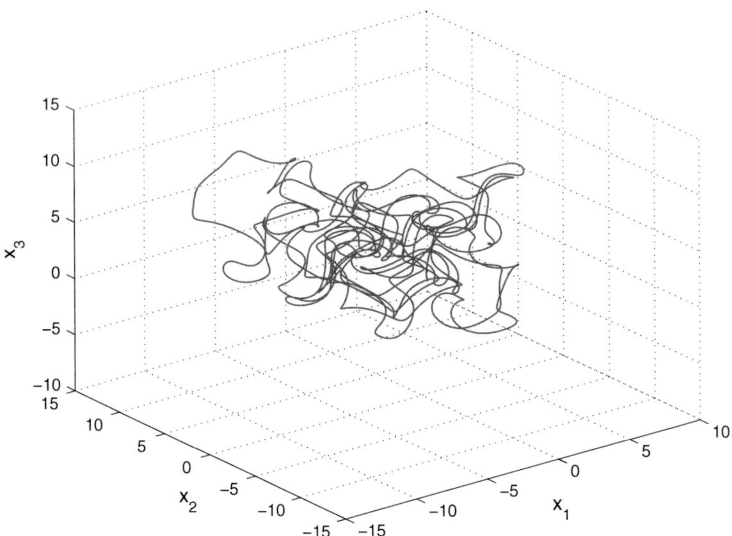

Fig. 1 3-D phase portrait of the Thomas circulant chaotic system with labyrinth chaos

$$E_\theta = \begin{bmatrix} \theta \\ \theta \\ \theta \end{bmatrix}, \tag{7}$$

where θ is a root of the transcendental equation

$$\sin\theta = b\theta \quad (b = 0.2082) \tag{8}$$

Since the Eq. (8) has infinitely many roots θ, it follows that the Thomas circulant system (2) has infinitely many equilibrium points E_θ given by (7).

In this research work, we announce a novel circulant chaotic system with labyrinth chaos, which is described by

$$\begin{cases} \dot{x}_1 = a(\sin x_2 + \cos x_2) - bx_1 \\ \dot{x}_2 = a(\sin x_3 + \cos x_3) - bx_2 \\ \dot{x}_3 = a(\sin x_1 + \cos x_1) - bx_3 \end{cases} \tag{9}$$

where a and b are constant, positive, parameters.

The novel circulant system (9) is *chaotic* when we take the parameter values as

$$a = 75, \quad b = 0.01 \tag{10}$$

For numerical simulations, we take the initial state of the circulant system (9) as

$$x_1(0) = 2.5, \quad x_2(0) = 2.7, \quad x_3(0) = 2.5 \tag{11}$$

The Lyapunov exponents of the novel circulant system (9) for the parameter values (10) and the initial state (11) are numerically found as

$$L_1 = 10.3755, \quad L_2 = 0, \quad L_3 = -10.4113 \tag{12}$$

Since the Maximal Lyapunov Exponent (MLE) of the novel circulant system (9) is $L_1 = 10.3755$, which is a large value, it follows that the novel circulant system (9) is highly chaotic.

Also, the Kaplan-Yorke dimension of the novel circulant system (9) is derived as

$$D_{KY} = 2 + \frac{L_1 + L_2}{|L_3|} = 2.9966, \tag{13}$$

which is close to three.

The Maximal Lyapunov Exponent (MLE) of the novel circulant chaotic system (9) is $L_1 = 10.3755$, which is much higher than the Maximal Lyapunov Exponent (MLE) of the Thomas circulant chaotic system (2), viz. $L_1 = 0.0179$.

Also, the Kaplan-Yorke dimension of the novel circulant chaotic system (9) is $D_{KY} = 2.9966$, which is much higher than the Kaplan-Yorke dimension of the Thomas circulant chaotic system (2), viz. $D_{KY} = 2.0281$.

This shows that the novel circulant chaotic system (9) exhibits more chaotic and complex behaviour than the Thomas circulant system (2). Also, the large value of D_{KY}, which is close to three, indicates that the novel circulant chaotic system (9) exhibits high complexity.

It is also easy to see that the novel circulant circulant system (9) has infinitely many equilibrium points given by

$$E_\theta = \begin{bmatrix} \theta \\ \theta \\ \theta \end{bmatrix}, \tag{14}$$

where θ is a root of the transcendental equation

$$\sin\theta + \cos\theta = \frac{b}{a}\,\theta, \quad (a = 75, \quad b = 0.01) \tag{15}$$

If we define

$$\mu = \frac{b}{a}, \tag{16}$$

then we can express (15) equivalently as

$$\sin\theta + \cos\theta = \mu\,\theta, \quad (\mu = 1.3333 \times 10^{-4}) \tag{17}$$

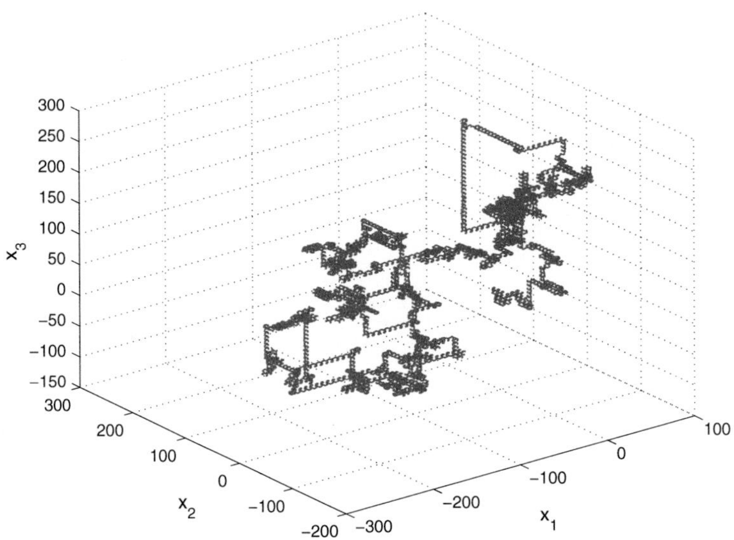

Fig. 2 3-D phase portrait of the novel circulant chaotic system with labyrinth chaos

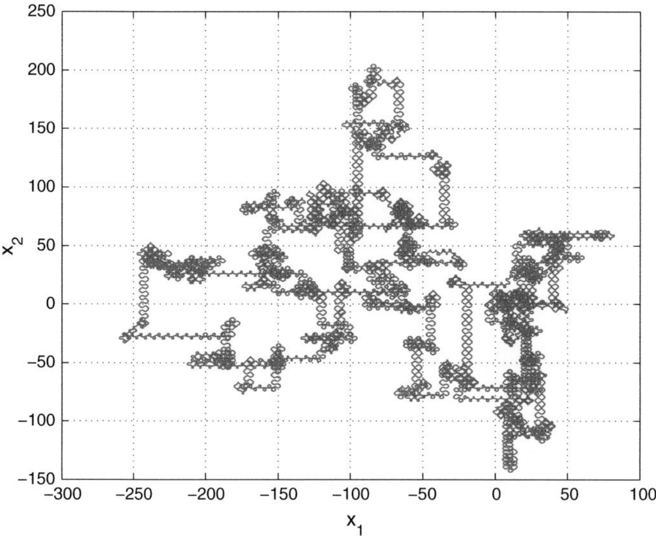

Fig. 3 2-D projection of the novel circulant chaotic system on the (x_1, x_2) plane

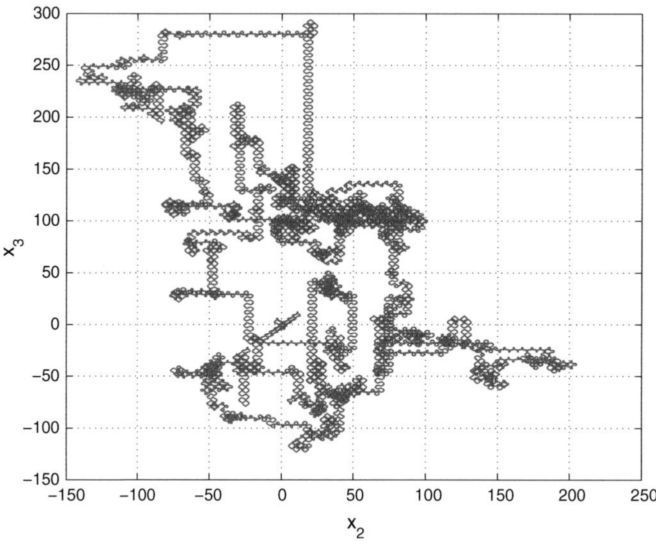

Fig. 4 2-D projection of the novel circulant chaotic system on the (x_2, x_3) plane

Since the Eq. (17) has infinitely many roots θ, it follows that the novel chaotic circulant system (9) has infinitely many equilibrium points E_θ given by (14).

In this work, we shall show that all the equilibrium points E_θ, $(\theta \in \mathbf{R})$ are saddle-focus points, which are unstable.

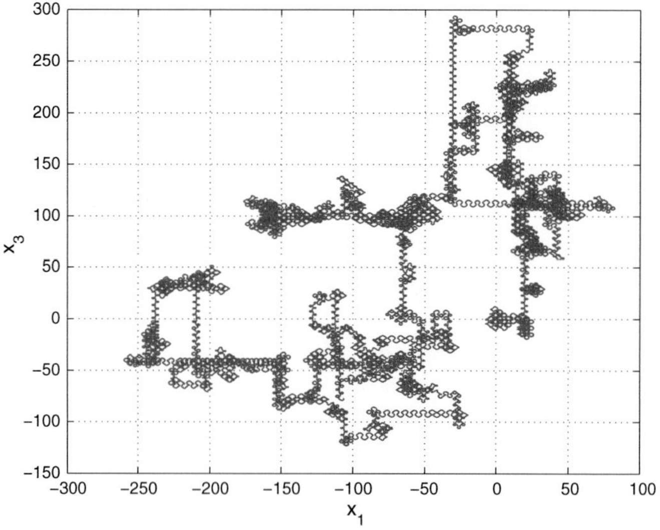

Fig. 5 2-D projection of the novel circulant chaotic system on the (x_1, x_3) plane

Figure 2 depicts the 3-D phase portrait of the novel circulant chaotic system (9) with labyrinth chaos.

Figures 3, 4 and 5 depict the 2-D projection of the novel circulant chaotic system (9) on the (x_1, x_2), (x_2, x_3) and (x_1, x_3) planes, respectively.

3 Analysis of the Novel 3-D Circulant Highly Chaotic System

In this section, we study the qualitative properties of the 3-D novel circulant chaotic system (9). We take the parameter values as in (10), viz. $a = 75$ and $b = 0.01$.

3.1 Dissipativity

In vector notation, the novel chaotic system (9) can be expressed as

$$\dot{\mathbf{x}} = f(\mathbf{x}) = \begin{bmatrix} f_1(x_1, x_2, x_3) \\ f_2(x_1, x_2, x_3) \\ f_3(x_1, x_2, x_3) \end{bmatrix}, \tag{18}$$

where

$$\begin{cases} f_1(x_1, x_2, x_3) = a(\sin x_2 + \cos x_2) - bx_1 \\ f_2(x_1, x_2, x_3) = a(\sin x_3 + \cos x_3) - bx_2 \\ f_3(x_1, x_2, x_3) = a(\sin x_1 + \cos x_1) - bx_3 \end{cases} \tag{19}$$

Let Ω be any region in $\mathbf{R}^3$ with a smooth boundary and also, $\Omega(t) = \Phi_t(\Omega)$, where Φ_t is the flow of f. Furthermore, let $V(t)$ denote the volume of $\Omega(t)$.

By Liouville's theorem, we know that

$$\dot{V}(t) = \int\limits_{\Omega(t)} (\nabla \cdot f)\, dx_1\, dx_2\, dx_3 \tag{20}$$

The divergence of the novel chaotic system (18) is found as

$$\nabla \cdot f = \frac{\partial f_1}{\partial x_1} + \frac{\partial f_2}{\partial x_2} + \frac{\partial f_3}{\partial x_3} = -3b = -\varepsilon < 0 \tag{21}$$

where $\varepsilon = 3b = 0.03 > 0$.

Inserting the value of $\nabla \cdot f$ from (21) into (20), we get

$$\dot{V}(t) = \int\limits_{\Omega(t)} (-\varepsilon)\, dx_1\, dx_2\, dx_3 = -\varepsilon V(t) \tag{22}$$

Integrating the first order linear differential equation (22), we get

$$V(t) = \exp(-\varepsilon t) V(0) \tag{23}$$

Since $\varepsilon > 0$, it follows from Eq. (23) that $V(t) \to 0$ exponentially as $t \to \infty$. This shows that the novel highly chaotic system (9) is dissipative.

Hence, the system limit sets are ultimately confined into a specific limit set of zero volume, and the asymptotic motion of the novel chaotic system (9) settles onto a strange attractor of the system.

3.2 Equilibrium Points

We take the parameter values as in the chaotic case (10), viz. $a = 75$ and $b = 0.01$.

It is easy to see that the novel circulant circulant system (9) has infinitely many equilibrium points given by

$$E_\theta = \begin{bmatrix} \theta \\ \theta \\ \theta \end{bmatrix}, \tag{24}$$

where θ is a root of the transcendental equation

$$\sin\theta + \cos\theta = \frac{b}{a}\,\theta, \quad (a = 75, \quad b = 0.01) \tag{25}$$

If we define

$$\mu = \frac{b}{a}, \tag{26}$$

then we can express (25) equivalently as

$$\sin\theta + \cos\theta = \mu\,\theta, \quad (\mu = 1.3333 \times 10^{-4}) \tag{27}$$

Since the Eq. (27) has infinitely many roots θ, the novel chaotic circulant system (9) has infinitely many equilibrium points E_θ given by (24).

Using MATLAB, some equilibrium points of the novel chaotic circulant system (9) can be listed as follows:

$$\ldots, \begin{bmatrix} -0.7855 \\ -0.7855 \\ -0.7855 \end{bmatrix}, \begin{bmatrix} 2.3560 \\ 2.3560 \\ 2.3560 \end{bmatrix}, \begin{bmatrix} 5.4983 \\ 5.4983 \\ 5.4983 \end{bmatrix}, \begin{bmatrix} 8.6386 \\ 8.6386 \\ 8.6386 \end{bmatrix}, \ldots \tag{28}$$

The Jacobian matrix of the novel circulant chaotic system (9) at any point $\mathbf{x} \in \mathbf{R}^3$ is obtained as

$$J(\mathbf{x}) = \begin{bmatrix} -b & a(\cos x_2 - \sin x_2) & 0 \\ 0 & -b & a(\cos x_3 - \sin x_3) \\ a(\cos x_1 - \sin x_1) & 0 & -b \end{bmatrix} \tag{29}$$

For all equilibrium points E_θ, $J(E_\theta)$ has the same eigenvalues, viz.

$$\lambda_1 = 106.06, \;\; \lambda_{2,3} = -53.04 \pm 91.86\,i \tag{30}$$

Thus, all the equilibrium points E_θ of the novel circulant chaotic system (9) are saddle-focus points, which are unstable.

3.3 Lyapunov Exponents and Kaplan-Yorke Dimension

We take the parameter values of the novel system (9) as in the chaotic case (10), i.e.

$$a = 75, \;\; b = 0.01 \tag{31}$$

We take the initial state of the novel system (9) as given in (11), i.e.

$$x_1(0) = 2.5, \quad x_2(0) = 2.8, \quad x_3(0) = 2.5 \tag{32}$$

Then the Lyapunov exponents of the system (9) are numerically obtained as

$$L_1 = 10.3755, \quad L_2 = 0, \quad L_3 = -10.4113 \tag{33}$$

Figure 6 shows the dynamics of the Lyapunov exponents of the novel circulant chaotic system (9). From Fig. 6, we see that the Maximal Lyapunov Exponent (MLE) of the novel circulant chaotic system (9) is $L_1 = 10.3755$, which is a large value. This shows that the novel circulant chaotic system (9) is *highly chaotic*.

Since the sum of the Lyapunov exponents of the novel circulant chaotic system (9) is negative, the system is dissipative.

Also, the Kaplan-Yorke dimension of the novel circulant chaotic system (9) is found as

$$D_{KY} = 2 + \frac{L_1 + L_2}{|L_3|} = 2.9966 \tag{34}$$

which is a very large value for a 3-D chaotic system and it is very close to three. This shows the high complexity of the novel circulant chaotic system (9). Hence, it is very suitable for many engineering applications such as cryptosystems, secure communications, etc.

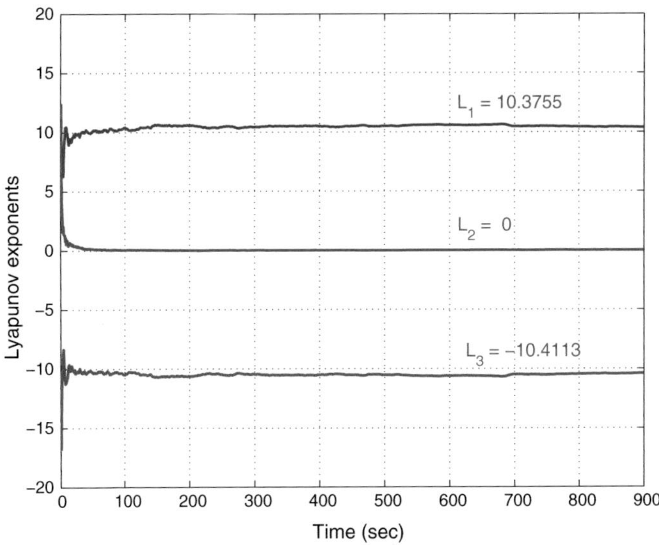

Fig. 6 Dynamics of the Lyapunov exponents of the novel circulant chaotic system

4 Adaptive Control of the Novel Circulant Highly Chaotic System

In this section, we use adaptive control method to derive an adaptive feedback control law for globally stabilizing the novel 3-D circulant highly chaotic system with unknown parameters.

Thus, we consider the novel circulant chaotic system given by

$$\begin{cases} \dot{x}_1 = a(\sin x_2 + \cos x_2) - bx_1 + u_1 \\ \dot{x}_2 = a(\sin x_3 + \cos x_3) - bx_2 + u_2 \\ \dot{x}_3 = a(\sin x_1 + \cos x_1) - bx_3 + u_3 \end{cases} \tag{35}$$

In (35), x_1, x_2, x_3 are the states and u_1, u_2, u_3 are the adaptive controls to be determined using estimates $\hat{a}(t)$, $\hat{b}(t)$ for the unknown parameters a, b, respectively.

To simplify the notation, we define

$$F(\alpha) = \sin\alpha + \cos\alpha \tag{36}$$

Using (36), we can represent (35) in a simple form as

$$\begin{cases} \dot{x}_1 = aF(x_2) - bx_1 + u_1 \\ \dot{x}_2 = aF(x_3) - bx_2 + u_2 \\ \dot{x}_3 = aF(x_1) - bx_3 + u_3 \end{cases} \tag{37}$$

We consider the adaptive feedback control law

$$\begin{cases} u_1 = -\hat{a}(t)F(x_2) + \hat{b}(t)x_1 - k_1x_1 \\ u_2 = -\hat{a}(t)F(x_3) + \hat{b}(t)x_2 - k_2x_2 \\ u_3 = -\hat{a}(t)F(x_1) + \hat{b}(t)x_3 - k_3x_3 \end{cases} \tag{38}$$

where k_1, k_2, k_3 are positive gain constants.

Substituting (38) into (37), we get the closed-loop plant dynamics as

$$\begin{cases} \dot{x}_1 = \left[a - \hat{a}(t)\right] F(x_2) - \left[b - \hat{b}(t)\right] x_1 - k_1x_1 \\ \dot{x}_2 = \left[a - \hat{a}(t)\right] F(x_3) - \left[b - \hat{b}(t)\right] x_2 - k_2x_2 \\ \dot{x}_3 = \left[a - \hat{a}(t)\right] F(x_1) - \left[b - \hat{b}(t)\right] x_3 - k_2x_3 \end{cases} \tag{39}$$

The parameter estimation errors are defined as

$$\begin{cases} e_a(t) = a - \hat{a}(t) \\ e_b(t) = b - \hat{b}(t) \end{cases} \tag{40}$$

In view of (40), we can simplify the plant dynamics (39) as

$$\begin{cases} \dot{x}_1 = e_a F(x_2) - e_b x_1 - k_1 x_1 \\ \dot{x}_2 = e_a F(x_3) - e_b x_2 - k_2 x_2 \\ \dot{x}_3 = e_a F(x_1) - e_b x_3 - k_2 x_3 \end{cases} \tag{41}$$

Differentiating (40) with respect to t, we obtain

$$\begin{cases} \dot{e}_a(t) = -\dot{\hat{a}}(t) \\ \dot{e}_b(t) = -\dot{\hat{b}}(t) \end{cases} \tag{42}$$

We consider the quadratic candidate Lyapunov function defined by

$$V(\mathbf{x}, e_a, e_b) = \frac{1}{2}\left(x_1^2 + x_2^2 + x_3^2\right) + \frac{1}{2}\left(e_a^2 + e_b^2\right) \tag{43}$$

Differentiating V along the trajectories of (41) and (42), we obtain

$$\begin{aligned} \dot{V} = & -k_1 x_1^2 - k_2 x_2^2 - k_3 x_3^2 + e_a\left[x_1 F(x_2) + x_2 F(x_3) + x_3 F(x_1) - \dot{\hat{a}}\right] \\ & + e_b\left[-x_1^2 - x_2^2 - x_3^2 - \dot{\hat{b}}\right] \end{aligned} \tag{44}$$

In view of (44), we take the parameter update law as

$$\begin{cases} \dot{\hat{a}}(t) = x_1 F(x_2) + x_2 F(x_3) + x_3 F(x_1) \\ \dot{\hat{b}}(t) = -x_1^2 - x_2^2 - x_3^2 \end{cases} \tag{45}$$

Next, we state and prove the main result of this section.

Theorem 1 *The novel 3-D circulant highly chaotic system (37) with unknown system parameters is globally and exponentially stabilized for all initial conditions by the adaptive control law (38) and the parameter update law (45), where k_1, k_2, k_3 are positive gain constants and $F(\alpha)$ is defined by (36).*

Proof We prove this result by applying Lyapunov stability theory [130].

We consider the quadratic Lyapunov function defined by (43), which is clearly a positive definite function on $\mathbf{R}^5$.

By substituting the parameter update law (45) into (44), we obtain the time-derivative of V as

$$\dot{V} = -k_1 x_1^2 - k_2 x_2^2 - k_3 x_3^2 \tag{46}$$

From (46), it is clear that $\dot{V}$ is a negative semi-definite function on $\mathbf{R}^5$.

Thus, we can conclude that the state vector $\mathbf{x}(t)$ and the parameter estimation error are globally bounded i.e.

$$\left[x_1(t) \; x_2(t) \; x_3(t) \; e_a(t) \; e_b(t) \right]^T \in \mathbf{L}_\infty.$$

We define $k = \min\{k_1, k_2, k_3\}$.

Then it follows from (46) that

$$\dot{V} \leq -k\|\mathbf{x}(t)\|^2 \tag{47}$$

Thus, we have

$$k\|\mathbf{x}(t)\|^2 \leq -\dot{V} \tag{48}$$

Integrating the inequality (48) from 0 to t, we get

$$k \int_0^t \|\mathbf{x}(\tau)\|^2 \, d\tau \;\leq\; V(0) - V(t) \tag{49}$$

From (49), it follows that $\mathbf{x} \in \mathbf{L}_2$.

Using (41), we can conclude that $\dot{\mathbf{x}} \in \mathbf{L}_\infty$.

Using Barbalat's lemma [130], we conclude that $\mathbf{x}(t) \to 0$ exponentially as $t \to \infty$ for all initial conditions $\mathbf{x}(0) \in \mathbf{R}^3$.

This completes the proof. ∎

For the numerical simulations, the classical fourth-order Runge-Kutta method with step size $h = 10^{-8}$ is used to solve the systems (37) and (45), when the adaptive control law (38) is applied.

The parameter values of the novel 3-D circulant chaotic system (37) are taken as in the chaotic case (10), i.e. $a = 75$ and $b = 0.01$.

We take the positive gain constants as $k_i = 5$ for $i = 1, 2, 3$.

Furthermore, as initial conditions of the novel highly chaotic system (37), we take

$$x_1(0) = 25.3, \quad x_2(0) = -14.7, \quad x_3(0) = 16.2 \tag{50}$$

Also, as initial conditions of the parameter estimates, we take

$$\hat{a}(0) = 12.3, \quad \hat{b}(0) = 10.8 \tag{51}$$

In Fig. 7, the exponential convergence of the controlled states of the 3-D novel highly chaotic system (37) is depicted. Figure 7 shows that the controlled states $x_1(t), x_2(t), x_3(t)$ converge to zero in 1 s.

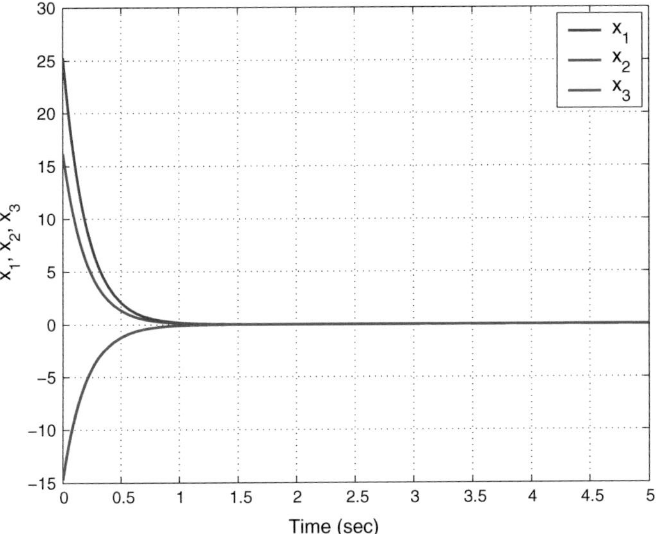

Fig. 7 Time-history of the controlled states x_1, x_2, x_3

5 Adaptive Synchronization of the Identical Novel Circulant Highly Chaotic Systems

In this section, we apply adaptive control method to derive an adaptive feedback control law for globally synchronizing identical 3-D novel circulant highly chaotic systems with unknown parameters.

To simplify the notation, we define

$$F(\alpha) = \sin\alpha + \cos\alpha \tag{52}$$

As the master system, we consider the novel circulant highly chaotic system given by

$$\begin{cases} \dot{x}_1 = aF(x_2) - bx_1 \\ \dot{x}_2 = aF(x_3) - bx_2 \\ \dot{x}_3 = aF(x_1) - bx_3 \end{cases} \tag{53}$$

In (53), x_1, x_2, x_3 are the states and a, b are unknown system parameters.

As the slave system, we consider the novel circulant highly chaotic system given by

$$\begin{cases} \dot{y}_1 = aF(y_2) - by_1 + u_1 \\ \dot{y}_2 = aF(y_3) - by_2 + u_2 \\ \dot{y}_3 = aF(y_1) - by_3 + u_3 \end{cases} \tag{54}$$

In (54), y_1, y_2, y_3 are the states and u_1, u_2, u_3 are the adaptive controls to be determined using estimates of the unknown system parameters.

The synchronization error between the novel chaotic systems is defined by

$$\begin{cases} e_1 = y_1 - x_1 \\ e_2 = y_2 - x_2 \\ e_3 = y_3 - x_3 \end{cases} \tag{55}$$

Then the error dynamics is obtained as

$$\begin{cases} \dot{e}_1 = a[F(y_2) - F(x_2)] - be_1 + u_1 \\ \dot{e}_2 = a[F(y_3) - F(x_3)] - be_2 + u_2 \\ \dot{e}_3 = a[F(y_1) - F(x_1)] - be_3 + u_3 \end{cases} \tag{56}$$

To simplify the notation, we define

$$G(\alpha, \beta) = F(\beta) - F(\alpha) \tag{57}$$

Then the error dynamics (56) can be simplified as

$$\begin{cases} \dot{e}_1 = a\, G(x_2, y_2) - be_1 + u_1 \\ \dot{e}_2 = a\, G(x_3, y_3) - be_2 + u_2 \\ \dot{e}_3 = a\, G(x_1, y_1) - be_3 + u_3 \end{cases} \tag{58}$$

We consider the adaptive feedback control law

$$\begin{cases} u_1 = -\hat{a}(t)\, G(x_2, y_2) + \hat{b}(t)e_1 - k_1e_1 \\ u_2 = -\hat{a}(t)\, G(x_3, y_3) + \hat{b}(t)e_2 - k_2e_2 \\ u_3 = -\hat{a}(t)\, G(x_1, y_1) + \hat{b}(t)e_3 - k_3e_3 \end{cases} \tag{59}$$

where k_1, k_2, k_3 are positive gain constants.

Substituting (59) into (58), we get the closed-loop error dynamics as

$$\begin{cases} \dot{e}_1 = \left[a - \hat{a}(t)\right] G(x_2, y_2) - \left[b - \hat{b}(t)\right] e_1 - k_1e_1 \\ \dot{e}_2 = \left[a - \hat{a}(t)\right] G(x_3, y_3) - \left[b - \hat{b}(t)\right] e_2 - k_2e_2 \\ \dot{e}_3 = \left[a - \hat{a}(t)\right] G(x_1, y_1) - \left[b - \hat{b}(t)\right] e_3 - k_3e_3 \end{cases} \tag{60}$$

The parameter estimation errors are defined as

$$\begin{cases} e_a(t) = a - \hat{a}(t) \\ e_b(t) = b - \hat{b}(t) \end{cases} \tag{61}$$

In view of (61), we can simplify the error dynamics (60) as

$$\begin{cases} \dot{e}_1 = e_a G(x_2, y_2) - e_b e_1 - k_1 e_1 \\ \dot{e}_2 = e_a G(x_3, y_3) - e_b e_2 - k_2 e_2 \\ \dot{e}_3 = e_a G(x_1, y_1) - e_b e_3 - k_3 e_3 \end{cases} \tag{62}$$

Differentiating (61) with respect to t, we obtain

$$\begin{cases} \dot{e}_a(t) = -\dot{\hat{a}}(t) \\ \dot{e}_b(t) = -\dot{\hat{b}}(t) \end{cases} \tag{63}$$

We consider the quadratic candidate Lyapunov function defined by

$$V(\mathbf{e}, e_a, e_b, e_c) = \frac{1}{2}\left(e_1^2 + e_2^2 + e_3^2\right) + \frac{1}{2}\left(e_a^2 + e_b^2\right) \tag{64}$$

Differentiating V along the trajectories of (62) and (63), we obtain

$$\begin{aligned} \dot{V} = & -k_1 e_1^2 - k_2 e_2^2 - k_3 e_3^2 \\ & + e_a \left[e_1 G(x_2, y_2) + e_2 G(x_3, y_3) + e_3 G(x_1, y_1) - \dot{\hat{a}} \right] \\ & + e_b \left[-e_1^2 - e_2^2 - e_3^2 - \dot{\hat{b}} \right] \end{aligned} \tag{65}$$

In view of (65), we take the parameter update law as

$$\begin{cases} \dot{\hat{a}}(t) = e_1 G(x_2, y_2) + e_2 G(x_3, y_3) + e_3 G(x_1, y_1) \\ \dot{\hat{b}}(t) = -e_1^2 - e_2^2 - e_3^2 \end{cases} \tag{66}$$

Next, we state and prove the main result of this section.

This result is proved by applying adaptive control theory and Lyapunov stability theory.

Theorem 2 *The novel circulant highly chaotic systems (53) and (54) with unknown system parameters are globally and exponentially synchronized for all initial conditions by the adaptive control law (59) and the parameter update law (66), where* k_1, k_2, k_3 *are positive gain constants and* F, G *are defined by the Eqs. (52) and (57), respectively.*

Proof We prove this result by applying Lyapunov stability theory [130].

We consider the quadratic Lyapunov function defined by (64), which is clearly a positive definite function on $\mathbf{R}^5$.

By substituting the parameter update law (66) into (65), we obtain

$$\dot{V} = -k_1 e_1^2 - k_2 e_2^2 - k_3 e_3^2 \tag{67}$$

From (67), it is clear that $\dot{V}$ is a negative semi-definite function on $\mathbf{R}^5$.

Thus, we can conclude that the error vector $\mathbf{e}(t)$ and the parameter estimation error are globally bounded, i.e.

$$\left[e_1(t) \; e_2(t) \; e_3(t) \; e_a(t) \; e_b(t) \right]^T \in \mathbf{L}_\infty. \tag{68}$$

We define $k = \min\{k_1, k_2, k_3\}$.

Then it follows from (67) that

$$\dot{V} \leq -k\|\mathbf{e}(t)\|^2 \tag{69}$$

Thus, we have

$$k\|\mathbf{e}(t)\|^2 \leq -\dot{V} \tag{70}$$

Integrating the inequality (70) from 0 to t, we get

$$k \int_0^t \|\mathbf{e}(\tau)\|^2 \, d\tau \; \leq \; V(0) - V(t) \tag{71}$$

From (71), it follows that $\mathbf{e} \in \mathbf{L}_2$.

Using (62), we can conclude that $\dot{\mathbf{e}} \in \mathbf{L}_\infty$.

Using Barbalat's lemma [130], we conclude that $\mathbf{e}(t) \to 0$ exponentially as $t \to \infty$ for all initial conditions $\mathbf{e}(0) \in \mathbf{R}^3$.

This completes the proof. ■

For the numerical simulations, the classical fourth-order Runge-Kutta method with step size $h = 10^{-8}$ is used to solve the systems (53), (54) and (66), when the adaptive control law (59) is applied.

The parameter values of the novel chaotic systems are taken as in the chaotic case (10), i.e. $a = 75$ and $b = 0.01$. We take the positive gain constants as $k_i = 5$ for $i = 1, 2, 3$. As initial conditions of the parameter estimates, we take $\hat{a}(0) = 6.4$ and $\hat{b}(0) = 10.6$.

As initial conditions of the master system (53), we take $x_1(0) = 6.2$, $x_2(0) = 3.5$ and $x_3(0) = -7.9$.

As initial conditions of the slave system (54), we take $y_1(0) = 12.6$, $y_2(0) = 4.7$ and $y_3(0) = 2.8$.

Figures 8, 9 and 10 describe the complete synchronization of the novel circulant chaotic systems (53) and (54). From these figures, it is clear that the states of the highly chaotic circulant systems are synchronized in 1 s. Figure 11 describes the time-history of the synchronization errors e_1, e_2, e_3. Figure 11 shows that the chaos synchronization errors converge to zero in 1 s.

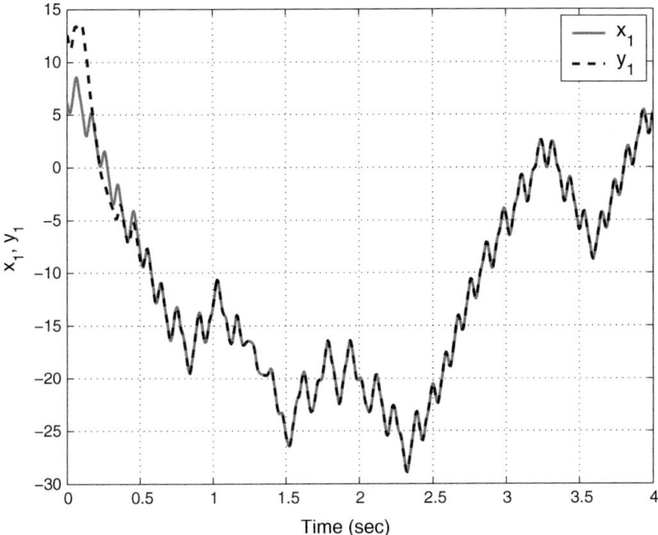

Fig. 8 Synchronization of the states x_1 and y_1

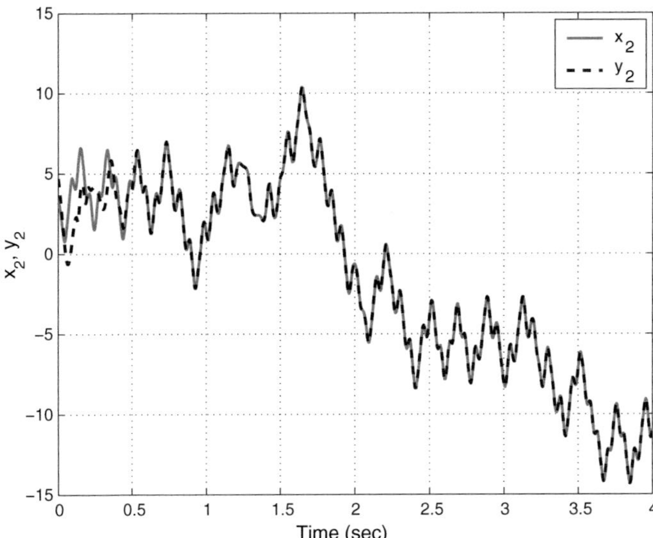

Fig. 9 Synchronization of the states x_2 and y_2

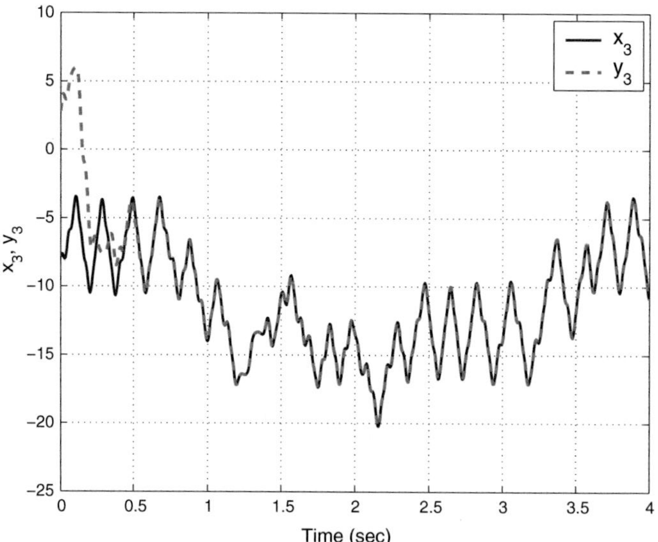

Fig. 10 Synchronization of the states x_3 and y_3

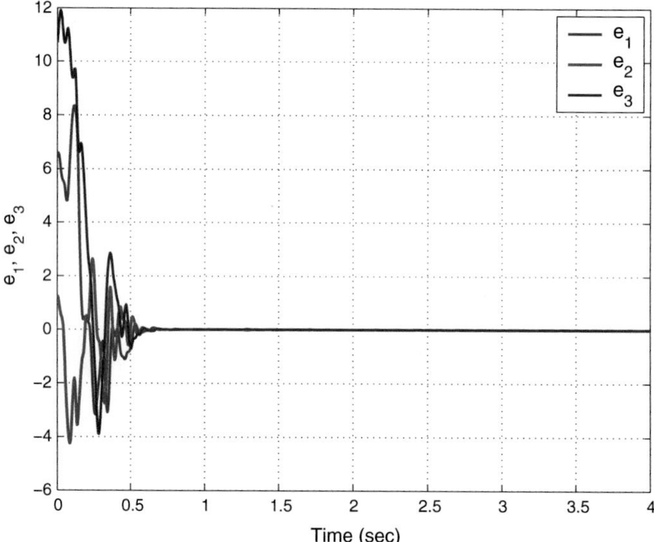

Fig. 11 Time-history of the synchronization errors e_1, e_2, e_3

6 Conclusions

In this work, we described a novel 3-D circulant highly chaotic system with labyrinth chaos. The novel chaotic system is a nine-term polynomial system with six sinusoidal nonlinearities. The phase portraits of the novel circulant chaotic system have been demonstrated and the qualitative properties of the novel circulant chaotic system were discussed in detail. We showed that novel circulant chaotic system has infinitely many equilibrium points and it exhibits labyrinth chaos. We also showed that all the equilibrium points of the novel circulant chaotic system are saddle-foci and hence they are unstable. The Lyapunov exponents of the novel circulant chaotic system have been obtained as $L_1 = 10.3755$, $L_2 = 0$ and $L_3 = -10.4113$. Thus, the Maximal Lyapunov Exponent (MLE) of the novel chaotic system is seen as $L_1 = 10.3755$, which is a large value. This shows that the novel 3-D circulant chaotic system is highly chaotic. Also, the Kaplan-Yorke dimension of the novel circulant highly chaotic system has been derived as $D_{KY} = 2.9966$. Since the Kaplan-Yorke dimension of the the novel circulant chaotic system has a large value and close to three, the novel circulant chaotic system with labyrinth chaos exhibits highly complex behaviour. Thus, the novel circulant chaotic system can be used for applications such as cryptosystems, secure communications, etc. Since the sum of the Lyapunov exponents is negative, the novel chaotic system is dissipative. Next, we derived new results for the global chaos control of the novel circulant highly chaotic system with unknown parameters using adaptive control method. We also derived new results for the global chaos synchronization of the identical novel circulant highly chaotic systems with unknown parameters using adaptive control method. The main control results have been established using Lyapunov stability theory. We showed MATLAB simulations to illustrate all the main results derived in this work.

References

1. Lorenz EN (1963) Deterministic periodic flow. J Atmos Sci 20(2):130–141
2. Rössler OE (1976) An equation for continuous chaos. Phys Lett A 57(5):397–398
3. Arneodo A, Coullet P, Tresser C (1981) Possible new strange attractors with spiral structure. Commun Math Phys 79(4):573–576
4. Sprott JC (1994) Some simple chaotic flows. Phys Rev E 50(2):647–650
5. Chen G, Ueta T (1999) Yet another chaotic attractor. Int J Bifurc Chaos 9(7):1465–1466
6. Lü J, Chen G (2002) A new chaotic attractor coined. Int J Bifurc Chaos 12(3):659–661
7. Cai G, Tan Z (2007) Chaos synchronization of a new chaotic system via nonlinear control. J Uncertain Syst 1(3):235–240
8. Tigan G, Opris D (2008) Analysis of a 3D chaotic system. Chaos Solitons Fractals 36: 1315–1319
9. Zhou W, Xu Y, Lu H, Pan L (2008) On dynamics analysis of a new chaotic attractor. Phys Lett A 372(36):5773–5777
10. Zhu C, Liu Y, Guo Y (2010) Theoretic and numerical study of a new chaotic system. Intell Inf Manag 2:104–109
11. Li D (2008) A three-scroll chaotic attractor. Phys Lett A 372(4):387–393

12. Sundarapandian V (2013) Analysis and anti-synchronization of a novel chaotic system via active and adaptive controllers. J Eng Sci Technol Rev 6(4):45–52
13. Sundarapandian V, Pehlivan I (2012) Analysis, control, synchronization, and circuit design of a novel chaotic system. Math Comput Model 55(7–8):1904–1915
14. Vaidyanathan S (2013) A new six-term 3-D chaotic system with an exponential nonlinearity. Far East J Math Sci 79(1):135–143
15. Vaidyanathan S (2013) Analysis and adaptive synchronization of two novel chaotic systems with hyperbolic sinusoidal and cosinusoidal nonlinearity and unknown parameters. J Eng Sci Technol Rev 6(4):53–65
16. Vaidyanathan S (2014) A new eight-term 3-D polynomial chaotic system with three quadratic nonlinearities. Far East J Math Sci 84(2):219–226
17. Vaidyanathan S (2014) Analysis and adaptive synchronization of eight-term 3-D polynomial chaotic systems with three quadratic nonlinearities. Eur Phys J: Special Topics 223(8): 1519–1529
18. Vaidyanathan S (2014) Analysis, control and synchronisation of a six-term novel chaotic system with three quadratic nonlinearities. Int J Modell Identif Control 22(1):41–53
19. Vaidyanathan S (2014) Generalised projective synchronisation of novel 3-D chaotic systems with an exponential non-linearity via active and adaptive control. Int J Model Identif Control 22(3):207–217
20. Vaidyanathan S (2015) A 3-D novel highly chaotic system with four quadratic nonlinearities, its adaptive control and anti-synchronization with unknown parameters. J Eng Sci Technol Rev 8(2):106–115
21. Vaidyanathan S (2015) Analysis, properties and control of an eight-term 3-D chaotic system with an exponential nonlinearity. Int J Model Identif Control 23(2):164–172
22. Vaidyanathan S, Azar AT (2015) Analysis, control and synchronization of a nine-term 3-D novel chaotic system. In: Azar AT, Vaidyanathan S (eds) Chaos modelling and control systems design. Studies in computational intelligence, vol 581. Springer, Germany, pp 19–38
23. Vaidyanathan S, Madhavan K (2013) Analysis, adaptive control and synchronization of a seven-term novel 3-D chaotic system. Int J Control Theory Appl 6(2):121–137
24. Vaidyanathan S, Pakiriswamy S (2015) A 3-D novel conservative chaotic system and its generalized projective synchronization via adaptive control. J Eng Sci Technol Rev 8(2): 52–60
25. Vaidyanathan S, Volos C (2015) Analysis and adaptive control of a novel 3-D conservative no-equilibrium chaotic system. Arch Control Sci 25(3):333–353
26. Vaidyanathan S, Volos C, Pham VT, Madhavan K, Idowu BA (2014) Adaptive backstepping control, synchronization and circuit simulation of a 3-D novel jerk chaotic system with two hyperbolic sinusoidal nonlinearities. Arch Control Sci 24(3):375–403
27. Vaidyanathan S, Rajagopal K, Volos CK, Kyprianidis IM, Stouboulos IN (2015) Analysis, adaptive control and synchronization of a seven-term novel 3-D chaotic system with three quadratic nonlinearities and its digital implementation in LabVIEW. J Eng Sci Technol Rev 8(2):130–141
28. Vaidyanathan S, Volos CK, Kyprianidis IM, Stouboulos IN, Pham VT (2015) Analysis, adaptive control and anti-synchronization of a six-term novel jerk chaotic system with two exponential nonlinearities and its circuit simulation. J Eng Sci Technol Rev 8(2):24–36
29. Vaidyanathan S, Volos CK, Pham VT (2015) Analysis, adaptive control and adaptive synchronization of a nine-term novel 3-D chaotic system with four quadratic nonlinearities and its circuit simulation. J Eng Sci Technol Rev 8(2):181–191
30. Vaidyanathan S, Volos CK, Pham VT (2015) Global chaos control of a novel nine-term chaotic system via sliding mode control. In: Azar AT, Zhu Q (eds) Advances and applications in sliding mode control systems. Studies in computational intelligence, vol 576. Springer, Germany, pp 571–590
31. Pehlivan I, Moroz IM, Vaidyanathan S (2014) Analysis, synchronization and circuit design of a novel butterfly attractor. J Sound Vib 333(20):5077–5096

32. Sampath S, Vaidyanathan S, Volos CK, Pham VT (2015) An eight-term novel four-scroll chaotic system with cubic nonlinearity and its circuit simulation. J Eng Sci Technol Rev 8(2):1–6
33. Pham VT, Vaidyanathan S, Volos CK, Jafari S (2015) Hidden attractors in a chaotic system with an exponential nonlinear term. Eur Phys J Special Topics 224(8):1507–1517
34. Azar AT (2010) Fuzzy systems. IN-TECH, Vienna, Austria
35. Azar AT, Vaidyanathan S (2015) Chaos modeling and control systems design. Studies in computational intelligence, vol 581. Springer, Germany
36. Azar AT, Vaidyanathan S (2015) Computational intelligence applications in modeling and control. Studies in computational intelligence, vol 575. Springer, Germany
37. Azar AT, Vaidyanathan S (2015) Handbook of research on advanced intelligent control engineering and automation. Advances in computational intelligence and robotics (ACIR). IGI-Global, USA
38. Azar AT, Zhu Q (2015) Advances and applications in sliding mode control systems. Studies in computational intelligence, vol 576. Springer, Germany
39. Zhu Q, Azar AT (2015) Complex system modelling and control through intelligent soft computations. Studies in fuzzines and soft computing, vol 319. Springer, Germany
40. Kengne J, Chedjou JC, Kenne G, Kyamakya K (2012) Dynamical properties and chaos synchronization of improved Colpitts oscillators. Commun Nonlinear Sci Numer Simul 17(7):2914–2923
41. Sharma A, Patidar V, Purohit G, Sud KK (2012) Effects on the bifurcation and chaos in forced Duffing oscillator due to nonlinear damping. Commun Nonlinear Sci Numer Simul 17(6):2254–2269
42. Li N, Pan W, Yan L, Luo B, Zou X (2014) Enhanced chaos synchronization and communication in cascade-coupled semiconductor ring lasers. Commun Nonlinear Sci Numer Simul 19(6):1874–1883
43. Yuan G, Zhang X, Wang Z (2014) Generation and synchronization of feedback-induced chaos in semiconductor ring lasers by injection-locking. Optik Int J Light Electr Optics 125(8): 1950–1953
44. Gaspard P (1999) Microscopic chaos and chemical reactions. Phys A: Stat Mech Appl 263(1–4):315–328
45. Petrov V, Gaspar V, Masere J, Showalter K (1993) Controlling chaos in Belousov-Zhabotinsky reaction. Nature 361:240–243
46. Vaidyanathan S (2015) Adaptive control of a chemical chaotic reactor. Int J PharmTech Res 8(3):377–382
47. Vaidyanathan S (2015) Adaptive synchronization of chemical chaotic reactors. Int J ChemTech Res 8(2):612–621
48. Vaidyanathan S (2015) Anti-synchronization of Brusselator chemical reaction systems via adaptive control. Int J ChemTech Res 8(6):759–768
49. Vaidyanathan S (2015) Dynamics and control of Brusselator chemical reaction. Int J ChemTech Res 8(6):740–749
50. Vaidyanathan S (2015) Dynamics and control of Tokamak system with symmetric and magnetically confined plasma. Int J ChemTech Res 8(6):795–803
51. Vaidyanathan S (2015) Synchronization of Tokamak systems with symmetric and magnetically confined plasma via adaptive control. Int J ChemTech Res 8(6):818–827
52. Das S, Goswami D, Chatterjee S, Mukherjee S (2014) Stability and chaos analysis of a novel swarm dynamics with applications to multi-agent systems. Eng Appl Artif Intell 30:189–198
53. Kyriazis M (1991) Applications of chaos theory to the molecular biology of aging. Exp Gerontol 26(6):569–572
54. Vaidyanathan S (2015) 3-cells cellular neural network (CNN) attractor and its adaptive biological control. Int J PharmTech Res 8(4):632–640
55. Vaidyanathan S (2015) Adaptive backstepping control of enzymes-substrates system with ferroelectric behaviour in brain waves. Int J PharmTech Res 8(2):256–261

56. Vaidyanathan S (2015) Adaptive biological control of generalized Lotka-Volterra three-species biological system. Int J PharmTech Res 8(4):622–631
57. Vaidyanathan S (2015) Adaptive chaotic synchronization of enzymes-substrates system with ferroelectric behaviour in brain waves. Int J PharmTech Res 8(5):964–973
58. Vaidyanathan S (2015) Adaptive synchronization of generalized Lotka-Volterra three-species biological systems. Int J PharmTech Res 8(5):928–937
59. Vaidyanathan S (2015) Chaos in neurons and adaptive control of Birkhoff-Shaw strange chaotic attractor. Int J PharmTech Res 8(5):956–963
60. Vaidyanathan S (2015) Lotka-Volterra population biology models with negative feedback and their ecological monitoring. Int J PharmTech Res 8(5):974–981
61. Vaidyanathan S (2015) Synchronization of 3-cells cellular neural network (CNN) attractors via adaptive control method. Int J PharmTech Res 8(5):946–955
62. Gibson WT, Wilson WG (2013) Individual-based chaos: extensions of the discrete logistic model. J Theor Biol 339:84–92
63. Suérez I (1999) Mastering chaos in ecology. Ecol Model 117(2–3):305–314
64. Lang J (2015) Color image encryption based on color blend and chaos permutation in the reality-preserving multiple-parameter fractional Fourier transform domain. Optics Commun 338:181–192
65. Zhang X, Zhao Z, Wang J (2014) Chaotic image encryption based on circular substitution box and key stream buffer. Signal Process: Image Commun 29(8):902–913
66. Rhouma R, Belghith S (2011) Cryptoanalysis of a chaos based cryptosystem on DSP. Commun Nonlinear Sci Numer Simul 16(2):876–884
67. Usama M, Khan MK, Alghatbar K, Lee C (2010) Chaos-based secure satellite imagery cryptosystem. Comput Math Appl 60(2):326–337
68. Azar AT, Serrano FE (2014) Robust IMC-PID tuning for cascade control systems with gain and phase margin specifications. Neural Comput Appl 25(5):983–995
69. Azar AT, Serrano FE (2015) Adaptive sliding mode control of the Furuta pendulum. In: Azar AT, Zhu Q (eds) Advances and applications in sliding mode control systems. Studies in computational intelligence, vol 576. Springer, Germany, pp 1–42
70. Azar AT, Serrano FE (2015) Deadbeat control for multivariable systems with time varying delays. In: Azar AT, Vaidyanathan S (eds) Chaos modeling and control systems design. Studies in computational intelligence, vol 581. Springer, Germany, pp 97–132
71. Azar AT, Serrano FE (2015) Design and modeling of anti wind up PID controllers. In: Zhu Q, Azar AT (eds) Complex system modelling and control through intelligent soft computations. Studies in fuzziness and soft computing, vol 319. Springer, Germany, pp 1–44
72. Azar AT, Serrano FE (2015) Stabilizatoin and control of mechanical systems with backlash. In: Azar AT, Vaidyanathan S (eds) Handbook of research on advanced intelligent control engineering and automation. Advances in computational intelligence and robotics (ACIR). IGI-Global, USA, pp 1–60
73. Feki M (2003) An adaptive chaos synchronization scheme applied to secure communication. Chaos Solitons Fractals 18(1):141–148
74. Murali K, Lakshmanan M (1998) Secure communication using a compound signal from generalized chaotic systems. Phys Lett A 241(6):303–310
75. Zaher AA, Abu-Rezq A (2011) On the design of chaos-based secure communication systems. Commun Nonlinear Syst Numer Simul 16(9):3721–3727
76. Mondal S, Mahanta C (2014) Adaptive second order terminal sliding mode controller for robotic manipulators. J Franklin Inst 351(4):2356–2377
77. Nehmzow U, Walker K (2005) Quantitative description of robot-environment interaction using chaos theory. Robot Auton Syst 53(3–4):177–193
78. Volos CK, Kyprianidis IM, Stouboulos IN (2013) Experimental investigation on coverage performance of a chaotic autonomous mobile robot. Robot Auton Syst 61(12):1314–1322
79. Qu Z (2011) Chaos in the genesis and maintenance of cardiac arrhythmias. Progr Biophys Mol Biol 105(3):247–257

80. Witte CL, Witte MH (1991) Chaos and predicting varix hemorrhage. Med Hypotheses 36(4):312–317
81. Azar AT (2012) Overview of type-2 fuzzy logic systems. Int J Fuzzy Syst Appl 2(4):1–28
82. Li Z, Chen G (2006) Integration of fuzzy logic and chaos theory. Studies in fuzziness and soft computing, vol 187. Springer, Germany
83. Huang X, Zhao Z, Wang Z, Li Y (2012) Chaos and hyperchaos in fractional-order cellular neural networks. Neurocomputing 94:13–21
84. Kaslik E, Sivasundaram S (2012) Nonlinear dynamics and chaos in fractional-order neural networks. Neural Netw 32:245–256
85. Lian S, Chen X (2011) Traceable content protection based on chaos and neural networks. Appl Soft Comput 11(7):4293–4301
86. Guégan D (2009) Chaos in economics and finance. Annu Rev Control 33(1):89–93
87. Sprott JC (2004) Competition with evolution in ecology and finance. Phys Lett A 325(5–6):329–333
88. Pham VT, Volos CK, Vaidyanathan S, Le TP, Vu VY (2015) A memristor-based hyperchaotic system with hidden attractors: dynamics, synchronization and circuital emulating. J Eng Sci Technol Rev 8(2):205–214
89. Volos CK, Kyprianidis IM, Stouboulos IN, Tlelo-Cuautle E, Vaidyanathan S (2015) Memristor: a new concept in synchronization of coupled neuromorphic circuits. J Eng Sci Technol Rev 8(2):157–173
90. Sundarapandian V (2010) Output regulation of the Lorenz attractor. Far East J Math Sci 42(2):289–299
91. Vaidyanathan S (2011) Output regulation of Arneodo-Coullet chaotic system. Commun Comput Inf Sci 133:98–107
92. Vaidyanathan S (2011) Output regulation of the unified chaotic system. Commun Comput Inf Sci 198:1–9
93. Sundarapandian V (2013) Adaptive control and synchronization design for the Lu-Xiao chaotic system. Lecture Notes in Electrical Engineering, vol 131, pp 319–327
94. Vaidyanathan S (2012) Adaptive controller and syncrhonizer design for the Qi-Chen chaotic system. Lecture notes of the institute for computer sciences, social-informatics and telecommunications engineering, vol 84, pp 73–82
95. Vaidyanathan S (2013) A ten-term novel 4-D hyperchaotic system with three quadratic nonlinearities and its control. Int J Control Theory Appl 6(2):97–109
96. Vaidyanathan S (2013) Analysis, control and synchronization of hyperchaotic Zhou system via adaptive control. Adv Intell Syst Comput 177:1–10
97. Vaidyanathan S (2014) Qualitative analysis and control of an eleven-term novel 4-D hyperchaotic system with two quadratic nonlinearities. Int J Control Theory Appl 7:35–47
98. Vaidyanathan S, Azar AT (2015) Analysis and control of a 4-D novel hyperchaotic system. Stud Comput Intell 581:3–17
99. Vaidyanathan S, Volos C, Pham VT (2014) Hyperchaos, adaptive control and synchronization of a novel 5-D hyperchaotic system with three positive Lyapunov exponents and its SPICE implementation. Arch Control Sci 24(4):409–446
100. Vaidyanathan S, Volos C, Pham VT, Madhavan K (2015) Analysis, adaptive control and synchronization of a novel 4-D hyperchaotic hyperjerk system and its SPICE implementation. Nonlinear Dynam 25(1):135–158
101. Vaidyanathan S, Volos CK, Pham VT (2015) Analysis, control, synchronization and SPICE implementation of a novel 4-D hyperchaotic Rikitake dynamo system without equilibrium. J Eng Sci Technol Rev 8(2):232–244
102. Vaidyanathan S (2012) Global chaos control of hyperchaotic Liu system via sliding control method. Int J Control Theory Appl 5(2):117–123
103. Vaidyanathan S (2012) Sliding mode control based global chaos control of Liu-Liu-Liu-Su chaotic system. Int J Control Theory Appl 5(1):15–20
104. Njah AN, Sunday OD (2009) Generalization on the chaos control of 4-D chaotic systems using recursive backstepping nonlinear controller. Chaos Solitons Fractals 41(5):2371–2376

105. Vaidyanathan S, Idowu BA, Azar AT (2015) Backstepping controller design for the global chaos synchronization of Sprott's jerk systems. Stud Comput Intell 581:39–58
106. Vincent UE, Njah AN, Laoye JA (2007) Controlling chaos and deterministic directed transport in inertia ratchets using backstepping control. Phys D 231(2):130–136
107. Karthikeyan R, Sundarapandian V (2014) Hybrid chaos synchronization of four-scroll systems via active control. J Electr Eng 65(2):97–103
108. Sarasu P, Sundarapandian V (2011) Active controller design for generalized projective synchronization of four-scroll chaotic systems. Int J Syst Signal Control Eng Appl 4(2):26–33
109. Sarasu P, Sundarapandian V (2011) The generalized projective synchronization of hyperchaotic Lorenz and hyperchaotic Qi systems via active control. Int J Soft Comput 6(5): 216–223
110. Vaidyanathan S, Rajagopal K (2011) Hybrid synchronization of hyperchaotic Wang-Chen and hyperchaotic Lorenz systems by active non-linear control. Int J Syst Signal Control Eng Appl 4(3):55–61
111. Vaidyanathan S, Rasappan S (2011) Global chaos synchronization of hyperchaotic Bao and Xu systems by active nonlinear control. Commun Comput Inf Sci 198:10–17
112. Vaidyanathan S, Azar AT, Rajagopal K, Alexander P (2015) Design and SPICE implementation of a 12-term novel hyperchaotic system and its synchronisation via active control. Int J Modell Identif Control 23(3):267–277
113. Sarasu P, Sundarapandian V (2012) Generalized projective synchronization of two-scroll systems via adaptive control. Int J Soft Comput 7(4):146–156
114. Sundarapandian V, Karthikeyan R (2011) Anti-synchronization of hyperchaotic Lorenz and hyperchaotic Chen systems by adaptive control. Int J Syst Signal Control Eng Appl 4(2):18–25
115. Sundarapandian V, Karthikeyan R (2011) Anti-synchronization of Lü and Pan chaotic systems by adaptive nonlinear control. Eur J Sci Res 64(1):94–106
116. Sundarapandian V, Karthikeyan R (2012) Adaptive anti-synchronization of uncertain Tigan and Li systems. J Eng Appl Sci 7(1):45–52
117. Vaidyanathan S (2012) Anti-synchronization of Sprott-L and Sprott-M chaotic systems via adaptive control. Int J Control Theory Appl 5(1):41–59
118. Vaidyanathan S, Pakiriswamy S (2013) Generalized projective synchronization of six-term Sundarapandian chaotic systems by adaptive control. Int J Control Theory Appl 6(2):153–163
119. Vaidyanathan S, Rajagopal K (2012) Global chaos synchronization of hyperchaotic Pang and hyperchaotic Wang systems via adaptive control. Int J Soft Comput 7(1):28–37
120. Sundarapandian V, Sivaperumal S (2011) Sliding controller design of hybrid synchronization of four-wing chaotic systems. Int J Soft Comput 6(5):224–231
121. Vaidyanathan S (2014) Global chaos synchronisation of identical Li-Wu chaotic systems via sliding mode control. Int J Model Identif Control 22(2):170–177
122. Vaidyanathan S, Sampath S (2012) Anti-synchronization of four-wing chaotic systems via sliding mode control. Int J Autom Comput 9(3):274–279
123. Vaidyanathan S, Sampath S, Azar AT (2015) Global chaos synchronisation of identical chaotic systems via novel sliding mode control method and its application to Zhu system. Int J Model Identif Control 23(1):92–100
124. Rasappan S, Vaidyanathan S (2013) Hybrid synchronization of n-scroll Chua circuits using adaptive backstepping control design with recursive feedback. Malays J Math Sci 73(1):73–95
125. Rasappan S, Vaidyanathan S (2014) Global chaos synchronization of WINDMI and Coullet chaotic systems using adaptive backstepping control design. Kyungpook Math J 54(1): 293–320
126. Suresh R, Sundarapandian V (2013) Global chaos synchronization of a family of n-scroll hyperchaotic Chua circuits using backstepping control with recursive feedback. Far East J Math Sci 7(2):219–246
127. Vaidyanathan S, Rasappan S (2014) Global chaos synchronization of n-scroll Chua circuit and Lur'e system using backstepping control design with recursive feedback. Arab J Sci Eng 39(4):3351–3364
128. Sprott JC (2010) Elegant chaos. World Scientific, Singapore

129. Thomas R (1999) Deterministic chaos seen in terms of feedback circuits: analysis, synthesis, 'labyrnth chaos'. Int J Bifurc Chaos 9:1889–1905
130. Khalil HK (2001) Nonlinear systems, 3rd edn. Prentice Hall, New Jersey

Dynamic Analysis, Adaptive Control and Synchronization of a Novel Highly Chaotic System with Four Quadratic Nonlinearities

Sundarapandian Vaidyanathan

Abstract In this work, we describe an eight-term novel highly chaotic system with four quadratic nonlinearities. The phase portraits of the novel highly chaotic system are illustrated and the dynamic properties of the highly chaotic system are discussed. The novel highly chaotic system has three unstable equilibrium points. We show that the equilibrium point at the origin is a saddle point, while the other two equilibrium points are saddle foci. The novel highly chaotic system has rotation symmetry about the x_3 axis. The Lyapunov exponents of the novel highly chaotic system are obtained as $L_1 = 11.0572$, $L_2 = 0$ and $L_3 = -28.0494$, while the Kaplan-Yorke dimension of the novel chaotic system is obtained as $D_{KY} = 2.3942$. Since the Maximal Lyapunov Exponent (MLE) of the novel chaotic system has a large value, viz. $L_1 = 11.0572$, the novel chaotic system is highly chaotic. Since the sum of the Lyapunov exponents is negative, the novel highly chaotic system is dissipative. Next, we derive new results for the global chaos control of the novel highly chaotic system with unknown parameters via adaptive control method. We also derive new results for the global chaos synchronization of the identical novel highly chaotic systems with unknown parameters via adaptive control method. The main adaptive control results are established using Lyapunov stability theory. MATLAB simulations are shown to depict the phase portraits of the novel highly chaotic system and also the adaptive control results derived in this work.

Keywords Chaos · Chaotic systems · Highly chaotic system · Chaos control · Adaptive control · Synchronization

---

S. Vaidyanathan (✉)
Research and Development Centre, Vel Tech University,
Avadi, Chennai 600062, Tamil Nadu, India
e-mail: sundarvtu@gmail.com

A.T. Azar and S. Vaidyanathan (eds.), *Advances in Chaos Theory and Intelligent Control*, Studies in Fuzziness and Soft Computing 337,
DOI 10.1007/978-3-319-30340-6_17

1 Introduction

Chaotic systems are defined as nonlinear dynamical systems which are sensitive to initial conditions, topologically mixing and with dense periodic orbits. Sensitivity to initial conditions of chaotic systems is popularly known as the *butterfly effect*.

Chaotic systems are either conservative or dissipative. The conservative chaotic systems are characterized by the property that they are *volume conserving*. The dissipative chaotic systems are characterized by the property that any asymptotic motion of the chaotic system settles onto a set of measure zero, i.e. a strange attractor. In this research work, we shall announce and discuss a novel 3-D dissipative highly chaotic circulant chaotic system with six sinusoidal nonlinearities.

The Lyapunov exponent of a chaotic system is a measure of the divergence of points which are initially very close and this can be used to quantify chaotic systems. Each nonlinear dynamical system has a spectrum of Lyapunov exponents, which are equal in number to the dimension of the state space. The largest Lyapunov exponent of a nonlinear dynamical system is called the *maximal Lyapunov exponent* (MLE).

In the last few decades, Chaos theory has become a very important and active research field, employing many applications in different disciplines like physics, chemistry, biology, ecology, engineering and economics, among others.

Some classical paradigms of 3-D chaotic systems in the literature are Lorenz system [1], Rössler system [2], ACT system [3], Sprott systems [4], Chen system [5], Lü system [6], Cai system [7], Tigan system [8], etc.

Many new chaotic systems have been discovered in the recent years such as Zhou system [9], Zhu system [10], Li system [11], Sundarapandian systems [12, 13], Vaidyanathan systems [14–30], Pehlivan system [31], Sampath system [32], Pham system [33], etc.

Chaos theory and control systems have many important applications in science and engineering [34–39]. Some commonly known applications are oscillators [40, 41], lasers [42, 43], chemical reactions [44–50], biology [51–60], ecology [61, 62], encryption [63, 64], cryptosystems [65, 66], mechanical systems [67–71], secure communications [72–74], robotics [75–77], cardiology [78, 79], intelligent control [80, 81], neural networks [82–84], finance [85, 86], etc.

The control of a chaotic system aims to stabilize or regulate the system. There are many methods available for controlling a chaotic system such as active control [87–89], adaptive control [90–98], sliding mode control [99, 100], backstepping control [101–103], etc.

Major works on synchronization of chaotic systems deal with the complete synchronization (CS) which has the design goal of using the output of the master system to control the slave system so that the output of the slave system tracks the output of the master system asymptotically with time.

There are many methods available for chaos synchronization such as active control [104–109], adaptive control [110–118], sliding mode control [119–122], backstepping control [123–126], etc.

In this research work, we announce an eight-term novel highly chaotic system with four quadratic nonlinearities. Using adaptive control method, we have also derived new results for the global chaos control of the novel highly chaotic system and global chaos synchronization of the identical novel highly chaotic systems when the system parameters are unknown.

This work is organized as follows. Section 2 describes the dynamic equations and phase portraits of the eight-term novel highly chaotic system. Section 3 details the dynamic analysis and properties of the novel highly chaotic system. The Lyapunov exponents of the novel chaotic system are obtained as $L_1 = 11.0572$, $L_2 = 0$ and $L_3 = -28.0494$, while the Kaplan-Yorke dimension of the novel chaotic system is obtained as $D_{KY} = 2.3942$. Since the maximal Lyapunov exponent of the novel chaotic system has a large value, viz. $L_1 = 11.0572$, the novel chaotic system is highly chaotic.

In Sect. 4, we derive new results for the global chaos control of the novel highly chaotic system with unknown parameters. In Sect. 5, we derive new results for the global chaos synchronization of the identical novel highly chaotic systems with unknown parameters. Section 6 contains a summary of the main results derived in this work.

2 A Novel 3-D Highly Chaotic System

In this section, we describe an eight-term novel chaotic system, which is given by the 3-D dynamics

$$\begin{cases} \dot{x}_1 = a(x_2 - x_1) + px_2x_3 \\ \dot{x}_2 = bx_1 - x_1x_3 \\ \dot{x}_3 = x_1x_2 - cx_3 + qx_2^2 \end{cases} \tag{1}$$

where x_1, x_2, x_3 are the states and a, b, c, p, q are constant, positive parameters.

The novel 3-D system (1) is an eight-term polynomial system with four quadratic nonlinearities.

The system (1) exhibits a *highly chaotic attractor* for the parameter values

$$a = 12, \quad b = 16, \quad c = 5, \quad p = 96, \quad q = 10 \tag{2}$$

For numerical simulations, we take the initial conditions as

$$x_1(0) = 0.2, \quad x_2(0) = 0.1, \quad x_3(0) = 0.2 \tag{3}$$

Figure 1 depicts the 3-D phase portrait of the novel highly chaotic system (1), while Figs. 2, 3 and 4 depict the 2-D projection of the novel highly chaotic system (1) on the (x_1, x_2), (x_2, x_3) and (x_1, x_3) planes, respectively. Figures 1, 2, 3 and 4 show that the novel highly chaotic system (1) exhibits a *two-scroll* chaotic attractor.

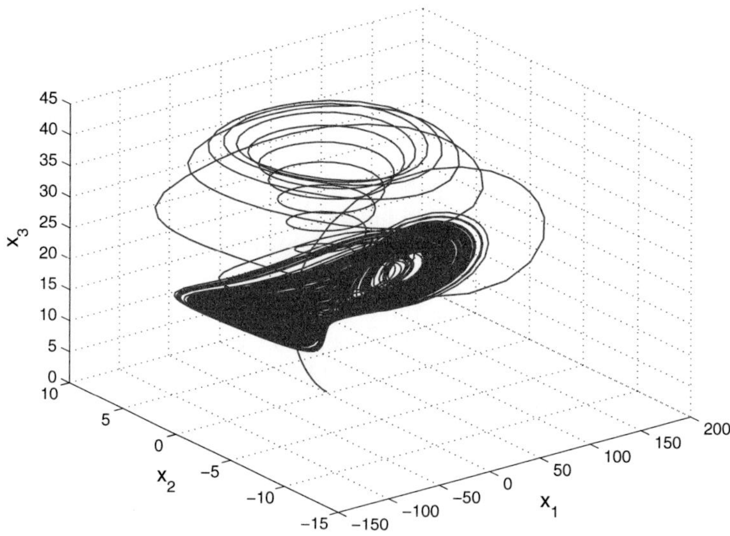

Fig. 1 3-D phase portrait of the novel highly chaotic system

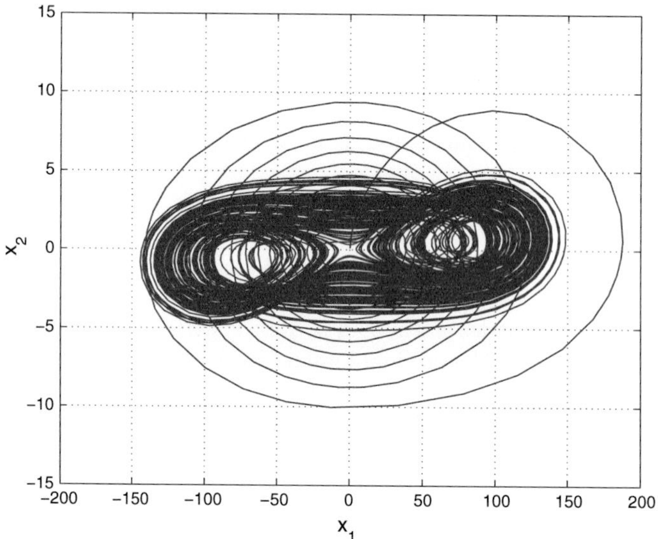

Fig. 2 2-D projection of the novel highly chaotic system on the (x_1, x_2) plane

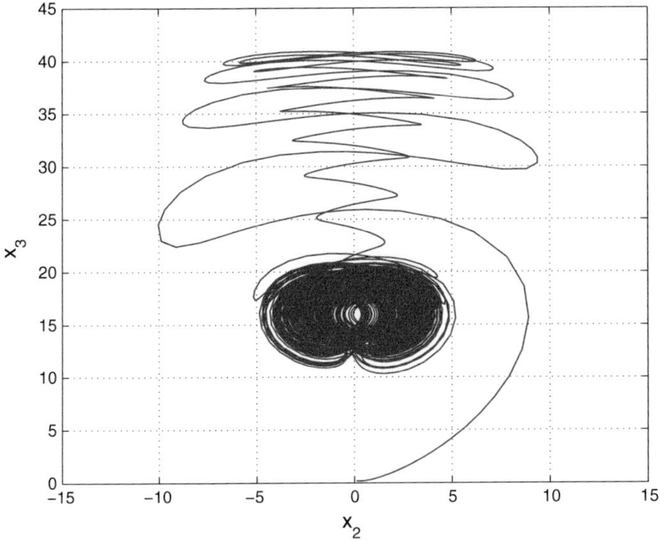

Fig. 3 2-D projection of the novel highly chaotic system on the (x_2, x_3) plane

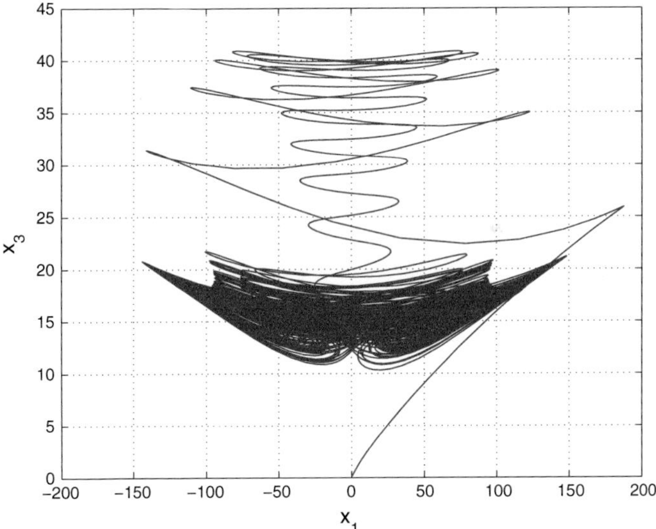

Fig. 4 2-D projection of the novel highly chaotic system on the (x_1, x_3) plane

3 Analysis of the Novel 3-D Highly Chaotic System

In this section, we give a dynamic analysis of the 3-D novel highly chaotic system (1). We take the parameter values as in the chaotic case (2), viz. $a = 12$, $b = 16$, $c = 5$, $p = 96$ and $q = 10$.

3.1 Dissipativity

In vector notation, the novel chaotic system (1) can be expressed as

$$\dot{\mathbf{x}} = f(\mathbf{x}) = \begin{bmatrix} f_1(x_1, x_2, x_3) \\ f_2(x_1, x_2, x_3) \\ f_3(x_1, x_2, x_3) \end{bmatrix}, \tag{4}$$

where

$$\begin{cases} f_1(x_1, x_2, x_3) = a(x_2 - x_1) + px_2x_3 \\ f_2(x_1, x_2, x_3) = bx_1 - x_1x_3 \\ f_3(x_1, x_2, x_3) = x_1x_2 - cx_3 + qx_2^2 \end{cases} \tag{5}$$

Let Ω be any region in $\mathbf{R}^3$ with a smooth boundary and also, $\Omega(t) = \Phi_t(\Omega)$, where Φ_t is the flow of f. Furthermore, let $V(t)$ denote the volume of $\Omega(t)$.

By Liouville's theorem, we know that

$$\dot{V}(t) = \int\limits_{\Omega(t)} (\nabla \cdot f)\, dx_1\, dx_2\, dx_3 \tag{6}$$

The divergence of the novel chaotic system (4) is found as

$$\nabla \cdot f = \frac{\partial f_1}{\partial x_1} + \frac{\partial f_2}{\partial x_2} + \frac{\partial f_3}{\partial x_3} = -(a + c) = -\mu < 0 \tag{7}$$

since $\mu = a + c = 17 > 0$.

Inserting the value of $\nabla \cdot f$ from (7) into (6), we get

$$\dot{V}(t) = \int\limits_{\Omega(t)} (-\mu)\, dx_1\, dx_2\, dx_3 = -\mu V(t) \tag{8}$$

Integrating the first order linear differential equation (8), we get

$$V(t) = \exp(-\mu t) V(0) \tag{9}$$

Since $\mu > 0$, it follows from Eq. (9) that $V(t) \to 0$ exponentially as $t \to \infty$. This shows that the novel chaotic system (1) is dissipative.

Hence, the system limit sets are ultimately confined into a specific limit set of zero volume, and the asymptotic motion of the novel chaotic system (1) settles onto a strange attractor of the system.

3.2 Equilibrium Points

We take the parameter values as in the chaotic case (2), viz. $a = 12, b = 16, c = 5$, $p = 96$ and $q = 10$.

It is easy to see that the system (1) has three equilibrium points, viz.

$$E_0 = \begin{bmatrix} 0 \\ 0 \\ 0 \end{bmatrix}, \quad E_1 = \begin{bmatrix} 97.8650 \\ 0.7586 \\ 16.0000 \end{bmatrix}, \quad E_2 = \begin{bmatrix} -97.8650 \\ -0.7586 \\ 16.0000 \end{bmatrix} \tag{10}$$

The Jacobian of the system (1) at any point $\mathbf{x} \in \mathbf{R}^3$ is calculated as

$$J(\mathbf{x}) = \begin{bmatrix} -a & a + px_3 & px_2 \\ b - x_3 & 0 & -x_1 \\ x_2 & x_1 + 2qx_2 & -c \end{bmatrix} = \begin{bmatrix} -12 & 12 + 96x_3 & 96x_2 \\ 16 - x_3 & 0 & -x_1 \\ x_2 & x_1 + 20x_2 & -5 \end{bmatrix} \tag{11}$$

The Jacobian of the system (1) at the equilibrium E_0 is obtained as

$$J_0 = J(E_0) = \begin{bmatrix} -12 & 12 & 0 \\ 16 & 0 & 0 \\ 0 & 0 & -5 \end{bmatrix} \tag{12}$$

We find that the matrix $J_0 = J(E_0)$ has the eigenvalues

$$\lambda_1 = -5, \quad \lambda_2 = -21.0997, \quad \lambda_3 = 9.0997 \tag{13}$$

This shows that the equilibrium point E_0 is a saddle-point, which is unstable.

The Jacobian of the system (1) at the equilibrium E_1 is obtained as

$$J_1 = J(E_1) = \begin{bmatrix} -12 & 1548 & 72.8256 \\ 0 & 0 & -97.8650 \\ 0.7586 & 113.0370 & -5 \end{bmatrix} \tag{14}$$

We find that the matrix $J_1 = J(E_1)$ has the eigenvalues

$$\lambda_1 = -22.15, \quad \lambda_{2,3} = 2.575 \pm 105.709i \tag{15}$$

This shows that the equilibrium point E_1 is a saddle-focus, which is unstable.

The Jacobian of the system (1) at the equilibrium E_2 is obtained as

$$J_2 = J(E_2) = \begin{bmatrix} -12 & 1548 & -72.8256 \\ 0 & 0 & 97.8650 \\ -0.7586 & -113.0370 & -5 \end{bmatrix} \tag{16}$$

We find that the matrix $J_2 = J(E_2)$ has the eigenvalues

$$\lambda_1 = -22.15, \quad \lambda_{2,3} = 2.575 \pm 105.709i \tag{17}$$

This shows that the equilibrium point E_2 is a saddle-focus, which is unstable.

3.3 Symmetry and Invariance

It is easy to see that the system (1) is invariant under the change of coordinates

$$(x_1, x_2, x_3) \mapsto (-x_1, -x_2, x_3) \tag{18}$$

Thus, it follows that the 3-D novel chaotic system (1) has rotation symmetry about the x_3-axis and that any non-trivial trajectory must have a twin trajectory.

Next, it is easy to see that the x_3-axis is invariant under the flow of the 3-D novel chaotic system (1).

The invariant motion along the x_3-axis is characterized by

$$\dot{x}_3 = -cx_3, \quad (c > 0) \tag{19}$$

which is globally exponentially stable.

3.4 Lyapunov Exponents and Kaplan-Yorke Dimension

We take the parameter values of the novel system (1) as in the chaotic case (2), i.e.

$$a = 12, \quad b = 16, \quad c = 5, \quad p = 96, \quad q = 10 \tag{20}$$

We take the initial state of the novel system (1) as given in (3), i.e.

$$x_1(0) = 0.2, \quad x_2(0) = 0.1, \quad x_3(0) = 0.2 \tag{21}$$

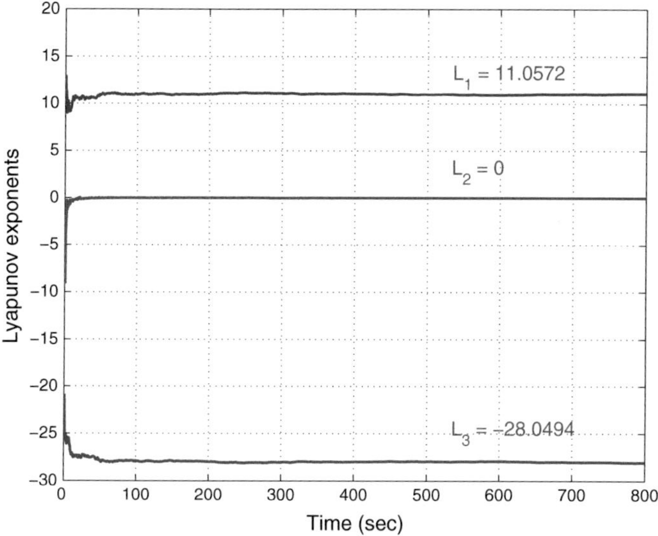

Fig. 5 Dynamics of the Lyapunov exponents of the highly chaotic system

Then the Lyapunov exponents of the system (1) are numerically obtained as

$$L_1 = 11.0572, \quad L_2 = 0, \quad L_3 = -28.0494 \tag{22}$$

Figure 5 shows the dynamics of the Lyapunov exponents of the novel system (1). From Fig. 5, we note that the Maximal Lyapunov Exponent (MLE) of the novel system (1) is given by $L_1 = 11.0572$, which is a large value. This shows that the novel system (1) is *highly chaotic*.

We also note that the sum of the Lyapunov exponents in (22) is negative, i.e.

$$L_1 + L_2 + L_3 = -16.9922 < 0 \tag{23}$$

This shows that the novel chaotic system (1) is dissipative.

Also, the Kaplan-Yorke dimension of the novel chaotic system (1) is found as

$$D_{KY} = 2 + \frac{L_1 + L_2}{|L_3|} = 2.3942, \tag{24}$$

which is fractional.

Also, the relatively large value of the Kaplan-Yorke dimension of the novel chaotic system (1), i.e. $D_{KY} = 2.3942$, indicates that the system exhibits highly complex behaviour. Hence, the novel chaotic system (1) has applications in cryptosystems, secure communication devices, etc.

4 Adaptive Control of the Novel Highly Chaotic System

In this section, we use adaptive control method to derive an adaptive feedback control law for globally stabilizing the novel 3-D highly chaotic system with unknown parameters.

Thus, we consider the novel highly chaotic system given by

$$\begin{cases} \dot{x}_1 = a(x_2 - x_1) + px_2x_3 + u_1 \\ \dot{x}_2 = bx_1 - x_1x_3 + u_2 \\ \dot{x}_3 = x_1x_2 - cx_3 + qx_2^2 + u_3 \end{cases} \tag{25}$$

In (25), x_1, x_2, x_3 are the states and u_1, u_2, u_3 are the adaptive controls to be determined using estimates for the unknown system parameters.

We consider the adaptive feedback control law

$$\begin{cases} u_1 = -\hat{a}(t)(x_2 - x_1) - \hat{p}(t)x_2x_3 - k_1x_1 \\ u_2 = -\hat{b}(t)x_1 + x_1x_3 - k_2x_2 \\ u_3 = -x_1x_2 + \hat{c}(t)x_3 - \hat{q}(t)x_2^2 - k_3x_3 \end{cases} \tag{26}$$

where k_1, k_2, k_3 are positive gain constants.

Substituting (26) into (25), we get the closed-loop plant dynamics as

$$\begin{cases} \dot{x}_1 = \left[a - \hat{a}(t)\right](x_2 - x_1) + \left[p - \hat{p}(t)\right]x_2x_3 - k_1x_1 \\ \dot{x}_2 = \left[b - \hat{b}(t)\right]x_1 - k_2x_2 \\ \dot{x}_3 = -\left[c - \hat{c}(t)\right]x_3 + \left[q - \hat{q}(t)\right]x_2^2 - k_3x_3 \end{cases} \tag{27}$$

The parameter estimation errors are defined as

$$\begin{cases} e_a(t) = a - \hat{a}(t) \\ e_b(t) = b - \hat{b}(t) \\ e_c(t) = c - \hat{c}(t) \\ e_p(t) = p - \hat{p}(t) \\ e_q(t) = q - \hat{q}(t) \end{cases} \tag{28}$$

In view of (28), we can simplify the plant dynamics (27) as

$$\begin{cases} \dot{x}_1 = e_a(x_2 - x_1) + e_px_2x_3 - k_1x_1 \\ \dot{x}_2 = e_bx_1 - k_2x_2 \\ \dot{x}_3 = -e_cx_3 + e_qx_2^2 - k_3x_3 \end{cases} \tag{29}$$

Differentiating (28) with respect to t, we obtain

$$\begin{cases} \dot{e}_a(t) = -\dot{\hat{a}}(t) \\ \dot{e}_b(t) = -\dot{\hat{b}}(t) \\ \dot{e}_c(t) = -\dot{\hat{c}}(t) \\ \dot{e}_p(t) = -\dot{\hat{p}}(t) \\ \dot{e}_q(t) = -\dot{\hat{q}}(t) \end{cases} \tag{30}$$

We consider the quadratic candidate Lyapunov function defined by

$$V(\mathbf{x}, e_a, e_b, e_c, e_p, e_q) = \frac{1}{2}\left(x_1^2 + x_2^2 + x_3^2\right) + \frac{1}{2}\left(e_a^2 + e_b^2 + e_c^2 + e_p^2 + e_q^2\right) \tag{31}$$

Differentiating V along the trajectories of (29) and (30), we obtain

$$\begin{aligned} \dot{V} = -k_1 x_1^2 - k_2 x_2^2 - k_3 x_3^2 + e_a\left[x_1(x_2 - x_1) - \dot{\hat{a}}\right] + e_b\left[x_1 x_2 - \dot{\hat{b}}\right] \\ + e_c\left[-x_3^2 - \dot{\hat{c}}\right] + e_p\left[x_1 x_2 x_3 - \dot{\hat{p}}\right] + e_q\left[x_2^2 x_3 - \dot{\hat{q}}\right] \end{aligned} \tag{32}$$

In view of (32), we take the parameter update law as

$$\begin{cases} \dot{\hat{a}}(t) = x_1(x_2 - x_1) \\ \dot{\hat{b}}(t) = x_1 x_2 \\ \dot{\hat{c}}(t) = -x_3^2 \\ \dot{\hat{p}}(t) = x_1 x_2 x_3 \\ \dot{\hat{q}}(t) = x_2^2 x_3 \end{cases} \tag{33}$$

Next, we state and prove the main result of this section.

Theorem 1 *The novel 3-D highly chaotic system (25) with unknown system parameters is globally and exponentially stabilized for all initial conditions by the adaptive control law (26) and the parameter update law (33), where k_1, k_2, k_3 are positive gain constants.*

Proof We prove this result by applying Lyapunov stability theory [127].

We consider the quadratic Lyapunov function defined by (31), which is clearly a positive definite function on $\mathbf{R}^8$.

By substituting the parameter update law (33) into (32), we obtain the time-derivative of V as

$$\dot{V} = -k_1 x_1^2 - k_2 x_2^2 - k_3 x_3^2 \tag{34}$$

From (34), it is clear that $\dot{V}$ is a negative semi-definite function on $\mathbf{R}^8$.

Thus, we can conclude that the state vector $\mathbf{x}(t)$ and the parameter estimation error are globally bounded i.e.

$$\left[x_1(t)\ x_2(t)\ x_3(t)\ e_a(t)\ e_b(t)\ e_c(t)\ e_p(t)\ e_q(t)\right]^T \in \mathbf{L}_\infty.$$

We define $k = \min\{k_1, k_2, k_3\}$.
Then it follows from (34) that

$$\dot{V} \leq -k\|\mathbf{x}(t)\|^2 \tag{35}$$

Thus, we have

$$k\|\mathbf{x}(t)\|^2 \leq -\dot{V} \tag{36}$$

Integrating the inequality (36) from 0 to t, we get

$$k \int_0^t \|\mathbf{x}(\tau)\|^2 \, d\tau \ \leq \ V(0) - V(t) \tag{37}$$

From (37), it follows that $\mathbf{x} \in \mathbf{L}_2$.
Using (29), we can conclude that $\dot{\mathbf{x}} \in \mathbf{L}_\infty$.
Using Barbalat's lemma [127], we conclude that $\mathbf{x}(t) \to 0$ exponentially as $t \to \infty$ for all initial conditions $\mathbf{x}(0) \in \mathbf{R}^3$.
Hence, the novel highly chaotic system (25) with unknown system parameters is globally and exponentially stabilized for all initial conditions by the adaptive control law (26) and the parameter update law (33).
This completes the proof. ∎

For the numerical simulations, the classical fourth-order Runge-Kutta method with step size $h = 10^{-8}$ is used to solve the systems (25) and (33), when the adaptive control law (26) is applied.

The parameter values of the novel 3-D highly chaotic system (25) are taken as in the chaotic case (2), i.e.

$$a = 12, \quad b = 16, \quad c = 5, \quad p = 96, \quad q = 10 \tag{38}$$

We take the positive gain constants as $k_i = 6$ for $i = 1, 2, 3$.
Furthermore, as initial conditions of the novel highly chaotic system (25), we take

$$x_1(0) = 12.4, \quad x_2(0) = -21.5, \quad x_3(0) = 16.2 \tag{39}$$

Also, as initial conditions of the parameter estimates, we take

$$\hat{a}(0) = 2.7, \quad \hat{b}(0) = 1.3, \quad \hat{c}(0) = 3.1, \quad \hat{p}(0) = 7.5, \quad \hat{q}(0) = 6.3 \tag{40}$$

In Fig. 6, the exponential convergence of the controlled states of the 3-D novel highly chaotic system (25) is depicted. From Fig. 6, we see that the controlled states $x_1(t), x_2(t), x_3(t)$ converge to zero in just 1 s. This shows the efficiency of the adaptive controller designed in this section for the novel highly chaotic system (25).

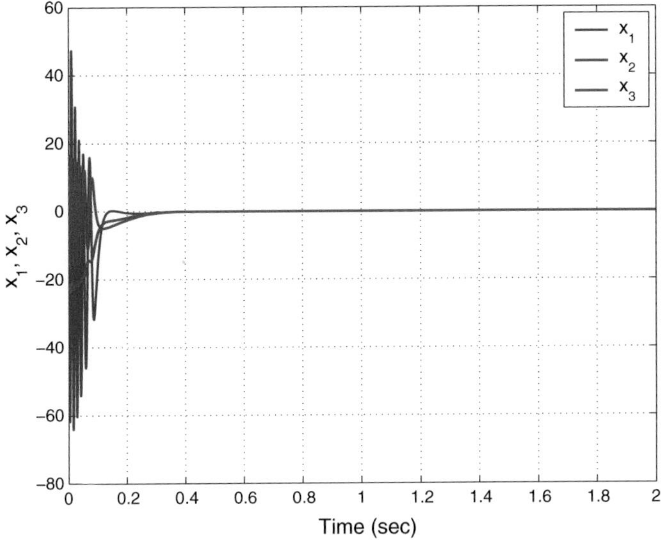

Fig. 6 Time-history of the controlled states x_1, x_2, x_3

5 Adaptive Synchronization of the Identical Novel Highly Chaotic Systems

In this section, we apply adaptive control method to derive an adaptive feedback control law for globally synchronizing identical 3-D novel highly chaotic systems with unknown parameters. The main result is established using Lyapunov stability theory.

As the master system, we consider the novel 3-D chaotic system given by

$$\begin{cases} \dot{x}_1 = a(x_2 - x_1) + px_2x_3 \\ \dot{x}_2 = bx_1 - x_1x_3 \\ \dot{x}_3 = x_1x_2 - cx_3 + qx_2^2 \end{cases} \tag{41}$$

In (41), x_1, x_2, x_3 are the states and a, b, c, p, q are unknown system parameters.
As the slave system, we consider the novel 3-D chaotic system given by

$$\begin{cases} \dot{y}_1 = a(y_2 - y_1) + py_2y_3 + u_1 \\ \dot{y}_2 = by_1 - y_1y_3 + u_2 \\ \dot{y}_3 = y_1y_2 - cy_3 + qy_2^2 + u_3 \end{cases} \tag{42}$$

In (42), y_1, y_2, y_3 are the states and u_1, u_2, u_3 are the adaptive controls to be determined using estimates of the unknown system parameters.

The synchronization error between the novel chaotic systems is defined by

$$\begin{cases} e_1 = y_1 - x_1 \\ e_2 = y_2 - x_2 \\ e_3 = y_3 - x_3 \end{cases} \tag{43}$$

Then the error dynamics is obtained as

$$\begin{cases} \dot{e}_1 = a(e_2 - e_1) + p(y_2 y_3 - x_2 x_3) + u_1 \\ \dot{e}_2 = b e_1 - y_1 y_3 + x_1 x_3 + u_2 \\ \dot{e}_3 = -c e_3 + y_1 y_2 - x_1 x_2 + q(y_2^2 - x_2^2) + u_3 \end{cases} \tag{44}$$

We consider the adaptive feedback control law

$$\begin{cases} u_1 = -\hat{a}(t)(e_2 - e_1) - \hat{p}(t)(y_2 y_3 - x_2 x_3) - k_1 e_1 \\ u_2 = -\hat{b}(t) e_1 + y_1 y_3 - x_1 x_3 - k_2 e_2 \\ u_3 = \hat{c}(t) e_3 - y_1 y_2 + x_1 x_2 - \hat{q}(t)(y_2^2 - x_2^2) - k_3 e_3 \end{cases} \tag{45}$$

where k_1, k_2, k_3 are positive gain constants.

Substituting (45) into (44), we get the closed-loop error dynamics as

$$\begin{cases} \dot{e}_1 = \left[a - \hat{a}(t)\right](e_2 - e_1) + \left[p - \hat{p}(t)\right](y_2 y_3 - x_2 x_3) - k_1 e_1 \\ \dot{e}_2 = \left[b - \hat{b}(t)\right] e_1 - k_2 e_2 \\ \dot{e}_3 = -\left[c - \hat{c}(t)\right] e_3 + \left[q - \hat{q}(t)\right](y_2^2 - x_2^2) - k_3 e_3 \end{cases} \tag{46}$$

The parameter estimation errors are defined as

$$\begin{cases} e_a(t) = a - \hat{a}(t) \\ e_b(t) = b - \hat{b}(t) \\ e_c(t) = c - \hat{c}(t) \\ e_p(t) = p - \hat{p}(t) \\ e_q(t) = q - \hat{q}(t) \end{cases} \tag{47}$$

In view of (47), we can simplify the error dynamics (46) as

$$\begin{cases} \dot{e}_1 = e_a(e_2 - e_1) + e_p(y_2 y_3 - x_2 x_3) - k_1 e_1 \\ \dot{e}_2 = e_b e_1 - k_2 e_2 \\ \dot{e}_3 = -e_c e_3 + e_q(y_2^2 - x_2^2) - k_3 e_3 \end{cases} \tag{48}$$

Differentiating (47) with respect to t, we obtain

$$\begin{cases} \dot{e}_a(t) = -\dot{\hat{a}}(t) \\ \dot{e}_b(t) = -\dot{\hat{b}}(t) \\ \dot{e}_c(t) = -\dot{\hat{c}}(t) \\ \dot{e}_p(t) = -\dot{\hat{p}}(t) \\ \dot{e}_q(t) = -\dot{\hat{q}}(t) \end{cases} \tag{49}$$

We consider the quadratic candidate Lyapunov function defined by

$$V(\mathbf{e}, e_a, e_b, e_c, e_p, e_q) = \frac{1}{2}\left(e_1^2 + e_2^2 + e_3^2\right) + \frac{1}{2}\left(e_a^2 + e_b^2 + e_c^2 + e_p^2 + e_q^2\right) \tag{50}$$

Differentiating V along the trajectories of (48) and (49), we obtain

$$\begin{aligned} \dot{V} = &-k_1 e_1^2 - k_2 e_2^2 - k_3 e_3^2 + e_a\left[e_1(e_2 - e_1) - \dot{\hat{a}}\right] + e_b\left[e_1 e_2 - \dot{\hat{b}}\right] \\ &+ e_c\left[-e_3^2 - \dot{\hat{c}}\right] + e_p\left[e_1(y_2 y_3 - x_2 x_3) - \dot{\hat{p}}\right] + e_q\left[e_3(y_2^2 - x_2^2) - \dot{\hat{q}}\right] \end{aligned} \tag{51}$$

In view of (51), we take the parameter update law as

$$\begin{cases} \dot{\hat{a}}(t) = e_1(e_2 - e_1) \\ \dot{\hat{b}}(t) = e_1 e_2 \\ \dot{\hat{c}}(t) = -e_3^2 \\ \dot{\hat{p}}(t) = e_1(y_2 y_3 - x_2 x_3) \\ \dot{\hat{q}}(t) = e_3(y_2^2 - x_2^2) \end{cases} \tag{52}$$

Next, we state and prove the main result of this section.

Theorem 2 *The novel highly chaotic systems (41) and (42) with unknown system parameters are globally and exponentially synchronized for all initial conditions by the adaptive control law (45) and the parameter update law (52), where* k_1, k_2, k_3 *are positive gain constants.*

Proof We prove this result by applying Lyapunov stability theory [127].

We consider the quadratic Lyapunov function defined by (50), which is clearly a positive definite function on $\mathbf{R}^8$.

By substituting the parameter update law (52) into (51), we obtain

$$\dot{V} = -k_1 e_1^2 - k_2 e_2^2 - k_3 e_3^2 \tag{53}$$

From (53), it is clear that $\dot{V}$ is a negative semi-definite function on $\mathbf{R}^8$.

Thus, we can conclude that the error vector $\mathbf{e}(t)$ and the parameter estimation error are globally bounded, i.e.

$$\left[\, e_1(t)\ e_2(t)\ e_3(t)\ e_a(t)\ e_b(t)\ e_c(t)\ e_p(t)\ e_q(t) \,\right]^T \in \mathbf{L}_\infty. \tag{54}$$

We define $k = \min\{k_1, k_2, k_3\}$.

Then it follows from (53) that

$$\dot{V} \le -k\|\mathbf{e}(t)\|^2 \tag{55}$$

Thus, we have

$$k\,\|\mathbf{e}(t)\|^2 \le -\dot{V} \tag{56}$$

Integrating the inequality (56) from 0 to t, we get

$$k \int\limits_0^t \|\mathbf{e}(\tau)\|^2 \, d\tau \;\le\; V(0) - V(t) \tag{57}$$

From (57), it follows that $\mathbf{e} \in \mathbf{L}_2$.

Using (48), we can conclude that $\dot{\mathbf{e}} \in \mathbf{L}_\infty$.

Using Barbalat's lemma [127], we conclude that $\mathbf{e}(t) \to 0$ exponentially as $t \to \infty$ for all initial conditions $\mathbf{e}(0) \in \mathbf{R}^3$.

This completes the proof. ■

For the numerical simulations, the classical fourth-order Runge-Kutta method with step size $h = 10^{-8}$ is used to solve the systems (41), (42) and (52), when the adaptive control law (45) is applied.

The parameter values of the novel chaotic systems are taken as in the chaotic case (2), i.e. $a = 12$, $b = 16$, $c = 5$, $p = 96$ and $q = 10$.

We take the positive gain constants as $k_i = 5$ for $i = 1, 2, 3$.

Furthermore, as initial conditions of the master system (41), we take

$$x_1(0) = 3.4, \quad x_2(0) = -12.7, \quad x_3(0) = 10.3 \tag{58}$$

As initial conditions of the slave system (42), we take

$$y_1(0) = 6.2, \quad y_2(0) = 3.5, \quad y_3(0) = 1.8 \tag{59}$$

Also, as initial conditions of the parameter estimates, we take

$$\hat{a}(0) = 2.1, \quad \hat{b}(0) = 6.3, \quad \hat{c}(0) = 17.4, \quad \hat{p}(0) = 10.9, \quad \hat{q}(0) = 8.5 \tag{60}$$

Figures 7, 8 and 9 describe the complete synchronization of the novel highly chaotic systems (41) and (42), while Fig. 10 describes the time-history of the synchronization errors e_1, e_2, e_3.

From Fig. 7, we see that the states x_1 and y_1 are synchronized in just 1 s. From Fig. 8, we see that the states x_2 and y_2 are synchronized in just 1 s. From Fig. 9, we see that the states x_3 and y_3 are synchronized in just 1 s. From Fig. 10, we see that the errors e_1, e_2, e_3 converge to zero in just 1 s. This shows the efficiency of the adaptive controller developed in this section for the synchronization of identical highly chaotic systems.

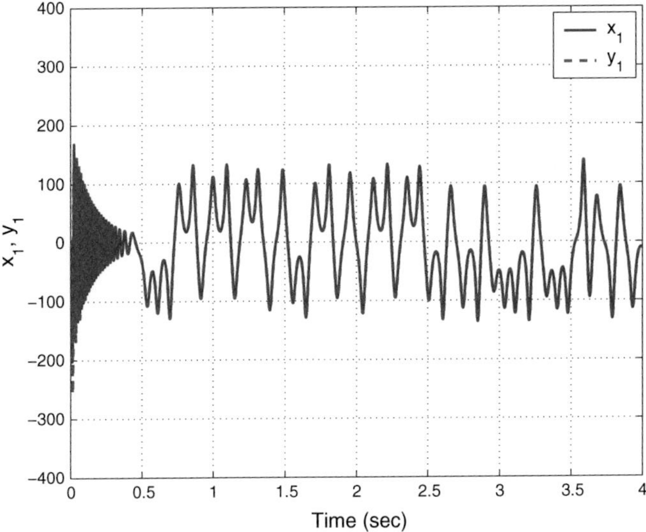

Fig. 7 Synchronization of the states x_1 and y_1

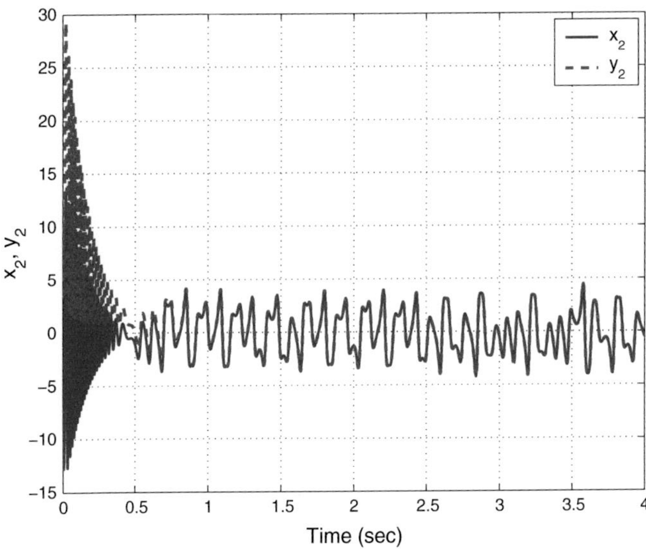

Fig. 8 Synchronization of the states x_2 and y_2

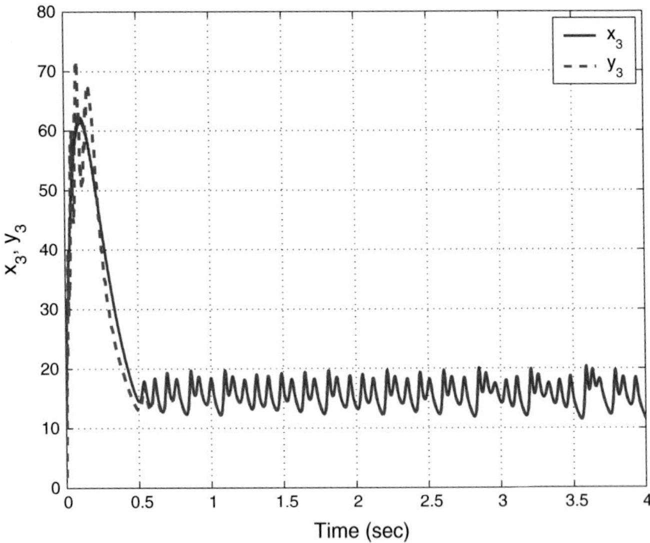

Fig. 9 Synchronization of the states x_3 and y_3

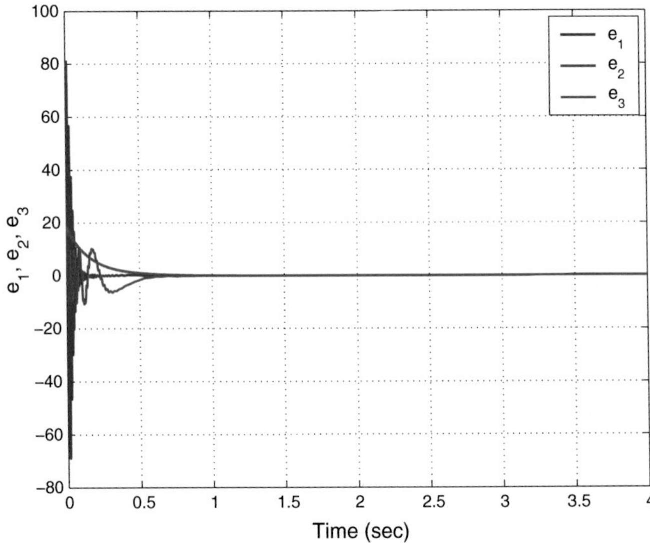

Fig. 10 Time-history of the synchronization errors e_1, e_2, e_3

6 Conclusions

This work announced an eight-term novel highly chaotic system with four quadratic nonlinearities. First, the qualitative properties of the highly chaotic system are discussed. We showed that the novel highly chaotic system has three unstable equilibrium points. The novel highly chaotic system has rotation symmetry about the x_3 axis. The Lyapunov exponents of the novel highly chaotic system have been obtained as $L_1 = 11.0572$, $L_2 = 0$ and $L_3 = -28.0494$, while the Kaplan-Yorke dimension of the novel chaotic system has been derived as $D_{KY} = 2.3942$. Since the Maximal Lyapunov Exponent (MLE) of the novel chaotic system has a large value, viz. $L_1 = 11.0572$, the novel chaotic system is highly chaotic. Since the sum of the Lyapunov exponents is negative, the novel highly chaotic system is dissipative. Then we derived new results for the global chaos control of the novel highly chaotic system with unknown parameters via adaptive control method. We also derived new results for the global chaos synchronization of the identical novel highly chaotic systems with unknown parameters via adaptive control method. The main control results have been proved using Lyapunov stability theory. MATLAB simulations were shown to illustrate all the results derived in this work.

References

1. Lorenz EN (1963) Deterministic periodic flow. J Atmos Sci 20(2):130–141
2. Rössler OE (1976) An equation for continuous chaos. Phys Lett A 57(5):397–398
3. Arneodo A, Coullet P, Tresser C (1981) Possible new strange attractors with spiral structure. Commun Math Phys 79(4):573–576
4. Sprott JC (1994) Some simple chaotic flows. Phys Rev E 50(2):647–650
5. Chen G, Ueta T (1999) Yet another chaotic attractor. Int J Bifurcat Chaos 9(7):1465–1466
6. Lü J, Chen G (2002) A new chaotic attractor coined. Int J Bifurcat Chaos 12(3):659–661
7. Cai G, Tan Z (2007) Chaos synchronization of a new chaotic system via nonlinear control. J Uncertain Syst 1(3):235–240
8. Tigan G, Opris D (2008) Analysis of a 3D chaotic system. Chaos, Solitons Fractals 36: 1315–1319
9. Zhou W, Xu Y, Lu H, Pan L (2008) On dynamics analysis of a new chaotic attractor. Phys Lett A 372(36):5773–5777
10. Zhu C, Liu Y, Guo Y (2010) Theoretic and numerical study of a new chaotic system. Intell Inform Manag 2:104–109
11. Li D (2008) A three-scroll chaotic attractor. Phys Lett A 372(4):387–393
12. Sundarapandian V (2013) Analysis and anti-synchronization of a novel chaotic system via active and adaptive controllers. J Eng Sci Technol Rev 6(4):45–52
13. Sundarapandian V, Pehlivan I (2012) Analysis, control, synchronization, and circuit design of a novel chaotic system. Math Comput Model 55(7–8):1904–1915
14. Vaidyanathan S (2013) A new six-term 3-D chaotic system with an exponential nonlinearity. Far East J Math Sci 79(1):135–143
15. Vaidyanathan S (2013) Analysis and adaptive synchronization of two novel chaotic systems with hyperbolic sinusoidal and cosinusoidal nonlinearity and unknown parameters. J Eng Sci Technol Rev 6(4):53–65

16. Vaidyanathan S (2014) A new eight-term 3-D polynomial chaotic system with three quadratic nonlinearities. Far East J Math Sci 84(2):219–226
17. Vaidyanathan S (2014) Analysis and adaptive synchronization of eight-term 3-D polynomial chaotic systems with three quadratic nonlinearities. Eur Phys J: Special Topics 223(8): 1519–1529
18. Vaidyanathan S (2014) Analysis, control and synchronisation of a six-term novel chaotic system with three quadratic nonlinearities. Int J Model Ident Control 22(1):41–53
19. Vaidyanathan S (2014) Generalised projective synchronisation of novel 3-D chaotic systems with an exponential non-linearity via active and adaptive control. Int J Model Ident Control 22(3):207–217
20. Vaidyanathan S (2015) A 3-D novel highly chaotic system with four quadratic nonlinearities, its adaptive control and anti-synchronization with unknown parameters. J Eng Sci Technol Rev 8(2):106–115
21. Vaidyanathan S (2015) Analysis, properties and control of an eight-term 3-D chaotic system with an exponential nonlinearity. Int J Model Ident Control 23(2):164–172
22. Vaidyanathan S, Azar AT (2015) Analysis, control and synchronization of a nine-term 3-D novel chaotic system. In: Azar AT, Vaidyanathan S (eds) Chaos modelling and control systems design, Studies in computational intelligence, vol 581. Springer, Germany, pp 19–38
23. Vaidyanathan S, Madhavan K (2013) Analysis, adaptive control and synchronization of a seven-term novel 3-D chaotic system. Int J Control Theory Appl 6(2):121–137
24. Vaidyanathan S, Pakiriswamy S (2015) A 3-D novel conservative chaotic system and its generalized projective synchronization via adaptive control. J Eng Sci Technol Rev 8(2): 52–60
25. Vaidyanathan S, Volos C (2015) Analysis and adaptive control of a novel 3-D conservative no-equilibrium chaotic system. Arch Control Sci 25(3):333–353
26. Vaidyanathan S, Volos C, Pham VT, Madhavan K, Idowu BA (2014) Adaptive backstepping control, synchronization and circuit simulation of a 3-D novel jerk chaotic system with two hyperbolic sinusoidal nonlinearities. Arch Control Sci 24(3):375–403
27. Vaidyanathan S, Rajagopal K, Volos CK, Kyprianidis IM, Stouboulos IN (2015) Analysis, adaptive control and synchronization of a seven-term novel 3-D chaotic system with three quadratic nonlinearities and its digital implementation in LabVIEW. J Eng Sci Technol Rev 8(2):130–141
28. Vaidyanathan S, Volos CK, Kyprianidis IM, Stouboulos IN, Pham VT (2015) Analysis, adaptive control and anti-synchronization of a six-term novel jerk chaotic system with two exponential nonlinearities and its circuit simulation. J Eng Sci Technol Rev 8(2):24–36
29. Vaidyanathan S, Volos CK, Pham VT (2015) Analysis, adaptive control and adaptive synchronization of a nine-term novel 3-D chaotic system with four quadratic nonlinearities and its circuit simulation. J Eng Sci Technol Rev 8(2):181–191
30. Vaidyanathan S, Volos CK, Pham VT (2015) Global chaos control of a novel nine-term chaotic system via sliding mode control. In: Azar AT, Zhu Q (eds) Advances and applications in sliding mode control systems, Studies in computational intelligence, vol 576. Springer, Germany, pp 571–590
31. Pehlivan I, Moroz IM, Vaidyanathan S (2014) Analysis, synchronization and circuit design of a novel butterfly attractor. J Sound Vib 333(20):5077–5096
32. Sampath S, Vaidyanathan S, Volos CK, Pham VT (2015) An eight-term novel four-scroll chaotic system with cubic nonlinearity and its circuit simulation. J Eng Sci Technol Rev 8(2):1–6
33. Pham VT, Vaidyanathan S, Volos CK, Jafari S (2015) Hidden attractors in a chaotic system with an exponential nonlinear term. Eur Phys J—Special Topics 224(8):1507–1517
34. Azar AT (2010) Fuzzy systems. IN-TECH, Vienna
35. Azar AT, Vaidyanathan S (2015) Chaos modeling and control systems design, Studies in computational intelligence, vol 581. Springer, Germany
36. Azar AT, Vaidyanathan S (2015) Computational intelligence applications in modeling and control, Studies in computational intelligence, vol 575. Springer, Germany

37. Azar AT, Vaidyanathan S (2015c) Handbook of Research on advanced intelligent control engineering and automation. Advances in computational intelligence and robotics (ACIR), IGI-Global, USA
38. Azar AT, Zhu Q (2015) Advances and applications in sliding mode control systems, Studies in computational intelligence, vol 576. Springer, Germany
39. Zhu Q, Azar AT (2015) Complex System modelling and control through intelligent soft computations, Studies in fuzzines and soft computing, vol 319. Springer, Germany
40. Kengne J, Chedjou JC, Kenne G, Kyamakya K (2012) Dynamical properties and chaos synchronization of improved Colpitts oscillators. Commun Nonlinear Sci Numer Simul 17(7):2914–2923
41. Sharma A, Patidar V, Purohit G, Sud KK (2012) Effects on the bifurcation and chaos in forced Duffing oscillator due to nonlinear damping. Commun Nonlinear Sci Numer Simul 17(6):2254–2269
42. Li N, Pan W, Yan L, Luo B, Zou X (2014) Enhanced chaos synchronization and communication in cascade-coupled semiconductor ring lasers. Commun Nonlinear Sci Numer Simul 19(6):1874–1883
43. Yuan G, Zhang X, Wang Z (2014) Generation and synchronization of feedback-induced chaos in semiconductor ring lasers by injection-locking. Optik—Int J Light Electr Optics 125(8):1950–1953
44. Gaspard P (1999) Microscopic chaos and chemical reactions. Phys A 263(1–4):315–328
45. Petrov V, Gaspar V, Masere J, Showalter K (1993) Controlling chaos in Belousov-Zhabotinsky reaction. Nature 361:240–243
46. Vaidyanathan S (2015) Adaptive synchronization of chemical chaotic reactors. Int J ChemTech Res 8(2):612–621
47. Vaidyanathan S (2015) Anti-synchronization of Brusselator chemical reaction systems via adaptive control. Int J ChemTech Res 8(6):759–768
48. Vaidyanathan S (2015) Dynamics and control of Brusselator chemical reaction. Int J ChemTech Res 8(6):740–749
49. Vaidyanathan S (2015) Dynamics and control of Tokamak system with symmetric and magnetically confined plasma. Int J ChemTech Res 8(6):795–803
50. Vaidyanathan S (2015) Synchronization of Tokamak systems with symmetric and magnetically confined plasma via adaptive control. Int J ChemTech Res 8(6):818–827
51. Das S, Goswami D, Chatterjee S, Mukherjee S (2014) Stability and chaos analysis of a novel swarm dynamics with applications to multi-agent systems. Eng Appl Artif Intell 30:189–198
52. Kyriazis M (1991) Applications of chaos theory to the molecular biology of aging. Exp Gerontol 26(6):569–572
53. Vaidyanathan S (2015) 3-cells cellular neural network (CNN) attractor and its adaptive biological control. Int J PharmTech Res 8(4):632–640
54. Vaidyanathan S (2015) Adaptive backstepping control of enzymes-substrates system with ferroelectric behaviour in brain waves. Int J PharmTech Res 8(2):256–261
55. Vaidyanathan S (2015) Adaptive biological control of generalized Lotka-Volterra three-species biological system. Int J PharmTech Res 8(4):622–631
56. Vaidyanathan S (2015) Adaptive chaotic synchronization of enzymes-substrates system with ferroelectric behaviour in brain waves. Int J PharmTech Res 8(5):964–973
57. Vaidyanathan S (2015) Adaptive synchronization of generalized Lotka-Volterra three-species biological systems. Int J PharmTech Res 8(5):928–937
58. Vaidyanathan S (2015) Chaos in neurons and adaptive control of Birkhoff-Shaw strange chaotic attractor. Int J PharmTech Res 8(5):956–963
59. Vaidyanathan S (2015) Lotka-Volterra population biology models with negative feedback and their ecological monitoring. Int J PharmTech Res 8(5):974–981
60. Vaidyanathan S (2015) Synchronization of 3-cells cellular neural network (CNN) attractors via adaptive control method. Int J PharmTech Res 8(5):946–955
61. Gibson WT, Wilson WG (2013) Individual-based chaos: Extensions of the discrete logistic model. J Theor Biol 339:84–92

62. Suérez I (1999) Mastering chaos in ecology. Ecol Model 117(2–3):305–314
63. Lang J (2015) Color image encryption based on color blend and chaos permutation in the reality-preserving multiple-parameter fractional Fourier transform domain. Optics Commun 338:181–192
64. Zhang X, Zhao Z, Wang J (2014) Chaotic image encryption based on circular substitution box and key stream buffer. Sig Process Image Commun 29(8):902–913
65. Rhouma R, Belghith S (2011) Cryptoanalysis of a chaos based cryptosystem on DSP. Commun Nonlinear Sci Numer Simul 16(2):876–884
66. Usama M, Khan MK, Alghatbar K, Lee C (2010) Chaos-based secure satellite imagery cryptosystem. Comput Math Appl 60(2):326–337
67. Azar AT, Serrano FE (2014) Robust IMC-PID tuning for cascade control systems with gain and phase margin specifications. Neural Comput Appl 25(5):983–995
68. Azar AT, Serrano FE (2015) Adaptive sliding mode control of the Furuta pendulum. In: Azar AT, Zhu Q (eds) Advances and applications in sliding mode control systems, Studies in computational intelligence, vol 576. Springer, Germany, pp 1–42
69. Azar AT, Serrano FE (2015) Deadbeat control for multivariable systems with time varying delays. In: Azar AT, Vaidyanathan S (eds) Chaos modeling and control systems design, Studies in computational intelligence, vol 581. Springer, Germany, pp 97–132
70. Azar AT, Serrano FE (2015) Design and modeling of anti wind up PID controllers. In: Zhu Q, Azar AT (eds) Complex system modelling and control through intelligent soft computations, Studies in fuzziness and soft computing, vol 319. Springer, Germany, pp 1–44
71. Azar AT, Serrano FE (2015) Stabilizatoin and control of mechanical systems with backlash. In: Azar AT, Vaidyanathan S (eds) Handbook of research on advanced intelligent control engineering and automation. Advances in computational intelligence and robotics (ACIR), IGI-Global, USA, pp 1–60
72. Feki M (2003) An adaptive chaos synchronization scheme applied to secure communication. Chaos, Solitons Fractals 18(1):141–148
73. Murali K, Lakshmanan M (1998) Secure communication using a compound signal from generalized chaotic systems. Phys Lett A 241(6):303–310
74. Zaher AA, Abu-Rezq A (2011) On the design of chaos-based secure communication systems. Commun Nonlinear Syst Numer Simul 16(9):3721–3727
75. Mondal S, Mahanta C (2014) Adaptive second order terminal sliding mode controller for robotic manipulators. J Franklin Inst 351(4):2356–2377
76. Nehmzow U, Walker K (2005) Quantitative description of robot-environment interaction using chaos theory. Robot Auton Syst 53(3–4):177–193
77. Volos CK, Kyprianidis IM, Stouboulos IN (2013) Experimental investigation on coverage performance of a chaotic autonomous mobile robot. Robot Auton Syst 61(12):1314–1322
78. Qu Z (2011) Chaos in the genesis and maintenance of cardiac arrhythmias. Prog Biophys Mol Biol 105(3):247–257
79. Witte CL, Witte MH (1991) Chaos and predicting varix hemorrhage. Med Hypotheses 36(4):312–317
80. Azar AT (2012) Overview of type-2 fuzzy logic systems. Int J Fuzzy Syst Appl 2(4):1–28
81. Li Z, Chen G (2006) Integration of fuzzy logic and chaos theory, Studies in fuzziness and soft computing, vol 187. Springer, Germany
82. Huang X, Zhao Z, Wang Z, Li Y (2012) Chaos and hyperchaos in fractional-order cellular neural networks. Neurocomputing 94:13–21
83. Kaslik E, Sivasundaram S (2012) Nonlinear dynamics and chaos in fractional-order neural networks. Neural Networks 32:245–256
84. Lian S, Chen X (2011) Traceable content protection based on chaos and neural networks. Appl Soft Comput 11(7):4293–4301
85. Guégan D (2009) Chaos in economics and finance. Ann Rev Control 33(1):89–93
86. Sprott JC (2004) Competition with evolution in ecology and finance. Phys Lett A 325(5–6):329–333

87. Sundarapandian V (2010) Output regulation of the Lorenz attractor. Far East J Math Sci 42(2):289–299
88. Vaidyanathan S (2011) Output regulation of Arneodo-Coullet chaotic system. Commun Comput Inform Sci 133:98–107
89. Vaidyanathan S (2011) Output regulation of the unified chaotic system. Commun Comput Inform Sci 198:1–9
90. Sundarapandian V (2013) Adaptive control and synchronization design for the Lu-Xiao chaotic system. Lecture notes in electrical engineering, vol 131, pp 319–327
91. Vaidyanathan S (2012) Adaptive controller and syncrhonizer design for the Qi-Chen chaotic system. Lecture notes of the institute for computer sciences, Social-informatics and telecommunications engineering, vol 84, pp 73–82
92. Vaidyanathan S (2013) A ten-term novel 4-D hyperchaotic system with three quadratic nonlinearities and its control. Int J Control Theory Appl 6(2):97–109
93. Vaidyanathan S (2013) Analysis, control and synchronization of hyperchaotic Zhou system via adaptive control. Adv Intell Syst Comput 177:1–10
94. Vaidyanathan S (2014) Qualitative analysis and control of an eleven-term novel 4-D hyperchaotic system with two quadratic nonlinearities. Int J Control Theory Appl 7:35–47
95. Vaidyanathan S, Azar AT (2015) Analysis and control of a 4-D novel hyperchaotic system. Studies in computational intelligence vol 581, pp 3–17
96. Vaidyanathan S, Volos C, Pham VT (2014) Hyperchaos, adaptive control and synchronization of a novel 5-D hyperchaotic system with three positive Lyapunov exponents and its SPICE implementation. Arch Control Sci 24(4):409–446
97. Vaidyanathan S, Volos C, Pham VT, Madhavan K (2015) Analysis, adaptive control and synchronization of a novel 4-D hyperchaotic hyperjerk system and its SPICE implementation. Nonlinear Dyn 25(1):135–158
98. Vaidyanathan S, Volos CK, Pham VT (2015) Analysis, control, synchronization and SPICE implementation of a novel 4-D hyperchaotic Rikitake dynamo system without equilibrium. J Eng Sci Technol Rev 8(2):232–244
99. Vaidyanathan S (2012) Global chaos control of hyperchaotic Liu system via sliding control method. Int J Control Theory Appl 5(2):117–123
100. Vaidyanathan S (2012) Sliding mode control based global chaos control of Liu-Liu-Liu-Su chaotic system. Int J Control Theory Appl 5(1):15–20
101. Njah AN, Sunday OD (2009) Generalization on the chaos control of 4-D chaotic systems using recursive backstepping nonlinear controller. Chaos, Solitons Fractals 41(5):2371–2376
102. Vaidyanathan S, Idowu BA, Azar AT (2015) Backstepping controller design for the global chaos synchronization of Sprott's jerk systems. Studies in computational intelligence, vol 581, pp 39–58
103. Vincent UE, Njah AN, Laoye JA (2007) Controlling chaos and deterministic directed transport in inertia ratchets using backstepping control. Phys D 231(2):130–136
104. Karthikeyan R, Sundarapandian V (2014) Hybrid chaos synchronization of four-scroll systems via active control. J Electr Eng 65(2):97–103
105. Sarasu P, Sundarapandian V (2011) Active controller design for generalized projective synchronization of four-scroll chaotic systems. Int J Syst Signal Control Eng Appl 4(2):26–33
106. Sarasu P, Sundarapandian V (2011) The generalized projective synchronization of hyperchaotic Lorenz and hyperchaotic Qi systems via active control. Int J Soft Comput 6(5): 216–223
107. Vaidyanathan S, Rajagopal K (2011) Hybrid synchronization of hyperchaotic Wang-Chen and hyperchaotic Lorenz systems by active non-linear control. Int J Syst Signal Control Eng Appl 4(3):55–61
108. Vaidyanathan S, Rasappan S (2011) Global chaos synchronization of hyperchaotic Bao and Xu systems by active nonlinear control. Commun Comput Inform Sci 198:10–17
109. Vaidyanathan S, Azar AT, Rajagopal K, Alexander P (2015) Design and SPICE implementation of a 12-term novel hyperchaotic system and its synchronisation via active control. Int J Model Ident Control 23(3):267–277

110. Vaidyanathan S, Pham VT, Volos CK, (2015) A 5-D hyperchaotic Rikitake dynamo system with hidden attractors. Eur Phys J: Special Topics 224(8):1575–1592
111. Sarasu P, Sundarapandian V (2012) Generalized projective synchronization of two-scroll systems via adaptive control. Int J Soft Comput 7(4):146–156
112. Sundarapandian V, Karthikeyan R (2011) Anti-synchronization of hyperchaotic Lorenz and hyperchaotic Chen systems by adaptive control. Int J Syst Signal Control Eng Appl 4(2):18–25
113. Sundarapandian V, Karthikeyan R (2011) Anti-synchronization of Lü and Pan chaotic systems by adaptive nonlinear control. Eur J Sci Res 64(1):94–106
114. Sundarapandian V, Karthikeyan R (2012) Adaptive anti-synchronization of uncertain Tigan and Li systems. J Eng Appl Sci 7(1):45–52
115. Vaidyanathan S (2012) Anti-synchronization of Sprott-L and Sprott-M chaotic systems via adaptive control. Int J Control Theory Appl 5(1):41–59
116. Vaidyanathan S (2015) Hyperchaos, qualitative analysis, control and synchronisation of a ten-term 4-D hyperchaotic system with an exponential nonlinearity and three quadratic nonlinearities. Int J Model Ident Control 23(4):380–392
117. Vaidyanathan S, Pakiriswamy S (2013) Generalized projective synchronization of six-term Sundarapandian chaotic systems by adaptive control. Int J Control Theory Appl 6(2):153–163
118. Vaidyanathan S, Rajagopal K (2012) Global chaos synchronization of hyperchaotic Pang and hyperchaotic Wang systems via adaptive control. Int J Soft Comput 7(1):28–37
119. Sundarapandian V, Sivaperumal S (2011) Sliding controller design of hybrid synchronization of four-wing chaotic systems. Int J Soft Comput 6(5):224–231
120. Vaidyanathan S (2014) Global chaos synchronisation of identical Li-Wu chaotic systems via sliding mode control. Int J Model Ident Control 22(2):170–177
121. Vaidyanathan S, Sampath S (2012) Anti-synchronization of four-wing chaotic systems via sliding mode control. Int J Autom Comput 9(3):274–279
122. Vaidyanathan S, Sampath S, Azar AT (2015) Global chaos synchronisation of identical chaotic systems via novel sliding mode control method and its application to Zhu system. Int J Model Ident Control 23(1):92–100
123. Rasappan S, Vaidyanathan S (2013) Hybrid synchronization of n-scroll Chua circuits using adaptive backstepping control design with recursive feedback. Malays J Math Sci 73(1):73–95
124. Rasappan S, Vaidyanathan S (2014) Global chaos synchronization of WINDMI and Coullet chaotic systems using adaptive backstepping control design. Kyungpook Math J 54(1): 293–320
125. Suresh R, Sundarapandian V (2013) Global chaos synchronization of a family of n-scroll hyperchaotic Chua circuits using backstepping control with recursive feedback. Far East J Math Sci 7(2):219–246
126. Vaidyanathan S, Rasappan S (2014) Global chaos synchronization of n-scroll Chua circuit and Lur'e system using backstepping control design with recursive feedback. Arab J Sci Eng 39(4):3351–3364
127. Khalil HK (2001) Nonlinear systems, 3rd edn. Prentice Hall, New Jersey

Analysis, Adaptive Control and Synchronization of a Novel 3-D Chaotic System with a Quartic Nonlinearity and Two Quadratic Nonlinearities

Sundarapandian Vaidyanathan

Abstract In this work, we announce a seven-term novel 3-D chaotic system with a quartic nonlinearity and two quadratic nonlinearities. The proposed chaotic system is highly chaotic and it has interesting qualitative properties. The phase portraits of the novel chaotic system are illustrated and the dynamic properties of the highly chaotic system are discussed. The novel 3-D chaotic system has three unstable equilibrium points. We show that the equilibrium point at the origin is a saddle point, while the other two equilibrium points are saddle foci. The novel 3-D chaotic system has rotation symmetry about the x_3 axis, which shows that every non-trivial trajectory of the system must have a twin trajectory. The Lyapunov exponents of the novel 3-D chaotic system are obtained as $L_1 = 8.6606$, $L_2 = 0$ and $L_3 = -26.6523$, while the Kaplan-Yorke dimension of the novel chaotic system is obtained as $D_{KY} = 2.3249$. Since the Maximal Lyapunov Exponent (MLE) of the novel chaotic system has a large value, viz. $L_1 = 8.6606$, the novel chaotic system is highly chaotic. Since the sum of the Lyapunov exponents is negative, the novel chaotic system is dissipative. Next, we apply adaptive control method to derive new results for the global chaos control of the novel chaotic system with unknown parameters. We also apply adaptive control method to derive new results for the global chaos synchronization of the identical novel chaotic systems with unknown parameters. The main adaptive control results are established using Lyapunov stability theory. MATLAB simulations are shown to illustrate all the main results derived in this work.

Keywords Chaos · Chaotic systems · Chaos control · Nonlinear control · Adaptive control · Chaos synchronization

---

S. Vaidyanathan (✉)
Research and Development Centre, Vel Tech University,
Avadi, Chennai 600062, Tamil Nadu, India
e-mail: sundarvtu@gmail.com

A.T. Azar and S. Vaidyanathan (eds.), *Advances in Chaos Theory and Intelligent Control*, Studies in Fuzziness and Soft Computing 337,
DOI 10.1007/978-3-319-30340-6_18

1 Introduction

Chaotic systems are defined as nonlinear dynamical systems which are sensitive to initial conditions, topologically mixing and with dense periodic orbits. Sensitivity to initial conditions of chaotic systems is popularly known as the *butterfly effect*.

Chaotic systems are either conservative or dissipative. The conservative chaotic systems are characterized by the property that they are *volume conserving*. The dissipative chaotic systems are characterized by the property that any asymptotic motion of the chaotic system settles onto a set of measure zero, i.e. a strange attractor. In this research work, we shall announce and discuss a novel 3-D dissipative highly chaotic circulant chaotic system with six sinusoidal nonlinearities.

The Lyapunov exponent of a chaotic system is a measure of the divergence of points which are initially very close and this can be used to quantify chaotic systems. Each nonlinear dynamical system has a spectrum of Lyapunov exponents, which are equal in number to the dimension of the state space. The largest Lyapunov exponent of a nonlinear dynamical system is called the *maximal Lyapunov exponent* (MLE).

In the last few decades, Chaos theory has become a very important and active research field, employing many applications in different disciplines like physics, chemistry, biology, ecology, engineering and economics, among others.

Some classical paradigms of 3-D chaotic systems in the literature are Lorenz system [31], Rössler system [45], ACT system [1], Sprott systems [52], Chen system [14], Lü system [32], Cai system [13], Tigan system [64], etc.

Many new chaotic systems have been discovered in the recent years such as Zhou system [128], Zhu system [129], Li system [27], Sundarapandian systems [57, 61], Vaidyanathan systems [72, 74, 76–79, 83, 90, 100, 101, 103, 109, 111, 114, 117, 118, 120], Pehlivan system [37], Sampath system [47], Pham system [39], etc.

Chaos theory and control systems have many important applications in science and engineering [2, 9–12, 130]. Some commonly known applications are oscillators [23, 51], lasers [28, 125], chemical reactions [17, 38, 87, 88, 91, 93, 94, 98], biology [15, 25, 82, 84–86, 89, 92, 96, 97], ecology [18, 54], encryption [26, 127], cryptosystems [44, 65], mechanical systems [4–8], secure communications [16, 34, 126], robotics [33, 35, 122], cardiology [41, 124], intelligent control [3, 29], neural networks [20, 22, 30], finance [19, 53], memristors [40, 123],etc.

The control of a chaotic system aims to stabilize or regulate the system with the help of a feedback control. There are many methods available for controlling a chaotic system such as active control [55, 66, 67], adaptive control [56, 68, 73, 75, 81, 99, 110, 116, 119], sliding mode control [70, 71], backstepping control [36, 113, 121], etc.

Major works on synchronization of chaotic systems deal with the complete synchronization (CS) which has the design goal of using the output of the master system to control the slave system so that the output of the slave system tracks the output of the master system asymptotically with time.

There are many methods available for chaos synchronization such as active control [21, 48, 49, 104, 106, 112], adaptive control [46, 50, 58–60, 69, 95, 102, 105], sliding mode control [62, 80, 108, 115], backstepping control [42, 43, 63, 107], etc.

In this research work, we announce a seven-term novel 3-D chaotic system with a quartic nonlinearity and two quadratic nonlinearities. Section 2 describes the 3-D dynamic equations and phase portraits of the seven-term novel 3-D chaotic system.

Section 3 details the qualitative analysis and properties of the novel 3-D chaotic system. The Lyapunov exponents of the novel chaotic system are obtained as $L_1 = 8.6606$, $L_2 = 0$ and $L_3 = -26.6523$, while the Kaplan-Yorke dimension of the novel chaotic system is obtained as $D_{KY} = 2.3249$. Since the maximal Lyapunov exponent of the novel chaotic system has a large value, viz. $L_1 = 8.6606$, the novel chaotic system is highly chaotic.

In Sect. 4, we derive new results for the global chaos control of the novel highly chaotic system with unknown parameters. In Sect. 5, we derive new results for the global chaos synchronization of the identical novel highly chaotic systems with unknown parameters. Section 6 contains a summary of the main results derived in this work.

2 A Novel 3-D Chaotic System

In this section, we describe a seven-term novel chaotic system, which is given by the 3-D dynamics

$$\begin{cases} \dot{x}_1 = a(x_2 - x_1) + qx_3 + x_2x_3 \\ \dot{x}_2 = bx_1 - x_1x_3 \\ \dot{x}_3 = -cx_3 + px_1^4 \end{cases} \tag{1}$$

where x_1, x_2, x_3 are the states and a, b, c, p, q are constant, positive parameters.

The novel 3-D system (1) is a seven-term polynomial system with a quartic nonlinearity and three quadratic nonlinearities.

The system (1) exhibits a *highly chaotic attractor* for the parameter values

$$a = 12, \quad b = 55, \quad c = 6, \quad p = 25, \quad q = 0.2 \tag{2}$$

For numerical simulations, we take the initial conditions as

$$x_1(0) = 2.6, \quad x_2(0) = 1.8, \quad x_3(0) = 2.5 \tag{3}$$

Figure 1 depicts the 3-D phase portrait of the novel 3-D chaotic system (1), while Figs. 2, 3 and 4 depict the 2-D projection of the novel 3-D chaotic system (1) on the (x_1, x_2), (x_2, x_3) and (x_1, x_3) planes, respectively. Figures 1, 2, 3 and 4 show that the novel 3-D chaotic system (1) exhibits a *highly chaotic* attractor.

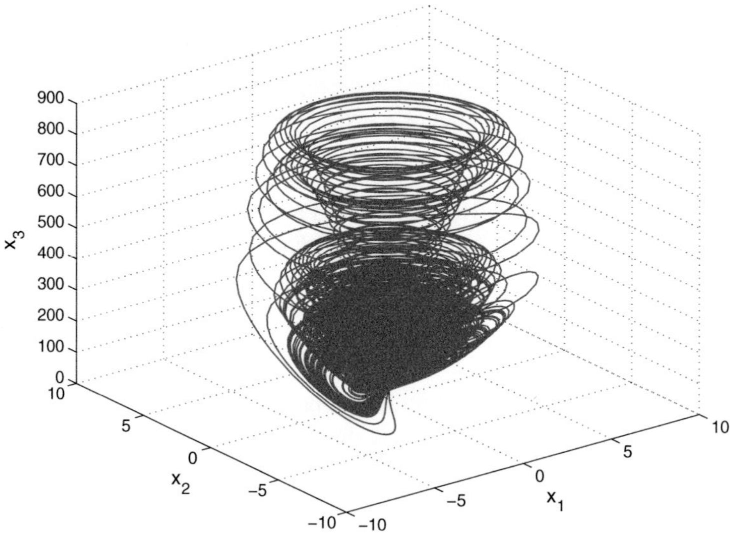

Fig. 1 3-D phase portrait of the novel 3-D chaotic system

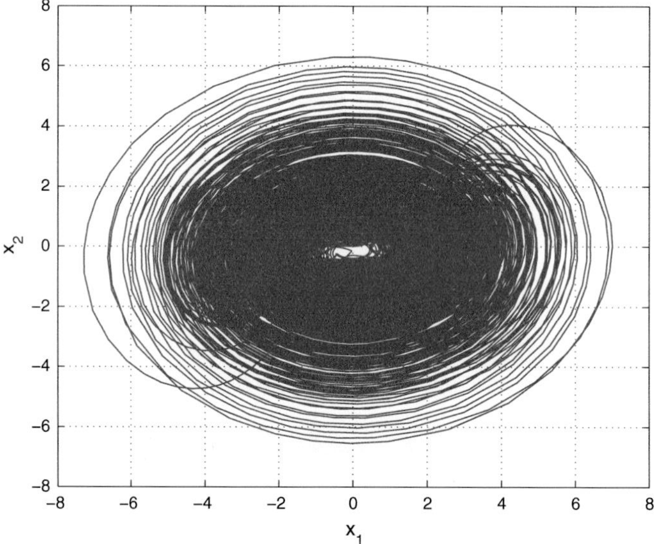

Fig. 2 2-D projection of the novel 3-D chaotic system on the (x_1, x_2) plane

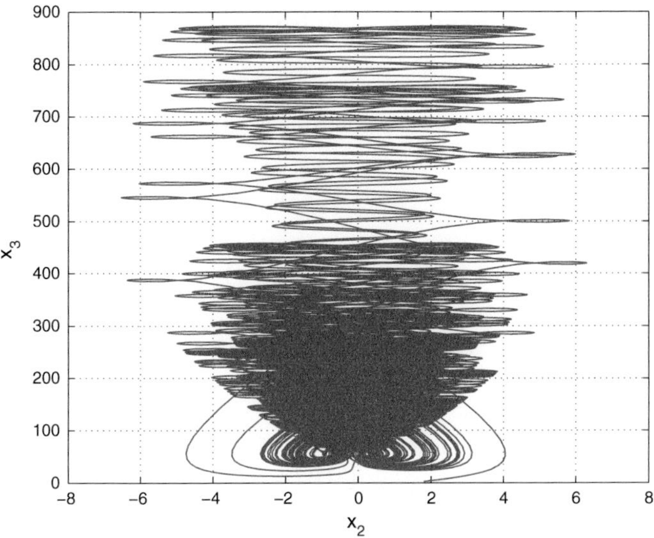

Fig. 3 2-D projection of the novel 3-D chaotic system on the (x_2, x_3) plane

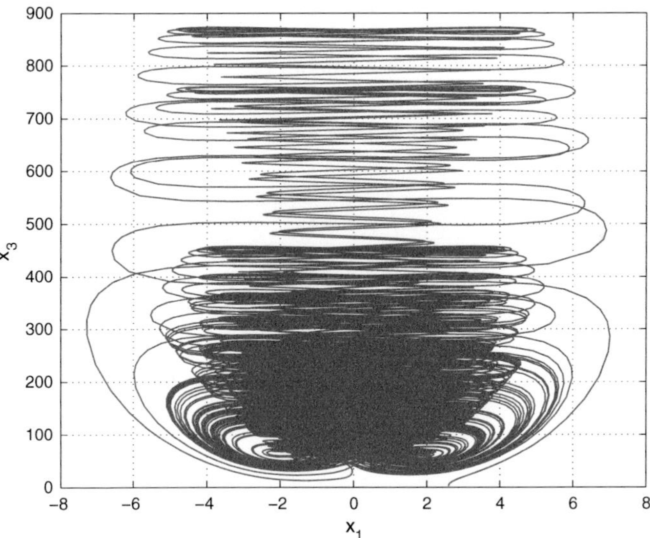

Fig. 4 2-D projection of the novel 3-D chaotic system on the (x_1, x_3) plane

3 Analysis of the Novel 3-D Chaotic System

In this section, we give a dynamic analysis of the 3-D novel highly chaotic system (1). We take the parameter values as in the chaotic case (2), i.e. $a = 12$, $b = 55$, $c = 6$, $p = 25$ and $q = 0.2$.

3.1 Dissipativity

In vector notation, the novel chaotic system (1) can be expressed as

$$\dot{\mathbf{x}} = f(\mathbf{x}) = \begin{bmatrix} f_1(x_1, x_2, x_3) \\ f_2(x_1, x_2, x_3) \\ f_3(x_1, x_2, x_3) \end{bmatrix}, \tag{4}$$

where

$$\begin{cases} f_1(x_1, x_2, x_3) = a(x_2 - x_1) + qx_3 + x_2x_3 \\ f_2(x_1, x_2, x_3) = bx_1 - x_1x_3 \\ f_3(x_1, x_2, x_3) = -cx_3 + px_1^4 \end{cases} \tag{5}$$

Let Ω be any region in $\mathbf{R}^3$ with a smooth boundary and also, $\Omega(t) = \Phi_t(\Omega)$, where Φ_t is the flow of f. Furthermore, let $V(t)$ denote the volume of $\Omega(t)$.

By Liouville's theorem, we know that

$$\dot{V}(t) = \int\limits_{\Omega(t)} (\nabla \cdot f)\, dx_1\, dx_2\, dx_3 \tag{6}$$

The divergence of the novel chaotic system (4) is found as

$$\nabla \cdot f = \frac{\partial f_1}{\partial x_1} + \frac{\partial f_2}{\partial x_2} + \frac{\partial f_3}{\partial x_3} = -(a + c) = -\mu < 0 \tag{7}$$

since $\mu = a + c = 18 > 0$.

Inserting the value of $\nabla \cdot f$ from (7) into (6), we get

$$\dot{V}(t) = \int\limits_{\Omega(t)} (-\mu)\, dx_1\, dx_2\, dx_3 = -\mu V(t) \tag{8}$$

Integrating the first order linear differential equation (8), we get

$$V(t) = \exp(-\mu t) V(0) \tag{9}$$

Since $\mu > 0$, it follows from Eq. (9) that $V(t) \to 0$ exponentially as $t \to \infty$. This shows that the novel chaotic system (1) is dissipative.

Hence, the system limit sets are ultimately confined into a specific limit set of zero volume, and the asymptotic motion of the novel chaotic system (1) settles onto a strange attractor of the system.

3.2 Equilibrium Points

We take the parameter values as in the chaotic case (2), i.e. $a = 12$, $b = 55$, $c = 6$, $p = 25$ and $q = 0.2$.

It is easy to see that the system (1) has three equilibrium points, viz.

$$E_0 = \begin{bmatrix} 0 \\ 0 \\ 0 \end{bmatrix}, \quad E_1 = \begin{bmatrix} 1.9061 \\ 0.1772 \\ 55.0000 \end{bmatrix}, \quad E_2 = \begin{bmatrix} -1.9061 \\ -0.5056 \\ 55.0000 \end{bmatrix} \tag{10}$$

The Jacobian of the system (1) at any point $\mathbf{x} \in \mathbf{R}^3$ is calculated as

$$J(\mathbf{x}) = \begin{bmatrix} -a & a + x_3 & q + x_2 \\ b - x_3 & 0 & -x_1 \\ 4px_1^3 & 0 & -c \end{bmatrix} = \begin{bmatrix} -12 & 12 + x_3 & 0.2 + x_2 \\ 55 - x_3 & 0 & -x_1 \\ 100x_1^3 & 0 & -6 \end{bmatrix} = \tag{11}$$

The Jacobian of the system (1) at the equilibrium E_0 is obtained as

$$J_0 = J(E_0) = \begin{bmatrix} -12 & 12 & 0.2 \\ 55 & 0 & 0 \\ 0 & 0 & -6 \end{bmatrix} \tag{12}$$

We find that the matrix $J_0 = J(E_0)$ has the eigenvalues

$$\lambda_1 = -6, \quad \lambda_2 = -32.3818, \quad \lambda_3 = 20.3818 \tag{13}$$

This shows that the equilibrium point E_0 is a saddle-point, which is unstable.

The Jacobian of the system (1) at the equilibrium E_1 is obtained as

$$J_1 = J(E_1) = \begin{bmatrix} -12 & 67 & 0.3772 \\ 0 & 0 & -1.9061 \\ 692.5275 & 0 & -6 \end{bmatrix} \tag{14}$$

We find that the matrix $J_1 = J(E_1)$ has the eigenvalues

$$\lambda_1 = -53.0246, \quad \lambda_{2,3} = 17.5123 \pm 36.8953i \tag{15}$$

This shows that the equilibrium point E_1 is a saddle-focus, which is unstable.
The Jacobian of the system (1) at the equilibrium E_2 is obtained as

$$J_2 = J(E_2) = \begin{bmatrix} -12 & 67 & -0.3056 \\ 0 & 0 & 1.9061 \\ -692.5275 & 0 & -6 \end{bmatrix} \tag{16}$$

We find that the matrix $J_2 = J(E_2)$ has the eigenvalues

$$\lambda_1 = -52.6091, \quad \lambda_{2,3} = 17.3045 \pm 37.1708i \tag{17}$$

This shows that the equilibrium point E_2 is a saddle-focus, which is unstable.

3.3 Symmetry

It is easy to see that the system (1) is invariant under the change of coordinates

$$(x_1, x_2, x_3) \mapsto (-x_1, -x_2, x_3) \tag{18}$$

Thus, it follows that the 3-D novel chaotic system (1) has rotation symmetry about the x_3-axis.

As a consequence, we conclude that any non-trivial trajectory $\begin{bmatrix} x_1(t) \\ x_2(t) \\ x_3(t) \end{bmatrix}$ of the system (1) must have a twin trajectory $\begin{bmatrix} -x_1(t) \\ -x_2(t) \\ x_3(t) \end{bmatrix}$ of the system (1).

3.4 Lyapunov Exponents and Kaplan-Yorke Dimension

We take the parameter values of the novel system (1) as in the chaotic case (2), i.e.

$$a = 12, \quad b = 55, \quad c = 6, \quad p = 25, \quad q = 0.2 \tag{19}$$

We take the initial state of the novel system (1) as given in (3), i.e.

$$x_1(0) = 2.6, \quad x_2(0) = 1.8, \quad x_3(0) = 2.5 \tag{20}$$

Then the Lyapunov exponents of the system (1) are numerically obtained as

$$L_1 = 8.6606, \quad L_2 = 0, \quad L_3 = -26.6523 \tag{21}$$

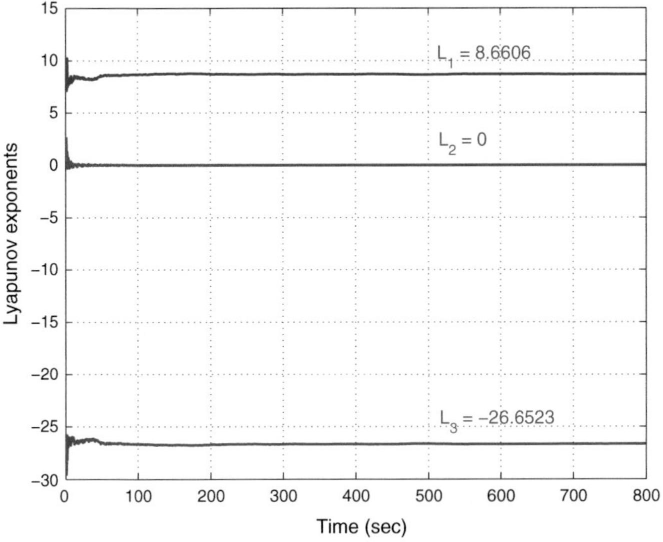

Fig. 5 Dynamics of the Lyapunov exponents of the novel chaotic system

Figure 5 shows the dynamics of the Lyapunov exponents of the novel system (1). From Fig. 5, we note that the Maximal Lyapunov Exponent (MLE) of the novel system (1) is given by $L_1 = 8.6606$, which is a large value. This shows that the novel system (1) is *highly chaotic*.

We also note that the sum of the Lyapunov exponents in (21) is negative, i.e.

$$L_1 + L_2 + L_3 = -17.9917 < 0 \tag{22}$$

This shows that the novel chaotic system (1) is dissipative.

Also, the Kaplan-Yorke dimension of the novel chaotic system (1) is found as

$$D_{KY} = 2 + \frac{L_1 + L_2}{|L_3|} = 2.3249, \tag{23}$$

which is fractional.

Also, the relatively large value of the Kaplan-Yorke dimension of the novel 3-D chaotic system (1), i.e. $D_{KY} = 2.3249$, indicates that the novel 3-D chaotic system exhibits highly complex behaviour. Hence, the novel chaotic system (1) has applications in cryptosystems, secure communication devices, etc.

4 Adaptive Control of the Novel 3-D Chaotic System

In this section, we use adaptive control method to derive an adaptive feedback control law for globally stabilizing the novel 3-D chaotic system with unknown parameters.

Thus, we consider the novel highly chaotic system given by

$$\begin{cases} \dot{x}_1 = a(x_2 - x_1) + qx_3 + x_2x_3 + u_1 \\ \dot{x}_2 = bx_1 - x_1x_3 + u_2 \\ \dot{x}_3 = -cx_3 + px_1^4 + u_3 \end{cases} \tag{24}$$

In (24), x_1, x_2, x_3 are the states and u_1, u_2, u_3 are the adaptive controls to be found using estimates for the unknown system parameters.

We consider the adaptive feedback control law

$$\begin{cases} u_1 = -\hat{a}(t)(x_2 - x_1) - \hat{q}(t)x_3 - x_2x_3 - k_1x_1 \\ u_2 = -\hat{b}(t)x_1 + x_1x_3 - k_2x_2 \\ u_3 = \hat{c}(t)x_3 - \hat{p}(t)x_1^4 - k_3x_3 \end{cases} \tag{25}$$

where k_1, k_2, k_3 are positive gain constants.

Substituting (25) into (24), we get the closed-loop plant dynamics as

$$\begin{cases} \dot{x}_1 = \left[a - \hat{a}(t)\right](x_2 - x_1) + \left[q - \hat{q}(t)\right]x_3 - k_1x_1 \\ \dot{x}_2 = \left[b - \hat{b}(t)\right]x_1 - k_2x_2 \\ \dot{x}_3 = -\left[c - \hat{c}(t)\right]x_3 + \left[p - \hat{p}(t)\right]x_1^4 - k_3x_3 \end{cases} \tag{26}$$

The parameter estimation errors are defined as

$$\begin{cases} e_a(t) = a - \hat{a}(t) \\ e_b(t) = b - \hat{b}(t) \\ e_c(t) = c - \hat{c}(t) \\ e_p(t) = p - \hat{p}(t) \\ e_q(t) = q - \hat{q}(t) \end{cases} \tag{27}$$

Using (27), we can simplify the plant dynamics (26) as

$$\begin{cases} \dot{x}_1 = e_a(x_2 - x_1) + e_qx_3 - k_1x_1 \\ \dot{x}_2 = e_bx_1 - k_2x_2 \\ \dot{x}_3 = -e_cx_3 + e_px_1^4 - k_3x_3 \end{cases} \tag{28}$$

Differentiating (27) with respect to t, we obtain

$$\begin{cases} \dot{e}_a(t) = -\dot{\hat{a}}(t) \\ \dot{e}_b(t) = -\dot{\hat{b}}(t) \\ \dot{e}_c(t) = -\dot{\hat{c}}(t) \\ \dot{e}_p(t) = -\dot{\hat{p}}(t) \\ \dot{e}_q(t) = -\dot{\hat{q}}(t) \end{cases} \tag{29}$$

We consider the quadratic candidate Lyapunov function defined by

$$V(\mathbf{x}, e_a, e_b, e_c, e_p, e_q) = \frac{1}{2}\left(x_1^2 + x_2^2 + x_3^2\right) + \frac{1}{2}\left(e_a^2 + e_b^2 + e_c^2 + e_p^2 + e_q^2\right) \tag{30}$$

Differentiating V along the trajectories of (28) and (29), we obtain

$$\begin{aligned} \dot{V} &= -k_1 x_1^2 - k_2 x_2^2 - k_3 x_3^2 + e_a\left[x_1(x_2 - x_1) - \dot{\hat{a}}\right] + e_b\left[x_1 x_2 - \dot{\hat{b}}\right] \\ &\quad + e_c\left[-x_3^2 - \dot{\hat{c}}\right] + e_p\left[x_1^4 x_3 - \dot{\hat{p}}\right] + e_q\left[x_1 x_3 - \dot{\hat{q}}\right] \end{aligned} \tag{31}$$

In view of (31), we take the parameter update law as

$$\begin{cases} \dot{\hat{a}}(t) = x_1(x_2 - x_1) \\ \dot{\hat{b}}(t) = x_1 x_2 \\ \dot{\hat{c}}(t) = -x_3^2 \\ \dot{\hat{p}}(t) = x_1^4 x_3 \\ \dot{\hat{q}}(t) = x_1 x_3 \end{cases} \tag{32}$$

Next, we state and prove the main result of this section.

Theorem 1 *The novel 3-D highly chaotic system (24) with unknown system parameters is globally and exponentially stabilized for all initial conditions by the adaptive control law (25) and the parameter update law (32), where k_1, k_2, k_3 are positive gain constants.*

Proof We prove this result by applying Lyapunov stability theory [24].

We consider the quadratic Lyapunov function defined by (30), which is clearly a positive definite function on $\mathbf{R}^8$.

By substituting the parameter update law (32) into (31), we obtain the time-derivative of V as

$$\dot{V} = -k_1 x_1^2 - k_2 x_2^2 - k_3 x_3^2 \tag{33}$$

From (33), it is clear that $\dot{V}$ is a negative semi-definite function on $\mathbf{R}^8$.

Thus, we can conclude that the state vector $\mathbf{x}(t)$ and the parameter estimation error are globally bounded i.e.

$$\left[x_1(t)\ x_2(t)\ x_3(t)\ e_a(t)\ e_b(t)\ e_c(t)\ e_p(t)\ e_q(t) \right]^T \in \mathbf{L}_\infty.$$

We define $k = \min\{k_1, k_2, k_3\}$.
Then it follows from (33) that

$$\dot{V} \leq -k\|\mathbf{x}(t)\|^2 \tag{34}$$

Thus, we have

$$k\|\mathbf{x}(t)\|^2 \leq -\dot{V} \tag{35}$$

Integrating the inequality (35) from 0 to t, we get

$$k \int_0^t \|\mathbf{x}(\tau)\|^2 \, d\tau \leq V(0) - V(t) \tag{36}$$

From (36), it follows that $\mathbf{x} \in \mathbf{L}_2$.
Using (28), we can conclude that $\dot{\mathbf{x}} \in \mathbf{L}_\infty$.
Using Barbalat's lemma [24], we conclude that $\mathbf{x}(t) \to 0$ exponentially as $t \to \infty$ for all initial conditions $\mathbf{x}(0) \in \mathbf{R}^3$.

Hence, the novel highly chaotic system (24) with unknown system parameters is globally and exponentially stabilized for all initial conditions by the adaptive control law (25) and the parameter update law (32).

This completes the proof. ■

For the numerical simulations, the classical fourth-order Runge-Kutta method with step size $h = 10^{-8}$ is used to solve the systems (24) and (32), when the adaptive control law (25) is applied.

The parameter values of the novel 3-D chaotic system (24) are taken as in the chaotic case (2), i.e.

$$a = 12, \quad b = 55, \quad c = 6, \quad p = 25, \quad q = 0.2 \tag{37}$$

We take the positive gain constants as $k_i = 8$ for $i = 1, 2, 3$.
Furthermore, as initial conditions of the novel highly chaotic system (24), we take

$$x_1(0) = 7.4, \quad x_2(0) = -10.5, \quad x_3(0) = 12.1 \tag{38}$$

Also, as initial conditions of the parameter estimates, we take

$$\hat{a}(0) = 3.2, \quad \hat{b}(0) = 12.3, \quad \hat{c}(0) = 13.4, \quad \hat{p}(0) = 17.8, \quad \hat{q}(0) = 16.7 \tag{39}$$

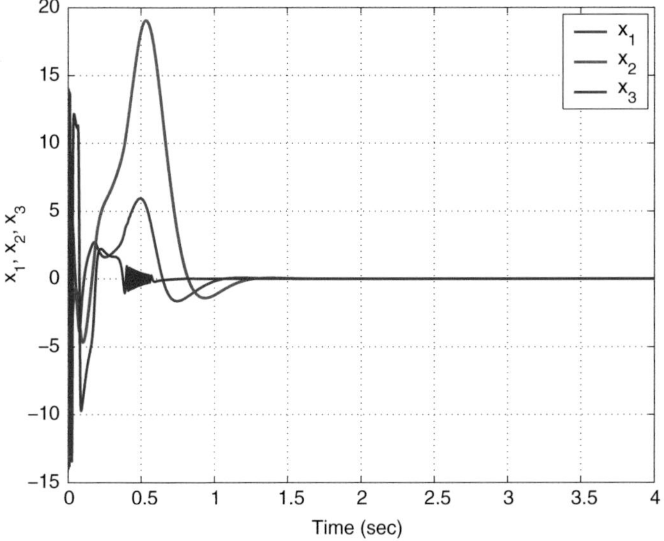

Fig. 6 Time-history of the controlled states x_1, x_2, x_3

In Fig. 6, the exponential convergence of the controlled states of the novel 3-D chaotic system (24) is depicted. From Fig. 6, we see that the controlled states $x_1(t)$, $x_2(t)$, $x_3(t)$ converge to zero in just two seconds.

This shows the efficiency of the adaptive controller designed in this section for the novel 3-D chaotic system (24).

5 Adaptive Synchronization of the Identical Novel 3-D Chaotic Systems

In this section, we apply adaptive control method to derive an adaptive feedback control law for globally synchronizing identical novel 3-D chaotic systems with unknown parameters. The main result is established using Lyapunov stability theory.

As the master system, we consider the novel 3-D chaotic system given by

$$\begin{cases} \dot{x}_1 = a(x_2 - x_1) + qx_3 + x_2x_3 \\ \dot{x}_2 = bx_1 - x_1x_3 \\ \dot{x}_3 = -cx_3 + px_1^4 \end{cases} \tag{40}$$

In (40), x_1, x_2, x_3 are the states and a, b, c, p, q are unknown system parameters.

As the slave system, we consider the novel 3-D chaotic system given by

$$\begin{cases} \dot{y}_1 = a(y_2 - y_1) + qy_3 + y_2y_3 + u_1 \\ \dot{y}_2 = by_1 - y_1y_3 + u_2 \\ \dot{y}_3 = -cy_3 + py_1^4 + u_3 \end{cases} \tag{41}$$

In (41), y_1, y_2, y_3 are the states and u_1, u_2, u_3 are the adaptive controls to be determined using estimates of the unknown system parameters.

The synchronization error between the novel chaotic systems is defined by

$$\begin{cases} e_1 = y_1 - x_1 \\ e_2 = y_2 - x_2 \\ e_3 = y_3 - x_3 \end{cases} \tag{42}$$

Then the error dynamics is obtained as

$$\begin{cases} \dot{e}_1 = a(e_2 - e_1) + qe_3 + y_2y_3 - x_2x_3 + u_1 \\ \dot{e}_2 = be_1 - y_1y_3 + x_1x_3 + u_2 \\ \dot{e}_3 = -ce_3 + p(y_1^4 - x_1^4) + u_3 \end{cases} \tag{43}$$

We consider the adaptive feedback control law

$$\begin{cases} u_1 = -\hat{a}(t)(e_2 - e_1) - \hat{q}(t)e_3 - y_2y_3 + x_2x_3 - k_1e_1 \\ u_2 = -\hat{b}(t)e_1 + y_1y_3 - x_1x_3 - k_2e_2 \\ u_3 = \hat{c}(t)e_3 - \hat{p}(t)(y_1^4 - x_1^4) - k_3e_3 \end{cases} \tag{44}$$

where k_1, k_2, k_3 are positive gain constants.

Substituting (44) into (43), we get the closed-loop error dynamics as

$$\begin{cases} \dot{e}_1 = \left[a - \hat{a}(t)\right](e_2 - e_1) + \left[q - \hat{q}(t)\right]e_3 - k_1e_1 \\ \dot{e}_2 = \left[b - \hat{b}(t)\right]e_1 - k_2e_2 \\ \dot{e}_3 = -\left[c - \hat{c}(t)\right]e_3 + \left[p - \hat{p}(t)\right](y_1^4 - x_1^4) - k_3e_3 \end{cases} \tag{45}$$

The parameter estimation errors are defined as

$$\begin{cases} e_a(t) = a - \hat{a}(t) \\ e_b(t) = b - \hat{b}(t) \\ e_c(t) = c - \hat{c}(t) \\ e_p(t) = p - \hat{p}(t) \\ e_q(t) = q - \hat{q}(t) \end{cases} \tag{46}$$

In view of (46), we can simplify the error dynamics (45) as

$$\begin{cases} \dot{e}_1 = e_a(e_2 - e_1) + e_q e_3 - k_1 e_1 \\ \dot{e}_2 = e_b e_1 - k_2 e_2 \\ \dot{e}_3 = -e_c e_3 + e_p(y_1^4 - x_1^4) - k_3 e_3 \end{cases} \tag{47}$$

Differentiating (46) with respect to t, we obtain

$$\begin{cases} \dot{e}_a(t) = -\dot{\hat{a}}(t) \\ \dot{e}_b(t) = -\dot{\hat{b}}(t) \\ \dot{e}_c(t) = -\dot{\hat{c}}(t) \\ \dot{e}_p(t) = -\dot{\hat{p}}(t) \\ \dot{e}_q(t) = -\dot{\hat{q}}(t) \end{cases} \tag{48}$$

We consider the quadratic candidate Lyapunov function defined by

$$V(\mathbf{e}, e_a, e_b, e_c, e_p, e_q) = \frac{1}{2}\left(e_1^2 + e_2^2 + e_3^2\right) + \frac{1}{2}\left(e_a^2 + e_b^2 + e_c^2 + e_p^2 + e_q^2\right) \tag{49}$$

Differentiating V along the trajectories of (47) and (48), we obtain

$$\begin{aligned} \dot{V} = &-k_1 e_1^2 - k_2 e_2^2 - k_3 e_3^2 + e_a\left[e_1(e_2 - e_1) - \dot{\hat{a}}\right] + e_b\left[e_1 e_2 - \dot{\hat{b}}\right] \\ &+ e_c\left[-e_3^2 - \dot{\hat{c}}\right] + e_p\left[e_3(y_1^4 - x_1^4) - \dot{\hat{p}}\right] + e_q\left[e_1 e_3 - \dot{\hat{q}}\right] \end{aligned} \tag{50}$$

In view of (50), we take the parameter update law as

$$\begin{cases} \dot{\hat{a}}(t) = e_1(e_2 - e_1) \\ \dot{\hat{b}}(t) = e_1 e_2 \\ \dot{\hat{c}}(t) = -e_3^2 \\ \dot{\hat{p}}(t) = e_3(y_1^4 - x_1^4) \\ \dot{\hat{q}}(t) = e_1 e_3 \end{cases} \tag{51}$$

Next, we state and prove the main result of this section.

Theorem 2 *The novel 3-D chaotic systems (40) and (41) with unknown system parameters are globally and exponentially synchronized for all initial conditions by the adaptive control law (44) and the parameter update law (51), where k_1, k_2, k_3 are positive gain constants.*

Proof We prove this result by applying Lyapunov stability theory [24].

We consider the quadratic Lyapunov function defined by (49), which is clearly a positive definite function on $\mathbf{R}^8$.

By substituting the parameter update law (51) into (50), we obtain

$$\dot{V} = -k_1 e_1^2 - k_2 e_2^2 - k_3 e_3^2 \tag{52}$$

From (52), it is clear that $\dot{V}$ is a negative semi-definite function on $\mathbf{R}^8$.

Thus, we can conclude that the error vector $\mathbf{e}(t)$ and the parameter estimation error are globally bounded, i.e.

$$\left[e_1(t)\ e_2(t)\ e_3(t)\ e_a(t)\ e_b(t)\ e_c(t)\ e_p(t)\ e_q(t) \right]^T \in \mathbf{L}_\infty. \tag{53}$$

We define $k = \min\{k_1, k_2, k_3\}$.

Then it follows from (52) that

$$\dot{V} \leq -k\|\mathbf{e}(t)\|^2 \tag{54}$$

Thus, we have

$$k\|\mathbf{e}(t)\|^2 \leq -\dot{V} \tag{55}$$

Integrating the inequality (55) from 0 to t, we get

$$k\int_0^t \|\mathbf{e}(\tau)\|^2\, d\tau \ \leq\ V(0) - V(t) \tag{56}$$

From (56), it follows that $\mathbf{e} \in \mathbf{L}_2$.

Using (47), we can conclude that $\dot{\mathbf{e}} \in \mathbf{L}_\infty$.

Using Barbalat's lemma [24], we conclude that $\mathbf{e}(t) \to 0$ exponentially as $t \to \infty$ for all initial conditions $\mathbf{e}(0) \in \mathbf{R}^3$.

This completes the proof. ∎

For the numerical simulations, the classical fourth-order Runge-Kutta method with step size $h = 10^{-8}$ is used to solve the systems (40), (41) and (51), when the adaptive control law (44) is applied.

The parameter values of the novel chaotic systems are taken as in the chaotic case (2), i.e. $a = 12$, $b = 55$, $c = 6$, $p = 25$ and $q = 0.2$.

We take the positive gain constants as $k_i = 8$ for $i = 1, 2, 3$.

Furthermore, as initial conditions of the master system (40), we take

$$x_1(0) = 12.5, \quad x_2(0) = 20.7, \quad x_3(0) = -5.3 \tag{57}$$

As initial conditions of the slave system (41), we take

$$y_1(0) = 6.8, \quad y_2(0) = 4.5, \quad y_3(0) = 11.4 \tag{58}$$

Also, as initial conditions of the parameter estimates, we take

$$\hat{a}(0) = 3.1, \quad \hat{b}(0) = 4.3, \quad \hat{c}(0) = 10.2, \quad \hat{p}(0) = 5.9, \quad \hat{q}(0) = 7.5 \tag{59}$$

Figures 7, 8 and 9 describe the complete synchronization of the novel 3-D chaotic systems (40) and (41), while Fig. 10 describes the time-history of the complete synchronization errors e_1, e_2, e_3.

From Fig. 7, we see that the states x_1 and y_1 are synchronized in just two seconds. From Fig. 8, we see that the states x_2 and y_2 are synchronized in just two seconds. From Fig. 9, we see that the states x_3 and y_3 are synchronized in just two seconds. From Fig. 10, we see that the errors e_1, e_2, e_3 converge to zero in just two seconds. This shows the efficiency of the adaptive controller developed in this section for the synchronization of identical 3-D chaotic systems.

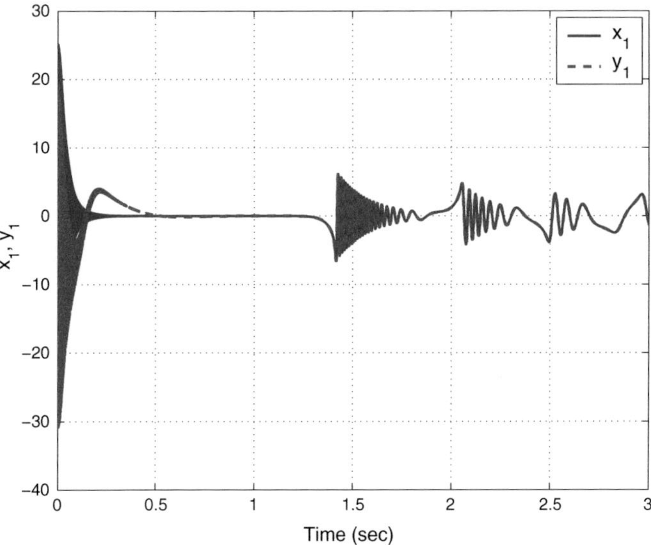

Fig. 7 Synchronization of the states x_1 and y_1

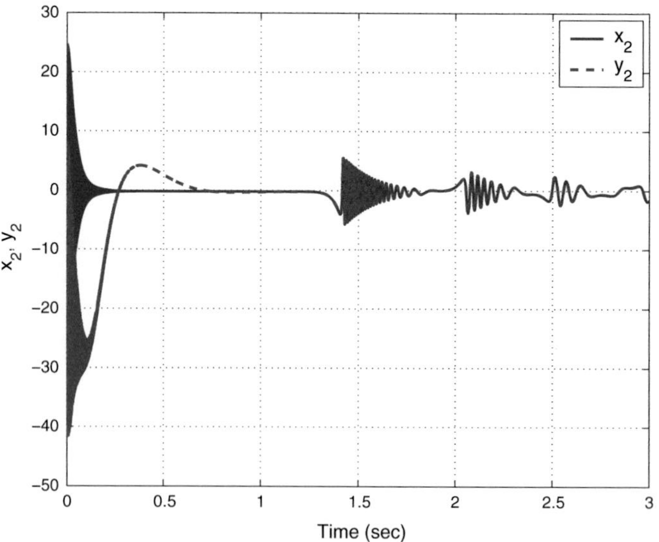

Fig. 8 Synchronization of the states x_2 and y_2

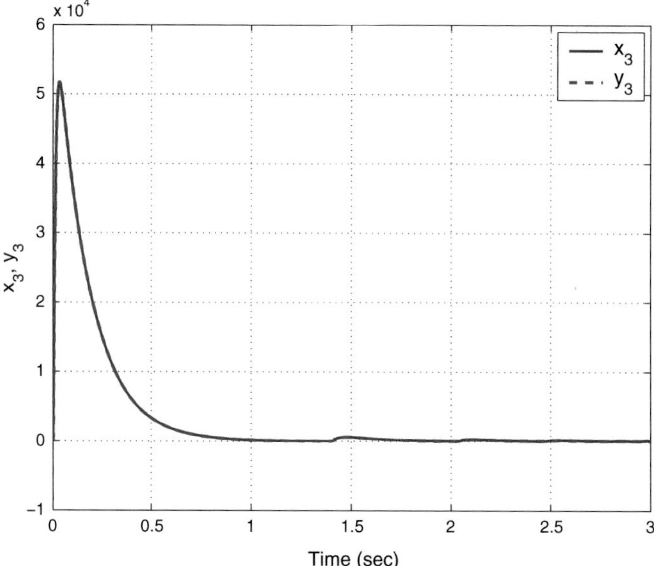

Fig. 9 Synchronization of the states x_3 and y_3

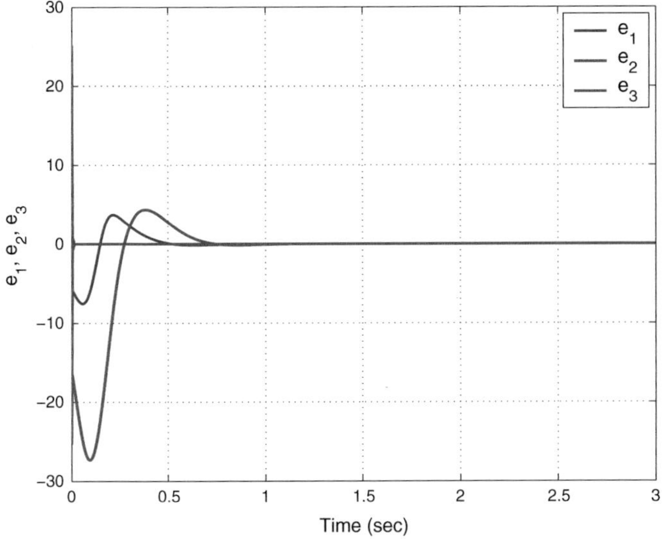

Fig. 10 Time-history of the synchronization errors e_1, e_2, e_3

6 Conclusions

In this work, we announced a seven-term novel 3-D chaotic system with a quartic nonlinearity and two quadratic nonlinearities. The dynamic properties of the novel 3-D chaotic system were discussed and phase portraits of the novel chaotic system were depicted. The novel 3-D chaotic system has three unstable equilibrium points. We showed that the equilibrium point at the origin is a saddle point, while the other two equilibrium points are saddle foci. The novel 3-D chaotic system has rotation symmetry about the x_3 axis, which shows that every non-trivial trajectory of the system must have a twin trajectory. The Lyapunov exponents of the novel 3-D chaotic system have been obtained as $L_1 = 8.6606$, $L_2 = 0$ and $L_3 = -26.6523$, while the Kaplan-Yorke dimension of the novel chaotic system has been obtained as $D_{KY} = 2.3249$. Since the Maximal Lyapunov Exponent (MLE) of the novel chaotic system has a large value, viz. $L_1 = 8.6606$, it follows that the novel chaotic system is highly chaotic. Since the sum of the Lyapunov exponents is negative, the novel chaotic system is dissipative. Next, we derived new results for the global chaos control of the novel chaotic system with unknown parameters via adaptive control method. We also derived new results for the global chaos synchronization of the identical novel chaotic systems with unknown parameters via adaptive control method. The main adaptive control results were proved using Lyapunov stability theory. MATLAB simulations have been shown to illustrate all the main results developed in this work.

References

1. Arneodo A, Coullet P, Tresser C (1981) Possible new strange attractors with spiral structure. Commun Math Phys 79(4):573–576
2. Azar AT (2010) Fuzzy systems. IN-TECH, Vienna, Austria
3. Azar AT (2012) Overview of type-2 fuzzy logic systems. Int J Fuzzy Syst Appl 2(4):1–28
4. Azar AT, Serrano FE (2014) Robust IMC-PID tuning for cascade control systems with gain and phase margin specifications. Neural Comput Appl 25(5):983–995
5. Azar AT, Serrano FE (2015) Adaptive sliding mode control of the Furuta pendulum. In: Azar AT, Zhu Q (eds) Advances and applications in sliding mode control systems. Studies in computational intelligence, vol 576. Springer, Berlin, pp 1–42
6. Azar AT, Serrano FE (2015) Deadbeat control for multivariable systems with time varying delays. In: Azar AT, Vaidyanathan S (eds) Chaos modeling and control systems design. Studies in computational intelligence, vol 576. Springer, Berlin, pp 97–132
7. Azar AT, Serrano FE (2015) Design and modeling of anti wind up PID controllers. In: Zhu Q, Azar AT (eds) Complex system modelling and control through intelligent soft computations. Studies in computational intelligence, vol 319. Springer, Berlin, pp 1–44
8. Azar AT, Serrano FE (2015) Stabilizatoin and control of mechanical systems with backlash. In: Azar AT, Vaidyanathan S (eds) Handbook of research on advanced intelligent control engineering and automation. Advances in computational intelligence and robotics (ACIR). IGI-Global, USA, pp 1–60
9. Azar AT, Vaidyanathan S (2015) Chaos modeling and control systems design. Studies in computational intelligence, vol 581. Springer, Berlin
10. Azar AT, Vaidyanathan S (2015) Computational intelligence applications in modeling and control. Studies in computational intelligence, vol 575. Springer, Berlin
11. Azar AT, Vaidyanathan S (2015) Handbook of research on advanced intelligent control engineering and automation. Advances in computational intelligence and robotics (ACIR). IGI-Global, USA
12. Azar AT, Zhu Q (2015) Advances and applications in sliding mode control systems. Studies in computational intelligence, vol 576. Springer, Berlin
13. Cai G, Tan Z (2007) Chaos synchronization of a new chaotic system via nonlinear control. J Uncertain Syst 1(3):235–240
14. Chen G, Ueta T (1999) Yet another chaotic attractor. Int J Bifurc Chaos 9(7):1465–1466
15. Das S, Goswami D, Chatterjee S, Mukherjee S (2014) Stability and chaos analysis of a novel swarm dynamics with applications to multi-agent systems. Eng Appl Artif Intell 30:189–198
16. Feki M (2003) An adaptive chaos synchronization scheme applied to secure communication. Chaos, Solitons Fractals 18(1):141–148
17. Gaspard P (1999) Microscopic chaos and chemical reactions. Physica A 263(1–4):315–328
18. Gibson WT, Wilson WG (2013) Individual-based chaos: Extensions of the discrete logistic model. J Theor Biol 339:84–92
19. Guégan D (2009) Chaos in economics and finance. Annu Rev Control 33(1):89–93
20. Huang X, Zhao Z, Wang Z, Li Y (2012) Chaos and hyperchaos in fractional-order cellular neural networks. Neurocomputing 94:13–21
21. Karthikeyan R, Sundarapandian V (2014) Hybrid chaos synchronization of four-scroll systems via active control. J Electr Eng 65(2):97–103
22. Kaslik E, Sivasundaram S (2012) Nonlinear dynamics and chaos in fractional-order neural networks. Neural Netw 32:245–256
23. Kengne J, Chedjou JC, Kenne G, Kyamakya K (2012) Dynamical properties and chaos synchronization of improved Colpitts oscillators. Commun Nonlinear Sci Numer Simul 17(7):2914–2923
24. Khalil HK (2001) Nonlinear systems, 3rd edn. Prentice Hall, New Jersey
25. Kyriazis M (1991) Applications of chaos theory to the molecular biology of aging. Exp Gerontol 26(6):569–572

26. Lang J (2015) Color image encryption based on color blend and chaos permutation in the reality-preserving multiple-parameter fractional Fourier transform domain. Opt Commun 338:181–192
27. Li D (2008) A three-scroll chaotic attractor. Phys Lett A 372(4):387–393
28. Li N, Pan W, Yan L, Luo B, Zou X (2014) Enhanced chaos synchronization and communication in cascade-coupled semiconductor ring lasers. Commun Nonlinear Sci Numer Simul 19(6):1874–1883
29. Li Z, Chen G (2006) Integration of fuzzy logic and chaos theory. Studies in fuzziness and soft computing, vol 187. Springer, Berlin
30. Lian S, Chen X (2011) Traceable content protection based on chaos and neural networks. Appl Soft Comput 11(7):4293–4301
31. Lorenz EN (1963) Deterministic periodic flow. J Atmos Sci 20(2):130–141
32. Lü J, Chen G (2002) A new chaotic attractor coined. Int J Bifurc Chaos 12(3):659–661
33. Mondal S, Mahanta C (2014) Adaptive second order terminal sliding mode controller for robotic manipulators. J Franklin Inst 351(4):2356–2377
34. Murali K, Lakshmanan M (1998) Secure communication using a compound signal from generalized chaotic systems. Phys Lett A 241(6):303–310
35. Nehmzow U, Walker K (2005) Quantitative description of robot-environment interaction using chaos theory. Robot Auton Syst 53(3–4):177–193
36. Njah AN, Sunday OD (2009) Generalization on the chaos control of 4-D chaotic systems using recursive backstepping nonlinear controller. Chaos, Solitons Fractals 41(5):2371–2376
37. Pehlivan I, Moroz IM, Vaidyanathan S (2014) Analysis, synchronization and circuit design of a novel butterfly attractor. J Sound Vib 333(20):5077–5096
38. Petrov V, Gaspar V, Masere J, Showalter K (1993) Controlling chaos in Belousov-Zhabotinsky reaction. Nature 361:240–243
39. Pham VT, Vaidyanathan S, Volos CK, Jafari S (2015) Hidden attractors in a chaotic system with an exponential nonlinear term. Eur Phys J: Spec Topics 224(8):1507–1517
40. Pham VT, Volos CK, Vaidyanathan S, Le TP, Vu VY (2015) A memristor-based hyperchaotic system with hidden attractors: dynamics, synchronization and circuital emulating. J Eng Sci Technol Rev 8(2):205–214
41. Qu Z (2011) Chaos in the genesis and maintenance of cardiac arrhythmias. Prog Biophys Mol Biol 105(3):247–257
42. Rasappan S, Vaidyanathan S (2013) Hybrid synchronization of n-scroll Chua circuits using adaptive backstepping control design with recursive feedback. Malays J Math Sci 73(1):73–95
43. Rasappan S, Vaidyanathan S (2014) Global chaos synchronization of WINDMI and Coullet chaotic systems using adaptive backstepping control design. Kyungpook Math J 54(1): 293–320
44. Rhouma R, Belghith S (2011) Cryptoanalysis of a chaos based cryptosystem on DSP. Commun Nonlinear Sci Numer Simul 16(2):876–884
45. Rössler OE (1976) An equation for continuous chaos. Phys Lett A 57(5):397–398
46. Vaidyanathan S, VTP, Volos CK, (2015) A 5-D hyperchaotic Rikitake dynamo system with hidden attractors. Eur Phys J: Spec Topics 224(8):1575–1592
47. Sampath S, Vaidyanathan S, Volos CK, Pham VT (2015) An eight-term novel four-scroll chaotic system with cubic nonlinearity and its circuit simulation. J Eng Sci Technol Rev 8(2):1–6
48. Sarasu P, Sundarapandian V (2011) Active controller design for generalized projective synchronization of four-scroll chaotic systems. Int J Syst Signal Control Eng Appl 4(2):26–33
49. Sarasu P, Sundarapandian V (2011) The generalized projective synchronization of hyperchaotic Lorenz and hyperchaotic Qi systems via active control. Int J Soft Comput 6(5): 216–223
50. Sarasu P, Sundarapandian V (2012) Generalized projective synchronization of two-scroll systems via adaptive control. Int J Soft Comput 7(4):146–156
51. Sharma A, Patidar V, Purohit G, Sud KK (2012) Effects on the bifurcation and chaos in forced Duffing oscillator due to nonlinear damping. Commun Nonlinear Sci Numer Simul 17(6):2254–2269

52. Sprott JC (1994) Some simple chaotic flows. Phys Rev E 50(2):647–650
53. Sprott JC (2004) Competition with evolution in ecology and finance. Phys Lett A 325(5–6):329–333
54. Suérez I (1999) Mastering chaos in ecology. Ecol Model 117(2–3):305–314
55. Sundarapandian V (2010) Output regulation of the Lorenz attractor. Far East J Math Sci 42(2):289–299
56. Sundarapandian V (2013) Adaptive control and synchronization design for the Lu-Xiao chaotic system. Lecture notes in electrical engineering, vol 131, pp 319–327
57. Sundarapandian V (2013) Analysis and anti-synchronization of a novel chaotic system via active and adaptive controllers. J Eng Sci Technol Rev 6(4):45–52
58. Sundarapandian V, Karthikeyan R (2011) Anti-synchronization of hyperchaotic Lorenz and hyperchaotic Chen systems by adaptive control. Int J Syst Signal Control Eng Appl 4(2):18–25
59. Sundarapandian V, Karthikeyan R (2011) Anti-synchronization of Lü and Pan chaotic systems by adaptive nonlinear control. Eur J Sci Res 64(1):94–106
60. Sundarapandian V, Karthikeyan R (2012) Adaptive anti-synchronization of uncertain Tigan and Li systems. J Eng Appl Sci 7(1):45–52
61. Sundarapandian V, Pehlivan I (2012) Analysis, control, synchronization, and circuit design of a novel chaotic system. Math Comput Model 55(7–8):1904–1915
62. Sundarapandian V, Sivaperumal S (2011) Sliding controller design of hybrid synchronization of four-wing chaotic systems. Int J Soft Comput 6(5):224–231
63. Suresh R, Sundarapandian V (2013) Global chaos synchronization of a family of n-scroll hyperchaotic Chua circuits using backstepping control with recursive feedback. Far East J Math Sci 7(2):219–246
64. Tigan G, Opris D (2008) Analysis of a 3D chaotic system. Chaos, Solitons Fractals 36: 1315–1319
65. Usama M, Khan MK, Alghatbar K, Lee C (2010) Chaos-based secure satellite imagery cryptosystem. Comput Math Appl 60(2):326–337
66. Vaidyanathan S (2011) Output regulation of Arneodo-Coullet chaotic system. Commun Comput Inf Sci 133:98–107
67. Vaidyanathan S (2011) Output regulation of the unified chaotic system. Commun Comput Inf Sci 198:1–9
68. Vaidyanathan S (2012) Adaptive controller and syncrhonizer design for the Qi-Chen chaotic system. Lecture notes of the institute for computer sciences. Social-informatics and telecommunications engineering, vol 84, pp 73–82
69. Vaidyanathan S (2012) Anti-synchronization of Sprott-L and Sprott-M chaotic systems via adaptive control. Int J Control Theory Appl 5(1):41–59
70. Vaidyanathan S (2012) Global chaos control of hyperchaotic Liu system via sliding control method. Int J Control Theory Appl 5(2):117–123
71. Vaidyanathan S (2012) Sliding mode control based global chaos control of Liu-Liu-Liu-Su chaotic system. Int J Control Theory Appl 5(1):15–20
72. Vaidyanathan S (2013) A new six-term 3-D chaotic system with an exponential nonlinearity. Far East J Math Sci 79(1):135–143
73. Vaidyanathan S (2013) A ten-term novel 4-D hyperchaotic system with three quadratic nonlinearities and its control. Int J Control Theory Appl 6(2):97–109
74. Vaidyanathan S (2013) Analysis and adaptive synchronization of two novel chaotic systems with hyperbolic sinusoidal and cosinusoidal nonlinearity and unknown parameters. J Eng Sci Technol Rev 6(4):53–65
75. Vaidyanathan S (2013) Analysis, control and synchronization of hyperchaotic Zhou system via adaptive control. Adv Intell Syst Comput 177:1–10
76. Vaidyanathan S (2014) A new eight-term 3-D polynomial chaotic system with three quadratic nonlinearities. Far East J Math Sci 84(2):219–226
77. Vaidyanathan S (2014) Analysis and adaptive synchronization of eight-term 3-D polynomial chaotic systems with three quadratic nonlinearities. Eur Phys J: Spec Topics 223(8): 1519–1529

78. Vaidyanathan S (2014) Analysis, control and synchronisation of a six-term novel chaotic system with three quadratic nonlinearities. Int J Model Ident Control 22(1):41–53
79. Vaidyanathan S (2014) Generalised projective synchronisation of novel 3-D chaotic systems with an exponential non-linearity via active and adaptive control. Int J Model Ident Control 22(3):207–217
80. Vaidyanathan S (2014) Global chaos synchronisation of identical Li-Wu chaotic systems via sliding mode control. Int J Model Ident Control 22(2):170–177
81. Vaidyanathan S (2014) Qualitative analysis and control of an eleven-term novel 4-D hyperchaotic system with two quadratic nonlinearities. Int J Control Theory Appl 7:35–47
82. Vaidyanathan S (2015) 3-cells cellular neural network (CNN) attractor and its adaptive biological control. Int J PharmTech Res 8(4):632–640
83. Vaidyanathan S (2015) A 3-D novel highly chaotic system with four quadratic nonlinearities, its adaptive control and anti-synchronization with unknown parameters. J Eng Sci Technol Rev 8(2):106–115
84. Vaidyanathan S (2015) Adaptive backstepping control of enzymes-substrates system with ferroelectric behaviour in brain waves. Int J PharmTech Res 8(2):256–261
85. Vaidyanathan S (2015) Adaptive biological control of generalized Lotka-Volterra three-species biological system. Int J PharmTech Res 8(4):622–631
86. Vaidyanathan S (2015) Adaptive chaotic synchronization of enzymes-substrates system with ferroelectric behaviour in brain waves. Int J PharmTech Res 8(5):964–973
87. Vaidyanathan S (2015) Adaptive control of a chemical chaotic reactor. Int J PharmTech Res 8(3):377–382
88. Vaidyanathan S (2015) Adaptive synchronization of chemical chaotic reactors. Int J ChemTech Res 8(2):612–621
89. Vaidyanathan S (2015) Adaptive synchronization of generalized Lotka-Volterra three-species biological systems. Int J PharmTech Res 8(5):928–937
90. Vaidyanathan S (2015) Analysis, properties and control of an eight-term 3-D chaotic system with an exponential nonlinearity. Int J Model Ident Control 23(2):164–172
91. Vaidyanathan S (2015) Anti-synchronization of Brusselator chemical reaction systems via adaptive control. Int J ChemTech Res 8(6):759–768
92. Vaidyanathan S (2015) Chaos in neurons and adaptive control of Birkhoff-Shaw strange chaotic attractor. Int J PharmTech Res 8(5):956–963
93. Vaidyanathan S (2015) Dynamics and control of Brusselator chemical reaction. Int J ChemTech Res 8(6):740–749
94. Vaidyanathan S (2015) Dynamics and control of Tokamak system with symmetric and magnetically confined plasma. Int J ChemTech Res 8(6):795–803
95. Vaidyanathan S (2015) Hyperchaos, qualitative analysis, control and synchronisation of a ten-term 4-D hyperchaotic system with an exponential nonlinearity and three quadratic nonlinearities. Int J Model Ident Control 23(4):380–392
96. Vaidyanathan S (2015) Lotka-Volterra population biology models with negative feedback and their ecological monitoring. Int J PharmTech Res 8(5):974–981
97. Vaidyanathan S (2015) Synchronization of 3-cells cellular neural network (CNN) attractors via adaptive control method. Int J PharmTech Res 8(5):946–955
98. Vaidyanathan S (2015) Synchronization of Tokamak systems with symmetric and magnetically confined plasma via adaptive control. Int J ChemTech Res 8(6):818–827
99. Vaidyanathan S, Azar AT (2015) Analysis and control of a 4-D novel hyperchaotic system. Studies in computational intelligence, vol 581, pp 3–17
100. Vaidyanathan S, Azar AT (2015) Analysis, control and synchronization of a nine-term 3-D novel chaotic system. In: Azar AT, Vaidyanathan S (eds) Chaos modelling and control systems design. Studies in computational intelligence, vol 581. Springer, Berlin, pp 19–38
101. Vaidyanathan S, Madhavan K (2013) Analysis, adaptive control and synchronization of a seven-term novel 3-D chaotic system. Int J Control Theory Appl 6(2):121–137
102. Vaidyanathan S, Pakiriswamy S (2013) Generalized projective synchronization of six-term Sundarapandian chaotic systems by adaptive control. Int J Control Theory Appl 6(2):153–163

103. Vaidyanathan S, Pakiriswamy S (2015) A 3-D novel conservative chaotic system and its generalized projective synchronization via adaptive control. J Eng Sci Technol Rev 8(2): 52–60
104. Vaidyanathan S, Rajagopal K (2011) Hybrid synchronization of hyperchaotic Wang-Chen and hyperchaotic Lorenz systems by active non-linear control. Int J Syst Signal Control Eng Appl 4(3):55–61
105. Vaidyanathan S, Rajagopal K (2012) Global chaos synchronization of hyperchaotic Pang and hyperchaotic Wang systems via adaptive control. Int J Soft Comput 7(1):28–37
106. Vaidyanathan S, Rasappan S (2011) Global chaos synchronization of hyperchaotic Bao and Xu systems by active nonlinear control. Commun Comput Inf Sci 198:10–17
107. Vaidyanathan S, Rasappan S (2014) Global chaos synchronization of n-scroll Chua circuit and Lur'e system using backstepping control design with recursive feedback. Arab J Sci Eng 39(4):3351–3364
108. Vaidyanathan S, Sampath S (2012) Anti-synchronization of four-wing chaotic systems via sliding mode control. Int J Autom Comput 9(3):274–279
109. Vaidyanathan S, Volos C (2015) Analysis and adaptive control of a novel 3-D conservative no-equilibrium chaotic system. Arch Control Sci 25(3):333–353
110. Vaidyanathan S, Volos C, Pham VT (2014) Hyperchaos, adaptive control and synchronization of a novel 5-D hyperchaotic system with three positive Lyapunov exponents and its SPICE implementation. Arch Control Sci 24(4):409–446
111. Vaidyanathan S, Volos C, Pham VT, Madhavan K, Idowu BA (2014) Adaptive backstepping control, synchronization and circuit simulation of a 3-D novel jerk chaotic system with two hyperbolic sinusoidal nonlinearities. Arch Control Sci 24(3):375–403
112. Vaidyanathan S, Azar AT, Rajagopal K, Alexander P (2015) Design and SPICE implementation of a 12-term novel hyperchaotic system and its synchronisation via active control. Int J Model Ident Control 23(3):267–277
113. Vaidyanathan S, Idowu BA, Azar AT (2015) Backstepping controller design for the global chaos synchronization of Sprott's jerk systems. Studies in computational intelligence, vol 581, pp 39–58
114. Vaidyanathan S, Rajagopal K, Volos CK, Kyprianidis IM, Stouboulos IN (2015) Analysis, adaptive control and synchronization of a seven-term novel 3-D chaotic system with three quadratic nonlinearities and its digital implementation in LabVIEW. J Eng Sci Technol Rev 8(2):130–141
115. Vaidyanathan S, Sampath S, Azar AT (2015) Global chaos synchronisation of identical chaotic systems via novel sliding mode control method and its application to Zhu system. Int J Model Ident Control 23(1):92–100
116. Vaidyanathan S, Volos C, Pham VT, Madhavan K (2015) Analysis, adaptive control and synchronization of a novel 4-D hyperchaotic hyperjerk system and its SPICE implementation. Nonlinear Dyn 25(1):135–158
117. Vaidyanathan S, Volos CK, Kyprianidis IM, Stouboulos IN, Pham VT (2015) Analysis, adaptive control and anti-synchronization of a six-term novel jerk chaotic system with two exponential nonlinearities and its circuit simulation. J Eng Sci Technol Rev 8(2):24–36
118. Vaidyanathan S, Volos CK, Pham VT (2015) Analysis, adaptive control and adaptive synchronization of a nine-term novel 3-D chaotic system with four quadratic nonlinearities and its circuit simulation. J Eng Sci Technol Rev 8(2):181–191
119. Vaidyanathan S, Volos CK, Pham VT (2015) Analysis, control, synchronization and SPICE implementation of a novel 4-D hyperchaotic Rikitake dynamo system without equilibrium. J Eng Sci Technol Rev 8(2):232–244
120. Vaidyanathan S, Volos CK, Pham VT (2015) Global chaos control of a novel nine-term chaotic system via sliding mode control. In: Azar AT, Zhu Q (eds) Advances and applications in sliding mode control systems. Studies in computational intelligence, vol 576. Springer, Berlin, pp 571–590
121. Vincent UE, Njah AN, Laoye JA (2007) Controlling chaos and deterministic directed transport in inertia ratchets using backstepping control. Physica D 231(2):130–136

122. Volos CK, Kyprianidis IM, Stouboulos IN (2013) Experimental investigation on coverage performance of a chaotic autonomous mobile robot. Robot Auton Syst 61(12):1314–1322
123. Volos CK, Kyprianidis IM, Stouboulos IN, Tlelo-Cuautle E, Vaidyanathan S (2015) Memristor: A new concept in synchronization of coupled neuromorphic circuits. J Eng Sci Technol Rev 8(2):157–173
124. Witte CL, Witte MH (1991) Chaos and predicting varix hemorrhage. Med Hypotheses 36(4):312–317
125. Yuan G, Zhang X, Wang Z (2014) Generation and synchronization of feedback-induced chaos in semiconductor ring lasers by injection-locking. Optik: Int J Light Electron Opt 125(8):1950–1953
126. Zaher AA, Abu-Rezq A (2011) On the design of chaos-based secure communication systems. Commun Nonlinear Syst Numer Simul 16(9):3721–3727
127. Zhang X, Zhao Z, Wang J (2014) Chaotic image encryption based on circular substitution box and key stream buffer. Sig Process Image Commun 29(8):902–913
128. Zhou W, Xu Y, Lu H, Pan L (2008) On dynamics analysis of a new chaotic attractor. Phys Lett A 372(36):5773–5777
129. Zhu C, Liu Y, Guo Y (2010) Theoretic and numerical study of a new chaotic system. Intell Inf Manag 2:104–109
130. Zhu Q, Azar AT (2015) Complex system modelling and control through intelligent, soft computations. Studies in fuzzines and soft computing, vol 319. Springer, Berlin

Qualitative Analysis and Properties of a Novel 4-D Hyperchaotic System with Two Quadratic Nonlinearities and Its Adaptive Control

Sundarapandian Vaidyanathan

Abstract A hyperchaotic attractor is typically defined as chaotic behavior with at least two positive Lyapunov exponents. Combined with one null Lyapunov exponent along the flow and one negative Lyapunov exponent to ensure the boundedness of the solution, the minimal dimension for an autonomous continuous-time hyperchaotic system is four. In this work, we announce an eleven-term novel 4-D hyperchaotic system with only two quadratic nonlinearities. The phase portraits of the eleven-term novel hyperchaotic system are depicted and the dynamic properties of the novel hyperchaotic system are discussed. We establish that the novel hyperchaotic system has three unstable equilibrium points. The Lyapunov exponents of the novel hyperchaotic system are obtained as $L_1 = 2.5112$, $L_2 = 0.3327$, $L_3 = 0$ and $L_4 = -24.7976$. The maximal Lyapunov exponent of the novel hyperchaotic system is found as $L_1 = 2.5112$. Also, the Kaplan-Yorke dimension of the novel hyperchaotic system is derived as $D_{KY} = 3.1147$. Since the sum of the four Lyapunov exponents is negative, the novel 4-D hyperchaotic system is dissipative. Next, an adaptive controller is designed to globally stabilize the novel hyperchaotic system with unknown parameters. Finally, an adaptive controller is also designed to achieve complete synchronization of the identical novel hyperchaotic systems with unknown parameters. The main adaptive control results for stabilization and synchronization of the novel hyperchaotic system are established using Lyapunov stability theory. MATLAB simulations are shown to illustrate all the main results derived in this work for the novel 4-D hyperchaotic system.

Keywords Chaos · Chaotic systems · Hyperchaos · Hyperchaotic systems · Adaptive control · Synchronization

S. Vaidyanathan (✉)
Research and Development Centre, Vel Tech University,
Avadi, Chennai 600062, Tamil Nadu, India
e-mail: sundarvtu@gmail.com

A.T. Azar and S. Vaidyanathan (eds.), *Advances in Chaos Theory and Intelligent Control*, Studies in Fuzziness and Soft Computing 337,
DOI 10.1007/978-3-319-30340-6_19

1 Introduction

In the last few decades, Chaos theory has become an active research field and it has important applications in different disciplines like physics, chemistry, biology, ecology, engineering and economics, among others.

Some classical paradigms of 3-D chaotic systems in the literature are Lorenz system [37], Rössler system [52], ACT system [1], Sprott systems [60], Chen system [15], Lü system [38], Cai system [13], Tigan system [72], etc.

Many new chaotic systems have been discovered in the recent years such as Zhou system [140], Zhu system [142], Li system [32], Wei-Yang system [132], Sundarapandian systems [65, 69], Vaidyanathan systems [80, 82, 84–87, 91, 98, 107, 108, 110, 116, 118, 121, 124, 125, 127], Pehlivan system [42], Sampath system [54], Pham system [45], etc.

The Lyapunov exponent is a measure of the divergence of phase points that are initially very close and can be used to quantify chaotic systems. It is common to refer to the largest Lyapunov exponent as the *Maximal Lyapunov Exponent* (MLE). A positive maximal Lyapunov exponent and phase space compactness are usually taken as defining conditions for a chaotic system.

When the evolution of a system occurs in a four-dimensional space, a second stretching direction can appear. A hyperchaotic system is defined as a chaotic system with at least two positive Lyapunov exponents. Since a hyperchaotic system must have at least two positive Lyapunov exponents, one null Lyapunov exponent along the flow and one negative Lyapunov exponent to ensure the boundedness of the solution, the minimal dimension for an autonomous continuous-time hyperchaotic system is four. Since a hyperchaotic system must have at least two positive Lyapunov exponents, the dynamics of a hyperchaotic system can expand in several different directions simultaneously.

The hyperchaotic systems have more complex dynamical behaviour and they have miscellaneous applications in engineering such as secure communications [18, 31, 134], cryptosystems [22, 50, 141], fuzzy logic [58, 138], electrical circuits [131, 136], etc.

The first 4-D hyperchaotic system was reported by Rössler [53]. Hyperchaotic Rössler system is a nine-term polynomial system with just one quadratic nonlinearity. In the recent decades, hyperchaotic systems have been reported in the chaos literature such as hyperchaotic Lorenz system [24], hyperchaotic Lü system [14], hyperchaotic Chen system [34], hyperchaotic Wang system [130], hyperchaotic Vaidyanathan systems [81, 89, 106, 117, 119, 123, 126], hyperchaotic Pham system [44], etc.

Chaos theory and control systems have many important applications in science and engineering [2, 9–12, 143]. Some commonly known applications are oscillators [27, 59], lasers [33, 135], chemical reactions [19, 43, 95, 96, 99, 101, 102, 105], biology [16, 29, 90, 92–94, 97, 100, 103, 104], ecology [20, 62], encryption [30, 139], cryptosystems [51, 73], mechanical systems [4–8], secure communications [17, 40, 137], robotics [39, 41, 128], cardiology [47, 133], intelligent control [3, 35], neural networks [23, 26, 36], finance [21, 61], memristors [46, 129], etc.

The control of a chaotic or hyperchaotic system aims to stabilize or regulate the system with the help of a feedback control. There are many methods available for controlling a chaotic system such as active control [63, 74, 75], adaptive control [64, 76, 83], sliding mode control [78, 79], backstepping control [120], etc.

The synchronization of chaotic systems aims to synchronize the states of master and slave systems asymptotically with time. There are many methods available for chaos synchronization such as active control [25, 55, 56, 111, 113], adaptive control [57, 66–68, 77, 109, 112], sliding mode control [70, 88, 115, 122], backstepping control [48, 49, 71, 114], etc.

In this research work, we announce an eleven-term novel 4-D hyperchaotic system with two quadratic nonlinearities. Section 2 describes the dynamic equations and phase portraits of the eleven-term novel 4-D hyperchaotic system.

Section 3 details the qualitative properties of the novel hyperchaotic system. The Lyapunov exponents of the novel hyperchaotic system are obtained as $L_1 = 2.5112$, $L_2 = 0.3327$, $L_3 = 0$ and $L_4 = -24.7976$, while the Kaplan-Yorke dimension of the novel hyperchaotic system is obtained as $D_{KY} = 3.1147$. We shall show that the novel hyperchaotic system has a symmetry about the x_3-axis, which will imply that every non-trivial trajectory of the novel 4-D hyperchaotic system has a twin trajectory. We shall also show that the novel hyperchaotic system has three unstable equilibrium points.

In Sect. 4, we design an adaptive controller to globally stabilize the novel hyperchaotic system with unknown parameters. In Sect. 5, an adaptive controller is designed to achieve global chaos synchronization of the identical novel hyperchaotic systems with unknown parameters. The main control results in Sects. 4 and 5 are established using Lyapunov stability theory. Section 6 summarizes the main results obtained in this work for the novel 4-D hyperchaotic system.

2 A Novel 4-D Hyperchaotic System

In this section, we describe a novel hyperchaotic system, which is given by the 4-D dynamics

$$\begin{cases} \dot{x}_1 = a(x_2 - x_1) + x_3 + x_4 \\ \dot{x}_2 = bx_1 - x_1x_3 + x_4 \\ \dot{x}_3 = x_1x_2 - cx_3 \\ \dot{x}_4 = -p(x_1 + x_2) \end{cases} \tag{1}$$

where x_1, x_2, x_3, x_4 are the states and a, b, c, p are constant positive parameters.

The novel 4-D system (1) is an eleven-term polynomial system with two quadratic nonlinearities.

The system (1) exhibits a *strange hyperchaotic attractor* for the parameter values

$$a = 18, \quad b = 125, \quad c = 4, \quad p = 6 \tag{2}$$

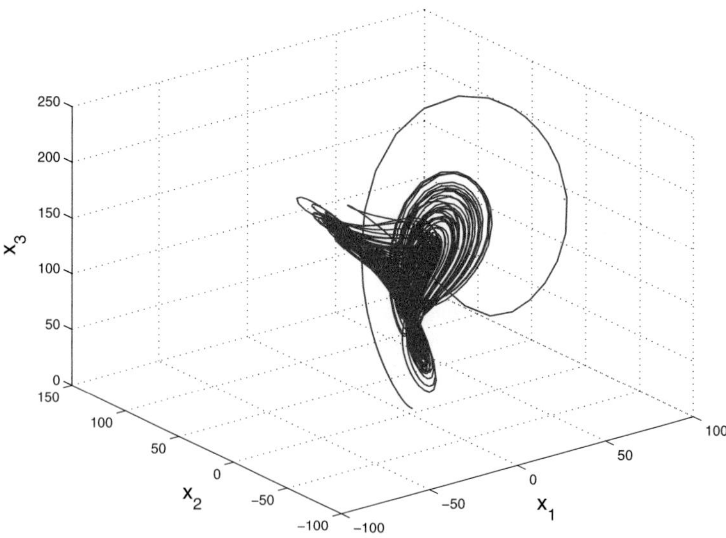

Fig. 1 3-D projection of the novel hyperchaotic system on the (x_1, x_2, x_3) space

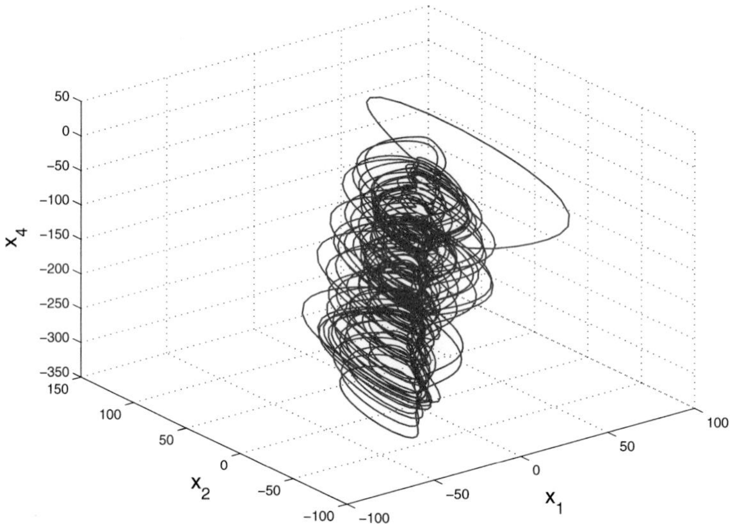

Fig. 2 3-D projection of the novel hyperchaotic system on the (x_1, x_2, x_4) space

For numerical simulations, we take the initial conditions as

$$x_1(0) = 2.8, \quad x_2(0) = 2.1, \quad x_3(0) = 2.5, \quad x_4(0) = 2.4 \tag{3}$$

The Lyapunov exponents for the novel 4-D system (1) for the parameter values (2) and the initial values (3) are numerically calculated as

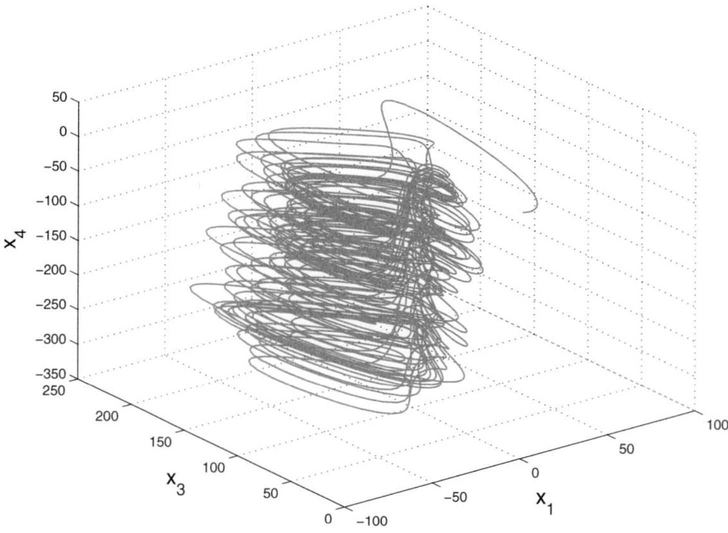

Fig. 3 3-D projection of the novel hyperchaotic system on the (x_1, x_3, x_4) space

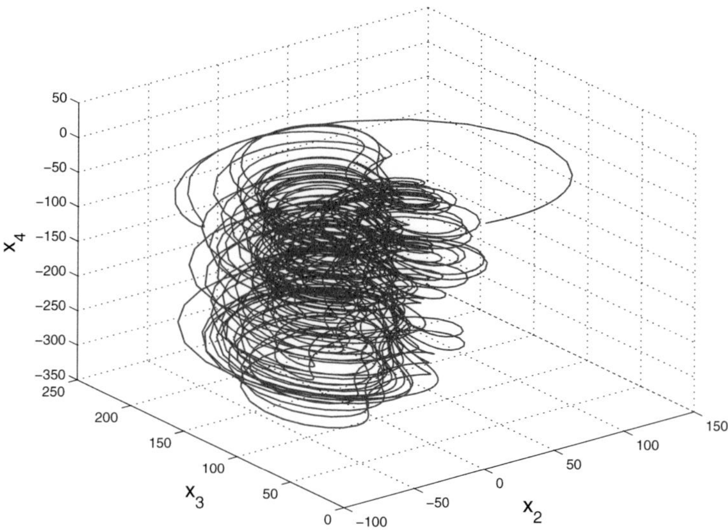

Fig. 4 3-D projection of the novel hyperchaotic system on the (x_2, x_3, x_4) space

$$L_1 = 2.5112, \quad L_2 = 0.3327, \quad L_3 = 0, \quad L_4 = -24.7976 \tag{4}$$

Since there are two positive Lyapunov exponents in the Lyapunov exponents spectrum (4), we conclude that the novel 4-D system (1) is hyperchaotic.

Figures 1, 2, 3 and 4 show the 3-D projection of the novel hyperchaotic system (1) on the (x_1, x_2, x_3), (x_1, x_2, x_4), (x_1, x_3, x_4) and (x_2, x_3, x_4) spaces, respectively. Figure 1 shows that the 3-D projection of the strange attractor on the (x_1, x_2, x_3) space of the novel hyperchaotic system (1) has the shape of a *two-scroll* attractor or a *butterfly* attractor.

3 Analysis of the Novel 4-D Hyperchaotic System

In this section, we study the qualitative properties of the novel 4-D hyperchaotic system (1). We take the parameter values as in the hyperchaotic case (2).

3.1 Dissipativity

In vector notation, the novel hyperchaotic system (1) can be expressed as

$$\dot{\mathbf{x}} = f(\mathbf{x}) = \begin{bmatrix} f_1(x_1, x_2, x_3, x_4) \\ f_2(x_1, x_2, x_3, x_4) \\ f_3(x_1, x_2, x_3, x_4) \\ f_4(x_1, x_2, x_3, x_4) \end{bmatrix}, \tag{5}$$

where

$$\begin{cases} f_1(x_1, x_2, x_3, x_4) = a(x_2 - x_1) + x_3 + x_4 \\ f_2(x_1, x_2, x_3, x_4) = bx_1 - x_1x_3 + x_4 \\ f_3(x_1, x_2, x_3, x_4) = x_1x_2 - cx_3 \\ f_4(x_1, x_2, x_3, x_4) = -p(x_1 + x_2) \end{cases} \tag{6}$$

Let Ω be any region in $\mathbf{R}^4$ with a smooth boundary and also, $\Omega(t) = \Phi_t(\Omega)$, where Φ_t is the flow of f.

Furthermore, let $V(t)$ denote the hypervolume of $\Omega(t)$.

By Liouville's theorem, we know that

$$\dot{V}(t) = \int\limits_{\Omega(t)} (\nabla \cdot f) \, dx_1 \, dx_2 \, dx_3 \, dx_4 \tag{7}$$

The divergence of the novel hyperchaotic system (5) is found as:

$$\nabla \cdot f = \frac{\partial f_1}{\partial x_1} + \frac{\partial f_2}{\partial x_2} + \frac{\partial f_3}{\partial x_3} + \frac{\partial f_4}{\partial x_4} = -(a + c) = -\mu < 0 \tag{8}$$

since $\mu = a + c = 18 + 4 = 22 > 0$.

Inserting the value of $\nabla \cdot f$ from (8) into (7), we get

$$\dot{V}(t) = \int\limits_{\Omega(t)} (-\mu)\, dx_1\, dx_2\, dx_3\, dx_4 = -\mu V(t) \tag{9}$$

Integrating the first order linear differential equation (9), we get

$$V(t) = \exp(-\mu t) V(0) \tag{10}$$

Since $\mu > 0$, it follows from Eq. (10) that $V(t) \to 0$ exponentially as $t \to \infty$. This shows that the novel hyperchaotic system (1) is dissipative. Hence, the system limit sets are ultimately confined into a specific limit set of zero hypervolume, and the asymptotic motion of the novel hyperchaotic system (1) settles onto a strange attractor of the system.

3.2 Equilibrium Points

We take the parameter values as in the hyperchaotic case (2), i.e.

$$a = 18, \quad b = 125, \quad c = 4, \quad p = 6 \tag{11}$$

The equilibrium points of the novel 4-D hyperchaotic system (1) are obtained by solving the system of equations (12):

$$a(x_2 - x_1) + x_3 + x_4 = 0 \tag{12a}$$
$$bx_1 - x_1x_3 + x_4 = 0 \tag{12b}$$
$$x_1x_2 - cx_3 = 0 \tag{12c}$$
$$-p(x_1 + x_2) = 0 \tag{12d}$$

Since $p \neq 0$, it follows from (12d) that $x_1 + x_2 = 0$ or

$$x_2 = -x_1 \tag{13}$$

Substituting (13) into (12c), we get

$$-x_1^2 - cx_3 = 0 \text{ or } x_1^2 = -cx_3 \tag{14}$$

Subtracting (12a) from (12b) and noting $x_2 = -x_1$, we obtain

$$-2ax_1 - bx_1 + x_1x_3 = 0 \text{ or } x_1(-2a + b + x_3) = 0 \tag{15}$$

If we suppose that $x_1 = 0$, then it is easy to see that $x_2 = 0$, $x_3 = 0$ and $x_4 = 0$.

Thus, the first equilibrium point of the novel hyperchaotic system (1) is the origin, viz.

$$E_0 = \begin{bmatrix} 0 \\ 0 \\ 0 \\ 0 \end{bmatrix}. \tag{16}$$

Next, we suppose that $x_1 \neq 0$. From (15), we must have

$$-2a + b + x_3 = 0 \;\text{ or }\; x_3 = 2a - b = 2(18) - 125 = -89 \tag{17}$$

Substituting $x_3 = -89$ in (14), we get

$$x_1^2 = -c(-89) = 89c = 89 \times 4 = 356 \tag{18}$$

Thus, we have

$$x_1 = \pm\sqrt{356} = \pm 18.868 \tag{19}$$

When $x_1 = 18.868$, we have the set of solutions $x_2 = -18.868$, $x_3 = -89$ and $x_4 = 768.248$.

Thus, we get the second equilibrium point of the novel hyperchaotic system (1) as

$$E_1 = \begin{bmatrix} 18.868 \\ -18.868 \\ -89 \\ 768.248 \end{bmatrix}. \tag{20}$$

When $x_1 = -18.868$, we have the set of solutions $x_2 = 18.868$, $x_3 = -89$ and $x_4 = -590.248$.

Thus, we get the third equilibrium point of the novel hyperchaotic system (1) as

$$E_2 = \begin{bmatrix} -18.868 \\ 18.868 \\ -89 \\ -590.248 \end{bmatrix}. \tag{21}$$

Since we have exhausted all cases, E_0, E_1, E_2 are the only equilibrium points of the novel hyperchaotic system (1).

To test the stability type of the equilibrium point $E_0 = \mathbf{0}$, we calculate the Jacobian matrix of the novel hyperchaotic system (1) at $\mathbf{x} = \mathbf{0}$:

We find that

$$J_0 \triangleq J(E_0) = \begin{bmatrix} -18 & 18 & 1 & 1 \\ 125 & 0 & 0 & 1 \\ 0 & 0 & -4 & 0 \\ -6 & -6 & 0 & 0 \end{bmatrix} \tag{22}$$

The matrix J_0 has the eigenvalues

$$\lambda_1 = -57.3306, \ \ \lambda_2 = -4, \ \ \lambda_3 = 0.4332, \ \ \lambda_4 = 38.8974 \tag{23}$$

This shows that the equilibrium point $E_0 = \mathbf{0}$ is a saddle-point, which is unstable. Next, we find that

$$J_1 \triangleq J(E_1) = \begin{bmatrix} -18 & 18 & 1 & 1 \\ 214 & 0 & -18.868 & 1 \\ -18.868 & 18.868 & -4 & 0 \\ -6 & -6 & 0 & 0 \end{bmatrix} \tag{24}$$

The matrix J_1 has the eigenvalues

$$\lambda_1 = -67.9969, \ \ \lambda_2 = -5.6662, \ \ \lambda_3 = 0.5329, \ \ \lambda_4 = 51.1302 \tag{25}$$

This shows that the equilibrium point E_1 is a saddle-point, which is unstable. Next, we find that

$$J_2 \triangleq J(E_2) = \begin{bmatrix} -18 & 18 & 1 & 1 \\ 214 & 0 & 18.868 & 1 \\ 18.868 & -18.868 & -4 & 0 \\ -6 & -6 & 0 & 0 \end{bmatrix} \tag{26}$$

The matrix J_2 has the eigenvalues

$$\lambda_1 = -69.3799, \ \ \lambda_2 = -3.6053, \ \ \lambda_3 = 0.8002, \ \ \lambda_4 = 50.1850 \tag{27}$$

This shows that the equilibrium point E_2 is a saddle-point, which is unstable.

Thus, all three equilibrium points of the novel hyperchaotic system (1) are saddle points, which are unstable.

3.3 Rotation Symmetry About the x_3-Axis

It is easy to see that the novel 4-D hyperchaotic system (1) is invariant under the change of coordinates

$$(x_1, x_2, x_3, x_4) \mapsto (-x_1, -x_2, x_3, -x_4) \tag{28}$$

Since the transformation (28) persists for all values of the system parameters, it follows that the novel 4-D hyperchaotic system (1) has rotation symmetry about the x_3-axis and that any non-trivial trajectory must have a twin trajectory.

3.4 Lyapunov Exponents and Kaplan-Yorke Dimension

We take the parameter values of the novel system (1) as in the hyperchaotic case (2), i.e.

$$a = 18, \quad b = 125, \quad c = 4, \quad p = 6 \tag{29}$$

We take the initial state of the novel system (1) as given in (3), i.e.

$$x_1(0) = 2.8, \quad x_2(0) = 2.1, \quad x_3(0) = 2.5, \quad x_4(0) = 2.4 \tag{30}$$

Then the Lyapunov exponents of the system (1) are numerically obtained using MATLAB as

$$\begin{cases} L_1 = 2.5112 \\ L_2 = 0.3327 \\ L_3 = 0 \\ L_4 = -24.7946 \end{cases} \tag{31}$$

Since there are two positive Lyapunov exponents in (31), the novel system (1) exhibits *hyperchaotic* behavior.

The maximal Lyapunov exponent of the novel hyperchaotic system (1) is easily seen as $L_1 = 2.5112$.

Since $L_1 + L_2 + L_3 + L_4 = -21.9537 < 0$, it follows that the novel hyperchaotic system (1) is dissipative.

Also, the Kaplan-Yorke dimension of the novel hyperchaotic system (1) is calculated as

$$D_{KY} = 3 + \frac{L_1 + L_2 + L_3}{|L_4|} = 3.1147, \tag{32}$$

which is fractional.

4 Adaptive Control of the Novel Hyperchaotic System

In this section, we apply adaptive control method to derive an adaptive feedback control law for globally stabilizing the novel 4-D hyperchaotic system with unknown parameters. The main control result in this section is established using Lyapunov stability theory.

Thus, we consider the controlled novel 4-D hyperchaotic system given by

$$\begin{cases} \dot{x}_1 = a(x_2 - x_1) + x_3 + x_4 + u_1 \\ \dot{x}_2 = bx_1 - x_1x_3 + x_4 + u_2 \\ \dot{x}_3 = x_1x_2 - cx_3 + u_3 \\ \dot{x}_4 = -p(x_1 + x_2) + u_4 \end{cases} \tag{33}$$

In (33), x_1, x_2, x_3, x_4 are the states and u_1, u_2, u_3, u_4 are the adaptive controls to be determined using estimates $\hat{a}(t)$, $\hat{b}(t)$, $\hat{c}(t)$, $\hat{p}(t)$ for the unknown parameters a, b, c, p, respectively.

We consider the adaptive feedback control law

$$\begin{cases} u_1 = -\hat{a}(t)(x_2 - x_1) - x_3 - x_4 - k_1x_1 \\ u_2 = -\hat{b}(t)x_1 + x_1x_3 - x_4 - k_2x_2 \\ u_3 = -x_1x_2 + \hat{c}(t)x_3 - k_3x_3 \\ u_4 = \hat{p}(t)(x_1 + x_2) - k_4x_4 \end{cases} \tag{34}$$

where k_1, k_2, k_3, k_4 are positive gain constants.

Substituting (34) into (33), we get the closed-loop plant dynamics as

$$\begin{cases} \dot{x}_1 = [a - \hat{a}(t)](x_2 - x_1) - k_1x_1 \\ \dot{x}_2 = [b - \hat{b}(t)]x_1 - k_2x_2 \\ \dot{x}_3 = -[c - \hat{c}(t)]x_3 - k_3x_3 \\ \dot{x}_4 = -[p - \hat{p}(t)](x_1 + x_2) - k_4x_4 \end{cases} \tag{35}$$

The parameter estimation errors are defined as

$$\begin{cases} e_a(t) = a - \hat{a}(t) \\ e_b(t) = b - \hat{b}(t) \\ e_c(t) = c - \hat{c}(t) \\ e_p(t) = p - \hat{p}(t) \end{cases} \tag{36}$$

In view of (36), we can simplify the plant dynamics (35) as

$$\begin{cases} \dot{x}_1 = e_a(x_2 - x_1) - k_1x_1 \\ \dot{x}_2 = e_bx_1 - k_2x_2 \\ \dot{x}_3 = -e_cx_3 - k_3x_3 \\ \dot{x}_4 = -e_p(x_1 + x_2) - k_4x_4 \end{cases} \tag{37}$$

Differentiating (36) with respect to t, we obtain

$$\begin{cases} \dot{e}_a(t) = -\dot{\hat{a}}(t) \\ \dot{e}_b(t) = -\dot{\hat{b}}(t) \\ \dot{e}_c(t) = -\dot{\hat{c}}(t) \\ \dot{e}_p(t) = -\dot{\hat{p}}(t) \end{cases} \tag{38}$$

We consider the quadratic candidate Lyapunov function defined by

$$V(\mathbf{x}, e_a, e_b, e_c, e_p) = \frac{1}{2}\left(x_1^2 + x_2^2 + x_3^2 + x_4^2\right) + \frac{1}{2}\left(e_a^2 + e_b^2 + e_c^2 + e_p^2\right) \tag{39}$$

Differentiating V along the trajectories of (37) and (38), we obtain

$$\begin{aligned} \dot{V} = &-k_1x_1^2 - k_2x_2^2 - k_3x_3^2 - k_4x_4^2 + e_a\left[x_1(x_2 - x_1) - \dot{\hat{a}}\right] \\ &+e_b\left[x_1x_2 - \dot{\hat{b}}\right] + e_c\left[-x_3^2 - \dot{\hat{c}}\right] + e_p\left[-x_4(x_1 + x_2) - \dot{\hat{p}}\right] \end{aligned} \tag{40}$$

In view of (40), we take the parameter update law as

$$\begin{cases} \dot{\hat{a}}(t) = x_1(x_2 - x_1) \\ \dot{\hat{b}}(t) = x_1x_2 \\ \dot{\hat{c}}(t) = -x_3^2 \\ \dot{\hat{p}}(t) = -x_4(x_1 + x_2) \end{cases} \tag{41}$$

Next, we state and prove the main result of this section.

Theorem 1 *The novel 4-D hyperchaotic system (33) with unknown system parameters is globally and exponentially stabilized for all initial conditions by the adaptive control law (34) and the parameter update law (41), where k_1, k_2, k_3, k_4 are positive gain constants.*

Proof We prove this result by applying Lyapunov stability theory [28].

We consider the quadratic Lyapunov function defined by (39), which is clearly a positive definite function on $\mathbf{R}^8$.

By substituting the parameter update law (41) into (40), we obtain the time-derivative of V as

$$\dot{V} = -k_1x_1^2 - k_2x_2^2 - k_3x_3^2 - k_4x_4^2 \tag{42}$$

From (42), it is clear that $\dot{V}$ is a negative semi-definite function on $\mathbf{R}^8$.

Thus, we can conclude that the state vector $\mathbf{x}(t)$ and the parameter estimation error are globally bounded, i.e.

$$\left[x_1(t)\ x_2(t)\ x_3(t)\ x_4(t)\ e_a(t)\ e_b(t)\ e_c(t)\ e_p(t)\right]^T \in \mathbf{L}_\infty.$$

We define $k = \min\{k_1, k_2, k_3, k_4\}$.

Then it follows from (42) that

$$\dot{V} \leq -k\|\mathbf{x}(t)\|^2 \tag{43}$$

Thus, we have

$$k\|\mathbf{x}(t)\|^2 \le -\dot{V} \tag{44}$$

Integrating the inequality (44) from 0 to t, we get

$$k\int_0^t \|\mathbf{x}(\tau)\|^2\, d\tau \le V(0) - V(t) \tag{45}$$

From (45), it follows that $\mathbf{x} \in \mathbf{L}_2$.

Using (37), we can conclude that $\dot{\mathbf{x}} \in \mathbf{L}_\infty$.

Using Barbalat's lemma [28], we conclude that $\mathbf{x}(t) \to 0$ exponentially as $t \to \infty$ for all initial conditions $\mathbf{x}(0) \in \mathbf{R}^4$.

Thus, the novel 4-D hyperchaotic system (33) with unknown system parameters is globally and exponentially stabilized for all initial conditions by the adaptive control law (34) and the parameter update law (41).

This completes the proof. □

For the numerical simulations, the classical fourth-order Runge-Kutta method with step size $h = 10^{-8}$ is used to solve the systems (33) and (41), when the adaptive control law (34) is applied.

The parameter values of the novel 4-D hyperchaotic system (33) are taken as in the hyperchaotic case (2), i.e.

$$a = 18, \quad b = 125, \quad c = 4, \quad p = 6 \tag{46}$$

We take the positive gain constants as

$$k_1 = 6, \;\; k_2 = 6, \;\; k_3 = 6, \;\; k_4 = 6 \tag{47}$$

Furthermore, as initial conditions of the novel 4-D hyperchaotic system (33), we take

$$x_1(0) = 17.5, \;\; x_2(0) = -12.7, \;\; x_3(0) = 27.8, \;\; x_4(0) = 22.3 \tag{48}$$

Also, as initial conditions of the parameter estimates, we take

$$\hat{a}(0) = 14.5, \;\; \hat{b}(0) = 20.3, \;\; \hat{c}(0) = 11.2, \;\; \hat{p}(0) = 15.4 \tag{49}$$

In Fig. 5, the exponential convergence of the controlled states of the novel hyperchaotic system (33) is shown. From Fig. 5, it is clear that the controlled states $x_1(t), x_2(t), x_3(4), x_4(t)$ converge to zero in 2 s. This shows the efficiency of the adaptive controller (41) designed in this section.

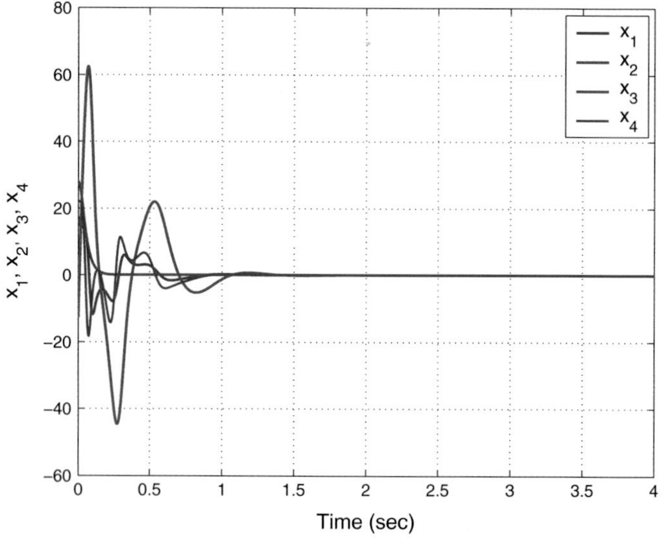

Fig. 5 Time-history of the controlled states x_1, x_2, x_3, x_4

5 Adaptive Synchronization of the Identical Novel Hyperchaotic Systems

In this section, we use adaptive control method to derive an adaptive feedback control law for globally synchronizing identical novel 4-D hyperchaotic systems with unknown parameters.

As the master system, we consider the novel 4-D hyperchaotic system given by

$$\begin{cases} \dot{x}_1 = a(x_2 - x_1) + x_3 + x_4 \\ \dot{x}_2 = bx_1 - x_1x_3 + x_4 \\ \dot{x}_3 = x_1x_2 - cx_3 \\ \dot{x}_4 = -p(x_1 + x_2) \end{cases} \tag{50}$$

In (50), x_1, x_2, x_3, x_4 are the states and a, b, c, p are unknown system parameters.

As the slave system, we consider the 4-D novel hyperchaotic system given by

$$\begin{cases} \dot{y}_1 = a(y_2 - y_1) + y_3 + y_4 + u_1 \\ \dot{y}_2 = by_1 - y_1y_3 + y_4 + u_2 \\ \dot{y}_3 = y_1y_2 - cy_3 + u_3 \\ \dot{y}_4 = -p(y_1 + y_2) + u_4 \end{cases} \tag{51}$$

In (51), y_1, y_2, y_3, y_4 are the states and u_1, u_2, u_3, u_4 are the adaptive controls to be determined using estimates $\hat{a}(t), \hat{b}(t), \hat{c}(t), \hat{p}(t)$ for the unknown parameters a, b, c, p, respectively.

The synchronization error between the novel hyperchaotic systems (50) and (51) is defined by

$$\begin{cases} e_1 = y_1 - x_1 \\ e_2 = y_2 - x_2 \\ e_3 = y_3 - x_3 \\ e_4 = y_4 - x_4 \end{cases} \tag{52}$$

Then the error dynamics is obtained as

$$\begin{cases} \dot{e}_1 = a(e_2 - e_1) + e_3 + e_4 + u_1 \\ \dot{e}_2 = be_1 + e_4 - y_1 y_3 + x_1 x_3 + u_2 \\ \dot{e}_3 = -ce_3 + y_1 y_2 - x_1 x_2 + u_3 \\ \dot{e}_4 = -p(e_1 + e_2) + u_4 \end{cases} \tag{53}$$

We consider the adaptive feedback control law

$$\begin{cases} u_1 = -\hat{a}(t)(e_2 - e_1) - e_3 - e_4 - k_1 e_1 \\ u_2 = -\hat{b}(t)e_1 - e_4 + y_1 y_3 - x_1 x_3 - k_2 e_2 \\ u_3 = \hat{c}(t)e_3 - y_1 y_2 + x_1 x_2 - k_3 e_3 \\ u_4 = \hat{p}(t)(e_1 + e_2) - k_4 e_4 \end{cases} \tag{54}$$

where k_1, k_2, k_3, k_4 are positive gain constants.

Substituting (54) into (53), we get the closed-loop error dynamics as

$$\begin{cases} \dot{e}_1 = \left[a - \hat{a}(t)\right](e_2 - e_1) - k_1 e_1 \\ \dot{e}_2 = \left[b - \hat{b}(t)\right] e_1 - k_2 e_2 \\ \dot{e}_3 = -\left[c - \hat{c}(t)\right] e_3 - k_3 e_3 \\ \dot{e}_4 = -\left[p - \hat{p}(t)\right](e_1 + e_2) - k_4 e_4 \end{cases} \tag{55}$$

The parameter estimation errors are defined as

$$\begin{cases} e_a(t) = a - \hat{a}(t) \\ e_b(t) = b - \hat{b}(t) \\ e_c(t) = c - \hat{c}(t) \\ e_p(t) = p - \hat{p}(t) \end{cases} \tag{56}$$

In view of (56), we can simplify the error dynamics (55) as

$$\begin{cases} \dot{e}_1 = e_a(e_2 - e_1) - k_1 e_1 \\ \dot{e}_2 = e_b e_1 - k_2 e_2 \\ \dot{e}_3 = -e_c e_3 - k_3 e_3 \\ \dot{e}_4 = -e_p(e_1 + e_2) - k_4 e_4 \end{cases} \tag{57}$$

Differentiating (56) with respect to t, we obtain

$$\begin{cases} \dot{e}_a(t) = -\dot{\hat{a}}(t) \\ \dot{e}_b(t) = -\dot{\hat{b}}(t) \\ \dot{e}_c(t) = -\dot{\hat{c}}(t) \\ \dot{e}_p(t) = -\dot{\hat{p}}(t) \end{cases} \tag{58}$$

We use adaptive control theory to find an update law for the parameter estimates. We consider the quadratic candidate Lyapunov function defined by

$$V(\mathbf{e}, e_a, e_b, e_c, e_p) = \frac{1}{2}\left(e_1^2 + e_2^2 + e_3^2 + e_4^2\right) + \frac{1}{2}\left(e_a^2 + e_b^2 + e_c^2 + e_p^2\right) \tag{59}$$

Differentiating V along the trajectories of (57) and (58), we obtain

$$\begin{aligned} \dot{V} = & -k_1 e_1^2 - k_2 e_2^2 - k_3 e_3^2 - k_4 e_4^2 + e_a\left[e_1(e_2 - e_1) - \dot{\hat{a}}\right] \\ & + e_b\left[e_1 e_2 - \dot{\hat{b}}\right] + e_c\left[-e_3^2 - \dot{\hat{c}}\right] + e_p\left[-e_4(e_1 + e_2) - \dot{\hat{p}}\right] \end{aligned} \tag{60}$$

In view of (60), we take the parameter update law as

$$\begin{cases} \dot{\hat{a}}(t) = e_1(e_2 - e_1) \\ \dot{\hat{b}}(t) = e_1 e_2 \\ \dot{\hat{c}}(t) = -e_3^2 \\ \dot{\hat{p}}(t) = -e_4(e_1 + e_2) \end{cases} \tag{61}$$

Next, we state and prove the main result of this section.

This main result for complete synchronization of the identical novel hyperchaotic systems is established using Lyapunov stability theory.

Theorem 2 *The novel hyperchaotic systems (50) and (51) with unknown system parameters are globally and exponentially synchronized for all initial conditions by the adaptive control law (54) and the parameter update law (61), where k_1, k_2, k_3, k_4 are positive gain constants.*

Proof We prove this result by applying Lyapunov stability theory [28].

We consider the quadratic Lyapunov function defined by (59), which is clearly a positive definite function on $\mathbf{R}^8$.

By substituting the parameter update law (61) into (60), we obtain

$$\dot{V} = -k_1 e_1^2 - k_2 e_2^2 - k_3 e_3^2 - k_4 e_4^2 \tag{62}$$

From (62), it is clear that $\dot{V}$ is a negative semi-definite function on $\mathbf{R}^8$.

Thus, we can conclude that the error vector $\mathbf{e}(t)$ and the parameter estimation error are globally bounded, i.e.

$$\left[e_1(t)\ e_2(t)\ e_3(t)\ e_4(t)\ e_a(t)\ e_b(t)\ e_c(t)\ e_p(t) \right]^T \in \mathbf{L}_\infty. \tag{63}$$

We define $k = \min\{k_1, k_2, k_3, k_4\}$.
Then it follows from (62) that

$$\dot{V} \le -k\|\mathbf{e}(t)\|^2 \tag{64}$$

Thus, we have

$$k\|\mathbf{e}(t)\|^2 \le -\dot{V} \tag{65}$$

Integrating the inequality (65) from 0 to t, we get

$$k \int_0^t \|\mathbf{e}(\tau)\|^2 \, d\tau \ \le \ V(0) - V(t) \tag{66}$$

From (66), it follows that $\mathbf{e} \in \mathbf{L}_2$.

Using (57), we can conclude that $\dot{\mathbf{e}} \in \mathbf{L}_\infty$.

Using Barbalat's lemma [28], we conclude that $\mathbf{e}(t) \to 0$ exponentially as $t \to \infty$ for all initial conditions $\mathbf{e}(0) \in \mathbf{R}^4$.

Thus, it follows that the novel hyperchaotic systems (50) and (51) with unknown system parameters are globally and exponentially synchronized for all initial conditions by the adaptive control law (54) and the parameter update law (61).

This completes the proof. □

For the numerical simulations, the classical fourth-order Runge-Kutta method with step size $h = 10^{-8}$ is used to solve the systems (50), (51) and (61), when the adaptive control law (54) is applied.

The parameter values of the novel hyperchaotic systems are taken as in the hyperchaotic case (2), i.e.

$$a = 18, \quad b = 125, \quad c = 4, \quad p = 6 \tag{67}$$

We take the positive gain constants as

$$k_1 = 6, \quad k_2 = 6, \quad k_3 = 6, \quad k_4 = 6 \tag{68}$$

Furthermore, as initial conditions of the master system (50), we take

$$x_1(0) = 2.3, \quad x_2(0) = 21.4, \quad x_3(0) = 12.7, \quad x_4(0) = 22.9 \tag{69}$$

As initial conditions of the slave system (51), we take

$$y_1(0) = 15.1, \quad y_2(0) = 7.5, \quad y_3(0) = 2.8, \quad y_4(0) = 7.4 \tag{70}$$

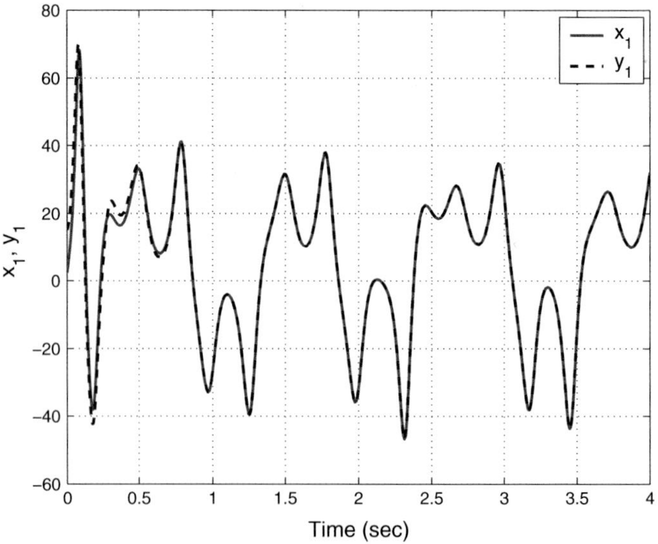

Fig. 6 Synchronization of the states x_1 and y_1

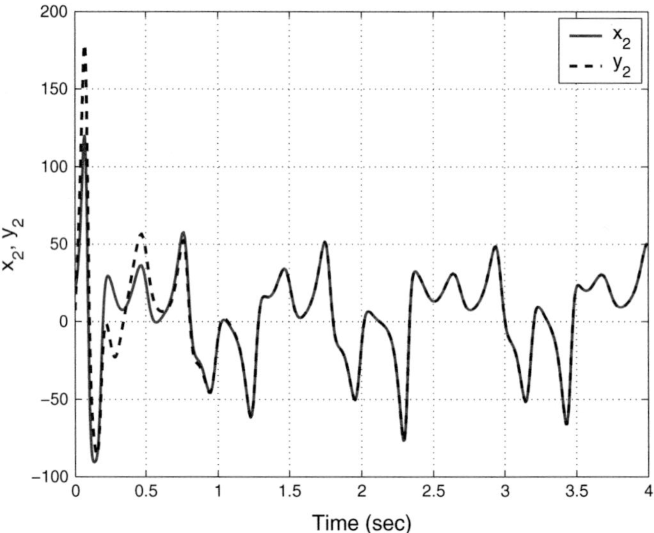

Fig. 7 Synchronization of the states x_2 and y_2

Also, as initial conditions of the parameter estimates, we take

$$\hat{a}(0) = 12.5, \quad \hat{b}(0) = 1.6, \quad \hat{c}(0) = 14.7, \quad \hat{p}(0) = 21.2 \tag{71}$$

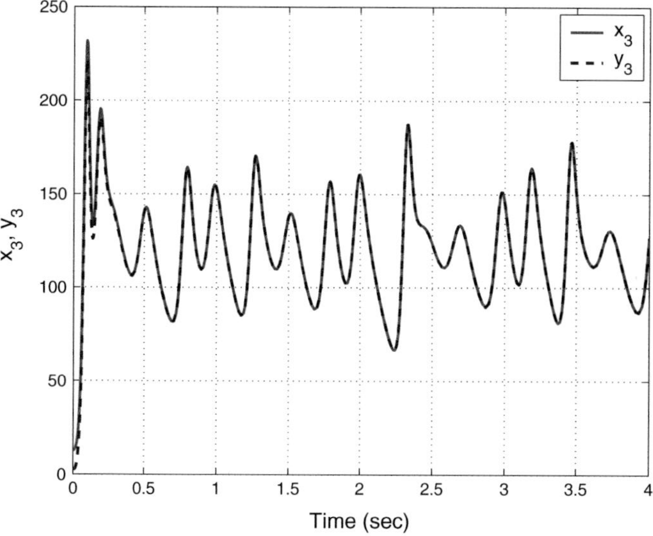

Fig. 8 Synchronization of the states x_3 and y_3

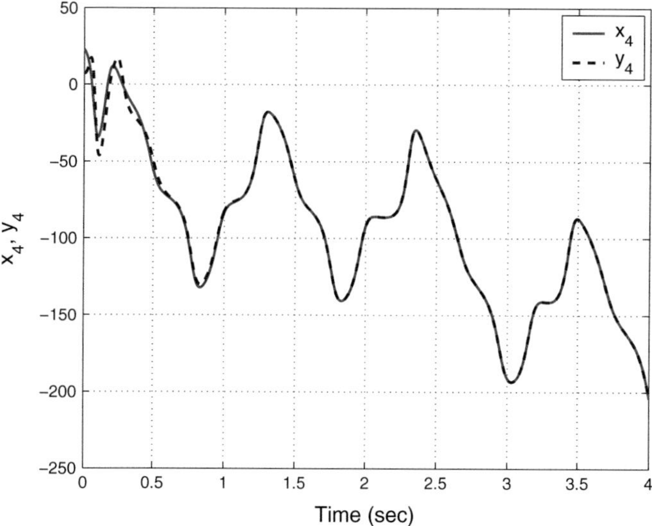

Fig. 9 Synchronization of the states x_4 and y_4

Figures 6, 7, 8 and 9 describe the complete synchronization of the novel hyperchaotic systems (50) and (51). From Figs. 6, 7, 8 and 9, it is clear that the respective states of the novel hyperchaotic systems are completely synchronized in 2 s.

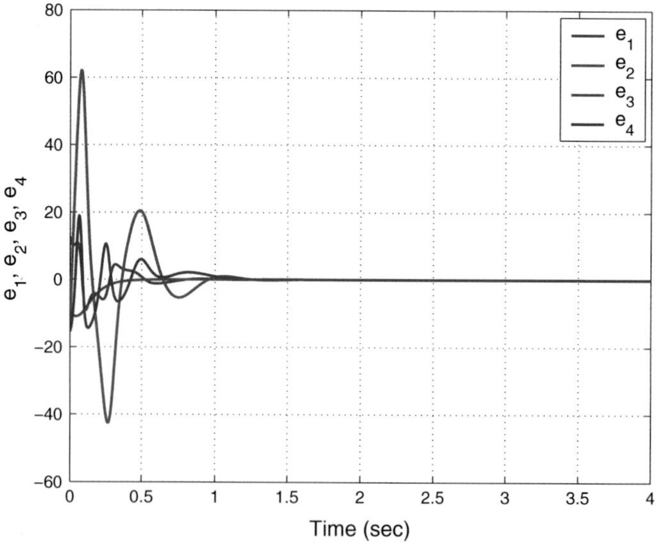

Fig. 10 Time-history of the synchronization errors e_1, e_2, e_3, e_4

Figure 10 describes the time-history of the synchronization errors e_1, e_2, e_3, e_4. From Fig. 10, it is clear that the synchronization errors e_1, e_2, e_3, e_4 converge to zero in 2 s. This shows the efficiency of the adaptive controller developed in this section for the complete synchronization of the novel hyperchaotic systems.

6 Conclusions

In this work, we described an eleven-term novel 4-D hyperchaotic system with only two quadratic nonlinearities. First, we detailed the qualitative properties of the novel hyperchaotic system. We established that the novel hyperchaotic system has three unstable equilibrium points. The Lyapunov exponents of the novel hyperchaotic system have been obtained as $L_1 = 2.5112$, $L_2 = 0.3327$, $L_3 = 0$ and $L_4 = -24.7976$. Thus, the maximal Lyapunov exponent of the novel hyperchaotic system is easily seen as $L_1 = 2.5112$. Also, the Kaplan-Yorke dimension of the novel hyperchaotic system has been derived as $D_{KY} = 3.1147$. Since the sum of the four Lyapunov exponents is negative, the novel 4-D hyperchaotic system is dissipative. Next, an adaptive controller was designed to globally stabilize the novel hyperchaotic system with unknown parameters. Finally, an adaptive controller was also designed to achieve complete synchronization of the identical novel hyperchaotic systems with unknown parameters. The main adaptive control results for stabilization and synchronization of the novel hyperchaotic system were established using Lyapunov stability theory. MATLAB simulations have been depicted to illustrate all the main results derived in this work for the novel 4-D hyperchaotic system.

References

1. Arneodo A, Coullet P, Tresser C (1981) Possible new strange attractors with spiral structure. Commun. Math. Phys. 79(4):573–576
2. Azar AT (2010) Fuzzy systems. IN-TECH, Vienna, Austria
3. Azar AT (2012) Overview of type-2 fuzzy logic systems. Int J Fuzzy Syst Appl 2(4):1–28
4. Azar AT, Serrano FE (2014) Robust IMC-PID tuning for cascade control systems with gain and phase margin specifications. Neural Comput Appl 25(5):983–995
5. Azar AT, Serrano FE (2015a) Adaptive sliding mode control of the Furuta pendulum. In: Azar AT, Zhu Q (eds) Advances and applications in sliding mode control systems. Studies in computational intelligence, vol 576. Springer, Germany, pp 1–42
6. Azar AT, Serrano FE (2015b) Deadbeat control for multivariable systems with time varying delays. In: Azar AT, Vaidyanathan S (eds) Chaos modeling and control systems design, Studies in computational intelligence, vol 581. Springer, Germany, pp 97–132
7. Azar AT, Serrano FE (2015c) Design and modeling of anti wind up PID controllers. In: Zhu Q, Azar AT (eds) Complex system modelling and control through intelligent soft computations, Studies in fuzziness and soft computing, vol 319. Springer, Germany, pp 1–44
8. Azar AT, Serrano FE (2015d) Stabilizatoin and control of mechanical systems with backlash. In: Azar AT, Vaidyanathan S (eds) Handbook of research on advanced intelligent control engineering and automation. Advances in Computational Intelligence and Robotics (ACIR). IGI-Global, USA, pp 1–60
9. Azar AT, Vaidyanathan S (2015a) Chaos modeling and control systems design. Studies in computational intelligence, vol 581. Springer, Germany
10. Azar AT, Vaidyanathan S (2015b) Computational intelligence applications in modeling and control. Studies in computational intelligence, vol 575. Springer, Germany
11. Azar AT, Vaidyanathan S (2015c) Handbook of research on advanced intelligent control engineering and automation. Advances in Computational Intelligence and Robotics (ACIR). IGI-Global, USA
12. Azar AT, Zhu Q (2015) Advances and applications in sliding mode control systems. Studies in computational intelligence, vol 576. Springer, Germany
13. Cai G, Tan Z (2007) Chaos synchronization of a new chaotic system via nonlinear control. J Uncertain Syst 1(3):235–240
14. Chen A, Lu J, Lü J, Yu S (2006) Generating hyperchaotic Lü attractor via state feedback control. Phys A 364:103–110
15. Chen G, Ueta T (1999) Yet another chaotic attractor. Int J Bifurcat Chaos 9(7):1465–1466
16. Das S, Goswami D, Chatterjee S, Mukherjee S (2014) Stability and chaos analysis of a novel swarm dynamics with applications to multi-agent systems. Eng Appl Artif Intell 30:189–198
17. Feki M (2003) An adaptive chaos synchronization scheme applied to secure communication. Chaos Solitons Fractals 18(1):141–148
18. Filali RL, Benrejeb M, Borne P (2014) On observer-based secure communication design using discrete-time hyperchaotic systems. Commun Nonlinear Sci NumerSimul 19(5):1424–1432
19. Gaspard P (1999) Microscopic chaos and chemical reactions. Phys A: Stat Mech Appl 263(1–4):315–328
20. Gibson WT, Wilson WG (2013) Individual-based chaos: extensions of the discrete logistic model. J Theor Biol 339:84–92
21. Guégan D (2009) Chaos in economics and finance. Annu Rev Control 33(1):89–93
22. Hammami S (2015) State feedback-based secure image cryptosystem using hyperchaotic synchronization. ISA Trans 54:52–59
23. Huang X, Zhao Z, Wang Z, Li Y (2012) Chaos and hyperchaos in fractional-order cellular neural networks. Neurocomputing 94:13–21
24. Jia Q (2007) Hyperchaos generated from the Lorenz chaotic system and its control. Phys Lett A 366:217–222
25. Karthikeyan R, Sundarapandian V (2014) Hybrid chaos synchronization of four-scroll systems via active control. J Electr Eng 65(2):97–103

26. Kaslik E, Sivasundaram S (2012) Nonlinear dynamics and chaos in fractional-order neural networks. Neural Netw 32:245–256
27. Kengne J, Chedjou JC, Kenne G, Kyamakya K (2012) Dynamical properties and chaos synchronization of improved Colpitts oscillators. Commun Nonlinear Sci Numer Simul 17(7):2914–2923
28. Khalil HK (2001) Nonlinear systems, 3rd edn. Prentice Hall, New Jersey, USA
29. Kyriazis M (1991) Applications of chaos theory to the molecular biology of aging. Exp Gerontol 26(6):569–572
30. Lang J (2015) Color image encryption based on color blend and chaos permutation in the reality-preserving multiple-parameter fractional Fourier transform domain. Opt Commun 338:181–192
31. Li C, Liao X, Wong KW (2005) Lag synchronization of hyperchaos with application to secure communications. Chaos Solitons Fractals 23(1):183–193
32. Li D (2008) A three-scroll chaotic attractor. Phys Lett A 372(4):387–393
33. Li N, Pan W, Yan L, Luo B, Zou X (2014) Enhanced chaos synchronization and communication in cascade-coupled semiconductor ring lasers. Commun Nonlinear Sci Numer Simul 19(6):1874–1883
34. Li X (2009) Modified projective synchronization of a new hyperchaotic system via nonlinear control. Commun Theor Phys 52:274–278
35. Li Z, Chen G (2006) Integration of fuzzy logic and chaos theory. Studies in fuzziness and soft computing, vol 187. Springer, Germany
36. Lian S, Chen X (2011) Traceable content protection based on chaos and neural networks. Appl Soft Comput 11(7):4293–4301
37. Lorenz EN (1963) Deterministic periodic flow. J Atmos Sci 20(2):130–141
38. Lü J, Chen G (2002) A new chaotic attractor coined. Int J Bifurcat Chaos 12(3):659–661
39. Mondal S, Mahanta C (2014) Adaptive second order terminal sliding mode controller for robotic manipulators. J Franklin Inst 351(4):2356–2377
40. Murali K, Lakshmanan M (1998) Secure communication using a compound signal from generalized chaotic systems. Phys Lett A 241(6):303–310
41. Nehmzow U, Walker K (2005) Quantitative description of robot-environment interaction using chaos theory. Robot Auton Syst 53(3–4):177–193
42. Pehlivan I, Moroz IM, Vaidyanathan S (2014) Analysis, synchronization and circuit design of a novel butterfly attractor. J Sound Vib 333(20):5077–5096
43. Petrov V, Gaspar V, Masere J, Showalter K (1993) Controlling chaos in Belousov-Zhabotinsky reaction. Nature 361:240–243
44. Pham VT, Volos C, Jafari S, Wang X, Vaidyanathan S (2014) Hidden hyperchaotic attractor in a novel simple memristive neural network. Optoelectron Adv Mater Rapid Commun 8(11–12):1157–1163
45. Pham VT, Vaidyanathan S, Volos CK, Jafari S (2015a) Hidden attractors in a chaotic system with an exponential nonlinear term. Eur Phys J Spec Top 224(8):1507–1517
46. Pham VT, Volos CK, Vaidyanathan S, Le TP, Vu VY (2015b) A memristor-based hyperchaotic system with hidden attractors: dynamics, synchronization and circuital emulating. J Eng Sci Technol Rev 8(2):205–214
47. Qu Z (2011) Chaos in the genesis and maintenance of cardiac arrhythmias. Prog Biophys Mol Biol 105(3):247–257
48. Rasappan S, Vaidyanathan S (2013) Hybrid synchronization of n-scroll Chua circuits using adaptive backstepping control design with recursive feedback. Malays J Math Sci 73(1):73–95
49. Rasappan S, Vaidyanathan S (2014) Global chaos synchronization of WINDMI and Coullet chaotic systems using adaptive backstepping control design. Kyungpook Math J 54(1):293–320
50. Rhouma R, Belghith S (2008) Cryptanalysis of a new image encryption algorithm based on hyper-chaos. Phys Lett A 372(38):5973–5978
51. Rhouma R, Belghith S (2011) Cryptoanalysis of a chaos based cryptosystem on DSP. Commun Nonlinear Sci Numer Simul 16(2):876–884

52. Rössler OE (1976) An equation for continuous chaos. Phys Lett A 57(5):397–398
53. Rössler OE (1979) An equation for hyperchaos. Phys Lett A 71:155–157
54. Sampath S, Vaidyanathan S, Volos CK, Pham VT (2015) An eight-term novel four-scroll chaotic system with cubic nonlinearity and its circuit simulation. J Eng Sci Technol Rev 8(2):1–6
55. Sarasu P, Sundarapandian V (2011a) Active controller design for generalized projective synchronization of four-scroll chaotic systems. Int J Syst Signal Control Eng Appl 4(2):26–33
56. Sarasu P, Sundarapandian V (2011b) The generalized projective synchronization of hyperchaotic Lorenz and hyperchaotic Qi systems via active control. Int J Soft Comput 6(5):216–223
57. Sarasu P, Sundarapandian V (2012) Generalized projective synchronization of two-scroll systems via adaptive control. Int J Soft Comput 7(4):146–156
58. Senouci A, Boukabou A (2014) Predictive control and synchronization of chaotic and hyperchaotic systems based on a $T - S$ fuzzy model. Math Comput Simul 105:62–78
59. Sharma A, Patidar V, Purohit G, Sud KK (2012) Effects on the bifurcation and chaos in forced Duffing oscillator due to nonlinear damping. Commun Nonlinear Sci Numer Simul 17(6):2254–2269
60. Sprott JC (1994) Some simple chaotic flows. Phys Rev E 50(2):647–650
61. Sprott JC (2004) Competition with evolution in ecology and finance. Phys Lett A 325(5–6):329–333
62. Suérez I (1999) Mastering chaos in ecology. Ecol Modell 117(2–3):305–314
63. Sundarapandian V (2010) Output regulation of the Lorenz attractor. Far East J Math Sci 42(2):289–299
64. Sundarapandian V (2013a) Adaptive control and synchronization design for the Lu-Xiao chaotic system. Lect Notes Electr Eng 131:319–327
65. Sundarapandian V (2013b) Analysis and anti-synchronization of a novel chaotic system via active and adaptive controllers. J Eng Sci Technol Rev 6(4):45–52
66. Sundarapandian V, Karthikeyan R (2011a) Anti-synchronization of hyperchaotic Lorenz and hyperchaotic Chen systems by adaptive control. Int J Syst Signal Control Eng Appl 4(2):18–25
67. Sundarapandian V, Karthikeyan R (2011b) Anti-synchronization of Lü and Pan chaotic systems by adaptive nonlinear control. Eur J Sci Res 64(1):94–106
68. Sundarapandian V, Karthikeyan R (2012) Adaptive anti-synchronization of uncertain Tigan and Li systems. J Eng Appl Sci 7(1):45–52
69. Sundarapandian V, Pehlivan I (2012) Analysis, control, synchronization, and circuit design of a novel chaotic system. Math Comput Modell 55(7–8):1904–1915
70. Sundarapandian V, Sivaperumal S (2011) Sliding controller design of hybrid synchronization of four-wing chaotic systems. Int J Soft Comput 6(5):224–231
71. Suresh R, Sundarapandian V (2013) Global chaos synchronization of a family of n-scroll hyperchaotic Chua circuits using backstepping control with recursive feedback. Far East J Math Sci 7(2):219–246
72. Tigan G, Opris D (2008) Analysis of a 3D chaotic system. Chaos Solitons Fractals 36:1315–1319
73. Usama M, Khan MK, Alghatbar K, Lee C (2010) Chaos-based secure satellite imagery cryptosystem. Comput Math Appl 60(2):326–337
74. Vaidyanathan S (2011a) Output regulation of Arneodo-Coullet chaotic system. Commun Comput Inf Sci 133:98–107
75. Vaidyanathan S (2011b) Output regulation of the unified chaotic system. Commun Comput Inf Sci 198:1–9
76. Vaidyanathan S (2012a) Adaptive controller and syncrhonizer design for the Qi-Chen chaotic system. Lect Notes Inst Comput Sci Soc-Inf Telecommun Eng 84:73–82
77. Vaidyanathan S (2012b) Anti-synchronization of Sprott-L and Sprott-M chaotic systems via adaptive control. Int J Control Theory Appl 5(1):41–59
78. Vaidyanathan S (2012c) Global chaos control of hyperchaotic Liu system via sliding control method. Int J Control Theory Appl 5(2):117–123

79. Vaidyanathan S (2012d) Sliding mode control based global chaos control of Liu-Liu-Liu-Su chaotic system. Int J Control Theory Appl 5(1):15–20
80. Vaidyanathan S (2013a) A new six-term 3-D chaotic system with an exponential nonlinearity. Far East J Math Sci 79(1):135–143
81. Vaidyanathan S (2013b) A ten-term novel 4-D hyperchaotic system with three quadratic nonlinearities and its control. Int J Control Theory Appl 6(2):97–109
82. Vaidyanathan S (2013c) Analysis and adaptive synchronization of two novel chaotic systems with hyperbolic sinusoidal and cosinusoidal nonlinearity and unknown parameters. J Eng Sci Technol Rev 6(4):53–65
83. Vaidyanathan S (2013d) Analysis, control and synchronization of hyperchaotic Zhou system via adaptive control. Adv Intell Syst Comput 177:1–10
84. Vaidyanathan S (2014a) A new eight-term 3-D polynomial chaotic system with three quadratic nonlinearities. Far East J Math Sci 84(2):219–226
85. Vaidyanathan S (2014b) Analysis and adaptive synchronization of eight-term 3-D polynomial chaotic systems with three quadratic nonlinearities. Eur Phys J: Spec Top 223(8):1519–1529
86. Vaidyanathan S (2014c) Analysis, control and synchronisation of a six-term novel chaotic system with three quadratic nonlinearities. Int J Modell Ident Control 22(1):41–53
87. Vaidyanathan S (2014d) Generalized projective synchronisation of novel 3-D chaotic systems with an exponential non-linearity via active and adaptive control. Int J Modell Ident Control 22(3):207–217
88. Vaidyanathan S (2014e) Global chaos synchronisation of identical Li-Wu chaotic systems via sliding mode control. Int J Modell Ident Control 22(2):170–177
89. Vaidyanathan S (2014f) Qualitative analysis and control of an eleven-term novel 4-D hyperchaotic system with two quadratic nonlinearities. Int J Control Theory Appl 7:35–47
90. Vaidyanathan S (2015a) 3-cells cellular neural network (CNN) attractor and its adaptive biological control. Int J PharmTech Res 8(4):632–640
91. Vaidyanathan S (2015b) A 3-D novel highly chaotic system with four quadratic nonlinearities, its adaptive control and anti-synchronization with unknown parameters. J Eng Sci Technol Rev 8(2):106–115
92. Vaidyanathan S (2015c) Adaptive backstepping control of enzymes-substrates system with ferroelectric behaviour in brain waves. Int J PharmTech Res 8(2):256–261
93. Vaidyanathan S (2015d) Adaptive biological control of generalized Lotka-Volterra three-species biological system. Int J PharmTech Res 8(4):622–631
94. Vaidyanathan S (2015e) Adaptive chaotic synchronization of enzymes-substrates system with ferroelectric behaviour in brain waves. Int J PharmTech Res 8(5):964–973
95. Vaidyanathan S (2015f) Adaptive control of a chemical chaotic reactor. Int J PharmTech Res 8(3):377–382
96. Vaidyanathan S (2015g) Adaptive synchronization of chemical chaotic reactors. Int J ChemTech Res 8(2):612–621
97. Vaidyanathan S (2015h) Adaptive synchronization of generalized Lotka-Volterra three-species biological systems. Int J PharmTech Res 8(5):928–937
98. Vaidyanathan S (2015i) Analysis, properties and control of an eight-term 3-D chaotic system with an exponential nonlinearity. Int J Modell Ident Control 23(2):164–172
99. Vaidyanathan S (2015j) Anti-synchronization of Brusselator chemical reaction systems via adaptive control. Int J ChemTech Res 8(6):759–768
100. Vaidyanathan S (2015k) Chaos in neurons and adaptive control of Birkhoff-Shaw strange chaotic attractor. Int J PharmTech Res 8(5):956–963
101. Vaidyanathan S (2015l) Dynamics and control of Brusselator chemical reaction. Int J ChemTech Res 8(6):740–749
102. Vaidyanathan S (2015m) Dynamics and control of Tokamak system with symmetric and magnetically confined plasma. Int J ChemTech Res 8(6):795–803
103. Vaidyanathan S (2015n) Lotka-Volterra population biology models with negative feedback and their ecological monitoring. Int J PharmTech Res 8(5):974–981

104. Vaidyanathan S (2015o) Synchronization of 3-cells cellular neural network (CNN) attractors via adaptive control method. Int J PharmTech Res 8(5):946–955
105. Vaidyanathan S (2015p) Synchronization of Tokamak systems with symmetric and magnetically confined plasma via adaptive control. Int J ChemTech Res 8(6):818–827
106. Vaidyanathan S, Azar AT (2015a) Analysis and control of a 4-D novel hyperchaotic system. Stud Comput Intell 581:3–17
107. Vaidyanathan S, Azar AT (2015b) Analysis, control and synchronization of a nine-term 3-D novel chaotic system. In: Azar AT, Vaidyanathan S (eds) Chaos modelling and control systems design. Studies in computational intelligence, vol 581. Springer, Germany, pp 19–38
108. Vaidyanathan S, Madhavan K (2013) Analysis, adaptive control and synchronization of a seven-term novel 3-D chaotic system. Int J Control Theory Appl 6(2):121–137
109. Vaidyanathan S, Pakiriswamy S (2013) Generalized projective synchronization of six-term Sundarapandian chaotic systems by adaptive control. Int J Control Theory Appl 6(2):153–163
110. Vaidyanathan S, Pakiriswamy S (2015) A 3-D novel conservative chaotic system and its generalized projective synchronization via adaptive control. J Eng Sci Technol Rev 8(2):52–60
111. Vaidyanathan S, Rajagopal K (2011) Hybrid synchronization of hyperchaotic Wang-Chen and hyperchaotic Lorenz systems by active non-linear control. Int J Syst Signal Control Eng Appl 4(3):55–61
112. Vaidyanathan S, Rajagopal K (2012) Global chaos synchronization of hyperchaotic Pang and hyperchaotic Wang systems via adaptive control. Int J Soft Comput 7(1):28–37
113. Vaidyanathan S, Rasappan S (2011) Global chaos synchronization of hyperchaotic Bao and Xu systems by active nonlinear control. Commun Comput Inf Sci 198:10–17
114. Vaidyanathan S, Rasappan S (2014) Global chaos synchronization of n-scroll Chua circuit and Lur'e system using backstepping control design with recursive feedback. Arab J Sci Eng 39(4):3351–3364
115. Vaidyanathan S, Sampath S (2012) Anti-synchronization of four-wing chaotic systems via sliding mode control. Int J Autom Comput 9(3):274–279
116. Vaidyanathan S, Volos C (2015) Analysis and adaptive control of a novel 3-D conservative no-equilibrium chaotic system. Arch Control Sci 25(3):333–353
117. Vaidyanathan S, Volos C, Pham VT (2014a) Hyperchaos, adaptive control and synchronization of a novel 5-D hyperchaotic system with three positive Lyapunov exponents and its SPICE implementation. Arch Control Sci 24(4):409–446
118. Vaidyanathan S, Volos C, Pham VT, Madhavan K, Idowu BA (2014b) Adaptive backstepping control, synchronization and circuit simulation of a 3-D novel jerk chaotic system with two hyperbolic sinusoidal nonlinearities. Arch Control Sci 24(3):375–403
119. Vaidyanathan S, Azar AT, Rajagopal K, Alexander P (2015a) Design and SPICE implementation of a 12-term novel hyperchaotic system and its synchronisation via active control. Int J Modell Ident Control 23(3):267–277
120. Vaidyanathan S, Idowu BA, Azar AT (2015b) Backstepping controller design for the global chaos synchronization of Sprott's jerk systems. Stud Comput Intell 581:39–58
121. Vaidyanathan S, Rajagopal K, Volos CK, Kyprianidis IM, Stouboulos IN (2015c) Analysis, adaptive control and synchronization of a seven-term novel 3-D chaotic system with three quadratic nonlinearities and its digital implementation in LabVIEW. J Eng Sci Technol Rev 8(2):130–141
122. Vaidyanathan S, Sampath S, Azar AT (2015d) Global chaos synchronisation of identical chaotic systems via novel sliding mode control method and its application to Zhu system. Int J Modell Ident Control 23(1):92–100
123. Vaidyanathan S, Volos C, Pham VT, Madhavan K (2015e) Analysis, adaptive control and synchronization of a novel 4-D hyperchaotic hyperjerk system and its SPICE implementation. Arch Control Sci 25(1):135–158
124. Vaidyanathan S, Volos CK, Kyprianidis IM, Stouboulos IN, Pham VT (2015f) Analysis, adaptive control and anti-synchronization of a six-term novel jerk chaotic system with two exponential nonlinearities and its circuit simulation. J Eng Sci Technol Rev 8(2):24–36

125. Vaidyanathan S, Volos CK, Pham VT (2015g) Analysis, adaptive control and adaptive synchronization of a nine-term novel 3-D chaotic system with four quadratic nonlinearities and its circuit simulation. J Eng Sci Technol Rev 8(2):181–191
126. Vaidyanathan S, Volos CK, Pham VT (2015h) Analysis, control, synchronization and SPICE implementation of a novel 4-D hyperchaotic Rikitake dynamo system without equilibrium. J Eng Sci Technol Rev 8(2):232–244
127. Vaidyanathan S, Volos CK, Pham VT (2015i) Global chaos control of a novel nine-term chaotic system via sliding mode control. In: Azar AT, Zhu Q (eds) Advances and applications in sliding mode control systems. Studies in computational intelligence, vol 576. Springer, Germany, pp 571–590
128. Volos CK, Kyprianidis IM, Stouboulos IN (2013) Experimental investigation on coverage performance of a chaotic autonomous mobile robot. Robot Auton Syst 61(12):1314–1322
129. Volos CK, Kyprianidis IM, Stouboulos IN, Tlelo-Cuautle E, Vaidyanathan S (2015) Memristor: a new concept in synchronization of coupled neuromorphic circuits. J Eng Sci Technol Rev 8(2):157–173
130. Wang J, Chen Z (2008) A novel hyperchaotic system and its complex dynamics. Int J Bifurcat Chaos 18:3309–3324
131. Wei X, Yunfei F, Qiang L (2012) A novel four-wing hyper-chaotic system and its circuit implementation. Procedia Eng 29:1264–1269
132. Wei Z, Yang Q (2010) Anti-control of Hopf bifurcation in the new chaotic system with two stable node-foci. Appl Math Comput 217(1):422–429
133. Witte CL, Witte MH (1991) Chaos and predicting varix hemorrhage. Med Hypotheses 36(4):312–317
134. Wu X, Zhu C, Kan H (2015) An improved secure communication scheme based passive synchronization of hyperchaotic complex nonlinear system. Appl Math Comput 252:201–214
135. Yuan G, Zhang X, Wang Z (2014) Generation and synchronization of feedback-induced chaos in semiconductor ring lasers by injection-locking. Optik—Int J Light Electron Opt 125(8):1950–1953
136. Yujun N, Xingyuan W, Mingjun W, Huaguang Z (2010) A new hyperchaotic system and its circuit implementation. Commun Nonlinear Sci Numer Simul 15(11):3518–3524
137. Zaher AA, Abu-Rezq A (2011) On the design of chaos-based secure communication systems. Commun Nonlinear Sci Numer Simul 16(9):3721–3727
138. Zhang H, Liao X, Yu J (2005) Fuzzy modeling and synchronization of hyperchaotic systems. Chaos Solitons Fractals 26(3):835–843
139. Zhang X, Zhao Z, Wang J (2014) Chaotic image encryption based on circular substitution box and key stream buffer. Signal Process: Image Commun 29(8):902–913
140. Zhou W, Xu Y, Lu H, Pan L (2008) On dynamics analysis of a new chaotic attractor. Phys Lett A 372(36):5773–5777
141. Zhu C (2012) A novel image encryption scheme based on improved hyperchaotic sequences. Opt Commun 285(1):29–37
142. Zhu C, Liu Y, Guo Y (2010) Theoretic and numerical study of a new chaotic system. Intell Inf Manage 2:104–109
143. Zhu Q, Azar AT (2015) Complex system modelling and control through intelligent soft computations. Studies in fuzzines and soft computing, vol 319. Springer, Germany

Global Chaos Synchronization of a Novel 3-D Chaotic System with Two Quadratic Nonlinearities via Active and Adaptive Control

Sundarapandian Vaidyanathan

Abstract In this research work, we announce a six-term novel 3-D dissipative chaotic system with two quadratic nonlinearities. First, this work describes the dynamic equations and qualitative properties of the novel chaotic system. We show that the novel chaotic system has three unstable equilibrium points. We also show that the novel chaotic system has a rotation symmetry about the x_3 axis. The Lyapunov exponents of the novel chaotic system are obtained as $L_1 = 1.2334$, $L_2 = 0$ and $L_3 = -4.7329$. Since the sum of the Lyapunov exponents is negative, the novel chaotic system is dissipative. Also, the Kaplan-Yorke dimension of the novel chaotic system is derived as $D_{KY} = 2.2606$. Next, this work describes the active synchronization of identical novel chaotic systems with known parameters. Furthermore, this work describes the adaptive synchronization of identical novel chaotic systems with unknown parameters. Both the active and adaptive synchronization results are established using Lyapunov stability theory. MATLAB simulations are depicted to illustrate all the main results derived in this work for the six-term novel 3-D novel chaotic system.

Keywords Chaos · Chaotic systems · Synchronization · Active control · Adaptive control

1 Introduction

Chaos theory is the study of how dynamical systems that are modelled by simple, straightforward, deterministic laws can exhibit very complicated and seemingly random long term behavior. A classic example of chaos is the chaos in weather patterns observed by Lorenz [38].

S. Vaidyanathan (✉)
Research and Development Centre, Vel Tech University,
Avadi, Chennai 600062, Tamil Nadu, India
e-mail: sundarvtu@gmail.com

A.T. Azar and S. Vaidyanathan (eds.), *Advances in Chaos Theory and Intelligent Control*, Studies in Fuzziness and Soft Computing 337,
DOI 10.1007/978-3-319-30340-6_20

A hallmark of chaotic systems is its sensitive dependence on the initial conditions [68]. This means that if two trajectories of the system start from initial conditions which are very close to each other, then after a relatively short period of time, the two trajectories will diverge and appear very different from each other. The "butterfly effect" is an example of this, alluding to the idea that the flap of a butterfly's wings in Africa can cause a cascade of events culminating in a tornado in Texas. This sensitive dependence on initial conditions also guarantees that weather forecasting will not be accurate for more than a few days in advance.

Some classical paradigms of 3-D chaotic systems in the literature are Lorenz system [38], Rössler system [56], ACT system [1], Sprott systems [66], Chen system [15], Lü system [39], Liu system [37], Cai system [13], Chen-Lee system [16], Tigan system [80], etc.

Many new chaotic systems have been discovered in the recent years such as Zhou system [151], Zhu system [152], Li system [32], Wei-Yang system [144], Sundarapandian systems [72, 77], Vaidyanathan systems [90, 91, 93–96, 99, 106, 116, 119, 121, 130, 132, 135, 137, 138, 140], Pehlivan system [45], Sampath system [58], Pham system [47], etc.

Chaos theory and control systems have many important applications in science and engineering [2, 9–12, 153]. Some commonly known applications are oscillators [28, 65], lasers [34, 147], chemical reactions [21, 46, 103, 104, 107, 109, 110, 114], biology [18, 30, 98, 100–102, 105, 108, 112, 113], ecology [22, 69], encryption [31, 150], cryptosystems [55, 81], mechanical systems [4–8], secure communications [19, 41, 148], robotics [40, 42, 142], cardiology [49, 145], intelligent control [3, 35], neural networks [24, 27, 36], finance [23, 67], memristors [48, 143], etc.

Control of chaotic systems aims to regulate or stabilize the chaotic system about its equilibrium points. There are many methods available in the literature such as active control method [70, 83, 84, 133], adaptive control method [43, 139], sliding mode control method [87, 89], backstepping control method [141], etc.

Synchronization of chaotic systems is a phenomenon that occurs when two or more chaotic systems are coupled or when a chaotic system drives another chaotic system. Major works on synchronization of chaotic systems deal with the complete synchronization of a pair of chaotic systems called the *master* and *slave* systems. The design goal of the complete synchronization is to apply the output of the master system to control the slave system so that the output of the slave system tracks the output of the master system asymptotically with time.

Thus, if $x(t)$ and $y(t)$ denote the states of the master and slave systems, then the design goal of complete synchronization is to satisfy the condition

$$\lim_{t \to \infty} \|x(t) - y(t)\| = 0, \quad \forall x(0), y(0) \in \mathbb{R}^n \tag{1}$$

Pecora and Carroll pioneered the research on synchronization of chaotic systems with their seminal papers [14, 44]. The active control method [26, 59, 60, 71, 76, 82, 88, 122, 123, 126] is typically used when the system parameters are available for measurement. Adaptive control method [57, 61–63, 73–75, 86, 92, 111, 115,

120, 124, 125, 131, 136] is typically used when some or all the system parameters are not available for measurement and estimates for the uncertain parameters of the systems. Sampled-data feedback control method [20, 33, 146, 149] and time-delay feedback control method [17, 25, 64] are also used for synchronization of chaotic systems.

Backstepping control method [50–54, 79, 127, 134] is also used for the synchronization of chaotic systems, which is a recursive method for stabilizing the origin of a control system in strict-feedback form. Another popular method for the synchronization of chaotic systems is the sliding mode control method [78, 85, 97, 117, 118, 128, 129], which is a nonlinear control method that alters the dynamics of a nonlinear system by application of a discontinuous control signal that forces the system to "slide" along a cross-section of the system's normal behavior.

In this research work, we announce a six-term novel 3-D dissipative chaotic system with two quadratic nonlinearities. First, this work describes the dynamic equations and qualitative properties of the novel chaotic system. We show that the novel chaotic system has three unstable equilibrium points. We also show that the novel chaotic system has a rotation symmetry about the x_3 axis. The Lyapunov exponents of the novel chaotic system are obtained as $L_1 = 1.2334$, $L_2 = 0$ and $L_3 = -4.7329$. Since the sum of the Lyapunov exponents is negative, the novel chaotic system is dissipative. Also, the Kaplan-Yorke dimension of the novel chaotic system is derived as $D_{KY} = 2.2606$.

Section 2 describes the dynamic equations and phase portraits of the novel chaotic system. Section 3 describes the qualitative properties of the novel chaotic system. Section 4 describes the active synchronization of the identical novel chaotic systems with known parameters. Section 5 describes the adaptive synchronization of the identical novel chaotic systems with unknown parameters. The active and adaptive synchronization results are established using Lyapunov stability theory [29]. MATLAB simulations are shown to illustrate all the main results derived in this work. Finally, Sect. 6 contains the conclusions of this work.

2 A Six-Term Novel 3-D Chaotic System

In this section, we describe the six-term novel 3-D chaotic system, which is given by the 3-D dynamics

$$\begin{cases} \dot{x}_1 = a(x_2 - x_1) \\ \dot{x}_2 = x_1 x_3 \\ \dot{x}_3 = 50 - b x_1^2 - c x_3 \end{cases} \tag{2}$$

where x_1, x_2, x_3 are the states and a, b, c are constant, positive, parameters.

The novel 3-D system (2) is *chaotic* when the parameter values are taken as

$$a = 2.9, \quad b = 0.7, \quad c = 0.6 \tag{3}$$

For numerical simulations, we use the classical fourth-order Runge-Kutta method with step-size $h = 10^{-8}$ to solve the system of differential equations (2) with the parameter values as in (3).

We take the initial values of the novel chaotic system (2) as

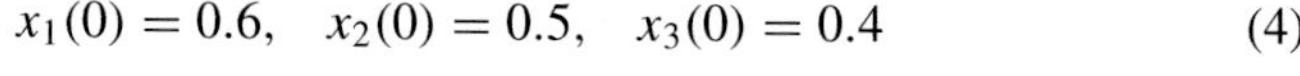

$$x_1(0) = 0.6, \quad x_2(0) = 0.5, \quad x_3(0) = 0.4 \tag{4}$$

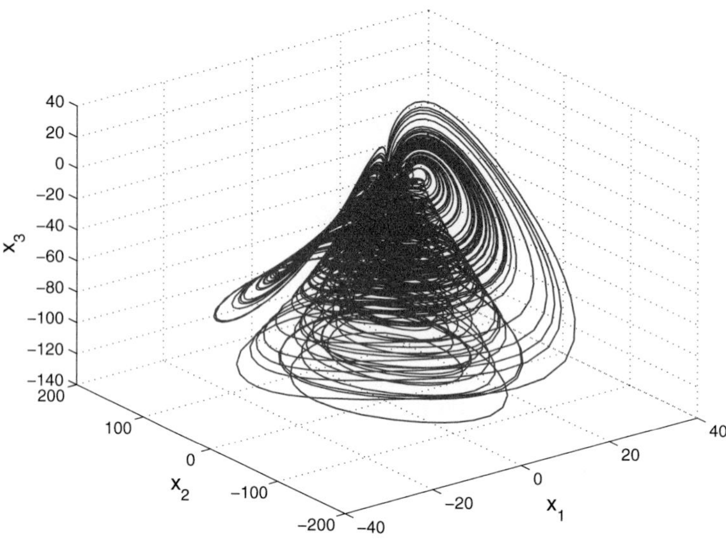

Fig. 1 Strange attractor of the novel chaotic system in $\mathbf{R}^3$

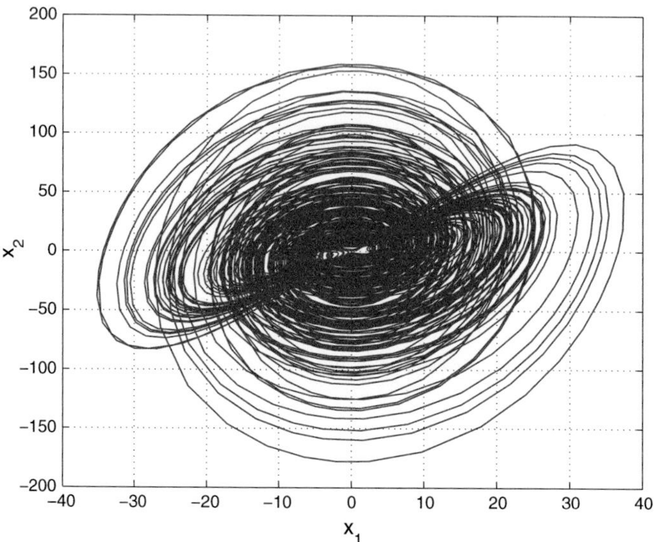

Fig. 2 2-D projection of the novel chaotic system on (x_1, x_2)-plane

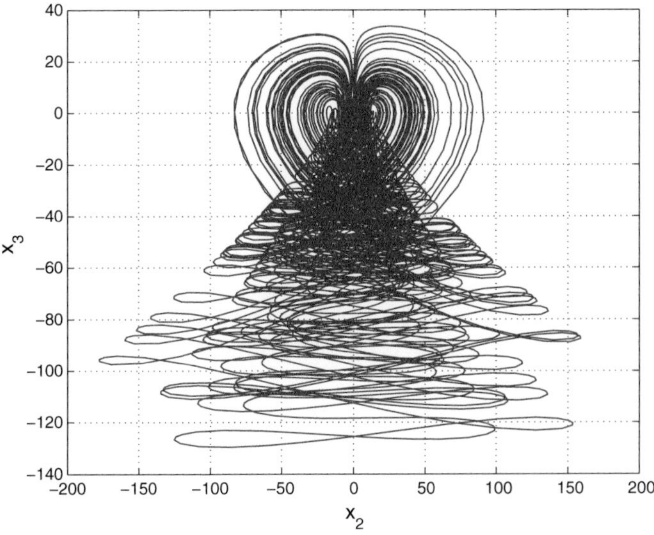

Fig. 3 2-D projection of the novel chaotic system on (x_2, x_3)-plane

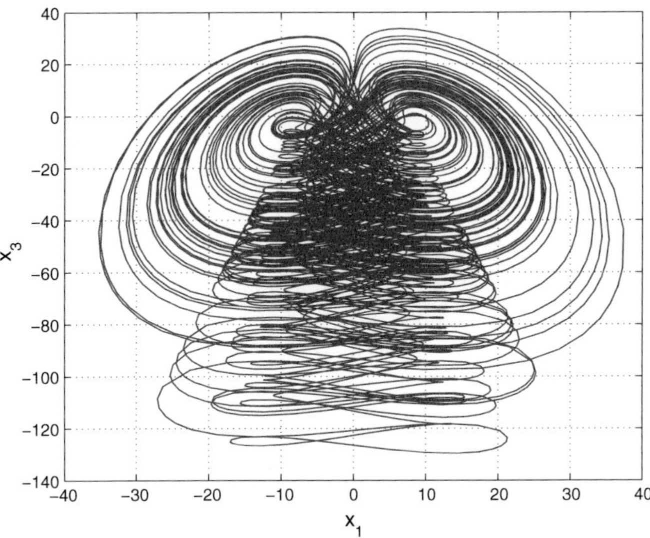

Fig. 4 2-D projection of the novel chaotic system on (x_1, x_3)-plane

Figure 1 describes the *strange attractor* of the novel chaotic system (2) in $\mathbb{R}^3$.

Figures 2, 3 and 4 describe the 2-D projection of the strange chaotic attractor of the novel chaotic system (2) on (x_1, x_2), (x_2, x_3) and (x_1, x_3) coordinate planes, respectively.

3 Properties of the Novel Chaotic System

This section describes the qualitative properties of the novel chaotic system (2). We take the parameter values as in the chaotic case (3), viz. $a = 2.9, b = 0.7$ and $c = 0.6$.

3.1 Dissipativity

We write the novel chaotic system (2) in vector notation as

$$\dot{x} = f(x) = \begin{bmatrix} f_1(x) \\ f_2(x) \\ f_3(x) \end{bmatrix}, \tag{5}$$

where

$$\begin{aligned} f_1(x) &= a(x_2 - x_1) \\ f_2(x) &= x_1 x_3 \\ f_3(x) &= 50 - bx_1^2 - cx_3 \end{aligned} \tag{6}$$

We take the parameter values as in the chaotic case (3).

The divergence of the vector field f on $\mathbb{R}^3$ is obtained as

$$\text{div } f = \frac{\partial f_1(x)}{\partial x_1} + \frac{\partial f_2(x)}{\partial x_2} + \frac{\partial f_3(x)}{\partial x_3} = -(a + 0 + c) = -\mu \tag{7}$$

where $\mu = a + c = 3.5 > 0$.

Let Ω be any region in $\mathbb{R}^3$ with a smooth boundary. Let $\Omega(t) = \Phi_t(\Omega)$, where Φ_t is the flow of the vector field f.

Let $V(t)$ denote the volume of $\Omega(t)$.

By Liouville's theorem, it follows that

$$\frac{dV(t)}{dt} = \int\limits_{\Omega(t)} (\text{div} f) dx_1 \, dx_2 \, dx_3 \tag{8}$$

Substituting the value of $\text{div} f$ in (8) leads to

$$\frac{dV(t)}{dt} = -\mu \int\limits_{\Omega(t)} dx_1 \, dx_2 \, dx_3 = -\mu V(t) \tag{9}$$

Integrating the linear differential equation (9), $V(t)$ is obtained as

$$V(t) = V(0) \exp(-\mu t), \quad \text{where } \mu = 3.5 > 0. \tag{10}$$

From Eq. (10), it follows that the volume $V(t)$ shrinks to zero exponentially as $t \to \infty$.

Thus, the novel chaotic system (2) is dissipative. Hence, any asymptotic motion of the system (2) settles onto a set of measure zero, i.e. a strange attractor.

3.2 Equilibrium Points

The equilibrium points of the novel chaotic system (2) for the chaotic case (3) are obtained by solving the system of equations

$$a(x_2 - x_1) = 0 \tag{11a}$$

$$x_1 x_3 = 0 \tag{11b}$$

$$50 - bx_1^2 - cx_3 = 0 \tag{11c}$$

From (11b), it follows that $x_1 = 0$ or $x_3 = 0$.

First, we take $x_1 = 0$. From (11a), we get $x_2 = x_1 = 0$.

Substituting $x_1 = x_2 = 0$ into (11c), we get $cx_3 = 50$ or

$$x_3 = \frac{50}{c} = \frac{50}{0.6} = 83.3333 \tag{12}$$

Thus, we get an equilibrium point of the system (2) as

$$E_1 = \begin{bmatrix} 0 \\ 0 \\ 83.3333 \end{bmatrix} \tag{13}$$

Next, we take $x_3 = 0$. From (11c), we get

$$x_1 = \pm\sqrt{\frac{50}{b}} = \pm\sqrt{\frac{50}{0.7}} = \pm 8.4515 \tag{14}$$

From (11a), we get $x_2 = x_1$.

Thus, we get two more equilibrium points of the system (2) as

$$E_2 = \begin{bmatrix} 8.4515 \\ 8.4515 \\ 0 \end{bmatrix}, \quad E_3 = \begin{bmatrix} -8.4515 \\ -8.4515 \\ 0 \end{bmatrix} \tag{15}$$

To test the stability type of the equilibrium points, we calculate the Jacobian of the system (2) at any $x \in \mathbb{R}^3$ as

$$J(x) = \begin{bmatrix} -a & a & 0 \\ x_3 & 0 & x_1 \\ -2bx_1 & 0 & -c \end{bmatrix} = \begin{bmatrix} -2.9 & 2.9 & 0 \\ x_3 & 0 & x_1 \\ -1.4x_1 & 0 & -0.6 \end{bmatrix} \tag{16}$$

We find that

$$J_1 = J(E_1) = \begin{bmatrix} -2.9 & 2.9 & 0 \\ 83.3333 & 0 & 0 \\ 0 & 0 & -0.6 \end{bmatrix} \tag{17}$$

The matrix J_1 has the eigenvalues

$$\lambda_1 = -17.0631, \quad \lambda_2 = -0.6, \quad \lambda_3 = 14.1631 \tag{18}$$

This shows that the equilibrium point E_1 is a saddle-point, which is unstable.
Next, we find that

$$J_2 = J(E_2) = \begin{bmatrix} -2.9 & 2.9 & 0 \\ 0 & 0 & 8.4515 \\ -11.8321 & 0 & -0.6 \end{bmatrix} \tag{19}$$

The matrix J_2 has the eigenvalues

$$\lambda_1 = -7.9123, \quad \lambda_{2,3} = 2.2061 \pm 5.6378i \tag{20}$$

This shows that the equilibrium point E_2 is a saddle-focus, which is unstable.
It is easy to verify that $J_3 = J(E_3)$ has the same set of eigenvalues as J_2.
Hence, E_3 is also a saddle-focus, which is unstable.

Thus, we conclude that all the three equilibrium points (E_1, E_2, E_3) of the novel chaotic system (2) are unstable.

3.3 Symmetry

It is easy to check that the novel chaotic system (2) is invariant under the change of coordinates

$$(x_1, x_2, x_3) \mapsto (-x_1, -x_2, x_3) \tag{21}$$

which persists for all values of the parameters a, b and c.

This shows that the novel chaotic system (2) has rotation symmetry about the x_3-axis. Hence, it follows that any non-trivial trajectory of the novel chaotic system (2) must have a twin trajectory.

3.4 Invariance

It is easy to check that the x_3-axis is invariant under the flow of the 3-D novel chaotic system (2). The invariant motion along the x_3-axis is characterized by the scalar dynamics

$$\dot{x}_3 = 50 - cx_3 \quad (c > 0) \tag{22}$$

which is unstable. This is because as $t \to \infty$, $x_3(t) \to \frac{50}{c} = 83.3333$. Since we cannot make $x_3(t)$ to stay within an ϵ-neighbourhood of $x_3 = 0$, it follows that the flow of the scalar dynamics (22) is unstable.

3.5 Lyapunov Exponents and Kaplan-Yorke Dimension

We take the parameter values of the novel chaotic system (2) as in the chaotic case (3), i.e.

$$a = 2.9, \quad b = 0.7, \quad c = 0.9 \tag{23}$$

We take the initial values of the novel chaotic system (2) as in (4).

Then the Lyapunov exponents of the novel chaotic system (2) are numerically obtained as

$$L_1 = 1.2334, \quad L_2 = 0, \quad L_3 = -4.7329 \tag{24}$$

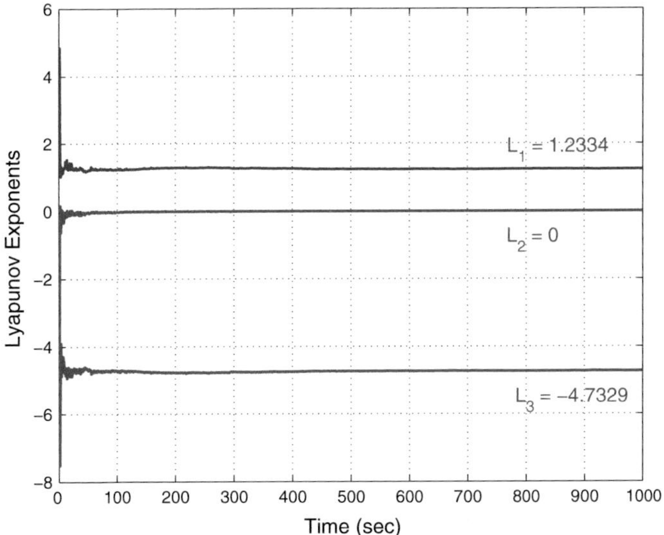

Fig. 5 Dynamics of the Lyapunov exponents of the novel chaotic system

Since $L_1 + L_2 + L_3 = -3.4995 < 0$, the novel chaotic system (2) is dissipative.

Figure 5 depicts the dynamics of the Lyapunov exponents of the novel chaotic system (2).

From Fig. 5, we note that the Maximal Lyapunov Exponent (MLE) of the novel chaotic system (2) is obtained as $L_1 = 1.2334$.

Also, the Kaplan-Yorke dimension of the novel chaotic system (2) is obtained as

$$D_{KY} = 2 + \frac{L_1 + L_2}{|L_3|} = 2.2606 \tag{25}$$

4 Active Synchronization of the Identical Novel Chaotic Systems

This section derives a new result for the active synchronization of the identical novel chaotic systems with known parameters. The main result is established using Lyapunov stability theory.

The master system is given by the novel chaotic system

$$\begin{aligned} \dot{x}_1 &= a(x_2 - x_1) \\ \dot{x}_2 &= x_1 x_3 \\ \dot{x}_3 &= 50 - b x_1^2 - c x_3 \end{aligned} \tag{26}$$

where x_1, x_2, x_3 are state variables and a, b, c are constant, known, parameters of the system.

The slave system is given by the controlled novel chaotic system

$$\begin{aligned} \dot{y}_1 &= a(y_2 - y_1) + u_1 \\ \dot{y}_2 &= y_1 y_3 + u_2 \\ \dot{y}_3 &= 50 - b y_1^2 - c y_3 + u_3 \end{aligned} \tag{27}$$

where y_1, y_2, y_3 are state variables and u_1, u_2, u_3 are active controls to be designed.

The synchronization error is defined as

$$\begin{aligned} e_1 &= y_1 - x_1 \\ e_2 &= y_2 - x_2 \\ e_3 &= y_3 - x_3 \end{aligned} \tag{28}$$

The error dynamics is easily obtained as

$$\begin{aligned} \dot{e}_1 &= a(e_2 - e_1) + u_1 \\ \dot{e}_2 &= y_1 y_3 - x_1 x_3 + u_2 \\ \dot{e}_3 &= -b(y_1^2 - x_1^2) - c e_3 + u_3 \end{aligned} \tag{29}$$

An active control law is taken as

$$\begin{aligned} u_1 &= -a(e_2 - e_1) - k_1 e_1 \\ u_2 &= -y_1 y_3 + x_1 x_3 - k_2 e_2 \\ u_3 &= b(y_1^2 - x_1^2) + c e_3 - k_3 e_3 \end{aligned} \tag{30}$$

where k_1, k_2, k_3 are positive gain constants.

The closed-loop control system is obtained by substituting (30) into (29) as

$$\begin{aligned} \dot{e}_1 &= -k_1 e_1 \\ \dot{e}_2 &= -k_2 e_2 \\ \dot{e}_3 &= -k_3 e_3 \end{aligned} \tag{31}$$

Theorem 1 *The identical novel chaotic systems (38) and (39) with known system parameters are globally and exponentially synchronized for all initial conditions* $x(0), y(0) \in \mathbb{R}^3$ *by the active control law (30), where* $k_i, (i = 1, 2, 3)$ *are positive constants.*

Proof We consider the quadratic Lyapunov function defined by

$$V(e_1, e_2, e_3) = \frac{1}{2}\left(e_1^2 + e_2^2 + e_3^2\right), \tag{32}$$

which is positive definite on $\mathbb{R}^3$.

Differentiating V along the trajectories of (31), we get

$$\dot{V} = -k_1 e_1^2 - k_2 e_2^2 - k_3 e_3^2 \tag{33}$$

which is a negative definite function on $\mathbb{R}^3$.

Thus, by Lyapunov stability theory [29], the error dynamics (31) is globally exponentially stable.

Hence, The identical novel chaotic systems (26) and (27) with known system parameters are globally and exponentially synchronized for all initial conditions $x(0), y(0) \in \mathbb{R}^3$ by the active control law (30).

This completes the proof. □

For numerical simulations, the parameter values of the novel systems (26) and (27) are taken as in the chaotic case, viz.

$$a = 2.9, \quad b = 0.7, \quad c = 0.6 \tag{34}$$

The gain constants are taken as

$$k_1 = 6, \quad k_2 = 6, \quad k_3 = 6 \tag{35}$$

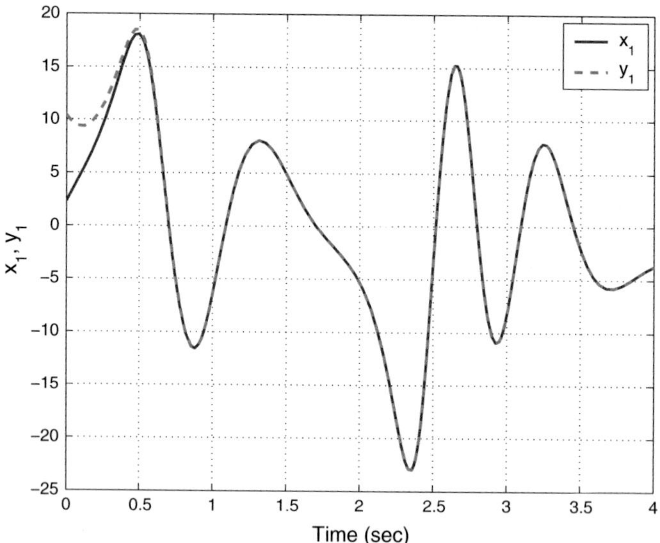

Fig. 6 Synchronization of the states x_1 and y_1

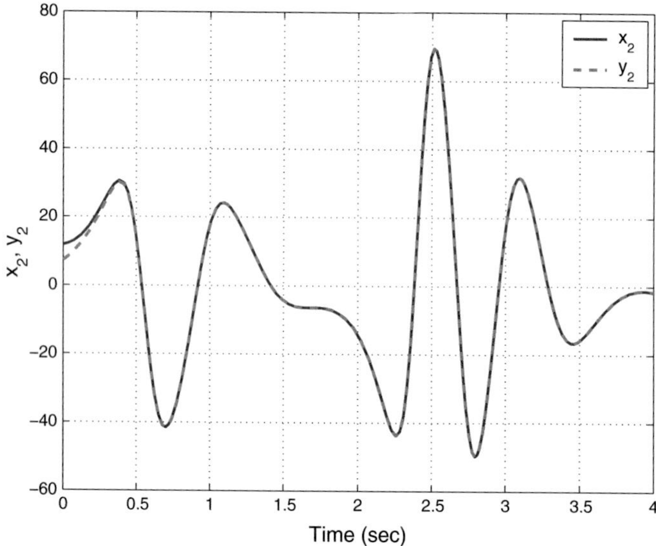

Fig. 7 Synchronization of the states x_2 and y_2

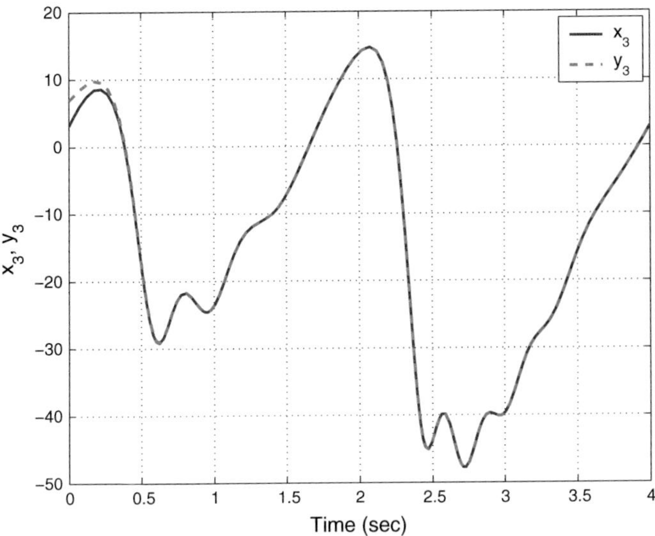

Fig. 8 Synchronization of the states x_3 and y_3

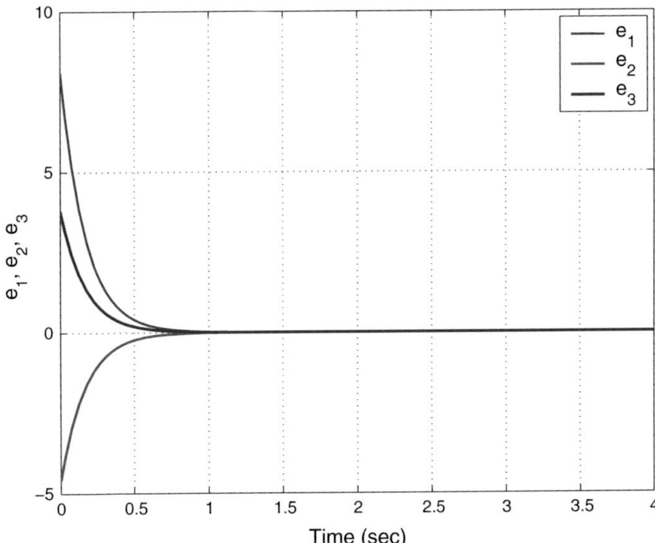

Fig. 9 Time-history of the synchronization errors e_1, e_2, e_3

The initial values of the master system (38) are taken as

$$x_1(0) = 2.3, \quad x_2(0) = 11.8, \quad x_3(0) = 3.1 \tag{36}$$

The initial values of the slave system (39) are taken as

$$y_1(0) = 10.4, \quad y_2(0) = 7.2, \quad y_3(0) = 6.9 \tag{37}$$

Figures 6, 7 and 8 show the complete synchronization of the identical chaotic systems (26) and (27).

Figure 6 shows that the states $x_1(t)$ and $y_1(t)$ are synchronized in 2 s (MATLAB).

Figure 7 shows that the states $x_2(t)$ and $y_2(t)$ are synchronized in 2 s (MATLAB).

Figure 8 shows that the states $x_3(t)$ and $y_3(t)$ are synchronized in 2 s (MATLAB).

Figure 9 shows the time-history of the synchronization errors $e_1(t)$, $e_2(t)$, $e_3(t)$. From Fig. 9, it is seen that the errors $e_1(t)$, $e_2(t)$ and $e_3(t)$ are stabilized in 2 s (MATLAB).

This shows the efficiency of the active control law (30) designed in this section.

5 Adaptive Synchronization of the Identical Novel Chaotic Systems

This section derives a new result for the adaptive synchronization of the identical novel chaotic systems with unknown parameters. The main result is established using Lyapunov stability theory.

The master system is given by the novel chaotic system

$$\begin{aligned} \dot{x}_1 &= a(x_2 - x_1) \\ \dot{x}_2 &= x_1 x_3 \\ \dot{x}_3 &= 50 - b x_1^2 - c x_3 \end{aligned} \tag{38}$$

where x_1, x_2, x_3 are state variables and a, b, c are constant, unknown, parameters of the system.

The slave system is given by the controlled novel chaotic system

$$\begin{aligned} \dot{y}_1 &= a(y_2 - y_1) + u_1 \\ \dot{y}_2 &= y_1 y_3 + u_2 \\ \dot{y}_3 &= 50 - b y_1^2 - c y_3 + u_3 \end{aligned} \tag{39}$$

where y_1, y_2, y_3 are state variables and u_1, u_2, u_3 are adaptive controls to be designed.

The synchronization error is defined as

$$\begin{aligned} e_1 &= y_1 - x_1 \\ e_2 &= y_2 - x_2 \\ e_3 &= y_3 - x_3 \end{aligned} \tag{40}$$

The error dynamics is easily obtained as

$$\begin{aligned}\dot{e}_1 &= a(e_2 - e_1) + u_1 \\ \dot{e}_2 &= y_1 y_3 - x_1 x_3 + u_2 \\ \dot{e}_3 &= -b(y_1^2 - x_1^2) - c e_3 + u_3\end{aligned} \tag{41}$$

An adaptive control law is taken as

$$\begin{aligned}u_1 &= -\hat{a}(t)(e_2 - e_1) - k_1 e_1 \\ u_2 &= -y_1 y_3 + x_1 x_3 - k_2 e_2 \\ u_3 &= \hat{b}(t)(y_1^2 - x_1^2) + \hat{c}(t) e_3 - k_3 e_3\end{aligned} \tag{42}$$

where $\hat{a}(t), \hat{b}(t), \hat{c}(t)$ are estimates for the unknown parameters a, b, c, respectively, and k_1, k_2, k_3 are positive gain constants.

The closed-loop control system is obtained by substituting (42) into (41) as

$$\begin{aligned}\dot{e}_1 &= \left[a - \hat{a}(t)\right](e_2 - e_1) - k_1 e_1 \\ \dot{e}_2 &= -k_2 e_2 \\ \dot{e}_3 &= -\left[b - \hat{b}(t)\right](y_1^2 - x_1^2) - \left[c - \hat{c}(t)\right] e_3 - k_3 e_3\end{aligned} \tag{43}$$

To simplify (43), we define the parameter estimation error as

$$\begin{aligned}e_a(t) &= a - \hat{a}(t) \\ e_b(t) &= b - \hat{b}(t) \\ e_c(t) &= c - \hat{c}(t)\end{aligned} \tag{44}$$

Using (44), the closed-loop system (43) can be simplified as

$$\begin{aligned}\dot{e}_1 &= e_a(e_2 - e_1) - k_1 e_1 \\ \dot{e}_2 &= -k_2 e_2 \\ \dot{e}_3 &= -e_b(y_1^2 - x_1^2) - e_c e_3 - k_3 e_3\end{aligned} \tag{45}$$

Differentiating the parameter estimation error (44) with respect to t, we get

$$\begin{aligned}\dot{e}_a &= -\dot{\hat{a}} \\ \dot{e}_b &= -\dot{\hat{b}} \\ \dot{e}_c &= -\dot{\hat{c}}\end{aligned} \tag{46}$$

We consider the quadratic Lyapunov function defined by

$$V(e_1, e_2, e_3, e_a, e_b, e_c) = \frac{1}{2}\left(e_1^2 + e_2^2 + e_3^2 + e_a^2 + e_b^2 + e_c^2\right), \tag{47}$$

which is positive definite on $\mathbb{R}^6$.

Differentiating V along the trajectories of (45) and (46), we get

$$\begin{aligned} \dot{V} = &-k_1 e_1^2 - k_2 e_2^2 - k_3 e_3^2 + e_a[e_1(e_2 - e_1) - \dot{\hat{a}}] \\ &+ e_b[-e_3(y_1^2 - x_1^2) - \dot{\hat{b}}] + e_c[-e_3^2 - \dot{\hat{c}}] \end{aligned} \tag{48}$$

In view of Eq. (48), an update law for the parameter estimates is taken as

$$\begin{aligned} \dot{\hat{a}} &= e_1(e_2 - e_1) \\ \dot{\hat{b}} &= -e_3(y_1^2 - x_1^2) \\ \dot{\hat{c}} &= -e_3^2 \end{aligned} \tag{49}$$

Next, we state and prove the main result of this section.
This main result is established using Lyapunov stability theory.

Theorem 2 *The identical novel chaotic systems (38) and (39) with unknown system parameters are globally and exponentially synchronized for all initial conditions* $x(0), y(0) \in \mathbb{R}^3$ *by the adaptive control law (42) and the parameter update law (49), where* $k_i, (i = 1, 2, 3)$ *are positive constants.*

Proof The result is proved using Lyapunov stability theory [29].

We consider the quadratic Lyapunov function V defined by (47), which is positive definite on $\mathbb{R}^6$.

Substitution of the parameter update law (49) into (48) yields

$$\dot{V} = -k_1 e_1^2 - k_2 e_2^2 - k_3 e_3^2, \tag{50}$$

which is a negative semi-definite function on $\mathbb{R}^6$.

Therefore, it can be concluded that the synchronization error vector $e(t)$ and the parameter estimation error are globally bounded, i.e.

$$\left[e_1(t) \; e_2(t) \; e_3(t) \; e_a(t) \; e_b(t) \; e_c(t) \right]^T \in \mathbf{L}_\infty. \tag{51}$$

Define

$$k = \min\{k_1, k_2, k_3\} \tag{52}$$

Then it follows from (50) that

$$\dot{V} \le -k\|e\|^2 \;\text{ or }\; k\|\mathbf{e}\|^2 \le -\dot{V} \tag{53}$$

Integrating the inequality (53) from 0 to t, we get

$$k \int_0^t \|\mathbf{e}(\tau)\|^2 \, d\tau \le -\int_0^t \dot{V}(\tau)\, d\tau = V(0) - V(t) \tag{54}$$

From (54), it follows that $\mathbf{e}(t) \in \mathbf{L}_2$.

Using (45), it can be deduced that $\dot{\mathbf{e}}(t) \in \mathbf{L}_\infty$.

Hence, using Barbalat's lemma, we can conclude that $\mathbf{e}(t) \to 0$ exponentially as $t \to \infty$ for all initial conditions $\mathbf{e}(0) \in \mathbb{R}^3$.

Since $\mathbf{e}(t) \to 0$ exponentially as $t \to \infty$ for all $\mathbf{e}(0) \in \mathbb{R}^3$, it is immediate that $\mathbf{x}(t) \to \mathbf{y}(t)$ as $t \to \infty$ for all $\mathbf{x}(0), \mathbf{y}(0) \in \mathbb{R}^3$.

This completes the proof. □

For numerical simulations, the parameter values of the novel systems (38) and (39) are taken as in the chaotic case, $a = 2.9$, $b = 0.7$ and $c = 0.6$.

The gain constants are taken as $k_i = 6$ for $i = 1, 2, 3$.

The initial values of the parameter estimates are taken as

$$\hat{a}(0) = 5.2, \quad \hat{b}(0) = 11.6, \quad \hat{c}(0) = 17.4 \tag{55}$$

The initial values of the master system (38) are taken as

$$x_1(0) = 5.2, \quad x_2(0) = 4.3, \quad x_3(0) = 12.1 \tag{56}$$

The initial values of the slave system (39) are taken as

$$y_1(0) = 16.9, \quad y_2(0) = 12.8, \quad y_3(0) = 4.7 \tag{57}$$

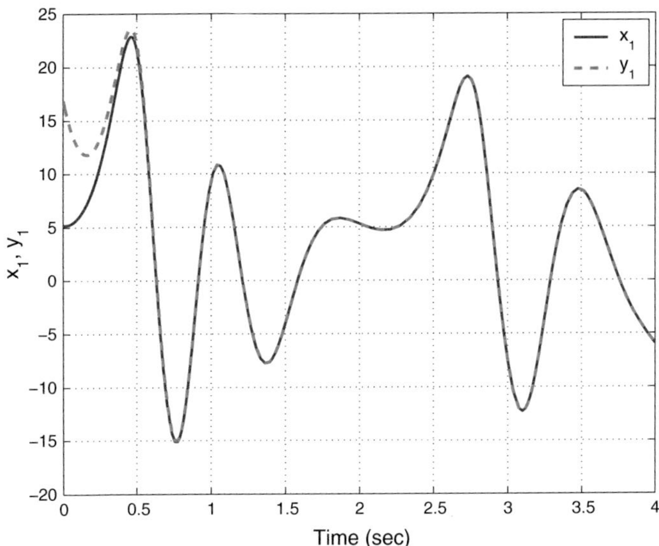

Fig. 10 Synchronization of the states x_1 and y_1

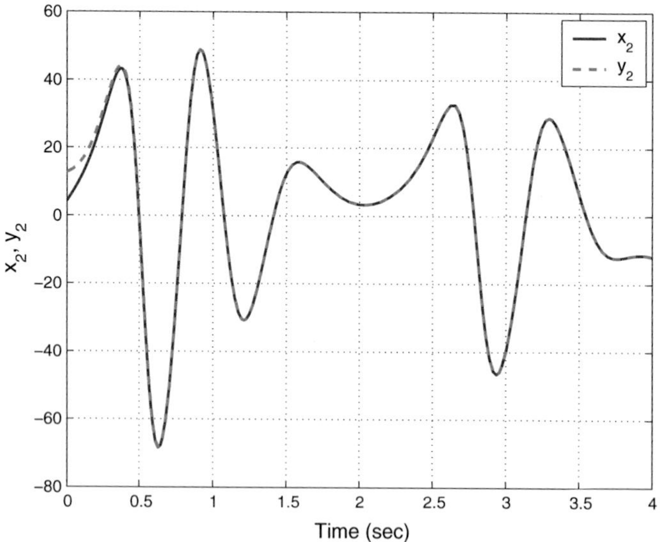

Fig. 11 Synchronization of the states x_2 and y_2

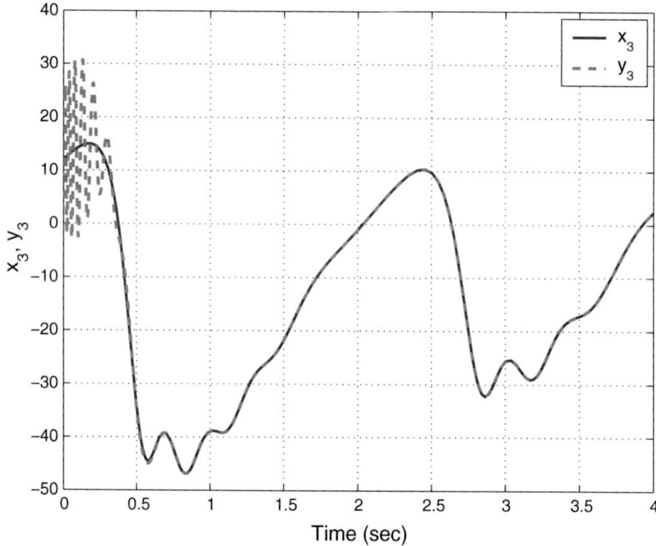

Fig. 12 Synchronization of the states x_3 and y_3

Figures 10, 11 and 12 show the complete synchronization of the identical chaotic systems (38) and (39).

Figure 10 shows that the states $x_1(t)$ and $y_1(t)$ are synchronized in 2 s (MATLAB).

Figure 11 shows that the states $x_2(t)$ and $y_2(t)$ are synchronized in 2 s (MATLAB).

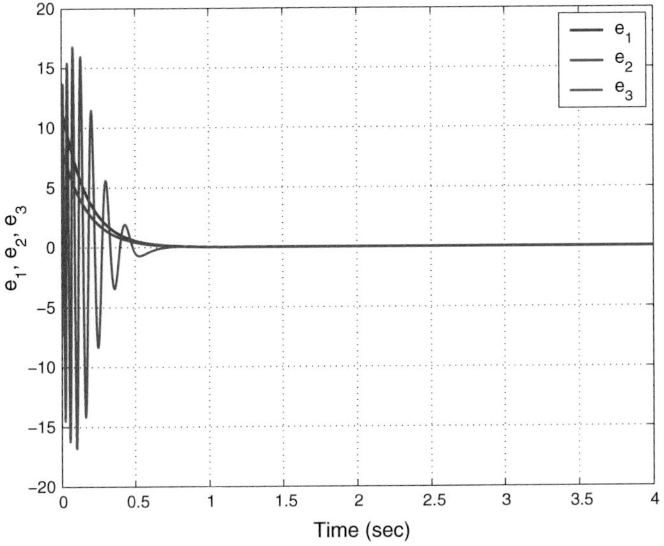

Fig. 13 Time-history of the synchronization errors e_1, e_2, e_3

Figure 12 shows that the states $x_3(t)$ and $y_3(t)$ are synchronized in 2 s (MATLAB).

Figure 13 shows the time-history of the synchronization errors $e_1(t), e_2(t), e_3(t)$. From Fig. 13, it is seen that the errors $e_1(t), e_2(t)$ and $e_3(t)$ are stabilized in 2 s (MATLAB).

This shows the efficiency of the adaptive control law (42) and the parameter update law (49) designed in this section.

6 Conclusions

In this research work, we described a six-term novel 3-D dissipative chaotic system with two quadratic nonlinearities. First, we detailed the dynamic equations and qualitative properties of the novel chaotic system. We showed that the novel chaotic system has three unstable equilibrium points. We also showed that the novel chaotic system has a rotation symmetry about the x_3 axis. The Lyapunov exponents of the novel chaotic system have been obtained as $L_1 = 1.2334$, $L_2 = 0$ and $L_3 = -4.7329$. Since the sum of the Lyapunov exponents is negative, the novel chaotic system is dissipative. Also, the Kaplan-Yorke dimension of the novel chaotic system has been derived as $D_{KY} = 2.2606$. Next, we described the active synchronization of identical novel chaotic systems with known parameters. We also described the adaptive synchronization of identical novel chaotic systems with unknown parameters. Both the active and adaptive synchronization results were established using Lyapunov stability theory. MATLAB simulations have been shown to demonstrate all the main results derived in this work for the six-term novel 3-D novel chaotic system.

References

1. Arneodo A, Coullet P, Tresser C (1981) Possible new strange attractors with spiral structure. Commun Math Phys 79(4):573–576
2. Azar AT (2010) Fuzzy systems. IN-TECH, Vienna, Austria
3. Azar AT (2012) Overview of type-2 fuzzy logic systems. Int J Fuzzy Syst Appl 2(4):1–28
4. Azar AT, Serrano FE (2014) Robust IMC-PID tuning for cascade control systems with gain and phase margin specifications. Neural Comput Appl 25(5):983–995
5. Azar AT, Serrano FE (2015a) Adaptive sliding mode control of the Furuta pendulum. In: Azar AT, Zhu Q (eds) Advances and applications in sliding mode control systems. Studies in computational intelligence, vol 576. Springer, Germany, pp 1–42
6. Azar AT, Serrano FE (2015b) Deadbeat control for multivariable systems with time varying delays. In: Azar AT, Vaidyanathan S (eds) Chaos modeling and control systems design. Studies in computational intelligence, vol 581. Springer, Germany, pp 97–132
7. Azar AT, Serrano FE (2015c) Design and modeling of anti wind up PID controllers. In: Zhu Q, Azar AT (eds) Complex system modelling and control through intelligent soft computations. Studies in fuzziness and soft computing, vol 319. Springer, Germany, pp 1–44
8. Azar AT, Serrano FE (2015d) Stabilizatoin and control of mechanical systems with backlash. In: Azar AT, Vaidyanathan S (eds) Handbook of research on advanced intelligent control engineering and automation. Advances in Computational Intelligence and Robotics (ACIR). IGI-Global, USA, pp 1–60
9. Azar AT, Vaidyanathan S (2015a) Chaos modeling and control systems design. Studies in computational intelligence, vol 581. Springer, Germany
10. Azar AT, Vaidyanathan S (2015b) Computational intelligence applications in modeling and control. Studies in computational intelligence, vol 575. Springer, Germany
11. Azar AT, Vaidyanathan S (2015c) Handbook of research on advanced intelligent control engineering and automation. Advances in Computational Intelligence and Robotics (ACIR). IGI-Global, USA
12. Azar AT, Zhu Q (2015) Advances and applications in sliding mode control systems. Studies in computational intelligence, vol 576. Springer, Germany
13. Cai G, Tan Z (2007) Chaos synchronization of a new chaotic system via nonlinear control. J Uncertain Syst 1(3):235–240
14. Carroll TL, Pecora LM (1991) Synchronizing chaotic circuits. IEEE Trans Circ Syst 38(4):453–456
15. Chen G, Ueta T (1999) Yet another chaotic attractor. Int J Bifurcat Chaos 9(7):1465–1466
16. Chen HK, Lee CI (2004) Anti-control of chaos in rigid body motion. Chaos Solitons Fractals 21(4):957–965
17. Chen WH, Wei D, Lu X (2014) Global exponential synchronization of nonlinear time-delay Lur'e systems via delayed impulsive control. Commun Nonlinear Sci Numer Simul 19(9):3298–3312
18. Das S, Goswami D, Chatterjee S, Mukherjee S (2014) Stability and chaos analysis of a novel swarm dynamics with applications to multi-agent systems. Eng Appl Artif Intell 30:189–198
19. Feki M (2003) An adaptive chaos synchronization scheme applied to secure communication. Chaos Solitons Fractals 18(1):141–148
20. Gan Q, Liang Y (2012) Synchronization of chaotic neural networks with time delay in the leakage term and parametric uncertainties based on sampled-data control. J Franklin Inst 349(6):1955–1971
21. Gaspard P (1999) Microscopic chaos and chemical reactions. Phys A: Stat Mech Appl 263(1–4):315–328
22. Gibson WT, Wilson WG (2013) Individual-based chaos: extensions of the discrete logistic model. J Theor Biol 339:84–92
23. Guégan D (2009) Chaos in economics and finance. Annu Rev Control 33(1):89–93
24. Huang X, Zhao Z, Wang Z, Li Y (2012) Chaos and hyperchaos in fractional-order cellular neural networks. Neurocomputing 94:13–21

25. Jiang GP, Zheng WX, Chen G (2004) Global chaos synchronization with channel time-delay. Chaos Solitons Fractals 20(2):267–275
26. Karthikeyan R, Sundarapandian V (2014) Hybrid chaos synchronization of four-scroll systems via active control. J Electr Eng 65(2):97–103
27. Kaslik E, Sivasundaram S (2012) Nonlinear dynamics and chaos in fractional-order neural networks. Neural Netw 32:245–256
28. Kengne J, Chedjou JC, Kenne G, Kyamakya K (2012) Dynamical properties and chaos synchronization of improved Colpitts oscillators. Commun Nonlinear Sci Numer Simul 17(7):2914–2923
29. Khalil HK (2001) Nonlinear Syst. Prentice Hall, New Jersey, USA
30. Kyriazis M (1991) Applications of chaos theory to the molecular biology of aging. Exp Gerontol 26(6):569–572
31. Lang J (2015) Color image encryption based on color blend and chaos permutation in the reality-preserving multiple-parameter fractional Fourier transform domain. Opt Commun 338:181–192
32. Li D (2008) A three-scroll chaotic attractor. Phys Lett A 372(4):387–393
33. Li N, Zhang Y, Nie Z (2011) Synchronization for general complex dynamical networks with sampled-data. Neurocomputing 74(5):805–811
34. Li N, Pan W, Yan L, Luo B, Zou X (2014) Enhanced chaos synchronization and communication in cascade-coupled semiconductor ring lasers. Commun Nonlinear Sci Numer Simul 19(6):1874–1883
35. Li Z, Chen G (2006) Integration of fuzzy logic and chaos theory. Studies in fuzziness and soft computing, vol 187. Springer, Germany
36. Lian S, Chen X (2011) Traceable content protection based on chaos and neural networks. Appl Soft Comput 11(7):4293–4301
37. Liu C, Liu T, Liu L, Liu K (2004) A new chaotic attractor. Chaos Solitons Fractals 22(5):1031–1038
38. Lorenz EN (1963) Deterministic periodic flow. J Atmos Sci 20(2):130–141
39. Lü J, Chen G (2002) A new chaotic attractor coined. Int J Bifurcat Chaos 12(3):659–661
40. Mondal S, Mahanta C (2014) Adaptive second order terminal sliding mode controller for robotic manipulators. J Franklin Inst 351(4):2356–2377
41. Murali K, Lakshmanan M (1998) Secure communication using a compound signal from generalized chaotic systems. Phys Lett A 241(6):303–310
42. Nehmzow U, Walker K (2005) Quantitative description of robot-environment interaction using chaos theory. Robot Auton Syst 53(3–4):177–193
43. Noroozi N, Roopaei M, Karimaghaee P, Safavi AA (2010) Simple adaptive variable structure control for unknown chaotic systems. Commun Nonlinear Sci Numer Simul 15(3):707–727
44. Pecora LM, Carroll TL (1990) Synchronization in chaotic systems. Phys Rev Lett 64(8):821–824
45. Pehlivan I, Moroz IM, Vaidyanathan S (2014) Analysis, synchronization and circuit design of a novel butterfly attractor. J Sound Vib 333(20):5077–5096
46. Petrov V, Gaspar V, Masere J, Showalter K (1993) Controlling chaos in Belousov-Zhabotinsky reaction. Nature 361:240–243
47. Pham VT, Vaidyanathan S, Volos CK, Jafari S (2015a) Hidden attractors in a chaotic system with an exponential nonlinear term. Eur Phys J—Spec Top 224(8):1507–1517
48. Pham VT, Volos CK, Vaidyanathan S, Le TP, Vu VY (2015b) A memristor-based hyperchaotic system with hidden attractors: dynamics, synchronization and circuital emulating. J Eng Sci Technol Rev 8(2):205–214
49. Qu Z (2011) Chaos in the genesis and maintenance of cardiac arrhythmias. Prog Biophys Mol Biol 105(3):247–257
50. Rasappan S, Vaidyanathan S (2012a) Global chaos synchronization of WINDMI and Coullet chaotic systems by backstepping control. Far East J Math Sci 67(2):265–287
51. Rasappan S, Vaidyanathan S (2012b) Hybrid synchronization of n-scroll Chua and Lur'e chaotic systems via backstepping control with novel feedback. Arch Control Sci 22(3):343–365

52. Rasappan S, Vaidyanathan S (2012c) Synchronization of hyperchaotic Liu system via backstepping control with recursive feedback. Commun Comput Inf Sci 305:212–221
53. Rasappan S, Vaidyanathan S (2013) Hybrid synchronization of n-scroll chaotic Chua circuits using adaptive backstepping control design with recursive feedback. Malays J Math Sci 7(2):219–246
54. Rasappan S, Vaidyanathan S (2014) Global chaos synchronization of WINDMI and Coullet chaotic systems using adaptive backstepping control design. Kyungpook Math J 54(1):293–320
55. Rhouma R, Belghith S (2011) Cryptoanalysis of a chaos based cryptosystem on DSP. Commun Nonlinear Sci Numer Simul 16(2):876–884
56. Rössler OE (1976) An equation for continuous chaos. Phys Lett A 57(5):397–398
57. Vaidyanathan S, VTP, Volos CK, (2015) A 5-D hyperchaotic Rikitake dynamo system with hidden attractors. Eur Phys J: Spec Top 224(8):1575–1592
58. Sampath S, Vaidyanathan S, Volos CK, Pham VT (2015) An eight-term novel four-scroll chaotic system with cubic nonlinearity and its circuit simulation. J Eng Sci Technol Rev 8(2):1–6
59. Sarasu P, Sundarapandian V (2011a) Active controller design for the generalized projective synchronization of four-scroll chaotic systems. Int J Syst Signal Control Eng Appl 4(2):26–33
60. Sarasu P, Sundarapandian V (2011b) The generalized projective synchronization of hyperchaotic Lorenz and hyperchaotic Qi systems via active control. Int J Soft Comput 6(5):216–223
61. Sarasu P, Sundarapandian V (2012a) Adaptive controller design for the generalized projective synchronization of 4-scroll systems. Int J Syst Signal Control Eng Appl 5(2):21–30
62. Sarasu P, Sundarapandian V (2012b) Generalized projective synchronization of three-scroll chaotic systems via adaptive control. Eur J Sci Res 72(4):504–522
63. Sarasu P, Sundarapandian V (2012c) Generalized projective synchronization of two-scroll systems via adaptive control. Int J Soft Comput 7(4):146–156
64. Shahverdiev EM, Shore KA (2009) Impact of modulated multiple optical feedback time delays on laser diode chaos synchronization. Opt Commun 282(17):3568–3572
65. Sharma A, Patidar V, Purohit G, Sud KK (2012) Effects on the bifurcation and chaos in forced Duffing oscillator due to nonlinear damping. Commun Nonlinear Sci Numer Simul 17(6):2254–2269
66. Sprott JC (1994) Some simple chaotic flows. Phys Rev E 50(2):647–650
67. Sprott JC (2004) Competition with evolution in ecology and finance. Phys Lett A 325(5–6):329–333
68. Sprott JC (2010) Elegant chaos: algebraically simple chaotic flows. World Scientific, Singapore
69. Suérez I (1999) Mastering chaos in ecology. Ecol Modell 117(2–3):305–314
70. Sundarapandian V (2010a) Output regulation of the Lorenz attractor. Far East J Math Sci 42(2):289–299
71. Sundarapandian V (2010b) Output regulation of the Lorenz attractor. Far East J Math Sci 42(2):289–299
72. Sundarapandian V (2013) Analysis and anti-synchronization of a novel chaotic system via active and adaptive controllers. J Eng Sci Technol Rev 6(4):45–52
73. Sundarapandian V, Karthikeyan R (2011a) Anti-synchronization of hyperchaotic Lorenz and hyperchaotic Chen systems by adaptive control. Int J Syst Signal Control Eng Appl 4(2):18–25
74. Sundarapandian V, Karthikeyan R (2011b) Anti-synchronization of Lü and Pan chaotic systems by adaptive nonlinear control. Eur J Sci Res 64(1):94–106
75. Sundarapandian V, Karthikeyan R (2012a) Adaptive anti-synchronization of uncertain Tigan and Li systems. J Eng Appl Sci 7(1):45–52
76. Sundarapandian V, Karthikeyan R (2012b) Hybrid synchronization of hyperchaotic Lorenz and hyperchaotic Chen systems via active control. J Eng Appl Sci 7(3):254–264
77. Sundarapandian V, Pehlivan I (2012) Analysis, control, synchronization, and circuit design of a novel chaotic system. Math Comput Modell 55(7–8):1904–1915

78. Sundarapandian V, Sivaperumal S (2011) Sliding controller design of hybrid synchronization of four-wing chaotic systems. Int J Soft Comput 6(5):224–231
79. Suresh R, Sundarapandian V (2013) Global chaos synchronization of a family of n-scroll hyperchaotic Chua circuits using backstepping control with recursive feedback. Far East J Math Sci 73(1):73–95
80. Tigan G, Opris D (2008) Analysis of a 3D chaotic system. Chaos Solitons Fractals 36:1315–1319
81. Usama M, Khan MK, Alghatbar K, Lee C (2010) Chaos-based secure satellite imagery cryptosystem. Comput Math Appl 60(2):326–337
82. Vaidyanathan S (2011a) Hybrid chaos synchronization of Liu and Lü systems by active nonlinear control. Commun Comput Inf Sci 204
83. Vaidyanathan S (2011b) Output regulation of Arneodo-Coullet chaotic system. Commun Comput Inf Sci 133:98–107
84. Vaidyanathan S (2011c) Output regulation of the unified chaotic system. Commun Comput Inf Sci 198:10–17
85. Vaidyanathan S (2012a) Analysis and synchronization of the hyperchaotic Yujun systems via sliding mode control. Adv Intell Syst Comput 176:329–337
86. Vaidyanathan S (2012b) Anti-synchronization of Sprott-L and Sprott-M chaotic systems via adaptive control. Int J Control Theory Appl 5(1):41–59
87. Vaidyanathan S (2012c) Global chaos control of hyperchaotic Liu system via sliding control method. Int J Control Theory Appl 5(2):117–123
88. Vaidyanathan S (2012d) Output regulation of the Liu chaotic system. Appl Mech Mater 110–116:3982–3989
89. Vaidyanathan S (2012e) Sliding mode control based global chaos control of Liu-Liu-Liu-Su chaotic system. Int J Control Theory Appl 1(2):15–20
90. Vaidyanathan S (2013a) A new six-term 3-D chaotic system with an exponential nonlinearity. Far East J Math Sci 79(1):135–143
91. Vaidyanathan S (2013b) Analysis and adaptive synchronization of two novel chaotic systems with hyperbolic sinusoidal and cosinusoidal nonlinearity and unknown parameters. J Eng Sci Technol Rev 6(4):53–65
92. Vaidyanathan S (2013c) Analysis, control and synchronization of hyperchaotic Zhou system via adaptive control. Adv Intell Syst Comput 177:1–10
93. Vaidyanathan S (2014a) A new eight-term 3-D polynomial chaotic system with three quadratic nonlinearities. Far East J Math Sci 84(2):219–226
94. Vaidyanathan S (2014b) Analysis and adaptive synchronization of eight-term 3-D polynomial chaotic systems with three quadratic nonlinearities. Eur Phys J: Spec Top 223(8):1519–1529
95. Vaidyanathan S (2014c) Analysis, control and synchronisation of a six-term novel chaotic system with three quadratic nonlinearities. Int J Modell Ident Control 22(1):41–53
96. Vaidyanathan S (2014d) Generalized projective synchronisation of novel 3-D chaotic systems with an exponential non-linearity via active and adaptive control. Int J Modell Ident Control 22(3):207–217
97. Vaidyanathan S (2014e) Global chaos synchronization of identical Li-Wu chaotic systems via sliding mode control. Int J Modell Ident Control 22(2):170–177
98. Vaidyanathan S (2015a) 3-cells cellular neural network (CNN) attractor and its adaptive biological control. Int J PharmTech Res 8(4):632–640
99. Vaidyanathan S (2015b) A 3-D novel highly chaotic system with four quadratic nonlinearities, its adaptive control and anti-synchronization with unknown parameters. J Eng Sci Technol Rev 8(2):106–115
100. Vaidyanathan S (2015c) Adaptive backstepping control of eEnzymes-substrates system with ferroelectric behaviour in brain waves. Int J PharmTech Res 8(2):256–261
101. Vaidyanathan S (2015d) Adaptive biological control of generalized Lotka-Volterra three-species biological system. Int J PharmTech Res 8(4):622–631
102. Vaidyanathan S (2015e) Adaptive chaotic synchronization of enzymes-substrates system with ferroelectric behaviour in brain waves. Int J PharmTech Res 8(5):964–973

103. Vaidyanathan S (2015f) Adaptive control of a chemical chaotic reactor. Int J PharmTech Res 8(3):377–382
104. Vaidyanathan S (2015g) Adaptive synchronization of chemical chaotic reactors. Int J ChemTech Res 8(2):612–621
105. Vaidyanathan S (2015h) Adaptive synchronization of generalized Lotka-Volterra three-species biological systems. Int J PharmTech Res 8(5):928–937
106. Vaidyanathan S (2015i) Analysis, properties and control of an eight-term 3-D chaotic system with an exponential nonlinearity. Int J Modell Ident Control 23(2):164–172
107. Vaidyanathan S (2015j) Anti-synchronization of Brusselator chemical reaction systems via adaptive control. Int J ChemTech Res 8(6):759–768
108. Vaidyanathan S (2015k) Chaos in neurons and adaptive control of Birkhoff-Shaw strange chaotic attractor. Int J PharmTech Res 8(5):956–963
109. Vaidyanathan S (2015l) Dynamics and control of Brusselator chemical reaction. Int J ChemTech Res 8(6):740–749
110. Vaidyanathan S (2015m) Dynamics and control of Tokamak system with symmetric and magnetically confined plasma. Int J ChemTech Res 8(6):795–803
111. Vaidyanathan S (2015n) Hyperchaos, qualitative analysis, control and synchronisation of a ten-term 4-D hyperchaotic system with an exponential nonlinearity and three quadratic nonlinearities. Int J Modell Ident Control 23(4):380–392
112. Vaidyanathan S (2015o) Lotka-Volterra population biology models with negative feedback and their ecological monitoring. Int J PharmTech Res 8(5):974–981
113. Vaidyanathan S (2015p) Synchronization of 3-cells cellular neural network (CNN) attractors via adaptive control method. Int J PharmTech Res 8(5):946–955
114. Vaidyanathan S (2015q) Synchronization of Tokamak systems with symmetric and magnetically confined plasma via adaptive control. Int J ChemTech Res 8(6):818–827
115. Vaidyanathan S, Azar AT (2015a) Analysis and control of a 4-D novel hyperchaotic system. In: Azar AT, Vaidyanathan S (eds) Chaos modeling and control systems designl Studies in computational intelligence, vol 581. Springer, Germany, pp 19–38
116. Vaidyanathan S, Azar AT (2015b) Analysis, control and synchronization of a nine-term 3-D novel chaotic system. In: Azar AT, Vaidyanathan S (eds) Chaos modelling and control systems design. Studies in computational intelligence, vol 581. Springer, Germany, pp 19–38
117. Vaidyanathan S, Azar AT (2015c) Anti-synchronization of identical chaotic systems using sliding mode control and an application to Vaidhyanathan-Madhavan chaotic systems. Stud Comput Intell 576:527–547
118. Vaidyanathan S, Azar AT (2015d) Hybrid synchronization of identical chaotic systems using sliding mode control and an application to Vaidhyanathan chaotic systems. Stud Comput Intell 576:549–569
119. Vaidyanathan S, Madhavan K (2013) Analysis, adaptive control and synchronization of a seven-term novel 3-D chaotic system. Int J Control Theory Appl 6(2):121–137
120. Vaidyanathan S, Pakiriswamy S (2013) Generalized projective synchronization of six-term Sundarapandian chaotic systems by adaptive control. Int J Control Theory Appl 6(2):153–163
121. Vaidyanathan S, Pakiriswamy S (2015) A 3-D novel conservative chaotic system and its generalized projective synchronization via adaptive control. J Eng Sci Technol Rev 8(2):52–60
122. Vaidyanathan S, Rajagopal K (2011a) Anti-synchronization of Li and T chaotic systems by active nonlinear control. Commun Comput Inf Sci 198:175–184
123. Vaidyanathan S, Rajagopal K (2011b) Global chaos synchronization of hyperchaotic Pang and Wang systems by active nonlinear control. Commun Comput Inf Sci 204:84–93
124. Vaidyanathan S, Rajagopal K (2011c) Global chaos synchronization of Lü and Pan systems by adaptive nonlinear control. Commun Comput Inf Sci 205:193–202
125. Vaidyanathan S, Rajagopal K (2012) Global chaos synchronization of hyperchaotic Pang and hyperchaotic Wang systems via adaptive control. Int J Soft Comput 7(1):28–37
126. Vaidyanathan S, Rasappan S (2011) Global chaos synchronization of hyperchaotic Bao and Xu systems by active nonlinear control. Commun Comput Inf Sci 198:10–17

127. Vaidyanathan S, Rasappan S (2014) Global chaos synchronization of n-scroll Chua circuit and Lur'e system using backstepping control design with recursive feedback. Arab J Sci Eng 39(4):3351–3364
128. Vaidyanathan S, Sampath S (2011) Global chaos synchronization of hyperchaotic Lorenz systems by sliding mode control. Commun Comput Inf Sci 205:156–164
129. Vaidyanathan S, Sampath S (2012) Anti-synchronization of four-wing chaotic systems via sliding mode control. Int J Autom Comput 9(3):274–279
130. Vaidyanathan S, Volos C (2015) Analysis and adaptive control of a novel 3-D conservative no-equilibrium chaotic system. Arch Control Sci 25(3):333–353
131. Vaidyanathan S, Volos C, Pham VT (2014a) Hyperchaos, adaptive control and synchronization of a novel 5-D hyperchaotic system with three positive Lyapunov exponents and its SPICE implementation. Arch Control Sci 24(4):409–446
132. Vaidyanathan S, Volos C, Pham VT, Madhavan K, Idowu BA (2014b) Adaptive backstepping control, synchronization and circuit simulation of a 3-D novel jerk chaotic system with two hyperbolic sinusoidal nonlinearities. Arch Control Sci 24(3):375–403
133. Vaidyanathan S, Azar AT, Rajagopal K, Alexander P (2015a) Design and SPICE implementation of a 12-term novel hyperchaotic system and its synchronisation via active control. Int J Modell Ident Control 23(3):267–277
134. Vaidyanathan S, Idowu BA, Azar AT (2015b) Backstepping controller design for the global chaos synchronization of Sprott's jerk systems. Stud Comput Intell 581:39–58
135. Vaidyanathan S, Rajagopal K, Volos CK, Kyprianidis IM, Stouboulos IN (2015c) Analysis, adaptive control and synchronization of a seven-term novel 3-D chaotic system with three quadratic nonlinearities and its digital implementation in LabVIEW. J Eng Sci Technol Rev 8(2):130–141
136. Vaidyanathan S, Volos C, Pham VT, Madhavan K (2015d) Analysis, adaptive control and synchronization of a novel 4-D hyperchaotic hyperjerk system and its SPICE implementation. Arch Control Sci 25(1):5–28
137. Vaidyanathan S, Volos CK, Kyprianidis IM, Stouboulos IN, Pham VT (2015e) Analysis, adaptive control and anti-synchronization of a six-term novel jerk chaotic system with two exponential nonlinearities and its circuit simulation. J Eng Sci Technol Rev 8(2):24–36
138. Vaidyanathan S, Volos CK, Pham VT (2015f) Analysis, adaptive control and adaptive synchronization of a nine-term novel 3-D chaotic system with four quadratic nonlinearities and its circuit simulation. J Eng Sci Technol Rev 8(2):174–184
139. Vaidyanathan S, Volos CK, Pham VT (2015g) Analysis, control, synchronization and SPICE implementation of a novel 4-D hyperchaotic Rikitake dynamo System without equilibrium. J Eng Sci Technol Rev 8(2):232–244
140. Vaidyanathan S, Volos CK, Pham VT (2015h) Global chaos control of a novel nine-term chaotic system via sliding mode control. In: Azar AT, Zhu Q (eds) Advances and applications in sliding mode control systems. Studies in computational intelligence, vol 576. Springer, Germany, pp 571–590
141. Vaidyanathan S, Volos CK, Rajagopal K, Kyprianidis IM, Stouboulos IN (2015i) Adaptive backstepping controller design for the anti-synchronization of identical WINDMI chaotic systems with unknown parameters and its SPICE implementation. J Eng Sci Technol Rev 8(2):74–82
142. Volos CK, Kyprianidis IM, Stouboulos IN (2013) Experimental investigation on coverage performance of a chaotic autonomous mobile robot. Robot Auton Syst 61(12):1314–1322
143. Volos CK, Kyprianidis IM, Stouboulos IN, Tlelo-Cuautle E, Vaidyanathan S (2015) Memristor: a new concept in synchronization of coupled neuromorphic circuits. J Eng Sci Technol Rev 8(2):157–173
144. Wei Z, Yang Q (2010) Anti-control of Hopf bifurcation in the new chaotic system with two stable node-foci. Appl Math Comput 217(1):422–429
145. Witte CL, Witte MH (1991) Chaos and predicting varix hemorrhage. Med Hypotheses 36(4):312–317

146. Xiao X, Zhou L, Zhang Z (2014) Synchronization of chaotic Lur'e systems with quantized sampled-data controller. Commun Nonlinear Sci Numer Simul 19(6):2039–2047
147. Yuan G, Zhang X, Wang Z (2014) Generation and synchronization of feedback-induced chaos in semiconductor ring lasers by injection-locking. Optik—Int J Light Electron Opt 125(8):1950–1953
148. Zaher AA, Abu-Rezq A (2011) On the design of chaos-based secure communication systems. Commun Nonlinear Sci Numer Simul 16(9):3721–3727
149. Zhang H, Zhou J (2012) Synchronization of sampled-data coupled harmonic oscillators with control inputs missing. Syst Control Lett 61(12):1277–1285
150. Zhang X, Zhao Z, Wang J (2014) Chaotic image encryption based on circular substitution box and key stream buffer. Signal Process: Image Commun 29(8):902–913
151. Zhou W, Xu Y, Lu H, Pan L (2008) On dynamics analysis of a new chaotic attractor. Phys Lett A 372(36):5773–5777
152. Zhu C, Liu Y, Guo Y (2010) Theoretic and numerical study of a new chaotic system. Intell Inf Manage 2:104–109
153. Zhu Q, Azar AT (2015) Complex system modelling and control through intelligent soft computations. Studies in fuzziness and soft computing, vol 319. Springer, Germany

A Novel 2-D Chaotic Enzymes-Substrates Reaction System and Its Adaptive Backstepping Control

Sundarapandian Vaidyanathan

Abstract In this research work, we announce a novel 2-D chaotic enzymes-substrates reaction system and discuss its adaptive backstepping control. First, this work describes the dynamic equations and qualitative properties of the novel 2-D biological chaotic system. Our novel chaotic system is obtained by modifying the equations of the 2-D enzymes-substrates reaction system with ferroelectric behaviour in brain waves obtained by Kadji et al. (Chaos Solitons Fractals 32:862–882, 2001, [27]). The Maximal Lyapunov Exponent (MLE) of the novel 2-D chaotic enzymes-substrates reaction system is obtained as $L_1 = 0.14425$. Next, this work describes the adaptive control of the novel 2-D chaotic enzymes-substrates reaction system via backstepping control method. Furthermore, this work describes the adaptive synchronization of identical novel 2-D chaotic enzymes-substrates reaction systems via backstepping control method. The main stabilization and synchronization results derived in this work are established via Lyapunov stability theory. MATLAB simulations are depicted to illustrate all the main results derived in this work for the novel 2-D chaotic enzymes-substrates reaction system.

Keywords Chaos · Chaotic systems · Enzymes-substrates system · Adaptive control · Synchronization · Backstepping control

1 Introduction

Chaos theory is the study of how dynamical systems that are modelled by simple, straightforward, deterministic laws can exhibit very complicated and seemingly random long term behavior. A classic example of chaos is the chaos in weather patterns observed by Lorenz [41].

S. Vaidyanathan (✉)
Research and Development Centre, Vel Tech University, Avadi, Chennai 600062, Tamil Nadu, India
e-mail: sundarvtu@gmail.com

A.T. Azar and S. Vaidyanathan (eds.), *Advances in Chaos Theory and Intelligent Control*, Studies in Fuzziness and Soft Computing 337,
DOI 10.1007/978-3-319-30340-6_21

A hallmark of chaotic systems is its sensitive dependence on the initial conditions [71]. This means that if two trajectories of the system start from initial conditions which are very close to each other, then after a relatively short period of time, the two trajectories will diverge and appear very different from each other. The "butterfly effect" is an example of this, alluding to the idea that the flap of a butterfly's wings in Africa can cause a cascade of events culminating in a tornado in Texas. This sensitive dependence on initial conditions also guarantees that weather forecasting will not be accurate for more than a few days in advance.

Some classical paradigms of 3-D chaotic systems in the literature are Lorenz system [41], Rössler system [59], ACT system [1], Sprott systems [69], Chen system [15], Lü system [42], Liu system [40], Cai system [13], Chen-Lee system [16], Tigan system [83], etc.

Many new chaotic systems have been discovered in the recent years such as Zhou system [154], Zhu system [155], Li system [35], Wei-Yang system [147], Sundarapandian systems [75, 80], Vaidyanathan systems [93, 94, 96–99, 102, 109, 119, 122, 124, 133, 135, 138, 140, 141, 143], Pehlivan system [48], Sampath system [61], Pham system [50], etc.

Chaos theory and control systems have many important applications in science and engineering [2, 9–12, 156]. Some commonly known applications are oscillators [31, 68], lasers [37, 150], chemical reactions [22, 49, 106, 107, 110, 112, 113, 117], biology [18, 33, 101, 103–105, 108, 111, 115, 116], ecology [23, 72], encryption [34, 153], cryptosystems [58, 84], mechanical systems [4–8], secure communications [19, 44, 151], robotics [43, 45, 145], cardiology [52, 148], intelligent control [3, 38], neural networks [25, 30, 39], finance [24, 70], memristors [51, 146], etc.

Control of chaotic systems aims to regulate or stabilize the chaotic system about its equilibrium points. There are many methods available in the literature such as active control method [73, 86, 87, 136], adaptive control method [46, 142], sliding mode control method [90, 92], backstepping control method [144], etc.

Synchronization of chaotic systems is a phenomenon that occurs when two or more chaotic systems are coupled or when a chaotic system drives another chaotic system. Major works on synchronization of chaotic systems deal with the complete synchronization of a pair of chaotic systems called the *master* and *slave* systems. The design goal of the complete synchronization is to apply the output of the master system to control the slave system so that the output of the slave system tracks the output of the master system asymptotically with time.

Pecora and Carroll pioneered the research on synchronization of chaotic systems with their seminal papers [14, 47]. The active control method [29, 62, 63, 74, 79, 85, 91, 125, 126, 129] is typically used when the system parameters are available for measurement. Adaptive control method [60, 64–66, 76–78, 89, 95, 114, 118, 123, 127, 128, 134, 139] is typically used when some or all the system parameters are not available for measurement and estimates for the uncertain parameters of the systems. Sampled-data feedback control method [21, 36, 149, 152] and time-delay feedback control method [17, 26, 67] are also used for synchronization of chaotic systems.

Backstepping control method [53–57, 82, 130, 137] is also used for the synchronization of chaotic systems, which is a recursive method for stabilizing the origin of a control system in strict-feedback form. Another popular method for the synchronization of chaotic systems is the sliding mode control method [81, 88, 100, 120, 121, 131, 132], which is a nonlinear control method that alters the dynamics of a nonlinear system by application of a discontinuous control signal that forces the system to "slide" along a cross-section of the system's normal behavior.

Coherent oscillations in biological systems are studied by Frohlich [20] and the following suggestions have been made which are taken as a physical basis for theoretical investigation of enzymatic substrate reaction with ferroelectric behaviour in brain waves model [28]:

(i) When metabolic energy is available, long-wavelength electric vibrations are very strongly and coherently excited in active biological system.
(ii) Biological systems have metastable states with very high electric polarization.

These long range interactions may lead to a selective transport of enzymes, and hence specific chemical reactions may become possible. From Frohlich ideas [20], it may be supposed that in large regions of the system of proteins, substrates, ions and structured water are activated by the chemical energy available from substrate enzyme reaction. Recently, Enjieu Kadji, CHabi Orou, Yamapi and Woafo have derived a chaotic enzymes-substrates reactions system with ferroelectric behaviour in brain waves [27].

In this work, by modifying the dynamics of Kadji chaotic system [27], we announce a novel 2-D chaotic enzymes-substrates reaction system. Section 2 describes the dynamic equations and phase portraits of the novel chaotic system. Section 3 describes the adaptive control of the novel 2-D chaotic enzymes-substrates reaction system with unknown parameters via backstepping control method. Section 4 describes the adaptive synchronization of the identical novel 2-D chaotic enzymes-substrates reaction systems with unknown parameters via backstepping control method. The adaptive control and synchronization results derived in this work are established using Lyapunov stability theory [32]. MATLAB simulations are shown to illustrate all the main results derived in this work. Finally, Sect. 5 contains the conclusions of this work.

2 Dynamics and Analysis of a Novel 2-D Biological Chaotic System

In [27], Enjieu Kadji, Chabi Orou, Yamapi and Woafo derived a 2-D enzymes-substrates reaction system with ferroelectric behaviour in brain waves, which is given by the differential equation

$$\ddot{x} - \mu \left(1 - x^2 + ax^4 - bx^6\right) \dot{x} + x = E \cos(\Omega t) \tag{1}$$

In (1), a, b are positive parameters, μ is the parameter of nonlinearity. Also, $f(t) = E\cos(\Omega t)$ is the external periodic force, where E and Ω are the amplitude and frequency of the external cosinusoidal excitation, respectively.

The enzymes-substrates reaction system (1) can be compactly put in system form as follows:

$$\begin{cases} \dot{x} = y \\ \dot{y} = \mu y\left(1 - x^2 + ax^4 - bx^6\right) - x + E\cos(\Omega t) \end{cases} \tag{2}$$

For the external excitation force $f(t) = E\cos(\Omega t)$, the constants are taken as

$$E = 8.27, \quad \Omega = 3.465 \tag{3}$$

The biological system (2) is *chaotic*, when the system parameters are chosen as

$$a = 2.55, \quad b = 1.70, \quad \mu = 2.001 \tag{4}$$

For numerical simulations, we take the initial conditions as

$$x(0) = 0.1, \quad y(0) = 0.1 \tag{5}$$

Figure 1 depicts the strange attractor of the Kadji biological chaotic system (2).

For calculating the Lyapunov exponents of the Kadji 2-D chaotic system (2), it is convenient to express it as a 3-D autonomous system given by

$$\begin{cases} \dot{x} = y \\ \dot{y} = \mu y\left(1 - x^2 + ax^4 - bx^6\right) - x + E\cos(\Omega z) \\ \dot{z} = 1 \end{cases} \tag{6}$$

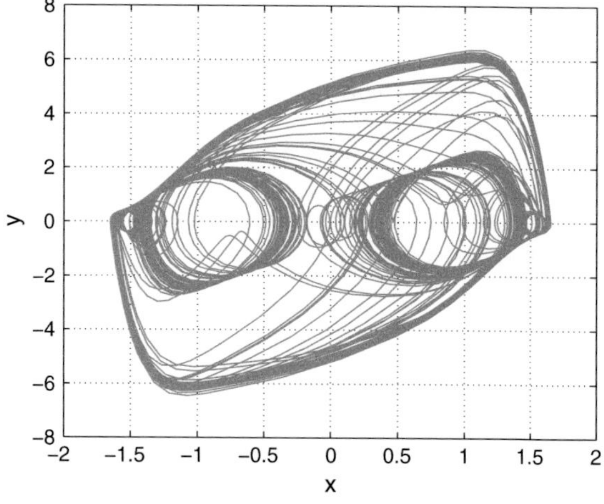

Fig. 1 The strange attractor of the Kadji biological chaotic system

Then the Lyapunov exponents of the Kadji biological system (6) are numerically obtained as

$$L_1 = 0.08320, \quad L_2 = 0, \quad L_3 = -9.92488 \tag{7}$$

Thus, the Kaplan-Yorke dimension of the Kadji biological system (6) is determined as

$$D_{KY} = 2 + \frac{L_1 + L_2}{|L_3|} = 2.0084 \tag{8}$$

In this research work, we announce a new biological chaotic system by modifying the dynamics (2) as

$$\begin{cases} \dot{x} = y \\ \dot{y} = \mu y \left(1 - x^2 + ax^4 - bx^6 + cx^8\right) - x + E\cos(\Omega t) \end{cases} \tag{9}$$

For the external excitation force $f(t) = E\cos(\Omega t)$, the constants are taken as

$$E = 8.4, \quad \Omega = 3.465 \tag{10}$$

The biological system (9) is *chaotic*, when the system parameters are chosen as

$$a = 2.55, \quad b = 1.70, \quad c = 0.0001, \quad \mu = 2.001 \tag{11}$$

For numerical simulations, we take the initial conditions as

$$x(0) = 0.2, \quad y(0) = 0.1 \tag{12}$$

Figure 2 depicts the strange attractor of the novel biological chaotic system (9).

Figures 3 and 4 describe the x and y waveforms of the novel biological chaotic system (9), respectively.

For calculating the Lyapunov exponents of the novel biological chaotic system (9), it is convenient to express it as a 3-D autonomous system given by

$$\begin{cases} \dot{x} = y \\ \dot{y} = \mu y \left(1 - x^2 + ax^4 - bx^6 + cx^8\right) - x + E\cos(\Omega z) \\ \dot{z} = 1 \end{cases} \tag{13}$$

Then the Lyapunov exponents of the novel biological system (13) are numerically obtained as

$$L_1 = 0.14425, \quad L_2 = 0, \quad L_3 = -9.85121 \tag{14}$$

Thus, the Kaplan-Yorke dimension of the novel biological system (13) is determined as

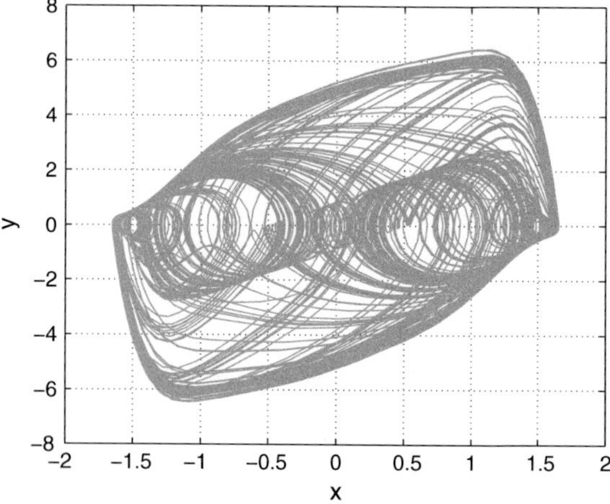

Fig. 2 The strange attractor of the novel biological chaotic system

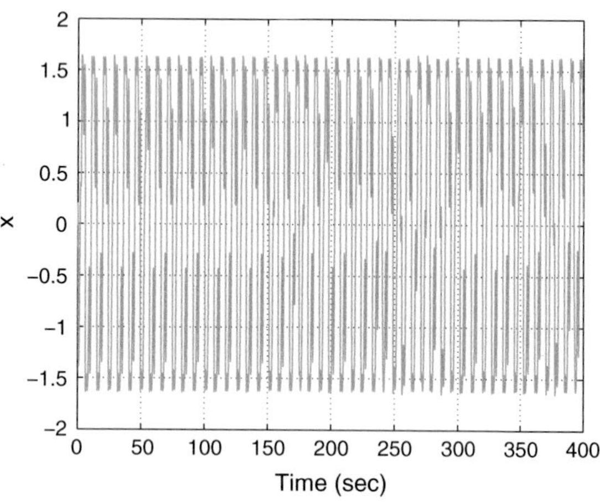

Fig. 3 The x-waveform of the novel biological chaotic system

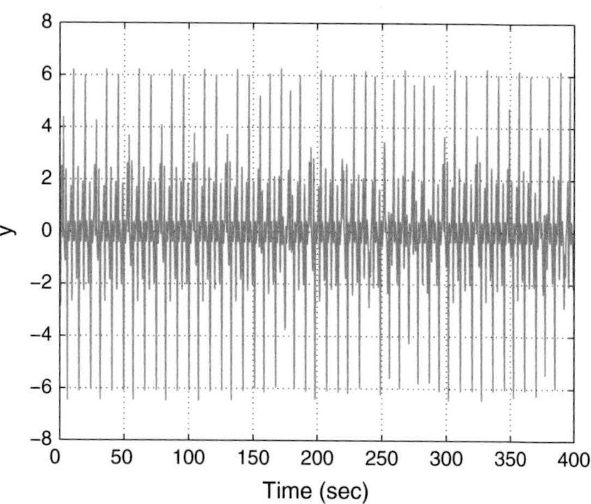

Fig. 4 The y-waveform of the novel biological chaotic system

$$D_{KY} = 2 + \frac{L_1 + L_2}{|L_3|} = 2.0146 \tag{15}$$

Thus, we observe that the Maximal Lyapunov Exponent (MLE) of the novel biological system (13) is found as $L_1 = 0.14425$, which is greater than the Maximal Lyapunov Exponent (MLE) of the Kadji biological system (6), which is found as $L_1 = 0.08320$.

Also, the Kaplan-Yorke dimension of the novel biological chaotic system (13) is found as $D_{KY} = 2.0146$, which is greater than the Kaplan-Yorke dimension of the Kadji biological system (6), which is found as $D_{KY} = 2.0084$.

These calculations show that the novel biological chaotic system (13) exhibits more chaotic behaviour than the Kadji biological chaotic system (6).

3 Adaptive Control of the Novel Biological Chaotic System

In this section, we design an adaptive backstepping feedback control law for globally stabilizing the novel enzymes-substrates reaction system with unknown parameters a, b and c.

It is supposed that the constants E and Ω associated with the external excitation force $f(t) = E\cos(\Omega t)$ are maintained at the constant values given in Eq. (10). It is also supposed that the parameter μ is maintained at the constant value given in Eq. (11).

Thus, we consider the novel chaotic enzymes-substrates reaction system given by the dynamics

$$\begin{cases} \dot{x} = y \\ \dot{y} = \mu y\left(1 - x^2 + ax^4 - bx^6 + cx^8\right) - x + E\cos(\Omega t) + u \end{cases} \tag{16}$$

In Eq. (16), x, y are the states and u is the adaptive backstepping controller to be found using estimates $\hat{a}(t), \hat{b}(t), \hat{c}(t)$ of the unknown parameters a, b, c, respectively.

Now, we define the parameter estimation errors as

$$\begin{cases} e_a(t) = a - \hat{a}(t) \\ e_b(t) = b - \hat{b}(t) \\ e_c(t) = c - \hat{c}(t) \end{cases} \tag{17}$$

Differentiating (17) with respect to t, we get

$$\begin{cases} \dot{e}_a = -\dot{\hat{a}} \\ \dot{e}_b = -\dot{\hat{b}} \\ \dot{e}_c = -\dot{\hat{c}} \end{cases} \tag{18}$$

Next, we state and prove the main result of this section.

Theorem 1 *The novel chaotic enzymes-substrates reaction system (16) with unknown parameters a, b and c is globally and exponentially stabilized for all initial conditions* $x(0), y(0) \in \mathbb{R}$ *by the adaptive control law*

$$u = -x - 2y - \mu y \left(1 - x^2 + \hat{a}(t)x^4 - \hat{b}(t)x^6 + \hat{c}(t)x^8\right) - E\cos(\Omega t) - kz_2 \tag{19}$$

where $k > 0$ *is a gain constant,*

$$z_2 = x + y, \tag{20}$$

and the update law for the parameter estimates $\hat{a}(t), \hat{b}(t), \hat{c}(t)$ *is given by*

$$\begin{cases} \dot{\hat{a}} = \mu z_2 x^4 y \\ \dot{\hat{b}} = -\mu z_2 x^6 y \\ \dot{\hat{c}} = \mu z_2 x^8 y \end{cases} \tag{21}$$

Proof We prove this result by applying backstepping control and Lyapunov stability theory [32].

First, we define a quadratic Lyapunov function

$$V_1(z_1) = \frac{1}{2}z_1^2, \tag{22}$$

where

$$z_1 = x \tag{23}$$

Differentiating V_1 along the dynamics (16), we get

$$\dot{V}_1 = z_1\dot{z}_1 = -z_1^2 + z_1(x + y) \tag{24}$$

Now, we define

$$z_2 = x + y \tag{25}$$

Substituting (25) into (24), we get

$$\dot{V}_1 = -z_1^2 + z_1z_2 \tag{26}$$

Next, we define a quadratic Lyapunov function

$$V(z_1, z_2, e_a, e_b, e_c) = V_1(z_1) + \frac{1}{2}z_2^2 + \frac{1}{2}e_a^2 + \frac{1}{2}e_b^2 + \frac{1}{2}e_c^2 \tag{27}$$

Differentiating (27) along the dynamics (16) and (18), we get

$$\dot{V} = -z_1^2 - z_2^2 + z_2 S - e_a \dot{\hat{a}} - e_b \dot{\hat{b}} - e_c \dot{\hat{c}} \tag{28}$$

where

$$S = z_1 + z_2 + \dot{z}_2 = x + 2y + \mu y(1 - x^2 + ax^4 - bx^6 + cx^8) + E\cos(\Omega t) + u \tag{29}$$

Substituting the feedback control law (19) into (29), we get

$$S = \mu[a - \hat{a}(t)]x^4 y - \mu[b - \hat{b}(t)]x^6 y + \mu[c - \hat{c}(t)]x^8 y - kz_2 \tag{30}$$

Using (17), Eq. (30) can be simplified as

$$S = \mu e_a x^4 y - \mu e_b x^6 y + \mu e_c x^8 y - kz_2 \tag{31}$$

Substituting the value of S from (31) into (28), we get

$$\dot{V} = -z_1^2 - (1+k)z_2^2 + e_a\left[\mu z_2 x^4 y - \dot{\hat{a}}\right] + e_b\left[-\mu z_2 x^6 y - \dot{\hat{b}}\right] + e_c\left[\mu z_2 x^8 y - \dot{\hat{c}}\right] \tag{32}$$

Substituting the parameter update law (21) into (32), we obtain

$$\dot{V} = -z_1^2 - (1+k)z_2^2 \tag{33}$$

which is negative semi-definite in $\mathbb{R}^5$.

From (33), it follows that the vector $z(t) = (z_1(t), z_2(t))$ and the parameter estimation error $(e_a(t), e_b(t), e_c(t))$ are globally bounded, *i.e.*

$$\left[z_1(t) \; z_2(t) \; e_a(t) \; e_b(t) \; e_c(t) \right] \in \mathbf{L}_\infty \tag{34}$$

Also, it follows from (33) that

$$\dot{V} \le -z_1^2 - z_2^2 = -\|\mathbf{z}\|^2 \tag{35}$$

That is,

$$\|\mathbf{z}\|^2 \le -\dot{V} \tag{36}$$

Integrating the inequality (36) from 0 to t, we get

$$\int_0^t \|\mathbf{z}(\tau)\|^2 \, d\tau \le V(0) - V(t) \tag{37}$$

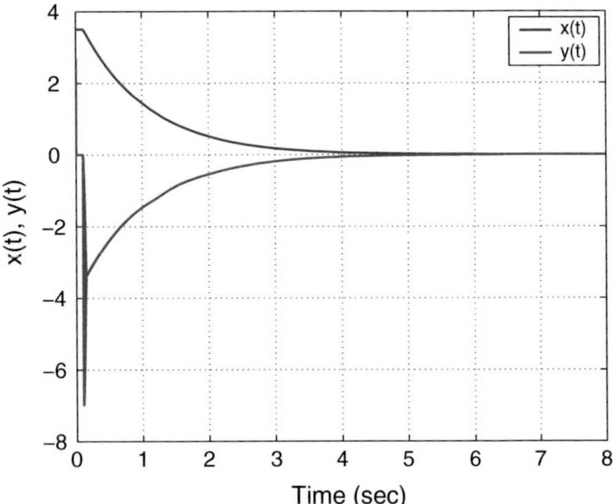

Fig. 5 Time-history of the controlled states of the novel biological system

From (37), it follows that $\mathbf{z} \in \mathbf{L}_2$. From Eq. (16), it can be deduced that $\dot{\mathbf{z}} \in \mathbf{L}_\infty$.

Thus, using Barbalat's lemma [32], we conclude that $\mathbf{z}(t) \to \mathbf{0}$ exponentially as $t \to \infty$ for all initial conditions $\mathbf{z}(0) \in \mathbb{R}^2$. Hence, it is immediate that $x(t) \to 0$ and $y(t) \to 0$ exponentially for all initial conditions $x(0), y(0) \in \mathbb{R}$.

This completes the proof. □

For numerical simulations, we use the classical fourth-order Runge-Kutta method in MATLAB with step-size $h = 10^{-8}$ for solving the novel enzymes-substrates reaction system (16) when the backstepping control law (19) and parameter update law (21) are implemented.

We take the positive gain constant as $k = 10$.

The parameter values are taken as in (10) and (11) for the chaotic case.

We take the initial values of the novel enzymes-substrates reaction system (16) as $x(0) = 3.5$ and $y(0) = 2.7$.

Also, we take $\hat{a}(0) = 5.2$, $\hat{b}(0) = 6.1$ and $\hat{c}(0) = 2.3$.

Figure 5 shows the time-history of the exponential convergence of the controlled states $x(t)$ and $y(t)$. From Fig. 5, it is clear that $x(t)$ and $y(t)$ converge to zero in approximately six seconds.

4 Adaptive Synchronization of the Identical Novel Biological Chaotic Systems

In this section, we design an adaptive backstepping feedback control law for globally synchronizing the trajectories of the novel enzymes-substrates reaction systems with unknown parameters a, b and c.

It is supposed that the constants E and Ω associated with the external excitation force $f(t) = E\cos(\Omega t)$ are maintained at the constant values given in Eq. (10). It is also supposed that the parameter μ is maintained at the constant value given in Eq. (11).

As the master system, we consider the novel chaotic enzymes-substrates reaction system given by the dynamics

$$\begin{cases} \dot{x}_1 = y_1 \\ \dot{y}_1 = \mu y_1 \left(1 - x_1^2 + a x_1^4 - b x_1^6 + c x_1^8\right) - x_1 + E\cos(\Omega t) \end{cases} \tag{38}$$

In Eq. (38), x_1, y_1 are the states and the parameters a, b, c are unknown.

As the slave system, we consider the novel chaotic enzymes-substrates reaction system given by the dynamics

$$\begin{cases} \dot{x}_2 = y_2 \\ \dot{y}_2 = \mu y_2 \left(1 - x_2^2 + a x_2^4 - b x_2^6 + c x_2^8\right) - x_2 + E\cos(\Omega t) + u \end{cases} \tag{39}$$

In Eq. (39), x_2, y_2 are the states and u is the adaptive controller to be determined using the estimates $\hat{a}(t), \hat{b}(t), \hat{c}(t)$ of the unknown system parameters a, b, c, respectively.

Now, we define the synchronization error as

$$\begin{cases} e_x = x_2 - x_1 \\ e_y = y_2 - y_1 \end{cases} \tag{40}$$

Then the synchronization error dynamics is obtained as

$$\begin{cases} \dot{e}_x = e_y \\ \dot{e}_y = \mu e_y - \mu(x_2^2 y_2 - x_1^2 y_1) + a\mu(x_2^4 y_2 - x_1^4 y_1) - b\mu(x_2^6 y_2 - x_1^6 y_1) \\ \qquad + c\mu(x_2^8 y_2 - x_1^8 y_1) - e_x + u \end{cases} \tag{41}$$

Next, we define the parameter estimation error as

$$\begin{cases} e_a(t) = a - \hat{a}(t) \\ e_b(t) = b - \hat{b}(t) \\ e_c(t) = c - \hat{c}(t) \end{cases} \tag{42}$$

Differentiating (42) with respect to t, we get

$$\begin{cases} \dot{e}_a = -\dot{\hat{a}} \\ \dot{e}_b = -\dot{\hat{b}} \\ \dot{e}_c = -\dot{\hat{c}} \end{cases} \tag{43}$$

Theorem 2 *The novel chaotic enzymes-substrates reaction systems (38) and (39) with unknown parameters a, b and c are globally and exponentially synchronized for all initial conditions $x_1(0), y_1(0), x_2(0), y_2(0) \in \mathbb{R}$ by the adaptive control law*

$$\begin{aligned} u = & -e_x - (2+\mu)e_y + \mu(x_2^2 y_2 - x_1^2 y_1) - \hat{a}(t)\mu(x_2^4 y_2 - x_1^4 y_1) \\ & + \hat{b}(t)\mu(x_2^6 y_2 - x_1^6 y_1) - \hat{c}(t)\mu(x_2^8 y_2 - x_1^8 y_1) - k z_2 \end{aligned} \tag{44}$$

where $k > 0$ is a gain constant,

$$z_2 = e_x + e_y, \tag{45}$$

and the update law for the parameter estimates $\hat{a}(t), \hat{b}(t), \hat{c}(t)$ is given by

$$\begin{cases} \dot{\hat{a}} = \mu z_2(x_2^4 y_2 - x_1^4 y_1) \\ \dot{\hat{b}} = -\mu z_2(x_2^6 y_2 - x_1^6 y_1) \\ \dot{\hat{c}} = \mu z_2(x_2^8 y_2 - x_1^8 y_1) \end{cases} \tag{46}$$

Proof We prove this result by applying backstepping control and Lyapunov stability theory [32].

First, we define a quadratic Lyapunov function

$$V_1(z_1) = \frac{1}{2} z_1^2, \tag{47}$$

where

$$z_1 = e_x \tag{48}$$

Differentiating V_1 along the dynamics (41), we get

$$\dot{V}_1 = z_1 \dot{z}_1 = -z_1^2 + z_1(e_x + e_y) \tag{49}$$

Now, we define

$$z_2 = e_x + e_y \tag{50}$$

Substituting (50) into (49), we get

$$\dot{V}_1 = -z_1^2 + z_1 z_2 \tag{51}$$

Next, we define a quadratic Lyapunov function

$$V(z_1, z_2, e_a, e_b, e_c) = V_1(z_1) + \frac{1}{2} z_2^2 + \frac{1}{2} e_a^2 + \frac{1}{2} e_b^2 + \frac{1}{2} e_c^2 \tag{52}$$

Differentiating (52) along the dynamics (41) and (43), we get

$$\dot{V} = -z_1^2 - z_2^2 + z_2 S - e_a \dot{\hat{a}} - e_b \dot{\hat{b}} - e_c \dot{\hat{c}} \tag{53}$$

where

$$\begin{aligned} S &= z_1 + z_2 + \dot{z}_2 \\ &= e_x + (2+\mu)e_y - \mu(x_2^2 y_2 - x_1^2 y_1) + a\mu(x_2^4 y_2 - x_1^4 y_1) \\ &\quad - b\mu(x_2^6 y_2 - x_1^6 y_1) + c\mu(x_2^8 y_2 - x_1^8 y_1) + u \end{aligned} \tag{54}$$

Substituting the feedback control law (44) into (54), we get

$$\begin{aligned} S &= [a - \hat{a}(t)]\mu(x_2^4 y_2 - x_1^4 y_1) - [b - \hat{b}(t)]\mu(x_2^6 y_2 - x_1^6 y_1) \\ &\quad + [c - \hat{c}(t)]\mu(x_2^8 y_2 - x_1^8 y_1) - k z_2 \end{aligned} \tag{55}$$

Using (42), Eq. (55) can be simplified as

$$S = e_a \mu(x_2^4 y_2 - x_1^4 y_1) - e_b \mu(x_2^6 y_2 - x_1^6 y_1) + e_c \mu(x_2^8 y_2 - x_1^8 y_1) - k z_2 \tag{56}$$

Substituting the value of S from (56) into (53), we get

$$\begin{aligned} \dot{V} &= -z_1^2 - (1+k)z_2^2 + e_a \left[\mu z_2 (x_2^4 y_2 - x_1^4 y_1) - \dot{\hat{a}} \right] \\ &\quad + e_b \left[-\mu z_2 (x_2^6 y_2 - x_1^6 y_1) - \dot{\hat{b}} \right] + e_c \left[\mu z_2 (x_2^8 y_2 - x_1^8 y_1) - \dot{\hat{c}} \right] \end{aligned} \tag{57}$$

Substituting the parameter update law (46) into (57), we obtain

$$\dot{V} = -z_1^2 - (1+k)z_2^2 \tag{58}$$

which is negative semi-definite in $\mathbb{R}^5$.

From (58), it follows that the vector $z(t) = (z_1(t), z_2(t))$ and the parameter estimation error $(e_a(t), e_b(t), e_c(t))$ are globally bounded, *i.e.*

$$\left[z_1(t)\ z_2(t)\ e_a(t)\ e_b(t)\ e_c(t) \right] \in \mathbf{L}_\infty \tag{59}$$

Also, it follows from (58) that

$$\dot{V} \le -z_1^2 - z_2^2 = -\|\mathbf{z}\|^2 \tag{60}$$

That is,

$$\|\mathbf{z}\|^2 \le -\dot{V} \tag{61}$$

Integrating the inequality (61) from 0 to t, we get

$$\int_0^t \|\mathbf{z}(\tau)\|^2 \, d\tau \ \leq \ V(0) - V(t) \tag{62}$$

From (62), it follows that $\mathbf{z} \in \mathbf{L}_2$. From Eq. (41), it can be deduced that $\dot{\mathbf{z}} \in \mathbf{L}_\infty$.

Thus, using Barbalat's lemma [32], we conclude that $\mathbf{z}(t) \to \mathbf{0}$ exponentially as $t \to \infty$ for all initial conditions $\mathbf{z}(0) \in \mathbb{R}^2$. Hence, it is immediate that $e_x(t) \to 0$ and $e_y(t) \to 0$ exponentially for all initial conditions $e_x(0), e_y(0) \in \mathbb{R}$.

This completes the proof. □

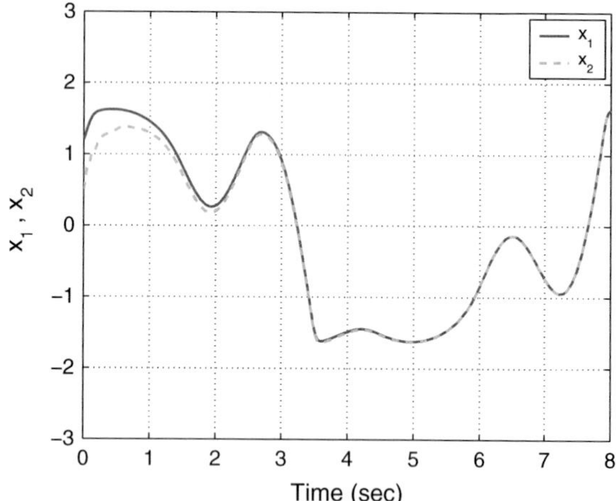

Fig. 6 Synchronization of the states x_1 and x_2

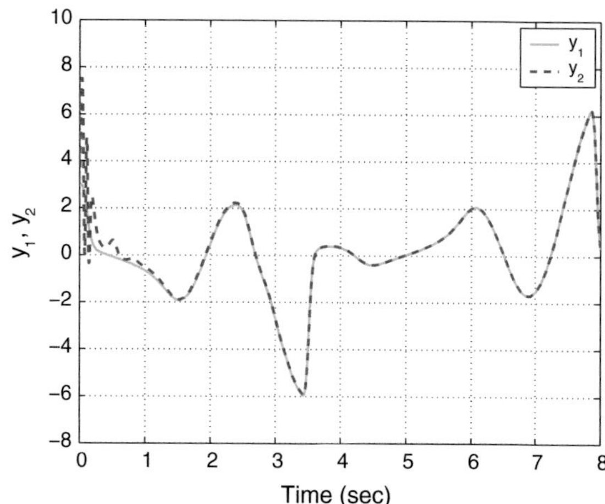

Fig. 7 Synchronization of the states y_1 and y_2

Fig. 8 Time-history of the synchronization errors e_x, e_y

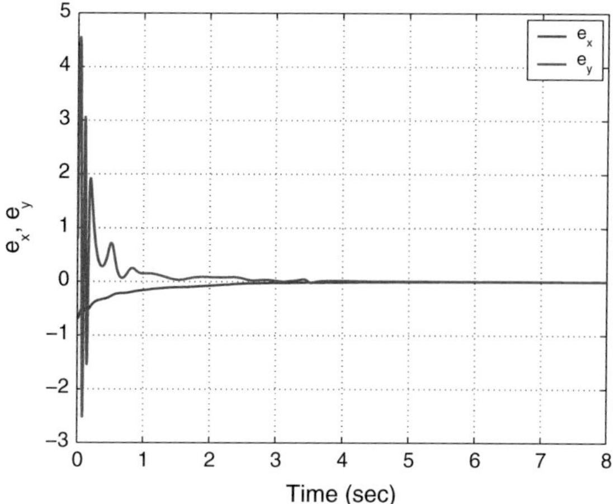

For numerical simulations, we take the positive gain constant as $k = 10$.

The parameter values are taken as in (10) and (11) for the chaotic case.

We take $\hat{a}(0) = 6.1$, $\hat{b}(0) = 2.2$ and $\hat{c}(0) = 4.8$.

We also take $x_1(0) = 1.2$, $y_1(0) = 2.9$, $x_2(0) = 0.5$ and $y_2(0) = 3.7$.

Figures 6 and 7 show the synchronization of the novel enzymes-substrates reaction systems (38) and (39). Figure 8 shows the time-history of the synchronization errors e_x, e_y.

From Figs. 6, 7 and 8, it is clear that the novel enzymes-substrates reaction systems (38) and (39) are synchronized in approximately six seconds.

5 Conclusions

In this research work, we described a novel 2-D chaotic enzymes-substrates reaction system and discuss its adaptive backstepping control. First, we discussed the dynamic equations and qualitative properties of the novel 2-D biological chaotic system. We derived our novel biological chaotic system by modifying the equations of the 2-D enzymes-substrates reaction system with ferroelectric behaviour in brain waves obtained by Kadji, Chabi Orou, Yamapi and Woafo (2007). The Maximal Lyapunov Exponent (MLE) of the novel 2-D chaotic enzymes-substrates reaction system is obtained as $L_1 = 0.14425$. Next, we derived new results for the adaptive control of the novel 2-D chaotic enzymes-substrates reaction system via backstepping control method. We also derived new results for the adaptive synchronization of identical novel 2-D chaotic enzymes-substrates reaction systems via backstepping control method. The main stabilization and synchronization results were established via Lyapunov stability theory. MATLAB simulations were shown to illustrate all the main results derived in this work for the novel 2-D biological chaotic system.

References

1. Arneodo A, Coullet P, Tresser C (1981) Possible new strange attractors with spiral structure. Commun Math Phys 79(4):573–576
2. Azar AT (2010) Fuzzy systems. IN-TECH, Vienna
3. Azar AT (2012) Overview of type-2 fuzzy logic systems. Int J Fuzzy Syst Appl 2(4):1–28
4. Azar AT, Serrano FE (2014) Robust IMC-PID tuning for cascade control systems with gain and phase margin specifications. Neural Comput Appl 25(5):983–995
5. Azar AT, Serrano FE (2015) Adaptive sliding mode control of the Furuta pendulum. In: Azar AT, Zhu Q (eds) Advances and applications in sliding mode control systems. Studies in computational intelligence, vol 576. Springer, Germany, pp 1–42
6. Azar AT, Serrano FE (2015) Deadbeat control for multivariable systems with time varying delays. In: Azar AT, Vaidyanathan S (eds) Chaos modeling and control systems design. Studies in computational intelligence, vol 581. Springer, Germany, pp 97–132
7. Azar AT, Serrano FE (2015) Design and modeling of anti wind up PID controllers. In: Zhu Q, Azar AT (eds) Complex system modelling and control through intelligent soft computations. Studies in fuzziness and soft computing, vol 319. Springer, Germany, pp 1–44
8. Azar AT, Serrano FE (2015) Stabilizatoin and control of mechanical systems with backlash. In: Azar AT, Vaidyanathan S (eds) Handbook of research on advanced intelligent control engineering and automation. Advances in computational intelligence and robotics (ACIR). IGI-Global, USA, pp 1–60
9. Azar AT, Vaidyanathan S (2015) Chaos modeling and control systems design. Studies in computational intelligence, vol 581. Springer, Germany
10. Azar AT, Vaidyanathan S (2015) Computational intelligence applications in modeling and control. Studies in computational intelligence, vol 575. Springer, Germany
11. Azar AT, Vaidyanathan S (2015) Handbook of research on advanced intelligent control engineering and automation. Advances in computational intelligence and robotics (ACIR). IGI-Global, USA
12. Azar AT, Zhu Q (2015) Advances and applications in sliding mode control systems. Studies in computational intelligence, vol 576. Springer, Germany
13. Cai G, Tan Z (2007) Chaos synchronization of a new chaotic system via nonlinear control. J Uncertain Syst 1(3):235–240
14. Carroll TL, Pecora LM (1991) Synchronizing chaotic circuits. IEEE Trans Circ Syst 38(4):453–456
15. Chen G, Ueta T (1999) Yet another chaotic attractor. Int J Bifurcat Chaos 9(7):1465–1466
16. Chen HK, Lee CI (2004) Anti-control of chaos in rigid body motion. Chaos Solitons Fractals 21(4):957–965
17. Chen WH, Wei D, Lu X (2014) Global exponential synchronization of nonlinear time-delay Lur'e systems via delayed impulsive control. Commun Nonlinear Sci Numer Simul 19(9):3298–3312
18. Das S, Goswami D, Chatterjee S, Mukherjee S (2014) Stability and chaos analysis of a novel swarm dynamics with applications to multi-agent systems. Eng Appl Artif Intell 30:189–198
19. Feki M (2003) An adaptive chaos synchronization scheme applied to secure communication. Chaos Solitons Fractals 18(1):141–148
20. Frohlich H (1968) Long range coherence and energy storage in biological systems. Int J Quantum Chem 2:641–649
21. Gan Q, Liang Y (2012) Synchronization of chaotic neural networks with time delay in the leakage term and parametric uncertainties based on sampled-data control. J Franklin Inst 349(6):1955–1971
22. Gaspard P (1999) Microscopic chaos and chemical reactions. Phys A: Stat Mech Appl 263(1–4):315–328
23. Gibson WT, Wilson WG (2013) Individual-based chaos: extensions of the discrete logistic model. J Theor Biol 339:84–92

24. Guégan D (2009) Chaos in economics and finance. Ann Rev Control 33(1):89–93
25. Huang X, Zhao Z, Wang Z, Li Y (2012) Chaos and hyperchaos in fractional-order cellular neural networks. Neurocomputing 94:13–21
26. Jiang GP, Zheng WX, Chen G (2004) Global chaos synchronization with channel time-delay. Chaos Solitons Fractals 20(2):267–275
27. Kadji HGE, Orou JBC, Yamapi R, Woafo P (2007) Nonlinear dynamics and strange attractors in the biological system. Chaos Solitons Fractals 32:862–882
28. Kaiser F (1978) Coherent oscillagions in biological systems, I. Bifurcation phenomena and phase transitions in an enzyme-substrate reaction with ferroelectric behavior. Z Naturforsch A 294:304–333
29. Karthikeyan R, Sundarapandian V (2014) Hybrid chaos synchronization of four-scroll systems via active control. J Electr Eng 65(2):97–103
30. Kaslik E, Sivasundaram S (2012) Nonlinear dynamics and chaos in fractional-order neural networks. Neural Netw 32:245–256
31. Kengne J, Chedjou JC, Kenne G, Kyamakya K (2012) Dynamical properties and chaos synchronization of improved Colpitts oscillators. Commun Nonlinear Sci Numer Simul 17(7):2914–2923
32. Khalil HK (2001) Nonlinear Syst. Prentice Hall, New Jersey
33. Kyriazis M (1991) Applications of chaos theory to the molecular biology of aging. Exp Gerontol 26(6):569–572
34. Lang J (2015) Color image encryption based on color blend and chaos permutation in the reality-preserving multiple-parameter fractional fourier transform domain. Opt Commun 338:181–192
35. Li D (2008) A three-scroll chaotic attractor. Phys Lett A 372(4):387–393
36. Li N, Zhang Y, Nie Z (2011) Synchronization for general complex dynamical networks with sampled-data. Neurocomputing 74(5):805–811
37. Li N, Pan W, Yan L, Luo B, Zou X (2014) Enhanced chaos synchronization and communication in cascade-coupled semiconductor ring lasers. Commun Nonlinear Sci Numer Simul 19(6):1874–1883
38. Li Z, Chen G (2006) Integration of fuzzy logic and chaos theory. Studies in fuzziness and soft computing, vol 187. Springer, Germany
39. Lian S, Chen X (2011) Traceable content protection based on chaos and neural networks. Appl Soft Comput 11(7):4293–4301
40. Liu C, Liu T, Liu L, Liu K (2004) A new chaotic attractor. Chaos Solitions Fractals 22(5):1031–1038
41. Lorenz EN (1963) Deterministic periodic flow. J Atmos Sci 20(2):130–141
42. Lü J, Chen G (2002) A new chaotic attractor coined. Int J Bifurcat Chaos 12(3):659–661
43. Mondal S, Mahanta C (2014) Adaptive second order terminal sliding mode controller for robotic manipulators. J Franklin Inst 351(4):2356–2377
44. Murali K, Lakshmanan M (1998) Secure communication using a compound signal from generalized chaotic systems. Phys Lett A 241(6):303–310
45. Nehmzow U, Walker K (2005) Quantitative description of robot-environment interaction using chaos theory. Robot Auton Syst 53(3–4):177–193
46. Noroozi N, Roopaei M, Karimaghaee P, Safavi AA (2010) Simple adaptive variable structure control for unknown chaotic systems. Commun Nonlinear Sci Numer Simul 15(3):707–727
47. Pecora LM, Carroll TL (1990) Synchronization in chaotic systems. Phys Rev Lett 64(8):821–824
48. Pehlivan I, Moroz IM, Vaidyanathan S (2014) Analysis, synchronization and circuit design of a novel butterfly attractor. J Sound Vib 333(20):5077–5096
49. Petrov V, Gaspar V, Masere J, Showalter K (1993) Controlling chaos in Belousov-Zhabotinsky reaction. Nature 361:240–243
50. Pham VT, Vaidyanathan S, Volos CK, Jafari S (2015a) Hidden attractors in a chaotic system with an exponential nonlinear term. Eur Phys J Spec Top 224(8):1507–1517

51. Pham VT, Volos CK, Vaidyanathan S, Le TP, Vu VY (2015b) A memristor-based hyperchaotic system with hidden attractors: dynamics, synchronization and circuital emulating. J Eng Sci Technol Rev 8(2):205–214
52. Qu Z (2011) Chaos in the genesis and maintenance of cardiac arrhythmias. Prog Biophys Mol Biol 105(3):247–257
53. Rasappan S, Vaidyanathan S (2012a) Global chaos synchronization of WINDMI and Coullet chaotic systems by backstepping control. Far East J Math Sci 67(2):265–287
54. Rasappan S, Vaidyanathan S (2012b) Hybrid synchronization of n-scroll Chua and Lur'e chaotic systems via backstepping control with novel feedback. Arch Control Sci 22(3):343–365
55. Rasappan S, Vaidyanathan S (2012c) Synchronization of hyperchaotic Liu system via backstepping control with recursive feedback. Commun Comput Inf Sci 305:212–221
56. Rasappan S, Vaidyanathan S (2013) Hybrid synchronization of n-scroll chaotic Chua circuits using adaptive backstepping control design with recursive feedback. Malays J Math Sci 7(2):219–246
57. Rasappan S, Vaidyanathan S (2014) Global chaos synchronization of WINDMI and Coullet chaotic systems using adaptive backstepping control design. Kyungpook Math J 54(1):293–320
58. Rhouma R, Belghith S (2011) Cryptoanalysis of a chaos based cryptosystem on DSP. Commun Nonlinear Sci Numer Simul 16(2):876–884
59. Rössler OE (1976) An equation for continuous chaos. Phys Lett A 57(5):397–398
60. Vaidyanathan S, VTP, Volos CK (2015) A 5-D hyperchaotic Rikitake dynamo system with hidden attractors. Eur Phys J Spec Top 224(8):1575–1592
61. Sampath S, Vaidyanathan S, Volos CK, Pham VT (2015) An eight-term novel four-scroll chaotic system with cubic nonlinearity and its circuit simulation. J Eng Sci Technol Rev 8(2):1–6
62. Sarasu P, Sundarapandian V (2011) Active controller design for the generalized projective synchronization of four-scroll chaotic systems. Int J Syst Signal Control Eng Appl 4(2):26–33
63. Sarasu P, Sundarapandian V (2011b) The generalized projective synchronization of hyperchaotic Lorenz and hyperchaotic Qi systems via active control. Int J Soft Comput 6(5):216–223
64. Sarasu P, Sundarapandian V (2012) Adaptive controller design for the generalized projective synchronization of 4-scroll systems. Int J Syst Signal Control Eng Appl 5(2):21–30
65. Sarasu P, Sundarapandian V (2012b) Generalized projective synchronization of three-scroll chaotic systems via adaptive control. Eur J Sci Res 72(4):504–522
66. Sarasu P, Sundarapandian V (2012c) Generalized projective synchronization of two-scroll systems via adaptive control. Int J Soft Comput 7(4):146–156
67. Shahverdiev EM, Shore KA (2009) Impact of modulated multiple optical feedback time delays on laser diode chaos synchronization. Opt Commun 282(17):3568–3572
68. Sharma A, Patidar V, Purohit G, Sud KK (2012) Effects on the bifurcation and chaos in forced Duffing oscillator due to nonlinear damping. Commun Nonlinear Sci Numer Simul 17(6):2254–2269
69. Sprott JC (1994) Some simple chaotic flows. Phys Rev E 50(2):647–650
70. Sprott JC (2004) Competition with evolution in ecology and finance. Phys Lett A 325(5–6):329–333
71. Sprott JC (2010) Elegant chaos: algebraically simple chaotic flows. World Scientific, Singapore
72. Suérez I (1999) Mastering chaos in ecology. Ecol Model 117(2–3):305–314
73. Sundarapandian V (2010a) Output regulation of the Lorenz attractor. Far East J Math Sci 42(2):289–299
74. Sundarapandian V (2010b) Output regulation of the Lorenz attractor. Far East J Math Sci 42(2):289–299
75. Sundarapandian V (2013) Analysis and anti-synchronization of a novel chaotic system via active and adaptive controllers. J Eng Sci Technol Rev 6(4):45–52

76. Sundarapandian V, Karthikeyan R (2011a) Anti-synchronization of hyperchaotic Lorenz and hyperchaotic Chen systems by adaptive control. Int J Syst Signal Control Eng Appl 4(2):18–25
77. Sundarapandian V, Karthikeyan R (2011b) Anti-synchronization of Lü and Pan chaotic systems by adaptive nonlinear control. Eur J Sci Res 64(1):94–106
78. Sundarapandian V, Karthikeyan R (2012a) Adaptive anti-synchronization of uncertain Tigan and Li systems. J Eng Appl Sci 7(1):45–52
79. Sundarapandian V, Karthikeyan R (2012b) Hybrid synchronization of hyperchaotic Lorenz and hyperchaotic Chen systems via active control. J Eng Appl Sci 7(3):254–264
80. Sundarapandian V, Pehlivan I (2012) Analysis, control, synchronization, and circuit design of a novel chaotic system. Math Comput Model 55(7–8):1904–1915
81. Sundarapandian V, Sivaperumal S (2011) Sliding controller design of hybrid synchronization of four-wing chaotic systems. Int J Soft Comput 6(5):224–231
82. Suresh R, Sundarapandian V (2013) Global chaos synchronization of a family of n-scroll hyperchaotic Chua circuits using backstepping control with recursive feedback. Far East J Math Sci 73(1):73–95
83. Tigan G, Opris D (2008) Analysis of a 3D chaotic system. Chaos Solitons Fractals 36:1315–1319
84. Usama M, Khan MK, Alghatbar K, Lee C (2010) Chaos-based secure satellite imagery cryptosystem. Comput Math Appl 60(2):326–337
85. Vaidyanathan S (2011) Hybrid chaos synchronization of Liu and Lü systems by active nonlinear control. In: Nagamalai D, Renault E, Dhanuskodi M (eds) Trends in computer science, engineering and information technology. Communications in computer and information science, vol 204. Springer, Berlin
86. Vaidyanathan S (2011b) Output regulation of Arneodo-Coullet chaotic system. In: Meghanathan N, Kaushik BK, Nagamalai D (eds) Advanced computing. Communications in computer and information, science, vol 133. Springer, Berlin, pp 98–107
87. Vaidyanathan S (2011c) Output regulation of the unified chaotic system. In: Wyld DC, Wozniak M, Chaki N, Meghanathan N, Nagamalai D (eds) Advances in computing and information technology. Communications in computer and information, science, vol 198, pp 10–17. Springer, Berlin
88. Vaidyanathan S (2012) Analysis and synchronization of the hyperchaotic Yujun systems via sliding mode control. In: Meghanathan N, Nagamalai D, Chaki N (eds) Advances in computing and information technology. Advances in intelligent systems and computing, vol 176. Springer, Berlin, pp 329–337
89. Vaidyanathan S (2012b) Anti-synchronization of Sprott-L and Sprott-M chaotic systems via adaptive control. Int J Control Theor Appl 5(1):41–59
90. Vaidyanathan S (2012c) Global chaos control of hyperchaotic Liu system via sliding control method. Int J Control Theor Appl 5(2):117–123
91. Vaidyanathan S (2012d) Output regulation of the Liu chaotic system. Appl Mech Mater 110–116:3982–39890
92. Vaidyanathan S (2012e) Sliding mode control based global chaos control of Liu-Liu-Liu-Su chaotic system. Int J Control Theor Appl 1(2):15–20
93. Vaidyanathan S (2013a) A new six-term 3-D chaotic system with an exponential nonlinearity. Far East J Math Sci 79(1):135–143
94. Vaidyanathan S (2013) Analysis and adaptive synchronization of two novel chaotic systems with hyperbolic sinusoidal and cosinusoidal nonlinearity and unknown parameters. J Eng Sci Technol Rev 6(4):53–65
95. Vaidyanathan S (2013) Analysis, control and synchronization of hyperchaotic Zhou system via adaptive control. In: Meghanathan N, Nagamalai D, Chaki N (eds) Advances in computing and information technology. Advances in intelligent systems and computing, vol 177. Springer, Berlin, pp 1–10
96. Vaidyanathan S (2014a) A new eight-term 3-D polynomial chaotic system with three quadratic nonlinearities. Far East J Math Sci 84(2):219–226

97. Vaidyanathan S (2014) Analysis and adaptive synchronization of eight-term 3-D polynomial chaotic systems with three quadratic nonlinearities. Eur Phys J Spec Top 223(8):1519–1529
98. Vaidyanathan S (2014c) Analysis, control and synchronisation of a six-term novel chaotic system with three quadratic nonlinearities. Int J Model Ident Control 22(1):41–53
99. Vaidyanathan S (2014d) Generalized projective synchronisation of novel 3-D chaotic systems with an exponential non-linearity via active and adaptive control. Int J Model Ident Control 22(3):207–217
100. Vaidyanathan S (2014e) Global chaos synchronization of identical Li-Wu chaotic systems via sliding mode control. Int J Model Ident Control 22(2):170–177
101. Vaidyanathan S (2015) 3-cells cellular neural network (CNN) attractor and its adaptive biological control. Int J PharmTech Res 8(4):632–640
102. Vaidyanathan S (2015) A 3-D novel highly chaotic system with four quadratic nonlinearities, its adaptive control and anti-synchronization with unknown parameters. J Eng Sci Technol Rev 8(2):106–115
103. Vaidyanathan S (2015) Adaptive backstepping control of enzymes-substrates system with ferroelectric behaviour in brain waves. Int J PharmTech Res 8(2):256–261
104. Vaidyanathan S (2015) Adaptive biological control of generalized Lotka-Volterra three-species biological system. Int J PharmTech Res 8(4):622–631
105. Vaidyanathan S (2015) Adaptive chaotic synchronization of enzymes-substrates system with ferroelectric behaviour in brain waves. Int J PharmTech Res 8(5):964–973
106. Vaidyanathan S (2015) Adaptive control of a chemical chaotic reactor. Int J PharmTech Res 8(3):377–382
107. Vaidyanathan S (2015) Adaptive synchronization of chemical chaotic reactors. Int J ChemTech Res 8(2):612–621
108. Vaidyanathan S (2015) Adaptive synchronization of generalized Lotka-Volterra three-species biological systems. Int J PharmTech Res 8(5):928–937
109. Vaidyanathan S (2015i) Analysis, properties and control of an eight-term 3-D chaotic system with an exponential nonlinearity. Int J Model Ident Control 23(2):164–172
110. Vaidyanathan S (2015) Anti-synchronization of Brusselator chemical reaction systems via adaptive control. Int J ChemTech Res 8(6):759–768
111. Vaidyanathan S (2015) Chaos in neurons and adaptive control of Birkhoff-Shaw strange chaotic attractor. Int J PharmTech Res 8(5):956–963
112. Vaidyanathan S (2015) Dynamics and control of Brusselator chemical reaction. Int J ChemTech Res 8(6):740–749
113. Vaidyanathan S (2015) Dynamics and control of Tokamak system with symmetric and magnetically confined plasma. Int J ChemTech Res 8(6):795–803
114. Vaidyanathan S (2015n) Hyperchaos, qualitative analysis, control and synchronisation of a ten-term 4-D hyperchaotic system with an exponential nonlinearity and three quadratic nonlinearities. Int J Model Ident Control 23(4):380–392
115. Vaidyanathan S (2015) Lotka-Volterra population biology models with negative feedback and their ecological monitoring. Int J PharmTech Res 8(5):974–981
116. Vaidyanathan S (2015) Synchronization of 3-cells cellular neural network (CNN) attractors via adaptive control method. Int J PharmTech Res 8(5):946–955
117. Vaidyanathan S (2015) Synchronization of Tokamak systems with symmetric and magnetically confined plasma via adaptive control. Int J ChemTech Res 8(6):818–827
118. Vaidyanathan S, Azar AT (2015) Analysis and control of a 4-D novel hyperchaotic system. In: Azar AT, Vaidyanathan S (eds) Chaos modeling and control systems design. Studies in computational intelligence, vol 581. Springer, Germany, pp 19–38
119. Vaidyanathan S, Azar AT (2015) Analysis, control and synchronization of a nine-term 3-D novel chaotic system. In: Azar AT, Vaidyanathan S (eds) Chaos modelling and control systems design. Studies in computational intelligence, vol 581. Springer, Germany, pp 19–38
120. Vaidyanathan S, Azar AT (2015c) Anti-synchronization of identical chaotic systems using sliding mode control and an application to Vaidhyanathan-Madhavan chaotic systems. In: Azar AT, Zhu Q (eds) Advances and applications in sliding mode control systems. Studies in computational intelligence, vol 576. Springer, Switzerland, pp 527–547

121. Vaidyanathan S, Azar AT (2015d) Hybrid synchronization of identical chaotic systems using sliding mode control and an application to Vaidhyanathan chaotic systems. In: Advances and applications in sliding mode control systems. Studies in computational intelligence, vol 576. Springer, Switzerland, pp 549–569
122. Vaidyanathan S, Madhavan K (2013) Analysis, adaptive control and synchronization of a seven-term novel 3-D chaotic system. Int J Control Theor Appl 6(2):121–137
123. Vaidyanathan S, Pakiriswamy S (2013) Generalized projective synchronization of six-term Sundarapandian chaotic systems by adaptive control. Int J Control Theor Appl 6(2):153–163
124. Vaidyanathan S, Pakiriswamy S (2015) A 3-D novel conservative chaotic system and its generalized projective synchronization via adaptive control. J Eng Sci Technol Rev 8(2):52–60
125. Vaidyanathan S, Rajagopal K (2011a) Anti-synchronization of Li and T chaotic systems by active nonlinear control. In: Wyld DC, Wozniak M, Chaki N, Meghanathan N, Nagamalai D (eds) Advances in computing and information technology. Communications in computer and information science, vol 198. Springer, Berlin, pp 175–184
126. Vaidyanathan S, Rajagopal K (2011b) Global chaos synchronization of hyperchaotic Pang and Wang systems by active nonlinear control. In: Nagamalai D, Renault E, Dhanuskodi M (eds) Trends in computer science, engineering and information technology. Communications in Computer and information science, vol 204. Springer, Berlin, pp 84–93
127. Vaidyanathan S, Rajagopal K (2011) Global chaos synchronization of Lü and Pan systems by adaptive nonlinear control. In: Nagamalai D, Renault E, Dhanuskodi M (eds) Advances in digital image processing and information technology. Communications in computer and information science, vol 205. Springer, Berlin, pp 193–202
128. Vaidyanathan S, Rajagopal K (2012) Global chaos synchronization of hyperchaotic Pang and hyperchaotic Wang systems via adaptive control. Int J Soft Comput 7(1):28–37
129. Vaidyanathan S, Rasappan S (2011) Global chaos synchronization of hyperchaotic Bao and Xu systems by active nonlinear control. Commun Comput Inf Sci 198:10–17
130. Vaidyanathan S, Rasappan S (2014) Global chaos synchronization of n-scroll Chua circuit and Lur'e system using backstepping control design with recursive feedback. Arab J Sci Eng 39(4):3351–3364
131. Vaidyanathan S, Sampath S (2011) Global chaos synchronization of hyperchaotic Lorenz systems by sliding mode control. Commun Comput Inf Sci 205:156–164
132. Vaidyanathan S, Sampath S (2012) Anti-synchronization of four-wing chaotic systems via sliding mode control. Int J Autom Comput 9(3):274–279
133. Vaidyanathan S, Volos C (2015) Analysis and adaptive control of a novel 3-D conservative no-equilibrium chaotic system. Arch Control Sci 25(3):333–353
134. Vaidyanathan S, Volos C, Pham VT (2014a) Hyperchaos, adaptive control and synchronization of a novel 5-D hyperchaotic system with three positive Lyapunov exponents and its SPICE implementation. Arch Control Sci 24(4):409–446
135. Vaidyanathan S, Volos C, Pham VT, Madhavan K, Idowu BA (2014b) Adaptive backstepping control, synchronization and circuit simulation of a 3-D novel jerk chaotic system with two hyperbolic sinusoidal nonlinearities. Arch Control Sci 24(3):375–403
136. Vaidyanathan S, Azar AT, Rajagopal K, Alexander P (2015a) Design and SPICE implementation of a 12-term novel hyperchaotic system and its synchronisation via active control. Int J Model Ident Control 23(3):267–277
137. Vaidyanathan S, Idowu BA, Azar AT (2015b) Backstepping controller design for the global chaos synchronization of Sprott's jerk systems. In: Azar AT, Vaidyanathan S (eds) Chaos modeling and control systems design. Studies in computational intelligence, vol 581. Springer, Switzerland, pp 39–58
138. Vaidyanathan S, Rajagopal K, Volos CK, Kyprianidis IM, Stouboulos IN (2015c) Analysis, adaptive control and synchronization of a seven-term novel 3-D chaotic system with three quadratic nonlinearities and its digital implementation in LabVIEW. J Eng Sci Technol Rev 8(2):130–141

139. Vaidyanathan S, Volos C, Pham VT, Madhavan K (2015d) Analysis, adaptive control and synchronization of a novel 4-D hyperchaotic hyperjerk system and its SPICE implementation. Arch Control Sci 25(1):5–28
140. Vaidyanathan S, Volos CK, Kyprianidis IM, Stouboulos IN, Pham VT (2015e) Analysis, adaptive control and anti-synchronization of a six-term novel jerk chaotic system with two exponential nonlinearities and its circuit simulation. J Eng Sci Technol Rev 8(2):24–36
141. Vaidyanathan S, Volos CK, Pham VT (2015f) Analysis, adaptive control and adaptive synchronization of a nine-term novel 3-D chaotic system with four quadratic nonlinearities and its circuit simulation. J Eng Sci Technol Rev 8(2):174–184
142. Vaidyanathan S, Volos CK, Pham VT (2015g) Analysis, control, synchronization and SPICE implementation of a novel 4-D hyperchaotic Rikitake dynamo system without equilibrium. J Eng Sci Technol Rev 8(2):232–244
143. Vaidyanathan S, Volos CK, Pham VT (2015h) Global chaos control of a novel nine-term chaotic system via sliding mode control. In: Azar AT, Zhu Q (eds) Advances and applications in sliding mode control systems. Studies in computational intelligence, vol 576. Springer, Germany, pp 571–590
144. Vaidyanathan S, Volos CK, Rajagopal K, Kyprianidis IM, Stouboulos IN (2015i) Adaptive backstepping controller design for the anti-synchronization of identical WINDMI chaotic systems with unknown parameters and its SPICE implementation. J Eng Sci Technol Rev 8(2):74–82
145. Volos CK, Kyprianidis IM, Stouboulos IN (2013) Experimental investigation on coverage performance of a chaotic autonomous mobile robot. Robot Auton Syst 61(12):1314–1322
146. Volos CK, Kyprianidis IM, Stouboulos IN, Tlelo-Cuautle E, Vaidyanathan S (2015) Memristor: a new concept in synchronization of coupled neuromorphic circuits. J Eng Sci Technol Rev 8(2):157–173
147. Wei Z, Yang Q (2010) Anti-control of Hopf bifurcation in the new chaotic system with two stable node-foci. Appl Math Comput 217(1):422–429
148. Witte CL, Witte MH (1991) Chaos and predicting varix hemorrhage. Med Hypotheses 36(4):312–317
149. Xiao X, Zhou L, Zhang Z (2014) Synchronization of chaotic Lur'e systems with quantized sampled-data controller. Commun Nonlinear Sci Numer Simul 19(6):2039–2047
150. Yuan G, Zhang X, Wang Z (2014) Generation and synchronization of feedback-induced chaos in semiconductor ring lasers by injection-locking. Optik Int J Light Electron Opt 125(8):1950–1953
151. Zaher AA, Abu-Rezq A (2011) On the design of chaos-based secure communication systems. Commun Nonlinear Syst Numer Simul 16(9):3721–3727
152. Zhang H, Zhou J (2012) Synchronization of sampled-data coupled harmonic oscillators with control inputs missing. Syst Control Lett 61(12):1277–1285
153. Zhang X, Zhao Z, Wang J (2014) Chaotic image encryption based on circular substitution box and key stream buffer. Signal Process Image Commun 29(8):902–913
154. Zhou W, Xu Y, Lu H, Pan L (2008) On dynamics analysis of a new chaotic attractor. Phys Lett A 372(36):5773–5777
155. Zhu C, Liu Y, Guo Y (2010) Theoretic and numerical study of a new chaotic system. Intell Inf Manage 2:104–109
156. Zhu Q, Azar AT (2015) Complex system modelling and control through intelligent soft computations. Studies in fuzzines and soft computing, vol 319. Springer, Germany

Analysis, Control and Synchronization of a Novel 4-D Highly Hyperchaotic System with Hidden Attractors

Sundarapandian Vaidyanathan

Abstract In this work, we announce a ten-term novel 4-D highly hyperchaotic system with three quadratic nonlinearities. The phase portraits of the ten-term novel highly hyperchaotic system are depicted and the qualitative properties of the novel highly hyperchaotic system are discussed. We shall show that the novel hyperchaotic system does not have any equilibrium point. Hence, the novel 4-D hyperchaotic system exhibits hidden attractors. The Lyapunov exponents of the novel hyperchaotic system are obtained as $L_1 = 13.67837$, $L_2 = 0.04058$, $L_3 = 0$ and $L_4 = -45.64661$. The Maximal Lyapunov Exponent (MLE) of the novel hyperchaotic system is found as $L_1 = 13.67837$, which is large. Thus, the novel 4-D hyperchaotic system proposed in this work is highly hyperchaotic. Also, the Kaplan-Yorke dimension of the novel hyperchaotic system is derived as $D_{KY} = 3.30055$. Since the sum of the Lyapunov exponents is negative, the novel hyperchaotic system is dissipative. Next, an adaptive controller is designed to globally stabilize the novel highly hyperchaotic system with unknown parameters. Finally, an adaptive controller is also designed to achieve global chaos synchronization of the identical novel highly hyperchaotic systems with unknown parameters. MATLAB simulations are depicted to illustrate all the main results derived in this work.

Keywords Chaos · Chaotic systems · Hyperchaos · Hyperchaotic systems · Adaptive control · Synchronization

1 Introduction

In the last few decades, Chaos theory has become a very important and active research field, employing many applications in different disciplines like physics, chemistry, biology, ecology, engineering and economics, among others.

S. Vaidyanathan (✉)
Research and Development Centre, Vel Tech University,
Avadi, Chennai 600062, Tamil Nadu, India
e-mail: sundarvtu@gmail.com

A.T. Azar and S. Vaidyanathan (eds.), *Advances in Chaos Theory and Intelligent Control*, Studies in Fuzziness and Soft Computing 337,
DOI 10.1007/978-3-319-30340-6_22

Some classical paradigms of 3-D chaotic systems in the literature are Lorenz system [35], Rössler system [50], ACT system [1], Sprott systems [58], Chen system [15], Lü system [36], Cai system [13], Tigan system [69], etc.

Many new chaotic systems have been discovered in the recent years such as Zhou system [138], Zhu system [139], Li system [31], Wei-Yang system [131], Sundarapandian systems [62, 66], Vaidyanathan systems [77, 79, 82–86, 96, 99, 100, 104, 111, 115, 117, 119, 125, 126], Pehlivan system [40], Sampath system [52], Pham system [43], etc.

The Lyapunov exponent is a measure of the divergence of phase points that are initially very close and can be used to quantify chaotic systems. It is common to refer to the largest Lyapunov exponent as the *Maximal Lyapunov Exponent* (MLE). A positive maximal Lyapunov exponent and phase space compactness are usually taken as defining conditions for a chaotic system.

A hyperchaotic system is defined as a chaotic system with at least two positive Lyapunov exponents. Thus, the dynamics of a hyperchaotic system can expand in several different directions simultaneously. Thus, the hyperchaotic systems have more complex dynamical behaviour and they have miscellaneous applications in engineering such as secure communications [18, 30, 133], cryptosystems [21, 48, 140], fuzzy logic [56, 136], electrical circuits [130, 134], etc.

The minimum dimension of an autonomous, continuous-time, hyperchaotic system is four. The first 4-D hyperchaotic system was found by Rössler [51]. Many hyperchaotic systems have been reported in the chaos literature such as hyperchaotic Lorenz system [23], hyperchaotic Lü system [14], hyperchaotic Chen system [32], hyperchaotic Wang system [129], hyperchaotic Vaidyanathan systems [78, 81, 88, 90, 94, 98, 101, 102, 116], hyperchaotic Pham system [42], etc.

Chaos theory and control systems have many important applications in science and engineering [6–10, 141]. Some commonly known applications are oscillators [26, 57], chemical reactions [19, 41, 89, 93, 108, 109, 112, 114], biology [16, 28, 91, 92, 103, 105–107, 110, 113], ecology [20, 59], encryption [29, 137], cryptosystems [49, 70], mechanical systems [2–5, 12], secure communications [17, 38, 135], robotics [37, 39, 127], cardiology [45, 132], intelligent control [11, 34], neural networks [22, 25, 33], memristors [44, 128], etc.

The control of a chaotic or hyperchaotic system aims to stabilize or regulate the system with the help of a feedback control. There are many methods available for controlling a chaotic system such as active control [60, 71, 72], adaptive control [61, 73, 80], sliding mode control [75, 76], backstepping control [95], etc.

The synchronization of chaotic systems aims to synchronize the states of master and slave systems asymptotically with time. There are many methods available for chaos synchronization such as active control [24, 53, 54, 120, 122], adaptive control [55, 63–65, 74, 118, 121], sliding mode control [67, 87, 97, 124], backstepping control [46, 47, 68, 123], etc.

This work is organized as follows. Section 2 describes the dynamic equations and phase portraits of the ten-term novel 4-D hyperchaotic system. Section 3 details the qualitative properties of the novel hyperchaotic system. In this section, we establish that the novel hyperchaotic system does not have any equilibrium point. Thus, it follows that the novel hyperchaotic system exhibits hidden attractors.

The Lyapunov exponents of the novel hyperchaotic system are obtained as $L_1 = 13.67837$, $L_2 = 0.04058$, $L_3 = 0$ and $L_4 = -45.64661$, while the Kaplan-Yorke dimension of the novel hyperchaotic system is obtained as $D_{KY} = 3.30055$. Since the Maximal Lyapunov Exponent (MLE) of the novel hyperchaotic system is $L_1 = 13.67837$, which is a large value, we conclude that the proposed novel hyperchaotic system is highly hyperchaotic. A novel contribution of this research work is the finding of a highly hyperchaotic 4-D system with hidden attractors.

In Sect. 4, we design an adaptive controller to globally stabilize the novel highly hyperchaotic system with unknown parameters. In Sect. 5, an adaptive controller is designed to achieve global chaos synchronization of the identical novel highly hyperchaotic systems with unknown parameters. MATLAB simulations have been shown to illustrate all the main results derived in this research work. Section 6 summarizes the main results of this research work.

2 A Novel 4-D Hyperchaotic System

In this section, we describe an eleven-term novel hyperchaotic system, which is given by the 4-D dynamics

$$\begin{cases} \dot{x}_1 = a(x_2 - x_1) + x_2 x_3 + x_4 \\ \dot{x}_2 = -c x_1 x_3 + p x_2 \\ \dot{x}_3 = x_1 x_2 - b \\ \dot{x}_4 = -q(x_1 + x_2) \end{cases} \tag{1}$$

where x_1, x_2, x_3, x_4 are the states and a, b, c, p, q are constant positive parameters.

The system (1) exhibits a *strange hyperchaotic attractor* for the parameter values

$$a = 60, \quad b = 27, \quad c = 160, \quad p = 0.3, \quad q = 2.8 \tag{2}$$

For numerical simulations, we take the initial conditions as

$$x_1(0) = 1.2, \quad x_2(0) = 0.4, \quad x_3(0) = 0.3, \quad x_4(0) = 1.4 \tag{3}$$

Figures 1, 2, 3 and 4 show the 3-D projection of the novel hyperchaotic system (1) on the (x_1, x_2, x_3), (x_1, x_2, x_4), (x_1, x_3, x_4) and (x_2, x_3, x_4) spaces, respectively.

3 Analysis of the Novel 4-D Highly Hyperchaotic System

In this section, we study the qualitative properties of the novel 4-D highly hyperchaotic system (1). We take the parameter values as in the hyperchaotic case (2).

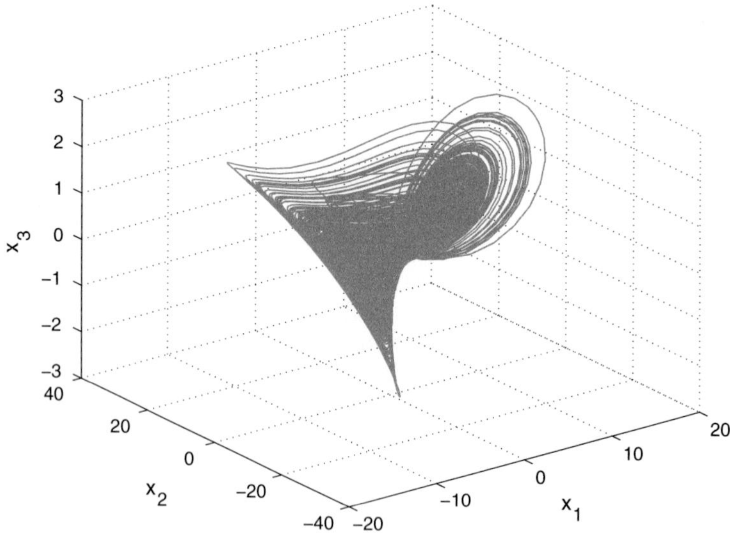

Fig. 1 3-D projection of the novel highly hyperchaotic system on the (x_1, x_2, x_3) space

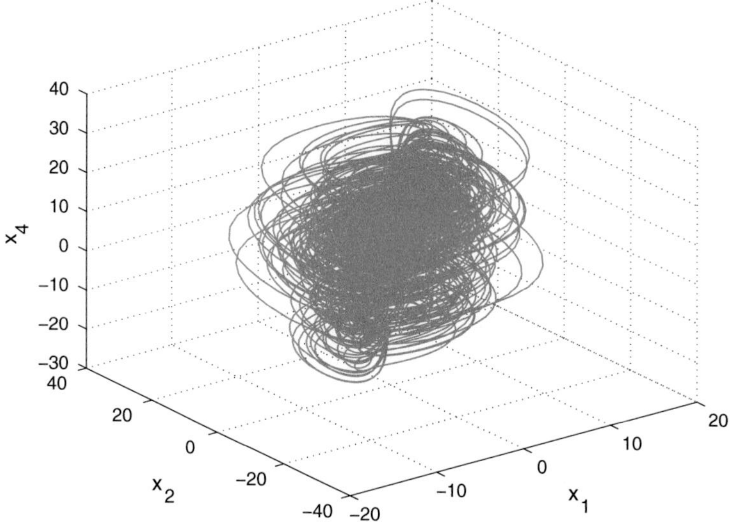

Fig. 2 3-D projection of the novel highly hyperchaotic system on the (x_1, x_2, x_4) space

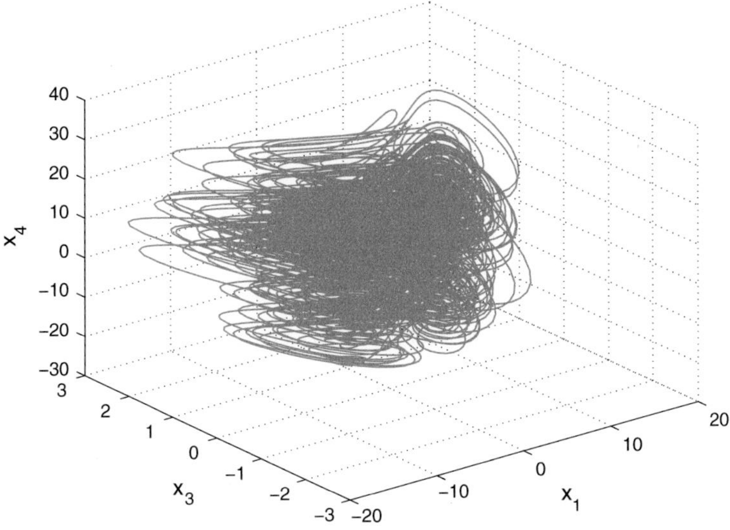

Fig. 3 3-D projection of the novel highly hyperchaotic system on the (x_1, x_3, x_4) space

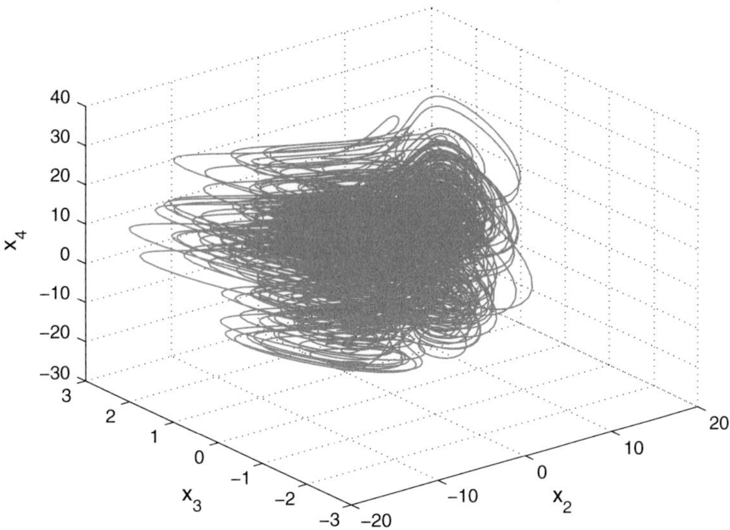

Fig. 4 3-D projection of the novel highly hyperchaotic system on the (x_2, x_3, x_4) space

3.1 Dissipativity

In vector notation, the novel highly hyperchaotic system (1) can be expressed as

$$\dot{\mathbf{x}} = f(\mathbf{x}) = \begin{bmatrix} f_1(x_1, x_2, x_3, x_4) \\ f_2(x_1, x_2, x_3, x_4) \\ f_3(x_1, x_2, x_3, x_4) \\ f_4(x_1, x_2, x_3, x_4) \end{bmatrix}, \tag{4}$$

where

$$\begin{cases} f_1(x_1, x_2, x_3, x_4) = a(x_2 - x_1) + x_2 x_3 + x_4 \\ f_2(x_1, x_2, x_3, x_4) = -c x_1 x_3 + p x_2 \\ f_3(x_1, x_2, x_3, x_4) = x_1 x_2 - b \\ f_4(x_1, x_2, x_3, x_4) = -q(x_1 + x_2) \end{cases} \tag{5}$$

Let Ω be any region in $\mathbf{R}^4$ with a smooth boundary and also, $\Omega(t) = \Phi_t(\Omega)$, where Φ_t is the flow of f.

Furthermore, let $V(t)$ denote the hypervolume of $\Omega(t)$.

By Liouville's theorem, we know that

$$\dot{V}(t) = \int\limits_{\Omega(t)} (\nabla \cdot f)\, dx_1\, dx_2\, dx_3\, dx_4 \tag{6}$$

The divergence of the novel hyperchaotic system (4) is found as:

$$\nabla \cdot f = \frac{\partial f_1}{\partial x_1} + \frac{\partial f_2}{\partial x_2} + \frac{\partial f_3}{\partial x_3} + \frac{\partial f_4}{\partial x_4} = -a + p = -\mu < 0 \tag{7}$$

since $\mu = a - p = 33 > 0$.

Inserting the value of $\nabla \cdot f$ from (7) into (6), we get

$$\dot{V}(t) = \int\limits_{\Omega(t)} (-\mu)\, dx_1\, dx_2\, dx_3\, dx_4 = -\mu V(t) \tag{8}$$

Integrating the first order linear differential equation (8), we get

$$V(t) = \exp(-\mu t) V(0) \tag{9}$$

Since $\mu > 0$, it follows from Eq. (9) that $V(t) \to 0$ exponentially as $t \to \infty$. This shows that the novel hyperchaotic system (1) is dissipative. Hence, the system limit sets are ultimately confined into a specific limit set of zero hypervolume, and the asymptotic motion of the novel hyperchaotic system (1) settles onto a strange attractor of the system.

3.2 Equilibrium Points

We take the parameter values as in the hyperchaotic case (2).

The equilibrium points of the 4-D system (1) are obtained by solving the system of equations

$$
\begin{aligned}
a(x_2 - x_1) + x_2 x_3 + x_4 &= 0 \qquad \text{(10a)} \\
-c x_1 x_3 + p x_2 &= 0 \qquad \text{(10b)} \\
x_1 x_2 - b &= 0 \qquad \text{(10c)} \\
-q(x_1 + x_2) &= 0 \qquad \text{(10d)}
\end{aligned}
$$

Since $q \neq 0$, it is immediate from (10d) that

$$x_1 + x_2 = 0 \ \text{ or } \ x_1 = -x_2 \tag{11}$$

Substituting $x_1 = -x_2$ in (10c), we get

$$x_2^2 = -b \tag{12}$$

which has no solutions since $b > 0$.

Thus, we conclude that the novel highly hyperchaotic system (1) does not have any equilibrium points. Hence, the novel highly hyperchaotic system (1) exhibits hidden attractors.

3.3 Rotation Symmetry About the x_3-Axis

It is easy to see that the novel 4-D highly hyperchaotic system (1) is invariant under the change of coordinates

$$(x_1, x_2, x_3, x_4) \mapsto (-x_1, -x_2, x_3, -x_4) \tag{13}$$

Since the transformation (13) persists for all values of the system parameters, it follows that the novel 4-D hyperchaotic system (1) has rotation symmetry about the x_3-axis and that any non-trivial trajectory must have a twin trajectory.

3.4 Invariance

It is easy to see that the x_3-axis is invariant under the flow of the 4-D novel hyperchaotic system (1).

The invariant motion along the x_3-axis is characterized by the scalar dynamics

$$\dot{x}_3 = -b, \tag{14}$$

which is unstable.

3.5 Lyapunov Exponents and Kaplan-Yorke Dimension

We take the parameter values of the novel system (1) as in the hyperchaotic case (2), i.e.

$$a = 60, \quad b = 27, \quad c = 160, \quad p = 0.3, \quad q = 2.8 \tag{15}$$

We take the initial state of the novel system (1) as given in (3), i.e.

$$x_1(0) = 1.2, \quad x_2(0) = 0.4, \quad x_3(0) = 0.3, \quad x_4(0) = 1.4 \tag{16}$$

Then the Lyapunov exponents of the system (1) are numerically obtained using MATLAB as

$$L_1 = 13.67837, \quad L_2 = 0.04058, \quad L_3 = 0, \quad L_4 = -45.64661 \tag{17}$$

Since there are two positive Lyapunov exponents in (17), the novel system (1) exhibits *hyperchaotic* behavior.

From the LE spectrum (17), we see that the maximal Lyapunov exponent of the novel hyperchaotic system (1) is $L_1 = 13.67837$, which is large.

We find that

$$L_1 + L_2 + L_3 + L_4 = -31.92766 < 0 \tag{18}$$

Thus, it follows that the novel highly hyperchaotic system (1) is dissipative.

Also, the Kaplan-Yorke dimension of the novel hyperchaotic system (1) is calculated as

$$D_{KY} = 3 + \frac{L_1 + L_2 + L_3}{|L_4|} = 3.30055, \tag{19}$$

which is fractional.

4 Adaptive Control of the Novel Highly Hyperchaotic System

In this section, we apply adaptive control method to derive an adaptive feedback control law for globally stabilizing the novel 4-D highly hyperchaotic system with unknown parameters. The main control result in this section is established using Lyapunov stability theory.

Thus, we consider the controlled novel 4-D highly hyperchaotic system given by

$$\begin{cases} \dot{x}_1 = a(x_2 - x_1) + x_2 x_3 + x_4 + u_1 \\ \dot{x}_2 = -c x_1 x_3 + p x_2 + u_2 \\ \dot{x}_3 = x_1 x_2 - b + u_3 \\ \dot{x}_4 = -q(x_1 + x_2) + u_4 \end{cases} \tag{20}$$

In (20), x_1, x_2, x_3, x_4 are the states and u_1, u_2, u_3, u_4 are the adaptive controls to be determined using estimates $\hat{a}(t), \hat{b}(t), \hat{c}(t), \hat{p}(t), \hat{q}(t)$ for the unknown parameters a, b, c, p, q, respectively.

We consider the adaptive feedback control law

$$\begin{cases} u_1 = -\hat{a}(t)(x_2 - x_1) - x_2 x_3 - x_4 - k_1 x_1 \\ u_2 = \hat{c}(t) x_1 x_3 - \hat{p}(t) x_2 - k_2 x_2 \\ u_3 = -x_1 x_2 + \hat{b}(t) - k_3 x_3 \\ u_4 = \hat{q}(t)(x_1 + x_2) - k_4 x_4 \end{cases} \tag{21}$$

where k_1, k_2, k_3, k_4 are positive gain constants.

Substituting (21) into (20), we get the closed-loop plant dynamics as

$$\begin{cases} \dot{x}_1 = [a - \hat{a}(t)](x_2 - x_1) - k_1 x_1 \\ \dot{x}_2 = -[c - \hat{c}(t)] x_1 x_3 + [p - \hat{p}(t)] x_2 - k_2 x_2 \\ \dot{x}_3 = -[b - \hat{b}(t)] - k_3 x_3 \\ \dot{x}_4 = -[q - \hat{q}(t)](x_1 + x_2) - k_4 x_4 \end{cases} \tag{22}$$

The parameter estimation errors are defined as

$$\begin{cases} e_a(t) = a - \hat{a}(t) \\ e_b(t) = b - \hat{b}(t) \\ e_c(t) = c - \hat{c}(t) \\ e_p(t) = p - \hat{p}(t) \\ e_q(t) = q - \hat{q}(t) \end{cases} \tag{23}$$

In view of (23), we can simplify the plant dynamics (22) as

$$\begin{cases} \dot{x}_1 = e_a(x_2 - x_1) - k_1 x_1 \\ \dot{x}_2 = -e_c x_1 x_3 + e_p x_2 - k_2 x_2 \\ \dot{x}_3 = -e_b - k_3 x_3 \\ \dot{x}_4 = -e_q(x_1 + x_2) - k_4 x_4 \end{cases} \tag{24}$$

Differentiating (23) with respect to t, we obtain

$$\begin{cases} \dot{e}_a(t) = -\dot{\hat{a}}(t) \\ \dot{e}_b(t) = -\dot{\hat{b}}(t) \\ \dot{e}_c(t) = -\dot{\hat{c}}(t) \\ \dot{e}_p(t) = -\dot{\hat{p}}(t) \\ \dot{e}_q(t) = -\dot{\hat{q}}(t) \end{cases} \tag{25}$$

We consider the quadratic candidate Lyapunov function defined by

$$V(\mathbf{x}, e_a, e_b, e_c, e_p, e_q) = \frac{1}{2}\left(x_1^2 + x_2^2 + x_3^2 + x_4^2\right) + \frac{1}{2}\left(e_a^2 + e_b^2 + e_c^2 + e_p^2 + e_q^2\right) \tag{26}$$

Differentiating V along the trajectories of (24) and (25), we obtain

$$\begin{aligned} \dot{V} = &-k_1x_1^2 - k_2x_2^2 - k_3x_3^2 - k_4x_4^2 + e_a\left[x_1(x_2 - x_1) - \dot{\hat{a}}\right] + e_b\left[-x_3 - \dot{\hat{b}}\right] \\ &+ e_c\left[-x_1x_2x_3 - \dot{\hat{c}}\right] + e_p\left[x_2^2 - \dot{\hat{p}}\right] + e_q\left[-(x_1 + x_2)x_4 - \dot{\hat{q}}\right] \end{aligned} \tag{27}$$

In view of (27), we take the parameter update law as

$$\begin{cases} \dot{\hat{a}}(t) = x_1(x_2 - x_1) \\ \dot{\hat{b}}(t) = -x_3 \\ \dot{\hat{c}}(t) = -x_1x_2x_3 \\ \dot{\hat{p}}(t) = x_2^2 \\ \dot{\hat{q}}(t) = -(x_1 + x_2)x_4 \end{cases} \tag{28}$$

Next, we state and prove the main result of this section.

Theorem 1 *The novel 4-D highly hyperchaotic system (20) with unknown system parameters is globally and exponentially stabilized for all initial conditions by the adaptive control law (21) and the parameter update law (28), where* k_1, k_2, k_3, k_4 *are positive gain constants.*

Proof We prove this result by applying Lyapunov stability theory [27].

We consider the quadratic Lyapunov function defined by (26), which is clearly a positive definite function on $\mathbf{R}^9$.

By substituting the parameter update law (28) into (27), we obtain the time-derivative of V as

$$\dot{V} = -k_1x_1^2 - k_2x_2^2 - k_3x_3^2 - k_4x_4^2 \tag{29}$$

From (29), it is clear that $\dot{V}$ is a negative semi-definite function on $\mathbf{R}^9$.

Thus, we can conclude that the state vector $\mathbf{x}(t)$ and the parameter estimation error are globally bounded, i.e.

$$\left[x_1(t)\ x_2(t)\ x_3(t)\ x_4(t)\ e_a(t)\ e_b(t)\ e_c(t)\ e_p(t)\ e_q(t)\right]^T \in \mathbf{L}_\infty.$$

We define $k = \min\{k_1, k_2, k_3, k_4\}$.

Then it follows from (29) that

$$\dot{V} \leq -k\|\mathbf{x}(t)\|^2 \tag{30}$$

Thus, we have

$$k\|\mathbf{x}(t)\|^2 \leq -\dot{V} \tag{31}$$

Integrating the inequality (31) from 0 to t, we get

$$k\int_0^t \|\mathbf{x}(\tau)\|^2 \, d\tau \;\leq\; V(0) - V(t) \tag{32}$$

From (32), it follows that $\mathbf{x} \in \mathbf{L}_2$.

Using (24), we can conclude that $\dot{\mathbf{x}} \in \mathbf{L}_\infty$.

Using Barbalat's lemma [27], we conclude that $\mathbf{x}(t) \to 0$ exponentially as $t \to \infty$ for all initial conditions $\mathbf{x}(0) \in \mathbf{R}^4$.

Thus, the novel 4-D highly hyperchaotic system (20) with unknown system parameters is globally and exponentially stabilized for all initial conditions by the adaptive control law (21) and the parameter update law (28).

This completes the proof. □

For the numerical simulations, the classical fourth-order Runge-Kutta method with step size $h = 10^{-8}$ is used to solve the systems (20) and (28), when the adaptive control law (21) is applied.

The parameter values of the novel 4-D hyperchaotic system (20) are taken as in the hyperchaotic case (2), i.e.

$$a = 60, \quad b = 27, \quad c = 160, \quad p = 0.3, \quad q = 2.8 \tag{33}$$

We take the positive gain constants as

$$k_1 = 6, \quad k_2 = 6, \quad k_3 = 6, \quad k_4 = 6 \tag{34}$$

Furthermore, as initial conditions of the novel 4-D highly hyperchaotic system (20), we take

$$x_1(0) = 8.5, \quad x_2(0) = -12.7, \quad x_3(0) = 4.3, \quad x_4(0) = -10.5 \tag{35}$$

Also, as initial conditions of the parameter estimates, we take

$$\hat{a}(0) = 3.2, \quad \hat{b}(0) = 7.6, \quad \hat{c}(0) = 2.7, \quad \hat{p}(0) = 15.8, \quad \hat{q}(0) = 9.4 \tag{36}$$

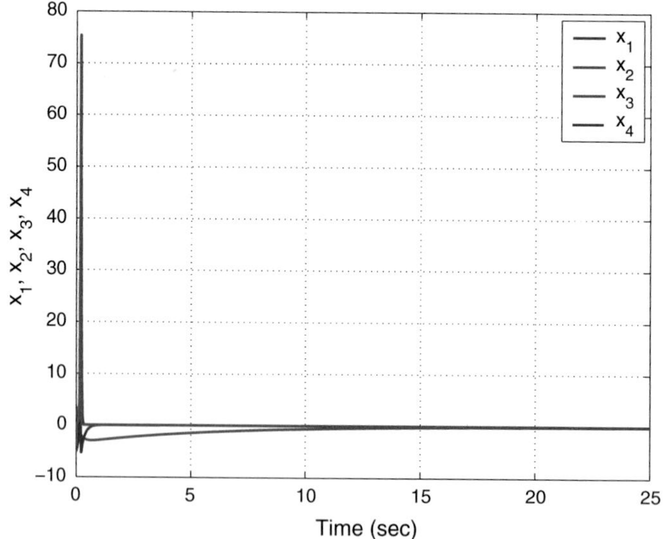

Fig. 5 Time-history of the controlled states x_1, x_2, x_3, x_4

In Fig. 5, the exponential convergence of the controlled states of the novel 4-D hyperchaotic system (20) is shown.

5 Adaptive Synchronization of the Identical Novel Hyperchaotic Systems

In this section, we use adaptive control method to derive an adaptive feedback control law for globally synchronizing identical novel 4-D highly hyperchaotic systems with unknown parameters.

As the master system, we consider the novel 4-D hyperchaotic system given by

$$\begin{cases} \dot{x}_1 = a(x_2 - x_1) + x_2 x_3 + x_4 \\ \dot{x}_2 = -c x_1 x_3 + p x_2 \\ \dot{x}_3 = x_1 x_2 - b \\ \dot{x}_4 = -q(x_1 + x_2) \end{cases} \tag{37}$$

In (37), x_1, x_2, x_3, x_4 are the states and a, b, c, p, q are unknown system parameters.

As the slave system, we consider the 4-D novel hyperchaotic system given by

$$\begin{cases} \dot{y}_1 = a(y_2 - y_1) + y_2 y_3 + y_4 + u_1 \\ \dot{y}_2 = -c y_1 y_3 + p y_2 + u_2 \\ \dot{y}_3 = y_1 y_2 - b + u_3 \\ \dot{y}_4 = -q(y_1 + y_2) + u_4 \end{cases} \tag{38}$$

In (38), y_1, y_2, y_3, y_4 are the states and u_1, u_2, u_3, u_4 are the adaptive controls to be determined using estimates $\hat{a}(t), \hat{c}(t), \hat{p}(t), \hat{q}(t)$ for the unknown parameters a, c, p, q, respectively.

The synchronization error between the novel hyperchaotic systems (37) and (38) is defined by

$$\begin{cases} e_1 = y_1 - x_1 \\ e_2 = y_2 - x_2 \\ e_3 = y_3 - x_3 \\ e_4 = y_4 - x_4 \end{cases} \tag{39}$$

Then the error dynamics is obtained as

$$\begin{cases} \dot{e}_1 = a(e_2 - e_1) + e_4 + y_2 y_3 - x_2 x_3 + u_1 \\ \dot{e}_2 = p e_2 - c(y_1 y_3 - x_1 x_3) + u_2 \\ \dot{e}_3 = y_1 y_2 - x_1 x_2 + u_3 \\ \dot{e}_4 = -q(e_1 + e_2) + u_4 \end{cases} \tag{40}$$

We consider the adaptive feedback control law

$$\begin{cases} u_1 = -\hat{a}(t)(e_2 - e_1) - e_4 - y_2 y_3 + x_2 x_3 - k_1 e_1 \\ u_2 = -\hat{p}(t) e_2 + \hat{c}(t)(y_1 y_3 - x_1 x_3) - k_2 e_2 \\ u_3 = -y_1 y_2 + x_1 x_2 - k_3 e_3 \\ u_4 = \hat{q}(t)(e_1 + e_2) - k_4 e_4 \end{cases} \tag{41}$$

where k_1, k_2, k_3, k_4 are positive gain constants.

Substituting (41) into (40), we get the closed-loop error dynamics as

$$\begin{cases} \dot{e}_1 = \left[a - \hat{a}(t)\right](e_2 - e_1) - k_1 e_1 \\ \dot{e}_2 = \left[p - \hat{p}(t)\right] e_2 - \left[c - \hat{c}(t)\right](y_1 y_3 - x_1 x_3) - k_2 e_2 \\ \dot{e}_3 = -k_3 e_3 \\ \dot{e}_4 = -\left[q - \hat{q}(t)\right](e_1 + e_2) - k_4 e_4 \end{cases} \tag{42}$$

The parameter estimation errors are defined as

$$\begin{cases} e_a(t) = a - \hat{a}(t) \\ e_c(t) = c - \hat{c}(t) \\ e_p(t) = p - \hat{p}(t) \\ e_q(t) = q - \hat{q}(t) \end{cases} \tag{43}$$

In view of (43), we can simplify the error dynamics (42) as

$$\begin{cases} \dot{e}_1 = e_a(e_2 - e_1) - k_1 e_1 \\ \dot{e}_2 = e_p e_2 - e_c(y_1 y_3 - x_1 x_3) - k_2 e_2 \\ \dot{e}_3 = -k_3 e_3 \\ \dot{e}_4 = -e_q(e_1 + e_2) - k_4 e_4 \end{cases} \tag{44}$$

Differentiating (43) with respect to t, we obtain

$$\begin{cases} \dot{e}_a(t) = -\dot{\hat{a}}(t) \\ \dot{e}_c(t) = -\dot{\hat{c}}(t) \\ \dot{e}_p(t) = -\dot{\hat{p}}(t) \\ \dot{e}_q(t) = -\dot{\hat{q}}(t) \end{cases} \tag{45}$$

We use adaptive control theory to find an update law for the parameter estimates. We consider the quadratic candidate Lyapunov function defined by

$$V(\mathbf{e}, e_a, e_c, e_p, e_q) = \frac{1}{2}\left(e_1^2 + e_2^2 + e_3^2 + e_4^2\right) + \frac{1}{2}\left(e_a^2 + e_c^2 + e_p^2 + e_q^2\right) \tag{46}$$

Differentiating V along the trajectories of (44) and (45), we obtain

$$\begin{aligned} \dot{V} = & -k_1 e_1^2 - k_2 e_2^2 - k_3 e_3^2 - k_4 e_4^2 + e_a\left[e_1(e_2 - e_1) - \dot{\hat{a}}\right] \\ & + e_c\left[-e_2(y_1 y_3 - x_1 x_3) - \dot{\hat{c}}\right] + e_p\left[e_2^2 - \dot{\hat{p}}\right] + e_q\left[-e_4(e_1 + e_2) - \dot{\hat{q}}\right] \end{aligned} \tag{47}$$

In view of (47), we take the parameter update law as

$$\begin{cases} \dot{\hat{a}}(t) = e_1(e_2 - e_1) \\ \dot{\hat{c}}(t) = -e_2(y_1 y_3 - x_1 x_3) \\ \dot{\hat{p}}(t) = e_2^2 \\ \dot{\hat{q}}(t) = -e_4(e_1 + e_2) \end{cases} \tag{48}$$

Next, we state and prove the main result of this section.

Theorem 2 *The novel hyperchaotic systems (37) and (38) with unknown system parameters are globally and exponentially synchronized for all initial conditions by the adaptive control law (41) and the parameter update law (48), where* k_1, k_2, k_3, k_4 *are positive gain constants.*

Proof We prove this result by applying Lyapunov stability theory [27].

We consider the quadratic Lyapunov function defined by (46), which is clearly a positive definite function on $\mathbf{R}^8$.

By substituting the parameter update law (48) into (47), we obtain

$$\dot{V} = -k_1 e_1^2 - k_2 e_2^2 - k_3 e_3^2 - k_4 e_4^2 \tag{49}$$

From (49), it is clear that $\dot{V}$ is a negative semi-definite function on $\mathbf{R}^8$.

Thus, we can conclude that the error vector $\mathbf{e}(t)$ and the parameter estimation error are globally bounded, i.e.

$$\left[\, e_1(t)\; e_2(t)\; e_3(t)\; e_4(t)\; e_a(t)\; e_c(t)\; e_p(t)\; e_q(t) \,\right]^T \in \mathbf{L}_\infty. \tag{50}$$

We define $k = \min\{k_1, k_2, k_3, k_4\}$.

Then it follows from (49) that

$$\dot{V} \leq -k\|\mathbf{e}(t)\|^2 \tag{51}$$

Thus, we have

$$k\|\mathbf{e}(t)\|^2 \leq -\dot{V} \tag{52}$$

Integrating the inequality (52) from 0 to t, we get

$$k \int_0^t \|\mathbf{e}(\tau)\|^2 \, d\tau \;\leq\; V(0) - V(t) \tag{53}$$

From (53), it follows that $\mathbf{e} \in \mathbf{L}_2$.

Using (44), we can conclude that $\dot{\mathbf{e}} \in \mathbf{L}_\infty$.

Using Barbalat's lemma [27], we conclude that $\mathbf{e}(t) \to 0$ exponentially as $t \to \infty$ for all initial conditions $\mathbf{e}(0) \in \mathbf{R}^4$.

This completes the proof. □

For the numerical simulations, the classical fourth-order Runge-Kutta method with step size $h = 10^{-8}$ is used to solve the systems (37), (38) and (48), when the adaptive control law (41) is applied.

The parameter values of the novel hyperchaotic systems are taken as in the hyperchaotic case (2), i.e.

$$a = 60, \quad b = 27, \quad c = 160, \quad p = 0.3, \quad q = 2.8 \tag{54}$$

We take the positive gain constants as

$$k_1 = 6, \quad k_2 = 6, \quad k_3 = 6, \quad k_4 = 6 \tag{55}$$

Furthermore, as initial conditions of the master system (37), we take

$$x_1(0) = -6.3, \quad x_2(0) = 12.4, \quad x_3(0) = 15.7, \quad x_4(0) = -12.9 \tag{56}$$

As initial conditions of the slave system (38), we take

$$y_1(0) = 5.1, \quad y_2(0) = -16.5, \quad y_3(0) = 4.8, \quad y_4(0) = 3.4 \tag{57}$$

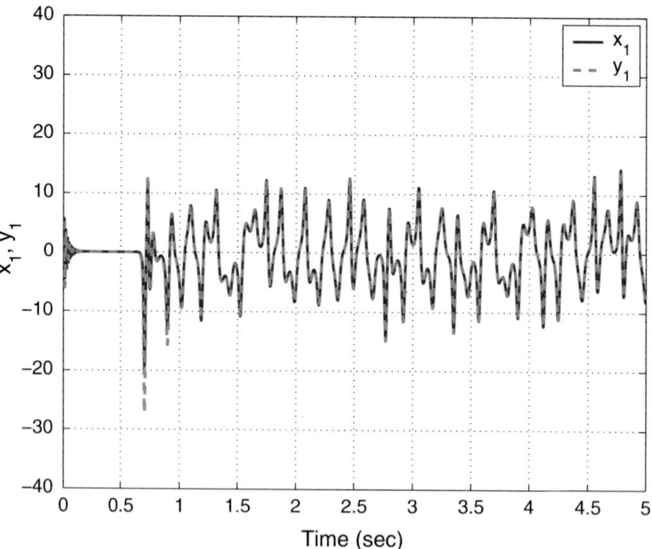

Fig. 6 Synchronization of the states x_1 and y_1

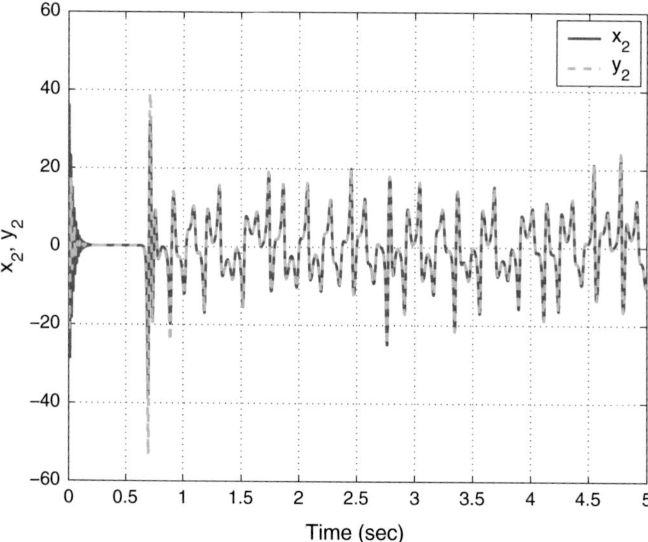

Fig. 7 Synchronization of the states x_2 and y_2

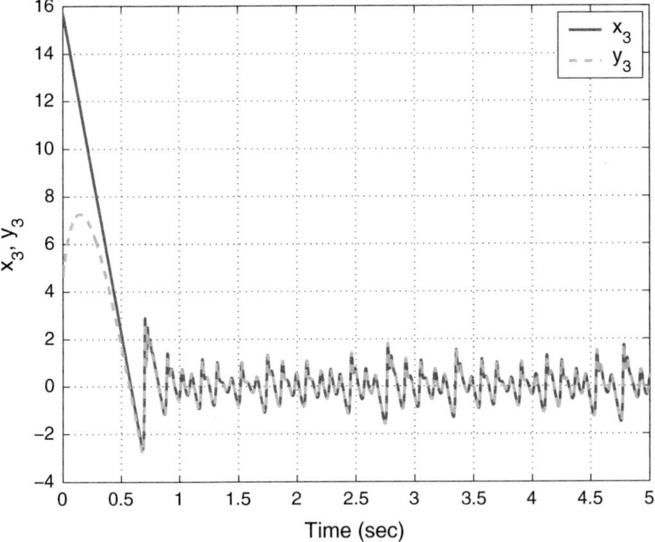

Fig. 8 Synchronization of the states x_3 and y_3

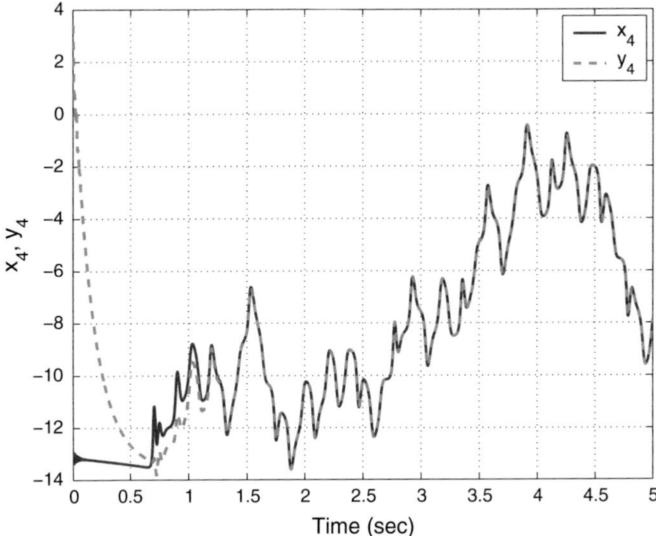

Fig. 9 Synchronization of the states x_4 and y_4

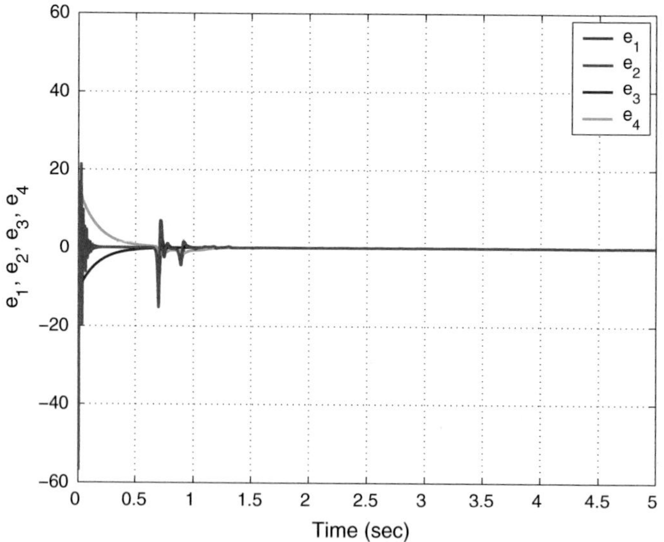

Fig. 10 Time-history of the synchronization errors e_1, e_2, e_3, e_4

Also, as initial conditions of the parameter estimates, we take

$$\hat{a}(0) = 2.8, \quad \hat{c}(0) = 1.4, \quad \hat{p}(0) = 12.7, \quad \hat{q}(0) = 5.8 \tag{58}$$

Figures 6, 7, 8 and 9 describe the complete synchronization of the novel hyperchaotic systems (37) and (38), while Fig. 10 describes the time-history of the synchronization errors e_1, e_2, e_3, e_4.

6 Conclusions

In this work, we described a ten-term novel 4-D highly hyperchaotic system with three quadratic nonlinearities. The phase portraits of the ten-term novel highly hyperchaotic system were depicted and the qualitative properties of the novel highly hyperchaotic system were discussed. We shall show that the novel hyperchaotic system does not have any equilibrium point. Hence, the novel 4-D hyperchaotic system exhibits hidden attractors. The Lyapunov exponents of the novel hyperchaotic system have been derived as $L_1 = 13.67837$, $L_2 = 0.04058$, $L_3 = 0$ and $L_4 = -45.64661$. The Maximal Lyapunov Exponent (MLE) of the novel hyperchaotic system is found as $L_1 = 13.67837$, which is large. Thus, the novel 4-D hyperchaotic system proposed in this work is highly hyperchaotic. Also, the Kaplan-Yorke dimension of the novel hyperchaotic system has been derived as $D_{KY} = 3.30055$. Since the sum of the Lyapunov exponents is negative, the novel hyperchaotic system is dissipative. Next, an adaptive controller was designed to globally stabilize the novel highly hyperchaotic system with unknown parameters. Finally, an adaptive controller was also designed to achieve global chaos synchronization of the identical novel highly hyperchaotic systems with unknown parameters. MATLAB simulations were depicted to illustrate all the main results derived in this work.

References

1. Arneodo A, Coullet P, Tresser C (1981) Possible new strange attractors with spiral structure. Commun Math Phys 79(4):573–576
2. Azar AT, Serrano FE (2015a) Adaptive sliding mode control of the Furuta pendulum. In: Azar AT, Zhu Q (eds) Advances and applications in sliding mode control systems. Studies in computational intelligence, vol 576. Springer, Germany, pp 1–42
3. Azar AT, Serrano FE (2015b) Deadbeat control for multivariable systems with time varying delays. In: Azar AT, Vaidyanathan S (eds) Chaos modeling and control systems design. Studies in computational intelligence, vol 581. Springer, Germany, pp 97–132
4. Azar AT, Serrano FE (2015c) Design and modeling of anti wind up PID controllers. In: Zhu Q, Azar AT (eds) Complex system modelling and control through intelligent soft computations. Studies in fuzziness and soft computing, vol 319. Springer, Germany, pp 1–44
5. Azar AT, Serrano FE (2015d) Stabilizatoin and control of mechanical systems with backlash. In: Azar AT, Vaidyanathan S (eds) Handbook of research on advanced intelligent control engineering and automation. Advances in computational intelligence and robotics (ACIR), IGI-Global, USA, pp 1–60
6. Azar AT, Vaidyanathan S (2015a) Chaos modeling and control systems design, Studies in computational intelligence, vol 581. Springer, Germany
7. Azar AT, Vaidyanathan S (2015b) Computational intelligence applications in modeling and control, Studies in computational intelligence, vol 575. Springer, Germany
8. Azar AT, Vaidyanathan S (2015c) Handbook of research on advanced intelligent control engineering and automation. Advances in computational intelligence and robotics (ACIR), IGI-Global, USA
9. Azar AT, Zhu Q (2015) Advances and applications in sliding mode control systems, Studies in computational intelligence, vol 576. Springer, Germany
10. Azar AT (2010) Fuzzy Systems. IN-TECH, Vienna
11. Azar AT (2012) Overview of type-2 fuzzy logic systems. Int J Fuzzy Syst Appl 2(4):1–28
12. Azar AT, Serrano FE (2014) Robust IMC-PID tuning for cascade control systems with gain and phase margin specifications. Neural Comput Appl 25(5):983–995
13. Cai G, Tan Z (2007) Chaos synchronization of a new chaotic system via nonlinear control. J Uncertain Syst 1(3):235–240
14. Chen A, Lu J, Lü J, Yu S (2006) Generating hyperchaotic Lü attractor via state feedback control. Phys A 364:103–110
15. Chen G, Ueta T (1999) Yet another chaotic attractor. Int J Bifurc Chaos 9(7):1465–1466
16. Das S, Goswami D, Chatterjee S, Mukherjee S (2014) Stability and chaos analysis of a novel swarm dynamics with applications to multi-agent systems. Eng Appl Artif Intell 30:189–198
17. Feki M (2003) An adaptive chaos synchronization scheme applied to secure communication. Chaos Solitons Fractals 18(1):141–148
18. Filali RL, Benrejeb M, Borne P (2014) On observer-based secure communication design using discrete-time hyperchaotic systems. Commun Nonlinear Sci Numer Simul 19(5):1424–1432
19. Gaspard P (1999) Microscopic chaos and chemical reactions. Phys A Stat Mech Appl 263(1–4):315–328

20. Gibson WT, Wilson WG (2013) Individual-based chaos: extensions of the discrete logistic model. J Theor Biol 339:84–92
21. Hammami S (2015) State feedback-based secure image cryptosystem using hyperchaotic synchronization. ISA Trans 54:52–59
22. Huang X, Zhao Z, Wang Z, Li Y (2012) Chaos and hyperchaos in fractional-order cellular neural networks. Neurocomputing 94:13–21
23. Jia Q (2007) Hyperchaos generated from the Lorenz chaotic system and its control. Phys Lett A 366:217–222
24. Karthikeyan R, Sundarapandian V (2014) Hybrid chaos synchronization of four-scroll systems via active control. J Electr Eng 65(2):97–103
25. Kaslik E, Sivasundaram S (2012) Nonlinear dynamics and chaos in fractional-order neural networks. Neural Netw 32:245–256
26. Kengne J, Chedjou JC, Kenne G, Kyamakya K (2012) Dynamical properties and chaos synchronization of improved Colpitts oscillators. Commun Nonlinear Sci Numer Simul 17(7):2914–2923
27. Khalil HK (2001) Nonlinear Syst, 3rd edn. Prentice Hall, New Jersey
28. Kyriazis M (1991) Applications of chaos theory to the molecular biology of aging. Exp Gerontol 26(6):569–572
29. Lang J (2015) Color image encryption based on color blend and chaos permutation in the reality-preserving multiple-parameter fractional Fourier transform domain. Opt Commun 338:181–192
30. Li C, Liao X, Wong KW (2005) Lag synchronization of hyperchaos with application to secure communications. Chaos Solitons Fractals 23(1):183–193
31. Li D (2008) A three-scroll chaotic attractor. Phys Lett A 372(4):387–393
32. Li X (2009) Modified projective synchronization of a new hyperchaotic system via nonlinear control. Commun Theor Phys 52:274–278
33. Lian S, Chen X (2011) Traceable content protection based on chaos and neural networks. Appl Soft Comput 11(7):4293–4301
34. Li Z, Chen G (2006) Integration of fuzzy logic and chaos theory. Studies in fuzziness and soft computing, vol 187. Springer, Germany
35. Lorenz EN (1963) Deterministic periodic flow. J Atmos Sci 20(2):130–141
36. Lü J, Chen G (2002) A new chaotic attractor coined. Int J Bifurc Chaos 12(3):659–661
37. Mondal S, Mahanta C (2014) Adaptive second order terminal sliding mode controller for robotic manipulators. J Frankl Inst 351(4):2356–2377
38. Murali K, Lakshmanan M (1998) Secure communication using a compound signal from generalized chaotic systems. Phys Lett A 241(6):303–310
39. Nehmzow U, Walker K (2005) Quantitative description of robot-environment interaction using chaos theory. Robot Auton Syst 53(3–4):177–193
40. Pehlivan I, Moroz IM, Vaidyanathan S (2014) Analysis, synchronization and circuit design of a novel butterfly attractor. J Sound Vib 333(20):5077–5096
41. Petrov V, Gaspar V, Masere J, Showalter K (1993) Controlling chaos in Belousov-Zhabotinsky reaction. Nature 361:240–243
42. Pham VT, Volos C, Jafari S, Wang X, Vaidyanathan S (2014) Hidden hyperchaotic attractor in a novel simple memristive neural network. Optoelectron Adv Mater Rapid Commun 8(11–12):1157–1163
43. Pham VT, Vaidyanathan S, Volos CK, Jafari S (2015a) Hidden attractors in a chaotic system with an exponential nonlinear term. Eur Phys J Spec Top 224(8):1507–1517
44. Pham VT, Volos CK, Vaidyanathan S, Le TP, Vu VY (2015b) A memristor-based hyperchaotic system with hidden attractors: dynamics, synchronization and circuital emulating. J Eng Sci Technol Rev 8(2):205–214
45. Qu Z (2011) Chaos in the genesis and maintenance of cardiac arrhythmias. Prog Biophys Mol Biol 105(3):247–257
46. Rasappan S, Vaidyanathan S (2013) Hybrid synchronization of n-scroll Chua circuits using adaptive backstepping control design with recursive feedback. Malays J Math Sci 73(1):73–95

47. Rasappan S, Vaidyanathan S (2014) Global chaos synchronization of WINDMI and Coullet chaotic systems using adaptive backstepping control design. Kyungpook Math J 54(1):293–320
48. Rhouma R, Belghith S (2008) Cryptanalysis of a new image encryption algorithm based on hyper-chaos. Phys Lett A 372(38):5973–5978
49. Rhouma R, Belghith S (2011) Cryptoanalysis of a chaos based cryptosystem on DSP. Commun Nonlinear Sci Numer Simul 16(2):876–884
50. Rössler OE (1976) An equation for continuous chaos. Phys Lett A 57(5):397–398
51. Rössler OE (1979) An equation for hyperchaos. Phys Lett A 71:155–157
52. Sampath S, Vaidyanathan S, Volos CK, Pham VT (2015) An eight-term novel four-scroll chaotic system with cubic nonlinearity and its circuit simulation. J Eng Sci Technol Rev 8(2):1–6
53. Sarasu P, Sundarapandian V (2011a) Active controller design for generalized projective synchronization of four-scroll chaotic systems. Int J Syst Signal Control Eng Appl 4(2):26–33
54. Sarasu P, Sundarapandian V (2011b) The generalized projective synchronization of hyperchaotic Lorenz and hyperchaotic Qi systems via active control. Int J Soft Comput 6(5):216–223
55. Sarasu P, Sundarapandian V (2012) Generalized projective synchronization of two-scroll systems via adaptive control. Int J Soft Comput 7(4):146–156
56. Senouci A, Boukabou A (2014) Predictive control and synchronization of chaotic and hyperchaotic systems based on a $T-S$ fuzzy model. Math Comput Simul 105:62–78
57. Sharma A, Patidar V, Purohit G, Sud KK (2012) Effects on the bifurcation and chaos in forced Duffing oscillator due to nonlinear damping. Commun Nonlinear Sci Numer Simul 17(6):2254–2269
58. Sprott JC (1994) Some simple chaotic flows. Phys Rev E 50(2):647–650
59. Suérez I (1999) Mastering chaos in ecology. Ecol Modell 117(2–3):305–314
60. Sundarapandian V (2010) Output regulation of the Lorenz attractor. Far East J Math Sci 42(2):289–299
61. Sundarapandian V (2013a) Adaptive control and synchronization design for the Lu-Xiao chaotic system. Lect Notes Electr Eng 131:319–327
62. Sundarapandian V (2013b) Analysis and anti-synchronization of a novel chaotic system via active and adaptive controllers. J Eng Sci Technol Rev 6(4):45–52
63. Sundarapandian V, Karthikeyan R (2011a) Anti-synchronization of hyperchaotic Lorenz and hyperchaotic Chen systems by adaptive control. Int J Syst Signal Control Eng Appl 4(2):18–25
64. Sundarapandian V, Karthikeyan R (2011b) Anti-synchronization of Lü and Pan chaotic systems by adaptive nonlinear control. Eur J Sci Res 64(1):94–106
65. Sundarapandian V, Karthikeyan R (2012) Adaptive anti-synchronization of uncertain Tigan and Li systems. J Eng Appl Sci 7(1):45–52
66. Sundarapandian V, Pehlivan I (2012) Analysis, control, synchronization, and circuit design of a novel chaotic system. Math Comput Model 55(7–8):1904–1915
67. Sundarapandian V, Sivaperumal S (2011) Sliding controller design of hybrid synchronization of four-wing chaotic systems. Int J Soft Comput 6(5):224–231
68. Suresh R, Sundarapandian V (2013) Global chaos synchronization of a family of n-scroll hyperchaotic Chua circuits using backstepping control with recursive feedback. Far East J Math Sci 7(2):219–246
69. Tigan G, Opris D (2008) Analysis of a 3D chaotic system. Chaos Solitons Fractals 36: 1315–1319
70. Usama M, Khan MK, Alghatbar K, Lee C (2010) Chaos-based secure satellite imagery cryptosystem. Comput Math Appl 60(2):326–337
71. Vaidyanathan S (2011a) Output regulation of Arneodo-Coullet chaotic system. Commun Comput Inf Sci 133:98–107
72. Vaidyanathan S (2011b) Output regulation of the unified chaotic system. Commun Comput Inf Sci 198:1–9
73. Vaidyanathan S (2012a) Adaptive controller and syncrhonizer design for the Qi-Chen chaotic system. Lect Notes Inst Comput Sci Soc Inf Telecommun Eng 84:73–82

74. Vaidyanathan S (2012b) Anti-synchronization of Sprott-L and Sprott-M chaotic systems via adaptive control. Int J Control Theory Appl 5(1):41–59
75. Vaidyanathan S (2012c) Global chaos control of hyperchaotic Liu system via sliding control method. Int J Control Theory Appl 5(2):117–123
76. Vaidyanathan S (2012d) Sliding mode control based global chaos control of Liu-Liu-Liu-Su chaotic system. Int J Control Theory Appl 5(1):15–20
77. Vaidyanathan S (2013a) A new six-term 3-D chaotic system with an exponential nonlinearity. Far East J Math Sci 79(1):135–143
78. Vaidyanathan S (2013b) A ten-term novel 4-D hyperchaotic system with three quadratic nonlinearities and its control. Int J Control Theory Appl 6(2):97–109
79. Vaidyanathan S (2013c) Analysis and adaptive synchronization of two novel chaotic systems with hyperbolic sinusoidal and cosinusoidal nonlinearity and unknown parameters. J Eng Sci Technol Rev 6(4):53–65
80. Vaidyanathan S (2013d) Analysis, control and synchronization of hyperchaotic Zhou system via adaptive control. Adv Intell Syst Comput 177:1–10
81. Vaidyanathan S, Volos C, Pham VT (2014a) Hyperchaos, adaptive control and synchronization of a novel 5-D hyperchaotic system with three positive Lyapunov exponents and its SPICE implementation. Arch Control Sci 24(4):409–446
82. Vaidyanathan S, Volos C, Pham VT, Madhavan K, Idowu BA (2014b) Adaptive backstepping control, synchronization and circuit simulation of a 3-D novel jerk chaotic system with two hyperbolic sinusoidal nonlinearities. Arch Control Sci 24(3):375–403
83. Vaidyanathan S (2014a) A new eight-term 3-D polynomial chaotic system with three quadratic nonlinearities. Far East J Math Sci 84(2):219–226
84. Vaidyanathan S (2014b) Analysis and adaptive synchronization of eight-term 3-D polynomial chaotic systems with three quadratic nonlinearities. Eur Phys J Spec Top 223(8):1519–1529
85. Vaidyanathan S (2014c) Analysis, control and synchronisation of a six-term novel chaotic system with three quadratic nonlinearities. Int J Model Identif Control 22(1):41–53
86. Vaidyanathan S (2014d) Generalized projective synchronisation of novel 3-D chaotic systems with an exponential non-linearity via active and adaptive control. Int J Model Identif Control 22(3):207–217
87. Vaidyanathan S (2014e) Global chaos synchronisation of identical Li-Wu chaotic systems via sliding mode control. Int J Model Identif Control 22(2):170–177
88. Vaidyanathan S (2014f) Qualitative analysis and control of an eleven-term novel 4-D hyperchaotic system with two quadratic nonlinearities. Int J Control Theory Appl 7:35–47
89. Vaidyanathan S (2015m) Dynamics and control of Tokamak system with symmetric and magnetically confined plasma. Int J ChemTech Res 8(6):795–803
90. Vaidyanathan S (2015n) Hyperchaos, qualitative analysis, control and synchronisation of a ten-term 4-D hyperchaotic system with an exponential nonlinearity and three quadratic nonlinearities. Int J Model Identif Control 23(4):380–392
91. Vaidyanathan S (2015o) Lotka-Volterra population biology models with negative feedback and their ecological monitoring. Int J PharmTech Res 8(5):974–981
92. Vaidyanathan S (2015p) Synchronization of 3-cells cellular neural network (CNN) attractors via adaptive control method. Int J PharmTech Res 8(5):946–955
93. Vaidyanathan S (2015q) Synchronization of Tokamak systems with symmetric and magnetically confined plasma via adaptive control. Int J ChemTech Res 8(6):818–827
94. Vaidyanathan S, Azar AT, Rajagopal K, Alexander P (2015a) Design and SPICE implementation of a 12-term novel hyperchaotic system and its synchronisation via active control. Int J Model Identif Control 23(3):267–277
95. Vaidyanathan S, Idowu BA, Azar AT (2015b) Backstepping controller design for the global chaos synchronization of Sprott's jerk systems. Stud Comput Intell 581:39–58
96. Vaidyanathan S, Rajagopal K, Volos CK, Kyprianidis IM, Stouboulos IN (2015c) Analysis, adaptive control and synchronization of a seven-term novel 3-D chaotic system with three quadratic nonlinearities and its digital implementation in LabVIEW. J Eng Sci Technol Rev 8(2):130–141

97. Vaidyanathan S, Sampath S, Azar AT (2015d) Global chaos synchronisation of identical chaotic systems via novel sliding mode control method and its application to Zhu system. Int J Model Identif Control 23(1):92–100
98. Vaidyanathan S, Volos C, Pham VT, Madhavan K (2015e) Analysis, adaptive control and synchronization of a novel 4-D hyperchaotic hyperjerk system and its SPICE implementation. Arch Control Sci 25(1):135–158
99. Vaidyanathan S, Volos CK, Kyprianidis IM, Stouboulos IN, Pham VT (2015f) Analysis, adaptive control and anti-synchronization of a six-term novel jerk chaotic system with two exponential nonlinearities and its circuit simulation. J Eng Sci Technol Rev 8(2):24–36
100. Vaidyanathan S, Volos CK, Pham VT (2015g) Analysis, adaptive control and adaptive synchronization of a nine-term novel 3-D chaotic system with four quadratic nonlinearities and its circuit simulation. J Eng Sci Technol Rev 8(2):181–191
101. Vaidyanathan S, Volos CK, Pham VT (2015h) Analysis, control, synchronization and SPICE implementation of a novel 4-D hyperchaotic Rikitake dynamo system without equilibrium. J Eng Sci Technol Rev 8(2):232–244
102. Vaidyanathan S, Pham VT, Volos CK (2015) A 5-D hyperchaotic Rikitake dynamo system with hidden attractors. Eur Phys J Spec Top 224(8):1575–1592
103. Vaidyanathan S (2015a) 3-cells cellular neural network (CNN) attractor and its adaptive biological control. Int J PharmTech Res 8(4):632–640
104. Vaidyanathan S (2015b) A 3-D novel highly chaotic system with four quadratic nonlinearities, its adaptive control and anti-synchronization with unknown parameters. J Eng Sci Technol Rev 8(2):106–115
105. Vaidyanathan S (2015c) Adaptive backstepping control of enzymes-substrates system with ferroelectric behaviour in brain waves. Int J PharmTech Res 8(2):256–261
106. Vaidyanathan S (2015d) Adaptive biological control of generalized Lotka-Volterra three-species biological system. Int J PharmTech Res 8(4):622–631
107. Vaidyanathan S (2015e) Adaptive chaotic synchronization of enzymes-substrates system with ferroelectric behaviour in brain waves. Int J PharmTech Res 8(5):964–973
108. Vaidyanathan S (2015f) Adaptive control of a chemical chaotic reactor. Int J PharmTech Res 8(3):377–382
109. Vaidyanathan S (2015g) Adaptive synchronization of chemical chaotic reactors. Int J ChemTech Res 8(2):612–621
110. Vaidyanathan S (2015h) Adaptive synchronization of generalized Lotka-Volterra three-species biological systems. Int J PharmTech Res 8(5):928–937
111. Vaidyanathan S (2015i) Analysis, properties and control of an eight-term 3-D chaotic system with an exponential nonlinearity. Int J Model Identif Control 23(2):164–172
112. Vaidyanathan S (2015j) Anti-synchronization of Brusselator chemical reaction systems via adaptive control. Int J ChemTech Res 8(6):759–768
113. Vaidyanathan S (2015k) Chaos in neurons and adaptive control of Birkhoff-Shaw strange chaotic attractor. Int J PharmTech Res 8(5):956–963
114. Vaidyanathan S (2015l) Dynamics and control of Brusselator chemical reaction. Int J ChemTech Res 8(6):740–749
115. Vaidyanathan S, Azar AT (2015b) Analysis, control and synchronization of a nine-term 3-D novel chaotic system. In: Azar AT, Vaidyanathan S (eds) Chaos modelling and control systems design. Studies in computational intelligence, vol 581. Springer, Germany, pp 19–38
116. Vaidyanathan S, Azar AT (2015a) Analysis and control of a 4-D novel hyperchaotic system. Stud Comput Intell 581:3–17
117. Vaidyanathan S, Madhavan K (2013) Analysis, adaptive control and synchronization of a seven-term novel 3-D chaotic system. Int J Control Theory Appl 6(2):121–137
118. Vaidyanathan S, Pakiriswamy S (2013) Generalized projective synchronization of six-term Sundarapandian chaotic systems by adaptive control. Int J Control Theory Appl 6(2):153–163
119. Vaidyanathan S, Pakiriswamy S (2015) A 3-D novel conservative chaotic system and its generalized projective synchronization via adaptive control. J Eng Sci Technol Rev 8(2): 52–60

120. Vaidyanathan S, Rajagopal K (2011) Hybrid synchronization of hyperchaotic Wang-Chen and hyperchaotic Lorenz systems by active non-linear control. Int J Syst Signal Control Eng Appl 4(3):55–61
121. Vaidyanathan S, Rajagopal K (2012) Global chaos synchronization of hyperchaotic Pang and hyperchaotic Wang systems via adaptive control. Int J Soft Comput 7(1):28–37
122. Vaidyanathan S, Rasappan S (2011) Global chaos synchronization of hyperchaotic Bao and Xu systems by active nonlinear control. Commun Comput Inf Sci 198:10–17
123. Vaidyanathan S, Rasappan S (2014) Global chaos synchronization of n-scroll Chua circuit and Lur'e system using backstepping control design with recursive feedback. Arab J Sci Eng 39(4):3351–3364
124. Vaidyanathan S, Sampath S (2012) Anti-synchronization of four-wing chaotic systems via sliding mode control. Int J Autom Comput 9(3):274–279
125. Vaidyanathan S, Volos CK, Pham VT (2015i) Global chaos control of a novel nine-term chaotic system via sliding mode control. In: Azar AT, Zhu Q (eds) Advances and applications in sliding mode control systems. Studies in computational intelligence, vol 576. Springer, Germany, pp 571–590
126. Vaidyanathan S, Volos C (2015) Analysis and adaptive control of a novel 3-D conservative no-equilibrium chaotic system. Arch Control Sci 25(3):333–353
127. Volos CK, Kyprianidis IM, Stouboulos IN (2013) Experimental investigation on coverage performance of a chaotic autonomous mobile robot. Robot Auton Syst 61(12):1314–1322
128. Volos CK, Kyprianidis IM, Stouboulos IN, Tlelo-Cuautle E, Vaidyanathan S (2015) Memristor: a new concept in synchronization of coupled neuromorphic circuits. J Eng Sci Technol Rev 8(2):157–173
129. Wang J, Chen Z (2008) A novel hyperchaotic system and its complex dynamics. Int J Bifurc Chaos 18:3309–3324
130. Wei X, Yunfei F, Qiang L (2012) A novel four-wing hyper-chaotic system and its circuit implementation. Procedia Eng 29:1264–1269
131. Wei Z, Yang Q (2010) Anti-control of Hopf bifurcation in the new chaotic system with two stable node-foci. Appl Math Comput 217(1):422–429
132. Witte CL, Witte MH (1991) Chaos and predicting varix hemorrhage. Med Hypotheses 36(4):312–317
133. Wu X, Zhu C, Kan H (2015) An improved secure communication scheme based passive synchronization of hyperchaotic complex nonlinear system. Appl Math Comput 252: 201–214
134. Yujun N, Xingyuan W, Mingjun W, Huaguang Z (2010) A new hyperchaotic system and its circuit implementation. Commun Nonlinear Sci Numer Simul 15(11):3518–3524
135. Zaher AA, Abu-Rezq A (2011) On the design of chaos-based secure communication systems. Commun Nonlinear Syst Numer Simul 16(9):3721–3727
136. Zhang H, Liao X, Yu J (2005) Fuzzy modeling and synchronization of hyperchaotic systems. Chaos Solitons Fractals 26(3):835–843
137. Zhang X, Zhao Z, Wang J (2014) Chaotic image encryption based on circular substitution box and key stream buffer. Signal Process Image Commun 29(8):902–913
138. Zhou W, Xu Y, Lu H, Pan L (2008) On dynamics analysis of a new chaotic attractor. Phys Lett A 372(36):5773–5777
139. Zhu C, Liu Y, Guo Y (2010) Theoretic and numerical study of a new chaotic system. Intell Inf Manag 2:104–109
140. Zhu C (2012) A novel image encryption scheme based on improved hyperchaotic sequences. Opt Commun 285(1):29–37
141. Zhu Q, Azar AT (2015) Complex system modelling and control through intelligent soft computations. Studies in fuzzines and soft computing, vol 319. Springer, Germany

A Novel Double Convection Chaotic System, Its Analysis, Adaptive Control and Synchronization

Sundarapandian Vaidyanathan

Abstract In this research work, we announce a novel 3-D double convection chaotic system and discuss its qualitative properties, adaptive control and synchronization. First, this work describes the dynamic equations and qualitative properties of the novel 3-D double convection chaotic system. We show that the novel 3-D double convection chaotic system has three unstable equilibrium points of which one equilibrium point is a saddle-point and the other two equilibrium points are saddle-foci. Our novel chaotic system is obtained by modifying the equations of the Rucklidge chaotic system (1992) for nonlinear double convection. The Lyapunov exponents of the novel 3-D double convection chaotic system are obtained as $L_1 = 1.03405$, $L_2 = 0$ and $L_3 = -4.03938$. Also, the Kaplan-Yorke dimension of the novel 3-D double convection chaotic system is derived as $D_{KY} = 2.2560$. Next, this work describes the global stabilization of the novel 3-D double convection chaotic system with unknown parameters via adaptive control method. Furthermore, this work describes the global chaos synchronization of identical novel 3-D double convection chaotic systems via adaptive control method. Our adaptive global stabilization and synchronization results are established via Lyapunov stability theory. MATLAB simulations are depicted to illustrate all the main results derived in this work for the novel 3-D double convection chaotic system.

Keywords Chaos · Chaotic systems · Nonlinear systems · Double convection system · Adaptive control · Synchronization

---

S. Vaidyanathan (✉)
Research and Development Centre, Vel Tech University,
Avadi, Chennai 600062, Tamil Nadu, India
e-mail: sundarvtu@gmail.com

A.T. Azar and S. Vaidyanathan (eds.), *Advances in Chaos Theory and Intelligent Control*, Studies in Fuzziness and Soft Computing 337,
DOI 10.1007/978-3-319-30340-6_23

1 Introduction

Chaos theory is the study of how dynamical systems that are modelled by simple, straightforward, deterministic laws can exhibit very complicated and seemingly random long term behavior. A classic example of chaos is the chaos in weather patterns observed by Lorenz [38].

A hallmark of chaotic systems is its sensitive dependence on the initial conditions [68]. This means that if two trajectories of the system start from initial conditions which are very close to each other, then after a relatively short period of time, the two trajectories will diverge and appear very different from each other.

Some classical paradigms of 3-D chaotic systems in the literature are Lorenz system [38], Rössler system [56], ACT system [1], Sprott systems [66], Chen system [17], Lü system [39], Liu system [37], Rucklidge system [57], Cai system [13], Chen-Lee system [16], Tigan system [80], etc.

Many new chaotic systems have been discovered in the recent years such as Zhou system [152], Zhu system [153], Li system [32], Wei-Yang system [145], Sundarapandian systems [72, 77], Vaidyanathan systems [90, 91, 94–98, 100, 107, 117, 119, 120, 126, 130, 132, 141, 142], Pehlivan system [45], Sampath system [58], Pham system [47], etc.

Chaos theory and control systems have many important applications in science and engineering [6–10, 154]. Some commonly known applications are oscillators [28, 65], lasers [34, 148], chemical reactions [21, 46, 104, 105, 108, 110, 111, 115], biology [18, 30, 101–103, 106, 109, 113, 114, 124], ecology [22, 69], encryption [31, 150], cryptosystems [55, 81], mechanical systems [2–5, 12], secure communications [19, 41, 149], robotics [40, 42, 143], cardiology [49, 146], intelligent control [11, 36], neural networks [24, 27, 35], finance [23, 67], memristors [48, 144], etc.

Control of chaotic systems aims to regulate or stabilize the chaotic system about its equilibrium points. There are many methods available in the literature such as active control method [70, 83, 84, 116], adaptive control method [43, 121], sliding mode control method [87, 89], backstepping control method [122], etc.

Synchronization of chaotic systems is a phenomenon that occurs when two or more chaotic systems are coupled or when a chaotic system drives another chaotic system. Major works on synchronization of chaotic systems deal with the complete synchronization of a pair of chaotic systems called the *master* and *slave* systems. The design goal of the complete synchronization is to apply the output of the master system to control the slave system so that the output of the slave system tracks the output of the master system asymptotically with time.

Pecora and Carroll pioneered the research on synchronization of chaotic systems with their seminal papers [14, 44]. The active control method [26, 59, 60, 71, 76, 82, 88, 133, 134, 137] is typically used when the system parameters are available for measurement. Adaptive control method [61–63, 73–75, 86, 92, 93, 112, 118, 123, 125, 131, 135, 136] is typically used when some or all the system parameters are not available for measurement and estimates for the uncertain parameters of the systems. Sampled-data feedback control method [20, 33, 147, 151] and time-delay feedback control method [15, 25, 64] are also used for synchronization of chaotic systems.

Backstepping control method [50–54, 79, 129, 138] is also used for the synchronization of chaotic systems, which is a recursive method for stabilizing the origin of a control system in strict-feedback form. Another popular method for the synchronization of chaotic systems is the sliding mode control method [78, 85, 99, 127, 128, 139, 140], which is a nonlinear control method that alters the dynamics of a nonlinear system by application of a discontinuous control signal that forces the system to "slide" along a cross-section of the system's normal behavior.

This paper starts with a discussion on the Rucklidge chaotic system [57] for nonlinear double convection. In fluid mechanics modelling, cases of two-dimensional convection in a horizontal layer of Boussinesq fluid with lateral constraints were studied by Rucklidge [57]. When the convection takes place in a fluid layer rotating uniformly about a vertical axis and in the limit of tall thin rolls, convection in an imposed vertical magnetic field and convection in a rotating fluid layer are both modelled by a new 3-D system of ordinary differential equations, which produces chaotic solutions like the Lorenz system [38].

The Lyapunov exponents of the Rucklidge chaotic system are obtained as $L_1 = 0.1877$, $L_2 = 0$ and $L_3 = -3.1893$. Thus, the Maximal Lyapunov Exponent (MLE) of the Rucklidge chaotic system is obtained as $L_1 = 0.1877$. Also, the Kaplan-Yorke dimension of the Rucklidge chaotic system is obtained as $D_{KY} = 2.0589$.

In this work, by modifying the dynamics of Rucklidge chaotic system [57], we derive a novel 3-D double convection chaotic system. In this work, we obtain the Lyapunov exponents of the novel double convection chaotic system as $L_1 = 1.03405$, $L_2 = 0$ and $L_3 = -4.03938$. Thus, the Maximal Lyapunov Exponent (MLE) of the novel double convection system is obtained as $L_1 = 1.03405$. Also, we derive the Kaplan-Yorke dimension of the novel chaotic system as $D_{KY} = 2.2560$.

Since the Maximal Lyapunov Exponent (MLE) of the novel double convection chaotic system is greater than the Maximal Lyapunov Exponent (MLE) of the Rucklidge chaotic system, it is evident that the novel double convection chaotic system is more chaotic than the Rucklidge chaotic system.

Section 2 describes the dynamic equations and phase portraits of the novel double convection chaotic system. We show that the novel 3-D double convection chaotic system has three unstable equilibrium points of which one equilibrium point is a saddle-point and the other two equilibrium points are saddle-foci. Section 3 describes the qualitative properties of the novel double convection chaotic system. Section 4 describes the adaptive control of the novel double convection chaotic system with unknown parameters via adaptive control method. Section 5 describes the adaptive synchronization of the identical novel double convection chaotic systems with unknown parameters via adaptive control method. The adaptive control and synchronization results derived in this work are established using Lyapunov stability theory [29]. MATLAB simulations are shown to illustrate all the main results derived in this work. Finally, Sect. 6 contains the conclusions of this work.

2 A Novel Double-Convection Chaotic System

In fluid mechanics modelling, cases of two-dimensional convection in a horizontal layer of Boussinesq fluid with lateral constraints were studied by Rucklidge [57]. When the convection takes place in a fluid layer rotating uniformly about a vertical axis and in the limit of tall thin rolls, convection in an imposed vertical magnetic field and convection in a rotating fluid layer are both modelled by a new 3-D system of ordinary differential equations, which produces chaotic solutions like the Lorenz system [38].

The Rucklidge chaotic system [57] for nonlinear double convection is described by the 3-D system

$$\begin{cases} \dot{x}_1 = -ax_1 + bx_2 - x_2x_3 \\ \dot{x}_2 = x_1 \\ \dot{x}_3 = -x_3 + x_2^2 \end{cases} \tag{1}$$

where x_1, x_2, x_3 are the states and a, b are constant, positive, parameters of the Rucklidge system.

The Rucklidge system (1) is chaotic when we take the parameter values as

$$a = 2, \quad b = 6.7 \tag{2}$$

For numerical simulations, we take the initial values of the Rucklidge system (1) as

$$x_1(0) = 1.2, \quad x_2(0) = 0.8, \quad x_3(0) = 1.4 \tag{3}$$

The Lyapunov exponents of the Rucklidge system (1) for the parameter values (2) and initial values (3) are numerically obtained using MATLAB as

$$L_1 = 0.1877, \quad L_2 = 0, \quad L_3 = -3.1893 \tag{4}$$

From the LE spectrum (4), it is clear that the Rucklidge system (1) is chaotic with the Maximal Lyapunov Exponent (MLE) as $L_1 = 0.1877$.

Also, the Kaplan-Yorke dimension of the Rucklidge chaotic system (1) is derived as

$$D_{KY} = 2 + \frac{L_1 + L_2}{|L_3|} = 2.0589 \tag{5}$$

Figure 1 depicts the strange chaotic attractor of the Rucklidge chaotic system (1) for nonlinear double convection.

In this research work, we announce a novel chaotic system for nonlinear double convection by modifying the Rucklidge dynamics (1) as

$$\begin{cases} \dot{x}_1 = -ax_1 + bx_2 - x_2x_3 \\ \dot{x}_2 = x_1 \\ \dot{x}_3 = -x_3 + x_2^4 \end{cases} \tag{6}$$

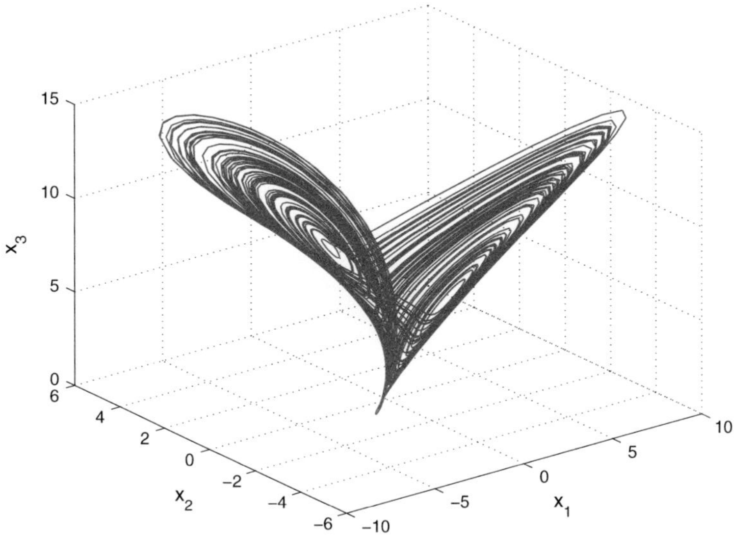

Fig. 1 The strange attractor of the Rucklidge chaotic system

The 3-D system (6) is *chaotic*, when the system parameters are chosen as

$$a = 2, \quad 5 \leq b \leq 120 \tag{7}$$

By fixing the value of a as $a = 2$, it is noted that the Maximal Lyapunov Exponent (MLE) of the novel chaotic system (6) occurs when $b = 105$.

Thus, in this work, we take the specific values of the parameters a and b for the chaotic case as

$$a = 2, \quad b = 105 \tag{8}$$

For numerical simulations, we take the initial values of the novel double convection chaotic system (6) as

$$x_1(0) = 1.2, \;\; x_2(0) = 0.8, \;\; x_3(0) = 1.4 \tag{9}$$

The Lyapunov exponents of the novel double convection chaotic system (6) for the parameter values (8) and the initial values (9) are numerically obtained using MATLAB as

$$L_1 = 1.03405, \quad L_2 = 0, \quad L_3 = -4.03938 \tag{10}$$

From the LE spectrum (10), it is clear that the novel 3-D system (6) is chaotic with the Maximal Lyapunov Exponent (MLE) as $L_1 = 1.03405$.

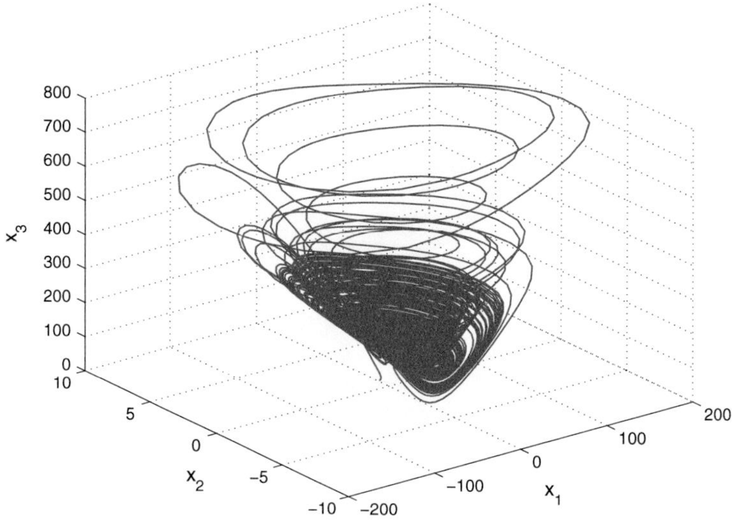

Fig. 2 The strange attractor of the novel chaotic system

Also, the Kaplan-Yorke dimension of the novel double convection chaotic system (6) is derived as

$$D_{KY} = 2 + \frac{L_1 + L_2}{|L_3|} = 2.2560 \tag{11}$$

Since the Maximal Lyapunov Exponent (MLE) and Kaplan-Yorke dimension of the novel double convection chaotic system (6) are greater than the Maximal Lyapunov Exponent (MLE) and Kaplan-Yorke dimension of the Rucklidge chaotic system (1), it is clear that the novel double convection chaotic system (6) is more chaotic than the Rucklidge system (1).

Figure 2 depicts the strange attractor of the novel double convection chaotic system (6). Figures 3, 4 and 5 describe the 2-D projections of the novel double convection chaotic system (6) on the (x_1, x_2), (x_2, x_3) and (x_1, x_3) coordinate planes, respectively.

3 Qualitative Properties of the Novel Double Convection Chaotic System

This section describes the qualitative properties of the novel double convection chaotic system (6). We take the parameter values as in the chaotic case (7), *viz.* $a = 2$ and $5 \leq b \leq 120$.

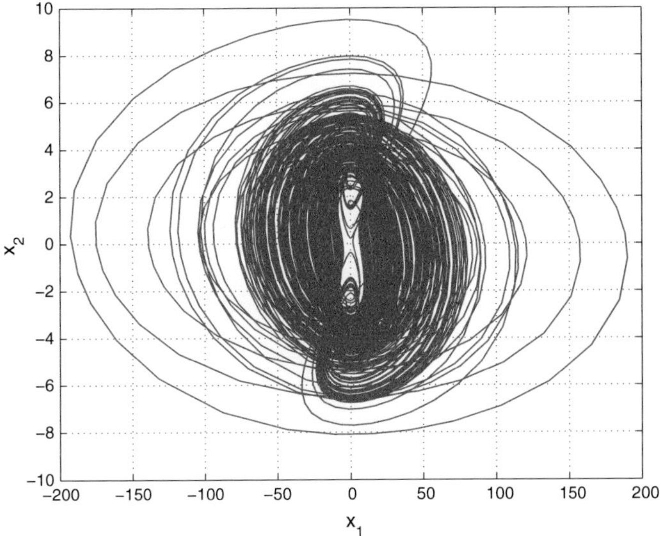

Fig. 3 2-D projection of the novel chaotic system on the (x_1, x_2) plane

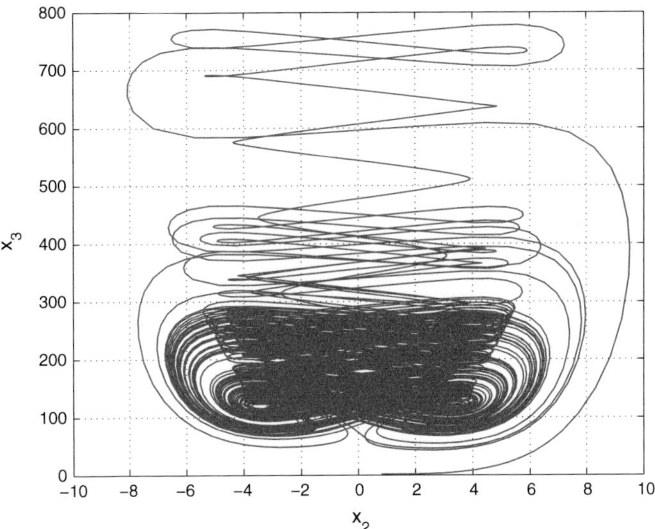

Fig. 4 2-D projection of the novel chaotic system on the (x_2, x_3) plane

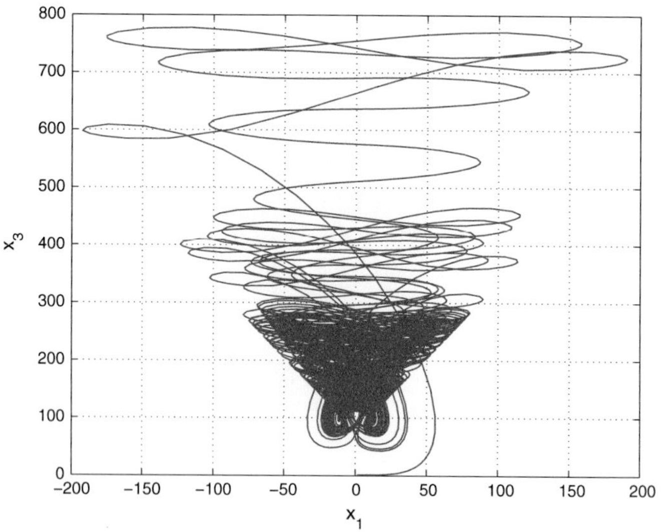

Fig. 5 2-D projection of the novel chaotic system on the (x_1, x_3) plane

3.1 Dissipativity

We write the novel chaotic system (6) in vector notation as

$$\dot{x} = f(x) = \begin{bmatrix} f_1(x) \\ f_2(x) \\ f_3(x) \end{bmatrix}, \tag{12}$$

where

$$\begin{aligned} f_1(x) &= -ax_1 + bx_2 - x_2x_3 \\ f_2(x) &= x_1 \\ f_3(x) &= -x_3 + x_2^4 \end{aligned} \tag{13}$$

We take the parameter values as in the chaotic case (7).

The divergence of the vector field f on $\mathbb{R}^3$ is obtained as

$$\operatorname{div} f = \frac{\partial f_1(x)}{\partial x_1} + \frac{\partial f_2(x)}{\partial x_2} + \frac{\partial f_3(x)}{\partial x_3} = -a - 1 = -\mu \tag{14}$$

where $\mu = a + 1 = 3 > 0$.

Let Ω be any region in $\mathbb{R}^3$ with a smooth boundary. Let $\Omega(t) = \Phi_t(\Omega)$, where Φ_t is the flow of the vector field f.

Let $V(t)$ denote the volume of $\Omega(t)$.

By Liouville's theorem, it follows that

$$\frac{dV(t)}{dt} = \int\limits_{\Omega(t)} (\operatorname{div} f) dx_1\, dx_2\, dx_3 \tag{15}$$

Substituting the value of div$f$ in (15) leads to

$$\frac{dV(t)}{dt} = -\mu \int\limits_{\Omega(t)} dx_1\, dx_2\, dx_3 = -\mu V(t) \tag{16}$$

Integrating the linear differential equation (16), $V(t)$ is obtained as

$$V(t) = V(0)\exp(-\mu t), \quad \text{where } \mu = 3.5 > 0. \tag{17}$$

From Eq. (17), it follows that the volume $V(t)$ shrinks to zero exponentially as $t \to \infty$.

Thus, the novel chaotic system (6) is dissipative. Hence, any asymptotic motion of the system (6) settles onto a set of measure zero, *i.e.* a strange attractor.

3.2 Equilibrium Points

The equilibrium points of the novel chaotic system (6) for the chaotic case (7) are obtained by solving the system of equations

$$-ax_1 + bx_2 - x_2 x_3 = 0 \tag{18a}$$

$$x_1 = 0 \tag{18b}$$

$$-x_3 + x_2^4 = 0 \tag{18c}$$

From (18b), it follows that $x_1 = 0$.

Substituting $x_1 = 0$ in (18a), we get

$$bx_2 - x_2 x_3 = 0 \ \text{ or } \ x_2(b - x_3) = 0 \tag{19}$$

Thus, we have two cases to consider: (A) $x_2 = 0$, and (B) $x_2 \neq 0$.

When $x_2 = 0$, we get the first equilibrium point of the system (6) as

$$E_0 = \begin{bmatrix} 0 \\ 0 \\ 0 \end{bmatrix} \tag{20}$$

Next, we suppose that $x_2 \neq 0$. From Eq. (19), we get $x_3 = b$.

Substituting $x_3 = b$ in (18c), we get

$$x_2^4 = b \text{ or } x_2 = \pm b^{\frac{1}{4}} \tag{21}$$

We take the value of b as $b = 105$. (In this case, the novel double convection chaotic system (6) has a maximum value of Lyapunov exponent).

Thus, we get two more equilibrium points of the system (6) as

$$E_1 = \begin{bmatrix} 0 \\ 3.2011 \\ 105 \end{bmatrix}, \quad E_2 = \begin{bmatrix} 0 \\ -3.2011 \\ 105 \end{bmatrix} \tag{22}$$

To test the stability type of the equilibrium points, we calculate the Jacobian of the system (6) at any $x \in \mathbb{R}^3$ as

$$J(x) = \begin{bmatrix} -a & b - x_3 & -x_2 \\ 1 & 0 & 0 \\ 0 & 4x_2^3 & -1 \end{bmatrix} = \begin{bmatrix} -2 & 105 - x_3 & -x_2 \\ 1 & 0 & 0 \\ 0 & 4x_2^3 & -1 \end{bmatrix} \tag{23}$$

We find that

$$J_0 = J(E_0) = \begin{bmatrix} -2 & 105 & 0 \\ 1 & 0 & 0 \\ 0 & 0 & -1 \end{bmatrix} \tag{24}$$

The matrix J_0 has the eigenvalues

$$\lambda_1 = -1, \quad \lambda_2 = -11.2956, \quad \lambda_3 = 9.2956 \tag{25}$$

This shows that the equilibrium point E_0 is a saddle-point, which is unstable.
Next, we find that

$$J_1 = J(E_1) = \begin{bmatrix} -2 & 0 & -3.2011 \\ 1 & 0 & 0 \\ 0 & 131.2072 & -1 \end{bmatrix} \tag{26}$$

The matrix J_1 has the eigenvalues

$$\lambda_1 = -8.5334, \quad \lambda_{2,3} = 2.7667 \pm 6.4470i \tag{27}$$

This shows that the equilibrium point E_1 is a saddle-focus, which is unstable.
We also find that

$$J_2 = J(E_2) = \begin{bmatrix} -2 & 0 & 3.2011 \\ 1 & 0 & 0 \\ 0 & -131.2072 & -1 \end{bmatrix} \tag{28}$$

The matrix J_2 has the eigenvalues

$$\lambda_1 = -8.5334, \quad \lambda_{2,3} = 2.7667 \pm 6.4470i \tag{29}$$

This shows that the equilibrium point E_2 is a saddle-focus, which is unstable. Hence, E_2 is also a saddle-focus, which is unstable.

Thus, we conclude that all the three equilibrium points (E_1, E_2, E_3) of the novel chaotic system (6) s are unstable.

3.3 Symmetry

It is easy to check that the novel chaotic system (6) is invariant under the change of coordinates

$$(x_1, x_2, x_3) \mapsto (-x_1, -x_2, x_3) \tag{30}$$

which persists for all values of the parameters a, b and c.

This shows that the novel chaotic system (6) has rotation symmetry about the x_3-axis. Hence, it follows that any non-trivial trajectory of the novel chaotic system (6) must have a twin trajectory.

3.4 Lyapunov Exponents and Kaplan-Yorke Dimension

We take the parameter values of the novel double convection chaotic system (6) as in the chaotic case (8), *i.e.*

$$a = 2, \quad b = 105 \tag{31}$$

We take the initial values of the novel chaotic system (6) as in (9).

Then the Lyapunov exponents of the novel chaotic system (6) are numerically obtained using MATLAB as

$$L_1 = 1.03405, \quad L_2 = 0, \quad L_3 = -4.03598 \tag{32}$$

Since $L_1 + L_2 + L_3 = -3.00533 < 0$, the novel chaotic system (6) is dissipative.

Figure 5 depicts the dynamics of the Lyapunov exponents of the novel chaotic system (6).

From Fig. 6, we note that the Maximal Lyapunov Exponent (MLE) of the novel chaotic system (6) is obtained as $L_1 = 1.03405$.

Also, the Kaplan-Yorke dimension of the novel chaotic system (6) is obtained as

$$D_{KY} = 2 + \frac{L_1 + L_2}{|L_3|} = 2.2560 \tag{33}$$

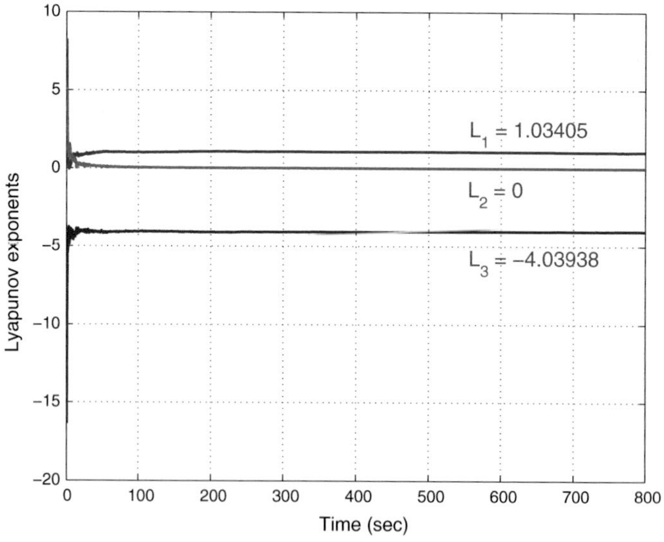

Fig. 6 Lyapunov exponents of the novel chaotic system

4 Adaptive Control of the Novel Double Convection Chaotic System

In this section, we design an adaptive feedback control law for globally stabilizing the novel double convection chaotic system with unknown parameters a and b.

Thus, we consider the novel double convection chaotic system given by the dynamics

$$\begin{cases} \dot{x}_1 = -ax_1 + bx_2 - x_2x_3 + u_1 \\ \dot{x}_2 = x_1 + u_2 \\ \dot{x}_3 = -x_3 + x_2^4 + u_3 \end{cases} \tag{34}$$

In Eq. (34), x_1, x_2, x_3 are the states and u is the adaptive controller to be found using estimates $\hat{a}(t), \hat{b}(t), \hat{c}(t)$ of the unknown parameters a, b, c, respectively.

We consider the adaptive controller defined by

$$\begin{cases} u_1 = \hat{a}(t)x_1 - \hat{b}(t)x_2 + x_2x_3 - k_1x_1 \\ u_2 = -x_1 - k_2x_2 \\ u_3 = x_3 - x_2^4 - k_3x_3 \end{cases} \tag{35}$$

where k_1, k_2, k_3 are positive gain constants.

Substituting (35) into (34), we get the closed-loop plant dynamics as

$$\begin{cases} \dot{x}_1 = -[a - \hat{a}(t)]x_1 + [b - \hat{b}(t)]x_2 - k_1 x_1 \\ \dot{x}_2 = -k_2 x_2 \\ \dot{x}_3 = -k_3 x_3 \end{cases} \tag{36}$$

Now, we define the parameter estimation errors as

$$\begin{cases} e_a(t) = a - \hat{a}(t) \\ e_b(t) = b - \hat{b}(t) \end{cases} \tag{37}$$

Using the definitions (37), we can simply the closed-loop plant dynamics (36) as

$$\begin{cases} \dot{x}_1 = -e_a x_1 + e_b x_2 - k_1 x_1 \\ \dot{x}_2 = -k_2 x_2 \\ \dot{x}_3 = -k_3 x_3 \end{cases} \tag{38}$$

Differentiating (37) with respect to t, we get

$$\begin{cases} \dot{e}_a = -\dot{\hat{a}} \\ \dot{e}_b = -\dot{\hat{b}} \end{cases} \tag{39}$$

Next, we consider the quadratic Lyapunov function defined by

$$V(x_1, x_2, x_3, e_a, e_b) = \frac{1}{2}\left(x_1^2 + x_2^2 + x_3^2 + e_a^2 + e_b^2\right) \tag{40}$$

which is positive definite on $\mathbb{R}^5$.

Differentiating V along the trajectories of (38) and (39), we get

$$\dot{V} = -k_1 x_1^2 - k_2 x_2^2 - k_3 x_3^2 + e_a\left(-x_1^2 - \dot{\hat{a}}\right) + e_b\left(x_1 x_2 - \dot{\hat{b}}\right) \tag{41}$$

In view of (41), we take the parameter update law as follows.

$$\begin{cases} \dot{\hat{a}} = -x_1^2 \\ \dot{\hat{b}} = x_1 x_2 \end{cases} \tag{42}$$

Next, we state and prove the main result of this section.

Theorem 1 *The novel double convection chaotic system (34) is globally and exponentially stabilized for all initial conditions* $\mathbf{x}(0) \in \mathbb{R}^3$ *by the adaptive control law (35) and the parameter update law (42), where* k_1, k_2, k_3 *are positive gain constants.*

Proof We prove this result by applying Lyapunov stability theory [29].

We consider the quadratic Lyapunov function V defined by Eq. (40), which is positive definite on $\mathbb{R}^5$.

Substituting the parameter update law (42) into (41), we obtain the time-derivative of V as

$$\dot{V} = -k_1x_1^2 - k_2x_2^2 - k_3x_3^2 \tag{43}$$

which is negative semi-definite on $\mathbb{R}^5$.

Thus, it follows that the state vector $\mathbf{x}(t) = (x_1(t), x_2(t), x_3(t))$ and the parameter estimation error vector $(e_a(t), e_b(t))$ are globally bounded, *i.e.*

$$\left[x_1(t) \;\; x_2(t) \;\; x_3(t) \;\; e_a(t) \;\; e_b(t) \right] \in \mathbf{L}_\infty \tag{44}$$

We define $k = \min\{k_1, k_2, k_3\}$. Then it follows from (43) that

$$\dot{V} \leq -kx_1^2 - kx_2^2 - kx_3^2 = -k\|\mathbf{x}\|^2 \tag{45}$$

or

$$\|\mathbf{x}\|^2 \leq -\dot{V} \tag{46}$$

Integrating the inequality (46) from 0 to t, we get

$$\int_0^t \|\mathbf{x}(\tau)\|^2 \, d\tau \;\leq\; V(0) - V(t) \tag{47}$$

From (47), it follows that $\mathbf{x} \in \mathbf{L}_2$. From Eq. (38), it can be deduced that $\dot{\mathbf{x}} \in \mathbf{L}_\infty$.

Thus, using Barbalat's lemma [29], we conclude that $\mathbf{x}(t) \to \mathbf{0}$ exponentially as $t \to \infty$ for all initial conditions $\mathbf{x}(0) \in \mathbb{R}^3$.

This completes the proof. ∎

For numerical simulations, we use the classical fourth-order Runge-Kutta method in MATLAB with step-size $h = 10^{-8}$ for solving the novel double convection chaotic system (34) when the adaptive control law (35) and parameter update law (42) are implemented.

We take the positive gain constants as $k_1 = 5$, $k_2 = 5$ and $k_3 = 5$.

The parameter values of the novel double convection chaotic system (34) are taken as in chaotic case, *i.e.* $a = 2$ and $b = 105$.

We take the initial values of the novel double convection chaotic system (34) as $x_1(0) = 6.3$, $x_2(0) = 12.4$ and $x_3(0) = 18.7$.

Also, we take the initial values of the parameter updates as $\hat{a}(0) = 23.4$ and $\hat{b}(0) = 10.8$.

Figure 7 shows the time-history of the exponential convergence of the controlled states $x_1(t)$, $x_2(t)$ and $x_3(t)$.

From Fig. 7, it is clear that $x_1(t)$, $x_2(t)$ and $x_3(t)$ converge to zero in two seconds.

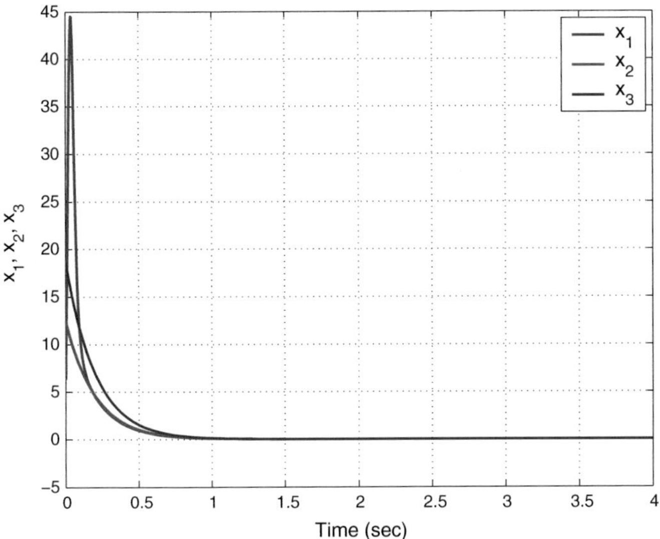

Fig. 7 Time-history of the controlled states of the novel double convection system

5 Adaptive Synchronization of the Identical Novel Double Convection Chaotic Systems

In this section, we design an adaptive feedback control law for globally synchronizing the states of the identical novel double convection chaotic systems with unknown parameters a and b.

As the master system, we consider the novel double convection chaotic system given by the dynamics

$$\begin{cases} \dot{x}_1 = -ax_1 + bx_2 - x_2x_3 \\ \dot{x}_2 = x_1 \\ \dot{x}_3 = -x_3 + x_2^4 \end{cases} \tag{48}$$

In Eq. (48), x_1, x_2, x_3 are the states and a, b are unknown system parameters.

As the slave system, we consider the novel double convection chaotic system given by the dynamics

$$\begin{cases} \dot{y}_1 = -ay_1 + by_2 - y_2y_3 + u_1 \\ \dot{y}_2 = y_1 + u_2 \\ \dot{y}_3 = -y_3 + y_2^4 + u_3 \end{cases} \tag{49}$$

In Eq. (49), y_1, y_2, y_3 are the states and u is the adaptive controller to be determined using estimates $\hat{a}(t), \hat{b}(t), \hat{c}(t)$ of the unknown parameters a, b, c, respectively.

The complete synchronization error between the novel double convection chaotic systems (48) and (49) is defined by

$$\begin{cases} e_1 = y_1 - x_1 \\ e_2 = y_2 - x_2 \\ e_3 = y_3 - x_3 \end{cases} \tag{50}$$

Then the error dynamics can be obtained as follows.

$$\begin{cases} \dot{e}_1 = -ae_1 + be_2 - y_2y_3 + x_2x_3 + u_1 \\ \dot{e}_2 = e_1 + u_2 \\ \dot{e}_3 = -e_3 + y_2^4 - x_2^4 + u_3 \end{cases} \tag{51}$$

We consider the adaptive controller defined by

$$\begin{cases} u_1 = \hat{a}(t)e_1 - \hat{b}(t)e_2 + y_2y_3 - x_2x_3 - k_1e_1 \\ u_2 = -e_1 - k_2e_2 \\ u_3 = e_3 - y_2^4 + x_2^4 - k_3e_3 \end{cases} \tag{52}$$

where k_1, k_2, k_3 are positive gain constants.

Substituting (52) into (51), we get the closed-loop plant dynamics as

$$\begin{cases} \dot{e}_1 = -[a - \hat{a}(t)]e_1 + [b - \hat{b}(t)]e_2 - k_1e_1 \\ \dot{e}_2 = -k_2e_2 \\ \dot{e}_3 = -k_3e_3 \end{cases} \tag{53}$$

Now, we define the parameter estimation errors as

$$\begin{cases} e_a(t) = a - \hat{a}(t) \\ e_b(t) = b - \hat{b}(t) \end{cases} \tag{54}$$

Using the definitions (54), we can simply the closed-loop plant dynamics (53) as

$$\begin{cases} \dot{e}_1 = -[a - \hat{a}(t)]e_1 + [b - \hat{b}(t)]e_2 - k_1e_1 \\ \dot{e}_2 = -k_2e_2 \\ \dot{e}_3 = -k_3e_3 \end{cases} \tag{55}$$

Differentiating (54) with respect to t, we get

$$\begin{cases} \dot{e}_a = -\dot{\hat{a}} \\ \dot{e}_b = -\dot{\hat{b}} \end{cases} \tag{56}$$

Next, we consider the quadratic Lyapunov function defined by

$$V(e_1, e_2, e_3, e_a, e_b) = \frac{1}{2}\left(e_1^2 + e_2^2 + e_3^2 + e_a^2 + e_b^2\right) \tag{57}$$

which is positive definite on $\mathbb{R}^5$.

Differentiating V along the trajectories of (55) and (56), we get

$$\dot{V} = -k_1 e_1^2 - k_2 e_2^2 - k_3 e_3^2 + e_a \left(-e_1^2 - \dot{\hat{a}} \right) + e_b \left(e_1 e_2 - \dot{\hat{b}} \right) \tag{58}$$

In view of (58), we take the parameter update law as follows.

$$\begin{cases} \dot{\hat{a}} = -e_1^2 \\ \dot{\hat{b}} = e_1 e_2 \end{cases} \tag{59}$$

Next, we state and prove the main result of this section.

Theorem 2 *The novel double convection chaotic systems (48) and (49) are globally and exponentially synchronized for all initial conditions* $\mathbf{x}(0), \mathbf{y}(0) \in \mathbb{R}^3$ *by the adaptive control law (52) and the parameter update law (59), where* k_1, k_2, k_3 *are positive gain constants.*

Proof We prove this result by applying Lyapunov stability theory [29].

We consider the quadratic Lyapunov function V defined by Eq. (57), which is positive definite on $\mathbb{R}^5$.

Substituting the parameter update law (59) into (58), we obtain the time-derivative of V as

$$\dot{V} = -k_1 e_1^2 - k_2 e_2^2 - k_3 e_3^2 \tag{60}$$

which is negative semi-definite on $\mathbb{R}^5$.

Thus, it follows that the error vector $\mathbf{e}(t) = (e_1(t), e_2(t), e_3(t))$ and the parameter estimation error vector $(e_a(t), e_b(t))$ are globally bounded, *i.e.*

$$\left[e_1(t) \;\; e_2(t) \;\; e_3(t) \;\; e_a(t) \;\; e_b(t) \right] \in \mathbf{L}_\infty \tag{61}$$

We define $k = \min\{k_1, k_2, k_3\}$.

Then it follows from (60) that

$$\dot{V} \leq -k e_1^2 - k e_2^2 - k e_3^2 = -k \|\mathbf{e}\|^2 \tag{62}$$

or

$$\|\mathbf{e}\|^2 \leq -\dot{V} \tag{63}$$

Integrating the inequality (46) from 0 to t, we get

$$\int_0^t \|\mathbf{e}(\tau)\|^2 \, d\tau \;\leq\; V(0) - V(t) \tag{64}$$

From (64), it follows that $\mathbf{e} \in \mathbf{L}_2$. From Eq. (55), it can be deduced that $\dot{\mathbf{e}} \in \mathbf{L}_\infty$.

Thus, using Barbalat's lemma [29], we conclude that $\mathbf{e}(t) \to \mathbf{0}$ exponentially as $t \to \infty$ for all initial conditions $\mathbf{e}(0) \in \mathbb{R}^3$.

This completes the proof. ∎

For numerical simulations, we use the classical fourth-order Runge-Kutta method in MATLAB with step-size $h = 10^{-8}$ for solving the novel double convection chaotic systems (48) and (49) when the adaptive control law (52) and parameter update law (59) are implemented.

We take the positive gain constants as $k_1 = 5$, $k_2 = 5$ and $k_3 = 5$.

The parameter values of the double convection chaotic systems (48) and (49) are taken as in the chaotic case, *i.e.* $a = 2$ and $b = 105$.

We take the initial values of the master system (48) as

$$x_1(0) = 2.3, \quad x_2(0) = 11.4, \quad x_3(0) = 7.5 \tag{65}$$

We take the initial values of the slave system (49) as

$$y_1(0) = 18.5, \quad x_2(0) = 6.9, \quad x_3(0) = 1.8 \tag{66}$$

Also, we take the initial values of the parameter updates as

$$\hat{a}(0) = 15.4, \quad \hat{b}(0) = 12.6 \tag{67}$$

Figures 8, 9 and 10 show the synchronization of the states of the novel double-convection chaotic systems (48) and (49).

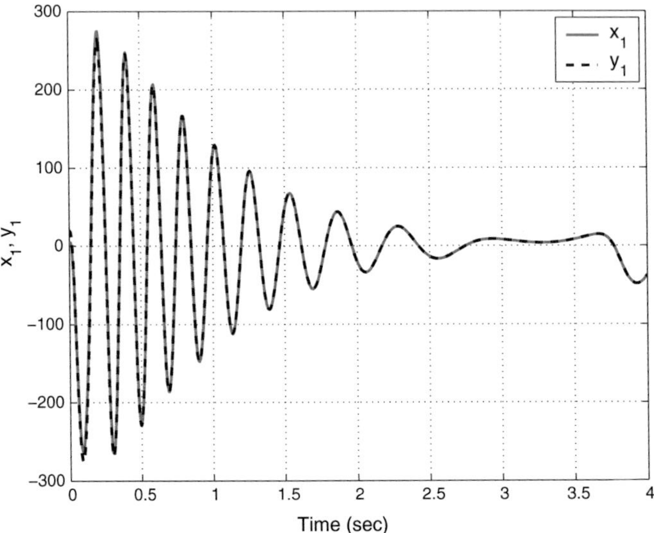

Fig. 8 Synchronization of the states x_1 and y_1

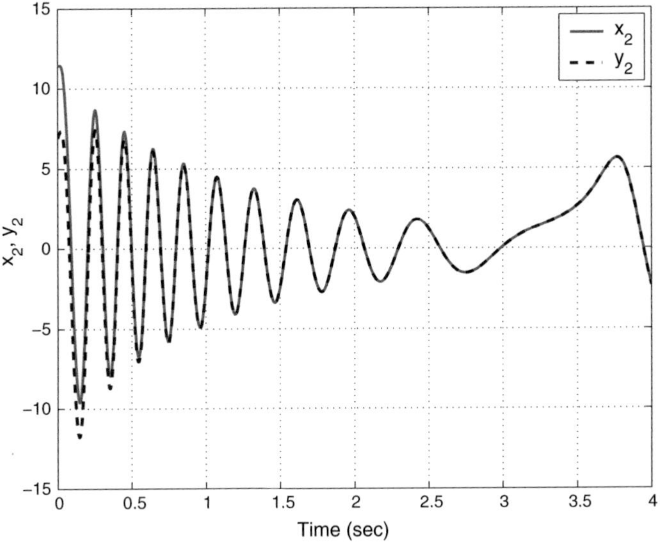

Fig. 9 Synchronization of the states x_2 and y_2

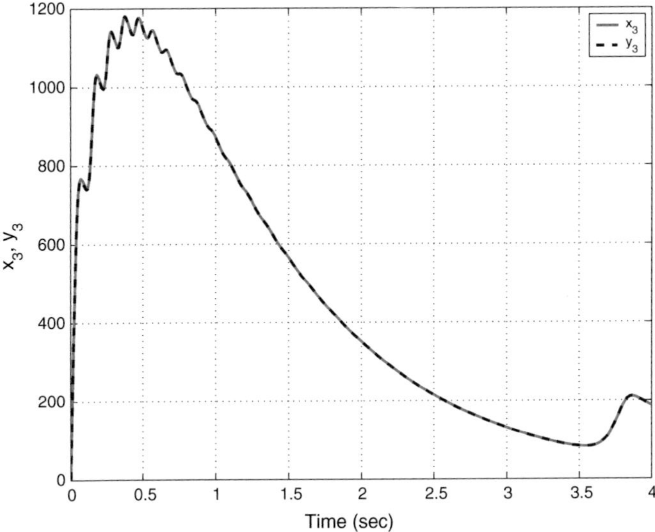

Fig. 10 Synchronization of the states x_3 and y_3

Figure 11 shows the time-history of the synchronization errors e_1, e_2, e_3.

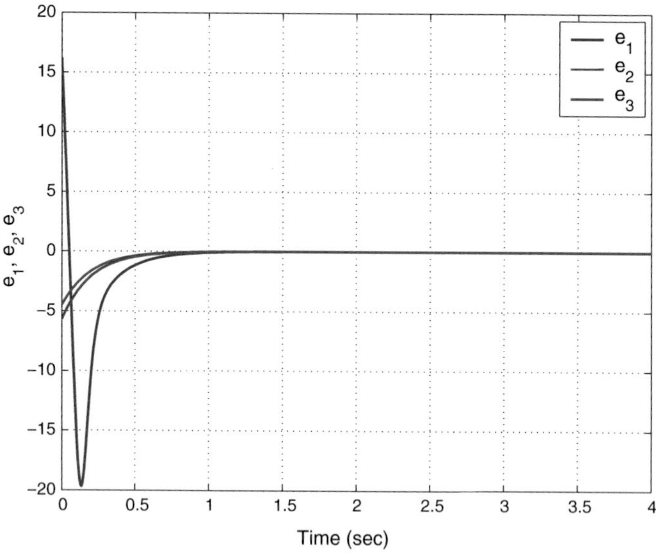

Fig. 11 Time-history of the synchronization errors e_1, e_2, e_3

6 Conclusions

In this research work, we detailed a novel 3-D double convection chaotic system and discuss its qualitative properties, adaptive control and synchronization. First, this work announced the dynamic equations and discussed the qualitative properties of the novel 3-D double convection chaotic system. We showed that the novel 3-D double convection chaotic system has three unstable equilibrium points of which one equilibrium point is a saddle-point and the other two equilibrium points are saddle-foci. Our novel chaotic system was obtained by modifying the equations of the Rucklidge chaotic system (1992) for nonlinear double convection. The Lyapunov exponents of the novel 3-D double convection chaotic system have been numerically obtained as $L_1 = 1.03405$, $L_2 = 0$ and $L_3 = -4.03938$. Also, the Kaplan-Yorke dimension of the novel 3-D double convection chaotic system has been calculated as $D_{KY} = 2.2560$. Next, this work described the adaptive controller design for the global stabilization of the novel 3-D double convection chaotic system. Furthermore, this work also described the adaptive synchronizer design for the global chaos synchronization of identical novel 3-D double convection chaotic systems. MATLAB simulations were shown to illustrate all the main results derived in this work for the novel 3-D double convection chaotic system.

References

1. Arneodo A, Coullet P, Tresser C (1981) Possible new strange attractors with spiral structure. Commun Math Phys 79(4):573–576
2. Azar AT, Serrano FE (2015a) Adaptive sliding mode control of the Furuta pendulum. In: Azar AT, Zhu Q (eds) Advances and Applications in Sliding Mode Control Systems. Studies in computational intelligence, vol 576. Springer, Germany, pp 1–42
3. Azar AT, Serrano FE (2015b) Deadbeat control for multivariable systems with time varying delays. In: Azar AT, Vaidyanathan S (eds) Chaos Modeling and Control Systems Design. Studies in computational intelligence, vol 581. Springer, Germany, pp 97–132
4. Azar AT, Serrano FE (2015c) Design and modeling of anti wind up PID controllers. In: Zhu Q, Azar AT (eds) Complex System Modelling and Control through Intelligent Soft Computations. Studies in fuzziness and soft computing, vol 319. Springer, Germany, pp 1–44
5. Azar AT, Serrano FE (2015d) Stabilizatoin and control of mechanical systems with backlash. In: Azar AT, Vaidyanathan S (eds) Handbook of Research on Advanced Intelligent Control Engineering and Automation. Advances in computational intelligence and robotics (ACIR), IGI-Global, USA, pp 1–60
6. Azar AT, Vaidyanathan S (2015a) Chaos Modeling and Control Systems Design. Studies in computational intelligence, vol 581. Springer, Germany
7. Azar AT, Vaidyanathan S (2015b) Computational Intelligence Applications in Modeling and Control. Studies in computational intelligence, vol 575. Springer, Germany
8. Azar AT, Vaidyanathan S (2015c) Handbook of Research on Advanced Intelligent Control Engineering and Automation. Advances in computational intelligence and robotics (ACIR), IGI-Global, USA
9. Azar AT, Zhu Q (2015) Advances and Applications in Sliding Mode Control Systems. Studies in computational intelligence, vol 576. Springer, Germany
10. Azar AT (2010) Fuzzy Systems. IN-TECH, Vienna
11. Azar AT (2012) Overview of type-2 fuzzy logic systems. Int J Fuzzy Syst Appl 2(4):1–28
12. Azar AT, Serrano FE (2014) Robust IMC-PID tuning for cascade control systems with gain and phase margin specifications. Neural Comput Appl 25(5):983–995
13. Cai G, Tan Z (2007) Chaos synchronization of a new chaotic system via nonlinear control. J Uncertain Syst 1(3):235–240
14. Carroll TL, Pecora LM (1991) Synchronizing chaotic circuits. IEEE Trans Circuits Syst 38(4):453–456
15. Chen WH, Wei D, Lu X (2014) Global exponential synchronization of nonlinear time-delay Lure systems via delayed impulsive control. Commun Nonlinear Sci Numer Simul 19(9):3298–3312
16. Chen HK, Lee CI (2004) Anti-control of chaos in rigid body motion. Chaos Solitons Fractals 21(4):957–965
17. Chen G, Ueta T (1999) Yet another chaotic attractor. Int J Bifurc Chaos 9(7):1465–1466
18. Das S, Goswami D, Chatterjee S, Mukherjee S (2014) Stability and chaos analysis of a novel swarm dynamics with applications to multi-agent systems. Eng Appl Artif Intell 30:189–198
19. Feki M (2003) An adaptive chaos synchronization scheme applied to secure communication. Chaos, Solitons Fractals 18(1):141–148
20. Gan Q, Liang Y (2012) Synchronization of chaotic neural networks with time delay in the leakage term and parametric uncertainties based on sampled-data control. J Frankl Inst 349(6):1955–1971
21. Gaspard P (1999) Microscopic chaos and chemical reactions. Phys A Stat Mech Appl 263 (1–4):315–328
22. Gibson WT, Wilson WG (2013) Individual-based chaos: extensions of the discrete logistic model. J Theor Biol 339:84–92
23. Guégan D (2009) Chaos in economics and finance. Annu Rev Control 33(1):89–93
24. Huang X, Zhao Z, Wang Z, Li Y (2012) Chaos and hyperchaos in fractional-order cellular neural networks. Neurocomputing 94:13–21

25. Jiang GP, Zheng WX, Chen G (2004) Global chaos synchronization with channel time-delay. Chaos, Solitons Fractals 20(2):267–275
26. Karthikeyan R, Sundarapandian V (2014) Hybrid chaos synchronization of four-scroll systems via active control. J Electr Eng 65(2):97–103
27. Kaslik E, Sivasundaram S (2012) Nonlinear dynamics and chaos in fractional-order neural networks. Neural Netw 32:245–256
28. Kengne J, Chedjou JC, Kenne G, Kyamakya K (2012) Dynamical properties and chaos synchronization of improved Colpitts oscillators. Commun Nonlinear Sci Numer Simul 17(7):2914–2923
29. Khalil HK (2001) Nonlinear Syst. Prentice Hall, New Jersey
30. Kyriazis M (1991) Applications of chaos theory to the molecular biology of aging. Exp Gerontol 26(6):569–572
31. Lang J (2015) Color image encryption based on color blend and chaos permutation in the reality-preserving multiple-parameter fractional Fourier transform domain. Opt Commun 338:181–192
32. Li D (2008) A three-scroll chaotic attractor. Phys Lett A 372(4):387–393
33. Li N, Zhang Y, Nie Z (2011) Synchronization for general complex dynamical networks with sampled-data. Neurocomputing 74(5):805–811
34. Li N, Pan W, Yan L, Luo B, Zou X (2014) Enhanced chaos synchronization and communication in cascade-coupled semiconductor ring lasers. Commun Nonlinear Sci Numer Simul 19(6):1874–1883
35. Lian S, Chen X (2011) Traceable content protection based on chaos and neural networks. Appl Soft Comput 11(7):4293–4301
36. Li Z, Chen G (2006) Integration of Fuzzy Logic and Chaos Theory. Studies in fuzziness and soft computing, vol 187. Springer, Germany
37. Liu C, Liu T, Liu L, Liu K (2004) A new chaotic attractor. Chaos Solitions Fractals 22(5): 1031–1038
38. Lorenz EN (1963) Deterministic periodic flow. J Atmos Sci 20(2):130–141
39. Lü J, Chen G (2002) A new chaotic attractor coined. Int J Bifurc Chaos 12(3):659–661
40. Mondal S, Mahanta C (2014) Adaptive second order terminal sliding mode controller for robotic manipulators. J Frankl Inst 351(4):2356–2377
41. Murali K, Lakshmanan M (1998) Secure communication using a compound signal from generalized chaotic systems. Phys Lett A 241(6):303–310
42. Nehmzow U, Walker K (2005) Quantitative description of robot-environment interaction using chaos theory. Robot Auton Syst 53(3–4):177–193
43. Noroozi N, Roopaei M, Karimaghaee P, Safavi AA (2010) Simple adaptive variable structure control for unknown chaotic systems. Commun Nonlinear Sci Numer Simul 15(3):707–727
44. Pecora LM, Carroll TL (1990) Synchronization in chaotic systems. Phys Rev Lett 64(8): 821–824
45. Pehlivan I, Moroz IM, Vaidyanathan S (2014) Analysis, synchronization and circuit design of a novel butterfly attractor. J Sound Vib 333(20):5077–5096
46. Petrov V, Gaspar V, Masere J, Showalter K (1993) Controlling chaos in Belousov-Zhabotinsky reaction. Nature 361:240–243
47. Pham VT, Vaidyanathan S, Volos CK, Jafari S (2015a) Hidden attractors in a chaotic system with an exponential nonlinear term. Eur Phys J Spec Top 224(8):1507–1517
48. Pham VT, Volos CK, Vaidyanathan S, Le TP, Vu VY (2015b) A memristor-based hyperchaotic system with hidden attractors: dynamics, synchronization and circuital emulating. J Eng Sci Technol Rev 8(2):205–214
49. Qu Z (2011) Chaos in the genesis and maintenance of cardiac arrhythmias. Prog Biophys Mol Biol 105(3):247–257
50. Rasappan S, Vaidyanathan S (2012a) Global chaos synchronization of WINDMI and Coullet chaotic systems by backstepping control. Far East J Math Sci 67(2):265–287
51. Rasappan S, Vaidyanathan S (2012b) Hybrid synchronization of n-scroll Chua and Lur'e chaotic systems via backstepping control with novel feedback. Arch Control Sci 22(3): 343–365

52. Rasappan S, Vaidyanathan S (2012c) Synchronization of hyperchaotic Liu system via backstepping control with recursive feedback. Commun Comput Inf Sci 305:212–221
53. Rasappan S, Vaidyanathan S (2013) Hybrid synchronization of n-scroll chaotic Chua circuits using adaptive backstepping control design with recursive feedback. Malays J Math Sci 7(2):219–246
54. Rasappan S, Vaidyanathan S (2014) Global chaos synchronization of WINDMI and Coullet chaotic systems using adaptive backstepping control design. Kyungpook Math J 54(1): 293–320
55. Rhouma R, Belghith S (2011) Cryptoanalysis of a chaos based cryptosystem on DSP. Commun Nonlinear Sci Numer Simul 16(2):876–884
56. Rössler OE (1976) An equation for continuous chaos. Phys Lett A 57(5):397–398
57. Rucklidge AM (1992) Chaos in models of double convection. J Fluid Mech 237:209–229
58. Sampath S, Vaidyanathan S, Volos CK, Pham VT (2015) An eight-term novel four-scroll chaotic system with cubic nonlinearity and its circuit simulation. J Eng Sci Technol Rev 8(2):1–6
59. Sarasu P, Sundarapandian V (2011a) Active controller design for the generalized projective synchronization of four-scroll chaotic systems. Int J Syst Signal Control Eng Appl 4(2):26–33
60. Sarasu P, Sundarapandian V (2011b) The generalized projective synchronization of hyperchaotic Lorenz and hyperchaotic Qi systems via active control. Int J Soft Comput 6(5):216–223
61. Sarasu P, Sundarapandian V (2012a) Adaptive controller design for the generalized projective synchronization of 4-scroll systems. Int J Syst Signal Control Eng Appl 5(2):21–30
62. Sarasu P, Sundarapandian V (2012b) Generalized projective synchronization of three-scroll chaotic systems via adaptive control. Eur J Sci Res 72(4):504–522
63. Sarasu P, Sundarapandian V (2012c) Generalized projective synchronization of two-scroll systems via adaptive control. Int J Soft Comput 7(4):146–156
64. Shahverdiev EM, Shore KA (2009) Impact of modulated multiple optical feedback time delays on laser diode chaos synchronization. Opt Commun 282(17):3568–3572
65. Sharma A, Patidar V, Purohit G, Sud KK (2012) Effects on the bifurcation and chaos in forced Duffing oscillator due to nonlinear damping. Commun Nonlinear Sci Numer Simul 17(6):2254–2269
66. Sprott JC (1994) Some simple chaotic flows. Phys Rev E 50(2):647–650
67. Sprott JC (2004) Competition with evolution in ecology and finance. Phys Lett A 325 (5–6):329–333
68. Sprott JC (2010) Elegant Chaos: Algebraically Simple Chaotic Flows. World Scientific, Singapore
69. Suérez I (1999) Mastering chaos in ecology. Ecol Model 117(2–3):305–314
70. Sundarapandian V (2010a) Output regulation of the Lorenz attractor. Far East J Math Sci 42(2):289–299
71. Sundarapandian V (2010b) Output regulation of the Lorenz attractor. Far East J Math Sci 42(2):289–299
72. Sundarapandian V (2013) Analysis and anti-synchronization of a novel chaotic system via active and adaptive controllers. J Eng Sci Technol Rev 6(4):45–52
73. Sundarapandian V, Karthikeyan R (2011a) Anti-synchronization of hyperchaotic Lorenz and hyperchaotic Chen systems by adaptive control. Int J Syst Signal Control Eng Appl 4(2):18–25
74. Sundarapandian V, Karthikeyan R (2011b) Anti-synchronization of Lü and Pan chaotic systems by adaptive nonlinear control. Eur J Sci Res 64(1):94–106
75. Sundarapandian V, Karthikeyan R (2012a) Adaptive anti-synchronization of uncertain Tigan and Li systems. J Eng Appl Sci 7(1):45–52
76. Sundarapandian V, Karthikeyan R (2012b) Hybrid synchronization of hyperchaotic Lorenz and hyperchaotic Chen systems via active control. J Eng Appl Sci 7(3):254–264
77. Sundarapandian V, Pehlivan I (2012) Analysis, control, synchronization, and circuit design of a novel chaotic system. Math Comput Model 55(7–8):1904–1915
78. Sundarapandian V, Sivaperumal S (2011) Sliding controller design of hybrid synchronization of four-wing chaotic systems. Int J Soft Comput 6(5):224–231

79. Suresh R, Sundarapandian V (2013) Global chaos synchronization of a family of n-scroll hyperchaotic Chua circuits using backstepping control with recursive feedback. Far East J Math Sci 73(1):73–95
80. Tigan G, Opris D (2008) Analysis of a 3D chaotic system. Chaos Solitons Fractals 36: 1315–1319
81. Usama M, Khan MK, Alghatbar K, Lee C (2010) Chaos-based secure satellite imagery cryptosystem. Comput Math Appl 60(2):326–337
82. Vaidyanathan S (2011a) Hybrid chaos synchronization of Liu and Lü systems by active nonlinear control. Commun Comput Inf Sci 204
83. Vaidyanathan S (2011b) Output regulation of Arneodo-Coullet chaotic system. Commun Comput Inf Sci 133:98–107
84. Vaidyanathan S (2011c) Output regulation of the unified chaotic system. Commun Comput Inf Sci 198:10–17
85. Vaidyanathan S (2012a) Analysis and synchronization of the hyperchaotic Yujun systems via sliding mode control. Adv Intell Syst Comput 176:329–337
86. Vaidyanathan S (2012b) Anti-synchronization of Sprott-L and Sprott-M chaotic systems via adaptive control. Int J Control Theory Appl 5(1):41–59
87. Vaidyanathan S (2012c) Global chaos control of hyperchaotic Liu system via sliding control method. Int J Control Theory Appl 5(2):117–123
88. Vaidyanathan S (2012d) Output regulation of the Liu chaotic system. Appl Mech Mater 110–116:3982–3989
89. Vaidyanathan S (2012e) Sliding mode control based global chaos control of Liu-Liu-Liu-Su chaotic system. Int J Control Theory Appl 1(2):15–20
90. Vaidyanathan S (2013a) A new six-term 3-D chaotic system with an exponential nonlinearity. Far East J Math Sci 79(1):135–143
91. Vaidyanathan S (2013b) Analysis and adaptive synchronization of two novel chaotic systems with hyperbolic sinusoidal and cosinusoidal nonlinearity and unknown parameters. J Eng Sci Technol Rev 6(4):53–65
92. Vaidyanathan S (2013c) Analysis, control and synchronization of hyperchaotic Zhou system via adaptive control. Adv Intell Syst Comput 177:1–10
93. Vaidyanathan S, Volos C, Pham VT (2014a) Hyperchaos, adaptive control and synchronization of a novel 5-D hyperchaotic system with three positive Lyapunov exponents and its SPICE implementation. Arch Control Sci 24(4):409–446
94. Vaidyanathan S, Volos C, Pham VT, Madhavan K, Idowu BA (2014b) Adaptive backstepping control, synchronization and circuit simulation of a 3-D novel jerk chaotic system with two hyperbolic sinusoidal nonlinearities. Arch Control Sci 24(3):375–403
95. Vaidyanathan S (2014a) A new eight-term 3-D polynomial chaotic system with three quadratic nonlinearities. Far East J Math Sci 84(2):219–226
96. Vaidyanathan S (2014b) Analysis and adaptive synchronization of eight-term 3-D polynomial chaotic systems with three quadratic nonlinearities. Eur Phys J Spec Top 223(8):1519–1529
97. Vaidyanathan S (2014c) Analysis, control and synchronisation of a six-term novel chaotic system with three quadratic nonlinearities. Int J Model Identif Control 22(1):41–53
98. Vaidyanathan S (2014d) Generalized projective synchronisation of novel 3-D chaotic systems with an exponential non-linearity via active and adaptive control. Int J Model Identif Control 22(3):207–217
99. Vaidyanathan S (2014e) Global chaos synchronization of identical Li-Wu chaotic systems via sliding mode control. Int J Model Identif Control 22(2):170–177
100. Vaidyanathan S (2015b) A 3-D novel highly chaotic system with four quadratic nonlinearities, its adaptive control and anti-synchronization with unknown parameters. J Eng Sci Technol Rev 8(2):106–115
101. Vaidyanathan S (2015c) Adaptive backstepping control of enzymes-substrates system with ferroelectric behaviour in brain waves. Int J PharmTech Res 8(2):256–261
102. Vaidyanathan S (2015d) Adaptive biological control of generalized Lotka-Volterra three-species biological system. Int J PharmTech Res 8(4):622–631

103. Vaidyanathan S (2015e) Adaptive chaotic synchronization of enzymes-substrates system with ferroelectric behaviour in brain waves. Int J PharmTech Res 8(5):964–973
104. Vaidyanathan S (2015f) Adaptive control of a chemical chaotic reactor. Int J PharmTech Res 8(3):377–382
105. Vaidyanathan S (2015g) Adaptive synchronization of chemical chaotic reactors. Int J ChemTech Res 8(2):612–621
106. Vaidyanathan S (2015h) Adaptive synchronization of generalized Lotka-Volterra three-species biological systems. Int J PharmTech Res 8(5):928–937
107. Vaidyanathan S (2015i) Analysis, properties and control of an eight-term 3-D chaotic system with an exponential nonlinearity. Int J Model Identif Control 23(2):164–172
108. Vaidyanathan S (2015j) Anti-synchronization of Brusselator chemical reaction systems via adaptive control. Int J ChemTech Res 8(6):759–768
109. Vaidyanathan S (2015k) Chaos in neurons and adaptive control of Birkhoff-Shaw strange chaotic attractor. Int J PharmTech Res 8(5):956–963
110. Vaidyanathan S (2015l) Dynamics and control of Brusselator chemical reaction. Int J ChemTech Res 8(6):740–749
111. Vaidyanathan S (2015m) Dynamics and control of Tokamak system with symmetric and magnetically confined plasma. Int J ChemTech Res 8(6):795–803
112. Vaidyanathan S (2015n) Hyperchaos, qualitative analysis, control and synchronisation of a ten-term 4-D hyperchaotic system with an exponential nonlinearity and three quadratic nonlinearities. Int J Modell Identif Control 23(4):380–392
113. Vaidyanathan S (2015o) Lotka-Volterra population biology models with negative feedback and their ecological monitoring. Int J PharmTech Res 8(5):974–981
114. Vaidyanathan S (2015p) Synchronization of 3-cells cellular neural network (CNN) attractors via adaptive control method. Int J PharmTech Res 8(5):946–955
115. Vaidyanathan S (2015q) Synchronization of Tokamak systems with symmetric and magnetically confined plasma via adaptive control. Int J ChemTech Res 8(6):818–827
116. Vaidyanathan S, Azar AT, Rajagopal K, Alexander P (2015a) Design and SPICE implementation of a 12-term novel hyperchaotic system and its synchronisation via active control. Int J Model Identif Control 23(3):267–277
117. Vaidyanathan S, Rajagopal K, Volos CK, Kyprianidis IM, Stouboulos IN (2015c) Analysis, adaptive control and synchronization of a seven-term novel 3-D chaotic system with three quadratic nonlinearities and its digital implementation in LabVIEW. J Eng Sci Technol Rev 8(2):130–141
118. Vaidyanathan S, Volos C, Pham VT, Madhavan K (2015d) Analysis, adaptive control and synchronization of a novel 4-D hyperchaotic hyperjerk system and its SPICE implementation. Arch Control Sci 25(1):5–28
119. Vaidyanathan S, Volos CK, Kyprianidis IM, Stouboulos IN, Pham VT (2015e) Analysis, adaptive control and anti-synchronization of a six-term novel jerk chaotic system with two exponential nonlinearities and its circuit simulation. J Eng Sci Technol Rev 8(2):24–36
120. Vaidyanathan S, Volos CK, Pham VT (2015f) Analysis, adaptive control and adaptive synchronization of a nine-term novel 3-D chaotic system with four quadratic nonlinearities and its circuit simulation. J Eng Sci Technol Rev 8(2):174–184
121. Vaidyanathan S, Volos CK, Pham VT (2015g) Analysis, control, synchronization and SPICE implementation of a novel 4-D hyperchaotic Rikitake dynamo System without equilibrium. J Eng Sci Technol Rev 8(2):232–244
122. Vaidyanathan S, Volos CK, Rajagopal K, Kyprianidis IM, Stouboulos IN (2015i) Adaptive backstepping controller design for the anti-synchronization of identical WINDMI chaotic systems with unknown parameters and its SPICE implementation. J Eng Sci Technol Rev 8(2):74–82
123. Vaidyanathan S, Pham VT, Volos CK (2015) A 5-D hyperchaotic Rikitake dynamo system with hidden attractors. Eur Phys J Spec Top 224(8):1575–1592
124. Vaidyanathan S (2015a) 3-cells cellular neural network (CNN) attractor and its adaptive biological control. Int J PharmTech Res 8(4):632–640

125. Vaidyanathan S, Azar AT (2015a) Analysis and control of a 4-D novel hyperchaotic system. In: Azar AT, Vaidyanathan S (eds) Chaos Modeling and Control Systems Design. Studies in computational intelligence, vol 581. Springer, Germany, pp 19–38
126. Vaidyanathan S, Azar AT (2015b) Analysis, control and synchronization of a nine-term 3-D novel chaotic system. In: Azar AT, Vaidyanathan S (eds) Chaos Modelling and Control Systems Design. Studies in computational intelligence, vol 581. Springer, Germany, pp 19–38
127. Vaidyanathan S, Azar AT (2015c) Anti-synchronization of identical chaotic systems using sliding mode control and an application to Vaidhyanathan-Madhavan chaotic systems. Studies in computational intelligence vol 576. pp 527–547
128. Vaidyanathan S, Azar AT (2015d) Hybrid synchronization of identical chaotic systems using sliding mode control and an application to Vaidhyanathan chaotic systems. Studies in computational intelligence, vol 576. pp 549–569
129. Vaidyanathan S, Idowu BA, Azar AT (2015b) Backstepping controller design for the global chaos synchronization of Sprott's jerk systems. Studies in computational intelligence, vol 581. pp. 39–58
130. Vaidyanathan S, Madhavan K (2013) Analysis, adaptive control and synchronization of a seven-term novel 3-D chaotic system. Int J Control Theory Appl 6(2):121–137
131. Vaidyanathan S, Pakiriswamy S (2013) Generalized projective synchronization of six-term Sundarapandian chaotic systems by adaptive control. Int J Control Theory Appl 6(2):153–163
132. Vaidyanathan S, Pakiriswamy S (2015) A 3-D novel conservative chaotic system and its generalized projective synchronization via adaptive control. J Eng Sci Technol Rev 8(2): 52–60
133. Vaidyanathan S, Rajagopal K (2011a) Anti-synchronization of Li and T chaotic systems by active nonlinear control. Commun Comput Inf Sci 198:175–184
134. Vaidyanathan S, Rajagopal K (2011b) Global chaos synchronization of hyperchaotic Pang and Wang systems by active nonlinear control. Commun Comput Inf Sci 204:84–93
135. Vaidyanathan S, Rajagopal K (2011c) Global chaos synchronization of Lü and Pan systems by adaptive nonlinear control. Commun Comput Inf Sci 205:193–202
136. Vaidyanathan S, Rajagopal K (2012) Global chaos synchronization of hyperchaotic Pang and hyperchaotic Wang systems via adaptive control. Int J Soft Comput 7(1):28–37
137. Vaidyanathan S, Rasappan S (2011) Global chaos synchronization of hyperchaotic Bao and Xu systems by active nonlinear control. Commun Comput Inf Sci 198:10–17
138. Vaidyanathan S, Rasappan S (2014) Global chaos synchronization of n-scroll Chua circuit and Lur'e system using backstepping control design with recursive feedback. Arab J Sci Eng 39(4):3351–3364
139. Vaidyanathan S, Sampath S (2011) Global chaos synchronization of hyperchaotic Lorenz systems by sliding mode control. Commun Comput Inf Sci 205:156–164
140. Vaidyanathan S, Sampath S (2012) Anti-synchronization of four-wing chaotic systems via sliding mode control. Int J Autom Comput 9(3):274–279
141. Vaidyanathan S, Volos CK, Pham VT (2015h) Global chaos control of a novel nine-term chaotic system via sliding mode control. In: Azar AT, Zhu Q (eds) Advances and Applications in Sliding Mode Control Systems. Studies in computational intelligence, vol 576. Springer, Germany, pp 571–590
142. Vaidyanathan S, Volos C (2015) Analysis and adaptive control of a novel 3-D conservative no-equilibrium chaotic system. Arch Control Sci 25(3):333–353
143. Volos CK, Kyprianidis IM, Stouboulos IN (2013) Experimental investigation on coverage performance of a chaotic autonomous mobile robot. Robot Auton Syst 61(12):1314–1322
144. Volos CK, Kyprianidis IM, Stouboulos IN, Tlelo-Cuautle E, Vaidyanathan S (2015) Memristor: A new concept in synchronization of coupled neuromorphic circuits. J Eng Sci Technol Rev 8(2):157–173
145. Wei Z, Yang Q (2010) Anti-control of Hopf bifurcation in the new chaotic system with two stable node-foci. Appl Math Comput 217(1):422–429

146. Witte CL, Witte MH (1991) Chaos and predicting varix hemorrhage. Med Hypotheses 36(4):312–317
147. Xiao X, Zhou L, Zhang Z (2014) Synchronization of chaotic Lure systems with quantized sampled-data controller. Commun Nonlinear Sci Numer Simul 19(6):2039–2047
148. Yuan G, Zhang X, Wang Z (2014) Generation and synchronization of feedback-induced chaos in semiconductor ring lasers by injection-locking. Optik Int J Light Electron Opt 125(8): 1950–1953
149. Zaher AA, Abu-Rezq A (2011) On the design of chaos-based secure communication systems. Commun Nonlinear Syst Numer Simul 16(9):3721–3727
150. Zhang X, Zhao Z, Wang J (2014) Chaotic image encryption based on circular substitution box and key stream buffer. Signal Process Image Commun 29(8):902–913
151. Zhang H, Zhou J (2012) Synchronization of sampled-data coupled harmonic oscillators with control inputs missing. Syst Control Lett 61(12):1277–1285
152. Zhou W, Xu Y, Lu H, Pan L (2008) On dynamics analysis of a new chaotic attractor. Phys Lett A 372(36):5773–5777
153. Zhu C, Liu Y, Guo Y (2010) Theoretic and numerical study of a new chaotic system. Intell Inf Manag 2:104–109
154. Zhu Q, Azar AT (2015) Complex System Modelling and Control through Intelligent Soft Computations. Studies in fuzzines and soft computing, vol 319, Springer, Germany

A Seven-Term Novel 3-D Jerk Chaotic System with Two Quadratic Nonlinearities and Its Adaptive Backstepping Control

Sundarapandian Vaidyanathan

Abstract In this research work, we announce a seven-term novel 3-D jerk chaotic system with two quadratic nonlinearities. First, we discuss the qualitative properties of the novel 3-D jerk chaotic system. We show that the novel jerk chaotic system has two unstable equilibrium points on the x_1-axis. We establish that the novel jerk chaotic system is dissipative. Next, we obtain the Lyapunov exponents of the novel jerk chaotic system as $L_1 = 0.11184$, $L_2 = 0$ and $L_3 = -0.61241$. Also, we derive the Kaplan-Yorke dimension of the novel jerk chaotic system as $D_{KY} = 2.18262$. Next, we design an adaptive backstepping controller to stabilize the novel jerk chaotic system with unknown system parameters. We also design an adaptive backstepping controller to achieve complete chaos synchronization of the identical novel jerk chaotic systems with unknown system parameters. The main control results are established using Lyapunov stability theory. The backstepping control method is a recursive procedure that links the choice of a Lyapunov function with the design of a controller and guarantees global asymptotic stability of strict feedback systems. MATLAB simulations are shown to illustrate all the main results developed in this work.

Keywords Chaos · Chaotic systems · Jerk systems · Backstepping control · Adaptive control · Synchronization

1 Introduction

Chaotic systems are defined as nonlinear dynamical systems which are sensitive to initial conditions, topologically mixing and with dense periodic orbits. Sensitivity to initial conditions of chaotic systems is popularly known as the *butterfly effect*.

S. Vaidyanathan (✉)
Research and Development Centre, Vel Tech University,
Avadi, Chennai 600062, Tamil Nadu, India
e-mail: sundarvtu@gmail.com

A.T. Azar and S. Vaidyanathan (eds.), *Advances in Chaos Theory and Intelligent Control*, Studies in Fuzziness and Soft Computing 337,
DOI 10.1007/978-3-319-30340-6_24

The Lyapunov exponent is a measure of the divergence of phase points that are initially very close and can be used to quantify chaotic systems. It is common to refer to the largest Lyapunov exponent as the *Maximal Lyapunov Exponent* (MLE). A positive maximal Lyapunov exponent and phase space compactness are usually taken as defining conditions for a chaotic system.

Some classical paradigms of 3-D chaotic systems in the literature are Lorenz system [38], Rössler system [56], ACT system [1], Sprott systems [66], Chen system [15], Lü system [39], Liu system [37], Cai system [13], Chen-Lee system [16], Tigan system [80], etc.

Many new chaotic systems have been discovered in the recent years such as Zhou system [151], Zhu system [152], Li system [32], Wei-Yang system [144], Sundarapandian systems [72, 77], Vaidyanathan systems [90, 91, 93–96, 99, 106, 116, 119, 121, 130, 132, 135, 137, 138, 140], Pehlivan system [45], Sampath system [58], Pham system [47], etc.

Chaos theory and control systems have many important applications in science and engineering [2, 9–12, 153]. Some commonly known applications are oscillators [28, 65], lasers [34, 147], chemical reactions [21, 46, 103, 104, 107, 109, 110, 114], biology [18, 30, 98, 100–102, 105, 108, 112, 113], ecology [22, 70], encryption [31, 150], cryptosystems [55, 81], mechanical systems [4–8], secure communications [19, 41, 148], robotics [40, 42, 142], cardiology [49, 145], intelligent control [3, 35], neural networks [24, 27, 36], finance [23, 68], memristors [48, 143], etc.

Control of chaotic systems aims to regulate or stabilize the chaotic system about its equilibrium points. There are many methods available in the literature such as active control method [71, 83, 84, 133], adaptive control method [43, 139], sliding mode control method [87, 89], backstepping control method [141], etc.

Synchronization of chaotic systems is a phenomenon that occurs when two or more chaotic systems are coupled or when a chaotic system drives another chaotic system. Major works on synchronization of chaotic systems deal with the complete synchronization of a pair of chaotic systems called the *master* and *slave* systems. The design goal of the complete synchronization is to apply the output of the master system to control the slave system so that the output of the slave system tracks the output of the master system asymptotically with time.

Pecora and Carroll pioneered the research on synchronization of chaotic systems with their seminal papers [14, 44]. The active control method [26, 59, 60, 71, 76, 82, 88, 122, 123, 126] is typically used when the system parameters are available for measurement. Adaptive control method [57, 61–63, 73–75, 86, 92, 111, 115, 120, 124, 125, 131, 136] is typically used when some or all the system parameters are not available for measurement and estimates for the uncertain parameters of the systems. Sampled-data feedback control method [20, 33, 146, 149] and time-delay feedback control method [17, 25, 64] are also used for synchronization of chaotic systems. Backstepping control method [50–54, 79, 127, 134] is also used for the synchronization of chaotic systems. Another popular method for the synchronization of chaotic systems is the sliding mode control method [78, 85, 97, 117, 118, 128, 129].

In the recent decades, there is some good interest in finding novel chaotic systems, which are of the jerk type [69]. Jerk chaotic systems are chaotic systems that can be expressed by an explicit third-order differential equation describing the time evolution of the single scalar variable x given by

$$\dddot{x} = j\left(x, \dot{x}, \ddot{x}\right). \tag{1}$$

The differential Eq. (1) is called *jerk system*, because the third order time derivative in mechanical systems is called *jerk*. Thus, in order to study different aspects of chaos, the ODE (1) can be considered instead of a 3-D system.

The simplest known jerk system was discovered by Sprott [67] and Sprott's jerk function contains just three terms with a quadratic nonlinearity:

$$j\left(x, \dot{x}, \ddot{x}\right) = -A\ddot{x} + \dot{x}^2 - x \quad (\text{with } A = 2.017) \tag{2}$$

In this research work, we consider jerk systems described by 3-D systems of differential equations.

Defining the phase variables

$$x_1 = x, \quad x_2 = \dot{x}, \quad x_3 = \ddot{x}, \tag{3}$$

we can rewrite the jerk differential Eq. (1) in system form as

$$\begin{cases} \dot{x}_1 = x_2 \\ \dot{x}_2 = x_3 \\ \dot{x}_3 = j(x_1, x_2, x_3) \end{cases} \tag{4}$$

In this research work, we propose a seven-term novel jerk chaotic system with two quadratic nonlinearities. We also derive new results for the adaptive control and synchronization of the novel jerk chaotic system with unknown parameters via backstepping control. The backstepping control method is a recursive procedure that links the choice of a Lyapunov function with the design of a controller and guarantees global asymptotic stability of strict feedback systems.

We obtain the Lyapunov exponents of the novel jerk system as $L_1 = 0.11184$, $L_2 = 0$ and $L_3 = -0.61241$. Also, we derive the Kaplan-Yorke dimension of the novel jerk chaotic system as $D_{KY} = 2.18262$.

Section 2 describes the dynamic equations and phase portraits of the novel jerk chaotic system. Section 3 describes the qualitative properties of the novel jerk chaotic system. Section 4 describes the adaptive control of the novel jerk chaotic system with unknown parameters. Section 5 describes the adaptive synchronization of the identical novel jerk chaotic system with unknown parameters. Finally, Sect. 6 contains the conclusions of this work.

2 A Novel 3-D Jerk Chaotic System

In this section, we announce a novel 3-D jerk chaotic system with two quadratic nonlinearities, which is described by the 3-D dynamics

$$\begin{cases} \dot{x}_1 = x_2 \\ \dot{x}_2 = x_3 \\ \dot{x}_3 = -x_1 - x_2 - ax_3 - bx_1^2 - cx_2^2 \end{cases} \tag{5}$$

where x_1, x_2, x_3 are the states and a, b, c are constant, positive, parameters.

The jerk system (5) is *chaotic* when the parameter values are taken as

$$a = 0.5, \quad b = 0.15, \quad c = 0.2 \tag{6}$$

For numerical simulations, we take the initial values of the novel jerk chaotic system (5) as

$$x_1(0) = 0.4, \quad x_2(0) = 0.8, \quad x_3(0) = 0.6 \tag{7}$$

Figure 1 describes the *strange chaotic attractor* of the novel jerk chaotic system (5) in $\mathbb{R}^3$.

Figures 2, 3 and 4 describe the 2-D projection of the strange chaotic attractor of the novel jerk chaotic system (5) on (x_1, x_2), (x_2, x_3) and (x_1, x_3) coordinate planes respectively.

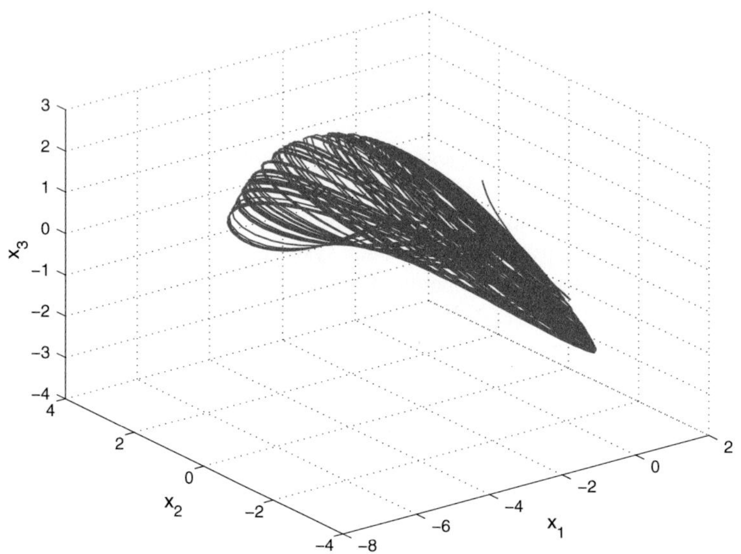

Fig. 1 Strange attractor of the novel jerk chaotic system in $\mathbf{R}^3$

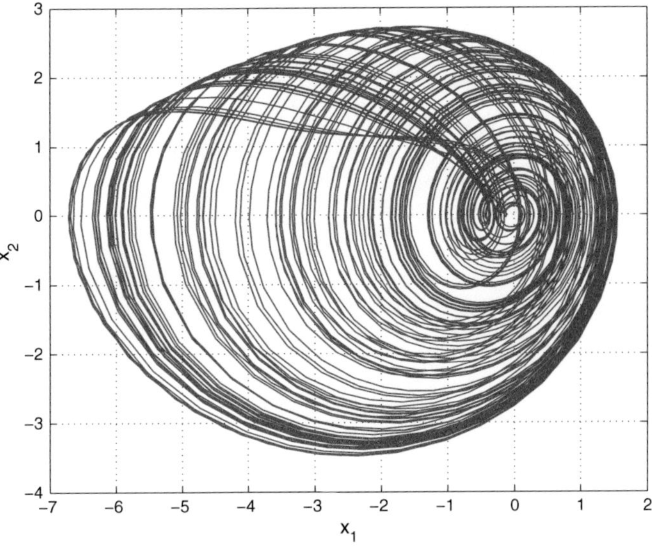

Fig. 2 2-D projection of the novel jerk chaotic system on (x_1, x_2)-plane

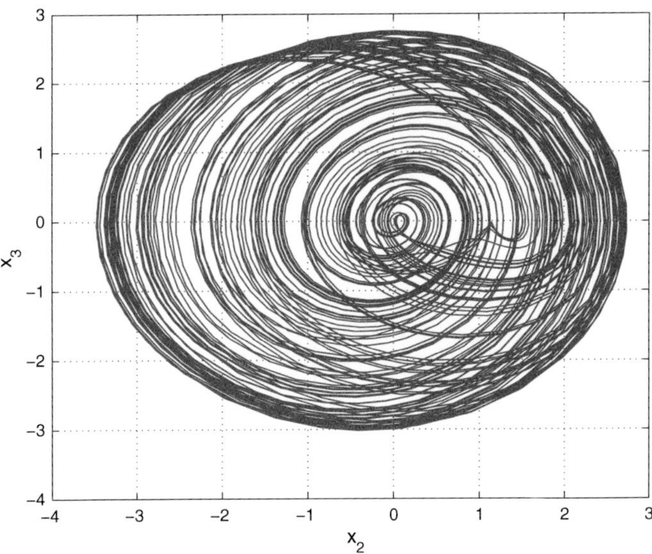

Fig. 3 2-D projection of the novel jerk chaotic system on (x_2, x_3)-plane

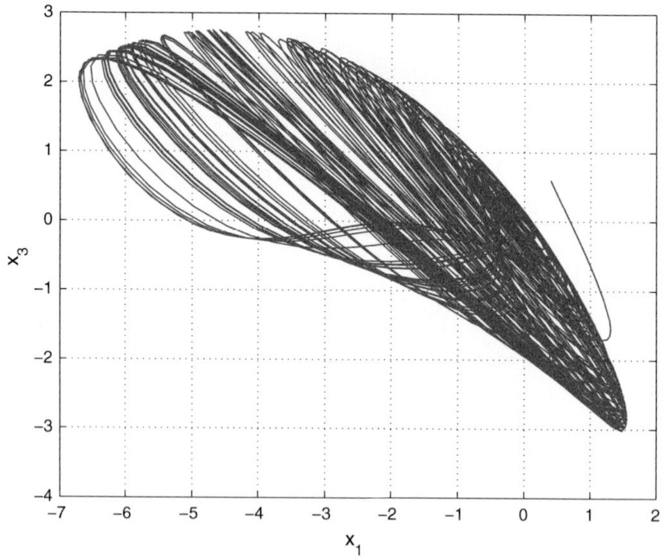

Fig. 4 2-D projection of the novel jerk chaotic system on (x_1, x_3)-plane

3 Properties of the Novel Jerk Chaotic System

This section describes the qualitative properties of the novel jerk chaotic system (5).

3.1 Dissipativity

We write the novel jerk chaotic system (5) in vector notation as

$$\dot{x} = f(x) = \begin{bmatrix} f_1(x) \\ f_2(x) \\ f_3(x) \end{bmatrix}, \tag{8}$$

where

$$\begin{aligned} f_1(x) &= x_2 \\ f_2(x) &= x_3 \\ f_3(x) &= -x_1 - x_2 - ax_3 - bx_1^2 - cx_2^2 \end{aligned} \tag{9}$$

We take the parameter values as in the chaotic case, *viz.* $a = 0.5$, $b = 0.15$ and $c = 0.2$.

The divergence of the vector field f on $\mathbb{R}^3$ is obtained as

$$\text{div } f = \frac{\partial f_1(x)}{\partial x_1} + \frac{\partial f_2(x)}{\partial x_2} + \frac{\partial f_3(x)}{\partial x_3} = 0 + 0 - a = -0.5 < 0 \tag{10}$$

Let Ω be any region in $\mathbb{R}^3$ with a smooth boundary. Let $\Omega(t) = \Phi_t(\Omega)$, where Φ_t is the flow of the vector field f.

Let $V(t)$ denote the volume of $\Omega(t)$.

By Liouville's theorem, it follows that

$$\frac{dV(t)}{dt} = \int_{\Omega(t)} (\text{div} f) dx_1\, dx_2\, dx_3 \tag{11}$$

Substituting the value of $\text{div} f$ in (11) leads to

$$\frac{dV(t)}{dt} = -\int_{\Omega(t)} (0.5) dx_1\, dx_2\, dx_3 = -0.5V(t) \tag{12}$$

Integrating the linear differential Eq. (12), $V(t)$ is obtained as

$$V(t) = V(0)\exp(-0.5t) \tag{13}$$

From Eq. (13), it follows that the volume $V(t)$ shrinks to zero exponentially as $t \to \infty$.

Thus, the novel jerk chaotic system (5) is dissipative. Hence, any asymptotic motion of the system (5) settles onto a set of measure zero, *i.e.* a strange attractor.

3.2 Equilibrium Point

The equilibrium points of the novel jerk chaotic system (5) are obtained by solving the system of equations

$$\begin{cases} x_2 = 0 \\ x_3 = 0 \\ -x_1 - x_2 - ax_3 - bx_1^2 - cx_2^2 = 0 \end{cases} \tag{14}$$

Solving the system (14), we obtain two equilibrium points of the novel jerk chaotic system (5) as

$$E_0 = \begin{bmatrix} 0 \\ 0 \\ 0 \end{bmatrix} \text{ and } E_1 = \begin{bmatrix} -6.6667 \\ 0 \\ 0 \end{bmatrix} \tag{15}$$

The Jacobian matrix of the novel jerk chaotic system (5) at any point $\mathbf{x} \in \mathbb{R}^3$ is obtained as

$$J(\mathbf{x}) = \begin{bmatrix} 0 & 1 & 0 \\ 0 & 0 & 1 \\ -1 - 0.3x_1 & -1 - 0.4x_2 & -0.5 \end{bmatrix} \tag{16}$$

Thus, it follows that

$$J_0 = J(E_0) = \begin{bmatrix} 0 & 1 & 0 \\ 0 & 0 & 1 \\ -1 & -1 & -0.5 \end{bmatrix} \tag{17}$$

The matrix J_0 has the eigenvalues

$$\lambda_1 = -0.8038, \quad \lambda_{2,3} = 0.1519 \pm 1.1050i \tag{18}$$

This shows that the equilibrium point E_0 is a saddle-focus, which is unstable. We also find that

$$J_1 = J(E_1) = \begin{bmatrix} 0 & 1 & 0 \\ 0 & 0 & 1 \\ 1 & -1 & -0.5 \end{bmatrix} \tag{19}$$

The matrix J_1 has the eigenvalues

$$\lambda_1 = 0.6015, \quad \lambda_{2,3} = -0.5507 \pm 1.1659i \tag{20}$$

This shows that the equilibrium point E_1 is a saddle-focus, which is unstable.

3.3 Lyapunov Exponents and Kaplan-Yorke Dimension

We take the parameter values of the novel jerk chaotic system (5) as in the chaotic case (6), *i.e.*

$$a = 0.5, \quad b = 0.15, \quad c = 0.2 \tag{21}$$

We take the initial values of the novel jerk chaotic system (5) as in (7), *i.e.*

$$x_1(0) = 0.4, \quad x_2(0) = 0.8, \quad x_3(0) = 0.6 \tag{22}$$

Then the Lyapunov exponents of the novel jerk chaotic system (5) are numerically obtained as

$$L_1 = 0.11184, \quad L_2 = 0, \quad L_3 = -0.61241 \tag{23}$$

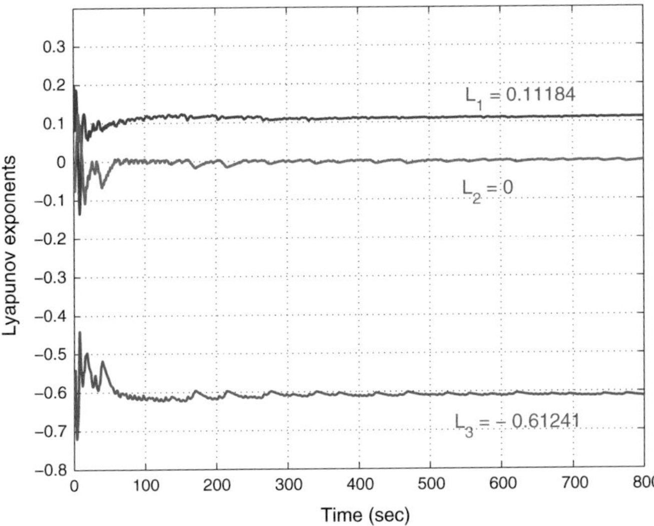

Fig. 5 Lyapunov exponents of the novel jerk chaotic system

Since $L_1 + L_2 + L_3 = -0.50057 < 0$, the novel jerk chaotic system (5) is dissipative.

Figure 5 shows the dynamics of the Lyapunov exponents of the novel jerk chaotic system (5). From Fig. 5, the Maximal Lyapunov Exponent (MLE) of the novel jerk chaotic system (5) is obtained as $L_1 = 0.11184$, which is a small value. Thus, the novel jerk chaotic system (5) is *mildly chaotic*.

Also, the Kaplan-Yorke dimension of the novel jerk chaotic system (5) is obtained as

$$D_{KY} = 2 + \frac{L_1 + L_2}{|L_3|} = 2.18262 \tag{24}$$

which is fractional.

4 Adaptive Control of Novel Jerk Chaotic System

In this section, we use backstepping control method to derive an adaptive control law for globally stabilizing the 3-D novel jerk chaotic system with unknown parameters.

Thus, we consider the 3-D novel jerk chaotic system given by

$$\begin{cases} \dot{x}_1 = x_2 \\ \dot{x}_2 = x_3 \\ \dot{x}_3 = -x_1 - x_2 - ax_3 - bx_1^2 - cx_2^2 + u \end{cases} \tag{25}$$

where a, b and c are unknown constant parameters, and u is a backstepping control law to be determined using estimates $\hat{a}(t)$, $\hat{b}(t)$ and $\hat{c}(t)$ for a, b and c, respectively.

The parameter estimation errors are defined as:

$$\begin{cases} e_a(t) = a - \hat{a}(t) \\ e_b(t) = b - \hat{b}(t) \\ e_c(t) = c - \hat{c}(t) \end{cases} \tag{26}$$

Differentiating (26) with respect to t, we obtain the following equations:

$$\begin{cases} \dot{e}_a(t) = -\dot{\hat{a}}(t) \\ \dot{e}_b(t) = -\dot{\hat{b}}(t) \\ \dot{e}_c(t) = -\dot{\hat{c}}(t) \end{cases} \tag{27}$$

Theorem 1 *The 3-D novel jerk chaotic system (25), with unknown parameters a and b, is globally and exponentially stabilized by the adaptive feedback control law,*

$$u(t) = -2x_1 - 4x_2 - [3 - \hat{a}(t)]x_3 + \hat{b}(t)x_1^2 + \hat{c}(t)x_2^2 - kz_3 \tag{28}$$

where $k > 0$ is a gain constant,

$$z_3 = 2x_1 + 2x_2 + x_3, \tag{29}$$

and the update law for the parameter estimates $\hat{a}(t)$, $\hat{b}(t)$, $\hat{c}(t)$ is given by

$$\begin{cases} \dot{\hat{a}}(t) = -x_3 z_3 \\ \dot{\hat{b}}(t) = -x_1^2 z_3 \\ \dot{\hat{c}}(t) = -x_2^2 z_3 \end{cases} \tag{30}$$

Proof We prove this result via backstepping control method and Lyapunov stability theory [29].

First, we define a quadratic Lyapunov function

$$V_1(z_1) = \frac{1}{2}z_1^2 \tag{31}$$

where

$$z_1 = x_1 \tag{32}$$

Differentiating V_1 along the dynamics (25), we get

$$\dot{V}_1 = z_1\dot{z}_1 = x_1x_2 = -z_1^2 + z_1(x_1 + x_2) \tag{33}$$

Now, we define

$$z_2 = x_1 + x_2 \tag{34}$$

Using (34), we can simplify the Eq. (33) as

$$\dot{V}_1 = -z_1^2 + z_1 z_2 \tag{35}$$

Secondly, we define a quadratic Lyapunov function

$$V_2(z_1, z_2) = V_1(z_1) + \frac{1}{2}\, z_2^2 = \frac{1}{2}\left(z_1^2 + z_2^2\right) \tag{36}$$

Differentiating V_2 along the dynamics (25), we get

$$\dot{V}_2 = -z_1^2 - z_2^2 + z_2(2x_1 + 2x_2 + x_3) \tag{37}$$

Now, we define

$$z_3 = 2x_1 + 2x_2 + x_3 \tag{38}$$

Using (38), we can simplify the Eq. (37) as

$$\dot{V}_2 = -z_1^2 - z_2^2 + z_2 z_3 \tag{39}$$

Finally, we define a quadratic Lyapunov function

$$V(z_1, z_2, z_3, e_a, e_b, e_c) = V_2(z_1, z_2) + \frac{1}{2}z_3^2 + \frac{1}{2}e_a^2 + \frac{1}{2}e_b^2 + \frac{1}{2}e_c^2 \tag{40}$$

which is a positive definite function on $\mathbb{R}^6$.

Differentiating V along the dynamics (25) and (27), we get

$$\dot{V} = -z_1^2 - z_2^2 - z_3^2 + z_3(z_3 + z_2 + \dot{z}_3) - e_a\dot{\hat{a}} - e_b\dot{\hat{b}} - e_c\dot{\hat{c}} \tag{41}$$

Equation (41) can be written compactly as

$$\dot{V} = -z_1^2 - z_2^2 - z_3^2 + z_3 S - e_a\dot{\hat{a}} - e_b\dot{\hat{b}} - e_c\dot{\hat{c}} \tag{42}$$

where

$$S = z_3 + z_2 + \dot{z}_3 = z_3 + z_2 + 2\dot{x}_1 + 2\dot{x}_2 + \dot{x}_3 \tag{43}$$

A simple calculation gives

$$S = 2x_1 + 4x_2 + (3 - a)x_3 - bx_1^2 - cx_2^2 + u \tag{44}$$

Substituting the adaptive control law (28) into (44), we obtain

$$S = -[a - \hat{a}(t)]x_3 - [b - \hat{b}(t)]x_1^2 - [c - \hat{c}(t)]x_2^2 - kz_3 \tag{45}$$

Using the definitions (27), we can simplify (45) as

$$S = -e_a x_3 - e_b x_1^2 - e_c x_2^2 - kz_3 \tag{46}$$

Substituting the value of S from (46) into (42), we obtain

$$\dot{V} = -z_1^2 - z_2^2 - (1+k)z_3^2 + e_a\left[-x_3 z_3 - \dot{\hat{a}}\right] + e_b\left[-x_1^2 z_3 - \dot{\hat{b}}\right] + e_c\left[-x_1^2 z_3 - \dot{\hat{c}}\right] \tag{47}$$

Substituting the update law (30) into (47), we get

$$\dot{V} = -z_1^2 - z_2^2 - (1+k)z_3^2, \tag{48}$$

which is a negative semi-definite function on $\mathbb{R}^6$.

From (48), it follows that the vector $\mathbf{z}(t) = (z_1(t), z_2(t), z_3(t))$ and the parameter estimation error $(e_a(t), e_b(t), e_c(t))$ are globally bounded, i.e.

$$\left[\, z_1(t)\ z_2(t)\ z_3(t)\ e_a(t)\ e_b(t)\ e_c(t) \,\right] \in \mathbf{L}_\infty \tag{49}$$

Also, it follows from (48) that

$$\dot{V} \leq -z_1^2 - z_2^2 - z_3^2 = -\|\mathbf{z}\|^2 \tag{50}$$

That is,

$$\|\mathbf{z}\|^2 \leq -\dot{V} \tag{51}$$

Integrating the inequality (51) from 0 to t, we get

$$\int_0^t |\mathbf{z}(\tau)|^2\, d\tau \leq V(0) - V(t) \tag{52}$$

From (52), it follows that $\mathbf{z}(t) \in \mathbf{L}_2$.

From Eq. (25), it can be deduced that $\dot{\mathbf{z}}(t) \in \text{Ł}_\infty$.

Thus, using Barbalat's lemma, we conclude that $\mathbf{z}(t) \to \mathbf{0}$ exponentially as $t \to \infty$ for all initial conditions $\mathbf{z}(0) \in \mathbb{R}^3$.

Hence, it is immediate that $\mathbf{x}(t) \to \mathbf{0}$ exponentially as $t \to \infty$ for all initial conditions $\mathbf{x}(0) \in \mathbb{R}^3$.

This completes the proof. □

For the numerical simulations, the classical fourth-order Runge-Kutta method with step size $h = 10^{-8}$ is used to solve the system of differential Eqs. (25) and (30), when the adaptive control law (28) is applied.

The parameter values of the novel jerk chaotic system (25) are taken as in the chaotic case, *i.e.* $a = 0.5$, $b = 0.15$ and $c = 0.2$.

We take the positive gain constant as $k = 10$.

Furthermore, as initial conditions of the novel jerk chaotic system (25), we take

$$x_1(0) = 8.2, \quad x_2(0) = -4.7, \quad x_3(0) = -3.6 \tag{53}$$

Also, as initial conditions of the parameter estimates $\hat{a}(t)$, $\hat{b}(t)$ and $\hat{c}(t)$, we take

$$\hat{a}(0) = 7.5, \quad \hat{b}(0) = 3.4, \quad \hat{c}(0) = 12.3 \tag{54}$$

In Fig. 6, the exponential convergence of the controlled states $x_1(t), x_2(t), x_3(t)$ is depicted, when the adaptive control law (28) and (30) are implemented.

From Fig. 6, it is clear that the controlled states $x_1(t), x_2(t), x_3(t)$ converge to zero exponentially in six seconds. This shows the efficiency of the adaptive backstepping controller derived in this section for the novel jerk chaotic system (25).

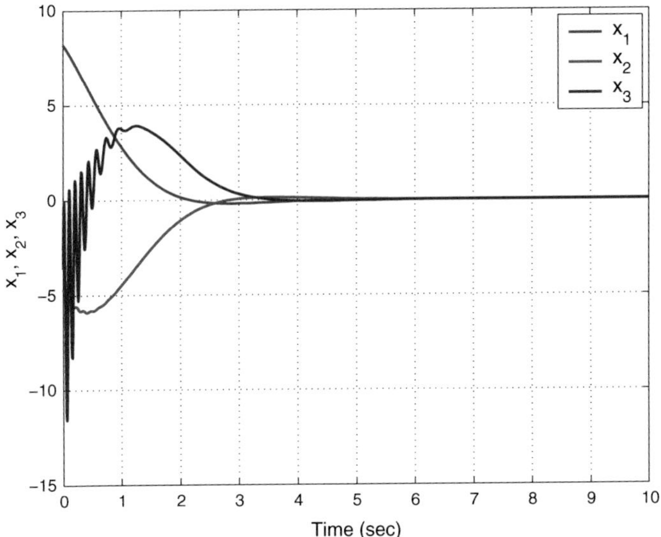

Fig. 6 Time-history of the controlled states $x_1(t), x_2(t), x_3(t)$

5 Adaptive Synchronization of Identical Novel Jerk Chaotic Systems

In this section, we use backstepping control method to derive an adaptive control law for globally and exponentially synchronizing the identical 3-D novel jerk chaotic systems with unknown parameters.

As the master system, we consider the 3-D novel jerk chaotic system given by

$$\begin{cases} \dot{x}_1 = x_2 \\ \dot{x}_2 = x_3 \\ \dot{x}_3 = -x_1 - x_2 - ax_3 - bx_1^2 - cx_2^2 \end{cases} \tag{55}$$

where x_1, x_2, x_3 are the states of the system, and a, b and c are unknown constant parameters.

As the slave system, we consider the 3-D novel jerk chaotic system given by

$$\begin{cases} \dot{y}_1 = y_2 \\ \dot{y}_2 = y_3 \\ \dot{y}_3 = -y_1 - y_2 - ay_3 - by_1^2 - cy_2^2 + u \end{cases} \tag{56}$$

where y_1, y_2, y_3 are the states of the system, and u is a backstepping control to be determined using estimates $\hat{a}(t), \hat{b}(t)$ and $\hat{c}(t)$ for a, b and c, respectively.

We define the synchronization errors between the states of the master system (55) and the slave system (56) as

$$\begin{cases} e_1 = y_1 - x_1 \\ e_2 = y_2 - x_2 \\ e_3 = y_3 - x_3 \end{cases} \tag{57}$$

Then the error dynamics is easily obtained as

$$\begin{cases} \dot{e}_1 = e_2 \\ \dot{e}_2 = e_3 \\ \dot{e}_3 = -e_1 - e_2 - ae_3 - b(y_1^2 - x_1^2) - c(y_2^2 - x_2^2) + u \end{cases} \tag{58}$$

The parameter estimation errors are defined as:

$$\begin{cases} e_a(t) = a - \hat{a}(t) \\ e_b(t) = b - \hat{b}(t) \\ e_c(t) = c - \hat{c}(t) \end{cases} \tag{59}$$

Differentiating (59) with respect to t, we obtain the following equations:

$$\begin{cases} \dot{e}_a(t) = -\dot{\hat{a}}(t) \\ \dot{e}_b(t) = -\dot{\hat{b}}(t) \\ \dot{e}_c(t) = -\dot{\hat{c}}(t) \end{cases} \tag{60}$$

Next, we shall state and prove the main result of this section.

Theorem 2 *The identical 3-D novel jerk chaotic systems (55) and (56) with unknown parameters a, b and c are globally and exponentially synchronized by the adaptive control law*

$$u(t) = -2e_1 - 4e_2 - \left[3 - \hat{a}(t)\right] e_3 + \hat{b}(t)\left(y_1^2 - x_1^2\right) + \hat{c}(t)\left(y_2^2 - x_2^2\right) - kz_3(t) \tag{61}$$

where $k > 0$ is a gain constant,

$$z_3 = 2e_1 + 2e_2 + e_3, \tag{62}$$

and the update law for the parameter estimates $\hat{a}(t), \hat{b}(t), \hat{c}(t)$ is given by

$$\begin{cases} \dot{\hat{a}}(t) = -e_3 z_3 \\ \dot{\hat{b}}(t) = -(y_1^2 - x_1^2) z_3 \\ \dot{\hat{c}}(t) = -(y_2^2 - x_2^2) z_3 \end{cases} \tag{63}$$

Proof We prove this result via backstepping control method and Lyapunov stability theory [29].

First, we define a quadratic Lyapunov function

$$V_1(z_1) = \frac{1}{2} z_1^2 \tag{64}$$

where

$$z_1 = e_1 \tag{65}$$

Differentiating V_1 along the error dynamics (58), we get

$$\dot{V}_1 = z_1 \dot{z}_1 = e_1 e_2 = -z_1^2 + z_1(e_1 + e_2) \tag{66}$$

Now, we define

$$z_2 = e_1 + e_2 \tag{67}$$

Using (67), we can simplify the equation (66) as

$$\dot{V}_1 = -z_1^2 + z_1 z_2 \tag{68}$$

Secondly, we define a quadratic Lyapunov function

$$V_2(z_1, z_2) = V_1(z_1) + \frac{1}{2} z_2^2 = \frac{1}{2} \left(z_1^2 + z_2^2\right) \tag{69}$$

Differentiating V_2 along the error dynamics (58), we get

$$\dot{V}_2 = -z_1^2 - z_2^2 + z_2(2e_1 + 2e_2 + e_3) \tag{70}$$

Now, we define

$$z_3 = 2e_1 + 2e_2 + e_3 \tag{71}$$

Using (71), we can simplify the equation (70) as

$$\dot{V}_2 = -z_1^2 - z_2^2 + z_2 z_3 \tag{72}$$

Finally, we define a quadratic Lyapunov function

$$V(z_1, z_2, z_3, e_a, e_b, e_c) = V_2(z_1, z_2) + \frac{1}{2} z_3^2 + \frac{1}{2} e_a^2 + \frac{1}{2} e_b^2 + \frac{1}{2} e_c^2 \tag{73}$$

Differentiating V along the error dynamics (58), we get

$$\dot{V} = -z_1^2 - z_2^2 - z_3^2 + z_3(z_3 + z_2 + \dot{z}_3) - e_a \dot{\hat{a}} - e_b \dot{\hat{b}} - e_c \dot{\hat{c}} \tag{74}$$

Equation (74) can be written compactly as

$$\dot{V} = -z_1^2 - z_2^2 - z_3^2 + z_3 S - e_a \dot{\hat{a}} - e_b \dot{\hat{b}} - e_c \dot{\hat{c}} \tag{75}$$

where

$$S = z_3 + z_2 + \dot{z}_3 = z_3 + z_2 + 2\dot{e}_1 + 2\dot{e}_2 + \dot{e}_3 \tag{76}$$

A simple calculation gives

$$S = 2e_1 + 4e_2 + (3 - a)e_3 - b(y_1^2 - x_1^2) - c(y_2^2 - x_2^2) + u \tag{77}$$

Substituting the adaptive control law (61) into (77), we obtain

$$S = -\left[a - \hat{a}(t)\right] e_3 - \left[b - \hat{b}(t)\right] (y_1^2 - x_1^2) - \left[c - \hat{c}(t)\right] (y_2^2 - x_2^2) - k z_3 \tag{78}$$

Using the definitions (60), we can simplify (78) as

$$S = -e_a e_3 - e_b(y_1^2 - x_1^2) - e_c(y_2^2 - x_2^2) - k z_3 \tag{79}$$

Substituting the value of S from (79) into (75), we obtain

$$\begin{cases} \dot{V} = -z_1 - z_2 - (1+k)z_3^2 + e_a\left[-e_3 z_3 - \dot{\hat{a}}\right] \\ \qquad +e_b\left[-(y_1^2 - x_1^2)z_3 - \dot{\hat{b}}\right] + e_c\left[-(y_2^2 - x_2^2)z_3 - \dot{\hat{c}}\right] \end{cases} \tag{80}$$

Substituting the update law (63) into (75), we get

$$\dot{V} = -z_1^2 - z_2^2 - (1+k)z_3^2, \tag{81}$$

which is a negative semi-definite function on $\mathbf{R}^6$.

From (81), it follows that the vector $\mathbf{z}(t) = (z_1(t), z_2(t), z_3(t))$ and the parameter estimation error $(e_a(t), e_b(t), e_c(t))$ are globally bounded, i.e.

$$\left[z_1(t)\ z_2(t)\ z_3(t)\ e_a(t)\ e_b(t)\ e_c(t) \right] \in \mathbf{L}_\infty \tag{82}$$

Also, it follows from (81) that

$$\dot{V} \leq -z_1^2 - z_2^2 - z_3^2 = -\|\mathbf{z}\|^2 \tag{83}$$

That is,

$$\|\mathbf{z}\|^2 \leq -\dot{V} \tag{84}$$

Integrating the inequality (84) from 0 to t, we get

$$\int_0^t |\mathbf{z}(\tau)|^2\, d\tau \leq V(0) - V(t) \tag{85}$$

From (85), it follows that $\mathbf{z}(t) \in \mathbf{L}_2$.

From Eq. (58), it can be deduced that $\dot{\mathbf{z}}(t) \in \mathbf{L}_\infty$.

Thus, using Barbalat's lemma, we conclude that $\mathbf{z}(t) \to \mathbf{0}$ exponentially as $t \to \infty$ for all initial conditions $\mathbf{z}(0) \in \mathbf{R}^3$.

Hence, it is immediate that $\mathbf{e}(t) \to \mathbf{0}$ exponentially as $t \to \infty$ for all initial conditions $\mathbf{e}(0) \in \mathbf{R}^3$.

This completes the proof. □

For the numerical simulations, the classical fourth-order Runge-Kutta method with step size $h = 10^{-8}$ is used to solve the system of differential Eqs. (55) and (56).

The parameter values of the novel jerk chaotic systems are taken as in the chaotic case, *i.e.* $a = 0.5$, $b = 0.15$ and $c = 0.2$.

We take the positive gain constant as $k = 10$.

Furthermore, as initial conditions of the master chaotic system (55), we take

$$x_1(0) = 1.1, \quad x_2(0) = 0.2, \quad x_3(0) = 2.1 \tag{86}$$

As initial conditions of the slave chaotic system (56), we take

$$y_1(0) = 0.9, \quad y_2(0) = 0.5, \quad y_3(0) = 1.4 \tag{87}$$

Also, as initial conditions of the parameter estimates, we take

$$\hat{a}(0) = 3.1, \quad \hat{b}(0) = 6.4, \quad \hat{c}(0) = 5.2 \tag{88}$$

In Figs. 7, 8 and 9, the complete synchronization of the identical 3-D jerk chaotic systems (55) and (56) is depicted.

From Fig. 7, it is clear that the states $x_1(t)$ and $y_1(t)$ are completely synchronized in six seconds.

From Fig. 8, it is clear that the states $x_2(t)$ and $y_2(t)$ are completely synchronized in six seconds.

From Fig. 9, it is clear that the states $x_3(t)$ and $y_3(t)$ are completely synchronized in six seconds.

Also, in Fig. 10, the time-history of the synchronization errors $e_1(t)$, $e_2(t)$, $e_3(t)$, is shown.

From Fig. 10, it is clear that the synchronization errors converge to zero in six seconds. This shows the efficiency of the adaptive backstepping controller designed in this section.

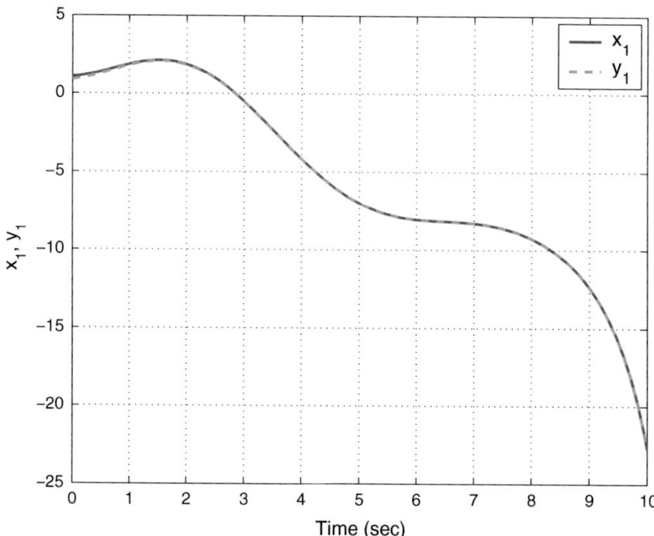

Fig. 7 Synchronization of the states $x_1(t)$ and $y_1(t)$

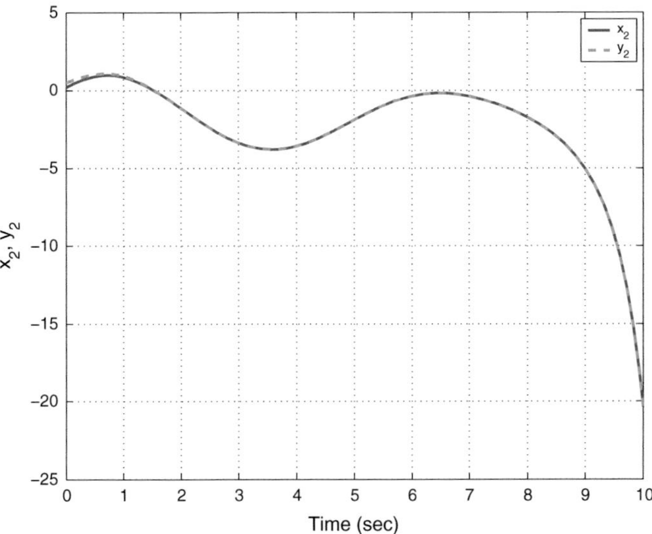

Fig. 8 Synchronization of the states $x_2(t)$ and $y_2(t)$

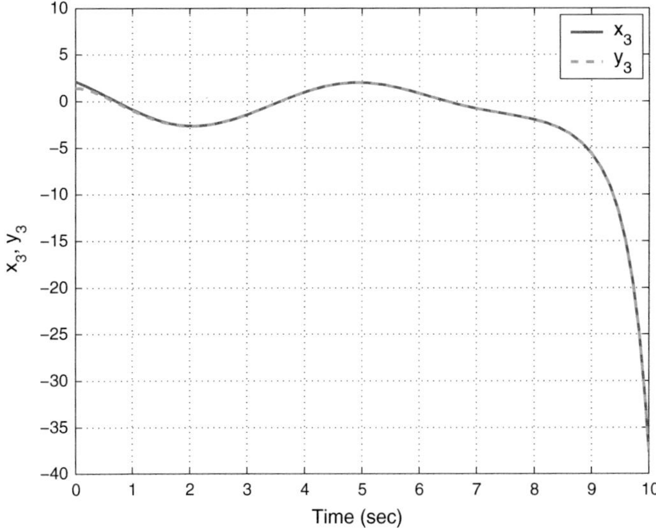

Fig. 9 Synchronization of the states $x_3(t)$ and $y_3(t)$

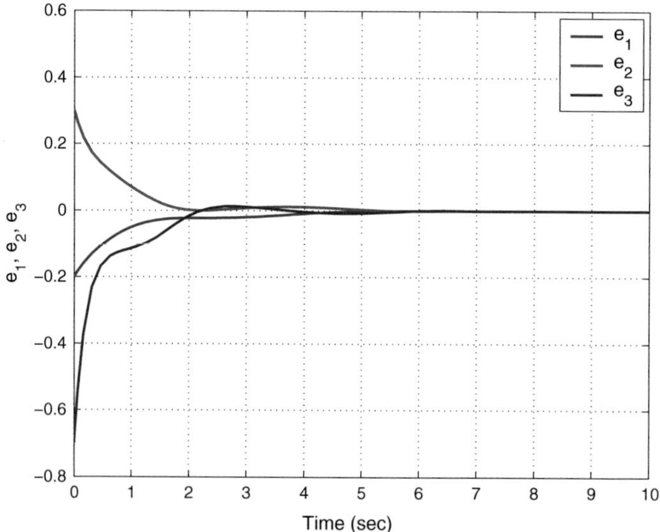

Fig. 10 Time-history of the synchronization errors $e_1(t)$, $e_2(t)$, $e_3(t)$

6 Conclusions

In this research work, we announced a seven-term novel jerk chaotic system with two quadratic nonlinearities. We derived new results for the adaptive controller and adaptive synchronizer design for the novel jerk chaotic system via backstepping control method. We first discussed the qualitative properties of the novel jerk chaotic system. The Lyapunov exponents of novel jerk chaotic system have been obtained as $L_1 = 0.11184$, $L_2 = 0$ and $L_3 = -0.61241$, while the Kaplan-Yorke dimension of the novel jerk chaotic system has been obtained as $D_{KY} = 2.18262$. The adaptive feedback control and synchronization results were proved using Lyapunov stability theory. MATLAB simulations have been depicted to illustrate all the main results for the novel jerk chaotic system.

References

1. Arneodo A, Coullet P, Tresser C (1981) Possible new strange attractors with spiral structure. Commun Math Phys 79(4):573–576
2. Azar AT (2010) Fuzzy systems. IN-TECH, Vienna
3. Azar AT (2012) Overview of type-2 fuzzy logic systems. Int J Fuzzy Syst Appl 2(4):1–28
4. Azar AT, Serrano FE (2014) Robust IMC-PID tuning for cascade control systems with gain and phase margin specifications. Neural Comput Appl 25(5):983–995
5. Azar AT, Serrano FE (2015) Adaptive sliding mode control of the Furuta pendulum. In: Azar AT, Zhu Q (eds) Advances and applications in sliding mode control systems. Studies in computational intelligence, vol 576. Springer, Germany, pp 1–42

6. Azar AT, Serrano FE (2015) Deadbeat control for multivariable systems with time varying delays. In: Azar AT, Vaidyanathan S (eds) Chaos modeling and control systems design. Studies in computational intelligence, vol 581. Springer, Germany, pp 97–132
7. Azar AT, Serrano FE (2015) Design and modeling of anti wind up PID controllers. In: Zhu Q, Azar AT (eds) Complex system modelling and control through intelligent soft computations. Studies in fuzziness and soft computing, vol 319. Springer, Germany, pp 1–44
8. Azar AT, Serrano FE (2015) Stabilizatoin and control of mechanical systems with backlash. In: Azar AT, Vaidyanathan S (eds) Handbook of research on advanced intelligent control engineering and automation. Advances in computational intelligence and robotics (ACIR). IGI-Global, USA, pp 1–60
9. Azar AT, Vaidyanathan S (2015) Chaos modeling and control systems design. Studies in computational intelligence, vol 581. Springer, Germany
10. Azar AT, Vaidyanathan S (2015) Computational intelligence applications in modeling and control. Studies in computational intelligence, vol 575. Springer, Germany
11. Azar AT, Vaidyanathan S (2015) Handbook of research on advanced intelligent control engineering and automation. Advances in computational intelligence and robotics (ACIR). IGI-Global, USA
12. Azar AT, Zhu Q (2015) Advances and applications in sliding mode control systems. Studies in computational intelligence, vol 576. Springer, Germany
13. Cai G, Tan Z (2007) Chaos synchronization of a new chaotic system via nonlinear control. J Uncertain Syst 1(3):235–240
14. Carroll TL, Pecora LM (1991) Synchronizing chaotic circuits. IEEE Trans Circ Syst 38(4):453–456
15. Chen G, Ueta T (1999) Yet another chaotic attractor. Int J Bifurcat Chaos 9(7):1465–1466
16. Chen HK, Lee CI (2004) Anti-control of chaos in rigid body motion. Chaos Solitons Fractals 21(4):957–965
17. Chen WH, Wei D, Lu X (2014) Global exponential synchronization of nonlinear time-delay Lur'e systems via delayed impulsive control. Commun Nonlinear Sci Numer Simul 19(9):3298–3312
18. Das S, Goswami D, Chatterjee S, Mukherjee S (2014) Stability and chaos analysis of a novel swarm dynamics with applications to multi-agent systems. Eng Appl Artif Intell 30:189–198
19. Feki M (2003) An adaptive chaos synchronization scheme applied to secure communication. Chaos Solitons Fractals 18(1):141–148
20. Gan Q, Liang Y (2012) Synchronization of chaotic neural networks with time delay in the leakage term and parametric uncertainties based on sampled-data control. J Franklin Inst 349(6):1955–1971
21. Gaspard P (1999) Microscopic chaos and chemical reactions. Phys A Stat Mech Appl 263(1–4):315–328
22. Gibson WT, Wilson WG (2013) Individual-based chaos: extensions of the discrete logistic model. J Theor Biol 339:84–92
23. Guégan D (2009) Chaos in economics and finance. Ann Rev Control 33(1):89–93
24. Huang X, Zhao Z, Wang Z, Li Y (2012) Chaos and hyperchaos in fractional-order cellular neural networks. Neurocomputing 94:13–21
25. Jiang GP, Zheng WX, Chen G (2004) Global chaos synchronization with channel time-delay. Chaos Solitons Fractals 20(2):267–275
26. Karthikeyan R, Sundarapandian V (2014) Hybrid chaos synchronization of four-scroll systems via active control. J Electr Eng 65(2):97–103
27. Kaslik E, Sivasundaram S (2012) Nonlinear dynamics and chaos in fractional-order neural networks. Neural Netw 32:245–256
28. Kengne J, Chedjou JC, Kenne G, Kyamakya K (2012) Dynamical properties and chaos synchronization of improved colpitts oscillators. Commun Nonlinear Sci Numer Simul 17(7):2914–2923
29. Khalil HK (2001) Nonlinear systems. Prentice Hall, New Jersey

30. Kyriazis M (1991) Applications of chaos theory to the molecular biology of aging. Exp Gerontol 26(6):569–572
31. Lang J (2015) Color image encryption based on color blend and chaos permutation in the reality-preserving multiple-parameter fractional Fourier transform domain. Opt Commun 338:181–192
32. Li D (2008) A three-scroll chaotic attractor. Phys Lett A 372(4):387–393
33. Li N, Zhang Y, Nie Z (2011) Synchronization for general complex dynamical networks with sampled-data. Neurocomputing 74(5):805–811
34. Li N, Pan W, Yan L, Luo B, Zou X (2014) Enhanced chaos synchronization and communication in cascade-coupled semiconductor ring lasers. Commun Nonlinear Sci Numer Simul 19(6): 1874–1883
35. Li Z, Chen G (2006) Integration of fuzzy logic and chaos theory. Studies in fuzziness and soft computing, vol 187. Springer, Germany
36. Lian S, Chen X (2011) Traceable content protection based on chaos and neural networks. Appl Soft Comput 11(7):4293–4301
37. Liu C, Liu T, Liu L, Liu K (2004) A new chaotic attractor. Chaos Solitions Fractals 22(5): 1031–1038
38. Lorenz EN (1963) Deterministic periodic flow. J Atmos Sci 20(2):130–141
39. Lü J, Chen G (2002) A new chaotic attractor coined. Int J Bifurcat Chaos 12(3):659–661
40. Mondal S, Mahanta C (2014) Adaptive second order terminal sliding mode controller for robotic manipulators. J Franklin Inst 351(4):2356–2377
41. Murali K, Lakshmanan M (1998) Secure communication using a compound signal from generalized chaotic systems. Phys Lett A 241(6):303–310
42. Nehmzow U, Walker K (2005) Quantitative description of robot-environment interaction using chaos theory. Robot Auton Syst 53(3–4):177–193
43. Noroozi N, Roopaei M, Karimaghaee P, Safavi AA (2010) Simple adaptive variable structure control for unknown chaotic systems. Commun Nonlinear Sci Numer Simul 15(3):707–727
44. Pecora LM, Carroll TL (1990) Synchronization in chaotic systems. Phys Rev Lett 64(8): 821–824
45. Pehlivan I, Moroz IM, Vaidyanathan S (2014) Analysis, synchronization and circuit design of a novel butterfly attractor. J Sound Vibr 333(20):5077–5096
46. Petrov V, Gaspar V, Masere J, Showalter K (1993) Controlling chaos in Belousov-Zhabotinsky reaction. Nature 361:240–243
47. Pham VT, Vaidyanathan S, Volos CK, Jafari S (2015) Hidden attractors in a chaotic system with an exponential nonlinear term. Eur Phys J Spec Top 224(8):1507–1517
48. Pham VT, Volos CK, Vaidyanathan S, Le TP, Vu VY (2015) A memristor-based hyperchaotic system with hidden attractors: dynamics, synchronization and circuital emulating. J Eng Sci Technol Rev 8(2):205–214
49. Qu Z (2011) Chaos in the genesis and maintenance of cardiac arrhythmias. Prog Biophys Mol Biol 105(3):247–257
50. Rasappan S, Vaidyanathan S (2012) Global chaos synchronization of WINDMI and Coullet chaotic systems by backstepping control. Far East J Math Sci 67(2):265–287
51. Rasappan S, Vaidyanathan S (2012) Hybrid synchronization of n-scroll Chua and Lur'e chaotic systems via backstepping control with novel feedback. Arch Control Sci 22(3):343–365
52. Rasappan S, Vaidyanathan S (2012) Synchronization of hyperchaotic Liu system via backstepping control with recursive feedback. Commun Comput Inf Sci 305:212–221
53. Rasappan S, Vaidyanathan S (2013) Hybrid synchronization of n-scroll chaotic Chua circuits using adaptive backstepping control design with recursive feedback. Malays J Math Sci 7(2): 219–246
54. Rasappan S, Vaidyanathan S (2014) Global chaos synchronization of WINDMI and Coullet chaotic systems using adaptive backstepping control design. Kyungpook Math J 54(1): 293–320
55. Rhouma R, Belghith S (2011) Cryptoanalysis of a chaos based cryptosystem on DSP. Commun Nonlinear Sci Numer Simul 16(2):876–884

56. Rössler OE (1976) An equation for continuous chaos. Phys Lett A 57(5):397–398
57. Vaidyanathan S, Pham VT, Volos CK (2015) A 5-D hyperchaotic Rikitake dynamo system with hidden attractors. Eur Phys J Spec Top 224(8):1575–1592
58. Sampath S, Vaidyanathan S, Volos CK, Pham VT (2015) An eight-term novel four-scroll chaotic system with cubic nonlinearity and its circuit simulation. J Eng Sci Technol Rev 8(2):1–6
59. Sarasu P, Sundarapandian V (2011) Active controller design for the generalized projective synchronization of four-scroll chaotic systems. Int J Syst Signal Control Eng Appl 4(2):26–33
60. Sarasu P, Sundarapandian V (2011) The generalized projective synchronization of hyperchaotic Lorenz and hyperchaotic Qi systems via active control. Int J Soft Comput 6(5): 216–223
61. Sarasu P, Sundarapandian V (2012) Adaptive controller design for the generalized projective synchronization of 4-scroll systems. Int J Syst Signal Control Eng Appl 5(2):21–30
62. Sarasu P, Sundarapandian V (2012) Generalized projective synchronization of three-scroll chaotic systems via adaptive control. Eur J Sci Res 72(4):504–522
63. Sarasu P, Sundarapandian V (2012) Generalized projective synchronization of two-scroll systems via adaptive control. Int J Soft Comput 7(4):146–156
64. Shahverdiev EM, Shore KA (2009) Impact of modulated multiple optical feedback time delays on laser diode chaos synchronization. Opt Commun 282(17):3568–3572
65. Sharma A, Patidar V, Purohit G, Sud KK (2012) Effects on the bifurcation and chaos in forced Duffing oscillator due to nonlinear damping. Commun Nonlinear Sci Numer Simul 17(6): 2254–2269
66. Sprott JC (1994) Some simple chaotic flows. Phys Rev E 50(2):647–650
67. Sprott JC (1997) Some simple chaotic jerk functions. Am J Phys 65(6):537–543
68. Sprott JC (2004) Competition with evolution in ecology and finance. Phys Lett A 325(5–6):329–333
69. Sprott JC (2010) Elegant chaos: algebraically simple chaotic flows. World Scientific, Singapore
70. Suérez I (1999) Mastering chaos in ecology. Ecol Model 117(2–3):305–314
71. Sundarapandian V (2010) Output regulation of the Lorenz attractor. Far East J Math Sci 42(2):289–299
72. Sundarapandian V (2013) Analysis and anti-synchronization of a novel chaotic system via active and adaptive controllers. J Eng Sci Technol Rev 6(4):45–52
73. Sundarapandian V, Karthikeyan R (2011) Anti-synchronization of hyperchaotic Lorenz and hyperchaotic Chen systems by adaptive control. Int J Syst Signal Control Eng Appl 4(2):18–25
74. Sundarapandian V, Karthikeyan R (2011) Anti-synchronization of Lü and Pan chaotic systems by adaptive nonlinear control. Eur J Sci Res 64(1):94–106
75. Sundarapandian V, Karthikeyan R (2012) Adaptive anti-synchronization of uncertain Tigan and Li systems. J Eng Appl Sci 7(1):45–52
76. Sundarapandian V, Karthikeyan R (2012) Hybrid synchronization of hyperchaotic Lorenz and hyperchaotic Chen systems via active control. J Eng Appl Sci 7(3):254–264
77. Sundarapandian V, Pehlivan I (2012) Analysis, control, synchronization, and circuit design of a novel chaotic system. Math Comput Model 55(7–8):1904–1915
78. Sundarapandian V, Sivaperumal S (2011) Sliding controller design of hybrid synchronization of four-wing chaotic systems. Int J Soft Comput 6(5):224–231
79. Suresh R, Sundarapandian V (2013) Global chaos synchronization of a family of n-scroll hyperchaotic Chua circuits using backstepping control with recursive feedback. Far East J Math Sci 73(1):73–95
80. Tigan G, Opris D (2008) Analysis of a 3D chaotic system. Chaos Solitons Fractals 36: 1315–1319
81. Usama M, Khan MK, Alghatbar K, Lee C (2010) Chaos-based secure satellite imagery cryptosystem. Comput Math Appl 60(2):326–337
82. Vaidyanathan S (2011) Hybrid chaos synchronization of Liu and Lü systems by active nonlinear control. Commun Comput Inf Sci 204:1–10

83. Vaidyanathan S (2011) Output regulation of Arneodo-Coullet chaotic system. Commun Comput Inf Sci 133:98–107
84. Vaidyanathan S (2011) Output regulation of the unified chaotic system. Commun Comput Inf Sci 198:10–17
85. Vaidyanathan S (2012) Analysis and synchronization of the hyperchaotic Yujun systems via sliding mode control. Adv Intell Syst Comput 176:329–337
86. Vaidyanathan S (2012) Anti-synchronization of Sprott-L and Sprott-M chaotic systems via adaptive control. Int J Control Theor Appl 5(1):41–59
87. Vaidyanathan S (2012) Global chaos control of hyperchaotic Liu system via sliding control method. Int J Control Theor Appl 5(2):117–123
88. Vaidyanathan S (2012) Output regulation of the Liu chaotic system. Appl Mech Mater 110–116:3982–3989
89. Vaidyanathan S (2012) Sliding mode control based global chaos control of Liu-Liu-Liu-Su chaotic system. Int J Control Theor Appl 1(2):15–20
90. Vaidyanathan S (2013) A new six-term 3-D chaotic system with an exponential nonlinearity. Far East J Math Sci 79(1):135–143
91. Vaidyanathan S (2013) Analysis and adaptive synchronization of two novel chaotic systems with hyperbolic sinusoidal and cosinusoidal nonlinearity and unknown parameters. J Eng Sci Technol Rev 6(4):53–65
92. Vaidyanathan S (2013) Analysis, control and synchronization of hyperchaotic Zhou system via adaptive control. Adv Intell Syst Comput 177:1–10
93. Vaidyanathan S (2014) A new eight-term 3-D polynomial chaotic system with three quadratic nonlinearities. Far East J Math Sci 84(2):219–226
94. Vaidyanathan S (2014) Analysis and adaptive synchronization of eight-term 3-D polynomial chaotic systems with three quadratic nonlinearities. Eur Phys J Spec Top 223(8):1519–1529
95. Vaidyanathan S (2014) Analysis, control and synchronisation of a six-term novel chaotic system with three quadratic nonlinearities. Int J Model Ident Control 22(1):41–53
96. Vaidyanathan S (2014) Generalized projective synchronisation of novel 3-D chaotic systems with an exponential non-linearity via active and adaptive control. Int J Model Ident Control 22(3):207–217
97. Vaidyanathan S (2014) Global chaos synchronization of identical Li-Wu chaotic systems via sliding mode control. Int J Model Ident Control 22(2):170–177
98. Vaidyanathan S (2015) 3-cells cellular neural network (CNN) attractor and its adaptive biological control. Int J PharmTech Res 8(4):632–640
99. Vaidyanathan S (2015) A 3-D novel highly chaotic system with four quadratic nonlinearities, its adaptive control and anti-synchronization with unknown parameters. J Eng Sci Technol Rev 8(2):106–115
100. Vaidyanathan S (2015) Adaptive backstepping control of enzymes-substrates system with ferroelectric behaviour in brain waves. Int J PharmTech Res 8(2):256–261
101. Vaidyanathan S (2015) Adaptive biological control of generalized Lotka-Volterra three-species biological system. Int J PharmTech Res 8(4):622–631
102. Vaidyanathan S (2015) Adaptive chaotic synchronization of enzymes-substrates system with ferroelectric behaviour in brain waves. Int J PharmTech Res 8(5):964–973
103. Vaidyanathan S (2015) Adaptive control of a chemical chaotic reactor. Int J PharmTech Res 8(3):377–382
104. Vaidyanathan S (2015) Adaptive synchronization of chemical chaotic reactors. Int J ChemTech Res 8(2):612–621
105. Vaidyanathan S (2015) Adaptive synchronization of generalized Lotka-Volterra three-species biological systems. Int J PharmTech Res 8(5):928–937
106. Vaidyanathan S (2015) Analysis, properties and control of an eight-term 3-D chaotic system with an exponential nonlinearity. Int J Model Ident Control 23(2):164–172
107. Vaidyanathan S (2015) Anti-synchronization of Brusselator chemical reaction systems via adaptive control. Int J ChemTech Res 8(6):759–768

108. Vaidyanathan S (2015) Chaos in neurons and adaptive control of Birkhoff-Shaw strange chaotic attractor. Int J PharmTech Res 8(5):956–963
109. Vaidyanathan S (2015) Dynamics and control of Brusselator chemical reaction. Int J ChemTech Res 8(6):740–749
110. Vaidyanathan S (2015) Dynamics and control of Tokamak system with symmetric and magnetically confined plasma. Int J ChemTech Res 8(6):795–803
111. Vaidyanathan S (2015) Hyperchaos, qualitative analysis, control and synchronisation of a ten-term 4-D hyperchaotic system with an exponential nonlinearity and three quadratic nonlinearities. Int J Model Ident Control 23(4):380–392
112. Vaidyanathan S (2015) Lotka-Volterra population biology models with negative feedback and their ecological monitoring. Int J PharmTech Res 8(5):974–981
113. Vaidyanathan S (2015) Synchronization of 3-cells cellular neural network (CNN) attractors via adaptive control method. Int J PharmTech Res 8(5):946–955
114. Vaidyanathan S (2015) Synchronization of Tokamak systems with symmetric and magnetically confined plasma via adaptive control. Int J ChemTech Res 8(6):818–827
115. Vaidyanathan S, Azar AT (2015) Analysis and control of a 4-D novel hyperchaotic system. In: Azar AT, Vaidyanathan S (eds) Chaos modeling and control systems design. Studies in computational intelligence, vol 581. Springer, Germany, pp 19–38
116. Vaidyanathan S, Azar AT (2015) Analysis, control and synchronization of a nine-term 3-D novel chaotic system. In: Azar AT, Vaidyanathan S (eds) Chaos modelling and control systems design. Studies in computational intelligence, vol 581. Springer, Germany, pp 19–38
117. Vaidyanathan S, Azar AT (2015) Anti-synchronization of identical chaotic systems using sliding mode control and an application to Vaidhyanathan-Madhavan chaotic systems. Stud Comput Intell 576:527–547
118. Vaidyanathan S, Azar AT (2015) Hybrid synchronization of identical chaotic systems using sliding mode control and an application to Vaidhyanathan chaotic systems. Stud Comput Intell 576:549–569
119. Vaidyanathan S, Madhavan K (2013) Analysis, adaptive control and synchronization of a seven-term novel 3-D chaotic system. Int J Control Theor Appl 6(2):121–137
120. Vaidyanathan S, Pakiriswamy S (2013) Generalized projective synchronization of six-term Sundarapandian chaotic systems by adaptive control. Int J Control Theor Appl 6(2):153–163
121. Vaidyanathan S, Pakiriswamy S (2015) A 3-D novel conservative chaotic system and its generalized projective synchronization via adaptive control. J Eng Sci Technol Rev 8(2): 52–60
122. Vaidyanathan S, Rajagopal K (2011) Anti-synchronization of Li and T chaotic systems by active nonlinear control. Commun Comput Inf Sci 198:175–184
123. Vaidyanathan S, Rajagopal K (2011) Global chaos synchronization of hyperchaotic Pang and Wang systems by active nonlinear control. Commun Comput Inf Sci 204:84–93
124. Vaidyanathan S, Rajagopal K (2011) Global chaos synchronization of Lü and Pan systems by adaptive nonlinear control. Commun Comput Inf Sci 205:193–202
125. Vaidyanathan S, Rajagopal K (2012) Global chaos synchronization of hyperchaotic Pang and hyperchaotic Wang systems via adaptive control. Int J Soft Comput 7(1):28–37
126. Vaidyanathan S, Rasappan S (2011) Global chaos synchronization of hyperchaotic Bao and Xu systems by active nonlinear control. Commun Comput Inf Sci 198:10–17
127. Vaidyanathan S, Rasappan S (2014) Global chaos synchronization of n-scroll Chua circuit and Lur'e system using backstepping control design with recursive feedback. Arab J Sci Eng 39(4):3351–3364
128. Vaidyanathan S, Sampath S (2011) Global chaos synchronization of hyperchaotic Lorenz systems by sliding mode control. Commun Comput Inf Sci 205:156–164
129. Vaidyanathan S, Sampath S (2012) Anti-synchronization of four-wing chaotic systems via sliding mode control. Int J Autom Comput 9(3):274–279
130. Vaidyanathan S, Volos C (2015) Analysis and adaptive control of a novel 3-D conservative no-equilibrium chaotic system. Arch Control Sci 25(3):333–353

131. Vaidyanathan S, Volos C, Pham VT (2014) Hyperchaos, adaptive control and synchronization of a novel 5-D hyperchaotic system with three positive Lyapunov exponents and its SPICE implementation. Arch Control Sci 24(4):409–446
132. Vaidyanathan S, Volos C, Pham VT, Madhavan K, Idowu BA (2014) Adaptive backstepping control, synchronization and circuit simulation of a 3-D novel jerk chaotic system with two hyperbolic sinusoidal nonlinearities. Arch Control Sci 24(3):375–403
133. Vaidyanathan S, Azar AT, Rajagopal K, Alexander P (2015) Design and SPICE implementation of a 12-term novel hyperchaotic system and its synchronisation via active control. Int J Model Ident Control 23(3):267–277
134. Vaidyanathan S, Idowu BA, Azar AT (2015) Backstepping controller design for the global chaos synchronization of Sprott's jerk systems. Stud Comput Intell 581:39–58
135. Vaidyanathan S, Rajagopal K, Volos CK, Kyprianidis IM, Stouboulos IN (2015) Analysis, adaptive control and synchronization of a seven-term novel 3-D chaotic system with three quadratic nonlinearities and its digital implementation in LabVIEW. J Eng Sci Technol Rev 8(2):130–141
136. Vaidyanathan S, Volos C, Pham VT, Madhavan K (2015) Analysis, adaptive control and synchronization of a novel 4-D hyperchaotic hyperjerk system and its SPICE implementation. Arch Control Sci 25(1):5–28
137. Vaidyanathan S, Volos CK, Kyprianidis IM, Stouboulos IN, Pham VT (2015) Analysis, adaptive control and anti-synchronization of a six-term novel jerk chaotic system with two exponential nonlinearities and its circuit simulation. J Eng Sci Technol Rev 8(2):24–36
138. Vaidyanathan S, Volos CK, Pham VT (2015) Analysis, adaptive control and adaptive synchronization of a nine-term novel 3-D chaotic system with four quadratic nonlinearities and its circuit simulation. J Eng Sci Technol Rev 8(2):174–184
139. Vaidyanathan S, Volos CK, Pham VT (2015) Analysis, control, synchronization and SPICE implementation of a novel 4-D hyperchaotic Rikitake dynamo system without equilibrium. J Eng Sci Technol Rev 8(2):232–244
140. Vaidyanathan S, Volos CK, Pham VT (2015) Global chaos control of a novel nine-term chaotic system via sliding mode control. In: Azar AT, Zhu Q (eds) Advances and applications in sliding mode control systems. Studies in computational intelligence, vol 576. Springer, Germany, pp 571–590
141. Vaidyanathan S, Volos CK, Rajagopal K, Kyprianidis IM, Stouboulos IN (2015) Adaptive backstepping controller design for the anti-synchronization of identical WINDMI chaotic systems with unknown parameters and its SPICE implementation. J Eng Sci Technol Rev 8(2):74–82
142. Volos CK, Kyprianidis IM, Stouboulos IN (2013) Experimental investigation on coverage performance of a chaotic autonomous mobile robot. Robot Auton Syst 61(12):1314–1322
143. Volos CK, Kyprianidis IM, Stouboulos IN, Tlelo-Cuautle E, Vaidyanathan S (2015) Memristor: a new concept in synchronization of coupled neuromorphic circuits. J Eng Sci Technol Rev 8(2):157–173
144. Wei Z, Yang Q (2010) Anti-control of Hopf bifurcation in the new chaotic system with two stable node-foci. Appl Math Comput 217(1):422–429
145. Witte CL, Witte MH (1991) Chaos and predicting varix hemorrhage. Med Hypotheses 36(4):312–317
146. Xiao X, Zhou L, Zhang Z (2014) Synchronization of chaotic Lur'e systems with quantized sampled-data controller. Commun Nonlinear Sci Numer Simul 19(6):2039–2047
147. Yuan G, Zhang X, Wang Z (2014) Generation and synchronization of feedback-induced chaos in semiconductor ring lasers by injection-locking. Optik Int J Light Electron Opt 125(8): 1950–1953
148. Zaher AA, Abu-Rezq A (2011) On the design of chaos-based secure communication systems. Commun Nonlinear Syst Numer Simul 16(9):3721–3727
149. Zhang H, Zhou J (2012) Synchronization of sampled-data coupled harmonic oscillators with control inputs missing. Syst Control Lett 61(12):1277–1285

150. Zhang X, Zhao Z, Wang J (2014) Chaotic image encryption based on circular substitution box and key stream buffer. Signal Process Image Commun 29(8):902–913
151. Zhou W, Xu Y, Lu H, Pan L (2008) On dynamics analysis of a new chaotic attractor. Phys Lett A 372(36):5773–5777
152. Zhu C, Liu Y, Guo Y (2010) Theoretic and numerical study of a new chaotic system. Intell Inf Manage 2:104–109
153. Zhu Q, Azar AT (2015) Complex system modelling and control through intelligent soft computations. Studies in fuzzines and soft computing, vol 319. Springer, Germany

Evidence of Chaos in EEG Signals: An Application to BCI

Kusuma Mohanchandra, Snehanshu Saha and K. Srikanta Murthy

Abstract The recent science and technology studies in neuroscience and machine learning have focused attention on investigating the functioning of the brain through nonlinear analysis. The brain is a nonlinear dynamic system, imparting randomness and nonlinearity in the EEG signals. The stochastic nature of the brain seeks the paramount importance of understanding the underlying neurophysiology. The nonlinear analysis of the dynamic structure may help to reveal the complex behavior of the brain signals. EEG signal analysis is helpful in various clinical applications to characterize the normal and diseased brain states. The EEG is used in predicting epileptic seizures, classifying the sleep stages, measuring the depth of anesthesia, and detecting the abnormal brain states. With the onset of EEG-based brain-computer interfaces, the characteristics of brain signals are used to control the devices through different mental states. Hence, the need to understand the brain state is important and crucial. In this chapter, the author introduces the theory and methods of chaos theory measurements and its applications in EEG signal analysis. A broad perspective of the techniques and implementation of the Correlation Dimension, Lyapunov Exponents, Fractal Dimension, Approximate Entropy, Sample Entropy, Hurst Exponent, Lempel-Ziv complexity, Hopf Bifurcation Theorem and Higher-order spectra is explained and their usage in EEG signal analysis is mentioned. We suggest that chaos theory provides not only potentially valuable diagnostic information but also a deeper understanding of neuropathological mechanisms underlying the brain in ways that are not possible by conventional linear analysis.

K. Mohanchandra (✉)
Department of Computer Science & Engineering, Medical Imaging Research Centre, Dayananda Sagar College of Engineering, Bangalore 560078, India
e-mail: kusumalak@gmail.com

S. Saha · K.S. Murthy
Department of Computer Science & Engineering, PESIT South Campus, Bangalore 560100, India

S. Saha
The Center for Basic Initiatives in Mathematical Modeling and Computation (CBIMMC), PESIT South Campus, Bangalore 560100, India

A.T. Azar and S. Vaidyanathan (eds.), *Advances in Chaos Theory and Intelligent Control*, Studies in Fuzziness and Soft Computing 337,
DOI 10.1007/978-3-319-30340-6_25

Keywords Chaos · EEG · Correlation dimension · Lyapunov exponents · Fractal dimension · Approximate entropy · Sample entropy · Hurst exponent · Lempel-Ziv complexity · Hopf bifurcation theorem · Higher-order spectra

1 Introduction

In the recent years, electroencephalography (EEG) based brain-computer interface (BCI) has emerged as a critical ammunition in the study of neuroscience, machine learning, and rehabilitation. The BCI is an interface in which a person uses his brain to control the machine needed for deployment to fulfill a particular task. The device can be a computer, wheelchair, robot, assistive or an alternative communication device. The human brain controls the devices and software directly by the mental activity [1] without any muscular movements. The non-invasive interface measures the brain's electrical activity through the scalp electrodes in the form of EEG signals. EEG signals record spontaneous electrical activity produced in the central nervous system. It includes information about the state and change of the neural system. Consequently, it is widely used in clinical as well as non-clinical research [18]. EEG signals reflect the dynamics of electrical activity generated by a large number of neurons in the brain. These dynamic properties are essential for understanding the EEG phenomena. There is a plethora of literature documenting the existence of high-frequency rhythmic activities in the human EEG under different circumstances, particularly in relation to the motor and cognitive functions. The complexity of understanding the phenomena calls for the use of mathematical models. The focus was mainly on the use of linear systems analysis in the past. Lately, contemporary EEG analysis incorporates analysis of nonlinear dynamic systems.

As nonlinear dynamics evolved in theory and practice, there is demonstrable indication that the brain is a nonlinear dynamic system, imparting randomness and nonlinearity in the EEG signal. Chaos means a state of inherent unpredictability in the behavior of the system. The chaos theory states that a small difference in the initial condition can yield widely diverging outcomes [2], rendering long-term prediction impossible. Analysis of EEG signal using methods of nonlinear dynamics and deterministic chaos theory may yield insight into the nonlinear dynamics and chaos in the brain.

Furthermore, studies on the synchronization in EEG signals associated with various brain disorders has been the focus of much attention. The synchronization process is expected to play a significant role in the real-time processing of chaotic signals. Synchronization of chaotic systems is a phenomenon that occurs when two or more chaotic systems are coupled or when a chaotic system drives another chaotic system [36]. The phenomenon of chaotic synchronization takes place when the phase space trajectories of two or more coupled chaotic systems, initially evolving on different attractors, eventually converge to a common trajectory of two or more chaotic systems. Many new types of synchronization have appeared in the last decade: global chaos synchronization [42], anti-synchronization [37], hybrid synchronization [38], adaptive synchronization [36], global chaos synchronization [41], to mention a few types used in the research. Synchronization is expected to play a significant role in establishing the communication between different regions of the brain.

The brain being a giant dynamic system is described as hyperchaotic. Any chaotic behavior with at least two positive Lyapunov exponents is defined as hyperchaotic [39, 40]. The hyperchaotic behavior in brain signals has only been studied to a limited extent. An exploration of the techniques to a greater extent may help understand the dynamics of the brain as a whole.

Small amplitude limit cycles, offsprings of the fixed point that loses stability in the local bifurcation, have the potential to localize EEG signals and thus could be used in source modeling. A study of the dynamical systems with EEG signals embedded is a technique that is fascinating and promising. The volume of literature is insignificant. There exist enough questions begging for answers. The manuscript is intended to explore the open areas and generate enough curiosity towards attaining momentum and sufficient traction among the BCI community. Real time applications [17] could be designed with the aid of dynamical modeling of the brain signals. Therefore in this chapter, the characteristics, dynamics and chaotic localization properties of EEG signals of the human brain have been presented. Accuracy and sensitivity of the interfaces thus designed would help the health professionals and patients, it is hoped. This book chapter is arranged as follows. In Sect. 2, a broad outlook of the techniques and applications of the Correlation Dimension, Lyapunov Exponents, Fractal Dimension, Approximate Entropy, Sample Entropy, Hurst Exponent, Lempel-Ziv complexity, Hopf Bifurcation Theorem and Higher-order spectra, and their usage in EEG signal analysis is described. The conclusion and the future work are explained in Sect. 3.

2 Theory and Methods

As EEG is considered highly noisy and nonlinear, the chaos theory is endowed with effective quantitative descriptors of EEG dynamics and the underlying chaos in the brain. However, the broad frequency spectrum of EEG makes the linear time-frequency analysis inefficient most of the time. Several nonlinear methods are developed to study the complexity, randomness and chaotic behavior of the EEG signal. Among those, the Correlation Dimension (D2), Lyapunov Exponents (LE), Fractal Dimension (FD), Approximate Entropy (ApEn), Sample Entropy (SampEn), Hurst Exponent (H), Lempel-Ziv complexity, and Hopf Bifurcation Theorem have been the most popular chaotic measures.

A chaotic system is illustrated by the strange attractors in the phase space [16]. Points that are initially close in the phase space become exponentially separated after some time lapse. Though they appear to be haphazard, they follow an unknown or hidden structure. This pattern called as an attractor can be visually observed through the plotting of data in phase space.

The brain is a complex nonlinear system that involves a large number of interrelated variables and is impossible to measure directly. With the limited number of variables available it is difficult to analyze the multi-dimensional dynamics of the brain. Takens [34] has demonstrated that with a single variable measured with sufficient accuracy for a long period, the underlying structure of the system can be reconstructed using delay coordinates. The information on the underlying brain dynamics can be extracted by analyzing a group of attractors reconstructed from the time series EEG signals.

The concept of phase space is a simple tool that helps to understand the chaotic behavior of the system. To understand phase space, it is first necessary to understand the concept of attractors. In chaos theory, an attractor is a pattern that forms when the behavior of the system is plotted in phase space [16]. When the line joins the points in chronological order, a pattern develops that can resemble a point, orbit, or some unusual pattern. The unusual pattern is referred as a strange attractor. The strange attractors have non-integer dimensions. Signals corresponding to strange attractors exhibit random structures. The two basic properties of the strange attractors are its sensitive dependence to initial conditions and unpredictability in the long run. The highly complex and dynamical nature of the neuronal interactions in the brain, calls for nonlinear methods to analyze the EEG signals.

Attractors range from being simple to greatly complex. Four types of attractors have been identified [6]: a point, a pendulum (limit cycle), a torus (a type of orbit), and a strange attractor. In phase space, a point attractor is a fixed point on a graph, there is no variation in the position and the dimension is zero. The point attractor occurs because the system behavior remains consistent over time. The pendulum (limit cycle) attractor, varies along a curve and resembles a narrow back-and-forth pattern when graphed in phase space; the dimension is one. The Torus attractor is a more complex pattern that forms an orbit, but also contains sub-orbits within the orbit, thus resembling a donut when graphed in phase space. Finally, the strange attractor, sometimes referred to as a fractal, is a complicated pattern that exists when the system is in chaos. The attractor is called strange because its shape may or may not resemble any known pattern.

For the nonlinear analysis of EEG signals, firstly the one-dimensional EEG data is transformed into multi-dimensional phase space using delay coordinates. In a hypothetical system with n variables, the phase space constructed will be n-dimensional. Each point in the phase space corresponds to the state of the system, and the n coordinates are the values assumed by the governing variables for that specific state. When the system is observed for a long period, the sequence of points in the phase space forms a trajectory. The trajectory fills a subspace of the phase space, called the attractor. The reconstruction of the attractor of the EEG data in the phase space is carried out through the technique of delay coordinates. Then the dynamical properties of the trajectories in the phase space are estimated by nonlinear parameters such as the fractal dimension (FD), correlation dimension (D2) and the first positive Lyapunov exponent [29].

2.1 Correlation Dimension (D2)

The Correlation Dimension is popularly used as a quantitative parameter to describe the attractors. It is a measure of the complexity of the system related to the number of degrees of freedom. It is a measure of the dimensionality of the space occupied by a set of random points. Computing the correlation dimension and study of its convergence for different input signals is a common practice. The correlation dimension converges to finite values for deterministic systems. However, it does not converge in the case of a random signal. Hence, correlation dimension is considered as a good parameter for evaluating the noisy nature of the EEG signals, inherent within the structure.

The correlation dimension is computed as follows: First, a phase space spanned by a set of embedding vectors is constructed. The n-dimensional vectors [34] are constructed in the following way:

$$\vec{x} = \{x(t), x(t+\tau), \ldots, x(t+(n-1)\cdot\tau)\} \tag{1}$$

where τ is a fixed time delay and n is the embedding dimension. Every instantaneous state of the system is represented by the vector $\vec{x}$ which denotes a point in the phase space. Then a correlation integral as a function of variable distances R (threshold distance) is computed as shown in Eq. (2).

$$C(R) = \lim_{N\to\infty} \frac{1}{N^2} \sum_{i \neq j} \theta(R - \|\vec{x}_i - \vec{x}_j\|), \qquad \vec{x}_i \in R^n \tag{2}$$

where N represents the number of data samples, $\|.\|$ is a norm and θ is the Heavyside function. The C(R) is a measure of the probability that two arbitrary points $\vec{x}_i - \vec{x}_j$ are separated by a distance less than R. The correlation dimension is defined as

$$D = \lim_{R\to 0} \frac{\log C(R)}{\log R} \tag{3}$$

Correlation dimension can be calculated from the slope of the plots of log C(R) versus log R. Low correlation dimension values correspond to simple systems and high correlation dimension values to more complex ones. In theory, these values approach infinity in the case of noise. The correlation dimension is proved to be very useful for characterizing the brain dynamics in different stages of sleep states [3]. Furthermore, correlation dimension is used to discriminate normal subjects and patients with different pathologies like epilepsy [30], Alzheimer [14], dementia and Parkinson [31] disease. The lower values of correlation dimension shows that the EEG signals are abnormal, indicating less complexity in the signals during the presence of an abnormality.

2.2 Lyapunov Exponents (LE)

Lyapunov exponent gives a quantitative measure of the sensitive dependence on the initial conditions. It is the averaged rate of divergence or convergence of two neighboring trajectories in the phase space. The measure is equal to the dimension of the phase space. To examine the behavior of an orbit around a point X*(t), the system is perturbed and represented as

$$X(t) = X^*(t) + U(t), \tag{4}$$

where U(t) represents the average deviation from the unperturbed trajectory at time t. In a chaotic region, the Lyapunov characteristic exponent σ is independent of X*(0). It is given by Oseledec theorem, which states that

$$\sigma_i = \lim_{t\to\infty} \frac{1}{t} \ln |U(t)| \tag{5}$$

For an n-dimensional mapping, the Lyapunov characteristic exponents are given by

$$\sigma_i = \lim_{N\to\infty} \ln |\lambda_i(N)| \tag{6}$$

For $i = 1, 2, \ldots, n$, and λ_i is the Lyapunov characteristic number. The larger the Lyapunov characteristic number greater is the rate of exponential divergence.

Say, two nearby trajectories have initial conditions (x_0, y_0) and $(x_1, y_1) = (x_0 + dx, y_0 + dy)$ respectively. At iteration k, the distance between trajectories is given by

$$d_k = |(x_1 - x_0, y_1 - y_0)| \tag{7}$$

The average, exponential rate of divergence of the trajectories is defined by

$$\sigma_1 = \lim_{k\to\infty} \frac{1}{k} \log\left(\frac{d_k}{d_0}\right) \tag{8}$$

For an n-dimensional phase space, there are n, Lyapunov characteristic exponents. Since, the largest exponent σ_1 dominates; this limit is useful only for finding the largest exponent. Given a Lyapunov characteristic exponent σ_i, the corresponding Lyapunov characteristic number λ_i is defined as

$$\lambda_i = e^{\sigma_i} \tag{9}$$

For an n-dimensional linear map,

$$X_{n+1} = MX_n \tag{10}$$

The Lyapunov characteristic numbers $\lambda_1, \ldots, \lambda_n$ are the eigenvalues of the map matrix. For an arbitrary map

$$x_{n+1} = f_1(x_n, y_n) \tag{11}$$

$$y_{n+1} = f_2(x_n, y_n) \tag{12}$$

the eigenvalues of the limit are the Lyapunov numbers

$$\lim_{n\to\infty} [J(x_n, y_n)J(x_{n-1}, y_{n-1}) \ldots J(x_1, y_1)]^{1/n} \tag{13}$$

where J(x, y) is the Jacobian. If $\lambda_i = 0$ for all i, the system is not chaotic. The determination of the existence of a positive Lyapunov exponent confirms the presence of chaos in the underlying dynamics of the time series. The Lyapunov exponents are used in the detection of epilepsy [33], sleep stages [3] and other pathologies but are used in comparison with the correlation dimension [26]. The largest Lyapunov exponent of EEG represents the degree of chaos in human brain activity [43].

2.3 *Fractal Dimension (FD)*

Fractal Dimension is used to estimate the dimensional complexity of biological signals. It can indicate the complete fractal cover of the space. The Higuchi algorithm calculates the fractal dimension of a time sequence x(1), x(2), ..., x(n). A new time series x(k, m) may be constructed as:

$$x(k, m) = \{x(m), x(m + k), x(m + 2k), \ldots, x(m + [(N - m)/k]k)\} \tag{14}$$

where $m = 1, 2, \ldots, k$. Here m is the initial time value, and k is the discrete time interval between the points. The length L(m, k) for each of the k time series or curves x(k, m) is computed as:

$$L(m, k) = \left\{ \frac{\sum_{i=1}^{(n-m)/k} |x(m + ik) - x[m + (i - 1)k]| (n - 1)}{\left[\frac{n-m}{k}\right] k} \right\} \frac{1}{k} \tag{15}$$

where n is the total length of data sequence x. L(k) denotes the mean value of the curve length. The fractal dimension is estimated as:

$$FD = \log(L(k))/\log(1/k) \tag{16}$$

The fractal dimension yields values between one and two since a simple curve has dimension one, and a plane is two-dimensional. Fractals are used to model the brain signals [9]. Although, the research on fractals in EEG is still in the nascent stages, there have been few breakthroughs relating fractals and chaos to the brain.

2.4 Approximate Entropy (ApEn)

Approximate Entropy is a measure of data regularity. It is defined as the logarithmic likelihood that the trends of the data patterns that are close to each other will remain close [3] in the next comparison with next pattern. Highly regular data results in producing smaller approximate entropy values and vice versa. The approximate entropy was proposed by Pincus [23]. The approximate entropy quantifies the predictability of subsequent amplitude values of data series based on the knowledge of the previous amplitude values. The approximate entropy value of an entirely irregular data series depends on the length of the epoch (m) and the number of previous values used for the prediction of the subsequent value. For an N sample time series {u(i): $1 \leq$ i $\leq N$}, given the length of the epoch (m), the approximate entropy is computed as shown in the following steps:

(i) Form a group of m-dimensional vector sequences X_1^m through X_{N-m+1}^m as: $X(i) = \{u(i), u(i+1), \ldots, u(i+m-1)\}, (i = 1, 2, \ldots, N-m+1)$.
(ii) Compute the distance between X(i) and other vectors X(j), where j $= 1, 2, \ldots,$ N $-$ m $+$ 1 and j $\neq$ i for every i value.
(iii) Given the threshold value r, count the number of $d[X(i), X(j)]$ which is smaller than r for every i value. Compute the ratio of this number to the total distance $N-m$:

$$C_i^m(r) = \frac{(number\ if\ X(j) such\ that\ d[X(i), X(j)] < r)}{(N-m+1)} \tag{17}$$

(iv) Define

$$\varphi^m(r) = (N-m+1)^{-1} \sum_{i=1}^{N-m+1} \ln C_i^m(r) \tag{18}$$

where ln is the natural logarithm.
(v) The approximate entropy is defined as:

$$\mathbf{ApEn\ (m, r) = \Phi^m(r) - \Phi^{m+1}(r)} \tag{19}$$

The approximate entropy is designed to work for small data samples and with less computational demand. The approximate entropy has been applied to classify EEG in psychiatric diseases such as schizophrenia [27], epilepsy [22], depth of anesthesia [11], and in classifying the sleep stages [7].

2.5 Sample Entropy (SampEn)

Sample entropy, was proposed by Richman and Moorman [25]. It is the negative natural logarithm of an estimate of the conditional probability. It states that the epoch of length m that match point-wise within a tolerance r also matches at the next point. Although sample entropy is also a measure of data regularity like approximate entropy, sample entropy is independent of record length. To calculate the sample entropy, points matching within the tolerance r are computed until there is no match. The count of matches are stored in counters A(k) and B(k) for all lengths k up to m. The Eq. (20) gives the sample entropy

$$\boldsymbol{SampEn} = -\boldsymbol{log}\frac{\boldsymbol{A(k)}}{\boldsymbol{B(k)}} \tag{20}$$

where $k = 0, 1, \ldots, m - 1$ and $B(0) = N$, the length of the input series.
$A =$ Number of all probable pairs having distance $([X_{m+1}(i), X_{m+1}(j)] < r)$ of length $m + 1$ and $i \neq j$.
$B =$ Number of all probable pairs having distance $([X_m(i), X_m(j)] < r)$ of length m.

A smaller value of sample entropy indicates more self-similarity in data set with less noise. The sample entropy is used for monitoring the depth of anesthesia [15] in detecting the state between awake and anesthesia. Furthermore, the sample entropy is used in detecting the EEG changes caused by epilepsy [44]. Data analysis results show that the values of both approximate entropy and sample entropy decrease significantly when the epilepsy is burst.

2.6 *Hurst Exponent (H)*

Hurst Exponent is a measure of self-similarity, predictability and the degree of long-range dependence in a time-series [19]. It is also a measure of the smoothness of a fractal time-series based on the asymptotic behavior of the rescaled range of the process. According to the Hurst's generalized equation of time series the hurst exponent, is defined as

$$H = \frac{\log\left(\frac{R}{S}\right)}{\log(T)} \tag{21}$$

where T is the duration of the sample of data and R/S is the corresponding value of rescaled range. R is the difference between the maximum and minimum deviation from the mean while S represents the standard deviation. In other words, the hurst exponent can be defined as

$$\left(\frac{R}{S}\right)_T = c^* T^H \tag{22}$$

Here, c is a constant and H is the hurst exponent. Hurst exponent is estimated by plotting (R/S) versus T in log–log axes. The slope of the regression line approximates the hurst exponent. Hurst exponent is used in identifying the different anesthetic states [20] of a patient during surgery.

2.7 The Lempel-Ziv Complexity (LZ)

Lempel-Ziv complexity is useful as a scalar metric to estimate the bandwidth of random processes and the harmonic variability in quasi-periodic signals. To measure the Lempel-Ziv complexity C, the EEG signal is transformed into a finite sequence of symbols. The discrete-time EEG signal x(i), i $= 1, 2, \ldots, n$, is converted into a binary sequence by comparing with the threshold T_d. The signal data is converted into a 0–1 sequence L as shown below:
$L = l_1, l_2, \ldots, l_r$ where

$$l_i = \begin{cases} 0, & \text{if } x(i) < T_d \\ 1, & \text{otherwise} \end{cases} \tag{23}$$

The median is used as the threshold T_d since it is robust to outliers. From the given finite symbol sequence L, the Lempel-Ziv complexity is computed. The sequence L is scanned from left to right, and the complexity counter C is increased by one, each time a new subsequence of consecutive characters is encountered. The complexity measure can be estimated using the following algorithm

1. Let ε and θ denote two sequences of L and let ε θ be the concatenation of ε θ. The sequence ε θ π is derived from ε θ after its last character is deleted (π denotes the operation of deleting the last character in the sequence). Let V(ε θ π) denote the vocabulary of all different subsequences of ε θ π. At the beginning $C = 1, \varepsilon = l_1$, and $\theta = l_2$. Therefore $\varepsilon\,\theta\,\pi = l_1$.
2. In general, $\varepsilon = l_1\ l_2\ l_3 \ldots l_r$ and $\theta = l_{r+1}$, so $\varepsilon\,\theta\,\pi = l_1\ l_2\ l_3 \ldots l_r$. If θ belongs to V(ε θ π), then θ is a subsequence of ε θ π; and not a new sequence.
3. Renew θ to be l_{r+1}, l_{r+2} and judge if θ belongs to V(ε θ π) or not.
4. Repeat the earlier steps until θ does not belong to V(ε θ π). Now $\theta = l_1\ l_2\ l_3 \ldots l_{r+i}$ is not a subsequence of $V(\varepsilon\,\theta\,\pi) = l_1\ l_2\ l_3 \ldots l_{r+i-1}$, so increase C by one.
5. Thereafter, ε is renewed to be $\varepsilon == l_1\ l_2\ l_3 \ldots l_{r+i}$ and $\theta = l_{r+i+1}$.
6. The above procedure is repeated until θ is the last character. Then the number of different subsequences in L is the measure of complexity C.

To make C, independent of the sequence length, it is normalized. If n is the length of the sequence and the number of different symbols in the symbol set is α, then the upper bound of C is given by,

$$c(n) < \frac{n}{(1-\delta_n)\,log_\alpha\,(n)} \tag{24}$$

where δ_n is a small number and $\delta_n \rightarrow 0$ as $n \rightarrow \infty$. In general, $n/\log_\alpha(n)$ is the upper bound of C, and the base of the logarithm is α, shown in Eq. (25),

$$\lim_{n\to\infty} c\,(n) = b\,(n) = \frac{n}{\log_\alpha(n)} \tag{25}$$

For a 0–1 sequence, $\alpha = 2$, therefore

$$b(n) = \frac{n}{log_2(n)} \tag{26}$$

and c(n) can be normalized via b(n)

$$C = \frac{c(n)}{b(n)} \tag{27}$$

where C is the normalized Lempel-Ziv complexity, which reflects the increasing rate of new patterns in the sequence.

The Lempel-Ziv complexity is a data compression technique [35]; it is also used in modeling and analyzing various biomedical signals. The Lempel-Ziv complexity is used to detect patterns in EEG signals during different mental tasks. These patterns are used in BCI applications [13] to help the disabled people to communicate and control external devices such as a robot or a wheelchair.

2.8 *Hopf Bifurcation Theorem*

A bifurcation is a radical change in the behavior of a one-dimensional dynamic system. The change occurs when one or several of its parameters pass through a critical value. Often it corresponds to the manifestation of limit cycles. Consider a planar system that depends on the parameter μ. Assume that the system has an equilibrium point at the origin for all μ.

$$\begin{cases} x' = f(x, y, \mu) \\ y' = g(x, y, \mu) \end{cases} \tag{28}$$

Suppose that the linearization at zero has two purely imaginary eigenvalues $\lambda_1(\mu)$ and $\lambda_2(\mu)$ while $\mu = \mu_c$. Suppose the real part of the eigenvalues satisfy the condition given in Eq. (29)

$$\frac{d}{d\mu}\left[Re\left(\lambda_{1,2}(\mu)\right)\right]_{|\mu=\mu c} > 0 \tag{29}$$

And the origin is asymptotically stable when $\mu = \mu_c$ then,

(i) $\mu = \mu_c$ is a bifurcation point of the system;
(ii) for some μ_1 such that $\mu_1 < \mu < \mu_c$ the origin is a stable focus;
(iii) for some μ_2 such that $\mu_c < \mu < \mu_2$ the origin is an unstable focus surrounded by a limit cycle whose size increases with μ.

At the bifurcation point, the system moves from a stable state to an unstable state surrounded by a stable limit cycle. Subsequently, the real part of the eigenvalues changes from negative to positive being zero at the bifurcation point. Then the Hopf bifurcation is said to be supercritical. The statistical parameters at and around the point of bifurcation are studied. The bifurcation point at the transition from an awake to a sleeping state and vice versa in normal and healthy subjects with sleep disorders is studied [32]. The phase transition at the bifurcation point during the prevalence of epileptic seizure activity is studied [28]. A change in neuronal activity from a resting, but excitable state to a periodic spiking behavior of a single neuron and neural populations is used in the detection of a seizure. The bifurcation diagram is widely used to describe Wendling's neuronal population model's behaviors [10] in EEG activity.

2.9 *Higher Order Spectra (HOS)*

The power spectrum estimation technique has been used in many signal processing applications, such as communications, image processing, speech processing and biomedical signal processing. The autocorrelation function and power spectrum work fine if the signal has a Gaussian or normal probability density function of known mean. However, for the EEG signals being inherently nonlinear and complex, one would need to obtain information regarding deviations from the Gaussianity and the presence of nonlinearities. In this case, the power spectrum (second-order spectrum) is of little help. Knowledge afar from the power spectrum or autocorrelation domain is required. The higher-order spectra of the order of three or higher, which are defined in terms of higher-order cumulants of the data, contains the information such as kurtosis and skewness. The third order spectrum is commonly referred to as bispectrum and the fourth-order as trispectrum. As higher-order spectra are a nonlinear function, it is reasonably helpful in the analysis of nonlinear systems operating under a random input. The higher-order correlations linking the input with the output has the ability to detect and characterize assured nonlinearities. Therefore, for this reason, several higher-order spectra based methods have been developed.

The cumulants and spectra estimation for the third-order case is presented in the following section. The EEG data of finite length is considered.

Let x(k), k = 1, ..., N be the available data.

1. Segment the data into P records of M samples each. Let $x^i(k)$, k = 0, ..., M − 1, represent the ith record.
2. Subtract the mean of each record.
3. Compute the discrete Fourier transform $X^i(k)$ of each segment, based on M points, i.e.,

$$X^i(k) = \sum_{n=0}^{M-1} x^i(n) e^{-j\frac{2\pi}{M}nk}, k = 0, 1, \ldots, M-1, i = 1, 2, \ldots, P \tag{30}$$

4. From each segment, the third-order spectrum is attained as

$$C_3^{x_i}(k_1,k_2) = \frac{1}{M} X^i(k_1)\, X^i(k_2)\, X^{i*}(k_1+k_2),\, i = 1, 2, \ldots, P \tag{31}$$

Due to the bispectrum symmetry properties, $C_3^{x_i}(k_1,k_2)$ has to be computed barely in the triangular region $0 \leq k_2 \leq k_1, k_1 + k_2 < M/2$.

5. A rectangular window of size ($M_3 \times M_3$) is used to perform additional smoothing around every frequency band, so as to reduce the variance of the estimate. The third order spectrum is assumed to be smooth.

$$\tilde{C}_3^{x_i}(k_1,k_2) = \frac{1}{M_3^2} \sum_{n_1=-M_3/2}^{M_3/2-1} \sum_{n_2=-M_3/2}^{M_3/2-1} C_3^{x_i}(k_1+n_1, k_2+n_2) \tag{32}$$

6. Finally, the third-order spectrum is given as the average overall third-order spectra, i.e.,

$$\hat{C}_3^{x}(k_1,k_2) = \frac{1}{P} \sum_{i=1}^{P} \tilde{C}_3^{x_i}(k_1,k_2) \tag{33}$$

The final bandwidth of this spectrum estimate is $\Delta = M_3/M$, which is the spacing between frequency samples in the bispectrum domain.

For large N, and as long as $\Delta \rightarrow 0$ and $\Delta^2 N \rightarrow \infty$, both the direct and the indirect methods produce asymptotically unbiased and consistent bispectrum estimates [24], with real and imaginary part variances:

$$\begin{aligned} var\left(Re\left[\hat{C}_3^x(k_1,k_2)\right]\right) &= var\left(Im\left[\hat{C}_3^x(k_1,k_2)\right]\right) \\ &= \frac{1}{\Delta^2 N} C_2^x(k_1)\, C_2^x(k_2)\, C_2^x(k_1+k_2) \\ &= \begin{cases} \frac{VL^2}{MK} C_2^x(k_1)\, C_2^x(k_2)\, C_2^x(k_1+k_2) & indirect \\ \frac{M}{KM_3^2} C_2^x(k_1)\, C_2^x(k_2)\, C_2^x(k_1+k_2) & direct \end{cases} \end{aligned} \tag{34}$$

where V is the energy of the bispectrum window [21]. From Eq. (34) it is evident that the bispectrum estimate variance can be reduced by increasing the number of records, or reducing the size of the region of support of the window in the cumulant domain (L), or by increasing the size of the smoothing window (M_3) [21].

Since, the EEG signals are nonlinear and dynamic in nature; the nonlinear higher-order spectral feature is used as a promising approach to predict the onset of seizures in the EEG recordings. It is reported that the higher-order spectra is a promising tool to differentiate between normal, pre-ictal and epileptic EEG signals [4, 8]. The results from the selected higher-order spectral based features achieve better classification accuracy compared to the power spectrum features. The higher-order spectral technique is used to determine the various sleep stages from the sleep EEG signal. Performing sleep staging by visual interpretation and linear techniques is difficult. A number of higher-order spectra based features are extracted from the various sleep stages—Wakefulness, Rapid Eye Movement (REM), Stage 1–4 Non-REM [5]. Thus, the nonlinear technique the HOS is used to extract hidden information in the sleep EEG signal. Hosseini [12] proposes an emotional stress recognition system with EEG signals using higher-order spectra to identify the emotional stress states.

3 Conclusion

The brain is one of the most complex parts of the human body and reveals a highly synergetic character. It is very intricate to understand and analyze the electrical signals emanating from the brain. The conventional linear methods are inadequate to understand the nonlinear dynamics of the EEG signals. However, chaos theory and dynamical equations may be used at the outset in understanding the potential brain signals. The chaos measurements have proven to generate new information that the conventional linear methods cannot reveal. Also, the chaos theory helps solve neurological diseases and help the neuroscience community in diagnosis. With the development of BCI technology, it calls for new methods of analysis of EEG signal to communicate and control the devices. With the advent of new analysis techniques and dynamical modeling of the brain signals, real-time applications could be designed to aid neuroscience community.

The theory and the application of different chaos measurements in the analysis of the EEG signal are presented in this work. A strong motivation is that a successful study on a complex EEG system may have a significant impact on our ability to forecast the behavior of the system and intervene to control the fatal crisis. Nonlinear dynamics has opened a new window for understanding the behavior of the brain. Nonlinear dynamic measures of complexity (D2, ApEn, SampEn) and stability (LE, FD) quantify the critical aspects of the dynamics of the brain as it evolves over time in its state space. To further understand and develop an intuition for these approaches and as a future work, simulations with well-defined data will be carried out. The study of synchronization and hyperchaotic behavior of EEG signals will also be considered in the future work.

More accurately designed experimental work is necessary to exploit the chaotic measurements and nonlinear analysis of the EEG signals. Unraveling the predicaments allows for deeper insight into the functioning of the human brain.

References

1. Azar AT, Balas VE, Olariu T (2014) Classification of EEG-based brain-computer interfaces. In: Springer international publishing in advanced intelligent computational technologies and decision support systems. Springer International Publishing, Switzerland, pp 97–106
2. Azar AT, Vaidyanathan S (2015) Chaos modeling and control systems design. In: Studies in computational intelligence, vol 581. Springer, Germany
3. Acharya R, Faust O, Kannathal N, Chua T, Laxminarayan S (2005) Non-linear analysis of EEG signals at various sleep stages. Comput Methods Programs Biomed 80(1):37–45
4. Acharya UR, Sree SV, Suri JS (2011) Automatic detection of epileptic EEG signals using higher order cumulant features. Int J Neural Syst 21(05):403–414
5. Acharya UR, Chua ECP, Chua KC, Min LC, Tamura T (2010) Analysis and automatic identification of sleep stages using higher order spectra. Int J Neural Syst 20(06):509–521
6. APICS magazine (2012) In the right space. http://media.apics.org/omnow/In%20the%20Right%20Space.pdf. Accessed 10 June 2015
7. Burioka N, Miyata M, Cornélissen G, Halberg F, Takeshima T, Kaplan DT, Shimizu E (2005) Approximate entropy in the electroencephalogram during wake and sleep. Clinical EEG Neurosci: Off J EEG Clin Neurosci Soc (ENCS) 36(1):21–24
8. Chua KC, Chandran V, Acharya UR, Lim CM (2011) Application of higher order spectra to identify epileptic EEG. J Med Syst 35(6):1563–1571
9. Eke A, Herman P, Kocsis L, Kozak LR (2002) Fractal characterization of complexity in temporal physiological signals. Physiol Meas 23(1):R1
10. Geng S, Zhou W, Yuan Q, Ma Z (2011) Bifurcation phenomenon of wendling's EEG model. In: The 2nd international IEEE symposium on bioelectronics and bioinformatics (ISBB), pp 111–114. Suzhou, China, 3–5 November 2011
11. Hans P, Dewandre PY, Brichant JF, Bonhomme V (2005) Effects of nitrous oxide on spectral entropy of the EEG during surgery under balanced anaesthesia with sufentanil and sevoflurane. Acta Anaesthesiol Belg 56(1):37–43
12. Hosseini SA, Khalilzadeh MA, Naghibi-Sistani MB, Niazmand V (2010) Higher order spectra analysis of EEG signals in emotional stress states. In: The IEEE second international conference on information technology and computer science (ITCS), 24–25 July 2010, Kiev, Ukraine, pp 60–63. doi:10.1109/ITCS.2010.21
13. Jahan IS, Prilepok M, Snasel V (2014) EEG data similarity using Lempel-Ziv complexity. In: AETA 2013: recent advances in electrical engineering and related sciences. Springer, Berlin, pp 289–295
14. Jelles B, Scheltens P, Van der Flier WM, Jonkman EJ, da Silva FL, Stam CJ (2008) Global dynamical analysis of the EEG in Alzheimer's disease: frequency-specific changes of functional interactions. Clin Neurophysiol 119(4):837–841
15. Jiang GJ, Fan SZ, Abbod MF, Huang HH, Lan JY, Tsai FF, Shieh JS (2015) Sample entropy analysis of EEG signals via artificial neural networks to model patients' consciousness level based on anesthesiologists experience. BioMed research international
16. Lorenz HW (1993) Nonlinear dynamical economics and chaotic motion, vol 334. Springer, Berlin
17. Mohanchandra K, Lingaraju GM, Kambli P (2013) Krishnamurthy V (2013) Using brain waves as new biometric feature for authenticating a computer user in real-time. Int J Biom Bioinf (IJBB) 7(1):49
18. Mohanchandra K, Saha S, Lingaraju GM (2015) EEG based brain computer interface for speech communication: principles and applications. In: Intelligent systems reference library, brain-computer interfaces: current trends and applications, vol 74. Springer-Verlag GmbH, Berlin/Heidelberg
19. Natarajan K, Acharya R, Alias F, Tiboleng T, Puthusserypady SK (2004) Nonlinear analysis of EEG signals at different mental states. Biomed Eng Online 3(1):7
20. Nguyen-Ky T, Wen P, Li Y (2014) Monitoring the depth of anaesthesia using Hurst exponent and Bayesian methods. IET Signal Process 8(9):907–917

21. Nikias CL, Petropulu AP (1993) Higher-order spectra analysis: a nonlinear signal processing framework. PTR Prentice Hall, Englewood Cliffs
22. Ocak H (2009) Automatic detection of epileptic seizures in EEG using discrete wavelet transform and approximate entropy. Expert Syst Appl 36(2):2027–2036
23. Pincus SM (1991) Approximate entropy as a measure of system complexity. Proc Natl Acad Sci 88(6):2297–2301
24. Rao TS, Gabr MM (2012) An introduction to bispectral analysis and bilinear time series models, vol 24. Springer Science & Business Media, Berlin
25. Richman JS, Moorman JR (2000) Physiological time-series analysis using approximate entropy and sample entropy. Am J Physiol-Heart Circ Physiol 278(6):H2039–H2049
26. Rosenstein MT, Collins JJ, De Luca CJ (1993) A practical method for calculating largest Lyapunov exponents from small data sets. Phys D: Nonlinear Phenom 65(1):117–134
27. Sabeti M, Katebi S, Boostani R (2009) Entropy and complexity measures for EEG signal classification of schizophrenic and control participants. Artif Intell Med 47(3):263–274
28. Scheffer M, Bascompte J, Brock WA, Brovkin V, Carpenter SR, Dakos V, Sugihara G (2009) Early-warning signals for critical transitions. Nature 461(7260):53–59
29. Shen Y, Olbrich E, Achermann P, Meier PF (2003) Dimensional complexity and spectral properties of the human sleep EEG. Clin Neurophysiol 114(2):199–209
30. Silva C, Pimentel IR, Andrade A, Foreid JP, Ducla-Soares E (1999) Correlation dimension maps of EEG from epileptic absences. Brain Topogr 11(3):201–209
31. Stam KJ, Tavy DL, Jelles B, Achtereekte HA, Slaets JP, Keunen RW (1994) Non-linear dynamical analysis of multichannel EEG: clinical applications in dementia and Parkinson's disease. Brain Topogr 7(2):141–150
32. Steriade MM, McCarley RW (2013) Brainstem control of wakefulness and sleep. Springer Science & Business Media, Berlin
33. Świderski B, Osowski S, Cichocki A, Rysz A (2007) Epileptic seizure prediction using Lyapunov exponents and support vector machine. In: Adaptive and natural computing algorithms. Springer, Berlin, Heidelberg, pp 373–381
34. Takens F (1981) Detecting strange attractors in turbulence. Springer, Berlin, Heidelberg, pp 366–381
35. Thakor NV, Tong S (2004) Advances in quantitative electroencephalogram analysis methods. Annu Rev Biomed Eng 6:453–495
36. Vaidyanathan S, Azar AT (2015) Analysis, control and synchronization of a nine-term 3-D novel chaotic system. Chaos modeling and control systems design. Springer International Publishing, Berlin, pp 19–38
37. Vaidyanathan S, Azar AT (2015) Anti-synchronization of identical chaotic systems using sliding mode control and an application to Vaidyanathan-Madhavan chaotic systems. In: Advances and applications in sliding mode control systems. Springer International Publishing, Berlin, pp 527–547
38. Vaidyanathan S, Azar AT (2015) Hybrid synchronization of identical chaotic systems using sliding mode control and an application to Vaidyanathan chaotic systems. In: Advances and applications in sliding mode control systems. Springer International Publishing, Berlin, pp 549–569
39. Vaidyanathan S, Azar AT (2015) Analysis and control of a 4-D novel hyperchaotic system. In: Chaos modeling and control systems design. Springer International Publishing, Berlin, pp 3–17
40. Vaidyanathan S, Azar AT, Rajagopal K, Alexander P (2015) Design and SPICE implementation of a 12-term novel hyperchaotic system and its synchronisation via active control. Int J Model Identif Control 23(3):267–277
41. Vaidyanathan S, Idowu BA, Azar AT (2015) Backstepping controller design for the global chaos synchronization of Sprott's jerk systems. In: Chaos modeling and control systems design. Springer International Publishing, Berlin, pp 39–58
42. Vaidyanathan S, Sampath S, Azar AT (2015) Global chaos synchronisation of identical chaotic systems via novel sliding mode control method and its application to Zhu system. Int J Model Identif Control 23(1):92–100

43. Wang X, Meng J, Tan G, Zou L (2010) Research on the relation of EEG signal chaos characteristics with high-level intelligence activity of human brain. Nonlinear Biomed Phys 4(1):2
44. Xiaobing BDQTL (2007) The sample entropy and its application in EEG based epilepsy detection [J]. J Biomed Eng 1:042

Circuit Realization of the Synchronization of Two Chaotic Oscillators with Optimized Maximum Lyapunov Exponent

V.H. Carbajal-Gómez, E. Tlelo-Cuautle and F.V. Fernández

Abstract The modeling, simulation and circuit realization of the synchronization of two optimized multi-scroll chaotic oscillators is described herein. The case of study is the master-slave synchronization of two multi-scroll chaotic oscillators generating four to seven scrolls, based on saturated function series. The maximum Lyapunov exponent (MLE) of the chaotic oscillator is optimized by applying meta-heuristics. We show the behavior on the synchronization for chaotic oscillators with low and high MLEs, while the synchronization is performed by generalized Hamiltonian forms and observer approach from nonlinear control theory. Numerical simulation results are given for the chaotic oscillators with and without optimized MLEs, and for their master-slave synchronization. Finally, we show the good agreement between theoretical results, SPICE simulations and the experimental results when the whole synchronized system is implemented with commercially available operational amplifiers.

Keywords Chaotic oscillator · Lyapunov exponent · Maximizing Lyapunov exponent · Particle swarm optimization · Chaos synchronization

1 Introduction

Numerous natural and artificial systems, completely described by deterministic dynamic laws and nonlinear differential equations without stochastic components, present an unpredictable and apparently random dynamical behavior extremely

V.H. Carbajal-Gómez · E. Tlelo-Cuautle (✉)
INAOE, Puebla, Mexico
e-mail: etlelo@inaoep.mx

V.H. Carbajal-Gómez
e-mail: vhcarbajal@inaoep.mx

F.V. Fernández
Instituto de Microelectronica de Sevilla, IMSE-CNM, CSIC, University of Seville, Seville, Spain
e-mail: pacov@imse-cnm.csic.es

A.T. Azar and S. Vaidyanathan (eds.), *Advances in Chaos Theory and Intelligent Control*, Studies in Fuzziness and Soft Computing 337,
DOI 10.1007/978-3-319-30340-6_26

sensitive to initial conditions. These systems present several interesting properties that have a great potential for commercial and industrial applications in several areas, such as engineering, computation, communication, chemistry, medicine, etc. Thus, the study of chaos in nonlinear electronic circuits has been a very active topic of research. Generation of multi-scroll chaotic attractors has received considerable attention for more than a decade; such interest is in both theoretical and practical issues [1, 2, 9, 13, 14, 23, 29]. For instance, Lyapunov exponents provide a means of ascertaining whether the behavior of a dynamical system is chaotic. Thus, the presence of at least one positive or maximum Lyapunov exponent (MLE) in a dynamical system can be taken as a confirmation of chaotic motion [24]. That way, the unpredictability of the chaotic oscillations increases when the MLE of the system also increases [11]. The Lyapunov exponent values that can be positive, zero or negative, directly depend on the nonlinear system parameter values [16, 17], like coefficients and perturbation functions. In this scenario, and since the number of combinations of values for the coefficients and nonlinear functions is quite huge, then evolutionary optimization algorithms are quite useful to find the best relation between the parameter values and unpredictability (MLE optimization).

One hot topic for chaotic oscillators is their synchronization [26, 27, 31], which has received increased interest. This property is supposed to have interesting applications in different fields, particularly in designing secure communication systems [19, 28–30]. For instance, private communication schemes are usually composed of chaotic systems [7, 15, 19, 22], where the confidential information is embedded into the transmitted chaotic signal by direct modulation, masking, or another technique. At the receiver end, if chaotic synchronization is achieved, then it is possible to extract the hidden information from the transmitted signal.

Better unpredictability is guaranteed if MLE is high. In this manner, the Particle Swarm Optimization (PSO) algorithm is applied herein to optimize MLE of a third order dynamical system in which a saturated nonlinear function (SNLF) series is used to model the nonlinear part. The chapter shows comparisons by plotting phase diagrams for generating four to seven scrolls, and highlighting that for low MLE values the chaotic attractors are well defined, whereas for optimized or high MLE values, the chaotic attractors are not well appreciated, thus improving the unpredictability of the dynamical system.

The chapter is structured as follows. In Sect. 2, the nonlinear system and the SNLF description are introduced. Here, the parameters of the multi-scroll chaotic oscillator and the voltage dynamic-range of commercially available operational amplifiers are used to implement the SNLF. Section 3 provides a review on the PSO algorithm. An approach to compute MLE is briefly introduced in Sect. 4, which is used in the PSO loop to optimize MLE. A summary on chaos synchronization via Hamiltonian systems and nonlinear observer approach from nonlinear control theory is provided in Sect. 5. Section 6 presents the synchronization of multi-scroll chaotic attractors. The circuit implementation is verified using the simulation program with integrated circuit emphasis (SPICE), and then it is realized with commercially available electronic circuits, for which experimental results on the synchronization are shown. Finally, the conclusions and future work are listed in Sect. 7.

2 Multi-scroll Chaotic Oscillator

A chaotic oscillator can be described by classifying a linear part and a nonlinear part [9]. In some cases, the nonlinear part can be approached to a piecewise-linear (PWL) function [9, 13, 18, 25], which is relatively easy to implement by using commercially available electronic devices.

In this chapter, the chaotic oscillator is based on SNLF series that can be modeled using PWL approximations. Figure 1 shows a SNLF series with 5 and 7 segments to generate 3 and 4-scrolls, respectively. In (1) a PWL approximation called series of a saturated function is described, where $k \geq 2$ is the slope of the saturated function and a multiplier factor to saturated plateaus, $plateau = \pm nk$, with n being an odd integer to generate an even number of scrolls and even integer to generate an odd number of scrolls. h is the saturated delay of the center of the slopes, and must agree with $h_i = \pm mk$, where $i = 1, \ldots, [(scrolls - 2)/2]$ and $m = 2, 4, \ldots, (scrolls - 2)$ to generate an even number of scrolls; and $i = 1, \ldots, [(scrolls - 1)/2]$ and $m = 1, 3, \ldots, (scrolls - 2)$ to generate an odd number of scrolls; p and q are positive integers. To generate n-scrolls attractors in a third order dynamical system, a controller (f) is added as shown in (2) [13], where $f(x1; k, h, p, q)$ is defined by (3), and a, b, c, d_1 are real and positive constants.

$$f(x_1; k, h, p, q) = \sum_{i=-p}^{q} f_i(x_1; h, k) \tag{1}$$

$$\begin{aligned}\dot{x_1} &= x_2 \\ \dot{x_2} &= x_3 \\ \dot{x_3} &= -ax_1 - bx_2 - cx_3 + d_1 f(x_1; k, h, p, q)\end{aligned} \tag{2}$$

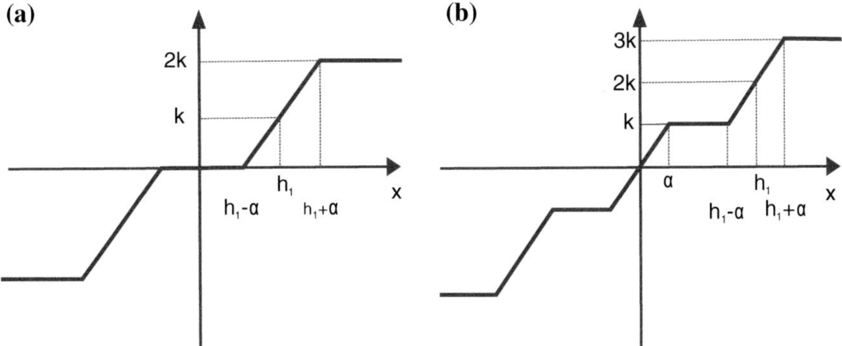

Fig. 1 PWL description of a SNLF with **a** 5 and **b** 7 segments

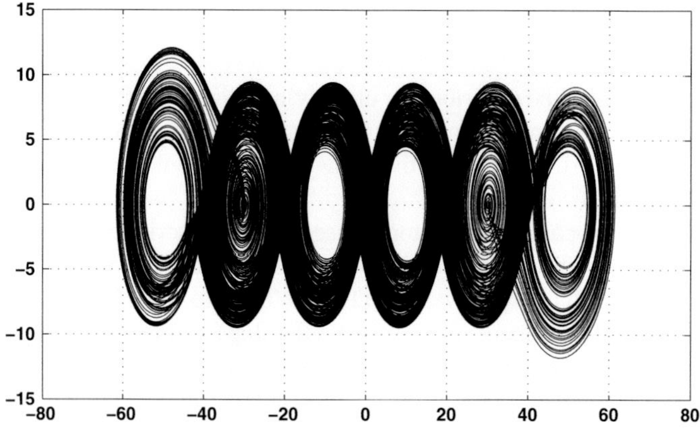

Fig. 2 6-scrolls attractor simulated in MATLAB

$$f(x_1; k, h, p, q) = \begin{cases} (2q+1)k & x_1 > qh+1 \\ k(x_1 - ih) + 2ik & \mid x_1 - ih \mid \leq 1 \\ & -p \leq i \leq q \\ (2i+1)k & ih+1 < x_1 < (i+1)h - 1 \\ & -p \leq i \leq q-1 \\ -(2p+1)k & x_1 < -ph - 1 \end{cases} \tag{3}$$

The simulation of a 6-scrolls chaotic attractor by using (2) and (3), was executed using package ODE45 in MATLAB. The parameter values were $a = b = c = d_1 = 0.7$, $k = 10$, $h = 20$, $p = q = 2$, and the simulation result is shown in Fig. 2, where two state variables are plotted, and where it can be appreciated that the dynamic ranges are very large for electronic devices, i.e. the horizontal axes goes from nearly -60 to 60 and the vertical one from -13 to 13, approximately. Since commercially available operational amplifiers work pretty well within ranges from -18 to 18V, then a scaling process is needed. Equation (3) cannot be implemented directly because $k \geq 2$ [13]. In addition, $h = 2k$ or $h = k$ should be accomplished to generate even or odd number of scrolls, respectively, to avoid superimposing the slopes because the plateaus can disappear. Henceforth, α is restricted to 1, so that to implement n-scrolls attractors using real operational amplifiers the dynamic range of the SNLF must be scaled [14]. In this case, the SNLF series is redefined by (4), where α allows that $k < 1$ because the chaos-condition now applies on $s = \frac{k}{\alpha}$, the new slope. In this manner, k and α can be selected to permit $k < 1$, so that the ranges in (3) can be scaled. As a result, 6-scrolls attractors are generated with $a = b = c = d_1 = 0.7$, $k = 1$, $\alpha = 0.1$, $s = 10$, $h = 2$, and $p = q = 2$, as shown in Fig. 3. Now, the ranges of the attractors are within the voltage dynamic ranges that can be processed by real amplifiers. Besides, it is possible to have smaller dynamic ranges by combining values of k and α.

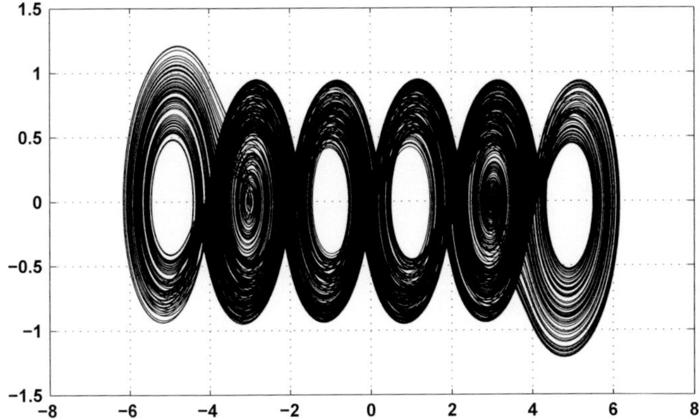

Fig. 3 6-scrolls attractor after scaling the original dynamical system

$$f(x_1; k, h, p, q) = \begin{cases} (2q+1)k & x_1 > qh + \alpha \\ \frac{k}{\alpha}(x_1 - ih) + 2ik & \mid x_1 - ih \mid \leq \alpha \\ & -p \leq i \leq q \\ (2i+1)k & ih + \alpha < x_1 < (i+1)h - \alpha \\ & -p \leq i \leq q-1 \\ -(2p+1)k & x_1 < -ph - \alpha \end{cases} \quad (4)$$

2.1 Circuit Implementation

The dynamical system in (2) has the block diagram representation shown in Fig. 4, which can be realized by using three integrators, one adder, a PWL function to implement $f(x)$, inverters and inverting amplifiers to tune the coefficient values. In electronics, each block can be realized with different kinds of active devices, namely: operational amplifiers, current-feedback operational amplifiers, current conveyors, unity-gain-cells, and so on. In this chapter, the realization of the dynamical system from (2) is done by using operational amplifiers, so that the electronic circuit is shown in Fig. 5, where it is easy to identify the integrators having one capacitor C.

To implement the SNLF block, the operational amplifier is saturated by using its finite-gain model sketched in Fig. 6, where its real limitations like gain, bandwidth, slew rate and saturation can be identified [21]. In this manner, to increase the number of saturation levels, a shift-voltage ($\pm E$) is added, so that the shifted-voltage equations determined by (5) for positive and negative shifts, respectively, as shown in Fig. 7, are obtained. From these descriptions, when realizing a PWL function like the SNLF block, the topology shown in Fig. 8 can be used, where the number of operational amplifiers equals the number of scrolls to be generated minus 1. In the same

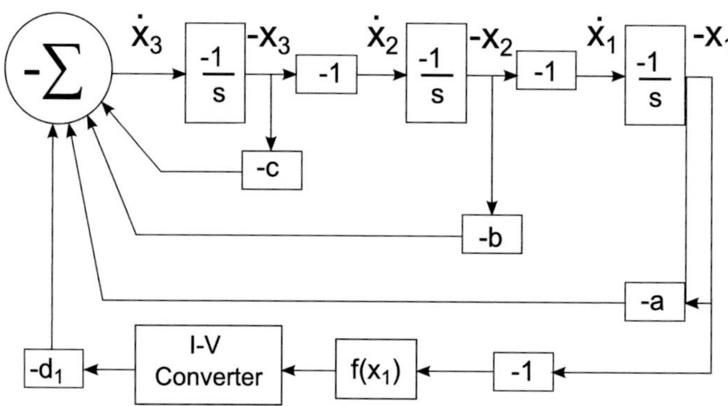

Fig. 4 Block diagram description of Eq. (2)

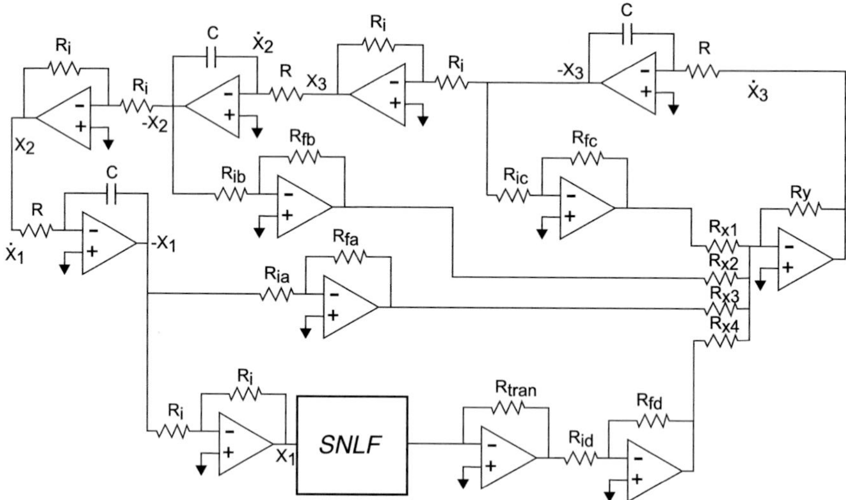

Fig. 5 Implementation of Eq. (2) by using operational amplifiers

Fig. 6 Finite gain model of the operational amplifier

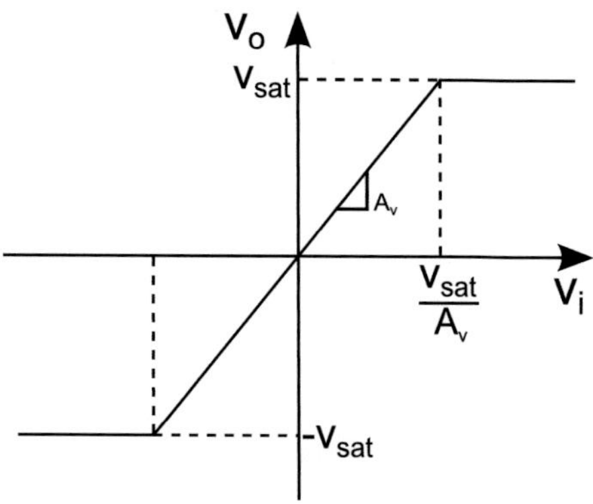

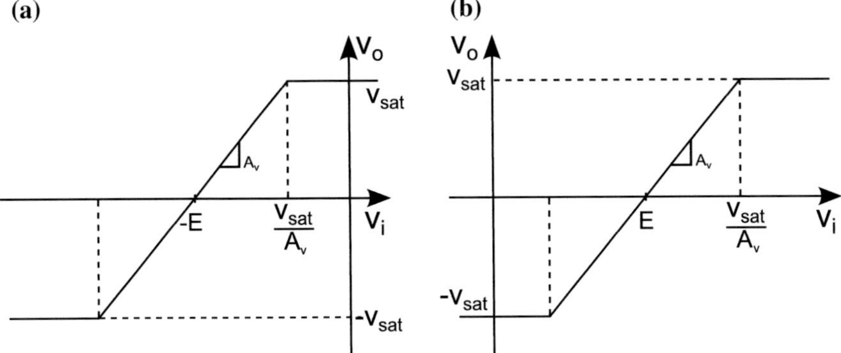

Fig. 7 Shift voltage to the **a** Negative, and **b** Positive sides

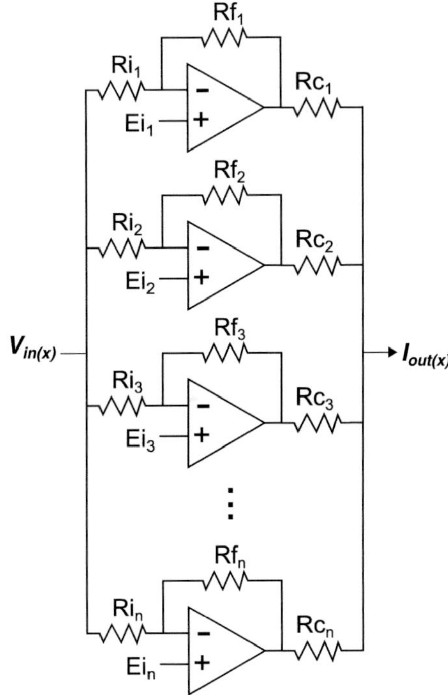

Fig. 8 Realization of the SNLF using operational amplifiers

manner, to generate different numbers of saturation levels, E takes different values in (5) to synthesize the required plateaus and slopes. In this chapter, the value of the plateaus k, in voltage and current, the breakpoints α, the slope s and the saturated delays h are evaluated by (6).

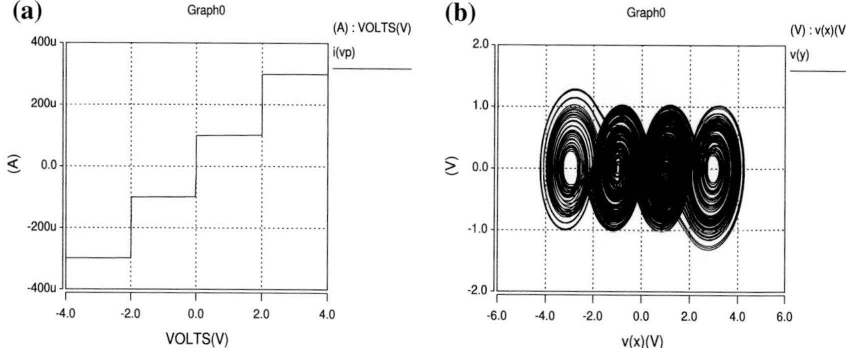

Fig. 9 Simulation results for generating 4-scrolls using SPICE. **a** SNLF. **b** 4-scrolls attractor chaotic oscillator

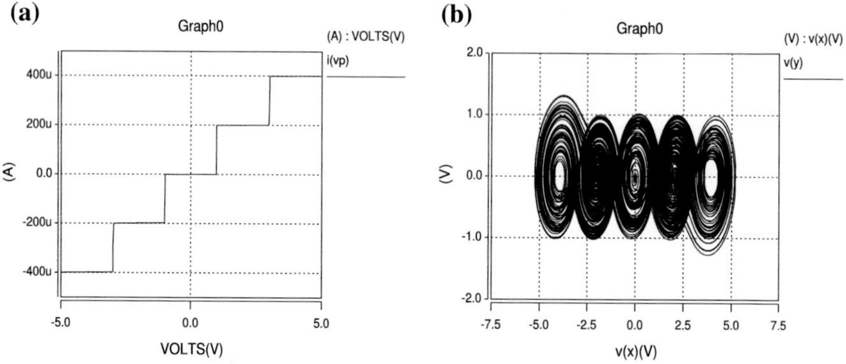

Fig. 10 Simulation results for generating 5-scrolls using SPICE. **a** SNLF. **b** 5-scrolls attractor chaotic oscillator

$$V_o = \frac{A_v}{2}(|V_i + \frac{V_{sat}}{A_v} - E| - |V_i - \frac{V_{sat}}{A_v} - E|) \tag{5}$$

$$V_o = \frac{A_v}{2}(|V_i + \frac{V_{sat}}{A_v} + E| - |V_i - \frac{V_{sat}}{A_v} + E|)$$

$$k = R_{ix} I_{sat}, \quad I_{sat} = \frac{V_{sat}}{R_c}, \quad \alpha = \frac{R_{iz}|V_{sat}|}{R_{fz}}, \quad h = \frac{E_i}{(1 + \frac{R_{iz}}{R_{fz}})} \tag{6}$$

Simulating (2) in the circuit simulator SPICE, using the model for the commercially available operational amplifier TL081, and the diagrams from Figs. 5 and 8, one can generate the attractors having 4 and 5 scrolls by selecting the functional specifications: $a = b = c = d = 0.7$, $k = 1$, $\alpha = 0.0165$, $s = 60.606$, $h_1 \simeq 1$, $I_{sat} = 100\,\mu\text{A}$, $R_{ix} = 10\,\text{k}\Omega$, $C = 1\,\mu\text{F}$, $R = 1\,\text{M}\Omega$, $R_{ia} = R_{ib} = R_{ic} = R_{id} = 10\,\text{k}\Omega$, $R_{fa} = R_{fb} = R_{fc} = R_{fd} = 7\,\text{k}\Omega$, $R_{x1} = R_{x2} = R_{x3} = R_{x4} = R_y = 10\,\text{k}\Omega$, $R_i = 10\,\text{k}\Omega$ and $V_{sat} = \pm 16\,V$, [21]. The simulation results obtained with SPICE are shown in Figs. 9 and 10.

3 Particle Swarm Optimization

Many global optimization problems can be formulated as the following form:

$$\min f(\mathbf{x}), \quad \mathbf{x} = [x_1, \ldots, x_d], \quad s.t. \; x_j \in [a_j, b_j], \quad j = 1, \ldots, d.$$

where f is the objective function, and $\mathbf{x}$ is a continuous variable vector in the d-dimensional space $\mathbb{R}^d$. The feasible domain of variables x_j is defined by specifying upper (b_j) and lower (a_j) limits of each component j.

Particle swarm optimization (PSO) was developed by Kennedy and Eberhart in 1995 [4], and was inspired on swarm behaviour observed in nature such as fish and bird schooling. Since then, PSO has generated a lot of attention, and now forms an exciting, ever-expanding research subject in the field of swarm intelligence. PSO has been applied to almost every area in optimization, computational intelligence, and design applications. There are at least two dozens of PSO variants, as well as hybrid algorithms obtained by combining PSO with other existing algorithms, which are also increasingly popular [3, 20].

PSO searches the space of an objective function by adjusting the trajectories of individual agents, called *particles*. Each *particle* traces a piecewise path that can be modeled as a time-dependent positional vector. Each particle is attracted toward the position of the current global best $pbest_i.pos_d$ and its own best known location $pbest_{gbest[i]}.pos$ in history.

When a particle finds a location that is better than any previously found locations, then it updates this location as the new current best for particle i. There is a current best for all N particles at any time t at each iteration. The aim is to find the global best among all the current best solutions until the objective no longer improves or after a certain number of iterations [8].

The pseudocode for PSO is shown in Algorithm 1. Each particle p_i has three associated values: position $p_i.pos$, velocity $p_i.vel$, and the value of the fitness function $p_i.fit$. Particle pbest has only position and fitness function value. *gbest*[] is a vector that stores indexes to reference pbest particles. *rand*() is a function that returns a random number greater or equal to zero and less that one. *evaluate*() is a function that calculates the value of the fitness for the problem to solve. This PSO version was inspired from [4] and [8]. Particles position p_i are initialized randomly and also their velocities (in lines 5–10 and 11–15 in Algorithm 1, respectively). Each particle is evaluated and $pbest_i$ particles are initialized equal to the p_i ones. For a given number of iterations the following process is applied: (1) three random numbers are calculated in $[1, N]$ (N = population size) with replacement; *gbest*[i] points to the best particle inside this cluster of three particles. (2) A new particle is calculated and its velocity is updated (line 22–23). If this new particle is better than its associated *pbest* then *pbest* particle takes the values of the new particle. The core of PSO is in the loop of lines 17–23. The update rules are

$$\begin{aligned} p_i.pos_d \leftarrow\; & wp_i.vel_d + \varphi_1 U_1(pbest_i.pos_d - p_i.pos_d) \\ & + \varphi_2 U_2(pbest_{gbest[i]}.pos_d - p_i.pos_d) \end{aligned}$$

Algorithm 1 Pseudocode of the particle swarm optimization algorithm

N is the number of particles
G is the number of iterations (generations)
Variable bounds $x_i \in [l_i, u_i]$, for $i = 1, 2, \ldots D$
Procedure PSO $(N, G, \{l_i\}, \{u_i\})$
for $i = 1 : N$ **do** ▷ Initialize particles positions
for $d = 1 : D$ **do**
$p_i.pos_d = l_d + (u_d - l_d) \cdot rand()$
$pbest_i.pos_d \leftarrow p_i.pos_d$
$p_i.fit \leftarrow evaluate(p_i.pos)$
$pbest_i.fit \leftarrow p_i.fit$
for $i = N : D$ **do** ▷ Initialize particles velocities
for $d = 1 : D$ **do**
$vmin = l_d - p_i.pos_d$
$vmax = u_d - p_i.pos_d$
$p_i.vel_d = vmin + (vmax - vmin) \cdot rand()$
for $g = 1 : G$ **do** ▷ Iterate G generations
for $i = 1 : N$ **do** ▷ For each particle
Let j_1, j_2 and j_3 be three random numbers in $\{1, N\}$
$gbest[i] = k|min(pbest_k.fit)$, for $k \in \{i, j_1, j_2, j_3\}$
for $i = N : D$ **do** ▷ For each particle
for $d = 1 : D$ **do** ▷ For each dimension
$p_i.pos_d \leftarrow w\ p_i.vel_d + \varphi_1 U_1(pbest_i.pos_d - p_i.pos_d) + \varphi_2 U_2(pbest_{gbest[i]}.pos_d - p_i.pos_d)$
$p_i.vel_d \leftarrow p_i.pos_d + p_i.vel_d$
If $p_i.pos_d < l_d$
$p_i.pos_d = l_d$; $p_i.vel_d = 0$
If $p_i.pos_d > L_d$ **then**
$p_i.pos_d = L_d$; $p_i.vel_d = 0$
$f = evaluate(p.pos_i)$
If $f < pbest.fit_i$ **then**
$pbest_i.pos \leftarrow p_{pos_i}$
$pbest_i.fit \leftarrow f$
search q $= pbest_k.pos - min(pbest_k.fit)$, for $k = 1, 2, \ldots, N$.
q is the solution at iteration g.

where w is a parameter called inertia weight, φ_1 and φ_2 are two parameters called acceleration coefficients, U_1 and U_2 are two random numbers uniformly distributed in the interval [0, 1).

4 Computing Lyapunov Exponents

The Lyapunov exponents give the most characteristic description of the presence of a deterministic nonperiodic flow. Therefore, Lyapunov exponents are asymptotic measures characterizing the average rate of growth (or shrinkage) of small perturbations to the solutions of a dynamical system. Lyapunov exponents provide quantitative measures of response sensitivity of a dynamical system to small changes in initial conditions [12]. The number of Lyapunov exponents equals the number of state variables, and if at least one is positive, this is an indication of chaos [12, 16, 24]. That way, an algorithm capable of computing the Lyapunov exponents in a simple fashion is very much in need to guarantee chaotic regime.

Lets us consider an n-dimensional dynamical system:

$$\dot{x} = f(x) \quad x \in \mathbb{R}^n \tag{7}$$

where x and f are n-dimensional vector fields, $t > 0$ and the initial conditions $x(0) = x_0$. To determine the n Lyapunov exponents of the system one have to find the long term evolution of small perturbations to a trajectory, which are determined by the variational equation from (7),

$$\dot{y} = \frac{\partial f}{\partial x}\big(x(t)\big)y = J\big(x(t)\big)y \tag{8}$$

where J is the $n \times n$ Jacobian matrix of f. A solution of (8) with a given initial perturbation $y(0)$ can be written as

$$y(t) = Y(t)y(0) \tag{9}$$

with $Y(t)$ as the fundamental solution satisfying

$$\dot{Y} = J\big(x(t)\big)Y \quad Y(0) = I_n \tag{10}$$

Here I_n denotes the $n \times n$ identity matrix. If we consider the evolution of an infinitesimal n-parallelepiped $[p_1(t), \ldots, p_n(t)]$ with the axis $p_i(t) = Y(t)p_i(0)$ for $i = 1, \ldots, n$, where $p_i(0)$ denotes an orthogonal basis of $\mathbb{R}^n$, then the i-th Lyapunov exponent, which measures the long-time sensitivity of the flow $x(t)$ with respect to the initial data $x(0)$ at the direction $p_i(t)$, is defined by the expansion rate of the length of the i-th axis $p_i(t)$ and is given by,

$$\lambda_i = \lim_{t\to\infty} \frac{1}{t} \ln \left\| p_i(t) \right\| \tag{11}$$

The Lyapunov exponents can be computed by applying the methods given in [16, 17, 24, 32]. In general, the method can be summarized as follows [16]:

1. Initial conditions of the system and the variational system are set to $\mathbf{X_0}$ and $\mathbf{I_{n\times n}}$, respectively.
2. The systems are integrated by several steps until the orthonormalization period TO is reached. The integration of the variational system $\mathbf{Y} = [y_1, y_2, y_3]$ depends on the specific Jacobian that the original system $\mathbf{X}$ is using in the current step.
3. The variational system is orthonormalized by using the standard Gram-Schmidt method [17], the logarithm of the norm of each Lyapunov vector contained in $\mathbf{Y}$ is obtained and accumulated in time.

4. The next integration is carried out by using the new orthonormalized vectors as initial conditions. This process is repeated until the full integration period T is reached.
5. The Lyapunov exponents are obtained by

$$\lambda_i \approx \frac{1}{T} \sum_{j=TO}^{T} \ln \|\mathbf{y}_i\|$$

For instance, in [24] the time step selection was made by using the minimum absolute value of all the eigenvalues of the system λ_{min}, and ψ was chosen well above the sample theorem as 50.

$$t_{step} = \frac{1}{\lambda_{min}\psi}$$

4.1 Optimizing the Maximum Lyapunov Exponent (MLE)

In most reported approaches using saturated nonlinear function series based chaotic oscillators [9, 13, 18, 25], the coefficients of the system are fixed and equal to 0.7, but the value of MLE is relatively small. The manner in which this MLE value can be increased is by performing optimization. However, since the search space is huge as the number of decimals increase for each coefficient a, b, c and d_1 in (2), then evolutionary algorithms are a good optimization option. In this section all coefficients are set to 0.7 and the corresponding MLE values are compared with the optimized ones generated by applying the PSO algorithm. The values are listed in Table 1. In this case, the optimization problem was established to vary the coefficients within the range 0.001 to 1.000, so that the search space has $10^3 \times 10^3 \times 10^3 \times 10^3$ potential solutions for the four coefficients, thus justifying the application of PSO algorithm [5, 10].

PSO is an stochastic algorithm, so that different results can be obtained among different runs and such results may also depend on the parameter settings [5, 10, 24]. In this work, PSO was executed with the inertia weight $w = 0.721$ and the acceleration coefficients $\varphi_1 = \varphi_2 = 1.193$. A population of 20 particles and 200 generations were used. Therefore, a total of 4000 fitness evaluations were allowed for this algorithm.

Figures 11, 12, 13 and 14 shows the phase diagram for the case of 4 to 7 optimized scrolls chaotic oscillator listed in Table 1. It can be appreciated that the dynamic behavior of the chaotic system is more complex as the positive Lyapunov exponent increases, because it achieves greater unpredictability of the chaotic behavior.

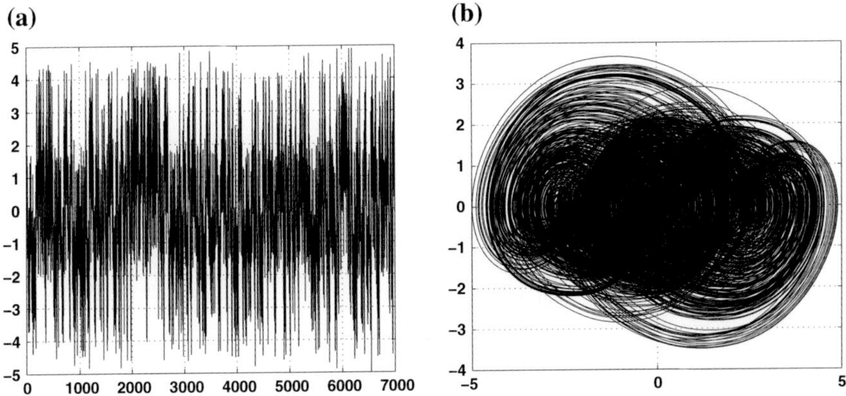

Fig. 11 Optimized 4-scrolls chaotic oscillator. **a** Time evolution, **b** phase space portraits

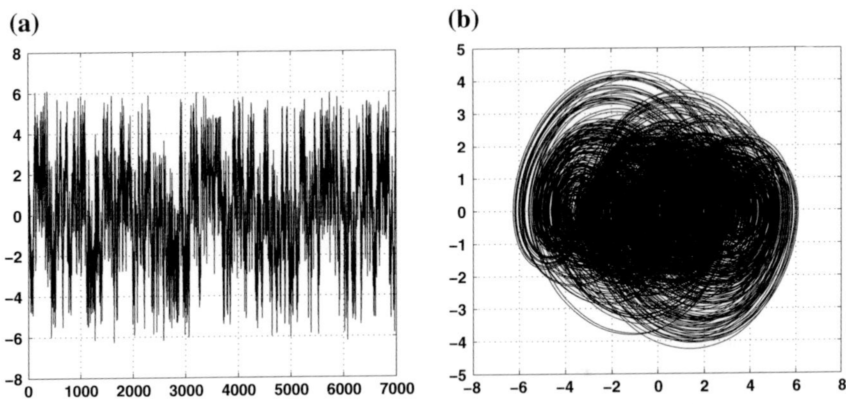

Fig. 12 Optimized 5-scrolls chaotic oscillator. **a** Time evolution, **b** phase space portraits

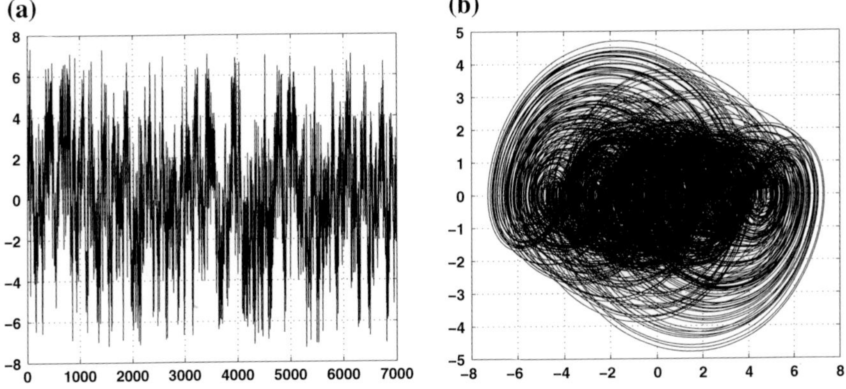

Fig. 13 Optimized 6-scrolls chaotic oscillator. **a** Time evolution, **b** phase space portraits

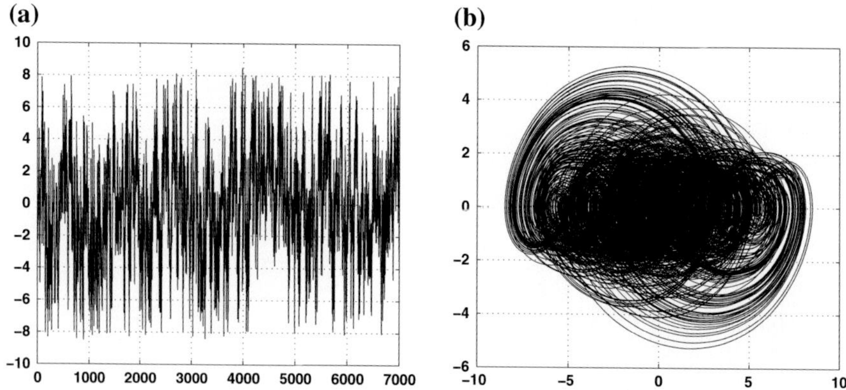

Fig. 14 Optimized 7-scrolls chaotic oscillator. **a** Time evolution, **b** phase space portraits

Table 1 MLE values optimized with PSO

Scrolls	MLE with fixed constants ($a = b = c = d_1$)	Best values for MLE and (a, b, c, d_1)
4	0.2652 (0.7, 0.7, 0.7, 0.7)	0.7001 (1.000 0.887 0.197 1.000)
5	0.2659 (0.7, 0.7, 0.7, 0.7)	0.7412 (0.990 0.836 0.150 1.000)
6	0.2781 (0.7, 0.7, 0.7, 0.7)	0.7918 (1.000 0.656 0.176 0.999)
7	0.2901 (0.7, 0.7, 0.7, 0.7)	0.8287 (1.000 0.591 0.178 1.000)

5 Hamiltonian Synchronization Approach

Multi-scrolls attractors can be synchronized by applying the Hamiltonian approach. This technique is well described in the seminal article [19]. We adopt it because of its suitability to automation [15, 25].

Lets us consider the dynamical system

$$\dot{x} = f(x) \tag{12}$$

where $x \in \mathbb{R}^n$ is the state vector and $f : \mathbb{R}^n \to \mathbb{R}^n$ is a nonlinear function. In [19], it is reported how system (12) can be written in the *Generalized Hamiltonian canonical form*:

$$\dot{x} = J(x)\frac{\partial H}{\partial x} + S(x)\frac{\partial H}{\partial x} + F(x), \quad x \in \mathbb{R}^n \tag{13}$$

where $H(x)$ denotes a smooth energy function which is globally positive definite in $\mathbb{R}^n$. The gradient vector of H, denoted by $\frac{\partial H}{\partial x}$, is assumed to exist everywhere. We use quadratic energy function $H(x) = \frac{1}{2}x^T Mx$ with M being a constant, symmetric positive definite matrix. In this case, $\frac{\partial H}{\partial x} = Mx$. The matrices $J(x)$ and $S(x)$ satisfy, for all $x \in \mathbb{R}^n$, the following properties: $J(x) + J^T(x) = 0$ and $S(x) = S^T(x)$. The vector field $J(x)\frac{\partial H}{\partial x}$ exhibits the conservative part of the system and it is also referred to as the workless part, or workless forces of the system, and $J(x)$ denotes the working or nonconservative part of the system.

For certain systems, $S(x)$ is *negative definite* or *negative semidefinite*. Thus, the vector field is referred to as the dissipative part of the system. If, on the other hand, $S(x)$ is positive definite, positive semidefinite, or indefinite, it clearly represents the global, semi-global, or local destabilizing part of the system, respectively. In the last case, we can always (although non-uniquely) decompose such an indefinite symmetric matrix into the sum of a symmetric negative semidefinite matrix $R(x)$ and a symmetric positive semidefinite matrix $N(x)$. Finally, $F(x)$ represents a locally destabilizing vector field.

In the context of observer design, we consider a special class of Generalized Hamiltonian forms with output $y(t)$, given by

$$\begin{aligned}\dot{x} &= J(y)\frac{\partial H}{\partial x} + (I + S)\frac{\partial H}{\partial x} + F(y), \quad x \in \mathbb{R}^n \\ y &= C\frac{\partial H}{\partial x}, \quad y \in \mathbb{R}^m\end{aligned} \tag{14}$$

where S is a constant symmetric matrix, not necessarily of a definite sign. I is a constant skew symmetric matrix, and C is a constant matrix.

We denote the estimate of the state $x(t)$ by $\xi(t)$, and consider the Hamiltonian energy function $H(\xi)$ to be the particularization of H in terms of $\xi(t)$. Similarly, we denote by $\eta(t)$ the estimated output, computed in terms of $\xi(t)$. The gradient vector $\frac{\partial H(\xi)}{\partial \xi}$ is, naturally, of the form $M\xi$ with M being a constant, symmetric positive definite matrix. A nonlinear state observer for the Generalized Hamiltonian form (14) is given by

$$\begin{aligned}\dot{\xi} &= J(y)\frac{\partial H}{\partial \xi} + (I + S)\frac{\partial H}{\partial \xi} + F(y) + K(y - \eta), \quad \xi \in \mathbb{R}^n \\ \eta &= C\frac{\partial H}{\partial \xi}, \quad \eta \in \mathbb{R}^m\end{aligned} \tag{15}$$

where K is the observer gain. The state estimation error, defined as $e(t) = x(t) - \xi(t)$, and the output estimation error, defined as $e_y(t) = y(t) - \eta(t)$, are governed by

$$\begin{aligned}\dot{e} &= J(y)\frac{\partial H}{\partial e} + (I + S - KC)\frac{\partial H}{\partial e}, \quad e \in \mathbb{R}^n \\ e_y &= C\frac{\partial H}{\partial e}, \quad e_y \in \mathbb{R}^m\end{aligned} \tag{16}$$

where $\frac{\partial H}{\partial e}$ actually stands, with some abuse of notation, for the gradient vector of the modified energy function, $\frac{\partial H(e)}{\partial e} = \frac{\partial H}{\partial x} - \frac{\partial H}{\partial \xi} = M(x - \xi) = Me$. We set, when needed, $I + S = W$.

Definition 1 (*Chaotic synchronization*) [19] The slave system (nonlinear state observer) (15) synchronizes with the chaotic master system in Generalized Hamiltonian form (14), if

$$\lim_{t \to \infty} \|x(t) - \xi(t)\| = 0 \tag{17}$$

no matter which initial conditions $x(0)$ and $\xi(0)$ have, where the state estimation error $e(t) = x(t) - \xi(t)$ corresponds to the synchronization error.

6 Synchronization of Multi-scroll Chaotic Attractors

The chaos generator model (12)–(14) in Generalized Hamiltonian form, according to (2) (master model) is given by

$$\begin{bmatrix} \dot{x}_1 \\ \dot{x}_2 \\ \dot{x}_3 \end{bmatrix} = \begin{bmatrix} 0 & \frac{1}{2b} & \frac{1}{2} \\ -\frac{1}{2b} & 0 & 1 \\ -\frac{1}{2} & -1 & 0 \end{bmatrix} \frac{\partial H}{\partial x} + \begin{bmatrix} 0 & \frac{1}{2b} & -\frac{1}{2} \\ \frac{1}{2b} & 0 & 0 \\ -\frac{1}{2} & 0 & -c \end{bmatrix} \frac{\partial H}{\partial x} + \begin{bmatrix} 0 \\ 0 \\ d_1 f(x) \end{bmatrix} \tag{18}$$

We take as Hamiltonian energy function

$$H(x) = \frac{1}{2}[ax_1^2 + bx_2^2 + x_3^2] \tag{19}$$

and as gradient vector

$$\frac{\partial H}{\partial x} = \begin{bmatrix} a & 0 & 0 \\ 0 & b & 0 \\ 0 & 0 & 1 \end{bmatrix} \begin{bmatrix} x_1 \\ x_2 \\ x_3 \end{bmatrix} = \begin{bmatrix} ax_1 \\ bx_2 \\ x_3 \end{bmatrix}$$

The destabilizing vector field calls for x_1 and x_2 signals to be used as the outputs, of the master model (18). We use $y = x_1$ in (18). The matrices C, S, and I are given by

$$C = \begin{bmatrix} \frac{1}{a} & 0 & 0 \end{bmatrix}$$

$$S = \begin{bmatrix} 0 & \frac{1}{2b} & -\frac{1}{2} \\ \frac{1}{2b} & 0 & 0 \\ -\frac{1}{2} & 0 & -c \end{bmatrix}$$

$$
I = \begin{bmatrix} 0 & \frac{1}{2b} & \frac{1}{2} \\ -\frac{1}{2b} & 0 & 1 \\ -\frac{1}{2} & -1 & 0 \end{bmatrix}
$$

The pair (C, S) is observable. Therefore, the nonlinear state observer for (18), to be used as the slave model, is designed according to (15) as

$$
\begin{bmatrix} \dot{\xi}_1 \\ \dot{\xi}_2 \\ \dot{\xi}_3 \end{bmatrix} = \begin{bmatrix} 0 & \frac{1}{2b} & \frac{1}{2} \\ -\frac{1}{2b} & 0 & 1 \\ -\frac{1}{2} & -1 & 0 \end{bmatrix} \frac{\partial H}{\partial \xi} + \begin{bmatrix} 0 & \frac{1}{2b} & -\frac{1}{2} \\ \frac{1}{2b} & 0 & 0 \\ -\frac{1}{2} & 0 & -c \end{bmatrix} \frac{\partial H}{\partial \xi} + \dots
$$
$$
+ \begin{bmatrix} 0 \\ 0 \\ d_1 f(\xi) \end{bmatrix} + \begin{bmatrix} k_1 \\ k_2 \\ k_3 \end{bmatrix} e_y \tag{20}
$$

with gain k_i, $i = 1, 2, 3$ to be selected in order to guarantee asymptotic exponential stability to zero of the state reconstruction error trajectories (i.e., synchronization error $e(t)$). From (18) and (20) we have that the synchronization error dynamics is governed by [15]

$$
\begin{bmatrix} \dot{e}_1 \\ \dot{e}_2 \\ \dot{e}_3 \end{bmatrix} = \begin{bmatrix} 0 & \frac{1}{2b} & \frac{1}{2} \\ -\frac{1}{2b} & 0 & 1 \\ -\frac{1}{2} & -1 & 0 \end{bmatrix} \frac{\partial H}{\partial e} + \begin{bmatrix} 0 & \frac{1}{2b} & -\frac{1}{2} \\ \frac{1}{2b} & 0 & 0 \\ -\frac{1}{2} & 0 & -c \end{bmatrix} \frac{\partial H}{\partial e} + \begin{bmatrix} k_1 \\ k_2 \\ k_3 \end{bmatrix} e_y \tag{21}
$$

When we have selected $K = (k_1, k_2, k_3)^T$ with $k_1 = 2$, $k_2 = 5$, $k_3 = 7$, and considering the initial condition $X(0) = [0, 0, 0.1]$, $\xi(0) = [1, -0.5, 3]$, we carry out the following numerical simulations by using *ode*45 integration algorithm in MATLAB with a full integration of $T = 2000$ for 4-scrolls chaotic oscillator and taking account the coefficients values of Table 1. Figures 15a, b and 16a, b show the state trajectories between the master and slave models (18) and (20), respectively. Figures 15c, d and 16c, d show the synchronization error (21) and the phase error between the master and slave chaotic oscillators, respectively.

6.1 Circuit Implementation

Our proposed scheme for the synchronization of multi-scroll chaotic oscillators of the form (2), by using operational amplifiers is shown in Fig. 17. The vector K in (20) is the observer gain and it is adjusted by selecting R_{io}, R_{fo}, R_{ko} according to the sufficient conditions for synchronization [15, 19].

By selecting values of the elements reported in [6, 7] on Fig. 17, the SPICE simulation of the synchronization is shown in Fig. 18 [7, 14]. The synchronization

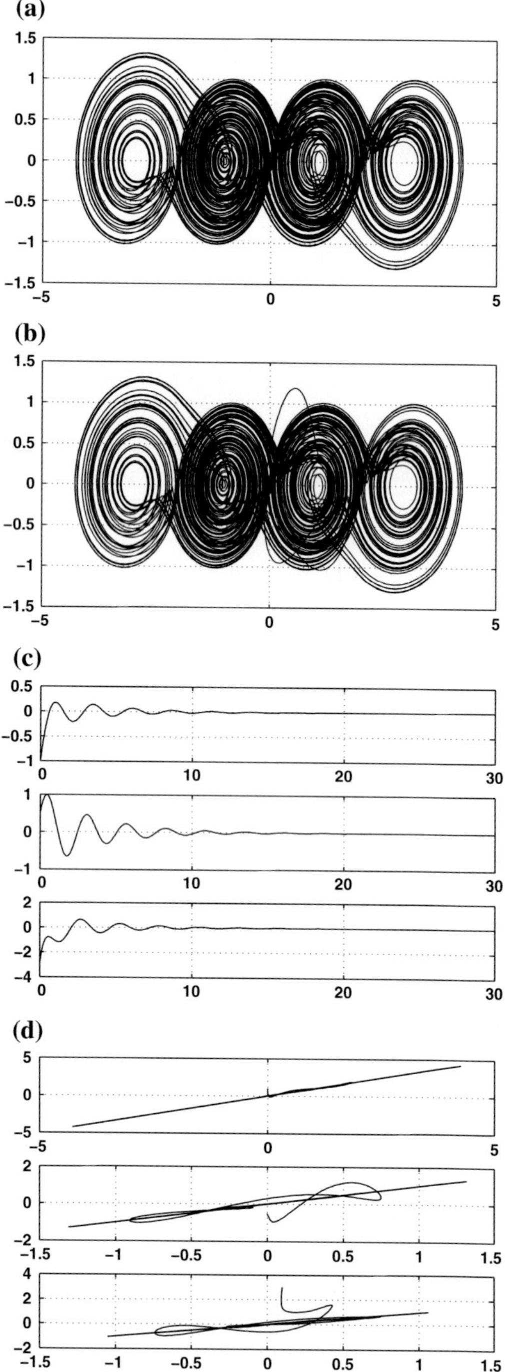

Fig. 15 4-scrolls chaotic oscillators master-slave synchronization with coefficients equal to 0.7. **a** Master. **b** Slave. **c** Error synchronization transient evolution. **d** Error phase diagram of the synchronized states x and ξ

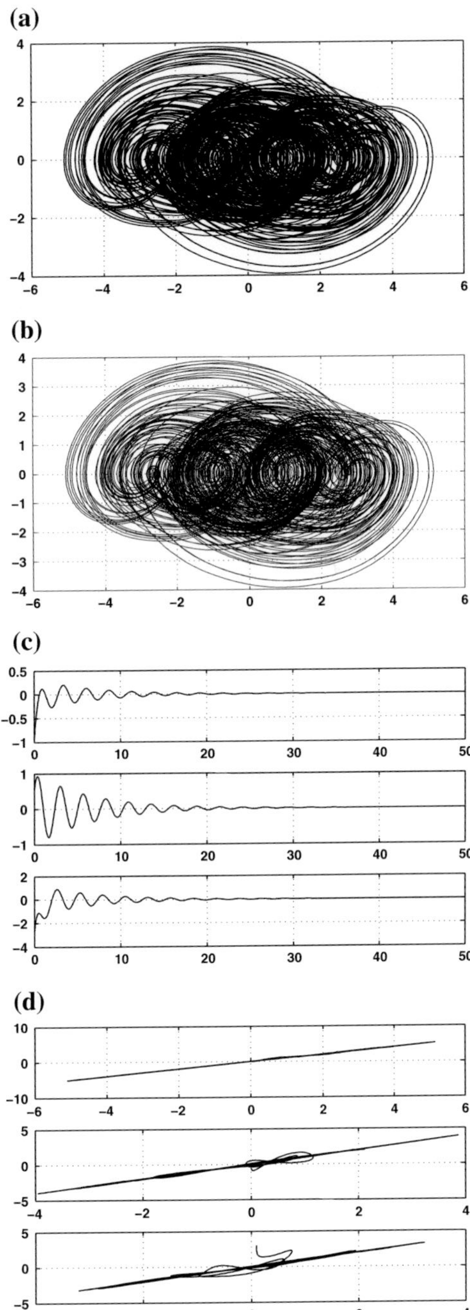

Fig. 16 Optimized 4-scrolls chaotic oscillator master-slave synchronization. **a** Master. **b** Slave. **c** Error synchronization transient evolution. **d** Error phase diagram of the synchronized states x and ξ

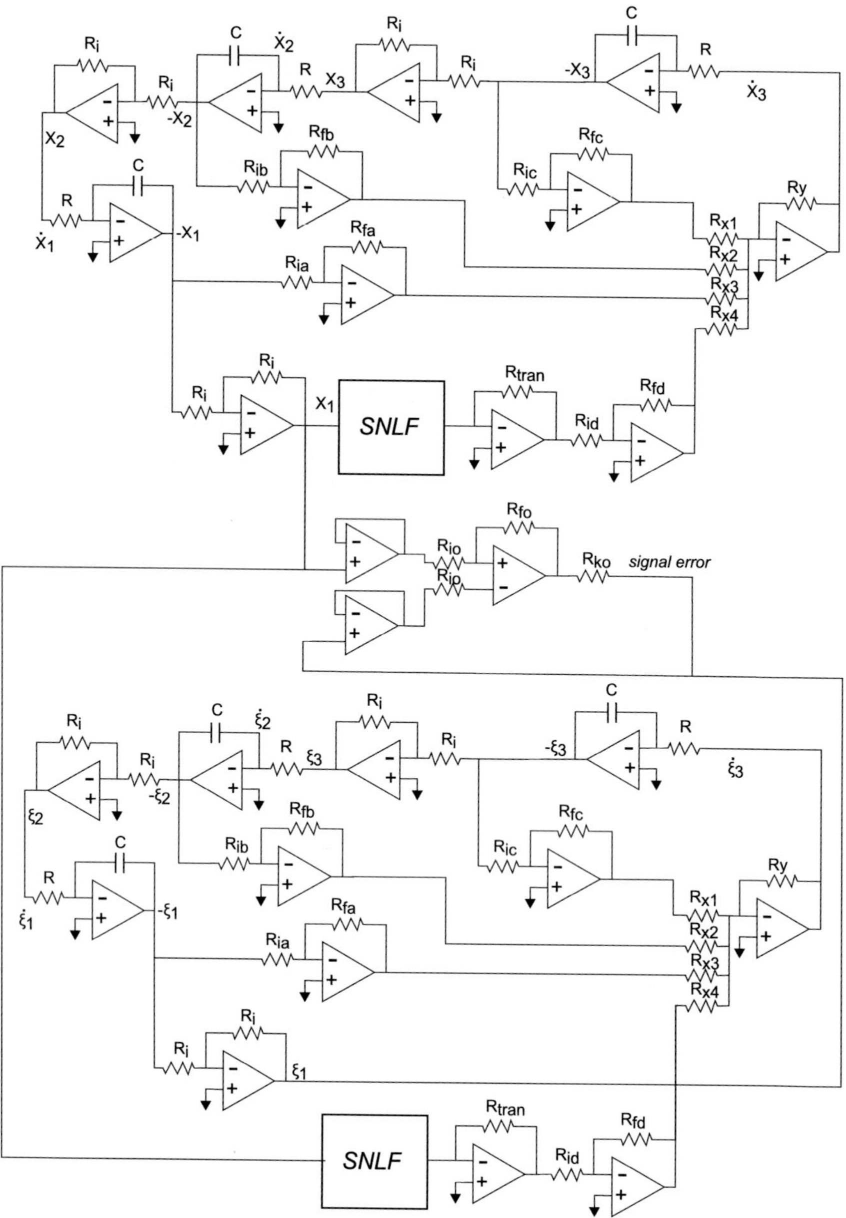

Fig. 17 Circuit implementation for the synchronization using operational amplifiers

error is shown in Fig. 19, which can be adjusted by varying the gain of the observer. The coincidence of the states is represented by a straight line with a unity slope (identity function) in the phase plane of each state.

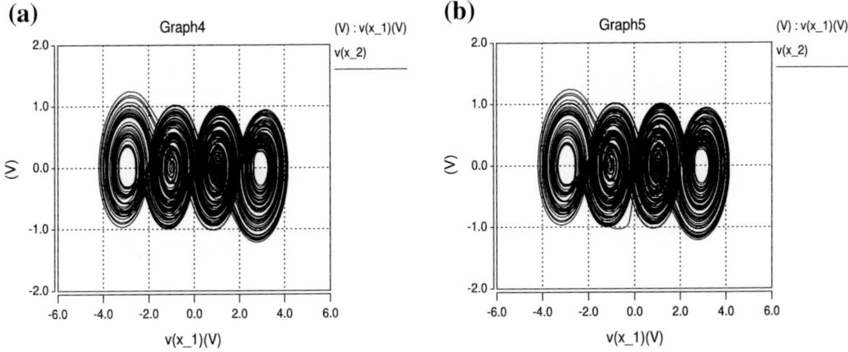

Fig. 18 4-scrolls attractor. **a** Master circuit. **b** Slave circuit

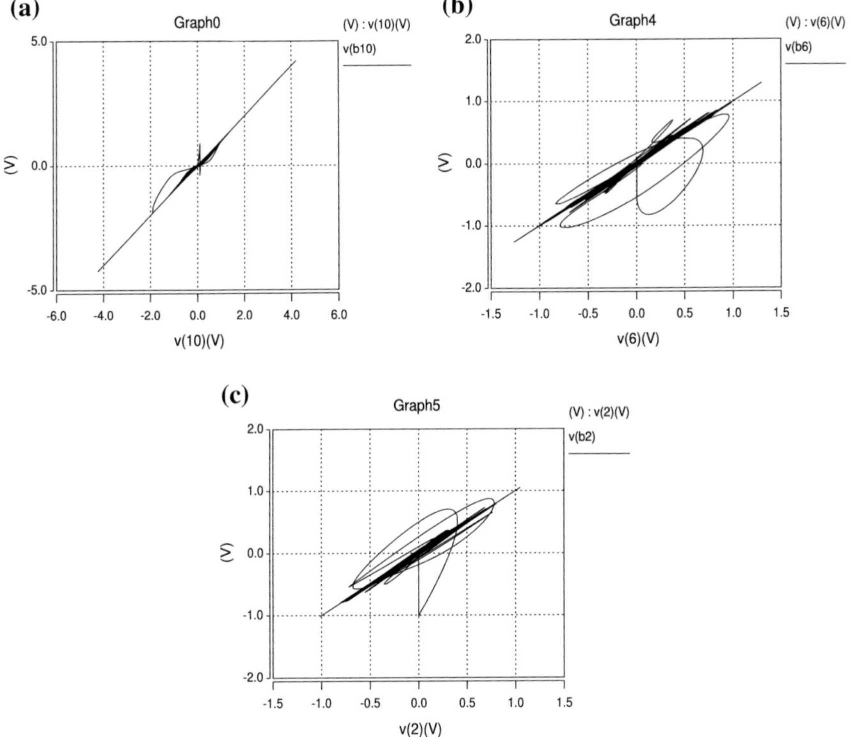

Fig. 19 Error phase diagram of **a** X_1 versus ξ_1, **b** X_2 versus ξ_2 and **c** X_3 versus ξ_3

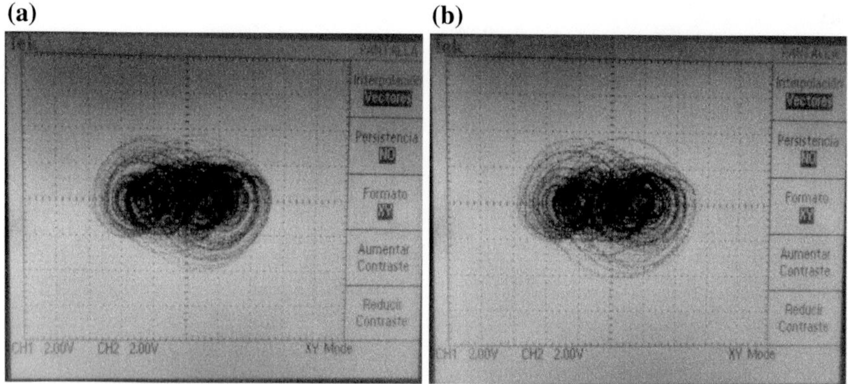

Fig. 20 5-scrolls optimized chaotic oscillator. **a** Master circuit. **b** Slave circuit

6.2 Experimental Synchronization Results

The circuit realization of Fig. 5 was performed by using the commercially available operational amplifier TL081. By selecting $a = 0.990, b = 0.836, c = 0.150, d_1 = 1$, $k = 1$, $\alpha = 0.0165$, $s = 60.606$, $h_1 \simeq 1$, $I_{sat} = 100\,\mu A$, $R_{ix} = 10\,k\Omega$, $C = 330$ pF, $R = 100\,k\Omega$, $R_{tran} = 10.54\,k\Omega$, $R_{ia} = R_{ib} = R_{ic} = R_{id} = 16\,k\Omega$, $R_{fa} = 15.84$ kΩ, $R_{fb} = 13.37\,k\Omega$, $R_{fc} = 2.4\,k\Omega$,$R_{fd} = 16\,k\Omega$, $R_{x1} = R_{x2} = R_{x3} = R_{x4} = R_y = 16\,k\Omega$, $R_i = 16\,k\Omega$ and $V_sat = \pm 18\ V$, it results in generating 5-scrolls as shown in Fig. 20.

The whole circuit used for the synchronization of two chaotic oscillators is shown in Fig. 17, where $R_{io} = 16\,k\Omega$, $R_{fo} = 48\,k\Omega$ and $R_{ko} = 3\Omega$. The synchronized results obtained by the physical realization are shown in Fig. 20, where the coincidence of the states is represented by a straight line with unitary slope in the phase plane for each state, as shown in Fig. 21.

7 Conclusions and Future Work

The MATLAB and SPICE simulations for generating multi-scrolls were described in detail. Further, the experimental realization of a chaotic oscillator and the synchronization of two of them generated results with good agreement with theoretical ones. Most important: the unpredictability of the chaotic oscillator was verified by optimizing MLE by applying PSO algorithm. This algorithm found different feasible solutions of the coefficient values for a, b, c and d_1 for generating 4 to 7-scrolls, which were listed in Table 1. SPICE simulations were presented in Figs. 9 and 10 to generate 4 and 5 scrolls, and experimental results were given for generating 5 scrolls

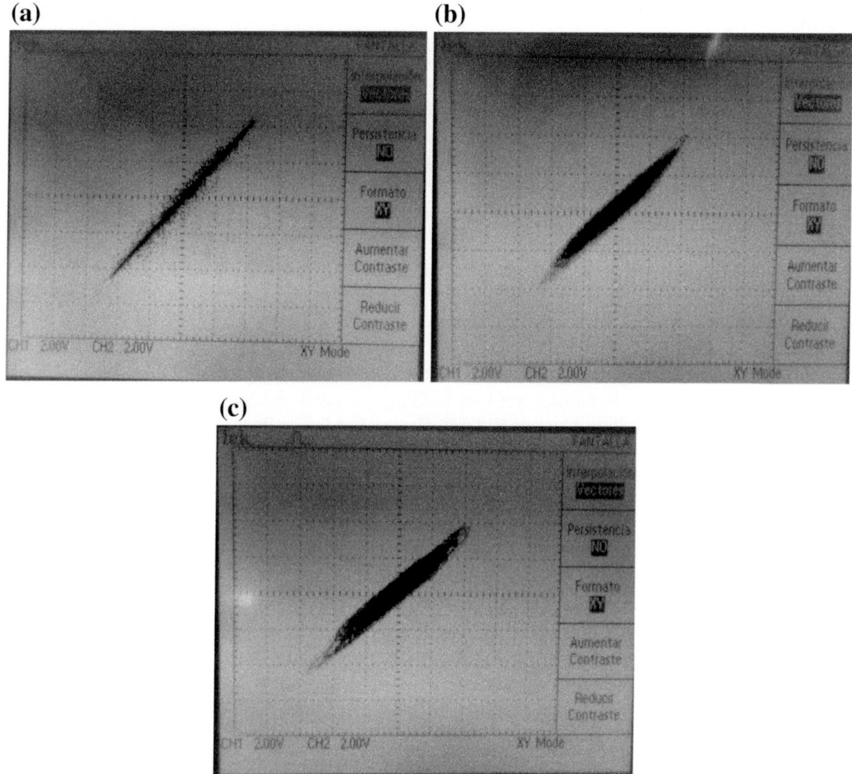

Fig. 21 Error phase diagram of **a** X_1 versus ξ_1, **b** X_2 verus ξ_2, and **c** X_3 verus ξ_3

shown in Fig. 20, and the synchronization of two oscillators with optimized MLE was shown in Fig. 21.

There are many opportunities for future work that spring from this document. In this work, no real attention has been given to the methods used to determine the chaotic regime. In particular, another measure of chaotic regimen that could be investigated is the entropy. Another important issue is how determining the optimal values of coefficients k to establish a better synchronization. Other kinds of active devices can be used to verify the response at different frequencies and levels of voltage and current signals.

Acknowledgments The first author wants to thank CONACyT-Mexico for the scholarship 331697. This work has been partially supported by CONACyT-Mexico under grant 237991-Y, in part by the TEC2013-45638-C3-3-R, funded by the Spanish Ministry of Economy and Competitiveness and ERDF, by the P12-TIC-1481 project, funded by Junta de Andalucia, and by CSIC project PIE 201350E058.

References

1. Azar AT, Vaidyanathan S (2015a) Chaos Modeling and Control Systems Design. Springer
2. Azar AT, Vaidyanathan S (2015b) Computational intelligence applications in modeling and control, vol 575. Springer
3. Blum C, Roli A (2003) Metaheuristics in combinatorial optimization: overview and conceptual comparison. ACM Comput Surv (CSUR) 35(3):268–308
4. Bratton D, Kennedy J (2007) Defining a standard for particle swarm optimization. In: IEEE swarm intelligence symposium, SIS 2007. IEEE, pp 120–127
5. Carbajal-Gómez V, Tlelo-Cuautle E, Fernández F, de la Fraga L, Sánchez-López C (2014) Maximizing lyapunov exponents in a chaotic oscillator by applying differential evolution. Int J Nonlinear Sci Numer Simul 15(1):11–17
6. Carbajal-Gómez V, Tlelo-Cuautle E, Trejo-Guerra R, Sánchez-López C, Munoz-Pacheco J (2011) Experimental synchronization of multiscroll chaotic attractors using current-feedback operational amplifiers. Nonlinear Sci Lett B: Chaos, Fractal and Synchron 1(1):37–42
7. Carbajal-Gomez V.H, Tlelo-Cuautle E, Trejo-Guerra R, Muñoz-Pacheco, JM (2013) Simulating the synchronization of multi-scroll chaotic oscillators. In: 2013 IEEE international symposium on circuits and systems (ISCAS). IEEE, pp 1773–1776
8. Clerc M (2011) From theory to practice in particle swarm optimization. In: Handbook of swarm intelligence. Springer, Berlin, pp 3–36
9. Tlelo-Cuautle, Pacheco JMM (2010) Electronic design automation of multi-scroll chaos generators. Bentham Science Publishers
10. de la Fraga LG, Tlelo-Cuautle E (2014) Optimizing the maximum lyapunov exponent and phase space portraits in multi-scroll chaotic oscillators. Nonlinear Dyn 76(2):1503–1515
11. de la Fraga LG, Tlelo-Cuautle E, Carbajal-Gómez V, Munoz-Pacheco J (2012) On maximizing positive lyapunov exponents in a chaotic oscillator with heuristics. Rev Mex Fis 58(3):274–281
12. Dieci L (2002) Jacobian free computation of lyapunov exponents. J Dyn Diff Equat 14(3):697–717
13. Lü J, Chen G (2006) Generating multiscroll chaotic attractors: theories, methods and applications. Int J Bifurcat Chaos 16(04):775–858
14. Muñoz-Pacheco J-M, Tlelo-Cuautle E (2008) Synthesis of n-scroll attractors using saturated functions from high-level simulation. In: Journal of physics: conference series, vol 96, p 012050. IOP Publishing
15. Muñoz-Pacheco JM, Zambrano-Serrano E, Félix-Beltrán O, Gómez-Pavón LC, Luis-Ramos A (2012) Synchronization of pwl function-based 2d and 3d multi-scroll chaotic systems. Nonlinear Dyn 70(2):1633–1643
16. Parker TS, Chua LO, Parker TS (1989) Practical numerical algorithms for chaotic systems. Springer
17. Rugonyi S, Bathe K-J (2003) An evaluation of the lyapunov characteristic exponent of chaotic continuous systems. Int J Numer Methods Eng 56(1):145–163
18. Sánchez-López C, Trejo-Guerra R, Munoz-Pacheco J, Tlelo-Cuautle E (2010) N-scroll chaotic attractors from saturated function series employing ccii+ s. Nonlinear Dyn 61(1–2):331–341
19. Sira-Ramirez H, Cruz-Hernández C (2001) Synchronization of chaotic systems: a generalized hamiltonian systems approach. Int J Bifurcat Chaos 11(05):1381–1395
20. Talbi E-G (2009) Metaheuristics: from design to implementation, vol 74. Wiley
21. Tlelo-Cuautle E, Pano-Azucena AD, Carbajal-Gomez VH, Sanchez-Sanchez M (2014) Experimental realization of a multiscroll chaotic oscillator with optimal maximum lyapunov exponent. Sci World J
22. Trejo-Guerra R, Tlelo-Cuautle E, Cruz-Hernández C, Sánchez-López C (2009) Chaotic communication system using chua's oscillators realized with ccii+ s. Int J Bifurcat Chaos 19(12):4217–4226
23. Trejo-Guerra R, Tlelo-Cuautle E, Jiménez-Fuentes J, Sánchez-López C, Muñoz-Pacheco J, Espinosa-Flores-Verdad G, Rocha-Pérez J (2012) Integrated circuit generating 3-and 5-scroll attractors. Commun Nonlinear Sci Numer Simul 17(11):4328–4335

24. Trejo-Guerra R, Tlelo-Cuautle E, Muñoz-Pacheco J, Sánchez-López C, Cruz-Hernández C (2010a) On the relation between the number of scrolls and the lyapunov exponents in pwl-functions-based η-scroll chaotic oscillators. Int J Nonlinear Sci Numer Simul 11(11):903–910
25. Trejo-Guerra R, Tlelo-Cuautle E, Sánchez-López C, Munoz-Pacheco J, Cruz-Hernández C (2010b) Realization of multiscroll chaotic attractors by using current-feedback operational amplifiers. Revista mexicana de física 56(4):268–274
26. Vaidyanathan S, Azar AT (2015a) Analysis, control and synchronization of a nine-term 3-d novel chaotic system. In: Azar AT, Vaidyanathan S (eds) Chaos modeling and control systems design. Studies in computational intelligence, vol 576. Springer International Publishing, pp 19–38
27. Vaidyanathan S, Azar AT (2015b) Anti-synchronization of identical chaotic systems using sliding mode control and an application to vaidyanathanmadhavan chaotic systems. In: Azar AT, Zhu Q (eds) Advances and applications in sliding mode control systems. Studies in computational intelligence, vol 576. Springer International Publishing, pp 527–547
28. Vaidyanathan S, Azar AT (2015c) Analysis and control of a 4-d novel hyperchaotic system. In: Chaos modeling and control systems design. Springer, pp 3–17
29. Vaidyanathan S, Azar AT, Rajagopal K, Alexander P (2015a) Design and spice implementation of a 12-term novel hyperchaotic system and its synchronisation via active control. Int J Model Ident Control 23(3):267–277
30. Vaidyanathan S, Idowu BA, Azar AT (2015b) Backstepping controller design for the global chaos synchronization of sprotts jerk systems. In: Chaos modeling and control systems design. Springer, pp 39–58
31. Vaidyanathan S, Sampath S, Azar AT (2015c) Global chaos synchronisation of identical chaotic systems via novel sliding mode control method and its application to zhu system. Int J Model Ident Control 23(1):92–100
32. Wolf A, Swift JB, Swinney HL, Vastano JA (1985) Determining lyapunov exponents from a time series. Physica D 16(3):285–317

Part II
Advances in Intelligent Control

Evolutionary Computational Technique in Automatic Generation Control of Multi-area Power Systems with Nonlinearity and Energy Storage Unit

K. Jagatheesan, B. Anand, K. Baskaran and Nilanjan Dey

Abstract In this study, a new meta-heuristic based evolutionary computational technique is reported for solving Automatic Generation Control (AGC) or Load Frequency Control (LFC) issue in multi-area power system with nonlinearity and an energy storage unit. Multi-area power system consists of two area equal reheat thermal power systems with Governor Dead Band (GDB) and Generation Rate Constraint (GRC) nonlinearity and boiler dynamics and energy storage element. During normal operating conditions, there no change in system parameters (Frequency and tie-line power flow) and stability. When sudden load demand occurs in any one of interconnected power, it affects system parameters and stability and system yield damping oscillation in their response with steady state error and settling time. In order to mitigate this biggest pose the proper selection of the controller is a major issue. In power system Automatic Voltage Regulator (AVR) loop is a primary control loop and in addition Proportional-Integral-Derivative (PID) controller is proposed as a secondary controller in AGC. The better performance of power system depends on proper selection of controller gain and also depends on the selected objective function for optimization of controller gain values. A new meta-heuristic based Ant Colony Optimization (ACO) evolutionary computational technique is used for tuning of PID controller with different operating conditions. Three different objective

K. Jagatheesan (✉)
Department of EEE, Mahendra Institute of Engineering and Technology, Namakkal, Tamilnadu, India
e-mail: jaga.ksr@gmail.com

B. Anand
Department of EEE, Hindusthan College of Engineering and Technology, Coimbatore, Tamilnadu, India
e-mail: b_anand_eee@yahoo.co.in

K. Baskaran
Department of EEE, Government College of Technology, Coimbatore, Tamilnadu, India
e-mail: drbaskaran@gct.ac.in

N. Dey
Department of ETCE, Jadavpur University, Kolkata, India
e-mail: neelanjandey@gmail.com

A.T. Azar and S. Vaidyanathan (eds.), *Advances in Chaos Theory and Intelligent Control*, Studies in Fuzziness and Soft Computing 337,
DOI 10.1007/978-3-319-30340-6_27

functions Integral Square Error (ISE), Integral Time Absolute Error (ITAE) and Integral Absolute Error (IAE) are used in ACO for tuning of controller gain. An electromechanical oscillation of power system is effectively damp out by introducing an energy storage unit in two area interconnected power system because of their inherent energy storage capacity with kinetic energy of the rotor. In this study Hydrogen generative Aqua Electroliser (HAE) with a fuel cell is incorporated into the investigated power system. The response of the proposed approach with different cost functions are obtained and compared with and without considering the effect of energy storage unit in LFC problem.

Keywords Ant Colony Optimization · Automatic Voltage Regulator (AVR) · Automatic Generation Control (AGC) · Energy storage unit · Interconnected power system · Load Frequency Control (LFC) · Nonlinearity · Proportional-Integral-Derivative (PID) controller

1 Introduction

The unpredicted actions and worries affect operating points of power system and system yield deviations in their parameters from its nominal value (Frequency and tie-line power flow between areas). The performance or efficiency of the system depends on effective power balance between total demands with total generation. By introducing the Automatic generation control scheme in power system the operation is controlled and it improves the quality of power supply by increasing reliability and generating sufficient power to consumers. AGC generates proper control signals to regulate power system operation. The name of the control signal is "Area Control Error" and it is established by linear combination of frequency and tie-line power flow between connected areas.

The main aim of LFC or AGC is to keep or retain power balance between generation and control in the power generating unit. Inorder to conquer this inadequacy many controllers was introduced and several optimization techniques also introduced to optimize controller gain values. The same time many energy storage units are developed and implemented in LFC/AGC of the interconnected power system.

Since, the system parameters and stability are affected by any sudden load demand occurs in any one of the interconnected power, which yield damping oscillation in their response with steady state error and settling time. In order to diminish this biggest case the proper selection of the controller is a major issue. Thus, the main contribution of the present work is to employ a new meta-heuristic technique for solving the Automatic Generation Control (AGC) or Load Frequency Control (LFC) issue in multi-area power system with nonlinearity and an energy storage unit. The ACO technique is used for tuning the PID controller with different operating conditions.

The work is organized as follows. The need of LFC/AGC and application of energy storage unit in the interconnected power system is discussed in literature review Sect. 1. In Sect. 3, investigated system is designed with different condition (Open loop model, closed loop model with PI/PID controller and PID controller with energy storage unit) and non linearity and an energy storage unit. The PI and PID controller is discussed and designed in Sect. 4 by considering two different optimization techniques with three objective functions. Section 5 gives the proposed optimization technique overview and its application in general purpose. The results are given in Sect. 6 and comparison tests are made to show the superiority of the proposed optimization technique. Finally, Sect. 7 gives the conclusion about the investigated work and proposed optimization technique.

2 Load Frequency Control\Automatic Generation Control Related Work

For the last few decades, many research articles have been dealing with the AGC or LFC in power system by using multi area interconnected power system (Nanda and kaul [70]; Tripathy et al. [103]; Nanda et al. [71]; Kothari and Nandha [60]; Pan and Lian [75]; Das et al. [25]; Demiroren et al. [31]; Shayeghi and Shayanfor [90]; Chidambaram and Paramasivam [21]; Ebrahim et al. [41]; Anand and Jeyakumar [5]; Nandha and Mishra [72]; Arivoli and Chidambaram [6]; Ali and Abd-Elazim [4]; Gozde et al. [48]; Gozde et al. [49]; Saikia et al. [86]; Omar et al. [73]; Samanta et al. [89]; Jagatheesan et al. 2014; Jagatheesan and Anand 2014; Dey et al. [34]; Dey et al. 2014b; Dey et al. [35]; Sahu et al. [84]; Dash et al. [27]; Francis and Chidambaram [44]; Sahu et al. [83]; Dash et al. [26, 28]; Padhan et al. [74]; Tripathy et al. [100]; Tam and Kumar [97]; Peterson et al. [77]; Mitani et al. [69]; Banerjee et al. [16]; Tam and Kumar [98]; Tripathy et al. [101]; Pothiye et al. [78]; Bhatt et al. [19]; Rajesh et al. [1]; Roy et al. [82]; Padhan et al. [79]; Chine et al. [22]; Beck et al. [18]; Kunisch et al. [61]; Banerjee et al. [16]; Salameh et al. [88]; Lu et al. [64]; Kunisch et al. [62]; Aditya et al. [2]; Kalyani et al. [57]; Tokuda [99]; Enomoto et al. [42]; Francis and Chidambaram [44]; Parmar [76]; Riberio et al. [81]; Ke'louwani et al. [58]; Little et al. [63]; Yu et al. [104]; Dimitris et al. [51]; Francis and Chidambaram [45]; Azar and Vaidyanathan [12]; Azar and Vaidyanathan [13]; Azar and Vaidyanathan [14]; Zhu and Azar [105]; Azar and Serrano [9]; Azar [10]; Azar [8]; Hassanien et al. [50]; Azar and Serrano [10]; Azar and Serrano [10]). The first LFC was discussed and published by Chon [23]. Table 1 report few noteworthy research works related to Load Frequency Control/Automatic Generation Control of interconnected power systems.

Literature survey about the LFC/AGC issue clearly shows that many energy storage devices are developed and introduced in the single/multi area power system. The Energy storage units are Superconducting Magnetic Energy Storage (SMES), Capacitive Energy Storage (CES), Battery Energy Storage (BES), Redox Flow

Table 1 Research works related to LFC/AGC of multi area interconnected power systems

Year	Control strategies	Authors and Years	References
1978	Parameter-plane technique	Nanda and kaul (1978)	[70]
1982	Lyapunov Technique	Tripathy et al. (1982)	[103]
1983	Continuous and Discrete mode Optimization	Nanda et al. (1983)	[71]
1988	Optimal Control theory	Kothari and Nandha (1988)	[60]
1989	Adaptive Controller	Pan and Lian (1989)	[75]
1991	Variable Structure Control (VSC)	Das et al. 1991	[25]
2006	Artificial Neural Network (ANN)	Demiroren et al. (2001) and Shayeghi and Shayanfor (2006)	[31, 90]
2009	Genetic Algorithm (GA)	Chidambaram and Paramasivam, 2009	[21]
2009	Particle Swarm Optimization (PSO)	Ebrahim et al. (2009)	[41]
2009	Fuzzy Logic Controller (FLC)	Anand and Jeyakumar 2009	[5]
2010	Classical controller	Nandha and Mishra (2010)	[72]
2011	Genetic Algorithm (GA)	Arivoli and Chidambaram (2010)	[6]
2011	Bacterial Foraging Optimization Algorithm (BFOA)	Ali and Abd-Elazim (2011)	[4]
2011	Craziness Based Particle Swarm Optimization (CRAZYPSO)	Gozde et al. (2011)	[48]
2012	Artificial Bee Colony (ABC)	Gozde et al. (2012)	[49]
2013	Bacterial Foraging (BF) technique	Saikia et al. (2013)	[86]
2013	Ant Colony Optimization (ACO)	Omar et al. (2013)	[73]
2013	Ant Weight lifting Algorithm	Samanta et al. (2013)	[89]
2014	Cuckoo Search (CS)	Dey et al. (2014a)	[34]
2014	Firefly Algorithm (FA)	Dey et al. (2014)	[35]
2014	Firefly Algorithm (FA)	Sahu et al. (2014)	[84]
2015	Bat inspired algorithm (BID)	Dash et al. (2015)	[27]

(continued)

Table 1 (continued)

Year	Control strategies	Authors and Years	References
2015	Beta Wavelet Neural Network (BWNN)	Francis and Chidambaram (2015)	[44]
2015	Teaching Learning Based Optimization (TLBO)	Sahu et al. (2015)	[83]
2015	Cuckoo Search (CS)	Dash et al. (2015)	[28]
2015	Firefly Algorithm (FA)	Padhan et al. (2015)	[74]

Battery (RFB) and Hydrogen Aqua Analyzer (HAE) and Fuel Cells (FC). Table 2 reports applications of energy storage units in LFC/AGC into the interconnected power system. The applications of energy storage units in power systems are reported in Table 2.

From the above literature review and discussions it is clearly evident that the proper controller is most essential requirement and also suitable optimization technique is more important to tune controller gain values and energy storage unit produce considerable effects in the system response. In this investigation Proportional-Integral-Derivative (PID) controller is implemented in two area interconnected power system with non linearity and boiler dynamics. The PID controller gain values are optimized by using AI based Ant Colony Optimization (ACO) technique.

3 Two Are Interconnected Tem Thermal Power System

The system under investigation consists of two equal areas interconnected open loop reheat thermal power system as shown in Fig. 1. Area 1 and area 2 comprisies reheater unit with the appropriate governor dead band and generation rate constraint nonlinearity and boiler dynamics. Each interconnected power system has three different inputs and two outputs. The inputs given to the systems are controlled input (u), load distrubace (SLP) and tie line power flow error between interconnected areas (delPtie). The outputs of systems are frequency (delF) and area control error (ACE). Figure 1 shows open loop transfer function model of to area power system considering one percent Step Load Perturbation (SLP) in area 2.

During nominal loading conditions there is no changes in the system parameters and each system carries its own load and maintain system power balance (Power generation is equal to load demand). When load demand occurs in any one of interconnected system, it affects system parameters and power balance. The system parameters get oscillations in their response and produce damping oscillations with large steady state error and the system take more time to settle with their scheduled values. The open loop response of the system is shown in figure. Figure 5 show open loop frequency deviations in area 1 and area2 and Fig. 6 shows the tie-line power flow

Table 2 Applications of energy storage unit in power system

Energy Storage Unit	Authors and Years	References
SMES	Tripathy et al. (1992)	[100]
	Tam and Kumar (1990)	[97]
	Peterson et al. (1975)	[77]
	Mitani et al. (1988)	[69]
	Banerjee et al. (1990)	[16]
	Tam and Kumar 1990	[98]
	Tripathy et al. (1992)	[101]
	Pothiye et al. (2007)	[78]
	Bhatt et al. (2010)	[19]
	Rajesh et al. (2011)	[1]
	Roy et al. (2014)	[82]
	Padhan et al. (2014)	[79]
	Chine et al. (2015)	[22]
BES	Beck et al. (1976)	[18]
	Kunisch et al. (1986)	[61]
	Banerjee et al. (1990)	[16]
	Salameh et al. (1992)	[88]
	Lu et al. (1995)	[64]
	Kunisch et al. (1996)	[62]
	Aditya et al. (2001)	[2]
	Kalyani et al. (2012)	[57]
RFB	Tokuda (1998)	[99]
	Enomoto et al. (2002)	[42]
	Francis and Chidambaram (2013)	[44]
	Parmar (2015)	[76]
HAE and FC	Riberio et al. (2001)	[81]
	Ke'louwani et al. (2007)	[58]
	Little et al. (2005)	[63]
	YU et al. (2009)	[104]
	Dimitris et al. (2009)	[51]
	Francis and Chidambaram (2015)	[45]

between area 1and 2. Table 3 gives the numerical values of open loop power system with 1 % SLP in area 2.

In order to overcome this big issue the design of proper control strategy is most essential thing. In this work two different controllers (PI and PID) and optimization techniques are used to solve above said issue. The closed loop model of thermal power system is shown in Fig. 2.

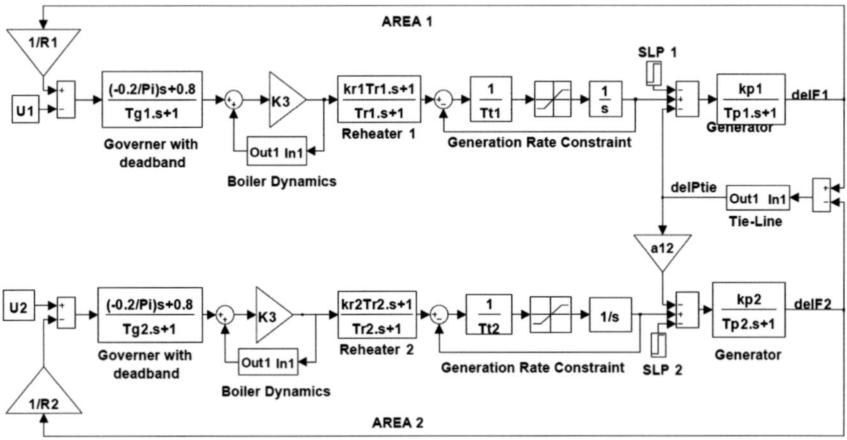

Fig. 1 Transfer function model of open loop system with GDB, GRC and boiler dynamics

Table 3 Open loop system parameters

Response	Steady state error
delF1	−0.0311 Hz
delF2	−0.03 Hz
delPtie12	−0.003 puMW

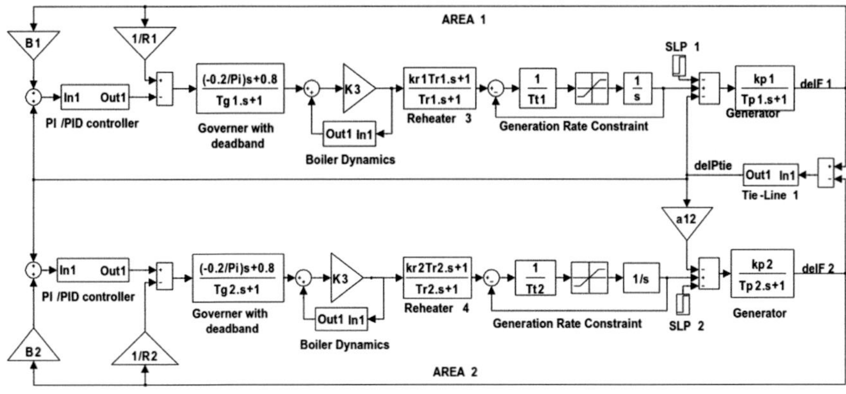

Fig. 2 Transfer function model of closed loop thermal power system

The design of proper controller in AGC gives the required control signal to improve system stability and overall performance of system response. The input of controller is Area Contol Error (ACE), it is defined as, “linear combination of changes in frequency deviation and tie-line power flow between the interconnected power system is called Area Control Error (ACE)” and it is given by.

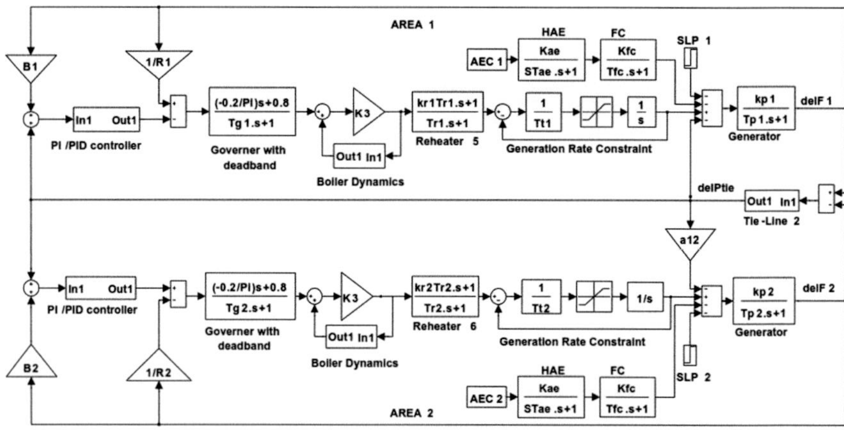

Fig. 3 Transfer function model of two area thermal power systems with nonlinearity, boiler dynamics and Energy storage unit

$$ACE_i = B_i \Delta f_i + \Delta P_{tie,ij} \tag{1}$$

Figures 7, 8 and 9, gives the comparison of open loop response with a conventional PI controller. Figures 7 and 8 shows frequency deviation comparison and Fig. 9 shows tie-line power flow deviation in between interconnected power systems. The transfer function model of two area interconnected power system with energy storage unit is shown in Fig. 3.

3.1 Governor Dead Band, Generation Rate and Boiler Dynamics

In the presence of GDB even for a small load perturbation, the system becomes highly non linear and more oscillatory. For the analysis, in this work backlash non linearity of about 0.05 % is considered for thermal system (Anand and Jeyakumar [5]; Kothari and Nandha [60]).

$$\text{Governor with the dead band } (\mathrm{G_g}) = \frac{-\frac{0.2}{\Pi}S + 0.8}{T_g S + 1} \tag{2}$$

When GRC is considered in the power system, it becomes non linear and linear control techniques are not fit for optimizing the controller gain. The GRC is considered in both areas of the systems. In this work GRC of 0.0017 p.u. MW $\sec^{-1}$ is considered. The maximum rate of valve opening or closing speed is controlled by the limiter. T_{sg}, g_{max} is the power rate limit imposed by valve or gate control.

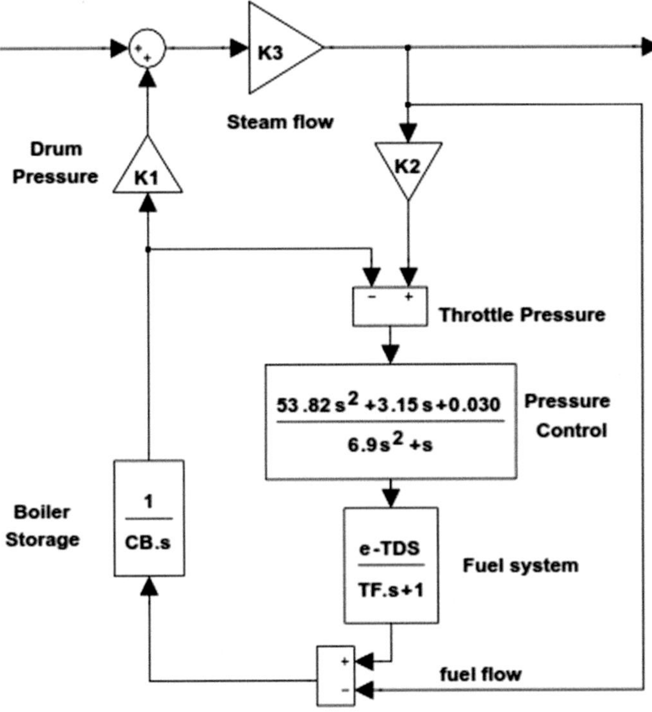

Fig. 4 Model boler dynamics

$$|\Delta \dot{Y}_E| < g_{\max} \quad (3)$$

In this work drum type boiler is considered for investigation and it is shown in Fig. 4 (Anand and Jeyakumar [5]; Kothari and Nandha [60]). It incorporated with long term dynamics of fuel, steam flow on boiler drum pressure and combustion controls. The boiler is a device, producing steam under pressure. There are three types of boiler available such as, gas or oil fired, coal fired well tuned and coal fired poorly tuned system. A gas or oil fired boiler is guaranteed for quick response compared to other system when the load demand occurs.

3.2 *Hydrogen Aqua Analyzer and Fuel Cell*

During nominal loading conditions the excessive power of the generating units are stored and it will be delivered during sudden load demand to keep power balance and maintain system stability. In this work excessive energy is stored in the form hydrogen energy and stored energy get back from the unit during the sudden load disturbance. Hydrogen energy storage unit comprises two more essential

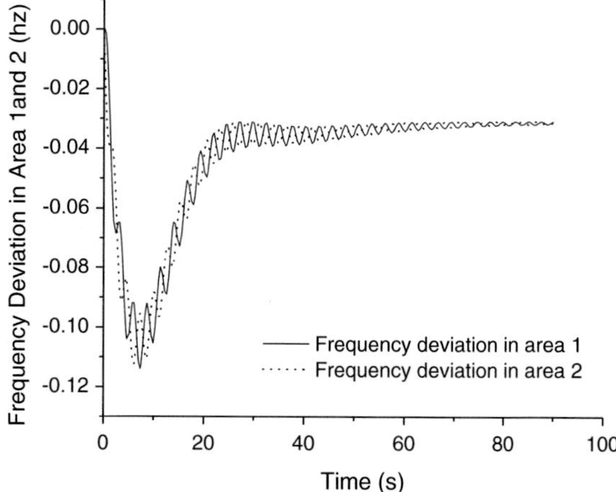

Fig. 5 Open loop response of system (delF1 and delF2)

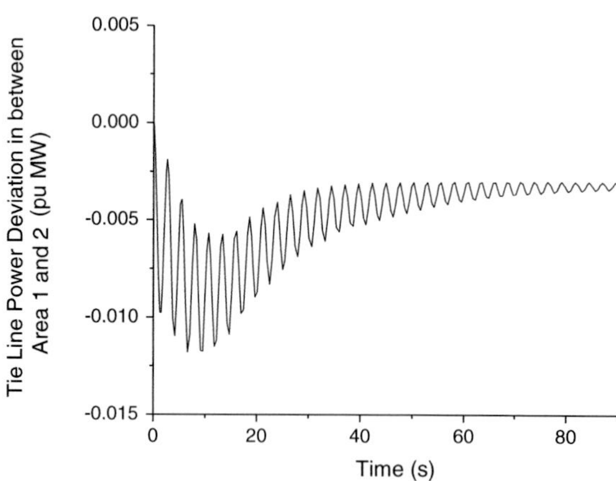

Fig. 6 Open loop response of system (delPtie12)

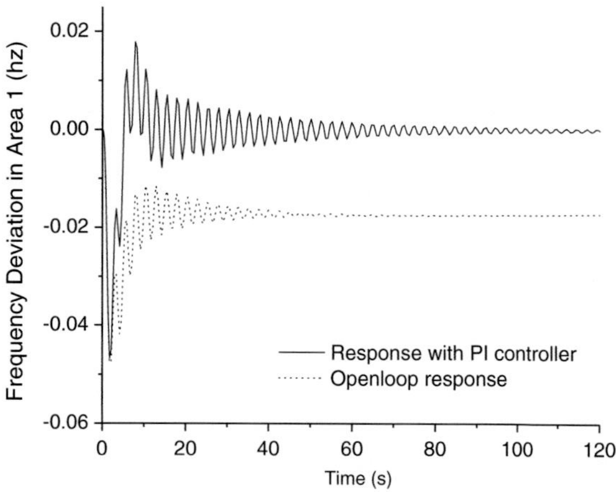

Fig. 7 Deviations of delF1 with open loop and conventional PI controller response

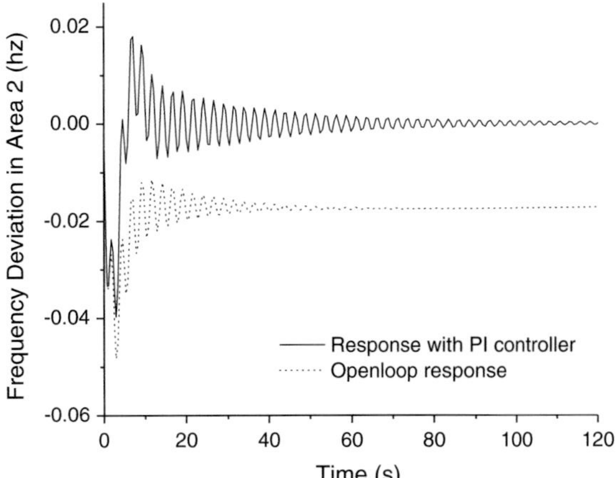

Fig. 8 Deviations of delF2 with open loop and conventional PI controller response

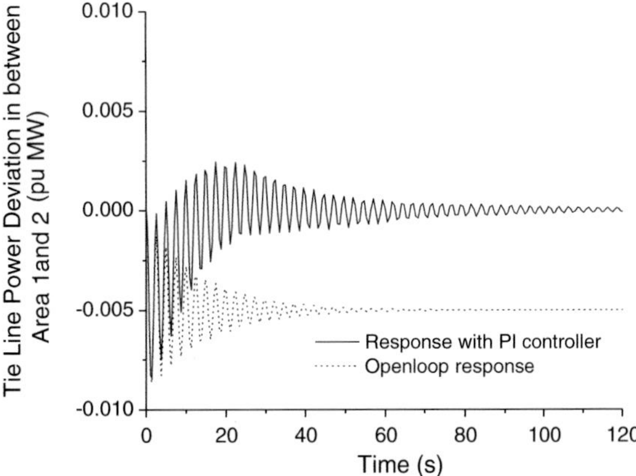

Fig. 9 Deviations of delPtie12 with open loop and conventional PI controller response

components, Such as, electrolyzer and fuel-cell. Electrolyze converts electrical power into hydrogen energy by decomposing after molecule and hydrogen is converted back into electrical energy. The fuel cell plays major role in the chemical energy conversion. The average conversion efficiency rate of water into hydrogen is 65–70 %, which improves up to 20–40 % of overall system efficiency (Riberio et al. [1]; Ke'louwani et al. [58]; Little et al. [63]; Yu et al. [104]; Dimitris et al. [51]; Francis and Chidambaram [45]). The hydrogen energy storage system is free harmful storage unit compare to others because of reducing the green house effect.

The transfer function of Aqua Electrolyzer is given by:

$$G_{AE}(s) = \frac{K_{AE}}{1 + ST_{AE}} \tag{4}$$

The transfer function of fuel cell is given by:

$$G_{FC}(s) = \frac{K_{FC}}{1 + ST_{FC}} \tag{5}$$

4 Proportional –Integral-Derivative Controller

The simple and easily implemented effective and powerful controller for modern industries is Proportional-Integral-Derivative (PID) controller and it is the most popular feedback controller. The characteristics of PID controllers are effective, robustness, easily understood and provide more superior performance under varied dynamic performance of process industries. From the name itself, it comprises the different operating modes. Such as: Proportional, Integral and Derivative modes. In AGC/LFC problem PID controller gives better control performance a secondary controller in order to eliminate the deviations in their system parameters (Frequency, tie-line power flow and area control error) and keep the values within the specified limit. The transfer function of PID controller is given by

$$U(S) = K_p E(S) + \frac{K_p}{T_i S} E(S) + K_d E(S) \tag{6}$$

The proportional mode in PID controller effectively shrinks the rise time in the system response, but steady state error never eliminated. Steady state error of system response is vanished by integral mode, but it makes system response shoddier. The overall stability of the system is improved by introducing derivative controller and reduces overshoots in the system parameters.

Before the selection of suitable controller as a secondary controller for better power system operation, the choice of controller gain values and choosing suitable objective functions are more crucial.

Integral Square Error (ISE)

$$J = \int_0^T \left\{ (\Delta f_i)^2 + \left(\Delta P_{tiei-j}\right)^2 \right\} dt \tag{7}$$

Integral Time Square Error

$$J = \int_0^\infty t \left\{ (\Delta f_i)^2 + \left(\Delta P_{tiei-j}\right)^2 \right\} dt \tag{8}$$

Integral Time Absolute Error

$$J = \int_0^{\infty} t \left|\left\{\Delta f_i + \Delta P_{tiei-j}\right\}\right| dt \tag{9}$$

The optimal controller gain values are obtained by using optimization technique. There are several optimization techniques are available to obtain the best optimal controller gain values (Trial and error method and Artificial Intelligence Technique, etc.).

5 Ant Colony Optimization (ACO)

The recent days swarms intelligence take more inspiration for solving optimization problem the help of social behaviors of insects and other animals. There are different types of swarm intelligence based optimization applications are available in the recent years. Such as Bacterial Foraging (BF), Particle Swarm Optimization (PSO), Impearilist competative Algorithm (ICA) and Genetic Algorithm etc.,

From this ant inspired by many methods and it as successfully implemented in the general and specific application in the name of Ant Colony Optimization (ACO) technique. The foraging behavior of ant species takes the inspiration for developing ACO. Some favorable paths are marked by ants by depositing pheromone chemical on the ground; it's helpful for other ants to identify some favorable and shortest path from the same colony. The same mechanism has been used for solving complex optimization problem. The table reports the some of noteworthy applications of ACO (Table 4).

In ACO algorithm three main major phases are available. Such as initialization, constructing ant solutions and updating pheromone concentration. The optimization process is stop when it reaches maximum iteration value. From the above said phses initialization take plays major role compared to other two phases. In this phase intial parameters of ACO is predetermined earlier and fed into the algorithm before it running. Such as: No of ants, number of nodes, number of iterations, number of variables and pheromone quantity. The quality and performace of optimization technique is depents on the choice of optimization parameters. In this study following parameters are considered:

Number of ants = 50
Pheromone (τ) = 0.6
Evaporation rate (ρ) = 0.95
Number of iterations = 100

Table 4 Applications of ACO algorithm grouped by year wise

Authors and Years	Problem	References
Dorigo et al. (1991)	Traveling salesman	[40]
Dorido et al. (1996)		[39]
Dorigo et al. (2002)		[38]
Stutzle and Hoos (1997)		[94]
Costa and Hertz (1997)	Graph coloring	[24]
Gambardella et al. (1999)	Vehicle routing	[47]
Maniezzo (1999)	Quadratic Assignment	[65]
Stutzle and Hoos (2000)		[95]
Gambardella and Dorigo (2000)	Sequential ordering problem	[46]
Den Besten et al. (2000)	Total weighted tardiness problem	[33]
Solnon (2000)	Constraint satisfaction problems	[93]
Merkle et al. (2002)	Project Scheduling	[68]
Solnon (2002)	Constraint satisfaction problems	[93]
Campos (2002)	Learning Bayesian networks	[29]
Campos (2002)		[30]
Socha et al. (2003)	Course timetabling	[92]
Merkle and Middendorf (2003)	Scheduling a single machine	[67]
Fenet and Solnon (2003)	Maximum clique	[43]
Blum (2005)	Open shop scheduling	[20]
Shmygelska and Hoos (2005)	Protein folding	[91]
Martens et al. (2006)	Knowledge fusion problem	[66]
Korb et al. (2006)	Docking	[49]

6 Results and Analysis

A two area interconnected open loop thermal power system as shown in Fig. 1 is considered for the first case. A thermal power system with PI and PID controller equipped power system is shown Fig. 2 is considered for the second case and energy storage unit is considered for the third case and transfermoel of power system shown in Fig. 3. The simulink model of investigating power plants is obtained with the help of MATLAB/SIMULINK environment. The performance of different cases and conditions are shown in Figs. 5, 6, 7, 8, 9, 10, 11, 12, 13, 14, 15, 16, 17 and 18.

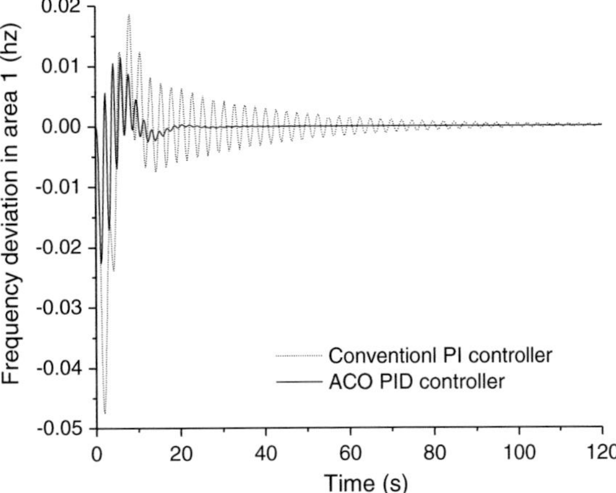

Fig. 10 Deviations of delF1 with conventional PI and ACO PID controller response

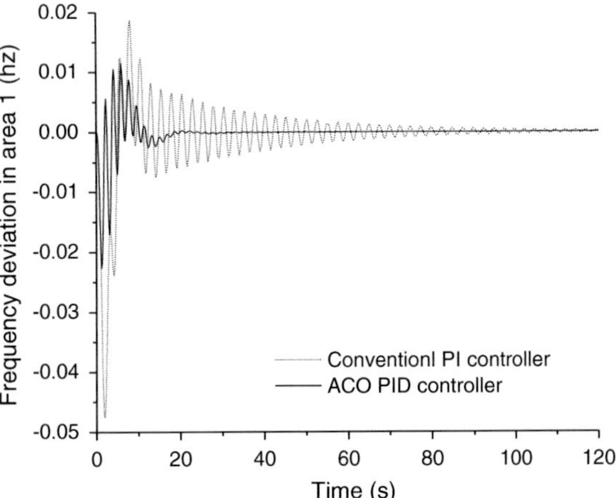

Fig. 11 Deviations of delF2 with conventional PI and ACO PID controller response

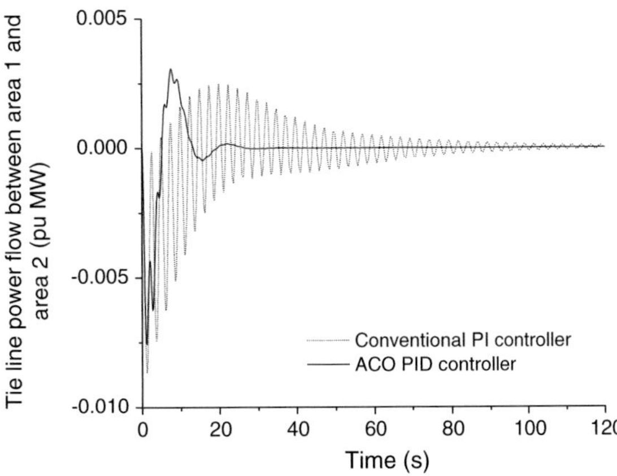

Fig. 12 Deviations of delPtie12 with conventional PI and ACO PID controller response

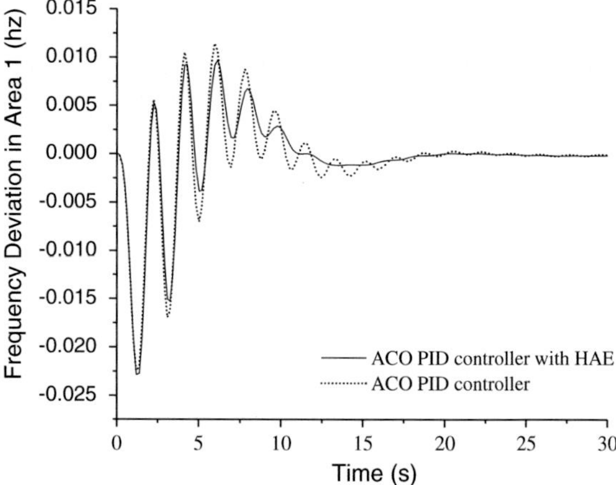

Fig. 13 Deviations of delF1 with and without considering HAE and FC unit

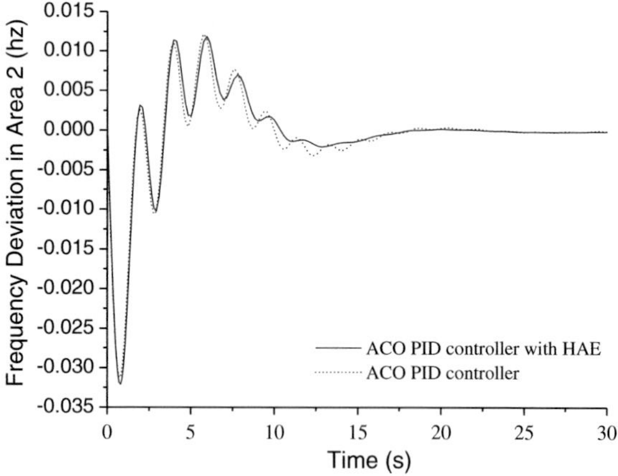

Fig. 14 Deviations of delF2 with and without considering HAE and FC unit

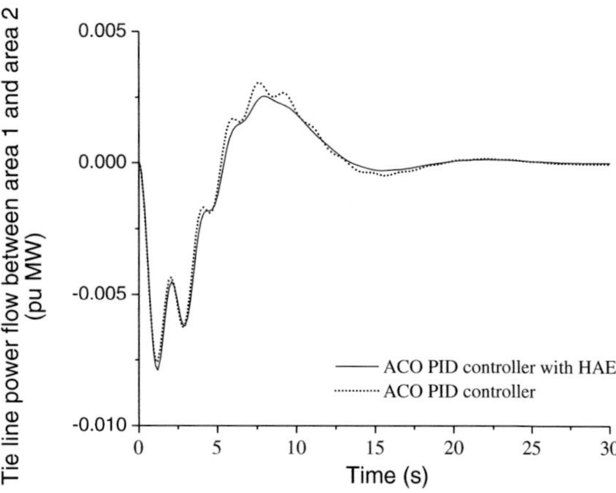

Fig. 15 Deviations of delPtie12 with and without considering HAE and FC unit

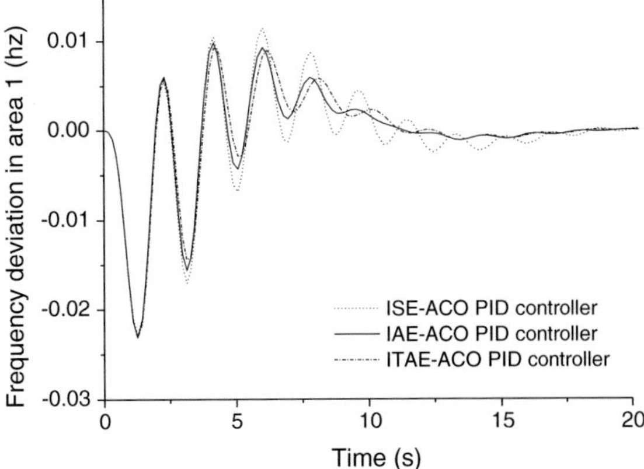

Fig. 16 Deviations of delF1 with ACO PID controller considering three diffent cost functions

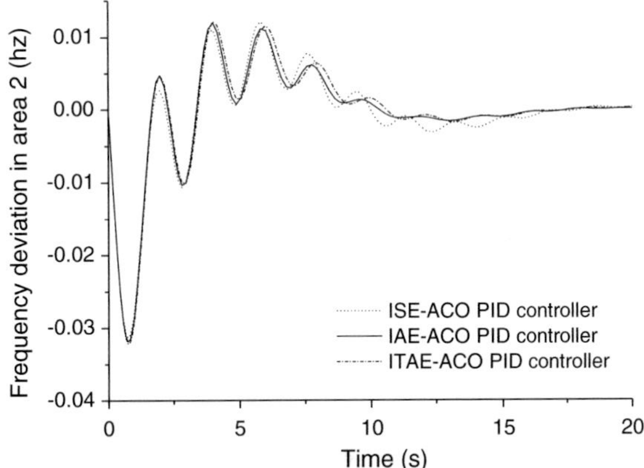

Fig. 17 Deviations of delF2 with ACO PID controller considering three different cost functions

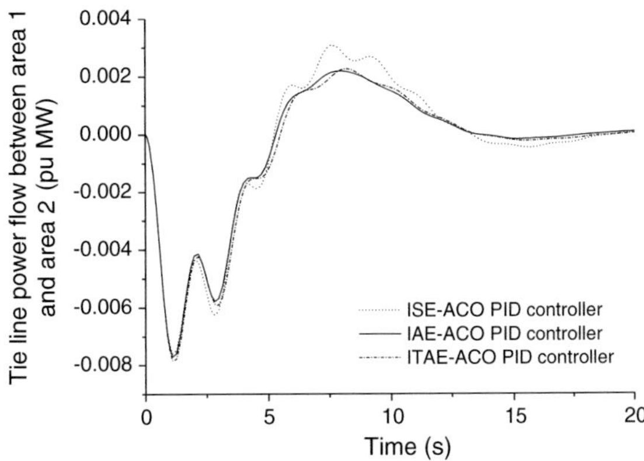

Fig. 18 Deviations of delPtie12 with ACO PID controller considering three different cost functions

6.1 Comparision Performance Analysis of Open Loop Response of System

A one percent step load perturbation (1 % SLP) is considered in thermal power system (area 2) to analyze open loop system performance. Frequency deviation deviations in area 1and are 2 shown in Figs. 5 and 6. It is clearly indicated that, when the load demand occurs in any one of interconnected power system, it produce more damping oscillations with large steady error and it takes more time to settle to it nominal value.

Tie-line power flow between interconnected power system response are shown in Fig. 6, it indicates that during sudden load disturbance or load demand affect power between connected power system and response yield more damping oscillations with large overshoots in their response. The values open loop parameters of investigated power system given in Table 3.

6.2 Comparision Performance Analysis of Open Loop Response of System with Conventional PI Controller

The closed loop transfer function model power system is discussed in system investigated section and system responses (delF1, delF2 and delPtie12) are shown in Figs. 7, 8 and 9. The solid line shows the response of conventional PI controller response and dotted line shows the response of open loop system without considering any controller. The performance of conventional Proportional-Integral (PI) controller is compared with the open loop response of the system. The conventional PI controller gain values are optimized by using trial and error method and discussed in Sect. 3.

The comparision response of system clearly decipate that, conventional PI controller effectively eliminate (nearly equal to zero) steady state error in frequency deviation and tie-line power flow between interconnected system. The response of the system yield large steady state error, peak over and under shoots during sudden load demand and without considering any controller.

6.3 Comparisons Performance Analysis of Conventional PI Controller and ACO Based PID Controller

Figures 10, 11 and 12 shows the frequency deviation in area 1 and area 2, tie-line power flow between area 1 and 2 of conventional PI and ACO-PID controller comparision performances in this case conventional PI controller equipped power system performance is compared with ACO-PID controller equipped system. In this response dotted line shows the response of conventional PI controller and solid line shows the response of ACO optimized PID controller in the same investigated power system.

Table 5 Settling time comparison of Conventional PI and ACO-PID

Response	Settling time (s)	
	Conventional PI	ACO-PID
delF1	100	59
delF2	116	30
delPtie12	120	120

Plot between frequency versus time and tie-line power flow versus time clearly indicated that, by the implementation of ACO-PID controller system settling time effectively reduced and peak overshoot, undershoot and damping oscillations also effectively reduced by implementing ACO based PID controller in investigated power system. Table 5 gives the numerical value comparison of conventional PI and ACO-PID controller in investigated power system.

6.4 *Comparisons of System Response with and Without Considering Energy Storage Unit Effect*

The characteristics of an energy storage unit are stores the energy during nominal or normal loading conditions and it delivers the stored energy during the sudden load disturbance. The different energy storage units and its applications are discussed in literature survey.

In this investigation Hydrogen Aqua Electrolyzer with Fuel Cell based energy storage unit is considered in each area of the power generating unit.

In the above response solid line shows the response of ACO-PID controller with Hydrogen Aqua Electrolyser (HAE) unit and dotted line shows the response of system with ACO PID controller and with out considering HAE energy storage unit. It is clearly evident that, system with HAE gives more superior response compared to system without HAE unit (Table 6).

Table 6 Settling time comparison of ACO-PID and ACO-PID with HAE

Response	Settling time (s)	
	ACO-PID	ACO-PID with HAE
delF1	59	17.07
delF2	30	17.54
delPtie12	120	18.12

Table 7 Settling time comparisions of different cost functions

Response	Settling time (s)		
	ACO-PID-ISE	ACO-PID-IAE	ACO-PID-ITAE
delF1	18.7	17.82	17.82
delF2	18.46	16.24	16.24
delPtie12	18.46	16.24	16.24

6.5 Comparison Analysis of System with a Proposed Approach Considering Different Objective Functions

In this case performance of investigating power system is analysed with PID controller considering non linearity and boiler dynamics and three different cost functions. The comparison performance of different objective functions is given the Figs. 16, 17 and 18, it shows frequency deviation in area 1, area 2 and tie-line power flow deviation between area 1 and 2 respectively. Where, dotted line shows the response of investigated power system with ISE objective function based PID controller, solid line shows the response of system response with IAE objective function based PID controller response and short dash dotted line shows the response of ITAE objective function based PID controller response.

It is clearly shown that peak over and undershoot values are effectively reduced and system settled very fast compare to the ISE optimized PID controller. And also performance of ITAE and IAE is same and numerical values are given in the Table 7.

Table 7 reports the settling comparisons of three different cost functions of investigated multi area thermal power system with PID controller. It is evident that performance of IAE and ITAE nearly same and it give mre superior performance over ISE optimized PID controller.

7 Conclusion

In this work performance of two area interconnected thermal power system is analyzed by considering generation rate constraint, governor dead band nonlinearities and boiler dynamics. Initially, performance of system is obtained without considering any supplementary controller and which is compared with proportional-Integral (PI) controller equipped system performance. The comparison clearly evident that supplementary controller effectively eliminates steady state error and it improves system stability.

In the second case system is equipped with Proportional-Integral-Derivative (PID) controller and controller gain values are optimized by using Artificial Intelligence (AI) based Ant Colony Optimization (ACO) technique. The performance of ACO-PID controller performance is compared with conventional tuned PI controller. The performance evalution reveals that ACO-PID controller based performance give more superior performance than the conventional PI controller and it effectively damping out the oscillations with good settling time and the overall stability of the system is increased.

The overall performance of the system is improved by adding energy storage unit in a power generating unit, which stores the energy during the normal loading condition and stored energy, is getting back during sudden load demand or disturbance. In this investigation Hydrogen Aqua Electrolyzer with Fuel Cell (HAE with FC) is considered in both areas and performances are obtained and compared with system performance without considering HAE and FC in generating units. Comparison result evident that considerable amount of performance is achieved by adding energy storage units.

The optimization of controller gain values is most important for better controller performance. At the same time choice or selection of objective function also more important. For this investigation three different objective functions are considered, such as, ITAE, IAE and ISE for selecting PID controller gain values. A performance comparison shows that IAE and ITAE give superior performance than ISE objective function. And also simulations performance numerical values of IAE and ITAE objective function optimized controller gives a nearly similar performance in terms of settling time, over and undershoots.

In future the same optimization technique was extended into (By increasing the size) multi-area interconnected power system. Also several new bio inspired algorithms are developed and implemented into the power system for solving same LFC/AGC issue in large interconnected power system.

References

1. Abraham RJ, Das D, Patra A (2011) Damping oscillations in tie-power and area frequencies in a thermal power system with SMES-TCPS combination. J Electr Syst 1(1):71–80
2. Aditya SK, Das D (2001) Battery energy storage for load frequency control of an interconnected power system. Electr Power Syst Res 58:179–185
3. Aldeeen M, Marsh JF (1991) Decentralized proportional-plus-integral design method for interconnected power systems. IEEE Proc C 138(4):263–274
4. Ali ES, Abd-Elazim SM (2009) Bacteria foraging optimization algorithm based load frequency controller for interconnected power system. Electr Power Energy Syst 33:633–638
5. Anand B, Jeyakumar AE (2009) Load frequency control with fuzzy logic controller considering non-linearities and boiler dynamics. ACSE 8:15–20
6. Arivoli R, Chidambaram IA (2011) Design of genetic algorithm (GA) based controller for load-frequency control of power systems interconnected with AC-DC TIE-LINE. Int J Sci Eng Tech 2:280–286
7. Azar AT (2010) Fuzzy systems. in-tech, Vienna
8. Azar AT (2012) Overview of type-2 fuzzy logic systems. Int J Fuzzy Syst Appl (IJFSA) 2(4):1–28
9. Azar AT, Serrano FE (2014) Robust IMC-PID tuning for cascade control systems with gain and phase margin specifications. Neural Comput Appl, 25(5):983–995. doi:10.1007/s00521-014-1560-x

10. Azar AT, Serrano FE (2015a) Design and modeling of anti wind up PID controllers. In: Zhu Q, Azar AT (eds) Complex system modelling and control through intelligent soft computations. Studies in fuzziness and soft computing, vol 319. Springer, Germany, pp 1–44. doi:10.1007/978-3-319-12883-2_1
11. Azar AT, Serrano FE (2015b) Deadbeat control for multivariable systems with time varying delays. In: Azar AT, Vaidyanathan S (eds) Chaos modeling and control systems design. Studies in computational intelligence, vol 581. Springer-Verlag GmbH, Berlin, pp 97–132. doi:10.1007/978-3-319-13132-0_6
12. Azar AT, Vaidyanathan S (2015a) Chaos modeling and control systems design. Studies in computational intelligence, vol 581. Springer, Germany
13. Azar AT, Vaidyanathan S (2015b) Handbook of research on advanced intelligent control engineering and automation. Advances in computational intelligence and robotics (ACIR) book series. IGI Global, USA
14. Azar AT, Vaidyanathan S (2015c) Computational intelligence applications in modeling and control. Studies in computational intelligence, vol 575. Springer, Germany. ISBN:978-3-319-11016-5
15. Balasundaram P, Chidambaram IA (2012) ABC algorithm based load frequency controller for an interconnected power system considering nonlinearities and coordinated with UPFC and RFB. Int J Eng Innovative Technol (IJEIT) 1(3):1–11
16. Banerjee S, Chatterjee JK, Tripathy SC (1990) Application of magnetic energy storage unit as load frequency stabilizer. IEEE Trans Energy Convers 5:46–51
17. Banerjee S, Chatterjee JK, Tripathy SC (1990) Application of magnetic energy storage unit as load-frequency stabilizer. IEEE Trans Energy Convers 5(1):46–51
18. Beck JW, Carrol DP, Gareis GE, Krause PC, Ong CM (1976) A computer study of battery energy storage and power conversion equipment operation. IEEE Trans Power Appar Syst 95(4):1064–1072
19. Bhatt P, Ghoshal SP, Roy R (2010) Load frequency stabilization by coordinated control of thyristor controlled phase shifters and superconducting magnetic energy storage for three types if interconnected two-area power systems. Electr Power Energy Syst 32:1111–1124
20. Blum C (2005) Beam-ACO–hybridizing ant colony optimization with beam search: an application to open shop scheduling. Comput Oper Res 32(6):1565–1591
21. Chidambaram IA, Paramasivam B (2009) Genetic algorithm based decentralized controller for load-frequency control of interconnected power systems with RFB considering TCPS in the tie-line. Int J Electr Eng Res 1:299–312
22. Chine S, Tripathy M (2015). Design of an optimal SMES for automatic generation control of two-area thermal power system using cuckoo search algorithm. J Electr Syst Inf Technol 2(1):1–13
23. Chon N (1957) Some aspects of tie-line biased control on interconnected power systems. Am Inst Electr Eng Trans 75:1415–1436
24. Costa D, Hertz A (1997) Ants can color graphs. J Oper Res Soc 48:295–305
25. Das S, Kothari ML, Kothari DP, Nanda J (1991) Variable structure control strategy to automatic generation control of interconnected reheat thermal system. IEEE Proc D 138(6):579–585
26. Dash P, Saikia LC, Sinha N (2014) Comparison of performances of several Cuckoo search algorithm based 2DOF controllers in AGC of multi-area thermal system. Electr Power Energy Syst 55:429–436
27. Dash P, Saikia LC, Sinha N (2015) Automatic generation control of multi area thermal system using Bat algorithm optimized PD-PID cascade controller. Electric Power Energy Syst 68:364–372
28. Dash P, Saikia LC, Sinha N (2015) Comparison of performance of several FACTS devices using cuckoo search algorithm optimized 2DOF controllers in multi-area AGC. Electr Power Energy Syst 65:316–324
29. De Campos LM, Ferna'ndez-Luna JM, Ga'mez JA, Puerta JM (2002) Ant colony optimization for learning Bayesian networks. Int J Approx Reason 31(3):291–311

30. DeCampos LM, Gamez JA, Puerta JM (2002) Learning bayesian networks by ant colony optimization: searching in the space of orderings. Mathware Soft Comput 9(2–3):251–268
31. Demiroren A, Sengor NS, Zeynelghi HL (2001) Automatic generation control by using ANN technique. Electr Power Compon Syst 29:883–896
32. Demiroren A, Zeynelgil AL, Sengor NS (2001) The application of ANN technique to load frequency control for three-area power systems. In: IEEE porto power tech conference on 10–13th September, Portugal
33. DenBesten ML, Stutzle T, Dorigo M (2000) Ant colony optimization for the total weighted tardiness problem. In: M. Schoenauer et al (eds) Proceedings of PPSN-VI. LNCS, vol 1917. Springer, Berlin, pp 611–620
34. Dey N, Samanta S, Chakraborty S, Das A, Chaudhuri SS, Suri JS (2014) Firefly algorithm for optimization of scaling factors during embedding of manifold medical information: an application in ophthalmology imaging. J Med Imaging Health Inform 4(3):384–394
35. Dey N, Samanta S, Yang XS, Chaudhri SS, Das A (2014) Optimization of scaling factors in electrocardiogram signal watermarking using cuckoo search. Int J Bio-Inspir Comput (IJBIC) 5(5):315–326
36. Dorigo M, Birattari M, Stiitzle T (2006) Ant colony optimization: artificial ants as a computational intelligence technique. IEEE Comput Intell Mag, 28–39
37. Dorigo M, Gambardella LM (1997) Ant colony system: a cooperative learning approach to the traveling salesman problem. IEEE Trans Evol Comput 1(1):53–66
38. Dorigo M, Gambardella LM (2002) Ants can solve constraint satisfaction problems. IEEE Trans Evol Comput 6(4):347–357
39. Dorigo M, Maniezzo V, Colorni A (1996) Ant system: optimization by a colony of cooperating agents. IEEE Trans Syst Man Cybern-Part B 26(1):29–41
40. Dorigo M, Maniezzo V, Colorni A (1991) Positive feedback as a search strategy. Dipartimento di Elettronica, Politecnico di Milano, Italy, Technical report, pp 91–016
41. Ebrahim MA, Mostafa HE, Gawish SA, Bendary FM (2009) Design of decentralized load frequency based-PID controller using stochastic particle swarm optimization technique. In: International conference on electric power and energy conversion system, pp 1–6
42. Enomoto K, Sasaki T, Shigematsu T, Deguchi H (2002) Evaluation study about redox flow battery response and its modeling. IEEE Trans Power Eng B 122(4):554
43. Fenet S, Solnon C (2003) Searching for maximum cliques with ant colony optimization. In: Raidl GR et al (eds) Applications of evolutionary computing, Proceedings of EvoWorkshops 2003. LNCS, vol 2611. Springer, Berlin, pp 236–245
44. Francis R, Chidambaram IA (2013) Load frequency control for an interconnected reheat thermal power systems with redox flow batteries using beta wavelet neural network controller. Int J Eng Innov Technol (IJEIT) 2(9):275–282
45. Francis R, Chidambaram IA (2015) Optimized PI+ load-frequency controller using BWNN approach for an interconnected reheat power system with RFB and hydrogen electrolyser units. Electr Power Energy Syst 67:381–392
46. Gambardella LM, Dorigo M (2000) Ant colony system hybridized with a new local search for the sequential ordering problem. INFORMS J Comput 12(3):237–255
47. Gambardella LM, Taillard ED, Agazzi G (1999) MACS-VRPTW: a multiple ant colony system for vehicle routing problems with time windows. In: Corne D et al (eds) New ideas in optimization. McGraw Hill, London, pp 63–76
48. Gozde H, Taplamacioglu MC (2011) Automatic generation control application with craziness based particle swarm optimization in a thermal power system. Electr Power Energy Syst 33:8–16
49. Gozde H, Taplamacioglu MC, Kocaarslan I (2012) Comparative performance analysis of Artificial Bee Colony algorithm in automatic generation control for interconnected reheat thermal power system. Electr Power Energy Syst 42:167–178
50. Hassanien AE, Tolba M, Azar AT (2014) Advanced machine learning technologies and applications: second international conference, AMLTA 2014, Cairo, Egypt, 28–30 Nov 2014. Proceedings, communications in computer and information science, vol 488. Springer-Verlag GmbH, Berlin. ISBN: 978-3-319-13460-4

51. Ipsakis D, Voutetakis S, Seferlisa P, Stergiopoulos F, Elmasides C (2009) Power management strategies for a stand-alone power system using renewable energy sources and hydrogen storage. Int J Hydrogen Energy 34:7081–7095
52. Jagatheesan K, Anand B (2012) Dynamic performance of multi-area hydro thermal power systems with integral controller considering various performance indices methods. IEEE international conference on emerging trends in science, engineering and technology, Tiruchirappalli, December 13–14
53. Jagatheesan K, Anand B (2014) Automatic generation control of three area hydro-thermal power systems considering electric and mechanical governor with conventional and ant colony optimization technique. Adv Nat Appl Sci 8(2):25–33
54. Jagatheesan K, Anand B (2014) Automatic generation control of three area hydro-thermal power systems considering electric and mechanical governor with conventional and ant colony optimization technique. Adv Nat Appl Sci 8(20):25–33
55. Jagatheesan K, Jeyanthi S, Anand B (2014) Conventional load frequency control of an interconnected multi-area reheat thermal power systems using HVDC link. Int J Sci Eng Res 5(4):88–92
56. Jagatheesan K, Anand B, Ebrahim MA (2014) Stochastic particle swarm optimization for tuning of PID controller in load frequency control of single area reheat thermal power system. Int J Electr Power Eng 8(2):33–40
57. Kalyani S, Nagalakshmi S, Marisha R (2012) Load frequency control using battery energy storage system in interconnected power system. In: Proceedings ICCCNT'12
58. Ke'louwani S, Agbossou K, Chahine R (2005) Model for energy conversion in renewable energy system with hydrogen storage. J Power Sources 140:392–399
59. Korb O, Stutzle T, Exner TE (2006) Application of ant colony optimization to structure-based drug design. In: Dorigo M et al (eds) Proceedings of ANTS 2006. LNCS, vol 4150. Springer, Berlin, pp 247–258
60. Kothari ML, Nanda J (1988) Application of optimal control strategy to automatic generation control of a hydrothermal system. IEEE Proc 135(4):268–274
61. Kunisch HJ, Kramer KG, Dominik H (1986) Battery energy storage, another option for load-frequency-control and instantaneous reserve. IEEE Trans Energy Convers 1(3):41–46
62. Kunisch HJ, Kramer KG, Dominik H (1996) Battery energy storage another option for load-frequency-control and instantaneous reserve. IEEE Trans Energy Convers 1(3):41–46
63. Little M, Thomson M, Infield I (2007) Electrical integration of renewable energy into stand-alone power supplies incorporating hydrogen storage. Int J Hydrogen Energy 32(10–11):1582–1588
64. Lu C, Liu C, Wu C (1995) Effect of battery energy storage system on load frequency control considering governor dead band and generation rate constraint. IEEE Trans Energy Convers 10(3):555–561
65. Maniezzo V (1999) Exact and approximate nondeterministic tree-search procedures for the quadratic assignment problem. INFORMS J Comput 11(4):358–369
66. Martens D, Backer MD, Haesen R, Baesens B, Mues C, Vanthienen J (2006) Ant-based approach to the knowledge fusion problem. In: Dori-go M et al (eds) Proceedings ANTS 2006. LNCS, vol 4150. Springer, Berlin, pp 84–95
67. Merkle D, Middendorf M (2003) Ant colony optimization with global pheromone evaluation for scheduling a single machine. Appl Intell 18(1):105–111
68. Merkle D, Middendorf M, Schmeck H (2002) Ant colony optimization for resource- constrained project scheduling. IEEE Trans Evol Comput 6(4):333–346
69. Mitani Y, Tsuji K, Murkami Y (1988) Application of superconducting magnet energy storage to improve power system dynamic performance. IEEE Trans Power Syst 3:1418–1425
70. Nanda J, Kaul BL (1978) Automatic generation control of an interconnected power system. IEEE Proc 125(5):385–390
71. Nanda J, Kothari ML, Satsangi PS (1983) Automatic generation control of an interconnected hydrothermal system in continuous and discrete modes considering generation rate constraints. IEEE Proc 130(1):17–27

72. Nandha J, Mishra S (2010) A novel classical controller for Automatic generation control in thermal and hydro thermal systems. In: PEDES, pp 1–6
73. Omar M, Soliman M, Abdel Ghany AM, Bendary F (2013) Optimal tuning of PID controllers for hydrothermal load frequency control using ant colony optimization. Int J Electr Eng Inform 5(3):348–356
74. Padhan S, Sahu RK, Panda S (2015) Application of firefly algorithm for load frequency control of multi-area interconnected power system. Electr Power Compon Syst 42(13):1419–1430
75. Pan CT, Liaw CM (1989) An adaptive controller for power system load-frequency control. IEEE Trans Power Syst 4(1):122–128
76. Parmar KP (2014) Load frequency control of multi-source power system with redox flow batteries: an analysis. Int J Comput Appl 88(8):46–52
77. Peterson HA, Mohan N, Boom RW (1975) Superconductive energy storage inductor-converter units for power systems. IEEE Trans Power Appar Syst 94:1337–1348
78. Pothiya S, Ngamroo I, Kongprawechnon W (2007) Design of optimal fuzzy logic-based PID controller using multiple Tabu search algorithm for AGC including SMES units. In: 8th international power engineering conference (IPEC 2007), pp 838–843
79. Padhan S, Sahu SK, Panda S (2014) Automatic generation control with thyristor controlled series compensator including superconducting magnetic energy storage units. Ain Shams Eng J 5:759–774
80. Dash P, Saikia LC, Sinha N (2014) Comparison of performances of several cuckoo search algorithm based 2DOF controllers in AGC of multi-area thermal system. Electr Power Energy Syst 55:429–436
81. Riberio PF, Johnson BK, Crow ML, Arsoy A, Liu Y (2001) Energy storage systems for advanced power applications. IEEE Proc 89(12):1744–1756
82. Roy A, Dutta S, Roy PK (2014) Automatic generation control by SMES-SMES controllers of two-area hydro-hydro system. Proceedings of 2014 1st international conference on non conventional energy (ICONCE 2014), pp 302–307
83. Sahu BK, Pati S, Mohanty PK, Panda S (2015) Teaching-learning based optimization algorithm based fuzzy-PID controller for automatic generation control of multi-area power system. Appl Soft Comput 27:240–249
84. Sahu RK, Panda S, Padhan S (2015) A hybrid firefly algorithm and pattern search technique for automatic generation control of multi area power systems. Electr Power Energy Syst 64:9–23
85. Sahu RK, Panda S, Rout UK (2013) DE optimized parallel 2-DOF PID controller for load frequency control of power system with governor dead-band nonlinearity. Electr Power Energy Syst 49:19–33
86. Saikia LC, Sinha N, Nanda J (2013) Maiden application of bacterial foraging based fuzzy IDD controller in AGC of a multi-area hydrothermal system. Electr Power Energy Syst 45:98–106
87. Saikia LC, Sinha N, Nanda J (2013) Maiden application of bacterial foraging based fuzzy IDD controller in AGC of a multi-area hydrothermal system. Electr Power Energy Syst 45:98–106
88. Salameh ZM, Casacca MA, Lynch WA (1992) A mathematical model of lead-acid batteries. IEEE Trans Energy Convers 7(1):93–98
89. Samanta S, Acharjee S, Mukherjee A, Das D, Dey N (2013) Ant weight lifting algorithm for image segmentation. In: 2013 IEEE international conference on computational intelligence and computing research (ICCIC), pp 1–5
90. Shayeghi H, Shayanfor HA (2006) Application of ANN technique based μ-synthesis to load frequency control of interconnected power and energy systems 28:503–511
91. Shmygelska A, Hoos HH (2005) An ant colony optimization algorithm for the 2D and 3D hydrophobic polar protein folding problem. BMC Bioinform 6(30):1
92. Socha K, Sampels M, Manfrin M (2003) Ant algorithms for the university course timetabling problem with regard to the state-of-the-art. In: Raidl GR et al (eds) Applications of evolutionary computing, proceedings of EvoWorkshops 2003. LNCS, vol. 2611, Springer, Berlin, pp 34–345
93. Solnon C (2000) Solving permutation constraint satisfaction problems with artificial ants. In: Proceedings of ECAI'2000. IOS Press, Amsterdam, pp 118–122

94. Stutzle T, Hoos HH (1997) The MAX -MIN ant system and local search for the traveling salesman problem. In: Back T et al (eds) Proceedings of 1997 IEEE international conference on evolutionary computation (ICEC'97). IEEE Press, Piscataway, pp 309–314
95. Stutzle T, Hoos HH (2000) MAX -MIN ant system. Future Gener Comput Syst 16(8):889–914
96. Taher SA, Fini MH, Aliabadi SF (2014) Fractional order PID controller design for LFC in electric power systems using imperialist competitive algorithm. Ain Shams Eng J 5:121–135
97. Tam K, Kumar P (1990) Application of SMES in an asynchronous link between power systems. IEEE Trans Energy Convers 5:436–444
98. Tam W, Kumar P (1990) Applications of superconductive magnetic energy storage in an asynchronous link between power systems. IEEE Trans Energy Convers 5:436–444
99. Tokuda N (1998) Development of a Redox Flow Battery System. In: Engineering conference IECEC-98-1074
100. Tripathy SC, Balasubramaniam R, Chandramohan Nair PS (1992) Adaptive automatic generation control with SMES in power systems. IEEE Trans Energy Convers 7:434–441
101. Tripathy SC, Balasubramaniam R, Chandramohan Nair PS (1992) Effect of superconducting magnetic energy storage on automatic generation control considering governor dead and boiler dynamics. IEEE Trans Power Syst 7(2):12661273
102. Tripathy SC, Bhatti TS, Jha CS, Malik OP, Hope GS (1984) Sampled data automatic generation control analysis with reheat steam turbines and governor dead-band effects. IEEE Trans Power Appar Syst 103(5):1045–1051
103. Tripathy SC, Hope GS, Malik OP (1982) Optimization of load-frequency control parameters for power systems with reheat steam turbines and governor dead band nonlinearity. IEEE Proc 129(1):10–16
104. YU S, Mays TJ, Dunn RW (2009) A new methodology for designing hydrogen energy storage in wind power systems to balance generation and demand. In: IEEE international conference (SUPERGEN 2011), Nanjing, pp 1–6
105. Zhu Q, Azar AT (2015) Complex system modeling and control through intelligent soft computations. Studies in fuzziness and soft computing, vol 319. Springer, Germany. ISBN: 978-3-319-12882-5

Fuzzy Adaptive Synchronization of Uncertain Fractional-Order Chaotic Systems

Abdesselem Boulkroune, Amel Bouzeriba, Toufik Bouden and Ahmad Taher Azar

Abstract In this chapter, a fuzzy adaptive controller for a class of fractional-order chaotic systems with uncertain dynamics and external disturbances is proposed to realize a practical projective synchronization. The adaptive fuzzy systems are used to online approximate unknown system nonlinearities. The proposed control law, which is derived based on a Lyapunov approach, is continuous and ensures the stability of the closed-loop system and the convergence of the underlying synchronization errors to a neighborhood of zero. Finally, a simulation example is provided to verify the effectiveness of the proposed synchronization method.

Keywords Intelligent adaptive control · Fuzzy systems · Projective synchronization · Fractional-order chaotic systems

1 Introduction

Fractional calculus is a generalization of ordinary differentiation and integration to arbitrary (non-integer) order. It can be traced to the early works of Leibniz and L'hospital in 1695. During the last decades, it has attracted a great deal of attention from chemists, physicians and engineers. In fact, it has been found that many systems in interdisciplinary fields can be accurately modelled by fractional-order

A. Boulkroune (✉) · A. Bouzeriba
LAJ Laboratory, University of Jijel, BP. 98, Ouled-Aissa, 18000 Jijel, Algeria
e-mail: boulkroune2002@yahoo.fr

A. Bouzeriba
e-mail: bouzeriba.amel@yahoo.fr

A. Bouzeriba · T. Bouden
NDT Laboratory, University of Jijel, BP. 98, Ouled-Aissa, 18000 Jijel, Algeria
e-mail: bouden_toufik@yahoo.com

A.T. Azar
Faculty of Computers and Information, Benha University, Banha, Egypt
e-mail: ahmad_t_azar@ieee.org; ahmad.azar@fci.bu.edu.eg

A.T. Azar and S. Vaidyanathan (eds.), *Advances in Chaos Theory and Intelligent Control*, Studies in Fuzziness and Soft Computing 337,
DOI 10.1007/978-3-319-30340-6_28

differential equations, such as dielectric polarization [40], viscoelastic systems [6], electromagnetic waves [20], electrode-electrolyte polarization [23], and so on.

Chaotic systems are nonlinear and deterministic systems. They are characterized by "sensitive dependence on initial conditions", a small perturbation eventually causes a large change in the system states [2–5, 54]. The main feature used to identify a chaotic behaviour is the well-known Lyapunov exponent criteria. In fact, a system that has one positive Lyapunov exponent is known as a chaotic system. It has been recently shown that many fractional-order systems can display chaotic behaviours, such as fractional-order Duffing system [18], fractional-order Lorenz system [53], fractional-order Chua's system [19], fractional-order Chen system [24], to name a few.

Synchronization problem consists in designing a slave system whose behavior mimics a master system. The latter drives the slave system via the transmitted signals. Various types of the chaos synchronization have been recently revealed, such as complete synchronization (CS) [12, 41] phase synchronization (PS) [38], generalized synchronization (GS) [52], generalized projective synchronization (GPS) [26], and so on. However, all these synchronization methods focus on integer-order chaotic systems (a special case of the fractional-order chaotic systems). In addition, it has been assumed in [12, 26, 38, 41, 52] that models of the chaotic systems are almost known. Therefore, it is interesting to extend these results to uncertain fractional-order chaotic systems by incorporating an adaptive fuzzy system to deal with model uncertainties [43–47].

Based on the universal approximation feature of the fuzzy systems [51], some adaptive fuzzy control schemes [8, 14, 22, 29, 36, 37, 50] have been designed for a class of unknown chaotic systems with integer-order. In these control schemes, the fuzzy systems are used to online approximate the unknown nonlinear functions. The stability analysis of the corresponding closed-loop control system has been carried out via a Lyapunov approach. The robustness issues with respect to the fuzzy approximation errors and the probable external disturbances have been improved by properly adding a robust control term to the fuzzy adaptive control term. This robust control term can be conceived by a sliding mode control approach [8, 14, 36, 37], an $H\infty$ control approach [22, 49, 50] and a quasi-sliding mode control approach [29]. However, it is should be noted that the above results [8, 14, 22, 29, 36, 37, 49, 50] are limited to chaotic integer-order systems.

The synchronization of fractional-order chaotic systems is certainly a challenging research topic [15, 21, 27, 28, 32, 33, 48] In [15], authors have researched the synchronization of fractional-order chaotic neural networks. In [48], a local stability criterion for synchronization of incommensurate fractional-order chaotic systems has been shown. In [32], an active pinning control for synchronization and anti-synchronization of uncertain unified chaotic systems with fractional-order has been reported. In addition, the author of [33] has designed a synchronization system of two identical fractional-order chaotic systems using a linear error feedback control. Hosseinnia et al. have designed a linear sliding surface with its switching control law for synchronization of two identical unknown fractional-order chaotic systems [21]. In [28], a fuzzy adaptive sliding mode control for synchronization of unknown fractional-order chaotic systems with time delay has been proposed. In [27], for achieving an H_∞ synchronizing for a class of uncertain fractional-order chaotic systems, a fuzzy adaptive controller has been designed. However, as stated in [1, 42], the results of [28] and [27] are already questionable, because the stability analysis has not been derived rigorously in mathematics.

The above considerations motivate our study work. In this chapter, a fuzzy adaptive controller for a class of fractional-order chaotic systems with uncertain dynamics and external disturbances is proposed to realize a practical projective synchronization. The adaptive fuzzy systems are used to online approximate unknown system nonlinearities. By using a coordinate transformation on the synchronization errors, the stability analysis as well as the control design are simplified. Furthermore, it is proved that all the resulting closed-loop system signals are bounded and the underlying synchronization errors converge to a neighborhood of zero. To show the effectiveness of the proposed synchronization system, an numerical illustrative example is presented. Compared with the existing results [15, 21, 27, 28, 32, 33, 48], the main contributions of this chapter are as follows:

(1) By designing an adaptive fuzzy control, a practical projective synchronization is properly achieved for a class of unknown fractional-order chaotic systems. To our best of knowledge, there are only few works dealing with the design of a fuzzy adaptive control for fractional-order systems. Some preliminaries works are given in [28, 28]. However, as stated in [1, 42], these works are already questionable.
(2) Unlike the previous results [15, 21, 27, 28, 32, 33, 48], the slave chaotic system is subject to unknown dynamic disturbances and the model of the master-slave structure is assumed to be completely unknown. In fact, adaptive fuzzy approximators are incorporated in the synchronization system to online estimate the uncertain nonlinear functions.
(3) Unlike the closely related works [27, 28], the stability analysis of the corresponding closed-loop system is strictly established in this chapter, by employing some properties of the Caputo fractional-order derivative [13, 17, 25, 34, 39].

This chapter is organized as follows. Basic definitions and preliminaries for fractional-order systems are given in Sect. 2. In Sect. 3, the problem statement and fuzzy logic systems are presented. The fuzzy adaptive controller is designed in Sect. 4. Simulation results to demonstrate the performances of the proposed projective synchronization scheme are provided in Sect. 5. Finally, Sect. 6 concludes this chapter.

2 Basic Definitions and Preliminaries for Fractional-Order Systems

There exist many definitions of fractional operators. As, the Caputo fractional operator is more consistent than another ones, then this operator will be used in this chapter. Also, a modification version of Adams-Bashforth-Moulton algorithm proposed in

[16] will be employed for computer numerical simulation of the Caputo fractional-order differential equations.

The Caputo fractional derivative of a function $x(t)$ with respect to t is defined as follows [34]:

$$D_t^{\alpha} x(t) = \frac{1}{\Gamma(m-\alpha)} \int_0^t (t-\tau)^{-\alpha+m-1} x^{(m)}(\tau) d\tau, \tag{1}$$

where m is the first integer which is not less than α, i.e. $m-1 \le \alpha \le m$, D_t^{α} is called the $\alpha - order$ Caputo differential operator, and $\Gamma(.)$ is the well-known Euler's gamma function:

$$\Gamma(s) = \int_0^{\infty} t^{s-1} e^{-t} dt; \tag{2}$$

The following three important properties of the Caputo fractional-order derivative will be employed in the sequent sections [13, 25, 34]:

Property 1 *Let* $0 < q < 1$, *then*

$$Dx(t) = D_t^{1-q} D_t^q x(t), \quad \text{where} \quad D = \frac{d}{dt}. \tag{3}$$

Property 2 *The Caputo fractional derivative operator is linear, i.e.*

$$D_t^q(\nu x(t) + \mu y(t)) = \nu D_t^q x(t) + \mu D_t^q y(t), \tag{4}$$

where ν *and* μ *are real constants.*

Especially, $D_t^q x(t) = D_t^q(x(t)+0) = D_t^q x(t) + D_t^q 0$, *then, we have* $D_t^q 0 = 0$.

Property 3 *Consider a Caputo fractional nonlinear system [17, 39]:*

$$D_t^q x(t) = f(x(t)), \quad \textit{with} \quad 0 < q < 1 \tag{5}$$

If $f(x(t))$ *satisfies the Lipschiz condition with respect to* x, *i.e.,*

$$\|f(x(t)) - f(x_1(t))\| \le \ell \|x(t) - x_1(t)\|, \tag{6}$$

where ℓ *is a positive constant. And without loss of generality, if one assumes that* $f(x)$ *satisfies* $f(x) = 0$ *at* $x = 0$.
It follows that:

$$\|f(x(t))\| \le \ell \|x(t)\|. \tag{7}$$

Remark 1 Compared to others definitions, the great advantage of the Caputo definition is that the initial conditions for fractional-order differential equations take on a similar form as for integer-order differential equations. In the literature, such a operator is sometimes referred as a smooth fractional derivative.

3 Problem Statement and Fuzzy Logic Systems

3.1 Problem statement

Our main motivation consists in designing a fuzzy adaptive control-based projective synchronization system for a class of fractional-order chaotic systems, as shown in Fig. 1, while ensuring that all the signals in the corresponding closed-loop system remain bounded.

Consider a class of fractional-order chaotic master systems described as

$$D_t^q X = F_1(X), \tag{8}$$

where $D_t^q = \frac{d^q}{dt^q}, 0 < q < 1$ is the fractional derivative order, with $X = [x_1, \ldots, x_n]^T \in R^n$ is the state vector of the master system which is assumed to be measureable, and $F_1(X) = [f_{11}(X), \ldots, f_{1n}(X)]^T \in R^n$ is a vector of unknown continuous nonlinear functions.

The corresponding fractional-order chaotic slave system is :

$$D_t^q Y = F_2(Y) + Gu + P(t, Y), \tag{9}$$

where $Y = [y_1, \ldots, y_n]^T \in R^n$ is its overall state vector which is also assumed to be available measurable. $F_2(Y) = [f_{21}(Y), \ldots .f_{2n}(Y)]^T \in R^n$ is a vector of smooth unknown nonlinear functions, $G = [g_{ii}] \in R^{n \times n}$ is an unknown constant control-gains matrix, $u = [u_1, \ldots, u_n]^T \in R^n$ is the control input, and $P(t, Y) = [P_1(t, Y), \ldots, P_n(t, Y)]^T \in R^n$ denotes the unknown dynamic disturbances.

Assumption 1 Without loss of generality, the matrix G is assumed to be symmetric positive-definite.

Remark 2 Note that many fractional-order chaotic systems belong to the class characterized by (8) or (9). Among these, we can cite: fractional-order Lorenz system, fractional-order unified chaotic system, fractional-order Chen system, so on.

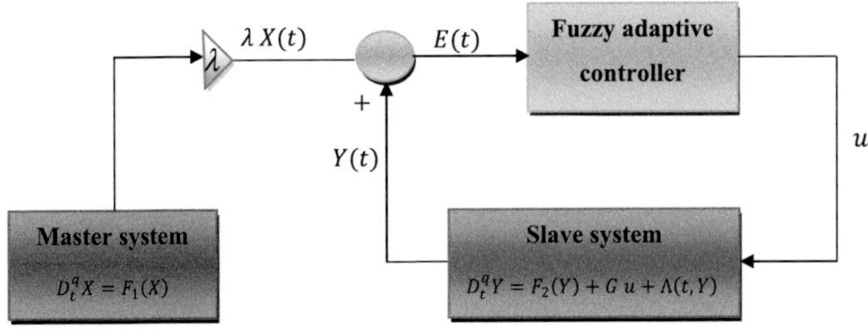

Fig. 1 The proposed synchronization scheme

Remark 3 The master-salve structure described by (8) and (9) has been considered in many studies [15, 17, 21, 32, 33, 48]. But, in these works, the model of such a structure is assumed to be known, partially known or without dynamic disturbances. Unlike in [15, 17, 21, 32, 33, 48], in this chapter, the model of the master-slave system is completely uncertain and with unknown dynamic disturbances.

Our main objective is to design an adaptive fuzzy controller u_i (for all $i = 1, \dots, n$) achieving a practical projective synchronization between the master system (8) and the slave system (9), while assuring the boundedness of all variables involved in the closed-loop system as well as the practical convergence of the corresponding synchronization errors to a neighborhood of zero.

Define the synchronization errors as

$$\begin{aligned} e_1 &= y_1 - \lambda x_1, \\ &\vdots \\ e_n &= y_n - \lambda x_n, \end{aligned} \tag{10}$$

where λ is a scaling factor that defines a proportional relation between the synchronized systems.

The vector of the synchronization errors is

$$E = [e_1, \dots, e_n]^T = Y - \lambda X. \tag{11}$$

In order to facilitate the design of the control system, we introduce a new variable $S = [S_1, \dots, S_n]^T$ as follows:

$$D_t^{1-q} S = E, \tag{12}$$

From Property 1 of the Caputo fractional derivative operator, we can rewrite (12) as:

$$D_t^q D_t^{1-q} S = \dot{S} = D_t^q E = F_2(Y) + G\,u - \lambda F_1(X) + p(t, Y). \tag{13}$$

Posing $G_1 = G^{-1}$, we have

$$G_1 \dot{S} = G_1[F_2(Y) - \lambda F_1(X)] + u + G_1 P(t, Y). \tag{14}$$

The dynamics of (14) can be expressed as

$$G_1 \dot{S} = \alpha(X, Y, \Lambda) + u, \tag{15}$$

where

$$\alpha(X, Y, \Lambda) = [\alpha_1(X, Y, \Lambda), \dots, \alpha_n(X, Y,)]^T = G_1[F_2(Y) - \lambda F_1(X)] + G_1 P(t, Y). \tag{16}$$

Later, (16) will be used in the development of the proposed fuzzy controller and the corresponding stability analysis.

Remark 4 Because the nonlinear function $\alpha(X, Y, \Lambda)$ is unknown, the design of a control system to practically stabilize the dynamics (15) is not easy. Later, to resolve such a problem, we will use an adaptive fuzzy system to approximate a functional upper-bound of $\alpha(X, Y, \Lambda)$.

Remark 5 From (12) and using Properties 1, 2 and 3 of the Caputo fractional derivative operator [17, 39], we can straightforwardly show the existence of a positive real number κ, such that $\|E\| \leq \kappa \|S\|$. So, $S = 0$ implies that $E = 0$, and the boundedness of S implies that of E.

3.2 *Fuzzy Logic Systems*

Typically, a fuzzy logic system (FLS) consists of four main parts: the knowledge base, the singleton fuzzifier, product inference, and the center average defuzzifier respectively [9, 10, 51].

The knowledge base for fuzzy logic system is constructed by a set of If-Then rules as follows:

$$R^{(i)} : if\ x_1\ is\ A_1^i\ and \ldots and\ x_n\ is\ A_n^i\ then\ \hat{f}\ is\ f^i, \tag{17}$$

where $A_1^i, A_2^i, \ldots..$ and A_n^i are fuzzy sets and f^i is a fuzzy singleton for the output in the ith rule.

Next, the FLS with a singleton fuzzifier, product inference, and center-average defuzzifier can be expressed as

$$\hat{f}(\underline{x}) = \frac{\Sigma_{i=1}^m f^i (\Pi_{j=1}^n \mu_{A_j^i}(x_j))}{\Sigma_{i=1}^m (\Pi_{j=1}^n \mu_{A_j^i}(x_j))} = \theta^T \psi(\underline{x}), \tag{18}$$

where $\underline{x} = [x_1, \ldots, x_n]^T \in R^n$, $\mu_{A_j^i}(x_j)$ is the membership function of the fuzzy set A_j^i, m is the number of fuzzy rules, $\theta^T = [f^1, \ldots, f^m]$ is the adjustable parameter (consequent parameters) vector, and $\psi^T = [\psi^1 \psi^2 \ldots \psi^m]$ with

$$\psi^i(\underline{x}) = \frac{(\Pi_{j=1}^n \mu_{A_j^i}(x_j))}{\Sigma_{i=1}^m (\Pi_{j=1}^n \mu_{A_j^i}(x_j))}, \tag{19}$$

being the fuzzy basis function (FBF).

Throughout the paper, the FBFs are appropriately selected so that [51]:

$$\Sigma_{i=1}^m (\Pi_{j=1}^n \mu_{A_j^i}(x_j)) > 0 \tag{20}$$

The fuzzy logic system (18) can estimate any nonlinear continuous function $f(\underline{x})$ defined on a compact operating space to an arbitrary accuracy [51]. Of particular importance, one assumes that the FBFs, i.e. $\psi(\underline{x})$, are properly specified in advance by designer. However, the consequent parameters, i.e. θ, are determined by some adaptation algorithms which will be designed later.

4 Design of Fuzzy Adaptive Controller

To facilitate the analysis, we need the following mild assumption.

Assumption 2 We assume that there exists an uncertain continuous positive function vector $\bar{\alpha}(Y)$ such that:

$$|\alpha(X, Y, \Lambda))| \leq \bar{\alpha}_n(Y), \tag{21}$$

with $\bar{\alpha}(Y) = [\bar{\alpha_1}(Y), \ldots, \bar{\alpha}_n(Y)]^T$.

Remark 6 Assumption 2 is not restrictive and used in many literatures [9, 10]. Note that this upper bound function $\bar{\alpha}(Y)$ always exist, as the state vector of the master system evolves in a compact set.

The uncertain nonlinear function $\bar{\alpha_i}(Y)$ can be approximated, over a compact set Ω_Y, by the FLS (18) as follows:

$$\hat{\bar{\alpha_i}}(Y, \theta_i) = \theta_i^T \psi_i(Y),\ with\ i = 1, \ldots, n, \tag{22}$$

where $\psi_i(Y)$ is the FBF vector and θ_i is the vector of the adjustable parameters of this FLS.

The ideal value of θ_i can be defined as [51]:

$$\theta_i^* = \arg\min_{\theta_i}[\sup_{Y \epsilon \Omega_Y} |\bar{\alpha}_i(Y) - \hat{\bar{\alpha}}_i(Y, \theta_i)|] \tag{23}$$

where θ_i^* is an artificial quantity required only for analytical purposes, as its value is not needed when implementing the control law.

Define

$$\tilde{\theta}_i = \theta_i - \theta_i^* \text{ and } \delta_i(Y) = \bar{\alpha}_i(Y) - \hat{\bar{\alpha}}_i(Y, \theta_i^*) = \bar{\alpha}_i(Y) - \theta_i^{*^T} \psi_i(Y) \tag{24}$$

as the parameter estimation errors and the approximation errors, respectively. $\delta_i(Y)$ satisfies $|\delta_i(Y)| \leq \bar{\delta}_i, \forall Y \in \Omega_Y$, with $\bar{\delta}_i$ is an unknown constant [7–11, 51].

Let us denote

$$\hat{\bar{\alpha}}(Y,\theta) = [\hat{\bar{\alpha}}_1(Y,\theta_1),\ldots,\hat{\bar{\alpha}}_n(Y,\theta_n)]^T = [\theta_1^T\psi_1(Y),\ldots,\theta_n^T\psi_n(Y)]^T,$$
$$\delta(Y) = [\delta_1(Y),\ldots,\delta_n(Y)]^T, \text{ and } \bar{\delta} = [\bar{\delta}_1,\ldots,\bar{\delta}_n]^T.$$

From the previous analysis, one can obtain

$$\begin{aligned}\hat{\bar{\alpha}}(Y,\theta) - \bar{\alpha}(Y) &= \hat{\bar{\alpha}}(Y,\theta) - \hat{\bar{\alpha}}(Y,\theta^*) + \hat{\bar{\alpha}}(Y,\theta^*) - \bar{\alpha}(Y),\\ &= \hat{\bar{\alpha}}(Y,\theta) - \hat{\bar{\alpha}}(Y,\theta^*) - \delta(Y) = \tilde{\theta}^T\psi(Y) - \delta(Y).\end{aligned} \quad (25)$$

with $\tilde{\theta}^T\psi(Y) = [\tilde{\theta}_1^T\psi_1(Y),\ldots,\tilde{\theta}_n^T\psi_n(Y)]^T$ and $\tilde{\theta}_i^T = \theta_i - \theta_i^*$, for $i = 1,\ldots,n$. For the master-slave system (8) and (9), The control input can be designed as

$$u_i = -\rho_i(t)Tanh(\rho_i(t)S_i/\varepsilon_i) \quad \text{for } i = 1,\ldots,n \quad (26)$$

with $\rho_i(t) = (\theta_i^T\psi_i(Y) + k_{0i} + k_{1i}|S_i|)$, where k_{0i} is an adaptive parameter will be designed later, and k_{1i} is a strictly positive design constant. *Tanh*(.) denotes the usual hyperbolic tangent function and ε_i is a strictly positive and small design constant.

Throughout this chapter, the following lemma with regard to function *Tanh*(.) will be used in the analysis [35]:

Lemma 1 *The following inequality holds for any $\varepsilon_i > 0$ and for any $z \in R$ [35]:*

$$0 \le |z| - zTanh(\frac{z}{\varepsilon_i}) \le \bar{\varepsilon}_i = \vartheta\varepsilon_i \quad (27)$$

where ϑ is a constant that satisfies $\vartheta = e^{-(1+\vartheta)}$, i.e. $\vartheta = 0.2785$.

The system (15) can be expressed as follows:

$$S^TG_1\dot{S} = S^T\alpha(X,Y,\Lambda) + S^Tu \le |S^T|\bar{\alpha}(Y) + S^Tu \quad (28)$$

Using the control law (26) and Lemma 1, (28) becomes:

$$S^TG_1\dot{S} \le |S^T|\bar{\alpha}(Y) - \sum_{i=1}^n \rho_i(t)|S_i| + \sum_{i=1}^n \rho_i(t)|S_i| - \sum_{i=1}^n \rho_i(t)S_iTanh(\rho_i(t)S_i/\varepsilon_i)$$

$$\begin{aligned}&\le -\Sigma_{i=1}^n|S_i|\tilde{\theta}_i^T\psi_i(Y) + \Sigma_{i=1}^n|S_i||\delta_i(Y)| - \Sigma_{i=1}^n k_{0i}|S_i|\\ &\quad - \Sigma_{i=1}^n k_{1i}s_i^2 + \Sigma_{i=1}^n\bar{\varepsilon}_i\end{aligned} \quad (29)$$

The adaptation laws associated to the control law (26) are

$$\dot{\theta}_i = \gamma_{\theta i}(|S_i|\psi_i(Y) - \sigma_{\theta i}\theta_i), \quad \theta_{ij}(0) > 0 \tag{30}$$

$$\dot{k}_{0i} = \gamma_{ki}(|S_i| - \sigma_{ki}k_{0i}), \quad k_{0i}(0) > 0 \tag{31}$$

where $\gamma_{\theta i}$, $\sigma_{\theta i}$, $\sigma_{ki}$and γ_{ki} are strictly positive design constants.
Now, we formulate the following theorem.

Theorem 1 *The closed-loop system consisting of the master-slave system* (8) *and* (9), *the control law* (26) *together with adaptation laws* (30) *and* (31) *has the following characteristics:*

- All variables in the closed-loop system remain bounded.
- Signals S and E exponentially converge to a residual set that can be made small by properly adjusting the design parameters.

Proof of Theorem 1 We define a Lyapunov function candidate as follows

$$V = \frac{1}{2}S^T G_1 S + \Sigma_{i=1}^n \frac{1}{2\gamma_{\theta i}}\tilde{\theta}_i^T \tilde{\theta}_i + \Sigma_{i=1}^n \frac{1}{2\gamma_{ki}}\tilde{k}_{0i}^2. \tag{32}$$

with $\tilde{k}_{0i} = k_{0i} - k_{0i}^*$, where $k_{0i}^* = \bar{\delta}_i$.

Taking the time derivative of V yields

$$\dot{V} = S^T G_1 \dot{S} + \Sigma_{i=1}^n \frac{1}{\gamma_{\theta i}}\tilde{\theta}_i^T \dot{\theta}_i + \Sigma_{i=1}^n \frac{1}{\gamma_{ki}}\tilde{k}_{0i}\dot{k}_{0i} \tag{33}$$

Substituting Eqs. (29)–(31) into (33), we have

$$\dot{V} \le -\Sigma_{i=1}^n k_{1i}S_i^2 + \Sigma_{i=1}^n \bar{\varepsilon}_i - \Sigma_{i=1}^n \sigma_{\theta i}\tilde{\theta}_i^T \theta_i - \Sigma_{i=1}^n \sigma_{ki}k_{0i}\tilde{k}_{0i}. \tag{34}$$

On the other hand, we can prove that

$$\begin{aligned} -\sigma_{\theta_i}\tilde{\theta}_i^T\theta_i &\le -\frac{\sigma_{\theta i}}{2}\left\|\tilde{\theta}_i\right\|^2 + \frac{\sigma_{\theta i}}{2}||\theta_i^*||^2 \\ -\sigma_{ki}k_{0i}\tilde{k}_{0i} &\le -\frac{\sigma_{ki}}{2}\tilde{k}_{0i}^2 + \frac{\sigma_{ki}}{2}k_{0i}^{*2} \end{aligned} \tag{35}$$

Using the previous inequality (35), (34) can be expressed in the following form

$$\begin{aligned} \dot{V} \le &-\Sigma_{i=1}^n k_{1i}S_i^2 - \Sigma_{i=1}^n \frac{\sigma_{\theta i}}{2}\left\|\tilde{\theta}_i\right\|^2 - \Sigma_{i=1}^n \frac{\sigma_{ki}}{2}\tilde{k}_{0i}^2 \\ &+ \Sigma_{i=1}^n \frac{\sigma_{ki}}{2}k_{0i}^{*2} + \Sigma_{i=1}^n \frac{\sigma_{\theta i}}{2}\left\|\theta_i^*\right\|^2 \end{aligned} \tag{36}$$

Consequently, (36) can be written as follows

$$\dot{V} \leq -\eta V + \mu \tag{37}$$

with $\mu = \Sigma_{i=1}^{n} \frac{\sigma_{ki}}{2} k_{0i}^{*2} + \Sigma_{i=1}^{n} \frac{\sigma_{\theta i}}{2} \left\| \theta_i^* \right\|^2 + \Sigma_{i=1}^{n} \bar{\varepsilon}_i$ and $\eta = min\{\frac{k_{1i}}{\sigma_{gmax}}, \sigma_{\theta i}\gamma_{\theta i}, \sigma_{ki}\gamma_{ki}\}$, where σ_{gmax} is the largest eigenvalue of the matrix G_1.

Multiplying (37) by $e^{\eta t}$, we can obtain

$$\frac{d}{dt}(e^{\eta t} V) \leq \mu e^{\eta t} \tag{38}$$

Integration (38) over $[0, t]$, it follows that

$$0 \leq V(t) \leq \frac{\mu}{\eta} + (V(0) - \frac{\mu}{\eta})e^{-\eta t} \tag{39}$$

Thus, it follows from (39) that $V(t)$ and hence all variables in the closed-loop system are bounded. Also, the error signal S_i converges to an adjustable residual set. According to definition of μ and η, it can be seen that the size of η depends on the control design parameters k_{1i}, γ_{ki}, $\gamma_{\theta i}$, σ_{ki}, $\sigma_{\theta i}$ (which should be selected strictly positive) and that of μ depends on the design parameters $\sigma_{ki}, \sigma_{\theta i}$. It is clear that if we increase k_{1i}, γ_{ki} , $\gamma_{\theta i}$ and decrease $\sigma_{ki}, \sigma_{\theta i}$, it will help to reduce the ultimate bound of S_i.

According to Remark 5 and Properties 1–3, the synchronization errors e_i are also stable and converge to an adjustable residual set.

5 Simulation Results

To test the performances of the proposed projective synchronization scheme, we consider two homogeneous fractional-order chaotic Chua's systems with a piecewise-linear nonlinearity [30]:

The master system:

$$\{ \begin{array}{l} D_t^q x_1 = \sigma[x_2 - x_1 - h(x_1)], \\ D_t^q x_2 = x_1 - x_2 + x_3, \\ D_t^q x_3 = -\beta x_2, \end{array} \tag{40}$$

The slave system:

$$\{ \begin{array}{l} D_t^q y_1 = \sigma[y_2 - y_1 - h(y_1)] + u_1 + P_1(t), \\ D_t^q y_2 = y_1 - y_2 + y_3 + u_2 + P_2(t), \\ D_t^q y_3 = -\beta y_2 + u_3 + P_3(t), \end{array} \tag{41}$$

where $h(y_1) = by_1 + 0.5(a - b)[|y_1 + 1| - |y_1 - 1|]$, $\sigma = 13$, $\beta = 14.87$, $a = -1.27$ and $b = -0.68$. The fractional-order q is set to 0.9. This value ensures the existence of chaos for this fractional-order chaotic Chua's systems [31]. The external dynamic disturbances are chosen as: $P_1(t) = P_2(t) = P_3(t) = 0.25sin(4t) - 0.15sin(5t)$.

The initial conditions of systems (41) and (40) are selected as: $X(0) = [x_1(0), x_2(0), x_3(0)]^T = [0.6, 0.1, -0.6]^T$ and $Y(0) = [y_1(0), y_2(0), y_3(0)]^T = [0.2, -0.1, 0.1]^T$.

The adaptive fuzzy systems, $\theta_i^T \psi_i(Y)$, with $i = 1, 2, 3$. For each input variable of these fuzzy systems, as in [11], we define three (one triangular and two trapezoidal) membership functions uniformly distributed on the intervals $[-4, 4]$. The design parameters are selected as follows: $k_{11} = k_{12} = k_{13} = 1$, $\gamma_{\theta 1} = \gamma_{\theta 2} = \gamma_{\theta 3} = 800$, $\sigma_{\theta 1} = \sigma_{\theta 2} = \sigma_{\theta 3} = 0.001$, $\gamma_{k1} = \gamma_{k2} = \gamma_{k3} = 5$, $\sigma_{k1} = \sigma_{k2} = \sigma_{k3} = 0.005$. The initial conditions of the adaptive laws are chosen as follows:$\theta_{1j}(0) = \theta_{2j}(0) = \theta_{3j}(0) = 0$, and $k_{0i}(0) = 0.001$.

The simulation results of a complete synchronization (i.e. with a scaling factor $\lambda = 1$) between the master system (40) and the slave system (41) are shown in Figs. 2 and 3. The time responses of both systems are given in Fig. 2a, b and c. It is clear from this figure that a complete synchronization is practically and quickly achieved. From Fig. 2d, one can see that the control signals are smooth, admissible and bounded. Figure 3a, b and c show the signal x_i versus y_i (respectively for

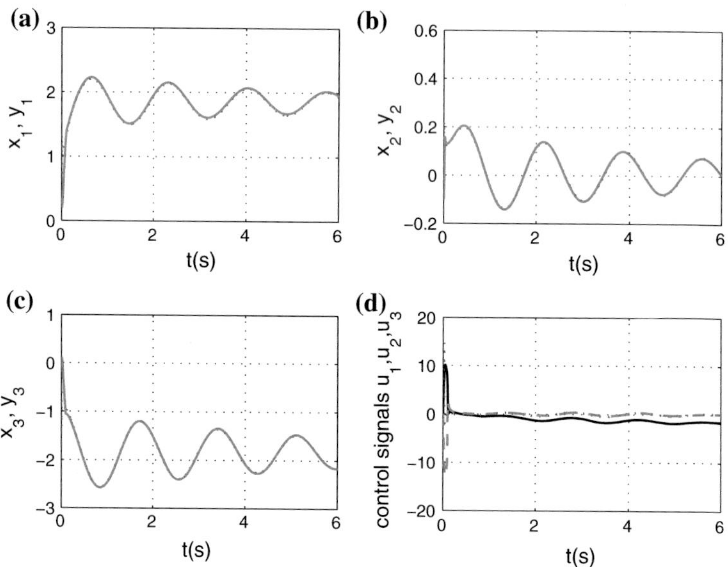

Fig. 2 Complete synchronization: **a** x_1 (*dotted line*) and y_1 (*solid line*). **b** x_2 (*dotted line*) and y_2 (*solid line*). **c** x_3 (*dotted line*) and y_3 (*solid line*). **d** Control signals: u_1 (*solid line*), u_2 (*dotted line*) and u_3 (*dashed line*)

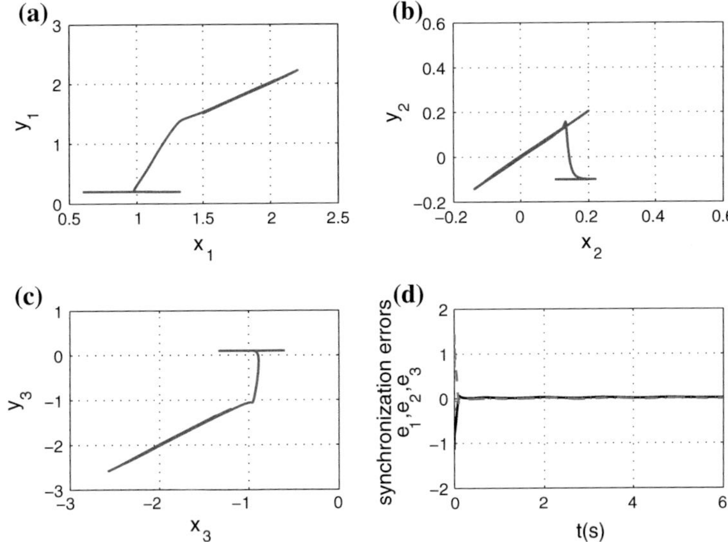

Fig. 3 Complete synchronization: **a**, **b** and **c**: The synchronization phase plots of $x_i - y_i$. **d**: The synchronization error curves: e_1 (solid line), e_2 (*dotted line*) and e_3 (*dashed line*)

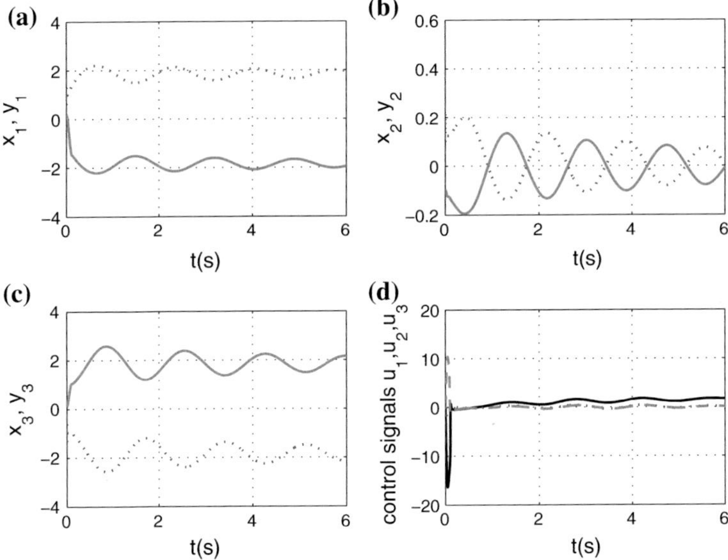

Fig. 4 Anti-phase synchronization: **a** x_1 (*dotted line*) and y_1 (*solid line*). **b** x_2 (*dotted line*) and y_2 (*solid line*). **c** x_3 (*dotted line*) and y_3 (*solid line*). **d** Control signals: u_1 (*solid line*), u_2 (*dotted line*) and u_3 (*dashed line*)

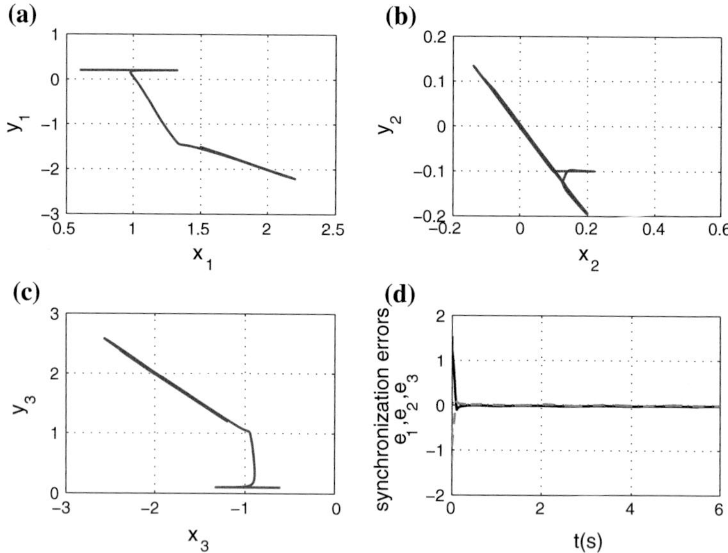

Fig. 5 Anti-phase synchronization: **a**, **b** and **c**: The anti-phase synchronization phase plots of $x_i - y_i$. **d**: The anti-phase synchronization error curves: e_1 (*solid line*), e_2 (*dotted line*) and e_3 (*dashed line*)

i = 1, 2, 3), Fig. 3d gives the time evolution of the corresponding synchronization errors. Hence, we can conclude from these results that the proposed adaptive fuzzy controller is robust and can effectively achieve a complete synchronization between the master system and the slave system, in spite of the presence of the external dynamic disturbances and uncertain dynamics.

Figures 4 and 5 show the simulation results of an anti-phase synchronization (i.e. with a scaling factor $\lambda = -1$). From these results, it is clear that the anti-phase synchronization is satisfactory achieved with proposed fuzzy adaptive controller, in spite of the presence of the unknown disturbances and uncertain dynamics.

6 Conclusion

In this chapter, we have designed a practical projective synchronization system based on a continuous adaptive fuzzy control for a class of fractional-order chaotic systems subject to both uncertain dynamics and external disturbances. the fuzzy approximators have been used to estimate unknown nonlinear functions. Of fundamental interest, a Lyapunov based analysis has been carried out to conclude about the practical stability of the closed-loop system as well as the convergence of the synchronization errors to a neighborhood of zero. Finally, an illustrative simulation example has been given to verify the effectiveness of the proposed projective synchronization system.

In our future work, we will focus on the design of an adaptive fuzzy controller to achieve a generalized function projective synchronization of a class of a class of fractional-order chaotic systems.

References

1. Aghababa MP (2012) Comments on H∞ synchronization of uncertain fractional order chaotic systems: adaptive fuzzy approach. ISA Trans 51:11–12
2. Azar AT, Vaidyanathan S (2015a) Handbook of research on advanced intelligent control engineering and automation. Advances in computational intelligence and robotics (ACIR) book series. IGI Global, USA
3. Azar AT, Vaidyanathan S (2015b) Computational intelligence applications in modeling and control. Studies in computational intelligence, vol 575. Springer, Germany. ISBN: 978-3-319-11016-5
4. Azar AT, Vaidyanathan S (2015c) Chaos modeling and control systems design. Studies in computational intelligence, vol 581. Springer, Germany. ISBN: 978-3-319-13131-3
5. Azar AT, Zhu Q (2015) Advances and applications in sliding mode control systems. Studies in computational intelligence, vol 576. Springer, Germany. ISBN: 978-3-319-11172-8
6. Bagley RL, Calico RA (1991) Fractional order state equations for the control of viscoelastically damped structures. J Guid Control Dyn 14:304–311
7. Boulkroune A, M'saad M (2012a) On the design of observer-based fuzzy adaptive controller for nonlinear systems with unknown control gain sign. Fuzzy Sets Syst 201:71–85
8. Boulkroune A, Msaad M (2012b) Fuzzy adaptive observer-based projective synchronization for nonlinear systems with input nonlinearity. J Vibr Control 18:437–450
9. Boulkroune A, Bouzeriba A, Hamel S, Bouden T (2014) A projective synchronization scheme based on fuzzy adaptive control for unknown multivariable chaotic systems. Nonlinear Dyn 78:433–447
10. Boulkroune A, M'saad M, Farza M (2011) Adaptive fuzzy controller for multivariable nonlinear state time-varying delay systems subject to input nonlinearities. Fuzzy Sets Syst 164:45–65
11. Boulkroune A, Tadjine M, M'saad M, Farza M (2008) How to design a fuzzy adaptive control based on observers for uncertain affine nonlinear systems. Fuzzy Sets Syst 159:926–948
12. Bowonga S, Kakmenib M, Koinac R (2006) Chaos synchronization and duration time of a class of uncertain systems. Math Comput Simulat 71:212–228
13. Carpinteri A, Mainardi F (1997) Fractals and fractional calculus. Springer, New York
14. Chen CS, Chen HH (2009) Robust adaptive neural-fuzzy-network control for the synchronization of uncertain chaotic systems. Nonlinear Anal: Real World Appl 10(3):1466–1479
15. Chen LP, Qu JF, Chai Y, Wu RC, Qi GY (2013) Synchronization of a class of fractional-order chaotic neural networks. Entropy 15:3265–3276
16. Diethelm K, Ford NJ (2002) Analysis of fractional differential equations. J Math Anal Appl 265:229–248
17. Dong-Feng W, Jin-Ying Z, Xiao-Yan W (2011) Synchronization of uncertain fractional-order chaotic systems with disturbance based on a fractional terminal sliding mode controller. Chin Phys B 20:110506
18. Gao X, Yu J (2005) Chaos in the fractional order periodically forced complex Duffing's oscillators. Chaos, Solitons Fractals 24(4):1097–1104
19. Hartley TT, Lorenzo CF, Qammer HK (1995) Chaos in a fractional order Chua's system. IEEE Trans. Circuits Syst I 42:485–490
20. Heaviside O (1971) Electromagnetic theory. Chelsea, New York
21. Hosseinnia SH, Ghaderi R, Ranjbar A, Mahmoudian M, Momani S (2010) Sliding mode synchronization of an uncertain fractional order chaotic system. Comput Math Appl 59:1637–1643

22. Hwang EJ, Hyun CH, Kim E, Park M (2009) Fuzzy model based adaptive synchronization of uncertain chaotic systems: robust tracking control approach. Phys Lett A 373(22):1935–1939
23. Ichise M, Nagayanagi Y, Kojima T (1971) An analog simulation of non-integer order transfer functions for analysis of electrode process. J Electroanal Chem Interfacial Electrochem 33:253–265
24. Li C, Peng G (2004) Chaos in Chen's system with a fractional order. Chaos, Solitons Fractals 22:443–450
25. Li Y, Chen YQ, Podlubny I (2010) Stability of fractional-order nonlinear dynamic systems: Lyapunov direct method and generalized Mittag-Leffler stability. Comput Math Appl 59:1810–1821
26. Li Z, Xu D (2004) A secure communication scheme using projective chaos synchronization. Chaos, Solitons Fractals 22:477–481
27. Lin TC, Kuo CH (2011) H_∞ synchronization of uncertain fractional order chaotic systems: adaptive fuzzy approach. ISA Trans 50:548–556
28. Lin TC, Lee TY (2011) Chaos synchronization of uncertain fractional-order chaotic systems with time delay based on adaptive fuzzy sliding mode control. IEEE Trans. Fuzzy Systems 19:623–635
29. Liu Y, Zheng Y (2009) Adaptive fuzzy approach to control unified chaotic systems. Nonlinear Dyn 57:431–439
30. Lu JG (2005) Chaotic dynamics and synchronization of fractional-order Chua's circuits with a piecewise-linear nonlinearity. Int J Modern Phys B 19(20):3249–3259
31. Lu JG, Chen G (2006) A note on the fractional order Chen system. Chaos, Solitons Fractals 27(3):685–688
32. Pan L, Zhou W, Fang J, Li D (2010) Synchronization and anti-synchronization of new uncertain fractional-order modified unified chaotic systems via novel active pinning control. Commun Nonlinear Sci Numer Simulat 15:3754–3762
33. Peng G (2007) Synchronization of fractional order chaotic systems. Phys Lett A 363:426–432
34. Podlubny I (1999) Fractional differential equations. Academic Press, New York
35. Polycarpou MM, Ioannou PA (1996) A robust adaptive nonlinear control design. Automatica 32:423–427
36. Poursamad A, Davaie-Markazi AH (2009) Robust adaptive fuzzy control of unknown chaotic systems. Appl Soft Comput 9(3):970–976
37. Roopaei M, Jahromi MZ (2008) Synchronization of two different chaotic systems using novel adaptive fuzzy sliding mode control. Chaos 18:033133
38. Rosenblum MG, Pikovsky AS, Kurths J (1996) Phase synchronization of chaotic oscillators. Phys Rev Lett 76:1804–1807
39. Ruo-Xun Z, Shi-Ping Y (2011) Adaptive stabilization of an incommensurate fractional order chaotic system via a single state controller. Chin Phys B 20:110506
40. Sun H, Abdelwahad A, Onaral B (1984) Linear approximation of transfer function with a pole of fractional power. IEEE Trans Autom Control 29:441–444
41. Sun J, Zhang Y (2004) Impulsive control and synchronization of Chua's oscillators. Math Comput Simul 66:499–508
42. Tavazoei MS (2012) Comments on Chaos synchronization of uncertain fractional-order chaotic systems with time delay based on adaptive fuzzy sliding mode control. IEEE Trans Fuzzy Syst 20:993–995
43. Vaidyanathan S, Azar AT (2015a) Anti-synchronization of identical chaotic systems using sliding mode control and an application to vaidyanathan-madhavan chaotic Systems. In: Azar AT, Zhu Q (eds) Advances and applications in sliding mode control systems. Studies in computational intelligence book series, vol 576. Springer, Berlin, pp 527–547. doi:10.1007/978-3-319-11173-5_19
44. Vaidyanathan S, Azar AT (2015b) Hybrid synchronization of identical chaotic systems using sliding mode control and an application to vaidyanathan chaotic systems. In: Azar AT, Zhu Q (eds) Advances and applications in sliding mode control systems. Studies in computational intelligence book series, vol 576. Springer, Berlin, pp 549–569. doi:10.1007/978-3-319-11173-5_20

45. Vaidyanathan S, Azar AT (2015c) Analysis, control and synchronization of a nine-term 3-D novel chaotic system. In: Azar AT, Vaidyanathan S (eds) Chaos modeling and control systems design. Studies in computational intelligence, vol 581. Springer, Berlin, pp 3–17. doi:10.1007/978-3-319-13132-0_1
46. Vaidyanathan S, Azar AT (2015d) Analysis and control of a 4-D novel hyperchaotic system. In: Azar AT, Vaidyanathan S (eds) Chaos modeling and control systems design. Studies in computational intelligence, vol 581. Springer, Berlin, pp 19–38. doi:10.1007/978-3-319-13132-0_2
47. Vaidyanathan S, Idowu BA, Azar AT (2015) Backstepping controller design for the global chaos synchronization of sprott's jerk systems. In: Azar AT, Vaidyanathan S (eds) Chaos modeling and control systems design. Studies in computational intelligence, vol 581. Springer, Berlin, pp 39–58. doi:10.1007/978-3-319-13132-0_3
48. Wang JW, Zhang YB (2009) Synchronization in coupled nonidentical incommensurate fractional-order systems. Phys Lett A 374:202–207
49. Wang J, Chen L, Deng B (2009) Synchronization of Ghostburster neuron in external electrical stimulation via H8 variable universe fuzzy adaptive control. Chaos, Solitons Fractals 39(5):2076–2085
50. Wang J, Zhang Z, Li H (2008) Synchronization of FitzHugh-Nagumo systems in EES via H8 variable universe adaptive fuzzy control. Chaos, Solitons Fractals 36:1332–1339
51. Wang LX (1994) Adaptive fuzzy systems and control: design and stability analysis. Prentice-Hall, Englewood Cliffs, NJ
52. Wang YW, Guan ZH (2006) Generalized synchronization of continuous chaotic systems. Chaos, Solitons Fractals 27:97–101
53. Yu Y, Li H, Wang S, Yu J (2009) Dynamic analysis of a fractional-order Lorenz chaotic system. Chaos, Solitons Fractals 42:1181–1189
54. Zhu Q, Azar AT (2015) Complex system modelling and control through intelligent soft computations. Studies in fuzziness and soft computing, vol 319, Springer, Germany. ISBN: 978-3-319-12882-5

Fuzzy Control-Based Function Synchronization of Unknown Chaotic Systems with Dead-Zone Input

Abdesselem Boulkroune, Sarah Hamel, Ahmad Taher Azar and Sundarapandian Vaidyanathan

Abstract This chapter deals with adaptive fuzzy control-based function vector synchronization between two chaotic systems with both unknown dynamic disturbances and input nonlinearities (dead-zone and sector nonlinearities). This synchronization scheme can be considered as a natural generalization of many existing projective synchronization systems (namely the function projective synchronization, the modified projective synchronization, the projective synchronization and so on). To effectively deal with the input nonlinearities, the control system is designed in a variable-structure framework. In order to approximate uncertain nonlinear functions, the adaptive fuzzy systems are incorporated in this control system. A Lyapunov approach is used to prove the boundedness of all signals as well as the exponential convergence of the corresponding synchronization errors to an adjustable region. The synchronization between two chaotic satellite systems is taken as an illustrative example to show the effectiveness of the proposed synchronization method.

Keywords Fuzzy adaptive control · Function vector synchronization · Chaotic satellite system · Uncertain chaotic system · Input nonlinearities

1 Introduction

Chaos is a nonlinear phenomenon encountered in science and mathematics. A chaotic system is a deterministic system behaves unpredictably and is characterized by the following special features: (1) It has an unusual sensitivity to initial states (therefore

---

A. Boulkroune (✉) · S. Hamel
LAJ, University of Jijel, BP. 98, Ouled-Aissa, 18000 Jijel, Algeria
e-mail: boulkroune2002@yahoo.fr

A.T. Azar
Faculty of Computers and Information, Benha University, Banha, Egypt
e-mail: ahmad_t_azar@ieee.org; ahmad.azar@fci.bu.edu.eg

S. Vaidyanathan
R & D Centre, Vel Tech University, Chennai, India

A.T. Azar and S. Vaidyanathan (eds.), *Advances in Chaos Theory and Intelligent Control*, Studies in Fuzziness and Soft Computing 337,
DOI 10.1007/978-3-319-30340-6_29

they are not predictable in the long run). (2) Its behaviour is not periodic. (3) It has a fractal structure. (4) It is governed by one or more control parameters. The main characteristic employed to identify a chaotic behavior is the well-known Lyapunov exponent criteria. In fact, a system that has one positive Lyapunov exponent is known as a chaotic system. However, a hyperchaotic system has more than one positive Lyapunov exponents. It should be noted that higher dimensional chaotic systems with more than one positive Lyapunov exponent can exhibit more complex dynamics. On the other hand, chaos synchronization is a phenomenon that may arise when two, or more, dissipative chaotic systems are coupled. Thus, the chaos synchronization has become the subject of several researchers in various fields [22, 28, 49]. Up to now, many types of synchronization have been reported for chaotic systems such as complete synchronization (CS) [22], anti-synchronization (AS) [35], lag-synchronization (LS) [23], generalized synchronization (GS) [48], projective synchronization (PS) [28, 49], and so on. In PS, the state vectors of two synchronized systems (master and slave systems) evolve in a proportional scale. When the responses of the synchronized dynamical states are synchronized up to a constant scaling matrix, the PS becomes the modified projective synchronization (MPS) [29, 30]. Recently, a new kind of PS called function projective synchronization (FPS) has been proposed in [24, 32, 38], where the master and slave systems can be synchronized up to a desired scaling function. It is evident that the MPS and PS are a special case of the FPS. In [50] a generalized function projective synchronization (GFPS) between two chaotic systems has been studied, while the responses of synchronized dynamical states can be synchronized up to a function matrix.

Several control methods [1, 5–8, 42, 43, 51] have been recently used to synchronize the chaotic systems. These include the sliding mode control [4, 9, 33, 40, 41], backstepping control [44], nonlinear control [36], adaptive control [32, 38, 46] and so on. To deal with the uncertainty problem in the chaotic systems, based on the universal approximation theorem [47], some adaptive fuzzy control systems have been integrated in the synchronization systems [26, 46]. It should be noted that most existing works were developed for the affine chaotic systems in which the control input appears linearly in the state equation. When the control input appears in a non-affine fashion, the control design becomes more complicated. In [19, 20], a fuzzy adaptive control-based projective synchronization has been proposed for a class of uncertain non-affine chaotic systems. In [15, 16], some adaptive fuzzy controllers have been also designed for the non-affine multivariable nonlinear systems.

On the other hand, the dead-zone is one of ubiquitous non-smooth nonlinear constraints that should be taken in account in the control design. If the dead-zone nonlinearity is ignored when designing the control system, the performances of the closed-loop are not guaranteed and can be even degraded, and the instability may occur. In the literature, there are extensive studies on control of systems with the actuator nonlinearities [2, 3, 12, 13, 17]. However, most existing results dealing with dead-zone nonlinearities were considered only for the uncertain affine chaotic systems. It should be noted that the extension of these control and synchronization approaches to non-affine chaotic systems is generally nontrivial. To the best of our knowledge, in the literature, there are already not results reported on the design of the control as well as control-based synchronization systems for uncertain non-affine chaotic systems with dead-zone nonlinearities. Recently, an adaptive fuzzy variable-structure control has been designed in [34] to achieve a function vector synchronization for affine chaotic systems with dead-zone nonlinearities. However, this work already suffers from the following limitations: (1) Disturbances considered in the model do not depend on the state. (2) The controller designed suffers from the well-known algebraic loop problem. (3) The slave system is affine in control.

Inspired by the above results, in this chapter, we propose a new approach for function vector synchronization of a class of unknown non-affine chaotic systems with dynamic disturbances and input nonlinearities (dead-zone with sector nonlinearities) using a suitable fuzzy adaptive variable-structure control. This synchronization scheme can be considered as a generalization of many existing projective synchronization systems (namely FPS, MPS, GFPS and so on) in the sense that the master and slave outputs are assumed to be some general function vectors. The main difficulties involved in designing such a synchronization scheme are: how to deal with uncertain nonlinear functions, non-affine multivariable control, the input nonlinearities and the combined effect of the unknown nonlinear dynamic disturbances, fuzzy approximation errors together with the higher order terms (HOT) issued from the use of the Taylor series expansion? In this chapter, these difficulties can be respectively solved by using the fuzzy approximation, the Taylor series expansion, the variable structure control and robust dynamic compensation. Compared to the existing results, main contributions of this chapter are as follows:

(1) Compared to [19, 20], a new function vector synchronization scheme based a fuzzy adaptive variable-structure control is designed a class of uncertain non-affine chaotic systems with dynamic disturbances and input nonlinearities. This synchronization scheme can be seen as a generalization of many previous synchronization systems (namely PS, CS, FPS, MPS, GFPS and so on).

(2) Compared to [22, 48, 50], the class of uncertain non-affine chaotic systems considered in this chapter is relatively large, as the slave systems are non-affine and with dynamic unknown disturbances and dead-zone in the input channel. To the our best knowledge, such a class of chaotic slave systems with all these features has not been considered in the literature.

(3) Unlike in [34], our proposed synchronization scheme has the following intrinsic aspects:

- The master-slave system is subject to state-depending disturbances.
- The considered class of chaotic systems is non-affine in control and relatively large.
- The proposed controller is free of the algebraic loop problem and rigorously derived in mathematics.

The rest of the chapter is organized in the following manner. Section 2 presents the system description and problem statement, followed by the design of fuzzy adaptive controller to practically achieve a function vector synchronization in Sect. 3. The simulation results are presented to demonstrate the effectiveness of the proposed synchronization scheme in Sect. 4. Finally, Sect. 5 contains the conclusions.

2 System Description and Problem Formulation

The chaotic master-slave systems considered can be described by:

$$\dot{y} = H_1(y) \tag{1}$$

$$\dot{x} = H_2(x, v) + p(x, t) \tag{2}$$

where $y = [y_1, \ldots, y_n]^T \in R^n$ and $x = [x_1, \ldots, x_n]^T \in R^n$ are the overall state vectors of the master and slave systems, respectively, which are assumed to be measurable. $H_1(Y) = [h_{11}(Y), \ \ldots, h_{1n}(Y)]^T \in R^n$ and $H_2(x, v) = [h_{21}(x, v), \ \ldots, h_{2n}(x, v)]^T \in R^n$ are vectors of smooth unknown nonlinear functions, with $v = \varphi(u) = [\varphi_1(u_1), \ldots, \varphi_n(u_n)]^T$ is the vector of input nonlinearities satisfying some properties which will be given later. And $P(x, t) = [P_1(x, t) \ , \ldots, P_n(x, t)]^T$ is the unknown dynamic disturbance vector.

Remark 1 Many chaotic systems can be expressed or transformed in the quite general form (1) and (2), such as unified chaotic systems, Chen's system, satellite system, *Lü* system and many others. Without input nonlinearities, this class of chaotic has been considered in our previous works [19, 20]. It is should be noted that the extension of these results to a function vector synchronization by considering the input nonlinearities in (2) is nontrivial.

Our objective consists in designing an adaptive fuzzy variable-structure law u, such that a function vector projective synchronization between the master system (1) and the slave one (2) is practically realized, and all involved signals in the closed-loop system remain bounded, as shown in Fig. 1.

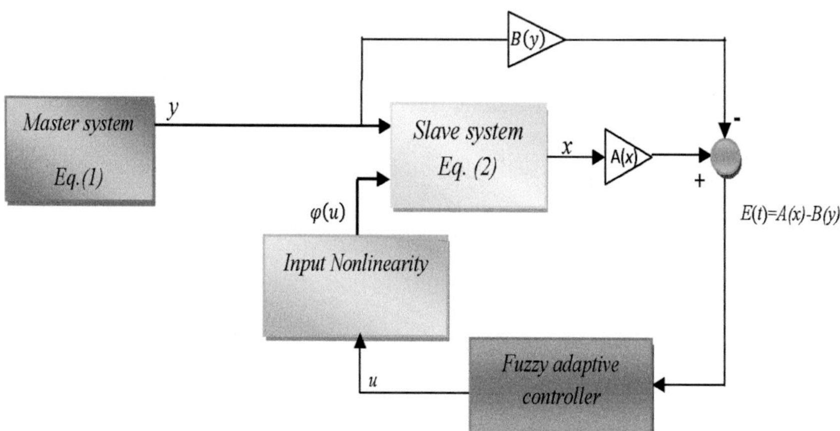

Fig. 1 Function vector synchronization scheme

Define the function vector synchronization error as

$$E = A(x) - B(y) \tag{3}$$

where $A(x)$ and $B(y)$ are continuous function vectors.

Remark 2 From (3), one can see that the proposed synchronization system as a simple generalization of many existing projective synchronization systems, such as PS, MPS, FPS, GFPS, so on.

Differentiating (3) with respect to time and using (1) and (2), we get

$$\dot{E} = \frac{\partial A(x)}{\partial x} H_2(x, \upsilon) + \frac{\partial A(x)}{\partial x} P(x, t) - \frac{\partial B(y)}{\partial y} H_1(y) \tag{4}$$

Let us define a Proportional-Integral (PI) sliding surface as follows:

$$S = [S_1, S_2, \ldots, S_n]^T = E + C_1 \int_0^t E(\tau) d\tau + C_2 \int_0^t Arctan(E(\tau)/\varepsilon) d\tau \tag{5}$$

where C_1 and C_2 are positive definite diagonal matrices defined respectively as: $C_1 = diag\,[C_{11}, \ldots, C_{1n}]$ and $C_2 = diag\,[C_{21}, \ldots, C_{2n}]$. ε is a small strictly positive constant. *Arctan* is the usual arctangent function.

Differentiating (5) yields

$$\dot{S} = \dot{E} + C_1 E + C_2\, Arctan(E/\varepsilon)$$

$$= \frac{\partial A(x)}{\partial x} H_2(x, \upsilon) + F_1(x, y) + \frac{\partial A(x)}{\partial x} P(x, t) \tag{6}$$

where $F_1(x, y) = -\frac{\partial B(y)}{\partial y} H_1(y) + C_1 E + C_2\, ArcTan(E/\varepsilon)$.

Assumption 1

(a) The matrix $\frac{\partial H_2(x,\upsilon)}{\partial \upsilon}$ is non-singular
(b) The matrix $\frac{\partial A(x)}{\partial x}$ is non-singular
(c) The matrix $\frac{\partial A(x)}{\partial x} \frac{\partial H_2(x,\upsilon)}{\partial \upsilon}$ has non-zero leading principal minors and their signs are known.

Remark 3 Assumption 1a can be seen as a controllability condition. This assumption is quite natural and not restrictive as it is satisfied by many physical systems: e.g. robotic systems, electric machine and chaotic systems. A similar assumptions has been made in [19, 20].

Remark 4 Assumption 1b is also a mild assumption. It ensures the invertibility of the matrix $\partial A(x)/\partial x$ involved in the control design. In general, it is implicitly considered (for controllability purposes) in the literatures dealing with function projective synchronization [24, 34, 50].

Remark 5 Assumption 1c is required to apply Lemma 1 given below and not restrictive as it is satisfied by many physical systems, for more details see [15, 16, 19].

Using the Taylor series expansion, the non-affine model (6) can be approximated as an affine model around an unknown optimal control $\upsilon^*(x)$, as follows:

$$H_2(x, \upsilon) = F(x) + G(x)\upsilon + HOT(x, \upsilon) \tag{7}$$

with $F(x) = [F_1(x), \ldots, F_n(x)]^T = H_2(x, \upsilon^*(X)) - \left[\frac{\partial H_2(x,\upsilon)}{\partial \upsilon}\right]_{\upsilon=\upsilon^*(x)} \upsilon^*(x)$, *and* $G(x) = \left[g_{ij}(x)\right] = \left[\frac{\partial H_2(x,\upsilon)}{\partial \upsilon}\right]_{\upsilon=\upsilon^*(x)}$

where $HOT(x, \upsilon)$ is the higher order terms (HOT) of this expansion, and $\upsilon = \upsilon^*(x)$ is an uncertain function minimizing the HOT.

Substituting (7) in (6), we obtain

$$\dot{S} = \frac{\partial A(x)}{\partial x} G(x)\upsilon + F_2(x, y) + \frac{\partial A(x)}{\partial x} HOT(x, \upsilon) + \frac{\partial A(x)}{\partial x} P(x, t) \tag{8}$$

with $F_2(x, y) = \frac{\partial A(x)}{\partial x} F(x) + F_1(x, y)$.

Since the matrix $\frac{\partial A(x)}{\partial x} G(x)$ is not generally symmetric, we should use the following lemma [15, 16, 20, 21]:

Lemma 1 *Any real matrix* $\frac{\partial A(x)}{\partial x} G(x) \in R^{n\times n}$ *with non-zero leading principal minors can be decomposed as follows [15, 16]:*

$$\frac{\partial A(x)}{\partial x} G(x) = G_{as}(x) D_a T_a(x) \tag{9}$$

with $G_{as}(x) \in R^{n\times n}$ *is a positive definite symmetric matrix,* $D_a \in R^{n\times n}$ *is a constant diagonal matrix with +1 or −1 on its diagonal, and* $T_a(x) \in R^{n\times n}$ *is a unity upper triangular matrix. The diagonal elements of* D_a *are nothing else than the ratios of the signs of the leading principal minors of* $\frac{\partial A(x)}{\partial x} G(x)$.

Considering Lemma 1, we have

$$\dot{S} = G_{as}(x) D_a T_a(x)\upsilon + F_2(x, y) + \frac{\partial A(x)}{\partial x} HOT(x, \upsilon) + \frac{\partial A(x)}{\partial x} P(x, t) \tag{10}$$

Assumption 2 $G_{as}(x)$ is of class C^1 and satisfies the following property:

$$\frac{1}{2}\left\|\frac{d\left(D_a^{-1} G_{as}^{-1}(x) D_a\right)}{dt}\right\| = \frac{1}{2}\left\|\frac{\partial\left(D_a^{-1} G_{as}^{-1}(x) D_a\right)}{\partial x}\dot{x}\right\| \le \bar{g}(x) \tag{11}$$

where $\bar{g}(x)$ is an unknown positive function.

Remark 6 Since in many synchronization literatures, the matrix $D_a^{-1}G_{as}^{-1}(x)D_a$ is considered to be constant, thus Assumption 2 can be seen as a mild assumption. It is should be noted that this assumption ensures that the time derivative of $D_a^{-1}G_{as}^{-1}(x)D_a$ depends only on the state vector x (i.e. it allows us to have a matrix $d\left(D_a^{-1}G_{as}^{-1}(x)D_a\right)/dt$ which does not depend on u).

2.1 Input Nonlinearity

The input nonlinearity considered in this chapter is a dead-zone with sector nonlinearities [14, 21]:

$$\varphi_i(u_i) = \begin{cases} \varphi_{i+}(u_i)(u_i - u_{i+}), & u_i > u_{i+} \\ 0, & -u_{i-} \le u_i \le u_{i+} \\ \varphi_{i-}(u_i)(u_i + u_{i-}), & u_i < -u_{i-} \end{cases} \tag{12}$$

where $\varphi_{i+}(u_i) > 0$ and $\varphi_{i-}(u_i) > 0$ are nonlinear functions of u_i, and $u_{i+} > 0$ and $u_{i-} > 0$.

The nonlinearity $\varphi_i(u_i)$ should satisfy the following properties:

$$(u_i - u_{i+})\varphi_i(u_i) \ge m_{i+}^*(u_i - u_{i+})^2, u_i > u_{i+}$$

$$(u_i + u_{i-})\varphi_i(u_i) \ge m_{i-}^*(u_i + u_{i-})^2, u_i < -u_{i-}, \tag{13}$$

where m_{i+}^* and m_{i-}^* are strictly positive constants, which are generally called *gain reduction tolerances* [14, 21, 39].

Assumption 3 Functions $\varphi_{i+}(u_i)$ and $\varphi_{i-}(u_i)$ and the constants m_{i+}^* and m_{i-}^* are uncertain. But, the constants u_{i+} and u_{i-} are known and strictly positive.

Remark 7 Unlike in [39], the so-called gain reduction tolerances m_{i+}^* and m_{i-}^* are assumed here to be uncertain.

2.2 Description of the Fuzzy Logic System

The configuration of a fuzzy logic system basically consists of a fuzzifier, some fuzzy IF–THEN rules, a fuzzy inference engine and a defuzzifier, as shown in Fig. 2.

The fuzzy inference engine is used to represent a non-linear relationship between an input vector $\underline{x}^T = [x_1, x_2, \ldots, x_n] \in R^n$ and an output $\hat{f} \in R$, this relationship is described by a set of fuzzy rules of the form:

$$if\, x_1\ is\ A_1^i\ and\ \ldots\ and\ x_n\ is\ A_n^i\ then\, \hat{f}\ is\, f^i \tag{14}$$

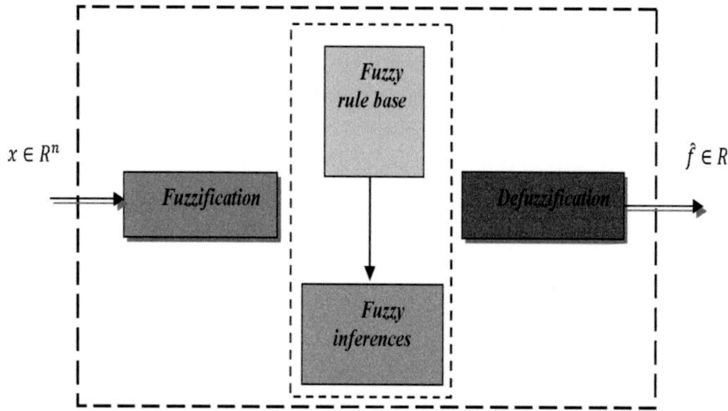

Fig. 2 Basic configuration of the used fuzzy logic system

where $A_1^i, A_2^i, \ldots, A_n^i$ are fuzzy sets and f^i is the fuzzy singleton for the output in the ith rule.

By using the singleton fuzzifier, product inference, and center-average defuzzifier, the output of the fuzzy system can be expressed as follows:

$$\hat{f}(\underline{x}) = \frac{\sum_{i=1}^{m} f^i (\prod_{j=1}^{n} \mu_{A_j^i}(x_j))}{\sum_{i=1}^{m} (\prod_{j=1}^{n} \mu_{A_j^i}(x_j))} = \theta^T \psi(\underline{x}) \tag{15}$$

where $\mu_{A_j^i}(x_j)$ is the degree of membership of x_j to A_j^i, m is the number of fuzzy rules, $\theta^T = [f^1, f^2, \ldots, f^m]$ is the adjustable parameter vector (which are the consequent parameters), and $\psi^T = [\psi^1 \psi^2 \ldots \psi^m]$ with

$$\psi^i(\underline{x}) = \frac{(\Pi_{j=1}^{n} \mu_{A_j^i}(x_j))}{\sum_{i=1}^{m} (\prod_{j=1}^{n} \mu_{A_j^i}(x_j))} \tag{16}$$

which is the fuzzy basis function (FBF).

Following the universal approximation theorem [47], the fuzzy system (15), which is commonly used in control, modeling and identification applications, is able to approximate any nonlinear smooth function $f(\underline{x})$ defined on a compact operating space to a given accuracy. It should be noted that the structure of the fuzzy system and the membership function parameters are appropriately specified beforehand. But, The vector of consequent parameters θ will be estimated online by using integral adaptation laws which will be designed later.

3 Control System Design and Stability Analysis

In this section, we will design our fuzzy adaptive control to practically achieve a function vector synchronization despite the presence of the input nonlinearities, uncertain dynamics and external disturbances.

Multiplying the Eq. (10) by $G_{as}^{-1}(X)$ and posing $\overline{S} = D_a^{-1}S$, we obtain

$$G_1(x)\dot{\bar{S}} = F_3(x, y, u) + \varphi(u) + R(x, u, t) \tag{17}$$

where $G_1(x) = D_a^{-1}G_{as}^{-1}(x)D_a$;

$$F_3(x, y, u) = D_a^{-1}G_{as}^{-1}(x)F_2(x, y) + [T_a(x) - I_n]\varphi(u);$$

$$R(x, u, t) = D_a^{-1}G_{as}^{-1}(x)\frac{\partial A(x)}{\partial x}HOT(x, v) + D_a^{-1}G_{as}^{-1}(x)\frac{\partial A(x)}{\partial x}P(x, t).$$

According to properties of D_a and $G_{as}(x)$, we can prove that the matrix $G_1(x)$ is also positive-definite symmetric.

The dynamics (17) can be written as follows

$$\frac{1}{2}\dot{G}_1(x)\bar{S} + G_1(x)\dot{\bar{S}} = \alpha(z) + \varphi(u) + \frac{1}{2}\dot{G}_1(x)\bar{S} - \bar{g}(x)\bar{S} + R(x, u, t) \tag{18}$$

with

$$\alpha(z) = [\alpha_1(z_1), \ldots, \alpha_n(z_n)]^T = \bar{g}(x)\bar{S} + F_3(x, y, u) \tag{19}$$

where $z = [z_1^T, \ldots, z_n^T]^T$.

Assumption 4 There exists an unknown continuous positive function $\bar{\alpha}_i(z_i)$ such that: $|\alpha_i(z_i)| \leq \eta\bar{\alpha}_i(z_i), \forall z_i \in \Omega_{z_i}$ with $\eta = min\left\{m_{i+}^*, m_{i-}^*\right\}$ for $i = 1, \ldots, n$, where the vectors z_i and their corresponding compact sets will be given below.

Remark 8 Assumption 4 is a mild assumption, as the function $\bar{\alpha}_i(z_i)$ is already assumed to be unknown. This assumption has been used in many literatures e.g. [14, 21].

As the state vector of the master system y always evolves in a compact set and the function vector $\alpha(z)$ is characterized by an "*upper triangular control structure*", the vectors z_i can be determined as follows:

$$\begin{aligned} z_1 &= [x^T, u_2, \ldots, u_n]^T \\ z_2 &= [x^T, u_3, \ldots, u_n]^T \\ &\vdots \\ z_{n-1} &= [x^T, u_n]^T \\ z_n &= x \end{aligned} \tag{20}$$

Let us define the corresponding operating compact sets as follows:

$\Omega_{z_i} = \left\{[x^T, u_{i+1}, \ldots, u_n]^T \,|x \in \Omega_x \subset R^n, u \in \Omega_u \right\}, for\ i = 1, \ldots, n-1$, and
$\Omega_{z_n} = \{x\,|x \in \Omega_x \subset R^n\}$.

$\bar{\alpha}_i(z_i)$ can be approximated, on the compact set Ω_{z_i}, by the fuzzy systems (15) as follows:

$$\hat{\bar{\alpha}}_i(z_i, \theta_i) = \theta_i^T \psi_i(z_i), \text{ with } i = 1, \ldots, n \tag{21}$$

where $\psi_i(z_i)$ is the FBF vector and θ_i is the adjustable parameter vector of this fuzzy system.

Let us define θ_i^* as the optimal vector of θ_i:

$$\theta_i^* = \arg\min_{\theta_i}\left[\sup_{z_i \in \Omega_{z_i}} \left|\bar{\alpha}_i(z_i) - \hat{\bar{\alpha}}_i(z_i, \theta_i)\right|\right] \tag{22}$$

Note that θ_i^* is mainly introduced for analysis purposes. Its value is not needed when implementing the control system.

Define the parameter estimation error and the fuzzy approximation error, respectively, as:

$$\tilde{\theta}_i = \theta_i - \theta_i^*, \quad \varepsilon_i(z_i) = \bar{\alpha}_i(z_i) - \bar{\alpha}_i(z_i, \theta_i^*) = \bar{\alpha}_i(z_i) - \theta_i^{*T}\psi_i(z_i) \tag{23}$$

As in [10, 11, 18, 47], it is assumed that the used fuzzy system does not violate the universal approximation property on the compact set Ω_{z_i}, which is taken large enough so that the input vector of those fuzzy systems remains in Ω_{z_i} under the closed-loop control system. It is therefore reasonable to suppose that the fuzzy approximation error is bounded for all $z_i \in \Omega_{z_i}$, i.e.: $|\varepsilon_i(z_i)| \leq \bar{\varepsilon}_i, \forall z_i \in \Omega_{z_i}$, where $\bar{\varepsilon}_i$ is an unknown positive constant.

Now, denote

$$\hat{\bar{\alpha}}(z, \theta) = \left[\hat{\bar{\alpha}}_1(z_1, \theta_1), \ldots, \hat{\bar{\alpha}}_n(z_n, \theta_n)\right]^T = \left[\theta_1^T\psi_1(z_1), \ldots, \theta_n^T\psi_n(z_n)\right]^T$$

$$\varepsilon(z) = [\varepsilon_1(z_1), \ldots, \varepsilon_n(z_n)]^T, \quad \bar{\varepsilon}(z) = [\bar{\varepsilon}_1, \ldots, \bar{\varepsilon}_n]^T.$$

Then, we have

$$\begin{aligned}\hat{\bar{\alpha}}(z, \theta) - \hat{\bar{\alpha}}(z) &= \hat{\bar{\alpha}}(z, \theta) - \hat{\bar{\alpha}}(z, \theta^*) + \hat{\bar{\alpha}}(z, \theta^*) - \hat{\bar{\alpha}}(z) \\ &= \hat{\bar{\alpha}}(z, \theta) - \hat{\bar{\alpha}}(z, \theta^*) - \bar{\varepsilon}(z) \\ &= \tilde{\theta}^T\psi(z) - \bar{\varepsilon}(z)\end{aligned} \tag{24}$$

where $\tilde{\theta}^T \psi(z) = \left[\tilde{\theta}_1^T \psi_1(z_1), \ldots, \tilde{\theta}_n^T \psi_n(z_n)\right]$ and $\tilde{\theta}_i = \theta_i - \theta_i^*$, for $i = 1, \ldots, n$.

Assumption 5 We assume that:

$$\varepsilon(z) \leq \eta \bar{R}(t, x, u)\kappa^* \tag{25}$$

with $\bar{R}(t, x, u) = 1 + \|x\| + \|u\|$, where $\kappa^* = [\kappa_1^*, \ldots, \kappa_n^*]^T$ is an unknown constant vector.

Remark 9 Following the same reasoning as in [15, 16], one can show that this assumption is logical and not restrictive.

To achieve our objective, we consider the following adaptive fuzzy variable-structure controller:

$$u_i = \begin{cases} -\rho_i(t) sign(\bar{S}_i) - u_{i-}, & \bar{S}_i > 0 \\ 0, & \bar{S}_i = 0 \\ -\rho_i(t) sign(\bar{S}_i) + u_{i+}, & \bar{S}_i < 0 \end{cases} \tag{26}$$

with

$$\rho_i(t) = u_{ri} Tanh\left(\frac{u_{ri}}{\varepsilon_t}\right) + k_{1i}\left|\bar{S}_i\right| + \theta_i^T \varphi_i(z_i), \forall i = 1, \ldots, n \tag{27}$$

$$\dot{\theta}_i(t) = -\gamma_{\theta_i} \sigma_{\theta i} \theta_i + \gamma_{\theta_i} |\bar{S}_i| \varphi_i(z_i), \ with \ \theta_i(0) > 0 \tag{28}$$

where $\gamma_{\theta i}$, $\sigma_{\theta i}$ and $k_{1i} > 0$ for $i = 1, \ldots, n$, are free positive design constants, and θ_i is the online estimate of θ_i^*. $u_r = [u_{r1}, \ldots, u_{rn}]^T$ is an adaptive dynamic control term given by:

$$\dot{u}_r = -\gamma_r u_r + \gamma_r \left[S_{ur} |\bar{S}| - \frac{Tanh(u_r/\varepsilon_t)}{u_r^t Tanh(u_r/\varepsilon_t) + \delta^2} \bar{R}(t, x, u) k^T |\bar{S}| \right], \tag{29}$$

$$\dot{\delta} = -\gamma_\delta \sigma_\delta \sigma - \gamma_\delta \frac{\delta}{u_r^T Tanh(u_r/\varepsilon_t) + \delta^2} \bar{R}(t, x, u) k^T |\bar{S}|, \quad \delta(0) > 0 \tag{30}$$

$$\dot{\kappa} = -\gamma_\kappa \sigma_\kappa \kappa + \gamma_\kappa \bar{R}(t, x, u) |\bar{S}|, \quad \kappa_i(0) > 0 \tag{31}$$

where $S_{ur} = diag\left[Tanh(u_{r1}/\varepsilon_t), \ldots, Tanh(u_r/\varepsilon_t)\right]$, $\kappa = [\kappa_1, \ldots, \kappa_n]^T$ is the estimate of κ^*. γ_κ, γ_δ, and γ_r are strictly positive design constants. σ_κ, σ_δ and ε_t are small strictly positive design constants.

Remark 10 Due to special structure (i.e. an upper triangular control structure) of the functions $\alpha_i(z_i)$ and dynamic feature of the robust dynamic controller u_{ri}, the proposed controller is free of the so-called algebraic-loop.

Using Assumptions 4 and 5 and the expressions (21), (27) and (23)–(25), (18) can be rewritten as follows

$$\frac{d}{dt}\left[\frac{1}{2\eta}\bar{S}^T G_1(x)\bar{S}\right] \leq \sum_{i=1}^{n}\left|\bar{S}_i\right|\overline{\alpha}_i(Z_i) + \frac{1}{\eta}\bar{S}^T\varphi(u) + \bar{R}(t,x,u)\kappa^{*T}\left|\bar{S}\right| - \bar{S}^T\varepsilon(z)$$
$$\leq -\sum\nolimits_{i=1}^{n}\left|\bar{S}_i\right|\left(u_{ri}Tanh(u_{ri}/\varepsilon_t) + k_{1i}\left|\bar{S}_i\right| + \tilde{\theta}_i^T\psi_i(z_i)\right) +$$
$$\sum\nolimits_{i=1}^{n}\left|\bar{S}_i\right|\left(u_{ri}Tanh\left(\frac{u_{ri}}{\varepsilon_t}\right) + k_{1i}\left|\bar{S}_i\right| + \theta_i^T\psi_i(z_i)\right) +$$
$$\frac{1}{\eta}\bar{S}^T\varphi(u) + \bar{R}(t,x,u)\kappa^{*T}\left|\bar{S}\right| \tag{32}$$

Theorem 1 *Consider the master system (1) and the slave system (2) subject to all the required modeling Assumptions 1–5. The proposed fuzzy adaptive variable structure controller, defined by (26)–(31), guarantees the following properties:*

(a) *The function vector synchronization between the systems (1) and (2) is practically achieved, while ensuring that all signals in the closed-loop system are semi-globally uniformly ultimately bounded (SGUUB)*
(b) *S_i exponentially converges to a residual set Ω_{Si} that can be made small by properly adjusting the design parameters.*

$$\Omega_{Si} = \left\{S_i|\,|S_i| \leq \left(\frac{2\pi}{\sigma_{g1}\mu}\right)^{1/2}\right\}. \tag{33}$$

where π, μ and σ_{g1} will be defined later.

Proof of Theorem 1 We define the Lyapunov function condidate as

$$V = \frac{1}{2\eta}\bar{S}^T G_1(x)\bar{S} + \frac{1}{2}\sum\nolimits_{i=1}^{n}\frac{1}{\gamma_{\theta i}}\tilde{\theta}_i^T\tilde{\theta}_i + \frac{1}{2\gamma_\kappa}\tilde{\kappa}^T\tilde{\kappa} + \frac{1}{2\gamma_\delta}\delta^2 + \frac{1}{2\gamma_r}u_r^T u_r \tag{34}$$

where $\tilde{\kappa} = \kappa - \kappa^*$.

Differentiating (34) with respect to time, we get

$$\dot{V} = \frac{1}{\eta}\bar{S}^T G_1(x)\dot{\bar{S}} + \frac{1}{2\eta}\bar{S}^T\dot{G}_1(x)\bar{S} + \sum\nolimits_{i=1}^{n}\frac{1}{\gamma_{\theta i}}\tilde{\theta}_i^T\dot{\theta}_i + \frac{1}{\gamma_\kappa}\tilde{\kappa}^T\dot{\kappa} + \frac{1}{\gamma_\delta}\delta\dot{\delta} + \frac{1}{\gamma_\delta}u_r^T\dot{u}_r \tag{35}$$

From (13) and (26), we obtain

$$u_i < -u_{i-} \text{ for } \bar{S}_i > 0 \Rightarrow (u_i + u_{i-})\varphi_i(u_i) \geq m_{i-}^*(u_i + u_{i-})^2 \geq \eta(u_i + u_{i-})^2 \tag{36}$$

$$u_i > u_{i+} \text{ for } \bar{S}_i < 0 \Rightarrow (u_i - u_{i+})\varphi_i(u_i) \geq m_{i+}^*(u_i - u_{i+})^2 \geq \eta(u_i - u_{i+})^2 \tag{37}$$

Considering (26) again, we can establish

$$\begin{aligned}&\bar{S}_i > 0 \Rightarrow \\ &\quad (u_i + u_{i-})\varphi_i(u_i) = -\rho_i(t)sign(\bar{S}_i)\varphi_i(u_i) \geq m_{i-}^*\rho_i^2(t)\left[sign(\bar{S}_i)\right]^2 \geq \eta\rho_i^2(t)\end{aligned} \tag{38}$$

$$\bar{S}_i < 0 \Rightarrow \\ (u_i - u_{i+})\varphi_i(u_i) = -\rho_i(t) sign(\bar{S}_i)\varphi_i(u_i) \geq m^*_{i+}\rho_i^2(t)\left[sign(\bar{S}_i)\right]^2 \geq \eta\rho_i^2(t) \tag{39}$$

Then, for $\bar{S}_i > 0$ and $\bar{S}_i < 0$, we have

$$-\rho_i(t) sign(\bar{S}_i)\varphi_i(u_i) \geq \eta\rho_i^2(t) \tag{40}$$

Using the fact that $\bar{S}_i sign(\bar{S}_i) = \left|\bar{S}_i\right|$, (40) can be expressed as

$$-\rho_i(t)\bar{S}_i^2\, sign(\bar{S}_i)\varphi_i(u_i) \geq \eta\rho_i^2(t)\bar{S}_i^2 = \eta\rho_i^2(t)\left|\bar{S}_i\right|^2 \tag{41}$$

Finally, because $\rho_i(t) > 0$, for all $\bar{S}_i$ we have

$$\bar{S}_i\varphi_i(u_i) \leq -\eta\rho_i(t)\left|\bar{S}_i\right| \tag{42}$$

By considering (27)–(32), $\dot{V}$ can be bounded as

$$\begin{aligned}\dot{V} &\leq -\sum_{i=1}^{n}\left|\bar{S}_i\right|\left(u_{ri} Tanh\left(u_{ri}/\varepsilon_t\right) + k_{1i}|\bar{S}_i|\right) - \sum_{i=1}^{n}\sigma_{\theta i}\tilde{\theta}_i^T\theta_i + \bar{R}(x,u)\left|\bar{S}\right|^T\kappa^* \\ &\quad + \frac{1}{\gamma_\kappa}\tilde{\kappa}^T\dot{\kappa} + \frac{1}{\gamma_\delta}\delta\dot{\delta} + \frac{1}{\gamma_r}u_r^T\dot{u}_r \\ &\leq -\sum\nolimits_{i=1}^{n}|\bar{S}_i|u_{ri} Tanh\left(\frac{u_{ri}}{\varepsilon_t}\right) - \sum\nolimits_{i=1}^{n}k_{1i}\bar{S}_i^2 - \sum\nolimits_{i=1}^{n}\sigma_{\theta i}\tilde{\theta}_i^T\theta_i - \sigma_\kappa\tilde{\kappa}^T\kappa + \\ &\quad \bar{R}(x,u)|\bar{S}|^T\kappa + \frac{1}{\gamma_\delta}\delta\dot{\delta} + \frac{1}{\gamma_r}u_r^T\dot{u}_r \\ &\leq -\sum\nolimits_{i=1}^{n}u_{ri}^2 - \sum\nolimits_{i=1}^{n}k_{1i}\bar{S}_i^2 - \sum\nolimits_{i=1}^{n}\sigma_{\theta i}\tilde{\theta}_i^T\theta_i - \sigma_\kappa\tilde{\kappa}^T\kappa - \sigma_\delta\delta^2\end{aligned} \tag{43}$$

On the other hand, we can established that

$$-\sigma_\kappa\tilde{\kappa}^T\kappa \leq -\frac{\sigma_\kappa}{2}\|\tilde{\kappa}\|^2 + \frac{\sigma_\kappa}{2}\|\kappa^*\|^2 \tag{44}$$

$$-\sigma_{\theta_i}\tilde{\theta}_i^T\theta_i \leq -\frac{\sigma_{\theta_i}}{2}\|\tilde{\theta}_i\|^2 + \frac{\sigma_{\theta_i}}{2}\|\theta_i^*\|^2 \tag{45}$$

By employing inequalities (44) and (45), (43) implies

$$\begin{aligned}\dot{V} &\leq -\sum\nolimits_{i=1}^{n}u_{ri}^2 - \sum\nolimits_{i=1}^{n}k_{1i}\bar{S}_i^2 - \sum\nolimits_{i=1}^{n}\frac{\sigma\theta_i}{2}\|\tilde{\theta}_i\|^2 + \sum\nolimits_{i=1}^{n}\frac{\sigma\theta i}{2}\|\tilde{\theta}\|^2 - \\ &\quad \frac{\sigma_\kappa}{2}\|\tilde{\kappa}\|^2 + \frac{\sigma_\kappa}{2}\|\kappa^*\|^2 - \sigma_\delta\delta^2\end{aligned} \tag{46}$$

According to properties of $G_{as}(x)$, there exists a positive scalar σ_{gs} such as $G_{as}(x) \geq \sigma_{gs} I_n$, thus

$$\bar{S}^T G_1(x)\bar{S} = S^T G_s^{-1} S \leq \frac{1}{\sigma_{gs}} \|\bar{S}\|^2 \tag{47}$$

And from (46) and (47), we obtain

$$\dot{V} \leq -\mu V + \pi \tag{48}$$

with $\pi = \sum_{i=1}^{n} \frac{\sigma_{\theta_i}}{2} \|\theta_i^*\|^2 + \frac{\sigma_\kappa}{2} \|\kappa^*\|^2$, and

$$\mu = \min \left\{ \min_i \left\{ 2\eta \sigma_{gs} k_{1i} \right\},\ \min_i \{\gamma_{\theta i} \sigma_{\theta i}\},\ 2\gamma_\delta \sigma_\delta,\ 2\gamma_r,\ \gamma_k \sigma_k \right\}$$

Multiplying (48) by $e^{\mu t}$, we get

$$\frac{d\left(V e^{\mu t}\right)}{dt} \leq \pi e^{\mu t} \tag{49}$$

And integrating (49) over $[0, t]$, we have

$$0 \leq V(t) \leq \frac{\pi}{\mu} + \left(V(0) - \frac{\pi}{\mu}\right) e^{-\mu t} \tag{50}$$

Therefore, all signals in the closed-loop control system, i.e. S_i, x, u_r, δ, $\tilde{\theta}_i$, and $\tilde{k}$, are SGUUB.

Considering (50) and (34), and because the matrix $G_1(x)$ is also positive definite symmetric, there exists an unknown positive constant σ_{g1} such that: $G_1(x) \geq \sigma_{g1} I_n$. Then, the following inequality results:

$$\left|\bar{S}_i\right| = |S_i| \leq \left(\frac{2}{\sigma_{g1}} \left(\frac{\pi}{\mu} + \left(V(0) - \frac{\pi}{\mu}\right) e^{-\mu t} \right) \right)^{1/2} \tag{51}$$

i.e. the solution of S_i exponentially converges to a bounded adjustable domain defined as follows: $\Omega_{Si} = \left\{ S_i \middle| \, |S_i| \leq \left(\frac{2\pi}{\sigma_{g1}\mu}\right)^{1/2} \right\}$. This completes the proof.

Remark 11 If $u_{i+} = u_{i-} = u_{i0}$, (26) can be simply rewritten as:

$$u_i = -\left(u_{ri} Tanh(u_{ri}/\varepsilon_t) + k_{1i} \left|\tilde{S}_i\right| + \theta_i^T \psi_i(z_i) + u_{i0} \right) Sign(\bar{S}_i) \tag{52}$$

Note that the sign function can be replaced by any equivalent smooth function to deal with the chattering effects.

4 Simulation Results

In this section, we will test our proposed controller for the synchronization of two chaotic satellite systems.

The dynamic equations of a typical satellite are given by [27, 37]:

$$\begin{cases} I_x\dot{w}_x = w_y w_z\left(I_y - I_z\right) + h_x + u_x \\ I_y\dot{w}_y = w_x w_z\left(I_z - I_x\right) + h_y + u_y \\ I_z\dot{w}_z = w_x w_y\left(I_x - I_y\right) + h_z + u_z \end{cases} \tag{53}$$

where I_x, I_y and I_z are the principal moments of inertia; w_x,w_y and w_z are the angular velocities of the satellite; h_x, h_y and h_z are perturbing torques and u_x, u_y and u_z are the three control torques. The authors in [27, 37] have taken $I_x = 3{,}000\,\text{kgm}^2$, $I_y = 2{,}000\,\text{kgm}^2$ and $I_z = 1{,}000\,\text{kgm}^2$ with the following perturbing torques:

$$\begin{bmatrix} h_x \\ h_y \\ h_z \end{bmatrix} = \begin{bmatrix} -1200 & 0 & 1000\sqrt{6}/2 \\ 0 & 350 & 0 \\ -1000\sqrt{6} & 0 & -400 \end{bmatrix}\begin{bmatrix} w_x \\ w_y \\ w_z \end{bmatrix} \tag{54}$$

It is worth noting that these torques are chosen such that the (non-controlled) satellite is forced into chaotic motion [25].

We consider the following two identical chaotic systems where the master system and the slave system are denoted with $y = [y_1, y_2, y_3]^T$ and $x = [x_1, x_2, x_3]^T$ respectively.

The master system:

$$\begin{cases} \dot{y}_1 = \frac{1}{3}y_2y_3 - 0.4y_1 + \frac{\sqrt{6}}{6}y_3 \\ \dot{y}_2 = -y_1y_2 + 0.175y_2 \\ \dot{y}_3 = y_1y_2 - \sqrt{6}y_1 - 0.4y_3 \end{cases} \tag{55}$$

The slave system controlled and subject to input nonlinearities and dynamical external disturbances is described by:

$$\begin{cases} \dot{x}_1 = \frac{1}{3}x_2x_3 - 0.4x_1 + \frac{\sqrt{6}}{6}x_3 + \varphi_1(u_1) + 0.2\varphi_1(u_1)^3 + P_1(x,t) \\ \dot{x}_2 = -x_1x_2 + 0.175x_2 + \varphi_2(u_2) + P_2(x,t) \\ \dot{x}_3 = x_1x_2 - \sqrt{6}x_1 - 0.4x_3 + \varphi_3(u_3) + 0.1\varphi_3(u_3)^3 + P_3(x,t) \end{cases} \tag{56}$$

Consider the following input nonlinearities $\varphi_i(u_i)$ for $i = 1, 2, 3$:

$$\varphi_i(u_i) = \begin{cases} (u_i - 3)\left(1.5 - \exp(0.3\,sin(u_i))\right), & u_i > 3 \\ 0, & -3 \le u_i \le 3 \\ (u_i + 3)\left(1.5 - \exp(0.3\,sin(u_i))\right), & u_i < -3 \end{cases} \tag{57}$$

The external disturbances are: $P_1(x,t) = 2x_1 + x_2$, $P_2(x,t) = x_2^2 + 0.5x_1$ and $P_3(x,t) = 0.1x_3^3 + \sin(t)$.

Three fuzzy adaptive systems have been used to online estimate the uncertainties. The first fuzzy system $\theta_1^T\psi_1(z_1)$ has the vector $z_1 = \left[x^T, u_2, u_3\right]^T$ as input, the second fuzzy system $\theta_2^T\psi_2(z_2)$ has $z_2 = \left[x^T, u_3\right]^T$ as input, but the third one $\theta_3^T\psi_3(z_3)$ has the state vector $z_3 = x$ as input. For each input variable of these fuzzy systems, as in [10], we define three (one triangular and two trapezoidal) membership functions uniformly distributed on the intervals $[-5, 5]$.

The initial conditions of master-slave systems and the adaptation laws are respectively selected as: $y(0) = [5, 3, -1]$, $x(0) = [3, 4.1, 2]$, $u_{r1}(0) = u_{r2}(0) = u_{r3}(0) = 1$, $\delta(0) = 2$, $\kappa_1(0) = \kappa_2(0) = \kappa_3(0) = 10$ and $\theta_{1_j}(0) = \theta_{2_j}(0) = \theta_{3_j}(0) = 0.01$, for $j = 1, \ldots, m$, where m is the number of the fuzzy rules. The design parameters are chosen as: $\gamma_{\theta_1} = \gamma_{\theta_2} = \gamma_{\theta_3} = 50$, $\sigma_{\theta 1} = \sigma_{\theta 2} = \sigma_{\theta 3} = 10^{-2}$, $\gamma_r = 100$, $\gamma_\kappa = 2$, $\sigma_\delta = 10^{-5}$, $\gamma_\delta = 10^{-4}$, $\sigma_\kappa = 10^{-2}$. Note that since $u_{i+} = u_{i-} = 3$, we can directly use the controller given by (52).

To demonstrate the effectiveness of the above fuzzy control-based function vector synchronization method, two simulation cases are considered:

Case 1 (Anti-phase synchronization): $b_1(y) = -y_1, b_2(y) = -y_2, b_3(y) = -y_3$, and $a_1(x) = x_1, a_2(x) = x_2, a_3(x) = x_3$. The simulation results with these function vectors are shown in Fig. 3.

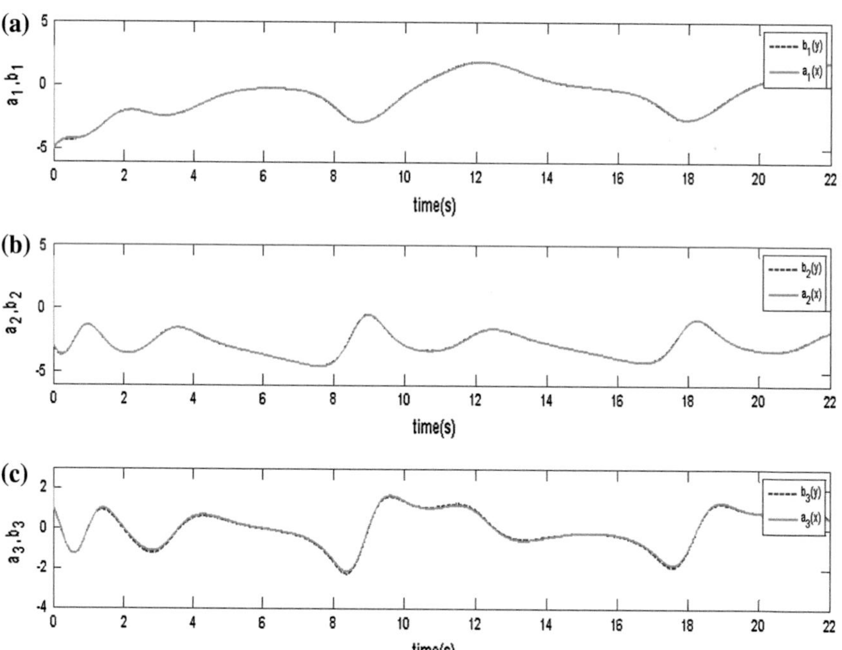

Fig. 3 Anti-phase synchronization with $b_1(y) = -y_1, b_2(y) = -y_2, b_3(y) = -y_3$. **a** $a_1(x)$ (*solid line*) and $b_1(y)$ (*dotted line*). **b** $a_2(x)$ (*solid line*) and $b_2(y)$ (*dotted line*). **c** $a_3(x)$ (*solid line*) and $b_3(y)$ (*dotted line*)

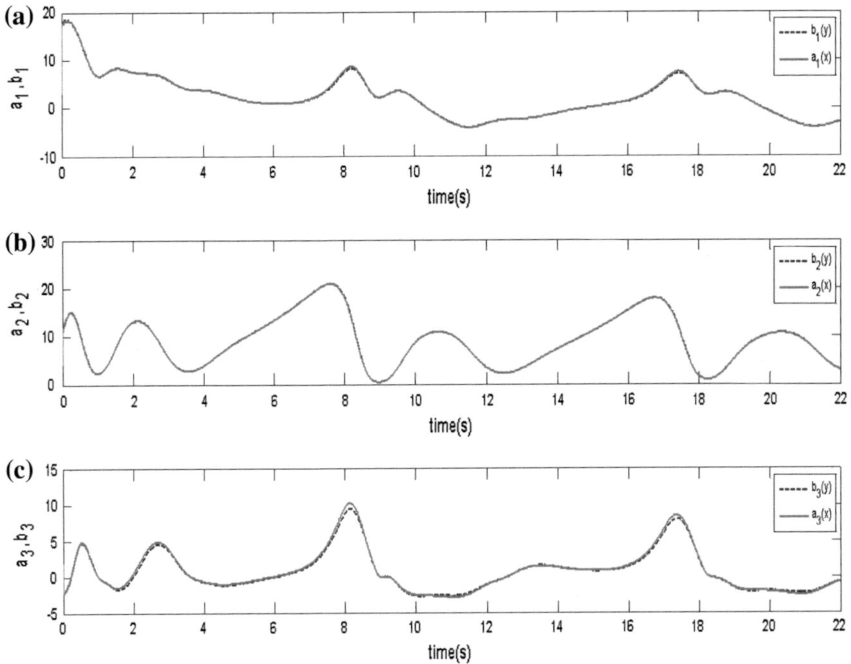

Fig. 4 GFPS with $b_1(y) = 0.1y_1^2 + y_1y_2, b_2(y) = y_2^2 + 0.1y_1y_2$ and $b_3(y) = sin(y_3)y_3 + y_2y_3$. **a** $a_1(x)$ (*solid line*) and $b_1(y)$ (*dotted line*). **b** $a_2(x)$ (*solid line*) and $b_2(y)$ (*dotted line*). **c** $a_3(x)$ (*solid line*) and $b_3(y)$ (*dotted line*)

Case 2 (GFP synchronization): $b_1(y) = 0.1y_1^2 + y_1y_2, b_2(y) = y_2^2 + 0.1y_1y_2$, $b_3(y) = sin(y_3) + y_2y_3$, and $a_1(x) = x_1, a_2(x) = x_2, a_3(x) = x_3$. The obtained results with these function vectors are given in Fig. 4.

From Figs. 3 and 4, we can easily remark in each plot that $a_i(x)$ converge quickly to $b_i(y)$. Then, the trajectories of the slave vector, $A(x) = [a_1(x), a_2(x), a_3(x)]^T$, practically converge to the trajectories of the master vector, $B(y) = \left[b_1(y), b_2(y), b_3(y)\right]^T$, in a short time. Hence, the function vector synchronization between the master and salve systems is effectively achieved, in spite of the presence of the unknown disturbances, the input nonlinearities (i.e. dead-zone input) and the uncertain dynamics.

5 Conclusion

The main contribution of this chapter is to design a fuzzy adaptive controller-based function vector synchronization of uncertain chaotic systems subject to input nonlinearities and dynamical disturbances. More precisely, on the basis of the Lyapunov stability theory, universal approximation capability of fuzzy logic systems and variable-structure control theory, a novel adaptive control system has been proposed to practically achieve a function vector synchronization, while ensuring that all involved signals in closed-loop system are bounded. In fact, the adaptive fuzzy systems have been used to estimate the unknown nonlinear functions. Unlike the previous works [34], the controller design and the synchronization error convergence to a small neighborhood of the origin have been rigorously derived. An illustrative simulation example has been given to verify our theoretical results. In the future works, we might hope to extend this proposed fuzzy adaptive variable-structure controller to handle other types of input nonlinearities such as saturation and hysteresis, etc. Furthermore, we look forward to apply this proposed fuzzy adaptive controller in the real application of chaos synchronization, where only the drive system output is available for measurement, using a state observer.

References

1. Azar AT, Serrano FE (2014) Robust IMC-PID tuning for cascade control systems with gain and phase margin specifications. Neural Comput Appl 25(5): 983–995. doi:10.1007/s00521-014-1560-x
2. Azar AT, Serrano FE (2015) Stabilization and control of mechanical systems with backlash. In: Azar AT, Vaidyanathan S (eds) Advanced intelligent control engineering and automation. advances in computational intelligence and robotics (ACIR) Book Series. IGI-Global, Hershey
3. Azar AT, Serrano FE (2015) Design and modeling of anti wind up PID controllers. In: Zhu Q, Azar AT (eds) Complex system modelling and control through intelligent soft computations. Studies in Fuzziness and Soft Computing, vol 319. Springer, Berlin, pp 1–44. doi:10.1007/978-3-319-12883-2_1
4. Azar AT, Serrano FE (2015) Adaptive sliding mode control of the Furuta pendulum. In Azar AT, Zhu Q (eds) Advances and applications in sliding mode control systems. Studies in computational intelligence, vol 576. Springer-Verlag GmbH, Berlin/Heidelberg, pp 1–42. doi:10.1007/978-3-319-11173-5_1
5. Azar AT, Serrano FE (2015) Deadbeat control for multivariable systems with time varying delays. In: Azar AT, Vaidyanathan S (eds) Chaos modeling and control systems design. Studies in computational intelligence, vol 581. Springer-Verlag GmbH, Berlin/Heidelberg, pp 97–132. doi:10.1007/978-3-319-13132-0_6
6. Azar AT, Vaidyanathan S (2015) Handbook of research on advanced intelligent control engineering and automation. In: Advances in computational intelligence and robotics (ACIR) Book Series. IGI Global, Hershey
7. Azar AT, Vaidyanathan S (2015) Computational intelligence applications in modeling and control. In: Studies in computational intelligence, vol 575. Springer, Berlin. ISBN: 978-3-319-11016-5
8. Azar AT, Vaidyanathan S (2015) Chaos modeling and control systems design. Studies in computational intelligence, vol 581. Springer, Berlin
9. Azar AT, Zhu Q (2015) Advances and applications in sliding mode control systems. In: Studies in computational intelligence, vol 576. Springer, Berlin. ISBN: 978-3-319-11172-8
10. Boulkroune A, Tadjine M, M'saad M, Farza M (2008) How to design a fuzzy adaptive control based on observers for uncertain affine nonlinear systems. Fuzzy Sets Syst 159:926–948
11. Boulkroune A, Tadjine M, M'saad M, Farza M (2009) Adaptive fuzzy controller for non-affine systems with zero dynamics. Int J Syst Sci 40(4):367–382

12. Boulkroune A, M'Saad M (2011) A fuzzy adaptive variable-structure control scheme for uncertain chaotic MIMO systems with sector nonlinearities and dead-zones. Expert Syst Appl 38(12):14744–14750
13. Boulkroune A, M'Saad M (2011) A practical projective synchronization approach for uncertain chaotic systems with dead-zone input. Commun Nonlinear Sci Numer Simul 16:4487–4500
14. Boulkroune A, M'Saad M, Farza M (2011) Adaptive fuzzy controller for multivariable nonlinear state time-varying delay systems subject to input nonlinearities. Fuzzy Sets Syst 164:45–65
15. Boulkroune A, M'Saad M, Farza M (2012) Adaptive fuzzy tracking control for a class of MIMO nonaffine uncertain systems. Neurocomputing 93:48–55
16. Boulkroune A, M'Saad M, Farza M (2012) Fuzzy approximation-based indirect adaptive controller for MIMO non-affine systems with unknown control direction. IET Control Theory Appl 17:2619–2629
17. Boulkroune A, M'Saad M (2012) Fuzzy adaptive observer-based projective synchronization for nonlinear systems with input nonlinearity. J Vib Control 18(3):437–450
18. Boulkroune A, M'Saad M (2012) On the design of observer-based fuzzy adaptive controller for nonlinear systems with unknown control gain sign. Fuzzy Sets Syst 201:71–85
19. Boulkroune A, Bouzeriba A, Hamel S, Bouden T (2014) Adaptive fuzzy control-based projective synchronization of uncertain non-affine chaotic systems. Complexity. doi:10.1002/cplx.21596
20. Boulkroune A, Bouzeriba A, Hamel S, Bouden T (2014) A projective synchronization scheme based on fuzzy adaptive control for unknown multivariable chaotic systems. Nonlinear Dyn 78(1):433–447
21. Boulkroune A, M'Saad M, Farza M (2014) State and output feedback fuzzy variable structure controllers for multivariable nonlinear systems subject to input nonlinearities. Int J Adv Manuf Technol 71:539–556
22. Bowonga S, Kakmenib M, Koinac R (2006) Chaos synchronization and duration time of a class of uncertain systems. Math Comput Simulat 71:212–228
23. Cailian C, Gang F, Xinping G (2005) An adaptive lag-synchronization method for time-delay chaotic systems. In: Proceedings of the American control conference, Portland, June 8–10, pp 4277–4282
24. Du HY, Zeng QS, Wang CH (2008) Function projective synchronization of different chaotic systems with uncertain parameters. Phys Lett A 372:5402–5410
25. Farid Y, Moghaddam TV (2014) Generalized projective synchronization of chaotic satellites problem using linear matrix inequality. Int J Dynam Control 2:577–586
26. Hwang E, Hyun C, Kim E, Park M (2009) Fuzzy model based adaptive synchronization of uncertain chaotic systems: robust tracking control approach. Phys Lett A 373:1935–1939
27. Kemih K, Kemiha A, Ghanes M (2009) Chaotic attitude control of satellite using impulsive control. Chaos Solitons Fractals 42:735–744
28. Li G (2006) Projective synchronization of chaotic system using backstepping control. Chaos Solitons Fractals 29:490–598
29. Li GH (2007) Generalized projective synchronization between Lorenz system and Chen's system. Chaos Solitons Fractals 32:1454–1458
30. Li GH (2007) Modified projective synchronization of chaotic system. Chaos Solitons Fractals 32:1786–1790
31. Li N, Xiang W, Li H (2012) Function vector synchronization of uncertain chaotic systems with nonlinearities and dead-zones. J Conmput Inf Syst 8:9491–9498
32. Luo RZ (2008) Adaptive function projective synchronization of Rössler hyperchaotic system with uncertain parameters. Phys Lett A 372:3667–3671
33. Mekki H, Boukhetala D, Azar AT (2015) Sliding modes for fault tolerant control. In: Azar AT, Zhu Q (eds) Advances and applications in sliding mode control systems. Studies in Computational Intelligence book Series, vol 576. Springer-Verlag GmbH, Berlin/Heidelberg, pp 407–433. doi:10.1007/978-3-319-11173-5_15
34. Ning L, Heng L, Wei X (2012) Fuzzy adaptive tracking control of uncertain chaotic system with input perturbance and nonlinearity. Acta Phys 61(23): 230505. doi:10.7498/aps.61.230505

35. Pikovsky AS, Rosenblum MG, Osipov GV, Kurths J (1997) Phase synchronization of chaotic oscillators by external driving. Phys D 104:219–238
36. Saaban AB, Ibrahim AB, Shahzad M, Ahmad I (2014) Identical synchronization of a new chaotic system via nonlinear control and linear active control techniques: a comparative analysis. Int J Hybrid Inf Technol 7(1):211–224
37. Sadaoui D, Boukabou A, Merabtine N, Benslama M (2011) Predictive synchronization of chaotic satellites systems. Expert Syst Appl 38(7):9041–9045
38. Sudheer KS, Sabir M (2009) Adaptive modified function projective synchronization between hyperchaotic Lorenz system and hyperchaotic Lu system with uncertain parameters. Phys Lett A 373:3743–3748
39. Shyu K-K, Liu W-J, Hsu K-C (2005) Design of large-scale time-delayed systems with deadzone input via variable structure control. Automatica 41:1239–1246
40. Vaidyanathan S, Azar AT (2015) Anti-synchronization of identical chaotic systems using sliding mode control and an application to Vaidyanathan-Madhavan chaotic systems. In: Azar AT, Zhu Q (eds) Advances and applications in sliding mode control systems. Studies in computational intelligence book series, vol 576. Springer-Verlag GmbH, Berlin/Heidelberg, pp 527–547. doi:10.1007/978-3-319-11173-5_19
41. Vaidyanathan S, Azar AT (2015) Hybrid synchronization of identical chaotic systems using sliding mode control and an application to Vaidyanathan chaotic systems. In: Azar AT, Zhu Q (eds) Advances and applications in sliding mode control systems. Studies in computational intelligence book series, vol 576. Springer-Verlag GmbH, Berlin/Heidelberg, pp 549–569. doi:10.1007/978-3-319-11173-5_20
42. Vaidyanathan S, Azar AT (2015) Analysis, control and synchronization of a nine-term 3-D novel chaotic system. In: Azar AT, Vaidyanathan S (eds) Chaos modeling and control systems design. Studies in computational intelligence, vol 581. Springer-Verlag GmbH, Berlin/Heidelberg, pp 3–17. doi:10.1007/978-3-319-13132-0_1
43. Vaidyanathan S, Azar AT (2015) Analysis and control of a 4-D novel hyperchaotic system. In: Azar AT, Vaidyanathan S (eds) Chaos modeling and control systems design. Studies in computational intelligence, vol 581. Springer-Verlag GmbH, Berlin/Heidelberg, pp 19–38. doi:10.1007/978-3-319-13132-0_2
44. Vaidyanathan S, Idowu BA, Azar AT (2015) Backstepping controller design for the global chaos synchronization of Sprott's jerk systems. In: Azar AT, Vaidyanathan S (eds) Chaos modeling and control systems design. Studies in computational intelligence, vol 581. Springer-Verlag GmbH, Berlin/Heidelberg, pp 39–58. doi:10.1007/978-3-319-13132-0_3
45. Vargas JA, Grzeidak E, Hemerly EM (2015) Robust adaptive synchronization of a hyperchaotic finance system. Nonlinear Dyn 80(1–2):239–248
46. Wang J, Chen L, Deng B (2009) Synchronization of ghostburster neuron in external electrical stimulation via H∞ variable universe fuzzy adaptive control. Chaos Solitons Fractals 39(5):2076–2085
47. Wang LX (1994) Adaptive fuzzy systems and control: design and stability analysis. Prentice-Hall, Englewood Cliffs
48. Wang YW, Guan ZH (2006) Generalized synchronization of continuous chaotic systems. Chaos Solitons Fractals 27:97–101
49. Yan J, Li C (2005) Generalized projective synchronization of a unified chaotic system. Chaos Solitons Fractals 26:1119–1124
50. Yu Y, Li H (2010) Adaptive generalized function projective synchronization of uncertain chaotic systems. Nonlinear Anal: Real World Appl 11:2456–2464
51. Zhu Q, Azar AT (2015) Complex system modelling and control through intelligent soft computations. In: Studies in fuzziness and soft computing, vol 319. Springer, Berlin. ISBN: 978-3-319-12882-5

Feature Selection and Recognition of Muzzle Point Image Pattern of Cattle by Using Hybrid Chaos BFO and PSO Algorithms

Santosh Kumar and Sanjay Kumar Singh

Abstract Recognition of cattle based on muzzle point image pattern (nose print) is a well study problem in the field of animal biometrics, computer vision, pattern recognition and various application domains. Missed cattle, false insurance claims and relocation at slaughter houses are major problems throughout the world. Muzzle pattern of cattle is a suitable biometric trait to recognize them by extracted features from muzzle pattern by using computer vision and pattern recognition approaches. It is similar to human's fingerprint recognition. However, the accuracy of animal biometric recognition systems is affected due to problems of low illumination condition, pose and recognition of animal at given distance. Feature selection is known to be a critical step in the design of pattern recognition and classifier for several reasons. It selects a discriminant feature vector set or pre-specified number of features from muzzle pattern database that leads to the best possible performance of the entire classifier in muzzle recognition of cattle. This book chapter presents a novel method of feature selection by using Hybrid Chaos Particle Swarm Optimization (PSO) and Bacterial Foraging Optimization (BFO) techniques. It has two parts: first, two types of chaotic mappings are introduced in different phase of hybrid algorithms which preserve the diversity of population and improve the global searching capability; (2) this book chapter exploited holistic feature approaches: Principal Component Analysis (PCA), Local Discriminant Analysis (LDA) and Discrete Cosine Transform (DCT) [28, 85] extract feature from the muzzle pattern images of cattle. Then, feature (eigenvector), fisher face and DCT feature vector are selected by applying hybrid PSO and BFO metaheuristic approach; it quickly find out the subspace of feature that is most beneficial to classification and recognition of muzzle pattern of cattle. This chapter provides with the stepping stone for future researches to unveil how swarm intelligence algorithms can solve the complex optimization problems and feature selection with helps to improve the cattle identification accuracy.

S. Kumar (✉) · S.K. Singh
Department of Computer Science & Engineering, Indian Institute of Technology, Banaras Hindu University, Varanasi, India
e-mail: santosh.rs.cse12@iitbhu.ac.in

S.K. Singh
e-mail: sks.cse@itbhu.ac.in

A.T. Azar and S. Vaidyanathan (eds.), *Advances in Chaos Theory and Intelligent Control*, Studies in Fuzziness and Soft Computing 337,
DOI 10.1007/978-3-319-30340-6_30

Keywords Animal biometrics · Muzzle point · Computer vision · Cattle recognition · Chaotic · PSO · BFO · Feature selection

1 Introduction

The chaos is a non-linear, deterministic dynamic behavioral based system. It is a bounded an unstable dynamic behavior that demonstrates sensitive dependence on initial conditions of the defined system. It also keeps the infinite unstable periodic motions in dynamic behavior. Although it seems to be a stochastic system and occur in a deterministic non-linear system under defined deterministic conditions. Since few years, growing interests from, biology, physics, chemistry, engineering and multidisciplinary based research have emphasized and stimulated the well studies of chaos theory and concept for control (Tot et al. 1990) [46] synchronization [67] and optimization [2, 56, 57] problems. Chaos based modeling, analysis, and control of dynamical systems have interested various scientists and engineers for a long time [10]. It has also various applications such synchronization of systems such as master and slave working [76, 77] and analysis of 3-D and 4-D dimension with help of hypochaotic algorithm [74, 75]. Backstepping Controller Design for the Global Chaos Synchronization of Sprott's Jerk Systems has been proposed by the author [78]. Due to the easy development of algorithms and it has a special ability to avoid being trapped in local optima, chaos has been a novel optimization technique and chaos-based searching algorithms have aroused deeply interests to solve the problems of different research fields [44, 48, 49, 79]. Chaos theory provides various applications in the variety of areas such as communications and numerical simulation [59, 80] engineering design [3], sound and vibration [23] and optimization algorithms [20] such genetic algorithms [31, 45], PSO [48], BFO and Black hole algorithm [52, 53].

Recently, a few techniques are focused on applications of chaos for optimization and finding the best solutions to defined problems in different research fields. The first one is the chaotic neural network (CNN) [2, 55] by incorporating chaotic dynamics into a neural network. Through the rich non-equilibrium dynamics with various concomitant attractors, chaotic neuron-dynamics can be used to continually search for the global optimum by following chaotic ergodic orbits. The other one is a chaotic optimization algorithm (COA) [56, 57, 68] based on the chaotic evolution of variables. The simple philosophy of the COA includes two main steps: firstly mapping from the chaotic space to the solution space, and then searching optimal regions using chaotic dynamics instead of random search in Genetic algorithms [4, 5, 27, 31, 37]. In random search, explorations of required search regions of problems are not meeting in giving criteria. Therefore, exploitation characteristics of given problems are not up to date for finding the optimal solutions. In the first step of the mapping process for chaotic, each has their own special property. Using distinct chaotic maps in BPSO yields different results [22, 70, 71, 86]. Chaos theory has successfully been applied to problems in the field of computer vision and pattern recognition for the selection of optimal features from the biometric characteristics such face, fingerprint iris and facial expression.

The cattle identification is an emerging research field in the animal biometrics based recognition systems. However, the traditional animal identification systems include ear tagging, tattooing, freeze branding, embedded microchip with ear tags, paint or dye for identification purpose of cattle. But ear tagging systems disintegrate the ear for long term usages and labels of ear tags can be lost easily. Therefore, such traditional animal identification systems are not able to provide a better security from the duplication and fraudulent, missed, swapped and reallocation at slaughter houses. The muzzle point pattern of cattle is a primary animal biometrics characteristic for recognition of cattle. It is similar to human fingerprint recognition systems [13, 41, 60]. The motivation of the propose approach is to mitigate the above issues related to livestock/cattle identification and provide an improved non-invasive—cost effective and robust recognition system based on muzzle point pattern for animal.

We propose an approach for cattle identification based muzzle point pattern that provides enhancement to traditional animal identification for registration, traceability and identification of false insurance claims, health management, control and outbreak of critical diseases.

The key contributions in the book chapter are as follows:

- In this book chapter, binary particle swarm optimization (BPSO) and bacterial foraging optimization (BFO) with chaotic maps proposed approach is used to determine the inertia weight values are used for optimization of feature values of images of muzzle pattern of cattle. The proposed approach for the optimization of features of muzzle images has been incorporated into the two kinds of chaotic maps (1) logistic maps and (2) tent maps, respectively and K-nearest neighbor (K-NN) classification approach [19] based on Euclidean distance calculations serves as a classifier to classify the selected feature for the cattle recognition based on the muzzle images database.
- Experimental results of proposed approach show that feature extraction techniques such PCA and LDA with Foraging Optimization (BFO) yields the recognition accuracy 94.73 % with processing time 245.7 ms. On other hand, Discrete Cosine Transform (DCT) features extraction technique with PSO provide the recognition accuracy of 96.25 % with processing time 164.46 ms. The Foraging Optimization (BFO) optimization technique may not only reduce the number of feature points, but also obtains higher classification accuracies.

The structure of book chapter is organized as follows. Section 2 presents the related work in the field of swarm intelligence and chaotic for the feature extraction and applications. Section 3 illustrates the standard Particle Swarm Intelligence (PSO) and Bacterial Foraging Optimization (BFO) algorithms have been overviewed. Section 4 discusses the proposed approach based framework biometric system for cattle recognition and Sect. 5 shows the features extraction and matching techniques. Section 6 presents the proposed approach for muzzle recognition of cattle and feature section and optimization with BFO-PSO approaches Sect. 7 demonstrates the experimental results and contains a discussion thereof. Results obtained by the proposed approach are compared with results of other feature selection methods and finally, conclusion and future direction is offered in Sect. 8.

2 Related Work

Chaos is a dynamic and a universal nonlinear phenomenon in nature [56]. The chaotic motion is not haphazard due to the three main characteristics of the chaotic variable, namely (1) ergodicity, (2) randomness and (3) sensitivity to initial conditions. Using these characteristics, a chaos optimization algorithm (COA) can solve complex and hard problems or function optimization with high efficiency of calculation. Nature-inspired based chaotic based metaheuristics approaches have been solving the most challenging problems of multidisciplinary since a few years. Numerous biological or bio-computations based systems have evolved with intriguing and better efficiency in maximizing their evolutionary objectives such as reproduction. There are many nature-inspired metaheuristics algorithms have been developed over the last few decades. For example, genetic algorithms (GAs) are working on the Darwinian evolution of biological systems and principle and particle swarm optimization (PSO) is based on the swarm behavior of birds and fish [11, 25, 81], which bat algorithm was based on the echolocation behavior of microbats and working principle of firefly algorithm was based on the flashing light patterns of tropic fireflies [25, 81]. All these algorithms have been applied to a wide range of applications.

Swarm intelligence approaches are highly applicable for the feature selection in the computer vision, pattern recognition medical images analysis and data mining. In general, medical datasets require more number of features to predict an activity for identifying the relevant and ignores the redundant features, consequently reducing the number of features to assess the fetal heart rate [7, 39]. The feature selection method has been hybridized with Rough Set Quick Reduct hybridized with Improved Harmony Search algorithm for dimession reduction and selection of optimal features [38]. Moreover, various application of metaheuristic approaches to identify the important features in the field of medical as well as data mining are getting more proliferations and provide better approaches to reducing the number of features to assess the fetal heart rate of individual with the help of medical heart recorded database of individual patients [6, 8]. The features are selected by using Unsupervised Particle Swarm Optimization (PSO) based Relative Reduct and are tested by using various measures of diagnostic accuracy. The Cloud security using Face Recognition along with feature exaction and selection has gained more significant for security and optimal load balancing of dynamic process [54].

Kumar et al. [52, 53] proposed a method based on Black Hole is a new bio-inspired metaheuristic approach to analysis and comprehensive study of black hole approach and its applications in different research fields like data clustering problem, image processing, data mining, computer vision, science and engineering.

Mpiperis et al. [63] proposed an recognition approach for 3D facial expression which is motivated by the advances of ant colony optimization (ACO) and particle swarm optimization (PSO) respectively in the field of computer vision. More Mirzayans et al. [62] proposed technique for object recognition based on the optimal feature from the object. The proposed approach is capable of extracting and exploiting the geometric properties of objects in images for fast and accurate recognition based on edges and corner of objects. In the direction of feature extraction and selection from the biometric databases such as face, iris and recognition of facial expression, Swarm Intelligence approaches optimal feature sets from the face [52–54].

Swarm intelligence is all regarding developing intelligent, smart and collective behaviors of self-contained and union cluster of independent people or group livings. In recent years the swarm intelligence has received widespread attention and most well-known paradigms in the research of swarm intelligence includes Particle Swarm Optimization (PSO) [49] Ant Colony Optimization (ACO) [24], Bacterial Foraging optimization (BFO) [21] Bee Colony Optimization (BCO) [72], Simulate Biological Immune Systems (SBIS) (Farmer et al. 1986) and Biogeography-Based Optimization are examples to this effect. These approaches are based on the simulation of the collective behavior of animals. The Advantages of Swarm Intelligence (SI) approaches are as follows:

- SI metaheuristics approach preserves information about the search space over the course of a number of iterations, whereas Evolutionary Algorithms (EA) discard the information of the previous generations.
- SI approaches often utilize memory to save the best solution obtained so far.
- SI approaches usually have fewer parameters to adjust.
- SI approaches have fewer operators compared to evolutionary approaches (crossover, mutation, elitism, and so on).
- It is easy to implement.

Based on recent survey, few metaheuristics algorithms are illustrated in the Table 1.

3 Methods and Materials

The section presents the working principle of PSO and BFO swarm intelligence approach in the brief and illustrates the proposed approach based framework biometric system for Muzzle Pattern Recognition of cattle.

3.1 Particle Swarm Optimization (PSO)

Particle Swarm Optimization (PSO) was initially proposed by James Kennedy and Russell Eberhart approach in 1995 which maintains a group of candidate solutions known as particles [48]. It is nature inspired algorithms and uses the metaphor of the flocking behavior of birds [1, 26, 33] fish schooling, ant colony and bee hives to solve optimization problems. The ability of PSO a global stochastic optimization technique is to solve many complex search problems efficiently and accurately has made it an interesting research area. Many autonomous particles are stochastically generated in the problem search space in particle swarm intelligence and each particle presents

Table 1 Description of metaheuristics algorithms

S. No.	Metaheuristics algorithms	Description of metaheuristics algorithms
1	Genetic Algorithms (GAs) (Tang et al. 1996)	Genetic algorithm is a search and optimization based techniques that evolve a population of candidate solutions to a given problem, using natural genetic variation and natural selection operators
2	Simulated Annealing (SA) Algorithm [51]	Simulated Annealing is developed by modeling the steel annealing process and gradually decreases the temperature (T)
3	Ant Colony Optimization (ACO) [24]	Ant Colony Optimization is inspired from the behavior of a real ant colony, which is able to find the shortest path between its nest and a food source (destination)
4	Particle Swarm Optimization (PSO) Algorithm [49]	Particle Swarm Optimization is developed based on the swarm behavior such as fish and bird schooling in nature
5	The Gravitational Search Algorithm (GSA) (Rashedi 2007)	It is constructed based on the law of gravity and the notion of mass interactions. In the GSA algorithm, the searcher agents are a collection of masses that interact with each other based on the Newtonian gravity and the laws of motion
6	Intelligent Water Drops Algorithm (Shah-Hosseini 2009)	It is inspired from observing natural water drops that flow in rivers and how natural rivers find almost optimal paths to their destination. In the IWD algorithm, several artificial water drops cooperate to change their environment in such a way that the optimal path is revealed as the one with the lowest soil on its links
7	Firefly Algorithm (FA) [25, 81]	The firefly algorithm (FA) was inspired by the flashing behavior of fireflies in nature. FA is nature inspired optimization algorithm that imitates or stimulates the flash pattern and characteristics of fireflies. It is used data analysis and to identify homogeneous groups of objects based on the values of their attributes
8	Honey Bee Mating Optimization (HBMO) Algorithm (Karaboga 2005)	It is inspired by the process of marriage in real honey bees
9	Bat Algorithm (BA) [81]	It is inspired by the echolocation behavior of bats. The capability of the echolocation of bats is fascinating as they can find their prey and recognize different types of insects even in complete darkness
10	Harmony Search Optimization Algorithm [30]	It is inspired by the improvising process of composing a piece of music. The action of finding the harmony in music is similar to finding the optimal solution in an optimization process

(continued)

Table 1 (continued)

S. No.	Metaheuristics algorithms	Description of metaheuristics algorithms
11	Big Bang–Big Crunch (Bb–By) Optimization [29]	It is based on one of the theories of the evolution of the universe. It is composed of the big bang and big crunch phases. In the big bang phase the candidate solutions are spread at random in the search space and in the big crunch phase a contraction procedure calculates a center of mass for the population
12	Cuckoo Search [83]	Cuckoo Search algorithm is based on the obligate brood parasitic behavior of some cuckoo species in combination with the Levy flight behavior of some birds and fruit flies
13	Firefly Algorithm (FA) [82]	Firefly algorithm is based on firefly behavior
14	Black Hole (BH) Algorithm [36]	It is inspired by the black hole phenomenon. The basic idea of a black hole is simply a region of space that has so much mass concentrated in it that there is no way for a nearby object to escape its gravitational pull. Anything falling into a black hole, including light, is forever gone from our universe
15	Flower Pollination Algorithm [84]	Flower Pollination Algorithm (FP) was founded by Yang in the year 2012. Inspired by the flow pollination process of flowering plants
16	Grey Wolf Optimizer (GWO) [61]	Grey Wolf Optimizer is new meta-heuristic which inspired by grey wolves (Canis lupus). The GWO algorithm mimics the leadership hierarchy and hunting mechanism of grey wolves in nature. Four types of grey wolves such as alpha, beta, delta, and omega are employed for simulating the leadership hierarchy

a candidate solution to a problem and is demonstrated by a velocity, location in the search space of problem it has a memory to preserve the its previous best position which helps it in remembering and provide best result to the problems.

3.1.1 Neighbourhood Selection Strategies in PSO

In PSO, swarm of N particles autonomous particles or entities is flying around in a hyper-dimensional (D-dimensional) search space. In addition, every swarm particle has numerous sort of topology (set or rules) which is useful to describe the interconnections among the particles. With each particle being attracted towards the best solution found by the particles neighborhood and the best solution found by the particle. The set of particles to which a particle j is topologically connected j neighborhood particle.

In PSO, two basic topologies for neighbor selection have been used to identify some other particle to influence the individual. These topologies are (1) global best (gbest) and (2) local best (lbest). In global best (gbest), the best neighbor in the entire initialized total population influence the target particle. On the other hand, local best (lbest) considers small number of swarm population and particles exchange information locally according to partial knowledge of the solution in space. Generally, in PSO the initialization phase, the positions and velocities of all individuals are randomly initialized. The position of each particle refers to a solution to a problem. Then the position moving process of a particle in the solution space related to a solution search process. The state of the position I is demonstrated by its current position, where D stands for the number of variable encountered in the optimization problem. The particle position i is updated during the evolutionary process.

The state of the position I is demonstrated by its current position $X_i = [x_{i1}, x_{i2}, \ldots x_{iD}]$ where D stands for the number of variable encountered in the optimization problem. The particle position i is updated during the evolutionary process.

$$v_{id}(t+1) = w \times v_{id}(t) + c_1 \times r_{id}(t) \times \left[p_{Bestid} - x_{id}\right] + c_2 \times r_{2d}(t) \times \left[g_{bestid} - x_{id}\right] \tag{1}$$

$$x_{id}(t+1) = x_{id}(t) + v_{id}(t+1) \tag{2}$$

where x_{id} represents the dth dimension of the next current position of the particle i and v_{id} demonstrates the dth variable of the next and current velocity of the particle. The p_{Bestid} shows the dth variable of the personal historical best position founded by particle I up to now. The g_{bestid} is variable of global best position founded by the overall particle so far c_1 and c_2 are the acceleration parameters which are commonly 2.0. r_{1d} and r_{2d} are two random numbers drawn for uniform distribution are (0, 1) and w is inertia weight which is used to set up the balance between the ability of global and local search feature of the particle swarm optimization [47]. In PSO, the particle behavior is demonstrated by the velocity and position update according to Eqs. (1) and (2) and the weight component of (1) models the tendency of the particle to continue in the same direction as before and second component of (1) is referred to as the particle "memory" and self-knowledge or remembrance. It represents the self-learning behavior of the particle. The third component in (1) is referred to as co-operation "*social knowledge*", "*group knowledge*" which reflects the social learning behavior of the particle. Equations (1) and (2) indicates the position of the particle in the solution space will be changed in the local of its current position and next velocity. After received each update, we check out the position and velocity of each particle to guarantee them being within predefine certain range of value. In order to keep the particles from flying out of the problem space, Kennedy [47] defined a clamping scheme to limit the velocity of each particle v_{id} therefore, that each component of is kept within the range [−Vmax, +Vmax]. The parameter choice for Vmax required some care since it appeared to influence the balance between exploration and exploitation characteristics of metaheuristic. As has been noted in [4, 5], the Vmax particle swarm succeeds at finding optimal regions of the search space, but has no feature that enables it to converge on optima. If the particle position and velocity are exceeded the range, they are modified as follow:

$$v_{id} = \min(v_d^{\max}, \max(v_d^{\min}, v_{id})) \tag{3}$$

$$x_{id} = p_{Best} \tag{4}$$

where $v_d^{\max}$, $v_d^{\min}$ maximum and minimum are values of the dth variable of the velocity respectively and p_{Best} is the mean of the dth variable of the personal historical best position of all particles calculated by Eqs. (1) and (2) respectively. The pseudo code for PSO is given as follows:

Pseudo code of Particle Swarm Optimization (PSO)

---

1. Initialize the population of particle (N) with random position and velocity in given (D-dimension) search space.
2. while terminating condition is not reached
3. do
4. for each particle i=1 to N Do
5. adapt velocity of the particle of the particle using
6. $v_{id}(t+1) = w \times v_{id}(t) + c_1 \times r_{id}(t) \times \left[p_{Bestid} - x_{id} \right] + c_2 \times r_{2d}(t) \times \left[g_{bestid} - x_{id} \right]$ ------- (5)
7. update the position of the particle using
8. $x_{id}(t+1) = x_{id}(t) + v_{id}(t+1)$
9. evaluate the fitness $f(x_i)$
10. $f(x_i) < f(p_i)$
11. $p \leftarrow x_i$
12. end
13. If $f(x_i) < f(p_g)$ then
14. $p_g \leftarrow x_i$
15. end
16. for end

An inertia weight w is a proportional agent that is related with the speed of last time and the formula for the change of the speed is given in Eq. 5. When w is bigger, bigger the PSO search ability for the whole while w is smaller, search ability is smaller and when $\omega = 1$, so at the later period of the several generations of PSO, there is a lack of the searching ability for the partial. Experimental results demonstrate that particle swarm optimization has the biggest speed of convergence when $w \in [0.8, 1.2]$. While experimenting, ω is confined from 0.9 to 0.4 according to the linear decrease, which makes PSO search for the bigger space at the beginning and locate the position quickly where there is the most optimist solution.

Rather than applying inertia to the velocity memory, Clerc and Kennedy [18] applied a constriction factor χ. The velocity update scheme proposed by Clerc and it can be expressed for the dth dimension of ith particle as by follows:

$$v_{id}(t+1) = \chi(v_{id}(t) + c_1(p_{id}(t) - x_{id}(t)) + c_2 p_2(p_{gd}(t) - x_{id}(t))) \tag{6}$$

$$\chi = \frac{2}{\varphi - 2 + \sqrt{\varphi^2 - 4\phi}}$$

where $\varphi = \phi_1 + \varphi_2 > 4$, this is known as hybrid cooperative PSO approach which improves the convergence rate of standard PSO (Eq. (6)). The example of a particle swarm optimization working is shown in Fig. 1.

In above Fig. 1, initially, the particles are scattered randomly in the search space. As they search for the optimum value, the particles balance the objectives of following the value gradient and random exploration. Over time they begin to congregate in the general area of the optimum value. Finally, the particles converge to the optimum value of a given hard or complex problem.

3.1.2 Major PSO Variants

The standard PSO exhibits some deficiencies including suffering from being premature and inefficient in solving complex multimodal optimization problem. On method to strengthen, the capacity of PSO is to dynamically adapt its parameters when the

Fig. 1 Example of a particle swarm optimization swarms progression in two-dimensions

particle evaluating process. In addition a fuzzy adaptive mechanism was used to the value of w. Kinnedy and Eberhart [48] recommendation that the proper value for acceleration parameter w is set to linearly decease over linearly.

- **Advantages and Disadvantages of PSO**

PSO is a population and intelligence based metaheuristics algorithm. PSO have no overlapping and mutation calculation and search can be carried out by speed of the particle deriving the development of the several generations, only the most PSO can transmit information on to the other particle and speed of the searching is very fast. Position and velocity of particle calculation in PSO is very simple compared with the other swarm intelligence method.

It occupies the biggest optimization ability and it can be completed easily. One of the limitations of the "*standard PSO*" is premature convergence and trapping in local optima of the problem. A great effort has been deployed to provide PSO convergence results through the stability analysis of the trajectories [18, 64].

These studies were aimed at understanding theoretically how PSO algorithm works and why under certain conditions it might fail to find a good solution. Considerable research has been also conducted into further refinement of the original formulation of PSO in both continuous and discrete problem spaces and areas such as dynamic environment, parallel implementation and multi-objective Optimization. The modified versions of PSO based on diversity, mutation, crossover and efficient initialization using different distribution. In 2008, Poli categorized a large number of publications dealing with PSO applications stored in the IEEE Xplore database. Therefore many papers related with the applications of PSO and its applications have been presented in the literature and several survey papers regarding these studies can be found in [10, 12, 14, 16, 17, 35, 39, 43, 48, 65, 89].

3.2 Bacterial Foraging Optimization (BFO)

Bacterial Foraging Optimization (BFO) is based on foraging strategy of bacteria *Escherichia coli*. After many generations, bacteria with poor foraging strategies are eliminated while; the individuals with good foraging strategy survive signifying survival of the fittest. The whole process can be divided into three subsections, namely—chemotaxis, reproduction and elimination and dispersal [66].

3.2.1 Chemotaxis

Chemotaxis is the process in which bacteria direct their movements according to certain chemical s in their environment. This is important for bacteria to find food by climbing up nutrient hills and at the same time avoids noxious substances. The sensors they use are receptor proteins which are very sensitive and possess high gain. That is a small change in the concentration of nutrients can cause a significant change in behavior (Liu and Passino 2002). Suppose that we want to find the minimum of J(θ), where $\theta \in R^D$ is the position of a bacterium in D-dimension al space and the cost function J (θ) is an attractant–repellant profile (where nutrients and noxious substances are located). Then J (h) ≤ 0 represents a nutrient rich environment J (θ) $= 0$ represents neutral medium and J (h) > 0 represents noxious substances.

- Let $\theta^i \in h(i, j, l)$ represent *i*th bacterium at *j*th chemotactic, *k*th reproductive and 1th elimination-dispersal step.
- The position of the bacterium at the $(j + 1)$th chemotactic step is calculated in terms of the position in the previous chemotactic step and the step size C(i) (termedas run length unit) applied in a random direction $\theta(i) : \theta^i(\mathrm{j} + 1, \mathrm{k}, 1) = \theta^i(\mathrm{j}, \mathrm{k}, 1) + \mathrm{C(i)} \times \phi(\mathrm{i})$.
- Where ϕ(i) is the unit random direction to discrete tumble and it is given by $\phi(\mathrm{i}) = \frac{\Delta(i)}{\sqrt{\Delta^T(i)\Delta(i)}}$. Where $\Delta(i) \in \mathrm{R}^D$ is a randomly generate d vector with elements within the interval $[-1, +1]$.
- The cost of each position is determined by the following equation:

$$J(i, j, k, l) = J(i, j, k, l) + \mathrm{J}_{cc}(\theta, \theta^i(j, k, l)). \tag{7}$$

It can be noticed that Eq. (7) the cost of a determined position J(i, j, k, l) is also affected by the attractive and repulsive forces existing among the bacteria of the population given by Jcc. If the cost at the location of the *i*th bacterium at $(j + 1\text{th})$ chemotactic step denoted by $\mathrm{J(i, j + 1, k, l)}$ is better (lower) than at the position $\theta^i(j, k, l)$ at the *j*th step, then the bacterium will take another chemotactic step of size C(i) in this same direction, up to a maximum number of permissible steps called Ns.

3.2.2 Swarming

Swarming is a general type of motility that is promoted by flagella and allows bacteria to move rapidly over and between surfaces and through viscous environments. Under certain conditions, cells of chemota ctic strains of *E. coli* excrete an attractant, aggregate in response to gradients of that attractant and form patterns of varying cell density. Central to this self-organization into swarm rings is chemotaxis. The cell-to-cell signaling in *E. coli* swarm may be represented by the following Eq. (8):

$$j_{cc}(\theta, \theta^i(j, k.l) = \sum_{i=1}^{S}\left[-d_{attractant}\exp\left(-w_{attractant}\sum_{m=1}^{D}\theta_m - \theta_m^i\right)^2\right] + \sum_{i=1}^{S}\left[-h_{repellant}\exp\left(-w_{repellant}\sum_{m=1}^{D}\theta_m - \theta_m^i\right)^2\right] \quad (8)$$

where h = $[h_1, h_2, \ldots, h_D]^T$ is a point in the D-dimensional search space, Jcc(h, h_i (j, k, 1)) is the objective function value that is to be added to the actual objective function and dattractant, wattractant, hrepellant, wrepellant are the coefficients which determine the depth and width of the attractant and the height and width of the repellant. These four parameters are to be chosen judiciously for a given problem, him is the *m*th dimension of the position of the *i*th bacterium hi in the population of the S bacteria.

3.2.3 Reproduction

The least healthy bacteria eventually die while each of the healthier bacteria (those yielding lower value of the objective function) asexually split into two bacteria, which are then placed in the same location. This keeps the swarm size constant.

3.2.4 Elimination and Dispersal

The gradual or sudden changes in the local environment where a bacterium population lives may occur due to various reasons e.g. a significant local rise of temperature may kill a group of bacteria that are currently in a region with a high concentration of nutrient gradients. Events can take place in such a fashion that all the bacteria in a region are killed or a group is dispersed into a new location. To simulate this phenomenon in BFOA some bacteria are liquidated at random with a very small probability while the new replacements are randomly initialized over the search space. In [21], the authors discussed some variations on the original BFOA algorithm and hybridizations of BFOA with other optimization techniques. They also provided an account of most of the significant applications of BFOA. However, experimentation with complex optimization problems reveal that the original BFOA algorithm possesses a poor converge nice behavior compared to other nature-inspired algorithms, like GAs and PSO and its performance also heavily decreases with the growth of the search space dimensionality. The pseudo code of BFO algorithm is given as follows:

Pseudo code of BFO Algorithm

1. Initialize parameters D, S, Nc, N_s, N_{re}, N_{ed}, P_{ed}, C(i) (i=1,. . . ,S) and θ^i where i$\in$ $(i, 2, 3......S)$.
2. while terminating condition is not reached do
3. /* Elimination dispersal loop *\
4. for i=1----N_{red} do
5. /* reproduction loop *\
6. for k=1……Nre Do
7. /* Chemotaxis loop *\
8. for j=1……Nc do
9. for each bacterium i=1…….S do
10. compute fitness function
11. $J(i,j,k,l) = J(i,j,k,l) + J_{CC}(\theta, \theta^i(j,k,l))$
12. J_{last}=J (I, j, k, l)
13. Tumble= generate a random vector $\Delta(i) \in R^D$
14. move: Compute the position of the bacterium $\theta^i(j+1,k,l) = \theta^i(j,k,l) + C(i) \times \phi(i)$ at j+1th characteristics step.
15. compute fitness function

    $$J(i,j,k,l) = J(i,j,k,l) + J_{CC}(\theta, \theta^i(j,k,l))$$

    swim: m=0
16. while M< N_s do
17. M=M+1
18. if j(I, j+1 ,k, l) < J_{last}
19. then
20. J_{last} =J(I, j+1 ,k, l)
21. move: Compute the position of bacterium $\theta^i(j+1,k,l)$ at J+1th chemotactic step using following step

22. $\theta^{i}(\mathrm{j}+1,\mathrm{k},\mathrm{l}) = \theta^{i}(\mathrm{j},\mathrm{k},\mathrm{l}) + \mathrm{C(i)} \times \phi(\mathrm{i})$
23. Compute fitness function

    $$J(i,j,k,\mathrm{l}) = J(i,j,k,\mathrm{l}) + \mathrm{J}_{CC}(\theta,\theta^{i}(j,k,l)) \tag{9}$$

    else
24. M=Ns
25. end
26. end
27. end
28. end
29. /* Re-production Process * /
30. for
31. i=1,2……S do
32. $j_{Health}(i) = \Sigma_{j=1}^{N_C+1} J(i,j,k,l)$
33. end
34. sort bacterium in order to ascending $J_{Health}(i)$ (highest cost means low health)
35. end
36. (The least healthy bacterium dies and the other heathier bacteria splt each into two bacteria which are placed in same location)
37. end
38. elimination-dispersal
39. for i=1…..S do
40. eliminate and disperse the ith bacterium with probability Ped
41. end
42. end

3.2.5 Chaotic Sequences for Inertia Weight

The inertia weight controls the balance between the global exploration and local search ability. A large inertia weight facilitates the global search, while a small inertia weight facilitates the local search. Proper adjustment of the inertia weight value is important. The inertia weight w is the key factor influencing the convergence and thus will greatly affect the BPSO search process, and through it the resulting classification accuracy. The BPSO process often suffers from entrapment of particles in a local optimum, which causes the premature convergence mentioned above. We employed chaotic binary particle swarm optimization (CBPSO) to prevent this early convergence, and thus achieve superior classification results.

Chaos is a deterministic dynamic system very sensitive to its initial conditions and parameters. The nature of chaos is apparently random and unpredictable; however it also possesses an element of regularity. A chaotic map is used to determine the inertia weight value at each iteration. Logistic maps and tent maps are the most frequently used chaotic behavior maps. They are similar to each other and both show unstable dynamic behavior. Chaotic sequences have been proven easy and fast to generate and store, as there is no need for storage of long sequences. In this book chapter, chaotic sequences are embedded in the BPSO and BFO. Logistic maps and tent maps determine the inertia weight values, respectively. The number of iterations in the CBPSO is given by t. The inertia weight value is modified by a logistic map according to Eq. (10):

$$w(t+1) = 4.0 \times w(t) \times w(1 - w(t)), w(t) \in (0, 1) \tag{10}$$

The inertia weight value modified by a tent map is given by:

$$\begin{aligned} &\mathit{if}((w(t) < 0.7) \\ &w(t = 1) = w(t)/0.7 \\ &\text{else} \\ &w(t + 10) = 10/3w(t)(1 - w(t)), w(t) \in (0, 1) \end{aligned}$$

When the inertia weight value is close to 1, BPSO increase the global search ability in given search space. For inertia weight values close to 0, PSO improves the local search ability. Further, in next section, we have discussed biometric system for recognition of cattle based on muzzle pattern.

4 Biometric System for Muzzle Pattern Recognition of Cattle

Recognition system for muzzle pattern of cattle is non-intrusive, invasive, and inexpensive by using muzzle pattern as common biometric identifier similar to human's fingerprint [40]. The attributes of muzzle images (beads and ridges) are probably the most similar biometric features used by biometric system to recognize them. In order that a recognition system for muzzle pattern of cattle works well in practice as follows:

(1) **Muzzle Pattern Detection**: It should automatically detect whether a muzzle image of cattle is present in the in acquired image.
(2) **Localization of muzzle images**: To locate the muzzle image if there is in the used muzzle pattern images database.
(3) **Recognition of muzzle of cattle**: To recognize the muzzle pattern of cattle from a general viewpoint from used muzzle images database.

Animal biometric system is basically a pattern recognition based system. It acquires biometric data from an individual, extracts a salient feature set from the data, compares feature set against the feature set(s) stored in the database and executes an action based on the result of the comparison [40]. For biometrics, different extracted features are incorporated for recognition such as—face, fingerprints, hand shape, calligraphy, iris, retinal, vein, and voice. Recognition of muzzle pattern images has a number of strengths to recommend it over other biometric modalities in certain circumstances. Muzzle recognition as a biometric derives a number of advantages from being the primary biometric that individuals use to recognize them. The muzzle recognition system is shown in Fig. 2.

1. **Acquisition module**: It is the module where the muzzle pattern image of cattle under consideration is presented to the system. It can request an image from several different environments: The muzzle image can be an image file that is located on storage (disk) or it can be captured by a frame grabber or can be scanned from paper with the help of a scanner.

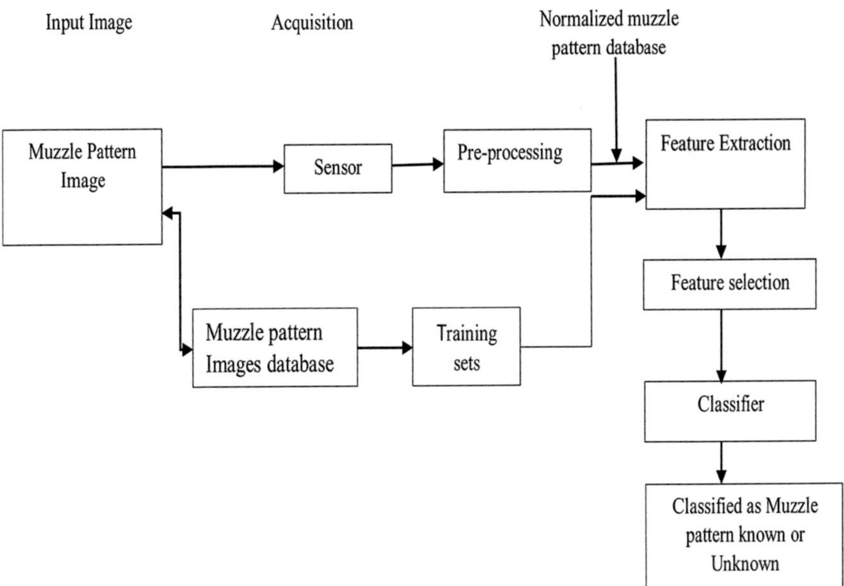

Fig. 2 Outline of typical recognition of muzzle pattern for cattle

2. **Pre-processing Process**: This is done to enhance the images to improve the recognition performance of the system:
    - **Image normalization**: It is done to change the acquired image size to a default size say, 128×128 on which the system can operate.
    - **Histogram equalization**: For images that are too dark or too light, it modifies the dynamic range and improves the contrast of the image so that facial features become more apparent.
    - **Median filtering**: For noisy images especially obtained from a camera or from a frame grabber, median filtering can clean the image without much loss of information.
    - **Background removal**: In order to deal primarily with facial information itself, muzzle image background can be removed. This is important for muzzle image recognition systems where entire information contained in the image is used. It is obvious that, for background removal, the pre-processing module should be capable of determining the muzzle outline.
    - **Translational and rotational normalizations**: In some cases, it is possible that in the muzzle image of subject the head is somehow shifted or rotated (pose problem). The head plays a major role in the determination of facial features. The pre-processing module determines and normalizes the shifts and rotations in the head position.
    - **Illumination normalization**: muzzle images taken under different illuminations can degrade recognition performance especially for muzzle recognition systems based on the principal component analysis in which entire muzzle information is used for recognition.
3. **Feature Extraction**: This module finds the key features in the muzzle pattern that may be used in classification of cattle. It is responsible for composing a feature vector that is well enough to represent the image.
4. **Feature Selection**: The feature selection seeks for the optimal set of d features out of m features obtained from feature extraction module. Several methods have been previously used to perform feature selection on training and testing data. In the developed muzzle recognition system for cattle. I have utilized evolutionary feature selection algorithms based on swarm intelligence called the Bacteria Foraging Optimization (BFO) and Particle Swarm Optimization (PSO).
5. **Classification Module**: In this module, the extracted features of the muzzle image are compared with the ones stored in the muzzle pattern images (database) library with the help of a pattern classifier.
6. **Training sets**: It is used during the learning phase of recognition process for muzzle pattern.
7. **Muzzle pattern image database**: On being classified as unknown by the classification module, the muzzle image database can be added to the database or library with their feature vectors for future comparisons.

5 Feature Extraction and Selection Methods

In this section, we have used computer vision based methods for extraction of feature from muzzle images database of cattle. These features basically have properties and geometric relations such as the areas, distances and angles between the feature points that are used as descriptors for muzzle recognition in this approach [88]. After feature extraction, feature selection is carried out for selection of discriminant feature vector from extracted feature sets of muzzle pattern.

Feature selection (FS) is a complex optimization problem in machine learning and computer vision that mitigate the number of extracted features and removes irrelevant noisy, redundant data from given database. Several universal approaches for feature selection are introduced briefly here. These approaches are (1) sequential forward selection approaches (SFS), (2) sequential backward selection approach (SBS) and sequential forward floating selection (SFFS). The sequential forward selection (SFS) approaches begins with a small feature subset and successively adds or removes feature values until some termination criterion is satisfied. However, SFS approach suffers from a nesting or looping affect. Since this algorithm does not check all possible feature subsets and they are not assured to give an optimal result [87]. In SFS, features discarded cannot be re-selected, and selected features cannot be removed later.

The sequential backward selection (SBS) approach is defined for the backward version of SFS. It chooses the plus-l-take-away-r process (PTA (l, r)) goes forward l stages (by adding l features via SFS) and then goes backward r stages (by deleting r features via SBS). This process is in repetitive nature several times. However, there is no theoretical guidance to determine the appropriate values of l and r [87].

In other method, sequential forward floating selection (SFFS) is the floating type of PTA (l, r). Unlike PTA (l, r), SFFS can be overturned an unlimited number of times as long as the reversal finds a better feature subset than the feature subset obtained so far at the same size. The SFFS suffers from entrapment in a local optimum when applied to a large-number feature problem. Simple genetic algorithms (SGA) generally have poor search ability in the near local optimum region, and premature convergence in SGA seems to be difficult to avoid. Therefore test parameters in SGA need to be carefully selected [87]. Although, feature selection is primarily performed to select relevant and informative features it can have other motivations including:

- General data reduction from high dimension to lower dimension, to limit storage requirements and increase algorithm speed.
- Feature set reduction, to save resources in the next round of data collection or during utilization.
- Performance improvement, to gain in predictive accuracy.
- Data understanding, to gain knowledge about the process that generated the data or simply visualize the data

In feature extraction phase of proposed muzzle recognition system for cattle, all the feature approaches exploited the top N principal components or most discriminative feature values or components of muzzle image database (or transform coefficients). These methods are used directly for dimension reduction into lower order. However, there may be some useful information in lower order principal components leading to a significant contribution in improving the recognition rate. This is known as curse of dimensionality.

Feature selection thus becomes an important step affecting the performance of a pattern recognition system. In this book chapter, search methods based on swarm intelligence algorithms are developed to select the appropriate principal components or transform coefficients from the set of extracted feature vectors. Feature selection approaches are Bacterial Foraging Optimization (BFO) and Particle Swarm Optimization (PSO) [48] which is used to select more discriminative feature values from is given dataset.

These feature extraction approaches try to define a muzzle images as a function and attempts to find a standard template of all the muzzle of cattle. The features can be defined independent of each other. Followings approaches for feature extraction are used.

5.1 Appearance-Based Recognition Approaches

Appearance-based methods consider the global properties of the image intensity pattern of an image [42]. Typically appearance-based face recognition algorithms proceed by computing basis vectors to represent the face data efficiently. In the next step, the faces are projected onto these vectors and the projection coefficients can be used for representing the face images. Popular algorithms such as Principal Component Analysis (PCA), Local Discriminant Analysis (LDA), Independent Component Analysis (ICA), LFA, Correlation Filters, manifolds and tensor faces are based on the appearance of the muzzle pattern. Appearance approaches to recognition of muzzle image have trouble dealing with pose variations. In general, appearance-based methods rely on techniques from statistical analysis and machine learning to find the relevant characteristics of muzzle images.

For feature extraction purpose, appearance-based methods include Principle Component Analysis (PCA) [73], Linear Discriminant Analysis (LDA) and Discrete Cosine Transform (DCT) have been used in this section. These approaches are described in detail in the next subsection.

5.1.1 Principal Component Analysis (PCA)

Principal Component Analysis (PCA) is unsupervised face recognition approach. It also used to reduce the dimensionality of large database. PCA [73] is an orthogonal linear transformation that transforms the data to a new coordinate system such that greatest variance by any projection of the data comes to lie on the first coordinate; the second greatest variance comes up in the second coordinate, and so on.

The basic working principal of PCA is illustrated as follows.

- Eigen faces also known as Principal Components Analysis (PCA) find the minimum mean squared error linear subspace that maps from the original N dimensional data space into an M-dimensional feature space.
- By doing this, Eigen faces (where typically M $\ll$ N) achieve dimensionality reduction by using the M eigenvectors of the covariance matrix corresponding to the largest eigenvalues. The resulting basis vectors are obtained by finding the optimal basis vectors that maximize the total variance of the projected data (i.e. the set of basis vectors that best describe the data) [73].
- Usually the mean x is extracted from the data, so that PCA is equivalent to Karhunen-Loeve Transform (KLT). So, let X n $\times$ m be the data matrix where $x_1 \ldots x_m$ are the image vectors (vector columns) and n is the number of pixels per image.
- The KLT basis is obtained by solving the Eigen value problem [50] as follows:

  Let $X = \{X_1, X_2, X_3, X_4, \ldots, X_m\}$, be a random vector with observations $X_i \in R^d$

1. Compute the mean μ

$$\mu = \frac{1}{n}\sum\nolimits_{i=1}^{m} X_i$$

2. Compute the Covariance Matrix S

$$S = \frac{1}{n}\sum\nolimits_{i=1}^{m} (X_i - \mu)(X_i - \mu)^{T_i}$$

3. Compute the eigenvalues λ_i and the eigenvectors of S

$$Sv_i = \lambda_i v_{i,i=1,2\ldots m}$$

Order the eigenvectors descending by their eigenvalues. The *k* principal components are the eigenvectors corresponding to the *k* largest Eigen values. The *k* principal components of the observed vector *k* are then given by:

$$y = w^T(x - \mu)$$
$$\text{where } W = (v_1, v_2 \ldots v_k)$$

4. The reconstruction from the PCA basis is given by:

$$X = Wy + \mu$$
$$W = (v_1, v_2 \ldots v_k)$$

5. The Eigen faces method then performs face recognition by:
   - Projecting all training samples into PCA subspace.
   - Projecting the query image into the PCA subspace.
   - Finding the nearest neighbor between the projected training images and the projected query image.

Still there's one problem left to solve. Imagine we are given 400 images sized 100 × 100 pixel. The PCA solves the Covariance matrix, $S = XX^T$ wherc $size(x) =$ 100 × 400 in our example. It would end up with 10000 × 10000 matrixes, roughly 0.8 GB. Solving this problem is not feasible. Therefore we will need to apply a trick. From your linear algebra lessons you know that an M × N matrix with M > N can only have (N − 1) non-zero eigenvalues. Therefore, it's possible to take the Eigenvalue decomposition of size N × N instead:

$X^T X v_i = \lambda_i v_i$ And get the original eigenvectors of $S = XX^T$ with a left multiplication of the data matrix:

$$XX^T(Xv_i) = \lambda(Xv_i)$$

The resulting eigenvectors are orthogonal; to get ortho-normal eigenvectors they need to be normalized to unit length.

5.2 *Linear Discriminant Analysis (LDA)*

Linear Discriminant Analysis (LDA) is a supervised learning based classification approach which is widely used to find linear combinations of features while preserving class separability. Unlike PCA, LDA tries to model the differences between classes. Classic LDA is designed to take into account only two classes. Specifically, it requires data points for different classes to be far from each other, while point from the same class is close. Consequently, LDA obtains differenced projection vectors for each class. Suppose we have m samples $x_1 \ldots x_m$ belonging to c classes; each class has m∗k elements. It can assume that the mean has been extracted from the samples, a sin PCA. The ratio of between-class scatter to the within-class scatter is calculated which the optimizing criterion in LDA is:

Let X be a random vector with samples drawn from c classes

$$X = \{X_{1,}X_1, \ldots X_c\}$$
$$X = \{X_{1,}X_1, \ldots X_i\}$$

The scatter matrices S_B and S_w are calculated as:

$$S_B = \sum\nolimits_{i=1}^{c} (\mu_i - \mu)(\mu_{i-\mu})^T$$
$$S_w = \sum\nolimits_{i=1}^{c} \sum\nolimits_{x_{i\in}X_j=1}^{c} (x_j - \mu_i)(x_j - \mu_i)^T$$

where μ is the total mean

$$\mu = \frac{1}{N}\sum_{i=1}^{n} X_i,\ where\ i \in (1, 2, 3 \ldots n)$$

And μ_i is the mean of class $\mu_i = \frac{1}{\|X_i\|}\sum_{x_j \in X_i}^{n}(x_j).$

Fisher's classic algorithm now looks for a projection W that maximizes the class reparability criterion:

$$W_{opt} = \arg\max_w \left| \frac{|W^T S_B W|}{|W^T S_w W|} \right|$$

The optimization problem can then be rewritten as:

$$W_{pca} = \arg\max_w |W^T S_T W|$$

$$W_{fld} = \arg\max_w \left| \frac{|W^T W_{pca}^T S_B W_{pca} W|}{|W^T W_{pca}^T S_w W_{pca} W|} \right|$$

The transformation matrix W that projects a sample into the $(c-1)$-dimensional space is then given by:

$$W = W_{fld}^T W_{pca}^T$$

The projection of two classes are shown in Fig. 3.

5.3 Discrete Cosine Transform (DCT)

Discrete Cosine Transform (DCT) has emerged as a popular transformation technique which is widely used in signal and image processing field due to its strong and well known "*energy compaction*" property: The most of information of signal tends to

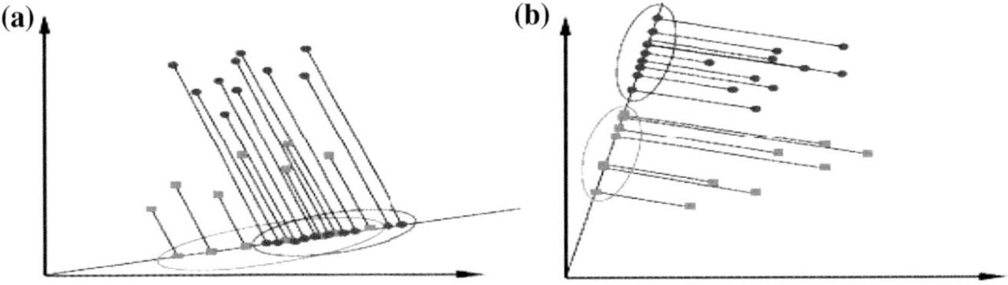

Fig. 3 Projection of two classes onto line

be concentrated in a few low-frequency components of the DCT. The exploitation of these DCT techniques for feature extraction in face recognition system has been described by several research groups [34, 58, 69] (Pan and Bolouri 1999).

DCT transforms a sequence of data points in terms of a sum of cosine functions oscillating at different frequencies based on energy compaction properties. DCT was found to be an effective method that yields high recognition rates with low computational complexity.

The DCT exploits inter-pixel redundancies to render excellent de-correlation for most natural images. After de-correlation each transform coefficient can be encoded independently without losing compression efficiency. The DCT helps separate the image into parts (or spectral sub-bands) of differing importance (with respect to the image's visual quality). Therefore, it can be used to transform images, compacting and allow an effective dimensionality reduction. They have been widely used for data compression. The DCT is based on the Fourier Discrete Transform (FDT), however, using only real numbers. When a DCT is performed over an image, the energy is compacted in the upper-left corner. Suppose an M $\times$ N image, where each image corresponds to a 2-D matrix, DCT coefficients are calculated as follows:

Let f(x, y) is M $\times$ N (2 D) image given as input. DCT coefficient of 2-D matrix of given input image is calculated. 2-D matrix can be truncated can be truncated retaining the upper-left area, which has the most information reducing the dimensionality of the problem.

$$f(u, v) = \frac{1}{\sqrt{MN}}\alpha(u) \times \alpha(v)\sum_{x=0}^{M-1}\sum_{y=0}^{N-1} \times F(x, y)$$

where $F(x, y) = f(x, y) \times \cos(\frac{(2x+1)\mu\prod}{2M}) \times \cos(\frac{(2y+1)\mu\prod}{2N})$ and u = 1, 2, . . .M and v = 1, … N.

$$\alpha(\mathrm{w}) = \begin{cases} 1 & \text{if w} = 0 \\ \sqrt{2}a & \text{Otherwise} \end{cases}$$

6 Proposed Muzzle Recognition Approach Using Hybrid Chaos Particle Swarm Optimization (ICPSO) and Bacterial Foraging Optimization (BFO) Algorithm

The proposed recognition system for muzzle pattern of cattle consists of two phases: (1) training phase and (2) testing phase. In training phase, the input muzzle image is used to extract the discriminant features by using algorithms PCA, LDA and DCT. The extracted feature values are stored in a matrix, known as feature matrix. This matrix represents the huge amount of extracted facial features values for each subject (cattle). The unique and most discriminative features are selected by using swarm intelligence approaches BFO and PSO and stored in muzzle pattern image database (gallery database).

6.1 Database Acquisition and Preparation

A database of muzzle pattern of cattle has been prepared from, Department of Animal Husbandry and Dairying (BHU), Institute of Agriculture Sciences Varanasi, India. In the acquisition of images, a 12 megapixel digital camera has been used to capture muzzle images. It took more than 5 months to keep adequate number of subject images for training and testing purpose. The preparation of cattle's muzzle pattern database is taken in two different sessions. The size of cattle's muzzle image database is 1200 (120 subject and 10 images per subject). Few sample images of muzzle database is shown from cattle muzzle database in Figs. 4 and 5 respectively.

In this experiment, the muzzle images of the muzzle image database are exploited to generate the training set and the test set. The training set is generated by 60 % of total image of muzzle pattern database and remaining 40 % image are used for testing phase.

In proposed approach, features are extracted from input images (training set) of muzzle pattern image database by using holistic approaches such as PCA, LDA and

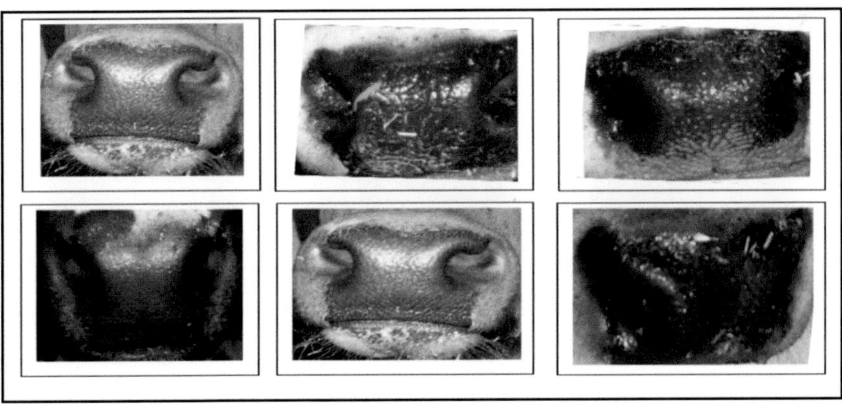

Fig. 4 Few cattle muzzle pattern image from cattle database

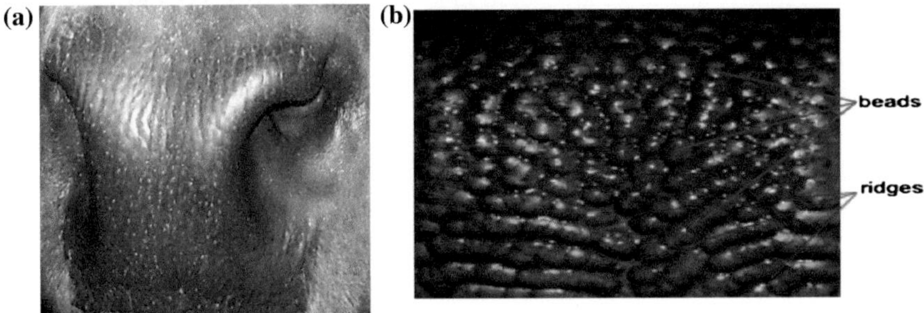

Fig. 5 **a** Original muzzle point image. **b** Beads and ridges in a muzzle pattern image

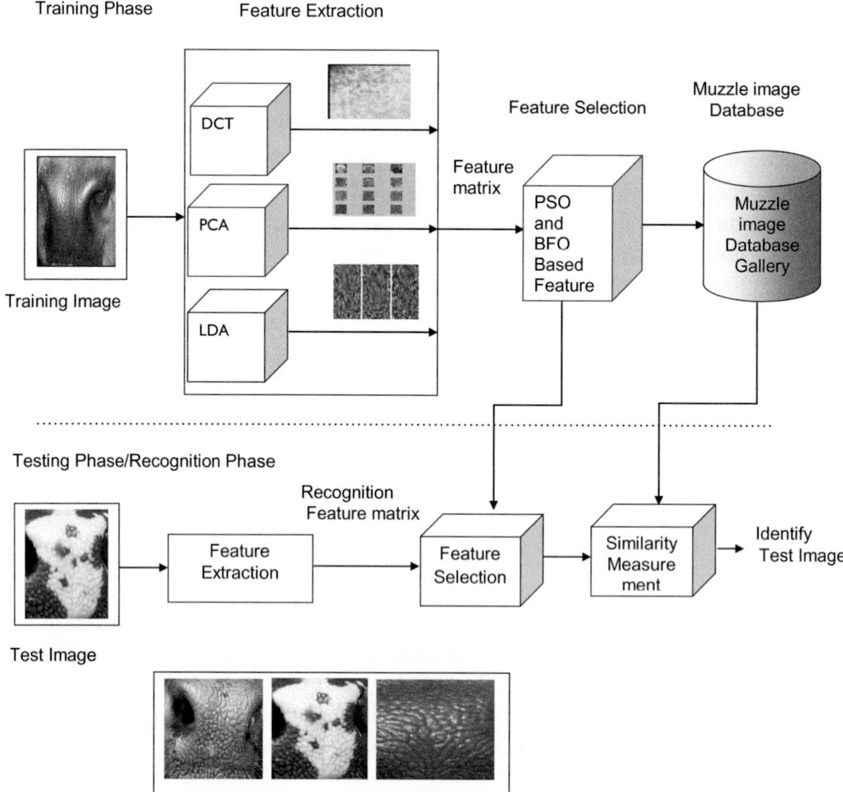

Fig. 6 Proposed approach for muzzle pattern recognition of cattle

DCT. These extracted features are stored in a define matrix is known as feature matrix. It holds huge number of feature vectors but these vectors are not more prominent to identifying the muzzle with better accuracy. Therefore, we have used swarm intelligence techniques such BFO [66] and PSO to select discriminant features vectors from feature matrix and these are stored in muzzle pattern database. The test set is generated by rest of the images in the given muzzle pattern image database with similarity measurement of match score of selected prominent features of test set. The schematic description of proposed muzzle image recognition system by swarm intelligence (SI) is shown in Fig. 6. In this experiment, two different methods are followed and a comparison of the two is made. They are described below in the Sects. 6.2 and 6.3 respectively.

6.2 BFO-Based Feature Selection Algorithm

In proposed muzzle recognition system, the Principal Component Analysis (PCA) has been used on the muzzle images of database to obtain the optimal bases before LDA. Then generate the eigenvectors as the feature vector set (which will be input to the BFO) through LDA. Feature selection applies the BFO algorithm on the feature vector sets and pick up the position of bacteria B with max (Jhealth) value. This position represents the best feature subset of the features defined in feature extraction step.

- **Classification**: Calculate the difference between the feature subset (obtained through feature selection) of each image of facial gallery and the test image with the help of Euclidean Distance defined below. The index of the image which has the smallest distance with the image under test is considered to be the required index. For an N-dimensional space, the Euclidean distance between two any points, p_i and q_i is given by:

$$D = \sum\nolimits_{i=1}^{n} sqrt(p_i - q_i)^2$$

where p_i and q_i are the co-ordinate values of p and q in given n space dimension.

6.3 PSO-Based Feature Selection Algorithm

The feature extraction obtained by applying Discrete Cosine Transformation (DCT) to image and stored into DCT array. Take the most representative features of size 50×50 from upper left corner of DCT array. Feature selection in this algorithm, each particle represents a range of possible candidate solutions. The evolution of each generation is accomplished through the fitness function. The fitness function is as follows: $F = \sqrt{\sum_{i=1}^{l} (M_i - M_o)(M_i - M_0)^T}$.

7 Experiment Results and Discussion

In order to construct the training set, 6 images per 120 subjects ($120 \times 6 = 720$ images) was used and the remaining 4 images ($120 \times 4 = 480$ images) were used for testing purpose. All 120 classes of subjects in the muzzle pattern images database were considered. The same training and testing data sets were used in both the approaches. The results are illustrated Table 2.

Table 2 Recognition accuracy (%) and training time (s) for muzzle pattern of cattle

S. No	Feature extraction approach	Feature selection approach	Muzzle recognition accuracy (%)	Training time (s)
1.	PCA and LDA	Foraging Optimization (BFO)	94.73	245.7
2.	Discrete Cosine Transform	Particle Swarm Optimization (PSO)	96.25	164.46

7.1 Comparative Analysis

On comparing BFO-based approach with PSO-based approach for feature selection approach, it is illustrated that the average muzzle recognition rate of BFO is better than that of PSO-based feature selection method. Also, on analysis it is demonstrated that number features required by BFO are less than that required for recognition using PSO (Table 2). However, in terms of computational time for training and testing process, PSO-based selection algorithm takes less training time than the BFO-based selection algorithm. Hence, BFO is computationally expensive than PSO. Therefore, the effectiveness of BFO in finding the optimal feature subset compared to PSO compensates its computational inefficiency.

8 Conclusion and Future Direction

In this book chapter, we have propose a novel BFO-based feature selection algorithm for muzzle recognition of cattle and the hybrid feature selection algorithm is applied to feature vectors which is extracted by applying the Discrete Cosine Transform (DCT), PCA and LDA feature extraction techniques. In the experimental results, the appearance based algorithms such as PCA and LDA yield the cattle recognition accuracy of 96.73 % for muzzle pattern which is better than Discrete Cosine Transform (DCT) using improved chaotic Particle Swarm Optimization (PSO) as feature selection method. It is notice that algorithm is exploited to search the feature space of extracted features) to obtained the optimal feature subset for cattle recognition based on muzzle point feature points. The evolution of experimental results has been driven by a fitness function which is defined in terms of class separation. The classifier performance and the length of selected feature vector were considered for performance evaluation using the own prepared database of muzzle pattern image of cattle.

Experimental results illustrate the superiority of the BFO-based feature selection algorithm in generating the good recognition accuracy with the minimal set of selected features. It founds out that the underlying bacterial foraging optimization (BFO) principle and the PSO swarm optimization integrated into evolutionary computational algorithms to provide a better search strategy for finding optimal feature vectors for muzzle image recognition. This proposed approach, in comparison to other object recognition systems so far is more advantageous and result-oriented because it cannot work on presumptions, it is unique and provides fast and contactless authentication. In the future work, this proposed approach can be applied for selecting optimal feature subset of biological database such as gene expression and protein sequence database. This proposed approach can also be prolonged to hybridize advanced swarm intelligence techniques such as Firefly black hole, Flower pollination, Gray wolf Optimizer (GWO), bees colony optimization, fish swarm, cuckoo search optimization. Moreover, these swarm intelligence algorithms can provide to improve the recognition accuracy of muzzle pattern of cattle by applying the different feature extraction techniques and swarm intelligence optimization with chaotic improved PSO and BFO algorithms. It can improve the recognition process using muzzle image to provide better solution to false insurance claimed problem through recognition system vastly. It can also help to animal biometrics by enhancement of cattle traceability, outbreak of diseases control and health management of animal. Furthermore, it may provide a better automatic animal system to recognize the animal across border with minimum false rejection rate (FRR) and better False Acceptance rate (FAR). Finally, swarm optimization methods namely bacterial foraging optimization (BFO) and particle swarm optimization (PSO) can provide a successful platform for the design and development of robust recognition system for muzzle pattern of cattle by selecting the discriminant features from given database.

References

1. Abbass HA (2001) MBO: marriage in honey bees optimization-A haplometrosis polygynous swarming approach. Proc IEEE Congr Evol Comput 1:207–214
2. Aihara K, Takabe T, Toyoda M (1990) Chaotic neural networks. Phys Lett A 144(6):333–340
3. Anderssen RS, Jennings LS, Ryan DM (1972) Optimization. Cvijovic', St. Lucia, Australia
4. Angeline PJ (1998) Evolutionary optimization versus particle swarm optimization: philosophy and performance differences. In: Evolutionary programming VII. Springer, Berlin, Heidelberg, pp 601–610
5. Angeline PJ (1998) Evolutionary optimization versus particle swarm optimization: philosophy and performance differences. In: Evolutionary programming VII. Springer, pp 601–610
6. Azar AT, Banu PKN, Inbarani HH (2013) PSORR—an unsupervised feature selection technique for fetal heart rate. In: 5th International conference on modelling, identification and control (ICMIC 2013), 31 Aug, 1–2 Sept 2013, Egypt
7. Azar AT, Hassanien AE (2015) Dimensionality reduction of medical big data using neural-fuzzy classifier. Soft Comput 19(4):1115–1127. doi:10.1007/s00500-014-1327-4
8. Azar AT (2014) Neuro-fuzzy feature selection approach based on linguistic hedges for medical diagnosis. Int J Modell Ident Control (IJMIC) 22(3):195–206. doi:10.1504/IJMIC.2014.065338
9. Azar AT, Vaidyanathan S (2015) Handbook of research on advanced intelligent control engineering and automation. Advances in Computational Intelligence and Robotics (ACIR) book series. IGI Global, USA

10. Azar AT, Vaidyanathan S (eds) (2015) Chaos modeling and control systems design. Springer International Publishing
11. Banks A, Vincent J, Anyakoha C (2007) A review of particle swarm optimization. Part I: background and development. Nat Comput 6(4):467–484
12. Banks A, Vincent J, Anyakoha C (2008) A review of particle swarm optimization. Part II: hybridisation, combinatorial, multicriteria and constrained optimization, and indicative applications. Nat Comput 7(1):109–124
13. Baranov AS, Graml R, Pirchner F, Schmid DO (1993) Breed differences and intrabreed genetic variability of dermatoglyphic pattern of cattle. J Anim Breed Genet 110(1–6):385–392
14. Blackwell T (2007) Particle swarm optimization in dynamic environments. In: Evolutionary computation in dynamic and uncertain environments. Springer, Berlin, Heidelberg, pp 29–49
15. Boussaïd I, Lepagnot J, Siarry P (2013) A survey on optimization metaheuristics. Inf Sci 237:82–117
16. Castillo O, Melin P (2012) Optimization of type-2 fuzzy systems based on bio-inspired methods: a concise review. Inf Sci Elsevier 205:1–19
17. Clerc M (2006) Particle swarm optimization. ISTE, London, UK
18. Clerc M, Kennedy J (2002) The particle swarm-explosion, stability, and convergence in a multidimensional complex space. IEEE Trans Evol Comput 6(1):58–73
19. Cover Thomas M, Hart Peter E (1967) Nearest neighbor pattern classification. IEEE Trans Inf Theory 13(1):21–27
20. Cvijovic D, Klinowski J (1995) Taboo search: an approach to the multiple-minima problem. Science 267:664–666 (University of Queensland Press)
21. Das S et al (2009) Bacterial foraging optimization algorithm: theoretical foundations, analysis, and applications. Foundations of computational intelligence, vol 3. Springer, Berlin, Heidelberg, pp 23–55
22. de Oca MAM, Stutzle T, Birattari M, Dorigo M (2009) Frankenstein's PSO: a composite particle swarm optimization algorithm. IEEE Trans Evol Comput 13:1120–1132
23. Dekkers A, Aarts E (1991) Global optimizations and simulated annealing. Math Program 50:367–93
24. Dorigo M, Maniezzo V, Colorni A (1996) Ant system: optimization by a colony of cooperating agents. IEEE Trans Syst Man Cybern Part B: Cybern 26(1):29–41
25. dos Santos Coelho L, Mariani VC (2008) Use of chaotic sequences in a biologically inspired algorithm for engineering design optimization. Expert Syst Appl 34:1905–1913
26. Eberhart RC, Shi Y (2000) Comparing inertia weights and constriction factors in particle swarm optimization. In: Proceedings of the 2000 congress on evolutionary computation, vol 1, pp 84–88
27. Eberhart RC, Shi Y (2001) Particle swarm optimization: developments, applications and resources. In: Proceedings of congress on evolutionary computation, Seoul, Korea, pp 81–86
28. Er MJ, Chen W, Wu S (2005) High-speed face recognition based on discrete cosine transform and RBF neural networks. IEEE Trans Neural Netw 16(3):679–691
29. Erol Osman K, Eksin I (2006) A new optimization method: big bang-big crunch. Adv Eng Softw 37(2):106–111
30. Geem ZW, Kim JH, Loganathan GV (2001) A new heuristic optimization algorithm: harmony search. Simulation 76(2):60–68
31. Goldberg DE (1998) Genetic algorithms in search, optimization, and machine learning. Addison Wesley, MA
32. Goldberg DE, Holland JH (1988) Genetic algorithms and machine learning. Mach Learn 3(2):95–99
33. Grosan C, Abraham A, Chis M (2006) Swarm intelligence in data mining. Springer, Berlin, Heidelberg
34. Hafed ZM, Levine MD (2001) Face recognition using discrete cosine transform. Int J Comput Vision 43(3):167–188
35. Hassanien AE, Azar AT, Snasel V, Kacprzyk J, Abawajy JH (2015) Big data in complex systems: challenges and opportunities. Studies in big data, vol 9. Springer-Verlag GmbH, Berlin/Heidelberg. ISBN: 978-3-319-11055-4

36. Hatamlou A (2013) Black hole: a new heuristic optimization approach for data clustering. Inf Sci 222:175–184
37. Ho S-Y, Lin H-S, Liauh W-H, Ho S-J (2008) OPSO: orthogonal particle swarm optimization and its application to task assignment problems. IEEE Trans Syst Man Cybern Part A: Syst Hum 38:288–298
38. Inbarani HH, Bagyamathi M, Azar AT (2015) A novel hybrid feature selection method based on rough set and improved harmony search. Neural Comput Appl 1–22. doi:10.1007/s00521-015-1840-0
39. Inbarani HH, Banu PKN, Azar AT (2014) Feature selection using swarm-based relative reduct technique for fetal heart rate. Neural Comput Appl 25(3–4):793–806. doi:10.1007/s00521-014-1552-x
40. Jain AK, Flynn P, Ross AA (2008) Handbook of biometrics. Springer Publication, New York. ISBN-13: 978-0-387-71040-2
41. Jain AK, Pankanti S, Prabhakar S, Hong L, Ross A (2004). Biometrics: a grand challenge. In Proceedings of the 17th IEEE International Conference on Pattern Recognition (ICPR, 2004), vol 2, pp 935–942
42. Jakhar R, Kaur N, Singh R (2011) Face recognition using bacteria foraging optimization-based selected features. Int J Adv Comput Sci Appl 1(3)
43. Jothi G, Inbarani HH, Azar AT (2013) Hybrid tolerance-PSO based supervised feature selection for digital mammogram images. Int J Fuzzy Syst Appl (IJFSA) 3(4):15–30
44. Kao Y-T, Zahara E, Kao IW (2008) A hybridized approach to data clustering. Expert Syst Appl 34:1754–1762
45. Kao Y-T, Zahara E (2008) A hybrid genetic algorithm and particle swarm optimization for multimodal functions. Appl Soft Comput 8:849–857
46. Kapitaniak T (1995) Continuous control and synchronization in chaotic systems. Chaos Solitons Fractals 6:237–244
47. Kennedy J (2010) Particle swarm optimization. In: Encyclopaedia of machine learning. Springer, US, pp 760–766
48. Kennedy J, Eberhart RC, Shi Y (2001) Swarm intelligence. Morgan Kaufmann Publishers, San Francisco
49. Kennedy J, Eberhart R (1995) Particle swarm optimization. Proc IEEE Int Conf Neural Netw 4(1995):1942–1948
50. Kirby M, Sirovich L (1990) Application of the Karhunen-Loeve procedure for the characterization of human faces. IEEE Trans Pattern Anal Mach Intell 12(1):103–108
51. Kirkpatrick S (1984) Optimization by simulated annealing: quantitative studies. J Stat Phys 34(5–6):975–986
52. Kumar S, Datta D, Singh SK (2015) Black hole algorithm and its applications. In: Computational intelligence applications in modeling and control. Springer International Publishing, pp 147–170
53. Kumar S, Datta D, Singh SK (2015) Swarm intelligence for biometric feature optimization. Handbook of research on swarm intelligence in engineering, vol 147
54. Kumar S, Sadhya D, Singh D, Singh SK (2014) Cloud security using face recognition. Handbook of research on securing cloud-based databases with biometric applications, vol 298
55. Lian Z, Gu X, Jiao B (2008) A novel particle swarm optimization algorithm for permutation flow-shop scheduling to minimize makespan. Chaos Solitons Fractals 35:851–861
56. Li B, Jiang WS (1998) Optimizing complex functions by chaos search. Cybern Syst 29: 409–419
57. Lu Z, Shieh LS, Chen GR (2003) On robust control of uncertain chaotic systems: a sliding-mode synthesis via chaotic optimization. Chaos Solitons Fractals 18:819–827
58. Matos FM, Batista LV, Poel J (2008) Face recognition using OCT coefficients selection. In: Proceedings of the ACM symposium on applied computing, pp 1753–1757
59. May R (1976) Simple mathematical models with very complicated dynamics. Nature 261:459–67

60. Minagawa H, Fujimura T, Ichiyanagi M, Tanaka K (2002) Identification of beef cattle by analysing images of their muzzle patterns lifted on paper. Publ Japan Soc Agric Inf 8: 596–600
61. Mirjalili SM, Lewis A (2014) Grey wolf optimizer. Adv Eng Softw 69:46–61
62. Mirzayans T, Parimi N, Pilarski P, Backhouse C, Wyard-Scott L, Musilek P (2005) A swarm-based system for object recognition. Neural Netw World 15(3):243–255
63. Mpiperis I, Malassiotis S, Petridis V, Strintzis MG (2008) 3D facial expression recognition using swarm intelligence. In: IEEE International Conference on Acoustics, speech and signal processing, 2008. ICASSP 2008. IEEE, pp 2133–2136
64. Ozcan E, Mohan CK (1999) Particle swarm optimization: surfing the waves. In: Proceedings of the 1999 IEEE congress on evolutionary computation (CEC 99), vol 3
65. Pant M, Thangaraj R, Abraham A (2009) Particle swarm optimization: performance tuning and empirical analysis. Foundations of computational intelligence, vol 3. Springer, Berlin, Heidelberg, pp 101–128
66. Passino KM (2002) Biomimicry of bacterial foraging for distributed optimization and control. IEEE Control Syst Mag 22:52–67
67. Pecora L, Carroll T (1990) Synchronization in chaotic systems. Phys Rev Lett 64:821–4
68. Qian W, Yang Y, Yang N, Li C (2008) Particle swarm optimization for SNP haplotype reconstruction problem. Appl Math Comput 196:266–272
69. Samra AS, El Taweel Gad Allah S, Ibrahim RM (2003) Face recognition using wavelet transform, fast Fourier transform and discrete cosine transform. In: IEEE 46th midwest symposium on circuits and systems, vol 1, pp 272–275
70. Shi Y, Eberhart RC (1998) A modified particle swarm optimizer. In: Proceedings of IEEE international conference on evolutionary computation. Anchorage, USA, pp 69–73
71. Shi Y, Eberhart RC (1999) Empirical study of particle swarm optimization. In: Proceedings of IEEE international congress on evolutionary computation, Washington, DC, pp 1945–50
72. Teodorović D (2009) Bee colony optimization (BCO). In: Innovations in swarm intelligence. Springer, Berlin, Heidelberg, pp 39–60
73. Turk MA, Pentland AP (1991) Face recognition using eigenfaces. In: Proceedings IEEE computer society conference on computer vision and pattern recognition (CVPR, 91), pp 586–591
74. Vaidyanathan S, Azar AT (2015) Analysis and control of a 4-D novel hyperchaotic system. In: Azar AT, Vaidyanathan S (eds) Chaos modeling and control systems design. Studies in computational intelligence, vol 581. Springer-Verlag GmbH, Berlin/Heidelberg, pp 19–38. doi:10.1007/978-3-319-13132-0
75. Vaidyanathan S, Azar AT (2015) Analysis, control and synchronization of a nine-term 3-D novel chaotic system. In: Azar AT, Vaidyanathan S (eds) Chaos modeling and control systems design. Studies in computational intelligence, vol 581, Springer-Verlag GmbH, Berlin/Heidelberg, pp 3–17. doi:10.1007/978-3-319-13132-0_1
76. Vaidyanathan S, Azar AT (2015) Anti-synchronization of identical chaotic systems using sliding mode control and an application to Vaidyanathan-Madhavan chaotic systems. In: Azar AT, Zhu Q (eds) Advances and applications in sliding mode control systems. Studies in computational intelligence book series, vol 576. Springer-Verlag GmbH, Berlin/Heidelberg, pp 527–547. doi:10.1007/978-3-319-11173-5_19
77. Vaidyanathan S, Azar AT (2015) Hybrid synchronization of identical chaotic systems using sliding mode control and an application to Vaidyanathan Chaotic systems. In: Azar AT, Zhu Q (eds) Advances and applications in sliding mode control systems. Studies in computational intelligence book series, vol 576. Springer-Verlag GmbH, Berlin/Heidelberg, pp 549–569. doi:10.1007/978-3-319-11173-5_20
78. Vaidyanathan S, Idowu BA, Azar AT (2015) Backstepping controller design for the global chaos synchronization of Sprott's Jerk systems. In: Azar AT, Vaidyanathan S (eds) Chaos modeling and control systems design. Studies in computational intelligence, vol 581. Springer-Verlag GmbH, Berlin/Heidelberg, pp 39–58. doi:10.1007/978-3-319-13132-0_3

79. Wang L, Zheng DZ, Lin QS (2001) Survey on chaotic optimization methods. Comput Technol Autom 20:1–5
80. Wang L (2001) Intelligent optimization algorithms with applications. Tsinghua University & Springer Press, Beijing
81. Yang XS (2010) A new metaheuristic bat-inspired algorithm. In: Nature inspired cooperative strategies for optimization (NICSO 2010). Springer, Berlin, Heidelberg, pp 65–74
82. Yang XS (2010) Firefly algorithm, stochastic test functions and design optimisation. Int J Bio-Inspired Comput. 2(2):78–84
83. Yang XS, Deb S (2010) Engineering optimisation by cuckoo search. Int J Math Model Numer Optim 1(4):330–343
84. Yang XS (2012) Flower pollination algorithm for global optimization. In: Unconventional computation and natural computation. Springer, Berlin, Heidelberg, pp 240–249
85. Yu M, Yan G, Zhu QW (2006) New face recognition method based on dwt/dct combined feature selection. In: Proceedings of IEEE international conference on machine learning and cybernetics, pp 3233–3236
86. Zhan ZH, Zhang J, Li Y, Chung HS (2009) Adaptive particle swarm optimization. IEEE Trans Syst Man Cybern Part B: Cybern 39:1362–1381
87. Zhang H, Sun G (2002) Feature selection using tabu search method. Pattern Recogn 35: 701–711
88. Zhao W, Chellappa R, Phillips PJ, Rosenfeld A (2003) Face recognition: a literature survey. ACM Comput Surv (CSUR) 35(4):399–458
89. Zhu Q, Azar AT (2015) Complex system modelling and control through intelligent soft computations. Studies in fuzziness and soft computing, vol 319. Springer-Verlag, Germany. ISBN: 978-3-319-12882-5

Control of Complex Systems Using Self Organizing Fuzzy Controller

Jitendra Kumar, Vineet Kumar and K.P.S. Rana

Abstract Robotic manipulators are complex multi-input multiple output systems finding lots of application in industries. Controlling such a complex system always has been an area of research owing to the inherent nonlinearities. In this work, a comparative study of Fuzzy Proportional Integral Derivative (FPID) and Self Organizing Fuzzy Controller (SOFC), applied for trajectory tracking and disturbance rejection to a two link planar rigid robotic manipulator with end-effector has been presented. Two layers of fuzzy logic controller (FLC) have been used to design SOFC in which second layer was used for adaptive mechanism and Takagi-Sugeno-Kang method has been used for inference mechanism in both the control schemes. Genetic algorithm (GA) has been used to optimize the gains of FPID and SOFC controllers for minimum Integral of Absolute Error (IAE) and Integral of Absolute Change in Controller Output. Simulation results revealed that SOFC outperformed FPID controller in both servo and regulatory mode. SOFC has offered 23.43 %, 60.50 % and 36.20 % improvements in IAE link-1, IAE link-2 and cost function for trajectory tracking and 35.21 %, 51.34 % and 39.63 % improvements in IAE link-1, IAE link-2 and cost function for disturbance rejection respectively.

Keywords Fuzzy PID controller · Self organizing fuzzy controller · Robotic manipulator · Genetic algorithm

---

J. Kumar (✉) · V. Kumar (✉) · K.P.S. Rana (✉)
Division of Instrumentation and Control Engineering,
Netaji Subhas Institute of Technology, New Delhi 110078, India
e-mail: singhjitendra86@gmail.com

V. Kumar
e-mail: vineetkumar27@gmail.com

K.P.S. Rana
e-mail: kpsrana1@gmail.com

A.T. Azar and S. Vaidyanathan (eds.), *Advances in Chaos Theory and Intelligent Control*, Studies in Fuzziness and Soft Computing 337,
DOI 10.1007/978-3-319-30340-6_31

1 Introduction

Intelligent techniques have offered new paradigm of research and challenges for scholars to deal with nonlinear, time-varying, and uncertain systems, which could not be properly catered by the conventional methods. These include approaches to design, optimization, and control of various systems without requiring exact mathematical models, in a way similar to how human works and involves many fields such as fuzzy logic, artificial neural network (ANN), evolutionary strategy like genetic algorithm (GA) and their hybrid combinations etc. Effective intelligent systems can be constructed by combining appropriate soft computing techniques for particular problem to be solved.

Classical adaptive control techniques like gain scheduling, self-tuning regulator, model reference adaptive control etc., were designed in 1960s to control complex systems. These classical control techniques highly depend on mathematical model of the system but if an accurate system analytical model is not available, not possible to obtain, or too complicated to use for control purposes, intelligent techniques are the best options. Out of different intelligent techniques, fuzzy logic control (FLC) scheme can be applied for efficient and systematic control to a good extent. Fuzzy logic was proposed by Zadeh [26–33] and FLC was firstly proposed by Mamdani [16, 17] and the related works have later been elaborated by Graham [8], Graham and Newell [9, 10]. In industrial application, firstly the FLC was used by King and Mamdani [13] for a dynamic complex and poorly defined processes where modeling difficulties and lack of suitable measurements make the control strategy manual. Later, Holmblad and Ostergaard [12] used FLC to control a cement kiln satisfactorily. In both these investigations, it was stated that to deal with uncertain and nonlinear system, FLC proved to be a better technique than conventional control methods. A survey on classical Proportional Integral and Derivative (PID) as well as fuzzy PID (FPID) controllers have been presented by Kumar, Nakra and Mittal [14] where it has been realized that classical PID controllers are effective for linear systems but not suitable for nonlinear systems.

Modeling and control of a chaotic system by applying different intelligent techniques have been presented by Azar and Vaidyanathan [4] where a feedback FLC has been designed and analyzed. Further to this perspective, applications areas of advanced intelligent systems in modeling and control of multi-disciplinary types of complex processes have been like electrical, electronic, chemical, mechanical, aerospace, as explained by Azar and Vaidyanathan [5, 6]. Windup control, waste management control, security issues control, biomedical applications and its control by using different soft computing methods have been presented by Zhu and Azar [34]. Different intelligent techniques which involve ANN, fuzzy logic, combination of ANN with fuzzy logic, combination of fuzzy with conventional controllers etc. have also been explained by Azar [1] where predictive modeling of dialysis variables have been found out in order to estimate the equilibrated dialysis dose. Azar and Zhu [7] presented nice works on control of nonlinear, uncertain and coupled systems like furuta pendulum, robot arms, internal combustion engines etc. using sliding mode control tuned by GA. Further, Hassanien, Tolba and Azar [11] presented use of intelligent techniques in machine learning to recognize the Arabic text, isolated printed Arabic character, sign language recognition in Arabic language etc. Mishra, Kumar and Rana [20] explored a novel and interesting application of fuzzy logic to compensate stiction in pneumatic control valves to suppress stiction based oscillations. The developed controller was compared with a conventional proportional integral controller for set point tracking and disturbance rejection on a real world plant for flow rate control and it was found that the proposed controller was able to suppress stiction based limit cycles in spite of varying nature of stiction magnitude in control valve. Based on the above facts, it could be easily concluded that intelligent methods like fuzzy logic is used not only in control applications but also in other fields.

On the other hand, there are some limitations of FLC, such as difficulty to obtain a reliable linguistic model of the operator's control strategy. Also some significant process details might be outside the operator's knowledge and due to these limitations FLC may fail to give desired performance and therefore an adaptive scheme needs to be incorporated to this so as to give a corrective action in run time to the plant and nullify the corresponding error. This scheme was initially proposed by Mamdani [17, 18], Mamdani, Procyk, and Baaklini, [19], Procyk, and Mamdani, [22] Yamazaki, and Mamdani [25] and the controller was called Self Organizing Fuzzy Controller (SOFC). A number of schemes for SOFC have been developed but the original attempt was made by Mamdani and his co-workers.

In control problems, main objective is to run the plant in such a way that it could track the reference set point as well as reject the undesired disturbances. Classical PID, Fuzzy PID (FPID) as well as SOFC are the type of controllers which are not model dependent i.e. the basic structure of controller remains the same even if the plant model is changed. But the gains associated with these controllers should be tuned properly for perfect control. Ziegler-Nichols and Cohen-Coon techniques are popularly known to tune the classical PID controllers, [24]. In the same context, Robust Internal Model Control PID for cascade control systems tuning with gain and phase margin specifications have been designed by Azar and Serrano [2]. To reduce the effect of final control element saturation, implementation and the tuning of anti-windup PID controller has been also presented by Azar and Serrano [3]. It may be noted that there is no well-defined technique to tune the intelligent controllers, even today. Also these classical tuning methods have designed to find the gains to satisfy a defined performance index at a time. These methods have not been defined to tune the classical controllers and to find the optimized gains for multi-objective cost function. Several efficient optimization techniques are available these days which could be used to tune the intelligent controllers for complex plant yielding customises performance. Controller tuning through optimization also helps the user to attain multiple performance criterion at the same time. Some of the recent applications of optimization techniques in controller tuning for customize performances are as follows. Sharma, Rana and Kumar [23] have controlled a two link planar robotic manipulator using fractional order PID and fractional order FPID controllers tuned by Cuckoo Search Algorithm (CSA). Mishra, Kumar and Rana [21] have applied GA to tune the parameters of a self-tuning FPID controller and its fractional order self-tuning FPID controller to control the composition of top and bottom products of a distillation column. Authors have shown that fractional order self-tuning FPID controller provided better set point tracking, disturbance rejection and uncertainty handling capability than simple integer order self-tuning FPID.

This chapter explores the capability of SOFC to control a 2-link planar rigid robotic manipulator with end-effector for reference trajectory tracking and disturbance rejection. A slight modification has been made in the internal structure of SOFC but main scheme has been kept as given by Mamdani et al. [19]. The considered plant is a highly nonlinear, time-varying, uncertain, coupled multi-input multi-output system. Classical adaptive controller design is a tedious task for this kind of plant and SOFC could be easily designed because of its design simplicity. The robustness of this control scheme has been tested and the results have also been compared to simple FPID controller.

This chapter has been organized as follows. In Sect. 2, brief description of SOFC is presented. Two layers of SOFC structure, FPID construction details, their membership functions (MF), and rule base for both the layers have also been explained in this section. The plant and its mathematical-model have been presented in Sect. 3. After that, simulation results have been shown graphically and their comparison has been also given in tabular form in Sect. 4, where the superiority of SOFC has been demonstrated over the FPID controller. Finally, conclusions are drawn in Sect. 5 and future research directions are also outlined.

2 SOFC Description

SOFC is a type of intelligent adaptive controller, having the capability of generating and modifying control rules based on the basis of existing error which occur due to variations of system's parameters, disturbance and noise. The generation and modification of rules can be achieved by different intelligent techniques like: fuzzy logic, ANN, GA etc. The control rules modification in each individual control action makes the major contribution to the present performance. A fuzzy logic system based SOFC has been explained and shown in Fig. 1 where two layers of FLC have been incorporated. The first layer of FLC is a simple structure of fuzzy PID controller. The output of fuzzy PID controller is the summation of output of fuzzy PD and fuzzy PI controller. The first layer fuzzy PID controller is described in Sect. 2.1. In Fig. 1, k_p and k_d are the gains of error and rate of change of error. Similarly k_{upd} and k_{upi} are the gains of FPD and FPI outputs of SOFC controller respectively. In this scheme, the robustness in SOFC comes from second layer. It changes the rule base of current control action on the basis of system performance so that the control action will be more aggressive. In this design strategy, the second layer is also designed by a FLC where rule base is different from first layer. The rule base for both the layers is given in the Tables 1 and 2.

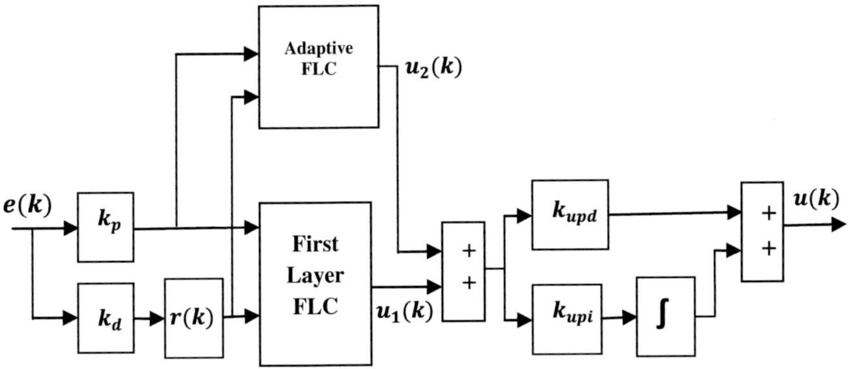

Fig. 1 SOFC block diagram

Table 1 Fuzzy rules for first layer

de	e						
	NB	NM	NS	Z	PS	PM	PB
NB	−1.0	−1.0	−0.8	−0.6	−0.4	−0.2	0
NM	−1.0	−0.8	−0.6	−0.4	−0.2	0	0.2
NS	−0.8	−0.6	−0.4	−0.2	0	0.2	0.4
Z	−0.6	−0.4	−0.2	0	0.2	0.4	0.6
PS	−0.4	−0.2	0	0.2	0.4	0.6	0.8
PM	−0.2	0	0.2	0.4	0.6	0.8	1.0
PB	0	0.2	0.4	0.6	0.8	1.0	1.0

Table 2 Fuzzy rules for second layer FLC

de	e						
	NB	NM	NS	Z	PS	PM	PB
NB	−1.0	−1.0	−0.8	−0.6	0	0	0
NM	−1.0	−0.8	−0.6	−0.3	0	0	0
NS	−0.8	−0.6	−0.3	−0.1	0	0	0
Z	−0.6	−0.3	−0.1	0	0.1	0.3	0.6
PS	0	0	0	0.1	0.3	0.6	0.8
PM	0	0	0	0.3	0.6	0.8	1.0
PB	0	0	0	0.6	0.8	1.0	1.0

2.1 First Layer of SOFC

The first layer of SOFC is a basic two–input and a single-output fuzzy PD + fuzzy PI i.e. FPID controller as shown in Fig. 1. If the adaptive layer is removed, it will become a simple fuzzy PID controller. The internal generic structure of FLC is shown in Fig. 2.

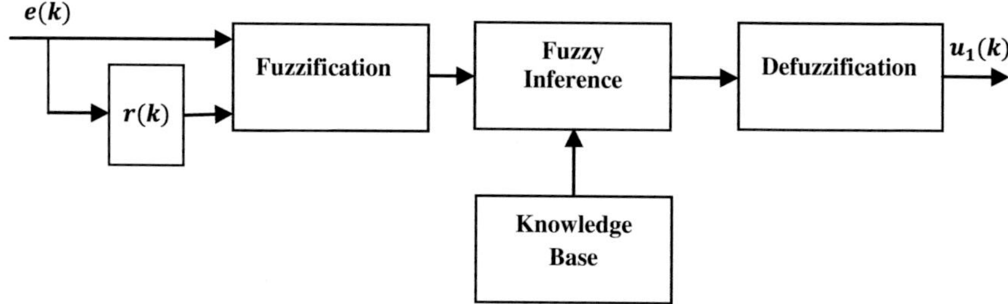

Fig. 2 Basic fuzzy logic control scheme

The error and rate of change of error are actually in the form of linguistic variables. The error and rate of change of error are fuzzified which can be further understood by the fuzzy control system as its inputs. Further, error and rate of change of error are fuzzified using seven equally distributed triangular MFs. The two inputs are scaled by multiplying optimized scaling factors for the customized system performance. In the Takagi-Sugeno-Kang (TSK) scheme, constant values are considered as consequent part of rule base. Scaling factors of error and rate of change of error can change the controller output abruptly. If the gains of controller are not tuned properly then it can make the system unstable. It is always preferred that the scaling gains of the controller should be optimized for optimum results. Initially manual tuning was used for this purpose, but now different optimization algorithms are being used. In this work, the optimizations of the scaling factors have been performed by GA because it is a nature based algorithm and its toolbox is available in MATLAB as a part of standard package. Error and rate of change of error have been defined as:

$$e(k) = ref(k) - y(k) \tag{1}$$

$$r(k) = \frac{e(k) - e(k-1)}{T} \tag{2}$$

where,
$ref(k)$ is the desired set point at time instant k.
$y(k)$ is the output of the plant.
$e(k)$ is the error.
$r(k)$ is the rate of change of error.
T is the sampling time.

After multiplying with the optimized gains, error and rate of change of error will transferred to the fuzzification block. The working of fuzzification is described in Sect. 2.1.1.

2.1.1 Fuzzification

Seven triangular MF have been defined over the full ranges of input space which linguistically describe the variable's universe of discourse as shown in Fig. 3. In this

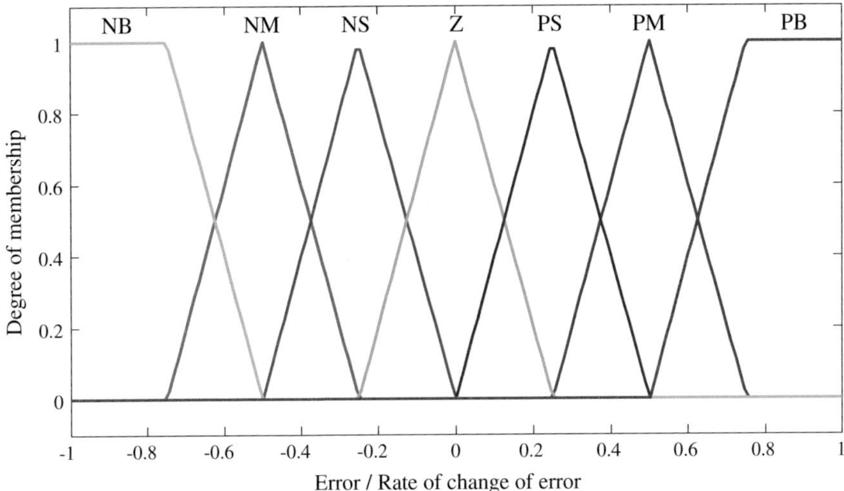

Fig. 3 Membership functions

control scheme, the universe of discourse of inputs have been taken between -1 to 1 and all the crisp inputs have been normalized by a corresponding optimized gain so that the crisp inputs always fall within this range. In case the input value is less than -1 or greater than 1 then the corresponding extreme MFs will be used to maximum strength. Name of MFs has been shown in Fig. 3 where NB stands for negative big, NM stands for negative medium, NS stands for negative small, Z stands for zero, PS stands for positive small, PM stands for positive medium, PB stands for positive big. The left and right halves of the triangular MFs for each linguistic variable are chosen to provide MFs with equal overlaps with adjacent MFs. MFs are defined to be symmetrical and equally spaced with an equal area so that each set can be described by its central value.

2.1.2 Rule Base

Selection of the number of MFs, type of MFs and their shapes are based on the knowledge of process and the intuition. As the number of MFs is increased, the resolution of controller will also be increase but number of rule base and the complexity of that FLC will be increased and which will make the controller mechanism relatively slow. So it's a task of control engineer to optimize the number and shape of MFs for a particular plant. As in this chapter, 7 MFs have been taken for both the inputs, so 49 rules have been generated. The main rule applied in the SOFC scheme is to define the fuzzy partitions over the entire operating ranges of the process variable and on the basis of that it could be understood that enough number of MFs have been taken here to provide adequate resolution. Fuzzy rules for first layer and second layer is shown

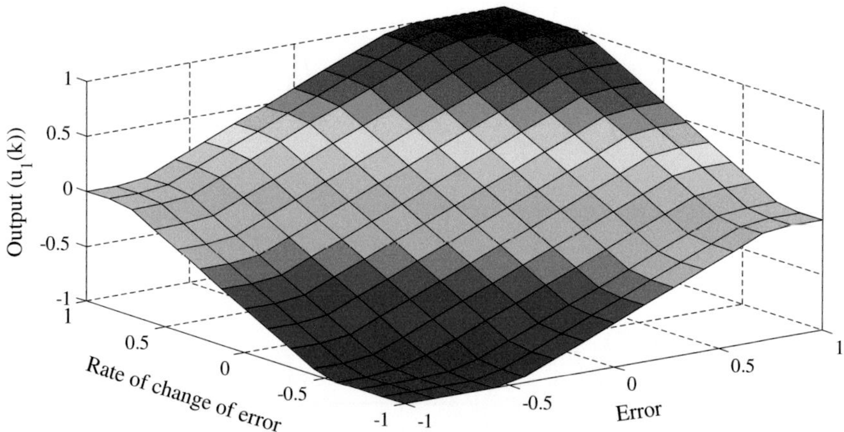

Fig. 4 Surface plot for first layer FLC

in Table 1 and Table 2, respectively. The three dimensional input-output surface plot for first layer is shown in Fig. 4. Fuzzy control rules are composed of a series of fuzzy if-then rules where the conditions are linguistic variables and the consequences are their corresponding values.

This collection of fuzzy rules simplifies input-output relation of the system in linguistic form as:

R: If $e(k)$ is E_i AND $r(k)$ is E_j, THEN $u_c(k)$ is $U_n(i,j)$.
where,

$e(k)$ is the error $e(k)$ at time instant k.

$r(k)$ is the rate of change of the error at time k.

$u_c(k)$ is the consequent part at time k.

E_i is the linguistic variable of $e(k)$.

E_j is the linguistic variable of $r(k)$.

$U_n(i, j)$ is the corresponding consequent part of $e(k)$ and $r(k)$.

2.1.3 Defuzzification

TSK fuzzy implication is adopted where fuzzy control law of the first layer fuzzy controller is shown as:

$$u_1(k) = \sum_{i,j} \frac{\{[\mu_{Ei}(e(k)) \bigwedge \mu_{Ej}(r(k))].U_{n(i,j)}\}}{[\mu_{Ei}(e(k)) \bigwedge \mu_{Ej}(r(k))]} \tag{3}$$

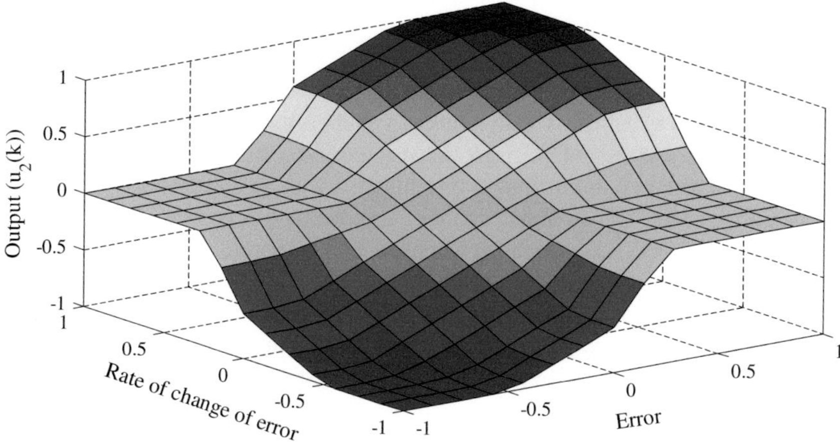

Fig. 5 Surface plot for second layer FLC

where, $\mu_{Ei}(e(k)) \bigwedge \mu_{Ej}(r(k))$ denotes the intersection of fuzzy sets $\mu_{Ei}(e(k))$ and $\mu_{Ej}(r(k))$, which is defined as the minimum operation between two fuzzy sets:
$\mu_{Ei}(e(k)) \bigwedge \mu_{Ej}(r(k)) = \text{minimum}\ [\mu_{Ei}(e(k)), \mu_{Ej}(r(k))]$.

2.2 *Second Layer of SOFC*

In order to compensate for the effects of model uncertainties, unexpected disturbances, time-varying parameters, and sensor noises etc., the rule base in the first layer needs to be updated in real time to achieve desired performance. An adaptive FLC is embedded in the second layer, and is also designed as a two-input and a single-output FPD controller in position form. The inputs and the scaling gains are same as the inputs of the first layer fuzzy controller. The shape of MFs and linguistic variables are kept same as those of the first layer fuzzy controller for easy computation. IF-THEN rules of adaptive FLC are also same as first layer fuzzy controller but rule Table of adaptive FLC is different from the first layer FLC. The fuzzy rules of the adaptive FLC are formulated in terms of linguistic statement and can be transformed in to a lookup decision table as shown in Table 2. The lookup table can be treated like a human expert who makes the appropriate corrections of the first layer FLC based on the current value of the error and rate of change of error. Based on the effectiveness of the last control action on the error reduction, the contribution of each activated fuzzy rule in the last step can be evaluated and modified accordingly. In this case, TSK fuzzy implication is also adopted as like in first layer FLC. The surface plot for second layer FLC is shown in Fig. 5.

3 Plant Description

A two link planar rigid robotic manipulator system with two degree of freedom is described in Fig. 6 as bellow. The first link of system is mounted on a rigid base with frictionless hinge and second link is mounted at the end of first link with a frictionless ball bearing. The end point of link two is attached with an end-effector and during the operation the mass of an object could be gripped by this end-effector. There are two inputs; torque τ_1 and τ_2 have been applied to this robotic system. First input is applied to link-1 and second input is applied to link-2 at joints. The outputs of the system are the joint position q_1 and q_2. Lagrangian equations in classical dynamics are used to develop the mathematical model of the system.

The dynamic equations of two link manipulator system have been taken from [15] which are as:

$$\begin{bmatrix} M_{11} & M_{12} \\ M_{21} & M_{22} \end{bmatrix} \begin{bmatrix} \ddot{q}_1 \\ \ddot{q}_2 \end{bmatrix} + \begin{bmatrix} -h\dot{q}_2 & -h\dot{q}_1 - h\dot{q}_2 \\ h\dot{q}_1 & 0 \end{bmatrix} \begin{bmatrix} \dot{q}_1 \\ \dot{q}_2 \end{bmatrix} + \begin{bmatrix} g_1 \\ g_2 \end{bmatrix} + \begin{bmatrix} v_1\dot{q}_1 \\ v_2\dot{q}_2 \end{bmatrix} + \begin{bmatrix} k_1 sgn(\dot{q}_1) \\ k_2 sgn(\dot{q}_2) \end{bmatrix} = \begin{bmatrix} \tau_1 \\ \tau_2 \end{bmatrix} \tag{4}$$

where,

$$M_{11} = I_1 + I_2 + m_1 l_{c1}^2 + m_2(l_1^2 + l_{c2}^2 + 2l_1 l_{c2}\cos(q_2)) + m_3(l_1^2 + l_2^2 + 2l_1 l_2 \cos(q_2)) \tag{5}$$

$$M_{12} = I_2 + m_2(l_{c2}^2 + l_1 l_{c2}\cos(q_2)) + m_3(l_2^2 + l_1 l_2 \cos(q_2)) \tag{6}$$

$$M_{21} = M_{12} \tag{7}$$

$$M_{22} = I_2 + m_2 l_{c2}^2 + m_3 l_2^2 \tag{8}$$

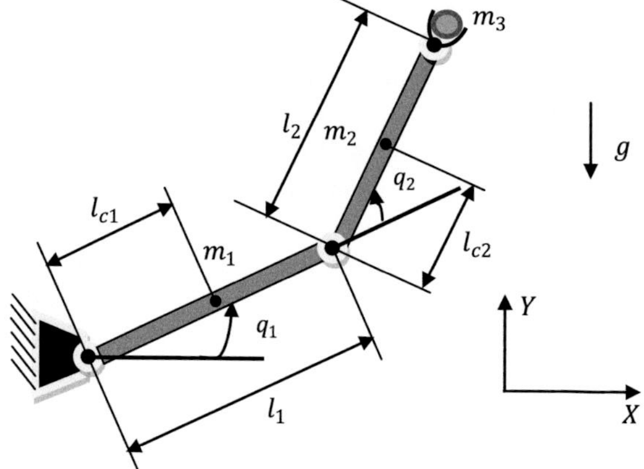

Fig. 6 Two link planar rigid robotic manipulator with end-effector

$$h = m_2 l_1 l_{c2} \sin(q_2) \tag{9}$$

$$g_1 = m_1 l_{c1} g \cos(q_1) + m_2 g(l_{c2} \cos(q_1 + q_2) + l_1 \cos(q_1 + q_2)) \tag{10}$$

$$g_2 = m_2 l_{c2} g \cos(q_1 + q_2) \tag{11}$$

Here, it can be shown that the matrix M is positive definite and due to that it will be always invertible. In this simulation work the following system parameters have been taken: $m_1 = 1.0$ kg, mass of link-1; $m_2 = 1.0$ kg, mass of link-2; $l_1 = 1.0$ m, length of link one; $l_2 = 1.0$ m, length of link two; $l_{c1} = 0.5$ m, distance from the joint of link-1 from its centre of gravity; $l_{c2} = 0.5$ m, distance from the joint of link-2 from its centre of gravity (COG); $I_1 = 0.2$ kgm^2, centroidal inertia of link one (lengthwise); $I_2 = 0.2$ kgm^2, centroidal inertia of link two (lengthwise); $g = 9.8$ m/sec^2; $v_1 = v_2 = 0.1$, coefficient of viscous friction; $k_1 = k_2 = 0.1$, coefficient of dynamic friction. The mass m_3 represents the mass of object holding by end-effector and this simulation study m_3 has been taken as 0.5 kg.

4 Simulation Studies

In this section result obtained for reference trajectory tracking and disturbance rejection has been shown for both SOFC and FPID controller. The simulation has been done in MATLAB (R2012a) on a personal computer having Intel core^{TM} i5 processor, 3.33 GHz frequency, 4 GB RAM with a 32-bit operating system. The plant model equations, made in SIMULINK, have been solved by fourth order Runge-Kutta method. The sampling time was taken as 0.001 s during the simulation. The torque limitation for both the link has been taken as [−100 100] Nm. The cost function has been chosen as:

$$F = w_1 f_1 + w_2 f_2 \tag{12}$$

$w_1 = 0.9999$ and $w_2 = 0.0001$ are chosen as the weights for f_1 and f_2.

$$f_1 = \int |e_1|\, dt + \int |e_1|\, dt \tag{13}$$

Here, f_1 is the summation of Integral Absolute Error (IAE) for both the links.

$$f_2 = \int |\Delta\tau_1|\, dt + \int |\Delta\tau_1|\, dt \tag{14}$$

f_2 is the summation of Integral of Absolute Change in Controller Output (IACCO) for both the links. The main motive to consider the IACCO in the objective function is that it minimizes the rate of change in controller output. Without considering this term in the cost function, IAE will be minimized but controller output may show highly oscillatory behavior. Due to this, it may accelerate the wear and tear of final control element. In the current trajectory tracking problem, the reference trajectory $q_{r_1}(t)$ for link-1 and $q_{r_2}(t)$ for link-2 have been chosen as:

Table 3 List of parameters used for GA optimization

S. no.	GA parameters	Variants
1	Population type	Double vector
2	Population size	20
3	Creation function	Uniform
4	Scaling function	Rank
5	Selection function	Tournament
6	Tournament size	4
7	Mutation function	Adaptive feasible
8	Crossover function	Arithmetic

$q_{r_1}(t)$ for link-1,

$$q_{r_1}(t) = (1.5 * t^2 - 0.5 * t^3) \quad \text{for } 0 \leq t \leq 2s \tag{15}$$

$$q_{r_1}(t) = (-5.5 + 9 * t - 3.375 * t^2 + 0.375 * t^3) \quad \text{for } 2 \leq t \leq 4s \tag{16}$$

and $q_{r_1}(t)$ for link-2,

$$q_{r_2}(t) = (2.25 * t^2 - 0.75 * t^3) \quad \text{for } 0 \leq t \leq 2s \tag{17}$$

$$q_{r_2}(t) = (13 - 12 * t + 4.5 * t^2 - 0.5 * t^3) \quad \text{for } 2 \leq t \leq 4s \tag{18}$$

There is no any thumb rule to effectively choose the weights $w_1$and w_2 for multi-objective GA optimization task but it should be considered that the value of $w_1 f_1$ and $w_2 f_2$ must be approximately equal to attract sufficient attention during the optimization processes for both the functions. It may be noted that the summation of the value of w_1 and w_2 is kept as 1. Parameters for GA toolbox, for both the controllers, have been listed in Table 3. Tuned values of all the eight parameters of fuzzy controllers

Table 4 Optimized SOFC and FPID controller parameters

	Parameters	SOFC	FPID
For link-1	k_p^1	4.7624	4.9763
	k_d^1	1.6652	3.0476
	k_{upd}^1	865.860	637.980
	k_{upi}^1	821.277	993.136
For link-2	k_p^2	4.5480	4.9467
	k_d^2	0.2480	0.6762
	k_{upd}^2	914.070	689.020
	k_{upi}^2	1191.510	806.796

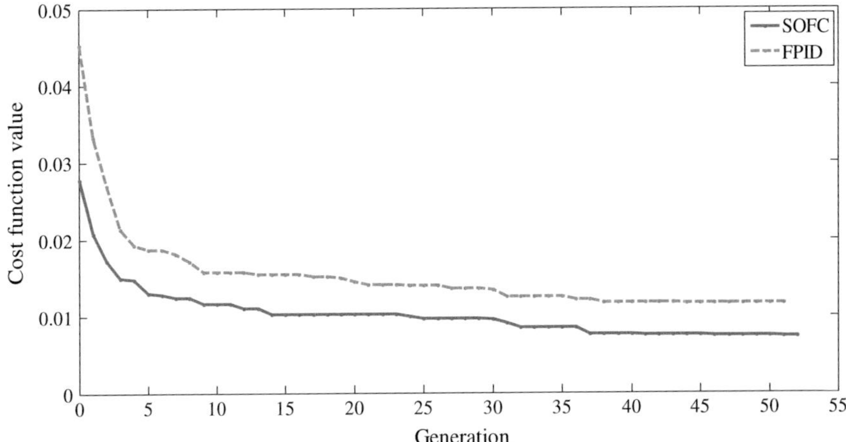

Fig. 7 Cost function value versus generation curve

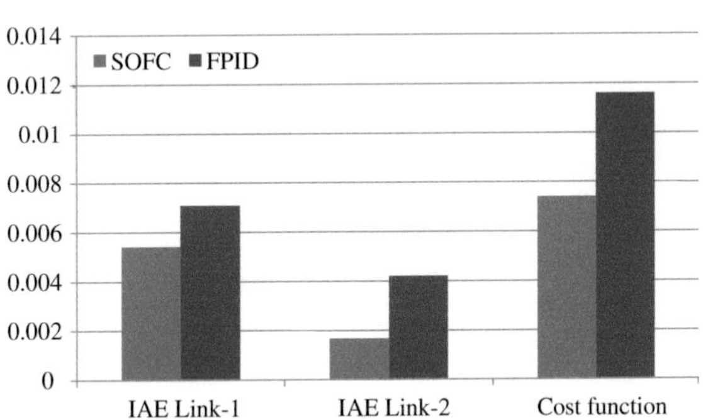

Fig. 8 Bar chart for reference trajectory tracking

are tabulated in Table 4. Figure 7 shows cost function value versus generation curve. In this Figure, it could be seen that the cost value for SOFC is lesser than FPID for all generations. The optimized cost value for SOFC is obtained as 0.0074 and 0.0116 for FPID. Both the controllers have been optimized for 100 generations but it could be seen from cost versus generation curve, Fig. 7, that the optimization task has been completed before 55 generations. The IAE of link-1 and link-2 and cost function are compared using barchart as shown in Fig. 8.

4.1 *Reference Trajectory Tracking*

Rigorous simulations were performed to evaluate the performance of both the controllers for reference trajectory tracking. Two trajectories were generated for link-1 and link-2 using Eqs. (15)–(18). The performance for trajectory tracking, for both

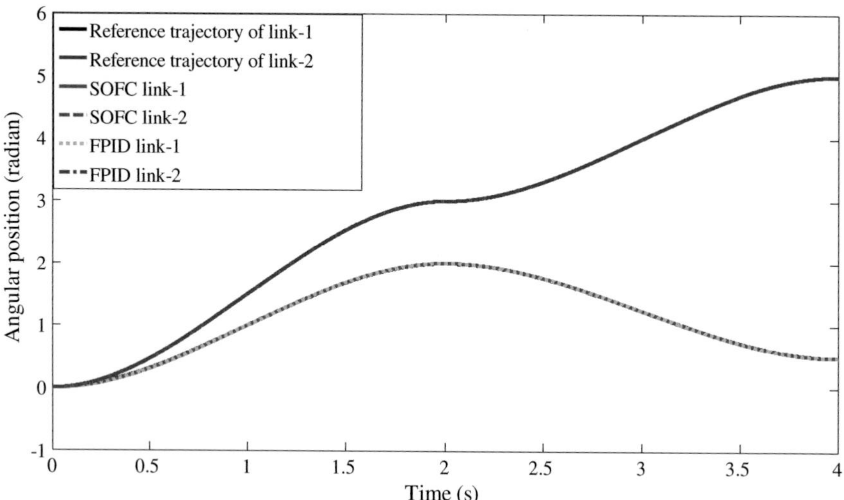

Fig. 9 Trajectory tracking performance

Table 5 IAE values for reference trajectory tracking

S. no.	Parameter	SOFC	FPID	Remark (% improvement in SOFC)
1.	IAE link-1	0.005423	0.007083	23.43
2.	IAE link-2	0.001659	0.004200	60.50
3.	Cost function	0.007400	0.011600	36.20

links and controllers, is shown in Fig. 9. Corresponding IAE values have been listed in Table 5 along with the bar chart in Fig. 8 for their relative assessment. Based on the cost function values it can be concluded that the performance of SOFC is 36.20 % much better than FPID controller. The variations of controller output, the resulting tracking errors and X-Y plot of the end-effector has been shown in Figs. 10, 11 and 12, respectively.

4.2 *Disturbance Rejection*

In most of the plant, there is always a chance of inescapable disturbance comes in the system. The main task of the controller is to reject the effect of these disturbances from the system so that the final output should not degrade. Here the disturbance rejection study has been performed by both the controller and the results have been shown.

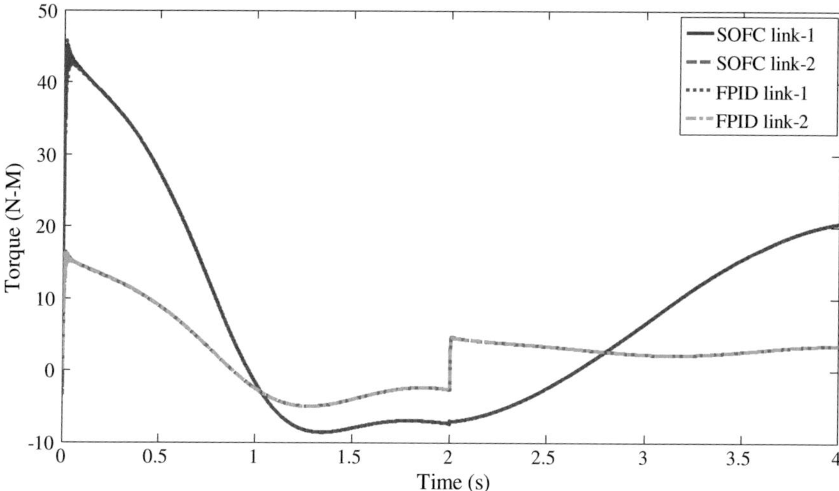

Fig. 10 Controller outputs

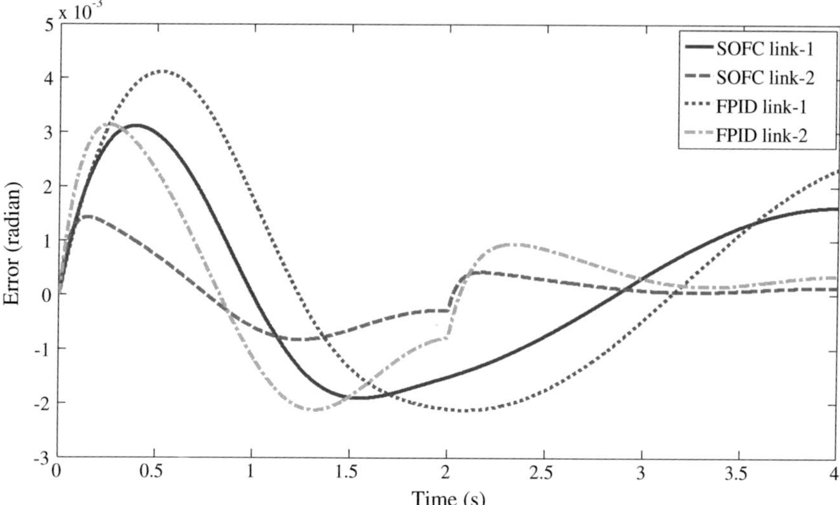

Fig. 11 Variations of error with time

A disturbance has been injected in the angular position of each link in the form of a pulse signal starting from $t = 1$ s and ending at $t = 2$ s. The amplitudes of these signals were kept as 0.13 and 0.18 for link-1 and link-2, respectively. Efficiency of the controllers to eliminate the effect of disturbance could be easily observed in Figs. 13, 14, 15 and 16 for trajectory tracking, controller output, tracking error

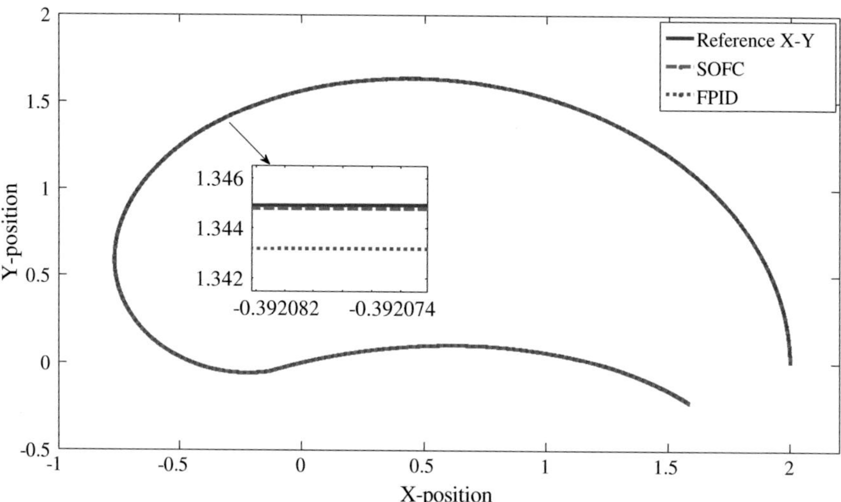

Fig. 12 X-Y curve for trajectory tracking

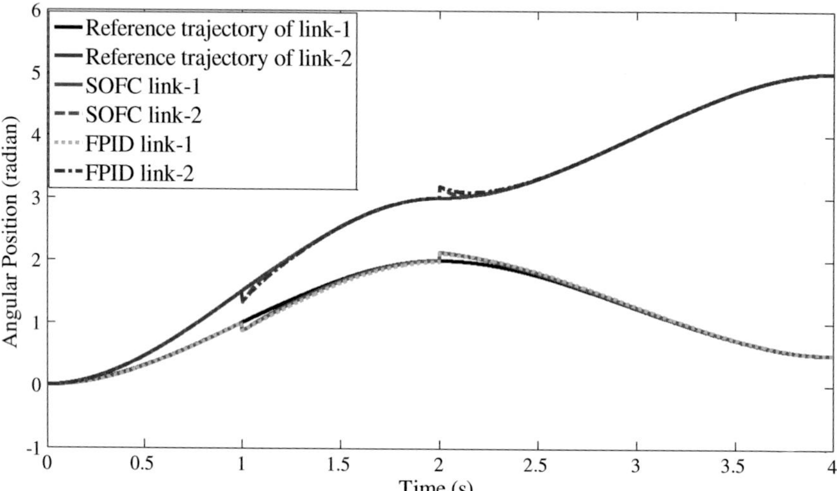

Fig. 13 Trajectory tracking for disturbance rejection

and X-Y curve. It can be observed that the trajectory has been tracked to a good accuracy. From the resulting IAE and cost function values presented in Table 6 and graphically shown in Fig. 17 in the form of bar-chart it can be inferred that the disturbance rejection performance of SOFC for IAE link-1, IAE link-2 and cost function is 35.21 %, 51.34 % and 39.63 % better than FPID controller. As seen in

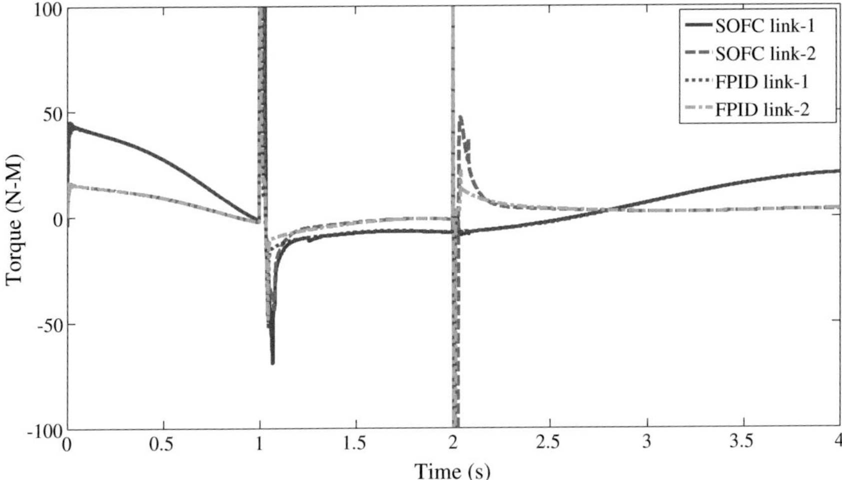

Fig. 14 Controller output for disturbance rejection study

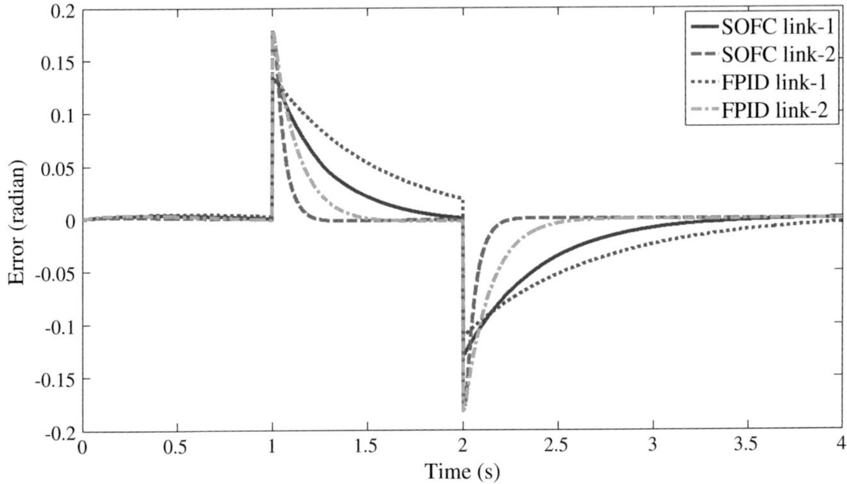

Fig. 15 Error profile for disturbance rejection study

Fig. 14, controller output signal shows a sudden change at $t = 1$ s and $t = 2$ s (the time at which the disturbance was incorporated) and saturates at 100 Nm as expected. Based on the present comparative study for reference trajectory tracking as well as disturbance rejection, the supremacy of SOFC on FPID controller could be easily seen.

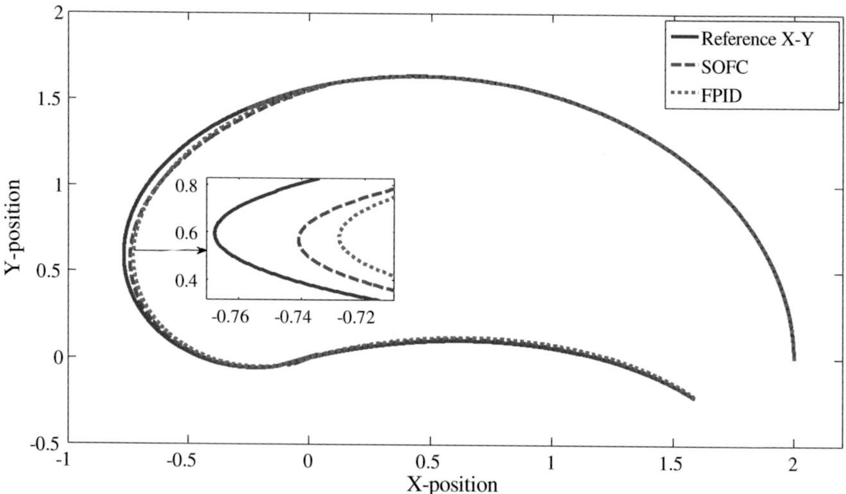

Fig. 16 X-Y curve for disturbance rejection study

Table 6 IAE values for disturbance rejection study

S. no.	Parameter	SOFC	FPID	Remark (% improvement in SOFC)
4.	IAE link-1	0.08642	0.1334	35.21
5.	IAE link-2	0.02519	0.05177	51.34
6.	Cost function	0.112100	0.185700	39.63

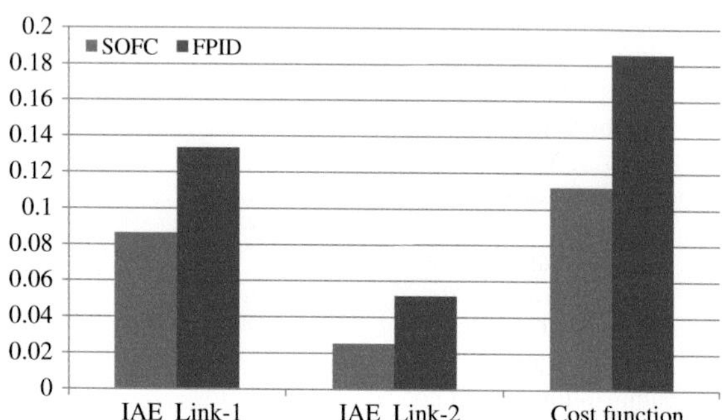

Fig. 17 Bar chart for disturbance rejection

5 Conclusion

In this chapter, two intelligent controllers namely, Self Organizing Fuzzy Controller (SOFC) and Fuzzy Proportional Integral and Derivative controller (FPID) have been implemented for a two link planar rigid robotic manipulator with end-effector for servo and regulatory control problem and a comparative study has been drawn. Both controllers employed Takagi-Sugeno-Kang (TSK) inference method. The major benefit of using TSK fuzzy system is that its rule base design is easy. The gains of the controllers were tuned by Genetic algorithm on the basis of a cost function comprising of a weighted summation of Integral of Absolute Error (IAE) and Integral of Absolute Change in Controller Output. Simulation studies show that SOFC gives better performance for trajectory tracking and for disturbance rejection problems i.e. the incorporation of second layer to the first layer FPID controller increases the efficacy of the controller for both types of problems. In the simulation studies, it has been found that SOFC has made enhancements 23.43 %, 60.50 % and 36.20 % in IAE link-1, IAE link-2 and cost function for trajectory tracking and 35.21 %, 51.34 % and 39.63 % enhancements in IAE link-1, IAE link-2 and cost function for disturbance rejection respectively.

This work can be further enhanced by designing the rule base of second layer of SOFC using other intelligent techniques like Artificial Neural Network (ANN) and the combination of Fuzzy as well as ANN. Furthermore, recent optimization techniques could also be used to efficiently tune the gains of controllers.

References

1. Azar AT (2010) Fuzzy Systems. IN-TECH, Vienna, Austria. ISBN: 978-953-7619-92-3
2. Azar AT, Serrano FE (2014) Robust IMC-PID tuning for cascade control systems with gain and phase margin specifications. Neural Comput Appl 25(5):983–995. doi:10.1007/s00521-014-1560-x
3. Azar AT, Serrano FE (2015) Design and modeling of anti-wind up PID controllers. In: Zhu Q, Azar AT (eds) Complex system modeling and control through intelligent soft computations. Studies in fuzziness and soft computing, vol 319. Springer, Germany, pp 1–44. doi:10.1007/978-3-319-12883-2_1
4. Azar AT, Vaidyanathan S (2015) Chaos modeling and control systems design. Studies in computational intelligence, vol 581. Springer, Germany. ISBN: 978-3-319-13131-3
5. Azar AT, Vaidyanathan S (2015) Computational intelligence applications in modeling and control. Studies in computational intelligence, vol 575. Springer, Germany. ISBN: 978-3-319-11016-5
6. Azar AT, Vaidyanathan S (2015) Handbook of research on advanced intelligent control engineering and automation. In: Advances in computational intelligence and robotics (ACIR) Book Series. IGI Global, USA
7. Azar AT, Zhu Q (2015) Advances and applications in sliding mode control systems. Studies in computational intelligence, vol 576. Springer, Germany. ISBN: 978-3-319-11172-8
8. Graham B (1986) Fuzzy identification and control. Ph.D. Thesis, University of Queensland, Australia
9. Graham B, Newell R (1988) Fuzzy identification and control of a liquid level rig. Fuzzy Sets Syst 26:255–273
10. Graham B, Newell R (1989) Fuzzy adaptive control of a first-order process. Fuzzy Sets Syst 26:47–65
11. Hassanien AE, Tolba M, Azar AT (2014) Advanced machine learning technologies and applications: second international conference, AMLTA 2014. Cairo, Egypt, 28–30 Nov 2014. In: Proceedings, communications in computer and information science, vol 488. GmbH, Springer, Heidelberg. ISBN: 978-3-319-13460-4

12. Holmblad LP, Ostergaard JJ (1982) Control of a cement kiln by fuzzy logic. In: Gupta MM, Sanchez E (eds) Fuzzy information and decision process. North-Holland, pp 389–399
13. King PJ, Mamdani EH (1977) The application of fuzzy control system to industrial processes. Automatica 13:235–242
14. Kumar V, Nakra BC, Mittal AP (2011) A review of classical and fuzzy PID controllers. Int J Intell Control Syst 16(3):170–181
15. Lee C-Y, Lee J-J (2003) Adaptive control of robot manipulators using multiple neural networks. In: Proceedings of the international conference on robotics and automation. Taipei, Taiwan, pp 1074–1079
16. Mamdani EH (1974) Application of fuzzy algorithms for control of simple dynamics plant. In: The proceedings of institute of electrical, control and science,vol 121, pp 1585–1588
17. Mamdani EH (1975) Prescriptive method for deriving control policy in a fuzzy logic controller. Electron Lett 11:625–626
18. Mamdani EH, Assilian S (1975) An experiment in linguistic synthesis with a fuzzy logic controller. Inf Sci 7:1–13
19. Mamdani EH, Procyk T, Baaklini N (1976) Application of fuzzy logic to controller design based on linguistic protocol. In: Mamdani EH, Gaines BR (eds) Discrete systems and systems and fuzzy reasoning, University of London, Queen Mary College, pp 125–149
20. Mishra P, Kumar V, Rana KPS (2014) A novel intelligent controller for combating stiction in pneumatic control valves. Control Eng Pract 33:94–104
21. Mishra P, Kumar V, Rana KPS (2015) A fractional order fuzzy PID controller for binary distillation column control. Expert Syst Appl 42(22):8533–8549
22. Procyk TJ, Mamdani EH (1979) A linguistic self-organizing process controller. Automatica 15(1):15–30
23. Sharma R, Rana KPS, Kumar V (2014) Performance analysis of fractional order fuzzy PID controllers applied to a robotic manipulator. Expert Syst Appl 41(9):4274–4289
24. Stephanopoulos G (1984) Chemical process control an introduction to theory and practice. Pearson Education, USA. ISBN: 81-7758-403-0
25. Yamazaki T, Mamdani EH (1982) On the performance of a rule-based self-organizing controller. In: Proceedings of IEEE conference on application of adaptive and multivariable control, Hull, England, pp 50–55
26. Zadeh LA (1965) Fuzzy sets. Inf Control 8:338–353
27. Zadeh LA (1968) Fuzzy algorithms. Inf Control 12:94–102
28. Zadeh LA (1970) Towards a theory of fuzzy systems. In: Kalman RE, Claris ND (eds) Aspects of networks and system theory. Rinehart and Winston, New York, pp 469–490
29. Zadeh LA (1972) A rationale for fuzzy control. J Dyn Syst Meas Contr 94(6):3–4
30. Zadeh LA (1973) Outline of a new approach to the analysis of complex systems and decision process. IEEE Trans Syst Man Cybern 3(1):28–44
31. Zadeh LA (1975) The concept of a linguistic variable and its application to approximate reasoning, part I. Inf Sci 8:199–249
32. Zadeh LA (1975) The concept of a linguistic variable and its application to approximate reasoning, part II. Inf Sci 8:199–249
33. Zadeh LA (1975) The concept of a linguistic variable and its application to approximate reasoning, part III. Inf Sci 8:199–249
34. Zhu Q, Azar AT (2015) Complex system modelling and control through intelligent soft computations. Studies in fuzziness and soft computing, vol 319. Springer, Germany. ISBN: 978-3-319-12882-5

Comparative Analysis of Different Nature Inspired Optimization Algorithms for Estimation of 3D Chaotic Systems

Sreejith S. Nair, K.P.S. Rana and Vineet Kumar

Abstract Among various nonlinear systems, parameter identification of chaotic systems turns out to be a very challenging task because of their complex and unpredictable nature. The control and synchronization of chaotic systems remains incomplete until the parameters of the chaotic systems are known. Traditionally, the trend has been to estimate the parameters using gradient based search methods which suffer from premature convergence and trapping in local minima. This chapter presents an optimization based scheme for estimation of the parameters of two chaotic systems namely Lorenz and Rossler, using two recently developed bio-inspired optimization algorithms, i.e. cuckoo search algorithm (CSA) and flower pollination algorithm (FPA). CSA is based on mimicking the breeding behavior and hostile reproduction strategies of cuckoo with the effective use of levy flight for providing global optimization while FPA is based on the natural process of flowering plants due to self and cross pollination using both levy flight strategies for global convergence and random walk for local convergence. The performance of these optimization algorithms, for efficient estimation of the parameters of chaotic system, is compared in terms of the resulting integral of absolute error (IAE). Simulation results demonstrated the effectiveness of CSA in offline 3D parameter estimation of the considered two chaotic systems over the FPA. The minimum fitness offered by FPA is 2.4E-03 and 5.03E-06 and by CSA it is 7.92E-06 and 1.31E-07 for the parameter estimation of the 3D Lorenz and Rossler chaotic system, respectively.

Keywords Cuckoo search algorithm · Flower pollination algorithm · Levy flight · Chaotic system

S.S. Nair (✉) · K.P.S. Rana · V. Kumar
Division of Instrumentation and Control Engineering,
Netaji Subhas Institute of Technology, Sector-3, Dwarka 110078, New Delhi, India
e-mail: sreejith336@gmail.com
URL: http://www.nsit.ac.in

K.P.S. Rana
e-mail: kpsrana1@gmail.com

V. Kumar
e-mail: vineetkumar27@gmail.com

A.T. Azar and S. Vaidyanathan (eds.), *Advances in Chaos Theory and Intelligent Control*, Studies in Fuzziness and Soft Computing 337,
DOI 10.1007/978-3-319-30340-6_32

1 Introduction

Nonlinear chaotic systems demonstrate a peculiar random collection of responses when they are excited by internal and external stimuli. Chaotic systems are also highly sensitive to initial conditions. They show 'butterfly effect' from similar initial conditions and turn out to totally different trajectories [1]. Lyapunov exponent is used to measure chaotic systems as it shows the divergence of phase points. Chaotic systems exhibit positive maximum Lyapunov exponent and phase space compactness. The chaotic systems finds many applications in various disciplines such as neural networks [11, 17, 20, 27], oscillators [24], lasers [23, 63], cryptosystems [39, 49], secure communication satellite [10, 33, 64], chemical reactions [13], biology [21], ecology [43, 45], finance [14, 43], fuzzy logic [15] and cardiology [37] etc.

First 3D chaotic system was discovered by Lorenz in 1963 for atmospheric convection studies in a weather model. In 1999, Chen coined another chaotic attractor which has similar 3D equation to that of Lorenz [30] but differs in topology [8, 31]. Later, researchers developed other chaotic systems such as Rossler system [40], Lu system and even some time delay systems such as Logistics and Mackey-Glass etc. [22, 28, 44, 47]. In 2015, Vaidyanathan and Azar proposed a nine-term "umbrella attractor" with four quadrant nonlinearities with its controlling and synchronizing done using adaptive controller and adaptive synchronizer, respectively [50, 52, 53]. They also proposed another interesting eleven term 4D hyperchaotic system with four quadrant nonlinearities which is dissipative. The work was supported by control and synchronization technique using adaptive controller [51, 53].

The control and synchronization of chaotic systems has been a topic of research for years [2–4, 6, 35, 53, 55, 56, 65]. Various methods used are: sample-data feedback method [10, 27], time delay feedback method [19, 42], backstopping method [34, 38, 48, 50, 51] active control method [29] and adaptive control method [6, 41]. For effective control and synchronization of chaotic systems parameter of the system plays an important role. In a real world scenario, there are difficulties in estimating parameters of a complex 3D chaotic system. Another major flaw is that the most of methods being gradient based methods are sensitive to initial conditions, and as a result they get trap in local minima solution. One of the potential solutions to these problems is to estimate the parameters by using soft computing techniques, particularly global optimization algorithms, evaluated at suitable cost function [9]. Several cases of parameter estimation of chaotic system using optimization algorithm like evolutionary programming approach [7], genetic algorithm (GA), particle swarm optimization (PSO) [16, 22], improved particle swarm optimization [32], drift particle swarm optimization [18], differential evolution (DE) [36, 46], chaotic gravitational search [5], cuckoo search [26] etc. have been reported in the literature. In this chapter two recently developed bio-inspired optimization algorithms, i.e. cuckoo search algorithm (CSA) and flower Pollination algorithm (FPA) are explored for estimation of the parameters of two chaotic systems namely Lorenz and Rossler. CSA functions based on mimicking the breeding behavior and hostile reproduction strategies of cuckoo with the effective use of levy flight for providing global optimization while FPA is based on the natural process of flowering plants due to self and cross pollination using both levy flight strategies for global convergence and random walk for local convergence. For performance comparison of these optimization algorithms, the resulting integral of absolute error (IAE) has been considered.

This chapter is organized as follows: following a brief introduction in Sect. 1, Sect. 2 outlines the general representation of the n-dimensional chaotic systems. Section 3 briefs about the used two optimization algorithms i.e. CSA and FPA. Intensive MATLAB simulation has been carried out in Sect. 4 which demonstrates the effectiveness of CSA and FPA for estimating the parameters of Lorenz and Rossler chaotic systems. The obtained results in the form of parameter values along with the resulting IAE values have been presented in the section. Finally, Sect. 5 draws the conclusions and recommends future scope in this area.

2 Problem Formulation

Chaotic systems are a special class of nonlinear systems that shows unique complex, random and noisy-like behaviors. This behavior depends on two major factors like initial conditions and the parameter variations. In a parameter estimation problem, using optimization techniques, the objective is to obtain the parameters of a chaotic system. This task is performed through minimization of a fitness function. A low value of the fitness function indicated a superior estimation result.

An n-dimensional continuous chaotic system can be represented as follows:

$$\dot{X} = f(X, X_0, \theta) \tag{1}$$

where, X_0 specifies the initial state and X is the state vector given by $X = (x_1, x_2, x_3 \ldots x_n) \in R_n$.

$\theta = (\theta_1, \theta_2, \theta_3, \ldots \theta_n) \in R_n$ is the unknown parameter vector. The estimated system can also be similarly stated as follows,

$$\dot{Y} = f(Y, Y_0, \hat{\theta}) \tag{2}$$

where, Y is the state vector of estimated system and are given by $Y = (y_1, y_2, y_3 \ldots y_n) \in R_n$

$$\hat{\theta} = (\hat{\theta}_1, \hat{\theta}_2, \hat{\theta}_3, \ldots \hat{\theta}_n) \tag{3}$$

$\hat{\theta}$ represents the estimated parameters of the chaotic system.

As mentioned earlier, the goal is to obtain such parameters for the chaotic system which yield minimum value of the objective function/fitness function defined in the form of a IAE between the actual system and estimated system. As given in Eq. 4 (Fig. 1),

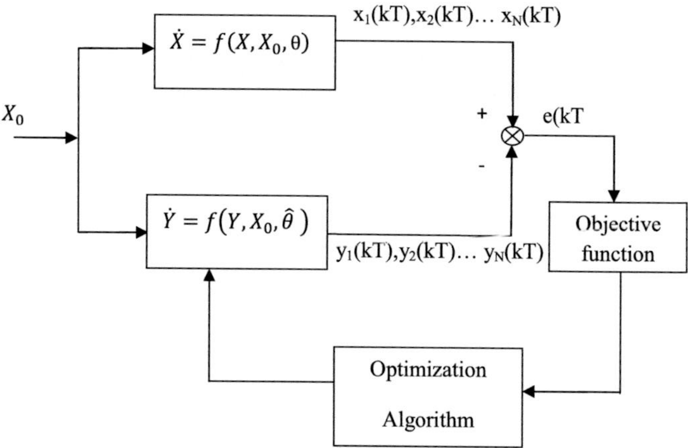

Fig. 1 The principle of parameter estimation for chaotic systems

$$\mathrm{J} = \sum_{K=1}^{N} |X_k - Y_k| \tag{4}$$

In the study, the parameters of chaotic system have been found by CSA and FPA, using IAE as cost function.

3 Review of CSA and FPA Bio-inspired Optimization Algorithms

Optimization being derivation free are simple, effective and adds flexibility to designs. There are different methods of optimization but are broadly classified into two categories namely deterministic and metaheuristic. Out of these metaheurestic algorithms such as GA, PSO and DE are very famous among the researchers. The benchmark studies revealed that CSA and FPA are much better in providing optimum solutions than other popular algorithm. It is also observed that levy flight based walking strategy is better for obtaining global optimization than most of the random walk based optimization. This is the main motivation to consider these two potential optimization algorithms for the problem of parameter estimation of chaotic system. A brief summary of these two algorithms are presented below.

3.1 Cuckoo Search Algorithm

CSA uses levy flight instead of isotropic random walk strategies to provide better global convergence. cuckoo fascinates us by its pleasing sound and its reproduction strategy. It has been found effective in solving multimodal problems compared to other popular metaheurestic algorithms. The algorithm imitates the breeding behavior of cuckoo and uses Lévy flight obtained by Mantegna's algorithm to generate new eggs [58]. The number of parameters used by CSA is lesser than other metaheuristic algorithms, therefore it is much faster. The elitism property enables to keep the best solution in the search space. The use of lévy flight, a complex random walking strategy, makes it prominent and effective compared to other optimization algorithms. The efficiency of an optimization algorithm depends on two factors namely diversification and intensification. The balance between these two factors is what makes the CSA superior and effective over other famous optimization algorithms such as GA, PSO and DE etc. A flowchart of CSA is given in Fig. 2.

There are three fundamental rules which were proposed by Yang and Deb in CSA [12, 58, 59]

1. Each cuckoo lays one egg (solution) at a time and dumps it in a random nest.
2. The best nests i.e. which contain the highest quality eggs passes to the next generations and others are discarded.
3. The number of available host nests is fixed and host can discover the alien with a probability $p_a \in [0, 1]$. If the host finds alien egg it can either throw away or abandon the nest so as to build a completely new nest in a new location.

The most important feature of CSA is the lévy Flight a complex random walk strategy that makes it stand-out with the other optimization algorithms. The CSA enables the use of Lévy (λ) flight to find the optima in a problem search space for multimodal function [60, 61].

The new solutions are generated by ith cuckoo using update equation,

$$X_i^{t+1} = X_i^t + \alpha \oplus \text{Lévy}(\lambda) \tag{5}$$

Lévy distribution is governed by random walk process which obeys a power-law step-size distribution with heavy tail and can be expressed as,

$$\text{Lévy} \sim u > t^{\lambda},\ 1 < \lambda < 3 \tag{6}$$

3.2 Flower Pollination Algorithm

In 2012, Yang developed FPA inspired by flowering plant by pollination process [61]. There are two kinds of pollination processes which occur in plants namely self pollination and cross pollination. Self pollination process happens when the pollens from the flower pollinate the flower or flowers of the same plant while cross pollination occurs when pollen grain are passed to flower from other plant. These characteristics of plants are considered for proposing the FPA for local and global convergence as seen in Fig. 3. Following four basic rules are formed for mimicking the flower pollination process [57].

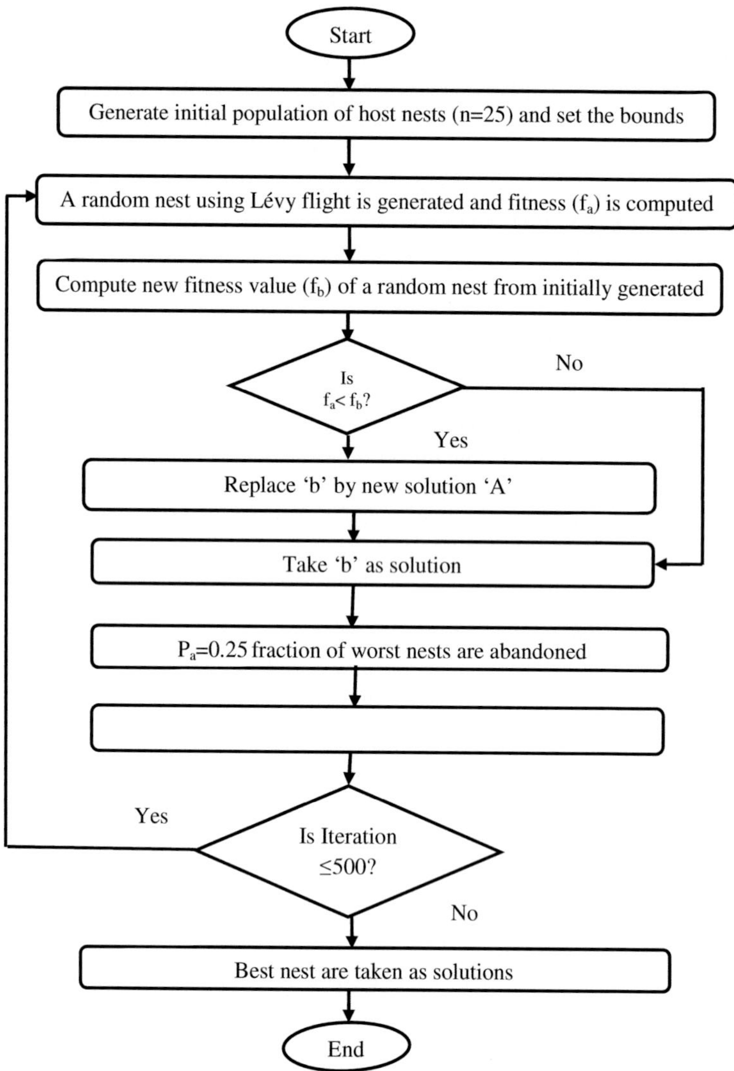

Fig. 2 Flow chart of Cuckoo search algorithm

a. Global pollination process is carried out by the process of biotic and cross pollination and the pollinators obeys the levy flight based walking strategy.
b. Abiotic and self pollination are used for self pollination.
c. Flower constancy is developed using pollinators such as insects which is equivalent to the reproduction probability that is proportional to the similarity of the two flowers involved.

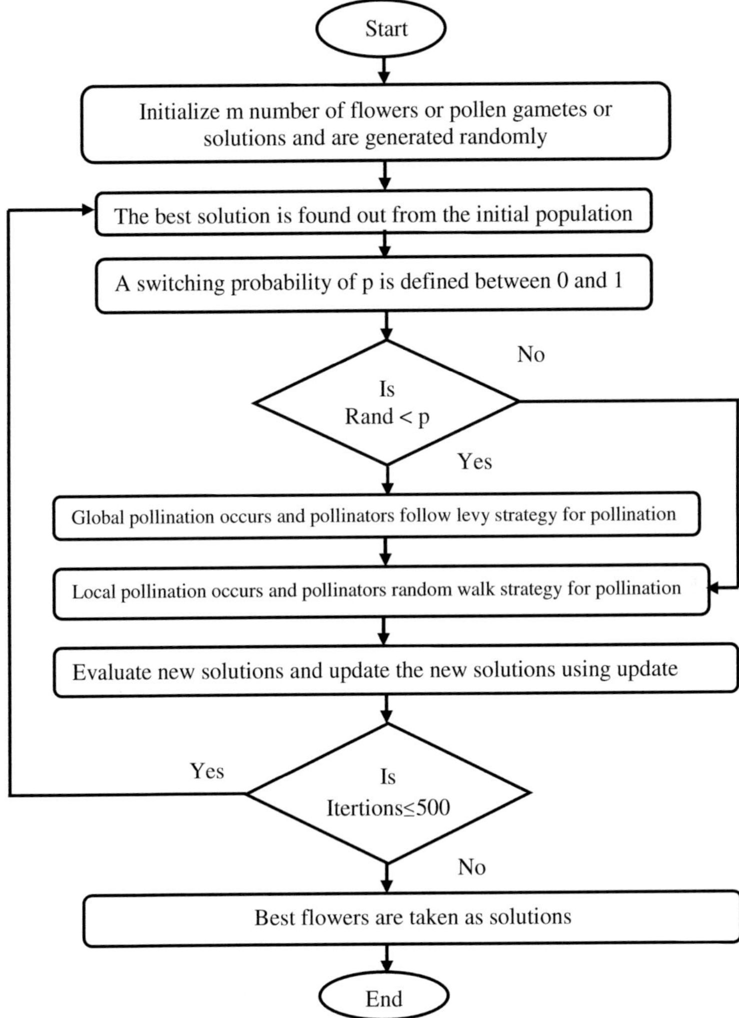

Fig. 3 Flow chart of Flower pollination algorithm

d. Switching probability of $p \in [0, 1]$ is used to switch between global pollination and local pollination with a slight bias towards local pollinations by comparing it with a random number (Rand) between 0 and 1.

These four rules are used to generate the updating equation for global and local optimization process.

The global pollination process is performed by pollinator such as insects by pollinating pollen gametes to a long distance with a longer range. The distance the insects can fly is governed by Lévy flight. Update equation can mathematically be represented as follows:

$$x_i^{n+1} = x_i^n + L(\lambda)(x_i^n - g_*) \tag{7}$$

where x_i^n is the pollen 'i' or solution vector x_i^n at iteration 'n' and g_* is the current best solution found among all the available solutions at the current iteration.

The local pollination process imitate the flower faithfulness to pollinatc in a limited locality and updation is done in a random walk strategies rather levy flight. The update equation can be stated as,

$$x_i^{n+1} = x_i^n + \varepsilon(x_j^n - x_k^n) \tag{8}$$

where, x_j^n and x_k^n are pollen. ε is the factor describes the random walk strategy for a uniform distribution in the range [0, 1].

4 Simulation Results

The comparative studies, between two bio-inspired optimization algorithms, CSA and FPA, for the parameter estimation of chaotic system i.e. Lorenz and Rossler are presented in this section. All the simulations were conducted using MATLAB software on an Intel core^{TM} i5 CPU 3.20 GHz with a 2 GB RAM. The performance evaluation has been done using IAE as the objective function which has been minimized using CSA and FPA. The statistical analysis has been done in terms of mean and standard deviation for best, average and worst results respectively.The ruggedness and repeatability of obtained solution is ensured with ten independent trials. Fourth order Runga-Kutta method has been used to solve the ordinary differential equations with a step-size of 0.01 s in order to simulate the 3D chaotic equation. The solutions having best fitness value is considered as best, those fall near to mean as the average results while those have the least fitness value is considered is the worst as seen from Tables 1 and 2. Tables 1 and 2 presents the estimated value of three parameters namely a, b and c obtained using CSA and FPA against the true value with the minimum fitness value, mean and standard deviation of best, average and worst results respectively. The tables clearly demonstrates that CSA better approximates the true value for both Lorenz and Rossler chaotic system than FPA.

The superiority of the optimization algorithms in identifying the well known chaotic systems such as Lorenz and Rossler system is considered. The searching range of the parameters of 3D Lorenz system was kept as $9 \leq a \leq 11$, $20 \leq b \leq 30$, $2 \leq c \leq 3$, while for Lorenz 3D chaotic system it is $0.01 \leq a \leq 0.5$, $0.01 \leq b \leq 0.5$, $2 \leq c \leq 10$, with the iterations kept at 500 for both the cases. The parameter sets for the optimization using CSA is defined as following, total number of nests (n) set to 25, number of dimension (nd) is taken as 3, probability of alien egg (p_a) is 0.25 with tolerance of 10^{-5} and for the flower pollination algorithm these are set as, population size as 20, and switching probability as 0.8.

Table 1 Results of 3D Lorenz system parameter estimation

Parameter	CSA			FPA		
	Best	Average	Worst	Best	Average	Worst
a = 10	10	9.9999	10.0000	9.9999	9.9998	10.0002
b = 28	28	28.0000	27.9999	27.9999	27.9999	28.0000
c = 8/3	2.6667	2.6667	2.6667	2.6666	2.6666	2.6666
Objective function	7.92E-06	1.31E-05	1.13E-05	2.4E-03	0.0721	0.1074
Mean (10 runs)	1.41E-05			2.34E-02		
Standard deviation	6.03946E-06			0.036084		

Table 2 Results of 3D Rossler system estimation

Parameter	CSA			FPA		
	Best	Average	Worst	Best	Average	Worst
a = 0.2	0.2000	0.2000	0.19999	0.1999	0.2000	0.1999
b = 0.2	0.2000	0.1999	0.2000	0.1999	0.2000	0.1999
c = 5.7	5.7000	5.7000	5.6999	5.6999	5.7001	5.6999
Objective function	1.31E-07	2.6E-06	3.9E-06	5.03E-06	3.5E-06	4.2E-06
Mean (ten runs)	4.65E-06			2.32E-06		
Standard deviation	2.45E-07			3.10E-06		

The first chaotic system considered is Lorenz system whose dynamic equation is stated below:

$$\begin{aligned} \dot{\mathrm{x}}(\mathrm{t}) &= \mathrm{a}(\mathrm{y}(\mathrm{t}) - \mathrm{x}(\mathrm{t})) \\ \dot{y}(t) &= bx(t) - x(t)z(t) - y(t) \\ \dot{z}(t) &= x(t)y(t) - cz(t) \end{aligned} \tag{9}$$

where a = 10, b = 28 and c = 8/3 are the system parameters and its chaotic behavior, shown in Fig. 4, depends upon the initial condition given by [x(0) y(0) z(0)] = [1 1 1].

Lorenz system is a set of ordinary differential equation which resembles a shape of butterfly or figure eight. This was first found out by Edward Lorenz when he conducted experiment for thermal variations in an air cell underneath a thunderhead. He observed that initial conditions with small differences leads to produce different solution. This has lead to the important property of the chaotic system that these are highly sensitive to initial conditions [30].

The Fig. 5 presents the Fitness versus iteration plots for the CSA and FPA. It can be observed that CSA takes least iterations and converge quickly to the true value and offers the minimum fitness value.

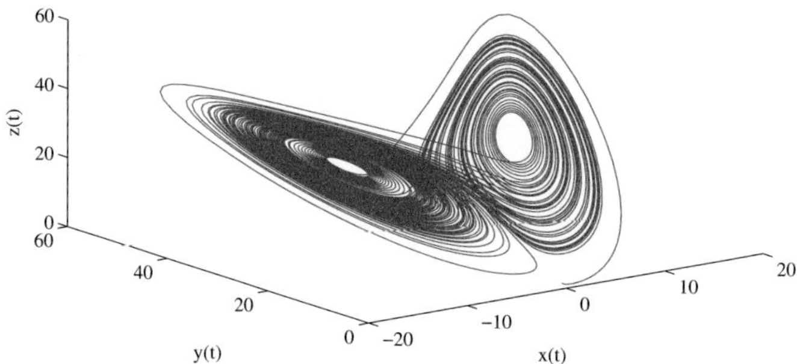

Fig. 4 Butterfly pattern of 3D Lorenz system

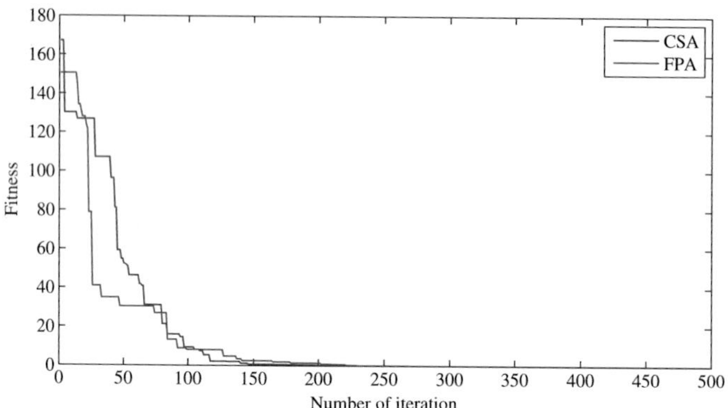

Fig. 5 Convergence profile of Lorenz chaotic system parameter estimation

The searching efficiency of the optimization algorithms is well illustrated by the parameter convergence profile. The parameter convergence profile shows the effectiveness of different optimization approaches in estimating the true values of chaotic system at faster rate or in minimum iteration. The profile clearly depicts that the parameter 'a' span between 9 and 10.7 against the true value of 10. The estimation of parameter in Fig. 6 shows the coefficient convergence profile and clearly suggests that CSA gives faster convergence for coefficients. CSA takes least iteration cycle compared to FPA to attain the true value.

The analysis of parameter variation with iterations is an interesting phenomenon and needs to be recorded in order to view the convergence of different parameters. In this case the parameter 'b' span between 27.1 and 28.4 as seen Fig. 7. Similar to the previous case here also the CSA quickly converges to the true value with least iteration cycle than FPA.

The profile shows the quality of estimated values of chaotic system obtained by the two different algorithms namely CSA and FPA. The parameter 'c' span between

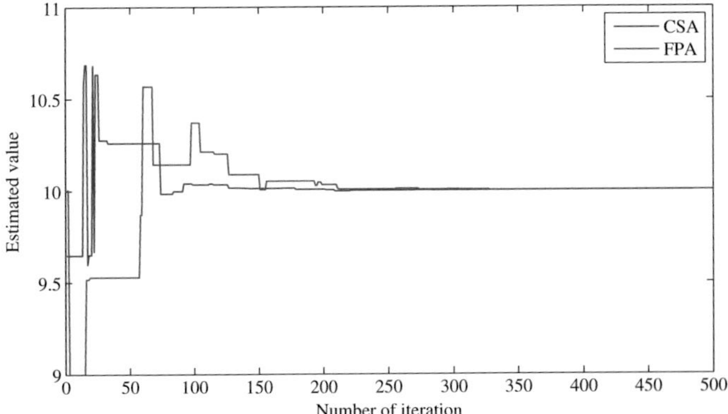

Fig. 6 Parameter convergence profile of ‘a’ for Lorenz chaotic system

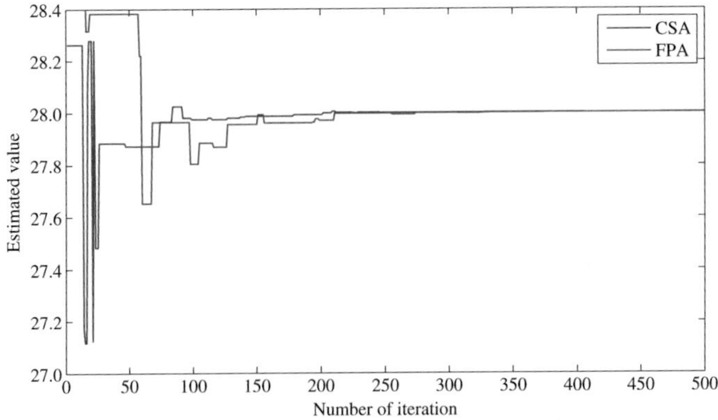

Fig. 7 Parameter convergence profile of ‘b’ for Lorenz chaotic system

2.61 and 2.76 as seen in Fig. 8. The parameter estimation using FPA oscillates more than CSA and CSA reaches to true value more easily.

The Table 1 shows the statistical results of 3D parameter estimation based on the fitness value of ten independent trials. Based on the fitness value, the solutions are organized into best, average and the least solutions. The table clearly depicts better estimation by CSA compared to FPA as even the worst solution given by CSA is better than best solution by FPA. Furthermore, the least value of standard deviation signifies the repeatability and the quality of solution offered by CSA when compared to FPA.

The second chaotic system considered for the study is the Rossler system whose dynamic equation is stated below:

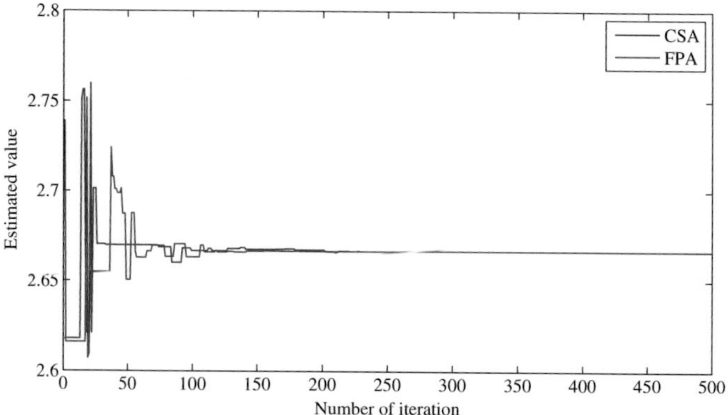

Fig. 8 Parameter convergence profile of 'c' for Lorenz chaotic system

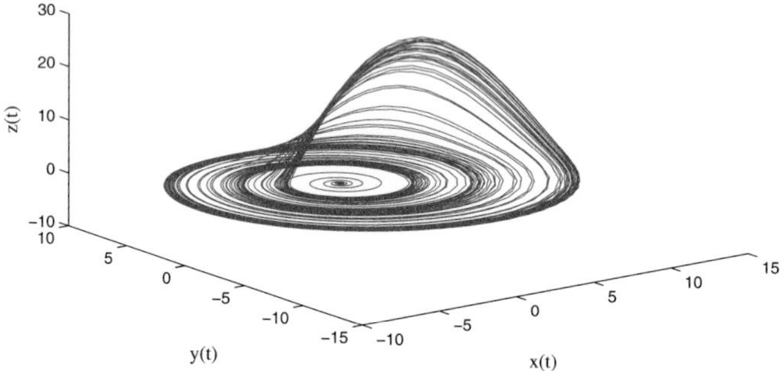

Fig. 9 Topological view of Rossler chaotic system

$$\begin{aligned} \dot{x}(t) &= -y(t) - z(t) \\ \dot{y}(t) &= x(t) + ay(t) \\ \dot{z}(t) &= b + z(t)x(t) - cz(t) \end{aligned} \tag{10}$$

where a = 0.2, b = 0.2 and c = 5.7 are the system parameters and its chaotic behavior depends upon the initial condition given by [x(0) y(0) z(0)] = [0.1 0.1 1.0] and are shown in Fig. 9.

The topological view of Rossler shows that it is a simple stretched and folded ribbon as seen in Fig. 9. This was first studied in 1970 by Otto Rossler by conducting experiments related to chemical kinematics [40]. This system does not reach a steady state value and with increase in the number of iterations the two close initial states will diverge.

The performance evaluation of Lorenz system can be visualized in the convergence profile as seen in Fig. 10. The fitness has been calculated with IAE as cost function

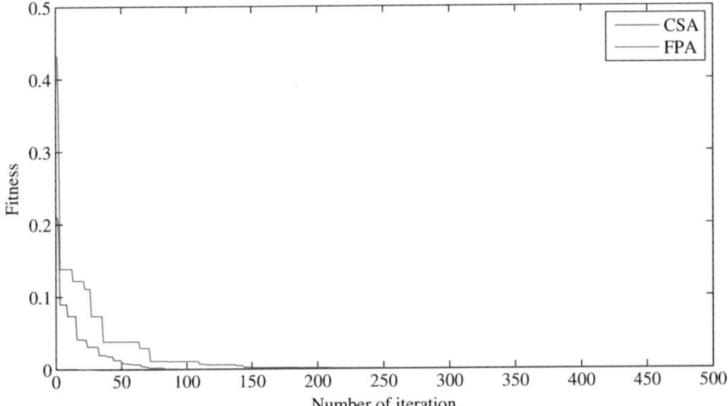

Fig. 10 Convergence profile of Lorenz chaotic system

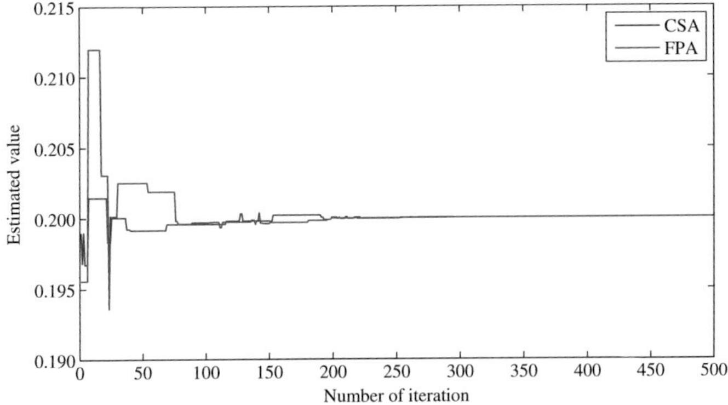

Fig. 11 Parameter convergence profile of ‘a’ for Rossler chaotic system

minimized using optimization algorithms. Both CSA and FPA converge to optimum solution in 100 and 180 iteration with minimum fitness of 1.31E-07 and 5.03E-06, respectively. This shows that CSA has been proved much effective in finding the optimum solutions of 3D Lorenz system than FPA.

The Fig. 11 demonstrates the effectiveness of CSA and FPA in estimating the true value of parameter ‘a’. It may be noted that parameter ‘a’ spans between 0.194 and 0.212 as seen in Fig. 11. Also there are more oscillations in case of FPA than CSA.

Figure 12 shows the variation of individual parameter with change in number of iterations. The parameter ‘b’ span between 0.12 and 0.26. It can be clearly inferred from the figure that FPA based parameter estimation oscillates more than CSA.

The convergence profile for parameter ‘c’ is provided in Fig. 13. The parameter ‘c’ span between 2.7 and 8. Again it can be seen from the figure that CSA oscillates less and converges quickly to the true value of 5.7.

The Table 2 shows the statistical analysis in terms of fitness values and the best, average and worst solutions of 10 independent trials calculated using optimization algorithms.

These values reflect the diversity of the performances given by the optimization algorithm for the purpose of estimating parameters of chaotic system. The best solutions provided by CSA are 1.31E-07 and FPA is 5.03E-06 respectively for identifying parameters of Rossler system.

5 Conclusion and Future Scope

In this research work, parameters of Lorenz and Rossler systems are estimated using two recently developed optimization algorithms namely cuckoo search algorithm (CSA) and flower pollination algorithm (FPA). The estimation of parameters through optimization rather than gradient based method proved to be an effective and simpler. The simulated experimental results demonstrated effectiveness of both the algorithms in the problem of parameter estimation in terms of Integral of absolute error (IAE). Comparative studies show CSA is better able to estimate the parameter for the Rossler and Lorenz chaotic system. As a future work, this research can be extended to hyperchaotic system and 5D system estimated using novel optimization algorithm. Fractional order chaotic system for both 3D chaotic systems and hyperchaotic sys-

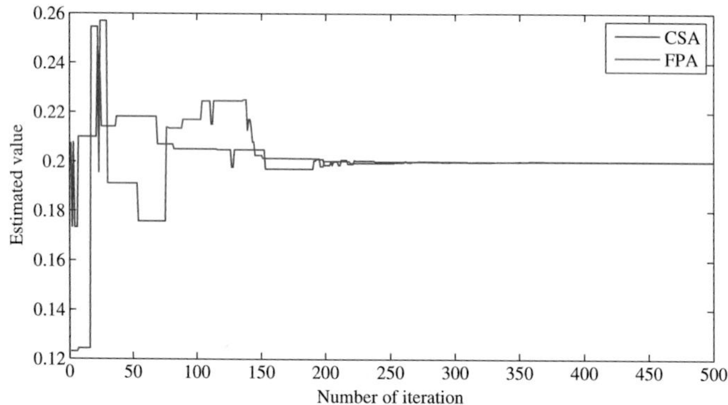

Fig. 12 Parameter convergence profile of ‘b’ for Rossler chaotic system

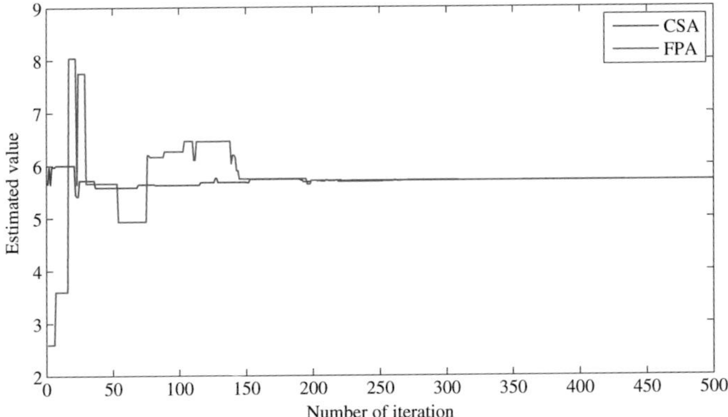

Fig. 13 Parameter convergence profile of ‘c’ for Rossler chaotic system

tem can also be considered for identification. The control and synchronization of both integer order and fractional order chaotic system will also be a potential topic of further research.

References

1. Alligood KT, Sauer T, Yorke JA (1997) Chaos: an introduction to dynamical systems, 1st edn. Springer, Berlin
2. Azar AT, Vaidyanathan S (eds) (2015a) Chaos modeling and control systems design. Studies in computational intelligence, vol 581. Springer, Germany. ISBN: 978-3-319-13131-3
3. Azar AT, Vaidyanathan S (eds) (2015b) Computational intelligence applications in modeling and control. Studies in computational intelligence, vol 575. Springer, Germany. ISBN: 978-3-319-11016-5
4. Azar AT, Vaidyanathan S (2014c) Chaos: handbook of research on advanced intelligent control engineering and automation. Advances in Computational Intelligence and Robotics (ACIR) Book Series, 1st edn. IGI Global, Hershey, PA. doi:10.4018/978-1-4666-7248-2
5. Li CS, Zhou JZ, Xiao JH (2012) Parameters identification of chaotic system by chaotic gravitational search algorithm. Chaos Solitons Fractals 45:539–547
6. Carroll TL, Pecora LM (1991) Synchronizing chaotic circuits. IEEE Trans Circ Syst 38(4):453–456
7. Chang et al (2008) Parameter identification of chaotic systems using evolutionary programming approach. Expert Syst Appl 35(4):2074–2079
8. Chen G, Ueta T (1999) Yet another chaotic attractor. Int J Bifurcat Chaos 9(7):1465–1466
9. Emary E et al (2014) Retinal vessel segmentation based on possibilistic fuzzy c-means clustering optimised with cuckoo search. In: IEEE 2014 international joint conference on neural networks (IJCNN 2014), 6–11 July. Beijing International Convention Center, Beijing, China
10. Feki M (2003) An adaptive chaos synchronization scheme applied to secure communication. Chaos Solitons Fractals 18(1):141–148
11. Gan Q, Liang Y (2012) Synchronization of chaotic neural networks with time delay in the leakage term and parametric uncertainties based on sampled-data control. J Franklin Inst 349(6):1955–1971

12. Gandomi AH, Yang XS, Alavi AH (2013) Cuckoo search algorithm. A metaheuristic approach to solve structural optimization problems. Eng Comput 29(1):17–35
13. Gaspard P (1999) Microscopic chaos and chemical reactions. Phys A: Stat Mech Appl 263(1–4):315–328
14. Guégan D (2009) Chaos in economics and finance. Annu Rev Control 33(1):89–93
15. Hassanien AE, Tolba M, Azar AT (2014) Advanced Machine Learning Technologies and Applications: Second International Conference, AMLTA 2014, Cairo, Egypt, 28–30 Nov 2014. Proceedings, Communications in computer and information science, vol 488. Springer-Verlag GmbH, Berlin/Heidelberg. ISBN: 978-3-319-13460-4
16. He Q, Wang L, Liu B (2007) Parameter estimation for chaotic systems by particle swarm optimization. Chaos Solitons Fractals 34:611–654
17. Huang X, Zhao Z, Wang Z, Li Y (2012) Chaos and hyperchaos in fractional-order cellular neural networks. Neurocomputing 94:13–21
18. Sun J et al (2010) Parameter estimation for chaotic systems with a Drift Particle Swarm Optimization method. Phys Lett A 374(28):2816–2822
19. Jiang G-P, Zheng WX, Chen G (2004) Global chaos synchronization with channel time-delay. Chaos Solitons Fractals 20(2):267–275
20. Kaslik E, Sivasundaram S (2012) Nonlinear dynamics and chaos in fractional-order neural networks. Neural Netw 32:245–256
21. Kyriazis M (1991) Applications of chaos theory to the molecular biology of aging. Exp Gerontol 26(6):569–572
22. Li D (2008) A three-scroll chaotic attractor. Phys Lett A 372(4):387–393
23. Li N et al (2014) Enhanced chaos synchronization and communication in cascade-coupled semiconductor ring lasers. Commun Nonlinear Sci Numer Simul 19(6):1874–1883
24. Li N, Zhang Y, Nie Z (2011) Synchronization for general complex dynamical networks with sampled-data. Neurocomputing 74(5):805–811
25. Li NQ, Pan W, Yan LS (2011) Parameter estimation for chaotic systems with and without noise using differential evolution-based method. Chin Phys B 20(6):060502(1–6)
26. Li XT, Yin MH (2012) Parameter estimation for chaotic systems using the cuckoo search algorithm with an orthogonal learning method. Chin Phys B 21(5):050507(1–6)
27. Lian S, Chen X (2011) Traceable content protection based on chaos and neural networks. Appl Soft Comput 11(7):4293–4301
28. Liu C et al (2004) A new chaotic attractor. Chaos Solitons Fractals 22(5):1031–1038
29. Liu L, Zhang C, Guo ZA (2007) Synchronization between two different chaotic systems with nonlinear feedback control. Chin Phys 16(6):1603–1607
30. Lorenz EN (1963) Deterministic periodic flow. J Atmos Sci 20(2):130–141
31. Lü J, Chen G (2002) A new chaotic attractor coined. Int J Bifurcat Chaos 12(3):659–661
32. Mondares H, Alfi A, Fateh MM (2010) Parameter identification of chaotic system through an improved particle swarm optimization. Expert Syst Appl 37(5):3714–3720
33. Murali K, Lakshmanan M (1998) Secure communication using a compound signal from generalized chaotic systems. Phys Lett A 241(6):303–310
34. Njah AN, Ojo KS, Adebayo GA, Obawole AO (2010) Generalized control and synchronization of chaos in RCL-shunted Josephson junction using backstepping design. Phys C 470(13–14):558–564
35. Pecora LM, Carroll TL (1990) Synchronization in chaotic systems. Phys Rev Lett 64(8):821–824
36. Peng B et al (2009) Differential evolution algorithm-based parameter estimation for chaotic systems. Chaos Solitons Fractals 39(5):2110–2118
37. Qu Z (2011) Chaos in the genesis and maintenance of cardiac arrhythmias. Prog Biophys Mol Biol 105(3):247–257
38. Rasappan S, Vaidyanathan S (2012) Global chaos synchronization of WINDMI and Coullet chaotic systems by backstepping control. Far East J Math Sci 67(2):265–287
39. Rhouma R, Belghith S (2011) Cryptoanalysis of a chaos based cryptosystem on DSP. Commun Nonlinear Sci Numer Simul 16(2):876–884

40. Rossler OE (1976) An equation for continuous chaos. Phys Lett A 57(5):397–398
41. Sarasu P, Sundarapandian V (2012) Generalized projective synchronization of two-scroll systems via adaptive control. Int J Soft Comput 7(4):146–156
42. Shahverdiev EM, Bayramov PA, Shore KA (2009) Cascaded and adaptive chaos synchronization in multiple time-delay laser systems. Chaos Solitons Fractals 42(1):180–186
43. Sprott JC (2004) Competition with evolution in ecology and finance. Phys Lett A 325(5–6):329–333
44. Sprott JC (1994) Some simple chaotic flows. Phys Rev E 50(2):647–650
45. Suerez I (1999) Mastering chaos in ecology. Ecol Modell 117(2–3):305–314
46. Tang YG, Guan XP (2009) Parameter estimation of chaotic system with tine-delay: a differential evolution approach. Chaos Solitons Fractals 42(5):3132–3139
47. Tigan G, Opris D (2008) Analysis of a 3D chaotic system. Chaos Solitons Fractals 36:1315–1319
48. Tu J, He H, Xiong P (2014) Adaptive back stepping synchronization between chaotic systems with unknown Lipchitz constant. Appl Math Comput 236:10–18
49. Usama M et al (2010) Chaos-based secure satellite imagery cryptosystem. Comput Math Appl 60(2):326–337
50. Vaidyanathan S, Azar AT (2015a) Analysis and control of a 4-D novel hyperchaotic system. In: Azar AT, Vaidyanathan S (eds) Chaos modeling and control systems design. Studies in computational intelligence, vol 581. Springer-Verlag GmbH, Berlin/Heidelberg, pp 19–38. doi:10.1007/978-3-319-13132-0_2
51. Vaidyanathan S, Azar AT, Rajagopal K, Alexander P (2015) Design and SPICE implementation of a 12-term novel hyperchaotic system and its synchronization via active control. Int J Modell Ident Control (IJMIC) 23(3):267–277
52. Vaidyanathan S, Azar AT (2015b) Anti-synchronization of identical chaotic systems using sliding mode control and an application to Vaidyanathan-Madhavan chaotic systems. In: Azar AT, Zhu Q (eds) Advances and applications in sliding mode control systems. Studies in computational intelligence book series, vol 576. Springer-Verlag GmbH, Berlin/Heidelberg, pp. 527–547. doi:10.1007/978-3-319-11173-5_19
53. Vaidyanathan S, Azar AT (2015c) Hybrid synchronization of identical chaotic systems using sliding mode control and an application to Vaidyanathan chaotic systems. In: Azar AT, Zhu Q (eds) Advances and applications in sliding mode control systems. Studies in computational intelligence book series, vol 576. Springer-Verlag GmbH, Berlin/Heidelberg, pp 549–569. doi:10.1007/978-3-319-11173-5_20
54. Vaidyanathan S, Azar AT, Rajagopal K, Alexander P (2015) Design and SPICE implementation of a 12-term novel hyperchaotic system and its synchronization via active control. Int J Modell Ident Control (IJMIC) 23(3):267–277
55. Vaidyanathan S, Idowu BA, Azar AT (2015) Backstepping controller design for the global chaos synchronization of Sprott's Jerk systems. In: Azar AT, Vaidyanathan S (eds) Chaos modeling and control systems design. Studies in computational intelligence, vol 581. Springer-Verlag GmbH, Berlin/Heidelberg, pp 39–58. doi:10.1007/978-3-319-13132-0_3
56. Vaidyanathan S, Sampath S, Azar AT (2015) Global chaos synchronisation of identical chaotic systems via novel sliding mode control method and its application to Zhu system. Int J Modell Ident Control (IJMIC) 23(1):92–100
57. Yang XS (2012) Flower pollination algorithm for global optimization. Unconventional Comput Nat Comput Lect Notes Comput Sci 7445:240–249
58. Yang XS, Deb S (2009) Cuckoo search via Levy Flights. In: Proceedings of world congress on nature and biologically inspired computing (NaBIC), pp 210–214
59. Yang XS et al (2013) Swarm intelligence and bio-inspired computation. Theory and applications. Elsevier, London
60. Yang XS (2010) Nature-inspired metaheuristic algorithms. Luniver Press, United Kingdom
61. Yang XS (2014) Nature-inspired optimization algorithms. Elsevier, London
62. Yang XS, Deb S (2010) Engineering optimisation by cuckoo search. Int J Math Modell Numer Optim 1(4):330–343

63. Yuan G, Zhang X, Wang Z (2014) Generation and synchronization of feedback-induced chaos in semiconductor ring lasers by injection-locking. Optik Int J Light Electron Opt 125(8):1950–1953
64. Zaher AA, Abu-Rezq A (2011) On the design of chaos-based secure communication systems. Commun Nonlinear Sci Numer Simul 16(9):3721–3727
65. Zhu Q, Azar AT (2015) Complex system modelling and control through intelligent soft computations. Studies in fuzziness and soft computing, vol 319. Springer-Verlag, Germany. ISBN: 978-3-319-12882-5

Swarm Intelligence PID Controller Tuning for AVR System

Naglaa K. Bahgaat and M.A. Moustafa Hassan

Abstract The voltage regulator is designed to automatically maintain a constant voltage level in the power system. It may be used to regulate one or more AC or DC voltages in power systems. Voltage regulator may be designed as a simple "feed-forward" or may include "negative feedback" control loops. Depending on the design, it may use an electromechanical mechanism, or electronic components. The role of an AVR is to keep constant the output voltage of the generator in a specified range. The PID controller can used to provide the control requirements.The chapter discusses the methods to get the best possible tuning controller parameters for an automatic voltage regulator (AVR) system of a synchronous generator. It was necessary to use PID controller to increase the stability margin and to improve performance of the system. Some modern techniques were defined. These techniques as Particle Swarm Optimization (PSO), also it illustrates the use of a Adaptive Weight Particle Swarm Optimization (AWPSO), Adaptive Acceleration Coefficients based PSO, (AACPSO), Adaptive Acceleration Coefficients based PSO (AACPSO). Furthermore, it introduces a new modification for AACPSO technique, Modified Adaptive Acceleration Coefficients based PSO (MAAPSO) is the new technique which will be discussed inside the chapter, A comparison between the results of all methods used will be given in this chapter. Simulation for comparison between the proposed methods will be displayed. The obtained results are promising.

Keywords Automatic Voltage Regulator · PID controller · Particle Swarm Optimization · Adaptive Weight Particle Swarm Optimization · Adaptive Acceleration Coefficients based Particle Swarm Optimization · Modified Adaptive Acceleration Coefficients based PSO

N.K. Bahgaat (✉)
Electrical Communication Department, Faculty of Engineering,
Canadian International College (CIC), 6 October City, Giza, Egypt
e-mail: nkahgaat@hotmail.com; n_mohamed2004@yahoo.com

M.A. Moustafa Hassan
Electrical Power Engineering Department, Faculty of Engineering,
Cairo University, Giza, Egypt
e-mail: mmustafa@eng.cu.edu.eg

A.T. Azar and S. Vaidyanathan (eds.), *Advances in Chaos Theory and Intelligent Control*, Studies in Fuzziness and Soft Computing 337,
DOI 10.1007/978-3-319-30340-6_33

1 Introduction

The main goal of the Automatic Voltage Regulator (AVR) loop is to control the magnitude of the terminal voltage V of the generator in the power system. The Dc signal, being proportional to |V|, is compared with a dc reference $|V|_{ref}$, the resulting "error voltage", after amplification and signal shaping, serves as the input to the exciter which finally delivers the voltage V_f to the generator field winding [11, 22, 37]. A simple AVR consists of amplifier, exciter, generator and sensor. There are many controllers used in practice in order to maintain the AVR in the power system in normal operating state,

In Previous works on AVR system with self-tuning control was initiated in the years of 1990s. Sweden bank and coworkers carried out the classical self-tuning control techniques to the AVR system in 1999 [13, 36]. After this study, Fitch used a generalized projecting control technique as a self-tuning control algorithm in the same year [18]. Since the conventional self-tuning control methods contains more mathematical calculation a conditions due to the complexity of the power systems such as nonlinear load characteristics and variable operating points. The usage of artificial intelligence based self-tuning controllers was preferred by researchers from the beginning of 2000 [14]. In particular, self-tuning PID type controllers which were tuned with theoptimization methods based on artificial intelligence have been initiated to carry out the AVR system since then. Gaing suggested a PSO based self tuning PID controller for AVR. In 2006, Kim and colleagues developed the hybrid method which contains genetic algorithm and bacterial foraging optimization technique in order to improve the performance of self-tuning PID controller in AVR system [26]. In 2007, Mukherjee andGhoshal reported the Sugeno fuzzy logic self-tuning algorithm based on crazy-PSO for PID controller [12, 23, 31].

This chapter discusses the application of some modern techniques used for tuning the controller parameters for an automatic voltage regulator (AVR) system. suchas Particle Swarm Optimization (PSO) as described in [19, 30, 32], Adaptive Weighted Particle Swarm Optimization techniques (AWPSO), Adaptive Accelerated Coefficients based on PSO (AACPSO) and Evolutionary Particle Swarm Optimization (MAACPSO). It will be used to determine the parameters of a PID controller according to the system dynamics [7, 9].

This chapter is prepared as follow: Sect. 1 introduces the chapter. The second section presents literature review of the study. Section 3 introduces the automatic voltage regulation equations, Particle Swarm Optimization (PSO), AWPSO, the Adaptive Accelerated Coefficients based PSO (AACPSO) and Modified Adaptive Accelerated Coefficients based PSO (MAACPSO). While Sect. 4 displays the case study and presents all the results and a comparative study between these methods, Furthermore, Sect. 5 concludes the chapter, and Finally discuss the future work. A list of references and Appendix of this chapter are given at the end of the chapter.

2 Literature Review

2.1 Linear Model of an AVR System

The role of an AVR is to keep constant output voltage of the generator in a specified range. A simple AVR consists of amplifier, exciter, generator and sensor. The model and the block diagram of AVR with PID controller is shown in Fig. 1.

The linear models of the elements of the AVR are given as the following [27]:

(a) PID Controller Model

The transfer function of PID controller is

$$G_c(S) = K_p + K_d S + \frac{K_i}{S} \tag{1}$$

where: kp, kd, and ki are the proportion coefficient, Differential coefficient,and integral coefficient, respectively.

(b) Amplifier Model

The transfer function of amplifier model is:

$$\frac{V_R(s)}{V_e(S)} = \frac{K_A}{1 + \tau_A S} \tag{2}$$

where K_A is an amplifier gain and τ_A is a time constant.

(c) Exciter Model

The transfer function of exciter model is:

$$\frac{V_F(s)}{V_R(S)} = \frac{K_E}{1 + \tau_E S} \tag{3}$$

where K_E is an amplifier gain and τ_E is a time constant.

(d) Generator Model

The transfer function of generator model is

$$\frac{V_t(s)}{V_F(S)} = \frac{K_G}{1 + \tau_G S} \tag{4}$$

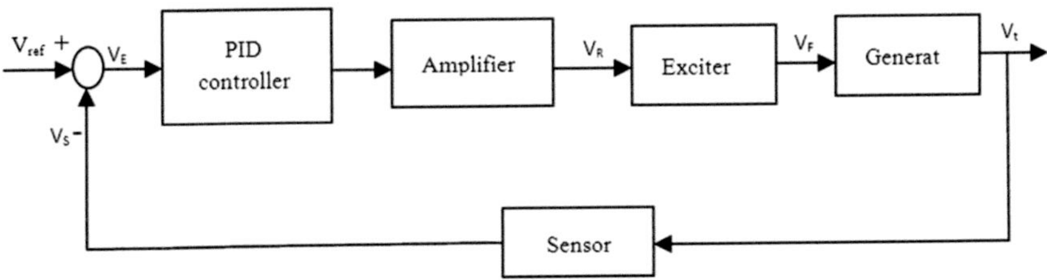

Fig. 1 The block diagram of the AVR

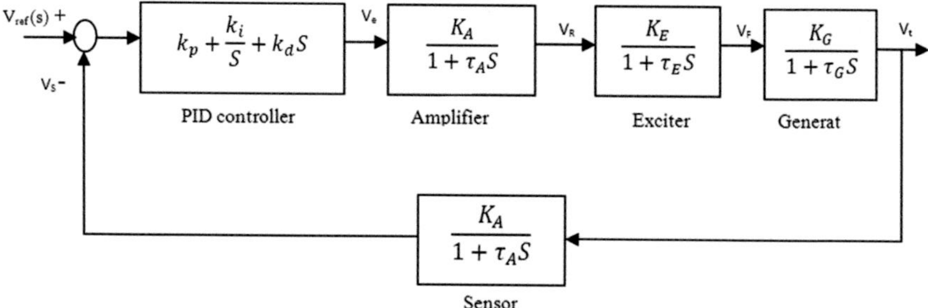

Fig. 2 Block diagram of an AVR system with a PID controller

where K_G is an amplifier gain and τ_G is a time constant.
(e) Sensor Model

The transfer function of sensor model is

$$\frac{V_S(s)}{V_t(S)} = \frac{K_R}{1 + \tau_R S} \tag{5}$$

where K_R is an amplifier gain and τ_R is a time constant.

Figure 2 presents the Block diagram of an AVR system with PID controller as the following:

2.2 PID controller

PID controller used more than 95 % in control applications. Also, they state that 30 % of the PID loops operate in the manual mode and 25 % of PID loops really operate under default factory settings. PID parameters can be achieved manually by trial and error, and by using many modern techniques as presented in [5, 9, 15, 41].

PID control is a linear control. Which consists of three parameters, these parameters operate on the error signal, which is the difference between the desired output and the actual output, and generate the actuating signal that drives the plant. They have three basic terms: the first one is the proportional action, in which the actuating signal is proportional to the error signal, the second on is the integral action, where the actuating signal is proportional to the time integral of the error signal; the last one is the derivative action, where the actuating signal is proportional to the time derivative of the error signal as presented in [7, 37]. The PID standard form is given by:

$$u = K_P \cdot e + K_i \int e \cdot dt + K_d \cdot \frac{de}{dt} \tag{6}$$

where:

e Is the error signal.
u Is the control action.
Kp Is the proportional gain
Ki Is the integral gain
Kd Is the derivative gain

Using conventional PD, PI, PID controllers does not provide enough control performance with the effect of governor dead band [34]. So many methods used since 1890 s till now to tune the controller. First method is the manual tuning. In this method the parameters (KP, Ki and Kd) have to be adjusted to arrive at acceptable performance.Second method is an automatic method called Ziegler–Nichols method which introduced by John G. Ziegler and Nathaniel B. Nichols in the 1940s [34]. In this method the control systems has an S-shaped curve called the process reaction curve and can be generated experimentally or from dynamic simulation of the plant as presented in [9, 16, 17, 37].

Some software can used to tuning the PID controller [28], such as PSO, AWPSO [30, 32], AACPSO and MAACPSO [2, 6, 9]. This chapter will present the design of the PID controller by using Particle Swarm Optimization(PSO), Adaptive Weighted Particle Swarm Optimization (AWPSO), Adaptive Accelerated Coefficients based PSO (AACPSO) and Modified Adaptive Accelerated Coefficients PSO (MAACPSO) by using the simulation based on MATLAB software program. This computer program which written on MATLAB and run many times until reaching to the best solution of the transfer function to have a value of PID parameters. These parameters produce the smallest value of settling time and over shoot. so, these values of PID parameters are the best values to reach. In the following there are a little review of these modern techniques and the equation of each one of them.

2.3 *Particle Swarm Optimization (PSO)*

PSO is a technique based on the behavior of the birds, as illustrated in [35]. particleacts individually and accelerates toward the best personal location (Pbest) while checking the fitness value of its current position. Fitness value of a position is obtained by evaluating the so-called fitness function at that location. If a particles' current location has a better fitness value than that of its current (Pbest), then the (Pbest) is replaced by the current location [8–10]. Each particle in the swarm has knowledge of the location with best fitness value of the whole swarm which is called the global best or (g best). At each point along their path, each particle also compares the fitness value of their (P best) to that of (g best). If any particle has a (P best) with better fitness value than that of current (gbest), then the current (g best) is replaced by that particle's (P best). The movement of particles is stopped once all particles reach close to the position with best fitness value of the swarm.

Let the particle of the swarm is represented by the N dimensional vector ith then the equations of the swarm as the following:

$$X_i = (X_1, X_2, X_3, \ldots X_N) \tag{7}$$

The previous best position of the Nth particles is recorded and represented as follows:

$$Pbest_i = (Pbest_1, Pbest_2, \ldots, Pbest_N) \tag{8}$$

where Pbest is Particle best position (m), N is the total number of iterations.

The best position of the particle among all particles in the swarm is represented by gbest the velocity of the particle is represented as follows:

$$Vi = (V1, V2, \ldots VN) \tag{9}$$

where V_i is the velocity of each i particle.

The modified velocity and position of each particle can be calculated from the current velocity and the distance from particle current position to particle best position P_{best} and to global best position g_{best} as shown in the following Equations [9]

$$V_i(t) = W \cdot V_i(t-1) + C_1 \cdot rand(0,1) \cdot (P_{best} - X_i(t-1)) + C_2 \cdot rand(0,1) \cdot (g_{best} - X_i(t-1)) \tag{10}$$

$$X_i(t) = X_i(t-1) + V_i(t) \tag{11}$$

$$i = 1, 2, 3 \ldots N \tag{12}$$

$$j = 1, 2, 3 \ldots D \tag{13}$$

where:

Vi(t) Velocity of the particle i at iteration t (m/s)
X i(t) The Current position of particle i at iteration t (m)
D The Dimension
C1 The cognitive acceleration coefficient and it is a positive number
C2 Social acceleration coefficient and it is a positive number
rand [0,1] A random number obtained from a uniform random distribution function in the interval [0,1]
gbest The Global best position (m)
W The Inertia weight

2.4 *Adaptive Weighted Particle Swarm Optimization*

Adaptive Weighted Particle Swarm Optimization (AWPSO) technique used is a modification of PSO in multi-objective optimization problems as presented in [1, 30]. AWPSO is consists of two terms which are: inertia weigh (W) and Acceleration factor (A) as illustrated in [10, 32]. The inertia weightformula is as follows which makes W value changes randomly from W_o to 1 as shown in [3, 4, 9].

$$W = W_o + \text{rand}(0, 1)(1 - W_o) \tag{14}$$

where:

Wo The initial positive constant in the interval chosen from [0, 1].

Particle velocity at ith iteration as follows:

$$V_i(t) = W \cdot V_i(t-1) + AC_1 \cdot \text{rand}(0,1) \cdot (P_{best} - X_i(t-1)) + AC_2 \cdot \text{rand}(0,1) \cdot (g_{best} - X_i(t-1)) \tag{15}$$

Additional term denoted by A called acceleration factor is added in the original velocity equation to improve the swarm search.

The acceleration factor formula is given as follows [32]:

$$A = A_o + \frac{i}{n} \tag{16}$$

where:

A_o: Is the initial positive constant in the interval [0.5, 1].
n: is the number of iteration.
C_1 and C_2: Are the constant representing the weighing of the stochastic acceleration terms that pull each particle towards P_{best} and g_{best} positions.

2.5 *Adaptive Accelerated Coefficients Based PSO*

This modification have emerged to improve AWPSO Algorithms, as Time-Varying Acceleration Coefficients (TVAC), where C_1 and C_2 which described in [20]; C_1 and C_2 values will be change linearly with time,This method studies how to deal with inertia weight and acceleration factors and how to change acceleration coefficients exponentially as presented in [2, 21, 38]. The parameters C_1 and C_2 vary adaptively according to the fitness value of G_{best} and P_{best}, as presented in [9, 20] becomes:

$$V_i^{(t+1)} = w^{(t)} V_i^{(t)} + C_1^{(t)} r_1 * \left(Pbest_i^{(t)} - X_i^{(t)}\right) + C_2^{(t)} r_2 * \left(Gbest^{(t)} - X_i^{(t)}\right) \tag{17}$$

$$w^{(t)} = w_o * \exp\left(-\alpha_w * t\right) \tag{18}$$

$$C_1^{(t)} = C_{1o} * \exp\left(-\alpha_c * t * k_c^{(t)}\right) \tag{19}$$

$$C_2^{(t)} = C_{2o} * \exp\left(\alpha_c * t * k_c^{(t)}\right) \tag{20}$$

$$\alpha_c = \frac{-1}{t_{max}} \ln\left(\frac{C_{2o}}{C_{1o}}\right) \tag{21}$$

$$k_c^{(t)} = \frac{\left(F_m^{(t)} - Gbest^{(t)}\right)}{F_m^{(t)}} \tag{22}$$

where:

$w^{(t)}$	The inertia weight factor
$C_1^{(t)}$	Acceleration coefficient at iteration t
i	Equal 1 or 2
t	The iteration number
ln	The neperian logarithm
α_w	Is determined with respect to initial and final values of ω with the same manner as αc described in [10, 24].
$k_c^{(t)}$	Determined based on the fitness value of Gbest and Pbest at iteration t
ω_o, c_{io}	c_{io} initial values of inertia weight factor and acceleration coefficients respectively with $i = 1$ or 2.
$F_m^{(t)}$	The mean value of the best positions related to all particles at iteration t

2.6 *Modified Adaptive Accelerated Coefficients PSO*

Modified Adaptive Accelerated Coefficients PSO will be described as presented in [25]. In this modification the first choice the value of C1 and C2 which described in the last section. Then C1 will be changes exponentially (with inertia weight) in the time, with respect to their minimal and maximal values.While, the other one C2 changes as a factor of the first coefficient. The choice of the exponential function is justified by the increasing or decreasing speed of such a function to accelerate the process of the algorithm and to get better search in the exploration s pace.

In this method the value of C1 which presented in the Eq. (20) the parameter $C_2(t)$ is suggested to be equal the following as described in [25]:

$$C_2(t) = 4 - C_1(t) \tag{23}$$

3 Cases Study

The objective of the system studies is to observe the performance of AVR in the power system. In this work the SIMULINK model was used to obtain the Simulation result using MATLAB program [33, 39, 40]. Figure 3 describes the SIMULINK of the model used [29].

The parameters value of the system presented in Table 1.

The load varies by 0.8 %; by using MATLAB program and choice the type of error which used in the program, there are three types of error as the following:

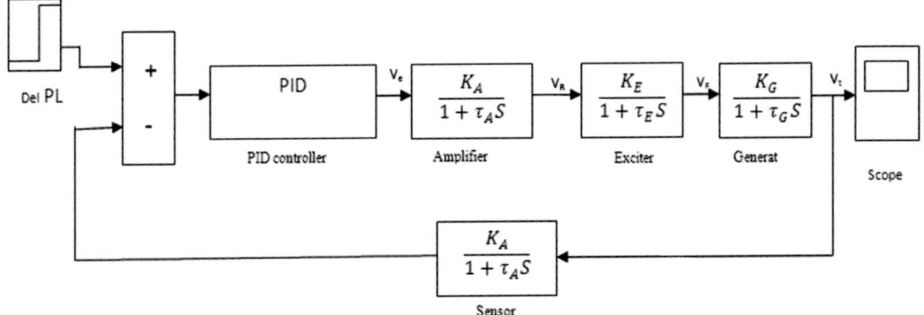

Fig. 3 SIMULINK model of AVR with PID Controller

Table 1 The value of the parameters in the system

Parameter	The value
Kg	1
Ka	10
Ke	1
Kr	1
Tg	0.02
Ta	0.1
Te	0.4
Tr	0.05

For Integral of Absolute Error (IAE):

$$\text{IAE} = \int_0^\infty |e(t)|\,dt \tag{24}$$

Integral of Squared Error (ISE)

$$\text{ISE} = \int_0^\infty e^2(t)\,dt \tag{25}$$

Integral of Time Weighted Absolute Error (ITAE)

$$\text{ITAE} = \int_0^\infty t\,|e(t)|\,dt \tag{26}$$

where:

e Is the error
f Is the objective function

In this study the ITAE error was used in the program.

Table 2 The results of the program using PSO, AWPSO, AACPSO and MAACPSO

Items of comparison	PSO	AWPSO	AACPSO	MAACPSO
Number of iterations	100	100	100	100
Error ITAE	0.0611	0.0252	0.0149	0.0267
Settling time (sec)	18.4281	13.9323	8.6514	3.7267
Over shoot	0.85	0.98	0.97	0.985
Kp	2.4	8.1	9.1	3.7
Ki	1.5	7.5	8.46	5.07
Kd	1.3	2.7	3.2	1.8

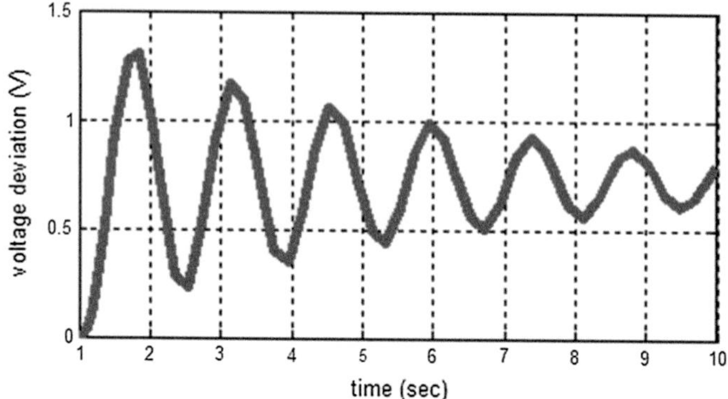

Fig. 4 The power system response before using the controller

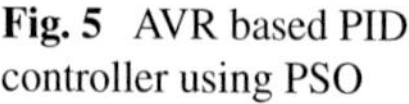

Fig. 5 AVR based PID controller using PSO

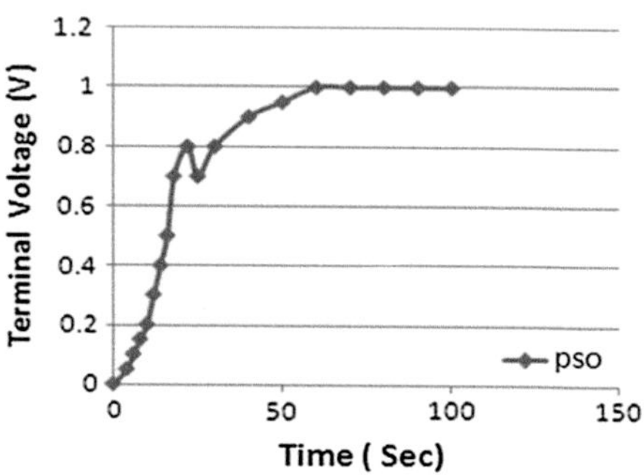

The program used the equations for PSO as shown above, after that the program was repeated using the equations of AWPSO, AACPSO and finally using EAACPSO, the results of the PID controller using different methods are shown in Table 2 and Fig. 5. Figure 4 presents the power system response before using the controller and Figs. 5, 6, 7, 8 and 9 below shows the terminal voltage response for change in load and regulation which obtained. It could be observed that the response settles in about

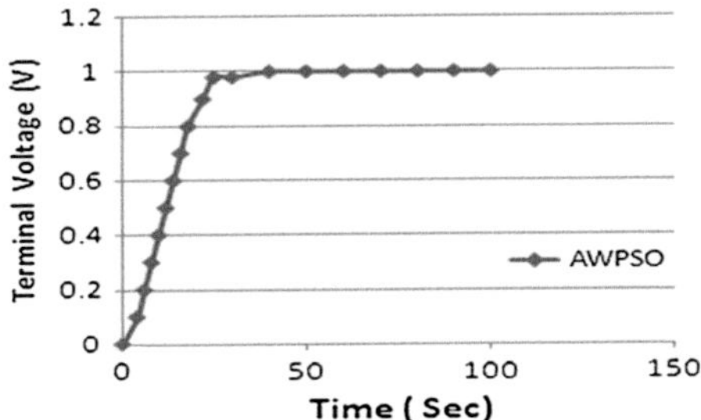

Fig. 6 AVR based PID controller using AWPSO

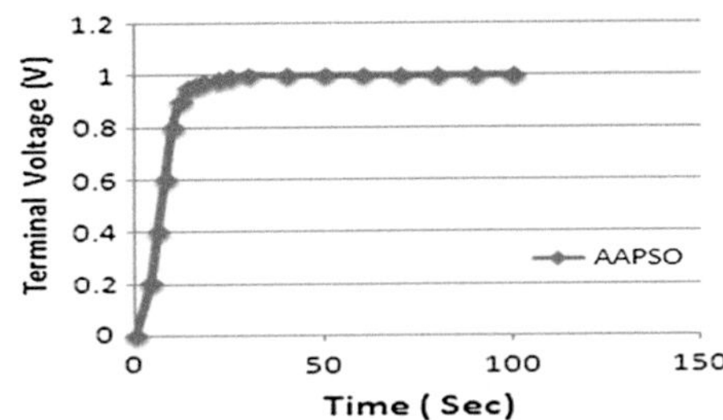

Fig. 7 AVR based PID controller using AAPSO

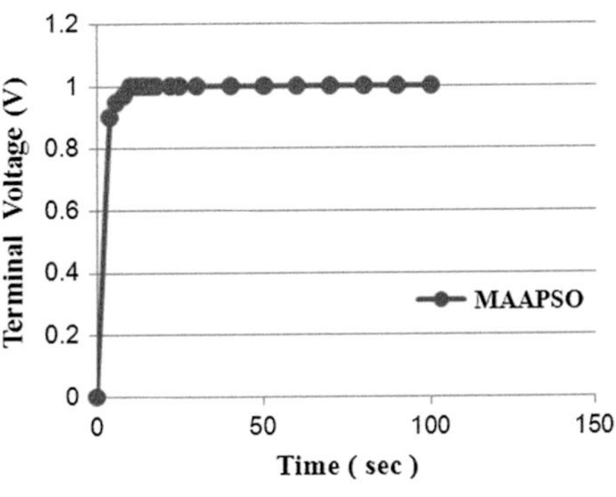

Fig. 8 AVR based PID controller using MAAPSO

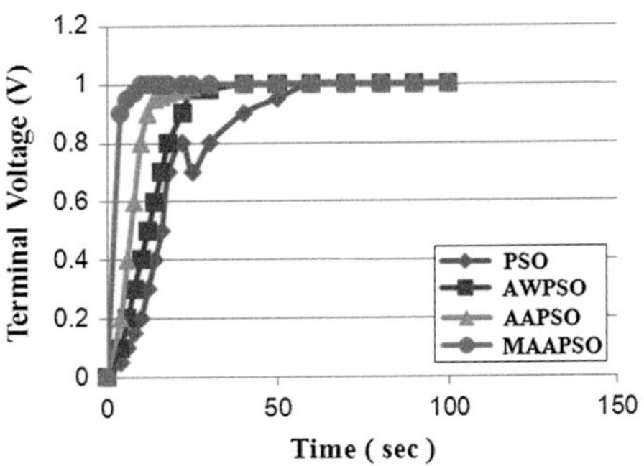

Fig. 9 AVR based PID controller using PSO, AWPSO, AAPSO and MAAPSO

3.7 s with very small overshoot. This results indicate that the PID using EAACPSO controller reduces the steady state error to zero. It is observed that the settling time of AVR with PID controller using MAAPSO is less and there is no transient peak overshoot. Other methods used give good results also, but the best method used is MAAPSO which give the smallest value of settling time and over shoot.

4 Discussion and Conclusion

The quality of controller used in power system is determined by constancy of frequency and voltage. Minimum frequency deviation and good terminal voltage level.the characteristic of a reliable power system. The AVR loop control is very important to be study to control the voltage and make the deviation of it more small as possible. It can be concluded that, the PID controllers provide a satisfactory stability between overshoot and transient oscillations with zero steady state error. The simulation results presented the difference between the methods used to tune the PID controller to reach to the best values of settling time and over shoot response when any changing of the load happen. From the results shown in Table 2 and Fig. 5 when evolutionary algorithms are applied to control system problems, their typical characteristics show a faster and smoother response.

An intelligent techniques have been proposed and can be implemented to improve The performance characteristics of the PID controller of the system. It is clear from the results that the proposed MAAPSO method can obtain higher quality solution with better computation efficiency than the other methods.

5 Future Work

AVR with load frequency control will be the future work to study the effect of changing the load frequency on the AVR control loop by using all methods as described before.

References

1. Allaoua B (2008) The efficiency of particle swarm optimization applied on fuzzy logic DC motor speed control. Serbian J Electr Eng 5(2):247–262
2. Azar AT, Serrano FE (2014) Robust IMC-PID tuning for cascade control systems with gain and phase margin specifications', Neural Computing and Applications, 25(5): 983–995 Springer. doi:10.1007/s00521-014-1560-x
3. Azar AT (2010) Fuzzy systems. IN-TECH, Austria, Vienna. ISBN: 978-953-7619-92-3
4. Azar AT (2012) Overview of type-2 fuzzy logic systems. Int J Fuzzy Syst Appl (IJFSA) 2(4): 1–28

5. Azar AT, Serrano FE (2015) Design and modeling of anti wind up PID controllers. In: Zhu Q, Azar AT (eds) Complex system modeling and control through intelligent soft computations. Studies in fuzziness and soft computing, vol 319. Springer, Germany, pp 1–44 doi:10.1007/978-3-319-12883-2_1.1
6. Azar AT, Vaidyanathan S (2015a) Handbook of research on advanced intelligent control engineering and automation. Advances in computational intelligence and robotics (ACIR). Book Series, IGI Global, USA
7. Azar AT, Vaidyanathan S (2015c) Computational intelligence applications in modeling and control studies in computational intelligence, vol 575. Springer, Germany. ISBN: 978-3-319-11016-5
8. Azar AT, Vaidyanathan S (2015b) Chaos modeling and control systems design. Studies in computational intelligence, vol 581. Springer, Germany. ISBN: 978-3-319-13131-3
9. Bahgaat NK, El-Sayed MI, Moustafa Hassan MA, Bendary FA (2014) Load frequency control in power system via improving PID controller based on particle swarm optimization and ANFIS techniques. Int J Syst Dyn Appl (IJSDA). IGI-Global, USA 3(3):1–24
10. Bahgaat NK, El-Sayed MI, Moustafa Hassan MA, Bendary FA (2015) Application of some modern techniques in load frequency control in power systems. In: Chaos Modeling and Control Systems Design, vol 581. Springer, Germany, pp 163–211
11. Bahgaat NK (2013) Artificial intelligent based controller for frequency control in power systems, Ph.D. Thesis, Faculty of Engineering at Al-Azhar University, Cairo, Egypt
12. Bevrani H (2009) Robust power system frequency control. Springer Science and Business Media LLC, Brisbane, Australia
13. Bhati S, Nitnawwre D (2012) Genetic optimization tuning of an automatic voltage regulator system. Int J Sci Eng Technol 1:120–124
14. Bhatt VK, Bhongade S (2013) Design of PID controller in automatic voltage regulator (AVR) system using PSO technique. Int J Eng Res Appl (IJERA) 3(4):1480–1485
15. Cheng G (1997) Genetic algorithms and engineering design. Wiley, New York
16. Coello CAC, Pulido GT, Lechuga MS (2004) Handling multiple objectives with particle swarm optimization. IEEE Trans Evol Comput 8(3):256–279
17. Eberhart RC, Shi Y (1998) Comparison between genetic algorithms and particle swarm optimization. In: Proceedings of IEEE International Conference on Evolution in Computing. Anchorage, AK, May, pp 611–616
18. Fitch JW, Zachariah KJ, Farsi M (1999) Turbo generator self-tuning automatic voltage regulator. IEEE Trans Energy Convers 14(3):843–848
19. Gaing ZL (2004) A particle swarm optimization approach for optimum design of PID controller in AVR system. IEEE Trans Energy Convers 19(2):384–391
20. Hamid A, Abdul-Rahman TK (2010) Short term load forecasting using an artificial neural network trained by artificial immune system learning algorithm in computer modeling and simulation (UKSim). In: 12th international conference on IEEE, pp 408–4131
21. Hassanien AE, Tolba M, Azar AT (2014) Advanced machine learning technologies and applications: second international conference, AMLTA 2014. Cairo, Egypt, 28–30 Nov 2014. In: Proceedings, communications in computer and information science, vol 488. GmbH, Springer, Heidelberg. ISBN: 978-3-319-13460-4
22. Ingemar EO (1983) Electric energy systems theory. McGrawhill Book Company, London
23. Ismail A (2006) Improving UAE power systems control performance by using combined LFC and AVR. In: 7th UAE University Research Conference on Engineering, pp 50–60
24. Kennedy J, Eberhart RC (1995) A new optimizer using particle swarm theory. In: Proceedings of 6th international symposium micro machine human science, Nagoya, Japan, pp 39–43
25. Khalifa F, Moustafa Hassan M, Abul-Haggag O, Mahmoud H (2015) The application of evolutionary computational techniques in medium term forecasting accepted for presentation at MEPCON'2015, Mansoura University, Mansoura, Egypt
26. Kim DH, Cho JH (2006) A biologically inspired intelligent PID controller tuning for AVR systems. Int J Control Autom Syst 4:624–636

27. Kumar A, Gupta R (2013) Compare the results of tuning of PID controller by using PSO and GA technique for AVR system. Int J Adv Res Comput Eng Technol (IJARCET) 2(6)
28. Kumar DV (1998) Intelligent controllers for automatic generation control. In: IEEE Region, 10 International Conference on Global Connectivity in Energy, Computer, Communication and Control, TENCON'98, vol 2, pp 557–574
29. Musa BU, Kalli BM, Kalli S (2013) Modeling and simulation of LFC and AVR with PID controller. Int J Eng Sci Invent 2:54–57
30. Naik RS, ChandraSekhar K, Vaisakh K (2005) Adaptive PSO based optimal fuzzy controller design for AGC equipped with SMES and SPSS. J Theor Appl Inf Technol 7(1):008–017
31. RamaSudha K, Vakula VS, Shanthi RV (2010) PSO based design of robust controller for two area load frequency control with nonlinearities. Int J Eng Sci 2(5):1311–1324
32. Rania HM (2012) Development of advanced controllers using adaptive weighted PSO algorithm with applications, M.Sc. Thesis, Faculty of Engineering, Cairo University, Cairo, Egypt
33. Shabib G, Abdel Gayed M, Rashwan AM (2010) Optimal tuning of PID controller for AVR system using modified particle swarm optimization. In: Proceedings of the 14th International Middle East Power Systems Conference (MEPCON'10), Cairo University, Egypt
34. Skogestad S (2003) Simple analytic rules for model reduction and PID controller tuning. J Process Control 13(4):291–309
35. Soundarrajan A, Sumathi S (2010) Particle swarm optimization based LFC and AVR of autonomous power generating system. IAENG Int J Comput Sci
36. Swidenbank E, Brown MD, Flynn D (1999) Self-tuning turbine generator control for power plant. Mechatronics 9:513–537
37. Tammam MA (2011) Multi objective genetic algorithm controllers tuning for load frequency control in electric power systems, M.Sc. Thesis, Faculty of Engineering at Cairo University, Cairo, Egypt
38. Vlachogiannis JG, Lee KY (2009) Economic load dispatch—acomparative study on heuristic optimization techniques with an improved coordinated aggregation based PSO. IEEE Trans Power Syst 24(2):991–1001
39. Wong CC, An Li S, Wang H (2009) Optimal PID controller design for AVR system. Tamkang J Sci Eng 12(3):259–270
40. Yoshida H, Kawata K, Fukuyama Y (2000) A particle swarm optimization for reactive power and voltage control considering voltage security assessment. IEEE Trans Power Syst 15: 1232–1239
41. Zhu Q, Azar AT (2015) Complex system modelling and control through intelligent soft computations. Studies in fuzziness and soft computing, vol 319. Springer, Germany. ISBN: 978-3-319-12882-5

Discrete Event Behavior-Based Distributed Architecture Design for Autonomous Intelligent Control of Mobile Robots with Embedded Petri Nets

Gen'ichi Yasuda

Abstract This chapter presents a design methodology of discrete event distributed control architecture for autonomous mobile robot systems. A modular, behavior-based distributed software architecture is presented on a hierarchical distributed microcontroller based hardware structure for intelligent control of mobile robots. Some intelligent behaviors, such as wall following, obstacle rounding, target seeking, and local environment mapping, have been implemented using sensor control modules such as multiple infrared range finding sensor modules and motion control modules to detect walls and obstacles in the surroundings of a mobile robot, based on environment features such as lines and corners estimated using a set of range sensors and a vision sensor. Upon these behavior modules, a Petri net based approach was applied to coordination of several concurrent activities of modules for the high-level tasks such as sensory navigation in unknown environments. Task specification implies the definition of a control program composed of behavior commands, which are not expressed in a sequential fashion but implicating parallel processing control. The net model can be directly obtained from the system requirements specification of each particular application. Thus, the remaining levels of the control structure are common to a wide range of applications. The Petri net based approach validates the implementation of synchronization and coordination in discrete event behavior-based control. Behavior modules are composed to design more complex modules according to applications. The detailed function of each control module is specialized according to the application, so that new control strategies can be easily embedded in the control modules for real-time performance of robotic actions. Compared to hand–written coding in robot program, because of explicit representation of robotic actions, behaviors and tasks, the system design procedure facilitates the understanding of the interaction among the different processes that might be present in the mobile robot control system. Consequently, it is easy and computationally inexpensive to design, write, and debug planned tasks. Besides it is possible to verify structural and behavioral properties of these programs owing to formal specification.

G. Yasuda (✉)
Nagasaki Institute of Applied Science, Nagasaki, Japan
e-mail: yasuda_genichi@pilot.nias.ac.jp; yasuda.genichi@gmail.com

A.T. Azar and S. Vaidyanathan (eds.), *Advances in Chaos Theory and Intelligent Control*, Studies in Fuzziness and Soft Computing 337,
DOI 10.1007/978-3-319-30340-6_34

Keywords Intelligent mobile robots · Behavior coordination · Task decomposition · Control system design · Discrete event control · Petri nets

1 Introduction

This chapter concerns the design and implementation of distributed control architectures for intelligent mobile robot systems. Focusing on building control systems for intelligent mobile robots, a behavior-based methodology for autonomous cooperative control is presented using discrete event net models as a formal approach. One of the main factors that have deeply restricted the development of intelligent robots is the problem of control system design. From the external viewpoint of autonomous robot behavior, the fundamental feature of the control system is that there are different programs, or processes, to be performed asynchronously and concurrently, such as locomotion control, image processing and/or rangefinder data processing for environment recognition and object detection, path planning, path execution with object avoidance, decision making for special operations in uncertain environments, etc. under a set of timing requirements [30]. At the lowest control level, actuators and sensors control, such as vehicle control and image capture, are prerequisite for task execution. Given techniques for controlling the robot motion and for integrating sensory information at some higher control levels, the principal problem is to develop a systematic methodology to combine all of these into one complete robot control system based on asynchronous, concurrent processes in a real-world application. Since the operations of a robotic system are basically processed in parallel, such as moving while path sensing, path tracking execution in parallel with local path planning, etc., it would be more efficient for the control system to process in parallel than in serial because of simplicity, load balancing, and real-time performance. In this context, modularity, localization of control and parallelism are critical issues in control system design.

Currently most of intelligent robots are application-specific and its programmability is limited and the reconfigurability is not embedded. So the coordination among robots demands a lot of machine-level programming to perform a specific task. According to the concept of the distributed autonomous robotic systems, some of these operations can be built into certain local robotic modules which can be also reorganized for different applications. A modular robotic control system should be decomposed to accommodate several independent simple jobs, tasks, operations at the same time, or be integrated to perform a more sophisticated task when it is needed. The essential concept of the modular robotic system is that the system can be considered as a set of multiple connected subsystems, where each subsystem has a minimum one degree of freedom, and its local intelligence is capable of executing some basic functions in sensing and actuation [14].

To accomplish a given task, all the subsystems would act in an aggregative manner to optimize the usage of available resources, such as multiple actuation and sensing devices. Thus, the optimization of the global-local interaction, where each modularized subsystem is governed by only one local center, is critical to the success of developing autonomous mobile robot systems to ensure the efficiency and the effectiveness, searching through multiple local optima to reach the global optimum. The essential purpose of utilizing parallel distributed processing such as neural networks and fuzzy logic is to build the robot autonomy through the establishment of the trainability and the self-learning capability of robot intelligence [3–6, 35]. Since, when the robot intelligence is localized, the concept of modularization would be meaningful, the hierarchically distributed intelligence, instead of predetermined precise commands, should enable a general command or task description to figure out how to execute the fuzzy command and accomplish the given task adequately in the unstructured environment through the adequate task planning. For intelligent robots with multiple redundant degrees of freedom, where unique solution to the associated inverse kinematics problem can not be obtained, the combination of modularization and parallel distributed controller design would adapt the command fuzziness with the redundancy of the controlled system, to support the system requirements for modularization of robot intelligence including the reconfiguration of global-local interaction. Thus, the distributed autonomous control system design would provide the necessary flexibility to overcome some of limitations of current mechatronic system design.

The main contribution of this chapter is to propose a novel, systematic method of the design and implementation of functionally distributed control architecture for real-time control of intelligent mobile robots. The architecture supports hierarchical task specification in terms of event driven state based Petri net modeling to provide flexible, efficient and high quality mission performance. All processes for the motion task have been integrated in hierarchical autonomous modular software of behavior and control modules designed to consider the real-time constraints of each control level of the system, especially to attain the synchronization of the various subsystems involved in the mapping, navigation and intelligent task operation in uncertain environments.

2 Related Works

For intelligent robot control, there are two extremes of control system, the centralized system and distributed system. In a centralized system, each subsystem or robot is only a collection of sensors and actuators with some local feedback loops and almost all tasks are processed in a host PC, single master subsystem or robot. The communication between the host and the other robots only involves sending data from sensors to the host and receiving detailed commands from the host. Because the processing of all of tasks is performed by the host, the defects such as the limitation of processing ability and reliability on fault tolerance might become more obvious as the system becoming larger. Conversely in a distributed system, each robot plans and solves a task problem independently while communicating information, which has been processed in each robot, with the other robots [28]. Interests in mobile robot systems can be dated back to the early 1970s [22, 31–33], when a non-centralized grouping robots system was first presented and shown as a feature of Okinawa International Exhibition [21]. To the best of our knowledge, our research is the first contribution addressing the cooperative formation control of multiple robots in obstacle environments under sensing and communication constraints, with guaranteed visibility and safety maintenance. The control structure is composed of several action modes with priorities: mainly electric barrier avoiding, team mate collision avoidance, and team mate following with one infrared light emitter and three infrared eyes, in the order of priority. Due to small differences of electronic sensibilities, the proposed electronic discrete event control ensures the safe navigation of the multiple robots and shows a line group formation in unknown cluttered environments, without the need for exchanging or estimating velocities between team mates.

In recent years, there have been many studies focusing on cooperation of distributed autonomous robot systems. In robotic applications, a robotic task can be thought of as a compound task composed of multiple primitive tasks with some constraints; one robot is required for each primitive task. A typical example is a two robotic arm system for assembly or welding operations, assisted with mobile robots for fetching parts, providing additional visual information, etc. Multirobot tasks are largely classified into two classes: individual behavior type and collective behavior type. In the former, each robot can be dynamically a master or slave according to situations. In the latter, all the robots are mutually activated, reactively and collectively as slaves, where one mastering entity is from the upper level. For the individual behavior type, mostly the cooperation strategy is based on distributed problem solving and distributed decision. For example, in collision avoidance among mobile robots and respective task accomplishment by multiple robots, each robot gets some information about other robots by communication, and then uses the information for making a distributed decision. Another illustrating example of individual type task is the task of transporting multiple discrete objects.

Compared with the above, for the collective behavior type cooperative tasks with dynamic constraints, such as object manipulation, should have more dynamic factors, and the increase in communication required for achieving harmonious cooperation between moving motion and manipulating motion in a robot makes the physical system difficult to realize in a pure distributed system. The overall collective type task of carrying one large and heavy object can be decomposed into multiple tasks of robots simultaneously holding with different places and moving with it. In contrast to collision avoidance among moving robots, the robots which are working on manipulation, interfere each other dynamically through the manipulated object. So, some information about the object and robots should be obtained for cooperation from sensors on each robot in real time [25]. Another example of collective behavior type cooperative task is the multiple synchronous sensing information collection, where multiple types of sensing information must be synchronously collected from each spatially distributed region by a group of robots. Because some tasks common to the whole system, such as tasks for planning for manipulation or global cooperation are suitable to be processed in a host robot or system controller than in each robot, if the processing of such cooperative tasks common to the whole system is done separately by each robot, the cost of unnecessary and unnatural cooperation due to excessive communication and tautological processing brings about a significant loss to the whole system [17]. On the other hand, since each robot is distributed physically, processing of some tasks is suited to be done separately rather than being concentrated at the host from the view points of flexibility, robustness and fault tolerant ability [12].

Nowadays, robotic cells in which two or more manipulator robots, mobile robots, and other active devices work simultaneously and in cooperation are not uncommon in industry. Although such cells are difficult to implement, they should become more prevalent as industrial application of robots continues to spread. Even in applications in which only one robot is used, there are often autonomous intelligent devices such as feeders, conveyors, or orienting tables whose motions occur simultaneously with robot motion, and they must be coordinated with that of the robot [13]. A central feature of such a cell is that it is a parallel processing system, where a user may define several processes which can then be run in parallel. For complex multirobot cells, the overall system should be naturally decomposed into several robotic subsystems, where each subsystem is coordinated by one or small number of robots. Generally, for a simple robotic cell, one process is defined for each active device in the cell; each robot or device would have its own process. Because tasks are processed concurrently and sequentially, a conflict may occur when two or more processes require an identical resource at the same time. Processes can interact through signal and wait primitives. This mimics the way the actual devices or human workers are likely to be coordinated on the factory floor. In fact, it turns out that initialization and communication instructions far outweigh robot motion commands. Planning and programming such a cell using only a teach pendant or even a robot programming language is tedious. Replacing the “teach by showing” method, robots are programmed by explicitly specifying the desired sequence of motions and actions to be performed with predefined motion and other primitives. Currently, by the development of task description languages, robot programs are created by specifying relatively high-level sub-goals, so that the programmer is freed from having to describe explicitly all robot actions. For example, when a high-level command like “grasp the workpiece” is given, the system plans a collision-free path to the workpiece, automatically determines a grasp location and performs all the necessary motions, although they encompass many general AI problems the solutions for which are often unknown or infeasible. Actually teaching mode is separated with programming mode in a hierarchical structure, such that at the higher level the execution order of operations is programmed and then for each operation or action in the execution program control parameters are specified by teaching. Anyway, it is important to be able to test out the complete interaction among devices in a cell, based on formal modeling tools such as Petri nets, in order to prove the viability of an application. Further, in programming robots and/or other devices which must cooperate in a single cell, it is clear that a planning system with a simulation capability is useful because such a graphic simulation system would allow the user to test several scenarios and investigate the details of synchronization and signaling between the various processes. Because a large amount of equipment cost is necessary for constructing real robot systems or manufacturing systems, it is crucial to analyze and evaluate the performance of systems in the system design stage. In evaluating the performance, simulations are popular as an experimental method, but it takes longer time to simulate the systems. The analytical method such as Markov analysis does not take longer time than simulation. However, as the scale of system increases, the state space also increases and it becomes difficult to analyze the performance. Therefore, a method to decrease the state space by decomposing and combining of analytical models is urgently required. Although the adoption of formal methods can greatly contribute to an improvement of software quality, such as reliability and readability, formal methods are still not used as much as they deserve because of their intrinsic complexity and difficulty for understanding and manipulating compared to a programming language. Thus Petri nets, like other formal methods, are not seen as an alternative to non-formal graphical design languages, such as UML, which are commonly used by the object oriented community.

Behavior-based robot control implies fully parallel control architecture composed of multiple behavior modules. In cases where control should be shared between multiple modules, the output would result from the arbitration of multiple conflicting module outputs. The subsumption architecture is an example of a switched parallel control through a linear suppression mechanism [1, 7, 8]. The architecture emphasizes the decomposition from a task to a set of pure reactive behaviors for real-time control. The subsumption means that a higher-level function of a behavior subsumes a subset of lower-level functions. The robot needs not to be programmed although it is limited to the behaviors implemented by its designer [11]. The behavior network is another example where a behavior is chosen by comparing and updating activation levels or bits for each behavior [2, 18]. Each behavior selects an action, compute run-time parameters, generate a bid describing the appropriateness of the behavior or action. The most appropriate behavior is determined in a distributed manner through inter-behavior bidding without any centralized mechanism [30]. Furthermore, because, in dynamic environment, a robot needs to quickly decide what to do and how to do it, the deliberative components must keep pace with changes in the environment to produce intelligent behavior where each behavior is responsible for planning the required actions [16], so that fixed computational resources need to be distributed among the behaviors.

3 Hierarchical Distributed Architecture for Behavior-Based Mobile Robot Control

The control system was implemented on a microcontroller-based multiprocessor network, where several microcontrollers are connected with one system controller through a shared serial bus. Each microcontroller with specialized interfaces for sensing and actuation is dedicated to intelligent control of a hardware device module. A view of an experimental mobile robot is shown in Fig. 1.

Fig. 1 View of an experimental mobile robot

Figure 2 shows the conceptual framework of hierarchical and distributed control software architecture for integrated deliberative planning and reactive execution for mobile robots. The distinction of planning and reactive control is based on the different time scales of interaction. The software architecture is constructed upon the distributed hardware structure, where each microcomputer-based intelligent device for sensing and actuation is connected via a serial bus. The proposed control architecture is not only functionally parallel/distributed but also hierarchically distinguished for effective development of a robot control system, considering the different abstraction levels that correspond to the different real-time constraints and the complex nature of its signals. It comprehends discrete event or sequential control at the top level and loop control at the lower levels. The upper layers are characterized by soft real time constraints whereas the lower layers are hard real time constrained. The components in all the layers run asynchronously to allow to interact continuously with the environment as well as to perform time consuming computations. The control architecture consists of five layers: device manager, control, network, coordination, and task organization layers. The device manager layer directly controls I/O peripherals such as motors, sensors, operator's communication devices, etc. to process raw sensory data.

The major subsystems of the hierarchical architecture are the coordination and control layers, which are implemented on the system controller. Each control module at the control layer acquires intermediate complex identifiable data computed from raw data by sensory device modules at the device manager layer and achieves primitive goal-directed action control. They are classified into environment recognition modules, including self-localization, self-referencing and landmark and obstacle detection, and motion execution modules such as line following and turning at a specified angle with robot based local environment information. For example, collision avoidance is performed through the cooperation between the obstacle detection and

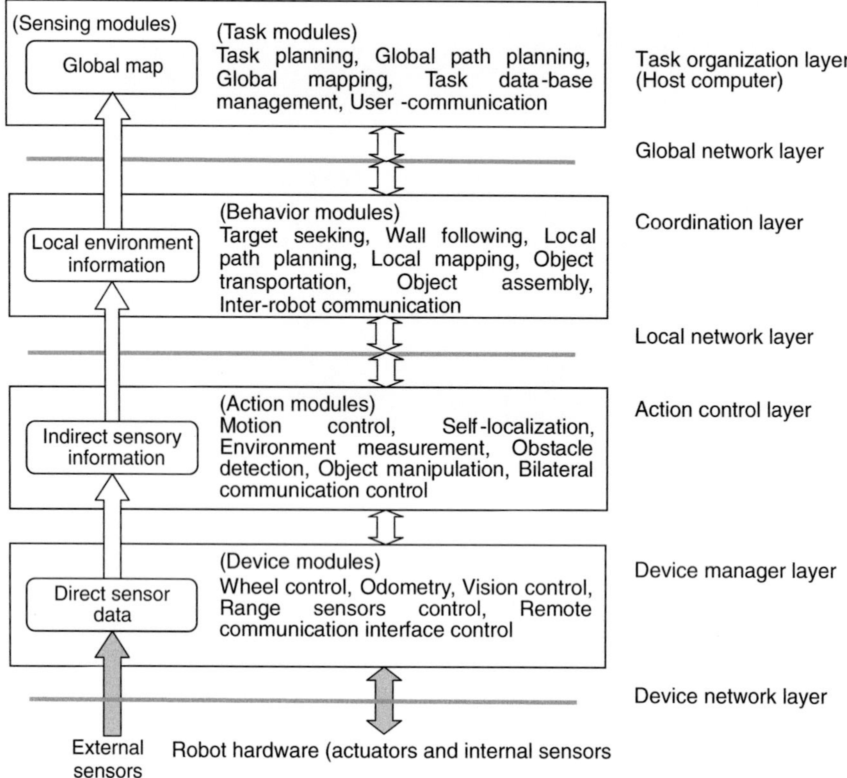

Fig. 2 Hierarchical modular software architecture for mobile robot control

motion control modules. Basically, for the same actuator system, such as arm, leg and hand, one motion exertion module can run at a time. It can deliver driving and steering velocities for the actuator device manager. According to the action groups, their parallel or sequential execution can be decided for efficient multi-agent system computation on a limited computer resource. At the network layer interrupts are used to communicate every event that occurs at the control layer. Each control module has an event handler and can carry out data transfer to suitable control modules or the system controller by only writing data, source address, and destination address in its own message buffer. Control modules communicate mutually using event data for emergency stop, remote control and other high-level information acquisition, such as human recognition.

The coordination layer consists of three parts: a behavior based controller, a local map manager, and a behavior modules database system for integrating behavior based robot control. All the communication relations between the neighboring layers are bidirectional real-time event flows. Control modules issue events to the behavior modules in order to report the current status and to notify the normal termination of an activity or unexpected condition. The coordination layer is in charge of the overall behavior control, coordinating the control modules by sending the activity start events and waiting for the incoming events and clock events. For this reason, the coordination layer is built as a behavior based controller which is a set of concurrent layers that embodies the different behaviors. The set of active behaviors and their resulting cooperation is controlled by the incoming events. Behavior modules have specific inputs and outputs: sensory and actuation information. They may share some of the input information with other modules. Sensory information is hierarchically classified into the raw sensor information, the local map from low-level sensor fusion, and the environmental feature from high-level sensor fusion and recognition, and the environmental model. The access to these data is managed by the shared information manager.

The task organization layer generates a task specification with a global planner for motion task applications. Its event based task description coordinates the system behaviors or subtasks: environment map building, global navigation, and particular manipulating operations. At the task organization layer, global planning, or long distance motion behavior planning, with a short term local map and a more strategic global map, is executed as one small part of the main controller's nominal cycle of activities. Because the global planning may usually meet too many unexpected events, it is difficult to build environment model. So, a geometrical map of the local environment is built in the coordination layer. Given a set of target locations, the coordination layer plans a detailed medium distance local path to the closest target position only and executes this plan. For simplicity, a local path consists of circular arcs and straight line segments. For a moving object, a local path is planned to the predicted future location of the object that is computed using an internal model of the object's dynamics. Upon reaching this target location, its local map will change based on the perceptual information extracted during motion. Only then the coordination layer triggers the local path planner to generate a path from the new location to the next target location. When a reference object is detected by a vision sensor, a trajectory planner is used to generate the short distance motion for docking or manipulation [34]. Figure 3 shows the generic diagram of the unified hardware and software structure for autonomous adaptive robot control.

For example, in wall following, the distance between the wall and the robot is kept constant using the range sensors. For straight-shaped wall segments with no abrupt turning, only the side sensors are used for the robot's quick motion control. The angle between the forward direction of the robot and the wall is estimated using the two range sensors. The reference angular velocity is computed using the angle, the error of distance between the robot and the wall, and the coefficients of feedback control. Thus the robot follows a smooth and close track beside the wall with real-time performance. In obstacle rounding, the robot first uses the wall following behavior to round the obstacle, and when the original direction can be found, the robot ends the behavior and the reference tangent and angular velocities of the mobile robot are determined to meet the target tracking.

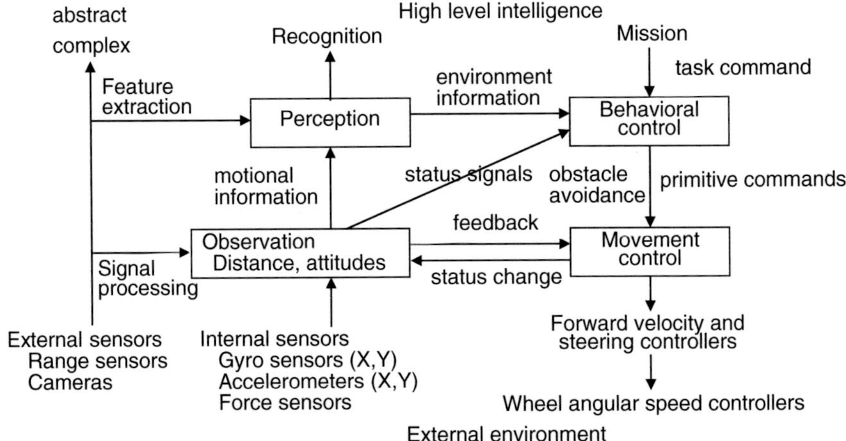

Fig. 3 Unified hardware and software structure for autonomous adaptive robot control

4 System Modeling and Communication of Robotic Activities

In this work, Petri nets are employed to model, design and evaluate a robotic task execution. Petri nets are marked bipartite graphs particularly suitable to discrete event control modeling. They can be seen as a state machine generalization that allows the modeling of resources, parallel and concurrent activities and their synchronization. A Petri net is defined as a directed bipartite graph having a structure and a marking. The structure has four components: a set of places, a set of transitions, a set of input arcs that connect places to transitions, and a set of output arcs that connect transitions to places. Each of these components has a graphical representation. The place is represented by a circle, the transition by a bar, and the input and output mapping by a set of directed arcs. The above structural features define the topology of the net. The marking is a way of showing the current state of the net, where the state of the net defines which places are active. Graphically, tokens, which reside at the places, are used to indicate the instantiation of a state. Formally, the marking vector is defined as a vector, where each element indicates the number of tokens at the corresponding place. A marking changes when a token moves from one place to another place. This is done by firing a transition, which may fire when it is enabled. Discrete event systems are characterized as nonlinear systems, because of nonlinear relationships between initial states and reachable states.

In comparison with finite state automata, combining Petri nets does not lead to the same state explosion as combining finite state automata. It seems more appropriate in a robotic environment to use a Petri net approach, such that each resource is modeled separately and the state of the robot has the state of all its components. The composition of separate Petri nets is done using place composition such that places with the same label are in fact the same place. Therefore, Petri nets can be used as basic modules of a robot system. Since the communication or connection of modules, including finite state machines, can be specified in the Petri net framework, the system net model in a single robot environment can be easily extended to multi-robot environments. Since the connection of finite state machines is no longer a finite state machine, the communication specification between modules can not be achieved using the finite state machine framework. Thus, the primary purpose of this chapter is to establish a control and communication mechanism for task execution of an autonomous intelligent robot system based on Petri net models.

The type of Petri nets considered here is the condition event net in which each place can contain not more than one token. Presence of arcs determines causality relations between conditions and events in the system. A token is placed in a place to indicate that the condition corresponding to the place is holding. An event occurrence is modeled by firing a transition which modifies the conditions. This allows a one-to-one semantic correspondence between places and conditions [23, 24]. The axioms of nets are as follows:

(1) A transition is enabled, if and only if, each of its input places has one token and each of its output places has no token;
(2) When an enabled transition fires, the marking is changed to the new one, where each of its input places has no token and each of output places has one token.

For the modeling of robotic activities, a place represents a condition or state of processes or resources. The global state of the system is defined by the marking of each place. A transition corresponds to an event of state change or robotic activity. For a robotic activity, the input places of the transition define the pre-conditions to the executions of the activity. The output places define the results or post-conditions of the activity [19, 20]. The Petri net model can describe precisely the causality relation of robotic activities in a system, such as sequencing, non-deterministic conflict, concurrency, and synchronization. A non-deterministic conflict between transitions corresponds to a fork, while a synchronization is represented by a join in the graph.

Conceptually, in a robotic activity, the start of the transition firing indicates the start of the process and the completion of the firing indicates the finishing of the process. Data from sensors is trapped by the transition firing and triggers actuators executing the control policy coded in the input tokens. Thus, transitions can be used for interaction with the environment as active elements, or agents for communication, in a Petri net, such that they read data from sensors and also send output data to actuators or other elements in the net. When a transition is enabled and waiting for the start signal, there is a condition that the transition cannot start firing until it receives the signal from an external sensor. When the start signal arrives, the transition sends command signals to the actuators and fires for a predefined period or until it receives a predefined status signal. When a transition starts firing, the tokens in the input places are marked as unavailable, and when the transition fires, the tokens are removed from the input places. Then, when it completes firing, tokens, the number of which depends on the arc weights of the output arcs, are deposited into the output places. The overall code in the transition can be implemented by a detailed ordinary Petri net model or in a general computer language such as C++. As the spectrum of robotic applications broadens toward the general purpose services, the rapid growth of computer technology helps the realization of many great ideas which previously would not be possible. With current C++ language, the object-oriented concept can be employed to develop the robot intelligence software, utilizing the traditional technique in spatial and kinematic planning, and dynamic control as the building blocks.

For real-time cooperative control of some robotic activities, a transition representing a robotic activity is decomposed into start and end transitions and a place representing the process, where tokens inside the place indicate that the process goes on and the processing time is determined by external conditions. Data from sensors trapped by an input transition is fed into the places as tokens, while a place loaded with tokens triggers actuators through an output transition as an ordinary place. In our work, for compact representation of interaction with the environment, the extended Petri net adopts the following elements as input and output interfaces which connect the net to its environment [15]: gate arcs and output arcs. A gate arc connects a transition with a status signal source (or places), and depending on the signal, it either permits or inhibits the occurrence of the event. An output arc connects a place with an external machine and sends a command signal to the machine.

Figure 4 shows an extended Petri net representation of robotic activity by an external robotic machine using transition firing with external and internal, permissive gate arcs. The net requests the resource and starts the action based on signals from an external device such as a switch. After the end of initialization, the external machine sends the acknowledge signal. Then the net sends the associated parameters and start signal through the output signal arc. At the end of activity, the external machine sends the signal status. After the ending activity (termination), the external machine sends the status to acknowledge the completion of the activity through the permissive gate arc, and finally the resource is released, where the ending activity is interlocked with another process in the system through the inhibitive gate arc.

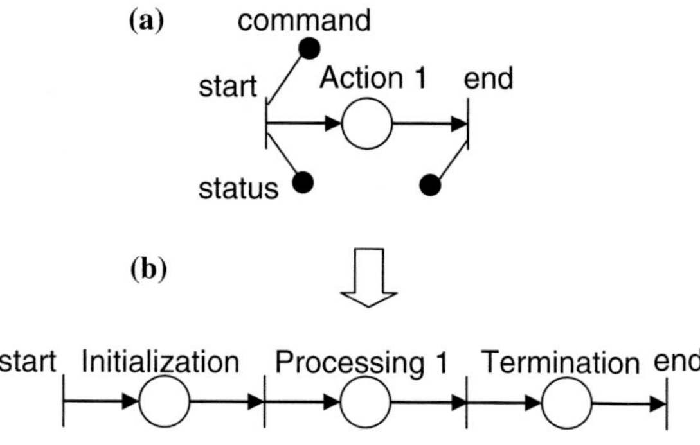

Fig. 4 Petri net model of general robotic task: **a** abbreviated form, **b** detailed representation

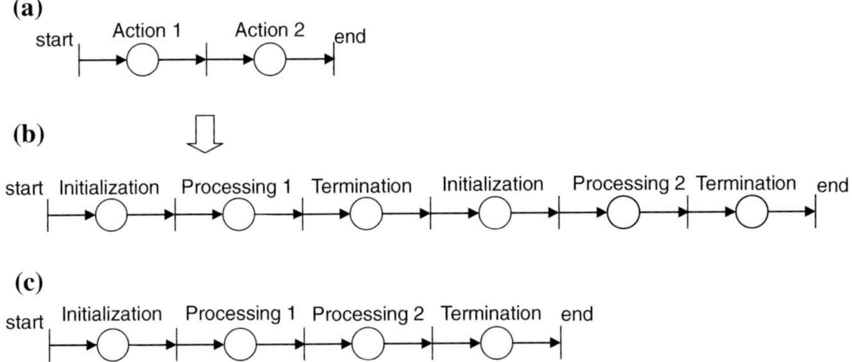

Fig. 5 Petri net model of sequential constructs of general robotic tasks: **a** abbreviated form, **b**, **c** two detailed representations

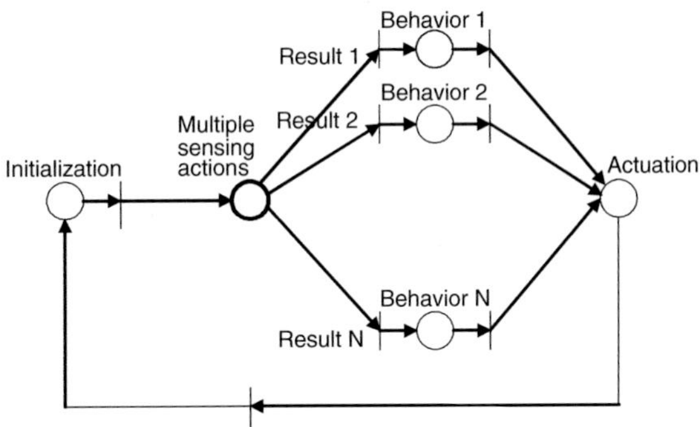

Fig. 6 Petri net model of sensor-based action selection

Basically, a robotic task can be subdivided in a sequence of subtasks or behaviors and these, in turn, will take place according to discrete event based state change dependent on the successive sensor data, as the basic modules of the task execution. All tasks can be executed by combining behaviors or reactions to objects in the environment. Each behavior has the common structure consisting of pre-conditions, post-conditions and the behavior action. The pre-conditions and the post-conditions represent sensory information of system status to determine respectively the conditions to start and end the behavior. The pre-conditions may include command input. Communication messages may be used as a pre-condition or a post-condition. In case of no uncertainties or no failures, the execution of a task will simply be the sequential execution of the constituent behaviors. From the view of behavior-based control, a behavior starts when there is a command or a set of preconditions, and stops upon success or failure of the pre-conditions and/or post-conditions.

The composition of the various behaviors will form robotic tasks, while behaviors can be described by combining different kinds of robotic actions using control structures such as sequential, conditional, iterative, and concurrent constructs [9, 10]. Figure 5 shows an example net representation of a sequential construct. Normally,

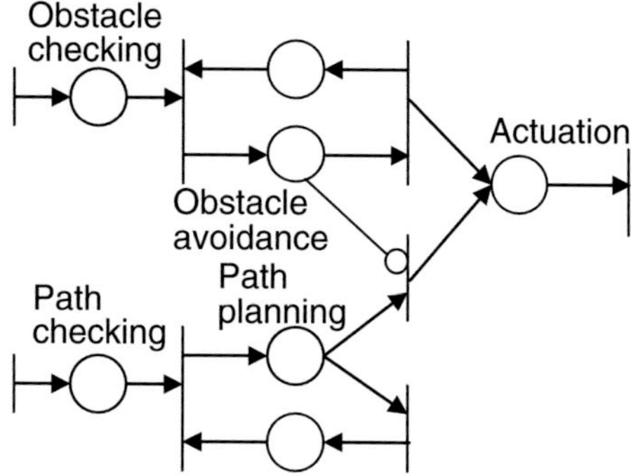

Fig. 7 Petri net representation of action priorities using inhibitive gate arcs, like subsumption architecture

after an action has completed the termination, the initialization of the next action starts. Following an action, the next action can be continuously executed without initialization. Figure 6 shows an example net representation of a conditional construct, or a sensing action, where the transitions ending the sensing action are in conflict and either of them is selected according to the sensing result. When some transitions which represent the end of sensing and the start of actuation are enabled, an arbitration mechanism is introduced to resolve conflict in firing. The execution of any of the behaviors, but only one at a time, is allowed. Figure 7 shows a net representation of fixed priorities according to activation of sensing subtask through direct coupling between perception and action based on the subsumption architecture, where the starting pre-conditions of each behavior are used as a condition for that behavior to be selectable. By constructing the coordinator, the commands sent by different behaviors can be executed in parallel, modified, combined and coordinated based on given goals, while producing continuously reactions based on sensor data for all the important objects in the environment. The complexity of the control system, which contains explicit representations for goal-oriented behaviors, is managed by decomposing it into subsystems such as producers and consumers of behaviors, and their interfaces.

Petri nets are decomposed by splitting the net into separate concurrent modules, called state machines, or finite state automata, which can be implemented independently in various devices with proper synchronization between them since the modules may work in different timing schemes. Several methods of net decomposition can be employed, using invariants computation, reachability set analysis, or concurrency graph analysis [29]. A p-invariant establishes a weighed ratio of a set of places which is maintained for any reachable marking. A shared transition represents a synchronous communication between two processes, while a shared place represents an asynchronous communication between the processes. A shared place whose destination is a single process is implemented with a buffer, or a mutual exclusive resource according to the number of destination processes. Figure 8 shows a net representation of synchronization for cooperative task execution with communication places, where a token indicates an associated message.

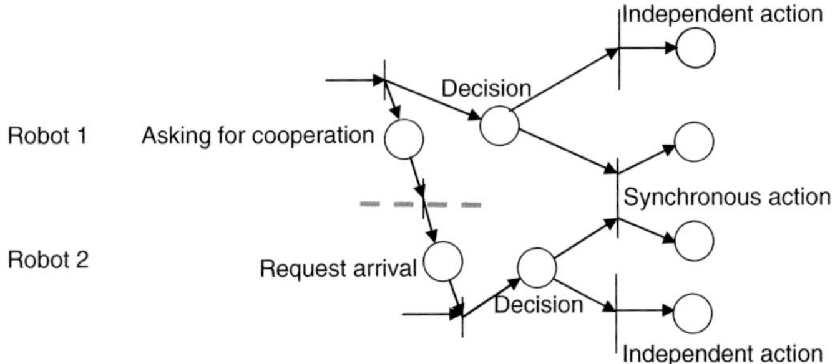

Fig. 8 Petri net representation of cooperative action selection

5 Architectural Design of Mobile Robot Tasks

Petri nets have been applied to a simple mobile robot task of sensory environment exploring and mapping, in which obstacle avoidance, target seeking and target following are specified as behaviors and sensory information leads the robot to change its own motion commands. The mobile robot has environment recognition, mapping, localization and communication capabilities, to sense the neighboring areas using range sensors, to build a local map, and to localize itself with regard to its own local map. Further, it upgrades the global map by integrating the previous local maps [27]. In a multirobot team, the communication capability enables each robot to talk to any other robot with very small time delay, to build up a global map by combining the local maps sent by other robots [26]. Thus, some intelligent behaviors, such as self-localization and environment mapping, have been implemented based upon primitive behavior modules and action control modules, such as motion control modules and multiple infrared range finding sensor control modules, to detect walls and obstacles and to build a local map in the surroundings of a wheeled mobile robot. Environment features of landmarks such as lines, corners and shaped or colored objects are estimated using a set of range sensors and a vision sensor. Upon these behavior modules, a Petri net based approach was applied to coordination of several concurrent activities of modules for the high-level behavior or task such as sensory exploration and mapping in unknown environments.

The Petri net based control architecture consists of four levels: task organization, behavior coordination, action control, and device execution, each corresponding to the hierarchical distributed software architecture. At the task organization level, a job goal is either assigned by a human operator or selected by the control system from a set of predetermined goals, with the organization of appropriate tasks into sequences of executions in order to accomplish the goal. At the coordinator level, the tasks are planned as a set of possible sequences of subtasks or behaviors and represented by a Petri net, with the conflict resolution in shared resources, such as task sharing and result sharing. A Petri net for behavior coordination specifies all the feasible sequences of behaviors for the task completion, with translating each input subtask into output commands to the associated control modules. Besides the

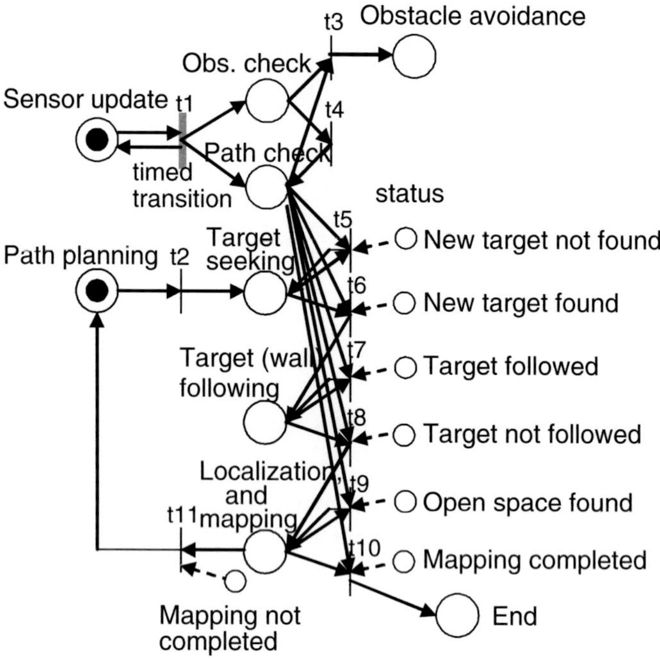

Fig. 9 Petri net model of a robotic task for environment exploration

translation mapping from the coordination level to the action control level, initial and final states, or markings, of the net are specified. At the action control level, a Petri net for action control defines all the possible sequences of primitive actions for the action execution, with translating each action into commands to the associated device modules. Basically, primitive actions are reactively defined as a mapping from sensor information to actuator execution. The major advantage of Petri nets over finite state machines is the capability to represent cooperation, parallelism, and conflict in shared resources and environments.

Figure 9 shows the resultant net model of the behavior based controller in the coordination layer, implemented in the system controller. The net model is configured through a sequential constraint of behavior modules based on the task specification sent from the task organization layer so that the behaviors of the whole system are brought about for achieving global control of a given job. The firing of transitions representing the sensing action models the updating of data from the range sensors. The places that receive tokens from the sensing module model data information checked and supplied about obstacle and wall distances and directions and target positions. Depending on these data, the firing of transitions defines which behavior is initiated. The activity of each behavior module is represented by a place in the net. Each behavior module is triggered by a particular sensing information, plans the corresponding actions, calculates the strength, or bid, of the behavior command

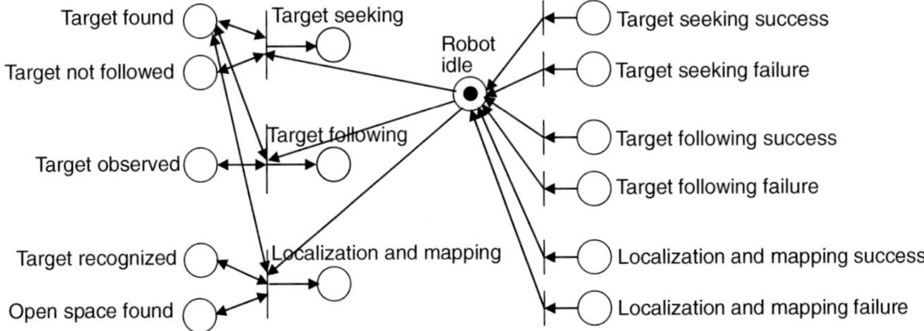

Fig. 10 A net representation of a mutual exclusive resource model for robotic behavior coordination

using the importance of the sensing information and the specified priorities among the different behaviors. The motor control mode models the motion states of the mobile robot according to the behavior decision, and is stored at the lowest level in the hierarchical shared memory. Figure 10 shows a net representation of a mutual exclusive resource model for robotic behavior coordination.

In Fig. 9, the robotic task is divided in a sequence of three behaviors, and the overall behavior will take place according to the behavior modules dependent on the successive sensor data. Initially, the robot is in standby. The firing of transition t1 means the start of the task. The global planner decides a target using the current global map. Then, behavior "Target seeking" starts when the robot starts environment search, that is, the corresponding transition t2 fires. Since a wall is detected, as indicated by the firing of transition t6, behavior "Target following" begins, that is, the robot keeps following the wall using range sensors and builds a local map. The end of the behavior and the start of behavior "Localization and mapping" take place once the wall is no longer present as indicated by the firing of transition t8. Thus, the robot behavior may oscillate between the exploring and mapping behaviors. At the end of "Localization and mapping", as indicated by the firing of transition t10, if the whole environment in the global map is searched, then the mobile robot stops as indicated by place "End", if not the next target is determined and the search and map behavior is continued.

Concurrently, the obstacle avoidance behavior is always checking whether an obstacle is present or not. If an obstacle is detected as indicated by the firing of transition t3, the avoid obstacle behavior takes place. Otherwise, if no obstacle is found as indicated by the firing of transition t4, behavior "Target seeking" begins. Figure 11 shows the detailed net model of the wall following behavior. When the robot is following a wall, it carries out distance measurements continuously. These concurrent processes must start and end simultaneously. So, the behavior control module requests wheel control and distance measurement modules with events. If the wall is not found, it terminates distance measurements and builds a local map estimating the direction of the wall. Then it localizes itself on its own local map and transfers the positional information to the global map builder.

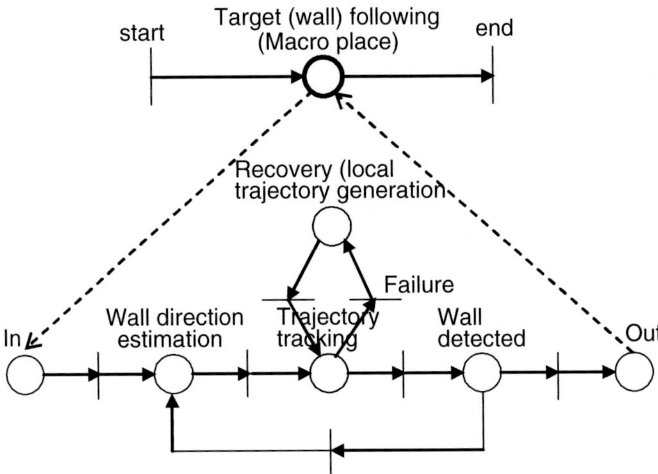

Fig. 11 Detailed Petri net model of wall following behavior in exploring and mapping behavior

The proposed distributed autonomous control framework has been implemented and tested to control the mobile robot. A Petri net simulator for robotic task execution has been implemented with a set of tools for designing and debugging tasks. Petri net models are executed reacting to the events occurring in the environment and to the state of the robot. During the real-time control execution, the simulator makes use of a set of functions that can access the internal state of the robot and return truth values about relevant properties for the execution of the robotic task. Figure 12 shows an experimental result indicating a sequence of events for steering control during wall following and obstacle avoidance. Considering the issue of the behavioral priority level, the net model reveals the concurrent procedure among the obstacle avoidance and any other possible behaviors, since concurrency can be easily implemented in a Petri net model. The sensed information of the robot's immediate surroundings is managed by the local map manager module, which generates appropriate events to allow sensor-based reaction when unexpected situations such as obstacles appear. It also can deal with the development of navigation control architecture with hierarchical control of a smooth planned path and sensor-based reactive control for an autonomous mobile robot.

The Petri net based control algorithm was implemented on a general microcomputer based network in accordance with the hardware architecture. The system controller, or the coordination layer, is implemented on one microcomputer, and the control modules are implemented on different microcontrollers, so that the control of the overall net is split into several subnets, each of which is sequentially implemented in a separate processor, concurrent with the others. Since each module integrates both control and operational parts of the subnet with inter-module connections through a communication mechanism like buffers or rendezvous, the actual concurrency is respected without the presence of the supervisor. The system controller coordinates

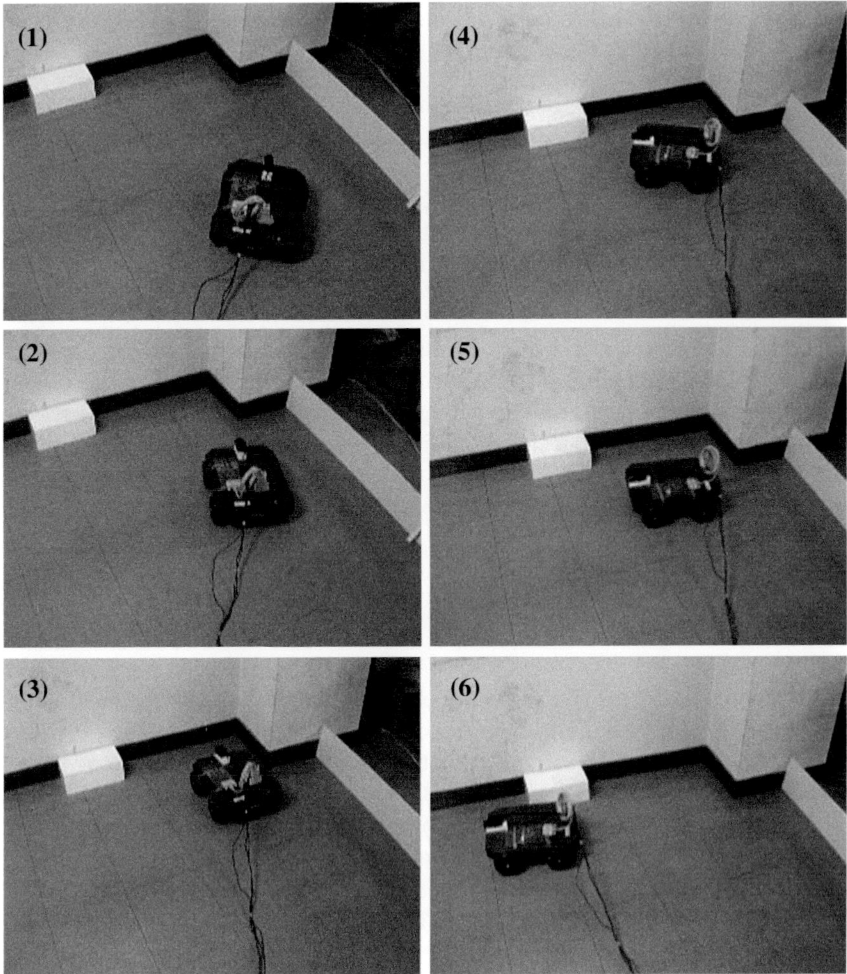

Fig. 12 An event sequence during wall following and obstacle avoidance behavior of a mobile robot

the activities of control modules, which control the sensors and actuators of the robot such that they are in accordance with transition firing in the hierarchical net model. When events occur at any control module, the module interrupts the system controller. The behavior-based system controller receives events and updates corresponding net model data, or enabled transitions, in the database system that manages the states of activity of all modules in the system using the event list table, where each entry of the table specifies the behavior module, the associated input and output event identifiers, and the parameter list. Using the database system and a map created by the global map manager for navigation, the system controller processes these events and generates global actions. Then, the controller requests tasks to the relevant control modules and these modules perform real-time control according to the task. The flow of the extended Petri net simulation consists in the following steps:

(1) Check all the transitions and update the enabled transition table;
(2) Check fireability of the enabled transitions in the table considering external gate conditions;
(3) Arbitrate fireable transitions in conflict using some arbitration rule;
(4) Execute transition firing and output corresponding signals to external machines;
(5) Change the marking to the new marking and remove the firing transition from the table.

Through the simulation steps, the transition vector table is efficiently used to extract enabled or fired transitions. Furthermore, the data structure of the net simulator is made up of several tables corresponding to the structure of the net specifying the robotic task. Macro transitions called substitution transitions, and fusion places which allow the use of the same place in different locations, are supported by the simulation software as a graphical mechanism to simplify the net drawing for user convenience.

6 Discussion

The overall net model for system control of a mobile robot was constructed through the integration of behavior and control modules in a bottom-up way. The Petri net based approach validates the implementation of synchronization and coordination in discrete event behavior based control. Behavior modules are composed to design more complex modules according to applications. The detailed function of each control module is specialized in a determined role according to the application, so that for real-time performance of robotic actions, new control strategies can be easily embedded in the control modules. The system design procedure facilitates the understanding of the interaction among the different processes that might be present in the mobile robot control system.

The vertical decomposition is based on a hierarchical dependency in terms of the activity of the nets. The upper layer constraints the activity of the lower layer, such that whenever a place in the net model at the upper layer is unmarked, the associated subnets at the lower layer are inactive. The execution of the overall net model is accomplished through the parallel execution of the global net model and the local subnet models in corresponding hierarchical layers. Using different levels of abstraction enables the construction of the model in an incremental way. The support for such specific model structuring mechanisms is of the most importance for the design of complex embedded systems to build up compact models. Even though modularity of the control system is helpful to make further improvement on the real-time performance, the amount of net elements, such as places and transitions, is increased considerably. This main disadvantage of using low-level Petri nets has been overcome with the use of hierarchical Petri nets. The proposed decomposition technique can be extended to some non-centralized system of maximally distributed agents, each equipped with at least one active effector, where each agent negotiates with neighboring active agents without any supervisor to asynchronously access intermediate passive elements between them.

7 Conclusions

This chapter presents a design methodology of discrete event distributed control architecture for autonomous mobile robot systems. A modular, behavior-based distributed software architecture is presented on a hierarchical distributed microcontroller based hardware structure for intelligent mobile robots. The system integration and implementation issues are discussed for the presented method. Because the robot has built-in intelligence of behaviors, remote control of complex tasks through the internet can be stably and efficiently performed using high-level behavior commands in unknown environments by human operators without a large amount of communication. Compared to direct remote control, the robot can autonomously avoid the potential collision with an obstacle so that the operator does not need to worry about too many motion details. Thus, high-level human command input is integrated with the coordination layer through the distributed autonomous control architecture.

As a future work, based on the concept of modular, behavior-based non-centralized control, self-learning abilities of behavior module synthesis will be implemented on multiple, reconfigurable LSI chips using agent-based autonomous Petri nets for exploring, mapping and rescue robots in unstructured, hazardous environments.

References

1. Arkin R (1987) Motor schema based navigation for a mobile robot: an approach to programming by behavior. In: IEEE international conference on robotics and automation, Raleigh, NC, pp. 264–271
2. Arkin R (1998) Behavior-based robotics. The MIT Press, Cambridge
3. Azar AT, Vaidyanathan S (2015a) Handbook of research on advanced intelligent control engineering and automation. Advances in computational intelligence and robotics (ACIR) book series, IGI Global, USA
4. Azar AT, Vaidyanathan S (2015b) Computational intelligence applications in modeling and control. In: Azar AT, Vaidyanathan S (eds) Studies in computational intelligence, vol 575. Springer, Germany
5. Azar AT, Vaidyanathan S (2015c) Chaos modeling and control systems design. Studies in computational intelligence, vol 581. Springer, Germany
6. Azar AT, Zhu Q (2015) Advances and applications in sliding mode control systems. Studies in computational intelligence, vol 576. Springer, Germany
7. Brooks RA (1986) A robust layered control system for a mobile robot. IEEE J Robot Autom 2:14–23
8. Brooks RA (1987) Asynchronous distributed control system for a mobile robot. In: SPIE conference on mobile robots, vol 0727, pp 77–84

9. Caccia M, Coletta P, Bruzzone G, Veruggio G (2005) Execution control of robotic tasks: a Petri net based approach. Control Eng Pract 13(8):959–971
10. Caloini A, Magnani GA, Pezze M (1998) A technique for designing robotic control systems based on Petri nets. IEEE Trans Control Syst Technol 6(1):72–86
11. Connell J (1992) A hybrid architecture applied to robot navigation. In: Proceedings of IEEE international conference on robotics and automation, pp 2719–2724
12. Doty KL, Van Aken RE (1993) Swarm robot material handling paradigm for a manufacturing workcell. In: Proceedings of the IEEE international conference on robotics and automation, vol 1, pp. 778–782
13. Duffie NA, Prabhu VV (1996) Heterarchical control of highly distributed manufacturing systems. Int J Comput Integr Manuf 9(4):270–281
14. Fleury S, Herrb M, Chatila R (1994) Design of a modular architecture for autonomous robot. In: Proceedings of the IEEE international conference on robotics and automation, pp 3508–3513
15. Freedman P (1991) Time, Petri nets and robotics. IEEE Trans Robot Autom 7(4):47–433
16. Gat E (1992) Integrating planning and reacting in a heterogeneous asynchronous architecture for controlling real-world mobile robots. In: Proceedings of the tenth national conference on artificial intelligence, pp 809–815
17. Liscano R, Manz A, Stuck ER, Fayek RE, Tigli JY (1995) Using a blackboard to integrate multiple activities and achieve strategic reasoning for mobile-robot navigation. IEEE Expert 10(2):24–36
18. Maes P (1990) Situated agents can have goals. In: Maes P (ed) Designing autonomous agents: theory and practice from biology to engineering and back. M.I.T. Press, pp 49–70
19. Milutinovic D, Lima P (2002) Petri net models of robotic tasks. In: Proceedings of the IEEE international conference on robotics and automation, vol 4, pp 4059–4064
20. Montano L, García FJ, Villarroel JL (2000) Using the time Petri net formalism for specification, validation, and code generation in robot-control applications. Int J Robot Res 19(1):59–76
21. Mori M (1975) The Three-eyed Beatles. Presented at the international ocean exposition, Okinawa, Japan
22. Mori M, Yasuda G (1973, 1974) A study on self-coordination mechanisms of systems with non-centralized neural nets 1st report, 2nd report, In: Reports of the meeting on neuro-sciences, sponsored by Ministry of Education, Science and Culture, p 177, Sep., 1973, p.168, Feb., 1974
23. Peterson JL (1981) Petri net theory and the modeling of systems. Prentice Hall
24. Murata T (1989) Petri nets: properties, analysis and applications. Proc IEEE 77:541–580
25. Osswald D, Martin J, Burghard C, Mikut R, Woern H, Bretthauer G (2003) Integrating a robot hand into the control system of a humanoid robot. In: Proceedings of the 2003 international conference on humanoid robots
26. Takai H, Mitsuoka J, Yasuda G, Tachibana K (2006) Feasibility study of sensing methods on cooperative localization for team operation of multiple mobile robots. In: Proceedings of the 3rd International conference on autonomous robots and agents, pp 393–398
27. Takai H, Mitsuoka J, Yasuda G, Tachibana K (2007) Cooperative workspace mapping for multi-robot team operations using ultrasonic sonar and image sensor. In: Proceedings of the 13th international conference on advanced robotics, pp 1129–1134
28. Takai H, Yasuda G, Tachibana K (2002) Construction of infrared wireless inter-robot communication networks for distributed sensing and cooperation of multiple autonomous mobile robots. In: Proceedings of the 15th IFAC world congress, Elsevier, pp 143–148
29. Wisniewski R, Barkalov A, Titarenko L Halang W (2011) Design of microprogrammed controllers to be implemented in FPGAs. Int J Appl Math Comput Sci 21(2):402–412
30. Yasuda G (1999) A multiagent architecture for sensor-based control of intelligent autonomous mobile robots. In: Proceedings of the 15th world congress of the international measurement confederation (IMEKO), ACTA IMEKO 1999, vol X (TC-17), pp 145–152
31. Yasuda G, Mori M (1971a) Application of graph theory to self-reproducing processes. In: Proceedings of the 10th SICE annual conference, in Japanese, pp 271–272
32. Yasuda G, Mori M (1971b) Construction of an artificial multi-molecular self-reproducing mechanism. In: Proceedings of the 14th joint automatic control conference of Japan, in Japanese, pp 65–66

33. Yasuda G, Mori M (1974) Future concepts on grouping robots. Presented at the meeting on neuro-sciences, sponsored by Ministry of Education, Science and Culture, Tokyo, Japan, Feb 1974
34. Yasuda G, Takai H (2001) Sensor-based path planning and intelligent steering control of non-holonomic mobile robots. In: Proceedings of the 27th annual conference of the IEEE industrial electronics society (IECON2001), pp 317–322
35. Zhu Q, Azar AT (2015) Complex system modelling and control through intelligent soft computations. Studies in fuzziness and soft computing, vol 319. Springer, Germany

Indoor Thermal Comfort Control Based on Fuzzy Logic

Lucio Ciabattoni, Gionata Cimini, Francesco Ferracuti, Gianluca Ippoliti and Sauro Longhi

Abstract Control and monitoring of indoor thermal conditions represent crucial tasks for people's satisfaction in working and living spaces. In the first part of the chapter we address thermal comfort issues in a working office scenario. Among all standards released, predicted mean vote (PMV) is the international index adopted to define users thermal comfort conditions in moderate environments. In order to optimize PMV index we designed a novel fuzzy controller suitable for commercial Heating, Ventilating and Air Conditioning (HVAC) systems. However in a residential scenario it would be extremely expensive to gather real time measures for PMV computation. Indeed in the second part of the chapter we introduce a novel approach for residential multi room comfort control based on humidex index. A fuzzy logic controller is introduced to reach and maintain comfort conditions in a living environment. Both control systems have been experimentally tested in the central east coast of Italy. Temperature regulation performances of both approaches have been compared with those of a classical PID based thermostat.

Keywords Fuzzy logic · Fuzzy control · Comfort control · Predicted mean vote · Humidex

---

L. Ciabattoni (✉) · G. Cimini · F. Ferracuti · G. Ippoliti · S. Longhi
Dipartimento di Ingegneria dell'Informazione, Università Politecnica delle Marche, Via Brecce Bianche, 60131 Ancona, Italy
e-mail: l.ciabattoni@univpm.it

G. Cimini
e-mail: g.cimini@univpm.it

F. Ferracuti
e-mail: f.ferracuti@univpm.it

G. Ippoliti
e-mail: gianluca.ippoliti@univpm.it

S. Longhi
e-mail: sauro.longhi@univpm.it

A.T. Azar and S. Vaidyanathan (eds.), *Advances in Chaos Theory and Intelligent Control*, Studies in Fuzziness and Soft Computing 337,
DOI 10.1007/978-3-319-30340-6_35

1 Introduction

During the last decade there has been an increasing demand by buildings occupants for the improvement of indoor comfort together with the reduction of energy consumption and CO_2 emissions [21, 30]. Indoor comfort plays a significant role and has a great impact on inhabitant's health, morale, productivity and satisfaction [48]. In this context, an energy and comfort management system (ECMS) must comprise an intelligent control system for buildings. One of the crucial aims of an ECMS is to fulfill the occupant's expected comfort [46].

A human being's comfort sensation is mainly related to the thermal balance of the body as a whole. This balance is influenced by physical activity, clothing and, obviously, environmental parameters namely air temperature, air velocity, air humidity and so on. Once these factors have been measured or estimated, the thermal sensation for the body can be predicted by calculating proper indexes. In extreme cases, people may decline to work or live in a particular environment with bad thermal sensation [16]. For this reason it is necessary to make efforts for the trade-offs between the maintenance of comfortable indoor environmental parameters and the reduction of energy consumption [44].

Especially in industrialized countries, several rules, indexes and standards have been recently released in order to provide to engineers and technicians the right tools to check of the indoor microclimate. Among all standards released, predicted mean vote (PMV) index [31] is adopted for assessing the thermal comfort condition for people exposed to moderate thermal environments [39].

In a thermal control scenario, various standard control schemes, such as an on/off switching thermostats, proportional-integral (PI) and proportional-integral-derivative (PID), have been extensively used in building engineering [45, 47]. Generally all these schemes do not have any direct knowledge of the system to be controlled and they are designed with constant parameters. Thus they provide poor control performance for noisy, disturbed and non-linear processes without taking into account users behavior [29].

In this context, control designers and engineers examined other control options to ensure thermal comfort and limit set-point overshoots, with consequent energy savings (see e.g. [7, 49]). Significant research was carried out in the last decades on optimal [32], predictive [10], distributed [11], adaptive [35] as well as artificial intelligence techniques [1, 36]. Most of the above control strategies did not consider the comfort factor but were only concerned with energy consumption savings [28]. Furthermore, since these are model-based control schemes, according to [27] their accuracy depend on the model they use. In this scenario, due to various complications and implementation challenges, there has been no industrial development based on this schemes. Furthermore this approaches makes it difficult to build a control algorithm taking weather effects explicitly into account.

Although some PMV based control methods exist, they treat all occupants the same, do not taking into account subjective adaptive parameters [37] (as the PMV itself). However the comfort temperature is a result of the interaction between the subjects and the building or other environment they are occupying. The main contextual variable to consider when referring to comfort is the climate. Climate has an overarching influence on the culture and thermal attitudes of any group of people and on the design of the buildings they inhabit. Whilst the basic mechanisms of the human relationship with the thermal environment may not change with climate, there are a number of detailed ways in which people are influenced by the climate they live in and these play a cumulative part in their response to the indoor climate. Another context is the time since the rate at which temperature changes occur is an important consideration if the conditions for comfort are to be properly specified. PMV comfort control approaches developed in literature, divided into model based, see e.g. [25], and model free, see e.g. [33], never consider an adaptation of PMV index. We developed a novel fuzzy controller for HVAC (Heating, Ventilating and Air Conditioning) systems based on an adaptive PMV index, taking into account the outdoor weather conditions as well as the time response of the system. Furthermore, considering all the control systems related issues we developed a model-free based controller and, at the same time, the adoption of fuzzy logic makes it easier to implement on a microcontroller.

In an office environment, metabolic and clothing related variables can be considered as constants due to the sedentary activities of the workers. Unfortunately the same variables can admit a wide range of values in a multiroom residential scenario, depending on the occupants activities. This issue could be solved by the installation of extremely expensive automation systems (e.g. thermal cameras or vision systems) to gather real time measures of all PMV variables. Another option could be to turn PMV equation into a non linear function of air temperature and relative humidity, making many strong assumptions as preliminarily reported in [12]. However, as stated in [41], this solution is unworkable in practice since we would lose the most important contributions to the PMV output. In the second part of this chapter we consider a novel approach for the indoor perceived temperature estimation according to the humidex index. This index number was introduced by Canadian meteorologists to describe the perceived thermal feeling of a person, by combining the effect of heat and humidity. Although humidex is mainly used in outdoor environments, recent studies promoted its feasibility as an indoor comfort predictor [40]. In this paper we investigate for the first time its use in a multiroom comfort control scenario. In particular a Fuzzy Inference System based control for different rooms of a house has been designed, considering the humidex value as the perceived temperature.

The performances of both approaches have been experimentally tested in residential and working buildings in the central East coast of Italy.

In this chapter, we will provide in Sect. 2 a brief introduction on the main indexes assessing human thermal comfort, including PMV index. The Fuzzy Logic approach is briefly presented in Sect. 3. The PMV optimization in an office environment scenario is introduced in Sect. 4, where input and output control variables as well as assumptions made and control rules formulation are presented. Experimental results of the control algorithm are reported as well, together with the comparison of its performances with those of a classical PID based thermostat. In Sect. 5 the design of the Fuzzy Logic based multiroom residential controller is presented.

Experimental setup and results of the control approach as well as a performance comparison with respect a traditional thermostat are then reported. Some remarks conclude the chapter.

2 Comfort Indexes

Several rules, indexes and standards have been recently released worldwide in order to provide the right tools to check indoor microclimate. Among all standards released, PMV index [31] is typically adopted for assessing the thermal comfort condition for people exposed to moderate thermal environments.

2.1 Predicted Mean Vote Index

The PMV model was developed by P.O. Fanger in the 70's using heat balance equations and empirical studies about skin temperature to define comfort. Standard thermal comfort surveys ask subjects about their thermal sensation on a seven point scale from cold (−3) to hot (+3). Fanger's equations are used to calculate the PMV of a large group of subjects for a particular combination of air temperature, mean radiant temperature, relative humidity, air speed, metabolic rate, and clothing insulation. The ideal value is 0, representing thermal neutrality, and the comfort zone is defined by the combinations of the six parameters for which the PMV is within the recommended limits ($-0.5 < \text{PMV} < +0.5$). This index is introduced in the third edition ISO-7730:2005. This third edition cancels and replaces the second edition (ISO-7730:1994), which has been technically revised. In particular, the equation of PMV index within an indoor environment is (Table 1):

$$\begin{aligned} PMV = &\left[0.303 \cdot e^{-0.036 \cdot M} + 0.028\right] \cdot \{(M-W) - \\ &3.05 \cdot 10^{-3} \cdot [5733 - 6.99 \cdot (M-W) - P_a] - \\ &0.42 \cdot (M - W - 58.15) - 1.710^{-5} \cdot M \cdot (5869 - P_a) - \\ &0.0014 \cdot M \cdot (34 - T_a) + \\ &\tfrac{1}{I_{cl}} \cdot (T_{cl} - 35.7 + 0.028 \cdot (M-W))\} \end{aligned} \quad (1)$$

where T_{cl} is given by:

$$\begin{aligned} T_{cl} = &\,35.7 - 0.028 \cdot (M-W) - I_{cl} \cdot \{3.96 \cdot 10^{-8} \cdot f_{cl} \cdot \\ &\left[(T_{cl}+273)^4 - (T_{mr}+273)^4\right] + f_{cl} \cdot h_c \cdot (T_{cl} - T_a)\} \end{aligned} \quad (2)$$

h_c can be obtained from:

$$h_c = \begin{cases} 2.38|T_{cl} - T_a|^{0.25} & if \quad 2.38|T_{cl} - T_a|^{0.25} \geq H \\ H & if \quad 2.38|T_{cl} - T_a|^{0.25} < H \end{cases} \quad (3)$$

Table 1 Nomenclature, description and measure units of the variables involved in the PMV equation

Variable	Measurement unit	Description
M	$\frac{W}{m^2}$	Metabolic rate
f_{cl}	–	Clothes area coefficient
I_{cl}	$m^2 \cdot \frac{K}{W}$	Clothes thermal insulation
T_a	°C	Air temperature
W	$\frac{W}{m^2}$	Effective mechanical power
T_{mr}	°C	Mean radiant temperature
Var	$\frac{m}{s}$	Air relative velocity
P_a	Pa	Water vapor partial pressure
h_c	$\frac{W}{K \cdot m^2}$	Convection heat transfer coefficient
T_{cl}	°C	Clothes external temperature

with $H = 12.1 \cdot \sqrt{Var}$. While f_{cl} results from:

$$f_{cl} = \begin{cases} 1.00 + 1.290 \cdot I_{cl} & if \quad I_{cl} \leq 0.078 \\ 1.05 + 0.645 \cdot I_{cl} & if \quad I_{cl} > 0.078 \end{cases} \tag{4}$$

2.1.1 Sensitivity Analysis

The Global Sensitivity Analysis (GSA) has seen a wide use in the field of model selection. Sensitivity analysis has the role to identify which uncertain factors contributes more on the output uncertainty. As the global sensitivity refers to an integrated sensitivity over the entire input parameter space it has been considered a valid approach for the aim of this work [9]. Indeed the normative provide useful information regarding both nominal and min-max values of the PMV equation's parameters. GSA has been solved using the well known Sobol' sensitivity, based on the ANalysis Of VAriance (ANOVA) decomposition. For the details about the approach the reader is invited to refers to [9, 42]. Consider the function $y = g(X)$ where $X = \{x_1, x_2, \ldots, x_n\}$ is the set of n parameters that we want to sort by importance. The total variance vector is obtained through Sobol' approach as $S^{\text{tot}} = \{S_1^{\text{tot}}, S_2^{\text{tot}}, \ldots, S_n^{\text{tot}}\}$. S_i^{tot} is the expected variance that remains in the output if all the parameters, except to the ith, are known. Please note that

$$0 \leq S_i^{\text{tot}} \leq 1 \tag{5}$$

$$\sum_{i=1}^{n} S_i^{\text{tot}} = 1. \tag{6}$$

The PMV Eqs. (1) and (2), suggest to fix the parameter set as

$$X = \{\mathrm{M, W, HR, T}_a, \mathrm{I}_{cl}\}. \tag{7}$$

The associated total variance vector for the set in Eq. (7), when the range of all the parameters are fixed as in [31], results to be:

$$S^{\text{tot}} = [0.7574, 0.2016, 0.0041, 0.0532, 0.0305]. \tag{8}$$

As expected, the parameters playing the most important role are the metabolic and the clothing related ones. The reason relies in the fact that they are allowed to vary in the whole space defined by the normative, which comprises different activities. Unfortunately these variables are very difficult to measure and they are often taken as given constant, considering the particular environment of operation. In this chapter we refer to an office environment and sedentary activities, thus it seems reasonable to fix the parameters that cannot be measured. In this way a reduced parameter set $X = \{\mathrm{HR, T}_a\}$ is considered and the associated total variance vector results to be:

$$\tilde{S}^{\text{tot}} = [0.062, 0.9395]. \tag{9}$$

This barely means that the humidity contributes to the 6.2 % of the PMV value when the other parameters are set.

2.1.2 Assumptions

Since it would be extremely expensive to gather real time measures of all the variables involved in PMV formula, we made some assumptions. According to a sensitivity analysis, as stated in [41], the most influencing variables are the metabolic rate, air temperature and humidity. Since the thermal comfort control tests have been performed in a sedentary working space scenario, human activities and clothing related variables have been estimated according to the European norm [31]. Concerning the metabolic rate M we choose a value of $M = 70\ \frac{W}{m^2}$, valid for sedentary activities such as office work or dwelling. The clothes thermal insulation I_{cl} has been chosen equal to $0.14\ m^2 \cdot \frac{K}{W}$ (for a typical medium combinations of indoor garments). The clothes area coefficient can be computed from Eq. (4) considering that $I_{cl} > 0.078$. Concerning the effective mechanical power we considered $W = 0$ for sedentary indoor activities. Mean radiant temperature will be considered equal to the air temperature thus allowing to compute from Eq. (3) the convection heat transfer coefficient as $h_c = 12.1 \cdot \sqrt{Var}$. Furthermore according to the federation of European heating, ventilation and air conditioning associations [8], the indoor air relative air velocity limit values range in summer from 0.15 to $0.30\ \frac{m}{s}$ and in winter from 0.15 to $0.25\ \frac{m}{s}$. We supposed a constant value of $0.2\ \frac{m}{s}$. Exploiting the above assumptions, clothes external temperature has as a linear relation with the air temperature in the range 16 to 25 °C, in the form $T_{cl} = 0.5760 \cdot Ta + 14.1810$. Water vapor partial pressure, a function of relative humidity HR, (see [31]) can be computed as:

$$\mathrm{Pa} = \mathrm{HR} \cdot 10 \cdot \mathrm{e}^{16.6536 - \frac{4030.183}{235 + \mathrm{Ta}}} \tag{10}$$

According to all these assumptions the final PMV formula will be a non linear function of the air temperature and relative humidity. Both variables can be measured to obtain a real time PMV estimation using wireless sensors, as will be described in Sect. 4.

2.2 *Humidex Index*

The humidex is an index number introduced by Canadian meteorologists [34] to describe the perceived thermal feeling of a person, by combining the effect of heat and humidity. Although humidex is mainly used in outdoor environments, recent studies promoted its feasibility as an indoor comfort predictor (as stated in [40]). In the second part of this chapter we investigate for the first time its use in a comfort control scenario. The typical formula to calculate humidex index is:

$$\text{humidex} = T + \frac{5}{9}(z - 10) \tag{11}$$

where T is the indoor temperature [°C] and z can be expressed as:

$$z = 6.112 \cdot 10^{(7.5 \cdot T/(237.7+T))} \cdot \frac{H}{100} \tag{12}$$

with H the relative humidity ([%]). Humidex is a unitless number, however, it roughly refers to the perceived temperature in Celsius degrees. As an example, with a temperature of 30 °C and a relative humidity of 70 %, the resulting humidex index is 35. It means that the perceived temperature by any average person will be close to 35 °C. In the remaining part of the chapter we will refer to "humidex index value" when talking about "perceived temperature".

2.3 *The Wet-Bulb Globe Temperature (WBGT) Index*

The wet-bulb globe temperature (WBGT) index is the most widely used and accepted index for the assessment of heat stress in industry. It has been published as British Standard BS EN 27243. The WBGT index is an empirical index representing the heat stress to which an individual is exposed. The index was developed specifically for use in industrial settings. Industrial technicians necessitated a compromise between the requirement for a precise index and the need to be able to easily take controlled measurements. WBGT, which emerged from the "corrected effective temperature" (CET), is a composite temperature used to estimate the effect of temperature, humidity, wind speed (wind chill), and visible and infrared radiation (usually sunlight) on humans. The WBGT formula consists of weighting of dry-bulb temperature (T_a) wet-bulb temperature (T_w) and black-globe temperature (T_g), in the following manner:

$$WBGT = 0.7T_w + 0.1T_a + 0.2T_g \tag{13}$$

For indoor conditions the index was modified as follows:

$$WBGT = 0.7T_w + 0.3T_g \tag{14}$$

The WBGT value is compared to the reference values provided in the standard for the appropriate metabolic rate and state of acclimation of the worker. This index is recommended by many international organizations for setting criteria for exposing workers to hot environment and was adopted as an ISO standard (ISO 7243). A limitation of this index is that the reference values are representative of the mean effect of heat, over a long period of work. It does not provide a reference for those instances where workers are exposed to harsh conditions for very short periods of time (e.g. few minutes). Another issue with WBGT is that reference values are chosen for workers physically fit, in good health, normally clothed, with adequate salt and water intake.

3 Fuzzy Logic Approach

Fuzzy rule-based systems (FRBS) have been successfully employed for system identification, control and modeling in many areas [3, 6, 13, 17] often overcoming the performances of other heuristic techniques (e.g. neural network based approaches [2, 14, 22–24]). Depending on the application, fuzzy research field can be divided into different approaches. The first is the linguistic fuzzy modeling (LFM) adopting the Mamdani rule structure. The other is the precise fuzzy modeling (PFM) where T-S and Tsukamoto fuzzy rule structures are generally used in the learned fuzzy model. Recently Type-2 Fuzzy Systems gained increasing attention, although not so many real time applications have been developed [4]. The approach considered for the PMV optimization based control is the linguistic fuzzy modeling (LFM) with Mamdani rule structure [18, 20] due to its capability to model human knowledge in an explicit way. The membership functions of the variables involved in both the fuzzy system presented consist of triangular asymmetric and trapezoidal functions chosen by experts in the field of thermal regulation. The trapezoidal fuzzy set A in the universe of discourse $U \in \mathbb{R}$ with the membership function μ_A is parameterized by four real scalar parameters: (a, b, c, d) with $a < b \leq c < d$. This representation can be interpreted as a mathematical membership function as follows [15]:

$$\mu_A(x) = \begin{cases} 0\,, & x < a \\ \frac{x-a}{b-a}\,, & a < x < b \\ 1\,, & b < x < c \\ \frac{d-x}{d-c}\,, & c < x < d \\ 0\,, & x > d \end{cases} \tag{15}$$

When $b = c$, the triangular function can be considered as a particular case of the trapezoidal one, as in [19].

Let A and B be two fuzzy sets in the universe of the discourse U and let the membership functions of the fuzzy sets A and B be μ_A and μ_B, respectively. The intersection between the fuzzy sets A and B is defined as follows:

$$\mu_{A\cap B}(x) = \min\{\mu_A(x), \mu_B(x)\}\,, \quad \forall x \in U \tag{16}$$

The union between the fuzzy sets A and B is defined as follows:

$$\mu_{A\cup B}(x) = \max\{\mu_A(x), \mu_B(x)\}\,, \quad \forall x \in U \tag{17}$$

The Max-Min-mCoA fuzzy inference algorithm is considered. The fuzzy logic controller uses the following equation to calculate the geometric center of the full area under the scaled membership functions:

$$\mathrm{mCoA} = \frac{\int \mathrm{f(x)} \cdot \mathrm{xdx}}{\int \mathrm{f(x)dx}} \tag{18}$$

where mCoA is the modified center of area and f(x) the output of the inference process. The interval of integration is between the minimum membership function value and the maximum membership function value (Fig. 1).

4 PMV Optimization in a Working Scenario

The PMV method, as it was originally developed, does not take into account location and adaptation to the outdoor thermal environment. It basically states that the indoor comfort temperature should not change according to the seasons. The aim of the fuzzy logic controller is to overcome this issue adapting the comfort conditions to the seasonality.

Input variables chosen for the FIS control are:

- the outdoor temperature (T_e)
- the value of the PMV index (PMV)
- the variation of the PMV index (Δ_{PMV})

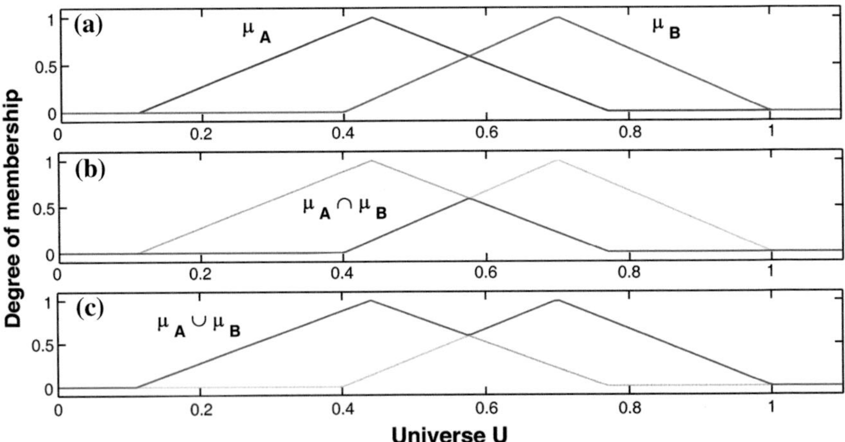

Fig. 1 Sample of two fuzzy sets (a) and their intersection (b) and union (c)

The output of the FIS engine is the Fan Coil Speed (U). Values chosen for the input and output variables fuzzy sets are reported in Table 2. Figures 2 and 3 depicts membership functions for the input variables PMV and Δ_{PMV} respectively, while Fig. 4 the output membership functions.

The defuzzification process computes the speed to drive the fan coil in order to maintain the desired indoor comfort level.

Figure 5 depicts the block diagram of the whole comfort control process. PMV index is computed in real time following the assumptions of Sect. 2.1.2 and measuring indoor air temperature and relative humidity. Values of PMV and its variation in a one minute interval as well as outdoor temperature are the inputs for the Fuzzy Inference System (FIS). The control value for the fan coil is its speed percentage (0–100 %) with respect the max speed. A saturation value has been chosen to stop the fan coil unit when the output value of the FIS is below 18 % of the max speed. The outdoor temperature, an input of the fuzzy controller, allows us to consider the seasonality when optimizing indoor comfort. Considering fuzzy sets of Table 2, if the outdoor temperature is "High" the control system acts to obtain and maintain "Neutral" indoor comfort conditions. On the contrary when the outdoor temperature is "Very High" the controller brings the environment to a "Summer Comfort" situation. Furthermore the presence of the PMV value and its variation in the FIS allow the control system to act as a regulator. We built 120 fuzzy rules and a sample is shown in Table 3.

4.1 Experimental Implementation

Sensors used in the experimental tests are built with the EnOcean technology, an energy harvesting wireless sensors technology used primarily in building automa-

Table 2 Considered fuzzy sets for the input variables: linguistic terms and their corresponding trapezoidal fuzzy sets

Input variables	Linguistic terms	Fuzzy sets (a, b, c, d)
Predicted mean vote (PMV)	Too cold	−3, −3, −0.9, −0.4
	Winter comfort	−0.6, −0.3, −0.1, 0
	Neutral	−0.2, 0, 0, 0.2
	Summer comfort	0, 0.2, 0.5, 0.7
	Too hot	0.5, 1, +3, +3
Outdoor temperature (T_e)	Very low	$-\infty, -\infty$, 0, 4
	Low	0, 4, 10, 14
	Medium	10, 14, 22, 26
	High	22, 26, 30, 34
	Very high	30, 34, $+\infty, +\infty$
PMV variation (Δ_{PMV})	Negative big	$-\infty, -\infty$, −0.5, −0.3
	Negative medium	−0.5, −0.3, −0.2, −0.1
	Negative small	−0.2, −0.1, −0.05, 0
	Zero	−0.05, −0.01, 0.01, 0.05
	Positive small	0, 0.05, 0.1, 0.2
	Positive medium	0.1, 0.2, 0.3, 0.5
	Positive big	0.3, 0.5, $+\infty, +\infty$
Output variable	**Linguistic terms**	**Fuzzy sets (a, b, c, d)**
Fan coil speed (U)	Zero	0, 0, 0.1, 0.2
	Low	0.1, 0.2, 0.3, 0.4
	Medium	0.3, 0.4, 0.6, 0.7
	High	0.6, 0.7, 0.8, 0.9
	Very high	0.8, 0.9, 1, 1

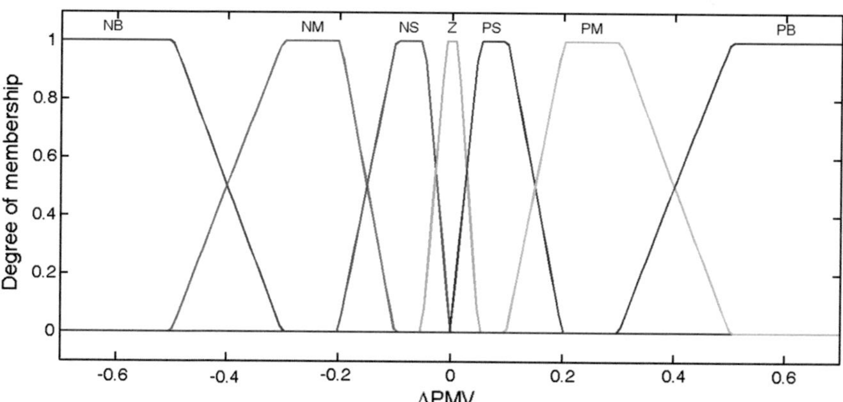

Fig. 2 Fuzzy sets for the input variable Δ_{PMV}. Labels for the sets are reported in Table 2

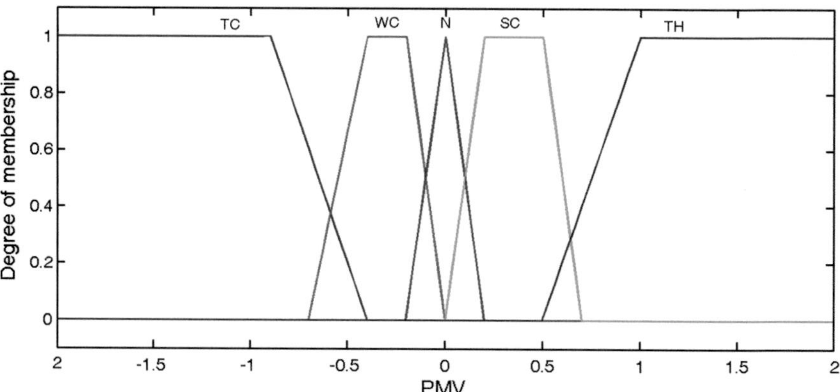

Fig. 3 Fuzzy sets for the input variable PMV. Labels for the sets are reported in Table 2

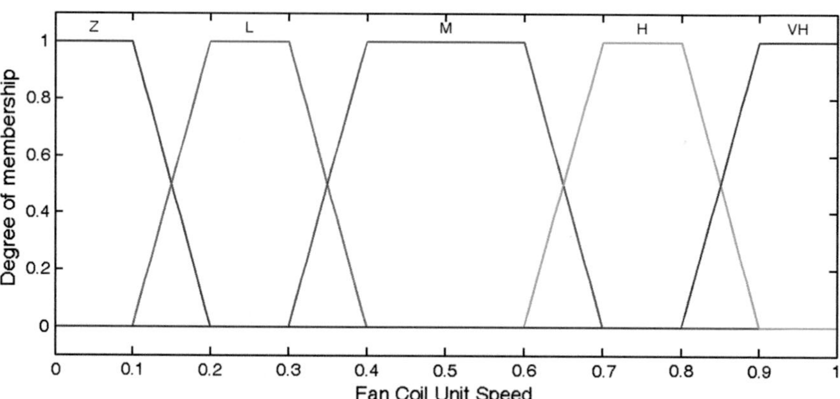

Fig. 4 Fuzzy sets for the output variable fan coil speed U. Labels for the sets are reported in Table 2

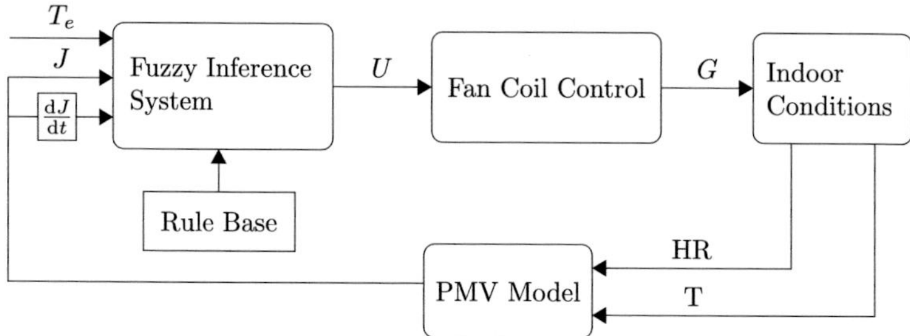

Fig. 5 Block diagram of the HVAC PMV based control. J and ΔJ represent the PMV and Δ_{PMV} values computed at time K. T_e is the outdoor temperature measured at time K

Table 3 Sample of the Fuzzy rules used for inference for "High" and "Very High" outdoor temperature

$T_e(K)$	PMV(K)	$\Delta_{PMV}(K)$	$U(K)$
VH	TH	NS	VH
VH	TH	NB	H
VH	SC	NM	Z
VH	SC	PS	M
H	TH	NB	H
H	SC	NM	S
H	N	NS	Z
H	N	Z	Z
H	N	PS	S

tion systems. These modules enable wireless communications between batteryless wireless sensors, controllers and gateways.

The external humidity and temperature sensor used is the model *FAFT*60 produced by Eltako company (the features of all sensors employed in the experimental tests can be found in [26]). It is an energy harvesting sensor with a micro solar panel to charge its battery. The internal humidity and temperature sensor is the model *FTF*55. The thermostat originally installed to control temperature is the *FTR*55*D*, allowing to preset temperature and acting as a PID temperature regulator. The microcontroller used to develop the proposed fuzzy controller is the model DIVO W, produced by the Italian company UMPI (technical data can be found in [43]). This module allows the management of both temperature and lighting control as well as the acquisition of EnOcean sensors values.

A fan coil unit has been used to actuate the indoor office comfort control signals and is part of a building HVAC system, thus allowing only to set up the fan coil speed. Indeed the air flow temperature is 19 °C during summer and 23 °C during winter periods. Only one unit was installed for the whole office environment and no structural configuration modifies have been performed.

The working space area was 42 m^2 and 2 outer walls of the office are directly exposed to sun (south and west oriented).

4.2 Experimental Results

The proposed fuzzy logic comfort controller has been tested in a working space (office) in the central east cost of Italy from May to August 2014. A series of experimental tests have been performed to verify if indoor comfort conditions were reached with different weather conditions. Figure 6 depicts the trend of the computed indoor PMV comfort index during a test in the summer. As it is possible to notice, starting from an extreme discomfort condition (PMV $>$ 1.2) in less than 45 minutes a summer comfort condition is reached (PMV $<$ 0.6). After 55 minutes the

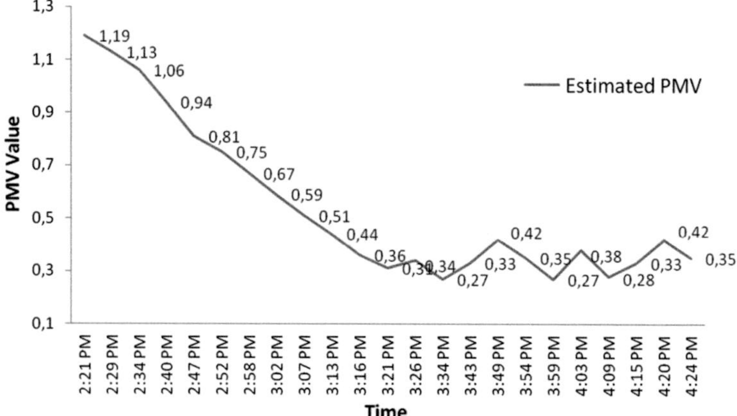

Fig. 6 July 25th 2014. Fuzzy controller performances. *Blue line* is the estimated PMV index

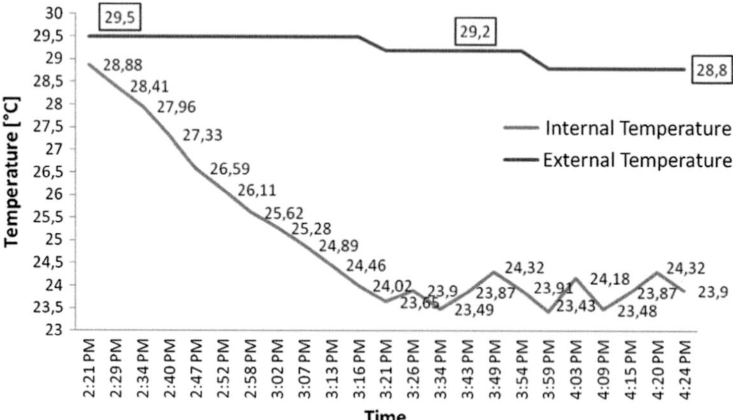

Fig. 7 July 25th 2014. Fuzzy controller performances. *Red line* is the external measured temperature while the *blue* one is the measured indoor temperature corresponding to the PMV index of Fig. 6

controller brings the room to the steady state comfort situation ($0.3 < \text{PMV} < 0.4$) and it is able to maintain that condition. In Fig. 7 is represented the trend of outdoor and indoor temperatures measured during the same test.

Since it is not possible to properly evaluate the controller performances by only transient and steady state error qualitative analysis, a quantitative performance comparison needs to be stated. In particular, problems of temperature regulation can extremely differ one to the other due to the configuration of the environment and its thermal properties. Thus, although from Figs. 6 and 7 the controller shows a good comfort optimization capability, we need to compare its transient and steady state error to a benchmark controller applied in the same experimental setup. We choose to compare our fuzzy controller with the PID based thermostat *FTR55D*, originally installed in the office.

However, since the aim of the fuzzy controller is to regulate indoor comfort, in order to compare its performances with those of the original thermostat, tests were performed according to the following protocol:

- test of the fuzzy controller for a certain outdoor and indoor starting temperature;
- computation of the steady state indoor temperature reached (considered as the set point when computing the error) and the related performance indexes. As an example, concerning the test of Fig. 7, the steady state temperature considered was of 23.9 °C;
- test of the PID controller in similar environmental conditions (starting indoor/outdoor temperature ± 0.2 °C) fixing as set point the previously computed steady state temperature;
- computation of the PID regulator performance indexes.

Three performance index measurements have been considered in the comparison process:

- the integral of the squared error (ISE):

$$\mathrm{ISE} = \int_0^T [e(t)]^2 dt \tag{19}$$

- the integral of the absolute value of error (IAE):

$$\mathrm{IAE} = \int_0^T |e(t)|\, dt \tag{20}$$

- the integral of time multiplied by the absolute value of error (ITAE), to penalize long duration transients as well as steady state oscillations:

$$\mathrm{ITAE} = \int_0^T t \cdot |e(t)|\, dt \tag{21}$$

Results of the experimental tests, carried out from May to August 2014, are summarized in Table 4. As reported in Table 4, the PMV Fuzzy based control shows better performances in terms of transient and steady state error with respect a PID based commercial thermostat.

To evaluate transient and steady state errors we computed the integral of the squared error (ISE) and the integral of time multiplied by the absolute value of error (ITAE). In particular, our fuzzy approach improves the performances by a 4.5 % according to the ISE index and the by an 10.4 % according to the ITAE index. Results are summarized in Table 4.

Table 4 Performance comparison of the proposed controller with the original built in PID based temperature control

	ISE	IAE	ITAE
PID based controller	16.884	2.561	104.725
Fuzzy logic based controller	16.122	2.492	93.806

5 Residential Multiroom Comfort Control

In a residential scenario it would be extremely expensive to gather real time measures of all the variables involved in PMV formula. Furthermore the most influential variables (i.e. the metabolic rate and clothes insulation, as stated in Sect. 2.1) have a wide range of admissible values in an household context. Thus it becomes necessary to make many strong assumptions to turn PMV equation into a non linear function of air temperature and relative humidity. In this context we consider another approach to measure the indoor perceived temperature. Fuzzy rule-based systems (FRBS) have been successfully employed for system identification, control and modeling in many areas [5, 38].

Figure 8 depicts the block diagram and the variables involved in the whole comfort control process. Input variables for the fuzzy inference system (FIS) control are the error between desired and measured perceived temperature (as shown in Fig. 9), its variation in a one minute interval and the outdoor temperature. The output of the FIS engine is the Fan Coil Speed percentage (0–100 %) with respect the max speed. A saturation value has been chosen to stop the fan coil unit when the output value of the FIS is below 18 % of the max speed. We built 140 Mamdani fuzzy rules of the following form:

IF **Error** is **Pos Big** AND **Error Derivative** is **Zero** AND **External Temperature** is **Low** THEN **Fan Coil Speed** is **Very High** (Table 5).

The defuzzyfication process computes the speed to drive the fan coil in order to maintain the desired indoor comfort level.

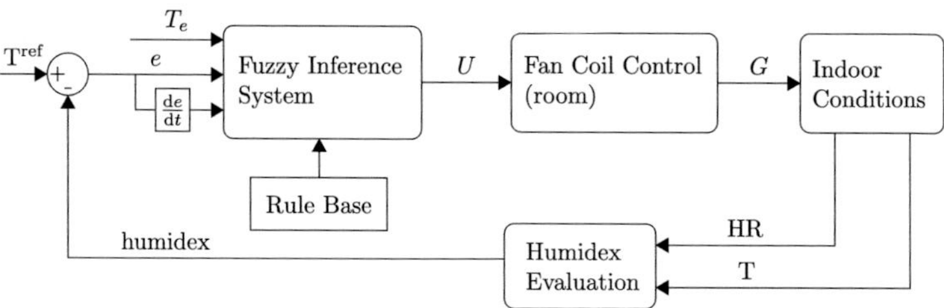

Fig. 8 Block diagram of a single room control

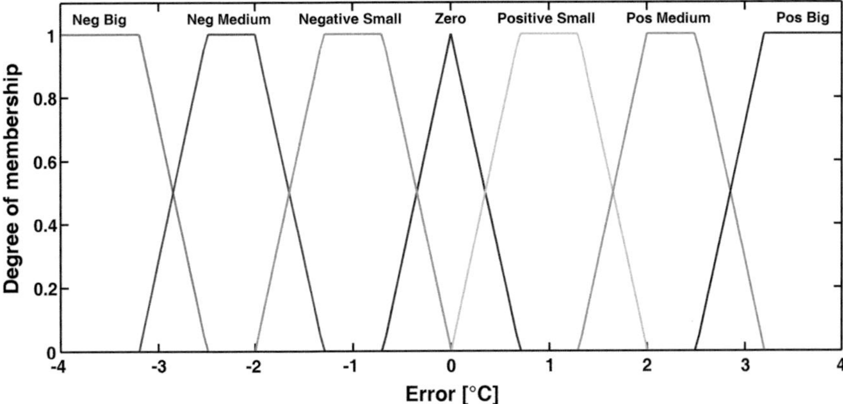

Fig. 9 Membership functions of the input variable error

Table 5 Considered fuzzy sets for the input variables: linguistic terms and their corresponding trapezoidal fuzzy sets

Input variables	Linguistic terms	Fuzzy sets (a, b, c, d)
Outdoor temperature (T_e)	Very Low	$-\infty, -\infty, 0, 4$
	Low	0, 4, 10, 14
	Medium	10, 14, 22, 26
	High	22, 26, 30, 34
	Very High	$30, 34, +\infty, +\infty$
Temperature error (e)	Negative Big	$-\infty, -\infty, -4, -2.5$
	Negative Medium	$-4, -2.5, -1.5, -0.8$
	Negative Small	$-1.5, -0.8, 0, 0$
	Zero	$-0.2, 0, 0, 0.2$
	Positive Small	0, 0, 0.8, 1.5
	Positive Medium	0.8, 1.5, 2.5, 4
	Positive Big	$2.5, 4, +\infty, +\infty$
Error variation (Δe)	Negative Medium	$-3, -3, -1, -0.5$
	Negative Small	$-0.7, -0.5, -0.2, 0$
	Zero	$-0.2, 0, 0, 0.2$
	Positive Small	0, 0.2, 0.5, 0.7
	Positive Medium	$0.5, 1, +3, +3$
Output variable	Linguistic terms	Fuzzy sets (a, b, c, d)
Fan coil speed (U)	Zero	0, 0, 0.1, 0.2
	Low	0.1, 0.2, 0.3, 0.4
	Medium	0.3, 0.4, 0.6, 0.7
	High	0.6, 0.7, 0.8, 0.9
	Very High	0.8, 0.9, 1, 1

5.1 Experimental Implementation

Sensors employed in the experimental tests are built with the EnOcean energy harvesting wireless technology. The external humidity and temperature sensor used is the model *FAFT*60 (the features of all sensors employed in the experimental tests can be found in [26]). The indoor humidity and temperature sensor is the Etalko model *FTF*55. The thermostat originally installed to control temperature is the *FTR*55*D*, allowing to preset temperature and acting as a PID temperature regulator. Fan coil units have been used to actuate the control signals. Since they are part of a building integrated HVAC system, we are only allowed to set up the fan coil speed. Indeed the air flow temperature is 19 °C during summer and 23 °C during winter periods. The test area is a 56 m^2 apartment composed by 3 rooms (kitchen, living room and bathroom). Each room under test has one fan coil unit installed and no configuration modifies have been performed. 2 outer walls of the area are directly exposed to sun (south and west oriented).

5.2 Experimental Results

Experimental tests have been performed to verify if multi room indoor comfort conditions were reached. Figure 10 depicts a test performed during a summer day. As it is possible to notice, varying the humidity of one of the rooms the controller acts to bring back the perceived temperature to the desired value (24.8 °C, set by the apartment occupants). A quantitative analysis has been performed comparing our fuzzy control approach with the PID based thermostat *FTR*55*D* originally installed in the building.

However, since the aim of the fuzzy controller is to regulate the perceived temperature, in order to compare its performances with those of the original thermostat, tests were performed according to the following protocol:

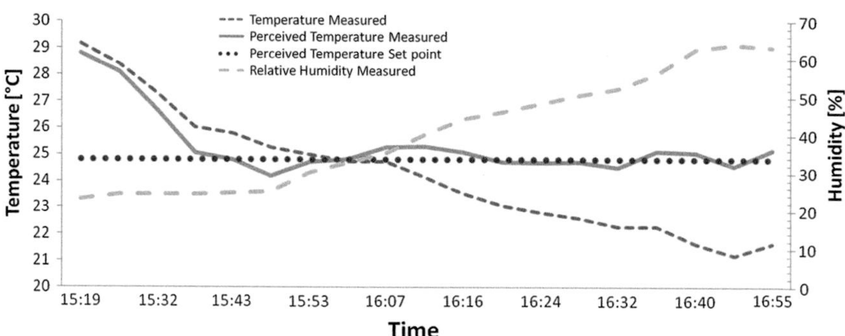

Fig. 10 June 21st 2014. Fuzzy controller performances

Table 6 Performance comparison of the multiroom Humidex based controller with the original built in PID based temperature control

	ISE	IAE	ITAE
PID based controller	15.91	2.74	108.55
Fuzzy logic based controller	14.10	2.32	94.81

- test of the humidex based fuzzy controller for a certain perceived temperature;
- computation of the corresponding air temperature;
- test of the PID controller in similar environmental conditions (starting indoor/outdoor temperature ± 0.2 °C) fixing as set point the previously computed air temperature;
- computation of the PID regulator performance indexes.

To evaluate transient and steady state errors we computed the integral of the squared error (ISE) and the integral of time multiplied by the absolute value of error (ITAE). In particular, our fuzzy approach improves the performances by a 5.7 % according to the ISE index and the by an 11.2 % according to the ITAE index. Results are summarized in Table 6.

6 Concluding Remarks

Due to the increasing attention on buildings indoor thermal comfort, many researchers focused their work in this area. In this chapter we proposed two different solutions for the comfort regulation for an office and a residential scenario. Among all international standards released, predicted mean vote (PMV) index is universally adopted for assessing users thermal comfort conditions. PMV index is a nonlinear function of various quantities, which limits its applicability, e.g. to the HVAC control problem.

A sensitivity analysis of the PMV showed that metabolic and clothing variables are the most influential in its formula. The result furthermore limits its applicability in dynamic and unpredictable living scenarios (e.g. an household). Furthermore PMV index treats all occupants in the same way do not taking into account specific location, outdoor condition and seasonality. Considering these issues we designed a novel fuzzy controller introducing external environmental parameters in order to build an Adaptive-PMV control system specific for office environments. Indeed in these scenario both metabolic and clothing parameters can be considered as constants without losing generality. Experimental tests have been carried out from May to August 2014 in a building in the central East coast of Italy. Results showed that the proposed control technique allows to avoid the use of a temperature set point for HVAC system, automatically regulating seasonal comfort conditions. Furthermore temperature regulation capabilities of the fuzzy controller have been compared with those of a classical PID regulator, showing better performances quantified in terms of IAE, ISE and ITAE indexes.

In the second part of the chapter we proposed a fuzzy logic approach for the indoor multi room comfort control. Considering the above mentioned issues about PMV use in dynamic scenarios, we investigated for the first time the use of the humidex index for comfort control purposes. The performances of the new approach have been experimentally tested in an apartment in Italy. Experimental tests have been carried out during summer 2014. Results showed that the proposed control technique allows to regulate the indoor perceived temperature for existent HVAC system in a multi room scenario. Temperature regulation capabilities of the fuzzy controller have been compared with those of a classical PID based thermostat, showing better performances (ranging from 5.7 % of ISE to 11.2 % of ITAE indexes).

References

1. Alcala R, Benitez JM, Casillas J, Cordon O, Perez R (2003) Fuzzy control of hvac systems optimized by genetic algorithms. Appl Intell 18(2):155–177
2. Azar AT (2010a) Adaptive neuro-fuzzy systems. In: Fuzzy systems. InTech, Vienna, Austria. ISBN: 978-953-7619-92-3
3. Azar AT (2010b) Fuzzy systems. InTech, Vienna, Austria
4. Azar AT (2012) Overview of type-2 fuzzy logic systems. Int J Fuzzy Syst Appl 2(4):1–28
5. Azar AT, Vaidyanathan S (2015a) Chaos modeling and control systems design. Studies in computational intelligence, vol. 581
6. Azar AT, Vaidyanathan S (2015b) Computational intelligence applications in modeling and control. Studies in computational intelligence, vol. 575
7. Azar AT, Vaidyanathan S (2015c) Handbook of research on advanced intelligent control engineering and automation. IGI Global
8. Brelih N, Seppanen O (2011) Ventilation rates and iaq in european standards and national regulations. In: 32nd AIVC conference and 1st TightVent conference
9. Cannavò F (2012) Sensitivity analysis for volcanic source modeling quality assessment and model selection. Comput Geosci 44:52–59
10. Castilla M, Alvarez J, Normey-Rico J, Rodriguez F (2014) Thermal comfort control using a non-linear mpc strategy: a real case of study in a bioclimatic building. J Process Control 24(6):703–713. Energy Efficient Buildings Special Issue
11. Chandan V, Alleyne AG (2014) Decentralized predictive thermal control for buildings. J Process Control 24(6):820–835. Energy Efficient Buildings Special Issue
12. Ciabattoni L, Cimini G, Ferracuti F, Grisostomi M, Ippoliti G, Pirro M (2015a). Indoor thermal comfort control through fuzzy logic pmv optimization. In: 2015 International joint conference on neural networks (IJCNN)
13. Ciabattoni L, Cimini G, Grisostomi M, Ippoliti G, Longhi S, Mainardi E (2013a) Supervisory control of PV-battery systems by online tuned neural networks. In: IEEE International conference on mechatronics (ICM), Vicenza, Italy, pp 99–104
14. Ciabattoni L, Corradini ML, Grisostomi M, Ippoliti G, Longhi S, Orlando G (2014a) A discrete-time verus controller based on rbf neural networks for pmsm drives. Asian J Control 16(2):396–408
15. Ciabattoni L, Ferracuti F, Grisostomi M, Ippoliti G, Longhi S (2015b) Fuzzy logic based economical analysis of photovoltaic energy management. Neurocomputing 170:296–305
16. Ciabattoni L, Ferracuti F, Ippoliti G, Longhi S, Turri G (2016). Iot based indoor personal comfort levels monitoring. In: 2016 IEEE International conference on consumer electronics (ICCE)

17. Ciabattoni L, Freddi A, Ippoliti G, Marcantonio M, Marchei D, Monteriu A, Pirro M (2013b) A smart lighting system for industrial and domestic use. In: 2013 IEEE international conference on mechatronics (ICM), pp 126–131
18. Ciabattoni L, Grisostomi M, Ippoliti G, Longhi, S (2013c) A fuzzy logic tool for household electrical consumption modeling. In: IECON 2013–39th Annual conference of the IEEE industrial electronics society, pp 8022–8027
19. Ciabattoni L, Grisostomi M, Ippoliti G, Longhi S (2014b) Fuzzy logic home energy consumption modeling for residential photovoltaic plant sizing in the new italian scenario. Energy 74:359–367
20. Ciabattoni L, Grisostomi M, Ippoliti G, Longhi S (2014c). Home energy management benefits evaluation through fuzzy logic consumptions simulator. In: 2014 International joint conference on neural networks (IJCNN), pp 1447–1452
21. Ciabattoni L, Ippoliti G, Benini M, Longhi S, Pirro M (2013d) Design of a home energy management system by online neural networks. In: 11th IFAC International workshop on adaptation and learning in control and signal processing. Caen, France, pp 677–682
22. Ciabattoni L, Ippoliti G, Longhi S, Cavalletti M (2013e) Online tuned neural networks for fuzzy supervisory control of PV-battery systems. In: IEEE PES innovative smart grid technologies conference (ISGT)
23. Ciabattoni L, Ippoliti G, Longhi S, Grisostomi M, Mainardi E (2012a) On line solar irradiation forecasting by minimal resource allocating networks. In: IEEE MED Conference 2012
24. Ciabattoni L, Ippoliti G, Longhi S, Grisostomi M, Rocchetti M (2012b) Online tuned neural networks for pv plant production forecasting. In: 2012 IEEE PVSC conference
25. Cigler J, Privara S, Vana Z, Zacekova E, Ferkl L (2012) Optimization of predicted mean vote index within model predictive control framework: computationally tractable solution. Energ Build 52:39–49
26. Etalko electronics (2014). EnOcean wireless sensors datasheets. http://www.eltako.com/en/the-wireless-building/2-active-wireless-sensors-and-transmitter-modules.html. Accessed 12 Jan 2015
27. Ferhatbegovic T, Zucker G, Palensky P (2012) An unscented kalman filter approach for the plant-model mismatch reduction in hvac system model based control. In: IECON 2012—38th Annual Conference on IEEE Industrial Electronics Society, pp 2180–2185
28. Ferreira P, Ruano A, Silva S, Conceicao E (2012). Neural networks based predictive control for thermal comfort and energy savings in public buildings. Energ Build 55(0):238–251. Cool Roofs, Cool Pavements, Cool Cities, and Cool World
29. Giantomassi A, Ferracuti F, Iarlori S, Longhi S, Fonti A, Comodi G (2014a) Kernel canonical variate analysis based management system for monitoring and diagnosing smart homes. In: 2014 international joint conference neural networks (IJCNN), pp 1432–1439
30. Giantomassi A, Ferracuti F, Iarlori S, Puglia G, Fonti A, Comodi G, Longhi S (2014b). Smart home heating system malfunction and bad behavior diagnosis by multi-scale pca under indoor temperature feedback control
31. ISO (International Standard Organization) (2007). ISO-7730:2006 norm. https://moodle.metropolia.fi/pluginfile.php/217631/mod_resource/content/1/EVS_EN_ISO_7730%3B2006_en.pdf. Accessed 18 Dec 2014
32. Jazizadeh F, Ghahramani A, Becerik-Gerber B, Kichkaylo T, Orosz M (2014) User-led decentralized thermal comfort driven hvac operations for improved efficiency in office buildings. Energ Build 70:398–410
33. Liang J, Du R (2005) Thermal comfort control based on neural network for hvac application. In: Proceedings of 2005 IEEE conference on control applications, CCA 2005, pp 819–824
34. Masterton J, Richardson F, service of atmospherique environnement C (1979) Humidex: a method of quantifying human discomfort due to excessive heat and humidity. 28cm. cli,1. Ministere de l'Environnement
35. Moon JW (2012) Performance of ann-based predictive and adaptive thermal-control methods for disturbances in and around residential buildings. Build Environ 48:15–26

36. Moon JW, Jung SK, Kim Y, Han S-H (2011) Comparative study of artificial intelligence-based building thermal control methods—application of fuzzy, adaptive neuro-fuzzy inference system, and artificial neural network. Appl Thermal Eng 31(14):2422–2429
37. Murakami S, Kato S, Kim T (2001). Coupled simulation of convicton, radiation, and hvac control for attaining a given pmv value. Build Environ 36(6):701–709. Building and Environmental Performance Simulation: Current State and Future Issues
38. Pepa L, Ciabattoni L, Verdini F, Capecci M, Ceravolo M (2014) Smartphone based fuzzy logic freezing of gait detection in parkinson's disease. In: 10th IEEE/ASME international conference on mechatronics and embedded systems and applications (MESA)
39. Pourshaghaghy A, Omidvari M (2012) Examination of thermal comfort in a hospital using pmv-ppd model. Appl Ergon 43(6):1089–1095
40. Rana R, Kusy B, Jurdak R, Wall J, Hu W (2013) Feasibility analysis of using humidex as an indoor thermal comfort predictor. Energ Build 64:17–25
41. Revel GM, Sabbatini E, Arnesano M (2012) Development and experimental evaluation of a thermography measurement system for real-time monitoring of comfort and heat rate exchange in the built environment. Meas Sci Technol 23(3)
42. Sobol I (2001) Global sensitivity indices for nonlinear mathematical models and their monte carlo estimates. Math Comput Simul 55(13):271–280. The second IMACS seminar on monte carlo methods
43. UMPI srl (2014) DIVO W module. http://www.umpi.it/en/simple-life-products-divo-w. Accessed 18 Dec 2014
44. Wang N, Fang F, Feng M (2014) Multi-objective optimal analysis of comfort and energy management for intelligent buildings. In: The 26th Chinese control and decision conference (2014 CCDC), pp 2783–2788
45. Wemhoff A (2012) Calibration of hvac equipment pid coefficients for energy conservation. Energ Build 45:60–66
46. Xiao J, Li J, Boutaba R, Hong J-K (2012) Comfort-aware home energy management under market-based demand-response. In: 2012 8th international conference on network and service management (cnsm) and 2012 workshop on systems virtualiztion management (svm), pp 10–18
47. Yang J-H, Bi X-Y (2010) High-precision temperature control system based on pid algorithm. In: 2010 International conference on computer application and system modeling (ICCASM), vol 12
48. Zhao Q, Cheng Z, Wang F, Jiang Y, Ding J (2014) Experimental study of group thermal comfort model. In: 2014 IEEE international conference on automation science and engineering (CASE), pp 1075–1078
49. Zhu Q, Azar AT (2015) Complex system modelling and control through intelligent soft computations. Studies in fuzziness and soft computing, vol 319

Load Frequency Control Based on Evolutionary Techniques in Electrical Power Systems

Naglaa K. Bahgaat, M.I. El-Sayed, M.A. Moustafa Hassan and F. Bendary

Abstract Load Frequency Control (LFC) used to regulate the power output of the electric generator within an area as the response of changes in system frequency and tie-line loading. Thus the LFC helps in maintaining the scheduled system frequency and tie-line power interchange with the other areas within the prescribed limits. Most LFCs are primarily composed of an integral controller. The integrator gain is set to a level that compromises between fast transient recovery and low overshoot in the dynamic response of the overall system. The disadvantage of this type of controllers that there are slow and does not allow the controller designer to take into account possible changes in operating conditions and non- linearities in the generator unit. Moreover, it lacks robustness. So there are many modern techniques used to tune the controller. This chapter discusses the application of evolutionary techniques in Load Frequency Control (LFC) in power systems. It gives introduction to evolutionary techniques. Then it presents the problem formulation for load frequency control with Evolutionary Particle Swarm Optimization (MAACPSO). It gives the application of Particle Swarm Optimization (PSO) in load frequency control, also it illustrates the use of a Adaptive Weight Particle Swarm Optimization (AWPSO), Adaptive Accelerated Coefficients based PSO, (AACPSO) Adaptive Accelerated Coefficients

N.K. Bahgaat (✉)
Electrical Communication Department, Faculty of Engineering,
Canadian International College (CIC), 6 October City, Giza, Egypt
e-mail: nkahgaat@hotmail.com; n_mohamed2004@yahoo.com

M.I. El-Sayed
Electrical Power Engineering Department Faculty of Engineering,
Al-Azhar University, Cairo, Egypt
e-mail: d_eng2009@yahoo.com

M.A. Moustafa Hassan
Electrical Power Engineering Department Faculty of Engineering,
Cairo University, Giza, Egypt
e-mail: mmustafa@eng.cu.edu.eg

F. Bendary
Electrical Power Engineering Department Faculty of Engineering,
Banha University, Cairo, Egypt
e-mail: fahmybendary10@gmail.com

A.T. Azar and S. Vaidyanathan (eds.), *Advances in Chaos Theory and Intelligent Control*, Studies in Fuzziness and Soft Computing 337,
DOI 10.1007/978-3-319-30340-6_36

based PSO (AACPSO). Furthermore, it introduces a new modification for AACPSO technique (MAACPSO). The new technique will be explained inside the chapter, it is abbreviated to Modified Adaptive Accelerated Coefficients based PSO (MAACPSO). A well done comparison will be given in this chapter for these above mentioned techniques. A reasonable discussion on the obtained results will be displayed. The obtained results are promising.

Keywords Modified Adaptive Accelerated Coefficients based PSO · Adaptive Accelerated Coefficients based Particle Swarm Optimization · Adaptive weight particle swarm optimization · Load Frequency Control · and Particle Swarm Optimization Technique

1 Introduction

Frequency is an important factor to describe the stability criterion in power systems [15, 23, 29]. To provide the stability of power system, active power balance and steady frequency are required. If any change occurs in active power demand or the generation in power systems, oscillations increase in both power and frequency. Frequency cannot be hold in its rated value because it depends on active power balance. Thus, system subjects to a serious instability problem. In electric power generation, system disturbances caused by load fluctuations result in changes to the desired frequency value. Automatic Generation Control (AGC) or Load Frequency Control (LFC) is an important issue in power system operation and control for supplying stable and reliable electric power with good quality [24, 27]. The principle aspect of Automatic Load Frequency Control is to maintain the generator power output and frequency within the prescribed limits [7].

There are many controllers used in practice in order to keep the power system in normal operating state [13, 30], one of the famous controllers used in power system are Proportional Integral (PI), Proportional Derivative (PD) and Proportional Integral Derivative (PID) controllers, PID will be used for the stabilization of the frequency in the load frequency control problems [4, 15, 23, 24, 29]. When changes of the loads occur each control area is responsible for individual load changes and scheduled interchanges with neighboring areas [25]. The changes of the loads and abnormal conditions leads to mismatches in frequency and tie line power interchanges which are to be kept in the allowable limits, for the strong operation of the power system. For simplicity, the effects of governor dead band are neglected in the LoadFrequency Control studies. To study the realistic analysis of the system performance, the governor dead band effect is to be incorporated. To improve the stability of the power networks, it is necessary to design LFC system that controls the power generation and active power at tie lines [7, 8].

There are many studies done in the past on this important issue in power systems, which is the load frequency control. As stated in some literature [9, 14, 20], its objective is to minimize the transient deviations in area frequency and tie-line power interchange and to ensure their steady state errors to be zeros. This chapter discusses the application of evolutionary techniques in Load Frequency Control (LFC) in power systems, such that Particle Swarm Optimization (PSO), Adaptive Weighted Particle Swarm Optimization techniques (AWPSO), Adaptive Accelerated Coefficients based on PSO (AACPSO) and Evolutionary Particle Swarm Optimization (MAACPSO). Will be used to determine the parameters of a PID controller according to the system dynamics. Using the same parameters of PID controller for the two different areas because it gives a better performance for the system frequency response than in case of using two different PID parameters for each different area [8, 19]. The main objectives of LFC in case of changes in system frequency, is to regulate the power output of the electric generator within an arranged area in response to, tie line loading so as to maintain the planned system frequency and interchange with the other areas within the prescribed limits.

In this chapter, the power systems contents of two area and load frequency control of this system is made based on PID controller. To choose best parameters of PID Controller many techniques are used, Particle Swarm Optimization and Adaptive Weight Particle Swarm Optimization Techniques (PSO) and (AWPSO) [11, 18, 21] and Also using Adaptive Accelerated Coefficients based PSO, (AACPSO), then a new modification for AACPSO technique will be discuss called evolutionary techniques based on Particle Swarm Optimization (MAACPSO).

This chapter is organized as follow: Sect. 1 introduces the chapter. The Sect. 2 presents literature review of the study. Section 3 introduces Particle Swarm (PSO), AWPSO, the Adaptive Accelerated Coefficients based PSO (AACPSO) and Modified Adaptive Accelerated Coefficients based PSO (MAACPSO). Section 4 displays the case study and a comparative study between these methods, while Sect. 5 concludes the chapter. Finally a list of references and Appendix of this chapter are given at the end of the chapter.

2 Literature Review

The PID controller was first described by Minorsky [2]. It has been confirmed that in control applications more than 95 % of the controllers are PID type. Also, they state that 30 % of the PID loops operate in the manual mode and 25 % of PID loops actually operate under default factory settings. The choice of appropriate PID parameters can be achieved manually by trial and error, using as guidelines the transient and steady response characteristic of each of the three terms. However, this procedure is very time consuming and requires certain skills [3, 8].

PID control is a linear control methodology. The structure of PID controllers as shown in Fig. 1 is very simple. They operate on the error signal, which is the difference between the desired output and the actual output, and generate the actuating signal that drives the plant. They have three basic terms: proportional action, in which the actuating signal is proportional to the error signal, integral action, where the actuating

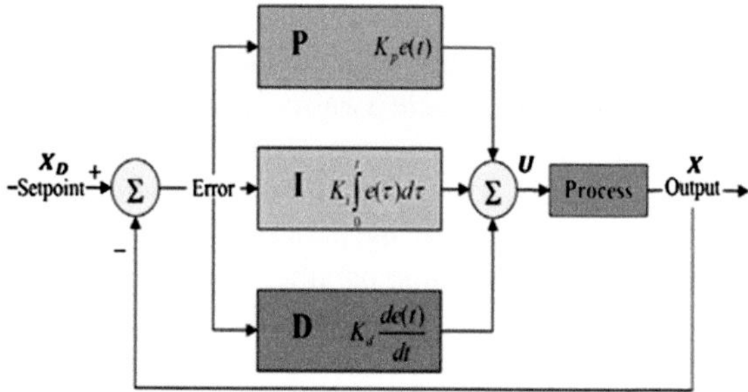

Fig. 1 Structure of PID controller

signal is proportional to the time integral of the error signal; and derivative action, where the actuating signal is proportional to the time derivative of the error signal as illustrated in [10, 22, 27].

With caution because it amplifies any existent noise in the signal. The PID standard form is given by:

$$u = K_P.e + K_i \int e.dt + K_d.\frac{de}{dt} \tag{1}$$

where:

e Is the error signal
u Is the control action
Kp Is the proportional gain
Ki Is the integral gain
Kd Is the derivative gain

In power system which consists of many neighboring areas there are mismatches in frequency and power transfer. The controller used in power systems should provide some degree of strength under different operating conditions. Using conventional PD, PI, PID controllers does not provide sufficient control performance with the effect of governor dead band [15, 23, 24, 29]. So many methods used since 1890s till now to tune the controller. First method is the manual tuning, this method.

To design a particular control loop, the three constants (KP, Ki, and Kd) have to be adjusted to arrive at acceptable performance; If the system must remain online, then first set Ki and Kd values to zero. Increase Kp until the output of the loop oscillates; then KP should be set to approximately half of that value for a "quarter amplitude decay" type response. Then increase Ki until any offset is correct in sufficient time for the process. However, increasing Ki will cause instability. Finally, increase Kd, if required, until the loop is acceptably quick to reach its reference after a load disturbance. However, too much Kd will cause excessive response and overshoot. A fast PID loop tuning usually overshoots slightly to reach the set point more quickly; however, some systems cannot accept overshoot, in which case an over damped closed-loop system is required, which will require a KP setting significantly less than half that of the KP setting causing oscillation. The effect of increasing each of the controller parameters KP, Ki and Kd can be summarized as illustrated in

[27]. Second method is an automatic method called Ziegler–Nichols method which introduced by John G. Ziegler and Nathaniel B. Nichols in the 1940s [24, 29]. It is recognized that the step response of most process control systems has an S-shaped curve called the process reaction curve and can be generated experimentally or from dynamic simulation of the plant. [8], the shape of the curve is characteristic of high order systems, and the plant behavior may be approximated by the following transfer function [27]:

$$\frac{Y(S)}{U(S)} = \frac{K.e^{-t_d.S}}{\tau.S + 1} \tag{2}$$

which is simply; a first order system plus a transportation lag. The constants in the above equation can be determined from the unit step response of the process. Ziegler and Nichols applied the PID controller to plants without integrator or dominant complex-conjugate poles, whose unit-step response resemble an S shaped curve with no overshoot. This S-shaped curve is called the reaction curve as shown in Fig. 1:

The following PID controller parameters were suggested:

$$K_P = 1.2T/L \tag{3}$$

$$K_i = K_P/2L \tag{4}$$

$$K_d = 0.5.L.\,K_P \tag{5}$$

Although the method provides a first approximation the response produced is under damped and needs further manual retuning. Some disadvantages of these control techniques for tuning PID controllers are:

(a) Excessive number of rules to set the gains.
(b) Inadequate dynamics of closed loop responses.
(c) Difficulty to deal with nonlinear processes.
(d) Mathematical complexity of the control design.

Therefore, it is interesting for academic and industrial communities the aspect of tuning for PID controllers, especially with a reduced number of parameters to be selected and a good performance to be achieved when dealing with complex processes.

The manual calculation methods no longer are used to tune loops inmost modern industrial facilities. Instead, PID tuning and loop optimization software are used to guarantee dependable results [14, 17, 25]. These software packages will gather the data, develop process models, and suggest optimal tuning.

Some software packages can even develop tuning by gathering data from reference changes, such as PSO, AWPSO [18, 19, 21], AACPSO [1, 8]. And this chapter will discuss the design of the PID controller by using modern method PSO, AWPSO, AACPSO and MAACPSO. These methods are simulated on MATLAB software program. This computer program which written on MATLAB had loops and run many times until reaching to a solution of the transfer function to have a value of PID parameters. These parameters lead to have the smallest value of settling time and over shoot. Therefore, these values of PID parameters (with these used methods) are the best values to reach to the best controller parameters. Moreover, a good comparison between the results of each used method will be done to choose the best one of them which will be suitable to use in the power system model used.

3 Overview on Practical Swarm Optimization Techniques

A Particle Swarm Optimization (PSO) is one of Artificial Intelligence (AI) Techniques. It's an optimization algorithm modeled. From the fields of AI with those of control engineering to design independent systems that can sense, reason, learn and act in an intelligent method. PSO depends on the simulation of the social behavior of bird and fish school [8, 21]. PSO is developed through the simulation of a bird flocking in two-dimension space by X-Y axis position where Vx and Vy express the velocity in X direction and Y direction. The flow chart described in Fig. 2, presented the steps of PSO. Modification of the agent position is realized by the position and velocity information [11, 18, 19, 21]. This information is analogy of personal experiences of each agent. Each agent knows its best value so far (Pbest) and its XY position; each agent knows the best value so far in the group (gbest) among Pbest s. This information is analogy of knowledge of how the other agents around them have performed. Namely, each agent tries to modify its position using the following information:

Let the particle of the swarm is represented by the N dimensional vector ith

$$X_i = (X_1, X_2, X_3, \ldots X_N) \tag{6}$$

The previous best position of the Nth particles is recorded and represented as follows:

$$\text{Pbesti} = (\text{Pbest1}, \text{Pbest2}, \ldots, \text{PbestN}) \tag{7}$$

where Pbest is Particle best position (m), N is the total number of iterations.

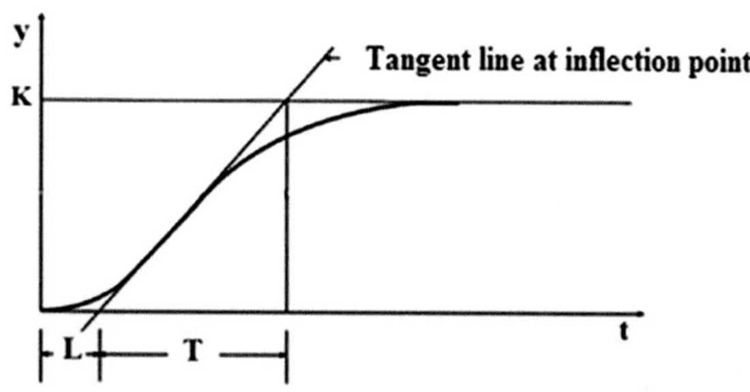

Fig. 2 Reaction curve used by Ziegler and Nichols

The best position of the particle among all particles in the swarm is represented by gbest the velocity of the particle is represented as follows:

$$Vi = (V1, V2, \ldots VN) \tag{8}$$

where V_i is the velocity of each i particle.

The modified velocity and position of each particle can be calculated from the current velocity and the distance from particle current position to particle best position P_{best} and to global best position g_{best} as shown in the following Equations [8]:

$$V_i(t) = W.V_i(t-1) + C_1.\text{rand}(0,1).(P_{best} - X_i(t-1)) + C_2.\text{rand}(0,1).(g_{best} - X_i(t-1)) \tag{9}$$

$$X_i(t) = X_i(t-1) + V_i(t) \tag{10}$$

$$i = 1, 2, 3.\ldots\ldots..N \tag{11}$$

$$j = 1, 2, 3.\ldots\ldots..D \tag{12}$$

where:

Vi(t)	Velocity of the particle i at iteration t (m/s)
Xi(t)	The Current position of particle i at iteration t (m)
D	The Dimension
C1	The cognitive acceleration coefficient and it is a positive number
C2	Social acceleration coefficient and it is a positive number
rand [0, 1]	A random number obtained from a uniform random distribution function in the interval [0, 1]
gbest	The Global best position (m)
W	The Inertia weight

3.1 Adaptive Weighted Particle Swarm Optimization

Adaptive Weighted Particle Swarm Optimization (AWPSO) technique has been anticipated for improving the performance of PSO in multi-objective optimization problems [18, 19]. AWPSO is consists of two terms which are: inertia weight (W) and Acceleration factor (A) [21]. The inertia weight (W) function is to balance global exploration and local exploration. It controls previous velocities effect on the new velocity. Larger the inertia weight, larger exploration of search space while smaller the inertia weights, the search will be limited and focused on a small region in the search space. The inertia weight formula is as follows which makes W value changes randomly from W_o to 1 [5, 6, 8].

$$W = W_o + \text{rand}(0, 1)(1 - W_o) \tag{13}$$

where:

Wo The initial positive constant in the interval chosen from [0, 1]

Particle velocity at its iteration as follows:

$$V_i(t) = W.V_i(t-1) + AC_1.\text{rand}(0, 1).(P_{best} - X_i(t-1)) + AC_2.\text{rand}(0, 1).(g_{best} - X_i(t-1)) \tag{14}$$

Additional term denoted by A called acceleration factor is added in the original velocity equation to improve the swarm search.

The iteration of the particle velocity described in [21] as the following:

$$A = A_o + \frac{i}{n} \tag{15}$$

where:

A_o: Is the initial positive constant in the interval [0.5, 1].
n: is the number of iteration.
C_1 and C_2: Are the constant representing the weighing of the stochastic acceleration terms that pull each particle towards P_{best} and g_{best} positions.

As shown in acceleration factor formula, that the acceleration term will increase as the number of iterations increases. This will increase the global search ability at the end of the run and help the algorithm to get far from the local optimum region. In this chapter, the term A_O is set at 0.5. Low values of C_1 and C_2 allow particles to roam far from the target region before being tugged back. However, high values result in abrupt movement toward, or past, target regions.

3.2 Adaptive Accelerated Coefficients Based PSO

In Sect. 3.1 the value of W can be located a good solution at a considerably faster rate but its ability to fine tune the optimum solution is weak, due to the lack of diversity at the end of the search. It has been observed by most researchers that in PSO, problem based tuning of parameters is a key factor to find the optimum solution accurately and efficiently [28]. New researches have emerged to improve PSO Algorithms, as Time-Varying Acceleration Coefficients (TVAC), where C_1 and C_2 in [12] change linearly with time, in the way that the cognitive component is reduced while the social component is increased as the search proceeds [1]. This method studies how to deal with inertia weight and acceleration factors and how to change acceleration coefficients exponentially (with inertia weight) in the time, with respect to their minimal and maximal values. The choice of the exponential function is justified by the increasing or decreasing speed of such a function to accelerate the convergence process of the algorithm and to get better search in the exploration s pace. Furthermore, C_1 and C_2 vary adaptively according to the fitness value of G_{best} and P_{best}, [8, 12] becomes:

$$V_i^{(t+1)} = w^{(t)} V_i^{(t)} + C_1^{(t)} r_1 * \left(Pbest_i^{(t)} - X_i^{(t)}\right) + C_2^{(t)} r_2 * \left(Gbest^{(t)} - X_i^{(t)}\right) \quad (16)$$

$$w^{(t)} = w_o * exp\,(- \propto_w * t) \quad (17)$$

$$C_1^{(t)} = C_{1o} * \exp\left(- \propto_c * t * k_c^{(t)}\right) \quad (18)$$

$$C_2^{(t)} = C_{2o} * \exp\left(\propto_c * t * k_c^{(t)}\right) \quad (19)$$

$$\propto_c = \frac{-1}{t_{max}} \ln\left(\frac{C_{2o}}{C_{1o}}\right) \quad (20)$$

$$k_c^{(t)} = \frac{\left(F_m^{(t)} - Gbest^{(t)}\right)}{F_m^{(t)}} \quad (21)$$

where:

$w^{(t)}$ The inertia weight factor
$C_1^{(t)}$ Acceleration coefficient at iteration t
i Equal 1 or 2
t The iteration number
ln The neperian logarithm
α_w Is determined with respect to initial and final values of ω with the same manner as αc described in [2].
$k_c^{(t)}$ Determined based on the fitness value of Gbest and Pbest at iteration t
ω_o, c_{io} c_{io} initial values of inertia weight factor and acceleration coefficients respectively with i = 1 or 2.
$F_m^{(t)}$ The mean value of the best positions related to all particles at iterationt

3.3 *Modified Adaptive Accelerated Coefficients PSO*

In this section, a new approach called Modified Adaptive Accelerated Coefficients PSO will be described as illustrated in [16]. A suggestion will be show how to choose the acceleration factors. The new approach will be make modification on the values of C1 and C2 which described in the last Sect. 3.2. The first coefficient changes exponentially (with inertia weight) in the time, with respect to their minimal and maximal values.While, the other one changes as a factor of the first coefficient. The choice of the exponential function is justified by the increasing or decreasing speed of such a function to accelerate the convergence process of the algorithm and to get better search in the exploration space.

Instead of the Eq. (18) the parameter $C_2(t)$ is suggested to be equal [16]:

$$C_2(t) = 4 - C_1(t) \tag{22}$$

The results of the program are shown in Tables 1, 2 and 3.

Table 1 Parameter description

Parameter	Description
Tg1, Tg2	Time constant for area 1 governor and area 2 governor in (seconds)
Tt1, Tt2	Turbine time delay between switching the valve and output turbine torque (seconds)
Tl1, Tl2	Generator 1 and generator 2 inertia constant
Kl1, Kl2	Power system gain constant (HZ/MW p.u)
R1, R2	Speed regulation constant of the governor (HZ/MW p.u)
B1, B2	Frequency bias p.u. MW/HZ
T12	Tie line synchronizing coefficient with area 2 MW p.u /HZ
a12	Gain
$\Delta f_1 or df_1$	Area 1 frequency deviation
$\Delta f_2 or df_2$	Area 2 frequency deviation
dPL1, dPL2	Frequency sensitive load change for area 1 and area 2
ΔPtie or dPtie	Net Tie line power flow
Vi	Area interface
ACE1	Area 1 control error
ACE2	Area 2 control error

Table 2 Parameters values

System parameters	Value
Tg1, Tg2	0.08 s
Tt1, Tt2	0.3 s
Tl1, Tl2	20 s
Kl1, Kl2	100 HZ/MW p.u
R1, R2	2.4 HZ/MW p.u
B1, B2	0.425 MW p.u/HZ
T12	0.05 MW p.u/HZ
a12	1

Table 3 The results of the program using PSO, AWPSO, AACPSO and MAACPSO

Items of comparison	PSO	AWPSO	AACPSO	MAACPSO
Number of iterations	500	500	500	500
Error IAE (Integrated error)	0.0611	0.0252	0.0149	0.0267
Settling time _Area 1 (sec)	5.4281	1.9323	1.6514	1.7267
Settling time _Area 2 (sec)	7.6946	4.1854	3.569	2.5288
Settling time _Tie line (sec)	7.7624	4.2082	3.6553	2.5696
Kp1	2.4283	8.1472	9.1995	3.7517
Ki1	1.5555	7.5774	9.4936	6.0754
Kd1	1.3753	2.7603	3.2393	0.8947
Kp2	2.9522	3.4998	4.7149	4.9802
Ki2	9.2078	1.6218	0.876	1.2982
Kd2	5.7955	8.6869	2.1397	8.4839

4 Cases Study

The model used as a cáse study is consists of two power system areas connected with each other's by tie transmission line as shown in Figs. 3 and 4 [8, 26]. Simulations are done by using MATLAB/SIMULINK for the case of the parameters of area 1 and area 2 are shown in the Appendix. Electric power system components are non-linear; therefore a linearization around a nominal operating point is usually performed to get a linearized system model which is used in the controller design process.

The operating conditions of power systems are continuously changing. Accordingly, the real plant usually differs from the assumed one. Therefore, classical algorithms to design an automatic generation controller using an assumed plant may not ensure the stability of the overall real system [8, 27]. The load frequency controller

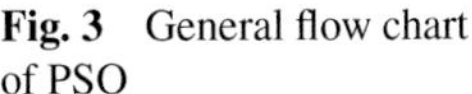

Fig. 3 General flow chart of PSO

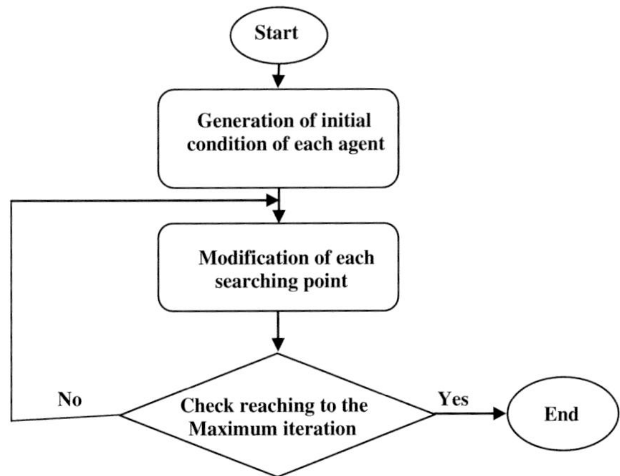

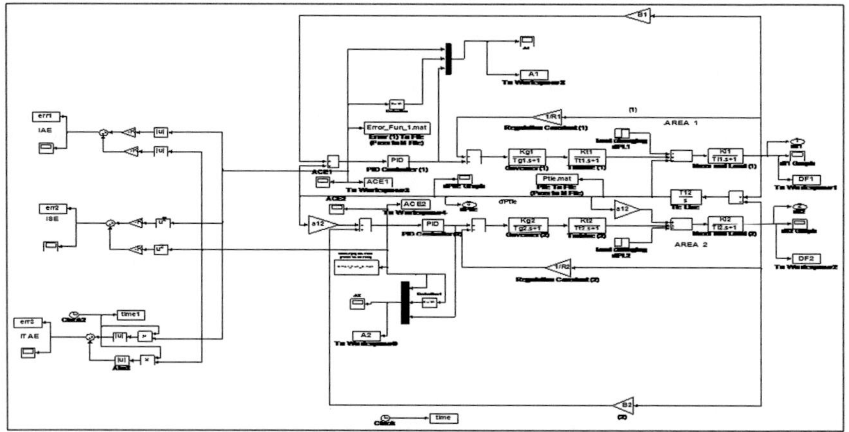

Fig. 4 Two-Area power system SIMULINK model using PID controller

function is to minimize the transient deviation of the frequency and maintains their values to steady state values and to restore the scheduled interchanges between different areas.

MATLAB programs are used for PSO, AWPSO, AACPSO and MAACPSO to make tuning of the PID controller's parameters. These parameters adjusted to have minimum integrated error value with shorted settling time. The objective function is defined as follows [8, 25]:

For Integral of Absolute Error (IAE):

$$\mathrm{IAE} = \int_0^\infty |e(t)|\, dt \tag{23}$$

$$f = \mathrm{IAE}_1 + \mathrm{IAE}_2 + \mathrm{IAE}_{\mathrm{Ptie}} \tag{24}$$

Integral of Squared Error (ISE)

$$\mathrm{ISE} = \int_0^\infty e^2(t)\, dt \tag{25}$$

$$f = \mathrm{ISE}_1 + \mathrm{ISE}_2 + \mathrm{ISE}_{\mathrm{Ptie}} \tag{26}$$

Integral of Time Weighted Absolute Error (ITAE)

$$\mathrm{ITAE} = \int_0^\infty t\, |e(t)|\, dt \tag{27}$$

$$f = \mathrm{ITAE}_1 + \mathrm{ITAE}_2 + \mathrm{ITAE}_{\mathrm{Ptie}} \tag{28}$$

where:

e Is the error
f Is the objective function
$IAE_1, IAE_2, IAE_{Ptie_1}$ The Integral of Absolute Error of area 1, area 2 and the tie line of the System
$ISE_1, ISE_2, ISE_{Ptie_1}$ The Integral of Squared Error of area 1, area 2 and the tie line of the System
$ITAE_1, ITAE_2, ITAE_{Ptie_1}$ Integral of Time Weighted Absolute Error of area 1, area 2 and the tie line of the System

For the two power system areas, step loading disturbance has been applied for each area, 0.07 p.u load throw has been withdrawn from the first area and 0.05 p.u loading added for the second area. The control objective is to control the frequency deviation for each area.

4.1 Steps of the Study

Using MATLAB/SIMULINK the steps of the study by using many intelligent techniques (PSO, AWPSO, AACPSO and finally using MAACPSO) areas the following:

(1) Using MATLAB/SIMULINK model of the system with its parameters.
(2) Choose the type of error used in the equations in the beginning of the MATLAB program (IAE, ISE, or ITAE).
(3) Using PSO program with the equations of the chosen type of error.
(4) Repeat using AWPSO program for the same type of error used.
(5) Repeat using AACPSO program for the same type of error used.
(6) Repeat using MAACPSO program for the same type of error used.
(7) Compare the results of the four methods used and determine the best which has a less value of settling time and frequency deviation.
(8) Assign the value of the PID controller for the best method results.
(9) Conclude the results.

The performance index selected by the user in the beginning of the program. Based on this performance index (f) optimization problem can be stated as: Minimize f the nominal system description and parameters are describing in the following:

4.2 Model Description and Parameters

The block diagram of the two areas power system model using PID controller presented at Fig. 3 as presented in [8, 27]. The description for the system parameters is displayed in Table 1 and the parameters values of the system is presented in Table 2.

So the transfer function of governors, turbine, mass and load becomes as given in [27]:

$$G_{h1}(S) = G_{h2}(S) = \frac{1}{0.08s + 1} \tag{29}$$

$$G_{t1}(S) = G_{t2}(S) = \frac{1}{0.3s + 1} \tag{30}$$

$$G_{y1}(S) = G_{y2}(S) = \frac{120}{20s + 1} \tag{31}$$

To optimize the performance of a PID controlled system, the PID gains K_P, K_i, and K_d of the two-area electric power system shown in Fig. 3 are adjusted to minimize a certain performance index. The performance index is calculated over a time interval; T, normally in the region of $0 < T < ts$ where t_s is the settling time of the system. By using different techniques in conjunction with Eqs. 22–29 the optimal controller parameters under various performance indices were obtained as shown in Tables 1, 2 and 3 show the results of the different methods used based PID controller.

4.3 *Results in Case of IAE Error*

A MATLAB code was written to carry out the PSO, AWPSO, AACPSO and MAACPSO algorithms. The Integral of Absolute Error (IAE) is considered as a choice in the run of the program. Table 3 illustrates The Results of the Program Using PSO, AWPSO, AACPSO and MAACPSO.

Figures 5 and 6 present the frequency deviation of area 1 and area 2 without using PID controller.

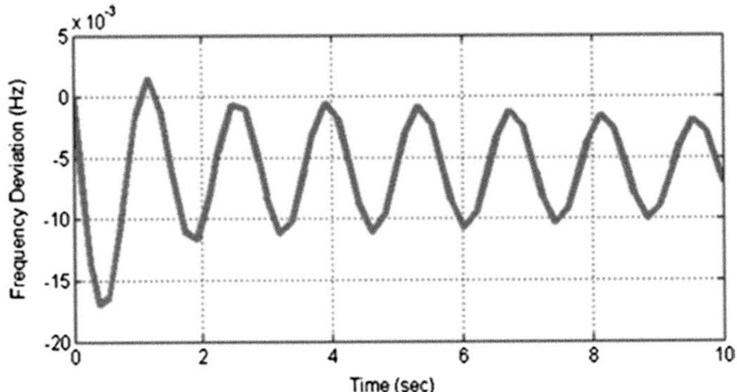

Fig. 5 The frequency deviation of area 1 without controller

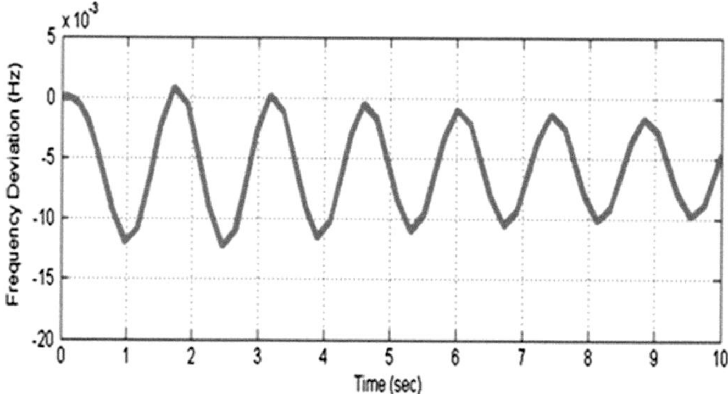

Fig. 6 The frequency deviation of area 2 without controller

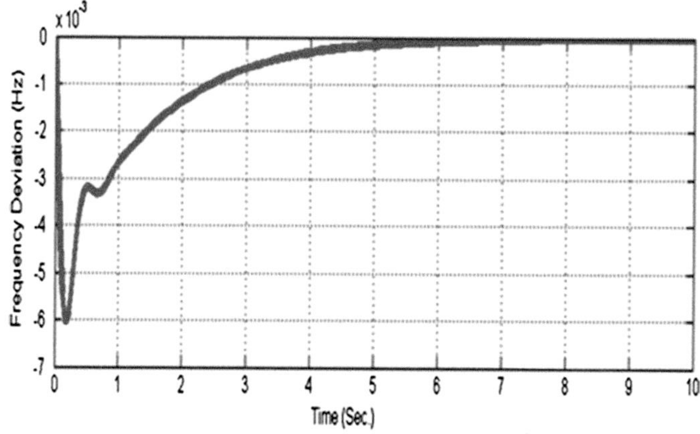

Fig. 7 The frequency deviation of area 1 with PSO based PID controller using IAE performance indices

Furthermore, there are the Figures describe the output of the system after controlling the error on area 1 and area 2. Figure 7 presents the frequency deviation of area 1 with PSO based PID Controller, Fig. 8 presents the frequency deviation of area 1 with AWPSO based PID controller and Fig. 9 illustrates the frequency deviation of area 1 with AACPSO based PID controller and finally Fig. 10 presented the frequency deviation of area 1using MAACPSO based PID controller.

The next Figures present the behavior of area 2 in different cases of Artificial Intelligence techniques.

Figure 11 presents the frequency deviation of area 2 with PSO based PID controller using IAE performance indices; Fig. 12 shows the behavior of the frequency deviation of area 2 in case of using AWPSO, while; Fig. 13 displays The frequency deviation of area 2 with AACPSO based PID controller, finally Fig. 14 presents the frequency deviation of area 2 with MAACPSO based PID controller using IAE performance indices.

From the results shown in Table 3 and also the above Figures from Figs. 7, 8, 9, 10, 11, 12, 13 and 14 all these show that:

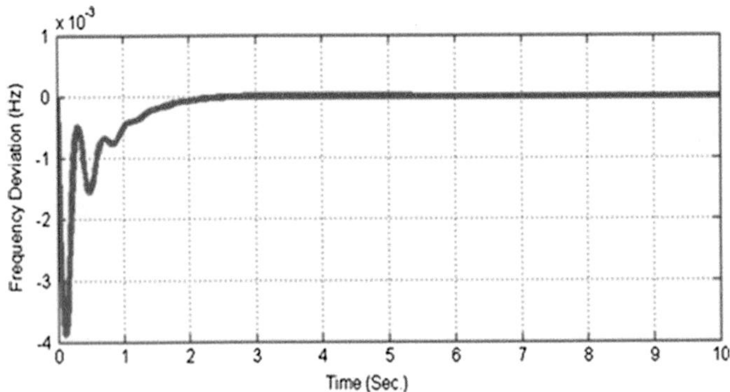

Fig. 8 The frequency deviation of area 1 with AWPSO based PID controller using IAE performance indices

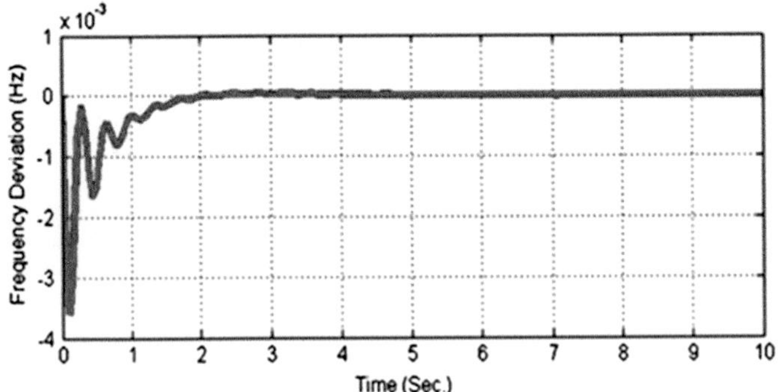

Fig. 9 The frequency deviation of area 1 with AACPSO based PID controller using IAE performance indices

Table 4 Tie line behavior at different types of control

Items of comparison	PSO	AWPSO	AACPSO	MAACPSO
Settling time_Tie line (s)	7.7624	4.2082	3.6553	2.5696
Maximum frequency of tie line power (Hz)	3.00E-07	4.24E-07	1.06E-06	4.98E-07
Time at maximum frequency of tie line power (s)	20.502	22.2727	4.8003	4.0135
Minimum frequency of tie line power (Hz)	−0.0011	−3.66E-04	−3.20E-04	−9.13E-04
Time at minimum frequency of tie line power (s)	1.0319	0.612	0.5743	0.442

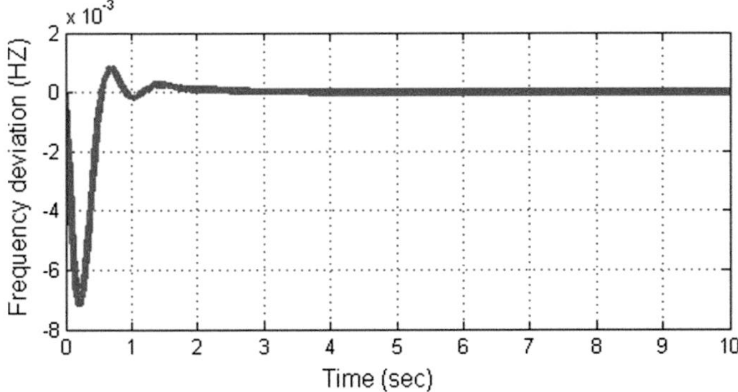

Fig. 10 The frequency deviation of area 1 with MAACPSO based PID controller using IAE performance indices

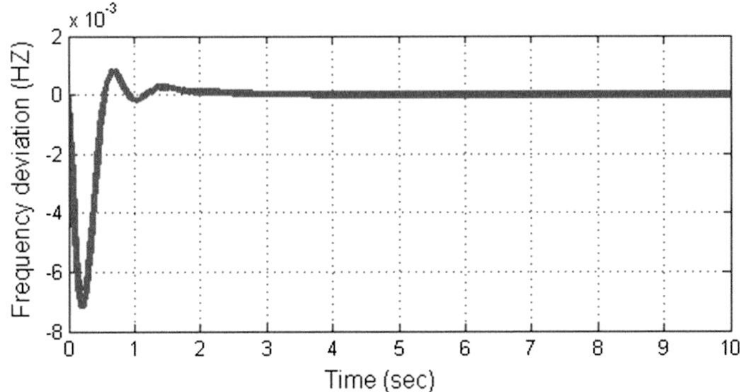

Fig. 11 The frequency deviation of area 2 with PSO based PID controller using IAE performance indices

(a) The settling time by using AACPSO is the smallest value of all the techniques used in the comparison. While, the settling time using MAACPSO comes next.
(b) The difference value between the settling time values using two methods “AACPSO, MAACPSO” is very small and equal approximately 0.08 s.
(c) All these results present that: the best method used to reach the minimum value of settling time in area 1 is AACPSO.
(d) The value of settling time of area 2 by using MAACPSO is less than all values using other methods.
(e) The difference between the value of settling time using MAACPSO and the nearest value using AACPSO is equal approximately 1.04 s.
(f) The value of settling time of the tie line using MAACPSO technique is the smallest compared to all methods used.

Table 5 Comparison between (MAACPSO) and (AACPSO)

Controller	Overshoot (Hz)	Settling time (s)
MAACPSO with IAE on area 1	5.5E-05	1.7267
AACPSO with IAE on area 1	4.10E-05	1.6514
MAACPSO with IAE on area 2	2.2E-03	2.5288
AACPSO with IAE on area 2	4.14E-06	3.6553
MAACPSO with IAE on tie line	4.98E-07	2.5696
AACPSO with IAE on tie line	1.06E-06	3.6553

(g) The difference between the values of settling time of the tie line using MAACPSO technique is less than the nearest value of settling time using AAPSO by approximately 1.08 s.

In the following sections there is Table 4 and Fig. 14 of Tie Line which describes the effects of using different techniques.

Figure 15 displays the Frequency Change Of The Tie Line Power With Using PSO, AWPSO And AAPSO Based PID Controller.

Table 5 shows Comparison of the value of Overshoot (Hz) and settling time (sec.) of the best two methods used MAACPSO and AACPSO.

The illustrated results in Table 5, Fig. 15 show that:

(a) Tables 4 and 5 indicate that on the Tie line power, the value of settling time in case of using MAACPSOis the best results and has a smaller value comparing with the other methods used (PSO, AWPSO and AACPSO).
(b) The settling time of Tie line in case of using MAACPSO is less than its value in case of using AACPSO by about 1.5 s, and less than its value when using AWPSO by about 1.6 s.
(c) Settling time by AWPSO is smaller than using PSO by 0.0359 s.

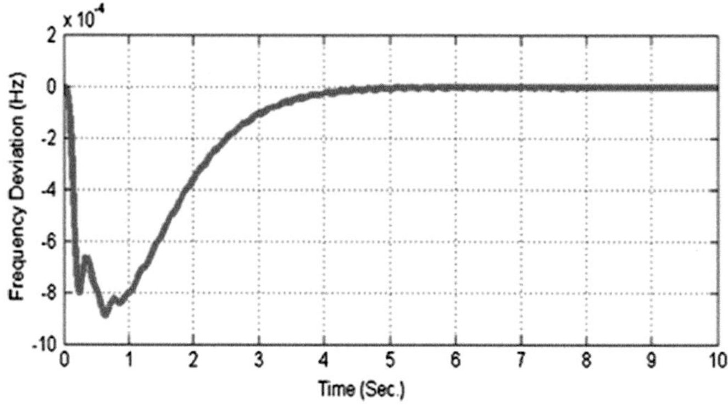

Fig. 12 The frequency deviation of area 2 with AWPSO based PID controller using IAE performance indices

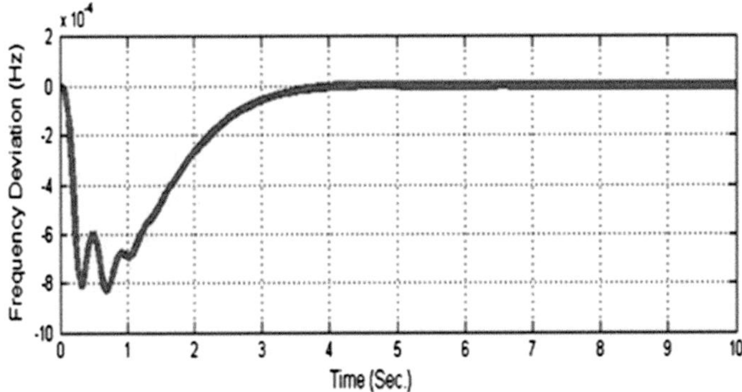

Fig. 13 The frequency deviation of area 2 with AACPSO based PID controller using IAE performance indices

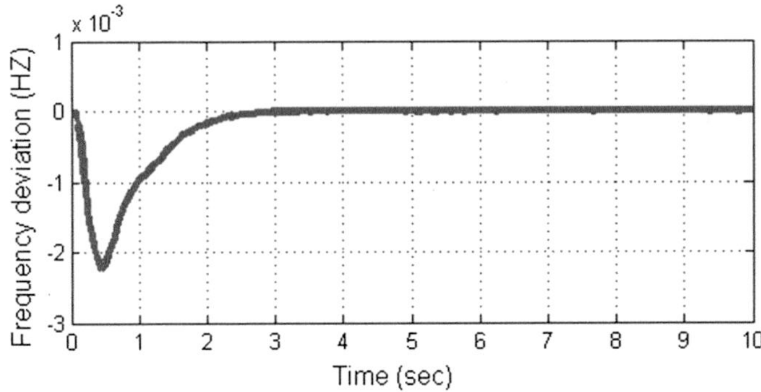

Fig. 14 The frequency deviation of area 2 with MAACPSO based PID controller using IAE performance indices

(d) The maximum frequency of Tie line power in case of using MAACPSO is less than its value of the other methods of controller used by a very small value.
(e) In general the maximum frequency of Tie line power is construed to be zero.
(f) Time at maximum power in case of using MAACPSO is less than the its value by using AACPSO by about 8 %, and the value of that time by using AACPSO is less than the other values of PSO and AWPSO. This value is less than the time of maximum power in case of using PSO by about 23.4 % and less than its value in case of using AWPSO by about 21.5 %.
(g) The minimum Tie line power in case of using MAACPSO is very small comparing with another methods, and its value in case of using AWPSO and AACPSO are almost equal and less than its value in case of using PSO.
(h) Time at minimum power in case of using MAACPSO is less than the other values of PSO, AWPSO and AACPSO.
(i) The Overshoot and settling time of area 1 by using MAACPSO is greater than that values by using AACPSO by a very small value.

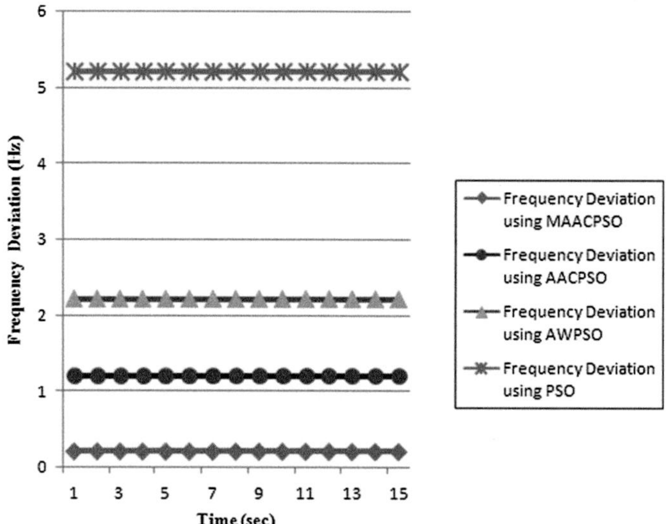

Fig. 15 Tie line power changes using PSO, AWPSO, AACPSO and MAACPSO based PID controller in case of using IAE error

(j) The Overshoot and settling time of area 2 by using MAACPSO is smaller than that values by using AACPSO.
(k) The Overshoot and settling time of tie line by using MAACPSO is very small than that values by using AACPSO.

All these results present that: the best method used to reach the minimum value of settling time is MAACPSO and AACPSO comes next.

5 Conclusions

The simulation of the proposed controllers explained in this chapter, indicate that:

Modified Adaptive Accelerated Coefficients based on PSO (MAACPSO) is the best method comparing with all methods used in this study. Then AACPSO comes next. As shown the settling time in area 1 using MAACPSO gives value near the value of settling time using AACPSO, the difference between the two methods was 0.07 s,and it's very small value.The settling time in area 2 using MAACPSO gives smaller than the value using AACPSO, the difference about 1.13 s, this is a very good result of MAACPSO. The settling time in tie line using MAACPSO gives very good value comparing with AACPSO, the difference was about 1.08 s., as presented in Tables 4 and 5 the frequency deviation values of area 1 of the best two methods MAACPSO and AACPSO was nearly equal, and the value using MAACPSO of tie tine is smaller than its value using AACPSO by about 5.6E-7 Hz.

6 Future Work

Studding the load frequency control using two different power system as a model and using MAACPSO technique will be a good study to examine the new technique, and make a comparison between the results of settling time and overshoot frequency using MAACPSO and some techniques like Fuzzy.

Appendix

Transmission line 1 parameters
Kg1 = 1
Kt1 = 1
Tg1 = 0.08
Tt1 = 20
R1 = 2.4
T11 = 20
Kl1 = 120
a12 = 1
Transmission line 2 parameters
Kg2 = 1
Kt2 = 1
Tg2 = 0.08
Tt2 = 0.33
R2 = 2.4
T12 = 20
Kl2 = 120

N = 25	Number of swarm beings
d = 6	Two dimensional problem
n = 500	Number of iterations
W0 = 0.15	Percentage of old velocity
A0 = 0.5	Acceleration factor constant between [0 1]
C1 = 2.05	Percentage towards personal optimum
C2 = 2.05	Percentage towards
x0range = [0 10]	Range of uniform initial distribution of positions
vstddev = 1	Std. deviation of initial velocities
C11 = 2	Percentage towards personal optimum used in ACC
C22 = 2.05	Percentage towards used in ACC

References

1. Ahmed S, Tarek B, Djemai N (2013) Economic dispatch resolution using adaptive accelerated coefficients based PSO considering generator constraints. In: International conference on control, decision and information technologies, (CoDIT'13)
2. Amjady N, Nasiri-Rad H (2009) Nonconvex economic dispatch with AC constraints by a new real coded genetic algorithm. IEEE Trans Power Syst 24(3):1489–1502
3. Ang KH, Chong GCY, Li Y (2005) PID: Control system analysis, design, and technology. IEEE Trans Control Syst Technol 13(4):559–576

4. Azar AT, Serrano FE (2015) Design and modeling of anti wind up PID controllers. In: Zhu Q, Azar AT, (eds) Complex system modelling and control through intelligent soft computations. Studies in fuzziness and soft computing, vol. 319, pp 1–44, Springer-Verlag, Germany. doi:10.1007/978-3-319-12883-2_1.1
5. Azar AT (2010) Fuzzy Systems. IN-TECH, Vienna. ISBN 978-953-7619-92-3
6. Azar AT (2012) Overview of type-2 fuzzy logic systems. Int J Fuzzy Syst Appl (IJFSA) 2(4):1–28
7. Bahgaat NK, El-Sayed MI, Moustafa Hassan MA, Bendary FA (2015) Application of some modern techniques in load frequency control in power systems, Chaos Model Control Syst Des. Springer-Verlag, Germany, 581: 163–211
8. Bahgaat NK, El-Sayed MI, Moustafa Hassan MA, Bendary FA (2014) Load frequency control in power system via improving PID controller based on particle swarm optimization and ANFIS techniques. Int J Syst Dyn Appl (IJSDA) 3(3):1–24
9. Bevrani H (2009) Robust power system frequency control. Springer Science and Business Media LLC, Brisbane
10. Darrell W (2005) A Genetic algorithm tutorial. Computer Science Department, Colorado State University, Colorado
11. Gaing ZL (2004) A particle swarm optimization approach for optimum design of PID controller in AVR system. IEEE Trans Energy Convers 19(2):384–391
12. Hamid A, Abdul-Rahman TK (2010) Short term load forecasting using an artificial neural network trained by artificial immune system learning algorithm in computer modeling and simulation (UKSim). In: 12th international conference on IEEE, pp. 408–413
13. Hassanien AE, Tolba M, Azar AT (2014) Advanced machine learning technologies and applications. In: Second international conference, AMLTA 2014, Cairo, Egypt, November 28–30, 2014. Proceedings, communications in computer and information science, vol. 488, Springer-Verlag GmbH Berlin/Heidelberg. ISBN: 978-3-319-13460-4
14. Ismail A (2006) Improving UAE power systems control performance by using combined LFC and AVR. In: 7th UAE university research conference, ENG, pp. 50–60
15. Ismail MM, Moustafa Hassan M (2012) Load frequency control adaptation using artificial intelligent techniques for one and two different areas power system. Int J Control Autom Syst 1(1):12–23
16. Khalifa F, Moustafa Hassan M, Abul-Haggag O, Mahmoud, H (2015) The application of evolutionary computational techniques in medium term forecasting. In: Accepted for presentation at MEPCON'2015, Mansoura University, Mansoura, Egypt
17. Kumar DV (1998) Intelligent controllers for automatic generation control. In: TENCON'98. IEEE region, 10 international conference on global connectivity in energy, computer, communication and control, vol. 2, pp. 557–574
18. Naik RS, ChandraSekhar K, Vaisakh K (2005) Adaptive PSO based optimal fuzzy controller design for AGC equipped with SMES and SPSS. J Theor Appl Inf Technol 7(1):008–017
19. Panigrahi BK, RavikumarPandi V, Das S (2008) Adaptive particle swarm optimization approach for static and dynamic economic load dispatch. Energy Convers Manag 49(6):1407–1415
20. RamaSudha K, Vakula VS, Shanthi RV (2010) PSO based design of robust controller for two area load frequency control with nonlinearities. Int J Eng Sci 2(5):1311–1324
21. Rania HM (2012) Development of advanced controllers using adaptive weighted PSO algorithm with applications, M.Sc thesis, Faculty of Engineering, Cairo University, Cairo, Egypt
22. Reeves CR, Rowe JE (2002) Genetic algorithm principles and perspective, A guide to GA theory, Kluwer Academic Publishers, ISBN: 1-4020-7240-6
23. Salami A, Jadid S, Ramezani N (2006) The Effect of load frequency controller on load pickup during restoration, power and energy conference, PECon'06. IEEE, international, pp 225–228
24. Skogestad S (2003) Simple analytic rules for model reduction and PID controller tuning. J Process Control 13(4):291–309
25. Tammam MA, Aboelela MAS, Moustafa MA, Seif AEA (2012a) Load Frequency Controller Design for Interconnected Electric Power System. In: 55th Annual Power Industry division Symposium POWID Austin. Texas, USA

26. Tammam MA, Aboelela MAS, Moustafa MA, Seif AEA (2012b) Fuzzy like PID controller tuning by multi-objective genetic algorithm for load frequency control in nonlinear electric power systems. Int J Adv Eng Technol 5(1):572–583
27. Tammam MA (2011) Multi objective genetic algorithm controllers Tuning for load frequency control in Electric power systems, M.Sc thesis, Faculty of Engineering at Cairo University, Cairo, Egypt
28. Vlachogiannis JG, Lee KY (2009) Economic load dispatch—a comparative study on heuristic optimization techniques with an improved coordinated aggregation based PSO. IEEE Trans Power Syst 24(2):991–1001
29. Wang Y, Zhou R, Wen C (1993) Robust load-frequency controller design for power systems. IEE Proc. C Gener. Trans. Distrib. IET Digit. Lib. 140(1):11–16
30. Zhu Q, Azar AT (2015) Complex system modelling and control through intelligent soft computations. Studies in fuzziness and soft computing, vol. 319, Springer, Germany. ISBN: 978-3-319-12882-5